ENCYCLOPEDIA
OF
FOOD SCIENCES
AND NUTRITION

SECOND EDITION

ENCYCLOPEDIA
OF
FOOD SCIENCES
AND NUTRITION

SECOND EDITION

Editor-in-Chief

BENJAMIN CABALLERO

Editors

LUIZ C TRUGO

PAUL M FINGLAS

ACADEMIC PRESS

An imprint of Elsevier Science

Amsterdam   Boston   London   New York   Oxford   Paris
San Diego   San Francisco   Singapore   Sydney   Tokyo

EDITORIAL ADVISORY BOARD

FOREWORD

There are no disciplines so all-encompassing in human endeavours as food science and nutrition. Whether it be biological, chemical, clinical, environmental, agricultural, physical – every science has a role and an impact. However, the disciplines of food science and nutrition do not begin or end with science. Politics and ethics, business and trade, humanitarian efforts, law and order, and basic human rights and morality all have something to do with it too.

As disciplines, food science and nutrition answer questions and solve problems. The questions and problems are diverse, and cover the full spectrum of every issue. Life span is one such issue, covered from the nutritional basis for fetal and infant development, to optimal nutrition for the elderly. Another such issue is the time span of the ancient and wild agro-biodiversity that we are working to preserve, to the designer cultivars from biotechnology that we are trying to develop. Still another is the age-old food preparation methods now honoured by the 'eco-gastronomes' of the world, to the high tech food product development advances of recent years.

As with most endeavours, our scientific and technological solutions can and do create new, unforeseen problems. The technologies that gave us an affordable and abundant food supply led to obesity and chronic diseases. The "green revolution" led to loss of some important agro-biodiversity. The technological innovation that gave us stable fats through hydrogenation, flooded the food supply with trans fatty acids. All these problems were identified through a multidisciplinary scientific approach and solutions are known. When technology created the problem and technology has found the solution, implementation is usually more successful. Reducing trans fatty acids in the food supply is case in point. Beyond the technologies, the solutions are more difficult to implement. We know how obesity can be reduced, but the solution is not directly technological. Hence, we show no success in the endeavour.

Of all the problems still confounding us in food science and nutrition, none is so compelling as reducing the number of hungry people in the world. FAO estimates that there are 800 million people who do not have enough to eat. The World Food Summit Plan of Action, the Millennium Development Goals and other international efforts look to food science and nutrition to provide the solution. Yet we only have part of the solution—the science part. The wider world of effort in food science and nutrition needs to be more conscientiously addressed by scientists. This is the world of advocacy and action: advocacy for food and nutrition as basic human rights, coupled with action to get food where it is needed.

But all those efforts would be futile if they are not based on sound scientific information. That is why works such as this Encyclopedia are so important. They provide to a wide readership, scientists and non-scientists alike, the opportunity to quickly gain understanding on specific topics, to clarify questions, and to orient to further reading. It is a pleasure to be involved in such an endeavour, where experts are willing to impart their knowledge and insights on scientific consensus and on exploration of current controversies. All the while, this gives us optimism for a brighter food and nutrition future.

Barbara Burlingame
25 February 2003

INTRODUCTION

There is no factor more vital to human survival than food. The only source of metabolic energy that humans can process is from nutrients and bioactive compounds with putative health benefits, and these come from the food that we eat. While infectious diseases and natural toxins may or may not affect people, everyone is inevitably affected by the type of food they consume.

In evolutionary terms, humans have increased the complexity of their food chain to an astounding level in a relatively short time. From the few staples of some thousand years ago, we have moved to an extraordinarily rich food chain, with many food items that would have been unrecognizable just some hundred years ago.

In this evolution, scientific discovery and technical developments have always gone hand in hand. The identification of vitamins and other essential nutrients last century, and the development of appropriate technologies, led to food fortification, and thus for the first time humans were able to modify foods to better fulfill their specific needs. As a result, nutritional deficiencies have been reduced dramatically or even eradicated in many parts of the world. This evolution is also yielding some undesirable consequences. The abundance of high-density, cheap calorie sources, and the market competition has facilitated overconsumption and promoted obesity, a problem of global proportions.

As the food chain grows in complexity, so does the scientific information related to it. Thus, providing accurate and integral scientific information on all aspects of the food chain, from agriculture and plant physiology to dietetics, clinical nutrition, epidemiology, and policy is obviously a major challenge.

The editors of the first edition of this encyclopedia took that challenge with, we believe, a great deal of success. This second edition builds on that success while updating and expanding in several areas. A large number of entries have been revised, and new entries added, amounting to two additional volumes. These new entries include new developments and technologies in food science, emerging issues in nutrition, and additional coverage of key areas. As always, we have made efforts to present the information in a concise and easy to read format, while maintaining rigorous scientific quality.

We trust that a wide range of scientists and health professionals will find this work useful. From food scientists in search of a methodological detail, to policymakers seeking update on a nutrition issue, we hope that you will find useful material for your work in this book. We also hope that, in however small way, the Encyclopedia will be a valuable resource for our shared efforts to improve food quality, availability, access, and ultimately, the health of populations around the world.

Benjamin Caballero
Luiz Trugo
Paul Finglas

GUIDE TO USE OF THE ENCYCLOPEDIA

Structure of the Encyclopedia

The material in the Encyclopedia is arranged as a series of entries in alphabetical order. Some entries comprise a single article, whilst entries on more diverse subjects consist of several articles that deal with various aspects of the topic. In the latter case the articles are arranged in a logical sequence within an entry.

To help you realize the full potential of the material in the Encyclopedia we have provided three features to help you find the topic of your choice.

1. Contents Lists

Your first point of reference will probably be the contents list. The complete contents list appearing in each volume will provide you with both the volume number and the page number of the entry. On the opening page of an entry a contents list is provided so that the full details of the articles within the entry are immediately available.

Alternatively you may choose to browse through a volume using the alphabetical order of the entries as your guide. To assist you in identifying your location within the Encyclopedia a running headline indicates the current entry and the current article within that entry.

You will find 'dummy entries' where obvious synonyms exist for entries or where we have grouped together related topics. Dummy entries appear in both the contents list and the body of the text. For example, a dummy entry appears for Allergy which directs you to Food Intolerance where the material is located.

Example
If you were attempting to locate material on Seafood via the contents list.

SEAFOOD *see* FISH: Introduction; Catching and Handling; Fish as Food; Demersal Species of Temperate Climates; Pelagic Species of Temperate Climates; Tuna and Tuna-like fish; Fish of Tropical Climates; Demersal Species of Tropical Climates; Pelagic Species of Topical Climates; Important Elasmobranch Species; Processing; Miscellaneous Fish Products; Spoilage of Seafood; Dietary importance of Fish and Shellfish; FISH FARMING; FISH MEAL; MARINE FOODS: Production and Uses of Marine Algae; Edible Animals Found in the Sea; Marine Mammals as Meat Sources: SHELLFISH: Characteristics of Crustacea; Commercially Important Crustacea: Characteristics of Molluscs; Commercially Important Molluscs; Contamination and Spoilage of Molluscs and Crustaceans; Aquaculture of Commercially Important Molluscs and Crustaceans

At the appropriate location in the contents list, the page numbers for these articles are given.

If you were trying to locate the material by browsing through the text and you looked up Seafood then the following information would be provided.

Seafood *see* **Fish**: Introduction; Catching and Handling; Fish as Food; Demersal Species of Temperate Climates; Pelagic Species of Temperate Climates; Tuna and Tuna-like fish; Fish of Tropical Climates; Demersal Species of Tropical Climates; Pelagic Species of Topical Climates; Important Elasmobranch Species; Processing; Miscellaneous Fish Products; Spoilage of Seafood; Dietary importance of Fish and Shellfish; **Fish Farming**; **Fish Meal**; **Marine Foods**: Production and Uses of Marine Algae; Edible Animals Found in the Sea; Marine Mammals as Meat Sources: **Shellfish**: Characteristics of Crustacea; Commercially Important Crustacea: Characteristics of Molluscs; Commercially Important Molluscs; Contamination and Spoilage of Molluscs and Crustaceans; Aquaculture of Commercially Important Molluscs and Crustaceans

Alternatively if you were looking up Shellfish the following information would be provided.

SHELLFISH

Contents

2. Cross References

All of the articles in the Encyclopedia have been extensively cross referenced.

The cross references, which appear at the end of an article, have been provided at three levels:

i. To indicate if a topic is discussed in greater detail elsewhere

> **JAMS AND PRESERVES/METHODS OF MANUFACTURE**
> *See also:* **Acids**: Properties and Determination; **Aluminum (Aluminium)**: Toxicology; **Apples; Ascorbic Acid**: Properties and Determination; **Cleaning Procedures in the Factory**: Overall Approach; **Colloids and Emulsions; Oxidation of Food Components; Pectin**: Food Use; **pH – Principles and Measurement; Raspberries and Related Fruits;** Ripening of Fruit; Storage Stability: Parameters Affecting Storage Stability; **Strawberries; Sucrose: Properties and Determination; Zinc**: Physiology

ii. To draw the reader's attention to a parallel discussions in other articles

> **JAMS AND PRESERVES/METHODS OF MANUFACTURE**
> *See also:* **Acids**: Properties and Determination; **Aluminum (Aluminium)**: Toxicology; **Apples; Ascorbic Acid**: Properties and Determination; **Cleaning Procedures in the Factory**: Overall Approach; Colloids and Emulsions; **Oxidation of Food Components; Pectin**: Food Use; pH – Principles and Measurement; **Raspberries and Related Fruits; Ripening of Fruit; Storage Stability**: Parameters Affecting Storage Stability; **Strawberries; Sucrose**: Properties and Determination; **Zinc**: Physiology

iii. To indicate material that broadens the discussion

> **JAMS AND PRESERVES/METHODS OF MANUFACTURE**
> *See also:* **Acids**: Properties and Determination; **Aluminum (Aluminium)**: Toxicology; **Apples; Ascorbic Acid**: Properties and Determination; **Cleaning Procedures in the Factory**: Overall Approach; **Colloids and Emulsions; Oxidation of Food Components; Pectin**: Food Use; **pH – Principles and Measurement; Raspberries and Related Fruits; Ripening of Fruit; Storage Stability**: Parameters Affecting Storage Stability; **Strawberries; Sucrose**: Properties and Determination; **Zinc**: Physiology

3. Index

The index will provide you with the volume number and page number of where the material is to be located, and the index entries differentiate between material that is a whole article, is part of an article or is data presented in a table. Detailed notes are provided on the opening page of the index.

4. Colour Plates

The colour figures for each volume have been grouped together in a plate section. The location of this section is cited in the contents list. Colour versions of black and white figures are cited in figure captions within individual articles.

5. Contributors

A full list of contributors appears in Volume 10.

CONTENTS

VOLUME 1

VOLUME 2

C

VOLUME 3

E

VOLUME 4

H

VOLUME 6

J

K

L

VOLUME 7

N

VOLUME 8

VOLUME 9

VOLUME 10

W

NATAMYCIN

L V Thomas and J Delves-Broughton, Danisco
Innovation, Beaminster, Dorset, UK

Introduction

Natamycin (also known as pimaricin) is a polyene macrolide antimycotic produced by *Streptomyces natalensis*. It is marketed commercially as the Natamax™ family of products by Danisco, and as the Delvocid® family of products by DSM. Natamycin is used worldwide as a food preservative; its primary application is surface treatment of cheese and processed meat (e.g., dry sausages) by dipping, spraying, or in emulsion coatings. It has several advantages as a preservative, including broad activity spectrum, efficacy at low concentrations, lack of resistance, and activity over a wide pH range. Due to its low solubility, natamycin does not migrate from the surface into the food, and thus does not affect the organoleptic properties. It has no effect on bacteria, including those used as starter cultures or to promote ripening. It is chemically stable and has prolonged effectiveness. Moreover, it is easy to apply and has a proven safety record. Its general characteristics are summarized in **Table 1**.

History

Natamycin was first discovered in 1955, when it was isolated from culture filtrates of an *Actinomycetes* bacterium isolated from soil near the town of Pietermaritzburg, in the Natal province of South Africa. The antimycotic was first called pimaricin, a name derived from this place of origin. Its current name natamycin, as well as the name of its producer organism *Streptomyces natalensis*, was similarly derived from Natal. Natamycin has also been produced from closely related *Streptomyces* strains isolated from other locations and has been given various names (e.g., tennecetin, antifungal A-5283) and trade names. Its potential as a food preservative was recognized shortly after its discovery, when its efficacy against molds and yeasts was investigated in foods including fruit juice, carbonated drinks, fresh strawberries and raspberries, dressed poultry, sausages, cottage cheese, and hard cheese.

Structure

Natamycin has the empirical formula $C_{33}H_{47}NO_{13}$ and a molecular weight of 665.7. The structure was first determined in 1958, and later revised. It belongs to the group of polyene macrolide antifungals, compounds characterized by possession of a macrocyclic ring of carbon atoms closed by lactonization, and four conjugated double bonds. This classes it as a tetraene antimycotic (**Figure 1**). The structure is closely related to other antimycotics such as nystatin, rimodicin, and amphotericin. The molecule is amphoteric due to one basic and one acid group and has an isoelectric point of 6.5.

Stability and Solubility

Natamycin is a white/cream-colored crystalline powder with no taste and little odor. It is usually found as the stable trihydrate formulation. No significant loss of activity occurs for several months if the powder is stored in the dark at room temperature. Aqueous suspensions are less stable, particularly if exposed to light, certain oxidants, and heavy metals, but remain sufficiently stable during practical use. Although solutions are more unstable in acid or alkaline conditions, the pH of food products is not normally at levels that would cause problems. For instance, natamycin stored at 30 °C for 21 days reportedly retained 100% activity at pH 5–7, retained c. 85% at pH 3.6, and 75% at pH 9.0. In acid solutions, the molecule undergoes hydrolysis at the glycosidic linkage, forming mycosamine and aponatamycin and subsequently other compounds. Aqueous suspensions at neutral pH are stable for 24 h at 50 °C and show little reduction in activity at this temperature for several days and for shorter periods at 100 °C.

Natamycin is poorly soluble in water (30–100 p.p.m. at room temperature) and almost insoluble in nonpolar solvents, but shows good solubility in strongly polar organic solvents (**Table 1**). To be active, natamycin must be in solution. However, solubility is not normally a limiting factor as natamycin is usually effective at relatively low concentrations. In solution, natamycin has an ultraviolet absorption spectrum with minima at 250, 295.5, and 311 nm, and maxima at 220, 290, 303, and 318 nm.

Table 1 General characteristics of natamycin

Characteristic	Description
Names	Natamycin, pimaricin (pimafucin), tennecetin, antifungal A-5283 (trade names: Natacyn, Myprozine)
Commercial products	Natamax™, Delvocid®
Producer organism	Streptomyces natalensis
EU number	E235
Formula	$C_{33}H_{47}NO_{13}$
Molecular weight	665.7 Da
Structure	Polyene macrolide antimycotic compound with a macrocyclic ring of carbon atoms closed by lactonization, and four conjugated double bonds
Properties	Amphoteric. Isoelectric point: 6.5
Solubility in different solvents	Water: 30–100 p.p.m.
	n-Butanol: 50–120 p.p.m.
	Glycerol: 15 000 p.p.m.
	Methylpyrrolidone: 120 000 p.p.m.
	Glacial acetic acid: 185 000 p.p.m.
Absorption spectrum	Maxima: 290, 303, 318 nm
	Minima: 250, 295.5, 311 nm
Antimicrobial spectrum	Most molds and yeasts (MIC: < 5–20 p.p.m.)
JECFA ADI	0.3 mg kg^{-1} body weight

MIC, minimum inhibitory concentration; JECFA, Joint Food and Agriculture Organization/World Health Organization Expert Committee on Food Additives; ADI, acceptable daily intake.

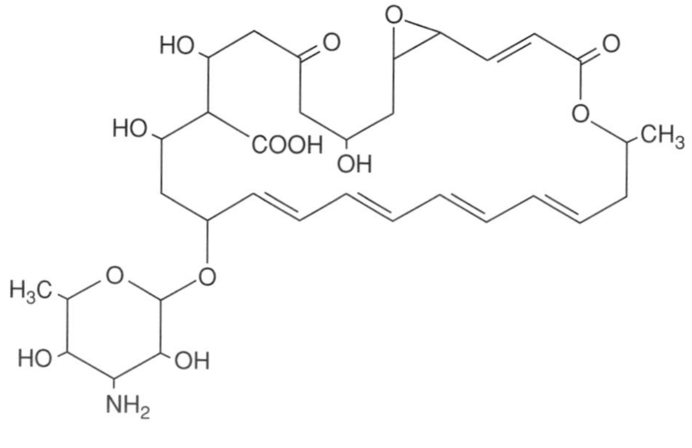

Figure 1 The structure of natamycin.

Mode of Action and Antimicrobial Effect

Mechanism of Action

Natamycin acts against yeasts and molds but is ineffective against bacteria. It combines with ergosterol and other sterols such as 24, 28-dehydroergosterol and cholesterol, compounds present in the cell membranes of yeast and molds but (with rare exceptions) not present in bacteria. The irreversible binding of natamycin to sterols, in particularl ergosterol, disrupts the cell membrane integrity and increases membrane permeability. This results in leakage of essential cellular constituents including cations, causing a rapid drop in intracellular pH and ultimately cell lysis. A study of laboratory-forced natamycin-resistant mutants of *Aspergillus flavus* has confirmed this theory. These either had reduced levels of ergosterol or none at all (and consequently much slower growth rates – critical to survival *in vivo*). A secondary mode of action is the inhibition of glycolysis and respiration.

Antimicrobial Spectrum

Natamycin is active against an extensive range of yeasts and molds (**Table 2**). The preservative is usually effective at concentrations between < 1 and 10 p.p.m. In general, yeasts are more sensitive than molds: the minimum inhibitory concentration (MIC) of yeasts is usually < 5 p.p.m., whereas that of molds can be at least 10 p.p.m. Examples of MICs are shown

Table 2 Antimicrobial spectrum: examples of yeast and molds sensitive to natamycin

Absidia	Penicillium camemberti
Alternaria	P. chrysogenum
Aspergillus chevalieri	P. digitatum
A. clavatus	P. expansum
A. flavus	P. glabrum
A. nidulans	P. islandicum
A. niger	P. notatum
A. ochraceus	P. roqueforti var. punctatum
A. oryzae	Rhizopus oryzae
A. penicilloides	Rhodotorula gracilis
A. roquefortii	Saccharomyces bailii
A. versicolor	S. bayanus
Botrytis cinerea	S. cerevisiae
Brettanomyces bruxellensis	S. exiguus
Candida albicans	S. ludwigii
C. guilliermondii	S. rouxii
C. vini	S. sake
Cladosporium cladosporioides	Sclerotina fructicola
Fusarium	Scopulariopsis asperula
Gloeosporium album	Torulopsis candida
Hansenula polymorpha	T. lactis var. condensi
Kloeckera apiculata	Wallenis sebii
Mucor mucedo	Zygosaccharomyces barkeri
M. racemosus	

Table 3 Examples of sensitivity of yeasts and molds to natamycin

Strain	Minimal inhibitory concentration (p.p.m.)
Aspergillus chevalieri 4298	0.1–2.5
Saccharomyces cerevisiae H	0.15
Penicillium chrysogenum	0.6–1.0
Aspergillus niger	1.0–1.8
Saccharomyces bailii	1.0
Candida albicans	1–2.5
Mucor mucedo	1.2–5.0
Penicillium notatum 4640	5.0
Saccharomyces rouxii 0562	5.0
Rhizopus oryzae 4758	10.0

in **Table 3**. Less susceptible species include *Verticillium cinnabarinum*, *Botrytis cinerea*, and *Penicillium discolour*, and also occur among the genera *Aspergillus*, *Fusarium*, *Penicillium*, and *Trichophyton*. The outcome of antifungal activity is usually cidal, in contrast to sorbate, which is fungistatic. Natamycin can inhibit aflatoxin synthesis, furthermore overall control of fungi leads to control of aflatoxins.

Resistance to polyene antimycotics such as natamycin does not seem to occur naturally and it has also proved difficult to generate resistant mutants in the laboratory. For example, in a survey of factories of natamycin-producers, no detectable difference in natamycin sensitivity was found for isolates compared to other factories. Similarly, in both a cheese warehouse and sausage factory where natamycin had been used for some time, no resistant strains were isolated. This lack of resistance may be partly explained by the fact that natamycin occurs as micelles in solution. Thus if a cell comes into contact with natamycin in solution, the antimycotic concentration is high and consequently lethal. Furthermore, the site of activity is an essential component of the cell membrane.

Factors Affecting Efficacy

The antimycotic activity of natamycin is affected by several factors, which may also affect its stability, such as pH, temperature, light, oxidants, and heavy metals. Natamycin is active over a wide pH range (pH 3–9); pH does not appear to affect its antimycotic activity but does affect stability. Compounds such as peroxides or chlorides, often used as cleaning/disinfecting agents, should be used with care in the proximity of natamycin.

Methods of Assay

The natamycin content of food products can be determined by microbiological, immunochemical, spectrophotometric, and liquid chromatographic (LC) procedures. The agar diffusion bioassay using *Saccharomyces cerevisiae* can also be used for quantitative assessment. Spectrophotometric and LC methods are commonly used for routine analysis. High-performance liquid chromatography (HPLC) with ultraviolet detection is considered one of the most sensitive and accurate methods, with a detection limit of $0.5\,mg\,kg^{-1}$. (*See* **Chromatography: High-performance Liquid Chromatography.**)

International Dairy Federation standard 140A: 1992 specifies the method for determining the natamycin contents of cheese rind and cheese adjacent to the rind. In this method a known quantity of sample is extracted with methanol, which is then diluted with water and cooled to between -15 and $-20\,°C$ to congeal and precipitate the majority of fat, followed by filtration. The natamycin content can then be determined in the filtrate by either a spectrometric or HPLC method. For the spectrometric measurement the spectrum of both a natamycin standard solution and the test sample is recorded in the range 300–340 nm. Absorbance is measured at the maximum of approximately 318 nm, at the minimum of approximately 311 nm, and then at 329 nm. For measurement by HPLC, ultraviolet detection should be at 308 nm and it is recommended that the mobile phase of the column should comprise methanol/water/acetic acid, $60+40+5$ (v/v/v). Before analysis of test samples, a standard natamycin solution is first injected to determine retention time and for calibration.

More recently, a rapid method using derivative spectrophotometry has been described. In this method cheese is extracted with acidified aqueous acetonitrile, and natamycin content is directly quantified in filtered extracts on the basis of the depth of the trough at 322.6 nm after third-derivative processing of the normal ultraviolet spectrum. An enzyme immunoassay involving the use of a natamycin–protein conjugate has also been reported, in which natamycin coupled to horseradish peroxidase was used as the labeled ligand. Detection limits were $200–2000\,pg\,ml^{-1}$, which enabled the determination of concentrations as low as $0.005\,mg\,dm^{-2}$ ($0.1\,mg\,kg^{-1}$) in cheese rinds.

Toxicology and Legislation

Toxicology studies have been undertaken using mice, rats, rabbits, and guinea pigs. Natamycin was least toxic if administered orally ($LD50 = 1500\,mg\,kg^{-1}$ in rats and mice) or subcutaneously ($LD50 = 5000\,mg\,kg^{-1}$ in rats), and most toxic if administered intravenously ($LD50 = 5–10\,mg\,kg^{-1}$). No natamycin was absorbed from the human intestinal tract after 7 days' feeding of up to a maximum of 500 mg day^{-1}. Feeding studies have been conducted in rats, rabbits, and dogs. The acceptable daily intake was set at 0.3 mg kg^{-1} body weight per day in 1976.

Specification of natamycin in the USA (21 CFR 172.155) requires purity of the anhydrous compound to be 97% $\pm$ 2%, containing < 1 p.p.m. arsenic and not more than 20 p.p.m. heavy metals.

Natamycin is approved for use as an antimycotic in various cheeses and processed meats in 32 countries worldwide. A more general use as a food additive is allowed in a few countries such as South Africa (Table 4). In the European Union (EU) natamycin (designated E235) is permitted for surface treatment of hard, semihard, and semisoft cheese as well as dry sausages. The maximum surface coverage permitted in the EU is 1 mg dm^{-2}, and penetration is restricted to 5 mm from the surface. In the USA, natamycin is permitted only if the cheese standard allows the use of 'safe and suitable' antimycotics. **Table 4** is a guide to food legislation. This list is not comprehensive – one should be aware that legislation for food additives is under constant review. The reader is advised to check the current legal situation with the appropriate authorities.

Table 4 Food legislation on the use of natamycin

Country	Food in which natamycin is permitted	Maximum permitted level
Algeria	Cheese rinds	Used in suspension at 2.5 g l^{-1}
Argentina	Surface treatment of hard and semihard paste cheeses	Limit of 1 mg dm^{-2}. Penetration limit of 2 mm
Australia	Surface treatment of cheese rind	Limit of 15 mg/kg
	Uncooked fermented manufactured meat products	Penetration limit of 3–5 mm
Bahrain	Permitted food preservative	
Brazil	Surface treatment of cheese	Limit of 2 mg dm^{-2}
Bulgaria	Cheese rind	500 mg kg^{-1}
Canada	Surface treatment of 47 listed cheeses	20 mg kg^{-1} based on total weight
	Grated/shredded cheese (0.5% sodium lauryl sulfate prohibited as dispersant)	10 mg kg^{-1}
Chile	Surface treatment of hard cheese (prohibited in wine)	
China	Surface treatment of cheese, processed meat products, moon cakes, baked goods, fruit juices, and processing utensils for easily moldy foods	Application by spraying or dipping in 200–300 mg kg^{-1}, to leave a residue of < 10 mg kg^{-1}
Colombia	Cheese	12.5 mg kg^{-1}
Czech Republic	Dairy and meat products – as EU regulations (contact authorities for further information)	Limit of 1 mg dm^{-2}. Penetration limit of 5 mm
Cyprus	Surface treatment of specified cheeses	
Egypt	Surface treatment of cooked cheese (dehydrated, semidehydrated, and semisoft cheese)	Limit of 2 mg 100 cm^{-2} (1 mg dm^{-2})
European Union (EU)	Surface treatment of specified cheese and sausage	Limit of 1 mg dm^{-2}. Penetration limit of 5 mm
Hungary	Surface treatment of hard and semihard cheese, dried, cured sausage	Limit of 1 mg dm^{-2}
Iceland	Surface treatment of ripened and whey cheese	Limit of 2 mg dm^{-2}
India	Surface treatment of hard cheese	Maximum application level: 2 mg dm^{-2}. Maximum residual level in finished cheese: 1 mg dm^{-2}. Penetration limit of 2 mm
Israel	Surface treatment of specified cheese	
Kuwait	Permitted food additive	
Lithuania	Surface treatment of hard, semihard, and semisoft cheese	Limit of 1 mg dm^{-2}. Penetration limit of 5 mm
Mauritius	Surface treatment of hard, semihard, and semisoft cheese and dried cured sausage	Limit of 1 mg dm^{-2}. Penetration limit of 5 mm

Continued

Table 4 Continued

Country	Food in which natamycin is permitted	Maximum permitted level
Mercosur	Surface treatment of cheese	Limit of 1 mg dm^{-2}. Maximum application of 5 mg kg^{-1}. Penetration limit of 2 mm
Mexico	Cheese surfaces	Limit of 0.002%
Norway		Maximum level of suspension: 2 g kg^{-1}
	Surface treatment of hard, firm, and semifirm cheese, dried, cured sausage	Limit of 1 mg dm^{-2}. Penetration limit of 5 mm
Oman	Surface treatment of specified cheese	
Philippines	Surface treatment of specified cheese	
Poland	Permitted in colored and uncolored soft wax and polyvinyl acetate for application to skin of hard cheese	No limits
	Surface treatment of smoked, dried sausage	Limit of 1 mg dm^{-2}
Saudi Arabia	Permitted as a mold inhibitor in foodstuffs but controlled by standards of composition	
Slovakia	Surface treatment of cheeses and dried, cured sausage	Limit of 1 mg dm^{-2}
South Africa	Wine, alcoholic fruit beverages, and grape-based liquors	30 mg l^{-1}
	Fresh fruit pulp	5 mg kg^{-1}
	Fruit juice (blackcurrant, pineapple, etc.)	5 mg kg^{-1}
	Fish sausages	6 mg kg^{-1} to be applied to the outer inedible casing only
	Manufactured fish products, fish pastes, fish roe and spawn, with the exception of frozen fish, salted snoek, and canned fish products	6 mg kg^{-1}
	Lobsters (quick frozen)	6 mg kg^{-1}
	Edam, Gouda, Tilsiter, Limburger, Cheddar, Cheshire	2 mg kg^{-1} in rind without plastic coating; 500 mg kg^{-1} in a plastic coating; 10 mg kg^{-1} for application to the surface of the cheese only
	Cottage cheese, cream cheese	Limit of 10 mg kg^{-1}
	Process or blended cheese, including cheese spread, process cheese preparations, and soft cheese	10 mg kg^{-1} for application to the surface of the cheese only
	Yogurt	10 mg kg^{-1}
	Canned foods	6 mg kg^{-1}
	Manufactured meat products	Limit of 500 mg kg^{-1} on casing or 6 mg kg^{-1} in contents
	Canned chopped meat, canned corned beef, cooked cured luncheon meat, cooked cured pork shoulder, Biltong, frozen cooked-meat pie fillings	Limit of 6 mg kg^{-1}
Tunisia	Surface treatment of hard, semihard and semisoft cheese; dried, cured sausage	Limit of 1 mg dm^{-2}
Turkey	Surface treatment of hard and semihard cheese, dried cured sausages, salami, and hot dogs	Limit of 1 mg dm^{-2}. Penetration limit of 5 mm
Ukraine	Surface treatment of cheese	Limit of 1 mg dm^{-2}. Penetration limit of 5 mm
United Arab Emirates	Permitted food additive	
USA	Surface treatment of cuts and slices of cheese	Limit of 20 mg kg^{-1}
	Nonstandard of identity yogurt	Limit of 7 mg kg^{-1}
	Nonstandard of identity cream cheese	Limit of 7 mg kg^{-1}
	Cottage cheese	Limit of 7 mg kg^{-1}
	Sour cream	Limit of 7 mg kg^{-1}
	Soft tortillas	Limit of 20 mg kg^{-1}
	Nonstandard of identity salad dressing	Limit of 20 mg kg^{-1}
Venezuela	Surface treatment of specified cheeses and sausages	Maximum 0.5% suspension

Preservation of Foods with Natamycin

Types of Food Suitable for Natamycin Use

The principal use of natamycin is on the surface of cheese and dry sausages. Due to its low solubility, natamycin remains effective on the surface for extended periods. When first applied, only 30–50 p.p.m. will be present in solution on the surface – the remainder is present in the more stable crystal formation. The preservative then gradually dissolves, insuring a slow release and prolonged effectiveness. An additional advantage of surface treatment is that natamycin does not penetrate far into the food; a limit of penetration of 1–4 mm has been reported in cheese rind. This is particularly useful for the preservation of blue cheese, especially in comparison to sorbate. Sorbate migrates into the cheese matrix where it inhibits the desired blue mold development inside the cheese. Natamycin remains on the surface, acting only where it is needed to prevent the growth of surface spoilage molds.

Much of the early work on natamycin investigated its ability to inhibit fungal growth on fruit. Strawberries, raspberries, and cranberries sprayed with 50 p.p.m. natamycin in the field prior to harvesting showed reduced spoiling during storage. Treatment by immersion in 10–100 p.p.m. proved more effective. A lecithin coating containing natamycin can be used to treat harder fruit, such as apples and pears. An addition level of 1–10 p.p.m. natamycin may be effective for a variety of beverages, including juices, beer, wine, and beverages containing tea or milk solids. Natamycin (at 20 p.p.m.) has proved effective against *S. cerevisiae* spoilage of orange juice. Further potential applications include ready-to-eat frostings, salad dressings, and mayonnaise susceptible to yeast spoilage. Natamycin control of *Aspergillus niger* and *S. cerevisiae* in cottage cheese has been investigated. Levels of 20–100 p.p.m. added to wash water or 1–5 p.p.m. added to the cheese dressing inhibited fungal growth. A level of 5–10 p.p.m. can be effective in controlling yeast and mold growth in yogurt. Natamycin has also been shown to be effective against molds isolated from bakery products. A level of 100 p.p.m. has been shown to inhibit the growth of yeasts and molds in fillings and icings. (*See* **Yeasts.**)

Natamycin can also be used to treat the surface of cured-meat products such as raw hams and investigations have been conducted into its ability to inhibit the spoilage microflora on raw cut chicken. Dipping in a solution of 10 p.p.m. inhibited yeast counts for 12–15 days at refrigeration temperature.

Mode of Application

To treat the surface of cheese and sausages, the food can be immersed, sprayed, or coated in an aqueous suspension or the suspension can be used in a plastic coating. Penetration into the food then depends on the food type, being greater in soft cheese compared to hard cheese.

Aqueous dipping solutions for cheese usually contain 1250 p.p.m. natamycin, but concentrations vary depending on cheese type. For example, dips for blue cheese may require dip concentrations as high as 2500 p.p.m. The cheese should be dipped into the natamycin slurry for a few seconds, whilst the slurry is kept constantly agitated to maintain the natamycin in suspension. The concentration of natamycin in the dipping solution becomes reduced after each dipping, at a rate dependent on the ratio of surface area to weight of each piece of cheese.

A suspension of 1250 p.p.m. is recommended for spraying cheese, applied at a rate of $6 \, l \, t^{-1}$ to achieve 7–15 p.p.m. natamycin on the surface. It is important that the natamycin is evently applied as it binds rapidly and tightly on contact with the cheese surface, so that subsequent mixing after initial application will not achieve better distribution of the preservative. A tumbler of at least 1 m in length is recommended with a space of at least 0.3 m between the natamycin and flow agent application.

Lower levels of natamycin (100–750 p.p.m.) can also be added to an emulsion of a polymer in water, mostly polyvinyl acetate, which can be applied as a cheese coating. Alternatively, 100–20 000 p.p.m. of natamycin can be mixed with 0.5–$50 \, g \, l^{-1}$ of a suitable thickener and approximately 20–$250 \, g \, l^{-1}$ salt and used on the surface of cheese or sausages. This method of application overcomes the problem of uneven distribution of the fungicide caused by the heterogeneity of the fat content of the product. Coatings can be applied by 'painting' multiple layers on to the surface with a sponge or soft cloth, or by a commercially available coating machine.

Natamycin can also be used in natural and fibrous casings of dry fermented sausage products, preventing mold growth during the ripening and storage process. Commonly, casings can be prepared by immersion for 2 h in a 1000-p.p.m. suspension. Alternatively the sausages can be dipped or sprayed in a 1000–5000-p.p.m. natamycin suspension.

Fermentation and Production

Natamycin is produced by fermentation of *Streptococous natalensis* in an aqueous nutrient medium containing a carbon source (e.g., starch, molasses, glycerol, glucose, lactose, maltose, sucrose, alcohols, organic acids), a fermentable organic and/or inorganic nitrogen source (e.g., corn steep liquor, casein, zein, lactalbumin, soya bean meal). The carbon source usually comprises 0.5–5% of the medium. Inorganic cations (potassium, sodium, or calcium) and anions (sulfate, phosphate, or chloride) may be needed as well as trace elements such as boron, molybdenum, or copper. Fermentation is aerobic and mechanical agitation and use of antifoaming agents can aid the process. The temperature is usually 26–30 °C and the pH range pH 6–8. The period of fermentation, varying with the medium, can be between 48 and 120 h.

Due to its low solubility in water, natamycin will accumulate mainly as crystals in the broth. Recovery commonly involves dissolving the natamycin using polar solvents with limited water miscibility such as butanol, methanol, and acetone and adjusting the medium to approximately pH 10. The fermentation broth is filtered to remove the mycelial biomass and impurities, and adjusted to a lower pH (*c*. pH 7) in order to crystallize the natamycin, which is then recovered and finally dried. An alternative process that does not use organic solvents involves the disintegration of the fermentation biomass by homogenization, high shear mixing, or ultrasonic techniques or treatment with heat, alkali, or enzymes.

See also: **Legislation**: History; **Mycotoxins**: Classifications; **Preservatives**: Classifications and Properties; Food Uses; Analysis; **Spoilage**: Molds in Spoilage; Yeasts in Spoilage; **Yeasts**

Further Reading

Anonymous (1992) Cheese and cheese rind determination of natamycin content method by molecular absorption spectrometry and by high-performance liquid chromatography. *IDF Standard 140A*. Brussels: International Dairy Federation.

Davidson PM and Doan CH (1993) Natamycin. In: Davidson PM and Branen AL (eds) *Antimicrobials in Foods*, pp. 395–407. New York: Marcel Dekker.

Hamilton-Miller JMT (1973) Chemistry and biology of the polyene macrolide antibiotics. *Bacteriological Reviews* 37: 166–196.

Raab WP (1972) *Natamycin (Pimaricin). Its Properties and Possibilities in Medicine*. Stuttgart: Georg Thieme.

Raghoenath D and Webbers JJP (1997) *Natamycin recovery*. AU patent application no. 1999717229 B2.

Stark J (1999) Permitted preservatives – natamycin. In: Robinson RK, Batt CA and Patel PD (eds) *Encyclopedia of Food Microbiology*, pp. 1776–1781. London: Academic Press.

Struyk AP and Waisvisz JM (1975) *Pimaricin and Process of Producing Same*. US patent 3892850.

Struyk AP, Hoette I, Drost G *et al.* (1957) Pimaricin, a new antifungal antibiotic. *Antibiotics Annual* 878–885.

Nectarines *See* **Peaches and Nectarines**

Neural Tube Defects *See* **Pregnancy**: Maternal Diet, Vitamins, and Neural Tube Defects; **Folic Acid**: Properties and Determination; Physiology

NIACIN

Contents
Properties and Determination
Physiology

Properties and Determination

K J Scott, Formerly of Institute of Food Research, Norwich, UK
P M Finglas, Institute of Food Research, Norwich, UK

Introduction

Niacin (nicotinic acid and nicotinamide) occurs widely in nature. Chemically, nicotinic acid is 3-pyridine carboxylic acid, and nicotinamide is 3-pyridine carboxylic acid amide (**Figure 1**). It should also be noted that in the USA niacin is used as a specific name for nicotinic acid, and niacinamide for nicotinamide.

Nicotinic acid (mol wt 123.11) occurs as colorless (white) nonhygroscopic needles or crystalline powder. It has a melting point of 235.5–236.5 °C and sublimes. It is soluble in water (1.67 g per 100 ml at 25 °C). It is also soluble in ethanol, acids, alkalis, and propylene glycol, but insoluble in ether. It has an absorption maximum at 261 nm, but the intensity is pH-dependent.

Nicotinamide (mol wt 122.11) also occurs as colorless (white) needles or crystalline powder but is slightly hygroscopic. It has a melting point of 128–131 °C and a distilling point of 150–160 °C at 5×10^{-4} mmHg. It is very soluble in water (100 g per 100 ml at c. 25 °C), ethanol (67 g per 100 ml) and glycerol (10 g per 100 ml) but is almost insoluble in ether. Like nicotinic acid, it has an absorption maximum at 261 nm. Both nicotinic acid and the amide are stable in the dry form and in neutral

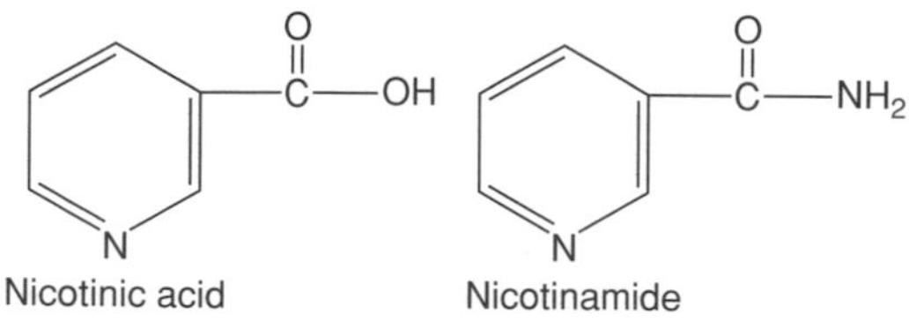

Figure 1 Structural formulae for nicotinic acid and nicotinamide. Reproduced from Niacin: Properties and Determination, *Encyclopaedia of Food Science, Food Technology and Nutrition*, Macrae R, Robinson RK and Salder MJ (eds), 1993, Academic Press.

aqueous solutions. Nicotinamide can be converted to nicotinic acid by treatment with acids or alkalis. Nicotinamide adenine dinucleotide (NAD) and nicotinamide adenine dinucleotide phosphate (NADP) are coenzyme forms. They combine with a wide variety of proteins and catalyze a large number of oxidation and reduction reactions and substrates in living organisms. (*See* **Coenzymes**.)

The human requirement of nicotinic acid or nicotinamide is dependent on the tryptophan intake (60 mg of tryptophan is equivalent to 1 mg of nicotinic acid). Tryptophan is converted via a complex pathway. This pathway from tryptophan to NAD^+ does not directly produce nicotinic acid or nicotinamide; NAD^+ is transformed to nicotinamide, which is further N-methylated, and reenters the pyridine nucleotide cycle or may be deaminated to nicotinic acid, which is further converted to nicotinic acid mononucleotide.

Occurrence

Nicotinic acid and nicotinamide are found in a wide variety of foods: yeast, liver, heart, and muscle meats are good sources (**Table 1**). Because of the tryptophan contribution, milk, milk products, and eggs are also considered to be sources of niacin activity, although their actual content of niacin itself is relatively low. In natural products, niacin is usually in the bound form. Thus, in order to measure the total content it is necessary to include a hydrolytic stage. It is important to realize that this measure of total niacin must not necessarily be equated with what is biologically available. Most nicotinic acid in maize is in the form of nicotinyl esters, which are thought not to be hydrolyzed in the gut. In wheat bran it is bound to polysaccharides, peptides, and glycopeptides. Around 50% of niacin in unenriched wheat and wheat products has been reported to be in a bound form. The treatment of maize with lime water has been reported to liberate bound niacin, making it biologically available to pigs. It has also been claimed that, in areas where corn is routinely steeped in lime water, there is an absence of pellagra compared to areas with an equal consumption of unsteeped corn where pellagra was endemic. However, the various claims for bound forms of

Table 1 Nicotinic acid, potential nicotinic acid from tryptophan, and niacin equivalents in various foods (mg 100 g^{-1})^a

Food	Nicotinic acid	Tryptophan/60	Niacin equivalents
Marmite	58	9	67
Liver			
Lamb (raw)	14.2	4.3	18.5
Lamb (fried)	19.9^b	4.9^b	24.8^b
Ox (raw)	13.4	4.5	17.9
Ox (stewed)	10.3^b	5.5^b	15.8^b
Heart			
Lamb (raw)	6.9	3.6	10.5
Ox (raw)	6.3	4.0	10.3
Ox (stewed)	4.7	6.7	11.4
Kidney			
Lamb (raw)	8.3	3.5	11.8
Lamb (fried)	9.6	5.3	14.9
Ox (raw)	6.0	3.4	9.4
Ox (stewed)	6.2^b	5.5^c	11.7^c
Meats (lean, raw)			
Beef	5.0^b	4.7^b	9.7^b
Lamb	5.4^b	3.9^b	9.3^b
Pork	6.9^b	4.5^b	11.4^b
Chicken	7.8	4.3^b	12.1^b
Milk (whole, average)	0.16^b	0.6^b	0.8^b
Eggs			
Chicken (whole, raw)	0.1	3.7	3.8
Chicken (boiled)	0.1	3.7	3.8
Cheese			
Cheddar	0.1	6.8^b	6.9^b
Danish blue	0.63^b	5.5^b	6.1^b

From: ^{a}McCance RA and Widdowson EM (1991) The Composition of Foods. In: Holland B, Welch AA and Unwin ID, Buss DH, Paul AA, Southgate DAT (eds) *The Composition of Foods*, 5th edn. Cambridge: Royal Society of Chemistry and Ministry of Agricultural, Fisheries and Food; ^{b}Food Standards Agency (2002) *McCance and Widdowson's The Composition of Foods*, Sixth Summary Edition. Cambridge; Royal Society of Chemistry. ^{c}Estimated.

niacin require thorough reinvestigation. (*See* **Bioavailability of Nutrients.**)

Biological Fluids (Blood) and Tissues

Nicotinamide is the primary circulating form of the vitamin. All tissues can incorporate nicotinamide into NAD. In the liver, nicotinic acid is incorporated into NAD, which is broken down into nicotinamide. The primary regulatory substance of the homeostasis in whole animals is nicotinamide itself. The levels in blood are buffered by the liver by being converted into a storage form of NAD.

Determination

Biological Assays

Historically, biological assays with animals such as the chick or the rat have been used as a baseline for other methods and to provide an evaluation of vitamin potency and availability of vitamins. However, biological assays are expensive to carry out and have an inherent lack of precision but, most importantly, are directly relevant only to the species used and

conditions of the assay procedure. Thus, 'animal' assays for the determination of niacin are generally no larger used.

Other Assays

Unlike a biological system, which responds to the bioavailable vitamin in a food material, nonbiological assays are usually geared to measuring the total vitamin content. However, it is possible, using differential hydrolysis in conjunction with the microbiological assay, to make some assessment of the distinction between the 'free' and 'bound' vitamins, although this information must be treated with caution. Hydrolysis to release bound forms is normally carried out with sulfuric acid, although in some methods the use of alkali has been recommended, particularly for cereals.

Chemical assays This assay is based on the formation of colored complexes formed as a result of the reaction between niacin and cyanogen bromide. The reaction is as follows: an α-unsubstituted pyridine forms a pyridinium salt with cyanogen bromide, which reacts with an amine under ring opening to form a derivative of glutaconaldehyde. The color produced can be measured spectrophotometrically. It has been reported that the colored products from the amide are not as stable or intense as from the acid. The amide is usually hydrolyzed to the acid. The Association of Official Analytical Chemists (AOAC) method employs sulfanilic acid as the chromogenic base in the manual and automated methods for nicotinic acid and nicotinamide in drugs, foods, feed, and cereal products, and barbituric acid in the analysis of the amide in multivitamin preparations. Foods and feeds are extracted with acid, and cereal products with calcium hydroxide. Factors which must be controlled for the method to produce reliable results include reaction temperature, pH, and preparation of blank corrections. However, this method lacks specificity, since all 3-pyridoxine derivatives react, and it requires the use of the highly toxic reagent cyanogen bromide. Thus, this method has been largely replaced by microbiological and high-performance liquid chromatography (HPLC) methods. (*See* **Spectroscopy**: Visible Spectroscopy and Colorimetry.)

Microbiological methods Although probably not suited for occasional use because of the specialist facilities and expertise required, the microbiological assay has been widely used for the determination of nicotinic acid in a variety of materials. It is more sensitive and specific than the chemical assay. The microbiological assay is based on the specific nutritional requirements in a defined medium for a particular vitamin(s) by a microorganism. The most widely used organism is *Lactobacillus plantarum* (ATCC 8014). In order to present the vitamin in a biologically active form to the organism the material to be assayed is normally subjected to acid hydrolysis. This procedure also converts any amide into nicotinic acid. This organism responds to nicotinic acid, nicotinamide, and nicotinuric acid (an inactive metabolite), and NAD, but not tryptophan, and it is also able to utilize bound nicotinic acid, present in cereals, to a considerable extent. It is specified in official methods for the determination of total niacin activity in food. The AOAC and AACC extraction procedures involve autoclaving the sample at 121–123 °C for 30 min with 1 mol l^{-1} sulfuric acid.

High-performance liquid chromatography HPLC offers an alternative to the chemical or microbiological assay, although the initial equipment cost and subsequent recurrent costs are relatively high. Analysis is commonly carried out after alkali, acid, or acid/enzyme hydrolysis, by separation on a reversed-phase column and ion pair reagents in the mobile phase and ultraviolet detection. A particular problem in the HPLC analysis of niacin in food

materials is that because of its relatively low ultraviolet absorption, interference from other compounds can make peak identification and quantification difficult. The application of this technique to food products often requires clean-up procedures, like cartridge extractions and column switching. The use of fluorescence detection increases specificity and sensitivity, but requires postcolumn derivatization, because niacin is not natively fluorescent. (*See* **Chromatography**: High-performance Liquid Chromatography.)

Dietary Requirements

As stated earlier, the evaluation of data obtained from the analysis of foods is complicated by the uncertainties surrounding the bioavailability of bound forms of the vitamin and the contribution of tryptophan. These factors must be considered when considering the nutritional implications of niacin intake. (*See* **Dietary Requirements of Adults**.)

The presently accepted conversion rate of tryptophan to niacin is 60:1. Using this conversion rate and the tryptophan requirement to maintain nitrogen balance, it is possible to establish the niacin requirement.

Recommended daily amounts or reference nutrient intake values are based on niacin equivalents. The UK figures are calculated in terms of resting metabolism, that is, 2.7 mg of niacin equivalents per MJ or 11.3 mg per 1000 kcal. The daily level for men is about 17 mg, and about 13 mg for women, and is independent of activity. The recommended level does not fall with increasing age. An additional 2 mg is recommended during lactation. Recommended levels for infants and children are dependent on age, ranging from 3 to 5 mg for infants under 1 year, about 11 mg at 5 years and about 12 mg at 10 years of age. US recommendations are very similar.

See also: **Bioavailability of Nutrients**; **Chromatography**: Thin-layer Chromatography; High-performance Liquid Chromatography; Gas Chromatography; **Coenzymes**; **Dietary Reference Values**; **Dietary Requirements of Adults**; **Spectroscopy**: Visible Spectroscopy and Colorimetry; **Vitamins**: Overview; Determination

Further Reading

AACC (2000) Niacin-microbiological method. AACC methods 86-51 and 86-50A. In: Grami B (ed.) *American Association of Cereal Chemists Methods*, 10th edn. St. Paul, MN, USA: American Association of Cereal Chemists.

AOAC (2002) Niacin and niacinamide. Methods 985.34, 944.13 and 975.41. In: Horwitz W (ed.) *Official Methods of Analysis*, 17th edn. Gaithersburg, MD: AOAC International.

Ball GFM (ed.) (1998) Niacin and tryptophan. In: *Bioavailability and Analysis of Vitamins in Foods*, pp. 319–359. London: Chapman & Hall.

Eitenmiller RR and DeSouza S (1985) Niacin. In: Augustin J, Klein B, Becker D and Venugopal P (eds) *Methods of Vitamin Assay*, 4th edn, pp. 385–398. New York: Wiley.

Lahely S, Bergaentzle M and Hasselmann C (1999) Fluorimetric determination of niacin in foods by high-performance liquid chromatography with post-column derivatization. *Food Chemistry* 65: 129–133.

Report of the Committee on Medical Aspects of Food Policy (1991) *Dietary Reference Values for Food Energy and Nutrients for the UK. DHS Report on Health and Social Subjects*, no. 41. Report of the Committee on Medical Aspects of Food Policy. London: Her Majesty's Stationery Office.

Shibata K and Taguchi H (2000) Nicotinic acid and nicotinamide. In: De Leenheer A, Lambert WE and Van Bocxlaer JF (eds), *Modern Chromatographic Analysis of Vitamins*, 3rd edn. pp. 325–364. New York: Marcel Dekker.

Physiology

D A Bender, University College London, London, UK

Introduction

The vitamin niacin (nicotinic acid and nicotinamide) forms the functional moiety of the nicotinamide nucleotide coenzymes, nicotinamide adenine dinucleotide (NAD) and nicotinamide adenine dinucleotide phosphate (NADP). These act as electron carriers in a wide variety of oxidation and reduction reactions. NAD is also the source of adenosine diphosphate (ADP)-ribose for the ADP-ribosylation of proteins, a mechanism of enzyme regulation, and the poly(ADP-ribosylation) of nucleoproteins controlling the deoxyribonucleic acid (DNA) repair mechanism. Nicotinic acid adenine dinucleotide and cyclic ADP-ribose both have a role in regulation of intracellular calcium concentration in response to hormone action.

Although niacin is generally regarded as a vitamin, it is not strictly a dietary essential, since the nicotinamide moiety of the coenzymes can be synthesized *in vivo* from the essential amino acid tryptophan. Under normal circumstances, the intake of tryptophan is probably adequate to meet niacin requirements without the need for any preformed niacin in the diet.

Metabolism of Niacin

Dietary Forms and Sources

Niacin is present in tissues, and therefore in foods, largely as NAD and NADP. The postmortem hydrolysis of NAD(P) is extremely rapid in animal tissues, so it is likely that much of the niacin of meat (a major dietary source of the vitamin) is free nicotinamide.

Coffee may provide significant amounts of nicotinic acid, formed as a result of the pyrolysis of trigonelline (N-methyl nicotinic acid) during roasting.

In the calculation of niacin intakes, the niacin content of cereals is normally ignored. Although chemical analysis reveals niacin in cereals (largely in the bran), this is mostly biologically unavailable, since it is bound as niacytin–nicotinoyl esters to a variety of polysaccharides and glycopeptides with molecular weights ranging between 1500 and 17 000. Although niacytin is not hydrolyzed by digestive enzymes, a small amount is hydrolyzed nonenzymically by gastric acid, and up to 10% of the niacytin in cereals may be biologically available.

Treatment of cereals with alkali (e.g., soaking overnight in calcium hydroxide solution, the traditional method for the preparation of tortillas in Mexico) or baking with alkaline baking powders releases much of the bound nicotinic acid. Roasting of whole-grain maize has a similar effect, since there is enough ammonia released from glutamine to form free nicotinamide by ammonolysis.

Digestion and Absorption

Nicotinamide nucleotides in the intestinal lumen are not absorbed as such, but undergo hydrolysis to free nicotinamide. A number of intestinal bacteria have nicotinamide deamidase activity, and a proportion of dietary nicotinamide may be deamidated in the intestinal lumen.

Both nicotinic acid and nicotinamide are absorbed from the small intestine by a sodium-dependent saturable process of active transport, although at unphysiologically high concentrations there is also passive diffusion across the intestinal mucosa.

Synthesis of the Nicotinamide Nucleotide Coenzymes

As shown in **Figure 1** NAD(P) can be synthesized from either of the niacin vitamers, or from quinolinic acid, which is a metabolite of the amino acid tryptophan.

In liver there is little utilization of preformed niacin for nucleotide synthesis. The enzymes for nicotinic acid and nicotinamide utilization are more or less saturated with their substrates at normal concentrations in the liver, and hence are unlikely to

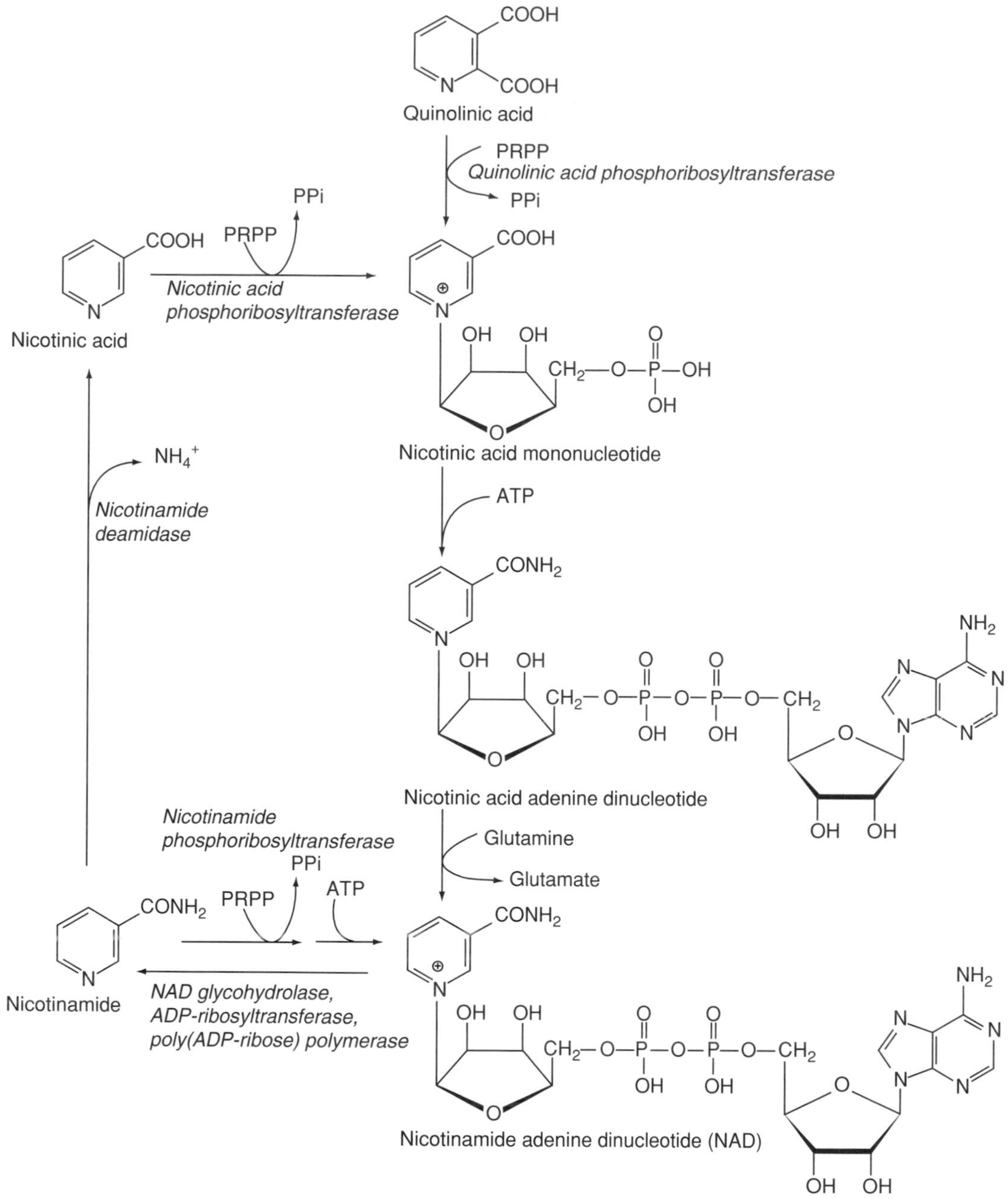

Figure 1 Synthesis of nicotinamide adenine dinucleotide (NAD) from nicotinamide, nicotinic acid, and quinolinic acid. Quinolinate phosphoribosyltransferase EC 2.4.2.19, nicotinic acid phosphoribosyltransferase EC 2.4.2.11, nicotinamide phosphoribosyltransferase EC 2.4.2.12, nicotinamide deamidase EC 3.5.1.19, NAD glycohydrolase EC 3.2.2.5, NAD pyrophosphatase EC 3.6.1.22, ADP-ribosyltransferases EC 2.4.2.31 and 2.4.2.36, poly(ADP-ribose) polymerase EC 2.4.2.30.

be able to use additional niacin for nucleotide synthesis. The liver synthesizes relatively large amounts of NAD(P) from tryptophan, followed by hydrolysis to release nicotinic acid and nicotinamide into the circulation for use by other tissues.

In most extrahepatic tissues, nicotinic acid is a better precursor of nucleotides than is nicotinamide. However, muscle and brain are able to take up

nicotinamide from the blood stream effectively, and apparently utilize it without prior deamidation.

Synthesis of Nicotinamide Nucleotides from Tryptophan

The oxidative pathway of tryptophan metabolism is shown in **Figure 2**. Under normal conditions, almost all of the dietary intake of tryptophan, apart from the

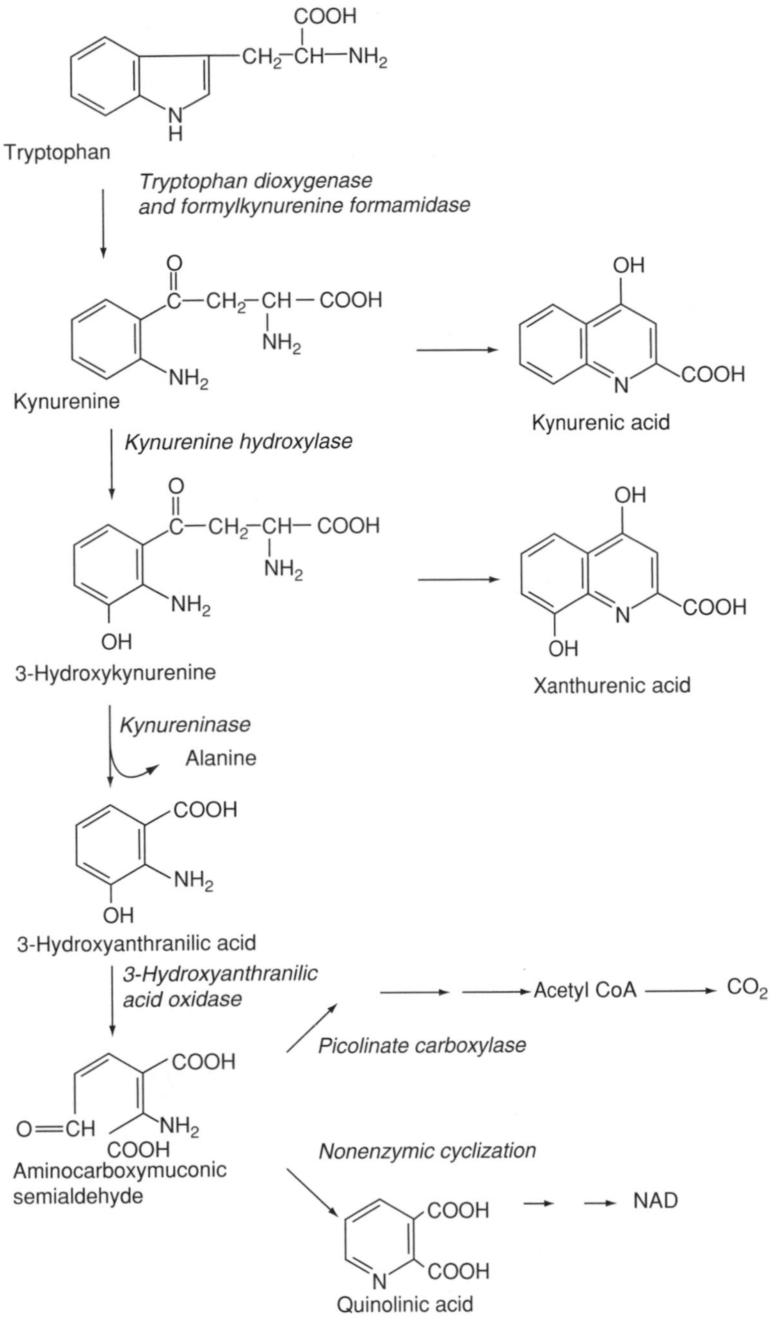

Figure 2 Tryptophan metabolism and the formation of quinolinic acid: tryptophan dioxygenase EC 1.13.11.11, formylkynurenine formamidase EC 3.5.1.9, kynurenine hydroxylase EC 1.14.13.9, kynureninase EC 3.7.1.3, 3-hydroxyanthranilate oxidase EC 1.10.3.5, picolinate carboxylase EC 4.1.1.45.

small amount that is used for net new protein synthesis, is metabolized by this pathway, and hence is potentially available for NAD synthesis. About 1% of tryptophan metabolism is by way of 5-hydroxylation and decarboxylation to form the neurotransmitter 5-hydroxytryptamine (serotonin).

The equivalence of dietary tryptophan and preformed niacin as precursors of the nicotinamide nucleotides has been assessed by determining the excretion of N^1-methylnicotinamide and methylpyridone carboxamide in response to test doses of the precursors, in subjects maintained on controlled diets. There is a considerable variation between subjects. It is generally accepted that in order to allow for individual variation it should be assumed that 60 mg of tryptophan is equivalent to 1 mg of preformed niacin – an overestimate of the average requirement of tryptophan for NAD synthesis. This is the basis for expressing niacin requirements and intake in terms of niacin equivalents – the sum of preformed niacin and 1/60 of the tryptophan.

The synthesis of NAD from tryptophan involves the nonenzymic cyclization of aminocarboxymuconic semialdehyde to quinolinic acid. The alternative metabolic fate of aminocarboxymuconic semialdehyde is decarboxylation, catalyzed by picolinate carboxylase, leading to the formation of acetyl coenzyme A. There is thus competition between an enzyme-catalyzed reaction, which has hyperbolic, saturable kinetics, and a nonenzymic reaction which has linear, first-order kinetics.

At low rates of tryptophan metabolism, most will be by way of the enzyme-catalyzed pathway, leading to oxidation, and there will be little accumulation of aminocarboxymuconic semialdehyde to undergo nonenzymic cyclization. As the rate of formation of aminocarboxymuconic semialdehyde increases, and picolinate carboxylase becomes saturated, there will be an increasing amount available to undergo cyclization to quinolinic acid, and hence onward metabolism to NAD. There is thus not a simple stoichiometric relationship between tryptophan and niacin, and the equivalence of the two coenzyme precursors will vary as the amount of tryptophan to be metabolized and the rate of metabolism vary.

The activities of three enzymes, tryptophan dioxygenase, kynurenine hydroxylase, and kynureninase, may all affect the rate of formation of aminocarboxymuconic semialdehyde, as may the rate of uptake of tryptophan into the liver.

Tryptophan dioxygenase Tryptophan dioxygenase (also known as tryptophan oxygenase or tryptophan pyrrolase) has a short half-life *in vivo* (of the order of 2 h) and is subject to regulation by three mechanisms: stabilization by its heme cofactor, hormonal induction, and feedback inhibition and repression by high concentrations of NADP.

The holoenzyme is more stable than the apoenzyme, and in the presence of relatively large amounts of heme both the activity and the total amount of enzyme protein in the liver are increased. Induction of heme synthesis thus results in increased oxidative metabolism of tryptophan. This is not induction of tryptophan dioxygenase apoenzyme, but the result of reduced catabolism of the enzyme protein.

Tryptophan and a number of tryptophan analogs have a similar effect by promoting conjugation of the apoenzyme with hematin, and thus stabilizing the holoenzyme.

Tryptophan dioxygenase is induced by glucocorticoid hormones and glucagon; the mechanisms involved are different, and the effects are at least partially additive.

Glucocorticoid hormones cause induction of the new messenger ribonucleic acid (mRNA) and protein synthesis, unlike the increase in activity observed in the presence of higher than normal amounts of tryptophan or heme. In response to the administration of the synthetic glucocorticoid, dexamethasone, there is increased transcription of the rat liver tryptophan dioxygenase gene, resulting in a 10-fold increase in tryptophan dioxygenase mRNA in the liver.

Glucagon, mediated by cyclic adenosine monophosphate (cAMP), increases the synthesis of tryptophan dioxygenase following the administration of glucocorticoids, although it has little effect in unstimulated animals. The effect of glucagon appears to be the result of an increase in the rate of translation of mRNA rather than an increase in transcription, and is antagonized by insulin.

Kynurenine hydroxylase and kynureninase The activities of kynurenine hydroxylase and kynureninase are only slightly higher than that of tryptophan dioxygenase under basal conditions. Impairment of the activity of either enzyme may reduce the rate of tryptophan metabolism, and so reduce the accumulation of aminocarboxymuconic semialdehyde, and the synthesis of NAD.

Kynurenine hydroxylase is a flavoprotein, and in riboflavin-deficient rats its activity is only 30–50% of that in control animals. Riboflavin deficiency may thus be a contributory factor in the etiology of pellagra when intakes of tryptophan and niacin are marginal.

In a number of studies, sexually mature women show a higher ratio of urinary kynurenine to hydroxykynurenine than do children, postmenopausal women, or men, suggesting impairment of kynurenine hydroxylase by endogenous estrogens or their metabolites. In experimental animals the administration of estrogens results in a very considerable reduction in kynurenine hydroxylase activity. The mechanism of this effect is unclear, but it is physiologically important; in most areas where pellagra was common, about twice as many women as men were affected.

Kynureninase is a pyridoxal phosphate-dependent enzyme; impairment of its activity in vitamin B_6 deficiency leads to accumulation of kynurenine and hydroxykynurenine, and their transamination products, kynurenic and xanthurenic acids. This is the basis of the tryptophan load test for assessing vitamin B_6

nutritional status. It is also inhibited by estrogen metabolites, and both vitamin B_6 deficiency and inhibition reduce the rate of metabolic flux through the oxidative pathway, hence reducing the formation of quinolinic acid and NAD from tryptophan.

Catabolism of the Nicotinamide Nucleotide Coenzymes

There is no evidence of any specific storage of niacin or the nicotinamide nucleotide coenzymes in the body. Free NAD(P), which is not associated with enzymes, is rapidly hydrolyzed, and the resultant nicotinamide is either used for resynthesis of nucleotides or is methylated and excreted. The catabolism of NAD is catalyzed by four enzymes:

1. NAD glycohydrolase, which catalyzes the hydrolysis of NAD(P)$^+$ at the N-glycoside linkage to yield nicotinamide and either ADP-ribose or ADP-ribose phosphate. This enzyme catalyzes hydrolysis of both NAD$^+$ and NADP$^+$. It is also important in the formation of nicotinic acid adenine dinucleotide and cyclic ADP-ribose
2. NAD pyrophosphatase, which releases nicotinamide mononucleotide. This can either be hydrolyzed by NAD glycohydrolase to release nicotinamide, or be a substrate for nicotinamide mononucleotide pyrophosphorylase, to form NAD
3. ADP-ribosyltransferase(s)
4. Poly(ADP-ribose) polymerase

ADP-ribosyltransferases and poly(ADP-ribose) polymerase normally transfer ADP-ribose on to acceptor proteins (see below), although both are also able to catalyze simple hydrolysis of NAD$^+$ in the absence of an acceptor protein.

Urinary Excretion of Niacin and Metabolites

There is normally little or no urinary excretion of either nicotinamide or nicotinic acid, because both vitamers are actively resorbed from the glomerular filtrate. It is only when the plasma concentration is so high that the resorption mechanism is saturated that significant excretion occurs.

The metabolites of niacin are shown in **Figure 3**, and the use of their urinary excretion as an index of niacin nutritional status in **Table 1**. The principal metabolite of nicotinamide is N^1-methylnicotinamide, which is actively secreted into the urine by the proximal renal tubule. N^1-Methylnicotinamide can also be metabolized further, to yield methylpyridone carboxamide. The extent to which this oxidation occurs, and the relative proportions of the 2-pyridone and 4-pyridone, varies from one species to another and also shows considerable variation between different strains of the same species. Aldehyde oxidase catalyzes the formation of both pyridones, and some additional 2-pyridone arises from the activity of xanthine oxidase.

Nicotinamide can also undergo oxidation to nicotinamide N-oxide. This is normally a minor metabolite in humans, unless large amounts of nicotinamide are ingested. At high levels of nicotinamide intake, some 6-hydroxynicotinamide may also be excreted.

Nicotinic acid can be conjugated with glycine to form nicotinuric acid (nicotinoyl glycine), or may be methylated to trigonelline (N^1-methylnicotinic acid). Small amounts of 6-hydroxynicotinic acid may also be formed.

It is not clear to what extent urinary excretion of trigonelline reflects endogenous methylation of nicotinic acid. There are significant amounts of trigonelline in foods, and some is formed by the intestinal bacterial metabolism of niacytin. Trigonelline is absorbed, but cannot be utilized as a source of niacin, and is excreted unchanged.

Nicotinic acid can give rise to N^1-methylnicotinamide and methyl pyridone carboxamide as a result of its incorporation into NAD(P) and subsequent hydrolysis to release nicotinamide. Similarly, nicotinamide can give rise to nicotinic acid, and hence nicotinuric acid and trigonelline, as a result of deamidation. However, test doses of nicotinamide are excreted mainly as N^1-methylnicotinamide, pyridones, and nicotinamide N-oxide, while doses of nicotinic acid are excreted mainly as nicotinuric acid.

Metabolic Functions of Niacin

Redox Function of NAD(P)

The nicotinamide nucleotide coenzymes are involved as proton and electron carriers in a wide variety of oxidation and reduction reactions. Before their chemical structures were known, NAD and NADP were known as coenzymes I and II respectively. Later, when the chemical nature of the pyridine ring of nicotinamide was discovered, they were called diphosphopyridine nucleotide, or DPN (NAD), and triphosphopyridine nucleotide, or TPN (NADP). These names will still be found in the literature, and the nicotinamide nucleotide coenzymes are sometimes referred to as the pyridine nucleotide coenzymes.

As shown in **Figure 4**, the oxidized coenzymes have a formal positive charge, and are represented as NAD$^+$ and NADP$^+$, while the reduced forms, carrying two electrons and one proton (and associated with an additional proton) are represented as NADH and NADPH. The two-electron reduction of NAD(P)$^+$ proceeds by way of a hydride (H$^-$) ion transfer to carbon-4 of the nicotinamide ring.

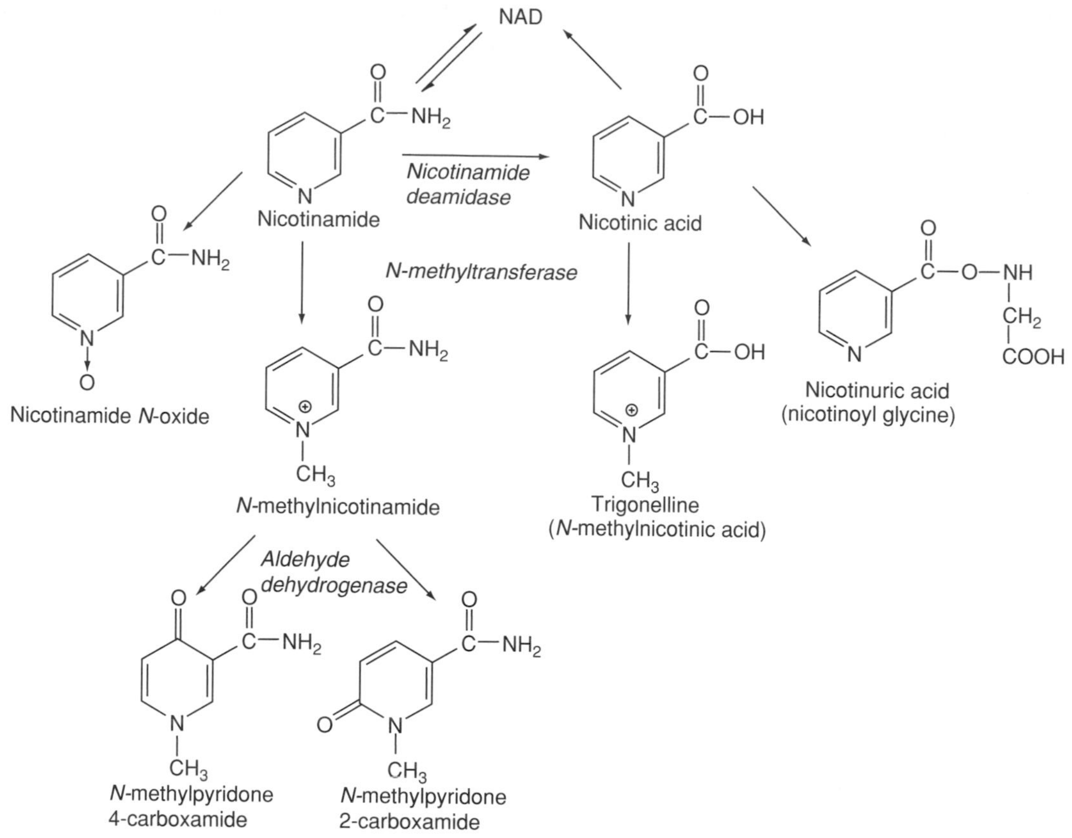

Figure 3 Metabolites of nicotinamide and nicotinic acid.

Table 1 Urinary excretion of niacin metabolites as an index of niacin nutritional status

	Elevated	Adequate	Marginal	Deficient
N^1-methylnicotinamide				
μmol 24 h^{-1}	> 48	17–47	5.8–17	< 5.8
mg g^{-1} creatinine	> 4.4	1.6–4.3	0.5–1.6	< 0.5
mmol mol^{-1} creatinine	> 4.0	1.3–3.9	0.4–1.3	< 0.4
Methylpyridone carboxamide				
μmol 24 h^{-1}		> 18.9	6.4–18.9	< 6.4
mg g^{-1} creatinine		> 4.0	2.0–3.9	< 2.0
mmol mol^{-1} creatinine		> 4.4	0.44–4.3	< 0.44
Ratio of methyl pyridone carboxamide to N^1-methylnicotinamide		1.3–4.0	1.0–1.3	< 1.0

In general, NAD$^+$ is involved as an electron acceptor in energy-yielding metabolism, being oxidized by the mitochondrial electron transport chain, while the major coenzyme for reductive synthetic reactions is NADPH. An exception here is the pentose phosphate pathway (hexose monophosphate shunt), which reduces NADP$^+$ to NADPH, and is the principal metabolic source of reductant for fatty acid synthesis.

Role of NAD in ADP-ribosylation of Proteins

ADP-ribosyltransferases catalyze the transfer of ADP-ribose from NAD$^+$ on to arginine, lysine, or asparagine residues in acceptor proteins, to form N-glycosides. In addition to endogenous ADP-ribosyltransferases, a number of bacterial toxins, including diphtheria and cholera toxins, *Escherichia coli* enterotoxin LT and *Pseudomonas aeruginosa* exotoxin A, also have ADP-ribosyltransferase activity, which is not subject to the same regulation as the enzymes in tissues.

ADP-ribosylation (**Figure 5**) is a reversible modification of proteins, and there are specific hydrolases which cleave the N-glycoside linkage. A variety of guanine nucleotide-binding protein (G protein)

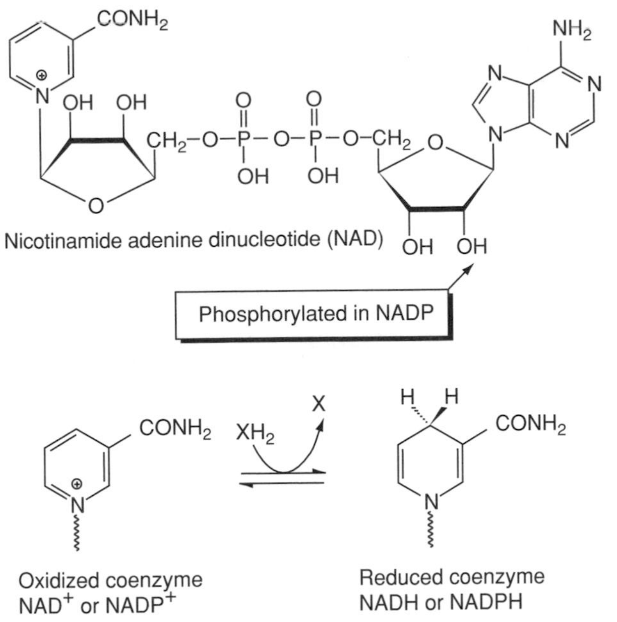

Figure 4 Oxidation and reduction of the nicotinamide coenzymes.

α-subunits involved with the regulation of adenylate cyclase activity are substrates for ADP-ribosylation, either activating the stimulatory G protein or inactivating the inhibitory G protein. The result of ADP-ribosylation by either mechanism is increased adenylate cyclase activity, and hence an increase in intracellular cAMP and the opening of membrane calcium channels. Other proteins that are regulated by ADP-ribosylation include the elongation factor(s) in protein synthesis and integrin and other cytoskeletal proteins.

Poly(ADP-ribose) polymerase (Figure 5) is primarily a nuclear enzyme, which is present in relatively large amounts – up to 1 mol per 1000 base pairs in DNA. It is a zinc finger protein that binds to a nick in DNA caused by strand breakage or excision of an incorrect base. When bound to DNA the enzyme undergoes an autocatalytic poly(ADP-ribosylation), adding branched chains of up to 200 ADP-ribose monomers at each of multiple sites on the enzyme. This poly(ADP-ribosylated) protein then interacts with the α-helical tail of histones, and displacing histones from DNA binding, so leaving the site clear for the DNA repair enzymes to act. There is then slow hydrolysis of the poly(ADP-ribose), permitting the histones to return to bind the repaired DNA.

There is some evidence that poly(ADP-ribose) polymerase may also be involved in DNA replication, since it copurifies with the replication fork and enzymes of DNA replication.

NAD Glycohydrolase and Intracellular Calcium Regulation

A cell membrane-bound NAD glycohydrolase catalyzes a base exchange between NAD^+ and nicotinic acid, yielding nicotinic acid adenine dinucleotide, as shown in Figure 6. The same enzyme also catalyzes exchange of the nicotinamide of NAD^+ with a variety of other compounds, including histamine; there is some evidence that this may provide a mechanism for rapid sequestration, and hence inactivation, of histamine.

The enzyme-bound ADP-ribose that is the intermediate in this base exchange reaction may also undergo either hydrolysis to yield ADP-ribose (so that the enzyme catalyzes overall hydrolysis of NAD^+ to nicotinamide and ADP-ribose), or internal cyclization to release cyclic ADP-ribose.

Both nicotinic acid adenine dinucleotide and cyclic ADP-ribose act to release intracellular stores of calcium. They appear to act as second messengers in transmembrane signal transduction, although the extracellular ligand has not been identified. There is evidence that all-trans-retinoic acid stimulates the cyclase activity of the enzyme, although it is not clear whether this is by a cell surface action or as a result of nuclear actions. In β-islet cells of the pancreas, glucose acts to stimulate the cyclase activity, and the resultant cyclic ADP-ribose causes a calmodulin-dependent mobilization of calcium and increased secretion of insulin.

Requirements and Recommendations

In view of the central role of the nicotinamide nucleotides in energy-yielding metabolism, and the fact that, at least under normal conditions, the nicotinamide released by ADP-ribosyltransferase and poly-(ADP-ribose) polymerase is available to be reutilized for nucleotide synthesis, niacin requirements are usually expressed per unit of energy expenditure.

Depletion and repletion studies suggest that, on the basis of urinary excretion of N^1-methylnicotinamide, the average niacin requirement is 5.5 mg per 1000 kcal (1.3 mg per MJ). Allowing for individual variation, recommended dietary allowances (RDAs) in most countries are set at 6.6 mg niacin equivalents (preformed niacin plus 1/60 of the dietary tryptophan) per 1000 kcal (1.6 mg per MJ). When energy intakes are very low, it is assumed that expenditure will not fall below about 2000 kcal, and this is the basis for the calculation of RDAs for adults with low energy intakes.

There is little requirement for preformed niacin in the diet, since average intakes of protein (at least in developed countries) will provide enough tryptophan

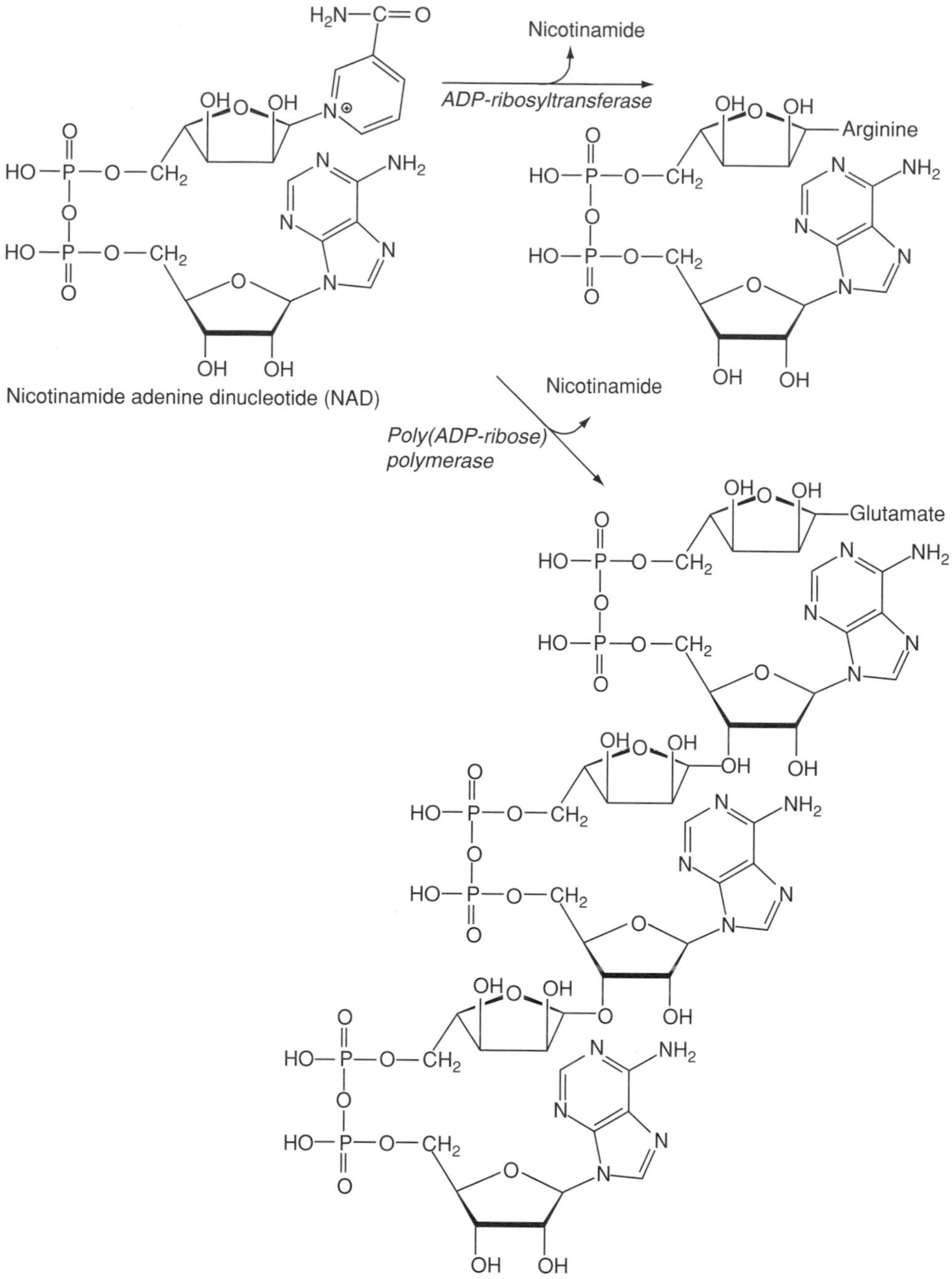

Figure 5 The reactions of adenosine diphosphate (ADP)-ribosyltransferase and poly(ADP-ribose) polymerase.

to meet requirements. Assuming that the diet provides some 15% of energy from protein, and this protein provides 14 g of tryptophan per kg, this implies an intake of 37.5 g of protein (525 mg of tryptophan) per 1000 kcal. Since 60 mg of tryptophan is equivalent to 1 mg of dietary niacin, this suggests that an average diet provides 8.75 mg niacin equivalents per 1000 kcal (2 mg per MJ) from tryptophan alone.

Groups at Risk of Niacin Deficiency (Pellagra)

The tryptophan–niacin deficiency disease pellagra has been a problem in areas of the world where maize or sorghum is the dietary staple, and intakes of both tryptophan and niacin are inadequate. The problem may be compounded by deficiency of riboflavin or vitamin B_6, both of which are required for the synthesis of NAD from tryptophan.

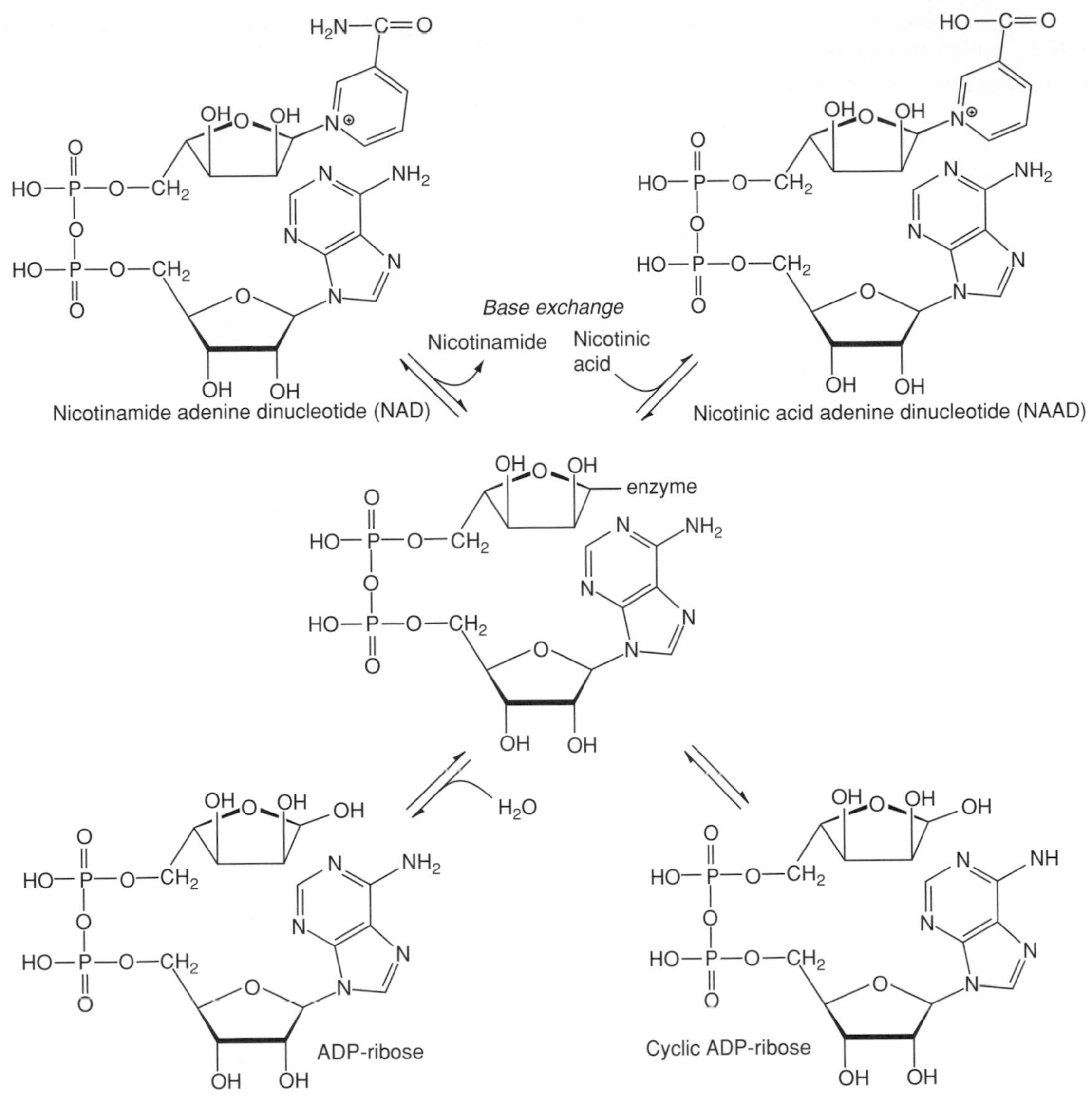

Figure 6 The reactions of nicotinamide adenine dinucleotide (NAD) glycohydrolase and formation of nicotinic acid adenine dinucleotide and cyclic adenosine diphosphate (ADP)-ribose.

A number of mycotoxins and chemotherapy agents used in the treatment of cancer activate poly(ADP-ribose) polymerase; exposure may result in significant depletion of NAD, and hence contribute to the development of pellagra.

Pellagra may arise as a result of drug-induced inhibition of tryptophan metabolism, e.g., by the anti-tuberculosis drug isoniazid (iso-nicotinic acid hydrazide) and the anti-parkinsonian drugs benserazide and carbidopa. Massively increased synthesis of 5-hydroxytryptamine, as occurs in the carcinoid syndrome, results in the development of pellagra, as a result of diversion of tryptophan away from NAD synthesis.

Pellagra also arises as a result of rare inborn errors of tryptophan metabolism or the impairment of tryptophan absorption (Hartnup disease).

Toxicity

Nicotinic acid has been used clinically in large doses (of the order of 1–3 g per day) as a hypolipidemic agent. It reduces both triglycerides and total cholesterol by about 20%, acting as an inhibitor of cholesterol synthesis. It has a more marked effect on cholesterol in low-density and very-low-density lipoproteins, and increases high-density lipoprotein cholesterol.

Nicotinic acid in such doses caused a marked vasodilatation, with flushing, burning, and itching of the skin; after a large dose there may be sufficient vasodilatation to cause hypotension. After the administration of 1–3 g of nicotinic acid daily for several days, the effect wears off to a considerable extent.

At intakes in excess of 1 g of niacin per day there are ultrastructural changes in the liver, and changes in liver function tests, carbohydrate tolerance, and uric acid metabolism, which are reversible on withdrawal of niacin. Sustained-release preparations are associated with more severe liver damage than simple preparations, and may cause clinical liver failure, presumably because they permit more prolonged maintenance of high blood and tissue concentrations of the vitamin.

Supplements of several grams of tryptophan per day have been used with some success in the treatment of depressive diseases, apparently without ill effect. However, a potentially fatal eosinophilia–myalgia syndrome associated with the use of tryptophan supplements has been reported, with more than 1200 cases reported to the US Center for Disease Control in 1989. It seems most likely that the problem was due to a trace contaminant (ethylidene *bis*-tryptophan) in a single batch of tryptophan, rather than toxicity of tryptophan *per se*.

See also: **Cereals**: Dietary Importance; **Coffee**: Analysis of Coffee Products; **Enzymes**: Functions and Characteristics; **Pellagra**; **Vitamins**: Overview; Determination

Further Reading

Bender DA (1983) Biochemistry of tryptophan in health and disease. *Molecular Aspects of Medicine* 6: 101–197.

Bender DA (2003) *Nutritional Biochemistry of the Vitamins*, 2nd edition. New York: Cambridge University Press.

Bender DA and Bender AE (1986) Niacin and tryptophan metabolism: the biochemical basis of niacin requirements and recommendations. *Nutrition Abstracts and Reviews (Series A)* 56: 695–719.

De Murcia G and de Murcia JM (1994) Poly(ADP-ribose) polymerase: a molecular nick-sensor. *Trends in Biochemical Sciences* 19: 172–176.

Desnoyers S, Shah GM, Broche G, Hoflack JC, Verreault A and Poirier GG (1995) Biochemical properties and function of poly(ADP-ribose) glycohydrolase. *Biochimie* 77: 433–438.

Frei B and Richter C (1988) Mono-(ADP-ribosylation) in rat liver mitochondria. *Biochemistry* 27: 529–535.

Galione A and White A (1994) Ca^{2+} release induced by cyclic ADP-ribose. *Trends in Cell Biology* 4: 431–436.

Moss J and Vaughan M (1988) ADP-ribosylation of guanyl nucleotide binding regulatory proteins by bacterial toxins. *Advances in Enzymology* 61: 303–379.

Okamoto H, Takasawa S and Tohgo A (1995) New aspects of the physiological significance of NAD, poly ADP-ribose and cyclic ADP-ribose. *Biochimie* 77: 356–363.

Rose DP (1972) Aspects of tryptophan metabolism in health and disease. *Journal of Clinical Pathology* 25: 17–25.

Ueda K and Hayaishi O (1985) ADP-ribosylation. *Annual Review of Biochemistry* 54: 73–100.

Nicotinic Acid *See* **Niacin**: Properties and Determination; Physiology

NISIN

G C Williams and J Delves-Broughton, Danisco Innovation, Beaminster, Dorset, UK

Introduction

Nisin is a polypeptide bacteriocin that exhibits antibacterial activity against a wide range of Gram-positive bacteria and is particularly effective against bacterial spores. It shows little or no activity against Gram-negative bacteria, yeasts, and molds. Nisin is produced by certain strains of *Lactococcus lactis* subsp. *lactis*. Commercial preparations of nisin are widely used as food preservatives throughout the world. It is recognized as both a toxicologically safe and natural substance.

History

Nisin was discovered in England in 1928 when problems arose in cheese-making. Batches of milk became contaminated with a nisin-producing strain of *L. lactis*, and as a result of nisin's inhibitory properties, the growth of the cheese starter cultures was inhibited. Nisin was subsequently isolated and characterized. Its name was derived from Group N (*Streptococcus*) inhibitory substance. The discovery of nisin predated that of penicillin, so therefore it is not surprising that initial research focused on its potential use for therapeutic purposes in medical and veterinary applications. It was found to be unsuitable for such purposes, mainly because of its limited antimicrobial spectrum in addition to its poor solubility and stability in body fluids. The first interest in its potential as a food preservative came in the 1950s when it was evaluated for the control of gas-producing *Clostridium* spp. in Swiss cheese. However, problems occurred due to the fact that the cheese starter cultures were inhibited, which adversely affected the cheese-ripening process. This problem has now been largely overcome by the development of nisin-resistant cheese starter cultures. The next milestone was its development as a preservative in processed cheese in the mid-1950s. During this period Aplin & Barrett Ltd were experiencing unacceptable spoilage problems due to clostridial growth in processed cheese spreads. This problem was eliminated by the inclusion of nisin-rich curd into the processed cheese mix. With this success the company went on to develop the commercial concentrated preparation of nisin termed Nisaplin. Nisaplin has a nisin content of $25 \, \text{mg g}^{-1}$ and is currently used worldwide as a food preservative. Recently Chr. Hansen from Denmark has introduced a similar product of the same potency, termed Chrisin.

Units of Activity

The first definition was that of a reading unit (later known as an international unit), which was defined as the amount of nisin necessary to inhibit one cell of *Streptococcus agalactiae* in 1 ml of broth. It was so termed in recognition that much of the early work on nisin was carried out at the National Institute for Research in Dairying, Shinfield (Reading), England. Since then an international reference preparation of nisin has been established by the World Health Organization Expert Committee in Biological Standardization. This reference preparation contains, like the commercial preparations Nisaplin and Chrisin, 25 mg (1 million international units) of pure nisin per gram. Throughout this article all levels of nisin are expressed as mg pure nisin per kilogram or liter.

Structure and Biosynthesis

Two natural nisin molecules exist, termed nisin A and nisin Z. The structure of the nisin A molecule was elucidated in 1971 and is presented in **Figure 1**. It is a 34-amino-acid polypeptide with amino and carboxyl endgroups, and five internal ring structures involving disulfide bridges. It possesses three unusual amino acids: dehydroalanine, lanthionine, and β-methyllanthionine. Lanthionine appears to be a common feature in a number of more recently characterized bacteriocins that are collectively known as lantibiotics. Nisin Z differs from nisin A by the substitution of asparagine for histidine at position 27. Nisin Z has a similar antimicrobial activity to nisin A, although nisin Z shows greater diffusion in agar gels. Japanese workers have successfully synthesized the nisin A molecule and confirmed its basic structure. Nisin A has a molecular weight of 3354 Da. There is evidence that nisin can exist as both dimers and tetramers.

A precursor molecule to nisin has been identified and its structural gene isolated and characterized. The

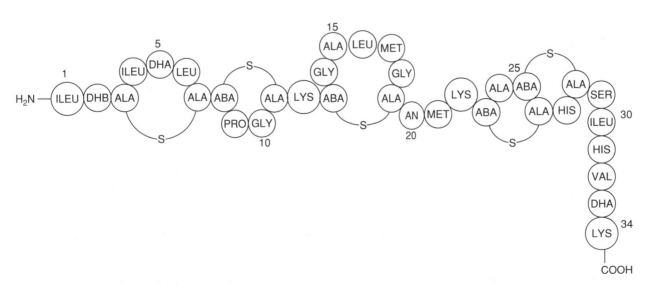

Figure 1　The structure of nisin A. ABA, aminobutyric acid; DHA, dehydroalanine; DHB, dehydrobutyrine (β-methyldehydroalanine); ALA-S-ALA, lanthionine; ABA-S-ALA, β-methyllanthionine.

nisin determinant of *L. lactis* is a component of a large transmissible gene block that also encodes for sucrose metabolism and nisin immunity. The gene block is located in the chromosome as opposed to being in a plasmid. The ability of *L. lactis* to synthesize nisin is a conjugally transmissible property that has been transferred to negative phenotype recipients.

Mode of Action and Antimicrobial Effect

Nisin works in a concentration-dependent manner both in terms of the amount of nisin applied and the number of vegetative cells or spores that need to be inhibited or killed. The primary site of action in sensitive Gram-positive cells is the cytoplasmic membrane where insertion of nisin into the membrane causes disruption of membrane function. An initial step is the docking of nisin on to lipid II, which is followed by the insertion of the nisin II lipid into the membrane where it forms pores. Through these pores, essential cell constituents leak from the inside to the outside of the cell and cause depletion of the proton motive force. Gram-negative bacteria are normally protected from nisin by their outer cell wall acting as an impermeable barrier to the nisin molecule. However, if the outer wall of Gram-negative cells is made permeable to nisin by partial or complete disruption, the cells can then become sensitive to nisin. This indicates that their cytoplasmic membranes are sensitive to nisin. Recent work has demonstrated that yeast protoplasts and spheroblasts are also sensitive to nisin. Gram-negative bacteria can be sensitized to nisin by sublethal heat treatment, freeze–thaw cycling, and exposure to chelating agents that remove divalent ions from their cell walls, thus making them more permeable to nisin. Good effects can be demonstrated against Gram-negative bacteria by the combined use of nisin and a chelating agent such as ethylenediaminetetraacetic acid (EDTA) in a simple buffer system. In food systems, however, such good results are not usually observed as the chelating agent will preferentially bind with the divalent ions that are more readily available in the food matrix.

The action of nisin against spores is predominantly sporostatic as opposed to sporicidal. Nisin affects the postgermination stages of spore development. It inhibits preemergent spore swelling, and thus the outgrowth and formation of vegetative cells. The active sites in spores are thought to be membrane-bound sulfhydryl groups. An important factor in relation to heat-processed foods is that progressive heat damaging of spores results in them becoming increasingly sensitive to nisin.

The nisin sensitivity of both cells and spores can vary between genera and even between strains of the same species. Nisin action against vegetative cells can either be bactericidal or bacteriostatic depending on a number of factors, such as nisin concentration, bacterial population size, physiological state of the bacteria, and the conditions of growth. Bactericidal effects against vegetative cells are enhanced under optimal growth conditions when the bacteria are in an energized state. In contrast, bacteriostatic effects are enhanced when nisin forms part of a multipreservation system in which growth conditions are nonoptimal and other inhibitory factors are exerted (hurdle technology). The fact that different conditions are required to insure either bacterial destruction or inhibition needs to be taken into consideration when applying nisin in a food preservation system. Nisin efficacy in foods is dependent on an effective level of nisin being maintained throughout the whole shelf-life of the food.

Solubility and Stability

Nisin is most soluble in acid substrates and becomes progressively less soluble as the pH increases. Thus at pH 2.2 the solubility is $56\,000\,\text{mg}\,l^{-1}$, at pH 5 it is $3000\,\text{mg}\,l^{-1}$ and at pH 7 it is $1000\,\text{mg}\,l^{-1}$. In practical food preservation situations the level of nisin treatment is unlikely to exceed $50\,\text{mg}\,l^{-1}$ and thus solubility is never a problem. Commercial powder preparations of nisin contain residual solids from the fermentation process that are insoluble. This can produce cloudy suspensions in water but has no detrimental effect on the efficacy of nisin.

Commercial preparations of nisin are remarkably stable. They show no loss of activity over a 2-year period providing they are stored under dry conditions, in the dark, and at temperatures below 25 °C. Nisin solutions are most stable to autoclaving (121 °C for 15 min) in the pH range 3.0–3.5 (< 10% activity loss). Values below and above this pH range result in a marked loss of activity, especially those furthest removed from the range (> 90% activity loss at pH 1 or pH 7 in a buffer system). Losses at typical pasteurization temperatures used in foods are significantly less and food components can protect nisin during heat processing.

The stability of nisin in a food system is dependent on three factors: incubation temperature, length of storage, and pH. In a nonheat-processed food a fourth factor will be the possible presence of protease enzymes. Nisin retention at warm ambient temperatures above 25 °C will be far less than at cooler temperatures. On a practical level this means that higher nisin addition levels are required for the preservation of foods in warmer climates. Retention of nisin activity in an acidic food product will be better than in a more neutral food stored under similar conditions. Nisin can be inactivated by many nonspecific proteases and by any proteolytic enzyme that can cleave the histidine-valine bond (residues 31–32) or the dehydroalanine bond (residues 33–34), e.g., thermolysin, chymotrypsin, subtilisin, ficin, papain, and bromelain. Elastase, pepsin, and leucine aminopeptidase have no action on nisin, and trypsin has a reversible action. A variety of bacteria can produce the enzyme 'nisinase' which specifically inactivates nisin. Some of the bacterial species reported as being able to produce nisinase include *Lactobacillus plantarum*, *Streptococcus thermophilus*, and *Bacillus cereus*.

The food additives, and related sulfiting agents sodium metabisulfite (an antioxidant, bleaching, and broad-spectrum antimicrobial agent) and titanium dioxide (a whitening agent) can also cause nisin degradation.

Methods of Assay

A number of bioassay methods have been devised. These include dye reduction methods using resazurin or methylene blue with a sensitive lactic acid bacterium in a milk-based medium, turbidometric growth measurement assay, horizontal agar plate diffusion assay, the bioluminescent measurement of released adenosine triphosphate (ATP) from *Lactobacillus casei*, and an enzyme-linked immunoabsorbent assay (ELISA). The method in most common use is the horizontal agar plate diffusion assay employing the test organism *Micrococcus luteus*. The lower limit of detection in this assay is about $0.025 \, \mathrm{mg \, l^{-1}}$. A recently described novel technique has been the development of a strain of *L. lactis* that can sense nisin and transduce the signal into bioluminescence with as little as $0.0125 \, \mathrm{ng \, ml^{-1}}$ nisin being detectable by this method.

Quantitative analysis of nisin can also be achieved by high-performance liquid chromatography (HPLC) analysis. A nisin-containing liquid is assayed chromatographically on a hydrophobic (C18) narrow-base HPLC column by gradient elution. Calculations are based on the peak height and quantification done by comparison with a standard nisin preparation. The lower limit of detection by HPLC is around $10 \, \mathrm{mg \, l^{-1}}$.

Toxicology and Legislation

Toxicity studies carried out in laboratory animals with levels of nisin far in excess of those used in foods have shown that nisin is nontoxic and is not carcinogenic. Nisin is rapidly inactivated in the intestine by digestive enzymes and cannot be detected in the saliva of human beings 10 min after consumption of liquid containing $5 \, \mathrm{mg \, l^{-1}}$ nisin. There is no evidence of sensitization (allergy problems) and microbiological studies have not shown any cross-resistance problems that may affect the efficacy of therapeutic antibiotics. It is important that nisin and other bacteriocins are not classified as antibiotics as this could hamper their future acceptance as food preservatives.

In 1969, the Joint Food and Agriculture Organization/World Health Organization (FAO/WHO) Expert Committee on Food Additives reviewed the toxicological data for nisin and recommended its use as a food preservative, with an acceptable daily intake (ADI) of $0.825 \, \mathrm{mg \, kg^{-1}}$ of body weight per day. At present its use is permitted in approximately 60 countries, including the US, the former USSR countries, and China. In the EU it has the food additive number E234.

Preservation of Foods using Nisin

Although nisin use as a food preservative originated with processed cheese products, other application areas have since been identified. Many of these are in products that, by their nature, are pasteurized during production but are not fully sterilized. In such foods Gram-negative bacteria, yeasts, and molds are destroyed by the heat treatment, and the surviving microflora consists of Gram-positive spore-forming bacteria that can be controlled by the use of nisin. The effectiveness of nisin in such products is only complete if postprocessing contamination is eliminated or minimized. Nisin is also used in canned vegetables to control thermophilic spoilage, in non-heated acidic products to control lactic acid bacteria, and in products that are at risk from the psychroduric food-poisoning bacterium, *Listeria monocytogenes*.

Table 1 summarizes the major categories of food in which nisin is used, and the typical spoilage or pathogenic bacteria in these products that are controlled by nisin.

Processed Cheese Products

Processed cheese products cover a wide range, including block cheese (approximately 44–46% moisture), slices (46–50% moisture), spreads (52–60% moisture), and sauces and dips (56–65% moisture). All are heat-processed and contain emulsifying salts. Product innovation in the industry is considerable and formulations can be of low fat or reduced sodium chloride content and may contain various flavor additives such as herbs, fish, shellfish, and meat. All these factors, along with bacterial quality of the raw ingredients, severity of the melt process, filling

Table 1 Typical addition levels of nisin (and Nisaplin) in examples of food applications

Type of food/application	Addition level of nisin ($mg\,kg^{-1}$ or $mg\,l^{-1}$)	Addition level of Nisaplin ($mg\,kg^{-1}$ or $mg\,l^{-1}$)	Typical target organisms
Processed cheese	5–15	200–600	Bacillus spp.
			Clostridium spp.
Pasteurized chilled dairy desserts	1.25–3.75	50–150	Bacillus spp.
Pasteurized milk and milk products	0.25–10.0	10–400	Bacillus spp.
			Clostridium spp.
Pasteurized liquid egg products	1.25–5.0	50–200	Bacillus spp.,
			e.g., B. cereus
Pasteurized chilled soups	2.5–5.0	100–200	Bacillus spp.
Crumpets	3.75–6.25	150–250	Bacillus cereus
Canned foods (low acid)	2.5–5.0	100–200	Bacillus stearothermophilus
			C. thermosaccharolyticum
Canned foods (high acid)	1.25–2.5		C. pasteurianum
			B. macerans
			B. coagulans
Canned lobster	25.0	1000	Listeria monocytogenes
Ricotta cheese	5.0	200	Listeria monocytogenes
Continental-type cooked sausage:			
Added to mix	1.25–6.25	50–250	
Dipping	5.0–25.0	200–1000	Lactic acid bacteria
Salad dressings	1.25–5	50–200	Lactic acid bacteria
Beer:			
Pitching yeast wash	25–37.5	1000–1500	Lactic acid bacteria, e.g.,
			Lactobacillus spp.
During fermentation	0.63–2.5	25–100	Pediococcus spp.
Postfermentation	0.25–1.25	10–50	

temperature, and shelf-life requirement can affect the microbial stability of processed cheese products and hence the requirement for nisin and the level at which it needs to be applied.

The ingredients used in the manufacture of these products are raw cheese, butter, skimmed milk powder, whey powder, phosphate, or citrate emulsifying salts and water. Spores of anaerobic clostridial species are often present in some of these ingredients, particularly the cheese, and are able to survive the heat process of 85–105 °C for 6–10 min commonly used in the heat process. The composition of processed cheese in terms of the relatively high pH (5.6–6.0) and moisture content combined with low redox potential (anaerobic conditions) can result in spore germination and growth, which may result in subsequent spoilage due to production of gas, off-odors, and digestion of the cheese. *Clostridium* spp. often associated with the spoilage of processed cheese are *C. sporogenes*, *C. butyricum*, and *C. tyrobutyricum*. Trials with processed cheese products have been carried out in the UK using a cocktail of spores of the aforementioned *Clostridium* spp. at inoculation levels of approximately 200 spores per gram. Spoilage was prevented during storage at 37 °C by 6.25 mg kg^{-1} nisin. Partial control was achieved with 2.5 mg kg^{-1} whilst control samples that did not contain nisin readily became spoiled.

The potential for growth and toxin production by *C. botulinum* in processed cheese products, particularly spreads, is of considerable significance. Trials in the US have indicated that, in processed cheese spreads, nisin is effective in delaying or preventing the growth and subsequent toxin production by inoculated spores of *C. botulinum* types A and B. These studies indicated that the use of nisin as an effective preservative in processed cheese spreads should form part of a multicomponent food preservation system. Levels of moisture, sodium chloride, phosphate emulsifier salt, and pH are all factors important in determining the necessary level of nisin to provide the required shelf-life. Facultative aerobic *Bacillus* spp. can also cause spoilage problems in processed cheese products and these organisms are also controllable using nisin. Levels used to prevent spoilage are 5–20 mg kg^{-1}, whereas levels used to provide protection against *C. botulinum* are 12.5 mg kg^{-1} and above.

Other Pasteurized Dairy Products

Other pasteurized dairy products, such as dairy desserts, cream, clotted cream, and mascarpone cheese, often cannot be subjected to full sterilization without damaging their organoleptic properties, texture, and/or appearance, and are thus sometimes preserved with nisin to extend their shelf-life. For example, tests on a chocolate dairy dessert resulted in a

20-day increase in shelf-life with 3.75 mg kg^{-1} nisin at $7\,^{\circ}\text{C}$. The same nisin level gave a 30-day increase in shelf-life for a crème caramel dessert stored at $12\,^{\circ}\text{C}$.

The addition of nisin to pasteurized milk is permitted in countries that may experience shelf-life problems due to high ambient temperatures, long-distance transport, and inadequate refrigeration. Trials at Reading University, UK, with nisin added at 1 mg l^{-1} before pasteurization at $72\,^{\circ}\text{C}$ for 15 s, $90\,^{\circ}\text{C}$ for 15 s, or $115\,^{\circ}\text{C}$ for 2 s all resulted in significant shelf-life extension of the milk when stored at $10\,^{\circ}\text{C}$. Similar benefits to shelf-life have also been demonstrated with pasteurized flavored milk.

Pasteurized Liquid Egg Products

Pasteurized liquid egg products (whole, yellow, and white) and value-added egg products (e.g., omelets, scrambled eggs, pancake mixes) receive heat treatments designed to insure the destruction of *Salmonella*. In the UK, for instance, liquid whole egg must be pasteurized for at least 2.5 min at a temperature of $64.4\,^{\circ}\text{C}$. However, such heat treatment is insufficient to kill bacterial spores and the more heat-resistant nonspore-forming Gram-positive bacteria such as *Enterococcus faecalis*. Many of these surviving bacteria are capable of growth at refrigerated temperatures and pasteurized egg products often have a short shelf-life. Application of nisin at levels of $2.5-5 \text{ mg l}^{-1}$ has been shown to act as an effective preservative, giving significant increases in shelf-life and providing protection against the growth of the psychroduric food-poisoning bacteria *B. cereus* and *L. monocytogenes*.

Pasteurized Soups

A recent trend in soup manufacture has been a move towards the production of fresh pasteurized products with a relatively limited chilled shelf-life. Heat-resistant spores of *Bacillus* spp. are able to survive the pasteurization treatment and may be capable of growing and causing spoilage under conditions of chill abuse storage. Trials with nisin at levels of $2.5-5.0 \text{ mg l}^{-1}$ have been found to be very effective at preventing or delaying the outgrowth of psychroduric spoilage *Bacillus* spp. during prolonged shelf-life storage of these pasteurized soup products.

High-moisture Hot Plate Products

Crumpets are high-moisture flour-based products that are popular in the UK, Australia, and New Zealand. Crumpets are produced on a hot plate from a flour batter and contain yeast, an aerating agent or both to give them a raised profile and open texture.

They are toasted before eating. Crumpets have a non-acid pH (pH 6), high moisture (48–54%) and high water activity (0.95–0.97). The product is sold at ambient temperature and has a shelf-life of 5 days. There have been a number of food-poisoning outbreaks due to the growth of *B. cereus* in crumpets, particularly in Australasia. Flour used in the manufacture of crumpets invariably contains low numbers of *B. cereus* spores that are not killed during the hot plate cooking process. During the 3–5-day ambient shelf-life of the product, levels of *B. cereus* can increase from undetectable levels to $> 10^5 \text{ cfu g}^{-1}$ – sufficient bacteria to cause food poisoning. Addition of nisin to the batter mix at 3.75 mg kg^{-1} prevents the growth of *B. cereus* to these potentially dangerous levels. Such use of nisin has received regulatory approval in Australia and New Zealand.

Canned Foods

Nisin is used in canned foods mainly for the control of thermophilic spoilage. It is mandatory in most countries that low-acid canned foods (pH > 4.5) receive a minimum heat process of $F_0 = 3$ to insure the destruction of *C. botulinum* spores, i.e., the minimum botulinum cook. Low-acid foods processed at F_0 of 3 are susceptible to spoilage from surviving heat-resistant spores of thermophilic bacterial species of *B. stearothermophilus* (cause of flat sour spoilage) and *C. thermosaccharolyticum* (cause of can swells). Thus nisin addition can facilitate prolonged storage of canned vegetables at warm ambient temperatures by inhibiting spore outgrowth of these thermophilic spoilage organisms. The use of nisin can also allow a reduction in the F_0 process down to the minimum of 3 without increasing the potential risk of thermophilic spoilage. Other advantages are reduced heat damage to the foods as well as potential saving in energy consumption. Nisin usage levels in low-acid canned vegetables are $2.5-5.0 \text{ mg kg}^{-1}$. Residual nisin levels in canned foods after high temperature processing can be as low as 2% of the addition level. However, the fact that heat-resistant thermophilic spores are highly sensitive to nisin combined with the heat damage enhancing their sensitivity means that extremely low levels of residual nisin can still be effective in this application. Preacidification of the brine with citric acid improves nisin retention with minimal effect on the pH of the vegetables after processing.

Examples of use are canned peas, carrots, peppers, potatoes, mushrooms, okra, baby sweetcorn, and asparagus. Nisin is also used in canned dairy puddings containing semolina and tapioca.

Bacterial spoilage of canned high-acid foods (pH below 4.5) is restricted to nonpathogenic spoilage species such as *C. pasteurianum*, *B. macerans*, and *B. coagulans*. Nisin addition levels of 1.25–2.50 mg kg^{-1} are used in high-acid tomato-based products.

Meat Products

Concern regarding the high levels of nitrite in cured meat has resulted in research investigating the use of nisin as a partial replacement for nitrite. Results indicated that only high (and uneconomic) levels of nisin achieved good control of *C. botulinum*. Further work is necessary before a case for such an application is demonstrated. More encouraging results have been obtained in vacuum-packed cooked continental-type sausages where lactic acid bacteria can cause spoilage by production of gas, off-odors, and slime. Addition of nisin into the sausage mix at levels of 1.25–6.25 mg kg^{-1} or dipping the cooked sausage into nisin solutions of 5.0–25.0 mg l^{-1} has proved effective in increasing shelf-life at storage temperatures of 6–12 °C. Investigations have shown that nisin is more inhibitory against lactic acid bacteria in sausages with lower fat levels. It was also shown that nisin had a greater effect in sausages containing diphosphate compared to those with orthophosphate.

Fish and Shellfish

Relatively few studies have been carried out using nisin in fresh fish mainly because the predominant flora tends to be Gram-negative bacteria. The potential hazard of botulism in both vacuum-packed and modified-atmosphere-packed fish has led to work at the Torry Research Station in the UK on the use of nisin as an antibotulinal agent. Application of nisin by spray to fillets of cod, herring, and smoked mackerel inoculated with *C. botulinum* type E spores resulted in a significant delay in toxin production at 10 and 26 °C. Another problem in smoked fish is the presence and growth of the psychroduric pathogen *L. monocytogenes*, especially in fresh and lightly preserved products. Nisin has been shown to be an effective antilisterial agent in smoked salmon, especially when packed in a carbon dioxide atmosphere.

L. monocytogenes can also be a problem in shellfish, particularly crabs and lobsters, as their meat can only be lightly heat-processed without significant product damage occurring. Nisin, in combination with a reduced heat process, that does not cause product damage of lobster meat, has been shown to achieve a *Listeria* kill that is significantly better than either heat or nisin used alone. An effective nisin level for this application is 25 mg kg^{-1}. Washing crabmeat with nisin has been shown to reduce levels of *L. monocytogenes*.

Salad Dressings

The development of cold blended salad dressings with reduced acidity can improve the flavor of many varieties that are considered to have an over-acid taste. Using reduced levels of acetic acid and raising the pH from 3.8 to 4.2 can make salad dressings prone to lactic acid bacterial spoilage during ambient storage. Such growth has been successfully controlled by the addition of nisin at 2.5–5.0 mg l^{-1}.

Natural Cheese

The first application of nisin was to prevent blowing problems in semihard ripe cheese such as Emmenthal and Gouda due to growth of *C. butyricum* and *C. tyrobutyricum*. Although promising results were obtained, a problem with inhibition of the starter cultures and consequent delay of the ripening process led to the work being discontinued. However, the increase in knowledge of lactic acid bacterial genetics and the need to devise methods for the control of *L. monocytogenes* has resulted in fresh interest into the use of nisin in natural cheese. To achieve success, nisin-resistant starters must be used in conjunction with nisin to insure successful development of the cheese. Natural nisin-producing, nisin-resistant strains tend to lack important properties associated with cheese starter cultures such as good flavor production, eye formation, acidifying characteristics, and bacteriophage resistance. Using the food-grade genetic transfer technique of conjugation it has been possible to develop nisin-producing, nisin-resistant starter cultures with the above-desired properties. Cheeses have been made with sufficient nisin content to provide protection against growth of *Clostridium* spp., *Staphylococcus aureus*, and *Listeria monocytogenes*.

Yogurt

The addition of nisin to stirred yogurt postproduction has an inhibitory effect on the starter culture (a mixture of *Lactobacillus delbrueckii* subsp. *bulgaricus* and *Streptococcus thermophilus* strains), thereby preventing subsequent overacidification of the yogurt. Thus an increase in shelf-life is obtained by maintaining the flavor of the yogurt (less sour) and preventing syneresis. Typical addition levels for this application are 0.5–1.25 mg kg^{-1}.

Alcoholic Beverages

Acid-tolerant lactic acid bacteria of the genera *Lactobacillus*, *Pediococcus*, and *Leuconostoc* can spoil beer and wine due to growth along with production of off-flavors, off-odors, slime, or haze. At levels of 0.25–2.5 mg l^{-1}, nisin is effective in preventing such spoilage. Yeasts are unaffected by nisin, thus the preservative can be added during the fermentation. Nisin can be added to fermenters to prevent or control contamination and can also be used to increase the shelf-life of unpasteurized and bottle-conditioned beers. Furthermore, nisin can be used in the pitching yeast wash as an alternative to acid washing for the control of lactic acid bacteria. Unlike nisin treatment, acid washing can have an adverse effect on yeast viability and performance. Typical levels for this application are $25.0–37.5\,mg\,l^{-1}$. Similar applications occur in the wine industry with the limitation that nisin cannot be used in wines that depend upon a desirable malolactic fermentation as the bacteria responsible are usually nisin-sensitive. This latter problem has been overcome by developing nisin-resistant strains of *Leuconostoc oenos* that can grow and maintain malolactic acid fermentation in the presence of nisin. In the production of distilled spirits, nisin can inhibit the lactic acid bacteria that compete with the yeast for substrate in the fermentation mash, thus resulting in increased alcohol yield in the final distillate. Alcohol yield has been increased by over 10% using this method.

Potential Future Applications

Research on nisin as a food additive continues as the demand increases for convenient long shelf-life and safe food preserved by safe preservatives with a natural connotation. Current research includes the use of nisin in combination with novel food preservation systems such as ultra high pressure, electroporation, pulsed electric fields, and nanothermosonication. It is evident that, like the classical use of heat preservation processes, bacterial spores are more resistant to these novel processes than bacterial vegetative cells and nisin in combination with these processes is synergistic against bacterial cells and surviving bacterial spores. Nisin in combination with other safe food additives that provide synergistic effects against Gram-positive bacteria or widen the antimicrobial spectrum to include Gram-negative bacteria, yeasts, and molds are also objectives of many research teams.

Incorporation of nisin into packaging and edible films may also provide new areas of application. Finally, emerging food spoilage bacteria such as *Alicyclobacillus acidoterrestris* (an acid-tolerant, spore-forming bacteria that can grow at pH 2.5 and is a potential problem in pasteurized fruit juice) and *B. sporothermodurans* (a mesophilic, spore-forming bacteria whose spores can survive ultra heat-treated (UHT) processing) have been shown to be nisin-sensitive. New applications to control such bacteria are currently being realized.

See also: **Alcohol**: Properties and Determination; **Canning**: Principles; **Cheeses**: Chemistry and Microbiology of Maturation; Processed Cheese; *Clostridium*: Occurrence of *Clostridium botulinum*; **Dressings and Mayonnaise**: The Products and Their Manufacture; Chemistry of the Products; **Eggs**: Use in the Food Industry; **Fish**: Processing; *Listeria*: Properties and Occurrence; **Meat**: Sausages and Comminuted Products; **Milk**: Processing of Liquid Milk; **Preservation of Food**; **Shellfish**: Contamination and Spoilage of Molluscs and Crustaceans; **Spoilage**: Bacterial Spoilage

Further Reading

Delves-Broughton J (1990) Nisin and its uses as a food preservative. *Food Technology* 44: 100–117.

Delves-Broughton J and Gasson MJ (1994) Nisin. In: Dillon VM, Board RG (eds) *Natural Antimicrobial Systems and Food Preservation*, p. 99. Wallingford: CAB International.

De Vuyst L and Vandamme EJ (1994) Nisin, a lantibiotic produced by *Lactococcus lactis* subsp. *lactis*: properties, biosynthesis, fermentation and applications. In: De Vuyst, Vandamme EJ (eds) *Bacteriocins of Lactic A Bacteria: Microbiology, Genetics and Applications*, p. 151. London: Blackie Academic & Professional.

Hurst A (1991) Nisin. *Advances in Applied Microbiology* 27: 85–123.

Hurst A and Hoover DG (1993) Nisin. In: Davidson PM, Branen AL (eds) *Antimicrobials in Foods*, p. 369. New York: Marcel Dekker.

Thomas LV, Clarkson MR and Delves-Broughton J (2000) Nisin. In: Naidu AS (ed) *Natural Food Antimicrobial Systems*. CRC Press, USA, pp. 463–524.

NITRATES AND NITRITES

M J Dennis and L A Wilson, Central Science Laboratory, Sand Hutton, York, UK

Background

Nitrate is a normal component of plant tissues, whereas nitrite is not normally found in significant quantities unless microbiological spoilage occurs. Both nitrate and nitrite are permitted food additives for use in curing meat. Their original function was to confer microbiological safety, but they were also found to contribute useful color and flavor characteristics, which, with the advent of refrigeration, are now of greater technological significance. There have been considerable concerns over the safety of nitrate and nitrite in foods. Nitrate is largely unreactive but can be reduced to nitrite, which can then react with secondary amines to form nitrosamines (many of which are carcinogens). Equally, it is becoming apparent that oxides of nitrogen play an important role in human physiology. Hence, in this review, we will cover the occurrence of nitrate and nitrite in food, the legislation governing their use as food additives, and legislation governing their occurrence in vegetables, their human dietary intake, their potential health benefits, and concerns.

Occurrence in Foods

Nitrate is found as a naturally occurring compound in foods such as vegetables, fruit, cereals, fish, milk, and dairy products, and is also found in water as a consequence of agricultural practices such as the use of nitrogen-containing fertilizers and from animal waste. Low levels are generally found from these sources, except in the case of some vegetables. Nitrate and nitrite are also permitted as food additives in some foods, primarily as protection against botulism.

Vegetables and Fruit

Nitrate in the soil is taken up by plants for use as a nitrogen source in the formation of proteins. Protein production occurs as a result of photosynthesis, but when light levels fall, the rate of photosynthesis decreases, and nitrate accumulates in cell fluids and sap. The levels of nitrate in vegetables grown under low light conditions are thus correspondingly higher than those grown under bright light, as shown in **Figure 1**. Overall, nitrate accumulation in plants is determined by genotype, growing conditions,

especially light levels and soil temperature, and nitrogen fertilization.

Nitrate content varies considerably according to species, with vegetables such as spinach and lettuce often containing up to $2500\,mg\,kg^{-1}$, whereas those such as asparagus have levels as low as $13\,mg\,kg^{-1}$. **Table 1** lists the concentrations of nitrate found in some vegetables and fruits, and compares the levels found in different countries. The natural levels of nitrate in vegetables generally are high when compared with other food groups. It is estimated that 75–80% of the total daily intake comes from vegetables, compared with only about 5–10% from drinking water. Cooking has been shown to decrease the concentration of nitrate in foods, dependent on the cooking technique used (see **Table 2**).

Nitrite levels in vegetables and fruit are low, usually below $2\,mg\,kg^{-1}$, except where there has been damage or improper storage leading to the microbiological reduction of nitrate to nitrite.

Meat and Meat Products

Fresh meat normally contains low levels of nitrate and nitrite, estimated at <4–$7\,mg\,kg^{-1}$ and <0.4–$0.5\,mg\,kg^{-1}$, respectively, in the UK in 1997, although higher concentrations of about 10–$30\,mg\,kg^{-1}$ of nitrate are found in cured meats such as ham and salami, where nitrite and nitrate have been incorporated as a permitted additive. Nitrite salts have

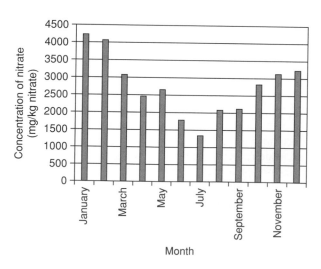

Figure 1 Seasonal variation in nitrate concentrations of lettuce (mg per kilogram of nitrate) (Denmark). From Petersen A and Stoltz S (1999) Nitrate and nitrite in vegetables on the Danish market: content and intake. *Food Additives and Contaminants* 16(7): 291–299, with permission.

Table 1 Comparison of average nitrate levels determined in vegetables and fruits (mg per kilogram of nitrate in fresh product) by different countries

Vegetable/fruit	Australia	Denmark[a]	Italy	UK	USA
Artichoke					16
Asparagus		13			60
Aubergine					370
Bean	265		450		466
Beetroot	2124	1490			1211	3288
Broccoli	310		400		1014
Brussels sprout	44		7–12	59	164
Cabbage	1062	342	240	338	712
Carrot	66		170–210	97	274
Cauliflower	221		37	86	658
Celery	1305				3151
Cucumber					151
Endive					1780
Kale					1096
Leek		308			700
Lettuce	943	2600		1051	2330
Melon					4932
Mushroom					219
Onion	22		80	48	235
Parsley	973				1380
Pea	66		57		40
Pepper (sweet)	88		78		165
Potato (sweet)	44				65
Potato	177	432	110	155	150
Pumpkin	177				550
Radish	1735				2600
Spinach (fresh)		1783	2100	1631	2470
Spinach (frozen)		680			
Tomato	44			17	80
Turnip (greens)					9040
Turnip (root)			970		535

[a]Average for 1993, 1994 and 1995/1996.
Data from Lyons DJ, Rayment GE, Nobbs PE and McCallum LE (1994) Nitrate and nitrite in fresh vegetables from Queensland. *Journal of the Science of Food and Agriculture* 64: 279–281; Petersen A and Stoltze S (1999) Nitrate and nitrite in vegetables on Danish market: content and intake. *Food Additives and Contaminants* 16(7): 291–299; Santamaria P (1997) Contributo degli ortaggi all'assunzione giornaliera di nitrato, nitrito e nitrosamine. *Industrie Alimentari* XXXVI (Novembre): 1329–1333; Ysort G, Miller P, Barrett G, Farrington D, Lawrance P and Harrison N (1999) Dietary exposures to nitrate in the UK. *Food Additives and Contaminants* 16(12): 521–532; Walker R (1990) Nitrates, nitrites and N-nitrosocompounds. A review of the occurrence in food and diet and the toxicological implications. *Food Additives and Contaminants* 7(6): 717–768. All data have been used with the author's permission.

Table 2 Effects of cooking on nitrate concentrations in vegetables (MAFF)

Vegetable	Cooking method	Mean nitrate concentration (mg per kilogram of nitrate)	Percentage change
Beetroot	Fresh	1211	
	Boiled	906	−25
Cabbage	Fresh	338	
	Boiled	114	−66
Carrot	Fresh	97	
	Boiled	71	−27
Cauliflower	Fresh	86	
	Boiled	46	−47
Onion	Fresh	48	
	Fried	44	−8
Potato	Fresh	167	
	Boiled	85	−49
	Fried	136	−19
	Baked	194	+16
Spinach	Fresh	1631	
	Boiled	468	−71
Sprout	Fresh	59	
	Boiled	24	−59
Swede	Fresh	118	
	Boiled	72	−39
Tomato	Fresh	17	
	Fried	27	+59

Data from: Ysart G, Miller P, Barrett G, Farrington D, Lawrance P and Harrison N (1999) Dietary exposures to nitrate in the UK. *Food Additives and Contaminants* 16(12): 521–532. All data used with the author's permission.

customarily been added to a range of meats for preservation purposes, especially as an antimicrobial agent with regard to *Clostridium botulinum*, to produce the pink color of cured meats via the formation of a nitric oxide–myoglobin complex and to give the traditional flavor expected from these products. Nitrite is preferred as a curing agent, as it reacts more quickly than nitrate, and less is required for color stabilization, but nitrate may be added to act as a reservoir in case the nitrite level is depleted during curing. Measures have been taken to reduce the amounts of nitrite used during meat curing processes owing to concern over the formation of nitrosamines. These restrictions have led to significant reductions in the nitrate and nitrite contents of cured meats. Nitrite is now frequently added to meat products at much lower levels, commonly $80–160\ \mathrm{mg\ kg^{-1}}$, in conjunction with sodium ascorbate or sodium erythorbate. It is considered that the benefits of adding nitrite to products of this type far outweigh the potential risks. (*See Clostridium*: Botulism.)

Cheese

Potassium nitrate is permitted as an additive in cheese manufacture in order to control microbiological contamination, which may produce early gassing, and thus to improve the quality of the finished product. Natural concentrations in cheese were found at $1–8\ \mathrm{mg\ kg^{-1}}$ of nitrate, with concentrations rising to $4–27\ \mathrm{mg\ kg^{-1}}$ of nitrate and $< 0.6–1\ \mathrm{mg\ kg^{-1}}$ of nitrite when nitrate was used as an additive.

Fish and Fish Products

Nitrite is permitted in some countries as an additive to some smoked and cured fish and fish products, when it is used as a preservative and color fixative. Levels of both nitrate and nitrite are low in the fresh product, but a recent Japanese survey (1998) analyzed nitrite levels in different salted and processed fish products and found that, although the mean concentrations were about 30–40% of the permitted limit, some products contained nitrite at levels of up to 32 mg kg^{-1}.

Beverages

Beverages make a significant contribution to nitrite intake due to high consumption. Although the concentrations found are low (< 0.4–0.8 mg l^{-1}), the mean nitrite intake in the UK from this source is 0.34 mg per person per day.

Water

Nitrate is ubiquitous in potable water. Its presence is due to a number of factors: significant sources are agricultural activity, for example, the use of nitrogenous fertilizers; sewage effluents and urban run-off and the geology of the area. It is formed when nitrogenous organic sources, such as urea and proteins, or fertilizers containing inorganic nitrogen are decomposed by microorganisms in water or soil to form ammonia, which is then oxidized to nitrate and nitrite. Up to 5 mg l^{-1} of nitrate have been found in rainwater in European urban areas, although levels are somewhat lower in rural areas. However, levels rise as a result of human activities, and groundwater with concentrations of up to 1500 mg l^{-1} have been found in an agricultural area in India. Nitrate is very soluble in water and easily moves through the soil into aquifers and into the drinking water supply. Nitrite is readily oxidized to nitrate and is consequently found in water at much lower levels, especially after water treatment involving chlorination. Water obtained from public water companies tends to contain lower levels of nitrate, usually less than 10 mg l^{-1} in most countries, but private wells throughout the world, especially in agricultural areas, may have concentrations often exceeding 50 mg l^{-1}. However, the exposure to water from the latter source is less common; for example, it is estimated that only about 2% of the European population draw their water from private wells.

Legislation

Maximum permitted levels of nitrate in potable water have been set at levels intended to prevent the occurrence of methaemoglobinemia in infants and other susceptible sections of the population. The second edition of the WHO Guidelines for drinking-water quality recommended a guideline value of 50 mg l^{-1} for nitrate and 3 mg l^{-1} for nitrite levels for water. The US Federal Government has set a similar maximum contaminant level of 45 mg l^{-1}.

Concern over the *in vivo* formation of nitrosamines from nitrites has led to a reduction in the levels permitted when they are used as food additives. In the UK, the current residual concentration of nitrite (as sodium nitrite) in cured bacon is limited to 175 mg kg^{-1} and 100 mg kg^{-1} for other cured meats. The maximum residual concentration of nitrate (as sodium nitrate) in all cured meats is 250 mg kg^{-1}. The use of potassium nitrate in cheese is limited to that giving a residual concentration of 50 mg kg^{-1}. Regulations in other countries are similar: Australia and New Zealand set a maximum of 125 mg kg^{-1} of sodium nitrite for cured meats and 50 mg kg^{-1} of total nitrate/nitrite in cheeses such as Gouda, Edam, and Havarti.

In Europe, Commission Regulation No. 194/97 sets the maximum levels for certain contaminants in foodstuffs, including nitrate in lettuce and spinach. By the use of specific measures intended to provide better control of the sources of nitrate and good agricultural practices, the levels in spinach should not exceed 2500 mg kg^{-1} (summer crop) and 3000 mg kg^{-1} (winter crop) and in lettuce, levels should not exceed 3500 mg kg^{-1} (summer crop except open-grown lettuce where the limit is 2500 mg kg^{-1} from May to September) and 4500 mg kg^{-1} (winter crop).

Dietary Intake of Nitrate and Nitrite

The acceptable daily intake (ADI) is defined as the maximum amount of a chemical that can be ingested daily over a lifetime with no appreciable health risk, and is based on the highest intake that does not give rise to observable adverse effects. The European Commission's Scientific Committee for Food set the ADI for nitrate at 0–3.65 mg per kilogram of body-weight (equivalent to an intake of 219 mg day^{-1} for a 60-kg person) and 0–0.06 mg per kilogram of body-weight (equivalent to an intake of 3.6 mg day^{-1} for a 60-kg person) for nitrite. These ADIs are similar to those set by the joint FAO/WHO experts committee. The ADIs apply to all sources of intake, although the use of nitrite as an additive in baby food intended for infants below the age of 3 months is not permitted. The nitrite ADI does not apply to infants under 3 months.

Dietary exposure to food components may be assessed by the use of several approaches, including the use of market basket studies, food diary records, *per capita* estimations using consumption data with the concentrations of nitrate and nitrite in foods, and urinary biomarkers. Intake studies have shown that the ADIs are not exceeded by the general population across the world, as shown in **Table 3**. The question of whether some subgroups of the population are consuming high levels of nitrate was addressed by the UK's Food Standards Agency, who carried out a study into the nitrate intake of vegetarians. This

Table 3 Comparison of daily dietary exposures to nitrate in different countries

Country	Dietary exposure (mg day^{-1})
UK	52
Finland	77
The Netherlands	52
The Netherlands	78–91
Basque Country, Spain	60
Belgium	154 (vegetables)
Egypt	296 (nitrate and nitrite)
USA	300
Poland	85 (duplicate hospital diets)
Poland	65 (estimate from hospital diets)
India	78
EU	18–131 (vegetables)

From Ysart G, Miller P, Barrett G et al. (1999) Dietary exposures to nitrate in the UK. *Food Additives and Contaminants* 16(12): 521–532.

showed that, although there was a slightly higher intake, dietary exposure was similar to other consumers and was well below the ADI.

In summary, food is the main source of nitrate, especially potatoes, green vegetables, and other vegetables. Water also makes a significant contribution. Estimates of nitrite intake suggest that 1–3 mg day^{-1} is typical and occurs mainly from beverages, due to high consumption.

Microbiological Nitrate Reduction

A detailed investigation of the reduction of nitrate to nitrite by microbiological contaminants of the brewing process has been undertaken. Concerns over the presence of nitrosamines in beer have led to detailed studies of the mechanism of formation of these compounds in order to prevent their formation as far as technologically feasible. Nitrosodimethylamine was formed in malt from nitrogen oxides present in kiln gases. Concentrations of this contaminant were greatly reduced by modifications to kilning conditions. Apparent total N-nitroso compounds (ATNC) provide an estimate of the total N-nitroso compounds present in a sample. The brewing industry sought to minimize ATNC and set a target of 20 µg (N-NO) kg^{-1}. It was established that the major source of nitrosating agent was from microbial reduction of nitrate. Hence, care in ensuring that microbiological contamination was minimized played an important role in meeting this target.

Microorganisms can be differentiated into assimilatory and dissimilatory nitrate reducers. The former reduce nitrate to nitrite and ammonia for use as their source of cell nitrogen. The latter use nitrate as a terminal electron acceptor for energy production (oxygen is the usual terminal electron acceptor under aerobic growth conditions). Nitrite, nitric oxide, nitrous oxide, and sometimes nitrogen may all be produced by dissimilatory nitrate reduction. A number of different bacteria and yeasts are capable of nitrate reduction. The effect is not merely to produce nitrite that subsequently undergoes the well-known acid-catalyzed chemical nitrosation reactions. Rather, the nitrosation reactions appear to be catalyzed by the nitrate and nitrite reductase enzymes. This mechanism enables nitrosation to occur at pH 6–9 which cannot happen in the chemical reaction at alkaline pH.

However, the significance of microbiological reduction of nitrate is wider than that caused by the microbiological contamination of food. Nitrate-reducing bacteria can be found in the mouth and the colon. Nitrate from the diet (or that produced metabolically) is transferred from the blood to the saliva in the salivary glands. It has been suggested that 25% of orally ingested nitrate may be secreted in saliva. Twenty per cent of this nitrate can then be reduced by oral bacteria to nitrite, which is swallowed. Hence, it is apparent that saliva makes the major contribution to dietary nitrite intake. Saliva nitrite levels rise at night and are higher in older people, especially men. Salivary nitrate and nitrite vary little from day to day, but vary more over a 5-year period.

Bacterial nitrosation may occur at any site in the body where there is infection, and these may be linked to cancers at these sites. However, the colon provides a site rich in microorganisms at all times. It is apparent from measurements of fecal ATNC that bacterial nitrosation reactions occur at this site. ATNC formation is influenced by the nature of the organisms present, the transit time of the fecal mass through the colon and to consumption of nitrate and (surprisingly) red meat. Since dietary nitrate is absorbed by the small intestine, it is difficult to appreciate how the nitrosation reaction may occur. It is possible that there may be diffusion of nitrate from the colonic epithelium. Alternatively, red meat might influence fecal ATNC by carrying nitric oxide into the colon as nitrosyl myoglobin, by favoring the growth of nitrate-reducing bacteria in the colon, or possibly by nitrosyl myoglobin acting as a nitrosating agent itself. These possibilities require further study.

Concerns over Nitrate and Health

A considerable body of evidence has been collected concerning the direct effects on health of the nitrate and nitrite ion. There is no evidence of carcinogenicity from nitrate consumption in experimental animals or through human epidemiological studies. Long-term animal studies on nitrite do not indicate carcinogenicity unless combined with amine intake leading to the formation of carcinogenic nitrosamines. The latter studies usually involved nitrite concentrations far higher than those commonly encountered in the diet. Nitrite has demonstrated mutagenic activity in *in vitro* assays. There is considerable evidence of the safety of substantially higher exposures to nitrite during clinical use than those normally encountered from the diet.

The most important direct toxic effect of nitrite (or nitrate after its reduction) is the condition known as methemoglobinemia. Absorbed nitrite ion is rapidly oxidized to nitrate by reaction with oxyhemoglobin to produce methemoglobin. The latter can not carry oxygen around the body, and although the reaction can be reversed enzymatically, if large quantities of methemoglobin are produced, fatality can result. Infants are deficient in this repair enzyme and hence are especially at risk if they receive unusually high intakes of nitrate, for example from contaminated well water. This may be exacerbated by concurrent infections resulting in endogenous nitrate synthesis.

Another concern is that consumption of nitrite (or nitrate reduced by oral bacteria to nitrite) may result in the nitrosation of amines to produce carcinogenic nitrosamines in the acid conditions of the stomach. Such an effect has been found in a human study after consuming a meal consisting of vegetables high in nitrate and fish (to provide a source of amines). Similar studies in animals have been shown to induce cancer. Such nitrosation reactions may be enhanced by bacterial nitrosation in the achlorhydric stomach.

Volatile nitrosamines are not normally found in urine (unless a urinary infection is present) or in feces. However, feces do contain ATNC, although the toxicological significance of this measurement is not understood. The formation of ATNC appears to require an active microbiological flora in the colon capable of nitrate reduction. Hence, ATNC would appear to be, at least, a biomarker of nitrosation reactions occurring in the colon. It is possible for the colonic bacteria to make a more subtle contribution to colon cancer than just the production of potential carcinogens. For example, nitrate-reducing enteric fermentative bacteria lead to the production of more acetate and less butyrate from substrate. Butyrate is considered to have a protective effect against cancer, since it promotes apoptosis (programmed cell death), so any selection for bacteria producing acetate may favor colonic epithelial cell proliferation.

There are a number of medical conditions other than cancer that have been associated with nitrate/nitrite intake. In Finland, a significant correlation has been found between the occurrence of type 1 diabetes and the nitrite intake of both the children and their mothers. No association was found for nitrate intake.

Since most nitrite intake arises from saliva, it seems likely that the association with nitrite is in fact a proxy for preformed dietary nitroso compounds, some of which (e.g., *N*-nitrosomethylurea) have well established toxic effects on pancreatic beta cells. Studies in Iceland and Sweden have also found similar correlations.

An association has been observed between maternal periconceptional exposure to nitrate and an increased risk of anencephalopathy (but not spina bifida). The lack of a quantitative association between nitrate intake and risk may indicate that something other than nitrate is the causative factor. It should be noted that not all studies have been able to demonstrate such a link, but nitrosocompounds have been shown to cause defects in the central nervous system of experimental animals.

Nitrate acts as a competitive inhibitor of iodine uptake by the thyroid and thyroid hypertrophy has been demonstrated experimentally in rats given drinking water containing large amounts of nitrate. There is some epidemiological evidence that populations exposed to high nitrate concentrations in drinking water show an average increase in thyroid volume, although it cannot be excluded that the effects are due to other compounds.

Potential Health Benefits of Nitrate

Nitric oxide plays a crucial role in human metabolism. It is synthesized from arginine by two related enzyme systems – one constitutive and one inducible. The constitutive form provides continuous vasodilation and inhibits platelet adhesion and aggregation; hence, it can help to prevent hypertension. The inducible isozyme is generated in response to bacterial lipopolysaccharide and hence is considered to play an important role in prevention of bacterial disease. However, NO production is also greatly increased in inflammatory bowel disease (e.g., Crohn's disease). NO synthesis also occurs in the central nervous system, but its physiological role in this tissue remains unexplained. These mechanisms for nitric oxide generation mean that man excretes more nitrate than is ingested.

Dietary nitrate is readily absorbed, transported in plasma, and concentrated 10-fold in the saliva by the salivary gland. Although nitrate is excreted in the urine, the kidney recovers some 80% of the nitrate. Hence, this renal salvage and salivary concentration suggest that nitrate is physiologically important. Nitric oxide has widespread antimicrobial activity, and hence, it is hypothesized that the salivary concentration of nitrate and its microbial conversion to nitrite are a symbiotic activity to generate quantities of nitric oxide in the stomach for protection against pathogens. Inhibition of gastric NO synthesis does not have clinical relevance, but this could be due to the high microbiological standards of contemporary diets. Nitrate is also expressed in sweat where significant amounts of nitrite and bacterially generated nitric oxide can also be found and where a protective effect against pathogens has also been postulated.

Gastric juices provide a rich source of nitrosatable substrates, which, some consider, far exceeds those provided by food. Hence, since nitrate can be endogenously formed, carried in the blood, and excreted in saliva, it would appear unavoidable that some gastric nitrosation takes place. However, much more gastric nitric oxide is produced than might be expected from acid-catalyzed dissociation of nitrite to nitric oxide and nitrogen dioxide. Thus, it would appear that another reductant (e.g., ascorbic acid) is produced in gastric juice in order to facilitate the generation of NO. Therefore, it may be that man is metabolically adapted to the possibility of gastric nitrosation. This view is supported by the failure of a number of epidemiological studies to demonstrate a link between nitrate intake and cancer in man.

See also: **Curing**; *Escherichia coli*: Occurrence and Epidemiology of Species other than *Escherichia coli*; **Legislation**: Contaminants and Adulterants; **Meat**: Preservation; **Nitrosamines**; **Smoked Foods**: Principles; Applications of Smoking; **Water Supplies**: Chemical Analysis

Further Reading

Benjamin N (2000) Nitrates in the human diet – good or bad? *Annales De Zootechnie* 49: 207–216.

Commission of the European Communities Commission Regulation (EC) No. 466/2001 (2001) Setting maximum levels for certain contaminants in foodstuffs. *Official Journal of the European Communities* No. L77: 1–13.

Croen LA, Todoroff K and Shaw GM (2001) Maternal exposure to nitrate from drinking water and diet and risk neural tube defects. *American Journal of Epidemiology* 153(4): 325–331.

Hill MJ (1996) Factors controlling endogenous N-nitrosation. *European Journal of Cancer Prevention* 5(supplement 1): 71–74.

Hughes R, Cross AJ, Pollock JRA and Bingham S (2001) Dose-dependent effect of dietary meat on endogenous colonic N-nitrosation. *Carcinogenesis* 22(1): 199–202.

Hunt J and Turner MK (1994) A survey of nitrite concentrations in retail fresh vegetables. *Food Additives and Contaminants* 11(3): 327–332.

Ishiwata H, Nishijima M, Fukasawa Y, Ito Y and Yamada T (1998) Evaluation of the inorganic food additive (nitrite, nitrate and sulfur dioxide) content of foods and estimation of daily intake based on the results of Official Inspection in Japan in the fiscal year 1994. *Journal of the Food Hygiene Society of Japan* 39(2): 78–88.

Lyons DJ, Rayment GE, Nobbs PE and McCallum LE (1994) Nitrate and nitrite in fresh vegetables from Queensland. *Journal of the Science of Food and Agriculture* 64: 279–281.

Ministry of Agriculture, Fisheries and Food (1998a) MAFF UK – Survey of nitrite and nitrate in bacon and cured meat products. *Food Surveillance Information Sheet No. 142*.

Ministry of Agriculture, Fisheries and Food (1998b) MAFF UK – 1997 Total Diet Study – Nitrate and nitrite. *Food Surveillance Information Sheet No. 163*.

Ministry of Agriculture, Fisheries and Food (1998c) MAFF UK – Duplicate diet study of vegetarians – nitrate analysis. *Food Surveillance Information Sheet No. 165*.

Mirvish SS, Reimers KJ, Kutler B *et al.* (2000) Nitrate and nitrite concentrations in human saliva for men and women at different ages and times of the day and their consistency over time. *European Journal of Cancer Prevention* 9: 335–342.

Parham NJ and Gibson GR (2000) Microbes involved in dissimilatory nitrate reduction in the human large intestine. *FEMS Microbiology Ecology* 31: 21–28.

Perner A and Rask-Madsen J (1999) Review article: the potential role of nitric oxide in chronic inflammatory bowel disorders. *Alimentary Pharmacology and Therapeutics* 13: 135–144.

Petersen A and Stoltze S (1999) Nitrate and nitrite in vegetables on the Danish market: content and intake. *Food Additives and Contaminants* 16(7): 291–299.

Rowland IR, Granli T, Bockman OC, Key PE and Massey RC (1991) Endogenous N-nitrosation in man assessed by measurement of apparent total N-nitroso compounds in faeces. *Carcinogenesis* 12(8): 1395–1401.

Santamaria P (1997) Contributo degli ortaggi all'assunzione giornaliera di nitrato, nitrito e nitrosamine. *Industrie Alimentari* XXXVI (Novembre): 1329–1334.

Smith NA (1994) Cambridge Prize Lecture: Nitrate reduction and N-nitrosation in brewing. *Journal of the Institute of Brewing* 100: 347–355.

van Maanen JMS, van Dijk A, Mulder K *et al.* (1994) Consumption of drinking water with high nitrate levels causes hypertrophy of the thyroid. *Toxicology Letters* 72: 365–374.

Virtanen SM, Jaakkola L, Rasanen L *et al.* (1994) Nitrate and nitrite intake and the risk for type 1 diabetes in Finnish children. *Diabetic Medicine* 11: 656–662.

Walker R (1990) Nitrates, nitrites and N-nitrosocompounds: a review of the occurrence in food and diet and the toxicological implications. *Food Additives and Contaminants* 7(6): 717–768.

World Health Organization (1998) Guidelines for drinking water quality, 2nd edn. *Addendum to vol. 2 Health criteria and other supporting information*, pp. 64–80.

Ysart G, Miller P, Barrett G, Farrington D, Lawrance P and Harrison N (1999) Dietary exposures to nitrate in the UK. *Food Additives and Contaminants* 16(12): 521–532.

NITROSAMINES

R A Scanlan, Oregon State University, Corvallis, OR, USA

Introduction

Investigations during the past 30 years on the occurrence of N-nitrosamines and the etiology of these compounds in human cancer began with two important discoveries. In 1956, the UK scientists David Barnes and Peter Magee reported N-nitrosodimethylamine (NDMA) as a carcinogen in experimental animals. In Norway in the early 1960s, NDMA proved to be the causative agent in the death of agricultural animals which had been fed nitrite-preserved fish meal. The nitrite had reacted with amines in the fish meal to produce lethal amounts of NDMA. Since nitrite is added to human foods, particularly cured meats, and amines occur commonly in most foods, this raised the question as to whether carcinogenic nitrosamines might also form in human foods. This article summarizes the results of the many investigations which have addressed this important issue.

Chemistry of N-Nitrosamine Formation

Basic Reactions

N-Nitrosamines are formed by a chemical reaction between a nitrosating agent and a secondary or a tertiary amine. Primary amines react with nitrosating agents to form unstable N-nitroso derivatives which degrade to olefins and alcohols. The structures of

several nitrosamines which occur in foods are shown in **Figure 1**.

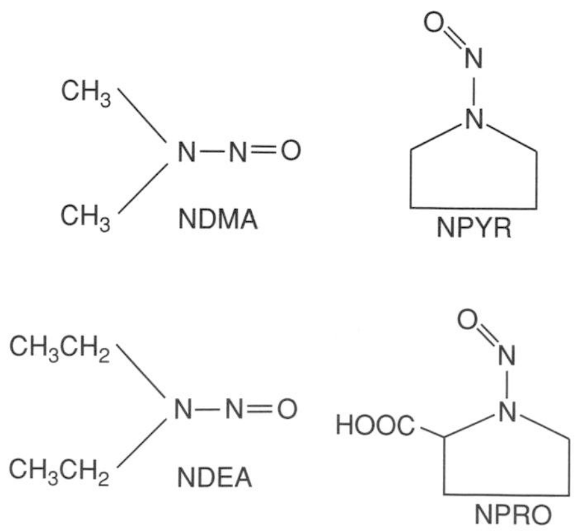

Oxides of nitrogen (NO_x) in which the nitrogen is in a $+3$ or $+4$ oxidation state can serve as nitrosating agents. A much studied nitrosating agent which participates in nitrosamine formation in foods is nitrous anhydride (N_2O_3). Nitrous anhydride forms readily from nitrite in aqueous acidic solution, as shown in eqn (1).

$$2NO_2^- + 2H^+ \rightleftharpoons N_2O_3 + H_2O \qquad (1)$$

Nitrous anhydride combines with the unshared pair of electrons on unpronated secondary amines through a nucleophilic substitution reaction to form N-nitrosamines, as depicted in eqn (2).

$$R_2NH + N_2O_3 \rightarrow R_2N-N=O + HNO_2 \qquad (2)$$

The rate of nitrosation is pH-dependent and is governed by the concentrations of amine and nitrite, as shown in eqn (3):

$$\text{Rate} = k[\text{amine}] \times [\text{nitrite}]^2 \qquad (3)$$

Since the nitrous anhydride reacts with unprotonated amine, the rate of nitrosation for secondary amines is inversely proportional to amine basicity. The pH optimum for nitrosation of most secondary amines is between 2.5 and 3.5. This is due to the counteracting effects of acidity on nitrous anhydride concentration, which increases at low pH, and the concentration of unprotonated amine, which increases at high pH.

The pH optimum has a number of implications. Although most foods are less acidic than pH 2.5–3.5, many foods are sufficiently acidic to allow nitrosation, albeit at rates slower than maximal. Furthermore, the 2.5–3.5 pH range is sufficiently close to the acidity of the human stomach to allow nitrosamine formation, providing that amines and nitrosating agents are present.

Tertiary amines react with nitrosating agents in acidic aqueous solution through a mechanism called nitrosative dealkylation to form nitrosamines. Nitrosative dealkylation involves conversion of a tertiary amine to a secondary amine, which subsequently reacts with nitrosating agent to form a nitrosamine.

Nitrosation Inhibitors

Nitrosation can be influenced by a wide range of catalysts and inhibitors. Ascorbic acid, α-tocopheral, and sulfur dioxide are used to inhibit nitrosamine formation in foods. Ascorbic acid reduces nitrous anhydride to nitric oxide, as shown in eqn (4).

Figure 1 Structure of some nitrosamines which occur in foods. NDMA, N-nitrosodimethylamine; NPYR, N-nitrosopyrrolidine; NDEA, N-nitrosodiethylamine; NPRO, N-nitrosoproline

Ascorbic acid + N_2O_3 → Dehydroascorbic acid
$$+ 2NO + H_2O \qquad (4)$$

Ascorbic acid is not completely effective in blocking nitrosation, since under oxidative conditions the nitric oxide can be reconverted to nitrous anhydride. Presumably α-tocopheral reacts with nitrous anhydride through a similar redox process to inhibit nitrosation. Both ascorbic acid and α-tocopheral are added to cured meats, and sulfur dioxide is used in processing barley malt to inhibit nitrosation.

Carcinogenicity

Although N-nitroso compounds have been shown to be acutely toxic, mutagenic, and teratogenic, their carcinogenic properties are of grave concern and have been studied extensively. Over 300 N-nitroso compounds, including many nitrosamines, have been demonstrated to be carcinogens in experimental animals. More than 50 animal species, including higher primates, are susceptible to N-nitroso compound-induced carcinogenesis. It has been observed that tumors induced by nitrosamines in experimental animals show similar morphological properties to tumors in corresponding human organs. Few scientists doubt that nitrosamines are capable of inducing cancer in humans. However, the amount of a given nitrosamine which is required to induce cancer in a human population is unknown. (See Cancer: Carcinogens in the Food Chain.)

Analytical Methodology

For purposes of analysis, nitrosamines are classified as either volatile or nonvolatile. Volatile nitrosamines are relatively nonpolar, low-molecular-weight compounds, such as NDMA and N-nitrosopyrrolidine (NPYR). They possess sufficient vapor pressure to allow removal from a food matrix by distillation and subsequent separation by gas chromatography. Chemiluminescent detectors designed to be relatively specific for nitrosamines allow quantitative analysis of volatile nitrosamines in foods and other biological materials at submicrogram per kilogram levels. Confirmation of nitrosamine identity is accomplished by mass spectrometry. (See Chromatography: Gas Chromatography; Mass Spectrometry: Principles and Instrumentation.)

Nonvolatile nitrosamines such as N-nitrosoproline (NPRO) tend to be of higher molecular weight, or more polar than volatile nitrosamines, and hence possess relatively low vapor pressures. These properties have impeded development of analytical methodology. Although procedures are available for determination of some nonvolatile nitrosamines, mainly N-nitrosated amino acids and amino acid derivatives, analytical methodology generally is less well developed for this class of compounds.

A procedure usually referred to as apparent total N-nitroso compound determination has been developed and used for analyses of foods and body fluids. Presumably when using this procedure all N-nitroso compounds in the sample are measured as a single entity. A limitation is that information regarding the identity of individual N-nitroso compounds is not provided by this methology. It is therefore impossible to access the carcinogenicity of compounds detected by this procedure.

Formation and Occurrence

Exogenous Formation

Foods During the past 25 years many foods have been analyzed for volatile nitrosamines. In general, foods in most western diets have been examined more extensively than foods from Asia, Africa, and South America. Although approximately 20 volatile nitrosamines have been identified in a variety of foods and beverages, NDMA and NPYR have been found most commonly. Most of the other volatile nitrosamines which occur in foods do so infrequently. Table 1 presents a condensation of many reports of the occurrence of volatile nitrosamines in foods from around the world. The summary is not comprehensive; rather, it is intended to show the foods in which nitrosamines occur most commonly.

Nitrosamines form in foods because under certain circumstances the precursors, amines and nitrosating agents, occur in foods. Amines occur commonly in food, and are formed biosynthetically and by

Table 1 Volatile nitrosamines which occur most commonly in foods

Food	Nitrosamine	Range[a] ($\mu g\,kg^{-1}$)	Occurrence
Fried bacon	NDMA, NPYR	1–100	Consistent
Cured meats	NDMA, NPYR, NPIP	ND–50	Sporadic
Beer	NDMA	ND–5	Sporadic
Cheese	NDMA	ND–5	Sporadic
Cooked fish	NDMA, NDEA	ND–50	Sporadic
Salt-dried fish	NDMA, NDEA	ND–1000	
Nonfat dry milk	NDMA	ND–1	Consistent in direct-fire dried product

[a]The range is intended to encompass data in most reports since 1980. In general, most positive samples contained nitrosamines at the lower end of the range. ND, not detectable.
NDMA, N-nitrosodimethylamine; NPYR, N-nitrosopyrrolidine; NDEA, N-nitrosodiethylamine; NPIP, N-nitrosopiperidine.

microbial activity. Nitrosating agents can be formed from certain compounds added to foods and as a result of specific processing conditions. In most instances, nitrosamines are found in foods in one of the three following categories.

Cured meats Nitrosamines are formed in cured meats because nitrite, and sometimes nitrate, are added to these products during processing. Nitrate is reduced to nitrite by the enzyme nitrate reductase, which occurs in a number of bacteria. As discussed earlier, nitrite is converted to nitrosating agents which subsequently react with amines in the meat during processing, storage, and cooking to form nitrosamines. (*See* **Curing.**)

Nitrite and nitrate have been added to cured meats for many years to prevent outgrowth and toxin formation by *Clostridium botulinum*. Nitrite, in combination with other curing ingredients such as sodium chloride, is particularly effective in inhibiting formation of the deadly botulism toxin. In addition, nitrite reacts with pigments in meat to impart the desirable pink color of cured meats and it prevents the development of off-flavors. (*See* **Nitrates and Nitrites.**)

Although not all cured meats have detectable amounts of nitrosamines, fried bacon has been shown consistently to contain these compounds. Typically, fried bacon contains $1-20\,\mu g\,kg^{-1}$ of NPYR and $1-3\,\mu g\,kg^{-1}$ of NDMA. The formation is related to the relatively high internal temperature of bacon during frying and the relatively low moisture content of bacon as compared to other cured meat products. When bacon is cooked by other methods, particularly in a microwave oven, considerably lower amounts of nitrosamines are found. The majority of evidence suggests that the free amino acid proline is first nitrosated and then decarboxylated to form NPYR during frying. Neither the precise chemical nature of the nitrosating agent nor the amine precursor for NDMA in bacon is known with certainty. However, the evidence suggests that the nitrosating agent is a reaction product of nitrite and lipids in the bacon.

Dried foods and ingredients A variety of processes and equipment are used to dry foods and food ingredients. During the direct-fire process, air used to dry the food is first heated by passing the air through the flames of burners. As a result, products of combustion, including oxides of nitrogen, are directly incorporated into the hot air used to dry the food. The oxides of nitrogen which include nitrosating agents such as nitrous anhydride can then react with amines in the food being dried to produce nitrosamines.

In 1979, scientists in Europe reported the occurrence of NDMA in beer. Soon after, reports from a number of countries confirmed that most beers contained $1-5\,\mu g\,kg^{-1}$ of NDMA. Although beer was found to contain NDMA, the NDMA was not formed during the brewing process. Investigation led to the discovery that direct-fire-dried malted barley, an ingredient used in the manufacture of beer, was the source of the NDMA.

A number of investigators have attempted to identify the amine precursors of NDMA in malted barley. The evidence suggests that dimethylamine and perhaps the tertiary amine alkaloids gramine and hordenine serve as precursors for NDMA in malted barley.

The discovery of nitrosamine formation in direct-fire-dried malted barley has led to the investigation of other dried foods. In the USA, certain dried dairy products, notably nonfat dry milk, are manufactured by the direct-fire drying process. Nonfat dry milk manufactured by this process consistently contains low amounts (less than $1\,\mu g\,kg^{-1}$) of NDMA. Since the direct-fire drying process is not used commonly in Europe for manufacture of nonfat dry milk, the product does not contain NDMA.

Seafood can contain nitrosamines as a result of either cooking or salt-drying. When fish is broiled with a gas flame, the oxides of nitrogen produced in the flame can cause nitrosamine formation in a manner analogous to nitrosamine formation in direct-fire drying. In certain areas of the world, particularly the Orient, fish is preserved by salt. Often sea salt is used which contains appreciable amounts of nitrate. The nitrate is reduced to nitrite with subsequent formation of nitrosating agents. In some cases, reaction with amines in the fish produces relatively large amounts of nitrosamines, principally NDMA. The potential for NDMA formation in fish is considerable since high levels of NDMA precursors, such as dimethylamine and trimethylamine, can occur in fish.

Migration from surfaces which contact foods Vulcanized rubber products, such as baby nursing nipples, have been shown to contain nitrosamines. It has been demonstrated that when baby bottle nipples are stored inverted in milk, the nitrosamines partially migrate from the nipples to the milk. Furthermore, studies have shown that nitrosamines migrate from rubber netting used to hold cured meats during the smoking process. In addition to rubber, nitrosamines in such substances as wax-treated wrapping paper and paperboard-based materials have been shown to migrate to foods.

Nonvolatile nitrosamines in foods The notion that nonvolatile nitrosamines might form in foods follows logically from the fact that the precursors, nonvolatile amine and nitrosating agents, occur in foods. Due to limitations in analytical methodology for nonvolatile nitrosamines, less is known about the occurrence of nonvolatile than volatile nitrosamines in foods. Recently developed methods for nonvolatile nitrosamines have been limited to the detection of N-nitrosated amino acids and amino acid derivatives. Most analyses have been on cured meats and include reports of compounds such as NPRO, N-nitrososarcosine (NSAR), N-nitroso-4-hydroxyproline, N-nitrosothiazolidine-4-carboxylic acid, N-nitroso-2-methylthiazolidine-4-carboxylic acid, and N-nitroso-2-hydroxyl-methylthiazolidine-4-carboxylic acid. Other than NSAR, which is a weak carcinogen, all the other N-nitrosated amino acids and amino acid derivatives which have been tested have not shown a carcinogenic response in animals.

Scientists conjecture that other nonvolatile nitrosamines such as nitrosated peptides and nitrosated amides occur in foods. Definitive information in this regard awaits further application of analytical methodology for nonvolatile nitrosamines to foods.

Reduction of nitrosamine formation Considerable efforts have been expended during the past 25 years to reduce nitrosamine formation in foods. Scientists have looked critically at the use of nitrite and nitrate in the manufacture of cured meats. Since nitrite is considered essential to guard against outgrowth and toxin production by C. botulinum, its use in the manufacture of cured meats has been retained. However, in many countries, permissible levels of nitrite have been reduced to the minimum necessary for control of botulism. Generally, nitrate is only permitted in a few fermented cured meat products where long-term inhibition of C. botulinum is required.

The nitrosation inhibitors ascorbic acid and α-tocopheral are either required or extensively used in processing cured meats. Alternatives and/or partial substitutes for nitrite in cured meats include use of lactic acid-producing organisms, potassium sorbate, sodium hypophosphite, fumarate esters, and ionizing radiation. To date, no adequate alternative for nitrite has been found. Consequently, nitrosamines continue to be found in cured meats, especially fried bacon, but generally at lower levels than occurred a number of years ago.

Efforts to reduce nitrosamine levels in barley malt, and hence in beer, have been very successful. This is largely due to conversion of direct-fired kilns to indirect-fired kilns for the manufacture of barley malt. With indirect-fired kilns, the products of combustion are not incorporated into the drying air and, therefore, nitrosamine formation in the barley malt is

greatly reduced. Interestingly, use of an indirect-fired kiln does not completely inhibit nitrosation, probably because ambient air which is drawn into the kiln usually contains trace levels of oxides of nitrogen and, therefore, some nitrosation occurs. In addition to the use of indirect-fired kilns, sulfur dioxide, which has been shown to be a nitrosation inhibitor, is sometimes used with both direct-fired kilns and indirect-fired kilns to reduce nitrosamine formation.

Due to these changes in processing, the NDMA levels in beer have been markedly reduced. In a recent survey, Canadian and US beers were found to contain on average $0.07\,\mu g\,kg^{-1}$ of NDMA whereas, prior to 1980, beer commonly contained $1-5\,\mu g\,kg^{-1}$ of NDMA.

Similarly, nitrosamine formation in rubber products, including baby nursing nipples, has been reduced. This has been accomplished by substituting nitrosatable with nonnitrosatable vulcanization accelerator compounds in the manufacture of rubber products.

Products other than food A variety of industrial, agricultural, and consumer items have been shown to contain nitrosamines. These include cosmetics, tobacco products, industrial cutting fluids, vehicle tires, and pesticides. In each case, the formation can be traced to use of amines and contact with nitrosating agents during product formulation, manufacture, or use.

Tobacco products are especially important in terms of nitrosamine occurrence and subsequent human exposure. During growth, tobacco plants biosynthesize a variety of alkaloids such as nicotine and nornicotine. These compounds are either secondary or tertiary amines. During curing, fermentation, and aging the alkaloids react with nitrosating agents formed from nitrate to produce a group of compounds commonly referred to as tobacco-specific nitrosamines. Tobacco-specific nitrosamines have been shown to occur in a wide range of tobacco products such as cigarettes, cigars, pipe tobacco, chewing tobacco, snuff, masheri, zarda, and nass. It is noteworthy that the tobacco-specific nitrosamines frequently occur in tobacco products at several orders of magnitude higher than volatile nitrosamine occurrence in foods and beverages.

Endogenous Formation

Convincing evidence exists for endogenous formation of nitrosamines in humans. Based on our knowledge of nitrosation in acidic, aqueous media, it is not surprising that nitrosamine formation has been demonstrated to occur in the stomach. Estimation of endogenous nitrosation in humans and experimental animals has been accomplished by measuring urinary

NPRO following oral administration of proline and nitrate. Inhibition of nitrosamine formation has been accomplished by the administration of nitrosation inhibitors such as ascorbic acid. Gastric nitrosation proceeds through reaction between amines and nitrosating agents derived from the diet. Nitrate, which occurs in substantial amounts in certain vegetables, is partially reduced to nitrite by nitrate reductase-containing bacteria in the oral cavity.

Recently, the existence of endogenous nitrosation pathways other than gastric nitrosation has been recognized. Evidence exists for mammalian cellular enzymatic conversion of arginine to nitric oxide, which in turn affects nitrosation. Scientists believe that this process occurs in several cell types, including macrophages. The extent and relevance of cellular nitrosation are yet to be determined.

Estimates of Exposure to Nitrosamines from Foods

Several groups have made estimates of human exposure to volatile nitrosamines from food and beverage consumption. Most of the estimates relate to exposure from consumption of foods and beverages in western Europe. Volatile nitrosamine exposure was predominately from NDMA and NPYR.

The estimation of daily exposure to volatile nitrosamines in most reports ranged from 0.1 to 1.0 µg per person. In comparison, a US National Academy of Sciences report in 1981 estimated exposure to nitrosamines from cigarette smoking to be 17 µg per person per day.

Several recent reviews reported an estimated daily exposure to nitrosamines of 10–120 µg per person from diet. Estimates in this range include, in addition to volatile nitrosamines, N-nitrosated amino acids and amino acid derivatives and some include apparent total N-nitroso compounds as well. It should be recognized that most N-nitrosated amino acids and amino acid derivates have failed to elicit a carcinogenic response in animals and the identity and carcinogenicity of the apparent total N-nitroso compounds are unknown.

Some cautionary comments are needed regarding estimates of exposure to nitrosamines. First, estimates should be based on nitrosamines with known identity and known carcinogenicity in animals. Second, a number of the estimates were conducted a decade ago or are based on reports of the volatile nitrosamine content of foods from over a decade ago. As discussed previously, the volatile nitrosamine content of foods, e.g., NDMA in beer, has been dramatically reduced in recent years. Therefore, some of the estimates in the literature may be higher than reflected by current levels of volatile nitrosamines. Third, information currently available does not allow reliable estimation of exposure from currently undetected carcinogenic nonvolatile nitrosamines, of nitrosamines formed endogenously, and of other N-nitroso compounds formed exogenously and endogenously. A fourth limitation of estimates is that they are based on average food consumption for relatively large populations. Food consumption patterns for subgroups and individuals within populations vary widely and, therefore, so does exposure to nitrosamines in the diet.

Role of Nitrosamines and Other N-Nitroso Compounds in Human Cancer

Based on current information on carcinogenicity in experimental animals, and on other pertinent information, most scientists believe that nitrosamines and other N-nitroso compounds are capable of inducing cancer in humans. However, at the present time we do not know what exposure to the various nitrosamines is required to induce cancer in a human population.

Convincing evidence exists for the role of tobacco-specific nitrosamines in cancer induction in people who use tobacco products. Particularly compelling evidence exists for the causative role of tobacco-specific nitrosamines in cancer of the oral cavity for people who engage in snuff dipping and chewing betel quid.

In order to assess more fully the part nitrosamines and other N-nitroso compounds play in human cancer, progress will be needed in the following areas: improvement and application of analytical methodology will be required for estimation of exposure to nonvolatile nitrosamines and other N-nitroso compounds in foods and in other materials; second, a better understanding of formation and exposure to endogenously formed nitrosamines and other N-nitroso compounds is needed; and, finally, methodology will need to be improved in order to allow reliable estimation of human cancer incidence from exposure to relatively low levels of carcinogens, including nitrosamines.

See also: **Amines**; **Cancer**: Carcinogens in the Food Chain; **Carcinogens**: Carcinogenic Substances in Food: Mechanisms; Carcinogenicity Tests; *Clostridium*: Occurrence of *Clostridium botulinum*; Botulism; **Curing**; **Drying**: Equipment Used in Drying Foods; **Nitrates and Nitrites**; **Smoking, Diet, and Health**

Further Reading

Forman D and Shuker D (eds) (1989) *Cancer Surveys, Advances and Prospects in Clinical, Epidemiological and Laboratory Oncology. Nitrate, Nitrite and*

Nitrosocompounds in Human Cancer, vol. 8, no. 2. Oxford: Oxford University Press (published for the Imperial Cancer Research Fund).

Gloria MBA, Barbour JF and Scanlan RA (1997) N-Nitrosodimethylamine in Brazilian, US domestic, and US imported beers. *Journal of Agriculture and Food Chemistry* 45: 814–816.

Gloria MBA, Barbour JF and Scanlan RA (1997) Volatile nitrosamines in fried bacon. *Journal of Agriculture and Food Chemistry* 45: 1816–1818.

Hill MJ and Giacosa A (eds) (1996) Proceedings of the thirteenth annual ECP symposium N-Nitroso compounds in human cancer: current status and future trends. *European Journal of Cancer Prevention* 5 (supplement 1): 1–163.

Hotchkiss JH (1989) Relative exposure to nitrite, nitrate and N-nitroso compounds from endogenous and exogenous sources. In: Taylor SL and Scanlan RA (eds)

Food Toxicology, a Perspective on the Relative Risks. IFT Basic Symposium Series. New York: Marcel Dekker.

O'Neill IK, Chen J and Bartsch H (eds) (1991) *Relevance to Human Cancer of N-Nitroso Compounds, Tobacco Smoke and Mycotoxins*. IARC Scientific Publication, no. 105. Lyon: International Agency for Research on Cancer.

Scanlan RA and Tannenbaum SR (eds) (1981) *N-Nitroso Compounds*. ACS Symposium Series, no. 174. Washington, DC: American Chemical Society.

Tricker AR (1997) N-nitroso compounds and man: sources of exposure, endogenous formation and occurrence in body fluids. *European Journal of Cancer Prevention* 6: 226–268.

Tricker AR and Preussmann R (1991) Carcinogenic N-nitrosamines in the diet: occurrence, formation, mechanisms and carcinogenic potential. *Mutation Research* 259: 277–289.

NMR Spectroscopy *See* **Spectroscopy**: Overview; Infrared and Raman; Near-infrared; Fluorescence; Atomic Emission and Absorption; Nuclear Magnetic Resonance; Visible Spectroscopy and Colorimetry

---

Nonstarch Polysaccharides *See* **Dietary Fiber**: Properties and Sources; Determination; Physiological Effects; Effects of Fiber on Absorption; Bran; Energy Value

---

NUCLEIC ACIDS

Contents
Properties and Determination
Physiology

Properties and Determination

D W Gruenwedel, University of California at Davis, Davis, CA, USA

This article is reproduced from *Encyclopaedia of Food Science, Food Technology and Nutrition*, Copyright 1993, Academic Press.

Presence of Nucleic Acids in Food

Nucleic acids are natural constituents of all foods derived from animal or plant sources, for they are intrinsic components of the cells making up food tissue. They occur usually in close association with basic proteins (histones, protamines), forming nucleoproteins. In general, their effect on texture, nutritive value, or sensory properties of foods is small in view of their low tissue concentrations (typically < 200 mg of nucleic acid phosphorus per 100 g of fresh tissue). These low tissue concentrations normally also preclude the occurrence of detrimental health effects. However, health consequences may arise with persons suffering from metabolic disorders such as impairment of purine excretion. In these instances, the dietary breakdown of the nucleic acids can generate unacceptably high levels of uric acid in blood as well as tissues, producing ultimately hyperuricemia, acute and chronic arthritis, or other symptoms of gout.

Properties of Nucleic Acids

Chemical Properties

Nucleic acids are macromolecules of high molecular weight, made up of heterocyclic pyrimidine (Py) and purine (Pu) bases, a five-membered (furanose-type) sugar ring, and orthophosphate. Bases, sugar and phosphate occur in equimolar ratios, i.e., Pu(Py): sugar: phosphate of 1:1:1. Thus, schematically, their chemical composition may be denoted as

$$[Pu(Py) - pentose - PO_4^-]n, \tag{1}$$

with n being a large number. Among naturally occurring nucleic acids, n may range from 4.5×10^3 (polyoma virus) to 7.0×10^7 (*Drosophila melanogaster* chromosome). Two broad categories exist, deoxyribonucleic acids (DNAs) and ribonucleic acids (RNAs), depending on whether the sugar moiety is β-D-2'-deoxyribose or β-D-ribose (note: primed entities refer to sugar-binding sites). Their elemental composition is roughly 15% nitrogen, 10% phosphorus, 36% carbon, 4% hydrogen, and 35% oxygen. Cellular DNA is localized almost exclusively in the nucleus, while RNA is found predominantly in the cytoplasm.

The principal bases found associated with DNAs are the purine derivatives adenine (6-aminopurine, Ade) and guanine (2-amino-6-hydroxypurine, Gua), and the pyrimidine derivatives cytosine (2-hydroxy-4-amino-pyrimidine, Cyt) and thymine (5-methyluracil, Thy). Occasional base variations (e.g., 5-methylcytosine or 5-hydroxymethylcytosine) are encountered. In RNAs, the base Thy is replaced by uracil (2,4-dihydroxypyrimidine, Ura). Certain RNAs, e.g., (transfer) tRNAs, show considerable base variations: they may contain a wide variety of methylated bases, including Thy, bases with thio groups, inosine, etc.

The Pu bases are bound to the sugar in an $N9 \rightarrow C1'$-glycosidic linkage; with Py bases, bonding occurs via $N1 \rightarrow C1'$. The resulting compounds are N-glycosides; they are called nucleosides. Some nucleosides are: adenosine (Ado or A), guanosine (Guo or G), cytidine (Cyd or C), uridine (Urd or U), and thymidine (dThd or dT). Attachment of orthophosphate in positions O3' (or O5') produces the corresponding 3' (or 5') (mono)nucleotides. 5'-Mononucleotides are the actual building blocks of DNAs and RNAs. Schematically, their polymerization takes place by $5' \rightarrow 3'$ phosphodiester bond formation, e.g., the 5'-phosphate group of one nucleotide links up with the free 3'-hydroxyl group of another, thereby forming a dinucleotide. It is customary to use short-hand notations such as 5'-pApCpGpT-3' or 5'-ApCpGpTp-3' to indicate 5'-terminal and 3'-terminal phosphate, respectively.

In both DNA and RNA, *in-vivo* diester bond formation proceeds solely in the $5' \rightarrow 3'$ direction, producing long-chain, unbranched structures (strands). Branching in RNA molecules can occur transiently during RNA splicing. Both DNA and RNA strands display polarity, meaning that their primary structure is to be read in the direction $\rightarrow O3' \rightarrow [PO_2 \rightarrow O5' \rightarrow C5' \rightarrow C4' \rightarrow C3' \rightarrow O3'] \rightarrow [PO_2 \rightarrow O5' \rightarrow C5' \rightarrow C4' \rightarrow C3' \rightarrow O3']$, etc. (the brackets define a DNA/RNA monomer subunit). The strand polarity is a consequence of furanose pucker as well as associated base stacking.

The chemical, i.e., base composition of nucleic acids is variable and differs from one species to another. Variation in base composition forms the basis of the genetic information content of DNAs. However, in DNAs, certain regularities apply (Chargaff's rules): (1) $\sum Pu = \sum Py$, (2) $\sum A = \sum T$ and $\sum G = \sum C$, (3) $\sum(A + C) = \sum(G + T)$. On occasion, 5-methylcytosine needs to be taken into consideration, too, in order to satisfy points (2) and (3). These rules hold because of the double-stranded (complementary) structure of DNAs brought about, in part, by Watson–Crick hydrogen bonding between opposite bases (A $=$ T and G $\equiv$ C base pair formation; the symbols $=$ and $\equiv$ denote that base pairing between A and T involves two hydrogen bonds, while three hydrogen bonds are formed in the base pairing between G and C). Chargaff's rules do not apply to single-stranded nucleic acids, whether they are RNAs (usually) or DNAs (occasionally).

Physiochemical Properties

At physiological pH values, the orthophosphate groups of the nucleic acids are negatively charged. Outward charge neutralization occurs through positively charged counterions aligning themselves with the negative charges on the polymers. Counterions can be Na^+, Mg^{2+}, basic proteins, polyamines, etc. Nucleic acids are therefore polyelectrolytes, a property that determines their solution behavior: they are extensively hydrated and form gel-like structures if present in higher concentrations. Even in dilute solutions they display non-Newtonian behavior, as noted, for instance, by their shear-rate-dependent viscosity. Once hydrated, they are readily soluble in water. Isolation of nucleic acids from tissues usually yields their sodium salt. How they really exist *in vivo* is unknown, although it is assumed that Na^+ also serves there as the principal counterion.

DNAs, with very few exceptions, are double-stranded helices of opposite polarity. By contrast, RNAs are usually single-stranded, although double- (as well as triple-) stranded RNAs are known to exist. There are also duplex helices made up of DNA and RNA single strands; they form, for example, *in vivo* during transcription. DNA double-strandedness, brought about by interstrand hydrogen bonding and intrastrand base stacking, is termed DNA secondary structure; this structure is salient to the preservation of genetic information by DNA. Hydrophobic bonding and London dispersion forces are the major factors contributing to base stacking. DNA secondary structure is quite stable over a wide range of pH (3–12), ionic strength, I (0.001–5), and temperature (up to 100 °C). The ranges given are approximate: not only are the effects of I and temperature interdependent (rule-of-thumb: the higher I is, the higher the thermal stability), but base composition (GC-rich DNAs are more heat-stable than AT-rich DNAs) contributes heavily to the stability of DNA secondary structure as well. Lastly, at elevated levels of I, individual cation/anion effects, often destabilizing in nature, also become noticeable. Transforming the DNA double helix into two single strands of DNA, a process occurring cooperatively, is called 'denaturation.' Synonymous terms are 'helix-to-random coil transition' or 'order–disorder transition.' 'Denaturation' represents a change in DNA conformation (conformational changes are those that rupture hydrogen bonds and/or hydrophobic bonds but leave covalent bonds intact). It is evident from the above that variations in pH, I, and temperature can be used to bring about changes in DNA conformation experimentally. While much is known regarding the subtleties of DNA secondary structure, very little information is available with respect to the structural properties of the single-stranded states of DNA (or RNA).

Nucleic acids are optically active molecules, i.e., they rotate the plane of (linearly) polarized light and, hence, display chirality. The major contribution to optical activity comes from base stacking, for it imparts molecular asymmetry. The optical activity of the asymmetric carbon atoms of deoxyribose or ribose does not contribute much to the overall effect. Double-stranded DNAs are usually of right-handed screwness or helicity. This information has been obtained from X-ray diffraction analysis of DNA fibers. Depending on the sign and magnitude of the parameters defining the orientation of a base pair in DNA relative to the helix axis (e.g., rotational twist, t, tilt, θ_T, roll θ_R, propeller twist, θ_p, axial rise, h, pitch, width, and depth of major and minor grooves), one can distinguish between conformational families such as A-, B-, C-, and Z-type DNA. They exist among naturally occurring DNAs, i.e., DNAs with random base sequences. Synthetic DNAs (polynucleotides with nonrandom sequences) are characterized by additional families such as B′, C″, D, E, S, Z′. Z-type DNA is a left-handed double helix.

Until quite recently, the consensus was that double-stranded DNA is rather inflexible ('rod' or 'worm-like chain') and that, apart from its exciting base-pairing capabilities, the structure of the sugar–phosphate backbone, in view of its seeming monotony, is of little consequence to its biological function. This perception was due to the fact that X-ray diffraction studies executed on DNA fibers gave structural information only at low levels of resolution. However, experimental evidence has shown that double-stranded DNA is surprisingly flexible locally, giving rise to what has been termed DNA 'polymorphism.' DNA polymorphism refers to seemingly minor, quite localized alterations in secondary structure that may very well be of crucial importance in biological processes such as replication and transcription. That there is such a thing as DNA polymorphism has been demonstrated by X-ray diffraction analysis of oligonucleotide crystals. Oligonucleotide crystals (formerly not available) yield high-resolution X-ray data down to atomic dimensions. It was found that DNA polymorphism resides in the nonplanar ring structure of β-D-2′-deoxyribose (or β-D-ribose in RNA) and that this structure is conformationally quite flexible. Its two main conformations (called sugar pucker) are envelope E (four atoms in a plane, the fifth out of plane) and twist T (three atoms in a plane, the two others out of plane on opposite sides). Out-of-plane atoms on the same side of C5′ are denoted *endo*, while those on the opposite side are denoted *exo*. Transitions between E and T are facile, giving rise to a pseudorotation cycle of the furanose ring in nucleosides. The cycle contains 10 different (*endo/exo*) T and E forms. In addition, O5′ can assume a number of orientations about the C4′–C5′ bond axis; they are known as *gauche, gauche* (+ *synclinal*); *gauche, trans* (*antiperiplanar*); and *trans, gauche* (− *synclinal*). Lastly, relative to the sugar moiety, a Pu can adopt two major orientations along the N—Cl′ bond; *anti* (the bulk of the base turns away from the sugar) and *syn* (it is over or toward the sugar). All orientations influence each other, giving rise to a multitude of sugar conformations – and, hence, structural DNA families – with important implications for biological function. Thus, the binding of enzymes (e.g., DNase I, restriction endonucleases) or regulatory proteins such as TFIIIA (zinc fingers) to DNA is now known to be greatly influenced by the geometry of the backbone. In fact, it has been said that while base pairing may be viewed as the 'brain' of DNA, the conformational flexibility of the sugar–phosphate backbone constitutes its 'heart.'

Determination and Characterization of Nucleic Acids

There exist literally hundreds of techniques that permit the detection, quantitative evaluation, structural, and biological characterization of nucleic acids. They are chemical, biochemical, physicochemical, or physical in nature; however, their description is beyond the scope of this article. 'Determination of nucleic acids' is therefore understood here to mean their detection (qualitatively as well as quantitatively) in cells and tissues in particular, and the determination of their chemical composition (average base composition) when in solution. Hence, techniques that serve their physical characterization in terms of structure, mass, or shape, or techniques used in DNA sequencing, in the enzymatic manipulation of DNA and RNA, in recombinant DNA research, in cloning, etc., are not presented. Techniques relating to their isolation from animal and plant tissues or from microbial sources are also not discussed here.

Detection of Nucleic Acids in Cells and Tissues

Histochemical approaches (*in-situ* staining) Most advantageous are histochemical staining techniques. Each constituent of DNA/RNA (cf. eqn (1)) can be identified histochemically: acidic phosphate groups are demonstrated with the aid of basic dyes, β-D-2'-deoxyribose with the help of a triphenylmethane dye, and Pu(Py) can be made visible via intercalation with fluorescent dyes. The specificity of the techniques can be improved further through judicious use of DNA (deoxyribonuclease) and RNA (ribonuclease) degrading enzymes prior to staining. This eliminates possible interference by the other polymer. Although staining techniques are usually used to obtain qualitative information, they can be adapted to furnish also quantitative data.

Techniques detecting DNA

Feulgen nucleal reaction DNA in tissue is depurinated by strong acid treatment. The resulting apurinic acid releases β-D-2' deoxyribose, which in turn is converted to its open-chain aldehyde form. The sugar aldehyde reacts with the colorless triphenylmethane derivative leucofuchsin (Schiff's reagent) to give a magenta color (fuchsin). The chemistry of the reaction is known. Since the ribose of RNA cannot be changed to an open-chain aldehyde form, RNA will not be stained; hence, this technique is very specific.

Kurnick's methyl green method Methyl green, a triphenylmethane dye possessing two positively charged amine groups, adds on to double-stranded DNA. Details of the process remain obscure but appear to be related to the secondary structure of DNA: the two positively charged amine groups of the dye are said to have just the correct distance to add on to phosphate groups separated from one another by one turn of the DNA double helix. DNA stains green or green-blue.

Fluorescence flow cytometry A number of macrocyclic dyes (acridine orange (derivative of acridines), Hoechst 33342, Hoechst 33258 (derivatives of piperazine), ethidium, propidium (derivatives of phenanthridine), etc.) bind to DNA bases. Some of them (e.g., acridine orange) will also react with RNA. Their commonality resides in the fact that they are planar, highly aromatic molecules that fluoresce when irradiated. Because of their flatness, they can bind to DNA through intercalation, i.e., they can sandwich themselves between bases. Intercalation increases their basic fluorescence, which makes the nucleic acids visible. Excitation sources are mercury arc lamp, argon ion laser, or krypton ion laser. Since DNA–dye emission spectra differ usually from those of the RNA–dye complexes, distinction between the two types of nucleic acids is possible. Determination of DNA (or RNA) is quantitative. The technique is used most frequently in cell-culture work or in studies in which free-flowing cellular or subcellular particles are available (e.g., red blood cells).

Techniques detecting DNA and RNA

Methyl green–pyronin method If proper experimental conditions are maintained, nuclei (DNA) will be stained green or blue (methyl green), and nucleoli and cytoplasm (RNA) red (pyronin). Since pyronin, a xanthene derivative, is also a fluorescent dye, the technique can be adopted for flow-cytometric determination of RNA.

Acridine orange method The dye acridine orange binds to DNA and RNA. The two dye–polymer complexes exhibit different fluorescence emission spectra: when illuminated with purple light, DNA fluoresces green–yellow to bright yellow and RNA reddish brown to orange; when illuminated with ultraviolet light, DNA fluoresces greenish yellow and RNA crimson red. (*See* **Spectroscopy**: Fluorescence.)

Nonhistochemical Approaches

Radioactive labeling method Nucleic acids can be further identified qualitatively or quantitatively in cells or tissues by labeling them radioactively. Frequently used radionuclides are ^{31}P, ^{14}C, and ^{3}H,

They are administered experimentally in the form of labeled precursors of DNA or RNA (e.g., [^{3}H]dT for DNA, [^{3}H]U for RNA, [α-^{32}P]dATP for DNA, [α-^{32}P]ATP for RNA, etc.) and are incorporated into the polymers enzymatically during macromolecular synthesis. Labeling can be performed *in vitro* or *in vivo*. The presence of the radionuclides and, hence, of the nucleic acids is noted by the radioactive decay of the nuclides in the form of α or β particles or γ rays.

Burton assay for DNA This assay is a colorimetric procedure for measuring the deoxyribose moiety of diphenylamine DNA. Color is produced by a secondary amine (phenylamine) reacting with the sugar. The assay is relatively specific, although RNA in high concentrations, or sucrose, should be absent. DNA does not have to be present in purified form; thus, determinations can be undertaken in crude cellular extracts. DNA is usually extracted from tissue or cells through treatment with strong acids (perchloric acid, trichloroacetic acid), although alkaline extractions have also been used. Extraction is undertaken to remove interfering substances, particularly proteins. Color readings are taken at 595 and 650 nm. A standard curve is prepared by using samples of pure DNA (commercially available). A number of modifications of the basic procedure exist. (*See* **Spectroscopy:** Visible Spectroscopy and Colorimetry.)

3,5-Diaminobenzoic acid fluorescence assay DNA is extracted from cellular material with acid treatment (usually perchloric acid) and the sample exposed to hydrochloric acid (1 N) for about 45 min at 55–57 °C. This produces apurinic acid and free β-D-2'-deoxyribose (see section *Feulgen nucleal reaction*). The C1' and C2' carbons of the sugar then react with 3,5-diaminobenzoic acid to produce a strongly fluorescent compound. Excitation occurs at 410 nm; fluorescence is measured at 510 nm. Calibration with pure DNA is required. Submicrogram quantities of DNA can be determined.

Detection of Nucleic Acids in Solution

These techniques assume DNA as well as RNA to exist in solution in essentially pure form, i.e., extraction from tissues or cells and purification occurring prior to analysis.

Spectrophotometric methods

Spectrophotometry in the ultraviolet (concentration determinations) The presence of stacked Pu(Py) bases in DNA/RNA makes the polymers absorb radiation in the ultraviolet region. With duplex DNA,

absorption starts at wavelengths below 300 nm and reaches a maximum near 260 nm and a minimum around 230 nm. The presence of a second peak around 200 nm has been suggested; however, its demonstration is not simple because most spectrophotometers cease functioning around 200 nm. Concentration determinations are usually performed at 260 nm by applying the Bouguer–Lambert–Beer law:

$$A^{260} = \varepsilon^{260} cd, \tag{2}$$

where A^{260} is the absorbance of DNA at 260 nm, ε^{260} the DNA molar absorptivity (1 (mol phosphate)$^{-1}$ cm^{-1}), c its molar concentration in solution (mol phosphate l^{-1}), and d the optical pathlength (cm). By measuring A^{260}, one can readily compute c, the desired quantity, ε^{260}, and these are available from the literature for almost all DNAs. They are usually around 6.6×10^3 l (mol phosphate)$^{-1}$ cm^{-1}. Because of the 1:1:1 molar relationship between phosphate, sugar, and base, c can also be expressed in terms of moles of base per liter or moles of monomer per liter, etc. Since the average molecular weight of a nucleic acid monomer unit is 330, it is easily derived from eqn (2) that for duplex DNA:

$$1 \, A^{260} \text{ unit} = 50 \, \mu\text{g DNA ml}^{-1}, \tag{3}$$

while for single-stranded DNA or for RNA

$$1 \, A^{260} \text{ unit} = 40 \, \mu\text{g RNA ml}^{-1}. \tag{4}$$

Spectrophotometric methods in the ultraviolet (base composition determinations) The secondary structure of DNA can be used for analytical purposes. Heating duplex DNA at a given pH and ionic strength ultimately yields single-stranded DNA ('denaturation'). The process can be followed spectrophotometrically since single-stranded DNA has a higher absorbance (by about 25% at room temperature) at, say, 260 nm than double-stranded DNA. The increase in absorbance is called 'hyperchromicity' (H) and is defined as follows:

$$H^{260} = \left(A_t^{260} - A_0^{260}\right)/A_0^{260}, \tag{5}$$

where A_t^{260} is the absorbance of DNA at 260 nm at a given temperature, t, while A_0 is its absorbance prior to heating (usually room temperature). H is thus the normalized absorbance increase of DNA due to heat-induced denaturation. Plotting H_{260} against temperature produces the so-called DNA 'melting curve.' The expression 'melting' refers to the fact that DNA 'helix-to-random coil transitions' resemble the infinitely sharp temperature-dependent phase transitions occurring in ice during melting. DNA 'melting curves' are of sigmoidal shape. The limiting value of A_t^{260} at

T_{max} is about 40% higher than A_0^{260}. The temperature at which H^{260} amounts to 50% of its limiting value H_{max}^{260} is called the 'melting temperature' (T_m in °C). T_m is linearly related to the (average) base composition of DNA. For the 'standard saline citrate' medium (0.15 M sodium chloride and 0.0015 M sodium citrate, pH 7.0):

$$\% \, GC = 2.44(T_m - 69.3). \qquad (6)$$

Since $\% \, GC = 100 X_{GC}$, with X_{GC}, representing the 'mole fraction' of GC base pairs in DNA, and since $X_{AT} = 1 - X_{GC}$, eqn (6) furnishes equally well information on %AT. Equations similar to eqn (6) have been established for other solvents. In conclusion, heating native DNA under controlled conditions (constant pH and ionic strength, application of a linear temperature gradient, automatic recording of hyperchromicity) enables the rapid determination of its chemical (base) composition. It is best to work at $A_0^{260} \approx 1$ ($\approx 50 \, \mu g$ DNA ml^{-1}).

See also: **Spectroscopy**: Fluorescence; Visible Spectroscopy and Colorimetry

Further Reading

Adams RLP, Knowler JT and Leader DP (1986) *The Biochemistry of the Nucleic Acids*, 10th edn. London: Chapman & Hall.

Alberts B, Bray D, Lewis J *et al.* (1989) *Molecular Biology of the Cell*, 2nd edn. New York: Garland.

Ausubel FM, Brent R, Kingston RE *et al.* (1989) *Current Protocols in Molecular Biology*, vols 1 and 2. New York: Greene, Wiley Interscience.

Chayen J and Bitensky L (1991) *Practical Histochemistry*, 2nd edn. Chichester, UK: John Wiley.

Saenger W (1984) *Principles of Nucleic Acid Structure*. New York: Springer-Verlag.

Shapiro HM (1985) *Practical Flow Cytometry*. New York: Alan R. Liss.

Walker JM (ed.) (1984) *Methods in Molecular Biology*, vol. 2. *Nucleic Acids*. Clifton: Humana Press.

Physiology

H A Simmonds, Purine Research Unit, Guy's Hospital, London, UK

Physiology

Nucleic acids are essential components of all cells and consequently are found in many foods. The term 'nucleic acid' commemorates the first isolation of this vital cell constituent, 'nuclein,' from the spermatic fluid of Rhine salmon and the nuclei of pus cells, by Miescher in 1868. Some 20 years later, it was demonstrated that uric acid, which had been recognized by the Swedish chemist Scheele in 1776 as a constituent of human urine and kidney stones, arose from nucleic acid degradation. Fischer and his school in Germany at the end of the nineteenth century (1895–1899), established the first chemical structure for uric acid, as well as that of other purine and pyrimidine bases. It was demonstrated subsequently that nucleic acids consisted of chains of these purine and pyrimidine bases linked to a pentose sugar, esterified with phosphoric acid (**Figure 1b**). Two types of nucleic acid have been identified – deoxyribonucleic acid (DNA) and ribonucleic acid (RNA).

Physiological Roles

The purine and pyrimidine bases of DNA carry the genetic information of all prokaryotic and eukaryotic organisms, with the sugar and phosphate groups performing a structural role. The human genome is considered to contain between 50 000 and 100 000 genes, each of which is composed of a linear polymer of DNA of varying length. In viruses, genes are made of either DNA or RNA. The infinite variation in genetic information is achieved by the sequence of the four bases that constitute DNA (**Figure 1b**) – the purines, *adenine* and *guanine*, and the pyrimidines, *thymine* and *cytosine*. DNA is double-stranded, and each nucleotide of the chain in one strand is linked by hydrogen bonding to a complementary nucleotide in the other. Complementary pairs of nucleotides are adenine and thymine, and guanine and cytosine. DNA is principally found in the nucleus and is considered to be relatively stable in most cell types. (*See* **Viruses**.)

Ribonucleic acid is essential for the transmission of the genetic message in the form of protein synthesis and must first be synthesized from DNA. In the case of RNA, one of the four bases differs from that in DNA – *uracil* replaces the pyrimidine base thymine – and the molecule is single-stranded, except in some viruses. In contrast to DNA, most of the RNA is in the cytoplasm. Cells contain three types of RNA: *messenger* RNA (mRNA; 5% of total RNA) provides the template for protein synthesis and is relatively labile; *transfer* RNA (tRNA; 15%) carries the message in the form of activated amino acids to the ribosome for the synthesis of specific polypeptides, as determined by the particular mRNA template; and *ribosomal* RNA (rRNA), the major RNA component (80%), is metabolically stable. (*See* **Protein**: Synthesis and Turnover.)

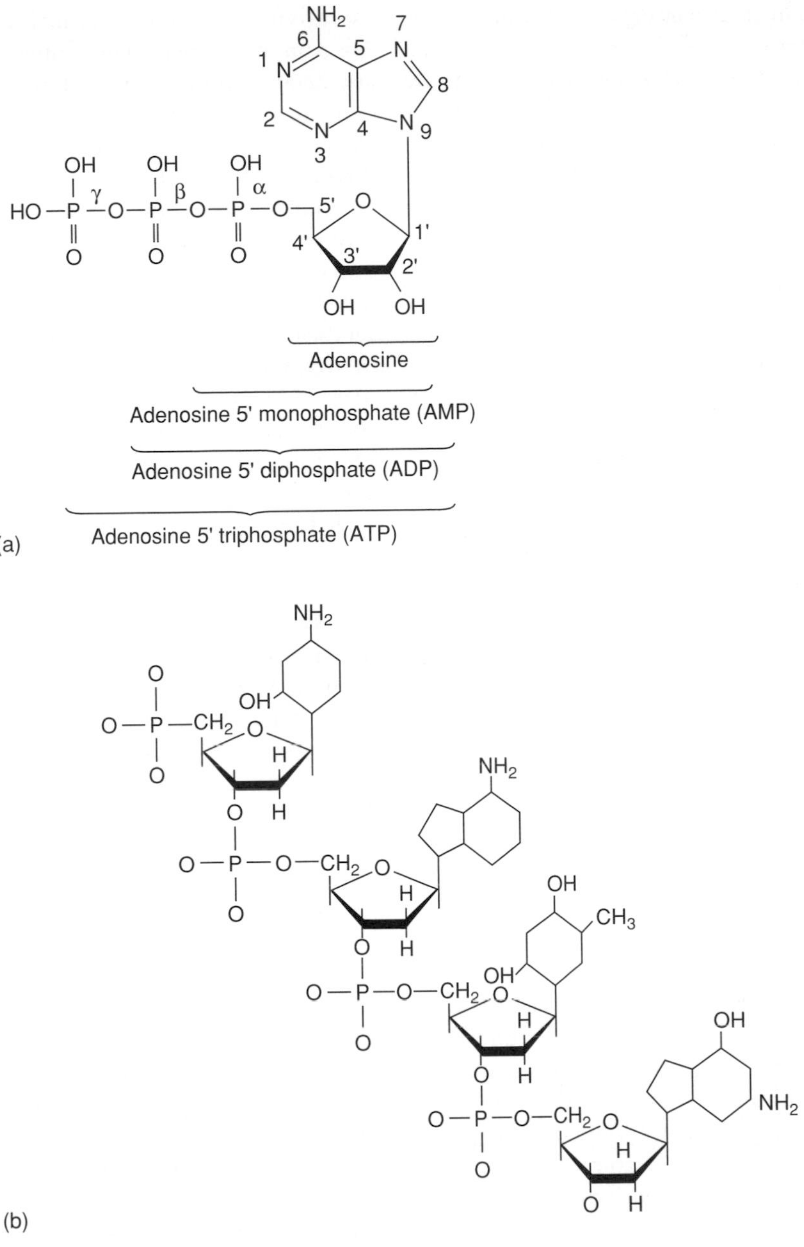

(a)

(b)

Figure 1 (a) Structural formula of adenosine 5′-triphosphate (ATP) indicating the numbering of the atoms on the ribose, as well as the purine ring, which consists of a six-membered pyrimidine ring fused to a five-membered imidazole ring. The position of attachment of the phosphate groupings for AMP, ADP, and ATP is also indicated. (b) Schematic representation of the structure of part of a single DNA strand made up of a sequence of the four bases, the pyrimidine cytosine, the purine adenine, the pyrimidine thymine, and purine guanine respectively, showing that the deoxyribose has an H group at the 2′ position on the pentose ring, instead of the OH group of ribose. These bases are linked via the 3′-OH group of the deoxyribose-phosphate moiety to the 5′-OH group of the next deoxyribose. Reproduced from Nucleic Acids: Physiology, *Encyclopaedia of Food Science, Food Technology and Nutrition*, Macrae R, Robinson RK and Sadler MJ (eds), 1993, Academic Press.

The important physiological roles played by the metabolic pathways responsible for sustaining these different nucleic acid pools in humans is demonstrated by the clinical manifestations when different steps in the synthesis, degradation, and repair of the constituent mononucleotides are defective or absent.

Nucleic Acid Metabolism in Humans

The building blocks of the nucleic acids, the purine and pyrimidine ribonucleotides, are of central importance to virtually all biological processes. Whereas cellular purines are derived exclusively from endogenous sources, cellular pyrimidines are not. As discussed in detail later, dietary purines are degraded by a battery of enzymes in the intestinal mucosa and enter the circulation as uric acid. By contrast, dietary pyrimidines can be absorbed in the nucleoside form and incorporated by salvage into the nucleotide pool, as evidenced by the lifelong treatment of patients with the pyrimidine *de-novo* salvage defect, hereditary oroticaciduria, with oral uridine.

Nucleotides in infants

Over the past two decades, there has been considerable interest in the role of dietary ribonucleotides in infant feeds, in which significant effects have been claimed. This debate stemmed from reports that ribonucleotides were present in human milk, but not in cows' milk or infant formulas. However, claims that nucleotide concentrations were much lower in cows' than in human milk were apparently related to the state of lactation, with concentrations higher during lactation. Infant formulas were confirmed as being devoid of nucleic acid.

These findings stimulated investigations to determine: (1) whether the nucleotides found in human milk result from degradation of nucleic acids, or are actively secreted as a response to a nutritional demand of the infant, and (2) the newborn infant's endogenous capacity to digest nucleic acids to absorbable products. Nucleotides were found in the greatest quantity in human milk followed by nucleic acids, nucleosides being a minor component. Importantly, the nucleotide/nucleosides were predominantly pyrimidines, purines being present only as uric acid. This specific profile must result from catalysis in the breast. Enzymes capable of degrading nucleotides, including xanthine dehydrogenase (XDH), are known to be present in human milk. Evaluation of the endogenous capability of infants to metabolize RNA and nucleotides confirmed that the intestine of infants, as in adults, digested RNA to cytidine, uridine, and uric acid *in vitro*. This study confirmed the extensive data derived from dietary purine loading in adults that dietary purine nucleotides (if not already degraded by gut bacteria) would be degraded to uric acid in the intestinal mucosa. However, uridine and cytidine would be absorbed (although the latter would be degraded rapidly to uridine by cells *in vivo*).

This research into the importance of human versus cows' milk, or formula feeding, was stimulated by animal studies that had indicated that dietary nucleotides may be required for maintenance of normal immune function in neonatal mice. A subsequent study in healthy term infants demonstrated that in those either breast-fed, or fed formula supplemented with nucleotides, some indices of immune function were indeed significantly higher (natural killer cell

cytotoxicity and interleukin-2 production by stimulated mononuclear cells) compared with nonsupplemented formula groups. However, this was found only in the newborn. The rate of growth and incidence and severity of infections did not differ significantly among these different dietary groups at 2 months. It was concluded that nucleotides may be a component of human milk that contributes to the enhanced immunity of the breast-fed infant. Another study implied that allergic diseases develop during feeding of cows' milk, but not human milk. It may be that the dearth of uridine nucleotides in cows' milk after 1 month of lactation could be implicated, since these effects were independent of any difference in the gut flora demonstrated in breast-fed, as distinct from formula-fed, infants.

This debate and the problems addressed arose from the use of animals. As indicated elsewhere in this chapter, rodents have both XDH and uridine phosphorylase everywhere. By contrast, in humans, uridine phosphorylase is confined to the liver, and XDH is confined to the liver, intestinal mucosa, and breast milk. This problem highlights the general lack of awareness of the two curious and unexplained differences in the synthesis of nucleotides in humans. First, pyrimidines derived from nucleotide degradation are salvaged at the nucleoside (uridine) level, purines as the base. Second, whereas purine nucleotides are derived exclusively from endogenous sources in humans, pyrimidine nucleotides can be formed from uridine orcytidine ingested in the diet as well. These important differences are underlined by the successful treatment with oral uridine of patients with uridine monophosphate synthase (UMPS) deficiency, which presents generally as macrocytic anemia. A few patients are also immunodeficient. UMPS is the complex catalyzing the last two steps of pyrimidine *de-novo* synthetic (DNS). Long-term follow-up of such patients confirms that humans cannot only survive without pyrimidine biosynthesis, but also reproduce. Thus, uridine certainly can restore immune function in humans as well.

Role of Endogenous Nucleotides, Nucleosides, and Bases in Cellular Metabolism

Purines and pyrimidines are effectively anchored inside the cell as the nucleotide by attachment to a pentose linked to a mono-, di- or triphosphate group (**Figure 1a and 1b**), principally the triphosphate. It was originally assumed that all reactions of biological significance took place intracellularly at the nucleotide level. Attention has been focused recently on the extracellular regulatory functions of purine nucleosides (base plus pentose), or the bases themselves. The pentose may be either ribose (ribonucleoside) or

2′-deoxyribose (deoxyribonucleoside) bound by the C1 atom through a glycosidic linkage to the N9 atom of the purine group, or to the N2 of the pyrimidine group (**Figure 1**).

The importance of purine and pyrimidine nucleotides in cellular metabolism is twofold. In addition to their role in the storage, transmission and translation of genetic information (as the polynucleotides DNA and RNA), as mononucleotides, they play an equally vital role in lipid and membrane synthesis (in the form of purine and pyrimidine sugars or lipids), signal transduction, and translation (in the form of guanosine triphosphate, cyclic adenosine monophosphate, and cyclic guanosine monophosphate), as well as providing the energy (adenosine triphosphate (ATP)) that drives many cellular reactions and forms the basis of the coenzymes (nicotinamide adenine dinucleotide, nicotinamide adenine dinucleotide phosphate, flavin adenine dinucleotide, etc.) (*See* **Coenzymes.**)

All cells require a balanced supply of purine and pyrimidine nucleotides for growth and survival, but this may vary from cell to cell, depending on function. Liver, for example, has a very complex nucleotide profile, compared with heart. These nucleotides may be built up by one of two routes: the energetically expensive multistep synthetic route, or the single-step so-called 'salvage' pathway. In normal circumstances, salvage predominates over synthesis. **Figure 2** illustrates the different metabolic pathways involved in the *de-novo* synthesis of these nucleotides, as well as the efficient recycling of the nucleotides or bases derived from them during the wear and tear of daily life (muscle work, wound healing, erythrocyte senescence, protein glycosylation, providing essential nourishment for the brain, etc.). Interestingly, whilst this recycling takes place at the base level for purines, it is the pyrimidine nucleosides that are actively recycled in humans, with only a small proportion being degraded further in either case (**Figure 2**). The importance of these different pathways to the overall control of nucleotide concentrations in the body is the subject of many excellent reviews and for this reason is summarized only briefly here.

Nucleotide Triphosphate Production and Nucleic Acid Synthesis

Purine and pyrimidine ribomononucleotides built up by either the *de-novo* or synthetic routes are phosphorylated via the correponding diphosphate to the triphosphate, which is the active intracellular form of most mononucleotides. In addition to the variety of vital individual cellular functions described above, these ribomononucleotides are essential intermediates in the synthesis of the polynucleotides RNA

and DNA respectively (**Figure 2**), being incorporated into DNA following the formation of the deoxyribonucleotide from the corresponding diphosphate by the enzyme ribonucleotide reductase. The latter is an allosteric enzyme; its activity and specificity are controlled in a complex manner by both purine and pyrimidine ribo and deoxyribonucleotides. This process is particularly active in cells and tissues with a high rate of turnover (e.g., lymphocytes, gut epithelium, skin, bone marrow, etc.). There is approximately five times as much RNA and DNA in the body.

Breakdown of Nucleic Acids

The polynucleotides DNA and RNA, although relatively stable in most tissues, turn over rapidly in dividing cells. Both DNA and RNA first must be degraded to the constituent mononucleotides, which are themselves degraded further. A variety of enzymes capable of hydrolyzing the phosphodiester bonds have been described and include ribonucleases specific for RNA and deoxyribonucleases for DNA as well as nonspecific nucleases, phosphorylases and phosphomonoesterases. The mononucleotides have the highest turnover rate, DNA the lowest. Further catabolism of the resulting monophosphate will differ depending on whether it is a pyrimidine or purine ribonucleotide of deoxyribonucleotide. For example, whereas adenine ribonucleotides are predominantly deaminated at the nucleotide level in humans by AMP-deaminase, deoxy-AMP is not a substrate for this enzyme and must first be degraded to deoxyadenosine and deaminated at the deoxynucleoside level (**Figure 2**).

Purine and pyrimidine (deoxy) nucleotides are degraded to the corresponding (deoxy) nucleosides by specific 5′ nucleotidases. Different purine endo- or ecto-5′ nucleotidases have been identified with different substrate specificities and may be of particular importance in providing bases for nucleotide resynthesis in tissues where there is rapid cell turnover and massive cell death (e.g., thymus, spleen, bone marrow). As mentioned above, whereas the normal metabolic route for pyrimidines is salvage at the nucleoside level, that for purine nucleosides and deoxynucleosides is degradation to the corresponding base by purine nucleoside phosphorylase, prior to salvage. This degradation is favored by the high intracellular inorganic phosphate and low ribose 1-phosphate levels in most tissues. Interestingly, these phosphorylases are not reactive toward either adenosine or cytidine, or their analogs, in human cells and first must be deaminated at the (deoxy) nucleoside (or nucleotide) level.

Salvage is an active process for both pyrimidines and purines. Consequently, only a small fraction of

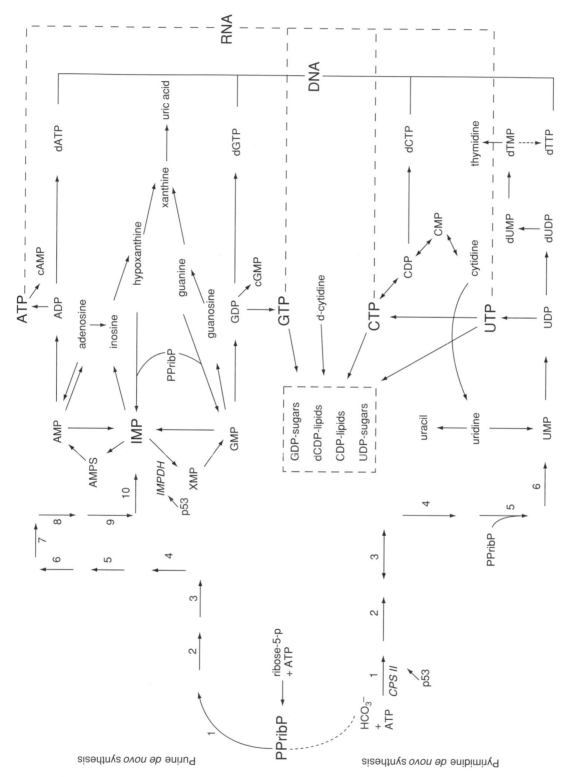

Figure 2 Multistep purine and pyrimidine nucleotide de novo synthetic (DNS) routes and single-step salvage (hypoxanthine, guanine, uridine, cytidine, D-cytidine) pathways, indicating the enzymes inosine monophosphate dehydrogenase (IMPDH) and carbamoyl phosphate synthetase 11 (CPS 11) targeted by analogs which deplete nucleotide pools, reportedly associated with a p53-dependent Go/G1 arrest. Note the involvement of 5-phosphoribosyl pyrophosphate (PPribP) in both purine DNS and salvage as well as in pyrimidine DNS. The role of ATP, GTP, UTP, and CTP in DNA and RNA synthesis, of ATP and GTP in 2nd messenger synthesis, of UTP, CTP, and GTP in the synthesis of the glycoconjugates (dashed box) essential for membrane synthesis, structure, and function and protein glycosylation, is also evident.

the nucleotides turned over daily is actually degraded and lost to the body. The pyrimidine bases, uracil and thymine, derived from nucleosides not recycled are degraded further to the β-amino acids, and there is thus no measurable end product. However, such loss is probably comparable with that for purines, the normal end product of which in humans is uric acid, formed from the precursor purine bases xanthine and hypoxanthine by the action of xanthine dehydrogenase (**Figure 2**).

Rates of Nucleotide Synthesis and Degradation in Different Cells

It has become apparent that the original concept of endogenous metabolism and its overall control, involving a complex interplay between *de-novo* synthesis and salvage, does not apply to all cells but is governed by a tissue- or cell-specific complement of enzymes and/or controls on them, depending on the function of that cell or tissue. The human erythrocyte, for instance, is anucleate and lacks the ability to use either salvage or *de-novo* synthesis to maintain its ΛTP levels, being dependent on adenosine scavenged from other tissues for this. In addition, the pyrimidine nucleotides found in nucleated cells are absent from mature erythrocytes, the only pyrimidines normally present being in the form of uridine diphosphate (UDP) sugars. ATP is also the most important purine in both skeletal and heart muscle, adenine nucleotides making up 95 and 90% of the total nucleotide complement, respectively. Although DNA in most tissues is considered relatively stable, it is evident from the two inherited disorders associated with immunodeficiency that cell death and rapid turnover of cells of the hemopoietic system (e.g., extrusion of the nucleus during erythrocyte maturation, or mounting an immune response) normally produces significant amounts of deoxyribonucleosides, as well as ribonucleosides, which must be degraded further. These disorders have highlighted the fact that removal of metabolic waste from DNA catabolism is vital to the normal immune response; failure to do so can result in the accumulation of deoxy-ATP and deoxy-GTP. These deoxynucleotide triphosphates are extremely toxic to T-lineage stem cells, resulting in severe combined immunodeficiency affecting both T- and B-cells in subjects deficient in adenosine deaminase, or a T-cell-specific immunodeficiency, in purine nucleoside phosphorylase deficiency (**Figure 2**).

Endogenous Nucleic Acid Synthesis in the Gut

It was originally reported from studies in guinea-pigs that the gastrointestinal tract was incapable of *de-novo* synthesis. However, subsequent workers have shown significant synthetic activity in rat intestine, as evaluated by the incorporation of radiolabeled glycine into RNA. The nucleic acid content of intestinal mucosa is high, as is the rate of cell turnover in the luminal villi, and it has been calculated in rat that about 30 mg of endogenous nucleic acid enters the lumen daily. This implies a considerable loss of both purines and pyrimidines, which can only be replaced by *de-novo* synthesis. Studies of nucleotide concentrations in rat intestine have shown both pyrimidine and purine nucleotides at concentrations equivalent to those in liver, supporting active nucleotide metabolism in the intestine.

Intestinal absorption is obviously related to intestinal motility and local blood flow. The presence of adenosine receptors on rat jejunal mucosal cells may be particularly important in the regulation of fluid and electrolyte absorption as well as the active transport of nutrients. Such adenosine may be generated slowly from nucleotides by nucleotidases physically or functionally coupled to the adenosine membrane translocator.

It is evident from the above that the pathways leading to the formation and degradation of nucleic acids are complex, and that many factors determine the origin as well as amount of endogenous pyrimidine or purine nucleotides turned over daily. Moreover, this will differ depending on the cell or tissue.

Role and Fate of Dietary Nucleic Acids

The metabolism of nucleic acids ingested in the diet differs from that of endogenous nucleic acids. The intestinal mucosa plays an important role in this. The above studies confirm that pyrimidine as well as purine nucleotides in intestinal mucosal cells are derived from endogenous sources. However, a growing body of evidence supports a role for dietary nucleotides derived from the gut in intestinal development, turnover, and repair. Suggested effects include enhancement of the host mucosal defense system and an influence on neonatal lipid metabolism and on iron bioavailability, implying a novel role for nutrition in the modulation of gut function.

Most of our knowledge relating to dietary nucleic acid metabolism in the intestine is derived from studies of the absorption of exogenous purine in different species – mouse, dog, rat, pig, etc., as well as humans – which date back as far as the latter half of the nineteenth century. Investigations in animals have shown that whereas the ribose attached to nucleosides derived from RNA was further metabolized, the phosphate was absorbed and excreted in the urine.

Recent studies using ^{13}C-labeled nucleic acid to supplement the diets of rats and chickens have provided further evidence for the incorporation of dietary pyrimidine nucleosides, but not purine nucleosides, directly into hepatic RNA. The successful lifelong treatment with uridine of patients with hereditary oroticaciduria – a defect in pyrimidine *de-novo* synthesis – confirms that dietary uridine is certainly absorbed and salvaged into mono- as well as polynucleotides in humans. The lack of effect of oral uracil in this disorder also confirms that uracil is a metabolic waste and that pyrimidine salvage occurs at the nucleoside level in humans.

Recent investigations have addressed the supposedly beneficial effects of dietary CDP-choline (citicholine) in patients with stroke or trauma, theoretically by increasing acetylcholine production in cholinergic neurons, as well as the amount of cell membrane. These studies confirmed that, when taken orally, CDP-choline elevates plasma levels of both choline and uridine (not cytidine) in humans, but no benefit was demonstrated in clinical trials.

Metabolism of Dietary Purines by Gut Bacteria

Animal studies have shown that the metabolism of dietary nucleic acid and the corresponding purine nucleotides, nucleosides, and bases in the gut is rapid. Isotope studies demonstrated that up to 50% of the radiolabeled purine is recovered as carbon dioxide within 30 min. The role of gut bacteria in this process was indicated by experiments using the XDH inhibitor, allopurinol, concomitantly, when the radiolabel was recovered *in toto* (in urine and feces only), presumably relating to inhibition of bacterial XDH. Allopurinol not only inhibited purine degradation but also decreased the absorption of dietary purine, a seemingly beneficial effect that may explain the reduction in total purine excretion noted in human subjects on a normal diet taking allopurinol.

Intestinal Mucosa Degrades Dietary Purines to Uric Acid

Interestingly, the above radiolabeling studies in pigs (where, as in humans, XDH activity is significant only in intestinal mucosa and liver), confirmed a lack of any radiolabel incorporation from dietary purine into tissue nucleotides. These studies demonstrated conclusively that dietary purine is degraded to a nonreutilizable form by the intestinal mucosa. Studies in other animal species have confirmed this, purine nucleotides, nucleosides, and bases absorbed from the gut lumen being largely converted to uric acid during passage across the mucosa and released as such in serosal secretions, prior to further degradation by urate oxidase (uricase) in the liver. Thus, in contrast to pyrimidines, humans have no apparent requirement for dietary purines. The intestine thus serves as

an effective barrier through the activity of a battery of enzymes capable of rapidly degrading purines already partly processed by gut bacteria to the nonreutilizable metabolic waste, uric acid. Such rapid degradation of purines by the gut presumably reflects an important evolutionary development to protect the integrity of the human genome. The uric acid produced daily in humans thus derives from two sources: catabolism of exogenous as well as endogenous mono and polynucleotides (**Figure 3**).

Excess Dietary Levels of Nucleic Acids and their Clinical Consequences

As mentioned above, pyrimidines have no measurable end product. The potential toxicity of dietary nucleic acids thus relates predominantly to a single mutational event that has resulted in the fact that the insoluble uric acid, and not the much more soluble allantoin as in other mammalian species, is the only measurable end product of nucleic acid degradation in humans. Although it had long been accepted that the enzyme uricase had been lost in the course of human evolution, more recent studies have established that the absence of uricase activity results from a lack of gene transcription, rather than loss of the gene itself. However, other factors play an equally important role in determining the pathogenesis of nucleic acids ingested in the diet.

Role of Exogenous Purine in Determining Circulating Uric Acid Concentrations

The ingestion of food now known to be rich in purines has been noted for millennia to be high in subjects with what has been designated 'primary' gout, the gout affecting predominantly the middle-aged male. This type of gout is a disorder of affluent societies consuming diets rich in purines; it is rare in women or children. During times of hardship, uric acid concentrations in the population fell considerably, and 'primary' gout almost vanished. The fact that gout was extremely prevalent among wealthy Englishmen for more than three centuries up to World War I is not surprising when the dietary habits of the day are examined. These affluent gentlemen habitually consumed vast meals comprising many courses and frequently including 16 different meats, the majority of which were rich in nucleic acids and other purines.

Type of Nucleic Acid and Toxicity

Detailed studies of dietary nucleic acid absorption in humans were carried out by Zöllner and coworkers in the late twentieth century and confirmed that nucleic acids ingested in the diet exert their major

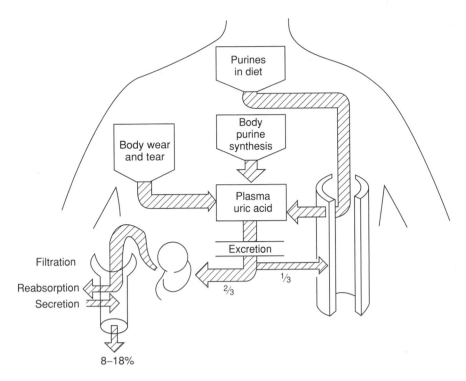

Figure 3 Schematic diagram showing the factors influencing the plasma uric acid, the end product of endogenous (body synthesis plus daily wear and tear) and exogenous (dietary) purine metabolism in humans. The different mechanisms by which this metabolic waste is eliminated (two-thirds by the kidney, one-third by the gut) in normal individuals are also shown. At the bottom left is a simplified version of the complex factors (involving filtration, reabsorption and secretion) interacting in the proximal tubule of the human kidney and resulting in the urinary excretion of only 8–18% of the filtered load depending on age and sex. Reproduced from Nucleic Acids: Physiology, *Encyclopaedia of Food Science, Food Technology and Nutrition*, Macrae R, Robinson RK and Sadler MJ (eds), 1993, Academic Press.

effect on uric acid levels. These studies showed that the nature of the nucleic acid ingested is equally important, with the effect of RNA being more than twice that of DNA (**Figure 4a**). This effect is also evident whether the purine is ingested in the form of nucleic acids, mononucleotides, nucleosides, or bases. Moreover, some forms of purine, e.g., guanosine – the principal RNA degradation product in yeast-rich beverages such as beer (especially real ale) – are absorbed and catabolized more readily than others. A controlled study of diet in primary gout patients compared with control males of comparable age demonstrated that the average daily intake of most nutrients, including purine nitrogen, was similar. However, gouty patients drank significantly more alcohol, predominantly in the form of beer (60 g per day, equivalent to 2.5 l of beer) and had a significantly higher mean plasma uric acid ($0.49 \, \text{mmol} \, \text{l}^{-1}$ compared with $0.39 \, \text{mmol} \, \text{l}^{-1}$).

Role of the Kidney or Drugs in the Genesis of Toxicity

The pathological changes that result from uric acid being the exclusive end product of purine metabolism in humans are due entirely to the insolubility of this metabolic waste, coupled with the other peculiarity that primates display, i.e., the renal tubule reabsorbs around 90% of the filtered urate (**Figure 3**). Net reabsorption is slightly higher in normal males (92%) than in females (88%) and is lower in children of either sex (80–85%). This explains the higher plasma uric acid in adult males and makes them more vulnerable to situations of overproduction, or overingestion. Clearly, sex and age are equally important contributory factors.

It is noteworthy that when normal subjects are fed yeast RNA, the plasma urate rises little with increased intake, the increase in the excretion of urate being dramatic when the rise in plasma urate is modest (**Figure 4b**). Thus, neither overingestion nor overproduction (unless sudden and massive) is likely to raise the concentration of uric acid in the plasma considerably, if the renal response is normal. It is obviously not so in the majority of gouty males with primary gout, where the striking change is that, for any plasma urate concentration, the excretion of urate is consistently less than in normal subjects. This defect in handling by the kidneys relates to a greatly reduced

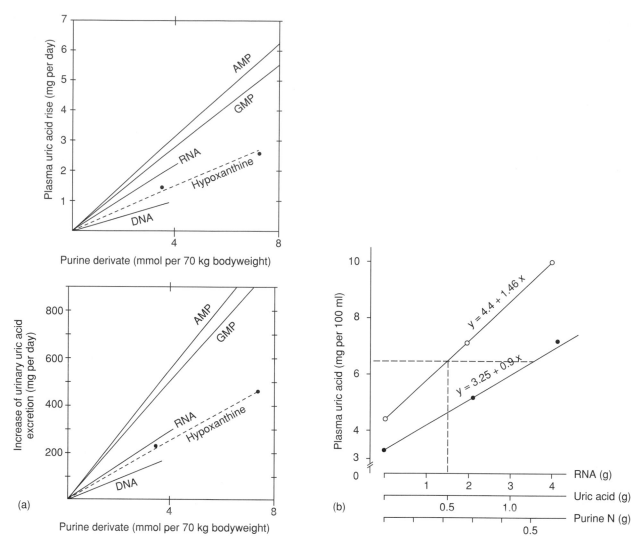

Figure 4 (a) Increase in plasma and urine uric acid (mg per day) in response to a purine load from the different sources indicated on a molar basis. (b) Influence of graded additions of RNA to a purine-free formula diet on plasma uric acid levels of either normouricemic (-•-) or hyperuricemic (-○-) individuals (some of whom had a positive family history of gout). Reproduced from Nucleic Acids: Physiology, *Encyclopaedia of Food Science, Food Technology and Nutrition*, Macrae R, Robinson RK and Sadler MJ (eds), 1993, Academic Press.

clearance of uric acid relative to the glomerular filtration rate (FE_{UR}); FE_{UR} is the fractional excretion of uric acid. The mean is 5.4% in gouty males compared with 8.1% for healthy males of comparable mean age (52 years).

The exact nature of this difference is not yet clear, since studies must be reevaluated in the light of recent knowledge of how complex the renal tubular handling of urate is. However, there is no doubt that it involves the transport of uric acid, which occurs predominantly in the proximal tubule, is bidirectional, and has both a secretory and reabsorptive component (**Figure 3**). Current opinion suggests that the majority of gouty patients underexcrete urate because of a defect in tubular secretion.

Confirmation must await identification at the molecular level of the different transporters involved in urate reabsorption and secretion. Clearly, a combination of events is needed to produce hyperuricemia and the clinical syndrome of gout. These are a large intake of readily absorbed purine coupled with a defect in the renal handling of uric acid at the kidney level, which means that the kidney cannot respond to a purine load without an abnormal rise in plasma urate concentration. (*See* **Renal Function and Disorders**: Kidney: Structure and Function.)

However, diet alone may not be the only culprit. Numerous other physiological and pathological agents such as lead are also capable of reducing urate excretion, and hence exacerbate the rise in

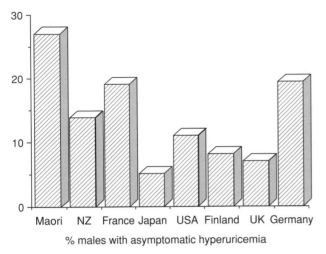

Figure 5 Percentage of Caucasian males, compared with a Polynesian race (the Maori), with asymptomatic hyperuricemia (defined as a plasma uric acid in excess of 0.42 mmol l^{-1}, or 7.0 mg dl^{-1}) in four EU countries, North America, Japan, and New Zealand (NZ).

plasma urate caused by diet. Any of these may lead to an acute attack of gout in a susceptible individual whose plasma urate is already elevated. The best known of the physiological substances are organic acids such as lactate; their overproduction may explain in part the hyperuricemia associated with excessive alcohol consumption coupled with inadequate food intake. Diuretics cause plasma volume contraction, and fractional urate reabsorption increases under these circumstances. The appearance of gout in patients treated with hypotensive agents over long periods currently accounts for over 50% of new presentations in gout clinics. Unusually, the number of elderly females in this group is high, and the mode of clinical presentation is frequently atypical.

Diet may also be important in drawing attention to an unusual subset of patients with juvenile onset of gout; a disorder affecting young males and, unusually, young females and children of either sex equally. Unlike primary gout, where renal function is normal for age, unrecognized, this dominantly inherited disorder is associated with progressive renal disease. Considerable variation in presentation has existed, and there has been no consensus as to whether the gout or the renal disease is the primary factor. However, it is now evident that there are two hallmarks for this disease: the first is hyperuricemia disproportionate to the degree of renal dysfunction; the second is that the degree of renal hypoexcretion of urate is generally extreme, which explains the associated tendency to gout. Moreover, the mean FE_{UR} in this group is only 5.1%, irrespective of age or sex: lower than that of the middle-aged gouty male (5.4%), and even

more remarkable considering the high percentage of children and young women in this group. This universally low clearance explains the added susceptibility to dietary purine leading to the isolated attack of gout in children as well as young adults, which has drawn attention to families with 'familial renal disease,' hitherto leading to death in the 30s. Recognition of the correct nature of the familial disorder has underlined the need to measure urate clearance in all kindred members and enabled early diagnosis and treatment of 42% of seemingly healthy children.

This is important because long-term follow-up (>20 years) in these patients treated with plasma uric acid lowering agents has shown that early recognition and treatment ameliorates the progression of the renal disease. These studies implicate uric acid in the etiology of the renal disease, but this has been disputed by others. However, very recent studies suggest that a reappraisal of the pathogenesis and consequences of hyperuricemia in hypertension, cardiovascular disease, and renal disease is warranted, which adds credence to the above hypothesis.

Role of Dietary Nucleic Acid in the Genesis of Urolithiasis

Although overingestion of purines by subjects with a normal renal response does not precipitate gout, it can predispose to uric acid lithiasis. Uric acid stones are relatively common in Australasia and similar countries, where the consumption of purine-rich beverages and food (e.g., beer, seafood, and meat) is high, and in individuals addicted to health foods such as yeast tablets. Vitamin C is also uricosuric, and acute uric acid nephropathy and sometimes urolithiasis have also been reported following therapy with a variety of uricosuric drugs. Perhaps less well recognized is the uricosuric effect of a high-protein diet and the fact that the intake of purine-rich foods also predisposes to calcium stone formation.

Some foods rich in nucleic acid – vegetarian diets consisting of pulses and grains and yeast extracts – have a particularly high adenine content. Although this is not a problem in the normal population, such diets can be potentially lethal to a rare group of subjects with a genetic defect leading to the inability to recycle adenine, overexcretion of adenine, and the even more insoluble uric acid analog 2,8-dihydroxyadenine (2,8-DHA). Patients with this defect generally present in childhood with kidney stones composed of 2,8-dihydroxyadenine. One such case from a commune consuming a macrobiotic diet presented in acute renal failure with severe renal damage, progressing to dialysis and transplantation. Diet can thus be an important precipitating factor in the pathogenesis of inherited disorders associated with the overexcretion of insoluble purines. (*See* Macrobiotic Diets; Vegetarian Diets.)

Reducing Dietary Levels of Nucleic Acids

Diet varies considerably in different countries and must be taken into account when investigating patients for suspected disorders of purine metabolism. Normal ranges for uric acid in plasma and urine differ greatly in the healthy population depending on the country (**Figure 5**). For example (and in contrast to the nineteenth century), until recently, the majority of subjects in the UK ingested a low purine diet consisting of one meat meal a day, and urinary uric acid excretion above 3.5 mmol per day was considered abnormal. By contrast, the upper limit of normal in Australia has been given as 7.00 mmol per day. However, the increasing affluence of some societies is now leading to an increase in gout, not only in adult Caucasian males, but also in countries such as Japan, where gout was once rare.

Turnover of Exogenous and Endogenous Uric Acid

A useful guide to the dietary purine consumption in different countries can be obtained from the percentage of males with asymptomatic hyperuricaemia. Using this yardstick, the highest consumption in the EU 20 years ago was in France (lovers of pâté and seafoods), the lowest in the UK (**Figure 5**). The statistics for Germany were from Bavaria where, as in New Zealand and Australia, beer consumption is high. Although similar statistics for diet-related differences are unavailable today, plasma uric acid will have risen along with the increase in obesity and blood pressure, especially in the UK. Race also plays a part, and Polynesians have a genetically low urate clearance compared with Caucasians, as exemplified by the New Zealand Maori (FE$_{UR}$ 4.9% in normouricaemic males, 3.9% in asymptomatic hyperuricemic males).

As we have seen, the body pool of urate, and hence the plasma urate concentration, is the result of a balance between production, ingestion, and excretion. The method for assessing the contribution of diet is to evaluate *de-novo* production of purines by placing the subject on a purine-free diet for 5–7 days and measuring the urinary excretion of urate, which will equal endogenous production. In this way, less than 1–5% of subjects in any country have been found to excrete abnormally large amounts of urate (>3 mmol per day). In such rare cases, an underlying genetic metabolic defect can be generally demonstrated in which the normal feedback controls on *de-novo* nucleotide synthesis and thus endogenous purine production are overridden, resulting in gross uric acid overproduction. Two such defects – both of which are X-linked and generally present in childhood, or adolescence, but sometimes neonatally – have been identified.

Purine Content of Foods

A knowledge of the purine content of specific foods is essential if dietary effects are to be reduced to a minimum. Until recently, such data have been difficult to find but are now available on the web. Most tables give only purine nitrogen, which, as demonstrated by the purine loading studies mentioned earlier, is not always a good guide because of the variation in absorption. Pâté is a particular culprit, as is most offal and organ meat (liver, kidney, heart, brains, sweetbreads), game (venison, pheasant, partridge, grouse), and the nucleic-acid-rich fish and seafoods – herring, kippers, sardines, smelts, sprats, anchovies, salmon, trout, mackerel, crustaceans (crab, lobster, prawns), shellfish (scallops, mussels), and caviar or roe. Purine nitrogen varies and ranges from 50 mg per 100 g in beef steak to 234 mg per 100 g in sardines. Many fresh vegetables, e.g., spinach, peas, beans, lentils, mushrooms, asparagus, and cauliflower, also have a considerable purine content, as have soya and other pulses and grains (porridge and oats, wheat and rye cereals). All meat extracts (Bovril, Oxo) or yeast extracts (Barmene, Tastex) are very rich in purine.

However, many studies have established that humans addicted to diets rich in nucleic acids are generally very reluctant to alter their dietary habits despite the strongest advice to do so. Consequently, therapy to reduce the pathological effects of dietary nucleic acids, namely the elevated uric acid levels, becomes essential.

Therapeutic Approaches to Reducing Uric Acid Levels

Numerous plasma uric acid-lowering drugs are in current use. Some act by increasing the renal elimination of uric acid (uricosuric drugs, e.g., benzbromarone), or restrict its formation (e.g., allopurinol). Allopurinol reduces urine uric acid levels as well. As mentioned earlier, studies in both humans and animals have shown that allopurinol has an additional beneficial effect in reducing dietary purine absorption. Allopurinol is usually a safe drug, but in rare instances, mostly in renal disease, other factors must be considered. The active metabolite of allopurinol, oxypurinol, is handled by the kidney in a fashion akin to uric acid and thus, even normally, has a long half-life. Since excretion of oxipurinol is reduced in renal failure, the allopurinol dose must be reduced to lower the plasma oxipurinol and minimize the risk of bone marrow depression, or other undesirable side-effects, which include epidermal necrolysis and hepatotoxicity. A recent study pinpointed the poor response to allopurinol in heavy drinkers and related this to the combined effect of ethanol in impairing urate excretion and increasing production. In rare instances, patients have had a severe allergic reaction to allopurinol. In such cases, the potent uricosuric drug, benzbromarone has proved beneficial, even in renal failure patients with kidney function as low as 25% of normal.

Allopurinol and Acute Renal Failure

It is important to note that in patients with genetic uric acid overproduction, allopurinol should be used with care. Its use to avert gout (or uric acid nephropathy) has precipitated acute or chronic renal failure due to xanthine nephropathy and sometimes xanthine stones instead. Xanthine is even more insoluble than uric acid, and as with uric acid, solubility cannot be improved by alkalinization of the urine. Xanthine nephropathy also may occur during massive endogenous nucleic acid breakdown in patients given allopurinol during aggressive therapy with cytotoxic drugs for malignant disorders. In such cases, much lower doses of allopurinol should be coupled with adequate hydration and alkalinization of the urine.

Role in Chemical Carcinogenesis

Substantial evidence from microorganisms and mammalian cells has implicated mutagenic events caused by damage to endogenous DNA as the initiating factor in carcinogenesis. Damage to critical regions of the genome of somatic cells can be produced by a variety of environmental mutagens. Examples of environmental factors affecting humans are cigarette smoke, asbestos, and ultraviolet (UV) irradiation. Recent putative additions to the list are radiation from microwaves and mobile phones.

A high correlation invariably exists between the mutagenic activity of different chemicals and their carcinogenic activity. Epidemiological evidence indicates a relationship between exposure to benzene and nonlymphocytic leukemia in humans, but the significance of DNA adduct formation in this is not clear. Normally, DNA possesses active repair systems to protect against such damage, which can be caused by a variety of chemical and physical agents, including ionizing, radiation, and UV light. Strands may become cross-linked, bases can be altered or lost, and phosphodiester bonds may be broken. UV light induces the formation of pyrimidine dimers, which are recognized and cleaved by specific endonucleases. Failure to repair the damage before DNA replication occurs results in the damaged region becoming the site of somatic mutations, chromosomal

rearrangements or amplifications, and aberrant DNA methylation.

In this context, it is of interest that the P53 gene is altered in many human cancers. Recent studies using analogs that target specific enzymes of either purine or pyrimidine nucleotide biosynthesis have identified P53 as a cellular regulator of both nucleotide synthetic pathways (**Figure 2**). These studies have led to the proposal that P53 is not only a sensor of DNA damage, inducing apoptosis in response to DNA strand breaks, but rather a sensor of nucleotide depletion. Upregulation of P53 expression, activated by nucleotide depletion or related processes, prevents cells entering the S phase when precursor nucleotide pools are low, thus avoiding replication of damaged DNA through a prolonged, but reversible, G0/G1 arrest. The significance of these findings is that elongation of the S phase during nucleotide depletion is known to be associated with increased chromosome breakage. (*See* **Carcinogens**: Carcinogenic Substances in Food: Mechanisms; Carcinogenicity Tests; **Mutagens**.)

See also: **Carcinogens**: Carcinogenic Substances in Food: Mechanisms; Carcinogenicity Tests; **Coenzymes**; **Infant Foods**: Milk Formulas; **Macrobiotic Diets**; **Mutagens**; **Protein**: Synthesis and Turnover; **Renal Function and Disorders**: Kidney: Structure and Function; **Vegetarian Diets**; **Viruses**

Further Reading

Diem K and Lentner C (eds) (1956 and 1970) *Scientific Tables – Chemical Composition of Foodstuffs*, 7th edn, pp. 230–243. Basle: Geigy.
Grahame RG, Simmonds HA and Carrey EA (eds) (2002) *Gont – an at your fingertips guide*. London: Class Publishing.
Grimble GK (1994) Dietary nucleotides and gut mucosal defence. *Gut* 35: S46–S51.
Henderson JF and Paterson ARP (1973) *Nucleotide Metabolism: An Introduction*. New York: Academic Press.
Linke SP, Clarkin KC, Di Leonardo A, Tson A and Wahl GM (1996) A reversible, p53-dependent G0/G1 cell cycle arrest induced by rNTP depletion in the absence of DNA damage. *Genes and Development* 10: 934–947.
Scriver CR, Beaudet AL, Sly WS and Valle D (eds) (2001) *The Metabolic and Molecular Basis of Inherited Disease*, 8th edn. New York: McGraw-Hill.
Secades JJ and Frontera G (1995) CDP-choline: Pharmacological and clinical review. *Methods and Findings in Experimental and Clinical Pharmacology* 17: 1–54.
Stone TW and Simmonds HA (1991) *Purines: Basic and Clinical Aspects*. London: Kluwer.
Zöllner N and Gresser U (eds) (1991) *Urate Deposition in Man and its Clinical Consequences*. Berlin: Springer-Verlag.

Nutraceuticals *See* Functional Foods

NUTRITION EDUCATION

S S Deshpande, Central Institute of Agricultural Engineering, Nabibagh, Bhopal, India

Background

Advances in nutrition research have provided means of prevention and control of some of the major crippling nutritional disorder with sever socioeconomic and health repercussions. The next logical step after the discovery of such solutions will be their propagation among the concerned community members. It is here that health and nutrition education has a role to play. The emphasis in these nutrition education programs ought to be on the proper use of natural dietary sources of nutrients and the need for full use of available health services and health and nutrition intervention program launched by government and other agencies.

Nutrition education programs, when properly implemented, have the potential to bring out desired behavior modifications among the communities. But for them to be successful, there are several obvious prerequisites such as the right approach suited to the community appropriately, motivated change agents, suitable educational strategies in terms of media, messages, etc. The ultimate impact of any nutritional education effort hinges on the extent to which the above criteria are satisfied.

Nutrition education has been defined as the process by which beliefs, attitudes, environmental influences, and understanding about food lead to practices that are scientifically sound, practical, and consistent with individual needs and available food resources; nutrition education should be available to all individuals and families. The fundamental philosophy of nutrition education is that efforts should focus on the establishment and protection of nutritional health rather than on crisis intervention. It is needed, regardless of income, location or cultural, social or economic practices, or level of education. Nutrition education must be a continuing process throughout the life cycle as new research brings additional knowledge.

Scope

Education means a change in behavior. It moves the individual from lack of interest and ignorance to increasing appreciation and knowledge and finally to action. Nutrition education offers a great opportunity to individuals to learn about the essentials of nutrition for health and to take steps to improve the quality of their diets and thus their well-being.

Nutrition education must continue throughout the individual's life in order to accommodate for developments in nutrition science and for changing economic circumstances, health requirements, and the new food products appearing in the nation's market. This requires a greatly expanded use of the mass media, and the involvement of governmental and private agencies and universities, as well as the food industries.

Concept

A concept is an idea around which the content of nutrition education curricula can be built.

What is Nutrition Education?

Nutrition education is the process of teaching the science of nutrition to an individual or group. Health professionals have a different role in educating an individual in the clinic, community, or long-term health-care facility. In these settings, the dietician, nutritionist, or nurse serves to assist or enable individuals to incorporate changes in eating patterns and behavior into their lives. The major focus of this type of nutrition is not knowledge and facts, but rather the development of permanent behavioral changes. This is the art of nutrition education – breaking down a large body of knowledge into small, individual components that are represented to a patient or client at a rate and level, at which they are able to absorb and use the information. Effective education is making nutrition information digestible and usable in an everyday setting.

Steps Involved in Nutrition Education

The development of an organized, coordinated nutrition education program requires a series of steps:

- Consumer research is needed to determine consumer use of existing information attitudes, knowledge, levels of awareness and concern, current nutrition practices, and type of education to which they should be most receptive.

- Nutritional surveillance data are needed to determine the most important public information problems for determining reasonable objectives to be used in formulating the nutrition education.
- Most appropriate target populations for nutrition education are determined on the basis of awareness, nutritional status, current knowledge, and receptivity to nutritional information.
- A nutrition education message should be designed that is appropriate for the needs of the selected population groups.
- The curricula for the nutrition education program must be well thought out and planned. At every step, the learner and the educator should work together to develop goals, objectives, and appropriate activities and resources to meet the goal.
- Evaluation is crucial to any successful educational effort.

Why Nutrition Education?

A number of forces have emerged to stimulate increased recognition of the need for a stronger, more comprehensive and creative effort at nutrition education. These factors include:

- Many serious as well as less serious health problems are the result of lifestyles and personal habits, including habits of food consumption. Individual modifications in lifestyles may improve health status.
- Food Patterns. Differing forms of food vary in nutritive value as well as in taste appeal. The average consumer is often mystified and unaware of how to make the wisest food choices from such a variety. Thus, there is a great need for education to enable consumers to select wisely an appropriate diet from the vast supply of food available.
- Lifestyles are drastically different from those of the turn of the century. The public must be knowledgeable about how to meet nutritional needs without overindulgence in foods.
- Nutrition education is also necessary to aid the consumer to make economical food choices, that is, to save money while getting optimal nutrition. This is especially important for the low-income consumer.
- On a global basis, there is a real possibility of food shortages. Therefore, the consumer should know how to provide nutritional substitutes for unavailable food items.

What to Emphasize in Nutrition Education

It is very difficult to define the basic knowledge that should be conveyed to the public about nutrition.

Much will depend upon what the learners already know, how they evaluate their present dietary habits and nutritional status, what their present eating habits actually are, how much they desire to know, and what changes they are willing or can afford to make.

Nutrition Education Strategies

In view of the multifactorial influences on food habits, there is no one way to persuade people to change their food choices. Thus, nutrition education must use a variety of techniques that must be a part of total family and community environments.

In other words, nutrition education involves information exchange as well as techniques to motivate and reinforce improved food habits. Successful nutrition education must include endeavors to make beliefs, attitudes, values, environmental factors, and individual ideas about food conducive to nutritionally sound, practical, and acceptable dietary habits. Nutritional education can be approached in various ways. What might be termed the rational–empirical curriculum design is the approach that is typically used by educators; it is logical and planned, and it involves behavioral objectives, activities that are designed to achieve objectives, and evaluation to determine the discrepancy between planned and achieved outcomes. The idea of self-responsibility for health suggests the basic notion that the individual possesses a dignity, worth, and responsibility to maximize these characteristics. Self-responsibility assumes that the individual has the potential and motivation to make wise judgments about factors that affect their health status. Another strategy has been termed the travel metaphor method. The idea is that there is a body of knowledge, facts, concepts, values and believes that are known and understood by the educator. It is the educator's responsibility to lead the learner through this body of knowledge and beliefs, and point out to the learner what is thought to be of importance. The educator is thus a travel guide whose job is to provide the student with a set of stimuli and the opportunity to take advantage and react positively. It then becomes the responsibility of the student to determine how they will make use of the contacts provided by the educator.

Another approach called the garden metaphor explains how humans grow, develop and flower naturally. The educator has the opportunity to nurture this environment so that the 'child' has the most opportunity to mature and to learn.

Another approach to health and nutrition education takes a strategy of social manipulation. The basic premise is that the matters affecting health are too urgent, too important, and beyond the individual's power to control on their own. Thus, the role of the health educator is seen as one of attempting to control behavior and to shape the desired responses. This approach is often employed by the mass media.

The manipulative approach is more concerned with the behavior response patterns rather than the manner in which the individual reaches these patterns. The approach is an attempt to reach the desired behavior through a short-cut method and does not take time to prepare the individual to make fully informed, wise choices.

The most appropriate and effective strategy for stimulating improved food habits will depend upon the knowledge, background, personal characteristics, and motivation of the audience. It may also depend upon the size of the target group. Ideally, the approach should be individualized to the target audience's needs, goals, objectives, and other characteristics. A mix of various approaches may be necessary.

Systems Approach to Nutrition Education

The 'systems' approach to nutrition education provides a useful and practical tool to aid the nutrition educator in deciding what should be taught and how it should be taught. Simply defined, the systems approach is a flow of steps needed to achieve the desired objectives. The approach consists of input, process, and output.

A successful nutrition education effort must incorporate managerial elements that are crucial to the success of all program planning. These elements, which could be termed the input, are essential to an effective nutrition education process. They include needs assessment and identification of problems; study of available resources and alternative strategies of intervention; determination of objectives and goals; assignment of resources to implement program; staff development to implement program, and program monitoring and evaluation. Crucial to the entire process is a firm organization base and administrative and budgetary support for nutrition education efforts.

Follow-up and Evaluation

Nutritional counseling follow-up, in order to evaluate learning, is an essential factor, which needs to be included in the plan. After an initial interview, follow-up is usually planned for some time during the next 1–2 weeks or perhaps at a later time in the same week. As clients progress toward a goal, follow-up visits are generally spaced further and further apart. In a follow-up session, the client's progress needs to be measured against the objectives, and the client needs to receive positive feedback about their progress. In

the event that no progress has been made or the client has slipped back into old habits, the counsellor needs to pursue the problem assertively. The problem needs to be identified and new goals or methods of achieving the goals established.

Summary

Nutrition education is a broad area; however, the nutritionists or nurses facilitating it are usually concentrated in the acute or long-term care facilities and community clinics. In all aspects of teaching and counseling, instructional objectives need to be established, the learner's knowledge assessed, behavioral objectives written, evaluation tools devised, information needs screened and organized, education materials selected, and instruction critiqued and revised. It is also essential to keep the learner's needs in the forefront and work to maintain a rapport with the client.

Education is not complete until the learning has been evaluated. Formal and informal tools may be used to measure learning against established objectives. Evaluation helps identify if further teaching or clarification of information is needed. To some degree, demonstrating successful learning at evaluation reinforces continuance of new behaviors, which is the goal of nutrition education.

See also: **Community Nutrition**; **Dietary Requirements of Adults**; **Famine, Starvation, and Fasting**; **Food Safety**

Further Reading

American Dietetic Association (1973) Position paper on nutrition education for the public. *Journal of the American Dietetic Association* 62: 429.

Caliendo MA (1986) *Nutrition Education. Definitions and Concepts in Nutrition and Preventive Health Care*, pp. 410–437. New York: Macmillan.

Deshpande SS (1999) Study of nutritional pattern of some selected villages around Bhopal with special reference to soybean consumption. Ph.D. Thesis, Barkatullah University, Bhopal.

Griffin GA (1974) Strategies in Curriculum Development In: *Conference on Education in Nutrition, Looking Forward from the Post*. New York: Columbia University, Teachers College.

Rau P (1982) Experiences in nutrition education. *Nutrition* 22(2): 3.

Robinson CH, Lawler MR, Chenowth WL and Garwick AE (1986) *Nutrition Problems and Programs in the Community. Normal & Therapeutic Nutrition*, 17th edn., pp. 355–357. New York: Macmillan.

SNE Concepts for Food & Nutrition Education (1982) *Journal of Nutrition Education* 14· 1

NUTRITION POLICIES IN WHO EUROPEAN MEMBER STATES

A Robertson, WHO Regional Office for Europe, Copenhagen, Denmark

Introduction

In December 1992, after more than 2 years of preparatory work by the member states and the joint efforts of the World Health Organization (WHO) and the Food and Agriculture Organization (FAO), the International Conference on Nutrition (ICN) was convened in Rome to:

- increase public awareness of the extent and seriousness of nutrition- and diet-related problems worldwide
- promote effective strategies and actions to address these problems
- encourage the political commitment necessary to do so

Each of the 159 participating countries and the European Community reaffirmed their determination to insure sustained nutritional well-being for all people and committed themselves to achieve this goal by unanimously adopting the World Declaration and Plan of Action for Nutrition. The nine action-oriented strategies of the World Declaration and Plan of Action for Nutrition are summarized in **Table 1**. These provide a framework for national plans of action.

In 1996 a follow-up to ICN (**Table 2**) was held in the European region through a FAO/WHO consultation in Warsaw, Poland. Its aim was to assess implementation of the World Declaration. Representatives from European member states and Canada, the USA, the European Commission and the Holy See attended the meeting along with representatives from the United Nations Development Program (UNDP), the United Nations Children's Fund (UNICEF), nongovernmental organizations (NGOs) and the private sector.

Methods

Member states were requested to prepare a summary report of their progress for the 1996 follow-up meeting in Poland. Based on the 35 reports received (response rate of 70%) the WHO Nutrition Unit compiled an overview of the situation.

Member States were grouped into eight geographic regions: Balkan, Baltic, Central Asian Republics (CAR), countries of central and eastern European (CCEE), western Europe, southern Europe, Commonwealth of Independent States (CIS), excluding CAR, and Nordic countries (**Table 3**). The aim of this is to facilitate comparative analysis and interpretation and assist in drawing conclusions and recommendations.

Results

The country data are presented in tables under the headings of the strategies of the Plan of Action for Nutrition adopted at the ICN.

Discussion

Developing Food and Nutrition Policy

Around the time of ICN (1992), some member states in Central Asia, the former Soviet Union and the Balkan region were not fully independent and so were unlikely to have information about the ICN. Thus food and nutrition specialists in some parts of the WHO European region were probably unaware of the ICN and so it is hardly surprising that many countries had not developed national plans of action according to the ICN objectives.

Moreover a few WHO member states are not members of FAO (1998) and others have only recently joined. This probably explains why there were more health than agriculture representatives present – 66% compared with 27% – at the consultation in Poland in 1996.

One of the difficulties when attempting to develop intersectoral policies is the level of commitment that

Table 1 The nine action-oriented strategies of the plan of action for nutrition

Incorporating nutritional objectives, considerations, and components into development policies and programs
Improving household food security
Protecting consumers through improved food quality and safety
Promoting breast-feeding
Caring for the socioeconomically deprived and nutritionally vulnerable
Preventing and controlling specific micronutrient deficiencies
Promoting appropriate diets and healthy lifestyles
Assessing, analyzing, and monitoring nutrition situations
Preventing and managing infectious diseases (N.B. not included in this analysis, since not mentioned in any country reports)

each stakeholder has to public health versus their own possibly diverging interests (**Figure 1**). It is improbable that intersectoral policies will be implemented, unless all stakeholders are committed to the process. However, the long-term interests of the agriculture sector, the food industry, wholesale and retail representatives are dependent on policies which have been developed in collaboration with the voluntary sector and consumers. Consumers, the customers of the food industry, should be more involved in the process whereby food policy is developed.

Intrasectoral Collaboration within the Health Sector

Good collaboration between nutrition and food safety is essential because the public and consumers perceive food in a holistic way. Consumers do not compartmentalize food or distinguish between food safety and nutrition – consumers want good wholesome food they can enjoy without fear. There appears to be far more collaboration between the nutritionists and food safety specialists working in central and eastern Europe compared with those in western Europe.

There may be several reasons for the stronger collaboration in central and eastern Europe. Nutrition is a relatively new science and in eastern parts of Europe its evolution seems to be closely linked to 'hygiene' and the sanitary–epidemiology system. In eastern Europe nutrition and food hygiene have evolved from the same postgraduate specialization, usually as a specialization of medicine. In former socialist countries food safety is traditionally under ministries of health, whereas in some western European

Table 2   Member states at the International Conference on Nutrition follow-up consultation, Poland, September 1996

Total no. of FAO/WHO member states invited	Total no. of countries present	Total no. of representatives	Percent of health and agriculture representatives from European region	Total no. of reports submitted (European region)
57	40	62	66% health 27% agriculture 7% others	35

FAO, Food and Agriculture Organization; WHO, World Health Organization.

Table 3   Member states of the World Health Organization European region, 1996

Balkan	Baltic	CAR and Turkey	CCEE	Western Europe	Southern Europe	CIS	Nordic

Albania
Bosnia and Herzegovina
Croatia
Federal Republic of
  Yugoslavia (Serbia
  and Montenegro)
Former Yugoslav Republic
  of Macedonia[a]
Slovenia[a]
	Estonia
	Latvia
	Lithuania
		Kazakhstan
		Kyrgyzstan[a]
		Tadjikistan[a]
		Turkey
		Turkmenistan[a]
		Uzbekistan
			Bulgaria[a]
			Czech Republic
			Hungary
			Poland
			Romania
			Slovakia
				Austria
				Belgium[a]
				France
				Germany
				Ireland
				Luxembourg
				Netherlands[a]
				Switzerland
				UK
					Greece
					Israel
					Italy
					Malta
					Monaco[a]
					Portugal[a]
					San Marino[a]
					Spain[a]
						Armenia
						Azerbaijan
						Belarus
						Georgia[a]
						Republic of
						  Moldova
						Russian
						  Federation[a]
						Ukraine
							Denmark
							Finland[a]
							Iceland
							Norway
							Sweden

[a]No report submitted (Finland was added later).
CAR, Central Asian Republics; CCEE, countries of central and eastern Europe; CIS, Commonwealth of Independent States.

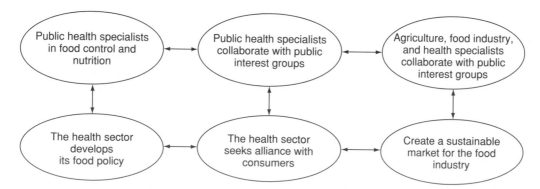

Figure 1   From intra- to intersectoral policy development food priorities for health.

countries only nutrition is under ministries of health and the responsibility for food legislation and enforcement may be with ministries of agriculture or trade in the west. However this is changing, especially in some European Union (EU) member states, where ministries of health are taking a much more proactive role in protecting consumers and their health.

Coordination between public health specialists working in food safety and nutrition is important for many other reasons. One is that from time to time food control authorities may issue warnings about food, e.g., chicken that contains *Salmonella*, or fish and vegetables which are contaminated. Meanwhile, as part of national campaigns, nutritionists promote the consumption of vegetables, fish, and poultry. Public health specialists must deliver consistent and reliable information to avoid public confusion. Closer cooperation between nutritionists and food safety specialists can prevent the promotion of conflicting messages.

Intersectoral Collaboration

Public health specialists dealing with food safety and nutrition should strengthen their collaboration with other sectors dealing with food, e.g., agriculture, food industry, wholesalers and retailers, voluntary sector and public interest groups representing consumer interests. Consumers have a vested interest in supporting the supply of safe food of good quality that is nutritionally healthy and the food industry has a vested interest in supplying healthy consumers. There is an opportunity for the health sector to strengthen health alliances with public interest groups working in the voluntary sector.

Surprisingly, in the country reports, there was very little mention about the importance of collaboration with consumer agencies except by Hungary, Norway, Poland, and Sweden. In these countries there is an understanding that the food industry can profitably increase the availability of vegetables and fruit, low-fat milk, and lean meat, on condition that it is assured a market.

Assessing the Nutrition Situation

Almost all countries reported a need to improve nutrition information systems. According to DAFNE (Network of the Pan-European Food Data Bank Based on Household Budget Surveys: countries participating are Belgium, Denmark, Greece, Germany, Hungary, Ireland, Poland, Portugal, Spain, and the UK), some had no information, while some use secondary data or access nutrition information from household budget surveys. Some had information from one-off surveys that were sponsored through

international cooperation. The Nordic countries and the UK seem to have established the most comprehensive surveillance systems that provide information on food consumption and nutritional status on a routine basis. Clearly it will take time for some countries to achieve this level of sophistication but ultimately these data are essential to facilitate food and nutrition policy development.

It seems unlikely that, at least in the near future, nationally representative nutrition surveys will be carried out on a regular basis; therefore other strategies have to be found: members of DAFNE extract nutrition-related data from household budget surveys; in Italy secondary data sources were used to develop a national policy. Mortality data are available from the WHO Health For All database which is accessible through the internet from the WHO website. These data allow comparison of mortality statistics between different countries. FAO food balance sheets provide data such as the percentage energy coming from carbohydrate, fat, protein, and alcohol and the quantity of vegetables and fruits available nationally. These data are directly accessible from the FAO website.

In the future the EU may encourage its member states, and those wishing to join, to develop standard information systems. This will allow comparison of data on food intake, nutritional status, and diet-related behavior in different EU member states and so facilitate the development of cost-effective strategies that enable countries to save money on treating preventable diseases.

Improving Food Security

After the Second World War the main aim in Europe was to increase the supply of food, especially animal products, e.g., meat and milk. However the postwar policy was too successful (**Table 4**) and in the 1980s until the present time Europe has surplus stocks of butter, meat, and milk.

Since the Second World War, new scientific evidence has led to the level of recommended daily protein intake being reduced. For example the level of protein intake recommended by the FAO, WHO, and the United Nations University (UNU) in 1986 was $0.75\,\mathrm{g\,kg^{-1}\,day^{-1}}$. This new level resulted in many countries reducing the level of their national recommended protein intakes for adults from $120\,\mathrm{g}$ to around $50\,\mathrm{g}$ per person per day. There is now some evidence suggesting that excess intake of some sources of protein, especially from red meat, could be associated with ill health.

Various hypotheses exist to explain how meat consumption may contribute to chronic diseases such as heart disease and certain cancers. Meat intake is associated with an increased intake of saturated fats

Table 4 Food policy in the twentieth century

1950s	1960s	1970s	1980s	21st century
Food insecurity and perceived nutrient deficiency, especially protein	Increased food production and consumption, especially of animal protein and fat	Surplus stocks of food, especially animal products	Links identified between mortality from NCDs and diets high in animal fats and low in vegetables and fruit	The opportunity to change food policy to meet health and environmental recommendations

NCDs, noncommunicable diseases.

which are linked to an increased risk of cardiovascular diseases and breast and colorectal cancer. Excessive protein intake may be associated with increased excretion of calcium and so possibly exacerbate the prevalence of osteoporosis. Excess intake of protein is associated with increased risk of renal disease, especially in infants, since the kidneys have to excrete excessive solute loads and so excess protein intake is perhaps linked to the risk of developing high blood pressure later in life. It is advisable that ministries of health review their national recommended intakes to insure that these are in line with current international recommendations.

Lessons learned over the past 50 years suggest future food policies should insure that:

- cereal and potato production should be geared to supply > 50% energy
- vegetables (excluding potatoes) and fruit production should be geared to supply a consumption of at least 400 g per day per person

Given the recommendations, contained in most dietary guidelines, 'to protect, promote and support the consumption of cereals, potatoes, vegetables and fruit,' it is surprising that there was no mention of the need to increase the production of these foods in the country reports. Without a food policy which guarantees food security in the form of cereals, potatoes, vegetables, and fruit, it is difficult to see how national targets for a healthy diet can be achieved.

One of the reasons why families living in countries undergoing economic transition (CEE and CIS) are not suffering from protein/energy deficiency is because many produce a large percentage of their own potatoes, vegetables, and fruit. In Russia, town dwellers produce 88% of their potatoes, 43% of their meat, 39% of their milk, and 28% of their eggs on urban household plots. This important share of production is generated on plots of 0.2–0.5 ha which together constitute only 4% of the total amount of agricultural land in Russia. In addition, imports from abroad have supplemented national food production and currently 50% of chickens and 30% of meat consumed in the Russian Federation is imported. In

the short term, imports are essential during times of hardship and economic transition, but in the long term agricultural polices should promote sustainable production and protect local food production to insure national food and nutrition security.

In eastern Europe the agriculture sector forms a much higher percentage of national gross domestic product compared with western Europe. Almost 30% of the population in some CEE countries are employed in agriculture compared with only 1 or 2% in western Europe. Many economists are concerned about the impact when more countries join the EU because the Common Agriculture Policy (CAP) is not sustainable. The CAP receives almost half the EU budget and in 1995 it cost EU taxpayers 39 billion ECU. During the CAP reform process, health recommendations and dietary guidelines should influence the reformed policy in addition to concerns about national employment and the national food security of the EU accession countries.

Protecting Health Through Food Safety and Quality

In some reports there was no mention of the importance of food safety, especially in the countries of southern Europe. As mentioned in the introduction, it is vital to strengthen the collaboration between food control and nutrition agencies. Some western governments are already attempting to bring these two disciplines under one food authority and three-quarters (27 out of 35 countries) did include food safety in their ICN reports. Close collaboration becomes even more essential as the line between food safety and nutrition becomes more and more blurred with the introduction of precooked foods; functional, novel, and special dietary foods; and supplements. Moreover, with the introduction of food into global trade, the public's need for information will be immense. To provide reliable information, both food safety specialists and nutritionists need data on food intake patterns. More effective use could be made of resources if joint surveillance and risk management of food intake are carried out together.

Foodborne diseases are among the most widespread health problems in the contemporary world.

In rich and poor countries alike they impose substantial health burdens, ranging in severity from mild indispositions to fatal illnesses. The emergence of new foodborne diseases is an ominous trend. Epidemics due to newly identified pathogens such as *Campylobacter jejuni, Listeria monocytogenes, Escherichia coli 157:H7* and bovine spongiform encephalopathy (BSE) have hit industrialized countries. *Salmonella* outbreaks are one of the main causes of epidemics in Europe. Global trade will make it more difficult to contain foodborne diseases within national borders and enforce national food laws. In addition, chemical contamination, toxic materials, pesticides, veterinary drugs, and other agrochemicals require constant surveillance to insure their safe use. Similarly, the use of food additives can improve the quality of the food supply but appropriate controls are necessary to insure their proper use.

Most regions included in their country reports the need to harmonize their national food legislation with the EU and World Trade Organization (WTO). Many countries stated they are lining up to join either the EU or WTO (12 countries) and so intend to harmonize their national legislation with international directives set by the EU or Codex Alimentarius. It is essential that health professionals, especially those working in public health, become more informed about international agreements on global food trade and take a full part in the WTO Committee on Sanitary and Phytosanitary Measures (SPS Committee). Only if public health specialists, from both food safety and nutrition, participate can they hope to influence future food policies.

Promoting Breast-Feeding

Clearly there is a lot to be done in the European region to promote breast-feeding. More than one-third (13) of the 35 member states made no reference to breast-feeding in their country report. This does not necessarily mean that there are no active breast-feeding programs in the country (e.g., Romania) but it does mean that there is probably little coordination between those responsible for nutrition policy and those responsible for promoting breast-feeding. Perhaps this is because breast-feeding is regarded as a clinical issue dealt with by midwives and primary health care workers. However, public health nutritionists and hygienists in CCEE and the CIS do have an important role and should be active in the development of national policies to promote breast-feeding.

National breast-feeding initiatives will only be successful if policies are developed and enforced according to guidelines from the Baby-Friendly Hospital Initiative (BFHI) and the International Code for the marketing of breast milk substitutes.

It is necessary to national and regional trends (marketing strategies and breast-feeding levels) to interpret the national situation and make recommendations to policy-makers.

The BFHI jointly launched in 1991 by WHO and UNICEF, aims to enable women to choose and practice breast-feeding as the primary source of nutrition up to about 6 months by minimizing obstacles which make it difficult or prevent women from exercising their right to breast-feed and to insure the cessation of free and low-cost infant formula supply to hospitals. The BFHI targets maternity services and hospitals, particularly health workers and those responsible for policies to help mothers succeed in breast-feeding. To become a baby-friendly hospital every facility providing maternity services and care for newborn infants should implement the Ten Steps to Successful Breastfeeding outlined in the joint WHO/UNICEF statement entitled *Protecting, Promoting and Supporting Breastfeeding: The Special Role of Maternity Services* and should end free and/or low cost supplies of breast milk substitutes.

There is clearly a vital need to increase the number of Baby-Friendly Hospitals in Europe and develop and enforce national legislation which is based on the International Code of Marketing of Breast Milk Substitutes.

In some countries there may be conflict between nutritionists, hygienists, and new breast-feeding policies. Some hygiene standards and decrees in the former Soviet countries, such as the recommendation that mothers should be separated from their babies after birth, contradict the current UNICEF and WHO recommendations to promote the 'rooming-in' of babies with their mothers. In addition, National Nutrition Institutes involved with research and development of breast milk substitutes and weaning foods may unintentionally interfere with breast-feeding promotion.

Complementary feeding (weaning) was not discussed in detail in any report and only mentioned very briefly in two country reports, those of Albania and Italy. Perhaps this is because the categories listed in the World Declaration of ICN do not specifically include complementary feeding, only breast-feeding. This was perhaps an oversight by the ICN committees since it is clearly important that all member states address their policies on infant and young child feeding. It is around the age of 6–24 months when mortality and morbidity levels are at their highest. It is likely that the high prevalence of anemia seen in some parts of the European region is due to poor feeding practices, such as the early introduction of foods and drinks, especially cows' milk and tea, before the age of 6 months.

Caring for the Deprived and Vulnerable

In the region around 70% of the population is suffering from physical and mental exhaustion as people cope with increasingly uncertain conditions related to very rapid economic and social change. This is especially true for the countries of the former Soviet Union where policy-makers are not used to developing safety nets for vulnerable groups. The transition to a market economy has resulted in soaring consumer prices due to a reduction in subsidies while salaries have remained low. In addition there is rising unemployment as a result of industrial restructuring and the new monetary polices have hit vulnerable groups in particular. Mechanisms must be developed to identify and protect the vulnerable groups in society.

Caring for vulnerable groups was mentioned in almost three-quarters (22) of the country reports. Many initiatives were related to the physiologically vulnerable, such as children, pregnant women, or the elderly. In addition, welfare safety nets, aimed at insuring a minimum family income by setting minimum wages, pensions, and unemployment benefits, were considered necessary for economically vulnerable groups such as the poor, low-income, unemployed, and refugees or immigrants and those with large or one-parent families.

The gap between the rich and poor is increasing in many countries in Europe, and as a phenomenon of the economic transition in central and eastern Europe, unemployment has appeared in many. If this trend continues, the food intake and nutritional status of vulnerable groups should be monitored closely in order to develop safety nets and prevent food insecurity. Among European industrialized nations the UK was exceptional in the pace and extent of the increase in inequality in the 1980s and by 1990 the level of inequality was almost back to 1930 levels.

Poverty is a growing concern in Europe, especially for those with incomes below societies' poverty lines. In 1994 it was estimated that 13.5% of the Polish population (about 5 million) had average monthly expenditures lower than the relative poverty line. In 1995, 4 million children in the UK lived in families with incomes below 50% of average earnings. Lack of food is associated with poverty, which can be measured by estimating the percentage of disposable income spent on food: in Romania the average figure is around 60%; and in Poland around 40%, compared with 22% in the EU.

Coping within a limited budget often means that healthier, safer foods are not affordable. In Lithuania increased food prices have caused socially deprived people to consume cheaper, less nutrient-dense foods which are more likely to be contaminated. In Poland the prevalence of nutrient deficiencies is highest amongst the unemployed and low-income families with many children. Poor people in the UK consume much lower quantities of vegetables and fruit.

There is growing evidence for the protective effect of vegetables and fruit against chronic diseases such as cancer and coronary heart disease. This suggests that the low intakes of vegetables by poor people may directly increase their risk of these diseases. Inadequate diets also have adverse consequences on a child's health, education, and future employability and there are demonstrable costs to society in each instance.

Micronutrient Deficiencies

There is no doubt that, in the European region of WHO, the most prevalent health problems related to diet are noncommunicable diseases such as cardiovascular diseases and cancers and not deficiency diseases. Nevertheless, micronutrient deficiency does exist in certain European countries under certain circumstances.

Iodine   One of the most prevalent nutrient deficiencies is iodine. Iodine deficiency disorders (IDD) are prevalent throughout Europe and only six countries have no IDD (Finland, Iceland, Norway, Sweden, Switzerland, and the UK). Although this problem has been largely eliminated in some countries in western Europe, it is reemerging in some countries in CEE and the CIS. For example, in Albania and Tadjikistan iodine deficiency is severe and in some other countries the prevalence of IDD is increasing.

The main public health strategy recommended to solve this problem is universal salt iodization. This means insuring that all the salt used by the food-processing industry, the mass catering sector, and households is iodized. In some countries there is no universal iodization and only table salt is iodized (Italy, Poland). This could be a problem, especially in western Europe where most of the population get their salt from processed food products and not from home-cooked food to which salt is added during cooking or eating. Therefore nutritional epidemiologists should monitor the source of a population's salt, and so iodine.

In addition to insuring that iodized salt is used in food processing and mass catering, policy-makers should consider the need to include a recommendation to feed iodized salt to animals (cows). In the Nordic countries and the UK iodine is given to cows in the form of 'salt-lick' and so milk and milk products supply around 40–70% of the iodine intake in humans. In addition to insuring elimination of iodine deficiency, this has the added benefit of limiting the amount of salt consumed by the population.

In CCEE and the CIS, cardiovascular diseases are responsible for 68% of all premature deaths, while in the rest of Europe this figure is only 43%. Moreover, proportionally, cerebrovascular diseases are markedly higher in CCEE and CIS. Because of the strong link between high salt intake and cerebrovascular disease, the WHO recommends a salt intake of no more than 6 g per capita per day. In Italy a campaign on the role of salt in the diet was launched. This campaign aims to eliminate IDD through the promotion of iodized salt, but at the same time preventing hypertension and vascular diseases through limitation of salt intake.

Iron   The other main nutrient likely to be deficient in European countries is iron. Iron deficiency was mentioned as a problem by one fourth (9) of 35 member states. Prevalence studies show that anemia is widespread in central Asia and the Caucasus. For example, in Kazakhstan and Azerbaijan up to 70% of children younger than 2 years of age have low hemoglobin levels. In addition, 20–30% of women of childbearing age also have low hemoglobin levels. These problems may not be related solely to iron deficiency in the diet. For example, it has already been mentioned that early introduction of cows' milk is a major cause of iron deficiency in the young. Anemia in adults is associated with the presence of iron-absorption inhibitors in the diet, such as tea and coffee, which is frequently consumed in eastern European countries; or lack of absorption enhancers, such as vitamin C from vegetables and fruit.

Folate   Folate deficiency is associated with neural tube defects. Denmark, Poland, and the UK mentioned national strategies related to solving this problem. Folate may also play an important role in the prevention of coronary heart disease by helping to reduce levels of homocysteine. The main food sources of folate are Brussels sprouts, asparagus, spinach, broccoli, cabbage, cauliflower, parsnip, iceberg lettuce, beans, peas, and beef and yeast extracts – yet another reason why national food and nutrition polices and dietary guidelines should promote vegetable production and consumption.

Other micronutrients   There may be situations when the only solution to solve micronutrient deficiency is by food fortification of bread or salt, for example (such as for iodine deficiency), or, more rarely, by supplements (during emergencies). However, generally, where possible, WHO promotes primary prevention strategies such as increasing vegetable consumption rather than advocating consumption of nutraceuticals, food supplements, or multivitamin tablets. Population strategies advocating the use of vitamin supplements are not the solution to micronutrient deficiencies in Europe. Vitamin supplements may have many disadvantages: side-effects; nutrient imbalances; toxicity; malabsorption; long-term dependence and lack of confidence in locally produced foods, and, finally, supplements are an unnecessary expense.

Appropriate Diets and Lifestyles

Countries of the CARs and the CIS must pay more attention to the link between diet and noncommunicable diseases (NCD), such as cardiovascular diseases, certain cancers, diabetes, hypertension, and obesity. Some of the country reports focused too much on deficiency of protein and micronutrients. In focusing on deficiency the link between premature mortality from NCD and a diet high in fat, salt, fatty red meat, and fatty/sugary foods and simultaneously low in vegetables and fruit is neglected.

Some countries in the CIS and central Asia mentioned the need to increase the teaching of dietology/dietetics. There is still a tendency in some countries to prescribe many different types of diets for different disorders. Many of these dietary prescriptions have not been scientifically proven. In addition, many of the individuals referred for dietary treatment are suffering from obesity, diabetes, heart disease, high blood pressure, and other conditions related to unhealthy lifestyles. These cases should be treated using the healthy nutrition principles developed for the population. Dental caries is also widespread in many European countries and there can be little doubt that this is related to frequent high intakes of sugar as well as poor oral hygiene.

Some countries, such as Albania, Israel, Kazakhstan, Latvia, Romania, Slovakia, the Republic of Moldova, Ukraine, and Uzbekistan made no mention of the link between diet and NCDs. Moreover, only one-third (12) of 35 member states mentioned obesity (Austria, Azerbaijan, Czech Republic, Denmark, Estonia, Hungary, Lithuania, Luxembourg, Malta, Poland, Turkey, and the UK), despite the fact that the prevalence of obesity is high and appears to be increasing in every country in Europe. For example, the Russian Federation has one of the highest prevalences, where 55% of the female population is overweight. Little mention of obesity probably reflects the lack of data on body weight and height and illustrates the need for countries to collect anthropometric data as part of their health information system.

Physical activity and obesity   Obesity is a chronic medical problem caused by a combination of an energy-dense diet leading to excess energy intake and lack of physical activity. There is abundant evidence that obesity is associated with a high risk of coronary heart disease, hypertension, diabetes mellitus, and gastrointestinal disorders. The risk of cancers of a number of sites (endometrial, renal, colon, gallbladder, and postmenopausal breast cancer) is also linked to obesity. As treatment of obesity is difficult, the need to readjust dietary energy intakes and/or physical activity permanently is essential. Other health advantages of high levels of physical activity include improved mental and psychological health.

Only 17% (six of 35) of member states made reference to the importance of increasing physical activity (Armenia, Estonia, Malta, Poland, Sweden, and the UK). Environments which support improved eating habits and more active living are needed. It is recommended that prevention efforts are focused on population-based public health strategies.

Conclusion

Cooperation and collaboration are needed throughout Europe and there is a need to share information and build alliances. In western Europe there exists the European Academy of Nutritional Sciences (EANS), the Federation of European Nutritional Sciences (FENS), and the Arbeitsgemeinschaft Ernährungsverhalten/Working Association for Nutritional Behavior (AGEV) and many countries have their Nutrition Society. In CIS and CCEE there seems to be a lack of coordination, and nutrition networks or societies should be established to facilitate sharing of information and developments in the area of food and nutrition.

See also: **Anemia (Anaemia)**: Iron-deficiency Anemia; **European Union**: European Food Law Harmonization; **Exercise**: Metabolic Requirements; **Folic Acid**: Properties and Determination; **Food Poisoning**: Statistics; **Food Safety**; **Infant Foods**: Milk Formulas; **Infants**: Breast- and Bottle-feeding; **Iodine**: Iodine-deficiency Disorders; **Obesity**: Epidemiology; **Quality Assurance and Quality Control**; *Salmonella*: Properties and Occurrence; **World Health Organization**; **Obesity**: Epidemiology **World Trade Organization**

Further Reading

Adams JS and Lee G (1997) Gains in bone mineral density with resolution of vitamin D intoxication. *Annals of Internal Medicine* 127(3): 203–206.

Agra Europe (1997) *East Europe Agriculture and Food.* No. 179. London: Agra Europe.

Bates CJ (1997) Plasma total homocysteine in a representative sample of 972 British men and women aged 65 and over. *Clinical Journal of Nutrition* 51: 691–697.

Braga C (1997) Homocysteine as a risk factor for vascular disease. *Nutrition Research Newsletter* 16: 7–8.

Borgdorff MW and Mortarjemi Y (1997) *Surveillance of Foodborne Diseases: What are the Options?* WHO/FSF/FOS/97.3. Geneva: WHO.

Brenner BM (1982) Dietary protein intake and the progressive nature of kidney disease: the role of hemodynamically mediated glomerular injury in the pathogenesis of progressive glomerular sclerosis in ageing, renal absorption, and intrinsic renal disease. *New England Journal of Medicine* 307: 652–659.

CARAK newsletter (1997). Geneva: WHO.

Consumers in Europe Group (1994) *The Common Agricultural Policy: How to Spend £28 Billion a Year Without Making Anyone Happy.* London: Consumers in Europe Group.

FAO statistical databases (FAOSTAT DATA): http://apps.fao.org/cgi-bin/nph-dp.pl.

Health Education Authority (1996) *Folic Acid – What all Women should Know.* UK: Health Education Authority.

Health Education Authority (1997) *Folic Acid and the Prevention of Neural Tube Defects: A Summary Guide for Health Professionals.*

Hecht K and Steinman E (1996) *Improving Access to Food in Low-income Communities: An Investigation of Three Bay Area Neighborhoods.* San Francisco: California Food Policy Advocates.

Itoh R, Nishiyama N and Suyama Y (1998) Dietary protein intake and urinary excretion of calcium: a cross-sectional study in a healthy Japanese population. *American Journal of Clinical Nutrition* 67: 438–444.

James WPT (1997) *An Interim Proposal on the Creation of a UK Food Standards Agency.* London: FSA.

James WPT, Nelson M, Ralph A and Leather S (1997) The contribution of nutrition to inequalities in health. *British Medical Journal* 314: 1545–1549.

Käferstein FK and Clugston GA (1995) Human health problems related to meat production and consumption. *Fleischwirtschaft* 75(7) 75–77.

Leather S (1996) *The Making of Modern Malnutrition: An Overview of Food Poverty in the UK.* London: Caroline Walker Trust.

Marriott BM (1997) Vitamin D supplementation: a word of caution (editorial). *Annals of Internal Medicine* 127(3): 231–232.

Motarjemi Y and Käferstein F (1996) *International Conference on Nutrition: A Challenge to the Food Safety Community.* 1996 WHO/FNU/FOS/96.4. Geneva: WHO.

National Food Alliance (1995) *Food and Low Income: A Conference Report.* London: National Food Alliance.

National Research Council (1997) *Premature Death in the New Independent States.* London: National Academy Press.

Proceedings of Regional Conference on Elimination of IDD in Central and Eastern Europe, the Commonwealth of Independent States, and the Baltic States (1997) Munich, Germany.

Report of a Working Party to the Chief Medical Officer for Scotland (1993) *Scotland's Health – A Challenge to us* all: *The Scottish Diet.* Scottish Office Home and Health Department.

Reports of the Scientific Committee for Food (1993) *Nutrient and Energy Intakes for the European Community.* Luxembourg: Commission of the European Communities, Directorate-General Industry.

Report of the Working Group on Diet and Cancer of the Committee on Medical Aspects of Food and Nutrition Policy (1998) *Nutritional Aspects of the Development of Cancer.* Department of Health report no. 48. London: Stationery Office.

Robertson A (1997) *Priorities for Eliminating IDD in CCEE and CIS.* Regional Conference on Elimination of IDD in Central and East Europe, the Commonwealth of Independent States, and the Baltic States. Munich, Germany.

SAFE Alliance (1997) *The Common Agricultural Policy Yesterday, Today and Tomorrow.* London: SAFE Alliance.

Third European Interdisciplinary Meeting and 20th Annual Scientific Meeting of AGEV (1997) *Public Health and Nutrition.* Abstract book. Copenhagen: AGEV/WZB.

USDA (1997) *Russian Food Consumption: Emerging Demand for Quality.* Newly Independent States and Baltics Update. Agriculture and Trade Report. Economic Research Service WRS-97-S1.

WHO (1990) *Diet, Nutrition, and the Prevention of Chronic Diseases: Report of a WHO Study Group.* Geneva: WHO Technical Report Series no. 797.

WHO (1996) *Guidelines for Strengthening a National Food Safety Programme.* WHO/FNU/FOS/96.2. Geneva: WHO.

WHO (1997) *Food Safety and Globalization of Trade in Food: A Challenge to the Public Health Sector.* WHO/FSF/FOS/97.8. Geneva: WHO.

WHO (1997) *Prevention and Management of the Global Epidemic of Obesity: Report of the WHO Consultation on Obesity.* WHO/NUT/NCD/97.2. Geneva: WHO.

WHO (1998) *Comparative Analysis of the Implementation of the Innocenti Declaration in WHO European Member States.* Copenhagen: WHO Regional Office.

WHO/FAO (1992) *World Declaration and Plan of Action for Nutrition.* International Conference on Nutrition, Rome: FAO/WHO.

WHO website: http://www.who.ch/ and Health for All: http://www.who.dk/www.who.dk/mainframe.htm.

Wilkinson RG (1996) *Unhealthy Societies; The Afflictions of Inequality.* London: Routledge.

Wiseman MJ (1987) Changing in renal function in response to protein restricted diet in type 1 (insulin-dependent)-diabetic patients. *Diabetologia* 30: 154–159.

World Cancer Research Fund in association with American Institute for Cancer Research (1997) *Food, Nutrition and the Prevention of Cancer: A Global Perspective.* Washington: American Institute for Cancer Research.

Wretlind A (1982) Standards for nutritional accuracy of the diet: European and WHO/FAO viewpoint. *American Journal of Clinical Nutrition* 36: 366–375.

NUTRITIONAL ASSESSMENT

Contents
Importance of Measuring Nutritional Status
Anthropometry and Clinical Examination
Biochemical Tests for Vitamins and Minerals
Functional Tests

Importance of Measuring Nutritional Status

F Fidanza, Universita degli Studi di Perugia, Perugia, Italy

Aims of Nutritional Status Assessment

Nutrition has been more and more implicated in various health problems. Consequently, improving nutritional status can play an important role in solving or preventing these problems.

Malnutrition – both undernutrition and overnutrition – is widespread both in single patients as well as in population groups. Surgical patients, for example, often suffer from protein-energy malnutrition, and subclinical blood levels of vitamins and minerals are present even in population groups of developed countries. (*See* **Malnutrition:** The Problem of Malnutrition.)

But scientific interest is now moving from deficiencies to a new role for some micronutrients in the maintenance of health and the prevention of degenerative diseases.

Efforts to obtain reliable information about the nutritional status of both patients and population groups encounter many difficulties. There are several reasons for these problems, the most important of which is the body's ability to adapt to adverse conditions, so that changes in biochemical or functional characteristics become evident only after substantial impairment has occurred. Methods of assessment must therefore be both highly specific and highly sensitive.

The complexity of the interactions between dietary inadequacy, disease, and personal and environmental variables also makes it particularly difficult to determine whether health impairment can be linked specifically to diet and whether or not any nutritional deficiency is secondary to some other defects. These problems can be solved only through continuing basic research to expand knowledge of the patterns of biochemical, physiological, pathological, and behavioral responses to deficiencies or excess of nutrients and to improve methods for applying this knowledge in practical situations.

Nutritional surveys can be cross-sectional (i.e., provide data on prevalence of malnutrition or deficiency) or longitudinal (i.e., provide data on incidence of malnutrition or deficiency and association with other factors). The serial collection of data at population level is a basic part of surveillance or of any intervention program.

A nutritional survey can be very general and comprehensive, in which case it includes the analysis not only of the prevailing nutritional problems of a given population but also of the responsible factors and the consequences. On the other hand, a survey can also be limited to some specific nutrients or selected population or patient groups. The selection is made according the objectives of the study and the funds, personnel, and time available. (See **Dietary Surveys**: Surveys of Food Intakes in Groups and Individuals.)

Malnutrition, whether under- or overnutrition, develops in various steps. In the prepathogenic period information on dietary inadequacy can be obtained from food balance sheets at national level and dietary surveys at individual or group level. These can provide an indication of potential nutritional problems of the population or group but do not assess the problems.

Late in the prepathogenic period some early changes in the body nutrient reserves can be present, but methods available are not sensitive enough to assess them.

In the pathogenic period, before the emergence of clear clinical signs, body tissue nutrient levels are modified to such a degree that metabolic and/or functional alterations appear. Biochemical and functional tests are then appropriate for status assessment and/or diagnosis.

Once the clinical horizon has been reached, non-specific clinical signs and symptoms are present. Anthropometric measurements can be of some use early at this stage, as can biochemical and functional tests.

When symptoms are present clinical nutrition assessment and studies of morbidity data are of great help. The final stage of the nutritional problem can be assessed from analysis of mortality rates and data from postmortem examination, but in this way only information of retrospective conditions is obtained.

Sampling Procedure

In population studies if all the subjects cannot be examined, a very careful sampling procedure has to be chosen, considering first study design and then sample size. Sampling details most appropriate for each study design can be found in epidemiology textbooks. (See **Epidemiology**.)

Collecting and Storing Samples for Biochemical and Immunological Tests

Once the population sample is selected it is necessary to consider in great detail the collection and storage of samples for biochemical and immunological tests.

The instructions for sample collection and preparation vary according to the specific assays to be performed. Only general rules are provided here.

Body tissues and excretions can rapidly deteriorate and/or be easily contaminated. In general, to obtain whole blood, serum, or plasma, appropriate syringes or vacuum collection tubes are used. If handled carefully, hemolysis can be avoided.

For some minerals and trace elements, in order to avoid contamination, one must use new plastic (polystyrene or polypropylene) syringes and collection tubes, and wear powder-free latex gloves when collecting or working with specimens.

Urine samples must be acidified and chilled on ice in polypropylene or polyethylene bottles.

Specimens of adipose tissue can be obtained by microbiopsy or needle aspiration by vacuum tube assembly.

The plasma or serum for protein determinations can be stored for 1 month at 2–6 °C and for several months at −20 °C.

Lipid fractions immediately separated from serum or plasma are stable for 1 year or more at −70 °C, if properly handled. Adipose tissue specimens for fatty acid determination can be stored at −20 °C without any other precaution up to 18 months.

For most vitamin determinations, plasma or serum is used and it can be stored at −20 °C for several months or, in a few cases, even for years. Samples should be protected from light for fat-soluble vitamins and riboflavin analysis. For the enzymatic assays of erythrocyte thiamin, riboflavin, and vitamin B_6 heparinized blood should be stabilized with proper ACD solution (ACD solution is a stabilizer made of citric acid, sodium citrate and D-glucose) for storage at 4–6 °C for 10 days. For the erythrocyte thiamin pyrophosphate (TPP) assay the determination should be carried out within 2 h of drawing blood because of TPP instability. For vitamin C determination, heparinized blood, after stabilization with metaphosphoric acid, can be stored frozen, preferably at −70 °C, for no more than 3 weeks. For vitamin C determination by high-performance liquid chromatography (HPLC), heparinized blood specimens can be stored for a few weeks at −80 °C after addition of reduced glutathione within 20 min of collection. The addition of dithiothreitol increases stability of plasma samples to more than 1 year at −70 °C.

Acidified urine specimens for thiamin, riboflavin, and vitamin B_6 determinations are stable at −20 °C for 3 months. The specimens should be protected from light. Longer storage is possible for biotin and pantothenic acid.

For mineral and trace element determination serum, plasma, whole blood, or blood cells are used with no great problems for storage, but glass containers should be avoided.

For cellular studies of immunocompetence the blood must be rapidly processed; in some cases the testing of frozen cells was successful.

Recommended Methods for Nutritional Status Assessment

Table 1 is a synopsis of some recommended methods. For a detailed description of them as well as for other methods, see the Further Reading section.

These methods can be used for the whole population as well as for single patients. In the clinical setting, nutritional assessment and support are best performed by a nutritional support team, including clinician, nutritionist, and clinical chemist. If needed, advanced methods for the assessment of body composition, nitrogen balance, and some prognostic indices are available to detect patient malnutrition and to predict outcome (sepsis, wound dehiscence, death).

Data Processing and Calculation

The most important stages of data processing are: coding (including a preliminary quality control); data input; quality control data; data bank; data analysis (using parametric and nonparametric techniques); and reporting. Specialized books describe in detail the above stages. For each type of variable, specific calculation methods may be used. Only general rules will be provided here.

For variables not normally distributed (some anthropometric data, ferritin, etc.), a logarithmic transformation can be applied. In some cases the use of nonparametric tests is preferred.

For the use and interpretation of anthropometric measurements an Expert Committee Report of the World Health Organization is highly recommended. Technical framework and detailed guidance on the use and interpretation of anthropometric measurements in pregnant and lactating women, newborn infants, infants and children, adolescents, overweight and thin adults, and adults aged 60 years and over are provided. An extensive series of reference data is included in an annex.

Computer programs are now commonly used to test specific hypotheses. Because of the highly specialized skills required for proper processing and analysis of data, it is recommended that specialists in these fields take part in the planning and execution of evaluation. To improve the quality of this work it is advisable that these specialists are also involved in the data recording – an essential preliminary step of data processing. This means that they must be involved in the design of the study and in the preparation and testing of various forms and/or questionnaires needed for the study.

For data analysis, a data stratification by age groups and sex or any other meaningful stratification like altitude (for hemoglobin), socioeconomic status, and health/disease data, must often be used. Results may be expressed in terms of means and standard deviations or standard errors or as percentages of values (prevalence) above or below an arbitrary cut-off point defining adequacy. Anthropometric indices can be expressed in terms of Z-score (or standard deviation score). If the sample size is large enough, the percentiles or the frequency distributions are very useful. This data presentation facilitates comparison between surveys in different countries; it permits at any time the selection of the percentile or the range of percentiles of reference, i.e., acceptability, and it can represent a combination of the descriptive and interpretative ways for analyzing and presenting collected data.

The final step is the interpretation of data. In general the prevalence of malnutrition or deficiency is obtained using cut-off from a reference population. This reference population and the population under study are often from different areas or the criteria used to specify the reference population are not carefully defined. However, since most reference values are population-specific, it is recommended that data from a clinically healthy reference population of the

Table 1 Nutritional status methodology

Item	Assessed variable	Method or apparatus
Anthropometry		
Stature	Height	Anthropometer or stadiometer
	Sitting height	Anthropometer
	Length	Measuring table
Body mass	Weight	Beam scale
Circumferences	Mid upper arm circumference (MAC)	Flexible nonelastic tape
	Head	Flexible nonelastic tape
	Chest	Flexible nonelastic tape
	Waist	Flexible nonelastic tape
	Hip	Flexible nonelastic tape
	Thigh	Flexible nonelastic tape
Skinfold	Triceps (TSF)[a]	Skinfold calliper
	Subscapular	Skinfold calliper
	Thigh	Skinfold calliper
Diameters	Biacromial breadth	Anthropometer
	Biiliac breadth	Anthropometer
	Bicondylar breadth	Sliding calliper
	Bistyloid breadth	Sliding calliper
Derived indices	Height for age	Z-score
	Weight for age	Z-score
	Weight for height	Z-score
	Body mass index	$Weight/height^2$
	Arm muscle circumference	$MAC - (\pi \times TSF)$
	Arm muscle area	$(MAC - (\pi \times TSF))^2/4\pi$
	Waist/hip ratio	
	Waist/thigh ratio	
Body composition		
Atomic level	Carbon	Inelastic neutron scattering
	Nitrogen	Neutron activation analysis
	Others	Neutron activation analysis
Molecular level	Water	Isotope dilution technique or bioelectrical impedance analysis
	Osseous minerals	Dual-photon absorptiometry
	Protein as total body nitrogen	Neutron activation analysis
	Lipid	Hydrodensitometry
Cellular level	Body cell mass	Total body potassium (TBK) or exchangeable TBK
	Extracellular fluid	Dilutometry
	Extracellular solids	Indirect methods (total body Ca by neutron activation analysis)
Tissue system	Subcutaneous and visceral adipose tissue	Computed axial tomography
	Skeletal muscle mass	Indirect methods (24-h urinary creatinine or TBK)
	Nitrogen	Indirect method (neutron activation analysis)
Whole body	Body mass	See anthropometry
	Stature	See anthropometry
	Circumferences	See anthropometry
	Skinfolds	See anthropometry
	Densitometry	Underwater weight
Physical working capacity and physical fitness	Cardiorespiratory efficiency	Treadmill or bicycle ergometer
	Muscle strength	Dynamometers
	Motor performance	Several tests
Energy balance	Energy intake	Individual dietary surveys
	Energy expenditure	Indirect calorimetry
		Doubly labeled water
Lipid		
Essential fatty acids	Serum, erythrocyte membranes, subcutaneous fat tissues	Gas chromatography
Protein		
Transport proteins	Plasma or serum	Immunological
Fibronectin	Plasma	Laser nephelometry
Albumin	Plasma or serum	Dye-binding

Continued

Table 1 Continued

Item	Assessed variable	Method or apparatus
Creatinine	Urine	Colorimetric (autoanalyser)
		HPLC
3-Methyl histidine	Urine	Colorimetric or fluorometric
		HPLC
Somatomedin C (IGF-1)	Plasma or serum	Radioimmunoassay
Amino acids	Plasma	Automated ion exchange
		HPLC
Vitamins		
Vitamin A (retinol)	Plasma or serum[b,c]	HPLC
Carotenoids	Plasma or serum[c]	HPLC
Vitamin E (α-tocopherol)	Plasma or serum[c]	HPLC
25 (OH)-vitamin D	Serum	Competitive protein-binding
		HPLC
		RIA
Vitamin K_1 (phylloquinone)	Serum or plasma	HPLC
Thiamin	Erythrocytes	Transketolase activity[d]
	Whole blood	HPLC
	Urine[e]	Fluorometric
Thiamin pyrophosphate	Erythrocytes	HPLC
Riboflavin	Erythrocytes	Glutathione reductase activity[d]
	Urine[e]	HPLC
Flavin adenine dinucleotide	Whole blood	HPLC
Niacin	Whole blood	Microbiological
Niacin metabolites	Urine[e]	HPLC
Vitamin B_6	Erythrocytes	Aspartate transaminase activity[d]
Pyridoxal 5′ phosphate	Whole blood	HPLC
Pyridoxal 5′ phosphate	Plasma	Radioenzymatic
4-Pyridoxic acid	Urine[e]	HPLC
Folate (5-Me-THF)	Serum or plasma[e]	Competitive protein-binding
	Erythrocytes	Radioassay
Homocysteine	Plasma	HPLC
Vitamin B_{12} (cobalamins)	Serum or plasma	Competitive protein-binding
Methylmalonic acid	Serum or urine	GC-MS
Biotin	Whole blood	Microbiological
	Plasma/urine	Radioimmunoassay
Pantothenic acid	Whole blood	Microbiological
Vitamin C	Plasma/buffy coat layer	Spectrophotometric
	Whole blood/plasma	HPLC-micromethod
Minerals and trace elements		
Total Ca	Plasma	Colorimetric
Ionized Ca	Plasma	Selective electrode
Bone density	Metacarpal indices	Radiographic or dual-photon absorptiometry
Hydroxyproline	Urine	Colorimetric
Osteocalcin	Serum	Radioimmunoassay
Phosphorus	Plasma or serum	Spectrophotometric
Magnesium	Serum or urine	AAS
	Serum	Colorimetric
Ferritin	Serum or plasma	ELISA
		RIA
		IRMA
Iron	Serum	Spectrophotometric or AAS
Iron binding capacity	Plasma or serum	Colorimetric or radioactive
Protoporphyrin	Erythrocytes	Hemofluorometer
		Fluorometric
Transferrin receptor	Serum	ELISA
Hemoglobin	Blood	Spectrophotometric or electronic counter
Hematocrit	Blood	Special centrifuge or electronic counter
Zinc	Plasma or serum	AAS
	Leukocyte and leukocyte subsets	AAS
	Hair	AAS
	Taste acuity	Threshold of test solutions
Copper	Plasma or serum	AAS or colorimetric
Ceruloplasmin	Serum	Spectrophotometric

Continued

Table 1 Continued

Item	Assessed variable	Method or apparatus
Superoxide dismutase	Erythrocytes	Spectrophotometric
Selenium	Whole blood or plasma	AAS or fluorometric
Glutathione peroxidase	Erythrocytes, plasma or platelets	Spectrophotometric
Chromium	Serum	Graphite furnace AAS
Manganese	Plasma or serum	Graphite furnace AAS
Immunocompetence		
Lymphocyte count	Blood	Blood cell counter
Delayed cutaneous hypersensitivity	Skin	Recall antigens (multitest)
Complement C_3 and factor B	Serum	Radial immunodiffusion or nephelometric
Secretory IgA	Saliva or tears	Radial immunodiffusion or nephelometric
T-lymphocyte and subset percentage	Blood mononuclear cells	Fluorescence microscopy or cytofluorometric
Lymphocyte proliferation	Blood mononuclear cells	Cell harvester and β-counter
Clinical examination		Various forms
Psychometrics		Various tests

[a]Percent body fat can be computed if population-specific equations are available.
[b]Micromethod available.
[c]Measurable simultaneously.
[d]Micromethod and automation available.
[e]Short-term status.
TSF, triceps skinfold; HPLC, high-performance liquid chromatography; IGF-1, insulin-like growth factor 1; RIA, radioimmunoassay; GC-MS, gas chromatography–mass spectrometry; AAS, atomic absorption spectrometry; ELISA, enzyme-linked immunosorbent assay; IRMA, immunoradiometric assay.

same country or even of the same area, are collected. This is not necessary for nutritional surveillance, for which international references should be used.

In addition, for the interpretation of data, confounding factors must be taken into consideration, because they can alter the status and needs of some nutrients. Good examples are smoking habits, alcohol intake, drug use and abuse, physical activity, and consumption of toxic and antinutritive substances.

Use of Data

The main use of nutritional status data is to assess prevalence of malnutrition or deficiency in cross-sectional studies and/or the study of the association of malnutrition and its determinants in longitudinal studies.

An additional and important use of these data is for surveillance or intervention programs. In this case the data collection is often limited to the most relevant indicators of major risk for the population surveyed or the most vulnerable groups inside the population, such as pregnant and lactating women, children, adolescents, and the elderly. Nutrition-related risk factors of chronic disease have recently been included in the surveillance programs.

The assessment of nutritional status of single patients can be used for predicting outcome, and for establishing the need for, and monitoring the effect of, nutritional support. In this case the measurement of special biochemical markers of stress is highly recommended

Limitations of Data

Most of the biochemical indices mentioned in **Table 1** are static indices (measures of the concentration of a nutrient or its metabolites in a suitable biochemical matrix) that do not give information on the functions of the body. It would be desirable to measure organ functions instead (they depend on the adequate availability of a nutrient or the response to the regulatory process to maintain body stores). However, few sensitive and reliable functional tests are available and some of them are not specific for a single nutrient or not easily applicable in the field.

Other limiting factors are intraindividual variance and the lack of harmonization and standardization of methodology. As yet, there are no official methods for nutritional status assessment and there are too few quality controls and interlaboratory comparisons. There are currently few reference laboratories and official standards from international or national agencies. This makes comparison of data from different studies difficult.

The reliability and validity of the methods used are not always assessed. It is true that in some cases assessment of true validity is almost impossible. But concurrent or other types of validity are available. This is an area that needs much more consideration.

Also important factors which are not always assessed in nutritional epidemiology include sensitivity, specificity, and variability of indicators. The sensitivity can be diagnostic (the capacity to detect the earliest stages of deficiency) and biochemical (reflecting every stage of the deficiency). The variability can be analytical (instrumental, preinstrumental), and biological (intraindividual, interindividual). The above items are limiting factors in data interpretation.

See also: **Dietary Surveys**: Measurement of Food Intake; Surveys of National Food Intake; Surveys of Food Intakes in Groups and Individuals; **Epidemiology**; **Malnutrition**: The Problem of Malnutrition; Malnutrition in Developed Countries

Further Reading

Fidanza F (ed.) (1991) *Nutritional Status Assessment – A Manual for Population Studies*. London: Chapman & Hall.

Gibson RS (1990) *Principle of Nutritional Assessment*. New York, NY: Oxford University Press.

Jelliffe DB and Jelliffe EFP (1989) *Community Nutritional Assessment*. Oxford: Oxford University Press.

Kinney JM, Jeejeeboy KN, Hill GL and Owens OE (eds) (1988) *Nutrition and Metabolism in Patient Care*. Philadelphia: Saunders.

Sauberlich HE (1999) *Laboratory Tests for the Assessment of Nutritional Status*, 2nd edn. Boca Raton: CRC Press.

van den Berg H, Heseker H, Lamand M *et al.* (1993) Flair concerted action no 10 status paper. *International Journal of Vitamin and Nutrition Research* 63: 247–316.

Wang ZM, Pierson RN and Heymsfield SB (1992) The five level model: a new approach to organizing body composition research. *American Journal of Clinical Nutrition* 56: 19–28.

WHO Expert Committee Report (1995) *Physical Status: The Use and Interpretation of Anthropometry*. Geneva: World Health Organization.

Anthropometry and Clinical Examination

J R Lustig and B J G Strauss, Monash Medical Centre, Clayton, Victoria, Australia

Introduction

A quick glance at a person will provide a gross estimation of nutritional status. Irrespective of training, we are all able to make these assessments: a Sumo wrestler is obese or overnourished, whilst a terminal AIDS patient is wasted and suffers from undernutrition. These assessments, however, are merely qualitative and contribute little to quantifying the nature and extent of these observations.

Table 1 Two-compartment model of body composition

Compartment	Anthropometric technique used
Fat mass/distribution	Body mass index
	Skinfold thicknesses
	Abdominal circumference
	Abdominal:gluteal ratio
Fat-free mass	Midarm circumference
	Midarm muscle circumference
	Midarm muscle area

Anthropometry comprises techniques that readily contribute to a more in-depth understanding of body composition and nutritional status, allowing the quantification of observations and the observation of changes with time.

Anthropometric methods are based on a model of body composition that consists of two distinct compartments: fat mass and fat-free mass (see **Table 1**). This two-compartment model of body composition defines the fat-free mass as a compartment consisting of body cells including skeletal muscle, extracellular water, the skeleton, and connective tissue, with the fat compartment consisting of the adipose tissue stores of the body, which has a very small cellular and water component. Anthropometric measurements can indirectly assess these two body compartments and thus provide an index of nutritional status. Alterations in body fat content generally are sensitive to changes in energy balance. Chronic malnutrition is reflected in changes to the protein stores found predominantly in skeletal muscle in the body.

Anthropometric techniques are rapid, portable, noninvasive, and inexpensive. The equipment required includes a tape measure, stadiometer (for measuring height), standardized weight scales, and skin-fold calipers. These techniques are frequently used in nutritional assessment and for monitoring changes in a diverse range of clinical settings. Such settings include tertiary referral centres, isolated rural practices, and population-based epidemiological studies. They are also used in gymnasia, infant welfare centers, schools, insurance assessments, sports clubs, and so on.

Anthropometric Assessment of Fat Mass

Body Mass Index (BMI)

$$\text{BMI} = \frac{\text{weight (kg)}}{\text{height (m}^2)}$$

Weight should be measured using standardized, good-quality weighing scales. The weight range of the scales should be known and should be appropriate for the subjects being weighed. This is particularly important for the larger weights. Many instruments are not designed for weighing above 120 kg and are inaccurate when approaching higher weights.

Height is measured in the standing position using a stadiometer. Alternatively, a measuring stick attached to a vertical surface with some form of right-angled headboard can be used. Portable versions have been developed for field work. Shoes and socks should be removed and clothing kept to a minimum in order to allow assessment of posture. The subject should stand upright with the feet, buttocks, and shoulder blades in contact with the vertical surface or stadiometer. The head is positioned so that the Frankfurt plane (the line between the top of the pinna and the outer canthus of the eye) is horizontal. The subject is asked to take a deep breath, and the movable headboard is lowered until it is touching the crown of the head. The height is measured at maximum inspiration with the examiner's eyes level with the headboard to avoid any errors of parallax. The measurement is read to the nearest millimeter, the lower value being used if the level falls between two values.

Because of kyphosis or other postural problems, or, in the case of a patient being confined to bed, knee height has been used as a surrogate for estimating stature, and regression formulae have been developed to estimate height from knee height.

The BMI or Quetelet's index is the most commonly used weight/height ratio in adult populations. The BMI has a reasonable correlation with relative fat mass, as measured by body-density techniques. The correlation between BMI and relative fat mass, as measured by underwater weighing, was found to be 0.82 for a sample of women and 0.70 for a male sample. Another limitation on the usefulness of BMI in predicting body fatness in a given individual relates to the weight nominator. As weight is influenced by muscle, bone, and water as well as by fat, an individual with a well-developed musculo-skeletal system relative to height may have a BMI in the obese range but will not be overfat. BMI should thus be considered in the light of an individual's physical activity level.

Classifications of BMI have been developed, and these define relative levels of adiposity (see **Table 2**). Such a classification is based on surveys of Caucasian populations and may not apply to other ethnic groups. In addition, there is a good correlation between BMI and morbidity and mortality from obesity, e.g., heart failure, osteoarthritis, and hypertension.

The BMI can also be used as an index of wasting. Population studies have shown an increase in morbidity and mortality, and alterations in immune function as the BMI falls below 18.5, irrespective of the type of population, from undernourished groups, to those

Table 2 Classifications for body mass index

Body mass index ($kg\,m^{-2}$)	Health association
< 16.0	Severe protein-energy malnutrition
16.0–16.9	Moderate protein-energy malnutrition
17.0–18.5	Mild protein-energy malnutrition
20.0–24.9	Apparently healthy, with low health risk
25.0–29.9	Overweight, with increased health risk
30.0–39.9	Obesity, increased macrovascular disease and diabetes mellitus
> 40.0	'Super obesity' with major morbidity

with clinical illnesses such as malignancies. The WHO has graded wasting on the basis of the BMI into mild (BMI, 18.5), moderate (BMI < 17.5) and severe (BMI < 16.5).

Abdominal:Gluteal Circumference Ratio

The ratio of the abdominal circumference to the maximal gluteal circumference defines the distribution of adipose tissue in the body. The waist is measured with the subject standing erect and undressed. The abdominal circumference is measured at the midpoint of the line between the rib or costal margin and the iliac crest in the midaxillary line. The maximal gluteal (buttock) circumference is also measured with the subject standing erect. It is important to define the sites measured and maintain consistency in these sites in order to compare sets of data.

A waist/hip ratio > 0.95 in males and > 0.85 females is consistent with abdominal obesity.

Abdominal fat distribution is associated with a range of adverse health consequences, including an increased risk of cardiovascular and cerebrovascular disease, impaired glucose tolerance, and hypertriglyceridemia, even when the BMI is in the 'healthy' range.

Skin-fold Thicknesses

Estimation of body fat content from skin-fold thicknesses assumes that measurements of subcutaneous fat represent total body fat. This assumption is not strictly true, as the thickness of subcutaneous fat does not necessarily reflect a constant proportion of total body fat. In addition, there are marked variations in the distribution of subcutaneous fat depending on age, sex, and race. None the less, this method provides useful information and is in routine use. Skin-fold thickness measurements may be taken at single or multiple sites. There is no universal agreement as to which skin-fold best represents total body fat. However, the triceps skin-fold is most frequently selected for a single-site assessment of body fat.

The measurement of the triceps skin-fold is taken at the midpoint of the upper arm. With the elbow bent to 90% the midpoint is located half-way between the acromium process of the shoulder and the olecranon process of the ulna. This point should be marked. The arm is then extended and allowed to hang loosely by the side. Using the thumb and forefinger, the skin-fold is grasped just above the midarm point. The calipers are applied at right angles to the skin at the midarm point and 1 cm in from the surface. The fold should remain held whilst the measurement is taken. Multiple skin-fold sites can be assessed and usually include both limb and trunk sites.

Body density can be estimated from skin-fold measurements. Using regression equations such as Durnin and Womersley's, skin-folds, sex, and age are used to derive body fat and fat-free mass. For example, for males aged 30–39 years:

1. $D \, (\text{gm}^{-3}) = 1.1422 - (0 - 0544 \times \log_{10}$ (Skinfold sum [mm])), where D = density; skinfold sum = biceps + triceps + subscapular + iliac.
2. Percentage body fat or relative fat mass can then be calculated. Several equations have been derived. The Siri equation assumes that the density of fat is $0.900 \, \text{g cm}^{-3}$ and that the density of the fat-free mass is $1.100 \, \text{g cm}^{-3}$: % fat = $[(4.95/D) - 4.50] \times 100$.
3. Total body fat is derived as follows: total body fat (kg) = [body weight (kg) × % body fat]/100.
4. Fat-free mass (kg) = body weight (kg) − body fat (kg).

Both the skill of the operator and the character of the subcutaneous fat in an individual can influence the precision of skin-fold measurements. In general, the error of this method is approximately 5%. Depending on the equation used and the population studied, this may range from 3 to 9% body fat mass. Durnin and Womersley derived values for calculation of fat mass for all age groups in both men and women.

Anthropometric Assessment of Fat-Free Mass

Mid-arm Circumference

The upper arm contains both muscle and subcutaneous fat, so that measurement of the circumference of the midupper arm may be used as an index of nutritional status. In underdeveloped countries where the population is often malnourished and with little fat reserves, a change in this measurement can reflect total body protein stores. Serial measurement of the upper-arm circumference may be used to monitor nutritional intervention. Measurement of this circumference is described in the section on skin-fold thicknesses.

Midarm Muscle Circumference

In practice, midarm muscle circumference is used in preference to midarm circumference as a measurement which reflects total body protein stores. The upper-arm circumference comprises central bone surrounded by a layer of skeletal muscle and subcutaneous fat. The midarm muscle circumference, derived from the midarm circumference and the triceps skin-fold thickness, takes into account an assessment of both fat and protein stores.

Small changes in muscle mass may not be detected by this technique. In addition, individual variations in the diameter of the humerus are not taken into account in this calculation. Midarm muscle circumference is derived using the following equation:

Midarm muscle circumference =
$$\text{midupper arm circumference} - (\pi \times TSF),$$

where TSF = triceps skinfold thickness (in cm).

Midarm Muscle Area

Midarm muscle area provides a higher correlation with body protein stores than does either of the midarm circumferences. Because it takes into account a two-dimensional assessment, it reflects more accurately small changes in muscle mass. None the less, one appraisal of the equation suggests that it may underestimate significantly the degree of muscle wasting. Midarm muscle area may be calculated using the following equation:

$$\text{arm muscle area} = [C - (\pi \times TSF)]^2/4\pi,$$

where C = midupper arm circumference, and TSF triceps skinfold thickness (in cm).

The coefficient of variation for the measurement of midarm muscle area has been estimated as approximately 7% when made by trained operators. This is therefore not a sensitive enough method by which to detect small changes in muscle mass.

Conclusions

Anthropometry provides a rapid methodology for the indirect quantification of fat and fat-free mass and, thus, the body compositional aspects of nutritional status. The techniques require little in the way of equipment and are readily portable and inexpensive. Although they are insensitive to small changes in nutritional status, they have a useful role to play in assessment of nutritional status and for monitoring progress over time. In addition, these techniques are useful in population-based field research of nutrition-related disorders.

See also: **Body Composition**; The Problem of Malnutrition; **Nutritional Assessment**: Importance of Measuring Nutritional Status; **Obesity**: Etiology and Diagnosis

Further Reading

Bjorntorp P (1987) Classification of obese patients and complications related to the distribution of surplus fat. *American Journal of Clinical Nutrition* 45: 1120–1125.

Durnin JVGA and Womersley J (1974) Body fat assessed from the total body density and its estimated skinfold thickness: measurements on 481 men and women aged from 16 to 72 years. *British Journal of Nutrition* 32: 77–97.

Forbes GB (1990) The abdomen:hip ratio. Normative data and observations on selected patients. *International Journal of Obesity* 14: 149–157.

Gibson RS (1990) *Principles of Nutritional Assessment.* London: Oxford University Press.

Heymsfield S, McManus CB, Smith J, Stevens V and Dixon DW (1982) Anthropometric measurement of muscle-mass: revised equations for calculating bone-free arm muscle area. *American Journal of Clinical Nutrition* 36: 680–690.

James WPT, Ferro-Luzzi A and Waterlow JC (1988) Definition of chronic energy deficiency in adults. *European Journal of Clinical Nutrition* 42: 909–981.

Lapidus L, Bengtsson Q, Larsson B *et al.* (1984) Distribution of adipose tissue and risk of cardiovascular death: a 12 year follow-up of participants in the population of women in Gothenburg, Sweden. *British Medical Journal* 289: 1257–1261.

Larsson L, Svardsudd K, Welin L *et al.* (1984) Abdominal adipose tissue distribution and risk of cardiovascular disease and death: a 13 year follow-up of participants in the study of men born in 1913. *British Medical Journal* 288: 1401–1404.

Lohman TG (1981) Skinfolds and body density and their relation to body fatness: a review. *Human Biology* 53: 181–225.

Lohman TG (1992) Advances in body composition assessment. *Current Issues in Exercise Science.* Monograph No. 3. Champaign, IL: Human Kinetics.

Siri WE (1961) Body composition from fluid spaces and density: Analysis of methods. *Techniques for Measuring Body Composition*, pp. 223–244. Washington, DC: National Academy of Sciences, National Research Council.

Smalley KJ, Knerr AN, Kendrick ZV, Colliver JA, and Owen OE (1990) Reassessment of body mass indices. *American Journal of Clinical Nutrition* 52: 405–408.

Physical Status. The Use and Interpretation of Anthropometry. (1995) WHO Technical Report 854. Geneva: WHO.

Biochemical Tests for Vitamins and Minerals

A V Lakshmi and M S Bamji[a], National Institute of Nutrition, Hyderabad, AP, India

Introduction

Increasing attention is being given to the definition of marginal malnutrition since the elimination of clinical signs of deficiency *per se* is not sufficient for optimal health. Biochemical measurements on easily available body fluids represent the most objective assessment of nutritional status and often provide subclinical information. It also provides information on micronutrient toxicity. During the development of any deficiency disease, biochemical changes precede clinical symptoms and hence biochemical tests help to identify the disease at the subclinical stage. They also help to confirm the clinical diagnosis.

An ideal biochemical test should be specific, sensitive, and indicative of tissue depletion at an early stage but not immediate dietary intake. However, we do not have an ideal biochemical test for most nutrients and the method selected depends on the situation in which it is applied. Often, more than one test is required. Classification of an individual as severely deficient, marginally deficient, normal, or intoxicated may require different tests. In this chapter biochemical tests used for the nutritional status assessment of fat- and water-soluble vitamins and of a few minerals – sodium, potassium, calcium, iron, iodine, zinc, copper, and selenium are discussed.

Types of Laboratory Tests

Laboratory tests for the assessment of vitamin or mineral nutritional status can be placed in the following categories:

1. Measurement of excretion of nutrient or its metabolite in urine, without or after a bolus load (load return test)
2. Measurement of nutrient level and/or its active metabolites in blood
3. Assay of activity of a vitamin-dependent enzyme and its *in vitro* stimulation with corresponding coenzyme in the blood. (*See* **Coenzymes.**)
4. Measurement of rise in the concentration of a metabolite in the blood or urine (resulting from inadequate intake of the vitamin) after administering a load of the precursor

[a]Current address: Dangoria Charitable Trust, 1–7–1074, Hyderabad 500020, India

Measurement of excretion of vitamin or its metabolite in urine can be carried out on a 24-h sample, random sample, or in the first voided morning sample of urine. Although a 24-h specimen provides the most accurate estimate, it is difficult to collect it in the free-living population. Measurements in random samples or a first voided urine sample, and expressing the value per gram of creatinine, may be useful for population studies but are not very reliable for individual assessment.

The limitation regarding the interpretation of urinary data is that it is influenced by immediate dietary intake and may not reflect tissue stores. Load return tests may yield better information regarding the state of tissues but they are seldom used now. Certain physiological and pathological conditions, as well as ingestion of drugs, influence urinary excretion of certain vitamins and minerals.

Either whole blood or different components such as plasma, erythrocytes, leukocytes, and platelets has been used for the assessment of vitamin and mineral status. The choice of the blood components is decided by the concentration of the nutrient and its sensitivity to vitamin or mineral depletion. High-performance liquid chromatography (HPLC) technique has revived interest in these measurements due to improved sensitivity. (See **Chromatography**: High-performance Liquid Chromatography.)

Details regarding enzymatic tests and metabolite tests will be discussed while dealing with the individual vitamins.

Guidelines for Interpretation of Values

Availability of standard reference material (serum and erythrocyte hemolysates), use of standardized procedures, and strict quality control measures are necessary for arriving at interpretative guidelines. Since reference quality control samples are not available for many vitamins, and different laboratory procedures are adopted by various workers, only tentative guidelines based on the values obtained with apparently healthy, well-nourished controls are available at present for interpretation of data. (See **Vitamins**: Determination.)

Fat-soluble Vitamins

Vitamin A

Serum vitamin A level is not a sensitive indicator of vitamin A stores. The levels are however indicative of vitamin A status when the stores are either fully replete or totally deplete and in the absence of infection. Serum levels below $0.35\ \mu mol\ l^{-1}$ are usually associated with

Table 1 Tentative guidelines for the interpretation of biochemical parameters of vitamin A status

Biochemical measurement	Acceptable	Medium risk	High risk
Serum ($\mu mol\ l^{-1}$)	> 70	35–70	< 35
Liver ($\mu mol\ g^{-1}$)	> 0.70	0.17–0.70	< 0.17
Relative dose–response (%)	< 20	> 20	
Modified relative dose–response (DR/R)	< 0.03	> 0.03	

DR, 3,4-didehydroretinol; R, retionol.

clinical signs of deficiency. Serum vitamin A levels above $0.70\ \mu mol\ l^{-1}$ are considered satisfactory and reflect adequate liver reserves (**Table 1**).

Marginal vitamin A status can be assessed by the relative dose–response (RDR) test or the modified relative dose–response (MRDR) test. The RDR test consists of determination of the plasma levels of vitamin A at baseline and 5 h after the administration of a small oral or intravenous dose of retinylpalmitate or retinyl acetate. An increase of more than 20% is indicative of inadequate hepatic reserves.

In the MRDR test, 3,4-didehydroretinol (vitamin A_2) is administered instead of retinol. This test requires only one blood sample for analysis. The ratio of dehydroretinol to retinol is used to assess liver vitamin A concentrations (**Table 1**).

Use of deuterated vitamin A for measuring the total body pool by means of isotope dilution is currently being refined.

Toxicity

High plasma levels of retinyl esters in the presence of high plasma levels of vitamin A is a sign of vitamin A toxicity. In conditions of normal vitamin A status, the plasma levels of retinyl esters is about 5%.

Spectrophotometric, HPLC fluorometric, and calorimetric techniques are used for measuring plasma vitamin A level. (See **Retinol**: Properties and Determination.)

Vitamin D

Measurement of serum 25-hydroxycholecalciferol, the major circulating metabolite of vitamin D, is the best indicator of vitamin D status. In healthy adults its concentration in serum varies from 20 to 130 nmol l^{-1} and less than 12.5 nmol l^{-1} is generally considered inadequate. It can be measured by HPLC or by competitive protein-binding assay. (See **Cholecalciferol**: Properties and Determination.)

Vitamin E

Vitamin E status is normally assessed by measuring levels of the vitamin in plasma or serum and other tissues (red cells, platelets, adipocytes) by spectrophotometric, fluorometric, or HPLC methods. Since the levels of vitamin E and lipids in the serum are directly correlated, their ratio is a better measure of status than serum vitamin E alone (**Table 2**). The ratio of plasma vitamin E to cholesterol + triglyceride or vitamin E to cholesterol has also been reported to be as good as that of vitamin E to total lipids. (*See* **Tocopherols**: Properties and Determination.)

In vitro hemolysis of red blood cells in the presence of hydrogen peroxide, quantitating the formation of malondialdehyde from hydroperoxides generated by the *in vitro* peroxidation of polyunsaturated fatty acids or erythrocytes exposed to hydrogen peroxide is considered a functional test for assessing vitamin E status. The latter tests, based on quantitation of malondialdehyde formed, are considered to be more sensitive than the hemolysis test.

Vitamin K

In the past, vitamin K status has been assessed by measuring prothrombin time and other clotting times. These measures are however not sensitive to detect subclinical vitamin K deficiency since clotting time does not change until the concentration of prothrombin decreases by about 50%.

Vitamin K is needed for the γ-carboxylation of certain glutamic acid residues in proteins. During vitamin K deficiency, the vitamin K-dependent proteins occur in an undercarboxylated form. Milder forms of vitamin K deficiency can be assessed by measuring urinary γ-carboxyglutamic acid excretion or circulating levels of undercarboxylated prothrombin and osteocalcin. The last parameter is considered to be a very sensitive marker of vitamin K status. Plasma level of phyloquinone has also been used as a measure of vitamin K status. Plasma phyloquinone and urinary γ-carboxyglutamic acid can be determined by HPLC separation followed by fluorometric detection. Undercarboxylated prothrombin and osteocalcin can be determined by radioimmunoassay. (*See* **Immunoassays**: Radioimmunoassay and Enzyme Immunossay.)

Thiamin

The biochemical tests used for the assessment of thiamin nutritional status include measurement of: (1) erythrocyte transketolase activation coefficient (ETK-AC); (2) urinary excretion of thiamin; and (3) measurement of whole blood or erythrocyte thiamin pyrophosphate (TPP) concentration.

TPP is the coenzyme for transketolase and determination of ETK-AC is the most widely accepted procedure for assessing thiamin status. Activity of this enzyme is measured in erythrocyte hemolysate without (basal) and with a saturating amount of TPP added *in vitro* (stimulated). The ratio of stimulated to basal activity is called the activity coefficient (ETK-AC). Basal activity falls and ETK-AC increases in thiamin deficiency (**Table 2**). Activity can be measured by determining the rate of disappearance of pentose or appearance of hexose. However interpretation of the results is complicated under certain disease conditions (some cancers, uremia, neuropathy, and diabetes) and drug treatment (diuretics, antacids) since ETK activity is affected under these conditions. (*See* **Enzymes**: Uses in Analysis.)

Normal adults receiving adequate dietary thiamin excrete more than 0.3 µmol per 24 h. Urinary thiamin can be measured by fluorometric method. (*See* **Thiamin**: Properties and Determination.)

Measurement of blood or erythrocyte TPP by HPLC appears to be a more sensitive and promising indicator of thiamin status and needs further verification.

Riboflavin

Measurement of urinary excretion of riboflavin, red blood cell riboflavin, and erythrocyte glutathione reductase activation coefficient (EGR-AC) are used for the biochemical assessment of riboflavin nutritional status.

The urinary excretion method is useful for population surveys. On an adequate dietary intake of riboflavin, normal adults excrete more than $0.21 \, \mu mol \, g^{-1}$ of creatinine. Riboflavin concentration

Table 2 Tentative guidelines for interpretation of blood vitamin levels and enzyme tests

Parameter measured	Acceptable	Medium risk	High risk
Serum vitamin E ($\mu mol \, l^{-1}$)	> 16.2	11.6–16.2	< 11.6
Serum vitamin E total lipid ratio	> 0.8		< 0.8
ETK-AC	< 1.15	1.15–1.25	> 1.25
EGR-AC	< 1.2	1.2–1.4	> 1.4
Serum folate ($nmol \, l^{-1}$)	> 11.0	6.7–11.0	< 6.7
Erythrocyte folate ($nmol \, l^{-1}$)	> 360	315–359	< 315
Plasma ascorbate ($\mu mol \, l^{-1}$)	17–85	11–17	< 11
Leukocyte ascorbate ($\mu mol \, 10^{-6}$ cells)	113–300		

ETK-AC, erythrocyte transketolase activation coefficient; EGR-AC, erythrocyte glutathione reductase activation coefficient in urine can be measured by fluorometric or microbiological assay or HPLC. (*See* **Riboflavin**: Properties and Determination.)

Red cell riboflavin level may be a better index of tissue status than urinary riboflavin. Fluorometric or HPLC technique can be used to estimate red cell riboflavin and the value can be expressed per gram hemoglobin or packed cell volume. However this test is not commonly used because tissue saturation is difficult to define and the value varies between individuals. Measurement of red cell flavin adenine dinucleotide (FAD) level using HPLC may improve the index.

At present, the EGR-AC test is the most accepted biochemical method for the assessment of riboflavin status. FAD, the coenzyme form of riboflavin, is necessary for the activity of glutathione reductase. The activity of this enzyme is measured in erythrocyte hemolysate or blood hemolysate without (basal) and with added FAD (stimulated). The ratio of stimulated to basal activity is called the activation coefficient (EGR-AC). This ratio increases in riboflavin deficiency (**Table 2**). EGR activity is assayed by rate reaction techniques. However the EGR-AC test fails to measure riboflavin status when glucose 6-phosphate dehydrogenase deficiency is also present.

During respiratory infections, urinary and blood levels of riboflavin are raised due to mobilization of the vitamin from the liver. There is a transient reduction in EGR-AC values also. Thus assessment of riboflavin status during such infection is difficult.

Niacin

Measurements of the urinary excretion of N^1-methylnicotinamide (NMN) and its oxidation product, N^1-myethyl-2-pyridone-5-carboxamide (2-pyridone) have been used to determine niacin status. Normal adults excrete more than $17 \, \mu mol$ NMN per day, whereas during deficiency its excretion is less than $5.8 \, \mu mol$ per day.

A more recent study suggested that erythrocyte nicotinamide adenine dinucleotide (NAD) concentration may serve as a sensitive indicator of niacin status and that a ratio of erythrocyte NAD to NAD phosphate (NADP) < 1.0 may suggest a risk of developing niacin deficiency. (*See* **Niacin**: Properties and Determination.)

Vitamin B$_6$

Several direct and indirect measures are used for the assessment of vitamin B$_6$ status.

Measurement of the plasma pyridoxal phosphate (PLP) level, the coenzyme form of the vitamin, is the most commonly used method and a concentration of $30 \, nmol \, l^{-1}$ is considered to be adequate. However, proper interpretation of plasma PLP data must take into account the factors which affect this value. High protein intake, pregnancy, age, smoking, and an increase in alkaline phosphatase activity are known to reduce PLP concentration. The measurement of both PLP and pyridoxal is suggested to be a better index of vitamin B$_6$ status during pregnancy. Pyridoxal phosphate levels in plasma can be measured by enzymatic or HPLC techniques.

The second direct measure is urinary excretion of 4-pyridoxic acid. It correlates well with dietary vitamin B$_6$ intake and this parameter is useful for population surveys. An excretion of $> 3.0 \, \mu mol$ per day is considered to be associated with adequate vitamin B$_6$ status and it can be measured by fluorometric method. (*See* **Vitamin B$_6$**: Properties and Determination.)

Functional tests for assessing vitamin B$_6$ status include measurement of PLP-dependent transaminases, such as erythrocyte aspartate aminotransferase (EAST) and alanine aminotransferase (EALT) and their activation coefficient (AC), and tryptophan or methionine load test. They reflect tissue levels of PLP.

EAST or EALT activity is measured without and with added PLP to give an activation coefficient (EAST-AC or EALT-AC) – an enzymatic test analogous to that used for thiamin and riboflavin. Although widely used for assessing vitamin B$_6$ status, the interpretation of EAST-AC value is not standardized. Values ranging from 1.7 to 2.0 have been used to differentiate between adequacy and deficiency. The EALT-AC value suggested for adequacy of vitamin B$_6$ status is less than 1.25. These two enzymes can be assayed by colorimetric or rate reaction technique.

The tryptophan load test has been used in population surveys as a functional index of B$_6$ status. An inadequate status results in an increased excretion of tryptophan metabolites such as kynurenine and xanthurenic acid, since vitamin B$_6$-dependent enzyme kynureninase is required for their metabolism. This test is usually carried out after administering an oral load of 2–5 g tryptophan and measuring the urinary excretion of xanthurenic acid in a 6–9-h period. However excretion of tryptophan metabolites is influenced by factors like protein intake and pregnancy.

Folic Acid and Vitamin B$_{12}$

Serum as well as red blood cell folate reflect folate status. While serum folate reflects the dietary intake and readily available tissue reserves, red blood cell folate reflects long-term folate status. Microbiological or competitive protein-binding radioassay is used to measure folate levels in serum and red blood cells. The guidelines for interpreting these values are given in **Table 2**. (*See* **Folic Acid**: Properties and Determination.)

The metabolism of histidine to glutamic acid is impaired in folate deficiency, leading to an increase in urinary excretion of formiminoglutamic acid (FIGLU). This occurs particularly after a loading dose of histidine. Normal adults excrete less than 35 mg FIGLU in 24 h after 5 g histidine load.

Deoxyuridine Suppression Test

In folic acid deficiency the activity of the enzyme thymidilate synthalase is impaired. This enzyme catalyzes the conversion of deoxyuridine to thymidine. In this test bone marrow cells or peripheral blood lymphocytes are preincubated with nonradioactive deoxyuridine and then with ^{3}H thymidine. The amount of radioactivity incorporated into the DNA of the cells is increased in folate as well as in vitamin B_{12} deficiency. These two deficiencies can be differentiated by *in vitro* incubation with respective vitamins.

Hypersegmentation of circulating neutrophils is a morphological change that occurs in folate deficiency due to slowed DNA synthesis. There are studies to suggest that this change occurs when plasma folate levels drop lower than $10 \, nmol \, l^{-1}$.

Measurement of plasma homocysteine level may also indicate the adequacy of folate and, to a lesser extent, vitamin B_{12} status. Vitamin B_6 deficiency also raises plasma homocysteine concentration.

Vitamin B_{12} status is generally assessed by measuring its concentration in serum by microbiological or competitive protein-binding radioassay. Radioassay using pure intrinsic factor measures true vitamin B_{12}, whereas microbiological assay measures a few analogs of vitamin B_{12} as well. The lowest acceptable serum level using microbiological assay is $150 \, pmol \, l^{-1}$, whereas for the radioassay specific for cobalamine, serum level $< 75 \, pmol \, l^{-1}$ is considered deficient. Measurement of plasma vitamin B_{12} bound to transcobalamine II is considered to be more sensitive than total vitamin B_{12} for the assessment of marginal vitamin B_{12} deficiency.

The urinary or plasma methylmalonic acid level, preferably after an oral load of valine (its precursor), is a metabolic test for detecting vitamin B_{12} deficiency. Normal adults excrete less than $100 \, \mu mol$ in 24 h.

Pantothenic Acid

Urinary excretion of pantothenic acid or blood levels of the vitamin can be used to assess pantothenic acid status. Adults taking an adequate amount of dietary pantothenate excrete between 9 and 24 μmol of the vitamin per gram creatinine and the blood levels

range between 1.2 and $1.8 \, \mu mol \, l^{-1}$. Pantothenic acid levels in blood and urine can be measured by microbiological or radioimmunoassay. (*See* **Pantothenic Acid**: Properties and Determination.)

Biotin

Biotin nutritional status can be assessed by using the following sensitive biochemical parameters: (1) urinary excretion of organic acids such as 3-hydroxyisovaleric, 3-hydroxypropionic, and methylcitric acids, or of biotin and its metabolite bisnorbiotin; (2) assay of the activity of biotin-dependent enzyme propionyl coenzyme A carboxylase with and without *in vitro* addition of biotin in lymphocytes; and (3) measurement of plasma biotin level. Plasma biotin is a less sensitive indicator compared to the other two methods. The normal reported range for plasma biotin is $140–356 \, pmol \, l^{-1}$. Biotin levels in blood and urine can be determined by microbiological or isotopic dilution assay. (*See* **Biotin**: Properties and Determination.)

Ascorbic Acid

Plasma and leukocyte ascorbate levels indicate ascorbic acid status. The simplicity and reliability of plasma ascorbic acid determination make it preferable for identifying individuals at risk of deficiency due to chronic low intake of the vitamin. However, plasma ascorbic acid levels may be less useful for defining the ascorbic acid status of individuals with adequate or high intake. Leukocyte ascorbic acid concentration reflects more accurately tissue stores of vitamin than plasma level.

Plasma and leukocyte ascorbic acid levels are determined by HPLC or colorimetric techniques. Thresholds for interpretation of ascorbic acid status are given in **Table 2**. (*See* **Ascorbic Acid**: Properties and Determination.)

Mineral Nutrition

In this section, biochemical methods used for the nutritional assessment of a few macro- and micromineral elements are given. However, the list is not comprehensive. Only those elements for which a laboratory test can give meaningful data from the nutritional point of view are included.

Sodium and Potassium

Dietary deficiency of sodium and potassium cations might occur after increased losses under certain conditions such as vomiting and diarrhea. Their status is generally assessed by measuring serum or plasma levels by atomic emission flame photometry or atomic absorption spectrophotometry (**Table 3**). (*See* **Sodium**: Properties and Determination.)

Calcium

In blood, calcium is present almost entirely in the plasma. As ionized calcium in the plasma is under strict homeostatic control ($1.2\,mmol\,l^{-1}$), it cannot serve as a sensitive parameter to assess nutritional deficiency of calcium. (*See* **Calcium**: Properties and Determination.)

Bone density is an indirect measure of calcium status, since 99% of it is present in the bones. It can be measured by radiographic, single- or dual-photon absorptiometry techniques.

Iron

Iron status can be assessed by three types of biochemical tests, reflecting different stages of deficiency. First, bone marrow iron and serum ferritin concentrations indicate iron stores. Serum ferritin levels less than $12\,\mu g\,l^{-1}$ suggest depletion. As an acute-phase reactant, serum ferritin is elevated by acute and chronic infections, inflammatory diseases, malignancies, and liver disorders.

In the second stage of iron deficiency, referred to as iron-deficiency erythropoiesis, the supply of iron to the erythroid marrow is reduced. This can be assessed by the transferrin saturation index. An increase in red cell protoporphyrin level also indicates restricted iron supply to the developing red cell (**Table 4**).

Recent studies suggest that serum transferrin receptor is a sensitive and quantitative index of tissue iron status (**Table 4**). The levels are elevated in iron-deficiency anemia. Inflammation or liver disease does not affect serum transferrin receptor levels.

The third stage of iron deficiency is anemia, which can be assessed by hemoglobin level or hematocrit. Hemoglobin levels less than $13\,g\,dl^{-1}$ of blood for adult males and $12\,g\,dl^{-1}$ blood for adult females are considered deficient, whereas during pregnancy hemoglobin concentration less than $11\,g\,dl^{-1}$ is considered to be anemia. (*See* **Anemia (Anaemia)**: Iron-deficiency Anemia.)

Iodine

Urinary excretion of iodine reflects the dietary intake. An excretion of more than $100\,\mu g$ per day suggests an adequate dietary intake of iodine. It can be measured by the kinetic method, based on iodide-catalyzed reduction of ceric to cerus by arsenic. (*See* **Iodine**: Properties and Determination.)

The only known requirement of iodine in humans is in the synthesis of thyroid hormones. Measurements of serum levels of thyrotropic hormone or triiodothyronine and thyroxine are sensitive functional tests for iodine status.

Zinc

Zinc levels in plasma, urine, hair and cellular compartments of blood, activity of zinc-dependent enzymes such as carbonic anhydrase in red cells and plasma levels of zinc-binding protein metallothionin have been used as indices of zinc nutrition status.

Plasma zinc appears to be the most widely used parameter. It decreases in cases of severe and moderate deficiency of zinc. Plasma zinc concentrations are susceptible to a number of pathophysiological influences. Since erythrocytes and hair have a longer half-life, their zinc content does not indicate acute zinc deficiency. Zinc levels in cells with rapid turnover (neutrophils and monocytes) can be used to detect mild zinc deficiency. Serum thymulin level without and with *in vitro* addition of zinc has been suggested as a sensitive method to assess mild zinc deficiency. Atomic absorption spectrometry is used for the analysis of zinc in urine and blood. **Table 5** gives the normal range of tissue zinc. (*See* **Zinc**: Properties and Determination.)

Table 3 Normal ranges of sodium, potassium, copper, and selenium in blood

Parameter	Range
Plasma sodium ($mmol\,l^{-1}$)	136–145
Plasma potassium ($mmol\,l^{-1}$)	3.5–5.0
Plasma copper ($mg\,l^{-1}$)	0.8–1.75
Serum ceruloplasmin ($\mu mol\,l^{-1}$)	1.7–2.9
Plasma selenium ($\mu g\,l^{-1}$)	60–120
Erythrocyte selenium ($\mu g\,l^{-1}$)	90–190

Table 4 Assessment of iron status (commonly used cut-off values)

Parameter	Value
Serum ferritin ($\mu g\,l^{-1}$)	12
Transferrin saturation (%)	16
Red cell protoporphyrin ($\mu mol\,l^{-1}$)	1.24
Serum transferrin receptor ($mg\,l^{-1}$: normal range)	2.5–8.5
Hemoglobin ($g\,l^{-1}$)	
Men	130
Women	120

Table 5 Normal range of tissue zinc

Parameter	Range
Plasma ($mg\,l^{-1}$)	0.7–1.4
Erythrocyte ($mg\,l^{-1}$)	10–14
Urine (μg per day)	400–600
Leukocytes (μg per 10^{10} cells)	80–130
Hair ($\mu g\,g^{-1}$)	124–320

Copper

Copper status is commonly assessed by estimating serum copper concentration or by assaying the activity of the copper-dependent enzyme ceruloplasmin in serum (**Table 3**). Serum copper levels tend to be higher in women. The values are influenced by certain physiological and pathological conditions. About 95% of copper in serum is in ceruloplasmin, which is an acute-phase reactant protein. Erythrocyte copper-zinc superoxide dismutase has been suggested to be a better index of copper status than serum copper or ceruloplasmin.

Serum copper can be estimated by atomic absorption spectrometry. (*See* **Copper**: Properties and Determination.)

Selenium

Selenium status can be assessed by measuring its levels in urine, plasma, erythrocytes, and blood or by assaying the selenium-dependent enzyme glutathione peroxidase in erythrocytes or platelets.

Plasma selenium is a sensitive index of short-term changes until the plasma level reaches a plateau. Erythrocyte selenium and glutathione peroxidase activity indicates long-term changes because of the long life span of these cells.

The commonly used techniques for measuring selenium in biological fluids include fluorimetry, atomic absorption spectrometry, and mass spectrometry. **Table 3** gives the frequently reported normal ranges in plasma and erythrocytes. (*See* **Selenium**: Properties and Determination.)

Selenium is an integral part of glutathione peroxidase and measurement of the activity of this enzyme is a functional test for the assessment of selenium status. The minimal erythrocyte selenium level required to reach a plateau of glutathione peroxidase activity has been estimated to be about $141 \, \mu g \, l^{-1}$.

Conclusions

The value of biochemical tests can be improved with a better understanding of their limitation and their relationship to functional consequences. What level of biochemical insult can an individual accommodate without any ill effects? Can generalized interpretative guidelines be applied to all population groups or are there substantial variations? Despite these dilemmas, biochemical tests have an important place in nutritional diagnosis.

See also: **Ascorbic Acid**: Properties and Determination; **Biotin**: Properties and Determination; **Calcium**: Properties and Determination; **Copper**: Properties and Determination; **Folic Acid**: Properties and Determination; **Iodine**: Properties and Determination; **Niacin**: Properties and Determination; **Riboflavin**: Properties and Determination; **Selenium**: Properties and Determination; **Sodium**: Properties and Determination; **Thiamin**: Properties and Determination; **Tocopherols**: Properties and Determination; **Vitamin B$_6$**: Properties and Determination; **Vitamins**: Determination; **Zinc**: Properties and Determination

Further Reading

Bamji MS (1981) *Vitamins in Human Biology and Medicine*, pp. 1–27, Boca Raton, Florida: CRC Press.

Flair EC Programme (1993) The measurement of micronutrient absorption and status. Flair concerted action no. 10 status papers. *International Journal of Vitamins and Nutrition Research* 63: 247–316.

Sauberlich HE (1984) Implications of nutritional status on human biochemistry, physiology and nutrition. *Clinical Biochemistry* 17: 132–142.

Shils ME and Young VR (eds) (1988) *Modern Nutrition in Health and Disease*, VII edn, pp. 142–446. Philadelphia: Lea & Febiger.

Sokoll LJ, Booth SL, O'Brien ME, Davidson KW, Tsaioin KI and Sadowski JA (1997) Changes in serum osteocalcin, plasma phylloquinone and urinary r-carboxyglutamic acid in response to altered intakes of dietary phylloquinone in human subjects. *American Journal of Clinical Nutrition* 65: 779–784.

Ziegler EE and Filer LJ Jr (eds) (1996) *Present Knowledge in Nutrition*, VII edn, pp. 120–236. Washington, DC: ILSI Press.

Functional Tests

N W Solomons, CeSSIAM, Guatemala City, Guatemala

Background

Reasons abound for assessing human nutritional status, ranging from the population level (epidemiology, public health) to assess risk of nutrient deficiency and excess, and to monitor changes and impacts of interventions to the clinical level, where a practitioner is interested in the patient's nutriture for routine maintenance, as a clue to disease, and in titrating nutritional and disease-specific therapies. (*See* **Epidemiology**.)

One of the options for assessing nutritional status is the approach of *functional assessment*. The purpose of this chapter is to discuss the history, origins, concepts, applications, advantages and disadvantages, and modern advances in the arena of functional tests of human nutriture.

Historical Perspective

Clinical assessment of nutritional status by examining signs and symptoms is the most venerable approach; in fact, the syndromes of scurvy, beriberi, pellagra, rickets, and anemia had been recognized clinically long before the chemical nature and etiological roles of ascorbic acid, thiamin, nicotinic acid, cholecalciferol, and the hematinic nutrients (iron, folic acid, vitamin B_{12}) were discovered. (*See* **Anemia (Anaemia): Iron-deficiency Anemia; Scurvy;** refer to individual nutrients.)

As *sensitivity* (the ability of a test to find all affected individuals) and *specificity* (the ability of a test to reject all unaffected individuals) of a diagnostic approach must always be considered, clinical examination has its limitations. In terms of sensitivity, clinical examination is a poor index of nutritional deficiency, as clinical manifestations emerge late in the process and only in those most deficient. They represent the tip of the iceberg. The specificity of clinical diagnosis is variable. Whereas scurvy and rickets are unmistakable, anemia has multiple nutritional bases and a host of causes unrelated to nutrition.

The twentieth century saw the chemical identification of the essential nutrients, producing reliable assays to measure nutrients or their metabolites in biological fluids and tissues. The first-line approach for the detection of deficiencies and excess of nutrients shifted from clinical examination to laboratory tests, with the realization that changes in body reserves and tissue content of nutrients decrease before the clinical manifestations of deficiency are expressed. Similarly, in terms of excessive accumulation, an increased burden of a nutrient occurs before overt toxic signs and symptoms develop. Most, but not all, biochemical procedures generate *static* indices of nutritional status. Assessment of individual status is pursued by constructing a normative reference distribution for a healthy, well-nourished population as the standard, and assigning cut-off criteria on the high and low ends that represent excess and deficiency, respectively. For a population assessment, the deviation of its biochemical distribution from that of the reference population is gauged, or the prevalence of individuals with values outside the criteria levels is tabulated.

The concept of *functional* assessment was added to the lexicon of nutritional status evaluation in the 1970s, although its origins go much further back to the Hess tourniquet test for capillary fragility for incipient scurvy of the early twentieth century, dark-adaptation tests for vitamin A status, and prothrombin- and coagulation-time tests for vitamin K adequacy of the middle of the century. However, in 1978 Study Team IX of the Committee on International Nutrition Programs of the National Academy of Sciences of the USA, chaired by the late Professor Doris Howes Calloway, issued a report arguing that the intactness of the physiological and behavioral functions that depended on nutrients was of more interest to policy-makers in governments and development assistant agencies than were the levels of nutrients in the body. In 1983, Noel Solomons and Lindsay Allen produced a *systematic* classification of tests of nutritional status based on measures of physiological performance or behavioral responses.

Definitions

This article compares and contrasts two types of indices that seek to determine the nutriture of individuals: *static* and *functional* indices.

Static Indices

Static indices are tests directed at assessing the *quantitative* content of a nutrient in the organism, either as whole-body reserves or as tissue concentrations. They represent chemically measurements of the nutrient itself, some active or inactive metabolite, or a complex, such as hemoglobin, that contains the nutrient. In the instance of iron status, for example, hemoglobin is a surrogate for iron in target tissues, whereas ferritin reflects its presence in body iron stores. Serum iron, transform saturation, and iron in hair or nails are all examples of static indices of iron status. (*See* **Anemia (Anaemia): Iron-deficiency Anemia.**)

Functional Indices

Functional indices of nutritional status are those behavioral, physiological, or biochemical functions of the organism dependent on the adequate availability of a nutrient or resulting from the homeostatic regulatory processes that maintain body stores and harmonic internal distribution of some nutrients.

As mentioned, perhaps the first example of a functional test of nutritional status was the capillary fragility test of Hess, followed in the 1930s by dark adaptation tests based on Wald's elucidation of the dependence of retinal cone function on vitamin A. With the advent of isolated radioisotopes and whole-body counting, tests of the absorption, retention, and distribution of several nutrients have been developed. With the molecular biology, genomic, and proteonomic revolutions, new and unexplored opportunities for functional tests have emerged. (*See* **Ascorbic Acid**: Physiology; **Retinol**: Physiology.)

Classification of Functional Indices

Classification by System

Functional tests can be conveniently classified according to the body system. The categories are as follows: (1) structure and structural integrity; (2) immunity and host defense; (3) transport; (4) hemostasis; (5) reproductive biology; (6) nervous system function; (7) hemodynamics and physical capacity. Some functional tests cannot be classified into any of these, and reside in a 'miscellaneous' catchment category.

Structure and structural integrity   Under 'structure and structural integrity' are classified all of those tests that relate to growth and development of children and adolescents. At the level of specific tissues, tests of the integrity of somatic membranes, red cell membranes, capillary walls, and collagen and osseous structures also pertain to this category. The destruction (lipid peroxidation) of membranes is presumably related to the adequacy of vitamin E and of other putative antioxidant nutrients such as selenium, zinc, and copper. (*See* **Growth and Development**; refer to individual nutrients.)

Immunity and host defense   In the category of immunity and host defense are included the multitude of tests of phagocytic, humoral, and cellular immunity, as well as the newer tests of messenger (hormonal) function of the cytokine mediators of the inflammatory response and of autoimmunity and autoimmune surveillance. Almost all macronutrients, and most of the micronutrients that relate to cellular proliferation or metabolism, influence the function at some level of host immunity. (*See* **Immunology of Food**.)

Transport   Tests of transport function begin at the intestinal level, specifically with its homeostatic regulation of the body reserves. Additionally, at the level of transmembrane movement of nutrients, we have different uptake potentials based on the nutrient status of the organism. In both cases, deficiency should enhance a net inward flux, and sufficiency should balance inward and outward movement. Among the nutrients with regulated uptake at the gut are iron and zinc. Cobalt ion can be used as a surrogate for iron in an isotopic absorption test. The conversion of carotene to vitamin A by the gut is also regulated by the level of body vitamin A stores, and theoretically could be measured by a clinical test

that measures the outcome of its uptake and intestinal metabolism. Isotopic turnover studies also fall into the category of tests of nutrient transport, as do the specific tests of mobilization from stores to the periphery, such as the relative dose–response and the modified relative dose–response for vitamin A status assessment. (*See* **Cobalt**.)

Hemostasis   Nutrition with respect to vitamin C, vitamin K, and all micronutrients that influence platelet aggregation can be reflected in tests of hemostasis related to capillary structure, bleeding time, or clotting capacity. It should be noted that capillary integrity can be classified under 'structural integrity' or under 'hemostasis.'

Reproductive biology   Spermatogenesis and gonadal–pituitary hormone axis function, as well as lactation and reproduction performance in the individual women and, at the population level, fertility and fecundity of the community, birthweights, and peripartum hemorrhage rates constitute functional tests of reproductive biology. (*See* **Menstrual Cycle: Nutritional Aspects; Premenstrual Syndrome: Nutritional Aspects; Lactation**: Physiology; **Lactic Acid Bacteria**.)

Nervous system function   From cognitive performance, social competence, scholastic performance, through sleep pattern to peripheral nerve transmission and, finally, to the specific sensory reception in eyes, ears, nose, and tongue, functional tests in the nervous system cover a ganut of nervous functions. Both macro- and micronutrient deficiencies have been associated with impairment of higher cerebral functions. Micronutrients, such as thiamin with respect to the sixth cranial nerve function, vitamin A, zinc, and vitamin E for retinal function, and zinc and vitamin A for olfactory acuity, are some of the nutrients that can be evaluated in tests of the special senses.

Hemodynamics and physical capacity   Functional tests of hemodynamics and physical capacity are based on techniques of noninvasive and work performance physiology. Doubly labeled water ($^2H_2^{18}O$) can be used as a long-term marker of energy expenditure. Protein, energy, and iron are the nutrients most regularly thought to be determinant of the performance of the individual in efforts of physical exertion. (*See* **Exercise**: Metabolic Requirements.)

Classification by Type of Test

The alternative form of classification of functional test of nutritional status is by type of test, i.e., by the manner in which the function is approached. These include four levels: *in vitro* tests of *in vivo* functions;

induced responses and load tests *in vivo*; spontaneous *in vivo* activities and responses of organs and tissues; and responses at the level of the individual or population. A selected array of functional tests are listed under the respective classifications described in the following four sections.

In vitro tests of *in vivo* functions

- erythrocyte transketolase activity coefficient;
- erythrocyte glutathione reductase activity coefficient;
- erythrocyte transaminase activity coefficient;
- circulating homocysteine concentration;
- granulocyte chemotactic response;
- lymphocyte chemotactic response;
- bactericidal capacity of granulocytes;
- leucocyte glycolysis;
- granulocyte reduction of nitroblue tetrazolium;
- serum opsonin activity;
- T-cell blastogenesis;
- monocyte interferon production;
- monocyte monokine (interleukin-1; tumor necrosis factor) production;
- erythrocyte fragility;
- ^{75}Se uptake by erythrocytes;
- ^{65}Zn uptake by erythrocytes;
- thymulin activity reconstitution by zinc *in vitro*;
- prothrombin time;
- descarboxy prothrombin levels;
- platelet aggregation;
- D(−)-uridine suppression test;
- ^{14}C-formate conversion in lymphocytes;
- tensile strength of skin strips.

This group of functional tests requires some form of biological sample, either blood to separate the various cellular elements or a skin biopsy. Where specific cellular elements such as platelets, lymphocytes, and granulocytes are concerned, the volumes of blood to be extracted can be considerable. The samples required for erythrocytes are usually small.

In vivo induced responses and load tests

- capillary fragility (Hess) test;
- experimental wound healing;
- collagen accumulation in subcutaneous sponge implant;
- mobilization of inflammatory cells (Rebuck skin window);
- delayed cutaneous hypersensitivity;
- *de novo* antibody formation;
- vasopressor response;
- total-body energy expenditure with $^{2}H_{2}^{18}$ metabolism;

- isotopic turnover studies with whole-body counting;
- radioiron absorption;
- radiocobalt absorption;
- zinc absorption tests;
- intravenous magnesium infusion tolerance (urinary excretion) test;
- retinol relative dose–response;
- dihydroxyretinol modified relative dose–response;
- β-carotene bioconversion to retinyl ester challenge
- thyroid radioiodine uptake;
- glucose tolerance test;
- postglucose plasma chromium response;
- postglucose urine chromium response;
- glucose load with exercise for lactate and pyruvate;
- urinary malonaldehyde excretion;
- mixed-function oxidase ($^{14}CO_2$) breath test;
- ^{14}C-histidine ($^{14}CO_2$) breath test;
- ^{14}C-serine ($^{14}CO_2$) breath test;
- histidine load for urinary formimino glutamic acid;
- histidine load test for urinary hydantoin propionic acid;
- purine load test for urinary xanthine;
- sulfur amino acid load test for abnormal sulfur metabolites;
- sodium bisulfite load test for abnormal sulfur metabolites;
- tryptophan load test for urinary xanthurenic acid;
- leucine load test for urinary 3-hydroxyisovaleric acid.

In this class of test, some form of substrate, either natural or artificial, and unlabeled, labeled with a stable isotope, or labeled with a radioisotope, is administered, and its metabolism is monitored. The metabolism is either governed by a nutrient-dependent function of the nutrient in question or related to the body's homeostatic regulation of nutrient reserves.

Spontaneous *in vivo* activities and responses of organs and tissues

- capillary bleeding tendency;
- dark adaptation;
- visual recovery time;
- central scotoma size;
- color discrimination;
- taste acuity;
- olfactory acuity;
- auditory acuity;
- abducens (sixth cranial nerve) function;
- nerve conduction time;
- skin conductivity;
- electroencephalography;
- sleep pattern;

- grip strength;
- weight gain as lean tissue;
- weight gain as fat tissue;
- maximal oxidative capacity;
- maximal work capacity;
- urinary malonaldehyde excretion;
- methyl selenol/methyl trimethyl selenomium ratio in urine;
- pulmonary excretion of volatile hydrocarbons (ethane or pentane);
- peripartum hemorrhage;
- sperm count.

The spontaneous function of organs and tissues, without provoking special handling of substrates or external stimuli, such as taste acuity and dark adaptation, constitute the least invasive and complex, and the most acceptable (to the person examined) of the range of functional tests of nutritional status.

Responses at the level of the whole individual or population

- linear growth velocity;
- total weight gain velocity;
- work productivity and spontaneous play activity;
- birthweight;
- Apgar score at birth;
- infant developmental scale performance;
- infant temperament;
- cognitive performance;
- scholastic performance;
- athletic performance;
- social competence;
- development of secondary sex characteristics;
- lactation performance;
- fertility rate;
- fecundity rate;
- infant mortality rate;
- disease resistance or susceptibility.

Finally, a series of functions of the whole individual or the average statistics for a community population can be used to assess nutritional status, as well as the impact of improvement or deterioration of health and dietary conditions on individual or collective nutriture. However, the more complex and integral the function, the greater is the number of factors – both nutritional and nonnutritional – that influence, mediate, and determine the performance. The quality of educational materials, the skill and motivation of teachers, and the state of civil tranquility are among the covariables that might influence the scholastic performance of children, in addition to the nutrition of the school population. If the first limiting factor in scholastic performance lies among these

variables, it is doubtful that any limitation to maximal output from malnutrition could be demonstrated or exposed.

Newer *In vitro* Approaches to Functional Assessment of Nutritional Status Based on Molecular Biological (Genomics, Proteomics) Techniques

The decade since the publication of the first edition of this volume has been marked by dramatic advances in the molecular biology of characterizing and expressing the genetic potential and regulation of the cell at the level of the genetic code information (genomics) and the expression of messenger RNA for specific protein (proteomics). Because nutrients can become limiting in the basic mechanisms, and can induce or repress genetic expression, genomics and proteomics protend a series of approaches to functional assessment:

- nutritional expression of nutrient-dependent or nutrient-containing enzymes or proteins;
- nutritional expression of homeostatic regulatory protein;
- nutritional modification of regulatory elements;
- nutritional pattern of upregulation (or downregulation) of gene expression.

In fact, several laboratories have begun to exploit the second of these methods, in tests for human zinc status based on the expression of mRNA for metallothionein, the intracellular binding protein of divalent metals, in peripheral nucleated blood cells.

Advantages and Disadvantages of Functional Assessment

Advantages

In terms of screening tests in epidemiology and clinical nutrition, functional indices of nutritional status have potential advantages of noninvasiveness, individual accuracy of diagnosis, timeliness and interpretability of nutritional interventions, and policy relevance.

Noninvasiveness   Almost by definition, static indices involve the sampling of blood or tissue. Many of the tests of performance are noninvasive in so far as no skin punctures or scrapings are involved. This factor may account for more compliance and cooperation on the part of potential clients or subjects. In an era of hepatitis B and human immunodeficiency virus viral endemicity, nonpenetrating procedures are also welcomed by investigators and laboratory personnel.

Individual accuracy of diagnosis   The supreme advantage of functional tests is their potential to provide an accurate diagnosis of status for the *individual* patient, client, or subject. However, this can be accomplished only through *serial* testing before and after a trial of physiological or therapeutic doses of the nutrient in question. Whereas static indices, or functional indices measured in a cross-sectional fashion, must be interpreted as normal or abnormal in a normative fashion, in relation to the distribution in a standard reference population, functional tests repeated in the same individual can be assessed with respect to the degree of incremental change. Thus, an accurate assessment of status in an individual is possible. Those with a nutrient-responsive improvement or normalization in a function are truly deficient. For example, consider two persons with the same dark adaptation threshold, in the highest 10th percentile of the reference population; both receive a course of vitamin A therapy, but only one subject improves. If there is no intrinsic retinal disease in the latter, we can conclude that the threshold is a normal variant. In the former, it is a manifestation of vitamin A deficiency.

Temporal sensitivity in detecting nutritional deficits   Another advantage of the functional test is the temporal sensitivity in detecting nutritional deficits. We generally recognize a hierarchy of *sequence* in the stage development of depletion. The earliest changes are those in nutrient-dependent function at the peripheral tissue level, often detectable before the nutrient levels in the tissues themselves have declined. Thus, between a functional deficit or a tissue-concentration decline, conventional wisdom favors the former as the earliest detectable alteration during nutritional depletion. As noted earlier, the signs and symptoms of the clinical deficiency syndrome appear much later in the course of depletion.

Interpretability of nutritional intervention   Demonstrating the efficacy and effectiveness, that is whether or not a nutritional intervention, such as the fortification of a food, the distribution of a supplement, or the application of a nutrient-sparing program such as deparasitization or immunizations, is actually having an impact, has been the bane of public health nutrition. Even when normative references are used to determine 'abnormals,' in a baseline survey, a marked shift in the distribution of a population toward a normal distribution after the implementation of an intervention can allow success to be interpreted. The failure to observe any change can be taken as tentative evidence for a null effect. However, in serial testing, one must account for any 'learning effect' or spontaneous improvement that would be expected from prior exposure to, or experience with, the test procedure.

Policy relevance   The policy relevance of functional assessment derives, in part, from the consequence of the foregoing discussion. For policy-makers, the demonstration that some aspect of *performance* was improved by a public health intervention generates more conviction and sustainability than simply showing that the level of the nutrient had increased in blood, hair, or urine. It provides a ready answer to the 'so what?' questions that those in charge of legislation or expenditures for health will ask.

Disadvantages

Functional assessment of nutritional status, however, is not a panacea; there are also notable disadvantages in their application. These include poor interpretability in cross-sectional studies, lack of specificity for nutritional status, and specific considerations in young children.

Poor interpretability in cross-sectional studies   Status and functional indices share a common limitation when the issue is the nutritional interpretation of a result from a single clinical examination or a cross-sectional study. This limitation is the difficulty in accurately separating those with suboptimal nutritional status from those with adequate status by simple reference to an arbitrary cut-off level. It is a *probabilistic* issue in which only a certain level of *risk* of pertaining to a truly deficient or truly sufficient population can be concluded from a single point-in-time evaluation.

Lack of specificity for nutritional status   Almost all of the functions that are measured to assess nutritional status can also be impaired by other (nonnutritional) causes. Thus, in a cross-sectional or single measurement assessment using a functional index of nutritional status, it is impossible to be sure that an impaired or abnormal function is related to nutritional deficiency or to some other pathology or a genetic abnormality. Moreover, even if a longitudinal study with supplementation is employed with a functional index, if the person has *multiple* nutrient deficiencies, and only a single nutrient is supplemented, the remaining deficits may impair a full recovery of the functional abnormality.

Considerations in young children   For obvious reasons, tests related to reproductive biology are not applicable to children. Moreover, the use of radioisotopes *in vivo* is relatively or absolutely contraindicated in young children. Stressful stimuli, such as vasopressors or electrical conductivity electrodes, or stressful performance challenges, such as maximal oxidative capacity, should be proscribed from the assessment of children.

Separating pharmacological from nutritional effects Even when serial or comparative measurements of a functional outcome are coupled to a nutrient-intervention format, the interpretation of the preexisting or prevalent nutritional state must be made with caution and judgment. What is a 'nutrient-responsive' situation, and what is a 'nutrient-deficient' condition? By way of analogy to personal esthetics, the application of make-up to a physically unattractive individual can make them 'presentable.' The appearance of a person already attractive by consensus standards can be enhanced with cosmetics to a state of 'ravishing,' however. Where are the valid boundaries to define an 'esthetic deficit'? Similarly, in household economy, what constitutes a 'deficient' income: when one cannot afford a Mercedes-Benz car? When one cannot afford any vehicle? Or when the funds do not allow for purchase of the basic necessities of food, clothes, shelter, and care? Additional wisdom and a grounded philosophical perspective are needed to differentiate enhancement (nutrient-responsive) from reversal of true deficits (nutrient-deficiency).

Important Applications of Functional Status Assessment

Assessment of Risk from Surgery and Other Major Therapies

Recently, grip strength, i.e., the development of contractive force by the hand musculature, has been shown to have some bearing on the prediction of survival or mortality, or the risks of complications in recovery from surgery or radiation or chemotherapy in association with static indices of anthropometry, immune status, and protein concentrations. Tests of active internal lipid peroxidation, such as the ethane or pentane breath tests, have theoretical promise as complementary indices for the prediction of risk of major therapy of precariously nourished patients.

Evaluation of Public Health Impact of Interventions

Faithful to the spirit of Study Team IX, the 1990s saw an increasing incorporation of functional assessment into the evaluation of public health interventions. Child growth in height and weight is the most commonly used variable. Other indicators of interest in public health nutrition are infection resistance, lactation performance, fertility, work output, and

mortality. (*See* **Infection, Fever, and Nutrition; Children**: Nutritional Requirements.)

Functional assessment can also be used to gain inferences for improving the balance of nutrients in traditional regional diets. As field intervention studies with supplements of protein and energy failed to show major functional improvements, but the association with animal protein did, the *quality* of the diet in terms of micronutrient (vitamin, mineral) malnutrition was focused on in the 1980s.

Assessment of Dietary Nutrient Intake Requirements

Functional assessment is gaining a place in the continued fine-tuning of our knowledge of human nutritional requirements and dietary recommendations. The dominant concept of where to set the recommendation for the daily intake of a nutrient is at the amount that will prevent the development of signs and symptoms of deficiency in the majority or a population. When possible, the concept of the amount that supports a specific pool size of total-body nutrient content has been introduced. Given that functional alterations are generally the earliest detectable changes with the onset of depletion, the logical approach to defining intake requirements of a nutrient is that level which prevents functional changes in the majority of a population. As experience is gained with interpreting and avoiding pitfalls in the application of functional tests, they will be used increasingly to refine our present understanding of what nutrient intakes to recommend to maintain adequate nutriture. (*See* **Dietary Requirements of Adults**.)

New Frontiers for Functional Assessment

Functional indicators of adverse interactions from excesses or toxicity associated with high doses of intakes of nutrients for use in nutritional toxicology have lagged behind the progress in diagnosis of deficient and marginal states. More thought and research need to go into this domain.

The late James Olson has made a pithy distinction between the functions and actions of constituents in the diet. The former relate to the *essential* functions that define a nutrient. However, at higher doses, these nutrients begin to have actions to favor health or prevent disease that go beyond their capacities to nourish. Bioactive (nutriceuticals, phytochemicals) compounds in foods such as flavonoids, carotenoids, etc. have assumed growing interest. Since the chemical nature of the active substances is often unknown, functional markers seem to be the only variables with which to assess the status of consumers.

Conclusions

The clinician, epidemiologist, investigator, and public health professional – all those involved with nutritional status diagnosis – should be aware of functional assessment, its potential, and its limitations. Any approach to nutritional assessment has its advantages, disadvantages, and limitations. The use of functional indices of nutritional status tends to be less invasive, and more client-friendly, than static assessment tests that seek to measure nutrients directly in body fluid or tissues.

Impairment of a peripheral nutrient-dependent function or the body's response to regulation often occurs before nutrient levels decline in blood and tissues, and well before clinical manifestations of deficiency emerge. When coupled to a trial of nutrient supplementation, the functional approach, unlike the static approach, can provide a definitive, retrospective assessment of the original nutritional status of the host. Functional tests can be classified by the physiological or anatomical system that is measured, or by the level of investigation of the test. As the nature of the metabolism and physiological and biochemical functions of nutrients becomes better understood, newer strategies to exploit the functional assessment of nutritional status will be suggested.

See also: **Anemia (Anaemia)**: Iron-deficiency Anemia; **Ascorbic Acid**: Physiology; **Children**: Nutritional Requirements; **Cobalt**; **Dietary Requirements of Adults**; **Epidemiology**; **Exercise**: Metabolic Requirements; **Growth and Development**; **Immunology of Food**; **Infection, Fever, and Nutrition**; **Lactation**: Physiology; **Lactic Acid Bacteria**; **Premenstrual Syndrome: Nutritional Aspects**; **Retinol**: Physiology; **Scurvy**

Further Reading

Calloway DH (1982) Functional consequences of malnutrition. *Reviews of Infectious Diseases* 4: 736–745.

Chandra RK (1981) Immunocompetence as a functional index of nutritional status. *British Medical Bulletin* 37: 89–94.

Chandra RK (2001) Effect of vitamin and trace-element supplementation on cognitive function in elderly subjects. *Nutrition* 17: 709–712.

Pollitt E (1996) Timing and vulnerability in research on malnutrition and cognition. *Nutrition Reviews* 54: 455–549.

Solomons NW (1985) Assessment of nutritional status: functional indicators of pediatric nutriture. *Pediatric Clinics of North America* 32: 319–334.

Solomons NW and Allen LH (1983) The functional assessment of nutritional status: principles, practices and potential. *Nutrition Reviews* 41: 33–50.

Nutritional Management *See* **Elderly**: Nutritional Management of Geriatric Patients; **Food Intolerance**: Types; Food Allergies; Milk Allergy; Lactose Intolerance; **HIV Disease and Nutrition**; **Hypertension**: Nutrition in the Diabetic Hypertensive; **Infection, Fever, and Nutrition**; **Multiple Sclerosis – Nutritional Management**; **Pregnancy**: Nutrition in Diabetic Pregnancy

---

Nutritional Policies *See* **Nutrition Policies in WHO European Member States**

NUTRITIONAL SURVEILLANCE

In Industrialized Countries

K F A M Hulshof, M R H Löwik and D C Welten,
TNO Nutrition and Food Research, The Netherlands

Introduction

Especially during the last two decades, evidence has accumulated that prevailing dietary patterns have adverse health effects. Nutritional assessment has become an important topic on the health-policy agenda. This article describes the aim of nutritional surveillance, the availability and usefulness of nutritional surveillance indicators (especially food consumption), and some trends, risk groups, and risk areas. Nutrition-related health problems and surveillance systems differ among countries, in particular between developed and developing countries. This article deals only with industrialized countries in Europe, Canada, Japan, Australia, and the USA. This paper reports the situation in the early 2000s.

Aim of Nutritional Surveillance

As stated by a Joint FAO/UNICEF/WHO Expert Committee, the objectives of nutritional surveillance are as follows: to describe the nutritional status of the population, with particular reference to groups at risk, to contribute to the analysis of causes for changes and differences, to promote decisions by governments on food and nutrition policy issues, to predict future trends, and to evaluate the effects of nutritional programs.

The international conference on Nutrition held in Rome in 1992, adopted a 'World declaration on Nutrition' and a 'Plan of actions for Nutrition.' In this plan, one of the nine action-oriented themes was 'Assessing, analysing and monitoring nutrition situations.' According to these themes, governments should, among other things: identify the priority of nutritional problems in their country, analyze their causes, plan and implement appropriate remedial situations, and monitor and evaluate efforts to improve the situation. Moreover, as a consequence of recent adopted directives of the European Union, the member states have to create systems that will monitor the use and intake of additives.

As illustrated in **Figure 1**, nutritional surveillance ideally provides information on a wide range of variables, from food availability, distribution and consumption and nutrient utilization (as reflected in nutritional status) to, ultimately, health status and mortality. This results in identification of public health problems that call for specific action and lead to nutrition research priorities (both applied and more fundamental). The data can be obtained from either existing sources, including administrative data, or surveys undertaken specifically for surveillance purposes.

Sources of Dietary Information

Insight into dietary patterns is a core target of nutritional surveillance, since this provides a comprehensive basis for nutritional risk assessment. In principle, three different types of data can be used: food supply data, data from household consumption surveys, and data from dietary surveys among individuals. Each type of data corresponds with a different stage in the food chain and is obtained by different methods.

Food-supply Data

Food-supply data provide information on the type and amount of food available for human consumption, to the country as a whole. The supply is calculated in food balance sheets (FBSs), which are accounts, on a national level, of annual production of food, changes in stocks, imports and exports, and agricultural use and industrial use. Food supply is usually expressed per head of the population in kilograms per year, or grams per day. The per-capita consumption of energy and some additional nutrients is calculated using food composition tables.

Food-supply data refer to food availability, which gives only a crude (overestimated) impression of potential consumption. Food and nutrient losses prior to consumption, owing to processing, spoilage, trimming and waste, may not be adequately accounted for. Furthermore, these data provide no information about the distribution of food among population groups or districts.

International FBSs are prepared and published by the FAO, the Organization for Economic Cooperation and Development (OECD) and the statistical office of the Commission of the European Union (EUROSTAT). The FAO has published FBSs since 1949, also covering the period 1934–1948. Since 1949, FBSs have been compiled on an annual basis from data supplied by about 200 countries. Information is available for all European countries, Australia,

NATIONAL FOOD SUPPLY ⟶ FOOD DISTRIBUTION ⟶ CONSUMPTION ⟶ NUTRIENT UTILIZATION ⟶ HEALTH OUTCOME

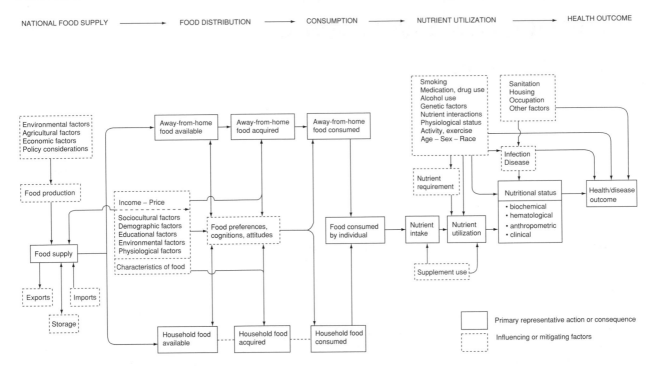

Figure 1   General conceptual model for food choice, food and nutrient intake, and nutritional and health status. Source: Life Sciences Research Office, FASEB (1989) Nutrition and Monitoring in the United States. An Update Report on Nutrition Monitoring. Prepared for the US Department of Agriculture and the US Department of Health and Human Services. DHS Publication No. (PHS) 89–1255. Public Health Service. Washington, US Government Printing Office, September 1989.

Canada, Japan, New Zealand, and the USA. Since 1971, the FAO has included its FBS data in the Interlinked Computer Storage and Processing System of Food and Agricultural Commodity Data (ICS). EUROSTAT publishes FBSs for the 15 member countries of the European Union. The OECD FBSs cover, besides several European countries, Australia, Canada, Japan, New Zealand, and the USA. Although FBSs are compiled in a similar way, they differ in coverage, food grouping and level of processing of commodities (e.g., FAO lists 300 food items classified into 17 food group categories; OECD 70 items in 13 categories) and in nutrient conversion. The FAO and OECD usually publish summaries of FBSs every 3–5 years, with a lag time of 3–4 years between data collection and publication. The ICS supplies figures on magnetic tape, on floppy disk and on Internet (http://apps.fao.org). EUROSTAT publishes supply balance sheets in the Agricultural Statistical Yearbook.

In addition to the international FBSs, many countries publish national FBSs, mostly in statistical yearbooks or special statistical publications. For example, US food-supply statistics are available from 1909 onwards. National FBSs tend to be more up to date. Owing to different methodologies underlying their compilation and presentation, these data can differ from international FBSs.

Despite their limitations, FBSs are useful in that they indicate the (in)adequate aspects of food supply, provide material for planning food supply (production, imports and exports) and give crude indications of (un)desirable changes in terms of potential (adverse) expected health impact. As a result of their long history, FBSs are especially used for assessing trends over time. In contrast to national FBSs, the international FBSs can be used for comparative studies, provided that the FAO and OECD data sets are used separately. **Figure 2** illustrates the use of FBS data (FAO) for comparison across countries and for trends over time. This table shows the consumption of meat in five selected countries. Only in the UK has the consumption of meat and meat products remained remarkably steady since 1965. In the other countries, the total consumption of meat has increased considerably, especially in Spain. The same tendency has been observed in other southern European countries and reflects one of the important changes in the Mediterranean diet over the past decades. Such comparisons implicitly assume that the demographic changes across the countries are similar.

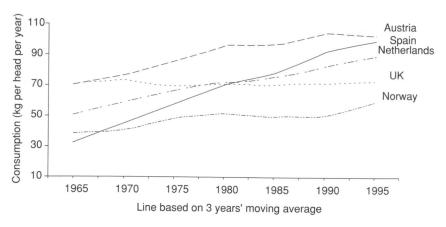

Figure 2   Available consumption of meat and meat products (kg per head per year) in five European countries. Source: FAO Statistical Databases: Food Balance Sheets. http://apps.fao.org. (10.03.2000).

Household Surveys

Food available at the household level may be estimated by budget surveys and by consumption surveys. The first type of survey gives information on the purchases of food in terms of expenditure and is used for economic policy. For example, weights for the construction of consumer price indices can be calculated. In household-consumption surveys, the amounts of foods and drinks brought into the household are also recorded. For the most part, only the expenditures of meals taken outdoors are noted. Some household surveys may even measure changes in food stocks, in addition to acquisition.

In general, household surveys do not provide information on how food is handled within the household, or on actual consumption by its members. Sometimes, the consumption data are converted to individual intake levels. The methods vary from simply dividing the total consumption by the number of people in the households, to assigning factors (consumer units) to persons weighed according to age and sex. In contrast to FBSs, household surveys can supply information on food (and nutrient) patterns in subgroups of households. These groups may be classified by economic, demographic, and other factors, which provide the opportunity for risk group identification.

In most countries, household surveys started in the 1940s or 1950s. Only few countries have a continuous system, some repeat surveys every 3–4 years, and others repeat surveys only every 5–10 years. In The Netherlands, the household budget surveys started in 1951, and since 1978, they have been conducted annually. In Europe, the best-known study is the specialized and ongoing household food consumption survey of the UK. Australia, Canada, and Japan have regularly conducted household consumption surveys. In the USA, the first national household

consumption survey was conducted in 1936–1937. Between 1942 and 1965, four nation-wide studies on household consumption were carried out. Since 1965, US household food-consumption surveys have also provided information on food intake at the individual level of household members. At present, a wide range of data on household surveys are available, as shown in the FAO Food and Nutrition Policy Papers and WHO publications. Since the dietary data are based on a variety of methods, the surveys are not very suitable for comparisons among countries. Differences exist in sampling procedures, food grouping, conversion to nutrients and period, frequency, and technique of data collection. Snacks, sweets, soft drinks, and alcoholic beverages are excluded from some studies. Data on the quantity of and/or expenditure on food may be collected by record keeping, by interviews or by both methods. Household accounts for nonfood items can cover a period of 4 weeks, but for foodstuffs, 2 weeks is more usual. The results of household surveys play an important role in nutritional surveillance at the national level, particularly when surveys are carried out annually, which reveals trends in food consumption.

To improve the possibilities for international comparison of these data in Europe, since 1993, an ongoing project of the European Union (DAFNE, DAta Food NEtworking) is harmonizing at the international level dietary exposure data from household budget surveys. DAFNE aims to create a pan-European food databank based on national household budget surveys by the development of the most appropriate way of using food and related data from these surveys. Methods were designed to calculate, for instance, the overall average availability per person per day of comparable food items or groups among DAFNE countries as well as average availability by degree of urbanization and educational level of

Table 1 Average availability of fruit and vegetables (fresh and processed) in DAFNE countries (grams per person per day)

	Greece	Ireland	Luxemburg	Norway	Spain	UK
Fruit	283	103	234	174	307	133
Vegetables	229	130	180	102	179	157

Source: DAFNE II project, 1998.

household head. Distribution and cumulative distribution functions of the availability of food items were estimated. **Table 1** shows the differences in availability of fruit and vegetables (fresh and processed) in Greece, Ireland, Luxembourg, Norway, Spain, and the UK.

Individual Dietary Surveys

In contrast to FBSs and household surveys, data from individual surveys provide information on average food and nutrient intake and their distribution over various well-defined groups of individuals. Data more closely reflect actual consumption and can provide additional information on meal patterns, etc.

To collect dietary intake data on an individual level, several methods can be used. Briefly, the methods can be divided into two categories: record and recall methods. Record methods collect information on current intake, keeping a record of all foods and drinks based on menu, household measures and/or weighing, over one or more days. Recall methods reflect past consumption, varying from intake over the previous day (24-h recall) to usual food intake (dietary history or food frequency). Each form has its own strengths and weaknesses, and there is no single ideal method. Details of the available methods for assessing food consumption of individuals are given in numerous reviews and manuals.

To characterize the average intake of food and nutrients and their distribution over various groups of individuals one 24-h recall or 1-day food record is appropriate, provided that the sample is representative of the population under study, and day-of-the-week and seasonal variations are taken into account. To determine the proportion 'at risk' of inadequate intake, the food consumption of each subject must be measured over more than one day, or retrospective information on intake over a longer period may be used (e.g., dietary history method). The appropriate period depends on the purpose of making an estimate, the precision desired, the food component(s) of interest, the intra- and interindividual variation components, and the period over which an intake has to be low or high before health risks are introduced.

Probably all (industrialized) countries, if not all, have carried out small-scale dietary surveys. These surveys provide valuable information, but owing to

samples of convenience and different food consumption methods, their usage in national nutrition policy and nutritional surveillance is often limited. Since the mid-1980s, the number of countries that have conducted individual dietary surveys is growing. The USA has one of the most comprehensive nutrition monitoring programs in the world and already has a longer tradition in the field of nutritional surveillance. The National Nutrition Monitoring and Related Research Program (NNMRRP) provides information about the dietary and nutritional status of the US population, conditions existing in the United States that affect the dietary and nutritional status of individuals, and relationships between diet and health. The Continuing Survey of Food Intakes by Individuals (CSFII) conducted by the US Department of Agriculture (USDA), and the National Health and Nutrition Examination Survey (NHANES) conducted by the Health and Human Services (HHS) are the two cornerstones of the NNMRRP surveys. They provide national estimates of food and nutrient intakes in the general US population and in subgroups. The CSFII emphasizes the food and nutrient intake defined by various socio-economic factors. In NHANES, information on dietary intake and health status in the same individuals is available. These surveys also provide the potential for assessing consumption of additives and pesticides in the diet.

Also Australia and several European countries have performed individual surveys on a national basis. **Table 2** presents examples from studies conducted since 1985. The surveys differ in coverage of population, methods used to collect dietary data, nutrition-related health indices, etc. In several countries, dietary data were collected using a record method, but the number of record days varied from 1 to 7. A 7-day weighed record is thought to be the most accurate method of dietary assessment. However, this method has a high respondent burden, which can have consequences for the response rate, representativeness of the sample and data quality. Response rates vary widely. Sometimes, weighing factors are used to adjust for sources of nonresponse. Some subgroups (such as ethnic minorities, pregnant or lactating women) do not occur in the population in sufficient numbers to appear in the survey with sufficient representation to allow for separate estimates of their diet and nutritional status that are reliable. Extra sampling can improve the precision of estimates in nutritional assessment in specific groups and is used in several studies, including those in the United States. Special (vulnerable) groups can also be examined in separate studies. For example, in 1998, in The Netherlands, a nation-wide survey was conducted among 8-year-old Turkish and Moroccan children and their

Table 2 Examples of nation-wide food-consumption surveys with nutrient-intake data on an individual level since 1985

Country	Year	Survey	Population Sex	Population Age	Sample size	Response rate (%)	Dietary method[a]
Australia	1985	National Dietary Survey of Schoolchildren	F+M	10–15	5 224	65	1d Rcd
Austria	1992	Austrian Study on Nutritional Status (ASNS)	F+M	6–18	2 173	na	7d Rcd
	1993–1997	Austrian Study on Nutritional Status (ASNS)	F+M	19–65	2 065	69	24h Rcl, DH
Denmark	1985	Dietary Habits in Denmark	F+M	15–80	2 242	na	DH
	1995	National Dietary Survey	F+M	1–80	3 098	65	7d Rcd
Finland	1992	Dietary Survey of Finnish Adults	F+M	25–64	1 861	60	3d Rcd
	1997	Dietary Survey of Finnish Adults	F+M	25–64	2 862	72	24h Rcl
			F+M	65–74	290		24h Rcl
France	1993–1994	National Food Consumption Survey (ASPCC Survey)	F+M	18+	1 229	na	7d Rcd
	1993–1994	National Food Consumption Survey (ASPCC Survey)	F+M	2–85	1 500	na	7d Rcd
	1998–1999	Individual National Food Consumption Survey (INCA)	F+M	3–14	1 018	na	7d Rcd
			F+M	15+	1 985		7d Rcd
Germany	1985–1989	National Nutrition Survey in Former West Germany	F+M	4–70+	24 632	71	7d Rcd
	1991–1992	National Health Survey in East Germany	F+M	18–80	1 897	71	DH
	1998	German Nutrition Survey	F+M	18–79	4 030	61	DH
Hungary	1985–1988	First Hungarian Representative Nutrition Survey	F+M	14–60+	16 641	61	2× 24h Rcl+FFQ
	1992–1994	Hungarian Randomized Nutrition Survey	F+M	18–60+	2 559	na	3× 24h Rcl+FFQ
Iceland	1990	Icelandic National Nutrition Survey	F+M	15–80	1 240	72	DH
Ireland	1990	Irish National Nutrition Survey (INNS)	F+M	8–18+	1 214	na	DH
	1998	North–South Food Consumption Survey	F+M	18–64	1 379	66	7d Rec
Italy	1994–1996	INN-CA1995	F+M		2 734	47	7d Rcd
Lithuania	1997	Baltic Nutrition and Health Survey	F+M	20–65	2 183	73	24h Rcl+FFQ
The Netherlands	1987–1988	Dutch National Food Consumption Survey (DNFCS-1)	F+M	1–79	5 898	79	2d Rcd
	1992	Dutch National Food Consumption Survey (DNFCS-2)	F+M	1–92	6 218	72	2d Rcd
	1997–1998	Dutch National Food Consumption Survey (DNFCS-3)	F+M	1–97	6 250	69	2d Rcd
Norway	1993–1994	National Dietary Survey Among Adults (Norkost)	F+M	16–79	3 144	63	FFQ
	1997	National Dietary Survey Among Adults (Norkost)	F+M	16–79	2 672	54	FFQ
	1999	National Dietary Survey (Norkost)	F+M	6+12 months	2 400	80	FFQ
	1999	National Dietary Survey (Norkost)	F+M	2 years	2 010	67	FFQ
Northern Ireland	1986–1987	Diet, Life-style and Health in Northern Ireland	F+M	16–64	616	na	7 d Rcd
Poland	1991–1994	Dietary Habits and Nutritional Status of Selected Populations	F+M	11–14	1 126	na	24h Rcl
				18	2 193		
				20–65	4 945		
Slovak Republic	1991–1994	Asessment of Food Habits and Nutritional Status, etc.	F+M	11–14	3 337	na	24h Rcl+FFQ
	1995–1998			15–18	4 556		
	1999–2001			19–88	4 807		
Sweden	1989	HULK	F+M	1–74	2 036	70	7d Rcd
	1997–1998	Riksmaten	F+M	18–74	1 215	60	7d Rcd
Switzerland	1992–1993	Swiss Health Survey	F+M	15–74	26 000	71	FFQ
UK	1986–1987	The Dietary and Nutritional Survey of British Adults	F+M	16–64	2 197	70	7d Rcd
	1992–1993	National Diet and Nutritional Survey: Children Aged 1½–4½ years	F+M	1½–4½	1 675	80	4d Rcd
	1994–1995	National Diet and Nutrition Survey: People Aged 65 Years and Over	F+M	65+	1 687	80?	4d Rcd
	1997	National Diet and Nutrition Survey: Young People Aged 4–18 Years	F+M	4–18	1 701	64	7d Rcd

[a]24h Rcl, 24-h recall; 1d Rcd, 1-day record; FFQ, food-frequency questionaire; DH, dietary history method; na, not available. See also: Löwik MRH and Brussaard JH (2002) EFCOSUM.

mothers, with a Dutch control group. Norway plans to carry out a survey among migrants in 2000–2001.

The presented national dietary surveys provide valuable information for usage in national policy and are central in nutritional surveillance, and, when repeated in a proper way, trends over time can be studied. However, for a detailed evaluation of dietary intake in Europe, there is a need for increasing comparability of sampling designs, dietary methods, and selected population descriptors. The establishment of a Health Monitoring Program in Europe should make it possible to measure health status, trends, and determinants throughout the Community; facilitate the planning, monitoring and evaluation of Community programmes and actions; provide Member States the appropriate health information to make comparisons; and support their national health policies. As part of this program, since the end of 1999, the project European Food Consumption Survey Method (EFCOSUM) has aimed to define a (minimum) set of dietary components that are relevant determinants of health. Moreover, the study aimed to define a method for the monitoring of food consumption in nationally representative samples of all age–sex categories in Europe in order to provide internationally comparable data. This method can be used alone, or as a calibration method for ongoing studies. The project made use of progress in relevant projects carried out until now, such as DAFNE and EPIC, and addressed the possibility for data fusion with other health monitoring studies. Fourteen EU Member States as well as eight other EU countries are participating in EFCOSUM.

Nutritional Status and Health Indices

The assessment of nutritional status includes, in addition to dietary intake, indicators of nutrition-related health status, such as anthropometric measurements, hematological and biochemical tests, clinical signs of deficiencies, and risk factors for diseases associated with diet (e.g., overweight). Furthermore, determinants of food and health-related behavior, such as nutritional knowledge and attitudes, may be studied as well. These indicators can be included in the surveys or studied in separate samples. Several national studies listed in **Table 2** studied both dietary intake and nutrition-related health-status indicators. In most surveys, anthropometric data were collected (sometimes self-reported data on body weight and body height); in some countries, also, medical examination and/or biochemical and hematological tests (e.g., Germany, Hungary, UK) were carried out, and information on physical activity was included (e.g., Norway, Sweden).

A major advantage of collecting comprehensive (broad oriented) information at the individual level is that interrelationships can be studied. In studying correlations between diet and nutritional status indicators, one of the characteristics of a cross-sectional study is that mostly low correlations are found. This is attributable to, among other things, intraindividual variation and inaccurate assessment of intake and status indicators. In a cross-sectional design, the observation that a particular dietary factor is positively or inversely associated with relevant variable is meaningful, even when there is a low P-value, since this provides suggestive evidence for diet–health relationships which should be studied in more detail. To establish a causal link between diet and health, both intervention and (semi)-longitudinal studies are necessary. Endpoints, such as morbidity and mortality data, provide valuable additional information on the role of nutritional factors in diseases.

Risk Areas and Risk Groups

Nutritional assessment includes a normative evaluation of dietary intake and nutritional status indicators in order to estimate, for instance, the proportion of the population at risk. Nutritional-status indices can be evaluated by comparing them with reference values mostly obtained from healthy adults. Alternatively, predetermined cut-off points (based on consensus reports) can be used. In evaluating dietary intake, the reference values applied in recommended dietary allowances (RDAs) or dietary guidelines are often used. However, the usage of cut-off values is prone to some misclassification owing to (biological) variation within and among individuals. Despite the weaknesses of cut-off points, these criteria are commonly used and often needed to evaluate dietary intake as well as nutritional status parameters.

In most industrialized countries, the principal nutrition-related health problems are related to unbalanced (mostly overconsumption) of some nutrients, particularly energy, fat, and saturated fatty acids. Although the mean intake of energy among adults is mostly lower than the recommendations, the data available from nutritional surveillance indicate a high prevalence of overweight and obesity in several countries. Obesity, defined as a body mass index greater than 30 kg m^{-2}, is a common condition in Europe and also in the USA. Although **Table 3** gives only a rough impression (age groups are not always comparable, the periods in which studies were conducted differ slightly, exclusion criteria might vary, etc.), the data show that the proportion of subjects classified as obese varies among countries. Despite the differences, however, in recent decades, the prevalence of obesity has increased in most

Table 3 Prevalence of obesity (BMI $\geq 30\,kg\,m^{-2}$) in some countries

Country	Year	Age (years)	Men (%)	Women (%)
Austria	1993	19–65	6	5
England	1994	16–64	15	16
Finland	1991	20–75	14	11
East Germany	1992	25–65	20.5	26.8
West Germany	1991	25–65	16.0	21.4
The Netherlands	1992	19–65	5.5	9.2
USA	1988–1994	20+	22.3	25.0

Table 4 Average difference (95% confidence intervals) in consumption of selected foods (grams per person per day) between the highest and lowest educational level in nine European countries

	Education (highest minus lowest level)	
	Men	Women
Fruit	+24 (+19.0; +29.0)	+26.7 (+21.7; +31.8)
Vegetables	+12.1 (+8.3; +15.8)	+17.5 (+13.7; +21.2)
Fats and oils (added)	−2.9 (−4.0; −1.9)	−3.1 (−3.9; −2.3)
Meat	−32.6 (−36.0; −29.1)	−24.3 (−26.9; −21.8)
Milk total	+46.9 (+38.0; +55.9)	+39.9 (+33.2; +46.6)
Cheese	+9.9 (+8.4; +11.4)	+10.6 (+9.3; +11.8)

Source: Roos and Prättälä (1999): Fair-97-3096 project.

European countries as well as in the USA in the past decade. For instance, in The Netherlands, among subjects aged 19 years and over, the proportion of obese subjects has increased during the last 10 years from 6.0% (1987–1988) to 10.4% (1997–1998). At the same time, in The Netherlands, a decrease in energy intake and fat consumption was observed. Data from DNFCS-1 (1987–1988) and DNFCS-3 (1997–1998) suggested that energy intake decreased from 2308 kcal day (9677 kJ) to 2190 kcal day (9191 kJ). This reduction of about 120 kcal day (~485 kJ) was attributable to a decrease in fat consumption (protein intake increased, and carbohydrate intake and alcohol consumption remained constant). This may imply that daily energy expenditure has decreased during the same period and may include an increased sedentary behavior. Findings of the statistical office confirmed a decrease in physical activity in leisure time.

Obesity is associated with several specific health risks, including an increased incidence of hypertension, increased noninsulin-dependent diabetes and high levels of cholesterol and other lipids. Corrected for these factors, obesity in itself has been reported to be an independent factor for cardiovascular disease. In most industrialized countries, a higher prevalence of overweight and obesity is observed among subjects with a lower socio-economic status.

Higher rates of mortality and morbidity have been found among lower socio-economic groups, as compared with higher socio-economic groups. In several studies, food-consumption patterns and nutrient intakes have been more consistent with current dietary guidelines among people with a higher socio-economic status. Recent analyses of 33 studies in 15 European countries, conducted within the framework of the European Union's FAIR program (FAIR-97-3096), showed that, particularly in the north and west of Europe, people with a higher education tend to consume more vegetables and fruits and less fats and oils. **Table 4** presents some selected results. Other studies also found that people with a higher socio-economic status have reported eating more wholemeal and brown bread and less whole milk, eggs,

and meat. A higher socio-economic status has been associated with a lower intake of fat, saturated fatty acids, and refined sugars, and a higher intake of fiber, although mostly, differences in nutrient intake levels have been quite small. In general, food disparities in relation to consumption levels might possibly explain some of the higher rates of mortality and morbidity among lower socio-economic groups.

Concerning micronutrients, in most countries, the average intake of most minerals and vitamins appears adequate for the population. In general, iron is an exception in that many subjects have a low iron intake in comparison with recommended values. In most countries, groups with a low intake are young children, adolescents and women of child bearing age. Moreover, the intake of vitamins (e.g., vitamin A and its precursors, vitamin B_6, vitamin C, folic acid), minerals, and/or trace elements (e.g., calcium, magnesium, zinc, iodine) might be not always adequate among certain population groups in several countries. However, it should be noted that the confirmation of nutritional risk obtained by biochemical data is essential for assessment of risk areas and risk groups.

Nutritional Surveillance in the Future

A growing awareness of potential relationships between diet (as part of lifestyle) and health is accompanied by an increasing demand for (comparable) data. Research in the field of molecular genetics and nutrient–gene interactions is promising. Monitoring systems may also provide more information on variations within the human genome. This will enable us increasingly to focus on the specific needs of the individual at potential risk.

Since it might be expected that in the coming years, bioactive components, additives, and contaminants will become increasingly a topic of interest, future food-consumption databases should allow also the assessment of these nonnutrients. Therefore, there will be a need for more descriptive specificity, such as brand names and more information on the food composition. The use of computer technology and advanced statistical methods may facilitate wider applications of survey data and cost-effective data fusions.

To improve valid comparisons between diet and health at the international level, a better harmonization of data collection, methodology and standardization of data analysis will be needed. Since 1990, this concept has already been implemented in the National Nutrition Monitoring and Related Research Program (NNMRRP) in the USA; in Europe, an increasing number of initiatives as to Pan-European projects are coming up with results.

See also: **Cholesterol**: Properties and Determination; **Coronary Heart Disease**: Etiology and Risk Factor; **Diabetes Mellitus**: Etiology; **Dietary Surveys**: Measurement of Food Intake; Surveys of National Food Intake; Surveys of Food Intakes in Groups and Individuals; **Food Composition Tables**; **Hypertension**: Physiology; **Nutrition Policies in WHO European Member States**; **Nutritional Assessment**: Importance of Measuring Nutritional Status; **Obesity**: Etiology and Diagnosis; **World Health Organization**

Further Reading

Becker W and Helsing E (1991) *Food and Health Data. Their Use in Nutrition-policy Making.* WHO Regional Publications, European Series, No. 34. Copenhagen: World Health Organization.

Briefel RR (1996) Nutrition monitoring in the United States. In: Ziegler EE and Filer LJ (eds) *Present Knowledge in Nutrition.* Washington, DC: ILSI Press.

FAO (1977, 1981, 1988) *Review of Food Consumption Surveys.* Nos 1, 27, 35, 44/Rome: Food and Agriculture Organization.

FAO/WHO (1992a) *International Conference on Nutrition. Major Issues for Nutrition Strategies. Promoting Appropriate Diets and Healthy Life-styles.* Theme Paper No. 5. Rome: Food and Agriculture Organization.

FAO/WHO (1992b) *International Conference on Nutrition. Nutrition and Development: a Global Assessment.* Rome: Food and Agriculture Organization.

Hulshof KFAM and Löwik MRH (1995) Nutritional problems in the industrialized world. In: Velden van der K, Ginneken van JKS, Velema JP, Walle de FB and Wijnen van JH (eds) *Health Matters. Public Health in North–South Perspective*, pp. 162–172. Bohn Stafleu Van Loghum: Houten/Diegem.

Kuczmarski RJ, Carroll MD, Flegal KM and Troiano RP (1997) Varying Body Mass Index cutoff points to describe overweight prevalence among U.A. adults: NHANES III (1988–1994). *Obesity Research* 5: 542–548.

Löwik MRH and Brussaard JH (eds) (2002) *EFCOSUM: European Food Consumption Methods. European Journal of Clinical Nutrition 56(Suppl. 2): 51–596.*

Nelson M and Bingham SA (1997) The assessment of food consumption and nutrient intake. In: Margetts BM and Nelson M (ed.) *Design Concepts in Nutritional Epidemiology.* Oxford: Oxford University Press.

Network for the Pan-European Food Data Bank based on Household Budget Surveys. DAFNE II (1998). Brussels: Commission of the European Union.

Roos G and Prättälä R (1999) *Disparities in Food Habits. Review of Research in 15 European Countries. FAIR-97-3096 Project.* Helsinki: National Public Health Institute.

Seidell JC and Flegal KM (1997) Assessing obesity: classification and epidemiology. *British Medical Bulletin* 53: 238–252.

Trichopoulou A and Lagiou P (1998) *Methodology for the Exploitation of HBS Food Data and Results on Food Availability in Six European Countries.* EUR 18357 EN. Brussels: Commission of the European Union, Research DG/B.I-I.

WHO (1976) *Methodology of Nutritional Surveillance. Report of a Joint FAO/UNICEF/WHO Expert Committee.* World Health Organization Technical Report Series 593.

NUTS

R A Stephenson, Maroochy Research Station, Nambour, Queensland, Australia

Introduction

Human consumption of nuts has been recorded from the earliest times and nuts continue as an important component of many diets throughout the world. Despite this, relatively few nut crops have become major commodities on world markets. In fact, most of the lesser known nuts, particularly those from warm climates (**Table 1**), are restricted to traditional diets, usually within relatively narrow regional boundaries. Consequently, information on them, including their composition and nutritional value, tends to be superficial. Only two of the lesser-known nuts of warm climates – macadamia (*Macadamia integrifolia*) and pistachio (*Pistacia vera*) – have developed

Table 1 Some lesser-known edible nuts of warm climates

Name and origin of nut	Features
Almondettes (*Buchanania lanzan* Spreng., family Anacardiaceae) Southern Asia	Related to pistachio nut; medium-sized tree; black, single-seeded fruits; pear-shaped kernels, 1 cm, eaten raw or roasted, delicious (combination of almond and pistachio flavor); 51.8% oil, 12.1% protein, 21.6% starch, 5% sugars
Bunya nut (*Araucaria bidwillii*, family Araucariaceae) Australia	Large pine tree; large cones bear starchy nuts *c.* 5 × 3 cm, usually roasted, flavor resembling that of chestnuts
Candle (Tung) nut (*Aleurites* spp., family Euphorbiaceae) China, South-east Asia, Indonesia	Attractive tree; fleshy fruit, containing a single nut, globose, *c.* 3 cm; hard, thick shell; soft, oily white kernel (*c.* 5 g); must be cooked, moderately poisonous raw, often used for oil
Cut nut (*Barringtonia procera*, *B. edulis*, *B. novae-hiberniae*, family Lecythidaceae, Brazil nut family) Pacific region	Confusion with taxonomy until recently, variable nut characteristics, trees are easy to propagate from seed or cuttings, common in coastal villages, *B. procera* is a more prolific bearer than the other two species – the fruits are a bit smaller (61 cf. 99 g) but have a higher kernel recovery (9%) and hence higher kernel production per tree (up to 50 kg fruit, 5 kg kernel), irregular bearing up to three times a year, high moisture content when fresh so shelf-life is limited, tasty, nutritious kernels, good nutty taste and long shelf-life when processed, relatively low oil content
Elaeocarpus spp., including silver, blue or bush quandong *E. grandis*, blueberry ash or blue oliveberry, *E. cyaneus*, and *E. dentatus*, family Elaeocarpaceae) Australia, New Zealand, Solomon Islands	Over 200 species, often attractive flowering trees producing blue berries from 12 to 25 mm, usually cooked before eating
Finschia nut (*Finschia chloroxantha*, family Proteaceae) Solomon Islands	Relative of macadamia
Galip nut (*Canarium indicum* L., family Burseraceae, also including related Java almond, *C. commune* and pili nut, *C. ovatum* Engl., adoa/bush ngali found only on Bougainville Island, *C. salmonense*) and *C. harveyi* from the Solomon Islands, Vanuatu, and Fiji Philippines, Moluccas (Indonesia), Papua New Guinea, Solomon Islands, Vanuatu	Tall, buttressed tree; roundish, dehiscent fruit bearing a single 5.5 × 2 cm nut, yield up to 300 kg nut in shell, hard, nonperishable shell, triangular in cross-section, *c.* 3 g, delicious, sweet almond-like flavor; 70–80% oil, 13% protein, 7% starch; remove testa before eating, raw or roasted, oil used for cooking and cosmetics, oleoresin burnt for lighting and incense or used to caulk boats, commercial production in Western Melanesia *c.* 100 000 tonnes nut-in-shell (16 000 t kernel, kernel recovery at 16%) from 2 million trees
Galo nut (*Anacolosa frutescens* (Blume) Blume, family Olacacae) South-east Asia, Philippines	Erect shrub or tree, hard wood used for building, fruit is a drupe (up to 18 g); the green-colored pulp and kernel can be eaten fresh but boiling brings out the delicious taste of the fruit; kernel often roasted; dry pulp and kernel contain respectively 9.5% and 10.7% protein, 4.5 and 7.5% fat, 70.8 and 75.5% carbohydrates, 7.2 and 3.7% fiber, and 7.9 and 2.9% ash; kernel recovery is high (73–85%)
Gnetum nut (*Gnetum gnemon*) Papua New Guinea, Solomon Islands	Some cultivation
Heritiera littoralis Solomon Islands	Common on seashore; some records of being eaten
Kauri pine nut (*Agathis* spp., family Araucariaceae) Pacific region	Attractive rainforest tree; overexploited for timber; limited use as a food source in the Pacific region only
Lotus seeds (*Nelumbo nucifera* Gaertn., family Nymphaceae) Asia, Papua New Guinea	Water lily revered by Buddhists; white, single-seeded carpers embedded in flat-topped, fleshy receptacle which dries when mature so that l-cm seeds rattle in their cavity; bitter green embryos removed before eating; eaten raw before fully ripe; nutty flavor, roasted or boiled when mature; rich in vitamin C, 58% carbohydrate (starchy), 17% protein, 2.5% fat
Macadamia nut (*Macadamia integrifolia* Maiden and Betche, *M. tetraphylla*, family Proteaceae) Australia	Evergreen tree adapted to the fringes of subtropical rainforests; spherical kernel enclosed by a thick, stony shell in a fibrous single-sutured husk; high oil content (> 72%); distinctive, delicate flavor
Moreton Bay chestnut (*Castanospernum australe*, family Fabaceae) Australia Pacific region	Fine spreading tree, large glossy fern-like leaves, deep green but lighter underneath, thick pea-pods with 5-cm 'chestnut' seeds; must be soaked and cooked before eating
Neisosperma oppositifolia	Seashore habitats; small kernel

Continued

Table 1 Continued

Name and origin of nut	Features
Okari nut (*Terminalia kaembachii* Warb., family Combretaceae, also *T. catappa* L., the tropical, Indian or sea almond, and *T. impedians*) Papua New Guinea, Vanuatu, South-east Asia, India	Tall tree easy to propagate; leaves clustered at twig tips; yields up to 100 kg large (up to 200 g) flattened, ellipsoid fruit with up to 10% kernel recovery; yields once (March to August) but sea almond bears sporadically throughout the year in low latitudes; stony endocarp encloses white kernel (up to 15 g) consisting of a leaf-like coiled cotyledon, *c.* 8 × 2 cm, 1.5–10 g; delicate almond flavor; contain about 50% sweet, colorless, nondrying edible oil; eaten raw or roasted, dry for long-term storage
Omphalea queenslandiae Solomon Islands	Woody climber; kernel needs processing
Oyster nuts (*Telfairia pedata* (sm. ex Sims) Hook, family Cucubitaceae) Tropical East Africa	Woody stemmed climbing vine, dioecious; large (< 15 kg), deeply ridged, dehiscent gourds containing up to 140 pale yellow seeds (nuts) enclosed in a strong, bitter-tasting, fibrous husk; nuts are flat and circular (3–4 × 15 cm thick); washed, sun-dried and dehusked before eating; raw or roasted; palatable, flavor similar to Brazil nuts; nutritious, *c.* 62% fat, 27% protein, rich edible oil
Pandanus nuts (various species, *Pandanus jiulianettii* Martelli, family Pandanaceae) Papua New Guinea, Pacific region	Wide ecological adaptation from coastal areas up to 1800 m, dioecious 'screw' pine; large, dense multiple fruit (syncarp) up to 30 cm diameter and 16 kg; individual fruits (*c.* 50 or more) separate easily, up to 10 cm long, 1.5 cm in diameter; contain two to four sweet seeds with a coconut taste; oily endosperm; normally roasted, fruitlet flesh of some species can also be eaten; dry seeds for long-term storage; oil content from 10 to 60%; depending on species, hardy stems and leaves also used for handicraft, furniture, and building materials
Pangi, Sis (*Pangium* edule, family Flacourtiaceae) Papua New Guinea, Solomon Islands	Berry an important food source in parts of Papua New Guinea
Paradise (Sapucaia) nut (*Lecythis usitata* Miers., family Lecythidaceae) Brazil, Guyana	Tall Amazon rainforest tree, related to Brazil nut but with superior, sweet delicate flavor (N.B.: some species poisonous); large, dehiscent woody fruits containing 30–40 irregular, oblong nuts resembling Brazil nuts but more rounded with a thinner, softer shell; white, creamy textured kernels; highly nutritious, 62% fat, 20% protein; eaten raw or roasted; edible oil
Pistachio nut (*Pistacia vera* L., family Anacardiaceae) Central Asia, the Middle East and Mediterranean basin	Small, deciduous, dioecious tree; resin ducts through all tree organs; bunches of nuts near shoot tips; harvesting at the correct stage of maturity is critical; fleshy hull surrounds dehiscent, bony shell which encloses the kernel; high in carbohydrates, mainly sucrose (16%), oil, largely unsaturated (55%), and essential amino acids (25%)
PNG oak (*Castanopis acuminatissima*) Papua New Guinea	Of minor importance; sometimes cooked before eating
Quandong (*Santalum acuminatum* Sprague and Summerhayes, family Santalaceae) Australia	Small, semiparasitic trees of desert regions; globular, edible, fleshy fruits; kernel enclosed in pitted, stony shell; oily; harsh aromatic flavor; nutritious, 60% fat, 25% protein; generally roasted
Sago plum (*Cycas revoluta* and other *Cycas* spp., family Cycadaceae) Australia, Papua New Guinea, Pacific region and East Asia	Glossy, fern-leaved plant with sturdy trunk, slow-growing, primitive seed plant. A sago-like meal is made from large seeds; must be soaked for a long time before eating
Souari nut (*Caryocar nuciferum* L., family Caryocaraceae) Brazil, Guianas	Attractive, large tree; fruits round, soft-wooded capsules, *c.* 15 cm diameter, containing two to five large, brown, kidney-shaped nuts up to 5 cm long; edible yellow pulp surrounding nut; kernel enclosed in hard, woody, warty shell up to 1 cm thick, hard to crack; kernel has soft, white, sweet, almond-like flavor; eaten raw or roasted; edible oil
Tahiti (Polynesian) chestnut (*Inocarpus fagiferus* (Parkinson) Fosberg., family Fabiaceae) Pacific Islands	Moderate-sized tree; stout, large (10 × 10 cm) green/brown kidney-shaped, nondehiscent, single-seeded pods borne in terminal clusters; fleshy 'nuts,' boiled or roasted when nearly ripe, taste like chestnuts, palatable, sometimes hard to digest; staple food on some islands; moderately nutritious, 80% carbohydrates (starch), 10% protein, 7% fat

Continued

Table 1 Continued

Name and origin of nut	Features
Tallow nuts (*Ximenia americana* L., family Olacaceae) Widespread throughout Tropics	Densely branched shrub, usually deciduous; egg-shaped, juicy, fleshy fruits contain a large, oily seed; white kernels; palatability varies; nutritious, rich in protein and oil; eaten raw or roasted; cooking oil
Water chestnut, Chinese water chestnut (*Trapa* spp., family Trapaceae) Tropical Africa, Central Europe, eastern Asia	Aquatic plant; not a true nut; hard-shelled, woody fruit with four woody, spiny horns contains a single large white starchy kernel; eaten raw, roasted, or boiled; *c.* 16% starch, 3% protein; not particularly nutritious

as commercial crops and will be discussed in more detail. Some commercial processing and marketing of local *Canarium*, *Terminalia*, and *Barringtonia* species has been initiated in the Solomon Islands, Vanautu, and Papua New Guinea in recent years. The pistachio is a temperate crop but has been included because of its requirement for long, hot, and dry summer and autumn seasons for commercial yields of acceptable quality.

The Macadamia Nut

The macadamia is considered to be one of the world's finest gourmet nuts because of its unique, delicate flavor, its fine crunchy texture and rich, creamy color. These features, together with relatively low volumes of production, have led to the successful promotion of macadamia as a luxury dessert nut. Macadamia nuts account for about 2.2% of the world's total tree nut production – approximately 17 800 t of kernel in 1998–99.

Origin

The macadamia nut is the only commercial food crop indigenous to Australia, originating along the fringes of rainforests in south-east Queensland and north-east New South Wales, from 25 °S to 28 °S latitude and within 24 km of the Pacific coast. Of the three species of macadamia, only two are edible – the smooth-shelled *M. integrifolia* and the rough-shelled *M. tetraphylla*. Only the former has been developed commercially. The rough-shelled macadamia, although producing a raw kernel of excellent eating quality, often contains a higher percentage of sugars which may caramelize on roasting, thus detracting from its appearance. The wild *M. ternifolia* produces a small, unpalatable, bitter kernel.

Botanical features

The evergreen macadamia tree is medium to large, attaining a height of up to 18 m and a spread of up to 15 m. It produces a number of vegetative growth flushes per year, with peaks of flushing in late summer and spring in Australia. The oblong to oblanceolate leaves are arranged in whorls of three and often have spiny, dentate margins, and short (5–15 mm) petioles. Three buds are arranged longitudinally in the axil of each leaf. Multiple branches and inflorescences may therefore be produced from each node. The pendulous racemes, 10–15 cm long and bearing up to 200 creamy white flowers, are borne on older wood. Although floral induction occurs in autumn, cool temperatures induce a period of dormancy. Racemes commence growing slowly in late winter and anthesis occurs in early spring. Fewer than 5% of flowers set fruit and many of these abscise 5–6 weeks after anthesis, coinciding with the stage of endosperm development and the commencement of rapid nut growth. Nuts take 6 months to mature.

Nut set is enhanced by cross-pollination. It is therefore recommended that at least two varieties be grown in orchards, often in alternate rows. Activity of pollinating insects, mainly native and domestic bees, is encouraged.

The fruit is a globose follicle in which only one of two ovules develops. In some varieties, however, a small percentage of the nuts produced are twins, resulting from the development of both ovules. Twins are undesirable because of the difficulty of extracting whole kernels. Mature fruits usually, but not always, abscise when the fibrous husk is still green. As the husk dries, it splits along a single suture to release the nut, consisting of a hard, thick, rough, stony, light-tan shell that encloses the kernel.

The rough-shelled macadamia is readily distinguished from the smooth-shelled species. The leaf margins are more serrated, with up to 40 spines on each side, and whereas new leaf growth of *M. integrifolia* is pale green in colour, young *M. tetraphylla* leaves are an attractive pink to red color. Racemes are longer (up to 30 cm) and bear up to 500 reddish-pink flowers.

Ecology

The macadamia occurs naturally in the fringes of subtropical Australian rainforests. Temperature is the major climatic variable determining growth and productivity, the optimum being 25 °C. Although the mature macadamia is capable of withstanding frosts as low as $-6\,°C$ for short periods, longer periods or lower temperatures may severely damage or kill mature trees. Developing inflorescences are particularly susceptible to frost damage. However, there is a low temperature requirement for flowering, the critical minimum above which flowering is suppressed being $20\,°C$. On the other hand, continuous and prolonged exposure to temperatures greater than $35\,°C$ often produces chlorotic and sometimes distorted growth.

The macadamia tree has several features suggesting adaptation to relatively harsh environments, including sclerophyllous leaves and dense clusters of fine, proteoid roots which develop to enhance nutrient uptake from poor soils, particularly those low in phosphorus. The conditions required for optimum production, however, are quite different from those for survival. Macadamia can be grown in a wide range of soils but not on heavy, impermeable clays and saline or calcareous soils. The trees are most suited to deep, well-drained soils with good organic matter content (3–4% carbon), medium cation exchange capacity with pH of 5.0–6.0.

Production

The commercial development of macadamia occurred in Hawaii where most of the crop is sold locally to tourists as 'suitcase exports' and, to a lesser extent, exported to the mainland USA domestic market where they comprise fewer than 3% of all tree nuts consumed. Australia and Hawaii dominate the world market, with 42% and 33% of the world's total production respectively. Australia produced 7500 t kernel from 12 200 ha in 1998–99, followed by Hawaii (5900 t), South Africa (1700 t), Kenya (950 t), Malawi (700 t), Costa Rica (450 t), Guatemala (450 t), and Brazil (180 t).

Over the last decade, macadamia production has more than doubled and will continue to increase as young plantings throughout the world come into production. The already intense competition from established nut crops has highlighted the need to develop new markets and promote more widespread consumption.

Composition

The quality of the macadamia nut is related to its oil content and composition. Mature nuts contain at least 72% oil (specific gravity, or SG < 1.0) for optimum eating and processing quality. Kernels with SG of 1.0–1.025 are classed as second-grade but can be used for lower-grade products. Third-grade kernels (SG > 1.025) are commercially unacceptable. Oil

accumulation does not commence until the nuts are fully grown and the shell hardens. It accumulates rapidly in the kernel during late summer when the reducing sugar content decreases.

The composition of mature, roasted, and salted macadamia nuts is shown in **Table 2**. As with many oil seeds, the protein is low in methionine. Fresh kernels contain up to 46% sugar, mostly nonreducing sugar. The oil consists of mainly unsaturated fatty acids and is similar in both species, although the proportion of unsaturated to saturated fatty acids appears to be slightly higher in *M. integrifolia* (62:1 compared with 48:1). Detailed fatty acid composition is shown in **Table 3**. The fatty acid composition and the absence of cholesterol may lead to the promotion of macadamias as a high-energy health food. The major volatile components in roasted macadamia kernels are apparently similar to those found in other roasted nuts, although little detailed information on these is available.

Table 2 Nutritive value per 100-g sample of macadamia nuts (roasted in oil and salted) and pistachio nuts (dried and shelled)

	Macadamia	Pistachio
Water (%)	2	4
Food energy (kJ)	3064 (c.732 kcal)	2465 (c.589 kcal)
Protein (g)	7.1	21.4
Fat (g)	78.6	50
Fatty acids (g)		
Saturated	11.4	6.1
Monounsaturated	61.1	33.2
Polyunsaturated	0.14	7.5
Cholesterol (mg)	0	0
Carbohydrate (g)	14.3	25
Calcium (mg)	46.4	135.7
Phosphorus (mg)	203.6	510.7
Iron (mg)	1.8	6.8
Potassium (mg)	332.1	1107.1
Sodium (mg)	264.3	
Sodium (unsalted raw kernel) (mg)	7.1	7.1
Vitamin A		
IU	Trace	250
RE	Trace	25
Thiamin (mg)	0.21	0.82
Riboflavin (mg)	0.11	0.18
Nicotinic acid (mg)	2.14	1.07
Ascorbic acid (mg)	0	Trace
Magnesium[a] (mg)	0.12	
Zinc[a] (mg)	1.4	
Manganese[a] (mg)	0.38	
Copper[a] (mg)	0.33	

[a]From Wenkam NS and Miller CD (1965) Composition of Hawaii Fruits. Hawaiian Agricultural Experiment Station bulletin no. 135. Honolulu. RE, retinal equivalents. International units, (IU) are used to express vitamin A activity based on a bioassay technique which is calibrated with actual concentration in mg g^{-1}.
Data from Gebhardt SE and Matthews RH (1989).

Table 3 Fatty acid composition (methyl esters) of macadamia nut oil

Fatty acid type	Content (%)
Oleate	67.14
Palmitoleate	19.11
Palmitate	6.15
Eicosenate	1.74
Stearate	1.64
Arachidate	1.59
Linoleate	1.34
Myristate	0.75
Laurate	0.62

Data from Cavaletto CG, Dela Cruz A, Ross E, and Yamamoto HY (1966) Factors affecting the stability of macadamia nuts 1. Raw Kernels. *Food Technology*. 20:108–111.

Harvesting and Storage

Macadamia nuts fall from the tree naturally when they are mature. They are either hand-harvested or picked up by machines; efficient machine harvesting requires a smooth, clean orchard floor.

Freshly fallen mature nuts may contain 25% moisture. The husk must therefore be removed as soon as possible to prevent overheating, mold development, and deterioration in quality. The dehusked nuts are initially dried, either artificially or air-dried, to 10% moisture or less before delivery to processors. The nuts are then further dried in silos to 10–15% moisture for longer-term storage, for most efficient cracking of the shell and thus more completed recovery of whole kernels. Drying is performed in stages (at 52 °C down to 45% moisture and then at 77 °C down to 15% moisture) to avoid adverse effects on kernel quality. After cracking and separation of shell from kernel, the product can then be lightly roasted and salted, or packaged raw in bulk in vacuum-filled, foil-laminate bags that help to prevent development of rancidity. Traditionally, kernel has been roasted in coconut oil, although dry-roasting techniques are now more popular. The packaged product is then kept in cold storage to prolong shelf-life. Under these conditions, kernel can be safely stored for at least a year.

The Pistachio Nut

The pistachio has a unique flavor and is popular as a gourmet nut, particularly in the USA. The crop has exacting environmental and cultural requirements for commercial production and these probably limit industry expansion.

Origin

Although the popularity of the pistachio has been recorded from the earliest times, its place of origin is uncertain. The genus *Pistacia* consists of 11 edible species which are widely distributed, including the USA, Africa, and central and eastern Asia. *P. vera* produces the largest nut and is the only species which has a dehiscent shell.

The pistachio appears to have originated on the slopes and plateau country of Iran, Turkestan, and Afghanistan south of the central Asian desert but is found growing wild as far east as India and Pakistan. It is widely cultivated in the Mediterranean basin, and more recently in Asia, the USA and, to a lesser extent, in Australia.

Botanical Features

The pistachio tree is small, dioecious and deciduous. It seldom exceeds 5 m in height and 10 m in diameter. The tree is hardy and can withstand extreme temperatures.

The leathery leaves with raised veins consist of one, three, or five leaflets. Juvenile leaves and those in environments with inadequate chilling have single leaflets, whereas mature leaflets in environments with adequate chilling have five. Resin ducts occur throughout tree tissues, and resin is exuded from any damaged parts.

In spring, inflorescent buds break first and there is often a delay of up to 7 days before vegetative buds break. Flowering may continue through to the latter part of spring, sometimes extending over a 2-week period, depending on variety and season. The transfer of pollen from male to female trees is by wind. Male pollinizer trees must be selected to flower at the same time as female trees in commercial orchards. One male tree per 10 female trees is required. The branched inflorescences are produced in clumps from several nodes just below the shoot tips. Bunches of nuts develop rapidly after pollination, the outer shell often attaining full size within a few weeks. The kernel, however, continues to develop over a 3–4-month period.

The nuts are mature when the outer skin of the husk changes from translucent to opaque. At this stage the husk is readily separated from the shell. The shells of good-quality nuts split to expose the edible kernel which is green or yellow in color with a green or reddish testa. Maturity of different varieties occurs at different times, up to 5 weeks apart, and each must be harvested at optimum maturity. Staining detracts from the appearance and the quality of the nut, resulting in much lower prices. Staining of shells is particularly a problem in humid or showery weather that also predisposes the development of aflatoxin-producing fungi. (*See* **Mycotoxins**: Occurrence and Determination.)

Apart from differences in time of flowering, maturity, and yield, varieties vary in the size and shape of nuts produced, the percentage of split, empty, and stained shells, and the proportion of shell to kernel.

Ecology

Although essentially a subtropical, semiarid Mediterranean species, the pistachio is grouped with warm climate nuts because of its requirement of long, hot, and dry summers and autumns with an average daily mean of 30 °C for 3 months for good commercial yields. It requires a frost-free period of about 200 days to insure that developing inflorescences are not damaged in spring or by very cold winters. It does, however, require chilling during winter for satisfactory flowering. A total of 1000 h per year below 7.5 °C is required.

Production

The world production of in-shell pistachios in 1998–99 was 232 000 t, of which 157 000 t were exported. The main producing countries are Iran, Turkey, and the USA. The pistachio tree commences bearing at 5 or 6 years and increases from only 1 or 2 kg per year to levels of up to 50 kg of fresh nuts at 15–20 years. There is a tendency for pistachio to be a biennial bearing crop. The pistachio is particularly popular in the USA where it accounts for 9% of total nut consumption. Pistachios are usually marketed in shell and less frequently as kernel, often lightly roasted and salted.

Composition

The pistachio has a similar energy value to many nutmeats such as almonds but is lower than that of macadamia (**Table 2**). It is high in carbohydrate, mainly sucrose (16%). The fatty acids are essentially monounsaturated, although its polyunsaturated fatty acid content is higher and its saturated fatty acid content lower than in macadamia nuts.

The pistachio is high in minerals, particularly in potassium. A 100-g portion of pistachio kernels provides 92% of the daily thiamin requirement. Refer to individual nutrients.

Harvesting, Handling, and Storage

Mature nuts hang on the tree well under normal conditions and may be left until most are ripe. Under some situations, however, if the harvest is delayed, the husk dries on to the nut. Consequently the nut is stained and quality is lowered. Ripe nuts are readily dislodged by light shaking. Young trees may be harvested by hand but mechanical harvesting is normally used in commercial orchards.

The nuts must be husked and dried as soon as possible after harvesting to maintain high quality and an unblemished appearance and to avoid spoilage. The husked nuts may be sun-dried but drying artificially in silos between 65 and 72 °C is preferred for large commercial orchards. Freshly harvested nuts can contain as much as 45% moisture. This is reduced to about 5% in 10 h.

Some cultivars produce up to 25% empty shells in which kernels have failed to develop. These blanks must be removed from sound nuts prior to drying by flotation separation.

The dried nuts are sorted to remove those with blemishes. They are graded according to size, roasted, salted, and packaged. Most pistachios are marketed in their shells in snack packs.

Future Prospects for Lesser-Known Nuts of Warm Climates

Both pistachio and macadamia nuts are expensive compared with the major nuts. Future expansion of these industries may result in reduced prices and perhaps on promotion of the health benefits of nuts which is now generating a lot of nutritional research activity. As the volume of these commodities entering world trade increases, greater emphasis will be needed on market development and promotion to insure that they can compete with other nut products. It is unlikely that any of the other lesser-known nuts will develop significantly beyond their present regional boundaries.

See also: **Carbohydrates**: Classification and Properties; **Fatty Acids**: Properties

Further Reading

Anonymous (1999) World nut production and exports. *The Cracker* 3: Edition 28, pp. 33–52.

Cavaletto CG, Dela Cruz A, Ross E and Yamamoto HY (1966) Factors affecting the stability of macadamia nuts I. Raw kernels. *Food Technology* 20: 108–111.

Doster MA and Michailides TJ (1999) Relationship between shell discoloration of pistachio nuts and incidence of fungal decay and insect infestation. *Plant Disease* 83: 259–264.

Gebhardt SE and Matthews RH (1991) *Nutritive Value of Foods. Pacific Northwest Extension bulletin no. 357.* Washington State: Pullman.

Henty EE (1985) Some Nut-Bearing Plants in Papua New Guinea. In: Anon (ed.) *West Australian Nut and Tree Crop Association Yearbook 10*, pp. 19–27. Perth: West Australian Nut and Tree Crops Association.

Howes FN (1948) *Nuts. Their Production and Everyday Uses.* London: Faber and Faber.

Maggs DH (1982) *An Introduction to Pistachio Growing in Australia.* Melbourne: Commonwealth Scientific and Industrial Research Organisation.

Maskan M and Karatas S (1999) Storage stability of whole-split pistachio nuts (*Pistachia vera* L.) at various conditions. *Food Chemistry* 66: 227–233.

Menninger EA (1977) *Edible Nuts of the World*. Florida: Stewart.

Rosengarten F (1984) *The Book of Edible Nuts*. New York: Walker.

Stephenson RA (1990) Macadamia nut. In: Bose TK and Mitra SK (eds) *Fruits: Tropical and Subtropical*, pp. 490–521. Calcutta: Naya Prokash.

Stevens ML, Bourke RM and Evans BR (eds) (1996) *South Pacific Indigenous Nuts*. Proceedings of a workshop held from 31 October to 4 November 1994 at Le Lagon Resort, Port Vila, Vanuatu. *ACIAR Proceedings no. 69*, 176pp.

Vallilo MI, Tavares M, Aued-Pimentel S, Campos NC and Neto JMM (1999) *Lecythis pisonis* Camb. Nuts: oil characterisation, fatty acids and minerals. *Food Chemistry* 66: 197–200.

Verheij EWM and Coronel RE (eds) (1991) *Plant Resources of South-East Asia, no. 2. Edible Fruits and Nuts*. Wageningen: Purdoc.

Wenkam NS and Miller CD (1965) *Composition of Hawaii Fruits. Hawaiian Agricultural Experiment Station bulletin no. 135*. Honolulu: Hawaii Agricultural Experiment Station.

OATS

G A Hareland and F A Manthey, North Dakota State
University, Fargo, ND, USA

Background

Oat is a multipurpose crop that has been grown throughout the world for centuries, generally in cool moist climates. Archeological discoveries have traced oat back to the Greeks, Romans, and Chinese from the first century, but the grain may have originated in areas surrounding the Mediterranean sea in countries of the Middle East. The world oat crop is diverse and includes thousands of commercial cultivars that are grown for multipurpose use. Oat has traditionally provided an inexpensive source of onfarm livestock feed, forage, and bedding. Approximately 70% of the world production of commercially grown oat is used for livestock feed, 20% for human food, and 5% for industrial usage. More than half of the oat crop never leaves the farm where it is produced. Oat is often-times grown as an alternate crop to break cycles of soilborne insects and crop diseases. Unlike most other cereal crops, oat has remained at a relatively low market cost and has been confined to growing on marginal soils associated with poor drainage and low fertility. Although oat production has decreased in recent years, high-quality oat is still in demand for human consumption. Studies have shown that oat is nutritious and contains physiologically active fiber components that aid the process of digestion. Oat bran has been reported to have positive effects on lowering serum cholesterol levels in humans. Compared with other cereal grains, however, the functional properties of oat have not been well defined. Breeders can now evaluate gene combinations by using genetic mapping methods to study the world oat collections from which the functional and nutritional merits of oat protein, oil, starch, and bran components will be better understood. This article presents an overview of oat production, breeding and genetics, anatomy, biochemical and nutritional quality, and utilization.

Oat Production

Oat is considered one of the major cereal grains that are grown throughout the world. Yet over the past 10 years there has been a substantial decline in world-wide oat production, primarily because of low price compared with other cereal grains. Between 1994 and 1999, the six-year average worldwide production was 29.5 million metric tonnes (**Table 1**) compared with an average of 42.8 million metric tonnes between 1986 and 1989. The Russian Federation recorded the highest percent, 40.7%, of the world total area harvested and the UK recorded the highest average yield of $5.81 \, t \, ha^{-1}$ between 1994 and 1999. The 20 countries listed in **Table 1** accounted for approximately 92.7% of the area harvested and 90.9% of the world production between 1994 and 1999.

Oat Breeding and Genetics

Oat (*Avena* spp. L.) belongs to the grass family, Gramineae. *Avena sativa* L. (**Figure 1**), often referred to as common spring or white oat, is a hexaploid containing three distinct diploid genomes, AACCDD, where $2n = 6x = 42$ chromosomes. *A. sativa* is the principal cultivated species grown throughout the world. *A. byzantina* Koch. is a red-oat type adapted to warmer climates where it is grown as a winter oat.

A. fatua L. (**Figure 2**) is a closely related wild oat hexaploid species that has little or no economic value and is considered one of the worst and ubiquitous cereal weeds in the world. A derivative of a wild/hull-less oat genetic cross can make an attractive ornamental decoration (**Figure 3**). Other wild oat species of *Avena* include the tetraploids ($2n = 4x = 28$ chromosomes) and diploids ($2n = 14$ chromosomes). The genetic variation among the approximately 21 000 oat genotypes, wild oat species included, has been explored only to limited extents.

Naked (hull-less) oat, *A. nuda* (**Figure 4**), is an agronomic variant of *A. sativa* and threshes free from its hulls (lemma and palea) during harvest. Naked oat is a species that originated in China

and has been grown for centuries. A single dominant gene controls the naked trait but the degree
of nakedness is affected by modifying genes and environmental factors. Naked oat is high in energy,
protein, and nutrition. The metabolizable energy
content of the groat is comparable to that of corn,
which makes naked oat suitable for animal feed
rations.

Oat-breeding efforts continue to focus on the production aspects of the crop, usually through hybridization, such as selecting for improved yield, winter
hardiness, lodging and disease resistance, plant
maturity, hull–groat relationships, seed size, test
weight, plant height, and protein content. Breeders
are characterizing the germplasm of the thousands of
accessions of oat in attempts to enhance specific traits
such as protein, oil, beta-glucans, and other nutritional qualities. Genetic maps are being developed
to identify molecular markers that are linked to
quality and disease resistance traits. The Genetic
Resources Information Network (GRIN) database
of the US Department of Agriculture disseminates
information on oat accessions to breeders for improving the overall quality of oat. Future breeding objectives will likely focus more on oat as a food, as
research reports tend to characterize the positive
nutritional and physiological properties of oat and
oat products in human diets. Future demands by

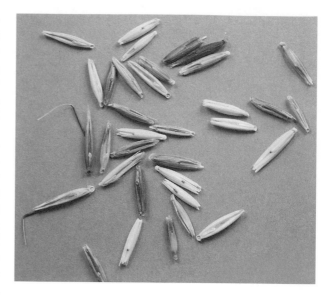

Figure 2 Whole kernels of the wild oat species, *Avena fatua*.
Note the characteristic long awns that are attached to two of the
kernels.

Figure 3 (see color plate 112) Oat panicle of *Avena* spp. From a
genetic cross between wild and hull-less oats. The oat derivative
can be used for decorative purposes.

Figure 1 (see color plate 111) Oat panicle of cultivated *Avena
sativa*.

Table 1 Six-year average area, yield, and production of oat (1994–99)[a]

Country	Area harvested ($\times 10^3$ ha)	Percent of world total area harvested	Average yield (t ha^{-1})	Average production ($\times 10^3$ t)	Percent of world total production
Argentina	245	1.5	1.46	368	1.2
Australia	941	5.8	1.50	1416	4.8
Belarus	324	2.0	2.07	672	2.3
Brazil	194	1.2	1.14	216	0.7
Canada	1482	9.1	2.47	3662	12.4
China	416	2.6	2.01	833	2.8
Finland	363	2.2	3.20	1157	3.9
France	141	0.9	4.36	614	2.1
Germany	308	1.9	4.86	1486	5.0
Italy	146	0.9	2.36	345	1.2
Kazakhstan	389	2.4	0.86	335	1.1
Poland	598	3.7	2.49	1485	5.0
Romania	246	1.5	1.55	380	1.3
Russian Federation	6637	40.7	1.17	7838	26.6
Spain	382	2.3	1.36	522	1.8
Sweden	310	1.9	3.66	1132	3.8
Turkey	154	0.9	1.78	276	0.9
Ukraine	550	3.4	1.78	989	3.4
UK	103	0.6	5.81	594	2.0
USA	1190	7.3	2.08	2476	8.4
% of world total		92.7			90.9
World total	16317		1.81	29465	

[a]Values from FAO Statistical Database, Food and Agriculture Organization of the United Nations (http://apps.fao.org/). Included are countries where average area harvested was greater than 100 $\times$ 10^3 ha.

food processors will encourage breeders to develop cultivars with improved milling qualities. Oat cultivars are likely to be adapted to different levels of protein, starch, lipids, and gums, and to changing seed size and shape for easier processing. Oat contains protein of superior quality, highly digestible carbohydrates, and an abundant source of non-digestible fiber. As the functional properties of the biochemical components become more evident, breeders will be encouraged to evaluate other genotypes for their agronomic, nutritional, and quality traits.

Anatomy of Oat Kernel

The compound inflorescence of an oat plant, referred to as a panicle (**Figures 1, 3** and **4**), is a continuation of the stem, and terminates in a single spikelet. Development of spikelets involves the formation of several florets, of which primary and secondary kernels develop to maturity (**Figure 5**). Occasionally tertiary kernels develop, but are smaller than the primary and secondary kernels. The hull of an oat kernel is normally attached loosely to the caryopsis. The groat is that portion of the oat kernel after removal of the hull (**Figure 6**). The major anatomical features of the oat kernel are the hull and groat (**Figure 7**). The hull contains two tissue layers, a lemma and palea. The

outer portion of the groat contains the bran layers, which consist of the pericarp, seed coat, and aleurone cells. The aleurone layer is developmentally and genetically part of the endosperm but adheres to the outer bran layers upon separation of bran from the endosperm during milling. The aleurone is usually a monolayer of cells that represent the thickest part of the bran fraction. Cells in the aleurone produce and secrete enzymes that degrade storage material in the endosperm. The embryo, also part of the groat, is made up of the scutellum and embryonic axis. The scutellum is located between the endosperm and embryo. Scutellum secretes enzymes during germination and is involved in food transfer from the endosperm to the embryo.

The proportional size of the hull and groat affects processing and impacts the economic value of oat. Oat kernels of uniform size and high groat percentages mill more efficiently and yield higher quantities of bran and endosperm than nonuniform and small kernels. Groat percentages range from 65% to 85% of the oat kernel and weigh from 10 to 40 mg in cultivated species. Groat weight of the primary kernel is higher than the groat weight of either secondary or tertiary kernels. Cultivated oat kernels are generally larger and have less hull content than kernels from wild oat species. Environmental growing conditions affect groat percentages. Thin kernels with high hull

Figure 4 (see color plate 113) Oat panicle of naked (hull-less) *Avena nuda*.

on differences in density. The efficiency of the dehulling process is affected by the quality of oat. Lightweight kernels that are thin or short, and double kernels in which the hull of the primary kernel envelops the secondary kernel, do not dehull as easily as larger and higher test-weight kernels. Whole oat is generally graded for size during the cleaning operation by separating thin and short kernels from larger kernels before dehulling.

The major fractions of the oat groat are the endosperm and bran layers. Their relative proportions are affected by both genotype and environment. Approximate proportions of kernel fractions from cultivated species have been estimated as follows: hull, 24–32%; groat, 65–85%; embryonic axis, 1–2%; bran layers, 27–41% with a total thickness of less than 0.1 mm; and starchy endosperm, 56–68%. The bran contains higher concentrations of vitamins, protein, lipid, minerals, and fiber while the endosperm contains higher concentrations of digestible carbohydrates. The ratio of bran to endosperm has an effect on the total chemical composition of the groat.

Biochemical and Nutritional Quality

Protein

Protein quantity varies within a single cultivar by 3–4% depending on environmental growing conditions and location. The approximate protein concentration in individual fractions of cultivated oat species is: groat, 12–25%; embryonic axis, 25–40%; scutellum, 24–32%, bran, 18–32% and starchy endosperm, 9–17%. Oat hulls contain less than 2% protein in cultivated species, but at higher concentrations in wild oat species. The embryonic axis and scutellum contain higher protein concentrations than other kernel fractions, but together make up only about 3% of the groat size. Thus, most of the protein is actually located in the bran and endosperm. The bran contains about twice the protein concentration as the endosperm, but only about half of the total groat protein because of the difference in relative size of bran and endosperm. Protein content in specific oat fractions varies among genotypes. This is most evident when comparing cultivated species with wild species, which normally have thicker hulls and smaller groats.

Oat groats contain protein of high quality. The amino acids of oat are important for their human nutritional quality; of special interest are the levels of essential amino acids, which the body cannot synthesize. Lysine, the first limiting amino acid in oat, averages about 4.2% of the groat protein, which is higher than other cereals but slightly below the

content and low test weight generally result from dry conditions after heading.

Oat hulls are cellulosic and fibrous materials that are high in crude fiber and low in caloric value. Hulls decrease bulk density during transportation, but function to keep the groat clean for processing purposes and to protect the groat from mechanical damage during thrashing. Hulls are removed during milling before processing groats into flour or rolled oats. Oat dehulling is accomplished by feeding whole kernels into the center of a high-speed rotor. The kernels are dehulled by the centrifugal force and impact against an abrasive rubber ring on the inside housing of the dehulling machine. The hulls are further separated from the groats by air aspiration based

Figure 5 Kernels of cultivated white oat, *Avena sativa* (left), and a derivative of a wild *Avena* sp. (right). The primary (large) and secondary (small) kernels are attached.

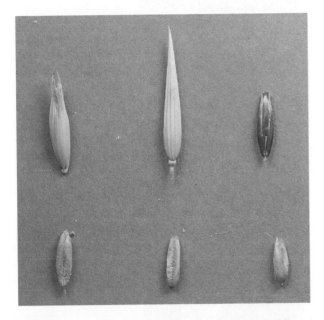

Figure 6 (see color plate 114) Kernels of cultivated white oat, *Avena sativa* (left), hull-less oat, *Avena nuda* (center), and a derivative of a wild *Avena* sp. (right). Whole oat kernels are pictured on the top row and the corresponding groats on the bottom row.

from approximately 2% to 12% in cultivated species. Estimated percentages of free lipids in oat fractions are: hull, 2%; endosperm, 5.2%; aleurone and bran, 6.4%; scutellum, 20.4%; and embryonic axis, 10.6%. The embryonic axis and scutellum together contain the highest concentration of lipids in the oat kernel, but because of its relative size compared with other oat fractions, lipid quantity is low. The endosperm contains low lipid concentration, but contains over 50% of the lipids in the groat. The aleurone layer is very rich in lipid content and constitutes the major source of bran lipids.

Oat lipids are nutritionally important because of the high concentration of polyunsaturated fatty acids, especially linoleic acid. Linoleic acid, an essential fatty acid, is utilized in the synthesis of prostaglandins, which function to regulate smooth muscles like the heart. The approximate percentages of fatty acids in oat lipids are: myristic, 0.4–4.9%; palmitic, 15.6–25.8%; stearic, 0.8–3.9%; oleic, 25.8–47.5%; linoleic, 31.3–46.2%; and linolenic, 0.9–3.7%.

Oat and oat product quality may decrease upon storage because of possible chemical changes in the polyunsaturated fatty acids. Whole oat or undamaged groats that are stored below room temperature and at low moisture levels (less than 10%) show little change in quality. Conversely, damaged oat, groats, or ground oat that are stored for extended periods of time above room temperature and at high moisture have increased levels of enzymatic (lipase) activity. Lipase activity causes the release of free fatty acids from triacylglycerols, which are the major abundant class of oat lipids. Subsequently, unbound polyunsaturated free fatty acids such as linoleic acid undergo oxidation by action of lipoxygenase activity, leading to product rancidity. Lipase activity is highest in the bran because of its association with the aleurone layer, but can be minimized by thermally treating whole oat or groats during processing.

Carbohydrates

Starch is the major carbohydrate in oat and is found primarily in the endosperm. The amylose and amylopectin starch components are present in a ratio of about 1:3, respectively, which is similar to that of wheat and corn. High amylopectin, or waxy oat, and high amylose oat have not been reported in oat genotypes.

Oat contains an abundant source of nondigestible polysaccharides, which are carbohydrates that make up dietary fiber. The nondigestible polysaccharides are classified as either cellulosic or noncellulosic carbohydrates. Cellulose is a linear polymer of beta-1,4-linked glucose units, unlike the alpha-1,4-linked

recommended Food and Agriculture Organization/World Health Organization reference standard of 5.5%. The embryo contains the greatest concentration of lysine. Oat ranks above other cereals in lysine content, including rice, because of the high protein concentration in the groat. Other essential amino acid concentrations associated within the groat are: methionine, 2.5%; valine, 6.4%; isoleucine, 3.9%; leucine, 7.4%; phenylalanine, 5.3%; and tryptophan, 1.7%.

Lipids

Oat contains relatively high lipid content compared with other cereal grains. Thus, oat serves as a unique source of energy as a livestock feed. Lipid concentration is genetically controlled and may vary

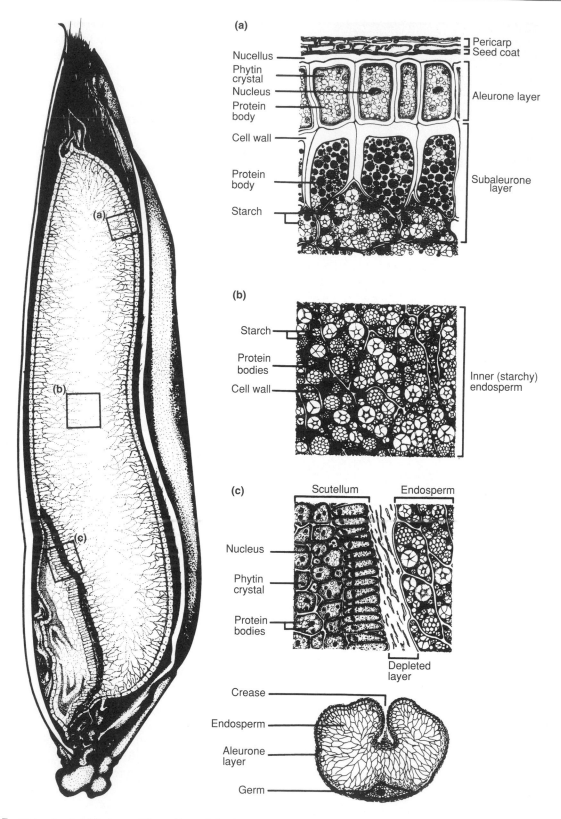

Figure 7 Major structural features of the oat kernel. Oat kernel on the left has been split longitudinally to reveal the approximate size and location of the major tissues. At the lower right is a cross-section view of the groat. (a), (b), and (c) are higher magnifications of portions of the bran, starchy endosperm, and germ. (Reproduced from Webster FH (ed.) (1986) *Oats: Chemistry and Technology*. St Paul, MN: American Association of Cereal Chemists, with permission.

glucose units that make up starch molecules. Noncellulosic polysaccharides include hemicelluloses, pectic substances, and gums and mucilage. Oat gum consists primarily of beta-glucans and pentosans. Beta-glucans are defined chemically as linear molecules of beta-1,3-and beta-1,4-linked D-glucopyranose units that are associated with cell wall structural components in both the bran and endosperm. Beta-glucans may be soluble or insoluble in water and contribute to the high viscosity of cooked rolled oats. Approximately 75% of oat beta-glucans are water-soluble. Pentosans are L-arabinofuranose residues linked as single-unit side chains to a backbone of beta-1,4-D-xylopyranose residues. Pentosans are present mostly in oat hulls and have been used in the commercial production of furfural. Upon acid hydrolysis of oat fiber, the polysaccharides yield sugars and sugar derivatives of D-xylose, L-arabinose, D-galactose, D-glucose, D-glucuronic acid, and 4-O-methyl-D-glucuronic acid.

Fiber-containing foods, upon ingestion, have a profound effect on food utilization from mouth to anus. Fiber directly affects gastric emptying time, rate of nutrient absorption from the intestine, fecal bulk, and frequency of bowel movements. Fiber indirectly affects pancreatic hormone secretions, hepatic glucose, and lipid metabolism. Under certain conditions, water-soluble beta-glucans have been reported to function in the small intestine as physiologically active, cholesterol-lowering ingredients. The hypocholesterolemic effects of beta-glucans tend to reduce the potential risk of ischemic heart disease. Oat hulls have little or no beta-glucan and are essentially water-insoluble fiber, which acts as fecal bulk in the large intestine. Total beta-glucan content in groats can vary from less than 2% to greater than 6%. Higher concentrations of beta-glucans are found in oat bran, which is obtained by separating endosperm and bran fractions by sieving. Beta-glucan content in oat bran is often higher than 7%.

Utilization of Oats

Oat is grown primarily for use as a forage crop and feed grain. Pliny, who reported on the human consumption of oat, stated in his *Natural History* that the Germanic tribes of the first century made 'their porridge of nothing else.' Oat later gained wide acceptance in Ireland and Scotland, where a variety of porridges were made. During the nineteenth century in America, oatmeal was sold almost exclusively in pharmacies and was suggested as a food for the infirm. In the late nineteenth century, Ferdinand Schumacher, a German immigrant, helped pioneer oatmeal into the largest selling cereal in the USA.

His innovations included large-scale domestic milling, movement of oatmeal from the pharmacy to the grocery, and the development of packaging, brand names, and promotional literature.

Oat and oat products are used extensively in a variety of commercial food products. Thermal processing of oat enhances a unique flavor and aroma that contribute to utilization in cereals and baked-food products. Whole groats and steel-cut (sectioned) groats are processed into flakes. The traditional rolled oats made from whole groats are thick flakes requiring long cooking times, whereas flakes made from steel-cut groats are thinner and require shorter cooking times. Steam applied to sectioned groats produces variations of quick and instant flakes, with cooking times of less than 3–5 min. Oat flour is utilized as an ingredient in a variety of food products, and contains antioxidant activity in products likely to undergo fat oxidation during storage. Oat bran, obtained by sieving coarsely ground groats, is utilized as an ingredient in a variety of hot and cold cereals, and in baked products as breads, cookies, and muffins. The high viscosity of oatmeal makes it useful as a thickener in soups, gravies, and sauces, and a meat extender in meat loaves and patties. Processed oat hulls are used in the production of low-calorie, high-fiber baked products. Studies have shown that oat can be included in diets of celiac patients. The prolamin storage proteins, which cause the chronic condition in the intestine of celiac patients, are denatured by thermally processing the groats. In Sweden, Mill Milk Oat Drink was developed as a supplement for people who lack the ability to produce the lactase enzyme for digesting lactose found in dairy milk. Nu-trim, a thermomechanically sheared product from oat fiber, was developed for its high soluble beta-glucan content and used as a fat substitute in certain low-calorie products.

Oat has numerous industrial applications and potential for other uses. Oat starch has adhesive properties as a glue extender. Oatmeal has been used for cosmetic purposes to absorb dirt and sebaceous secretions from the skin. Oat hulls have cariostatic properties to prevent dental caries. Oat hulls have been used in the production of furfural and other derivatives as components in petroleum extractions, resins, plasticizers, insecticides, and pharmaceuticals; they may be used as a brewery filter aid; a component for making linoleum; and a component for making antiskid tire treads.

See also: **Carbohydrates**: Classification and Properties; **Cereals**: Contribution to the Diet; Dietary Importance; **Fats**: Classification; **Protein**: Sources of Food-grade Protein

Further Reading

Coffman FA (1977) *Oat History, Identification and Classification*. Technical bulletin 1516. Washington, DC: US Department of Agriculture, ARS.

Webster FH (ed.) (1986) *Oats: Chemistry and Technology*. St Paul, Minnesota: American Association of Cereal Chemists.

Welch RW (ed.) (1995) *The Oat Crop: Production and Utilization*. London, England: Chapman and Hall.

Youngs VL and Forsberg RA (eds) (1987) Oat. In: *Nutritional Quality of Cereal Grains: Genetic and Agronomic Improvement*. Agronomy monograph no. 28. Madison, Wisconsin: American Society of Agronomy.

Youngs VL, Peterson DM and Brown CM (1982) Oats. In: Pomeranz Y (ed.) *Advances in Cereal Science and Technology*, vol. 5, pp. 49–105. St Paul, Minnesota: American Association of Cereal Chemists.

OBESITY

Contents
Etiology and Diagnosis
Fat Distribution
Treatment
Epidemiology

Etiology and Diagnosis

H Pande, Johns Hopkins Bayview Medical Center, Baltimore, MD, USA
L J Cheskin, Johns Hopkins University, Baltimore, MD, USA

Background

The prevalence of obesity world-wide is increasing rapidly. The WHO has designated obesity as a major health problem throughout the world. In 1995, there were an estimated 200 million adults classified as overweight and obese world-wide. By 2000, the number of overweight and obese adults had increased to over 300 million. The obesity epidemic is not restricted to industrialized societies. In developing countries, it is estimated that over 115 million people suffer from obesity-related problems.

In the USA, the incidence of obesity among adults increased by 50% from 1976 to 1994. The most recent data demonstrate a prevalence of overweight (BMI 25+) or obesity (BMI 30+) of 64% and of obesity (BMI 30+) of 30.5%. Nearly 80% of some groups, e.g. middle-aged black women, are overweight or obese.

In most countries of Western Europe, the prevalence of obesity in adults is 15–25%. In the UK, over 20% of women and 17% of men are obese. Obesity is also increasing in prevalence in Latin America and the Caribbean. In South-east Asia, the Middle East, China, Japan, and the Pacific region, a marked rise in obesity is being seen in all populations and is now being recognized as a major public-health problem.

Obesity is clearly associated with increased morbidity and mortality. It is estimated to cause between 280 000 and 325 000 deaths per year in the USA. In 1995, the direct obesity-related health care costs in the USA exceeded $50 billion – nearly 6% of the entire national expenditure for health care.

Etiology

A simple and straightforward definition of obesity is an excess of body fat. However, obesity is more than just an accumulation of excessive amount of fat and its cosmetic consequences. It is a chronic disease with serious health consequences, is a major cause of morbidity and mortality worldwide, and is now considered the most common disorder of nutrition in developed countries.

The etiology of obesity is complex, multifactorial, and incompletely understood, despite significant advances in recent decades. It is thought to result from a complex interaction between genetic, environmental, metabolic, dietary, and behavioral factors.

The pathogenesis of obesity can be considered the sum of those biological factors, which increase the predisposition towards expansion of adipose tissue mass together with consequences of adaptation to an environment that promotes and possibly rewards increased food intake and decreased physical activity.

When daily energy intake is equal to daily energy expenditure, a state of energy balance exists, and the body weight stays constant. When energy intake exceeds energy expenditure, a state of positive energy balance exists, and body weight increases. Overweight and obesity are usually the result of months and years of positive energy balance, resulting in enlargement of fat cells.

Genetic Factors

It has long been appreciated that obesity runs in families. Familial studies confirm that BMI is highly correlated among first-degree relatives.

Twin studies have shown a similarity in BMI between twins, which is stronger in monozygotic than in dizygotic twins. This correlation was found to persist even when the twins were raised separately.

Adoption studies have shown a significant correlation between the BMI of biological parents and adoptees, but not between the BMI of adoptive parents and adoptees. These studies suggest that the heritability of obesity may be as low as 10% (adoption studies) to as high as 80% (twin studies).

Some of the most compelling evidence for the genetic basis of obesity comes from the discovery of five mutations that cause spontaneous obesity and diabetes in mice. The Agouti (Ay), obese (ob), fat, diabetes (db) and tubby (tub) genes have all been cloned. In 1994, the discovery of the ob gene and its product leptin in ob/ob mice was a significant advancement in our understanding of the genetics of body-weight regulation. Since then, substantial progress has been made in characterizing genes related to human obesity. Genes responsible for previously existing mouse obesity mutations, Ay, db, ob, fat, tub, have been identified and have led to an understanding of the physiological pathways underlying each mutation.

Leptin, for example, is a 16-kDa hormone produced by adipose tissue, which acts by binding to receptors in the hypothalamus, which in turn alters the expression of several neuropeptides that regulate neuroendocrine function, appetite, and energy expenditure. A decrease in body fat leads to a decreased level of this hormone, which in turn stimulates food intake. In addition, decreased leptin levels activate a hormonal response that is characteristic of the starved state, including reductions in the resting metabolic rate. Increased body fat is associated with increased levels of leptin, which act to reduce food intake. By this mechanism, weight is maintained within a relatively narrow range physiologically.

Mutation of the leptin gene, leading to deficiency of leptin, results in the development of hyperphagia, hypogonadotrophic hypogonadism, and severe obesity with childhood onset (functionally equivalent to the ob mouse model). So far, only two kindreds with defects in leptin have been reported. In humans, mutations have also been identified in the leptin receptor in the hypothalamus (equivalent of the db mouse model), in the melanocortin 4 receptor (MC4R) and proopiomelanocortin (POMC), which is a precursor for the natural ligand of MC4R (Agouti model). Mutation in the prohormone and neuropeptide processing enzyme carboxypeptidase E in the fat mouse model and its equivalent in human peptidase PC1 have also been described.

Although monogenic mouse models have played an important role in our understanding of the biology of weight control, these mutations are rare and do not explain the common genetic predisposition to obesity seen in most human populations. They help us in understanding why, within a relatively homogenous environment, some individuals are lean and others are obese. However, only a small fraction of human obesity is due to a deficiency of leptin; instead, human obesity is associated with increased plasma concentrations of leptin. Thus, human obesity is characterized by a central and/or peripheral resistance or decreased responsiveness to leptin. The biological effects of the daily peripheral administration of recombinant human leptin in obese humans are currently being determined in clinical trials but are not expected to be dramatic, considering the nonresponsiveness to the satiety effects of high physiologic levels of leptin in obese individuals.

So far, more than 30 syndromes with a Mendelian pattern of inheritance associated with obesity have been identified. Most of these are pleiotropic syndromes in which obesity is one of the several features.

Prader–Willi syndrome (PWS) is the most common and best-characterized human obesity syndrome with an estimated prevalence of 1:25 000 and autosomal dominant pattern of inheritance. In addition to obesity, it is characterized by hypotonia at birth, small hands and feet, hyperphagia that usually develops between 12 and 18 months of age, mental retardation, short stature, and hypogonadism. Although familial inheritance of PWS is sometimes described, the vast majority of cases are sporadic. It has been established that PWS is most often caused by a deletion of the paternal 15q11.2–12-chromosome segment.

A second, less common disorder, Bardet–Biedl syndrome, which was previously incorrectly named Laurence–Moon–Bardet–Biedl syndrome, is an autosomal recessive disorder. It shares many characteristics with PWS, including an increased risk of obesity. Studies of affected families have identified several different chromosomal loci (chromosomes 16, 11, 3, and 5) responsible for the syndrome, suggesting that it is a heterogeneous condition, with obesity being one of the common features.

The more common forms of obesity are, however, polygenic, and do not display a Mendelian pattern of inheritance. It is a quantitative phenotype that is probably influenced by numerous susceptibility genes, accounting for variations in energy requirements, energy utilization and storage, and the metabolic characteristics of the muscles, as well as being strongly influenced by environmental factors.

The strong environmental pressure to consume calories in excess of the energy expended exceeds the capacity for homeostatic adaptation in genetically predisposed persons, leading to an energy imbalance favoring fat storage. Genetic determinants of interindividual variation in obesity and related phenotypes are likely to be multiple and interacting, with most single variants producing only a modest effect. Genome wide linkage studies in obese families and in different ethnic groups have led to the establishment of the 'human obesity gene map,' which contains entries for more than 120 genes and 16 chromosomal regions in which published studies indicate a possible relationship to excess adiposity or related phenotypes.

Landmark discoveries of leptin, uncoupling proteins, and neuropeptides involved in body-weight regulation have highlighted the importance of molecular genetic factors in determining an individual's susceptibility to obesity. These factors alone, however, cannot explain the rapidly rising prevalence of obesity in the past few decades and the current obesity epidemic, as our genes have not changed substantially over this period. It is believed that our genes simply permit us to become obese; the environment, however, may determine whether or not we do, indeed, become obese.

Metabolic Rate

Metabolic rate is the function of genetic, medical and voluntarily modifiable factors. The total daily energy expenditure (TDEE) is the sum of resting metabolic rate (RMR), the thermic effect of food (TEF), and the energy expenditure in physical activity (EEact). EEact can be divided into in activities of daily living and energy expended in additional physical activity.

TDEE has been found to correlate with body weight. The contribution of TEF and RMR to TDEE is on average 10 and 60%, respectively. The available evidence suggests that there is no association, or at best a very small association, between RMR and TEF and a tendency to gain weight. Activity-related energy expenditure contributes to 30% of the TDEE, is the most variable component, and may be the component that predisposes to obesity and can be most readily modified during efforts in weight control. Recent evidence suggests that incidental physical activity, such as large muscle contractile activity, which is, unfortunately largely involuntary, is a significant predictor of TDEE.

Decreases in TDEE can lead to obesity only if energy intake is not reduced to match the new level of energy expenditure. Humans have an effective homeostatic system with the ability to finely match the energy intake to energy expenditure to avoid positive or negative energy balance and changes in body weight. The effectiveness of this system is illustrated by the fact that adults on average consume 1 million calories every year, and a persistent mismatch of even 1% (about 10 000 calories with energy intake more than expenditure) would lead to an accumulation of fat and an increase in body weight of about 30 pounds in a decade. However, usually the body weight of an adult human increases only slightly over the course of a decade, which means that the homeostatic system operates with a remarkable degree of precision.

Endocrine Causes

Endocrine disorders can upset the precise physiologic regulation of metabolic rate and intake normally operating to maintain a stable body weight. These disorders include hypothyroidism, growth-hormone deficiency, Cushing's syndrome, insulinoma, polycystic ovary disease, and primary empty sella syndrome: all can lead to increased body fat. With treatment, patients who are deficient, for example, in either testosterone or growth hormone show a reduction in visceral adiposity when their hormone levels are normalized.

Studies have shown that patients with nonendocrine precipitants of visceral obesity have increased cortisol and insulin secretion combined with low production of sex steroids, such as testosterone in men, and a low rate of secretion of growth hormone. However, these altered endocrine functions seem secondary to the obese state and at least partially responsible for the development of complications related to the obesity. Weight reduction in general is followed by normalization of endocrine function.

Environmental Factors

Although genetic and metabolic-endocrine factors may explain some of the variability between individuals within a given environment, changes in the environment must be responsible for the increase in the prevalence of overweight and obesity in the population in the past few decades. The rapid increase in obesity that occurred from the 1970s through to the late 1990s suggests that the environment has changed to one that is now strikingly obesity-promoting.

An increase in body weight would occur only when environmental factors overwhelm the ability of the regulatory process to adjust energy intake to energy expenditure. Several changes in the environment have contributed directly to the inability to match energy intake to energy expenditure.

The availability and consumption of energy-dense, high-fat foods is one such factor. Numerous studies in laboratory animals and humans indicate that high-fat diets increase the risk of overeating and development of obesity, and reducing dietary fat results in reduced total caloric intake by reducing the probability of overeating. Moreover, the appetite-suppressant effect of leptin may be overridden by access to high-calorie foods. Body-fat storage also occurs at a greater rate when excess energy comes from fat than when it comes from carbohydrate or protein.

It is becoming clearer that, unlike total energy balance, fat balance is not as closely regulated. A positive energy balance most often results from disruptions in fat balance. It is known that the presence of excess dietary fat does not acutely increase the rate of oxidation of fat for energy. Also, unlike carbohydrates and protein, the capacity for fat storage in humans is virtually unlimited. Excess calories in the form of fat are readily stored as triglyceride in adipose tissue, with a very high efficiency.

However even with a high-fat diet, most, but not all, individuals will become obese. The ultimate amount of weight gained will depend upon the genetic makeup and nongenetic factors like voluntary physical activity.

Another environmental factor is the consumption of more total calories in food as a result of an essentially unlimited supply of convenient, high-energy-density, relatively inexpensive, highly palatable foods, along with an increase in the portion sizes. This increases the likelihood of overeating.

The fast-food and restaurant industry has played a major role in this by using enticing food advertise-ments, serving large portions of high-calorie foods, and encouraging social eating behaviors. Studies have demonstrated that many of the highly palatable food items that have been introduced into the market by the food companies are optimized to elicit a highly positive reinforcing response, ensuring their continued ingestion and incorporation into our menu of familiar foods.

The relationship between dietary fat consumption and obesity, though, is not universally accepted. The evidence for this association includes the close correlation across nations between the percentage of energy in the diet derived from fat and the prevalence of obesity in that country, as well as laboratory studies in animals and humans, which demonstrate that consumption of a high-fat diet *ad libitum* tends to cause excessive weight gain. However, the USDA nationwide food consumption surveys show that the average fat intake, and the average total calorie intake had decreased between 1977 and 1988. Similar surveys in UK from 1970 and 1990 and studies from France have shown that fat consumption has decreased, yet paradoxically, the prevalence of obesity in USA and other countries has increased.

By contrast, self-reported food intake surveys (National Health and Nutrition Examination Survey) suggests that while dietary fat intake in absolute terms has remained relatively constant over the period, since the total energy intake has increased, the proportional intake from fat has declined. It would therefore be misleading to conclude that dietary fat intake is not likely to play a role in the increase in prevalence of obesity. Obesity prevalence might have risen even faster, if not for the apparent dietary-fat restraint and availability of reduced-calorie food products.

Physical Activity

It is well recognized and appreciated that increased energy intake and decreased physical activity, independent of genetics, would inevitably promote the development of obesity in individuals.

Various studies have shown that there is, indeed, an inverse relationship between physical activity and obesity. Cross-sectional, population, and cohort studies have shown that a low level of physical activity predisposes to obesity, while a high level is protective. In addition, normal-weight, postobese women who report being nonexercisers gained more than twice as much weight over 4 years of follow-up as regular exercisers.

Our increasingly inactive lifestyle may play an important and perhaps dominating role in the increasing prevalence of obesity. There are data to support that work-related physical activity and energy expended in activities of daily living has declined in the past few decades. In the USA, however, leisure-time physical activity has not changed much since 1985, with 60% of the adults reporting that they are not regularly active, and 25% reporting that they are not active at all.

The amount of energy required for daily living has also decreased, owing to an increase in sedentary activities such as watching television, VCR, remote control, computer interactions, listening to music, etc.

Data from the national personal transportation survey in the USA have shown that the number of annual walking trips has declined significantly, whereas the number of daily car trips has increased by nearly an identical amount.

Decreases in TDEE as a consequence of the above factors would lead to obesity if energy intake were not reduced to match the new level of energy expenditure, which seems to be happening in developed countries. Even additional bouts of exercise, if followed by a high-fat and high-energy diet, can totally negate the energy expenditure of the exercise.

An individual's predisposition to engage in physical activity may be influenced by their genotype. Variations in muscle metabolic characteristics, like oxidative capacity, may play a role in the pathogenesis of obesity. The increased proportion of fast-twitch (type II B) as opposed to slow (type 1) muscle fibers in obese individuals may be responsible for decreased exercise tolerance and increased perception of fatigue by some of these individuals, predisposing them to be less active and gain weight.

Social and Cultural Factors

Social and cultural factors also make important contributions to the pathogenesis of obesity. The range of BMI of a population varies significantly according to the stage of transition to market economy and associated industrialization of a country, urbanization, changing social structures, and socioeconomic status. In the initial stages of the transition, the wealthier sections of society show an increase in the proportion of people with a high BMI, whereas in the later phases of transition, a high BMI among the poor shows increasing prevalence. This is usually accompanied by an increase in childhood and adolescent obesity and obesity-related disorders.

Cultural factors like cuisine, entertaining, and food selection, fat content of diet, attitude toward health, energy intake, body image, and physical activity also play a role in influencing the range of BMI in a population.

Fetal Undernutrition

Maternal malnutrition causing undernutrition of the fetus during intrauterine development may determine the later onset of obesity, independent of genetic factors. In support of the hypothesis is the finding of an inverse relationship between birthweight and adult obesity, blood pressure, and type 2 diabetes, in both men and women, in later life, with the highest BMI, glucose concentrations, and systolic blood pressures being observed in those with the lowest birthweight. The mechanism of this is unclear.

Diagnosis

The diagnosis of obesity is generally not difficult, but is is important to both quantify its degree and determine the relative risk of complications based on body-fat distribution, and other risk modifiers such as personal or family history of complicating conditions.

Body Mass Index

The body mass index (BMI) is accepted as the standard measure of relative body weight, i.e., weight adjusted for height. The BMI was found to correlate significantly with total fat content in the body. It can be used to assess overweight and obesity and to monitor changes in body weight and determine the efficacy of a treatment program.

BMI is calculated as weight in kilograms divided by height in meters squared.

$$BMI = weight(kg)/height(m)^2.$$

The formula using nonmetric measurements is

$$BMI = 703 \times (weight\ (lb.)/height\ (in)^2).$$

These values are independent of age and sex. Calculating BMI is simple, rapid, and inexpensive, and can be applied generally to adults. Although there are a number of accurate methods by which to assess body fat (e.g., total body water, total body potassium, bioelectrical impedance, and dual-energy X-ray absorptiometry), the measurement of BMI is the most practical approach in a clinical setting. This is especially true because no trial data exist to indicate that one measure of fatness is better than any other for monitoring overweight and obese patients during treatment.

Based on BMI, a standard classification of the degree of overweight and obesity has been established (**Table 1**). A BMI of > 25 confers increased risk for a number of health conditions. There is strong evidence that weight loss in overweight and obese individuals reduces risk factors for diabetes, coronary

Table 1 Classification of obesity according to the BMI

Classification	BMI $(kg\,m^{-2})$
Underweight	< 18.5
Normal	18.5–24.9
Overweight	25–29.9
Obesity	
Class 1	30–34.9
Class 2	35–39.9
Class 3	> 40

Table 2 Relationship between waist circumference and the risk of obesity-related complications

		Risk		
	Waist circumference	Low	Moderate	High
Men	(in)	< 37	37–40	> 40
	(cm)	< 94	94–102	> 102
Women	(in)	< 32	32–35	> 35
	(cm)	< 80	80–88	> 88

The high-risk cut off for men is a waist circumference of > 102 cm (40 in), and the high-risk cut off for women is a waist circumference of > 88 cm (35 in).

artery disease, strokes etc. While the risk of medical complications generally increases with obesity class, an important modifier is the distribution of excess adipose tissue, as estimated most conveniently by the waist circumference.

Waist Circumference

Body fat may be preferentially located in the abdomen (android obesity pattern) or surrounding the hips and thighs (gynoid obesity pattern). The android pattern often reflects an accumulation of fat surrounding the abdominal visceral organs. The presence of excess fat in the abdomen out of proportion to total body fat is an independent predictor of morbidity. Waist circumference is a clinically acceptable measurement for assessing an individual's abdominal fat content, and they positively correlate with each other. The sex-specific cut-off associated with increased relative risk for the development of obesity-associated risk factors, when the BMI is 24.9–34.9, is shown in **Table 2**. When the BMI is more than 35, the waist circumference is usually more than the cut-off and thus does not have any additional predictive value. The waist circumference is a better marker of abdominal fat content than the waist-to-hip ratio and is the most practical anthropometric measurement for estimating a patient's abdominal fat content before and during weight-loss treatment.

Weight-for-Height Tables

Historically, weight-for-height tables like the Metropolitan Life Insurance Company table were used to define the normal weight range. However, such tables have major limitations, such as a reliance on primarily white reference populations, use of unvalidated estimates of frame size, and derivation of the table from insurance company mortality data.

Assessment of Risk Status

Adults with a BMI of 30 or more have a 1.5–2.0 times relative risk of death from all causes compared with individuals with a BMI of 20–25, with most of the increase being due to cardiovascular causes. It is therefore reasonable to assess an individual's absolute risk status by looking for the presence of the following conditions:

Coronary heart disease Obesity is associated with coronary heart disease (CHD), primarily via its impact on the CHD risk factors of hypertension, dyslipidemia, and type 2 diabetes mellitus. Obesity may also directly confer some increased risk of coronary artery disease (CAD) though. Besides BMI, excessive abdominal fat has also been shown to have a positive correlation with morbidity and mortality from CAD. Body weight, independent of risk factors, is also directly related to the development of left ventricular hypertrophy and congestive cardiac failure, perhaps via increased cardiac workload, and changes in vascular resistance and blood return. Obese individuals should also be evaluated for other cardiovascular risk factors, especially smoking and family history of CAD, as these may magnify the obesity-associated risk.

Type 2 diabetes mellitus Excess body fat is associated with increased insulin resistance and risk of developing type 2 diabetes mellitus. This correlation is strengthened by the fact that a significant majority of people with type 2 diabetes mellitus are obese. Moreover, the dramatic increase in obesity during the past decade has been accompanied by a dramatic increase in the prevalence of type 2 diabetes mellitus. In fact, overweight and obesity, especially in those with a more central body fat distribution, are the major risk factors for the development of type 2 diabetes mellitus.

Obstructive sleep apnea Obstructive sleep apnea syndrome (OSA) is characterized by occlusion of the upper airways and cessation of airflow for at least 10 s during sleep. Obesity is believed to change upper-airway anatomy through increased deposition of periluminal fat. This leads to apnea, arterial hypoxemia, recurrent arousals from sleep, excessive daytime somnolence, and snoring. OSA is more common among the obese, with those having a BMI 30 or more being at greatest risk. Patients with OSA can develop complications like neuropsychiatric and behavioral disturbances, systemic and pulmonary hypertension, right heart failure, myocardial infarction and stroke. There are data indicating that the symptoms of sleep apnea improve with weight loss.

Hypertension Obesity may influence the blood pressure and predispose to hypertension. There is strong and consistent evidence that weight loss produced by lifestyle modifications reduces blood-pressure levels.

Dyslipidemia Obesity is associated with increased total and low-density lipoprotein cholesterols, increased serum triglycerides, and reduced levels of protective high-density lipoprotein cholesterol. All these lipid abnormalities are associated with an increased risk of developing CAD.

Other disease conditions associated with obesity Other conditions include ischemic stroke, nonalcoholic steatohepatitis, gallstones and cholecystectomy, osteoarthritis, menstrual irregularities, and infertility. Increased body weight is associated with an increased risk for certain forms of cancer, including colon cancer, endometrial cancer, and postmenopausal (but not premenopausal) breast cancer. Maternal obesity is a significant risk factor for the development of gestational diabetes mellitus, and neural tube defects.

Physical examination and laboratory evaluation

The physical exam should include, in addition to anthropometry, a pulse and blood pressure measurement, examination of the thyroid, auscultation of heart and lungs, palpation of the liver and examination of weight-bearing joints. A laboratory evaluation will help further in quantifying the risk of obesity-related complications and in monitoring weight loss. It should include serum chemistry, lipid profile and thyroid-function tests. In certain high-risk individuals, an electrocardiogram, stress test, chest X-ray, sleep study, pulmonary-function tests, ultrasound of gallbladder, liver, and ovaries, etc., may also be appropriate.

Conclusion

The etiology of obesity has been discussed from several perspectives. The concept of a genetic predisposition to weight gain, the biological mechanisms involved in the regulation of body weight, and its individual variation have been discussed. This was followed by a review of our current appreciation of the regulation of energy balance, presented from the point of view that small daily errors in positive energy balance integrated over months and years

increasingly lead people to obesity. The complexity of the interaction between genetics, behavior, and the environment was also outlined.

Most available evidence indicates that higher levels of body weight and body fat are associated with an increased risk for the development of numerous adverse health consequences. Medical history should evaluate etiological factors that may have played a role in the development of obesity. Physical examination and laboratory tests should extend this evaluation to risk factors associated with obesity.

See also: **Cholesterol**: Factors Determining Blood Cholesterol Levels; **Coronary Heart Disease**: Etiology and Risk Factor; **Diabetes Mellitus**: Etiology; **Energy Metabolism**; **Exercise**: Metabolic Requirements; **Hormones**: Steroid Hormones; **Hypertension**: Physiology; **Metabolic Rate**; **Obesity**: Fat Distribution; Treatment; Epidemiology

Further Reading

Anonymous (1998) *Clinical Guidelines on the Identification, Evaluation, and Treatment of Overweight and Obesity in Adults – The Evidence Report.* National Institutes of Health (published erratum appears in *Obesity Research* 1998 6: 464).

Astrup A (1998) The American paradox: the role of energy-dense fat-reduced food in the increasing prevalence of obesity. *Current Opinion in Clinical Nutrition & Metabolic Care* 1: 573–577.

Bray GA (1999) Etiology and pathogenesis of obesity. *Clinical Cornerstone* 2: 1–15.

Campfield LA and Smith FJ (1999) The pathogenesis of obesity. *Best Practice & Research Clinical Endocrinology & Metabolism* 13: 13–30.

Chagnon YC, Perusse L, Weisnagel SJ, Rankinen T and Bouchard C (2000) The human obesity gene map: the 1999 update. *Obesity Research* 8: 89–117.

Flegal KM, Carroll MD, Kuczmarski RJ and Johnson CL (1988) Overweight and obesity in the United States: prevalence and trends, 1960–1994. *International Journal of Obesity & Related Metabolic Disorders* 22: 39–47.

Hill JO and Peters JC (1998) Environmental contributions to the obesity epidemic. *Science* 280: 1371–1374.

Hill JO, Wyatt HR and Melanson EL (2000) Genetic and environmental contributions to obesity. *Medical Clinics of North America* 84: 333–346.

Kopelman PG (2000) Obesity as a medical problem. *Nature* 404: 635–643.

Lonnqvist F, Nordfors L and Schalling M (1999) Leptin and its potential role in human obesity. *Journal of Internal Medicine* 245: 643–652.

Mokdad AH, Serdula MK, Dietz WH *et al.* (1999) The spread of the obesity epidemic in the United States, 1991–1998. *Journal of the American Medical Association* 282: 1519–1522.

National Task Force on the Prevention and Treatment of Obesity (2000) Overweight, obesity, and health risk. *Archives of Internal Medicine* 160: 898–904.

Sheehan MT and Jensen MD (2000) Metabolic complications of obesity. Pathophysiologic considerations. *Medical Clinics of North America* 84: 363–385.

Tremblay A (1999) Physical activity and obesity. *Best Practice & Research Clinical Endocrinology & Metabolism* 13: 121–129.

WHO (1998) *Obesity: Preventing and Managing the Global Epidemic.* Geneva: World Health Organization.

Fat Distribution

F Deeb, Johns Hopkins Bayview Medical Center, Baltimore, MD, USA

L J Cheskin, Johns Hopkins University, Baltimore, MD, USA

Background

While obesity is easily diagnosed in most cases, and is well known to increase the risk of a variety of disease conditions, less well appreciated is the basic physiology of adipose tissue deposition and the effects of different distributions of adipose tissue on heath risks.

Types of Human Obesity

Obesity is defined as an excess of body weight as a result of an excessive accumulation of fat and results from interaction between the environment and genes. A historically high-fat diet, in concert with decreasing levels of physical activity, has recently resulted in an epidemic of obesity and overweight in much of the world, which now affects at least 64% of Americans. Human adipose tissue can be classified by site into four types: subcutaneous, yellow marrow, interstitial, and visceral. Subcutaneous adipose tissue lies directly under the dermis of the skin and superficial to the fascia overlying the skeletal muscles. More than half of total body fat is generally found in this site. Subcutaneous adipose tissue can be further divided into a deep layer and a superficial layer. One clinical study found that during weight loss, the deep layer decreased to a greater extent than the superficial layer; the significance of this finding is unknown.

Interstitial adipose tissue is found between cells and usually accounts for only a small proportion of total body fat. Yellow marrow is the fat within the bone marrow and is not normally subject to much increase with weight gain.

Variation in the amount visceral adipose tissue (VAT) has the greatest impact on health risks. VAT can be subdivided into retroperitoneal and intraperitoneal adipose tissue. The intraperitoneal adipose tissue can be further divided into mesenteric and omental depots, whose veins drain into the portal vascular system. The liver, too, may accumulate fat, primarily within the hepatocytes. This may occur as a benign result of weight gain (hepatic steatosis or benign fatty liver) or as a manifestation of the often-progressive disease known as nonalcoholic steatohepatitis (NASH), which is often accompanied by obesity and/or type 2 diabetes.

A variety of terms have been used to define differences in adipose tissue accumulation patterns including android/gynoid, apple/pear, central/peripheral, and upper body/lower body. Of note, typical body-fat distribution differs between men and women. Women tend to accumulate more lower-body fat during weight gain, whereas men tend to have upper-body abdominal accumulation of fat, as well as variable lower-body fat accumulation. Purely lower-body weight gain is not generally seen in non-castrated males, though the typical male pattern of predominantly upper-body weight gain is not infrequent in women.

The metabolic consequences of obesity are determined to some extent by the distribution of fat. A number of animal and human studies have found that central obesity poses a greater risk of, among other conditions, cardiovascular disease, type-2 diabetes mellitus, hypertension, gallstones, and gout compared with the risk associated with peripheral obesity. The amount of abdominal visceral fat appears to be the most important corollary of health risk for the individual with obesity. In fact, even in the absence of overall obesity, an excess of VAT appears to be associated with excess health risks.

Regulation of Adipose Tissue Distribution in Humans

In view of the close epidemiologic and metabolic associations between central obesity and disease, the regulation of adipose tissue growth and distribution is important to understand. Age and gender are two important, nonmodifiable determinants of the size of the VAT depot compared to subcutaneous fat. Increasing age and male gender tend to increase VAT magnitude at a given level of total body fat. That is, weight gain in general increases VAT, but to a lesser degree in younger people and in women. However, genetic factors account for a good deal of variability in the amount of fat deposited in VAT or other sites. In addition, some environmental (modifiable) factors appear important, and include smoking and lack of exercise. A number of endocrine abnormalities have been variably detected in association with central obesity, including low testosterone secretion in men, elevated cortisol and androgens in women, and low levels of growth hormone in men and women. Hypersensitivity of the hypothalamo–pituitary–adrenal (HPA) axis may be the mechanism underlying these abnormalities. Insulin resistance in peripheral tissues, leading to hyperinsulinemia, also plays a key role. All of these hormonal changes have important influences on adipose tissue metabolism and distribution.

At the adipocyte level, insulin and cortisol contribute to lipid accumulation by decreasing lipoprotein lipase (LPL) activity. Growth hormone (GH) and testosterone increase LPL activity, so that the decrease in these hormones often seen in central obesity would also tend to decrease LPL activity and favor further adipose tissue deposition. Because of the higher blood flow, cellularity, and innervation in visceral compared with subcutaneous adipose tissues, the consequences of the hormonal abnormalities described above are believed to be expressed more in visceral adipose tissue. Another factor accounting for the disproportionate health effects of VAT is that the number of androgen and cortisol receptors is higher in VAT than other adipose tissue deposits. Thus, endocrine abnormalities in VAT, e.g., elevated levels of insulin and cortisol, which favor accumulation of fat, and decreased levels of GH and sex hormones, which mobilize fat, are magnified by the higher receptor density, leading to further accumulation of fat in VAT. In various physiological conditions associated with increased visceral fat mass, the balance between the lipid accumulation hormones (cortisol and insulin) and the hormones that prevent lipid accumulation and instead activate lipid mobilization (sex hormones and GH) is shifted to favor the lipid-enhancing hormones. Such conditions include polycystic ovary syndrome, Cushing's syndrome, menopause, aging, GH deficiency, HIV lipodystrophy, and excess alcohol intake. Finally, local synthesis of steroid hormones in adipose tissue may play a role in regulating deposition of fat. Regional measurement of fat turnover reveals higher rates of lipid mobilization in upper-body (visceral) adipose tissue than in lower-body adipose tissue. The order in men is as follows: omental visceral = retroperitoneal > subcutaneous abdominal > subcutaneous femoral adipose tissue.

Effects of Fat Distribution on Health

As noted, body fat distribution is now recognized as an important predictor and modifier of many of the adverse health consequences of obesity. Upper body obesity and abdominal visceral fat are associated with a variety of metabolic complications, including hypertension, insulin resistance, type-2 diabetes mellitus, dyslipidemia, gout, and premature coronary death.

Evidence suggests that abnormally high adipose tissue turnover in VAT, mediated by high levels of LPL and/or increased sensitivity to catecholamines or decreased sensitivity to insulin's effect on inhibiting lipolysis, results in elevated free fatty acid availability and may explain some aspects of the metabolic consequences of upper-body obesity. Overall, though, the precise mechanism(s) by which excess VAT lead to adverse health consequences is poorly understood.

Heart Disease

Obesity is an independent risk factor for the development of coronary artery disease (CAD). Obesity also increases the risk of CAD indirectly through its adverse effects on insulin resistance, lipid metabolism, and blood pressure. Central obesity is the pattern most highly correlated with an adverse coronary risk profile. Treatment of obesity results in an improved coronary risk profile. Despite the high prevalence of obesity in coronary populations, the effect of weight loss on hard cardiovascular outcomes such as myocardial infarction and death has received relatively little attention.

Endocrine Abnormalities

The endocrine and metabolic complications most frequently seen in obesity are insulin resistance, type 2 diabetes mellitus, dyslipidemia, menstrual abnormalities, and infertility.

Insulin resistance is quite common in severe obesity, and occurs especially with central obesity. Insulin resistance is a central pathophysiologic feature of multiple conditions, the conglomeration of which is referred to as 'Syndrome X.' Syndrome X includes hyperglycemia, hyperinsulinemia, dyslipidemia, and hypertension, which are associated with an increased risk of CAD and stroke. HPA axis hypersensitivity, as evidenced by an increased response to challenges at all levels of the HPA axis, from the adrenals to the central regulatory centers, appears to be associated with obesity, in particular, central obesity. Sex steroid and GH secretions are blunted, despite heightened HPA reactivity. The hyperandrogenicity frequently seen in obese women may be related to increased testosterone production in the ovaries and adrenals.

Obesity in girls is associated with early menarche and puberty. The effect of childhood obesity on boys is more variable; obesity can lead to either early or delayed puberty. Pubertal gynecomastia is a common problem in obese boys.

Gastrointestinal Conditions

A gastrointestinal disorder for which obesity is a classical risk factor is cholelithiasis. The magnitude of risk is directly related to the severity of obesity but probably not to fat distribution. Since, for each kilogram of fat gained, an additional 20 mg of cholesterol is synthesized and excreted in the bile, this increased cholesterol turnover is the probable mechanism by which obesity increases the likelihood of precipitation of cholesterol gallstones. The risk for gallbladder stones increases during weight loss because of unfavorable changes in bile composition (lithogenicity), increases in the flux of cholesterol through the bile ducts as fat is mobilized, and decreased gallbladder contractility. New gallstone formation especially occurs in subjects who experience rapid weight loss after gastric surgery for obesity or after dieting with a very-low-calorie (< 800 kcal per day) and low-fat diet (as many as 25–35% of patients in some studies develop usually asymptomatic gallstones). The risk for gallstones appears to be lower in patients consuming a low calorie diet (800 kcal per day) containing 15–25 g of fat per day (4% after 10 weeks of dieting).

NASH, a liver disease with frequent sequelae of fibrosis and cirrhosis, occurs most commonly among obese patients and those with type 2 diabetes, and thus is likely to be more common in central obesity. Alanine and aspartate aminotransferase and alkaline phosphatase are the most commonly elevated liver enzyme levels in NASH but are usually no more than twice the upper limit of normal and do not correlate well with histological abnormities. The pathogenesis of liver injury in obese individuals with NASH is in part secondary to the proinflammatory cytokine, tumor necrosis factor α (TNF-α), which has emerged as a key factor in various forms of liver disease. Data from animal and clinical studies suggest that TNF-α mediates not only the early stage of fatty liver but also the transition to more advanced stages of liver damage. A modest weight loss (> 10% of body weight) can normalize liver tests, decrease liver fat content, and decrease liver size.

Another common gastrointestinal condition associated with obesity is gastroesophageal reflux disease (GERD). Abdominal obesity (and pregnancy) are especially associated with GERD, as these conditions result in increased intra-abdominal pressure, which is a central mechanism for reflux of gastric contents into the acid-sensitive esophagus. Abdominal hernias are also associated in particular with central obesity.

Pulmonary Conditions

Pulmonary conditions associated with obesity include obstructive sleep apnea (OSA) and obesity-hypoventilation syndrome (OHS). In OSA, patients suffer from repeated attacks of upper-airway obstruction during sleep. OSA is fat distribution-sensitive because it is believed to be associated with increased fat deposition in the tongue, uvula, pharynx, and hypopharynx. OHS is a condition associated with serious obesity in which an awake patient suffers from hypercarbia and hypoxemia. Abdominal obesity is a mechanical factor in the etiology of OHS, since increased abdominal fat tends to raise the diaphragm, as well as increase fat deposition within the diaphragm and intercostal muscles, and also increase chest-wall weight. Leptin, a protein produced by adipose cells, serves as a pulmonary growth factor and as a modulator of the central respiratory control center. Leptin deficiency or, more likely, leptin resistance is believed to play a role in OHS and OSA.

Assessment of Human Obesity

In order to assess an individual's body fat content and fat topography, a comparison of measures obtained in the individual with age- and gender-specific norms is necessary. This procedure has proved to be challenging in both clinical and research settings. At least three body-fat measures (total body fat content, upper-body fat, abdominal visceral fat) are very important, and must be evaluated not just in a laboratory environment but also in the physician's office and in large-scale population studies. **Table 1** lists the current methods available to obtain a direct or predicted measure of total body fat content.

Anthropometry is the cheapest, most widely used method of assessing human-body composition. Anthropometry measurements can be used both clinically and in epidemiological studies to grade the degree of adiposity in either individuals or groups. The measurements are also used to describe the anatomic distribution of adipose tissue and thus classify individuals and groups with regard to the type of obesity (central or peripheral). The anthropometric measures considered most useful in assessing obesity include weight, stature, skinfold thickness, and circumference of the trunk and limbs.

Weight

Body weight is best measured to the nearest $\pm$ 0.1 kg using a beam-balanced scale. Despite the requirement of frequent and careful calibration, spring scales or electronic scales may be used instead. Patients should

Table 1 Methods of estimating body fat and its distribution

Method	Cost	Ease of use	Accuracy	Measure of regional fat
Height and weight	$	Easy	High	No
Skinfolds	$	Easy	High	No
Circumferences	$	Easy	Low	Yes
Ultrasound	$$	Easy	Moderate	Yes
Density/immersion	$	Moderate	Moderate	Yes
Heavy water		Moderate	High	No
Tritiated	$$	Moderate	High	No
Deuterium oxide or heavy oxygen	$$$	Moderate	High	No
Potassium isotope (40K)	$$$$	Difficult	High	No
Total body electrical conductivity	$$$	Moderate	High	No
Bioelectric impedance	$$	Easy	High	No
Absorptiometry (dual-energy X-ray absorptiometry)	$$$	Easy	High	Yes
Computed tomography	$$$$	Difficult	High	Yes
Magnetic resonance imaging	$$$$	Difficult	High	Yes
Neutron activation	$$$$	Difficult	High	No

Adapted from Bray GA and Gray DS (1988) Obesity. Part I – Pathogenesis. *Western Journal of Medicine* 149(4): 429–441.

be measured either nude or wearing light clothing of known weight.

Stature

The preferred method of measuring stature is by using a wall-mounted stadiometer, measuring to the nearest ± 1 cm. Plastic stadiometers are available for clinical use that have acceptable accuracy. The position of the patient is very important and should be standardized, the patient standing erect with the head, shoulders, and buttocks against the stadiometer wall. Body weight and stature can be used to calculate the body-mass index (BMI) (or Quetelet's index) or weight corrected for height and to compare this value to percentiles for the distribution of this index by sex, age, and race tabulated from large surveys conducted by the National Center for Health Statistics. The BMI is calculated as the weight (in kilograms) divided by the square of height (in meters) or as the weight (in pounds) multiplied by 704 and divided by the square of the height (in inches). It is important to remember that age and body proportion (leg/trunk length ratio) influence the correlation of BMI with height and that BMI is correlated with lean as well as fat mass. This means that a very muscular individual will have a measured BMI that puts them in the obese category, although they do not necessarily have an abnormal percentage of body weight composed of fat.

Skinfold Thickness

This is a measurement of a double thickness or 'fold' of skin, including the underlying fascia and subcutaneous adipose tissue, which is taken using calipers at standard locations on the body. It is performed by pinching and elevating a skinfold at a specific anatomical site using the thumb and forefingers, and measuring the thickness of the fold with a specially designed caliper. The anatomic sites commonly measured are the triceps (midway between the shoulder and the elbow), the subscapular, suprailiac and paraumbilical area, and the medial thigh and calf. These measurements correlate with, but do not accurately measure, the actual mean thickness of the subcutaneous adipose tissue. This has been illustrated by comparisons of caliper skinfold thickness measurements with radiographic and ultrasound measurements of the true subcutaneous adipose tissue thickness at different anatomic sites. Reproducible skinfold measurement requires training. A major criticism of skinfold thickness methods is that they ignore variations in the amounts of internal adipose tissues, especially the visceral adipose tissue that is associated with so many of the increased health risks of obesity.

Circumferences

Body circumferences can be measured more easily in severely obese patients than skinfold thickness. Circumferences have an additional advantage over skinfold thicknesses in that they can measure internal as well as subcutaneous adipose tissue. However, unlike skinfold measures, circumferences are influenced by variation in the size of underlying muscles and bone. The most useful circumferences for grading or predicting body fat and for describing adipose tissue distribution are the upper arm, chest, waist or abdomen, hip or buttocks, and the thigh (proximal or midthigh). In men, the waist circumference is highly correlated with total body fat. In women, the hip or thigh circumference is more predictive of total body fat. Many circumference indices have been devised in an effort to capture the association between fatty-tissue distribution and morbidity and mortality. The most popular circumference index is the 'waist/hip ratio' (WHR), followed by the 'waist/thigh ratio' (WTR). The WHR is an independent predictor of metabolic disturbance including insulin resistance, hyperlipidemia, hypertension, and arteriosclerosis. Similar associations have been reported for WTR, as well as for skinfold thickness indices. Recent evidence suggests, however, that the waist circumference alone is as reliable as the WHR in predicting VAT and its associated health risks.

Ultrasound

Ultrasound has the advantages of measuring subcutaneous adipose tissue thickness more accurately than skinfold measures and of being able to measure otherwise inaccessible sites. The two major disadvantages of ultrasound are that an experienced technician is usually required and that the costs are high.

Hydrodensitometry

An obese person floats higher in water than a non-obese person. This is because the density of fat mass is lower than that of fat-free mass (FFM). Underwater weighing is one of the oldest *in vivo* methods of analyzing human-body composition as a two-compartment (fat and fat-free body mass) model. It held the status of being the 'gold standard' for body-fat composition analysis for many years but is both inconvenient and not as precise a measure of body composition as other methods.

Bioimpedance and Conductivity

Total body electrical conductivity and bioelectrical impedance are techniques for predicting body composition based on the measurement of the electrical resistance in the body to a tiny current. This electrical resistance is proportional to body shape and the volume of conductive tissue (body components with high water concentrations such as the FFM and skeletal-muscle mass). As a result, these methods do not estimate body fatness but instead predict FFM; fat must be derived secondarily as the difference between the predicted FFM and the body weight. These tests have the advantage of being simple, quick, and inexpensive, and can be readily used in population studies. Their disadvantages are the variability in the measurement and its dependence on the hydration level of the patient.

Absorptiometry

Dual-energy X-ray absorptiometry (DEXA) was primarily designed to measure the amount of mineral within bone, but can also be used to estimate the amount of fat in soft tissues. DEXA is an outgrowth of dual-photon absorptiometric systems, having been introduced in the early 1990s as faster and more accurate. DEXA is quick and simple to carry out, yields information about regional adipose tissue distribution, because it is a whole body scan, but is costly.

Imaging

Imaging studies, such as computed tomography and magnetic resonance imaging (MRI), are widely considered the most accurate means currently available for assessing body-fat distribution, but the limitations are the high cost, expertise needed, radiation exposure, and limited utility in a research setting. These are the only methods available for precise, accurate quantification of visceral and other internal depots of adipose tissue. MRI is now the method of choice for calibration of field methods of measuring body fat and skeletal muscles *in vivo*.

Isotope Dilution

Dilutional methods include a group of specialized research techniques for measuring body composition of fluid compartments (e.g., total body water (TBW), extracellular fluid, and plasma volume) and the electrolytes within those compartments, such as exchangeable potassium, sodium, and chlorine. They are costly, require specialized training, and are not widely available. Examples are heavy water and neutron activation measures.

Heavy Water

Water labeled with either two isotopes of hydrogen (deuterium, $2H_2O$; tritium, $3H_2O$) or oxygen ($H_2^{18}O$) has been used to quantitate TBW by a dilution method in healthy and diseased individuals.

Neutron Activation

Whole-body counting/*in vivo* neutron activation analysis can be used to estimate adiposity components. This method enables all main anatomic elements to be quantified at the anatomic level *in vivo*. Once the elements have been measured, the proportion of fat, protein, water, and mineral can be calculated by applying simultaneous equations. These techniques are not widely available but are precise.

In summary, an increasingly well-recognized and understood modifier of health risk in obesity is regional fat distribution. While the ideal method of measuring fat distribution has not been defined, efforts should be made to identify those individuals who are at high risk because of an unfavorably high VAT mass. As a marker of health risk, regional distribution of fat is arguably more important than obesity *per se*, and should be afforded the prominence it warrants in the evaluation and treatment of obesity.

See also: **Body Composition**; **Coronary Heart Disease**: Etiology and Risk Factor; **Diabetes Mellitus**: Etiology; **Hormones**: Pituitary Hormones; **Nutritional Assessment**: Importance of Measuring Nutritional Status; Anthropometry and Clinical Examination; **Obesity**: Etiology and Diagnosis

Further Reading

Bjorntorp P (1996) The regulation of adipose tissue distribution in humans. *International Journal of Obesity and Related Metabolic Disorders* 20(4): 291–302.

Bray GA (1996) Health hazards of obesity. *Endocrinology and Metabolism Clinics of North America* 25(4): 907–919.

Goran MI, Gower BA, Nagy TR and Johnson R (1998) Developmental changes in energy expenditure and physical activity in children: evidence for a decline in physical activity in girls before puberty. *Pediatrics* 101(5): 887–891.

Lo JC, Mulligan K, Tai VW, Algren H and Schambelan M (1998) 'Buffalo hump' in men with HIV-1 infection. *Lancet* 351(9106): 867–870.

Pietrobelli A, Wang Z and Heymsfield SB (1998) Techniques used in measuring human body composition. *Current Opinion in Clinical Nutrition and Metabolic Care* 1(5): 439–448.

Smith SR (1996) The endocrinology of obesity. *Endocrinology and Metabolism Clinics of North America* 25(4): 921–942.

Smith SR, Lovejoy JC, Greenway F *et al.* (2001) Contributions of total body fat, abdominal subcutaneous adipose tissue compartments, and visceral adipose tissue to the metabolic complications of obesity. *Metabolism* 50(4): 425–435.

Tilg H and Diehl AM (2000) Cytokines in alcoholic and nonalcoholic steatohepatitis. *New England Journal of Medicine* 343(20): 1467–1476.

Treatment

L F Donze and L J Cheskin, Johns Hopkins University, Baltimore, MD, USA

Introduction

The purpose of this chapter is to review critically the current state of knowledge about the treatment of obesity. We will discuss various treatment approaches, including the recommended multidisciplinary assessment and treatment; popular, commercial, and self-help approaches; and more aggressive methods such as very-low-calorie-diets (VLCDs), pharmacologic treatment, and surgery. We conclude with recommendations for improving the likelihood of long-term maintenance after weight loss.

Multidisciplinary Assessment and Treatment

Components of a comprehensive or multidisciplinary approach to weight loss typically include medical management, behavior modification, dietary modification, exercise modification, and long-term follow-up. All too often, only one or two elements of a comprehensive approach are considered or implemented in practice. Ideally, all five treatment components will be present in one program, so that overweight individuals will not need to coordinate their own care. True multidisciplinary treatment is most likely to occur in university- or hospital-based programs, with a treatment team consisting of physicians, psychologists, dietitians, and exercise physiologists. A typical treatment process is described below.

Multidisciplinary Assessment

Before treatment can begin, it is important for the treatment team to gain a good understanding of the patient's situation, both in regards to the weight problem and the individual as a whole. Such understanding is typically accomplished via comprehensive evaluations with the patient, including medical, behavioral, nutritional, and exercise/fitness assessments.

Medical assessment   Medical evaluation is strongly recommended for obese patients with a body mass index (BMI) of 30 or more, those who have medical comorbidities, and/or individuals who have not seen a medical care provider in a year or longer. This assessment is critical for understanding the etiology of the disorder in each patient and for establishing a reference point for response to therapy. Medical assessment usually consists of a focused medical history, a careful physical examination, and appropriate laboratory tests.

The medical history should include an assessment of patient and family medical history, including obesity, cardiovascular disease (CVD), hypertension, cancer, diabetes, thyroid and other endocrine diseases, and dyslipidemia. Patient history should include a detailed weight history (lowest and peak adult weight, weight changes and their precipitants, and current weight); history of previous weight-loss attempts; use of tobacco, alcohol, drugs, and medications; and level of motivation. Patients should also be screened for conditions that contraindicate or compromise participation in an exercise program, such as recent myocardial infarction, angina pectoris, disabling osteoarthritis (especially of the knees), severe obesity with restricted mobility, pulmonary disease, or traumatic injury.

The medical care provider should conduct a comprehensive physical examination, with special attention paid to signs of potential comorbidities such as type 2 diabetes mellitus, hypertension, dyslipidemia, and sleep apnea, as well as to possible causes of weight gain such as hypothyroidism, polycystic ovarian syndrome (PCOS), and other endocrine conditions. Blood pressure should be measured with a sufficiently large cuff to give an accurate reading, with the patient in a relaxed state.

Laboratory evaluations can serve as screening tests for certain complications associated with obesity. Blood chemistries should include fasting serum glucose, cholesterol, triglycerides, and liver profile tests. An electrocardiogram (ECG) and complete blood count and urinalysis should be performed to establish a baseline prior to treatment. Thyroid-stimulating hormone (TSH) levels should be checked if there is any suspicion of thyroid dysfunction. Other endocrine and metabolic tests can be performed, if indicated.

Behavioral assessment A clinical psychologist typically conducts the initial behavioral assessment of the patient. This indepth interview covers the patient's psychosocial history, including childhood experience; any trauma, abuse, or unusual events as a child or adult; perceived reasons for weight gain; educational, occupational, and relationship history; and any current stressors that may interfere with the patient's ability to prioritize or adhere to treatment. The patient's diet and weight history are also assessed, including periods of weight gain and losses, the patient's experience on previously attempted diets, patterns of weight maintenance or relapse, and motivation for weight loss. The patient's current eating behavior is assessed by discussing a 'typical day' with regard to eating (such as time of day, foods eaten, and reported hunger or other motivations for eating). Binge-eating behavior and other symptoms of clinical eating disorders are also assessed. Finally, the patient's psychiatric history and current mental status, including history of depression or other disorders, previous and current treatment with psychotherapy or medication, and current depressive (or other) symptoms, are explored.

Dietary assessment A formal dietary assessment is best done by a registered dietitian with training and experience in weight management. An initial nutrition evaluation may consist of a 24-h recall of food intake, a food frequency to determine adequacy and composition of the diet, anthropometric assessment (i.e., waist and hip measurements), and an interview assessing meal patterns, beverage consumption, food preferences, grocery shopping/cooking pattern, restaurant dining, and any religious or cultural customs that may affect the patient's diet. The nutrition assessment is necessary to evaluate the initial nutritional status of the patient and to make appropriate nutritional and treatment recommendations. Although many aspects of diet may be characterized as behaviors, understanding patients' taste preferences and the macronutrient composition of their usual array of food choices is useful in suggesting dietary and behavioral changes that are consistent with the patient's preferences and lifestyle.

The results of the dietary assessment should be interpreted cautiously, because both underreporting and restrained eating (while under observation) are common. Despite these shortcomings, the information gathered can be quite valuable. For example, if an individual reports drinking almost 1000 calories per day in juice and soda, simply switching to non-caloric beverages will likely promote a significant weight loss.

Exercise assessment The exercise/fitness assessment should be performed by a trained exercise physiologist and should explore the patient's usual degree of physical activity, any limiting factors such as joint disease or injuries, preferred types of activity, and measurement of the patient's current level of fitness. A fitness test may consist of a Harvard step test and a test of flexibility, and bioelectrical impedance testing may be used to assess body composition. A formal stress test is not required unless active cardiovascular disease is suspected. The exercise physiologist may also explore lifestyle factors such as usual work hours, social support, and nearby facilities or other resources that could promote increased activity level.

Metabolic testing Metabolic testing of the patient's resting metabolic rate (RMR) via a metabolic cart (indirect calorimetry) is recommended if the necessary equipment and staffing are available. A metabolic test may be performed by an exercise physiologist, dietitian, nurse, or technician. Resting metabolic rate (also called resting energy expenditure,

REE), obtained either by metabolic cart or estimation formulas, is then used to determine daily caloric needs by multiplying RMR by an activity factor ranging from 1.3 (for very sedentary) to 2.0 or more (for very physically active individuals). This estimate of total daily energy expenditure is used by the dietitian to determine the patient's initial caloric requirements and an estimated rate of weight loss, given a specific reduction in energy intake in the diet.

Summary   The entire multidisciplinary assessment can be an emotional and draining experience for the patient, who may not have previously talked about (or even thought about) his or her life in such an indepth way, and to a team of clinicians. We have found, however, that only after a comprehensive evaluation by the multidisciplinary treatment team can an optimized and individualized treatment plan be recommended.

Multidisciplinary Treatment

After multidisciplinary evaluations are complete, treatment can begin. The patient may participate in individual and/or group sessions with members of the treatment team. Weekly treatment sessions are recommended initially (for at least the first month of treatment), to provide the support and education necessary for the patient to stabilize his/her eating behavior and implement the suggested dietary and physical activity changes. Once weight loss is under way, less frequent appointments may suffice.

Medical monitoring   Ongoing medical management, including vital signs, brief review of systems, discussion of any adverse effects of the diet or weight loss, and periodic repeated blood tests, is recommended for patients on low-calorie diets (< 1200 calories per day) and/or those with serious medical comorbidities. Medication dosage or scheduling (for example, of orally administered glucose-lowering agents or insulin) may need to be adjusted as a result of energy restriction and weight loss.

Behavior modification and psychological treatment
Behavior modification in the treatment of obesity is a methodology for systematically modifying a patient's eating, exercise, or other behaviors that may be contributing to the obesity. It is typically the psychologist on the treatment team who suggests or implements behavioral techniques, although these tools may be useful in all components of treatment. The six principles used in behavior modification typically include self-monitoring, stimulus control, contingency management, stress management, cognitive restructuring, and social support.

Self-monitoring is the detailed, daily recording of food intake (or other behaviors) and the circumstances under which the behavior occurs. Food and exercise diaries (**Figure 1**) are used to assess the patient's eating and activity habits, to increase the patient's awareness of his or her own behavior, and to promote positive behavior change. Patients may be encouraged to record time of eating episodes, type and amount of food eaten, feelings of hunger and satiety, place, context, and/or comments (such as, 'Out with friends at happy hour' or 'Not hungry but feeling deprived'). These records are reviewed in treatment sessions, so that specific feedback can be provided to the patient. Research has consistently demonstrated that self-monitoring is associated with improved treatment outcomes, and patients often report that it is one of the most helpful obesity management tools.

Stimulus control involves identifying the environmental cues associated with unhealthy eating and inactivity. Modifying these cues involves strategies such as only eating at the kitchen table and not in front of the television; keeping tempting, high-calorie foods out of the house (or at least, out of sight); and laying out exercise clothes the night before as a reminder to exercise in the morning. Research has suggested that obese individuals are more strongly influenced by external cues to eating than are nonobese people. Therefore, modifying or eliminating these external cues can reduce eating in response to them. For this reason, patients are encouraged to 'turn off all distractions' while eating, in order to reduce environmental and behavioral cues to eating, such as watching television, reading the newspaper, or driving.

Contingency management is the use of rewards for appropriate behavior changes, such as increasing frequency of exercise or reducing intake of fast foods. The rewards can be granted by the patient, the patient's family, and/or the weight management center; however, rewards must be of value and reinforcing to the patient in order to be effective. At some weight-loss programs, patients can earn monetary 'vouchers' towards treatment sessions and products for regular attendance and compliance with treatment recommendations.

Stress management involves the use of problem-solving strategies to reduce or cope with stressful events. Meditation, relaxation procedures, and regular exercise are effective techniques to reduce feelings of stress. However, stress management also involves helping the patient understand the cause(s) of his/her stress in order to make the necessary life or environmental changes to prevent ongoing, chronic stress, which can interfere with weight-loss efforts or exacerbate the weight problem.

SELF-MONITORING RECORD

Name Susan   Date 1/17/01   Weight 205

Water (Please check)   Each ○ equals one 240-milliliter glass.

☑ ☑ ☑ ☑ ☑ ☑ ○ ○ ○ ○

Eating Log (Please specify food & beverages consumed, time, and amount)

Time	Food or Beverage	Amount	Calories	Fat (g)	Context or Comments
8am	banana	1	105		
	yogurt	1	90		
	toast	1 slice	80		
10:30	donuts	2	410		at the office
12pm	large salad	1	20		light lunch
	dressing	1 T	50		because of donut binge
	diet coke	1	0		
7pm	crackers	a lot	2,300		starved — while making dinner
8pm	chicken	3oz	140		
	rice	½ C	100		
	beans (green)	1 C	35		
9pm	popcorn	4c.	160		watching TV bored

Daily Caloric Intake = 1490
Daily Fat Intake = _____

Physical Activity

Time	Activity	Duration
	no time today!	

Figure 1   Example of a self-monitoring record.

Cognitive restructuring techniques are used to identify and modify a patient's dysfunctional attitudes and beliefs about weight regulation and body image. Examples include the use of affirmations (such as positive self-statements) and visual imagery (such as seeing oneself eating and exercising appropriately or visualizing a realistic body size). Many overweight patients adopt 'all or none' or dichotomous thinking patterns, which need to be combated with more flexible thoughts such as, 'Even though I ate two donuts this morning, the day is not ruined, and I will recommit to sticking to my diet plan, starting now!'

The final behavioral principle is social support. Such support, usually from the patient's family or a support group, is used to maintain motivation and provide reinforcement for appropriate behavior changes. Facilitated support groups are often an integral part of comprehensive weight-loss treatment for these reasons, and they further serve to 'normalize' the ups and downs of the treatment process.

These six behavioral principles are immensely useful in helping patients adhere to a healthy diet and exercise program. However, behavior modification alone may not be enough to address all of the psychological complexities of obesity and associated eating problems. Many patients seeking weight-loss treatment are clinically depressed in addition to being obese. For patients who are introspective and amenable to therapy, psychotherapy may be recommended, either as part of their weight-loss program or with a therapist outside the program. Depressed patients who are not interested in psychotherapy, cannot afford it, or are severely depressed may be prescribed an antidepressant by the physician on the treatment team or referred to a psychiatrist. Although depressive symptoms often improve with weight loss, adherence to treatment may be easier if the patient is not actively depressed. Individual or group psychotherapy may also be beneficial to address other issues related to obesity, such as low self-esteem, body image, and relationship problems. Binge eating or other eating disorders often coexist with obesity and may require separate treatment by a specialist.

Nutritional management   Nutritional management of obesity entails an individualized approach, based on the dietary assessment, within a multidisciplinary setting whenever possible. Diet prescription for weight management involves a caloric deficit to promote weight reduction. For patients with mild to moderate obesity, a caloric deficit of at most 500–750 calories per day is recommended to promote a 0.5–0.7 kg weight loss per week. A low-calorie, individualized food-based diet that is either balance-deficit (reducing the total number of calories while keeping proportions from carbohydrate, fat, and protein basically the same as before) or a fat-deficit diet, with most of the caloric reduction resulting from restriction of fat, can be prescribed. The latter approach is preferable for Americans, whose typical diet is too high in fat. Also, a greater volume of food can be eaten on a diet that emphasizes complex carbohydrates and reduces fat intake to 20–25% of calories consumed. In either case, the focus of a calorie- reduced food-based plan should be on nutritional balance, with calories distributed appropriately among carbohydrates, protein, and fat, based on recommendations outlined in the US Department of Agriculture Dietary Guidelines for Americans or similar World Health Organization guidelines.

The diet must be realistic – that is, based on dietary modification and practical changes in eating habits. Nutritional recommendations should be determined by the patient's current eating habits, lifestyle, ethnicity and culture, other coexisting medical conditions, and potential nutrient–drug interactions. The patient should be advised to drink at least 1.5–2.0 l of water daily, unless contraindicated, e.g., by congestive heart failure, edema, or renal insufficiency. Patients should also be encouraged by the dietitian to self-monitor their food intake, which may include measuring portion sizes and recording and calculating calories, fat grams, and/or carbohydrate grams. If energy intake

is prescribed below 1200 calories per day, daily supplementation is usually indicated to ensure adequate vitamin and mineral intake.

Suggesting gradual changes is helpful in altering diet composition. Depending on the initial quality of the patient's diet, the dietitian may focus on revamping one meal at a time, so as not to overwhelm the patient. For example, if a patient's usual breakfast consists of biscuits with gravy, bacon, and sausage from a fast-food restaurant, an alternative of home-prepared oatmeal, yogurt, and fruit may be suggested. Once the patient has incorporated this change, the dietitian may then move on to improving the lunch meal. It is better to recommend dietary changes that are feasible (and achievable) for a patient, rather than prescribing a diet that the patient will reject or not be able to follow. Similarly, reducing the fat content of milk or meats in stages (e.g., 2% to 1% to skim milk) gives the patient a chance to adjust to the new taste before further reducing to a lower-fat version.

Exercise treatment   Although regular, moderate physical activity alone results in limited weight loss over the long term, it is an essential and high-priority component of any weight management program. Regular physical activity is thought to be the most important predictor of long-term weight maintenance. Research has shown that patients who diet and exercise regularly are much more likely to maintain weight loss than those treated with diet alone. When performed in conjunction with caloric restriction, regular, moderate physical activity achieves the following effects: increases 24-h energy expenditure; maintains (or minimizes loss of) lean body mass; reduces cardiovascular risk by producing beneficial changes in the lipid profile; has positive psychological effects; improves insulin sensitivity; and may provide other health benefits independent of weight loss.

In determining an appropriate, individualized exercise program, the exercise physiologist and the patient together should consider a plan that: (1) fitsinto the patient's schedule and lifestyle; (2) considers the patient's likes/dislikes; (3) makes use of the patient's resources; and (4) is based on the patient's current level of fitness. The exercise physiologist should ensure that, before beginning a fitness program, all patients can recognize and dealwith abnormal physical responses to physical activity.

Eventually, the exercise goal of any weight management program should be 30–60 min of continuous, moderate-intensity physical activity five to seven times per week. Until a patient can tolerate

30 min of continuous activity, several 10-min periods of physical activity throughout the day can help build the patient's endurance while still burning calories. Additionally, short exercise and rest intervals can be alternated. The intensity, duration, and frequency of activity should be progressively increased, according to the patient's increasing ability. Patients should be encouraged to self-monitor their daily physical activities, including type of activity, time performed, and duration, so that progress can be appreciated and the goal reset a bit higher from time to time. The exercise physiologist may serve as a personal exercise trainer for those who desire additional motivation and accountability.

Popular Weight-Loss Approaches

Many overweight and obese individuals do not seek or have access to the type of multidisciplinary treatment described above, so instead, they turn to popular or fad diets, commercial weight-loss programs, and/or self-help groups. These three types of approaches will be briefly reviewed in the following section.

Popular Diets

Numerous commercially advertised diets are available to consumers today that promise quick results and a 'magic bullet' approach to weight reduction. We believe that it is best to be skeptical, and instill such skepticism in overweight individuals, about diets that are not part of a comprehensive approach to weight management. Many popular diets are based on very limited menus, because monotony may help curtail consumption. Any severely reduced-calorie diet will initially cause diuresis, which makes the diet seem efficacious at first. This diuresis usually results in approximately 2–4% loss of body weight during the first 7–10 days, although most of this weight will be regained when the period of severe caloric restriction ends.

Many popular diets are based on the dieter counting and limiting calories or some other constituent (such as fat, sugar, carbohydrates), with the end result that energy intake is decreased. Another type of popular diet is a fixed-energy-level diet, which limits intake by controlling portion sizes, menu choice, and composition. Popular diets of this type include prepackaged and portion-controlled foods available in supermarkets or from commercial and nonprofit weight management programs. Other popular diets may use an energy-deficit approach (described previously) or prescribe moderate hypocaloric plans, such as 1800 calories per day for men and 1200 calories for women.

A current trend in the USA is that of high-protein, low-carbohydrate plans, such as the Atkins diet, Sugar Busters, and the Carbohydrate-Lover's diet (to name a few). These diets are not nutritionally balanced and have not been scientifically demonstrated to be effective or healthful in the long term. They promote the metabolic state of ketosis, resulting in a large initial diuresis, as described above, which will be reversed once carbohydrates are reintroduced into the diet.

Commercial and Self-Help Approaches

Self-help and commercial groups have the potential to offer the personal involvement and comprehensive approach needed to promote behavior changes in complex problems such as obesity. This potential, combined with the large numbers of people who can be reached, make these groups the most influential approaches to weight management currently available. On the negative side, these groups often generate expectations greater than can be reasonably achieved, sometimes through misleading advertising. In addition, evaluations and outcome statistics of these programs (if existing) have not been made public.

Commercial programs Many commercial (for-profit) programs provide information on dieting and nutrition and may include elements of physical activity, behavioral techniques, the provision of food, and group support. In the USA, we have seen the recent emergence of local, commercial weight-loss centers, run by physicians, in which the primary treatment is medication that is either prescribed or dispensed at the center. Often, preprinted diet plans are given to patients with limited, if any, nutritional or behavioral counseling.

Self-help groups Self-help groups gather people with weight and/or eating problems and operate at little or no cost, generally without professional intervention. These groups offer considerable social support but vary in their philosophy. A group such as TOPS (Take Off Pounds Sensibly) includes more behavioral principles, whereas Overeaters Anonymous is based on the 12-Step model used in addiction programs and focuses on 'compulsive overeating' as the core problem.

Lack of evaluation It is striking how little is known about the effectiveness of commercial and self-help approaches. A few scattered papers have been published on such groups, but the focus has been on attrition rates rather than on the weight loss and maintenance of the participants. It is possible that the lack of evaluation is due to problems with confidentiality in the self-help groups and with protecting the financial interests of the commercial programs. The closed atmosphere that now surrounds these programs will probably remain so unless regulatory action forces a scrutiny of treatment results.

In light of the lack of evaluation of commercial and self-help programs, and the similar lack of research on the safety of many popular diets, we recommend that all self-initiated weight loss attempts be approved and monitored by the individual's primary physician or an expert in weight management, if the primary physician is not trained or experienced in this area.

Aggressive Approaches to Weight Loss

More aggressive weight management approaches are available under medical supervision and are appropriate for individuals who are severely overweight (generally BMI $\geq$ 35) and/or those with serious medical conditions related to their obesity. Obese patients who have failed multiple weight reduction attempts through traditional programs may also be considered for aggressive strategies. Examples of aggressive approaches include supervised VLCDs, pharmacologic treatment, and surgery.

Very-Low-Calorie-Diets

A medically supervised VLCD (fewer than 800 calories per day) may consist of food, commercially available liquid supplements, or a combination of the two. The first incarnation of these diets was in the form of egg- or milk-based protein combined with water. The use of these severely calorie-restricted diets in medically unsupervised individuals led to at least 58 highly publicized deaths, often resulting from electrolyte imbalances and sudden cardiac death. However, with experience and careful patient selection and monitoring, VLCDs have evolved into comparatively safe weight-loss techniques. It is recommended that patients being considered for VLCDs undergo a comprehensive medical examination and an ECG. Absolute contraindications include the presence of untreated or severe cardiac, hepatic, renal, or thromboembolic disease; type 1 diabetes; cancer; and eating disorders. Weekly medical supervision is recommended while on the VLCD. With full adherence, weight loss averages 1.1–1.8 kg per week, depending on body mass and physical activity level. The initial diuretic phase may be pronounced and accompanied

by symptoms such as lightheadedness, headache, or fatigue. Later symptoms may include constipation and cold intolerance. The diet period is followed by progressive refeeding. Throughout the course of dieting, education on nutrition, exercise, and behavior modification will help the patient maintain his/her weight loss during and beyond the refeeding phase. For this reason, multidisciplinary treatment is strongly recommended in combination with the VLCD. Without this education and support, weight gain will likely occur almost as rapidly as weight loss.

Pharmacologic Treatment

Adjuvant anorectic medications may be useful in select obese patients (BMI $\geq$ 30 or BMI = 27–29 with at least one major comorbidity) who have been unsuccessful in previous coordinated weight-loss efforts. Commonly used medications in the treatment of obesity are listed in **Table 1**. Among the safest and most effective medications currently available are phentermine and sibutramine, both centrally active appetite suppressants.

Phentermine is thought to increase the release and inhibit the reuptake of norepinephrine (noradrenaline) and dopamine from nerve terminals in the brain, thereby suppressing feelings of hunger. The Food and Drug Administration (FDA) has approved phentermine for short-term use (that is, a few weeks to 3 months). Sibutramine, a serotonin and norepinephrine reuptake inhibitor, appears to decrease food intake and may increase energy expenditure as well. Sibutramine has been approved for use for up to 1 year, to aid in weight loss or maintenance of weight loss.

An entirely different mechanism of action is employed by the recently FDA-approved agent, orlistat tetrahydrolipstatin. Orlistat is an inhibitor of pancreatic lipase; thus, it blocks absorption of fat, resulting in about 30% of ingested fat being

Table 1 Medications currently used for treatment of obesity and example trade names

Centrally acting anorexic agents
Phentermine (Fastin, Adipex-P, Ionamin)
Sibutramine (Meridia)
Mazindol (Mazanor, Sanorex)
Diethylpropion HCl (Tenuate)
Phenylpropanolamine (Dexatrim, Acutrim, decongestants)
Ephedrine (decongestants, herbal weight-loss supplements)
Locally acting agents
Orlistat tetrahydrolipstatin (Xenical)

HCl, hydrochloride.

excreted in the stools, with predictable side-effects. Its efficacy in weight loss is probably not only due to the decreased absorption of calories, but also to the Antabuse-like effect when a high-fat eating episode occurs. Orlistat is to be used in obese adults three times per day before meals and has been shown in studies of up to 2 years' duration to improve weight loss and maintenance. Because the malabsorption of fat induced by orlistat also causes malabsorption of fat-soluble vitamins, use of a vitamin supplement is recommended.

Incremental weight loss in controlled trials of anorexic agents tends to be in the range of 0.2 kg per week greater than placebo. This is a definite benefit, but the risks of drug therapy (as illustrated by the fen-phen controversy) must be carefully balanced against the benefits. There is the potential for medications to be prescribed inappropriately, to individuals who do not meet criteria for use or beyond the duration of continued efficacy. Physiologic or behavioral tolerance to these medications, resulting in decreased effects, may occur over time, and some are not recommended, and have not been studied, for longer than 3 months. At best, antiobesity agents may provide certain obese patients with a 'jump-start' to weight loss, or they may be used to improve dietary adherence during a particularly difficult time in the course of treatment.

Surgery

Surgical treatment of obesity may be considered only in carefully selected patients who meet the following criteria: (1) a very high medical risk exists (BMI $\geq$ 40 or BMI $=$ 35–39 with life-threatening or disabling comorbidities); (2) obesity has been present for at least 5 years; (3) no history of or current untreated alcoholism, drug abuse, or major psychiatric disorder is noted; and (4) the patient is between 18 and 65 years of age. In addition, patients should have failed previous attempts at medically supervised weight reduction and should have realistic expectations about the long-term outcome achievable with surgery. A multidisciplinary team approach to screening patients preoperatively, including evaluation with the surgeon, a dietitian, a psychologist, and the patient's primary physician, is critical to the patient's long-term success.

Two proven surgical options are available for the treatment of morbid obesity: purely restrictive operations such as the vertical banded gastroplasty (gastric stapling), and gastric bypass operations such as the Roux-en-Y gastric bypass. These procedures primarily involve either the mechanical restriction of caloric intake by creating a small gastric reservoir, or the induction of malabsorption by bypassing variable lengths of the small intestine. The vertical banded gastroplasty (**Figure 2**) achieves weight loss by creating a small pouch with a narrowed outlet in the stomach by using a band of synthetic material. With the gastric bypass procedure (**Figure 3**), a small pouch is created of 20–30 ml, which limits oral intake, and a moderate degree of malabsorption is created by a Roux-en-Y loop of bowel anastomosed to this small gastric remnant.

The Roux-en-Y gastric bypass produces more substantial weight loss than vertical banded gastroplasty and may have longer-lasting effects. The gastric bypass results in weight loss averaging 50–60% of excess body weight, with good weight control documented for up to 10 years. This procedure is not without risks, including post surgical complications, food intolerance, nutritional deficiencies, extreme psychiatric distress, and even death. However, in the majority of patients, obesity-related comorbidities, such as diabetes and hypertension, are either reversed or prevented after obesity surgery, as long as weight loss is maintained.

Maintenance of Weight Loss

Maintaining weight loss seems to be inherently more difficult than losing weight, particularly for

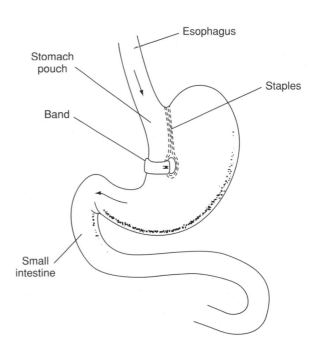

Figure 2 Vertical banded gastroplasty. A vertical staple line reduces the size and accommodation of the stomach. Mesh banding of the distal portion of the newly created gastric pouch slows the exit of food, leading to early satiety. Reproduced with permission from Schaefer DC and Cheskin LC (1998) Update on obesity treatment. *Gastroenterology* 6(2): 136–145.

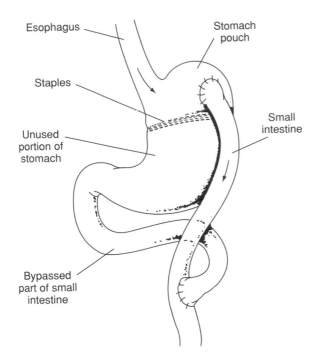

Figure 3 Gastric bypass. A horizontal staple line reduces the size and accommodation of the stomach. A jejunal limb is attached to the new gastric pouch, thus allowing bypass of the duodenum. Reproduced with permission from Schaefer DC and Cheskin LC (1998) Update on obesity treatment. *Gastroenterology* 6(2): 136–145.

individuals who used or were treated with caloric restriction. Unfortunately, the long-term success rate for all weight-loss methods (excluding surgery), defined as losing weight and keeping the majority of it off for at least 5 years, is low – perhaps 5–15%. Even after the most state-of-the-art, medically monitored comprehensive treatment, it is likely that patients will gradually return to their baseline weight within 3–5 years after treatment termination. Given this very high relapse rate, obesity can be considered a chronic condition, much like cigarette smoking or other substance abuse. As such, only the continuous-care model of management, which views obesity as a chronic disease requiring continuous support or contact after the end of formal treatment, has been found to produce significant maintenance of weight loss.

It is probable that some form of continuous or punctuated long-term care will be necessary for many (if not all) obese patients to maintain the dietary and lifestyle changes that produced their weight loss. With ongoing support from a physician or other health care providers, patients will be better able to maintain their changes in eating and activity behavior and the resultant weight loss. Regrettably, optimal strategies for weight maintenance are in a state of

evolution and represent the key limitation of our current treatment of this highly prevalent, medically hazardous, and chronic condition. Research has identified continued program contact (e.g. monthly), regular physical activity, nutritional sophistication, and self-monitoring as four predictors of weight maintenance. It not surprising that the strategies helpful in promoting weight loss, such as self-monitoring, physical activity, regular weigh-ins, and attendance at a multidisciplinary program, will be the same strategies necessary for long-term maintenance of weight loss.

See also: **Behavioral (Behavioural) Effects of Diet**; **Body Composition**; **Bulimia Nervosa**; **Exercise**: Muscle; Metabolic Requirements; **Fat Substitutes**: Use of Fat-replaced Foods in Reducing Fat and Energy Intake; **Hunger**; **Metabolic Rate**; **Nutrition Education**; **Nutritional Assessment**: Importance of Measuring Nutritional Status; **Obesity**: Etiology and Diagnosis; Fat Distribution; Epidemiology; **Satiety and Appetite**: Food, Nutrition, and Appetite; **Slimming**: Slimming Diets; Metabolic Conseqences of Slimming Diets and Weight Maintenance

Further Reading

AACE/ACE Obesity Task Force (1998) AACE/ACE position statement on the prevention, diagnosis, and treatment of obesity (1998 revision). *Endocrine Practice* 4(5): 297–330.

Bray GA (1998) *Contemporary Diagnosis and Management of Obesity*. Newtown, PA: Handbooks in Health Care.

Brownell KD and Fairburn CG (1995) *Eating Disorders and Obesity: A Comprehensive Handbook*. New York: Guilford Press.

Cheskin LJ (1997) *Losing Weight for Good: Developing Your Personal Plan of Action*. Baltimore, MD: Johns hopkins University Press.

Foreyt JP and Poston WSC (1998) What is the role of cognitive-behavior therapy in patient management? *Obesity Research* 6 (suppl. 1): 18S–22S.

National Institutes of Health Consensus Development Panel (1991) Gastrointestinal surgery for severe obesity. *Annals of Internal Medicine* 115: 956–961.

Perri MG, Nezu AM and Viegener BJ (1992) *Improving the Long-term Management of Obesity: Theory, Research, and Clinical Guidelines*. New York: John Wiley.

Schaefer DC and Cheskin LJ (1998) Update on obesity treatment. *Gastroenterologist* 6(2): 136–145.

Thomas PR (ed.) (1995) *Weighing the Options: Criteria for Evaluating Weight-Management Programs*. Washington, DC: National Academy Press.

Wadden TA and Van Itallie TB (1992) *Treatment of the Seriously Obese Patient*. New York: Guilford Press.

Epidemiology

N E Sherwood and M Story, University of Minnesota, Minneapolis, MN, USA

Introduction

Obesity is a major public health problem in the USA. The dramatic increase in the prevalence of obesity over the last few decades has raised concerns about associated health risks for children, adolescents, and adults. Obesity is considered the second leading cause of preventable death in the USA. Recent data suggest that the estimated number of annual deaths in the USA due to obesity approaches 300 000. A continuation of current trends could lead to further increases in the number of people affected by obesity-related health conditions and premature mortality. Prominent among the health risks associated with obesity are hypertension, type 2 diabetes mellitus, dyslipidemia, stroke, gallbladder disease, osteoarthritis, sleep apnea, respiratory problems, and certain cancers (e.g., endometrial, breast, prostate, and colon). Obesity is also associated with psychosocial problems such as binge-eating disorder and depression. Furthermore, individuals who are obese are adversely impacted by social bias and discrimination.

As the prevalence of obesity has been rising, the economic burden of obesity has been a topic of considerable discussion. The economic costs of obesity have been categorized into three areas, the cost to individuals because of the impact of obesity on health, costs to society due to lost productivity and premature mortality associated with obesity, and costs to individuals and service providers in treating obesity. Recent estimates suggest that obesity-related mortality may account for 7% of US health care costs. Medical care costs alone attributable to obesity in the US were estimated to toal almost $46 billion in 1990. Although no study to date has taken into account the entire range of costs associated with obesity, estimates of the economic burden are as high as $100 billion per year.

Monitoring the prevalence of obesity and associated health behaviors is critical for understanding and documenting the scope of the problem. The goal of this chapter is to review definitions of overweight and obesity in adults and children, present available data regarding the epidemiology of obesity and related risk factors, and discuss implications for future public health research and intervention.

Definitions of Overweight and Obesity

The terms obesity and overweight are often used interchangeably, but they are distinct conditions. Obesity refers to excess adipose tissue, whereas overweight refers to excess weight for height. Because it is difficult to obtain direct measures of obesity and since body weight tends to be highly correlated with adiposity, weight-for-height measures are generally used to classify overweight and obesity in adults and youths.

Adults

Definitions of obesity and overweight have varied over time and in different countries. In the USA, many studies have used the 1959 or 1983 Metropolitan Life Insurance tables of desirable weights for heights and defined overweight as a weight exceeding the midpoint of these standards by a certain percentage. In 1985, a NIH Consensus Conference endorsed the use of body mass index (BMI; $kg\,m^{-2}$) and adopted the 85th percentile of BMI from the second National Health and Nutrition Examination Survey (NHANES II) as the standard for defining overweight. The 85th percentile corresponded well to approximately 20% overweight according the 1983 Metropolitan Life tables. Using this standard, overweight was defined as BMI ≥ 27.8 for men and ≥ 27.3 for women. In 1995, a World Health Organization (WHO) expert committee recommended an alternative classification system for overweight and obesity; overweight was defined as a BMI of 25–$29.9\,kg\,m^{-2}$ and obesity as a BMI $\geq 30\,kg\,m^{-2}$. Although the previous definition of overweight in the US used a cut-off of about $27\,kg\,m^{-2}$, epidemiological data show increases in mortality with BMIs above $25\,kg\,m^{-2}$. The category of obesity is further subdivided into three classes: obese class I (BMI 30–34.9), obese class II (BMI 35.0–39.9), and obese class III (BMI ≥ 40).

BMI has been accepted by numerous groups as the most practical and valid screening tool for assessing weight status in individuals and the most useful population-level measure of overweight and obesity. BMI provides meaningful prevalence estimates within and between populations and is useful for identifying individuals and groups at increased health risk and for evaluating the efficacy of obesity interventions. However, it is important to recognize that BMI does not provide information on body fat or body-fat distribution. Body-fat distribution has been identified as an independent predictor of health risk. Individuals with excess abdominal fat are at increased risk of negative health consequences of obesity. In adults, the waist-to-hip ratio, or waist circumference, is often used as a marker of intraabdominal adipose tissue.

Children and Adolescents

Defining overweight or obesity in children and adolescents is complicated by the normal processes of growth, pubertal development, and body-composition changes. No standard definition exists for child or adolescent obesity. Classifications that have been used include weight-for-height percentiles, relative weight, percent of ideal body weight, skinfold measures, and BMI. BMI is currently the most accepted and widely used measure. For adults, BMI criteria have been based on mortality or mortality outcome research, but no risk-based criteria have been established for youths, as it is difficult to link youth weight status to chronic disease outcomes. Adult BMI criteria also utilize a single cut-off value for both sexes and all ages, which is inappropriate for children and adolescents who are experiencing growth and body-composition changes. Because BMI changes dramatically with age during childhood and adolescence, BMI needs to be assessed using age-specific reference curves. Gender-specific values are also needed for adolescents because of differences in body composition and timing of puberty. During puberty, males increase their lean body mass and decrease the amount of body fat, whereas girls experience rapid fat accretion.

The release of the revised National Center for Health Statistics (NCHS) and the Centers for Disease Control (CDC) growth charts incorporating smoothed gender- and age-specific BMI percentiles based on data from National Health Examination Survey (NHES) and NHANES will provide a useful reference for describing weight status in children and adolescents. It has been suggested by US expert panels that for clinical evaluation of obesity and epidemiologic application, overweight in youths should be defined as a BMI greater than the 95th percentile for age using national reference population data. Youths between the 85th and 95th percentile are considered at risk of becoming overweight.

Prevalence of Obesity: National Trends

Adults

Comprehensive data of trends in the prevalence of obesity provided by national surveys (NHES I (1960–1962); NHANES I (1971–1974); NHANES II (1976–1980); NHANES III (1988–1994)) show that the percentage of obese persons has increased over time. Although there was a small increase in the estimated prevalence of obesity across the first three periods (1960–1980), a much larger increase was observed between the third and the fourth surveys. The percentage of individuals classified as obese increased from 15% in NHANESII to 23% in NHANES III. An additional 33% of US adults in NHANES III were classified as overweight according to the BMI > 25.0 standard. According to these data, an estimated 97.1 million adults (55% of US adults) are classified as overweight.

Data from the Behavioral Risk Factor Surveillance System (BRFSS) also show that the prevalence of obesity has increased in recent years. The BRFSS is a cross-sectional telephone survey of noninstitutionalized adults aged 18 and older conducted by the Centers for Disease Control and Prevention and state health departments. Data from the BRFSS indicate that the prevalence of obesity increased from 12% in 1991 to 18% in 1998. The prevalence of obesity increased across both genders, all socio-demographic groups and in all states. These data are derived from self-reports of individuals; therefore, true rates of overweight and obesity are likely underestimated since individuals tend to underreport their body weight.

Children and Adolescents

Regardless of the method used to classify overweight or obesity, studies have shown high prevalence estimates among children and adolescents, and that rates have increased dramatically since the mid-1960s and continue to rise. NHANES data indicate that the prevalence of overweight (BMI at or above the 95th percentile) among youths doubled from 1976–1980 to 1988–1994, increasing from 8 to 14% for 6–11 year olds and from 6 to 12% for 12–17 year olds. Currently, about 11% of US children and adolescents are overweight, and an additional 14% have a BMI between the 85th and 95th percentiles, indicating that they are at risk for becoming overweight.

Gender Differences in Obesity

Data suggest that the prevalence of obesity is increasing across all populations in the USA, but epidemiological data show that certain subgroups of individuals are at a particularly high risk of obesity. Although the prevalence of obesity has increased over time in both men and women, the prevalence of overweight has remained relatively stable. Data show that men have higher rates of overweight compared with women, but women have higher rates of obesity than men. A major contributor to the high rates of obesity in US adults is weight gain with age. About two-thirds of obese adults became obese in adulthood. Between the ages of 20 and 50 years, weight gain tends to occur at a rate of 0.45–0.91 kg per year, and major weight gain is most likely to occur between the ages of 25 and 34 among both men and women. Prospective epidemiological data also show that women in the US appear to be at twice the risk of major weight gain as men. The prevalence of overweight and obesity increases with age across both sexes, peaking around the 50–59-year-old age range. After age 70, the prevalence of obesity is lower in adult men and women. This lower prevalence may be due to both obesity-related mortality and decreases in BMI with age.

Among children and adolescents, national data show few gender differences in obesity rates. Males tend to have slightly higher rates of overweight compared with females (e.g., 15% for males compared with 14% for females).

Racial/Ethnic Differences in Obesity

Significant racial/ethnic differences in the prevalence of overweight and obesity have also been observed. The most recent NHANES III data show that among women, African American and Hispanic women are at highest risk, with about two-thirds of women in these ethnic groups meeting BMI criteria for over-weight or obesity. Ethnic specific data from studies such as Hispanic HANES (HHANES), which includes Mexican Americans, Cuban Americans, and Puerto Rican Americans, and smaller studies of American Indians and Alaskan Natives, show high rates of overweight and obesity in these populations. In contrast, ethnic-specific data show that Asian American women have the lowest rates of overweight and obesity. Among men, fewer ethnic differences in overweight and obesity have been observed. According to NHANES III data, Hispanic men are at highest risk of overweight and obesity compared with non-Hispanic white men and African American men. According to data from the BRFSS, the prevalence of obesity in Hispanic men almost doubled over the last decade (10% in 1991; 18% in 1997). Data from smaller studies of American Indian and Alaskan Native men also show elevated rates of overweight and obesity compared with all-race data in US men.

As with adults, differences in overweight prevalence by race and ethnicity have been observed among children and adolescents. In NHANES III, the overweight prevalence among youths was considerably higher for male and female Hispanics, and African American girls compared with whites. For example, among 6–11 year olds, 16% of black girls, 14% of Hispanic girls, and 9% of white girls were overweight. Among males aged 6–11, 17% of Hispanic boys were overweight, compared with 10% of white and 12% of black boys. Other studies have shown excessively high prevalence estimates of obesity in American Indian youth. A recent survey of over 12 000 American Indian youths aged 5–17 years of age in the Northern Plains area found that 22% of the males and 18% of the females were overweight (BMI > 95th percentile).

Socio-economic Differences in Obesity

An inverse relationship between socio-economic status and obesity has been observed in the USA and is particularly pronounced for women. Data from NHANES III show that collapsing across all races, women with lower incomes are more likely to be overweight or obese compared with those with higher incomes. Examination of NHANES III all-race data for men shows that the prevalence of overweight does not vary across income groups. The relationship between socio-economic status and obesity, however, appears to vary across ethnic groups and according to gender. There is a strong inverse relationship between income and weight status among nonHispanic White women and a moderate inverse relationship between income and weight status among Hispanic American women. However, among African American women, weight status and income are only weakly associated. A similar trend has been observed in other studies. Although nonHispanic white men did not show any difference in weight status across income groups, examination of similar data in men shows that there is a positive relationship between income and obesity for Hispanic American and African American men.

Compared with adults, studies with children have shown a weak and less consistent relationship between SES and overweight. In NHANES III data, there was an inverse relationship between SES and overweight only for white adolescents. Overweight prevalence was not related to family income in African American or Hispanic youths.

Persistence of Childhood Obesity

Whether childhood-onset obesity leads to an increased likelihood of obesity in later life is an important issue with clinical implications. The likelihood of persistence of obesity from childhood to adulthood is related to the degree and duration of obesity, family adiposity, and age of the child. The likelihood of an overweight infant becoming an overweight adolescent or adult is small. Less than 15% of overweight infants and only about 25% of overweight preschool children will remain overweight into adulthood. Obesity is more likely to persist if it is present during the adolescent years. In general, the later into adolescence overweight persists, the severity of the obesity and the presence of parental obesity increase the likelihood that obesity will persist in adulthood. A recent Washington state study tracked 850 infants over 21–29 years and found that among obese 6 years olds, about 50% remained obese. By the age of 10–14 years, 80% of obese children with at least one obese parent remained obese.

Global Trends in Obesity

Both childhood and adult obesity are becoming global concerns. Other industrialized countries such as England, Singapore, Japan, Canada, and Australia are also experiencing increasing rates of obesity in both adults and children. Similar to data from the US, women from other countries tend to have higher rates of obesity compared with men, although men may have higher rates of overweight. The WHO reports that the prevalence of obesity is also increasing in countries undergoing economic transition and in developing countries where obesity coexists with undernutrition.

Risk Factors for Obesity in Adults and Children

It is widely accepted that overweight and obesity are multidetermined chronic problems resulting from complex interactions between genes and environments characterized by energy imbalance due to sedentary lifestyles and ready access to a wide variety of foods. Research suggests obesity runs in families and that some individuals are more vulnerable than others to gaining weight and becoming obese. Researchers also agree that there is no single gene that causes obesity, but that multiple genes and gene mutations associated with weight gain are involved. Various mechanisms through which genetic susceptibility to weight gain have been proposed including a low resting metabolic rate, low level of lipid oxidation rate, low fat-free mass, and poor appetite control. Although genetic research is a promising approach for understanding the development of obesity and identifying those at risk for obesity, it should be noted that the rapid increases in rates of obesity and overweight that have been observed have occurred over too brief a time period for there to have been significant genetic changes in the population. Therefore, the following sections briefly review environmental factors associated with the obesity epidemic.

Environmental Influences: Dietary Intake

Increased population levels of energy intake is a likely contributor to the epidemic of obesity. However, research on trends in energy intake over the time period in which the prevalence of obesity has increased has not consistently supported this hypothesis. Data from national surveys show a decrease in energy intake over time, whereas, nutrient intake data from NHANES III show an increase in energy intake over time. A recent review of ecological data regarding the amounts and types of food available in the USA during the past two decades provides interesting data that may shed light on some of the discrepancies found in previous survey data. Specifically, although the availability and purchasing of reduced fat and reduced energy foods has increased over the past few decades, people are eating more meals away from home, eating larger portions, and consuming more convenience foods. These behaviors are more conducive to higher energy intakes and may be contributing to the increased prevalence of obesity.

Environmental Influences: Physical Activity

Low levels of energy expenditure are increasingly recognized as important contributors to obesity and related health conditions. An abundance of cross-sectional research shows that heavier individuals are less active than lighter individuals, and prospective research indicates that changes in physical activity level are associated with changes in body weight in the direction predicted by the energy balance equation. The past century has produced dramatic changes in physical activity patterns in the USA. Machines and labor-saving devices have become so commonplace that the energy expenditure now required for daily life is a fraction of what it was a generation or two ago. Voluntary or leisure-time physical activity has thus assumed central importance in meeting physical-activity requirements. Unfortunately, voluntary physical activity is not very popular. Less than 25% of all adults, adolescents, and children report that they engage in regular physical activity (i.e., a minimum of 30 min of moderate to vigorous activity on most days of the week), the activity level most recently recommended by health experts. The prevalence of regular physical activity also varies according to demographic characteristics. Decreases in physical activity are observed during adolescence, with the decline particularly pronounced among girls. Women are less physically active than men, older adults are less active than younger adults, and African American and Hispanic adults are less active than whites. A lower socio-economic status is also associated with lower levels of physical activity.

Although low levels of leisure-time physical activity likely contribute to the epidemic of obesity, it is noteworthy that among adults, leisure-time activity has remained stable or increased since the mid-1980s, the time period during which the prevalence of obesity increased. Given this paradoxical finding, it is likely that increases in sedentary activities such as television watching and computer use and decreases in lifestyle, household, and occupational activity that have been less carefully measured have contributed to reductions in overall energy expenditure at the population level. Data from the Americans' Use of Time study show that the amount of free time spent watching television increased from about 4 h per week in 1965 to about 15 h in 1985. Documenting the prevalence of sedentary behavior among children and adults and targeting the reduction of sedentary activity as a health-behavior goal have received considerable attention in recent years.

Future Directions

Understanding and addressing the epidemic of obesity has become a global public-health priority. A recent 1997 WHO report on obesity, entitled 'Obesity: Preventing and Managing the global Epidemic,' noted that overweight and obesity represent a rapidly growing threat to the health of populations world-wide. Both children and adults in developing and developed countries are affected. The report concluded that the spectrum of problems seen across the world is of such magnitude that obesity should be regarded as the principally neglected public-health problem.

The recently released goals of Healthy People 2010 are to reduce the prevalence of obesity among adults from 23 to 15% and to reduce the prevalence of obesity among children and adolescents from 11 to 5%. The etiology of obesity is complex and encompasses a wide variety of social, behavioral, cultural, environmental, physiological, and genetic factors. To achieve these ambitious goals, considerable effort must be focused on helping individuals at the population level modify their diets and increase their physical activity levels, key behaviors involved in the regulation of body weight. A challenge to public-health professionals is to develop educational and environmental interventions that support diet and exercise patterns associated with a healthy body weight. Prevention of obesity should begin early in life and involve the development and maintenance of healthy eating and physical activity patterns. These patterns need to be reinforced at home, in schools, and throughout the community. Communities, government, health organizations, the media, and the food and health industry must form alliances if we are to combat obesity.

A coherent and standard international system for classifying overweight and obesity in children and adults is also needed. The WHO and US agencies recommend using BMI for adults, with BMI ≥ 25 denoting 'overweight' and BMI > 30 denoting 'obesity.' Criteria and methods for assessing and documenting obesity in children and adolescents need to be similarly developed. A common international standard would allow meaningful comparisons of obesity within and across populations and document obesity trends over time.

See also: **Adolescents**; **Children**: Nutritional Problems; **Nutritional Assessment**: Anthropometry and Clinical Examination; **Obesity**: Etiology and Diagnosis

Further Reading

Allison DB, Fontaine KR, Manson JE, Stevens J and Van-Itallie TB (1999) Annual deaths attributable to obesity in the United States. *Journal of the American Medical Association* 282(16): 1530–1538.

Clinical Guideline on the Identification, Evaluation, and Treatment of Overweight and Obesity in Adults. The Evidence Report (1999) National Institutes of Health. National Heart, Lung, and Blood Institute.

Flegal KM, Carroll MD, Kuczmarski RJ and Johnson CL (1998) Overweight and obesity in the United States: prevalence and trends, 1960–1994. *International Journal of Obesity* 22: 39–47.

Harnack L, Jeffery R and Boutelle K (in press) Temporal trends in energy intake in the U.S.: an ecological perspective. *American Journal of Clinical Nutrition.*

Kuczmarski RJ, Flegal KM, Campbell SM and Johnson CL (1994) Increasing prevalence of overweight among US adults. *Journal of the American Medical Association* 272 (3): 205–211.

Mokdad AH, Serdula MK, Dietz et al. (1999) The spread of the obesity epidemic in the United States, 1991–1998. *Journal of the American Medical Association* 282(16): 1519–1522.

Must A, Spadano J, Coakley EH, Field AE, Colditz G and Foryx WD (1999) The disease burden associated with overweight and obesity. *Journal of the American Medical Association* 282(16): 1523–1529.

Obesity: Preventing and Managing the Global Epidemic. Report of a WHO Consultation on Obesity. 3–5 June 1997. Division of noncommunicable diseases. Geneva: WHO.

Strauss R (1999) Childhood obesity. *Current Problems in Pediatrics* 29: 5–29.

Troiano RP and Flegal KM (1998) Overweight children and adolescents: description, epidemiology, and demographics. *Pediatrics* 101(supplement 3): 497–504.

US Department of Health and Human Services (1996) *Physical Activity and Health: A Report of the Surgeon General.* Atlanta, GA: US Department of Health and Human Services, Centers for Disease Control and

Prevention, National Center for Chronic Disease Prevention and Health Promotion.

Williamson DF (1991) Epidemiologic analysis of weight gain in U.S. adults. *Nutrition* 7(4): 285–286.

Williamson DF, Kahn HK, Remington PL and Anda RF (1990) The 10-year incidence of overweight and major weight gain in U.S. adults. *Archives of Internal Medicine* 150: 665–672.

Odors *See* **Sensory Evaluation**: Sensory Characteristics of Human Foods; Food Acceptability and Sensory Evaluation; Practical Considerations; Sensory Difference Testing; Sensory Rating and Scoring Methods; Descriptive Analysis; Appearance; Texture; Aroma; Taste

---

OFFAL

Types of Offal

W F Spooncer, CSIRO, University of Western Sydney, Richmond, NSW, Australia

This article is reproduced from *Encyclopaedia of Food Science, Food Technology and Nutrition*, Copyright 1993, Academic Press.

Introduction

Offals are all the noncarcass parts of slaughtered animals with the exception of hide and skin. Most offals have the potential to be used as human food but this potential can only be realized if the offals are collected hygienically, inspected, and passed fit for human consumption, then cleaned and prepared in an appropriate manner. Offals which are not collected for human use, either because there is no demand for them, or because they are diseased or contaminated, are inedible offals. Inedible offals are either destroyed by burning or burying, rendered to produce tallow and meat meal or, if suitable, used as pet food. (*See* **Meat**: Slaughter.)

Definition of Offals

Because of the variety of different types of offals, both edible and inedible, some attempts have been made to group offals together in categories. Many texts refer to offal as being derived from the 'off fall' from carcass preparation. Some of this 'off fall' may be attractive as edible product while some of it may not be edible at all. To avoid the connotation that offals include unattractive and inedible material, the terms 'variety meats' and 'fancy meats' have been introduced as categories of offals which are edible and prepared and consumed in a recognizable form. Variety meats and fancy meats are items such as

heart, tongue, kidney, sweetbread, liver, brain, tail, and tripe. In the USA, where the term variety meat is common, products which are not consumed in a recognizable form but could be used in processed meats are referred to as offal meat. Examples of offal meat are lungs, spleen, udders, and head meat. In this case the word 'offal' is reserved for inedible material.

Other ways in which offals are categorized include the distinction between white and red offals, and between organ offal and muscle offal.

The categories of white and red offal are used to distinguish stomach and intestinal offals from all other offals. This distinction is made because white offals require careful cleaning and preparation to remove intestinal and stomach contents before they can be used as edible material. Intestinal offal should be cleaned in an area separated from the preparation of other offals and carcass meat. Separate facilities for handling white and red offals minimize the chance of red offal being contaminated by intestinal contents.

The distinction between offals which are mainly organ tissue and those which are muscular tissue is made because of the difference in the potential keeping quality of the two types of offals. Organ offals, such as liver, kidney, lungs, spleen, and brain, are not as stable as muscular offals, such as tail, tongue, and diaphragm. In chilled storage, physiological activity in organ tissues can result in deterioration of the flavor and texture. (*See* **Meat**: Preservation.)

Specific Offals

Table 1 lists the items of offal available from slaughtered animals, alternative names, and possible categorization of the offals. **Table 2** lists typical weights of offals from different carcass types. (*See* **Beef; Pork; Poultry:** Chicken; Ducks and Geese; Turkey.)

According to American surveys, the more popular offals are liver, pigs' feet, oxtail, heart, tongue, pork maw (stomach), brains, sweetbread, tripe, and kidney. These offal items are described below. Other offals listed in **Table 1** do not appeal to western tastes but are popular in other markets, particularly in Asian countries.

Liver

Liver is the largest organ in meat-producing species. Its color varies from light red to dark red brown and even black in old animals. The liver is attached by ligament to the anterior abdominal wall and to the stomach by the lesser omentum. During evisceration these attachments are torn or cut, and the liver is removed attached by ligament to the diaphragm and the rest of the pluck (heart, liver, lung, and diaphragm). The liver is separated from the pluck by cutting through the attachment to the diaphragm. The gallbladder and duct are then pulled or cut off the liver.

Liver is composed of specialized liver cells with a network of blood vessels and epithelia-lined ducts between the cells. The cells are held together by a network of connective tissue.

The texture and flavor of liver are affected by species and animal age. Livers from young animals are lighter in color, have more delicate flavor, and are more tender than those from older animals. Lamb and calf liver, and liver from young beef are more suitable for table dishes. Liver from mutton, older bulls, and cows are used in manufactured or processed meats. Pig livers have a strong flavor which is

Table 1 Summary of offal types

Offal type	Alternative name	Category of offal	
		Variety meat	Red or white offal
Organ or glandular offal			
Liver	Lamb's fry	✓	Red
Heart } Pluck		✓	Red
Lung }	Lights		Red
Brain		✓	Red
Thymus	Sweetbread, heartbread, neckbread	✓	Red
Kidney		✓	Red
Pancreas	Gutbread		Red
Spleen	Melt		Red
Stomach	Pig's maw	✓	White
Rumen	Blanket tripe	✓	White
Reticulum	Honeycomb tripe	✓	White
Omasum	Bible		White
Abomasum	Reed (or vell in calves)		White
Testes	Fries		Red
Udder			Red
Uterus			White
Muscular offals			
Head meat			Red
Cheek			Red
Tail	Oxtail	✓	Red
Diaphragm	Skirt	✓	Red
Tongue (also considered glandular)		✓	Red
Esophagus	Weasand		Red
Others			
Head			Red
Feet			Red
Intestines	Chitterlings (pig only)		White
Ear			Red
Bone marrow			Red

Table 2 Typical weight of offals

	Ovine		Porcine		Bovine	
	Lamb (g)	Mutton (g)	Porker (g)	Baconer (g)	Veal (kg)	Beef (kg)
Liver	300–600	650–720	700–1200	1200–2000	0.75–3.0	2.5–8.0
Heart	90–150	180–215	150–250	200–300	0.25–1.0	0.8–2.0
Lung	200–350	400–500	500–850	750–1100	0.65–1.5	1.5–6.0
Brain	75–110	100–130	NA	NA	0.15–0.23	0.23–0.35
Thymus	50–100	NA	NA	NA	0.03–0.125	NA
Spleen	40–65	70–110	70–130	100–150	0.1–0.6	0.6–1.0
Stomach						
Maw	NA	NA	450–600	500–800	NA	NA
Rumen and reticulum	250–500	500–800	NA	NA	0.5–1.5	2.0–10.0
Omasum	30–400	400–500	NA	NA	0.2–1.0	1.0–5.0
Abomasum	80–140	150–200	NA	NA	0.35–0.75	0.7–3.0
Uterus	NA	NA	100–250	200–600	NA	NA
Cheek	NA	NA	1000–1800	1500–2000	0.3–0.7	0.7–1.2
Tail	NA	NA	NA	NA	0.2–0.6	0.55–1.2
Diaphragm	NA	NA	170–250	200–280	0.2–0.6	0.55–1.1
Tongue (short-cut)	50–100	80–150	120–200	150–220	0.15–0.7	0.7–1.6
Head	NA	NA	2000–3000	2800–3800	3.0–5.5	NA
Feet	NA	NA	230–280	300–400	0.2–0.4	NA
Ear	NA	NA	100–300	100–500	NA	NA

NA, not available.

appreciated in Asian countries, but in the west pig livers are used in processed meats rather than in table dishes.

To distinguish between the classes of bovine liver, they may be graded by weight. Livers in the range 1–3 kg are classed as calf livers. Larger livers from beef or ox may be graded into weight ranges of 3–4 kg, 4–5.5 kg, and over 5.5 kg. Ovine livers are not strictly graded but lamb liver should not exceed 700 g.

Heart

Hearts can be collected and trimmed in several styles. Whole hearts are removed from the pericardium and separated from the lungs and remainder of the pluck by cutting through the aorta and pulmonary vein. Whole hearts may be trimmed by cutting off the auricles along with the ossa cordis bones, aorta, and pulmonary vein. Even if the auricles remain in place, whole hearts are trimmed to remove remnants of the aorta, pulmonary vein, and the fibrous rings of the ossa cordis. Trimmed hearts comprise ventricles with associated fat, arteries, and veins, and may or may not include the auricles. They should be well washed to remove blood clots. Large hearts may be split open by cutting through the ventricle wall from the base to the apex.

The thick walls of the ventricles are composed of cardiac muscle, which is tough even in young animals. Cardiac muscle is supported by a network of connective tissue, which also contributes to the toughness of heart meat from older animals.

Tongue

Full tongues consist of the blade of the tongue with the roots attached, including the hyoid bones except the stylohyoid, the larynx, three tracheal rings, lymph nodes, salivary glands, and associated muscle and connective tissue. Short-cut tongues are a more common style of trim and are derived from full tongues by removing the larynx and roots by cutting behind the hyoid bones. Other trims of tongue are prepared with the hyoid bones removed.

The tongue blade is a core of skeletal muscle and connective tissue covered with an epithelial mucous membrane.

Brain

Brains include the cerebrum, cerebellum, and a portion of spinal cord. They are collected after opening the skull by splitting the occipital and maxillary bones. The brain is removed from the skull cavity, leaving the outer skin (the dura mater) in the skull but with a fine membrane (the pia mater and arachnoid meninges) covering the brain tissue. This membrane can be removed when the brain is cooked.

Brain tissue is not supported by a network of connective tissue and, consequently, has a soft, delicate texture.

Sweetbread

Sweetbread is the thymus gland and is only available from young animals. As animals mature, the gland degenerates into a mass of connective tissue and fat. The sweetbread is collected in two separate portions, although it is a single gland. The main portion, also called neckbread, is found in the neck, either side of the trachea. The other portion is in the chest cavity, close to the heart, and may be called heartbread. After collection, fat and connective tissue is trimmed off the gland.

The thymus gland consists mainly of lymphocytes and is covered by a capsule of connective tissue. The capsule penetrates the gland and divides it into lobules. The texture of sweetbread is delicate but may be tough if sweetbreads are collected from older animals in which the thymus gland has started to degenerate. Sweetbreads are soaked in water to remove blood, blanched to firm the texture, and skinned to remove the capsule before they are prepared for the table.

Kidney

Kidneys are contained in a capsule of fat and remain in the carcass after evisceration. The common way to handle kidneys is to remove them from the capsule for inspection, trim off the ureter and renal blood vessels close to their entry to the kidney and, for beef kidneys, cut out the core of fat and connective tissue from the renal hilus.

Kidneys are made up of renal tubules and small veins and arteries. The structural components are the endothelial and epithelial cells of these vessels, and a network of connective tissue. There is a fibrous capsule of connective tissue around the kidney which is usually removed before kidneys are cooked.

Kidneys from sheep and pigs have the characteristic bean shape. Beef and calf kidneys are divided into lobules.

Pigs' Feet

Forefeet are collected from pig carcasses by cutting the foreleg between the carpal and metacarpal bones. Similarly, hindfeet are collected by cutting between the tarsal and metatarsal bones. Pigs' feet comprise the bones, tendons, skeletal muscle, connective tissue, fat and skin of the trotters, but toenails and all hair should be removed.

The edible part of pigs' feet is fat, muscle, and connective tissue. Prolonged moist cooking of the feet will soften and at least partially gelatinize the collagenous connective tissue and make the flesh tender.

Oxtail

Tails are collected from beef and veal carcasses by cutting off the tail at the sacrococcygeal junction. Tails consist of the coccygeal (tail) vertebrae with associated skeletal muscle, connective tissue, and fat. The last two or three vertebrae may be removed, and some or all of the subcutaneous fat at the anterior end of the tail may be trimmed off.

Tripe

Cattle and sheep stomachs have four compartments, all of which can be used to make tripe. Tripe from the rumen and reticulum is the most common. Rumen tripe is known as blanket tripe and reticulum tripe is known as the honeycomb. The internal surface of the rumen has densely packed papillae, while the reticulum has ridges in the shape of a honeycomb structure. In addition, there are thickened folds in the rumen wall. These folds contain a core of smooth muscle and are not covered by papillae. The fold can be trimmed out of the rumen to produce pillae tripe, known commercially as pillar or mountain chain.

The third compartment of the ruminant stomach is the omasum. The internal wall is in the form of deep, thin folds like the pages of a book. This appearance accounts for the popular name for the omasum, which is 'bible.' Although the omasum has a delicate flavor and texture, it is not commonly used as tripe because of the difficulty of cleaning the stomach contents from between the folds. The fourth part of the ruminant stomach is the abomasum, sometimes called the reed.

The four parts of the stomach are collected in one piece and the omental fat (caul fat) and spleen are removed. The neck of the omasum is cut to separate the rumen and reticulum from the omasum and abomasum. The rumen is cut open and the contents of the rumen and reticulum washed out. The rumen and reticulum are then trimmed into the different tripes, and external fat is trimmed off. Mountain chain tripe does not receive further processing, but blanket and honeycomb tripes may be further cleaned, scalded, and bleached with hydrogen peroxide.

The walls of ruminant stomachs are composed of smooth muscle and connective tissue. The papillae in the rumen are composed of collagen and elastin fibers covered with cornified epithelia, while the ridges and folds in the reticulum and omasum contain smooth muscle as well as connective tissue. There are small, cornified papillae on the folds of the omasum, and the surfaces of the folds are covered by a keratinized mucous membrane. The abomasum has a thick epithelial lining.

Tripes are generally tough because of the high connective tissue content. They contain about 35 g of collagen per 100 g of protein. They require prolonged, moist cooking to tenderize them. Bleached tripe is treated with caustic soda and has a pH of about 7–9, which increases the water-holding capacity and helps to tenderize the tripe.

Maw

Pig's stomach or maw is a similar organ to the ruminant abomasum. The maw is separated from the rest of the viscera by pulling off the omental fat and spleen from their attachments to the stomach and cutting through the esophagus and duodenum. The maw is then split open and the contents washed out. After cleaning the maw, it is usually scalded in water at about 90 °C to remove the mucous membrane lining.

In common with the abomasum, the maw is composed of a wall of smooth muscle and connective tissue, with an inner lining of thick epithelia and mucous membrane.

Offal for Human Use

Offal consumption varies from about 1.7 kg per person per year in Canada to about 16 kg per person per year in Ireland. **Table 3** shows per capita consumption of offal in western countries.

As shown in **Table 1**, almost all noncarcass parts can be used as food after the carcass and offal have been inspected and found fit for human consumption. (Exceptions are uteri from pregnant animals, and fetuses.) After collection, offals are trimmed and washed to remove contamination by blood, saliva,

Table 3   Per capita consumption of offals by countries

	Consumption (kg per head per year)
Australia	4.7
Austria	4.5
Belgium and Luxembourg combined	8.0
Canada	1.7
Denmark	7.0
Republic of Ireland	16.7
Finland	8.1
France	6.6
Greece	6.0
Italy	3.8
Japan	2.5
The Netherlands	2.6
New Zealand	6.0
Norway	2.7
Portugal	5.0
Spain	4.0
Sweden	2.5
Switzerland	5.9
Turkey	2.1
UK	4.3
USA	4.1
Germany	5.7
Yugoslavia	2.9

mucus, and ingesta. They may be distributed frozen, chilled, or vacuum-packed and chilled. Frozen offals held at − 18 °C have a useful shelf-life of 6 months; chilled offals have a shelf-life of 2–3 days; and vacuum-packed tail, tongue, skirt, heart, kidney, and liver have a shelf-life of at least 3 weeks at 1 °C.

Offals are consumed both as table dishes and in manufactured meat products such as sausages and pâté. There are many options for preparing and cooking offals, commensurate with the wide ranges of offal types. Some offals require initial treatment such as soaking and precooking. Brains, sweetbreads, and testicles are soaked in cold water to remove blood pigment and improve color. They are then blanched by brief immersion in hot water to firm the texture before they are cooked by other methods. Liver and kidney, particularly from older animals, may also be soaked to leach out the strong flavor associated with these offals. Stomachs and intestines are precooked by prolonged immersion in hot water to remove the mucosal membrane, and soften connective tissues. Tongues are also precooked, after which the thick skin of epithelia on the tongue blade can be peeled off. (*See* **Meat**: Sausages and Comminuted Products.)

Regulations may restrict the way in which offals are used in manufactured and comminuted meat products. In the EC and Australia, offals are considered to be meat but are distinguished from meat of skeletal muscle origin by using terms such as carcass meat and meat flesh, which exclude offal meat. In the USA, heart and tongue are considered to be meat, and other offals have the status of meat byproducts. The greatest restrictions on the use of offals is in uncooked meat products such as fresh sausages.

In Australia, no offals are permitted in fresh sausage and in the USA only tongue and heart can be used in fresh sausage. In the UK the designated list of permitted offals which can be used in uncooked products includes heart, kidney, liver, and tongue. In general, cooked meat products may contain offals. The Codex Alimentarius Commission has standards which restrict the use of offals in specific products. In summary, these standards are as follows:

1. Canned corned beef. Heart meat is the only offal permitted.
2. Luncheon meat. Luncheon meat with binder may contain offals, with the exception of lungs that have been collected from animals scalded by immersion in water. Only heart and tongue are permitted in luncheon meat without binder.
3. Cooked cured chopped meat. Only heart and tongue are permitted in products which do not contain binder. Offals (except udders, lungs, and

organs and glands from the genital system) may be used in products that contain binder.

Other Uses of Offals

Collection of offals for edible use requires – at least – inspection, washing, and trimming. The cost of these procedures may not be justified by the value of some offals. In this case the offals are disposed of or used in other ways. The main alternatives are to render offals to produce tallow and meat meal, or to use them as pet food. Inspected offal which is not edible because of contamination or disease status is rendered along with other offals which are not required for edible use or pet food, e.g., heads, hooves, and the residues of gut material after collecting sausage casing and tripe.

Offals collected for pet food may be subject to less rigorous inspection, preparation, and trimming than offals for human use and can be handled at a lower cost. The main use of offals is in canned pet foods. These products receive severe heat treatment, and offals which retain form and texture, such as lung and tripe, are most suitable. Liver, heart, and spleen are also used in canned pet food.

There are specialized uses for some offals. Pig and beef pancreas glands are used to produce insulin for the treatment of diabetes. However, insulin from pancreas glands is being replaced by insulin produced from bacterial fermentation.

The mucosa from the lining of lungs and intestines is used to produce the anticoagulant, heparin. A long-standing use of calf vells (abomasum) is in the production of rennet.

See also: **Beef**; **Meat**: Slaughter; Preservation; Sausages and Comminuted Products; **Pork**; **Poultry**: Chicken; Ducks and Geese; Turkey

Further Reading

Codex Alimentarius Commission (1981) *Codex Standards for Processed Meat and Poultry Products and Soups and Broths*. Codex Alimentarius, vol. IV. Rome: Food and Agriculture Organization and World Health Organization.

Gerrard F and Mallion FJ (eds) (1977) *Complete Book of Meat*, pp. 407–413. London: Virtue.

Levie A (1970) *The Handbook of Meat*, pp. 296–301. Westport, Connecticut: AVI.

Patterson JT and Gibbs PA (1979) Vacuum packing of bovine edible offal. *Meat Science* 3: 209.

Spooncer WF (1978) *Yield of Organs and Glands in Relation to Carcass Weight*. Meat Research Report 8/78. Cannon Hill, Queensland: CSIRO Division of Food Research.

Spooncer WF (1988) Organs and glands as human food. In: Pearson AM and Dutson TR (eds) *Edible Meat Byproducts*, pp. 197–217. Essex: Elsevier Science.

Off-flavors (Off-flavours) in Food *See* **Taints**: Types and Causes; Analysis and Identification

---

Oils *See* **Fats**: Production of Animal Fats; Uses in the Food Industry; Digestion, Absorption, and Transport; Requirements; Fat Replacers; Classification; Occurrence; **Vegetable Oils**: Types and Properties; Oil Production and Processing; Composition and Analysis; Dietary Importance

---

Okra *See* **Vegetables of Temperate Climates**: Commercial and Dietary Importance; Cabbage and Related Vegetables; Leaf Vegetables; Oriental Brassicas; Carrot, Parsnip, and Beetroot; Swede, Turnip, and Radish; Miscellaneous Root Crops; Stem and Other Vegetables

---

Olestra *See* **Fat Substitutes**: Use of Fat-replaced Foods in Reducing Fat and Energy Intake

OLIVE OIL

M Tsimidou, G Blekas and D Boskou, Aristotle University of Thessaloniki, Thessaloniki, Greece

Introduction

Poseidon and Athena had a violent disagreement as to who should govern Attica (Greece). To settle this dispute, Zeus declared that he would give the country to whoever offered him the most useful present for mankind. Poseidon then presented a salty spring. Athena presented a twig. This twig would become a staunch tree, capable of living for untold centuries, whose fruit would be edible and from which an extraordinary liquid would be extracted for the preparation of man's food, alleviation of his wounds, strengthening of his system, and lighting his nights. Needless to say, Athena won, and the city, ever since, is named Athens (**Figure 1**).

The cultivation of the olive tree (**Figure 2**) dates back before recorded history. The origin of the tree is the area of Mesopotamia. The Minoans (2000–1450 BC) were quite advanced in the cultivation of the olive tree and the production of olive oil, which was also exported. One valuable source of information is the Linear B tablets from the royal archives of Knossos (Crete) and Pylos (Peloponnese) (see **Figure 3**). Greek mythology, the Old Testament, and Roman literature are full of information regarding the role of the olive tree and its products in religion and everyday life. The cultivation of the olive tree is an essential element of the civilizations developed in the Mediterranean Basin, closely related to their prosperity and cultural achievements.

Olive oil is obtained from the fruits of *Olea europaea* L. The evergreen tree has a life span of several centuries (300–600 years). Its longevity is partially

Figure 2 (see color plate 116) Olive tree in Rethimnon, Crete (Greece), courtesy of V. Zambounis, Athens.

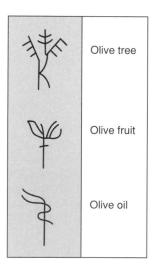

	Olive tree
	Olive fruit
	Olive oil

Figure 3 Symbols in the written language Linear B. Reproduced from Psillakis N and Kastanas E (1999) *The civilization of olive: olive oil*, 2nd edn. Greek Academy of Taste, Karmanor, Iraklion, Crete (Greece), with permission.

Figure 1 (see color plate 115) Picture of a fossilized olive leaf. Reproduced from Psillakis N and Kastanas E (1999) *The civilization of olive: olive oil*, 2nd edn. Greek Academy of Taste, Karmanor, Iraklion, Crete (Greece), with permission.

explained by the characteristic ability to send out shoots and roots from temporary buds at the lower part of the trunk. The height of an old tree may even be as much as 25 m, but trees in systematic plantations have an optimum height of 4–5 m to facilitate manual picking of fruits. The tree is resistant to unfavorable conditions (dryness, infertile soil, etc.) but requires a mild climate (not below −9 °C in winter – a moderate to high temperature in summer). Many species of olive tree are grown all over the world, some of them (> 35) suitable for olive oil production. Cultivars used for oil production have a medium-sized fruit containing 17–30% oil when ripe. Among the most important oil-producing cultivars are: *Koroneiki, Megaritiki,* and *Tsounati* (Greece), *Leccino* and *Frantoio* (Italy), *Picual, Hojiblanca,* and *Lechin de Sevilla* (Spain). (*See* **Olives.**)

Olive Oil Trade and Consumption

Olive oil is an important component of the Mediterranean diet, which is now also gaining interest among consumers of northern Europe, USA, Canada, and other countries as a component of healthy eating. Olive oil is almost exclusively produced in Spain, Italy, Greece, Tunisia, Turkey, Morocco, and Portugal (*c.* 31, 31, 18, 4, 5, 2.5, and 2% of the world's production, respectively) (see also **Table 1**). More than 800 million trees, covering a total area of 25 million acres, yield approximately 2 000 000 tonnes of oil (**Figure 4**). Though olive oil production accounts for no more than 2% of the volume of edible oils traded world-wide, its market value has a significant share (15%) in international trade because of its high price.

Table 1 World olive oil production

Country	1998/1999 (tonnes)	Percentage share	1999/2000 (tonnes, estimated)	Percentage share
European Union	1 698 500	71.5	1 563 000	77
Tunisia	215 000	9	200 000	10
Turkey	170 000	7	70 000	3
Syria	115 000	5	80 000	4
Morocco	65 000	3	40 000	2
Algeria	39 500	1.5	25 000	1
Other countries	70 600	3	55 500	3
Total	2 373 600	100	2 033 500	100

Data from IOOC (1999) Trade Standards Applying to Olive Oil and Olive Pomace Oil. COI/T. 15/NC, Doc. No. 2; rev. 9. Madrid: International Olive Oil Council.

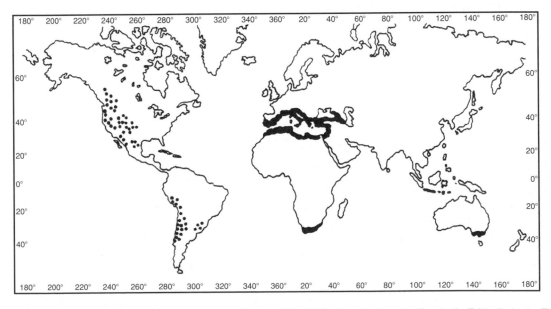

Figure 4 Olive tree growing areas of the world. From Kiritsakis KA (1998) *Olive Oil: From the Tree to the Table,* 2nd edn. Trumbull, CT: Food and Nutrition Press, with permission.

The International Olive Oil Council (IOOC), established in 1959, is the intergovernmental organization responsible for administering the International Agreement on Olive Oil and Table Olives. This agreement has been negotiated at the United Nations and lays down the policy that members should take for standardization of the market for olive oil, olive-pomace oil, and table olives. The Agreement is renewed regularly (for the time being, the 1986 Agreement has been prolonged until 31 December 2000). Principles and provisions set are compulsory in international trade, recommended in domestic trade, and incorporated in the trade standards for olive oil and olive-pomace oil.

According to recent IOOC reports, the increasing production of olive oil in Europe, as well as in the rest of the world, has led to a drop in prices and to a concomitant rise in consumption. Data concerning consumption in 1998/1999 and estimated consumption for 1999/2000 are given in **Table 2**. Similar trends stand for the year 2001/2002. Olive oil consumption is practically concentrated in the producing countries. The average annual per-capita consumption of olive oil reflects the economic policy and culinary traditions of each producing country. Thus, consumption that is reported for Greek consumers (19.5 kg) is twofold higher than that in Italy or Tunisia. In France and in the countries of central and northern Europe, consumption of olive oil is very low. In the last decade, there has been a significant increase in olive oil consumption. In countries such as Germany, Denmark, UK, and Ireland, the estimated per-capita per-year consumption in the last 2 years has been around 0.5–0.6 kg. In the USA and Canada, similar estimates are approximately 0.8 kg. The results of the 'Seven Countries Study' related the low rate of coronary heart disease among the Cretan population to the regular consumption of virgin olive oil. This was the first scientific evidence for the nutritional role of this 'high-in-monounsaturates' juice. Currently, the interest is focused on the role of olive oil in the prevention of bone and nervous-system functions, digestion, antiageing properties at cell and mitochondrial levels, diabetes, and breast tumors, and also in the preventive role of some of its minor constituents in the oxidation of low-density lipoprotein (LDL). (*See* **Coronary Heart Disease:** Etiology and Risk Factor; Antioxidant Status; Prevention.)

Commercial Types of Olive Oil

Olive oil is marketed in accordance with the designations of **Table 3**. Virgin olive oil is obtained by only mechanical or other physical means under conditions, particularly thermal, which do not lead to alterations in the oil. This oil has not undergone treatment other than washing, decantation, centrifugation, and filtration. *Extra, fine,* and *ordinary* virgin olive oils are different edible grades of virgin olive oil. *Lampante* is a nonedible grade of virgin olive oil suitable for refining or other technical uses. Refined olive oil is the oil refined by the methods used for other vegetable oils. These include neutralization, decolorization with bleaching earth, and deodorization. Olive oil is an edible blend of virgin olive oil and refined olive oil. Olive-pomace oil is obtained by solvent extraction of the cakes derived from olive milling. The triacylglycerol composition is similar to that of virgin olive oil, but some of the nonsaponifiable compounds differ significantly. Due to the high content of waxes, the oil has to be winterized before refining. *Olive-pomace oil* can be used as a mixture with virgin olive under the designation 'olive-pomace oil.' This oil cannot be called 'olive oil.' Virgin olive oils may also be traded as 'appelation d'origin' products in accordance with the existing legislation (e.g., EC Regulation 2081/92). (*See* **European Union:** European Food Law Harmonization.)

Table 2   World olive oil consumption

Country	1998/1999 (tonnes)	Percentage share	1999/2000 (tonnes, estimated)	Percentage share
European Union	1 669 000	70	1 679 000	71
USA	157 000	7	160 500	7
Turkey	97 000	4	60 000	2.5
Syria	88 000	4	85 000	3.5
Tunisia	49 000	2	65 000	3
Morocco	55 000	2	50 000	2
Algeria	35 000	1	29 000	1
Other producing countries	142 000	6	139 000	6
Other nonproducing countries	92 500	4	92 500	4
Total	2 385 200	100	2 359 900	100

Data from IOOC (1999) Trade Standards Applying to Olive Oil and Olive Pomace Oil. COI/T. 15/NC, Doc. No. 2; rev. 9. Madrid: International Olive Oil Council.

Table 3 Quality criteria for olive oils and olive-pomace oils

	Extra virgin olive oil	Virgin olive oil	Ordinary virgin olive oil[b]	Lampante virgin olive oil	Refined olive oil	Olive oil[d]	Crude olive-pomace oil	Refined olive-pomace oil	Olive-pomace oil
1. Organoleptic characteristics:									
Panel test score (scale 1–9)	≥ 6.5	≥ 5.5	≥ 3.5	≤ 3.5					
Odor and taste					Acceptable	Good		Acceptable	Good
Color					Light yellow	Light yellow to green		Light yellow to brownish yellow	Light yellow to green
Aspect at 20 °C for 24 h					Limpid	Limpid		Limpid	Limpid
2. Free acidity % m/m (expressed as oleic acid)	$\leq 1.0^a$	≤ 2.0	≤ 3.3	> 3.3	≤ 0.3	$\leq 1.5^e$	No limit	≤ 0.3	≤ 1.5
3. Peroxide value (mEq peroxide oxygen per kg oil)	≤ 20	≤ 20	≤ 20	No limit	≤ 5.0	≤ 15	No limit	≤ 5	≤ 15
4. Absorbance in ultraviolet ($K_{1cm}^{1\%}$)									
270 nm	≤ 0.25	≤ 0.25	$\leq 0.30^c$	No limit^c	≤ 1.10	≤ 0.90		≤ 2.00	≤ 1.70
ΔK	≤ 0.01	≤ 0.01	≤ 0.01		≤ 0.16	≤ 0.15		≤ 0.20	≤ 0.18

$^a \leq 0.8$; ^{b}Banning of the category agreed in the EC; ^{c}After passage of the sample through activated alumina, absorbance at 270 nm shall be equal to or less than 0.11; ^{d}commercial name is discussed; $^e \leq 1$ proposed.
Data from IOOC (1999) Trade Standards Applying to Olive Oil and Olive Pomace Oil. COI/T. 15/NC, Doc. No. 2; rev. 9. Madrid: International Olive Oil Council.

Olive-oil Extraction

There are three extraction procedures currently used to separate the oil from the other phases in the paste, liquid or solid. These are *pressure*, *centrifugation*, and *percolation*. Specialized manufacturers provide suitable machinery for the proper extraction of olive oil.

Pressure

This is the oldest method but is still in use. In this system, the paste is pressed to release an oily must (olive + vegetation water) (**Figure 5**). The liquid phase separates from the solid phase with the help of the drainage effect of the mats and the stone fragments. A cake is formed between the mats, while the two liquid phases (oil and vegetation water) are separated by centrifugation.

Centrifugation or Three-phase Systems

The crushed olives are mixed with water. A horizontal centrifuge (**Figure 6**) separates the mass into pomace and must, which is further separated into oil and vegetation water. In the 1970s and 1980s, three-phase centrifuges replaced pressure systems to a great extent to cut processing costs and to reduce olive storage time. A recent innovation to reduce polyphenol losses is the so-called two-phase system, which is able to separate the oily phase without the addition of

Figure 5 (see color plate 117) Hydraulic pressure unit (courtesy of ELAIS SA, Piraeus, Greece).

water. The oily phase is separated from the paste to give oil and water plus husks.

Percolation

A steel plate is plunged into olive oil paste. When it is withdrawn, it will be coated with oil because of the different surface tensions of the liquid phases (the metal phase is coated with a skin of oil). The system

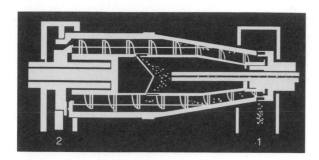

Figure 6 Horizontal centrifuge (decanter): 1, exit of solid phase. 2, exit of liquid phase (oil and water). From Kiritsakis KA (1998) *Olive Oil: From the Tree to the Table*, 2nd edn. Trumbull, CT: Food and Nutrition Press, with permission.

is combined with a continuous horizontal centrifuge to increase the capacity.

Cold Pressing

Cold pressing is the extraction process of olive oil from olive paste at a temperature of less than 25 °C.

Milling Conditions and Quality of Olive Oil

Traditional pressure-extraction systems yield good-quality oil when olives are in good condition and good manufacturing practice is applied. Percolation (selective filtration) also provides good-quality oil. Centrifugation plants have a larger hourly processing capacity and significant practical advantages. In such systems, however, the paste has to be thinned with warm water. This lowers the level of polar natural antioxidants, which have been associated with the storage stability of the oil and its beneficial health effects. Depending on the variety and stage of maturity of olives, different crushing systems are necessary to obtain the optimum polyphenol content and better aroma characteristics. Stone mills yield oils with a good aroma and less intense bitter taste than metal-disk crushers because of a lower polyphenol content. Hammer crushers usually yield oil with a characteristically high content of polyphenols. The malaxation time and temperatures are also critical for the aroma and quality.

Other methods to improve the quality and organoleptic scores during milling are based on the use of pectolytic and cellulolytic enzymes. The International Olive Oil Council, however, does not accept such techniques because olive oil is by definition obtained from the fruits of olive tree only by *'physical means.'*

Olive Oil Composition

The composition of olive oil is primarily triacylglycerols and secondarily free fatty acids, and some 0.5–1.5% of nonglyceridic constituents. The Codex Alimentarius Commission, the IOOC, and the Commission of the European Union provide identity and compositional characteristics for olive oil and olive-pomace oil.

EEC Regulation 2568/91 and subsequent amendments (EC 2632/94, EC 656/95, and EC 2472/97) set the limits for fatty acid composition and content in sterols, waxes, saturated fatty acids in position-2 of triacylglycerols, erythrodiol + uvaol, trilinolein and *trans* unsaturated fatty acids and also for specific absorption at various wavelengths. These limits are very strict and aim at protecting good-quality olive oil, and especially virgin olive oil, from adulteration. The same directives also define analytical methods. (*See* **Fatty Acids**: Properties; Analysis.)

Extended information for the composition of olive oil and other aspects of olive oil chemistry and technology are provided in reviews and books authored mainly by Spanish, Italian, and Greek authors.

Fatty Acids and Triacylglycerols

The fatty acid composition of olive oil ranges from 7.5 to 20.0% palmitic acid, 0.5 to 5.0% stearic acid, 0.3 to 3.5% palmitoleic acid, 55.0 to 83.0% oleic acid, 3.5 to 21.0% linoleic acid, 0.0 to 1.5% linolenic acid, 0.0 to 0.8% arachidic acid, 0.0 to 0.2% behenic acid, and 0.0 to 1.0% lignoceric acid. The olive oil composition may differ from sample to sample, depending on the zone of production, the latitude, the climate, the variety, and the stage of maturity of olives when collected. Olive oil has a fatty acid composition similar to that of hazelnut, almond and high-oleic sunflower oils. This composition differs significantly from any other type of edible oil and fat. (*See* **Nuts**; **Sunflower Oil**; **Vegetable Oils**: Types and Properties.)

The triacylglycerol profile of olive oil, as determined by reversed-phase liquid chromatography, is also different from the profiles of maize, cottonseed, sunflower soybean, and rapeseed oil, and resembles that of hazelnut oil. The main triacylglycerols found in olive oil are: OOO (40–59%), POO (12–20%), OOL (12.5–20%), POL (5.5–7%), SOO (3–7%) (O = oleic acid, P = palmitic acid, S = stearic acid, L = linoleic acid).

Tocopherols

The dietary benefits of olive oil are partly attributed to the presence of α-tocopherol and other natural antioxidants. α-Tocopherol acts not only as a free radical trapping agent but also as a singlet oxygen quencher. This protective effect against photooxidation is enhanced by the presence of β-carotene. The main homolog of vitamin E forms present in olive oil is α-tocopherol, which makes up approximately 95% of total tocopherols. The other 5% are β- and γ-tocopherols. Good-quality, fresh virgin olive oils have a considerable content of α-tocopherol (200–300 mg kg^{-1}). Refined, bleached, and deodorized olive oils have markedly reduced contents because of the losses of tocopherols during processing. (*See* **Tocopherols: Properties and Determination.**)

Squalene

Squalene ($C_{30}H_{50}$), an intermediate in the biosynthesis of sterols in plant and animal world, is the major olive oil hydrocarbon. Squalene is found in olive oil at a concentration ranging from 0.7 to 12 g per kilogram of oil. The squalene content depends on the olive cultivar and oil-extraction technology, and it is dramatically reduced during the process of refining. There are claims that squalene can enhance the quality of life, if taken continuously, and that its consumption is beneficial for patients with heart disease, diabetes, arthritis, hepatitis, and other diseases. It has been claimed that the high squalene content of olive oil, as compared with that of other human foods, is a major factor in the cancer-risk-reducing effect of this oil.

Carotenoids

The main carotenoids present in olive oil are β-carotene and lutein. Trace amounts of neoxanthin, violaxanthin, cryptoxanthin, and other xanthophyls may be found. The total carotenoids content may range between 1 and 20 mg per kilogram of oil; normal values do not exceed 10 mg kg^{-1}. Oils have been reported that have a higher content of lutein than of β-carotene and *vice versa*. So far, studies have indicated an inverse correlation between β-carotene intake and incidence of cardiovascular disease, but no clear effect on low-density lipoprotein (LDL) oxidation has been demonstrated with β-carotene. (*See* **Carotenoids: Occurrence, Properties, and Determination.**)

Sterols

Phytosterols are functional ingredients. They have an absorption level 20 times lower than that of cholesterol. Phytosterols inhibit the absorption of cholesterol in the body during digestion. Four classes of sterols occur in olive oil, common sterols (4-desmethylsterols), 4-α-methylsterols, 4,4-dimethylsterols, and triterpene dialcohols (erythrodiol and uvaol) (**Figure 7**). The major class is 4-desmethylsterols. Usual values reported for these classes are 1000–2000 mg per kilogram of oil. Part of the total sterols is present as esters with fatty acids. β-Sitosterol makes up 60–90% of the total sterol fraction. Other sterols found in considerable amounts are δ^5-avenasterol (**Figure 8**; 5–36% of the total sterol fraction) and campesterol (approx. 3% of the total sterol fraction). The rest of the 4-desmethylsterols present in olive oil are found in trace or very small amounts. Certain sterols are probably responsible for the resistance of olive oil to rapid deterioration at elevated temperatures. (*See* **Functional Foods.**)

Phenolic Compounds

Virgin olive oil contains phenolic substances that affect its stability and flavor. Tyrosol (4-hydroxyphenethyl alcohol) and hydroxytyrosol (3,4-dihydroxyphenethyl alcohol) are usually mentioned as the major constituents. Other phenolic compounds are caffeic acid, o-coumaric acid, p-coumaric acid, ferulic acid, gallic acid, homovanillic acid, p-hydroxybenzoic acid, p-hydroxyphenylacetic acid, protocatechuic acid, sinapic acid, syringic acid, tyrosol glucoside, and vanillic acid. Aglycons of oleuropein and ligstroside, the esters of hydroxytyrosol and tyrosol with

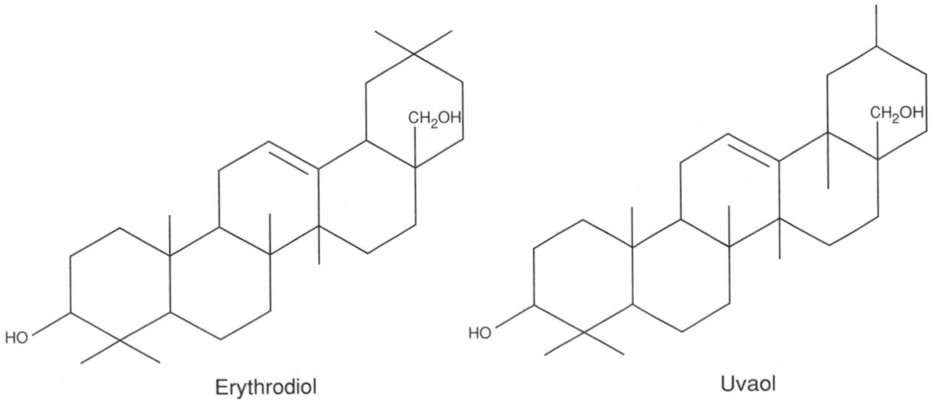

Figure 7 Two triterpene dialcohols used to check purity of virgin olive oil.

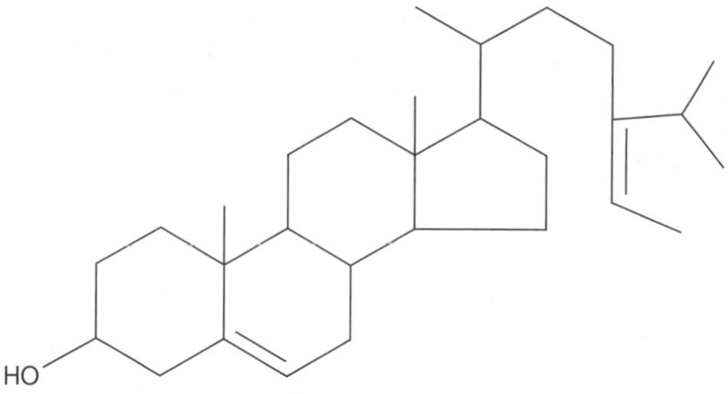

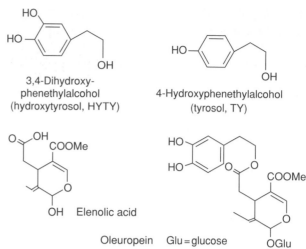

Figure 8 Chemical structure of Delta ^{5}-Avenasterol.

Figure 9 Major phenolic and related compounds in olive fruit and olive oil.

elenolic acid, diacetoxy and dialdehydic forms of these aglycons, elenolic acid, and flavonoids have been also reported to be present in the polar fraction of virgin olive oil (**Figure 9**).

The content in phenolic compounds differs from oil to oil (a few to more than 400 mg per kilogram of oil, expressed as caffeic acid). When the level exceeds $300 \, mg \, kg^{-1}$, the oil may have a bitter taste. However, a high polyphenol content appears to be beneficial for the shelf-life of the oil, and there is a good correlation of stability and total phenol or o-diphenol content. Among the various phenolic compounds tested for their contribution to the stability, hydroxytyrosol and caffeic acid were found to be the most potent antioxidants.

Dietary antioxidants present in exra virgin olive oil were found to increase the resistance of low-density lipoproteins to *in-vitro* or *in-vivo* oxidation experiments. According to a large number of reports, hydroxytyrosol is the most important biophenol of olive oil similar to the phenolic compounds encountered in green tea and in red wine. Oleuropein also presents valuable functional properties, but it occurs in olive oil only in trace amounts. It is found in abundance in olives and olive leaves. (*See* **Phenolic Compounds.**)

Chlorophyls

The color of olive oil is mainly due to pheophytins α and b. Traces of chlorophyls α and b may be found only in fresh oils. The chlorophyl content in virgin olive oil ranges from trace amounts to more than $30 \, mg \, kg^{-1}$. The level of chlorophyls (and carotenoids) depends on genetic factors, degree of fruit ripening, and extraction technology. The level decreases as the fruit ripens. In the absence of light, chlorophyls probably act as weak antioxidants. In the presence of light, it is accepted that chlorophyls act as strong oxidation promoters. Chlorophyls

are responsible for the yellow/green hues of the oil. (*See* **Chlorophyl.**)

Volatile and Aroma Compounds

More than 100 constituents have been identified in virgin olive oil, mainly hydrocarbons, alcohols, aldehydes, esters, and ketones. However, only a few of these compounds contribute significantly to its aroma. The most important odorants are volatiles with six carbon atoms, arising from the enzymic oxidation of both linoleic and linolenic acids during olive crushing and malaxation. Genetic, agronomic, and technological parameters, such as olive variety, degree of fruit ripening, crushing, and malaxation conditions or extraction system affect the enzymic activity and alter the quality and intensity of the aroma. Certain compounds were found to be responsible for desirable flavor attributes, e.g., (Z)-3-hexenal for the 'green, apple-like' odor, hexanal for the 'grassy' odor, (E)-2-hexenal for the 'green, bitter almond-like' odor, ethyl 2-methylbutanoate for the 'fruity' odor. Other compounds such as octanal, nonenals, decadienals, 1-octen-3-one, acetic acid, 3-methylbutanol, 2-phenylethanol, etc., are responsible for some negative attributes. (*See* **Flavor (Flavour) Compounds:** Structures and Characteristics.)

Storage and Packing

The oxidative stability of virgin olive oil is mainly related to its characteristic pattern of triacylglycerols (low unsaturation, *iodine value* < 90) and also to the considerable levels of polar phenolic antioxidants and the presence of α-tocopherol. Since the early 1970s, it was found that there was a good correlation between polyphenols and the oxidative stability of the oil, and of chlorophyls in the presence of light. Storage in cool places protected from air access is suggested to prevent autoxidation. Therefore, stainless steel containers are ideal for storage. Glass bottles are resistant to fat and oxygen permeation, but only colored bottles offer light protection.

Traditional and Modern Use

Olive oil is part of a cuisine that is simple, light, and placid, with defined tastes and full of harmony, the so-called Mediterranean cuisine. Olive oil resistance to the development of rancidity is combined with a vast array of flavor and color hues and distinct features due to differences in cultivars of olives from which the oil is extracted. A good-quality olive oil blends perfectly with the greens. The exquisite taste of olive oil is very often complemented by the sharp taste of vinegar, lemon, or tomato. In salads or in cooking, olive oil is usually mixed with herbs and spices, which are also an important element of the Mediterranean diet. Olive oil shows a remarkable resistance during domestic deep-frying of potatoes or in other uses at frying temperatures due to its low unsaturation. It is, therefore, recommended not only as a salad oil but also for cooking and frying.

See also: **Carotenoids**: Occurrence, Properties, and Determination; **Chlorophyl**; **Coronary Heart Disease**: Etiology and Risk Factor; Antioxidant Status; Prevention; **European Union**: European Food Law Harmonization; **Fatty Acids**: Properties; Analysis; **Flavor (Flavour) Compounds**: Structures and Characteristics; **Functional Foods**; **Nuts**; **Olives**; **Phenolic Compounds**; **Sunflower Oil**; **Tocopherols**: Properties and Determination; **Vegetable Oils**: Types and Properties

Further Reading

Boskou D (ed.) (1996) *Olive Oil, Chemistry and Technology*. Champaign, IL: AOCS Press.
Boskou D and Elmadfa I (eds) (1999) *Frying of Food*. Lancaster, PA: Technomic Publishing.
Codex Alimentarius Commission (1993) Revised Norm for Olive Oil, C1 1993/15-FO.
Commission of the European Communities (1991) Regulation No. 2568/91: On the characteristics of olive oil and olive-residue oil and on the relevant methods of analysis. *Official Journal of the European Commission* No. L248.
Fedeli E (1977) Lipids of olives. *Progress in the Chemistry of Fats and other Lipids* 15: 57–74.
Gracian Tous J (1968) The chemistry and analysis of olive oil. In: Boekenoogen HP (ed.) *Analysis and Characterisation of Oils, Fats and Fat Products*, pp. 315–606. London: Interscience Publishers.

Figure 10 (see color plate 118) Pottery from Knossos Palace. Reproduced from Psillakis N and Kastanas E (1999) *The civilization of olive: olive oil*, 2nd edn. Greek Academy of Taste, Karmanor, Iraklion, Crete (Greece), with permission.

all the researchers agreed that the level of polyphenols or of specific phenols, e.g., tyrosol level, coincided well with the overall oil quality. Thus, oils with a high initial peroxide value may have an unexpectedly long shelf-life due to a high content of natural antioxidants. In the last decade, some researchers focused their interest on the contribution of individual phenols present in virgin olive oil to its oxidative stability. Recent publications indicate that there is an interest in all natural antioxidants of olive oil that are studied in parallel. The importance of polyphenols for olive oil stability explains the efforts of investigators to develop an appropriate technology to recover them from olive oil wastewater.

Bottled virgin olive oil has a shelf-life of around 18 months. This period is very long for a fat product. The resistance of the oil to oxidation may be supported by special storage and packing conditions (see **Figure 10**).

Storage in dark places is essential for the keepability of the product to prevent the prooxidant activity

Harwood J and Aparicio R (eds) (2000) *Handbook of Olive Oil, Analysis and Properties*. Gaithersburg, MD: Aspen.

IOOC (1996) *International Encyclopedia of Olive Tree*, 1st edn. Madrid: International Olive Oil Council.

IOOC (1999) Trade Standards Applying to Olive Oil and Olive Pomace Oil. COI/T. 15/NC, Doc. No. 2; rev. 9. Madrid: International Olive Oil Council.

Kiritsakis KA (1998) *Olive Oil: From the Tree to the Table*, 2nd edn. Trumbull, CT: Food and Nutrition Press.

Kumpulainen JT and Salonen JT (eds) (1996) *Natural Antioxidants and Food Quality in Atherosclerosis and Cancer Prevention*. Cambridge: The Royal Society of Chemistry.

Newmark HL (1997) Squalene, olive oil and cancer risk: A review and a hypothesis. *Cancer Epidemiology, Biomarkers and Prevention* 6: 1101–1103.

Perrin JL (1992) Les composes mineurs et les antioxygenes naturels de l'olive et de son huile. Mise au point. *Revue Française des Corps Gras* 39: 25–32.

Ryan D and Robards K (1998) Phenolic compounds in olives (a review). *Analyst* 123: 31R–44R.

Tsimidou M (1997) Polyphenols and quality of virgin olive oil in retrospect (a review). *Italian Journal of Food Science* 10: 99–116.

OLIVES

B L Raina, Regional Research Laboratory (CSIR), Jammu, India

Introduction

Olive (*Olea europaea*, family Oleaceae) is an evergreen xerophytic tree grown for its drupes, which yield oil and are also marketed as table or pickled olives. The olive seems to have been cultivated since prehistoric times in the eastern Mediterranean region, and has been closely associated with human religious, sociocultural, medicinal, and nutritional needs. Today, olive cultivation has attained commercial crop status in every country with a Mediterranean climate. Spain, Italy, and Greece are leading producers of olives. Other important olive-producing countries are Turkey, Tunisia, Portugal, France, Morocco, Algeria, Syria, Yugoslavia, Jordan, the USA, Cyprus, Israel, Argentina, and Libya.

Production and Trade

There are about 800 million olive trees in the world, covering about 9 million hectares (ha), with the annual olive fruit production being about 13.7 million metric tonnes (t). Approximately 720 000 t of the production are consumed as table olives, and the remaining 8 680 000 t are used for the extraction of oil (**Table 1**). Nearly 2.3 million t of olive oil and 230 000 t of olive residue oil is produced globally. About 260 000 t of olive oil and 200 000 t of table olives is traded in international markets. Olive oil is exported by 35 countries to more than 100 countries in five continents. Spain is the largest exporter and

Italy buys nearly half of the entire quantity traded in the international market. Of the 300 000 t of olive oil produced in Greece, around 65% is consumed on the domestic market, whilst 100 000 t is exported, mostly to the European Union (EU). Seventy-five percent of all the olive oil produced in Greece is edible virgin olive oil, for which the global market is expanding. With its 4.2% share of world olive oil production and 8.7% share of world exports. Turkey ranks fourth and in some years fifth after Tunisia. The annual increase in olive oil production is 1.4% as compared to 4% in respect of other edible oils.

As the world's largest producer, Spain produces between 200 000 and 250 000 t of table olives per year. This is over 60% of EU production. Half is for the domestic market and half is for export. Greece is the second largest producer of table olives and exports of the typical table varieties are greater than

Table 1 Major production of olives and olive oil

Country	Fruit production (1000 MT)			Oil production (1000 MT)		
	1996	1997	1998	1996	1997	1998
Spain	4506	5686	3564	1037	1164	777
Italy	2147	3591	2680	420	698	497
Greece	1950	1879	2068	470	457	427
Turkey	1800	520	1550	197	56	56F
Tunisia	1550	500	1000F	329	96	210
Morocco	908	518	650	87	50	66
Syria	648	403	763	140	85	121F
Portugal	284	267	287F	46	43	43
Algeria	313	319	124	50	49	47F
Argentina	92	92F	92F	4	12	12F
World	15 308	14 702	13 757	2848	2770	2316

F, Food and Agriculture Organization estimate.
Source: www.fao.org.

sales on the home market. The production of table olives in Italy is around 86 397 t per year. Table olives account for about 3.4% of the annual crop, and barely 10% of world production. Turkey is one of the world's biggest producers and consumers of table olives, and lies sixth in the export ranking. It is the world's top producer of black olives (29%) and their second biggest exporter (19%).

The crop is very labor-demanding, provides employment to millions, and accounts in many countries for a major share of agricultural income. Almost 25% of the farming income in the Mediterranean basin is from olive products. The olive products, which have been consumed by the Mediterranean people for generations, are increasingly becoming a luxury in countries with a high standard of living.

Botany

Olive is an evergreen tree 3–12 m or even more in height and, in some instances, bears fruit for 1000 years or even more; branches are numerous; leaves are opposite, leathery, lanceolate, dark green above, silvery beneath; flowers are small, white, dimorphic (male and perfect), appearing in early spring in loose axillary clusters. Fruit setting in olive is erratic in certain areas, especially where irrigation is insufficient, the plantation is dense, the pests and diseases are not controlled and, above all, a single cultivar is planted over a wide area, reducing the chances for cross-pollination.

The olive fruit is a globular, oblong, or sometimes crescent-shaped drupe. The pericarp is composed of epicarp or skin, the mesocarp or pulp, and the endocarp or stone (pit) which contains the seed. The fruit reaches its maximum weight 6–8 months after flowering late in spring, and passes through successive shades of straw, pink, and red, before finally turning purplish-black at full maturity. The ripe fruit weighs from 1.5 to 13 g. The pit with its hard shell makes up 13–30% of the fruit weight and the skin 1.5–3.5%. The seed does not exceed 3% of the weight of the fruit. At full maturity, the mesocarp contains 6–10% soluble solids and 15–40% or even more oil, depending on variety. The pericarp contains 96–98% of the total amount of oil, while the remaining 2–4% of the oil is in the kernel. The characteristic bitter glycoside, oleuropein, is more concentrated close to the peel.

Varieties

The principal varieties cultivated for oil have medium-sized fruits and are harvested when ripe. More important among these are Arauco, Bouquetier, Corfolia, Dafnolia, Fratoio, Koroneiki, Lechin, Manzanillo, Morcal, Nevadillo, Picual, Rozzola, Rougette, Smertolia, Taggiasea, Tsounati, and Zorzalena. The varieties preferred for table olive production, with less than 8% oil, include Ascolano, Calamata, Chalkidikis, Conservolea, Gordal, Hojiblanca, Manzanillo, Megariticci, Mission, Sevillano, Throumbolia, Verdale, and Volou.

Physiology and Biochemistry

Fruit Development and Ripening

Along with an increase in fruit weight, various changes in constituents occur during the development of the fruit. In the cultivar Gordal, there is a continuous increase in the oil content, a brief rise in reducing sugar content, followed by a decline till maturity, a sharp initial drop in crude fiber content and a gradual one afterwards, and a low protein and ash content. There are no qualitative changes in chlorophyl and carotenoid pigments in Hojiblanca and Manzanillo olives during growth and ripening. The pigment content in the virgin olive oil varies according to the degree of ripeness of the fruits. Initially the major component is pheophytin, while in oil obtained at the end of the season lutein is the most abundant. The loss of chlorophyl pigments in the fresh fruits and virgin oil is not due to activation of chlorophyllase during oil processing. The β-carotene content and its provitamin A value diminish with ripeness (13.2 μg β-carotene per g oil is reduced to 1.27 μg g^{-1} in Sevillano variety). The considerable range in the lutein/β-carotene ratio (between 1.3 and 5.1 depending on variety) makes this ratio a differentiator of single-variety oils. The amount of coumaric acid, syringic acid, and flavonol varies with maturity. The quantitative differences in the composition of flavonol and flavone glycosides during ripening may be useful in the biochemical characterization of olive cultivars. Cultivar and stage of maturity influence tocopherol and tocotrienol contents. The concentrations of the volatile compounds during three stages of ripeness in four different cultivars are reported to decline with increasing ripeness, although hexyl acetate is highest and hexan-1-ol lowest at the ripe stage. During maturation of the fruit, the size of oil droplets within the olive increases, raising the extractability of the oil from about 20% in the early stage to as much as 90% in fully mature fruit. Furthermore, small fruits are characterized by high oleuropein and low verbascoside contents. The gradual loss of firmness and of anhydrogalacturonic acid content of the pectic chain is associated with increasing enzyme activities. Pectinesterase activity appears during ripening and polygalacturonase activity mainly during storage. The activity of cellulolytic enzymes increases with ripening of the fruit. (*See* **Colorants** (**Colourants**): Properties and Determination of Natural Pigments; **Ripening of Fruit**.)

Constituents of Olive Fruit

The moisture and oil content form 85–90% of the weight of pulp, while the rest comprises organic matter and minerals. The major monosaccharides in fruit pulp are glucose, mannose, xylose, galactose, and arabinose, accompanied by mannitol and rhamnose in certain cultivars. The pulp is rich in potassium. Small amounts of organic acids, such as citric, malic, oxalic, malonic, fumaric, tartaric, lactic, acetic, and tricarballylic (1,2,3 propanetricarboxylic acid) are present. (*See* **Acids**: Natural Acids and Acidulants; **Carbohydrates**: Classification and Properties.)

A variety of phenolic compounds are found in olive pulp; caffeic and ferulic acids are among the simpler of them. The major orthodiphenolic compound typical of olives is oleuropein, which is responsible for the bitter taste of immature olives. Two linear compounds are reported to be isolated from the ethyl acetate extract of residues resulting from Spanish olive oil processing. These compounds are identified as 3-(1-(hydroxymethyl)-(E)1-propenyl) glutaric acid and 3-(1-(formyl)-(E)1-propenyl) glutaric acid. These products are structural components of oleuropein. Oxidation products of oleuropein and other phenolics impart the black color to the fruit as a result of complex interactions between diphenol oxidase and oleuropein. Anthocyanins, specifically the glycosides of cyanidin and peonidin, are responsible for the purple and black color of ripe olives. The highest concentration of anthocyanin pigments has been observed in the epicarp ($0.55\,g\,kg^{-1}$); a complex mixture of flavonoids is found in the mesocarp; derivatives of luteolin and apigenin are found in the endocarp. (*See* **Phenolic Compounds**.)

The skin, pulp, and seed contain different lipid fractions and fatty acids, sterols, triterpene alcohols, dialcohols, and hydrocarbon fractions. Among the differences between the parts of the olive drupe are the presence of erythrodiol and traces of uvaol, oleanoic acid, traces of ursolic acid, and oleanoic aldehydes in the skin only. The distribution of unsaponifiable hydroxy compounds fractions within the drupe and their concentration in oils from various parts of Lingurian olives reveal the presence of sterols, 4-methylsterols, triterpene alcohols (TTA), triterpene dialcohols (TTDA), linear saturated alcohol (LSA), and phytol. Most of the sterols, 4-methylsterols,

TTA, and phytol are found in the pulp, while LSA and TTDA are found in the peel. The main hydroxy components are LSA in the peel, TTA in the pulp, and sterols in stone and seed. The kernel oil is characterized by the presence of estrone ester ($8\,\mu g$ per 100 ml oil).

Fifty-six volatile compounds are reported to be present in the leaves, flowers, epicarp, and mesocarp of Lucca and Mission varieties, and 29 compounds, including 3-vinylpyridine, methylvinylpyridine, and *cis*-jasmone, have been identified in these varieties. Higher concentrations of hexanal and *trans*-2-hexanal are found in the Lucca variety, while numerous long-chain hydrocarbons are found in flowers of both varieties. The peroxidase activity and lipid peroxidation intensity in leaves may be of potential use in selecting olive cultivars for yield.

Soil and Climate

Olive does well on moderately saline soils not suitable for most other crops. The soil should be deep and well-drained, so that the roots penetrate deeply to take advantage of underground moisture even in relatively arid areas. The olives can also grow on calcareous slopes provided the fertility of the soil is built up.

Olive can be grown commercially in warm-temperate and subtropical countries located between latitudes 30° and 45°N and 30° and 45°S, at 300–400 m altitude in the north to 1000–1220 m above sea level in favorable sections of the south. The production of olives under dry conditions needs better resource management. This includes use of special tools and traditional water conservation measures. It cannot be grown where the winter temperature falls below $-9\,°C$, as this temperature will kill the trees. A certain degree of chilling, differing considerably among cultivars, is necessary for floral induction. The olive blooms relatively late in the spring. A long, relatively hot growing season is required for the fruit to develop to the appropriate size. A consistently clear and dry atmosphere is necessary for maximum yield of fruit. Adequate soil moisture must be present throughout the year to produce heavy yields of the required size of fruit.

Cultural Practices

Propagation

Olives are propagated by hardwood cuttings, or by small softwood cuttings rooted under mist sprays, or by 'ovule' protuberances formed at the base of the trunk. The treatment with IBA+ putrescine HCl and placing the cuttings in the polyethylene tunnel improve the rooting ability. Although propagation by seeds provides a stronger root system and more fruitful trees, the seedlings produced require grafting and thus a longer time to reach fruiting. The variable nature of the rootstock may bring about variability in the growth and size of the trees. The nursery trees are planted 8–12 m apart in irrigated orchards or 12–22 m apart in unirrigated groves.

IRTA-i.18, a clone selection of the olive Arbequina, is superior to the cultivar for fruit color, fruit yield at a density of 300 trees h^{-1}, crown volume, fruit weight, flesh/stone ratio, and oil content, and identified as suitable for commercialization. The topical pollen application can be used instead of interplanting of pollinizer trees to reduce shotberry incidence in Manzanillo olive groves.

Irrigation and Drainage

For production of a good crop of large fruit size, the trees should have adequate soil moisture throughout the year, especially before flowering time in late winter and early spring and during the summer, when fruit growth takes place and new fruiting wood develops. In the Mediterranean region both oil and pickling olives are largely grown without irrigation. Under rain-fed conditions the olives require 40–50 years to reach full productivity, necessitating the use of many intercrops to provide income during the interim period. The yield of full-bearing olive trees in Spain is about 825 kg ha^{-1}, and water is the main limiting factor for low yield.

In California, the olive groves are grown under irrigated conditions. To insure adequate tree growth and profitable yields, supplementary summer irrigation is necessary. Olive is very sensitive to excessive soil moisture. Poor drainage at any time, if prolonged, will kill the tree.

Fertilizers

The olive tree can obtain the required amount of potassium and phosphorus from the soil. However, nitrogen is utilized by the plants more rapidly, necessitating the application of nitrogenous fertilizers. Under irrigated conditions in California, the growers apply 450 g nitrogen per tree per year just before onset of the winter season. In gravelly, shallow, foothill types of soil, the trees respond well to potassium and boron. In Greece, a fertilizer mixture containing nitrogen, phophorus, and potassium is used.

The organic method of cultivation using *Hedysarum coronarium* cover crop inoculated with a granular formulation of *Rhizobium hedysarii* to supply nitrogen to olive significantly increases the total and organic nitrogen in the soil and has an impact on olive productivity.

Pruning and Thinning

Pruning methods for the bearing trees vary greatly. In the Mediterranean region, heavy pruning is practiced. Heavy pruning under irrigated conditions leads to reduced yields owing to the removal of potential fruit-bearing surfaces, and results in fruit bearing in alternate years. A moderate amount of annual pruning, involving removal of injured, dead, or ill-placed branches, is all that is necessary to spread the fruiting surface uniformly, so that every part of the bearing wood is adequately exposed to light and air. The height of plants in regular plantation is generally maintained at up to 5 m. In some countries, thinning of fruits is achieved by spraying the young fruits with 150 p.p.m. naphthaleneacetic acid. Spray-thinned trees may be expected to bear fruit the following year. The girdling with GA3 plant regulator application (50 or 100 p.p.m.) increases the fruit yield.

Pests and Diseases

Olives have many pest problems, some of them affecting the tree or the fruit itself. Olive fruit fly (*Dacus oleae* Gmel.), olive kernel borer (*Prays oleae* Bern.), olive scale (*Parlatoria oleae* Colvee.), Oleander or ivy scale (*Aspidiotus hederae* (Vallot), greedy scale (*A. camelliae* Signoret), California red scale (*Aonidiella aurantii* (Maskell), and black scale (*Saissetia oleae* Bern.) are the important pests of olive. Two recent reports of beetles injurious to the olive tree involve *Nericonia nigra* and *Noemia semirufa*.

The olive fruitfiy (*Dacus oleae* Gmel.) is a major pest in the Mediterranean region. The insect attacks at early fruit maturity, causing fermentation and premature fruit dropping. Sometimes, owing to adverse climatic conditions, the production loss due to this fly can be as high as 30%. The infestation adversely affects the oil quality by increasing acidity, especially when infestation exceeds 30%. The pest is best controlled by proteinaceous attractant sprays. (E)-2-hexanal emitted by the crushed olives has repellent property to the fruitfly. Chrysopids are the most active and effective predators on olive kernel borer (*Prays oleae* Bern). The pesticides deltamethrin, fenthion, and formothion are also used to control *Dacus oleae*. The use of green immature olives for pickling has become popular in Spain as a means of escape from the olive fly.

The principal disease of the olive tree is 'olive-knot,' caused by *Bacterium savastanoi* Stevens, which leads to knots or tumors on the leaves, twigs, branches, and trunk. There is no proper control for this disease, except to avoid olive varieties which are susceptible to cold injury and to this organism, or by quarantine. The root-knot, citrus and reniform nematodes can be suppressed by less irrigation frequency and predacious nematode in olive.

Peacock spot (*Cycloconium oleaginum*) and *Verticillium* rot (*Verticillium alboatrum*) are the fungal diseases of olive, occasionally causing heavy losses in production and fruit quality. The problem is serious in humid areas, but can be controlled by various fungicides such as Bordeaux mixture or lime sulfur. Soil solarization of individual diseased trees, when combined with chemical weed control, is effective in controlling *Verticillium* wilt in olive trees. (*See* **Fungicides**.)

Harvesting

Traditionally, the olives are harvested by hand into a receptacle in the autumn or early winter at the appropriate green-yellow stage. Olives for oil extraction are collected by beating the branches of the tree with long poles so that the fruit falls on to cloth sheets laid on the ground. This is done after the fruit has turned black. Beating of the trees tends to promote the spread of olive-knot disease through injury to the fruit and tree. In Greece and elsewhere, the scarcity of farm labor has resulted in the introduction of plastic nets for collection of naturally falling fruit. To obtain better-quality oil, the olives should not be left on the nets for more than 15 days. Chemical sprays to induce fruit abscission are used to aid the mechanical harvesting of olives. Mechanical shakers with stationary cloth, canvas-covered, or mechanized collecting frames are used to a limited extent, mostly for harvesting olives for oil. A 'beating-down' machine for mechanical picking of olives without damaging the trees has been tested in Italy. The machine is particularly suitable for high-yielding trees with well-developed foliage, and for old trees, for which vibrating picking machines are not recommended.

For high-quality table olives, where uniformly ripe fruit is a requisite, the fruit is handpicked using ladders. The finest and most highly prized oil is made from olives picked just before they begin to soften and darken. The time of maturity may vary within the cultivated area, among the trees within the same grove, or within the fruits on the same tree.

The measurement of Brix values allows for rapid, inexpensive estimation of the oil content of olive fruits at the harvest site and is useful in determining optimum harvest time. The respiration rate is another new ripening index for drupes and the harvesting of olives is suggested to be conducted during the climacteric phase of the fruits

Postharvest Handling and Storage of Fruit

Ideally, the extraction of oil should follow the harvesting of the olives without delay. As this is impractical, the olives must be stored for a period before processing. During storage, several chemical and biochemical changes may occur that lead to the deterioration of the oil. The most serious damage is caused by enzymes indigenous to the olive tissue and those produced by the microflora on it. The pathogenic and saprophytic fungi isolated from stored olives, except *Saccharomyces* spp., increase the acidity of the extracted oil, and no aflatoxins are reported to be detected in samples infected by *Aspergillus flavus*. As the olive tissue respires, heat is produced, which, if not effectively dissipated, accelerates the enzymatic actions. Lipolysis, lipid oxidation, and other undesirable reactions occur, leading to a lowering of the oil quality.

To minimize the effect of these activities, the olives are stored in 25-cm-thick layers in cool buildings. The use of perforated trays for spreading and stacking saves space. The olives are better stored under water in tanks containing mild preservatives, such as 3% salt or 0.03% citric acid and 3% salt, or 2% metabisulfite. Aerobic storage with a low concentration of acetic acid gives stable color and texture even after 3 months; bulk storage can be done in 50% ethanol. Aerobic storage and the maintenance of a CO_2 concentration of 10–15 mg per 100 ml throughout the storage period eliminates shriveling. Fruit maturation is delayed by refrigeration compared with storage at room temperature and the firmness of the fruits stored at room temperature decreases much more rapidly. The sensory and chemical qualities of the oil produced from fruits stored at 5 °C can be maintained for fruit storage durations up to 45 days, but with storage at 8 °C, these qualities are maintained for storage durations up to 15 days. For fruits stored at room temperature, oil quality deteriorates after 7 days of fruit storage. The processing period of oil mills can be extended by freeze-storing of olives at −18 °C for 90 days. This also helps to increase oil extraction. (*See* **Preservation of Food**.)

Processing and Utilization

Oil is extracted traditionally by simple equipment involving crushing, pressing, and separation of oil from the liquor. It is now being replaced by modern mechanical equipment of widely varying designs, either as separate units or combined as a single-stage process. The olive oil recovery processes have been patented in many countries.

The oil obtained from sound fruit by pressing without further treatment is called virgin olive oil. After the first pressing, the residual pulp is still rich in oil and is usually recrushed and repressed with or without the addition of warm water. Oil obtained from the second pressing tends to have a more intense color and higher acid content, as well as a weaker aroma. This latter oil, together with inferior virgin oil, is further subjected to refining treatments such as neutralization, deodorization, bleaching, and winterization; this helps in removing acid, color, and odor. The oil so obtained is called refined olive oil, which is largely used for mixing with the first extraction to produce edible grades. Refining procedures decontaminate oil containing fenthion residues. The remaining press cake still retains about 4–5% oil, but more than 50% of the total oil in the cake can be obtained in a single step by direct extraction in a continuous centrifugation system, after adding about 1% Na_2CO_3 by weight of cake. It is then necessary to dry and further deoil the cake for better stability towards autoxidation. The dried cake is extracted with hexane and the resulting 'sulfur oil' is repeatedly rectified. The extraction with hexane results in an oil with higher acidity and moisture content, but a lower percentage of impurities.

Olive oil has been classified by the United Nations Conference on Olive Oil, held in Geneva in 1961, into four groups according to the method of preparation and acid content:

1. Virgin olive oil. Oil extracted by pressing, free of admixtures, called 'extra' when the oleic acid does not exceed 1 g per 100 g, and 'fine' if the acid content does not exceed 1.5 g per 100 g and the flavor is perfect; 'ordinary' olive oil may contain acid up to 3.0 g per 100 g and have a slight off-flavor; if the oil has a definite off-flavor it is classified as 'lampante.'
2. Refined olive oil. This oil may be called 'pure' when it is refined from virgin oil, and 'second-quality' when it is refined from solvent-extracted oil.
3. Blended olive oil. Blended oil can be called 'pure' when it consists of a blend of virgin and refined oil, and 'blended' when it contains a blend of virgin and second-quality refined oil.
4. Industrial oils. These oils are obtained by extraction of olive residue with solvents.

Storage and Packing of Oil

Olive oil undergoes deterioration during storage. It is important to store the oil extracted from olives of different varieties separately. The storage drums and tanks should be of inert material, impermeable to oil, and lined with epoxy resins, enameled tiles, or glass; and the oil should be stored at a constant temperature of about 15 °C.

Olive oil is packed in several types of retail containers including bottles made of glass, polyvinyl chloride (PVC) or polyethylene, tin-plated cans, and in tetra-briks. The containers should provide protection of the oil from light, and leave a minimum volume of head space. Packing of oil under vacuum and inert gases is also recommended.

Table Olives

Many different methods of converting olives into olive products exist in different olive-growing countries. The traditional method of preparing table olives by removing their bitterness through alkaline treatment, which grew up in the environs of Seville, Spain, is rapidly being replaced by technically advanced methods. Production has been increased and costs have been cut, and the industry has remained highly competitive. Sevillian green olives are the predominant type used (70%), followed by oxidation-darkened olives (25%), and untreated black olives (5%). The chief innovations in the last 25 years have been mass fermentation, development of conservation in an aerobic medium, development of automatic stoning and stuffing machines, developments of pastes, fixing of pack conditions and treatments, and general mechanization of all handling, packing, and storage of the end product.

The major commercial methods are as follows.

Green Fermented Olives (Spanish Method)

The fruits are picked when they are still firm and light green in color. The principal varieties processed in Spain are the Sevillano and Manzanillo. The fruit is covered in large containers with 1.8% lye solution and kept at 28 °C for 4–8 h. The treated olives are frequently washed with water for 24–36 h.

The washed fruits are then covered with brine in fermentation tanks. This changes the color of the olives from deep green to the typical 'olive-green' shade. The product, on attaining equilibrium, contains 6–8% NaCl. Spices may now be added to the fermented olives according to consumer preference. Closed containers help prevent the development of surface yeasts and molds.

The fermented olives, before marketing, are graded according to shape, size, and color. Sometimes the olives are pitted and stuffed with pimentos, onions, almonds, anchovies, or other products. The fermented brine is filtered (or replaced by fresh brine) for covering the olives in bottles or cans prior to pasteurization.

Olives at different stages of ripeness suffer major losses during the pitting process; green unripe olives are excessively firm, and the cell wall of changing-color olives is too disorganized to give good industrial results. The treatment with high concentration of lye and its 100% penetration into the flesh, irrespective of fruit size, results in significant textural decline and changes in the composition of polysaccharides that are most important to the cell wall structure.

Among the major spoilage organisms, zapatera – a butyric acid producer – is important.

Canned Ripe Olives (American Style)

The principal commercial varieties packed are the Ascolano, Manzanillo, Mission, and Sevillano. Straw-yellow to cherry-red olives are graded according to color and size to insure uniform lye penetration. The fruits are placed in 5.8% brine for 6 weeks or longer in large wooden or concrete tanks, and the brine concentration is gradually raised to 10%. The fruit is then transferred into shallow vats and treated 4–8 times with gradually decreasing strength of lye (3.0–0.5% NaOH). Each treatment is followed by exposure to air for 1–5 days, either with frequent turning in the presence of air or by bubbling air through the immersed fruit. The olives are then leached by frequent changes of water for 5–7 days until all traces of NaOH are removed. (*See* **Canning: Principles.**)

The washed olives are pasteurized, cured in 2–3% NaCl for 2–6 days, graded, and packed into enameled cans covered with 2.5–3.5% NaCl solution, sealed and sterilized at 116 °C for 1 h. Pitting and stuffing of olives are often practiced.

Black (Naturally Ripe) Olives (Greek Method)

The Greek black olive industry uses naturally ripe, fully matured and dark-purple fruits of mainly Calamata, Conservolea, and Magaritici varieties. The fruits are covered with brine, and the concentration is gradually increased to prevent shriveling. The covering brine concentration is kept at 10% NaCl during winter and increased to 15–16% during summer to prevent spoilage. The bitterness is lost within 3–6 months, and the lactic acid formed during fermentation does not exceed 0.5%. The dark-purple anthocyanins of the olives turn light red during fermentation.

The olives, while being prepared for marketing, are exposed to air until the dark color is regained. The fruits are carefully sorted and graded according to international standards, and repacked in fresh brine containing about 8% NaCl and 0.5–0.75% acetic acid.

Antifungal treatment trials of 'Greek-style' black olives indicate that sorbic acid 0.075% is the most efficient preservative, followed by 0.075% benzoic acid and 0.032% calcium proprionate. In the olives

inoculated with *Aspergillus parasiticus*, the preservatives totally inhibit aflatoxin formation, while the toxin can be detected at low concentrations in all controls.

The spoilage organisms encountered are zapatera, galazoma, and film-forming pectolytic organisms which disintegrate the 'meat' of the fruit. This last problem can be prevented by means of a thin layer of paraffin oil on the surface of the brine.

Utilization of Byproducts

Olive Cake or Meal

The press residue may be used as a fuel, manure, animal feed, soil conditioner, fiber (cellulose and lignin) for food use, or in single-cell protein production. Sludge obtained from olive-processing plants enriched with nitrogen fertilizer can be applied to olive orchards growing on alkaline alluvial soils and as manure for maize. The spent form of olive oil cake can be incorporated into the diets of rabbits at 30% level. Mixtures of organic acids of low molecular weight and fatty acids can be recovered from the waste gases produced during drying of cake. *Pleurotus eryngii* can be grown on a medium containing olive husks.

Olive Stones

The olive stones find use in molded products, plastics, and furfural manufacture. Activated carbon produced from olive stones has a high adsorption capacity against lead.

Olive Vegetation Water

Tocopherols, flavor compounds, antioxidants, and anthocyanins can be recovered from olive vegetation water. Olive oil deodorization distillate contains squalene in a concentration range of 10–30%. Squalene in high purity with 90% yield can be recovered by supercritical carbon dioxide extraction. The application of 100–200 m^3 waste waters from olive oil-processing plants per hectare, 45 days after sowing, markedly increased barley and sunflower yields and was reported to have no adverse effects on soil microorganisms. (*See* **Antioxidants:** Natural Antioxidants; **Flavor (Flavour) Compounds:** Structures and Characteristics; **Tocopherols:** Properties and Determination.)

Nutritional and Health Aspects

The use of olive oil, whether in medicine or as source of nutrition, has its roots in ancient history. During the last few decades, renewed interest has been generated in the nutritional and health aspects of the olive. The oil is used largely for culinary purposes, in wool combing, the manufacture of toilet preparations, and in the pharmaceutical industry.

Besides its physiological advantage in being rapidly and completely digested, the oil also has clinical importance e.g., antiulcer activity, effectiveness in combating gallbladder disease, and in lowering plasma cholesterol level. Omega-6-polyunsaturated fatty acid in olive oil may exacerbate adjuvant-induced arthritis. Olive pollen extract can induce an asthmatic response. (*See* **Fatty Acids**: Dietary Importance.)

See also: **Acids**: Natural Acids and Acidulants; **Antioxidants**: Natural Antioxidants; **Canning**: Principles; **Carbohydrates**: Classification and Properties; **Colorants (Colourants)**: Properties and Determination of Natural Pigments; **Fats**: Classification; **Fatty Acids**: Properties; Dietary Importance; **Flavor (Flavour) Compounds**: Structures and Characteristics; **Fungicides**; **Phenolic Compounds**; **Preservation of Food**; **Ripening of Fruit**; **Tocopherols**: Properties and Determination

Further Reading

Alves Mdo CFGP and Pinheiro Alves Mdo CFG (1991) Rationalization of national olive oil production

capacity: a critical review, from harvesting to storage of olive oil. *Revista de Ciencias Agrarias* 14(4): 59–71.

Codex Alimentarius WHO/FAO (1970) Recommended international standard for olive oil, virgin and refined olive residue oil. CAS/RS-33.

Damania AB (1995) Olive, the plant of peace reigns throughout Mediterranean. *Diversity* 11(1–2): 131–132.

di Giovacchino L and Di Giovacchino L (1997) From olive harvesting to origin olive oil production. *OCL Oleagineux, Corps Gras, Lipides* 4(5): 359–361.

FAO (1998) *Production Year Book*, vol. 52. Rome: Food and Agriculture Organization. (www.fao.org)

Fernandez Diez MJ (1971) The Olive. In: Hulme AC (ed.) *Biochemistry of Fruits and Their Products*. New York: Academic Press.

Katsoyannos P (1992) *The Olive Pests and their Control in the Near East*. FAO Plant Production and Protection Paper no. 115, ix + 178 pp. Rome: FAO.

Kiritsakis A and Markakis P (1987) Olive oil: a review. *Advances in Food Research* 31: 453–482.

Mangold HK and Fedeli E (1997) Olives, olive oils and the Mediterranean diet. *Rivista Italiana delle Sostanze Grasse* 74(8): 349–352.

Santos Antunes AF (1980) Technology of olive oil production. *Boletim Instituto de Azeite e Productos Oleaginosis* 8(1): 23107.

ONIONS AND RELATED CROPS

S Khokar, University of Leeds, Leeds, UK
G R Fenwick, Institute of Food Research, Norwich, UK

Introduction

The genus, *Allium* (family Alliaceae), is large and geographically diverse, comprising over 500 species, most of which have little commercial importance. Alliums are distributed across warm temperate and temperate regions and into the subArctic belt. The major centers of genetic diversity are Afghanistan, Iran, and Pakistan with the Mediterranean basin being regarded as a secondary center. In addition to the onion (*A. cepa* L.), cultivated vegetable Alliums include garlic (*A. sativum* L.), leek (*A. ampeloprasum* L.), Japanese bunching onion (*A. fistulosum* L.), rakkyo (*A. chinense* G. Don), Chinese chives (*A. tuberosum* Rottl. ex Spr.), chives (*A. schoenoprasum* L.), and shallot (*A. cepa* L.).

The onion is by far the most important of the cultivated Alliums. Production is mainly in the Northern Hemisphere, with relatively little in unfavorable tropical regions (where shallots are grown). Cultivation experience, cultural attitudes, social beliefs, and culinary practices have resulted in specific Alliums being prized in particular regions; for example, leek and shallots in Western and Northern Europe; kurrat in Egypt and the Eastern Mediterranean; rakkyo, Japanese bunching onion; and garlic in China and Japan. The dramatic increase in global travel, interest in exotic and ethnic foods, and associated culinary practices, allied to improvements in agronomic practices and varietal- and postharvest characteristics have combined to encourage demand for fresh and processed vegetable Alliums. This has been intensified by prevailing enthusiasms for both 'healthy' eating and naturally occurring health remedies and medicines.

Onions and other cultivated Alliums have long been prized for their culinary properties, flavor, and taste and for their use as natural medicines. More recently, their nutritional and health-enhancing properties have attracted much scientific investigation and consumer interest, thereby encouraging primary production and stimulating postharvest activities and opportunities. This brief article addresses the composition, flavor, postharvest handling, and processing of Alliums, emphasizing their significance for human health and well-being. The focus is mainly, but not exclusively, on onion and garlic.

History

The onion has been cultivated for over 4000 years; the earliest records are found in Ancient Egypt, including carvings dated to the third and fourth dynasties (2700 BC). In addition to being eaten, onions had significance in funeral rituals and embalming. During their Exodus (1500 BC), the Israelites recalled the foods of Egypt, including onion, leek, and garlic; mention was made of onions in Indian manuscripts dating from the sixth century BC. Locally grown onions were among the fermented vegetables given to the builders of the Great Wall of China (third century BC). Greek and Roman authors, notably Theophrastus and Columella, described the botany and cultivation of the onion, whereas others, such as Hippocrates and Dioscorides, focused on its medicinal properties. These culminated in the writings of Pliny, which addressed the cultivation, uses, and history of onions, garlic, and leek, and, together with the work of the military physician Galen, these had a great influence on the practice and writings of herbalists in medieval times and later. More extensive information has been published in the reviews of Fenwick and Hanley.

Production

Onion production in Asia accounts for more than 60% of the global production; together, India and China account for more than a third; in comparison, European production is less than 10%. Asian production of garlic exceeds 80% of the global volume, with China alone producing almost two-thirds. The world production of dry onions and garlic in 1999 is summarized in **Table 1**.

Composition

Onions and garlic are found in kitchens all over the world, being used as cooked vegetables, in salads, and as a flavoring in meat, poultry, and vegetarian dishes. Onions additionally possess excellent processing characteristics and may be canned, pickled, frozen, and fried. Dehydrated onion and garlic powders and

Table 1 Production of dry bulb onions and garlic ($\times 10^3$ tonnes)

Country	Onion	Garlic
World	41 527	9 042
Asia	25 511	7 585
Africa	2 780	178
North and Central America	3 370	301
South America	2 796	286
Europe	3 454	266
People's Republic of China	11 290	5 990
India	4 429	451
USA	2 995	224
Turkey	2 300	106
Iran	1 210	
Japan	1 240	
Republic of Korea	936	484
Brazil	921	63
Spain	985	170

From FAO Production Data (1999).

flavor extracts also serve to enhance the value of these Alliums in the home and in the food industry. The proximate composition of selected Alliums is listed in **Table 2**. The cultivar, agronomic and environmental conditions during growth, and postharvest conditions all affect the composition. Protein, fat, and fiber contents are low in comparison with other vegetables, and starch is absent. Carbohydrates include sucrose, glucose, fructose, and fructose polymers. Green onion tops have also been found to be rich in β-carotene and to contain significant amounts of vitamin C. Alliums are rich in amino acids and γ-glutamyl peptides (including precursors of the characteristic pungent flavoring compounds), as well as biologically active secondary metabolites such as anthocyanins, flavonols, and phenolics.

Flavor

Alliums are rich in sulfur-containing compounds that are broken down by endogenous enzymes to yield a number of more volatile, less stable chemical compounds. These are considered to be responsible for much of the biological activity found in Allium extracts and oils, as well as determining the particular flavor characteristics of individual members of the family. (*See* **Flavor (Flavour) Compounds**: Production Methods.)

Flavor Biochemistry

Intact alliums have no pungency, since the volatile products are only released following the interaction of the enzyme, alliinase, with the S-alk(en)ylcysteine sulfoxide (alliin, I) which occurs when tissue is damaged or disrupted (**Figure 1**). The initial products of this enzymic hydrolysis are ammonia, pyruvate, and

Table 2 Composition of cultivated alliums (fresh-weight basis)

	Water (g)	Energy (kcal)	Protein (g)	Fat (g)	Carbohydrates (g)	Sodium (mg)	Fiber (g)	Carotene (µg)	Sodium (mg)	Iodine (µg)	Folate (µg)
Onion, fried	65.7	164	2.3	11.2	14.1	10.0	3.2	40	4	6	38
Onion, raw	89.0	36	1.2	0.2	7.9	5.6	1.5	10	3	3	17
Garlic	64.3	98	7.9	0.6	16.3	1.6	4.1[a]	0	4	3	5
Leeks	92.2	21	1.2	1.2	2.6	2.0	2.4	575	6		40
Chives	91.0	23	2.8	0.6	1.7	1.7	1.9[a]	2300	5	1.6	45
Rakkyo	86.0	50	2.2	0.3	12.0						

[a]Nonstarch polysacchrides.
From Bender AE and Bender DA (1999) *Food Tables & Labelling*. Oxford: Oxford University Press.

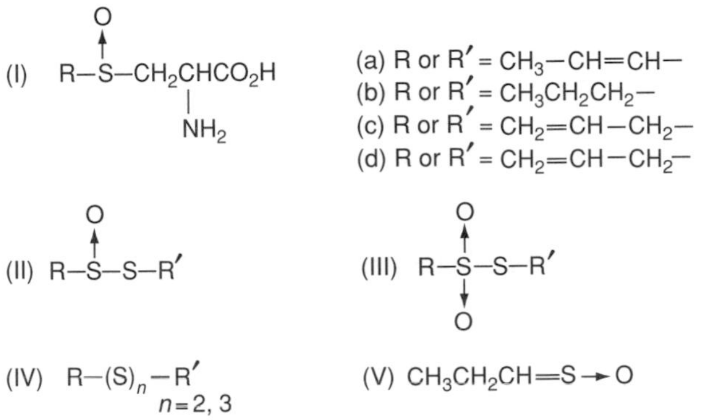

Figure 1 Sulfur-containing flavor volatiles from alliums.

an alk(en)ylthiosulphinate (allicin, II). The latter, which possesses odor characteristics typical of the freshly cut tissue, can undergo further nonenzymic reactions to yield a variety of compounds, including thiosulfinate [III] and di- and trisulfides [IV]. These compounds have slightly differing flavors and odors, and may impart a cooked note to steam-distilled onion or leek oils.

Not all alliums contain the same substituted cysteine sulfoxides; onion contains primarily the S-(1-propenyl), S-propyl, and, to a lesser extent, S-methyl derivatives (Ia–c, respectively). Thus, the typical flavor of onion is due to the presence of propyl- and 1-propenylthiosulphinates and corresponding di- and tri-sulfides. In contrast, garlic contains predominantly S-(2-propenyl)cysteine sulfoxide (S-allylcysteine sulfoxide, Id) and, and none of the 1-propenyl isomer. Its flavor is consequently due to the presence of 2-propenylthiosulphinate and related di- and tri-sulfides. Hive and leek contain relatively more S-methylcysteine sulfoxide, which yields methylthiolsulfinate (possessing a 'cabbagy' note), dimethyldisulfide, and related compounds.

Since it is possible for both symmetrical (II, R = R') and mixed (II, R ≠ R') thiolsulfinates to be formed during the enzymic breakdown of alliins, and since these can yield all of the breakdown products mentioned above and which can subsequently enter into secondary chemical reactions, the composition of the flavor volatiles of freshly chopped onion, leek, and garlic can become very complex. Almost 100 compounds have been identified in freshly cut onion and garlic, and processing – especially that involving thermal treatment – causes additional chemical reactions to occur. These give rise to thiphenes, polysulfides and the products of amino acid–sugar (Maillard) reactions.

Lachrymatory Factor

Alone amongst the common alliums, the onion contains large amounts of S-(1-propenyl) cysteine sulfoxide (Ia). This compound is unique because the action of alliinase does not give rise to allicin, but rather to a mixture of unstable isomers of thiopropanal-S-oxide (V), which comprise the lachrymatory principles of the onion. The mode of formation of these compounds and their properties suggests that the problems of peeling onions may be reduced if carried out under running water, or if the onion is first chilled.

Estimation of Flavor Strength

The development of simple and reliable methods for the assessment and comparison of strength of flavor in onions and garlic, which may be required to identify cultivars with advantageous processing characteristics, has been the subject of much investigation. Since the enzymic cleavage of flavor precursors by alliinase generates a range of breakdown products, these have been used to measure pungency. The lachrymatory principle has been measured colorimetrically and spectrophotometrically, but these determinations are necessarily confined to onions. A robust method for pyruvate employs the color reaction with 2,4-dinitrophenylhydrazine; however, this cannot be employed for dehydrated products due to the presence of interfering by-products formed during dehydration and storage. Gas-chromatographic methods, with initial flavor concentration, have also been used to measure the headspace volatiles. The amount and composition of the volatile fraction is largely dependent upon the conditions (degree of tissue disruption, temperature, time), so that all of these variables must be kept constant if reliable comparisons are to be conducted. Alternative approaches involve measurement of alliinase or γ-glutamyltranspeptidase activity. It is possible to monitor the flavor precursors, including γ-glutamyl peptides and alliins; however, such methods provide an indication of *potential* flavor, since the presence of the appropriate enzymes is necessary to release the volatile compounds.

Factors Determining Flavor

Amongst the many factors that have been found to have an effect on flavor are genotype, physiological age and condition, storage, and agronomic and environmental conditions. Thus, the range of flavor intensity in onions may vary 10-fold, with the most pungent being selected for processing. Sprouting of onions during storage leads to an increase in flavor as a result of a rise in transpeptidase levels and the conversion of γ-glutamyl peptides to alliins. Alliums grown under dry or arid conditions are smaller and visually less attractive than those grown in an abundance of water, but generally possess stronger flavors. The application of sulfur to the soil during growth significantly increases flavor, pungency and (in onions) lachrymatory character. The available sulfur is used preferentially for plant growth; only when these growing requirements are met is the secondary metabolic pathway, leading to alliin and allicin, enhanced.

Nutritional and Health Aspects

Garlic bulb and onions, two members of the Allium family, have been widely reported to possess a significant antioxidant activity which has been ascribed to allicin, a known scavenger of peroxyl radicals. Recent studies have indicated that, in addition to this scavenger, other compounds could also be involved. The onion contains a number of quercitin, isorhamnetin, and kaempferol conjugates and, indeed, is one of the major sources of flavonols in European diet. A relatively small range in content of quercitin conjugates for red ($110-295$ mg kg^{-1}) and yellow ($119-286$ mg kg^{-1}) cultivars have been reported. Flavonoids have been shown to offer protection against degenerative conditions such as cardiovascular disease, cancer, and aging.

Many alliums, including noncultivated wild types, have been extensively used for their therapeutic and medicinal properties. Most traditional medical practices include references to onion, garlic, and other alliums in the form of extracts, decoctions, concoctions, and poultices. Onions and garlic oils have proved effective in the treatment of a number of conditions, including intestinal worms, stomach ulcers, eye disorders, gastrointestinal disturbances, hypertension, and malarial fevers. Beneficial antitumor, hypoglycemic, hypolipemic, antiatherosclerotic, antiplatelet aggregation effects have also been reported. The active principles in many cases have been shown to be the sulfur-containing alliin derivatives.

A number of 'odor-free' garlic products, including capsules, extracts, and tablets, are commercially available. Since the biologically active principles also possess pungency, some care is needed to ensure that the treatment employed to remove odor has not at the same time reduced or removed the medicinally active compounds. (*See* **Garlic.**)

Postharvest Storage

Postharvest losses of onions may range from 10 to 30%, with even higher losses being reported from tropical regions. A number of factors have been shown to contribute to these figures; as a result, cultivars have been selected for improved storage characteristics, methods for sprout control developed, and storage practices and technologies optimized. Postharvest losses may be minimized if damage and bruising are reduced during harvest, storage, and packing; adequate ventilation of storage areas is vital if moisture accumulation on tissue surfaces, and subsequent disease proliferation, is to be avoided.

Dormancy, Sprouting, and Storage

Bulb dormancy is controlled by a complex interaction of growth inhibitors and promoters. Natural dormancy commences with harvest maturity and continues for 4–9 weeks, during which period, the bulbs will neither sprout nor continue to grow due to the influence of inhibitors translocated from the green leaves. However, these inhibitors are gradually destroyed with time. Once dormancy has passed, roots may emerge and leaf shoots appear; sprouting is optimal at 10–15 °C. Onions may be stored without sprouting if maintained at 0–5 °C and 65–70% relative humidity (RH). Sprouting has also been inhibited by controlled-atmosphere storage atmosphere storage at reduced oxygen levels (for example, 5–10% CO_2, 3% O_2, 5 °C), γ-irradiation or preharvest foliar treatment with maleic anhydride (MH). Onions treated with MH and held between −2 and 0 °C and 65–70% RH can be stored for 6–7 months without any deterioration of quality.

Garlic bulbs intended for consumption are held at ambient temperatures and can remain in good condition for several months. However, for extended storage of up to 8 months, bulbs should be stored at -2 to $0\,°C$ and 60% RH. Sprouting most readily occurs between 5 and $10\,°C$, and inhibition may be achieved, as described for onions. Unlike onions and garlic, leeks are in active growth when harvested; they can have a postharvest life of up to 8 weeks at $0\,°C$ and 95% RH. Controlled atmosphere treatment (for example, $10\% \ CO_2 + 1\% \ O_2$) can extend the storage period further. Leeks are stored upright to avoid curvature of the pseudostems.

Storage Diseases

Postharvest diseases of onions, which will reduce the quality and, thereby, the value of the crop, are largely determined by the growing and storage conditions. The major disease of stored onions in temperate regions is neck rot, caused by *Botrytis allii*. This may be controlled by seed and set treatment with benomyl/thiram and effective postharvest drying. Other soil-borne diseases (notably, white rot, pink rot, basal rot and smudge) are best controlled by combinations of field treatment and good husbandry.

Processing

Dehydration

Onions were included in the first dehydration experiments carried out in the late eighteenth century. A major program of onion and garlic dehydration was conducted in California in the 1920s, and dehydrated onion was already extensively used in food processing by the end of the next decade. It was the onset of World War II that provided the necessary impetus for the development of commercial food dehydration techniques. Onions for dehydration are generally white or yellow and have a high solids content (18–20% and above) and strong pungency. After preliminary cleaning, peeling, and slicing/chopping, onions are dehydrated at $75–60\,°C$, the temperature being reduced as the moisture content decreases. The final moisture content of 4% is achieved through warm air circulation. Discoloration may be minimized by effective curing and conditioning whilst the application of fluidized-bed techniques may overcome problems of scorching and agglomeration. In general, 8–10 kg of raw onions produce 1 kg of dehydrated product. Garlic may be dehydrated under similar conditions, with 5 kg of fresh bulbs producing 1 kg of dried product. Experiences of hot-air drying of chives have revealed problems such as an undesirable flavor, texture, and color. Flash-freezing and drying have proved effective in overcoming these problems. Dehydrated leek products are available for use in soups and other products, and dry green onions have been used to replace more expensive shallots and chives. Onion and garlic powders are used in the manufacture of onion and garlic salt. Both dehydrated onion and garlic are prone to discoloration, a problem that can be minimized by optimizing the drying conditions or employing approved additives. (*See* **Drying**: Theory of Air-drying; Physical and Structural Changes.)

Onion and Garlic Oils

Onion oil is obtained by the distillation of minced onions. The oil, obtained in a 0.002–0.03% yield, depending upon the source and processing conditions, has a flavor enhancement 500 times that of the dehydrated product. Garlic oil comprises 0.1–0.25% of the fresh weight of the clove, 1 g being equivalent to 200 g of dried garlic. The intense pungency of these products makes them difficult to use directly, and they are usually diluted in vegetable oil, or encapsulated.

Onion and Garlic Juices

Onion and garlic juices contain both flavor and aroma-active compounds, their precursors and sugars. They may be blended with volatile oils to restore a 'rounder' flavor profile. The juices, viscous and dark brown in color, can be mixed with a support (such as propylene glycol, lecithin or glucose) to provide an oleoresin with a flavor intensity 10 times greater than that of the dehydrated powder.

Others

Canned and bottled onions are used by the catering industry, especially in North America. Onion rings, coated in batter, are widely used in fast food outlets; these are supplied frozen, usually being extruded rather than machine-cut. Pickled onions, popular in the UK, are prepared from both 'brown' and pearl/cocktail onions (28–45 mm and 10–28 mm, respectively). Typically, brown onions are steeped in brine (10%) for 1–4 days, and lactic acid is added to control fermentation. They are then washed, covered in vinegar, and pasteurized; bisulfite may be added to maintain the color. Silverskin onions tend not to be pasteurized since this adversely affects the texture and flavor. Fermentation of Alliums has been described for thousands of years; onions, garlic, and leek are included in a variety of fermented products such as kimchi, ragi, sauces, and pickles. Many varieties of salamis and sausages are also strongly flavored with garlic before drying.

See also: **Drying**: Theory of Air-drying; Physical and Structural Changes; Chemical Changes

Further Reading

Bender AE and Bender DA (1999) *Food Tables & Labelling*. Oxford: Oxford University Press.

Fenwick GR and Hanley AB (1985) The genus *Allium* – parts 1,2 and 3. *CRC Critical Reviews in Food Science and Nutrition* 22: 199–271; 22: 273–376; 23: 1–73.

Rabinowitch HD and Brewster JL (eds) (1990) *Onions and Allied Crops*. Boca Raton, FL: CRC Press.

Rubatzky VE and Yamaguchi M (1997) *World Vegetables: Principles, Production and Nutritive Values*. Florence, KY: ITP Publishing.

Yamaguchi M (1983) *World Vegetables: Principles, Production and Nutritive Values*. Westport, CT: AVI Publishing.

Oranges *See* **Citrus Fruits**: Types on the Market; Composition and Characterization; Oranges; Processed and Derived Products of Oranges; Lemons; Grapefruits; Limes

---

ORGANICALLY FARMED FOOD

S Stolton, Elm Farm Research Centre, Berkshire, UK

Introduction

As our food system becomes ever more complex, foods become more highly processed, and the variety of foods available at any one time overcomes geographical and seasonal restraints, some consumers have been drawn towards alternative forms of production such as organically farmed food that are seen as more 'natural' and 'healthier' and perhaps of higher quality. The scientific evidence substantiating such claims remains limited, however, and is often the subject of intense debate. In practice, any analysis of the differences between organic and conventional food needs to encompass a wide range of different issues relating to health, environment, society, and the innate preferences of consumers.

What is Organic Farming?

Organic food is produced by farming methods that rely to a large extent on locally available resources and depend on maintaining ecological balances and optimizing the benefits from naturally occurring biological processes, rather than on manipulating the ecosystem through use of agrochemicals and fossil fuels. Organic agriculture aims to encompass agricultural systems that promote environmentally, socially, and economically sound production. By respecting the natural capacity of plants, animals, and the landscape, organic agriculture aims to maximize the quality of food whilst maintaining a sustainable agriculture and environment.

The principal aims of organic agriculture are summarized by the International Federation of Organic Agriculture Movements (IFOAM) in its *Basic Standards for Organic Agriculture and Food Processing*. These aims are:

- to produce food of high nutritional quality in sufficient quantity;
- to interact in a constructive and life-enhancing way with all natural systems and cycles;
- to encourage and enhance biological cycles within the farming system, involving microorganisms, soil flora and fauna, plants, and animals;
- to maintain and increase long-term fertility of soils;
- to use, as far as possible, renewable resources in locally organized agricultural systems;
- to work, as far as possible, within a closed system with regard to organic matter and nutrient elements;
- to work, as far as possible, with materials and substances which can be reused or recycled, either on the farm or elsewhere;
- to give all livestock life conditions which allow them to perform the basic aspects of their innate behavior;
- to minimize all forms of pollution that may result from agricultural practice,

- to maintain the genetic diversity of the agricultural system and its surroundings, including the protection of plant and wildlife habitats;
- to allow agricultural producers a life according to United Nation human rights, to cover their basic needs, and obtain an adequate return and satisfaction from their work, including a safe working environment;
- to consider the wider social and ecological impact of the farming system.

The IFOAM *Basic Standards* have provided a framework for almost all the national regulations and for the international World Health Organization/Food and Agriculture Organization Codex Alimentarius on organic agriculture. They are used by organic farmer organizations worldwide as a common platform. In many countries organic food and farming are regulated by legislation. In the European Union, for example, organic standards and certification procedures are enshrined in the European Community (EC) Regulation 2092/91, which provides a legal definition of organic production in Europe and for organic produce imported into Europe.

What Makes Organically Farmed Food Different?

There have been persistent calls from within the food industry for research that distinguishes clearly what, if anything, differs between organic and conventionally produced foods. Studies have been increasing in both number and scale, although results remain ambiguous. This is a highly political subject and there are regular well-publicised claims to 'prove' the issue one way or another. However, in reality many questions remain to be answered.

The majority of the scientific studies carried out to determine what, if any, special characteristics can be attributed to organically farmed food, as opposed to food produced under other production regimes, have been comparative, using one or two crops to determine general trends.

A comprehensive literature review of over 150 comparative studies of this sort was published in 1997. The review highlighted a number of inherent problems with comparative studies to date, including that there was no clear and consistent definition of 'conventional' agriculture (although primarily this term denotes a system where farming methods attempt to substitute natural processes – for instance, by using inorganic fertilizers or pesticides) and that sample sizes tended to be small – both of which make trend analysis difficult. Most studies reviewed are physicochemical evaluations of concentrations of

desirable or undesirable elements in food. Other analytical methods used in comparative studies include electrochemical, plant physiological, and multivariable physicochemical methods, susceptibility to infection, and postharvest behavior. More 'alternative/holistic' methods to determine the quality of organic food have also been developed but these are not generally recognized by conventional scientists. Such methods include the ascending-imaging method, round filter chromatography, copper chloride crystallization, and the measurement of ultraweak photon emissions. The studies carried out using these methods claim to have shown a trend towards organic foods having more 'vital activity' than other food from other production systems.

Other researchers argue that focusing solely on physicochemical evaluations misses many of the wider issues relating to organic food and also misses the reasons why most consumers buy organic food. Just as the organic farming system cannot be fully understood by studying one crop in the rotation, the qualities of organic food cannot be determined by one or two comparative studies of individual food products or the effects of an organic diet on laboratory-reared animals. A more valid approach to understanding the characteristics that may make organic food different is to look at a wider range of attributes that together could be said to describe 'food quality.'

A new framework, or definition, of food quality, combining six distinct criteria was drawn up by a group of international experts involved in the food industry, at a colloquium organized by the UK-based organic research organization, Elm Farm Research Centre, in 1989. Like all such concepts, this can never hope to cover all aspects of quality. However, the definition does incorporate the many issues that have been the focus for research into the quality of organic food and is thus a useful tool for reviewing research findings. In many areas quality characteristics can be measured quantitatively and comparisons between organic and conventional practices can be made; however, our choice of food is often led more by the subjective non-quantifiable values, which can make the judging of quality more difficult, but no less valid.

The Six Aspects of Food Quality

The six criteria of food quality (authentic, functional, biological, nutritional, sensual, and ethical) identify the range of issues that consumers and producers recognize as important with respect to food production and composition. They thus provide a framework within which a more comprehensive analysis of what makes organic food different can be carried out. Each criterion is briefly examined below.

Authentic

> Food that is authentic, traditional or natural and has not been adulterated in production, processing or storage, including the use of genetically modified organisms or their derivatives.

Consistently, research into consumers' attitudes shows that the main perceptions of organic food are linked to phrases such as 'no chemicals/additives/pesticides,' 'natural' and 'healthy.' Although there is an increasing range of processed organic foods being produced, these processes are themselves bound by standards for organic production. According to the IFOAM *Basic Standards*, the general principles of organic food processing and handling are that: 'any handling and processing of organic products should be optimized to maintain the quality and integrity of the product.' Furthermore, the processing methods should ensure that: 'the vital quality of an organic ingredient should be maintained throughout each step of its processing.'

Measuring authenticity is almost impossible because it relates to such a large extent to individual consumer preferences. However, issues like the avoidance of most agrochemicals, the fact that organic food aspires to be locally produced, and the generally reduced processing appeals to a substantial part of the food market. The fact that organic food is also free of genetically modified (GM) food also meshes well with consumer preferences in Europe and beyond. More generally, the less processed nature of most organic food helps fulfill the needs of those consumers who are consciously trying to move away from the highly processed foods that became a dominant part of the industry in the 1980s and 1990s.

Functional

> How appropriate food is to its specific purpose – i.e., food that produces, stores, or cooks well. 'Functionality' is also used by food scientists and technologists to describe ingredients that blend well, or which have a good shelf-life.

Concern about the world's diminishing biological diversity (or biodiversity) has been growing in recent years. Although biodiversity loss tends to be linked in people's minds with threats to rainforests or to rare animals like giant pandas, genetic diversity is also being lost within food production. This loss inevitably affects our choice of the variety of food and the functionality of crops produced.

Loss of variety means loss of potential to adapt to different conditions. A distinction between organic and conventional agriculture is that in the former, encouragement of greater variety is an important part of the overall system, because reduced reliance on agrochemicals to change conditions in the soil means that the plants must themselves be better adapted to local conditions. Organic standards encourage the expansion of varieties grown to gain benefit from those suited to particular growing conditions. For example, conventional wheat grown in an organic system will show a distinct reduction in protein content, whereas selection of a variety suitable for the particular conditions on the farm can allow baking-quality wheat to be grown.

This encouragement of diverse species use is threatened by the general reduction in crop diversity. A survey of some 75 crop species, carried out by the Rural Advancement Fund International, found that about 97% of the varieties given on old US Department of Agriculture lists are now extinct (assuming that varieties that are not stored in US seed banks are probably extinct). This loss is not unique to the USA, for example, indigenous wheat, rice, and sorghum varieties have virtually disappeared from many centers of diversity in areas of Europe, the Middle East, Africa, and Asia. Of the 2500 or so apple varieties that have been grown in the UK, just nine now dominate the retail outlets. Within the EC, this erosion of diversity in crops has become legally obligatory with the introduction of a common catalogue. Varieties not listed are seen as inferior and cannot be legally sold by seed companies.

Organic growing conditions can also contribute to the functionality of food. A review of research on storage degradation by the Soil Association shows that organic crops can have better storage quality and postharvest behavior than nonorganically grown produce. Organic plant production is characterized by a relatively low level of nutrient supply compared to conventional crops. This leads to earlier completion of vegetative growth and an early onset of maturity, which may contribute to a longer shelf-life. Respiration rates and enzyme activity, which can also result in lower storage losses, also tend to be lower in organically farmed foods.

Biological

> How food interacts with the body's functioning. This includes both negative and positive interactions, such as the detrimental effects of some food additives on allergies and beneficial effects of foods in stimulating immunity to disease. The concept also refers to the pharmacological qualities of foods and herbs.

Many consumers feel that organic food can have positive effects on health and this is a major reason for purchase. A comprehensive review of over 400 published papers by the Soil Association of existing research on the differences between organically and nonorganically grown foods has revealed significant differences to key areas of food quality important to the promotion of good health – food safety, nutritional content, and the observed health effects of organic food. The review notes that, while there has been little research carried out on humans, a number of feeding trials on animals have shown significant improvements in the growth, reproductive health, and recovery from illness of animals fed organically produced food.

Reproductive health is seen as a particularly sensitive indicator of environmental conditions and has been the focus of some research into the effects of organic food on health. It is known that occupational exposure to pesticides can impair reproductive health and it is hypothesized that long-term low-dose exposure to residues through food intake may also have adverse effects, so food with less pesticide residues might be expected to have beneficial results. In the mid-1990s, two small studies in Denmark indicated that semen quality was improved among consumers of organic food and this gained worldwide publicity – however, a larger study could not confirm these results. Several feeding experiments have also been carried out, but over a range of studies of mice, rats, pigeons, rabbits, and hens the reproductive effects of diet have shown no clear trend. In feed experiments with mice and rats the share of stillborn animals and animals that died shortly after birth was significantly higher in litters fed on conventionally produced feed.

The routine use of antibiotics in intensive farming has been linked to the development of 'superbugs' resistant to antibiotics used to combat human illness. As organic standards only allow the use of antibiotics in animals to cure specific problems, it is reasonable to suggest that an organic diet will reduce the buildup of antibiotic residues.

Nutritional

How food contributes to a balanced diet. This recognizes individual food values, such as vitamins, protein, and trace elements, and undesirable elements such as fats and sugar and nitrate and sodium, which in large amounts can be harmful.

Public concern about the presence of pesticide residues, additives, and nitrates in food is the reason many consumers purchase organically produced food. The organic sector has, however, been careful to stress to consumers that, given the state of the environment, and the presence of pesticide residues in soils, air pollution from spray drift, and industrial pollution, no food production method can guarantee a 100% chemical-free crop. There is however evidence that organic foods do have lower levels of

pesticide residues than comparable conventional crops, although these results should not be overestimated given the small number of studies.

Concern has also been expressed over increasing nitrate intake from dietary sources, notably from the consumption of vegetables, although the possible health impacts of this remain the subject of debate. There is considerable variation in the levels of nitrate in foods, and in the consumption patterns of different foodstuffs amongst people; for example, the nitrate content of leafy vegetables is generally higher in winter. Comparative studies on farming systems and the related accumulation of nitrate in vegetables have been the subject of several research projects in recent years. In the UK, trials showed significantly lower levels of nitrate accumulation when using composted farmyard manure compared to the use of the soluble compound fertilizer – although fertilization with either type of material did lead to an increase in the level of nitrate in the plants. This trend is confirmed over a range of research studies (the Soil Association reviewed 16 studies, of which 14 showed a trend toward significantly lower nitrate contents in organically grown food), with nitrate content of conventional produce as opposed to organic production being particularly high in nitrophilic leaf, root, and tuber vegetables.

The presence or absence of harmful or potentially harmful substances in food only represents one aspect of the nutrition equation, as our food is made up of a range of nutrients such as vitamins and minerals. The Soil Association reviewed 99 papers that compared the nutritional quality of organically and conventionally grown crops. These studies were assessed for validity in terms of agricultural practice and scientific analysis. Of those studies deemed valid, the Soil Association concluded that vitamin C and dry-matter contents are higher, on average, in organically grown crops. Mineral contents were also higher, on average, in organically grown crops, although the small number and heterogeneous nature of the studies included mean that more research is needed to confirm this finding. Research also indicates a clear long-term decline in the trace mineral content of fruit and vegetables, but to understand the reason for this the influence of farming practices requires further investigation.

As well as having effects on individual elements of crops, the choice of an organic diet may also have an effect on overall dietary habits. A survey of the diet of UK organic consumers concluded that those committed to the consumption of organic food have a different diet from the average consumer – a diet that is more in line with the productive capacity of the organic system. By collecting detailed diet diaries and comparing them with data from the Office of

Population Censuses and Surveys, organic consumers showed a significant swing towards the consumption of fruit and vegetables, with a reduction in the protein food 'centerpieces.' Protein consumption was less dependent on meat, being derived from other sources. Staples showed a greater diversity, with a greater intake of pasta, grains, and cereals, with less concentration on potatoes. The survey data also showed an increased consumption of fibrous foods and a decreased consumption of animal-based as well as sweet and fatty foods.

Sensual

> Food which appeals to the senses; the industry uses the term 'organoleptic' to refer to sensual quality of food, i.e. its look (esthetic appeal), taste, feel (as in mouth feel), smell, and sound.

The determination of whether organic food tastes better than food produced from other forms of production has been the subject of considerable research. Although it is widely believed by consumers that organically farmed food tastes better than the alternatives, there is no conclusive scientific evidence to prove this. Partly this is due to individual preferences in taste, which makes the determination of quality through taste alone totally subjective, and is complicated by a whole range of regional factors, such as soils and climate, which make direct comparisons between growing systems difficult (unless trial plots are extremely well matched). The Soil Association also suggests that other influences on taste could be the higher water content in conventionally grown food diluting its flavor, different choices of cultivars, and the fact that organic produce has sometimes been shown to have higher natural sugar content, which may be perceived as a better or worse taste, depending on the product.

Review of feeding experiments with animals has however drawn one unexpected, and so far unexplained, conclusion. When given the choice, animals almost always prefer organic produce to conventionally produced feed.

Ethical

> This concept has four related but distinct meanings: environmental (the effects on our environment of food production methods); social (the conditions in which food producers are treated – on the farm, in the factory, on the shop floor); ethical (the morality of the way specific foods are produced – for example, the conditions of animals raised for meat); and political (the effects of the food industry on specific countries).

Research into consumer preferences suggests that there has been a shift in emphasis amongst purchasers of organic food in the last decade, away from people primarily driven by personal health concerns to those with a wider interest in the environmental and social impacts of the food that they buy.

There is a large literature describing the environmental problems of conventional agriculture and the benefits of organic agriculture. For example, long-term research projects have accumulated substantial evidence that organic systems are beneficial to biodiversity. An overview of research findings from 23 of these projects by the Soil Association concluded that in most studies there were important differences between the biodiversity on organic and conventional farms, with generally greater levels of both abundance and diversity of species on organic farms. On average, the research found five times as many wild plants in arable fields and 57% more species, 25% more birds at the field edge, and 44% more species in fields in the autumn and winter, and three times as many nonpest butterflies in crop areas.

In a review of the environmental impacts of organic farming in 18 European countries, an assessment of a series of environmental indicators found that organic farming has more floral and faunal diversity than conventional farms, tends to conserve soil fertility and system stability better and on a per-hectare scale has between 40 and 60% lower CO_2 emissions and is likely to have lower emissions of N_2O, CH_4, and NH_3.

Sociological data comparing organic and conventional farming systems are not so well advanced. The IFOAM began a project on data collection and farm system comparison in 1997 and has developed a methodology which collects data on 13 indicators that provide information on subjects such as production of main and byproducts, recycling, male and female hired and own labor, financial results, energy consumption and production, water use, family expenditure, and education. The methodology is now being tested in Guatemala, India, and Nicaragua. Results to date have shown that self-reliance, which is defined as a percentage of output value that is consumed on the farm, is higher in the organic farms.

Conclusions

Many of the issues described above are not limited to organic production: for example, fresh food is available at the farm gate from many conventional farms and there is a growing fair-trade movement that insures equitable returns for food produced in developing countries. However, organic farming is the only system where all such issues are addressed together and within a legal framework through the organic standards, insuring that consumers can be reasonably confident that a 'package' of health, environmental, and social issues has been addressed in the production of the organic food that they purchase. When provided with the whole picture, across a range of issues such as those given above, there does seem to be sufficient research to conclude that organically farmed food does have some unique characteristics.

See also: **Antibiotics and Drugs**: Uses in Food Production; **Food and Agriculture Organization of the United Nations**; **Legislation**: Codex; **Quality Assurance and Quality Control**; **World Health Organization**

Further Reading

Alföldi T, Lockeretz W and Niggli U (2000) *Proceedings of the 13th International IFOAM Scientific Conference.* Germany: IFOAM.

Azzeez G (2000) *The Biodiversity Benefits of Organic Farming.* Bristol, UK: Soil Association.

Finesilver T (1989) *Comparison of Food Quality of Organically versus Conventionally Grown Plant Foods.* Ecological Agriculture Projects. Quebec, Canada: McGill University.

Gold M (2000) *Organically Produced Foods: Nutritive Content. Special Reference Briefs Series no. SRB 2000–03.* Beltsville, USA: Alternative Farming Systems Information Center, National Agricultural Library, US Department of Agriculture.

Heaton S (2001) *Organic Farming, Food Quality and Human Health: A Review of the Evidence.* Bristol, UK: Soil Association.

Lampkin N (1990) *Organic Farming.* Ipswich, UK: Farming Press.

Pretty JN, Brett C, Gee D *et al.* (2000) An assessment of the total external costs of UK agriculture. *Agricultural Systems* 65(2): 113–136.

Stolze M, Piorr A, Häring A and Dabbert S (2000) *The Environmental Impacts of Organic Farming in Europe. Organic Farming in Europe: Economics and Policy,* vol. 6. Hohenheim, Germany: University of Hohenheim.

Woese K, Lange D, Boess C and Bögl K-W (1997) A comparison of organically and conventionally grown foods – results of a review of the relevant literature. *Journal of the Science of Food and Agriculture* 74: 281–293.

Woodward L, Stolton S and Dudley N (1992) *Food Quality: Concepts and Methodology.* Hamstead Marshall, Berkshire, UK: Elm Farm Research Centre.

Organoleptic Evaluation *See* **Sensory Evaluation**: Sensory Characteristics of Human Foods; Food Acceptability and Sensory Evaluation; Practical Considerations; Sensory Difference Testing; Sensory Rating and Scoring Methods; Descriptive Analysis; Appearance; Texture; Aroma; Taste

Osmosis *See* **Membrane Techniques**: Principles of Reverse Osmosis; Applications of Reverse Osmosis; Principles of Ultrafiltration; Applications of Ultrafiltration

Osteomalacia *See* **Rickets and Osteomalacia**

OSTEOPOROSIS

J J B Anderson, University of North Carolina, Chapel Hill, NC, USA

Background

Osteoporosis is considered a chronic multifactorial disease because many variables contribute to the development of low bone mass and low bone mineral density accompanied by an increased risk of fracture. This disease usually occurs in late adulthood, following the menopause in women, and a decade or so later in men, but it may appear rarely earlier in life depending on the status of reproductive steroid hormones.

Both hereditary and environmental factors contribute to the multifactorial nature of this disease. Individual differences in constitutional factors as well as lifestyle/environmental determinants influence bone. In terms of lifestyle, both regular physical activities and a healthy diet remain two of the more important contributors to bone health and the maintenance of its functions into late life. The major function, sound ambulation, needs to be preserved as long as possible through exercise and diet. When calcium intake is not adequate, the adaptational role of the calcium regulatory system operates in an attempt to preserve skeletal structure for its functional uses. If the adaptation is not sufficient, bone deterioration, i.e., loss of both mass and density, will lead to fragility fractures. These and related topics are covered in this paper. (Although drugs that help conserve bone are widely used today, this review focuses on dietary risk factors and nutritional therapies.)

Bone Structure

Bone tissue consists of two types within the same specific bone, e.g., a vertebra of the spinal column: trabecular (cancellous) and cortical (compact). Trabecular tissue is the more metabolically active because it has about eight to 10 times more total surface area than cortical tissue, and these surfaces are all largely covered by bone cells that are responsible for new bone formation and old bone resorption (degradation). In the body of a vertebra most of the bone tissue is trabecular, but near the surfaces of the entire vertebra, cortical bone predominates. Therefore, each specific bone or organ contains both types of bone tissue but typically in different locations within the bone. For example, long bones, such as the femur, contain much more trabecular tissue at either end near the hip joint or knee joint and a much greater proportion of cortical bone in the shaft that connects the two ends. This distinction is important because most of the fractures of the bones (organs) occur where more metabolically active trabecular bone tissue exists.

Bone Gain and Bone Loss through the Life Cycle

In early life, during the growth periods of the skeleton, bone acquisition through bone formation dominates. This phase represents the making of the skeletal model, i.e., modeling, and it typically ends by 16–18 years in females and 18–22 years in males. Modeling is characterized by greater formation than resorption and results in a net gain of bone mass or bone mineral content (BMC) by the end of the growth phase. Modeling is completed at the end of linear growth. In remodeling of the skeleton, resorption of bone equals formation; so, the net amount of bone mass remains fairly constant, though some modest gain may still occur until the mid-20s or beyond. During the beginning of the remodeling, the increase in pubertal hormones is responsible for the cessation of growth in length, or height increase. At this time, bone mineral density (BMD) attains a maximum value that becomes the 'healthy norm' for an individual for the rest of their life. (In the definitions of osteopenia and osteoporosis, the healthy means of a population of 20–29-year-old males or females are taken as the standard values for determining osteoporosis late in life (see below.)

In late life, i.e., after the menopause in women and a decade or so later in men, imbalances in the remodeling of the skeleton result in bone losses, so that reductions in both BMC and BMD occur. The loss of estrogens after the menopause and probably the later decline of androgens by men contribute to the increase in resorption, the reduction of formation, or both. The increase in resorption is triggered by increased activity of osteoclasts, and the decline in formation is directly related to decreased activity of osteoblasts. An increase in bone turnover, i.e., increased rates of resorption and formation with resorption dominating formation, greatly accelerates the rate of bone loss. Most individuals are slow or moderate losers of bone, whereas only a small fraction are 'fast losers.' Bone turnover is assessed by measurement of chemical markers of degraded matrix proteins, such as collagen, resulting from bone resorption and of hormones, especially parathyroid hormone, involved in calcium homeostasis.

The changes in the skeleton over the life cycle reflect early-life gains and late-life losses. When the losses become sufficient to deteriorate to the state of osteoporosis, an individual becomes at risk of fragility fractures of bones such as the vertebrae and proximal femurs (hips). The hip fractures are the most severe and debilitating, and many individuals never recover.

Measurement of BMC and BMD

BMC and BMD are measured today using a technique called dual energy X-radiographic analysis (DXA). Osteopenia, or too little bone, is followed by osteoporosis, or even greater bone loss, which places an individual at great risk of a fracture. Quantitative definitions of osteopenia and osteoporosis have been established by the World Health Organization as follows: osteopenia is between minus 1 and 2.5 standard deviations (SDs) below the 20–29-year-old mean values for women or men; and osteoporosis is greater than −2.5 SDs below the 20–29-year-old means. (These values of individuals at any adult age compared with means of healthy young adults are known as T-scores.) According to this definition, a postmenopausal 60-year-old woman with −3.0 SDs below the mean at different skeletal measurement sites does not necessarily have any clinically diagnosed fracture, but she is likely to have one over the next few years without any drug (or drug and diet) therapy.

In addition to the DXA determination of bone status, biochemical markers may also be used to assess the severity of bone turnover and the potential risk of fracture. Therefore, both clinical observations, e.g., loss of height and pain in the lower back, and biochemical marker data may increase the diagnostic significance of DXA measurements that classify an individual in the osteoporotic range or >2.5 SDs below the healthy means.

Types of Osteoporosis

Type I osteoporosis uses the postmenopausal woman as the prototype, although men also rarely may suffer from the abrupt loss of sex steroids that impact greatly on the retention of bone tissue. Type II osteoporosis is age-related and typically occurs in both genders in the later decades of life; the causation of Type II is poorly understood, but it accelerates when the musculoskeletal system functions decline.

Although, for simplicity, only two types of osteoporosis are defined here, other variants clearly exist.

One that is of increasing importance is corticosteroid-induced osteoporosis because of the widespread use of prednisone and related drugs in therapy of diseases other than osteoporosis. This type of osteoporosis is called secondary. Other drugs may also induce secondary osteoporosis.

Nondietary Risk Factors for Osteoporosis

Several nondietary risk factors – all potentially adverse or harmful – have been identified that promote the loss of bone and the onset of fractures. Seven environmental/lifestyle factors are listed here:

- thinness with low LBM;
- cigarette smoking;
- excessive alcohol consumption;
- insufficient physical activity;
- drugs – over-the-counter and prescription;
- decline of sensory perceptions;
- falls.

Each of these factors has a risk associated with it, but when two or three exist together at the same time, the risk of an osteoporotic fracture may increase exponentially rather than additively. For example, the small-framed older postmenopausal woman who smokes a pack of cigarettes a day, drinks two or three servings of alcohol a day, and who has little physical activity in her daily life will typically be at great risk of an early hip fracture, i.e., by age 70 years or younger. When the old-old, i.e., greater than 80, suffer declines of acuity of their senses, such as vision and equilibrium, or take medications that result in the same effects, they are much more likely to fall and break their hip.

The most important of these adverse factors, in general, may be the decline in regular physical activity that results in the loss of lean body mass (LBM), characterized by declines in muscle strength and tone, because bone loss follows closely the loss of LBM.

Many elderly take several drugs, i.e., polypharmacy, and some of them become confused or have difficulty maintaining equilibrium. Each of these factors may contribute to falls and consequent hip fractures.

Dietary Risk Factors

In addition to adverse environmental/lifestyle factors, numerous dietary factors may also have adverse effects on skeletal tissue. These deleterious factors are thought to operate throughout the life cycle, not just during late life. The major variables are given here, but this list is not exhaustive:

- low calcium;
- high phosphorus;
- low vitamin D;
- high-animal protein and acid load;
- high-sodium snacks;
- vegetarian diet;
- poor diet in general.

The most common dietary problem associated with the development of osteoporosis has been an inadequate consumption of calcium, at least among Caucasians in Western nations, including pediatric populations. Low calcium intakes in these nations are typically accompanied by high phosphate consumption because of both the ingestion of naturally occurring phosphorus in foods, especially animal proteins, and phosphorus from food fortification. Phosphorus fortification is fairly common in Western nations because of the widespread use of processed foods – foods modified with the many applications that utilize phosphate salts of one type or another. Cola-type soft drink beverages also fit in this category, although, technically, they are not foods. The net result of consuming phosphate-rich foods may be a low calcium (Ca) to high phosphate (P) intake that is worsened when individuals consume little milk or cheese in their usual diets.

If the Ca:P ratio declines to approximately 1 to 4, the serum parathyroid hormone (PTH) concentration becomes elevated (but remains within the normal range). A constantly elevated PTH leads to bone loss and a gradual decline in BMD that eventually becomes osteoporosis. How many years bone takes to get into the osteoporotic range, according to WHO definitions, is not clear, but a long-term dietary pattern of low calcium–high phosphate may even contribute to low bone mass in females before they reach 20 years of age, if fractures among girls and pubertal females are a valid index of low bone mass.

Coupled with limited skin exposure to vitamin D-promoting sunlight, a low intake of vitamin D from foods, especially fortified dairy products in the USA and various deep-sea fish species in much of the world, is now considered a risk factor for low bone mass. The mechanism for this condition is not entirely clear, but a low circulating concentration of 25-hydroxyvitamin D is linked with both a decline in intestinal calcium absorption and an increase in bone turnover with a resultant loss of bone.

A usual high-animal protein intake has been reported by some investigators to contribute to bone loss and an increase in risk for osteoporosis. The mechanism is considered to reside in the increased production of acids, i.e., phosphoric and sulfuric,

from the degradation of phosphorus- and sulfur-containing amino acids that are considerably greater in animal than plant proteins, such as those derived from soy. The net effect is an increased loss of calcium ions in urine. This loss has been shown acutely and in short-term experiments, but it has not been established in long-term studies.

High-sodium snack foods have become very popular in Western nations, and they contribute additional sodium to an already high intake that derives from so many foods processed with sodium or salt. Calcium renal losses increase on high-sodium intakes because the kidneys favor sodium reabsorption at the expense of calcium ions. The net loss of calcium comes from bone, and therefore a loss of bone mass may be associated with excessive consumption of sodium-rich snack foods.

Vegetarian diets may also compromise bone health through a number of possible mechanisms, but the low calcium and vitamin D intakes from a vegan dietary pattern may be largely responsible for lower bone mass among vegetarians.

Finally, a poor diet, especially one based on a limited intake of fruits and vegetables, may be deficient in other bone-essential nutrients, such as magnesium, vitamin K, zinc, antioxidant nutrients, and probably a dozen others. Any one or a combination of these deficits may inhibit efficient bone formation. Adverse effects include loss of protection against oxidants, and poor regulation of acid–base balance, thus impacting bone as a major buffering store. Whatever the mechanisms may be, bone health requires a wide variety of nutrients that are best supplied by a varied diet consisting of the recommended numbers of servings of foods each day.

Adaptation to Low Calcium Intakes

Several adaptations to dietary intakes occur after each meal in healthy individuals; a number of these adaptations directly affect calcium homeostasis by impacting on serum PTH and 1,25-dihydroxyvitamin D concentrations. For example, a low calcium intake, especially when coupled with a high phosphate consumption pattern, stimulates PTH secretion via a calcium sensor on the membranes of parathyroid gland cells. Elevated PTH, in turn, stimulates the removal of calcium from bone (see above).

In addition, a low calcium intake stimulates the vitamin D regulatory mechanism by increasing the renal production of the hormonal form of vitamin D, i.e., 1,25-dihydroxyvitamin D, which leads to an increase in intestinal calcium absorption and bone utilization of calcium. The potential problem in older women, and perhaps men, is that the intestinal adaptation declines, and less calcium can be absorbed with the hormonal stimulus.

Finally, excessive, continuous PTH secretion that may develop late in life with a low Ca:high P ratio diet and low circulating 25-hydroxyvitamin D concentration contributes to bone loss almost across the 24 h of a day. It is now recognized that a persistent and continuous treatment with PTH causes significant bone loss, whereas a discontinuous treatment with PTH, i.e., once daily or once weekly, increases bone mass and density by stimulating osteoblasts to make new bone tissue. Thus, any type of low calcium intake that generates a continuous secretion and elevation of PTH in blood will have serious negative consequences on the maintenance of bone, especially of trabecular bone tissue.

Dietary Prevention and Treatment – Foods and Supplements

The principles of both primary and secondary prevention of osteoporosis are similar: increase calcium in the diet through foods and supplements to 1000 mg or more per day in order to suppress PTH secretion (see above), and assure that individuals obtain sufficient amounts of vitamin D through both foods and supplements: a dosage of 800 IU or more is recommended. Sunshine exposure is also encouraged, depending on geographic latitude and time of the year. Finally, a healthy diet containing virtually all nutrients from foods makes sense from a nutritional perspective. For the elderly, a daily supplement that contains a wide range of nutrients at recommended intake levels is both safe and inexpensive. The current dietary reference intakes should be used to guide consumers in maintaining appropriate amounts of nutrients each day (1997–2001).

Public Health Implications and Conclusions

Since the incidence and prevalence of osteoporosis are increasing as our populations are aging, greater expenditures are anticipated for direct care of hip fracture patients and then their subsequent rehabilitation. Many victims will not survive the first post-fracture year. Preventive strategies, either primary or secondary, that are cheap and effective, must be identified and implemented. Improving the diet may be one of the most easily modifiable approaches for the prevention of osteoporosis, but of course, if individuals cannot ambulate, the dietary changes will have little impact with regard to osteoporosis. Walking and maintaining activities of daily living by the elderly are critical for the prevention of osteoporosis. Any type of minimal exercise program should also yield some benefit to the retention of musculoskeletal function.

See also: **Bone**; **Calcium**: Properties and Determination; Physiology; **Cholecalciferol**: Properties and Determination; Physiology; **Hormones**: Thyroid Hormones; **Smoking, Diet, and Health**; **Sodium**: Properties and Determination; Physiology

Further Reading

Anderson JJB (1999) Plant-based diets and bone health: Nutritional implications. *American Journal of Clinical Nutrition* 70: 539S–542S.

Anderson JJB, Sell ML, Garner SC and Calvo MS (2001) Phosphorus. In: Russell R and Bowman B (eds) *Present Knowledge in Nutrition*, 8th edn. Washington, DC: ILSI Press.

Committee on Nutrition (1999) Calcium requirements of infants, children, and adolescents. *Pediatrics* 104: 1153–1157.

Dietary Reference Intakes (DRIs) (1997–2001) *Several Reports Including: Institute of Medicine (IOM), Food and Nutrition Board. Dietary Reference Intakes for Calcium, Phosphorus, Magnesium, Vitamin D, and Fluoride*. Washington, DC: National Academy Press.

Goulding A, Cannan R, Williams SM et al. (1998) Bone mineral density in girls with forearm fractures. *Journal of Bone and Mineral Research* 13: 143–148.

New SA, Robins SP, Campbell MK et al. (2000) Dietary influences on bone mass and bone metabolism: Further evidence of a positive link between fruit and vegetable consumption and bone health? *American Journal of Clinical Nutrition* 71: 142–151.

Wyshak G (2000) Teenaged girls, carbonated beverage consumption, and bone fractures. *Archives of Pediatrics and Adolescent Medicine* 154: 610–613.

OXALATES

S C Morrison, New Zealand Institute for Crop and Food Research, Lincoln, Canterbury, New Zealand
G P Savage, Lincoln University, Canterbury, New Zealand

Introduction

Oxalic acid and its salts occur as end products of metabolism in a number of plant tissues. When these plants are eaten they may have an adverse effect on mineral bioavailability because oxalates bind calcium and other minerals. While oxalic acid is a normal end product of mammalian metabolism, the consumption of additional oxalic acid may cause stone formation in the urinary tract when it is excreted in the urine. The mean daily intake of oxalate in the diet in the UK has been calculated to be 70–150 mg; tea, rhubarb, spinach, and beet are common high oxalate-containing foods. Soaking and cooking foodstuffs high in oxalate will reduce the oxalate content by leaching.

The consumption of high-oxalate foods is more likely to pose health problems in those who have an unbalanced diet or those with intestinal malfunction. A diet high in oxalate and low in essential minerals, such as calcium and iron, is not recommended. Vegans and lactose-intolerant persons may have a high-oxalate and low-calcium diet unless their diet is supplemented. Vegetarians who consume greater amounts of vegetables will have a higher intake of oxalates, which may reduce calcium availability. This may be an increased risk factor for women, who require greater amounts of calcium in the diet. Persons with an increased absorption rate of oxalate are advised to avoid or eat fewer high-oxalate foods to prevent kidney stone formation. In healthy individuals, the occasional consumption of high-oxalate foods as part of a balanced diet does not pose any particular problem.

Oxalates in Plants

Oxalate can be found in small amounts in many plants. Oxalate-rich foods are usually minor components in human diets but are present in higher quantities in seasonal diets in certain areas of the world (particularly in the tropics) as a component of grains, tubers, nuts, vegetables, and fruits. The highest levels of oxalates are found in the foods listed in **Table 1**.

In general, oxalate content is highest in the leaves, then in the seeds, and lowest in the stems. Reports show that the stems or stalks of plants, such as amaranth, rhubarb, spinach, and beet, contain significantly lower levels of oxalates than the leaves. In the buckwheat family (e.g., rhubarb, sorrel), there is almost twice as much oxalic acid in the leaves as in the stalk. However, in the goosefoot family (e.g., beet, spinach), oxalic acid is more abundant in the stalk than in the petiole of the leaf. It must be noted that the leaves of rhubarb are rarely eaten and therefore the oxalate content of its leaves is of no concern in human nutrition.

Oxalic acid concentration tends to be higher in plants than in meats, which may be considered oxalate-free when planning low-oxalate diets. Meats, fats, and dairy products contain very low levels of oxalates.

The levels of oxalates in fungi are low when compared to the levels found in spinach and rhubarb. The giant mushroom (*Tricholoma giganteum*), a large, edible fungi, is reported to contain $89 \, \text{mg} \, 100 \, \text{g}^{-1}$ dry weight (DW) oxalic acid while tropical species of mushrooms, including termite and ear mushrooms, contain between 80 and $220 \, \text{mg}$ oxalate $100 \, \text{g}^{-1}$ DW.

High oxalate levels in tropical plants are of some concern. Taro (*Colocasia esculenta*) and sweet potato (*Ipomoea batatas*) were reported to contain $278–574 \, \text{mg} \, 100 \, \text{g}^{-1}$ fresh weight (FW) and $470 \, \text{mg} \, 100 \, \text{g}^{-1}$, respectively. Total oxalate levels in yam (*Dioscorea alata*) tubers were reported in the range $486–781 \, \text{mg} \, 100 \, \text{g}^{-1}$ DW, but may not be of nutritional concern since 50–75% of the oxalates were present in the water-soluble form and therefore may leach out during cooking. Oca or New Zealand yam (*Oxalis tuberosa* Mol.) contains $80–221 \, \text{mg} \, 100 \, \text{g}^{-1}$ FW soluble oxalate. Higher levels of oxalates are usually found in the leaves and highest concentrations of oxalate have been found in the skin of these tropical root crops. Peanut greens, commonly consumed in tropical climates, are reported to contain $407 \, \text{mg} \, 100 \, \text{g}^{-1}$. Coriander leaf (*Coriandrum sativum*) contains $1268 \, \text{mg} \, 100 \, \text{g}^{-1}$; horsegram (*Macrotyloma uniflorum*) and santhi (*Boernavia diffusa*) contain $508 \, \text{mg} \, 100 \, \text{g}^{-1}$ and $3800 \, \text{mg} \, 100 \, \text{g}^{-1}$ respectively. Nuts such as peanuts, pecans, and cashews are relatively high in oxalates. Sesame seeds have been reported to contain high quantities of oxalate, ranging from 350 to $1750 \, \text{mg} \, 100 \, \text{g}^{-1}$ FW.

Beverages with a high-oxalate concentration include Indian black tea, cocoa drinks, Ovaltine, cola, and certain types of beer.

The oxalic acid content is variable within some species; some cultivars of spinach (Universal, Winter

Table 1 Common foods exhibiting high levels of oxalates

Spinach	Beet	Rhubarb
Swiss chard	Yam, oca, sweet potato	Gooseberries, strawberries
Kale	Peanuts, pecans	Black tea
Purslane	Soybean	Cocoa, chocolate, Ovaltine
Collards	Wheat germ, wheat bran	Cola
Mustard and turnip greens	Sorrel, parsley, amaranth	Beer

Giant) contain $400–600 \, \text{mg} \, 100 \, \text{g}^{-1}$, while others range from 700 to $900 \, \text{mg} \, 100 \, \text{g}^{-1}$. Oxalic acid accumulates in plants, especially during dry conditions. A study comparing two cultivars of spinach, Magic (cv. summer) and Lead (cv. autumn), revealed that the summer cultivar contained greater amounts of oxalate ($740 \, \text{mg} \, 100 \, \text{g}^{-1}$ FW) than the autumn cultivar ($560 \, \text{mg} \, 100 \, \text{g}^{-1}$ FW). Oxalate content has been reported to increase as the plant ages and becomes overripe. The proportion of oxalic acid in the leaves of the goosefoot family can double during ripening. However, in tomatoes, oxalic acid content has been reported to decrease during ripening.

Absorption and Metabolism in Mammals

Calcium can combine with oxalate to form calcium oxalate in the intestinal lumen, making the calcium unavailable for absorption; calcium oxalate is then excreted in the feces. Free or soluble oxalate is absorbed by passive diffusion in the colon in humans; comparative studies between healthy individuals and those with ileostomies indicate that the colon is the principal site for oxalate absorption. However, it is suggested that the small intestine may be the major absorptive site rather than the colon. The absorption of oxalates from individual foods varies depending on their dietary conditions and source; in general, the absorption is quite limited. It has been estimated that 2–5% of administered oxalate is absorbed in humans. Experiments have shown that more oxalate is absorbed when consumed while fasting (12%) compared to only 7% oxalate absorption when consumed with a normal diet. The percentage of oxalate absorption varied markedly from 1% for rhubarb and spinach to 22% from tea, but generally absorption was higher at low doses.

Oxalate is an end product of ascorbate, glyoxylate, and glycine metabolism in mammals. Thirty-three to fifty percent of urinary oxalate is derived from ascorbate, 40% from glycine and 6–33% from minor metabolic pathways and dietary oxalate; dietary oxalate appears to account for only 10–15% of excreted oxalates.

Chemical Properties and Toxic Effects

Oxalic acid forms water-soluble salts with Na^+, K^+, and NH_4^+ ions; it also binds with Ca^{2+}, Fe^{2+}, and Mg^{2+}, rendering these minerals unavailable to animals. However, Zn^{2+} appears to be relatively unaffected. Calcium oxalate ($Ca(COO)_2$) is insoluble at a neutral or alkaline pH, but freely dissolves in acid.

Ingestion of 4–5 g of oxalate is the minimum dose capable of causing death in an adult, but reports have shown that 10–15 g is the usual amount required to cause fatalities. Oxalic acid ingestion results in corrosion of the mouth and gastrointestinal tract, gastric hemorrhage, renal failure, and hematuria. Other associated problems include low plasma calcium, which may cause convulsions and high plasma oxalates. Most fatalities from oxalate poisoning are apparently due to the removal of calcium ions from the serum by precipitation. High levels of oxalate may interfere with carbohydrate metabolism, particularly by succinic dehydrogenase inhibition; this may be a significant factor in death from oxalate toxicity induced in animals grazing pastures containing high levels of *Halogeton glomeratus*. Halogeton (*H. glomeratus*) and wood sorrel (*Oxalis cernua*) are high in oxalates and are known to cause injury to grazing cattle and sheep.

Although garden sorrel is a herb and not normally consumed in large quantities, there has been one report of fatal oxalate poisoning after a man consumed an estimated 6–8 g oxalate in vegetable soup containing 500 g sorrel. Both fatal and nonfatal poisoning by rhubarb leaves is thought to be caused by toxic anthraquinone glycosides rather than oxalates. Experiments involving the consumption of more than $30–35 \, \text{g} \, \text{day}^{-1}$ of cocoa, a high-oxalate foodstuff, by eight women provoked symptoms of intoxication including loss of appetite, nausea, and headaches. However, cocoa contains theobromine ($1500–2500 \, \text{mg} \, 100 \, \text{g}^{-1}$) and tannic acid ($4000–6000 \, \text{mg} \, 100 \, \text{g}^{-1}$), both of which are more toxic than the oxalic acid present ($500–700 \, \text{mg} \, 100 \, \text{g}^{-1}$). There appears to be a great deal of confusion as to what was responsible for these poisonings and it would be unwise to assume that only one factor was responsible.

Effect on Bioavailability of Minerals

High-oxalate foods have been known to inhibit calcium and iron absorption. Even though vegetables such as spinach, rhubarb, and Swiss chard are high in calcium, the calcium cannot be absorbed due to the presence of oxalates in these vegetables. When calcium absorption from spinach, a high-oxalate and high-calcium food, was compared with calcium absorption from milk, a high-calcium food, the results showed that the calcium from spinach is not readily available (only 5.1% absorbed), probably due to the high content of oxalates. The adverse effect of oxalates is greater if the oxalate-to-calcium ratio exceeds 9:4 (or approximately 2). The oxalate-to-calcium ratio in a food varies widely and can be classified into three groups, as summarized in **Table 2**.

Oxalate and calcium levels and the oxalate-to-calcium ratio of specific foods are detailed in **Table 3**.

Foods that have a ratio greater than two, as well as containing no utilizable calcium, have excess oxalate, which can bind calcium in other foods eaten at the same time. Foodstuffs with a ratio of about one do not encroach on the utilization of calcium provided by other products and therefore do not exert any demineralizing effects. However, these foods are not good sources of calcium. Although parsley (*Petroselinum sativum*) contains average levels of oxalate (140–200 mg 100 g^{-1}), its high calcium levels (180–290 mg 100 g^{-1}) reduce the oxalate-to-calcium ratio to a low level.

Oxalate appears to interfere only slightly with zinc absorption. A counteracting or protective mechanism may prevent the precipitation of zinc by oxalates. Increasing the proportion of magnesium ions in solution was reported to inhibit the binding of calcium and zinc oxalates. This observation explains the minor effect oxalates have on zinc absorption from some leafy vegetables, such as spinach, which has high levels of calcium and zinc, and relatively high levels of magnesium.

Oxalic acid may cause greater decreases in mineral availability if consumed with a high-fiber diet but the decrease may be only temporary. Negative calcium, magnesium, zinc, and copper balances were detected in males consuming a diet containing fiber and oxalates. When spinach was replaced by cauliflower, a low-oxalate vegetable, fiber had no effect on the minerals studied, indicating that the apparent negative balances obtained were due to the presence of oxalic acid.

Adverse Effects of Oxalates

Acute

A number of plants contain calcium oxalate crystals (measured as insoluble oxalate). When ingested, they are not absorbed into the blood stream and remain largely undissolved within the digestive tract, so they have no systemic toxicity, but the sharp raphide crystals can penetrate the tissues of the mouth and the tongue, causing considerable discomfort. Most of the plants that contain calcium oxalate crystals are members of the arum family. It has been suggested that calcium oxalate crystals are responsible for the irritating sensation in kiwi fruit (*Actinidia* sp.) and soluble oxalates are thought to account for the bitter taste present in some oca (*O. tuberosa* Mol.). Conophor seeds (*Tetracarpidium conophorum*) are a popular Nigerian snack, which have a bitter taste when raw but are palatable when cooked. This observation was correlated with a 73% decrease in total oxalate concentration after cooking.

Chronic

Oxalate is poorly absorbed under nonfasting conditions. Once absorbed free oxalate binds to calcium ions to form insoluble calcium oxalate, it remains in the insoluble form.

Free oxalate and calcium can precipitate in the urine and may form kidney stones. These stones consist mainly of calcium oxalate (80%), which is relatively insoluble in urine, and calcium phosphate (5%). Oxalate crystallizes with calcium in the renal vasculature and infiltrates vessel walls causing renal tubular obstruction, vascular necrosis, and hemorrhage, which lead to anuria, uremia, electrolyte disturbances, or even rupture. Kidney stones are becoming more common in men between the ages 30 and 50 years in industrialized countries. The risk factors involved in stone formation are a low volume of urine, increased urinary excretion of oxalate, calcium, or uric acid, a persistently low or high urinary

Table 2 Examples of plants with varying oxalate to calcium ratios

Oxalate-to-calcium ratio	Examples
Group 1: Plants with a ratio greater than 2	Spinach, rhubarb, beet, sorrel, cocoa
Group 2: Plants with a ratio of approx. 1	Potatoes, amaranth, gooseberries, currants
Group 3: Plants with a ratio less than 1	Lettuce, cabbage, cauliflower, green beans, peas

Table 3 Oxalate, calcium, and oxalate-to-calcium ratio (Ox:Ca) of some common foods

Foodstuff	Oxalate (mg 100 g^{-1} FW)		Calcium (mg 100 g^{-1} FW)		Ox:Ca ratio
	Range	Mean	Range	Mean	(mmol l^{-1})
Group 1					
Rhubarb (*Rheum rhaponticum*)					
cv. Victoria, forced, stewed		260		12.4	9.32
raw	275–1336	805	40–50	45	7.95
Common sorrel (*Rumex acetosa*)	270–730	500	35–45	40	5.56
Red beetroot (*Beta vulgaris*)	121–450	275	121–450	275	5.09
Garden sorrel (*Rumex patientia*)	300–700	500	40–50	45	4.94
Pig spinach (*Chenopodium* spp.)		1100		99	4.94
Purslane (*Portulaca oleracea*)	910–1679	1294	13–236	125	4.60
Spinach (*Spinacia oleracea*)	320–1260	970	80–122	101	4.27
Garden orach (*Atriplex hortensis*)	300–1500	900		100	4.00
New Zealand spinach (*Tetragonia expansa*)		890		100	3.96
Coffee (*Coffea arabica*)	50–150	100	10–15	12	3.70
Cashew (*Anacardium occidentale*)		231		41	2.50
Cocoa (*Theobroma cacao*)	500–900	700	100–150	125	2.49
Beet leaves (*Beta vulgaris* var. *cicla*)	300–920	610	100–120	110	2.46
Rhubarb (*Rheum rhaponticum*)					
cv. Crimson, end of season, stewed		460		91.5	2.23
Group 2					
Potato (*Solanum tuberosum*)	20–141	80	10–34	22	1.62
Amaranth (*Amaranthus polygonoicles*)		1586		595	1.18
Tea (*Thea chinensis*)	300–2000	1150	400–500	450	1.14
Amaranth (*Amaranthus tricolor*)		1087		453	1.07
Rhubarb (*Rheum rhaponticum*)					
cv. Victoria, end of season, stewed		620		266	1.04
Group 3					
Apple (*Malus* spp.)	0–30	15	5–15	10	0.67
Blackcurrant (*Ribes nigrum*)	2–90	50	19–50	35	0.63
Tomato (*Lycopersicum esculentum*)	5–35	20	10–20	15	0.58
Parsley (*Petroselinum sativum*)	140–200	170	180–290	235	0.32
Cabbage (*Brassica oleracea*)	0–125	60	200–300	250	0.11
Lettuce (*Lactuca sativa*)	5–20	12	73–90	81	0.07

FW, fresh weight.

Adapted from Zarembski PM and Hodgkinson A (1962) The oxalic acid content of English diets. *British Journal of Nutrition* 16: 627–634; Gontzea I and Sutzescu P (1968) *Natural Antinutritive Substances in Foodstuffs and Forages*, pp. 84–108. Basel: S Karger; Meena BA, Umapathy KP, Pankaja N and Prakash J (1987) Soluble and insoluble oxalates in selected foods. *Journal of Food Science and Technology* 24: 43–44; Noonan SC and Savage GP (1999) Oxalates and its effects on humans. *Asia Pacific Journal of Clinical Nutrition* 8: 64–74 with permission.

pH, and a low concentration of urinary inhibitors, such as magnesium, citrate, and high-molecular-weight polyanions. Normal urine is usually supersaturated with calcium oxalate. The normal urinary excretion of oxalate is less than 40–50 mg day^{-1} with less than 10% coming from the diet. Intakes of oxalate exceeding 180 mg day^{-1} lead to a marked increase in the amount excreted. Small increases in oxalate excretion have pronounced effects on the production of calcium oxalate in the urine, implying that foods high in oxalate can promote hyperoxaluria (high oxalate excretion) and increase the risk of stone formation. Rhubarb, spinach, beet, nuts, chocolate, tea, coffee, parsley, celery, and wheat bran cause significant increases in urinary oxalate excretion in healthy individuals and have been identified as the main dietary sources in the risk of kidney stone formation. It

has been reported that black tea increased oxalate excretion by only 7.9%, compared with increases of 300% and 400% for spinach and rhubarb, respectively. Therefore 2–3 cups a day of black tea would have little effect on the risk of urinary stone formation when compared to spinach and rhubarb. It appears that tea is a significant source of oxalate intake in UK diets.

The main reason for the strong relationship between the risk of calcium stones and urinary oxalate excretion appears to be the effect that the latter has on the supersaturation of urine with calcium oxalate. The amount of oxalate excreted in the urine was higher in individuals with kidney stones than in healthy individuals, suggesting that those with kidney stones absorb more oxalate, consume more oxalate or oxalate-producing substances such as ascorbate, or metabolize more oxalate precursors. Excessive or increased absorption of oxalate from normal diets is the result of intestinal abnormalities or malfunction. This is termed 'enteric hyperoxaluria' and is the commonest cause of increased renal oxalate excretion. It has been indicated that people with abnormal gastrointestinal absorption are at greater risk for hyperoxaluria and, as a result, kidney stone formation, than healthy individuals and should reduce their intake of oxalate and its precursors, such as ascorbate. A low-oxalate diet has prevented stone formation in some cases involving gastrointestinal disorders associated with hyperoxaluria.

An increase in calcium intake should be accompanied by a lower oxalate consumption, because a low-calcium and high-oxalate diet enhances oxalate absorption and excretion, which carries an even greater risk of stone formation than high calcium excretion. An increase in calcium intake may reduce urinary oxalate excretion by binding to more oxalate in the gut, thus reducing the risk of stone formation. Varying the amounts of calcium does not significantly alter levels of urinary calcium. From experimental work, it has been concluded that hypercalciuria plays, at most, a secondary role in the formation of calcium stones compared with mild hyperoxaluria.

Excessive ascorbic acid (vitamin C) intake may increase urinary oxalate output with an increased risk of forming kidney stones. An excess dose is considered to be 2000 mg of vitamin C per day. However, ascorbic acid doses greater than $500 \, mg \, day^{-1}$ were reported to induce a significant increase in urinary oxalate, and doses of $1000 \, mg \, day^{-1}$ would increase urinary oxalate excretion by $6–13 \, mg \, day^{-1}$. The recommended daily intake in many countries is in the region of 80 mg.

Effects of Processing

Oxalates may be removed from food by leaching in water but this is not the most effective method as it removes only the soluble oxalate. Although the amount of oxalate in raw soybean (Glycine max) is relatively low, soaking and germination of the seed reduced the oxalate concentration. Cooking germinated soybeans reduced oxalate concentration below that in uncooked germinated soybeans. Soaking followed by cooking also proved to be effective, although not as effective as germination. Oxalate content in horsegram seeds (M. uniflorum) decreased by 38% when seeds were dehulled (508 and 315 mg $100 \, g^{-1}$, for seed and dehulled seed, respectively). Roasting was found to be the least effective method. Roasting chicory roots was reported to increase oxalate content. Roasting oca (New Zealand yam,

O. tuberosa Mol.) also increased oxalate levels by 10–26%. This may be caused by the decrease in moisture content, a hypothesis supported by reports of dry tropical leafy vegetables having higher oxalate concentrations than fresh vegetables.

A 40–50% loss of total oxalates by leaching was reported when yam tubers (D. alata and D. esculenta) were boiled, compared to steaming (20–25%) and baking (12–15%). Cooking proved most effective in reducing total oxalates. However, it must be noted that water-soluble minerals also leach out at the same time. Mineral leaching appears to vary between plant species. Blanching has been reported to decrease the oxalic acid content in spinach. However blanching, by conventional and microwave methods, reduced the oxalic acid content of sweet potato, peanut, and collard leaves only slightly whereas other antinutritional factors such as tannic and phytic acid were reduced significantly. Spinach, orach, and silverbeet are generally eaten after being boiled. However, rhubarb, cocoa, and common and garden sorrel may be consumed in the raw state and therefore should be eaten in smaller quantities.

Fermentation, frequently used in Asian countries, has been reported to decrease the oxalate content of foods. A marked decrease in oxalic acid content was reported in Icacinia manni (a starch tuber) after fermentation. Oxalic acid was observed to decrease by 37% (86 to $54 \, mg \, 100 \, g^{-1}$ FW) during souring of poi (a cooked taro paste) at $20 \, °C$.

Recommendations

Foods high in oxalates should be consumed in moderation to insure optimum intake of minerals from the diet. Although some foods are reported to be high in calcium and other essential minerals, the amount available may be limited due to the presence of oxalates. For instance, spinach is a high-calcium food, yet because of its high oxalate content, the calcium availability is almost negligible. The availability of magnesium, iron, sodium, potassium, and phosphorus may also be restricted.

High-oxalate foods should be cooked to reduce the oxalate content. Soaking raw foods will also reduce the oxalate content but other useful nutrients such as water-soluble vitamins and minerals may also be lost at the same time. Oxalates tend to occur in higher concentrations in the leafy parts of vegetables rather than in roots or stalks.

For the general population, the occasional consumption of high-oxalate foods as part of a balanced diet does not pose any health problems. However, there are some groups of people who may be at risk from oxalate-induced side-effects.

Vegans and vegetarians should be aware that some foods contain high levels of oxalates. The diets of vegans and those persons with lactose intolerance may be low in calcium as a result of the exclusion of dairy products, unless their diet is supplemented by some other high-calcium food products. It is recommended that high-oxalate foods should be accompanied by calcium-rich foods such as dairy products and shellfish. If high-oxalate foods are consumed in conjunction with a low-calcium diet, then the consumer may be at risk of hyperoxaluria, which may lead to kidney stone formation.

Women tend to be more susceptible to calcium and iron deficiencies than men. Osteoporosis is a concern amongst females, especially after menopause. People suffering from fractures should also be aware of the potential effects of oxalates on mineral availability, as high calcium levels are required for bone repair. Once again, consumption of high-oxalate foods with an adequate-to-high calcium intake should pose no health problems. It must also be noted that calcium is only absorbed and used when there are adequate levels of vitamin D in the body, either obtained via the diet or synthesized by the body when exposed to sunlight. Women should eat red meats, which are low in oxalate, to satisfy their iron intake. Adequate levels of vitamin C are required for the absorption of iron, but excess amounts are not advised because ascorbic acid is converted into oxalate.

The risk of stone formation is three times greater in males and they should avoid eating excess amounts of high-oxalate foods. Sufferers of hyperoxaluria and kidney stones are advised to restrict their diet to foods containing low or medium levels of oxalates, as although urinary oxalate arises predominantly from endogenous sources, it can be influenced by dietary intake. Excess vitamin C intake is not recommended in these patients.

Inhabitants of tropical countries should be aware that leafy tropical plants and tropical root crops tend to contain higher levels of oxalates than plants from temperate climates. People living in these areas are at possible risk of stone formation due to hyperoxaluria, and mineral deficiencies if sufficient minerals are not consumed.

See also: **Ascorbic Acid**: Properties and Determination Physiology; **Calcium**: Properties and Determination; Physiology; **Iron**: Properties and Determination; Physiology; **Plant Antinutritional Factors**: Characteristics; **Renal Function and Disorders**: Nutritional Management of Renal Disorders; **Toxins in Food – Naturally Occurring**

Further Reading

Concon JM (1988) *Food Toxicology – Principles and Concepts*, pp. 416–419. New York: Marcel Dekker.

Dobbins JW and Binder HJ (1977) Importance of the colon in enteric hyperoxaluria. *New England Journal of Medicine* 296: 298–301.

Gontzea I and Sutzescu P (1968) *Natural Antinutritive Substances in Foodstuffs and Forages*, pp. 84–108. Basel: S Karger.

Hagler L and Herman RH (1973) Oxalate metabolism I. *American Journal of Clinical Nutrition* 26: 758–765.

Hanson CF, Frankos VH and Thompson WO (1989) Bioavailability of oxalic acid from spinach, sugar beet fibre and a solution of sodium oxalate consumed by female volunteers. *Food and Chemical Toxicology* 27: 181–184.

Linder MC (1991) *Nutritional Biochemistry and Metabolism with Clinical Applications*, 2nd edn. New York: Elsevier.

Massey LK, Roman-Smith H and Sutton RAL (1993) Effect of dietary oxalate and calcium on urinary oxalate and risk of formation of calcium oxalate kidney stones. *Journal of the American Dietetic Association* 93: 901–906.

Meena BA, Umapathy KP, Pankaja N and Prakash J (1987) Soluble and insoluble oxalates in selected foods. *Journal of Food Science and Technology* 24: 43–44.

Noonan SC and Savage GP (1999) Oxalates and its effects on humans. *Asia Pacific Journal of Clinical Nutrition* 8: 64–74.

Ross AB, Savage GP, Martin RJ and Vanhanen L (1999) Oxalates in oca (New Zealand yam) (*Oxalis tuberosa* Mol.). *Journal of Agricultural and Food Chemistry* 47: 5019–5022.

Sangketkit C, Savage GP, Martin RJ, Mason SL and Vanhanen L (1999) Oxalates in oca: a negative feature? In: Jenson J and Savage GP (eds) *Second South West Pacific Nutrition and Dietetic Conference Proceedings*, pp. 44–50. Auckland, New Zealand.

Strenge A, Hesse A, Bach D and Vahlensieck W (1981) Excretion of oxalic acid following the ingestion of various amounts of oxalic acid-rich foods. In Smith LH, Robertson WG and Finlayson B (eds) *Urolithiasis: clinical and basic research*, pp. 789–794. New York: Plenum Press.

Wanasundera JPD and Ravindran G (1994) Nutritional assessment of yam (*Dioscorea alata*) tubers. *Plant Foods for Human Nutrition* 46: 33–39.

Zarembski PM and Hodgkinson A (1962) The oxalic acid content of English diets. *British Journal of Nutrition* 16: 627–634.

OXIDATION OF FOOD COMPONENTS

K L Parkin and S Damodaran, University of Wisconsin-Madison, Madison, WI, USA

Scope

Antoine Lavoisier (1743–94) recognized oxidation as a chemical process, concluding that oxygen was the element responsible for the formation of acidic residues, or oxides, upon combustion of certain substances. A contemporary definition of oxidation is the process by which oxygen is added, or hydrogen or electrons are withdrawn. For a component to be oxidized, another has to be reduced, and reduction can be defined as the withdrawal of oxygen, or the addition of hydrogen or electrons. The component that is oxidized and loses electrons is the reductant and the component that is reduced and gains electrons is the oxidant. Oxidation is distinct from oxygenation; the latter is a noncovalent coordination of oxygen with a component, as in the case where hemoglobin and myoglobin bind oxygen to facilitate oxygen transport in blood and muscle, respectively. In food systems, oxygen is the most common oxidant, although other endogenous and added chemicals can also serve as oxidants. The principal negative effect of oxidation in foods is that flavor quality is lost, giving rise to the defect often referred to as oxidative rancidity. In addition, functional, color, and nutritional qualities of food components can be lost as a consequence of oxidation in foods. However, there are also some oxidative processes in foods that have beneficial effects on quality.

The Basic Process of Oxidation

Oxidation–Reduction Potentials

The potential or thermodynamic favorability for two components to be involved in an oxidation–reduction (redox) reaction can be predicted from the corresponding half-reactions of oxidation and reduction. **Table 1** provides a selective list of some standard reduction half-reactions, using the hydrogen half-cell at pH 7.0 as a standard. As the reduction potential (voltage) becomes more positive, the tendency for that half-reaction to take place increases. Thus, the most powerful oxidants in **Table 1** are hydrogen peroxide and oxygen. For each component, oxidation half-reactions take place in the reverse direction of what appears in **Table 1** and

have voltages of the opposite sign of the same magnitude.

The redox potential (E_h) of food systems is dependent on the concentration and redox states of the components of that system. One of the most important components is oxygen and, at limited dissolved oxygen components, E_h is strongly dependent on the oxygen content. 'Reducing' conditions or a very negative E_h (e.g., $-400\,mV$) exist when dissolved oxygen is poised at near-anaerobic levels.

Transition metals such as copper and iron are believed to be involved in oxidation in foods via a redox cycling mechanism. Since there is very little 'free' iron and copper in biological systems, the types and concentrations of chelators present have a marked effect on the redox behavior and thus oxidative activity of these transition metals.

Oxidation in foods is often caused by free radical reactions. There are three stages of free radical oxidation, also referred to as autoxidation when the oxidant is oxygen. The first step, or 'initiation,' involves the formation of a free radical species ($X^{\cdot}$) from a biological component (XH), usually by the abstraction of a hydrogen atom ($H^{\cdot}$) by active oxygen or high-energy irradiation (eqn (1)). 'Propagation' of free radical oxidation processes occurs by chain reactions that consume oxygen and yield new free radical species (peroxy radicals, $XOO^{\cdot}$) or peroxides (XOOH), as in eqns (2) and (3). The products ($X^{\cdot}$ and XOOH, see also eqn (6)) can further propagate free radical reactions. 'Termination' of free radical oxidative reactions occurs when two radical species react with each other to form a nonradical adduct, as in eqn (4).

$$XH \rightarrow X^{\cdot} + H^{\cdot} \qquad (1)$$

$$X^{\cdot} + {}^3O_2 \rightarrow XOO^{\cdot} \qquad (2)$$

Table 1 Standard electrode potentials of selected reduction half-reactions

Reaction	Volts
$H_2O_2 + 2H^+ + 2e^- \rightarrow 2H_2O$	1.77
$O_2 + 4H^+ + 4e^- \rightarrow 2H_2O$	1.23
$Cu^{2+} + e^- \rightarrow Cu^+$	0.15
$Fe^{3+} + e^- \rightarrow Fe_2^+$	0.11
Dehydroascorbate $+ H^+ + 2e^- \rightarrow$ Ascorbate	0.054
$2H^+ + 2e^- \rightarrow H_2$	0.00
$RSSR + 2H^+ + 2e^- \rightarrow 2RSH$	-0.39^a

[a]Estimated value for oxidized disulfide (RSSR) conversion to reduced thiol (RSH) such as for oxidized/reduced glutathione and cystine/cysteine couples.
Standard Conditions: pH 7.0, 1 mol l^{-1} for each component

$$XOO \cdot + XH \to XOOH + X \cdot \qquad (3)$$

$$X \cdot + X \cdot \to X - X \qquad (4)$$

Activation of Oxygen

In food systems, molecular oxygen (dioxygen; O_2) is generally the source of oxidizing power. Other strong oxidants include the food additives hydrogen peroxide (H_2O_2), calcium and benzoyl peroxides, and bromates ($KBrO_3$). All of these compounds are conjugates of oxygen. However, not all strong oxidants are composed of oxygen, as fluorine and bromine are also strong oxidants.

Although the reaction of ground state oxygen ($^3O_2$, triplet oxygen) with organic compounds is thermodynamically favorable, it is kinetically slow due to the high energy of activation required for oxygen to react. The electron configuration of $^3O_2$ includes two unpaired electrons in the outer shell, yielding a triplet signal in a magnetic field. All known organic compounds are in a singlet state, having all electrons paired with another. Consequently, facile reaction between $^3O_2$ and organic molecules is forbidden due to the incompatibility of their electron 'spin' states.

'Activation' of $^3O_2$ overcomes much of the energy barrier to its reactivity as an oxidant. One means of activation is the excitation of $^3O_2$ to yield singlet molecular oxygen ($^1O_2$); the latter has the two outer electrons paired in a single orbital. Other forms of 'activated' or 'reactive' oxygen result from the first three one-electron reductions to $^3O_2$ in the process of reducing $^3O_2$ to water. These reactive species of oxygen include the superoxide anion radical (O_2^-) and its conjugate acid (HO_2), hydrogen peroxide (H_2O_2), and the hydroxyl radical ($\cdot OH$). The standard reduction potentials of each of these steps is provided in **Table 2**. The electronic structure of these activated forms of oxygen facilitate their reactivity with biological compounds. The strongest electrophiles (electron seekers), $\cdot OH$ and $^1O_2$ are the most reactive forms of 'active' oxygen, followed in reactivity by O_2^- and then H_2O_2.

Table 2 Standard electrode potentials for univalent reductions of O_2 to H_2O

Reaction	Volts
$O_2 + e^- \to O_2^-$	$-0.16\ (-0.33)$
$O_2^- + e^- + 2H^+ \to H_2O_2$	0.89
$H_2O_2 + e^- + H^+ \to \cdot OH + H_2O$	0.38
$\cdot OH + e^- + H^+ \to H_2O$	2.32

Standard conditions: pH 7.0, 1 mol l^{-1} for each component. For O_2, electrode potential also provided, in parentheses, at 10^5 Pa (0.987 atm).

Some of these active oxygen species can be interconverted, and these processes can be facilitated by the presence of specific catalysts. Activated forms of oxygen can also be formed by γ irradiation and by photosensitization of pigments in foods.

Catalysts of Oxidation Reactions

Catalysts of oxidation reactions can be enzymatic (protein) or nonenzymatic. Transition metals (M^n, reduced form; M^{n+1}, oxidized form) can participate in redox reactions with $^3O_2$ to yield O_2^-/HO_2, as in eqn. (5). The resulting O_2^- can initiate oxidation reactions. Another manner by which transition metals can cause oxidation reactions is by breaking down lipid hydroperoxides (LOOH) (eqn (6)), and the alkoxy radical (LO·) so formed can cause further oxidative reactions. Since there are often small quantities of LOOH in food systems this process is probably important. Transition metals can also take part in the interconversion of active oxygen species, as in the Haber–Weiss reaction (eqn (7)). This reaction can be mediated by three partial reactions (eqns (8)–(10)). In the first reaction (eqn (8)), O_2^- acts as a reductant and donates an electron to an oxidized transition metal (e.g., iron). In the second step (eqn (9)), O_2^-, acting as both an oxidant and reductant, undergoes dismutation to form H_2O_2 and $^3O_2$. In the third step (eqn (10)), also called the Fenton reaction, the reduced transition metal donates an electron to H_2O_2 to form the extremely reactive $\cdot OH$, and the transition metal reverts back to its oxidized state to allow another cycle. In food systems, several endogenous components, such as ascorbic acid and thiol compounds, can replace O_2^- as a reductant. This set of reactions also illustrates the participation in oxidative reactions of all activated oxygen species generated by univalent electron reductions of $^3O_2$.

$$M^n + {}^3O_2 \to M^{n+1} + O_2^- \qquad (5)$$

$$M^n + LOOH \to M^{n+1} + LO \cdot + OH^- \qquad (6)$$

$$O_2^- + H_2O_2 \to {}^3O_2 + OH^- + \cdot OH \qquad (7)$$

$$O_2^- + M^{n+1} \to {}^3O_2 + M^n \qquad (8)$$

$$2O_2^- + 2H^+ \to {}^3O_2 + H_2O_2 \qquad (9)$$

$$H_2O_2 + M^n \to M^{n+1} + \cdot OH + OH^- \qquad (10)$$

Other nonenzymatic catalysts include photosensitive pigments in foods. Photosensitive pigments become elevated to an excited triplet state upon the absorption of light energy, and can transfer that energy to $^3O_2$ or other biological components. Some pigments favor transmission of energy to organic compounds (type I process) which ultimately yield O_2^- and H_2O_2 from $^3O_2$. Other pigments favor transmission of energy directly to $^3O_2$ to yield $^1O_2$ (type II process). Examples of photosensitizers of each of these types in foods are riboflavin and chlorophyll, respectively. (*See* **Chlorophyl**; **Riboflavin**: Properties and Determination.)

Enzymatic catalysts of oxidative reactions usually cause oxidations of specific biological compounds. For example, the enzymes lipoxygenase, polyphenoloxidase, sulfhydryl oxidase and xanthine oxidase are common to foods and cause the specific oxidation of unsaturated fatty acids, mono- and diphenolic acids, protein thiol (cysteine) residues, and xanthine, respectively. Glucose oxidase converts glucose to gluconic acid and also produces H_2O_2. Xanthine oxidase and peroxidase can produce H_2O_2 and O_2^- and $^1O_2$, respectively, and this is dependent on which substrates are being utilized and the level of oxygen present. These active oxygen species may cause the oxidation of other biological compounds, leading to losses in food quality.

Oxidation of Food Components

Lipids

Polyunsaturated fatty acids with 1,4-pentadiene functional units are particularly sensitive to oxidative reactions. Using linoleic acid as an example, oxidation can be initiated by two basic mechanisms, abstraction (autoxidation) and 'ene' addition (**Figure 1**). Abstraction is when an electron (or hydrogen atom) is removed from the fatty acid by reaction with an electrophilic species such as ·OH or X·, or by interaction with high-energy radiation. The initial abstraction step yields a fatty acid free radical (initiation step) which can then undergo addition of $^3O_2$ (propagation step) and then abstract an electron from another biological compound. The methylene or 'allylic' hydrogen atoms of the pentadiene structure are most readily abstracted. The resulting free radical (L·) can be stabilized by resonance along the original pentadiene structure, and the fatty acid radical tends to undergo addition of $^3O_2$ when the unpaired electron is most 'delocalized' or located at the terminal sites, or C9 and C13, resulting in the formation of first the linoleic acid 9- and 13-hydroperoxyl radicals, and then the 9- and 13-OOH (hydroperoxides) isomers. Further oxidative processes can be initiated by interaction of these hydroperoxides with transition metals as previously described in eqn (6). (*See* **Fatty Acids**: Properties.)

The 'ene' addition reaction that can initiate lipid oxidation is caused by the highly electrophilic $^1O_2$,

which will add directly to the double bond since this is where the highest electron density can be found. Thus, a mixture of 9-, 10-, 12-, and 13-OOH isomers are produced by $^1O_2$ reaction with linoleic acid.

Once the fatty acid hydroperoxides and hydroperoxyl radicals are formed, additional initiation reactions can take place for these initial products, being unstable, can be subject to secondary reactions, as shown for linolenic acid oxidation (**Figure 2**). Oxidation of any remaining double bonds can take place, and in some cases the fatty acid radicals can attack adjacent intramolecular double bonds, forming cyclic structures. Alternatively, hydroperoxyl fatty acids can react with adjacent fatty acids to yield polymerized oxidation products. 'Scission' reactions lead to fracture of the fatty acid chain and result in the emanation of reduced molecular weight ketones and aldehydes. These latter secondary products, being fairly volatile, give rise to the off-flavors and odors that are associated with oxidized foods or oxidative rancidity. One product that can be formed by secondary reactions of oxidizing lipids is malondialdehyde (MDA). MDA is often used by food scientists as an indicator of the degree of oxidation of lipids in foods. In cases where specific secondary products are formed by enzyme reactions, such as by lipid hydroperoxide lyases in freshly cut cucumber and tomato fruits, the resulting volatile compounds are pleasant and contribute desirable aromatic qualities.

Thermally induced oxidation reactions can occur in both saturated and unsaturated lipids at temperatures encountered during processes such as deep-fat frying. Oxidation generally proceeds via the initial formation of hydroperoxides. The high temperatures can cause many isomerization and scission reactions to take place, producing a myriad of secondary or breakdown products such as epoxides, dihydroperoxides, cyclized fatty acids, dimers, and aldehydes and ketones resulting from scission reactions.

Proteins

Proteins, peptides, and amino acids in foods undergo several oxidative changes during food processing. The amino acids that are most susceptible to oxidative degradation are methionine, cysteine (cystine), histidine, and tryptophan. Under severe oxidizing conditions tyrosine, serine, and threonine are also oxidized to some extent. Oxidation of proteins and amino acids is caused by several agents, such as light, γ irradiation, peroxidizing lipids, metal ions, the products of enzymatic and nonenzymatic browning reactions, and food additives such as hydrogen peroxide, benzoyl peroxide, bromates ($KBrO_3$) and azodicarbonamide. (*See* **Amino Acids**: Properties and Occurrence; **Protein**: Chemistry.)

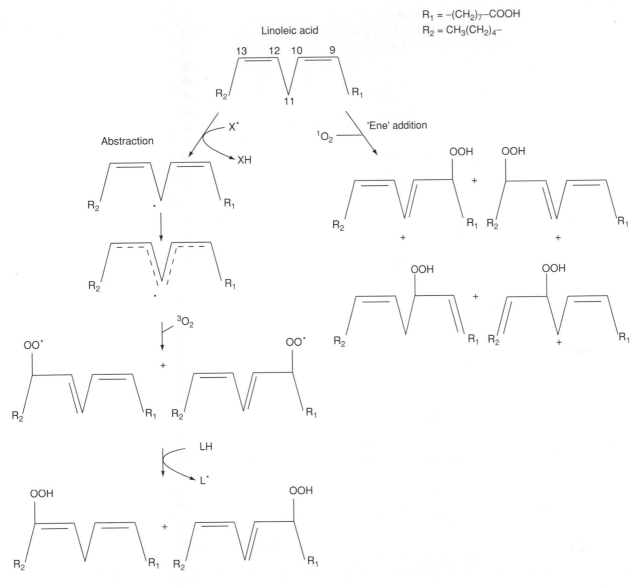

Figure 1 Initiation reactions for the oxidation of linoleic acid. Reproduced from Oxidation of Food Components, *Encyclopaedia of Food Science, Food Technology and Nutrition*, Macrae R, Robinson RK and Sadler MJ (eds), 1993, Academic Press.

Treatment of proteins with hydrogen peroxide or calcium peroxide causes oxidation of methionine sulfoxide (reversible), which can be further oxidized to methionine sulfones (irreversible) (eqn (11)). Cysteine residues can be oxidized by peroxides or other forms of activated oxygen to yield the sulfenic (Cy-SOH), sulfinic (Cy—SO_2H) and sulfonic (Cy—SO_3H), acid derivatives. Oxidation of cystine residues in proteins results in the formation of mono-, di-, tri-, and tetrasulfoxides.

Free thiol groups in proteins are readily oxidized by atmospheric oxygen to form disulfide cross-links. Free thiol groups also catalyze thiol-disulfide interchange reactions, which often lead to polymerization of proteins. Oxidizing agents such as $KBrO_3$ and azodicar-bonamide are often used as additives in wheat flour in order to improve dough formation. These additives are believed to oxidize and block the free thiol groups of protein and nonprotein constituents, and thus prevent the occurrence of thiol–disulfide interchange reactions in the dough. The modulation of oxidation–reduction behavior in dough systems may also be controlled by ascorbic acid, dehydroascorbic acid, and glutathione.

When foods containing photosensitive substances, such as riboflavin and chlorophyl, are exposed to light, the amino acids histidine, cysteine, methionine, tryptophan, and tyrosine are oxidized by the activated oxygen species O_2^-, H_2O_2 and $^1O_2$. γ Irradiation of foods results in the formation of H_2O_2

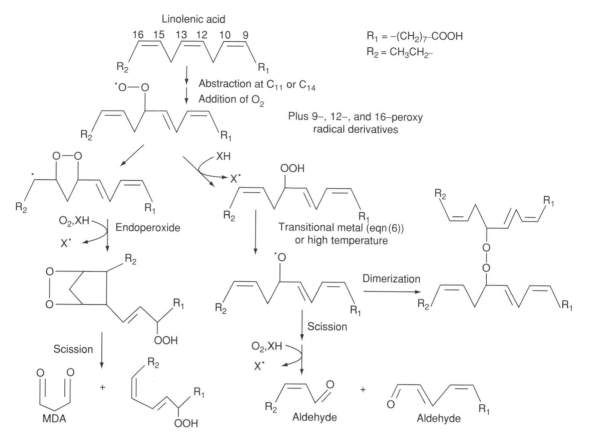

Figure 2 Initiation and secondary reactions for the oxidation of linolenic acid. Reproduced from Oxidation of Food Components, *Encyclopaedia of Food Science, Food Technology and Nutrition*, Macrae R, Robinson RK and Sadler MJ (eds), 1993, Academic Press.

$$-CH_2-CH_2-S-CH_3 \rightleftharpoons -CH_2-CH_2-\overset{O}{\overset{\|}{S}}-CH_3 \rightarrow -CH_2-CH_2-\overset{O}{\overset{\|}{\underset{\|}{S}}}-CH_3 \qquad (11)$$

through radiolysis of water in the presence of oxygen, which in turn causes oxidative changes in proteins. Tryptophan residues can also be oxidized upon exposure of proteins to acidic conditions.

Substantial oxidation of free amino acids and amino acid residues in proteins occurs in the presence of peroxidizing lipids. Methionine, cysteine, histidine, and lysine are the most susceptible amino acids/residues. Two types of mechanisms, one involving the alkoxy (LO·) and peroxy (LOO·) free radicals, and the other involving malondialdehyde and other carbonyl compounds, are believed to be involved in the oxidation of proteins by peroxidizing lipids. In the first case, the lipid free radicals react with proteins (P) and induce formation of protein free radicals (P·), followed by polymerization of protein molecules (eqns (12)–(16)). In addition to the free radical-induced polymerization of protein molecules, the lipid peroxides formed during the reactions oxidize methionine, cysteine, histidine, and tryptophan residues. The highly reactive malondialdehyde formed in peroxidizing lipids reacts with amino groups of lysyl residues, resulting in intermolecular cross-links.

$$LOO· + P \rightarrow ·LOOP \underset{O_2}{\rightarrow} ·OOLOOP \underset{P}{\rightarrow} POOLOOP \qquad (12)$$

or

$$LO· + P \rightarrow LOH + P· \qquad (13)$$
$$LOO· + P \rightarrow LOOH + P· \qquad (14)$$
$$P· + P \rightarrow P-P· \qquad (15)$$
$$P-P· + P \rightarrow P-P-P· \qquad (16)$$

Heat treatment of proteinaceous foods causes several oxidative changes in proteins. While mild heat treatment results in protein denaturation and loss of functionality, severe heat treatment often causes undesirable chemical changes in amino acid residues and complex reactions of proteins with other food components, such as carbohydrates and lipids. (*See* **Browning**: Nonenzymatic.)

When protein is heated at temperatures above 300 °C, as commonly encountered during broiling and grilling, several amino acid residues undergo thermal decomposition and pyrolysis. Several of these pyrolysis products have been isolated, identified, and shown to be highly mutagenic. The most carcinogenic/mutagenic products are formed from the decomposition of tryptophan, glutamate and lysyl residues. (*See* **Carcinogens**: Carcinogenic Substances in Food: Mechanisms; **Mutagens**.)

Carbohydrates

Carbohydrates are not as sensitive to oxidation reactions as are lipids and proteins. In addition, since many oxidation products are not volatile, the practical consequences of carbohydrate oxidations in foods are limited. Oxidation of food carbohydrates can take place, especially at high temperatures, resulting in caramelization reactions. (*See* **Caramel**: Methods of Manufacture; **Carbohydrates**: Interactions with Other Food Components.)

Some industrial processes employ oxidation reactions to prepare functional derivatives of monosaccharides for the chemical industries. However, the carbohydrate oxidations of most relevance to foods involve enzymatic reactions. Glucose oxidase oxidizes glucose to gluconic acid while simultaneously reducing O_2 to H_2O_2. The enzyme is added to foods to reduce glucose levels (to prevent nonenzymatic browning in eggs to be dried) or oxygen tension (to stabilize beverages and salad dressings from oxidative deterioration).

Carbohydrates can be oxidized by the same free radical mechanisms as described for lipids. Low-molecular-weight carbohydrates such as glucose, mannitol, and deoxyribose are known to react with ·OH and produce oxidized derivatives. Again, these derivatives, when present, have little impact on food quality and are thus of little practical significance.

Minor Food Components

Oxidation of minor food components can also influence food quality. Oxidation of ascorbic acid by enzymatic (ascorbic acid oxidase) or nonenzymatic means can compromise nutritional quality. Ascorbic acid, and other organic acids, can be degraded by active oxygen species and by reactions initiated by transition metals. Polyphenol oxidase and tyrosinase are enzymes found in plant and crustacean foods that oxidize phenolic acids and initiate secondary nonenzymatic reactions that are responsible for darkening and often loss of color quality. (*See* **Ascorbic Acid**: Properties and Determination.)

Environmental Factors

Temperature

Generally, as the temperature increases, the rate of oxidation reactions also increases. Rate increases usually follow a $Q_{10} = 2$ relationship, or a doubling in rate for every rise of 10 °C, provided no change in the mechanism of reaction occurs with a corresponding change in temperature and as long as competing reactions have little impact on the reactants. For oxidative reactions caused by enzymes, an optimum temperature exists. This is because enzymes, being proteins, are denatured above a characteristic temperature and will lose biological activity. Another factor is that, as temperature increases, oxygen solubility in water decreases and this could attenuate the temperature activation of oxidative reactions if oxygen was a limiting component in the process.

Moisture

Lipid oxidation reactions generally have rate minima at intermediate water activities (a_w) of about 0.3. As a_w increases above 0.3, rates of oxidation increase, probably due to increased mobility and activity of catalysts. At a_w below 0.3, rates of oxidation increase, perhaps due to solvation and removal of catalysts and reaction intermediates from lipids and into the aqueous phase. (*See* **Water Activity**: Effect on Food Stability.)

Chemical Composition of Food

The relative concentrations of various prooxidants and antioxidants in the food, and their identities and relative reactivities, will greatly influence the rate of oxidation reactions. Antioxidants, such as butylated hydroxytoluene, butylated hydroxyanisole, propyl gallate, ascorbic acid, and sodium bisulfite can be added to impede oxidative reactions. On the other hand, the inadvertent addition of transition metals, such as iron and copper, from processing equipment can hasten oxidative reactions in foods. In addition, the nature of the substrate sensitive to oxidative reactions is important. For example, foods rich in polyunsaturated fatty acids (vegetable and fish oils) are more sensitive to oxidation than those rich in monounsaturated fatty acids (animal depot fats). (*See* **Antioxidants**: Natural Antioxidants; Synthetic Antioxidants.)

Exogenous Factors

Packaging materials and strategies can have an influence on rates of oxidation. Foods containing photosensitive pigments are often packed in opaque or translucent containers to minimize photooxidative processes by preventing the activation of oxygen by these pigments. The head space of products in containers can also be controlled to inhibit oxidative processes. For example, foods can be sealed in containers after the head space has been flushed with an inert gas such as nitrogen or under vacuum. Both approaches serve to minimize the oxygen available to support oxidative reactions. Products may also be coated, such as with sugar syrups on fruits to be frozen, to minimize the availability of oxygen for oxidative reactions.

Examples in Foods

Many foods are susceptible to oxidative reactions. Plant and fish oils, due to their high levels of unsaturation, can deteriorate in flavor very rapidly if oxidation is allowed to take place. Lipid oxidation gives rise to the 'fishy' flavor and aroma in frozen fish and this limits acceptable storage life. Refining procedures for vegetable and seed oils are partly designed to remove chlorophyl which can cause flavor deterioration by initiating photooxidative reactions. Another light-sensitive food is milk. The endogenous riboflavin can cause photooxidation of lipids and proteins and yield the undesirable 'light-activated flavor.' Early recognition of this problem gave rise to the domestic delivery of milk in opaque containers, in many countries, to prevent its exposure to light. (*See* **Fish Oils**: Composition and Properties; **Milk**: Processing of Liquid Milk; **Vegetable Oils**: Dietary Importance.)

Nuts and high-fat products, such as potato crisps, can be packaged in containers having a modified atmosphere or head space. Nitrogen flushing of containers and packing under vacuum are processes designed to limit the amount of oxygen available for oxidation reactions. (*See* **Chilled Storage**: Use of Modified-atmosphere Packaging.)

Most plant tissues, particularly fresh fruits, brown excessively upon cutting or bruising. This is often due to the presence of polyphenol oxidase, which acts on phenolic acids in these foods. After the initial enzymatic oxidations, a series of subsequent nonenzymatic oxidations convert these phenolic acids into polymers that are responsible for the brown color. (*See* **Phenolic Compounds**.)

Not all oxidative reactions in foods are undesirable. In the extrusion processing of foods at alkaline pH, the oxidative polymerization of proteins results in the desirable texturization in simulated meat products. Protein and lipid oxidation are also recognized for their beneficial effects on dough strengthening in the baking industries. A third example is the emanation of characteristic flavors and aromas upon slicing of fresh cucumbers and tomatoes. The sources of these flavors are unsaturated fatty acids which have been initially oxidized by endogenous lipoxygenase activity and have been transformed further by other enzymes. The chemical 'fermentation' of tea leaves and the 'ripening' of olives are achieved by potentiating polyphenol oxidase enzyme activity on endogenous phenolic acids, causing the development of desirable coloration in these products.

Generally, oxidative reactions that can be controlled can be manipulated to yield beneficial effects, whereas those that cannot often yield detrimental effects on food quality.

See also: **Antioxidants**: Natural Antioxidants; Synthetic Antioxidants; **Ascorbic Acid**: Properties and Determination; **Browning**: Nonenzymatic; **Chilled Storage**: Use of Modified-atmosphere Packaging; **Chlorophyl**; **Fatty Acids**: Metabolism; **Fish Oils**: Composition and Properties; **Milk**: Processing of Liquid Milk; **Phenolic Compounds**; **Riboflavin**: Properties and Determination; **Vegetable Oils**: Dietary Importance; **Water Activity**: Effect on Food Stability

Further Reading

Buettner GR (1993) The pecking order of free radicals and antioxidants: lipid peroxidation, α-tocopherol, and ascorbate. *Archives of Biochemistry and Biophysics* 300: 535–543.

Clark WM (1960) *Oxidation–Reduction Potentials of Organic Systems*. Baltimore: Williams and Wilkins.

Frankel EN (1998) *Lipid Oxidation*. Dundee: Theory Press.

Kanner J, German JB and Kinsella JE (1987) Initiation of lipid peroxidation in biological systems. *CRC Critical Reviews in Food Science and Nutrition* 25: 317–364.

Packer L (ed.) (1984) Oxygen radicals in biological systems. *Methods in Enzymology* 105.

Richardson T and Finley JW (eds) (1985) *Chemical Changes in Food During Processing*. Westport, CT: AVI.

OXIDATIVE PHOSPHORYLATION

D A Bender, University College London, London, UK

Background

The total body content of ATP is of the order of 10 g, whereas the daily turnover of ATP is equal to the body weight, some 70 kg. A small number of metabolic reactions involve direct transfer of phosphate from a phosphorylated substrate on to ADP, forming ATP – substrate-level phosphorylation. Under normal conditions, almost all of the phosphorylation of ADP to ATP occurs in the mitochondria, by the process of oxidative phosphorylation – the oxidation of reduced coenzymes linked to the reduction of oxygen to water and (under normal conditions) obligatorily linked to phosphorylation of ADP $\rightarrow$ ATP. This obligatory linkage of substrate oxidation, reoxidation of reduced coenzymes, and reduction of oxygen to water with the phosphorylation of ADP mean that the availability of ADP controls the rate at which substrates are oxidized. In turn, the availability of ADP to be phosphorylated is dependent on the rate of utilization of ATP in performing physical and chemical work. Thus, energy expenditure in physical and chemical work controls the rate at which metabolic fuels are oxidized, rather than being used to form reserves of (mainly) adipose tissue triacylglycerol.

With the exception of glycolysis and the pentose phosphate pathway, most of the reactions in the oxidation of metabolic fuels occur inside the mitochondria and lead to the reduction of nicotinamide nucleotide and flavin coenzymes. Within the inner membrane of the mitochondrion, there is a series of coenzymes that are able to undergo reduction and oxidation. The first coenzyme in the chain is reduced by reaction with NADH, and is then reoxidized by reducing the next coenzyme. In turn, each coenzyme in the chain is reduced by the preceding coenzyme, and then reoxidized by reducing the next coenzyme. The final step is the oxidation of a reduced coenzyme by oxygen, resulting in the formation of water. **Figure 1** shows an overview of this mitochondrial electron transport chain.

Experimentally, the electron transport chain can be dissected into four complexes of coenzymes, which catalyze:

1. Oxidation of NADH leading to the reduction of ubiquinone to ubiquinol. This complex is associated with the phosphorylation of ADP $\rightarrow$ ATP.

2. Oxidation of reduced flavoproteins and reduction of ubiquinone to ubiquinol. This complex is not associated with phosphorylation of ADP.

3. Oxidation of ubiquinol, leading to reduction of cytochrome c. This complex is associated with the phosphorylation of ADP $\rightarrow$ ATP.

4. Oxidation of reduced cytochrome c leading to the reduction of oxygen to water. This complex is associated with the phosphorylation of ADP $\rightarrow$ ATP.

This means that there are three sites in the electron transport chain between NADH and oxygen that are linked to the phosphorylation of ADP $\rightarrow$ ATP, but only two between reduced flavoproteins and oxygen. Experimentally, this is seen as a ratio of phosphate esterified:oxygen consumed (the P:O ratio) of approximately 3 when substrates that reduce NAD^+ are oxidized, and approximately 2 when substrates that reduce flavoproteins are oxidized.

Experimentally, mitochondrial metabolism is measured using the oxygen electrode, in which the percentage saturation of the buffer with oxygen is measured electrochemically as the mitochondria oxidize substrates and reduce oxygen to water. **Figure 2** shows the oxygen electrode traces for oxidation of malate (which is linked to reduction of NAD^+) and succinate (which is linked to reduction of a flavoprotein). The greater consumption of oxygen for oxidation of succinate compared with the same amount of malate reflects the lower P:O ratio for succinate oxidation.

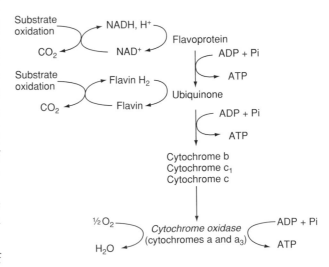

Figure 1 Overview of the mitochondrial electron transport chain.

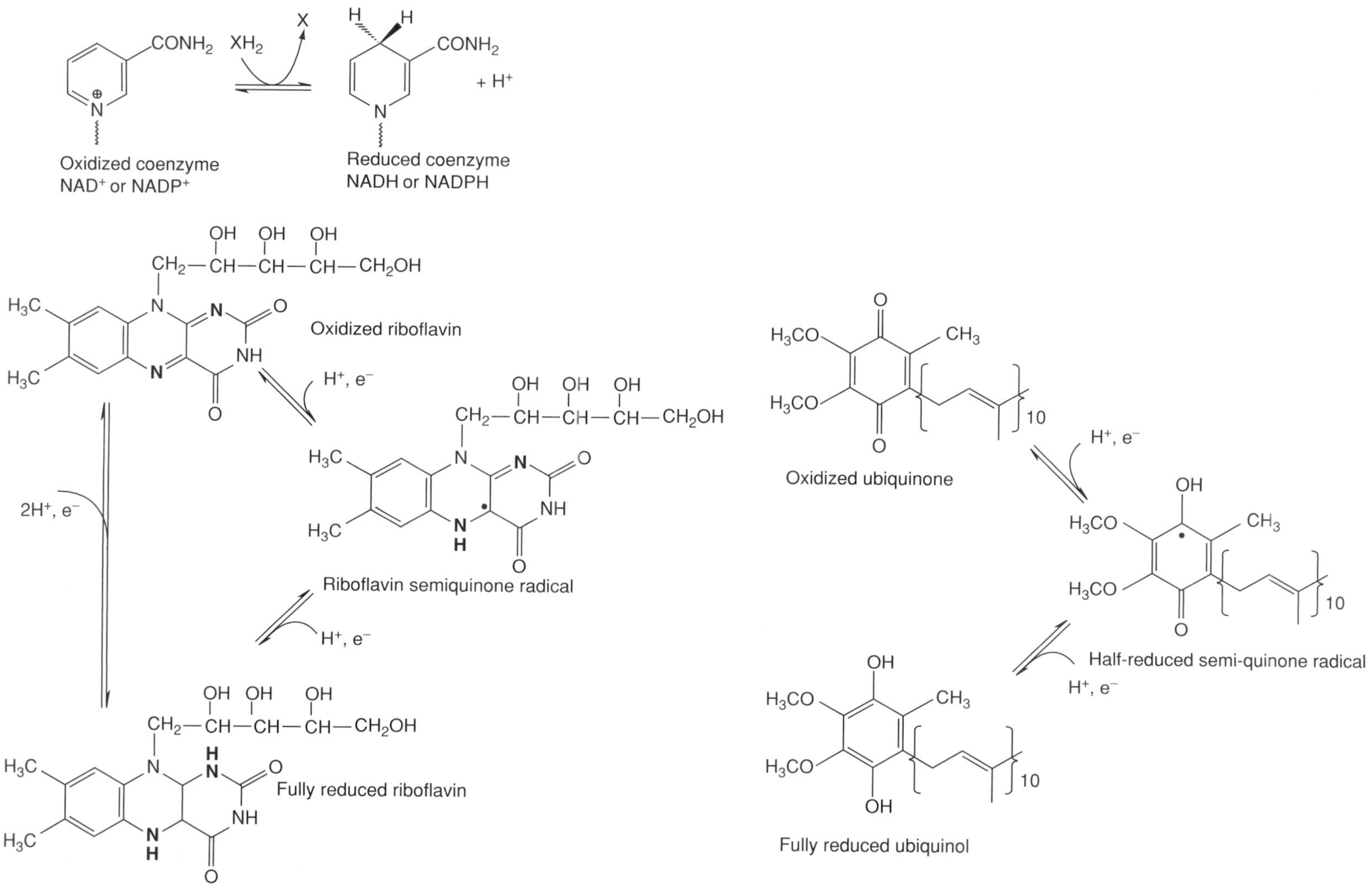

Figure 5 Oxidation and reduction of the hydrogen carriers of the electron transport chain: the nicotinamide nucleotide coenzymes (NAD and NADP), flavins and ubiquinone.

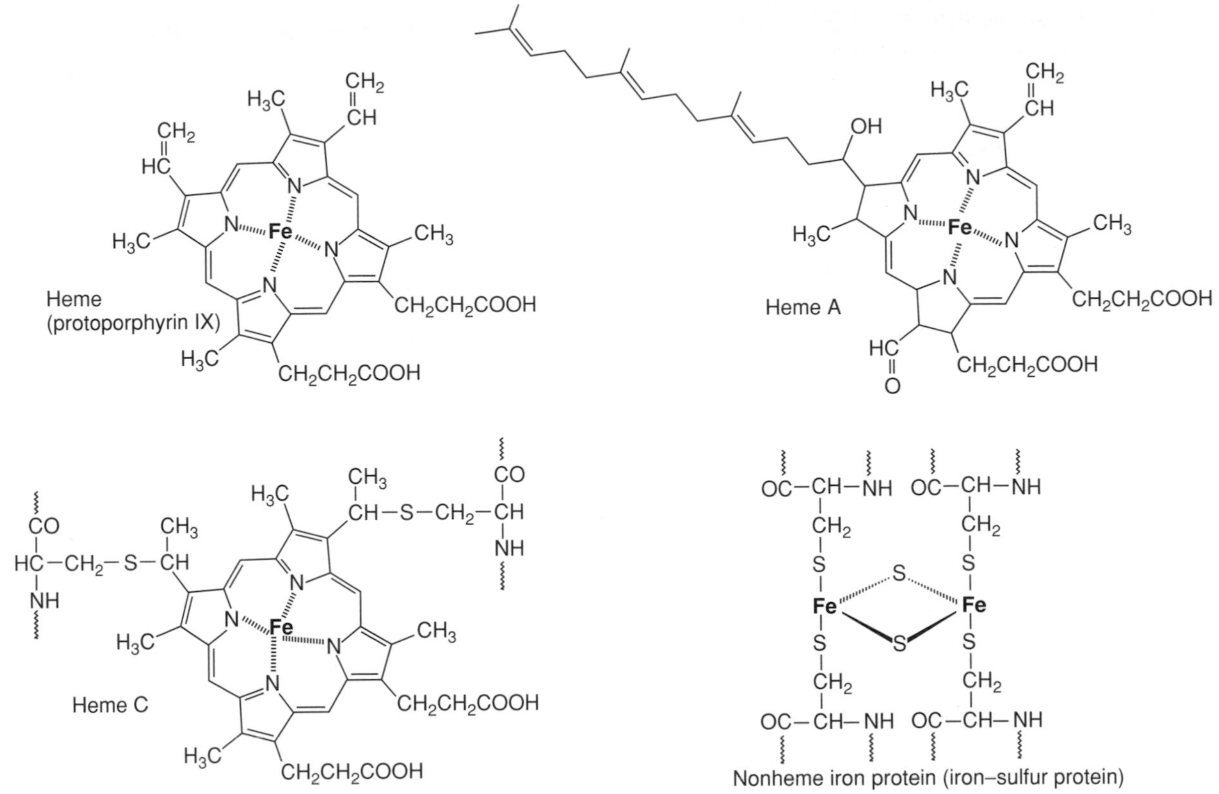

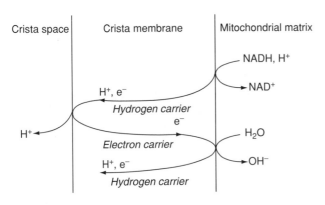

Figure 6 Types of heme in cytochromes and the binding of iron in nonheme iron proteins (iron–sulfur proteins).

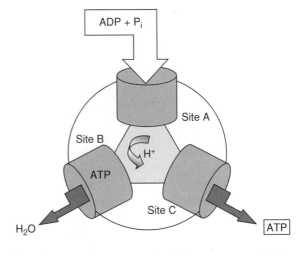

Figure 7 Proton pumping across the crista membrane in complex I.

Figure 8 Three active sites of the ATP synthase complex in the mitochondrial primary particle.

ATP. If ADP is not available to bind, rotation cannot occur – and if rotation cannot occur, protons cannot flow through the stalk from the crista space into the matrix.

Coupling of Electron Transport, Oxidative Phosphorylation, and Fuel Oxidation

The processes of oxidation of reduced coenzymes and the phosphorylation of ADP → ATP are normally tightly coupled:

- ADP phosphorylation cannot occur unless there is a proton gradient across the crista membrane resulting from the oxidation of NADH or reduced flavins.
- If there is little or no ADP available, the oxidation of NADH and reduced flavins is inhibited, because the protons cannot cross the stalk of the primary

particle, and so the proton gradient becomes large enough to inhibit further transport of protons into the crista space. Indeed, experimentally, it is possible to force reverse electron transport, and reduction of NAD^+ and flavins by creating a proton gradient across the crista membrane.

Metabolic fuels can only be oxidized when NAD^+ and oxidized flavoproteins are available. Therefore, if there is little or no ADP available in the mitochondria (i.e., it has all been phosphorylated to ATP), there will be an accumulation of reduced coenzymes and hence a slowing of the rate of oxidation of metabolic fuels. This means that substrates are only oxidized when there is a need for the phosphorylation of ADP to ATP, and ADP is available. In turn, the availability of ADP is dependent on the utilization of ATP in performing physical and chemical work.

It is possible to uncouple electron transport and ADP phosphorylation, by adding a weak acid such as dinitrophenol that transports protons across the crista membrane. As shown in **Figure 9**, in the presence of such an uncoupler, the protons extruded during electron transport do not accumulate in the crista space, but are transported into the mitochondrial matrix, where they react with hydroxyl ions, forming water. Under these conditions, ADP is not phosphorylated to ATP, and the oxidation of NADH and reduced flavins can continue unimpeded until all the available substrate or oxygen has been consumed. **Figure 10** shows the oxygen electrode trace in the presence of an uncoupler – there is more or less complete consumption of oxygen regardless of the amount of ADP present.

The result of uncoupling electron transport from the phosphorylation of ADP is that a great deal of substrate is oxidized, with little production of ATP, although heat is produced. This is one of the physiological mechanisms for heat production to maintain body temperature without performing physical work – nonshivering thermogenesis. There are a number of proteins in the mitochondria of various tissues that act as proton transporters across the crista membrane when they are activated.

The first such uncoupling protein to be identified was in brown adipose tissue, and was called thermogenin because of its role in thermogenesis. Brown adipose tissue is anatomically and functionally distinct from the white adipose tissue that is the main site of fat storage in the body. It has a red–brown color because it is especially rich in mitochondria. Brown adipose tissue is especially important in the maintenance of body temperature in infants, but it remains

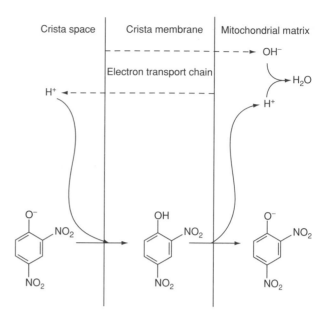

Figure 9 Uncoupling – discharge of the proton gradient by a weak acid.

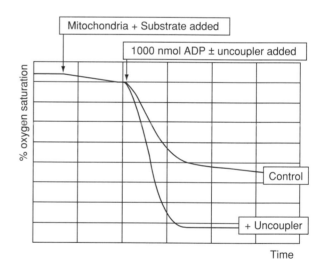

Figure 10 Oxygen electrode traces in the presence and absence of an uncoupler.

active in adults, although its importance compared with uncoupling proteins in muscle and other tissues is unclear.

In addition to maintenance of body temperature, uncoupling proteins are important in overall energy balance and body weight control. The hormone leptin, secreted by (white) adipose tissue, increases the expression of uncoupling proteins in muscle and adipose tissue, so increasing energy expenditure and the utilization of adipose tissue fat reserves.

Respiratory Poisons and other Inhibitors

Much of our knowledge of the processes involved in electron transport and oxidative phosphorylation has come from studies using inhibitors.

1. Rotenone, the active ingredient of derris powder, an insecticide prepared from the roots of the leguminous plant *Lonchocarpus nicou*. It is an inhibitor of complex I. The same effect is seen in the presence of amytal (amobarbital), a barbiturate sedative drug, which again inhibits complex I. These two compounds inhibit oxidation of malate, which requires complex I, but not succinate, which reduces ubiquinone directly. The addition of an uncoupler has no effect on malate oxidation in the presence of these two inhibitors of electron transport, but results in uncontrolled oxidation of succinate.

2. Antimycin A, an antibiotic produced by *Streptomyces* spp. that is used as a fungicide against fungi that are parasitic on rice. It inhibits complex III, and thus inhibits the oxidation of both malate and succinate, since both require complex III, and the addition of the uncoupler has no effect.

3. Cyanide, azide, and carbon monoxide bind irreversibly to the iron of cytochrome a_3, and thus inhibit complex IV. Again, these compounds inhibit oxidation of both malate and succinate, since both rely on cytochrome oxidase, and, again, the addition of the uncoupler has no effect.

4. Oligomycin, a therapeutically useless antibiotic produced by *Streptomyces* spp. Oligomycin inhibits the transport of protons across the stalk of the primary particle. This results in inhibition of oxidation of both malate and succinate, since, if the protons cannot be transported back into the matrix, they will accumulate and inhibit further electron transport. In this case, addition of the uncoupler permits reentry of protons across the crista membrane, and hence uncontrolled oxidation of substrates.

5. Atractyloside (a plant glycoside) and bongkrekic acid (a toxic antibiotic formed by *Pseudomonas cocovenans* growing on coconut – this is named after bongkrek, a mold-fermented coconut product in Indonesia, that becomes highly toxic when *Ps. cocovenans* outgrows the mold). Both compounds inhibit the transport of ADP and ATP across the mitochondrial membrane. Bongkrekic acid fixes nucleotides to the transport protein at the matrix side of the membrane, so that they cannot be released, whereas atractyloside has a higher affinity for the transport protein than does ADP, and so out-competes it for transport into the matrix.

See also: **Adaptation – Nutritional Aspects**; **Adipose Tissue**: Structure and Function of White Adipose Tissue

Further Reading

Bender DA (2002) *Introduction to Nutrition and Metabolism*, 3rd edn. London: Taylor & Francis.

Boyer PD (1997) The ATP synthase – a splendid molecular motor. *Annual Reviews of Biochemistry* 66: 717–749.

Murray RK, Granner DK, Mayes PA and Rodwell VW (2000) *Harper's Biochemistry*, 25th edn. New York: McGraw-Hill.

Oysters *See* **Shellfish**: Characteristics of Crustacea; Commercially Important Crustacea; Characteristics of Molluscs; Commercially Important Molluscs; Contamination and Spoilage of Molluscs and Crustaceans; Aquaculture of Commercially Important Molluscs and Crustaceans

PACKAGING

Contents

Packaging of Liquids

S D Deshpande, Central Institute of Agricultural
Engineering, Nabi Bagh, Bhopal, India

Background

In modern times, packaging has been identified as an integral part of processing in the food industry. Packaging is a technique of using the most appropriate packaging media for the safe delivery of the contents from the centers of production to the site of consumption. Packaging serves as the vital link in the long line of production, storage, transportation, distribution, and marketing. The package must ensure the same high quality of the product to the consumer, as they are used to getting, in freshly manufactured products. It is important that all products should reach the consumer in a usable condition.

Modern packaging systems for liquid foods are products from a synthesis of demands from producers, distributors, and consumers. The need for hygiene is the primary reason for retail packaging of perishable liquid food products like milk. Although this was realized more than a century ago, packaging techniques for liquid milk were slow in developing. The advent of pasteurization in the 1920s made retail packaging of liquid essential, and the returnable glass bottle was soon to become universal.

The commercial development of plastic materials, starting with polyethylene (PE) in the 1940s, opened up new possibilities for improving hygiene in liquid packaging. PE ultimately became the most frequently used thermoplastic in paper and carton board coating processes, also finding uses in inplant manufacture of packages from reel stock by form–fill–seal techniques. In current efforts to make retailing still more efficient, the focus is on standardized packages and transport wrappings, the aim being to simplify routines and cut costs. One-way cartons are suitable for these requirements. These developments have gradually led to a change in retail patterns in many countries, and the replacement of returnable glass bottles by single-service paper/plastic containers has been seen in many countries.

Today, a product distinction can be made between milk and non-milk on the one hand, and fresh and long-life products on the other. The main products retailed in one-way cartons are still milk and milk products, holding approximately 80% of the carton demand in liquid packaging, but a steady increase in market share for fruit juices, mineral water, sports drinks, vegetable oils and juices, soft drinks, and wine is observed. This trend is likely to continue, ensuring a further potential for the paper bottle.

Functional Packaging

Functional packaging of products contributes to the industrial prosperity of a country through optimal utilization of resources. The packaging has to protect the contents against hazards such as the vagaries of climate and transportation. During the course of the journey, the package would be exposed to varying climatic conditions, often resulting in evaporation and condensation of water of the contents inside the package. Also, atmospheric gases like O_2 and SO_2 contribute to the deterioration of the products by the oxidation of fat-rich products or corrosion of the metal containers. While this is the case in bulk packages, the unit container, which comes directly into contact with the product, must have requisite barriers and protective properties. A scientifically designed package should, therefore, afford protection against egress or ingress of moisture, flavor loss or odor pickup, light, oxygen, microbial and fungus attack, as well as being compatible with the food packaged. The package must preserve the quality, freshness, and functional performance of the product, afford the requisite shelf-life, and make it possible for the product to reach the consumer in prime condition.

Principles of Production

Filling and sealing machines for paper-based packages for liquids form two groups: those that work from a roll of packaging material, and those that work from premanufactured blanks, the difference reflecting the basic machine philosophy or concepts of companies like Tetra Pak and Elopak. The basic idea of Tetra Pak is that the package should be formed from a roll of packaging material, filled, and sealed in a continuous, closed process. The basic idea of Elopak is that as much of the package production as possible should be included in the converting process. Consequently, the production of blanks, being a highly specialized process, is therefore considered best performed when separated from the food-packaging plant.

Converting

In the converting process, the basic paper is coated on both sides, printed, and provided with scorelines to facilitate creasing when finally forming the package.

Filling and Sealing

When choosing a carton-based packaging system for liquids, there are three differently shaped packages presently predominating the market, namely the gable top, the tetrahedron, and the brick.

The filling and sealing procedure of a gable-top-type package starts with a blank being fed from a magazine. The lay-flat tube is then unfolded and enters a mandrel where the bottom is heated with hot air. The bottom is then folded in accordance with scorelines, and pressure is applied for finishing the bottom sealing. Now an open rectangular box, it is removed from the mandrel on to a conveyer, filled with liquid and the top sealed. The top seal is made using hot air and pressure.

The most striking feature of the tetrahedral package is the shape. The tetrahedral shape requires less packaging material than other designs, as it offers the most favorable ratio of area to volume.

The production of Tetra Brik-type packages from roll-fed machines follows basically the same principles as those for Tetra Standard, but the transverse seams are sealed parallel. The characteristic brick shape is formed after cutting off individual packages from the tube, by folding in the flaps and heat-sealing them.

Materials

The sandwich construction of the two common paper-based laminates used in liquid packaging is shown in **Figure 1**. If no high-gas barrier is required, the material consists of paper with a polyethylene coating on both sides. The paper layer may consist of unbleached, bleached, or semibleached sulfate pulp or laminates of these. The paper layer, being responsible for much of the machinability and mechanical properties of the package, requires a high and stable quality.

Additional barrier properties are usually provided by aluminum foil, laminated to the board, but the contact surface against the food remains polyethylene.

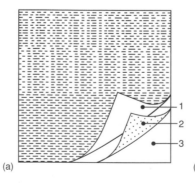 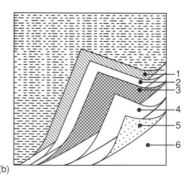

Figure 1 Sandwich construction of two common laminates for carton containers. (a) Typical laminate for short-life products like fresh milk consisting of (1) exterior PE, (2) paper, and (3) interior PE; (b) typical laminate for long-life products consisting of (1) exterior PE, (2) paper, (3) Surlyn, (4) Al-foil, (5) Surlyn, and (6) interior PE.

Packaging materials should provide the following properties:

1. high hygiene standards;
2. sufficient mechanical strength and internal bond;
3. liquid proofness;
4. interness to product;
5. light barrier;
6. low gas permeability;
7. seal ability;
8. machinability in converting and filling processes.

Canning

The process of canning is sometimes called sterilization because the heat treatment of the food eliminates all microorganisms that can spoil the food and those that are harmful to humans, including directly pathogenic bacteria and those that produce lethal toxins. Most commercial canning operations are based on the principle that bacteria destruction increases tenfold for each 10°C (18°F) increase in temperature. Food exposed to high temperatures for only minutes or seconds retains more of its natural flavor. Recent developments include the use of cans made of aluminum, very thin steel, and coated and uncoated plastic. Can openers are unnecessary for cans that have a pullable metal tab or ring attached at the top. Despite the widespread popularity of canned foods, the major limitation of canning is in the quality of the final product. Since food is not a good conductor of heat, excess heat needs to be applied to the container's surface for a period of time to guarantee sufficient heat at the center, or 'cold spot,' in order to destroy all organisms causing spoilage and disease. This method of preserving causes foods to lose juices, texture, flavor, and nutrients. The retort pouch, developed in the 1970s to alleviate this problem, is a 3-layered laminate with flexible plastic films as the outer and inner layers and aluminum foil in the middle. The pouch, which is approximately 19 mm (0.75 in) thick, is filled and sealed under vacuum. Because the pouch has a large surface-to-volume ratio, heat needs to penetrate less than 10 mm (0.38 in) from the surface to the 'cold spot,' thereby yielding greatly improved products.

Selection Criteria for Packaging Materials

Some of the important packaging considerations, which influence the choice of packaging materials, are given below.

Product Protection

The choice of packaging material has to be made depending upon the nature of the product, i.e., susceptibility of the ingredients used to undergo deterioration in contact with water, moisture, gases, light, etc. For products containing high fats, the packaging material must have high oxygen barrier properties. For such applications, material based on nylon, nitrile, or polyester may be considered. However, for products that are likely to deteriorate only by moisture gain or loss, a packaging material with sufficient moisture barrier properties may be adequate. For light-sensitive products, an opaque packaging material may be necessary. The shelf-life is another important criterion that influences packaging material specifications.

Convenience

The convenience for consumers and amenability of the packaging material to provide such features is a predominant factor in the final choice. Features such as reclosability, dispensing capability, stand-up facility, etc. are some examples of packages preferred by the end users.

Sales Appeal and Package Decoration

For some products, it may be necessary for the package to communicate effectively and visually attract the consumer while it is displayed on the shelf. Thus, the choice of packaging material is influenced by such properties as gloss, transparency, and ability to accept multicolor printing.

Product Package Compatibility

This aspect is critical, particularly for edible products. Product package compatibility testing is carried out to assess the possibility of any migration from and to the product. Any loss of product flavor or product tainting with unacceptable odor is also checked.

Packaging Machinery

For products that are mass-produced, the speed of packaging is very important, and the packaging material chosen must run smoothly on the packaging machine used. Here, the mechanical properties of the films and laminates such as tensile strength, rigidity, coefficient of friction, slip properties of the film and laminates, and their susceptibility to develop a static charge need to be considered.

Package Sealing Efficiency

However good a packaging material may be for a given product, its ability to protect the product is determined by its seal strength. Often, the product is spilled around the sealing area while filling. In such cases, sealing of the package with product contamination on the sealing surface is necessary.

Package Strength

Product distribution over longer distances invariably involves multiple handling and long storage periods. These are generally associated with both mechanical and climatic hazards. The choice of material, therefore, has to be made, based on the severity of hazards and the capacity of the material to withstand these in a given distribution situation.

Statutory Requirements

Often, there are certain regulations governing the choice of packaging material, for example, in the packaging of pesticides.

Material Availability

Any user would like to have multiple sources of procuring their packaging material. Therefore, the sources of supply and the availability of material become predominant considerations.

Cost

Sometimes, marketing success depends on how low the product is priced. The ultimate choice of packaging material is often influenced by the cost at which the functions indicated above are achieved.

The criteria discussed above are some of the more important considerations. Depending upon the type of product and the market, many additional considerations may become important. However, by and large, if the above criteria are used while selecting the packaging material, the user is likely to arrive at an optimum choice of packaging material.

Developments in Packaging of Liquid Foods

Liquid food products have been packed, preserved, and transported over the ages by a variety of methods. Wooden containers, earthen pots (glazed pottery or jars), and glass continue to be the major materials, even in the present day. All these had the benefits of being nonreactive to the food material and also efficient to contain the material resisting deformation besides being leak-proof (if properly closed). Over the years, with man's quest for newer technologies, packaging of liquid food products also gained greater importance. The need for developing alternative materials and methods for liquid food packaging has arisen for several reasons, such as to reduce the cost of packaging, improve the shelf-life, provide convenience in using the product, improve the handling during transport or in retail outlets, and also promote consumer acceptance of the product. It was further necessitated due to changed distribution

systems, competition, changing consumer needs, and availability of new packaging materials and/or methods.

For transporting in bulk, liquid material can be carried in wooden drums or metal drums. Alternatively, containers can be packed in wooden crated, or fiber mold cases, e.g., glass carboys in crates, polythene, or synthetic plastic carboys can be considered. A new development in this area is the bag-in-box type of packaging for liquids. In examining the shelf-life requirements of the product, microbial spoilage in the case of fresh food products such as milk and fruit juices, due to their high susceptibility to microbial contamination and hence requirement for sterilization/pasteurization, needs to be considered. Packaging material should withstand rigorous treatments during such packaging stages.

As the industry grows, and the demand for convenience increases, the need for quality packaging undoubtedly increases. Today, packaging is cost-effective in the sense that the package is pivotal to the marketing philosophy of the urban markets offering a wide choice at a low cost. Thus, the model food package has to keep down marketing costs and offer the consumer tangible price benefit. The increased speeds on machine lines and use of flexible laminates in place of tin containers may be considered as a development.

As outlined above, the selection of the packaging material and the packaging machines depends upon the food to be packed, and the methods together can be called as a packaging system. Today, fruit juices, wines, water, oil, soup, concentrated vegetable, purées, etc. are packed in a variety of systems such as prepac, Blockpak, Zupack, Purepak, Ceka, Tetrapak, thermoforming systems, and bag-in-box systems. Facilities such as asceptic packaging, hot filling, nitrogen flushing, etc. for the product are also available.

There are the range of cartoning systems and form–fill–seal machines now available and are replacing the glass and tin containers for liquid food packaging. Among these, asceptic packaging is yet another development.

Requirements with respect to hygiene and impermeability are stringent for liquid food products. Sealing must be tight. There are basically three types of machines for the packaging of liquids, viz. vertical form–fill–seal machines horizontal form–fill–seal machines and sachet-forming machines.

All the types can be used for liquid food product packaging. They can have such secondary facilities as gas flushing, vacuum zing, etc. introduced. Standi-packs and gusseted sachet packs are quite acceptable for liquid food products for consumer convenience.

Resources and Energy

Presently, one-way carton packages for milk, juices, and other noncarbonated beverages coexist with one-way and returnable alternatives like the glass or plastic bottle. At first sight, the returnable alternatives seem preferable from an energy and ecological point of view. But considering the entire process, from extraction of raw materials, through production and distribution to the handling of waste, this is not the case when compared to one-trip cartons. Investigations show that there are only minor differences, ecologically speaking, between a returnable bottle with a realistic trippage for modern retailing and a plastic-coated paper bottle.

The constituent materials of the dominant packages for the noncarbonated liquid foods, for glass bottles, cans, plastic bottles, and cartons, differ in practically every aspect: raw material consumption, energy consumption, impact on water and air, and amount and quality of waste. The overall energy requirements for the different packages are of the same magnitude (**Figure 2**), water and air pollutants are comparable, and a large amount of waste is generated in the use of returnable bottles if a bottle does fewer than 20 trips.

Carton-based packages using wood as the basic raw material exploit a renewable resource. Present planting of trees exceeds harvesting, so there will not be any deforestation as a result of the manufacture of paper containers.

Litter and waste may be regarded as minor problems as far as PE-coated cartons are concerned. If a plastic-coated cartonboard is left in nature, complete degradation will usually take several years, the main problem being the degradation of the polymer. Biodegradable PE has been discussed as an alternative outer coating. Although this solution would increase the rate of degradation considerably, it has never found applications. The waste problem is the more important issue when it comes to disposabilities. Nevertheless, the waste from PE-coated cartons is easily handled, does not require separate collection, and may be used as a source of energy when incinerating.

Marketing Aspects and Competition

The main alternatives in liquid food packaging are glass and plastic bottles, cans, and plastic-coated cartons. A packaging system based on cartons is facilitated by low overall costs. In fact, the one-way

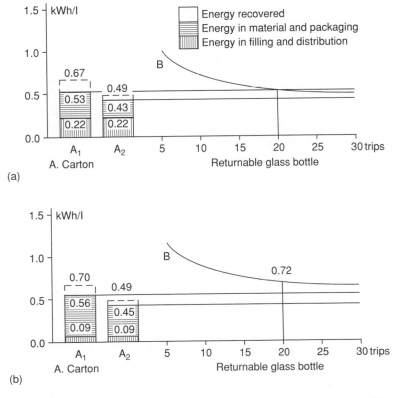

Figure 2 Comparison of energy counts for milk (a) and juice (b) packaged in cartons and returnable glass bottles. A_1 and A_2 represent the energy counts for cartons including and excluding wood energy, respectively. The figures are reproduced from data given in Sundstrom G (1982) *Milk Packages and Energy*. Malmo, Sweden: Sundstrom AB and Sundstrom G and Lundholm MF (1982) *Juice Packages and Energy*. Malmo, Sweden: Sundstrom AB.

carton container is found to be a cheaper alternative than cans and returnable glass bottles. The profitability is largely due to low distribution costs, storage efficiency, maximum use of shelf-space, and low labor costs.

For the working environment, systems based on one-way cartons are regarded as superior. Different products with very different processing qualities can be handled, including mineral water, nectars or fruit juices with a high pulp content, wine, milk, preserved dairy products, and others.

In modern retailing, the package serves the purpose of being an important source of information. Being printable and having four sides available for commercials and content declaration, the carton is its own sale promoter, offering free advertisement displayed on shelves and the consumer's table.

In Western European countries, milk represented somewhat below 80% of liquid carton demand at the beginning of the 1980s, and presently the cartons hold about 60% of the total milk market. There is currently a decline in the total milk market but an increased consumption of cartons for milk, thought to be due to the commercialization of long-life milk and the replacement of returnable bottles in the UK. Cartons have achieved a high penetration of the milk market in most Western countries, so the scope for further progress may be regarded as limited outside the UK. Furthermore, fresh milk (i.e., pasteurized milk) is believed to appeal to popular taste more than sterilized milk, and the market may well turn in favor of fresh milk where proper refrigeration conditions may be achieved.

The second largest product group is for fruit juices, holding about 15% and steadily increasing, and the third largest product group is soft drinks, holding about 5%. Other products for which the cartons are likely to increase in importance are wine, mineral water, vegetable oil, and juices.

Overview

Liquid food packaging has been vital in assuring man of a year-round supply of a variety of foods and has changed us from an agrarian to a cosmopolitan culture.

Significant changes have taken place over the last three decades, and packaging is becoming recognized as a definable and essential technology. It is considered earlier in the food product development cycle and a more organized approach to actual package development is taken.

Currently, package designs are first tested under laboratory conditions (vibration, drop tests, inclined impact) and then confirmed by actual field

use, sometimes with little agreement. Improvements in agreement between field experience and laboratory testing will be improved by such developments such as:

1. Shipping hazards recorders, such as that developed by the US Army Natick Development Center (**Figure 3**), to measure and record on tape such events as number and height of drops and environmental conditions such as temperature and relative humidity.
2. In estimating shelf-lives, recognition of diurnal temperature and humidity changes, and the unsteady state nature of oxygen and water vapor permeabilities.
3. In devising more relevant tests to replace archaic techniques, such as the Muller burst test for fiberboard.

Computers can be of great help in the statistical evaluation of the massive environmental and hazards data required, defining any distribution system sufficiently to permit trade-off decisions on product loss versus packaging cost. This will also be essential in handling laboratory test data and will be useful in the graphic mode, for actual package design.

On-line, nondestructive test procedures such as the infrared seal defect technique, the sonic vacuum test technique for cans, and magnetic metal contamination detection procedures will be more plentiful. Energy costs will govern package design to a great degree and will necessitate frequent substitutions for critical materials.

There will be greater emphasis on the nonprotective functions of a packaging system. The flexible

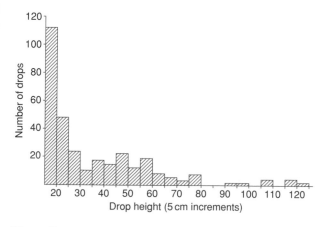

Figure 3 Frequency histogram of recorded drops. Distribution of 312 recorded drops in 5-cm increments. The system activation threshold enables drops of less than 15 cm to be recorded. This represents 1500 days and 80 000 km of shipments. From Barca FD (1975) *Acquisition of Drop Height Data during Package Handling Operations*. Natick, MA: Aeromechanical Engineering Laboratory, US Army Natick Development Center

packaging system for thermoprocessed foods extends the boil-in-bag heating capability to nonrefrigerated foods. Untended vending systems using microwave energy for in-package product heating appear to be feasible. Liquid foods are being studied with the aim to use the flat configuration to achieve a high product quality and container designs for heating for serving and serving from the container itself.

See also: **Milk**: Liquid Milk for the Consumer; Processing of Liquid Milk; **Pasteurization**: Pasteurization of Liquid Products

Further Reading

Athalye AS (1992) Packaging as a system. In: *Plastics in Packaging*, pp. 15–17. New Delhi: Tata McGraw – Hills.

Barca FD (1975) *Acquisition of Drop Height Data during Package Handling Operations*. Natick, MA: Aeromechanical Engineering Laboratory, US Army Natick Development Center.

Goyal GK (1991) Practice and research related to packaging of indigenous dairy products – A review. *Journal of Indian Dairyman* 43(11): 489.

Gupta TR (2000) Bulk aseptic packaging of fruit pulp and concentrate. 14th Indian Convention of Food Scientists and Technologists Souvenir, pp. 78. Mysore, India: Central Food Technological Research Institute.

Iversen A (1987) Cartons for liquids. In: *Modern Processing, Packaging and Distribution Systems for Food*, pp. 86–90, 105–107. Westport, CT: AVI.

Kumar KR (2000) Global standards for food packaging. 14th Indian Convention of Food Scientists and Technologists Souvenir, pp. 26–32. Mysore, India: Central Food Technological Research Institute.

Lampi RA (1978) Packaging of food – Overview. In: *Encyclopedia of Food Technology and Food Science Series*, vol. 3, pp. 591–597. Westport, CT: AVI.

Menon CPS (1986) Developments in packaging of liquid foods. In: *Packaging of Food Products*, pp. 187–192. Bombay: Indian Institute of Packaging.

Narayanan CK (2000) Trends in health food drinks packaging. 14th Indian Convention of Food Scientists and Technologists Souvenir, pp. 78. Mysore, India: Central Food Technological Research Institute.

Nunn DW (1980) Alternative packaging of milk – a consequence analysis. *CMI Report No. 790306–1*. Fantoft, Norway: Chr. Michelsens Institute.

Raj Baldev (2001) Food and packaging interaction – migration concepts and regulations. *Indian Food Industry* 20(5): 67–74.

Rangarao GCP (2001) Computer simulation of packaging and storage environments and prediction of shelf life of moisture sensitive foods. *Indian Food Industry* 20(6): 65–70.

Sarkar PC and Gupta PC (2002) Shellac-based tin can lacquers for the packaging industry. *Indian Food Packer* 56(2): 69–71.

Sogi DS, Shergill RS and Bawa AS (2000) Effect of packaging materials on storage of lemon juice beverage concentrate. *Journal of Food Science and Technology* 37(3): 296–299.

Sundstrom G (1982) *Milk Packages and Energy*. Malmo, Sweden: Sundstrom AB.

Sundstrom G and Lundholm MP (1982) *Juice Packages and Energy*. Malmo, Sweden: Sundstrom AB.

Sundstrom G and Lundholm MP (1985) *Resource and Environmental Impact of Tetra Brik Aseptic Carton and of Refillable and Nonrefillable Glass Bottles*. Malmo, Sweden: Sundstrom AB.

Packaging of Solids

M Mathlouthi, Université de Reims Champagne-Ardenne, Reims Cedex, France

Background

The aim of packaging solid foods is to protect them against spoilage and to preserve their quality. Selecting a suitable packaging system for this category of products requires a good knowledge not only of the factors affecting their stability but also of the extent to which they are deformed when submitted to mechanical constraints. Both stability and resistance to being deformed are influenced by the water content and water activity (a_w). Most of the physical, biochemical, and microbiological changes of solid foods also depend on a_w. Also, lipid oxidation and other causes of decrease in quality originating from the permeation of gases, water vapor, aroma substances, and other chemicals, or the transfer of energy (light, heat) to an acceptor phase (moisture, fat, etc.) in the product must be minimized through the choice of packaging material, which should be a barrier to matter and or energy transfer. Such a barrier should not be completely tight to take into account a possible release of water vapor from the product in the inner space of the package, which might increase the relative humidity to a limit value of a_w for which the caking of powdery solid foods can occur. In addition, as solid foods are stored and distributed in their packages, one of the objectives of packaging, besides the protection of the product, is to meet the requirements for successful marketing. (*See* **Oxidation of Food Components; Water Activity**: Effect on Food Stability.)

Classification

This classification of solid foods is based on the degree to which they are deformed when submitted to compression or agglomerated and compacted as a result of moisture increase. This depends on whether they are soft, rigid, or powdery products, and may influence the protectivity of the packaging material as well as its acceptance by the consumer. A special class is devoted to food powders for which flowability, density, morphology of particles, and their crystallinity play a major part. Three categories of products are distinguished:

- nondeformable solid foods like toasts, biscuits, crackers, hard cheeses, confectionery, chocolates, etc.;
- deformable products like sandwich loafs, industrial pastry, soft cheeses, meat, and meat products, etc.;
- food powders like refined granulated sugar, wheat flour, spray-dried milk, dairy-based infant formula powders, etc.

The choice of packaging solution depends on the objective sought, namely the nonperception by the consumer of any degradation of the product. As with heat treatment, absolute protection of a packaged foodstuff against alteration is not possible. Thus, protection in this context is only a commercial notion comparable to that of commercial sterilization. It was decided to classify solid foods into deformable and nondeformable products in order to discriminate between those products (deformable) for which the main origin of degradation is the barrier characteristic of the packaging material. In the case of deformation, the appearance of the product becomes the dominant criterion as regards acceptability by the consumer. For nondeformable solid foods, the product may be the origin of a change in structure of the packaging material, leading to a reduction of thickness, pinholes or tears, which modify the transfer of matter or radiation, and provoke biochemical alterations, reduction in shelf-life and, finally, rejection by the consumer. For solid foods, much more than for liquid foods, the concept of an efficient packaging system should be based on the interaction between the product, the packaging material, and the machinery involved.

To generalize for solid foods as a whole, baked goods are taken as a model for this class of food products. Indeed, it is easy to find among baked goods representation examples on nondeformable products like toasts, biscuits, or crackers, and on deformable products like bread or sponge cake. For these products, as for the majority of solid foods, a good criterion of differentiation is the a_w. For rigid baked goods, the a_w is low, about 0.05–0.30, and its value for soft products is around 0.70–0.96. Besides acting as a good reference mark for the aptitude for deforming on which we base our classification, the a_w is associated with most alterations that occur in solid foods.

As a representative product of food powders, white sugar can be taken as an example. Besides the sensitivity to moisture and temperature, the stability of packaged sugar, either granulated or cubes, depends on the stability of crystallization at the surface of the crystals, absence of amorphous or fine particles, and the minimization of the fraction of water content called 'free' water or solvent water on the water vapor sorption curve.

Causes of Alterations

Changes in packaged solid foodstuffs are perceived as being due to a mass or energy transfer across the package providing that there exists a donor (environment, packaging material, etc.) and an acceptor (mobile phase of the solid food). This approach may seem to be complicated, but it has the advantage of relying on classical equations of diffusion. The required characteristics of the packaging material are approached in a dynamic way and deduced from mathematical models of the prediction of shelf-life.

Instead of reviewing all possible matter and energy transfers and their effects on the degradation of solid foods, three major causes of alterations are described: the transfer of water vapor, the transfer of radiation (light), and the transfer of chemicals during storage.

Quality Changes originating from the Transfer of Water Vapor

Nondeformable Products

For this category of foods, an increase in a_w is generally the origin of the alteration and leads to a loss of crispness. However, an increase in moisture may be protective against lipid oxidation in low-moisture foods. In most cases, there remains in the headspace, after sealing the container, enough oxygen to initiate oxidation where an acceptor, e.g., ethylenic bonds in unsaturated fatty acids, exists. It may be noted that, in rigid baked goods with an a_w of 0.30, the risks of biochemical or microbiological alterations are absent.

Deformable Products

It may be recalled that the a_w for these foods is > 0.70, which corresponds to an increase in availability of degradation reaction sites due to an enhanced mobility of water. The risks, in this case, are due to water vapor transfer (gain or loss of weight), chemical and enzymatic reactions, and microbial spoilage. It is also observed that, in this zone of a_w values, a structural rearrangement of one of the constituents (starch) of the product may lead to an alteration known as staling of the baked goods. (*See* **Spoilage**: Chemical and Enzymatic Spoilage; Bacterial Spoilage.)

The kinetics of alteration reactions depend on the components of food. This is the case, for example, for the oxidation of lipids. If fats are sequestered by the starchy phase, as in a dry cake (sablé), lipid oxidation does not occur, and the shelf-life of the cake is prolonged. If lipids are also on the surface of the product, as with extruded snacks, lipid oxidation becomes a limiting factor in the quality assurance of packaged solid foods. The availability of water for enzymatic reactions or microbial growth may be limited by the addition of a_w depressors, 'humectants' like sugars, polyols, glycerol, or proteins. To minimize nonenzymatic browning, it is important to reduce the duration of exposure to water vapor pressure at the optimum value ($a_w = 0.65$–0.75) for the Maillard reaction. However, at least for baked goods, the appearance of the product, which is appreciated by the consumer, is partly due to the Maillard reaction. (*See* **Browning**: Nonenzymatic.)

Food Powders

Caking is the major alteration of free-flowing food powders ingredients and results in the transformation of a pulverulent product into lumps, then into agglomerated solid, and finally into a sticky material that has lost its functionality. Such a transformation depends on the temperature and moisture content. Different stages are involved amid which are the bridging of particles, their compaction, and liquefaction. For sugar crystals, the dynamic evolution of a_w and its increase to reach 0.80 can provoke the agglomeration of particles, especially when fine crystals with an average size less than $250\,\mu m$ are present. When the amount of fine particles increases (above 10%) the a_w value for caking is lowered to around 0.45.

Another origin of instability of food powders is their amorphous structure, which allows their existence below the glass transition temperature (T_g) in an out-of-equilibrium state. It is possible to reach T_g at normal temperatures if the relative humidity is increased. This might be at the origin of a situation where the viscosity is decreased at such a value that allows the product to be transformed into a rubbery material, which sticks to the package and loses its quality.

Alterations Originating from Exposure to Light

Radiation, particularly of ultraviolet light, has a catalytic effect on the quality changes in foods. Deteriorative changes like the destruction of lipid-soluble vitamins, the loss of riboflavin and other vitamins, and changes in proteins and food pigments are accelerated by light. The presence of oxygen is intimately linked to these light catalyzed reactions.

These alterations lead to rancid foods with a changed color and reduced nutritive value. The odor and color changes of the product are obviously perceived by the consumer. In order to limit the rate of light-induced deteriorative reactions, the choice of packaging material for solid nondeformable foods should insure a low oxygen partial pressure and a nontransparent package.

Alterations Originating from Food–Package Interactions

For this group of alterations, the notion of reciprocal mobility of a donor and a receptor acquires all its importance. Indeed, whatever the migration (from the product to the packaging material or from the package to the food), it is essential to have a carrier for the transfer of the migrant chemical. It is also necessary to distinguish between hydrophilic and lipophilic chemicals. For nondeformable products ($a_w \leq 0.30$), the migration of some constituents of the packaging material (low-molecular-weight polyolefins, plasticizers, lubricants, etc.) into the mobile fatty phase of the food may be observed. Likewise, the adsorption of fats by the material (polyethylene, polypropylene, etc.) in contact with solid foods may occur.

For deformable products ($a_w \geq 0.70$), the preponderant mobile phase is water, and only hydrophilic chemicals are implicated in migration, i.e., acrylic derivatives of varnish, glycol derivatives, urea, formaldehyde, etc. In this category of deformable foods, it may be that both the aqueous and fatty phases are mobile, as in some sponge cakes, and both migrations are then observed. The structure and composition of the product help to determine the preponderant mobile phase in contact with the packaging material. Although fat may be fixed to starch and water immobilized by 'humectants,' this category of solid foods is more difficult to control in terms of migration.

Stains of syrup are sometimes observed at the top of paper board package of sugar lumps. This is observed when the lumps are packed at a high temperature (above 40 °C) and not cured in a ventilated room. Such conditions (40 °C and relative humidity above 85%) lead to the diffusion of sucrose in the film of syrup surrounding the crystals. After crystallization of sugar in the film of syrup, and the release of hydration water, the syrup becomes sufficiently mobile to stick to the package. As granulated sugar may act as an adsorption medium, odors from the environment or other volatiles from the packaging material may be temporarily adsorbed by the sugar.

Preservation

In order to protect a solid food against alteration, it is necessary to place a barrier between it and the environment. This barrier should be adapted to the capacity of adsorption of the food for the factors responsible for the deteriorative reactions. The barrier properties of the packaging material are determined by its permeability to the degradation agents, mainly gas or vapor permeating from outside.

Permeability of Packaging Material

The mass transfer of oxygen, carbon dioxide, water vapor or aroma components, or heat or radiant energy (ultraviolet, infrared, β, γ or microwave radiation) across the packaging material is generally ruled by a relation derived from Fourier or Fick's laws. In the case of unidirectional transfer, under a steady state and at equilibrium, the quantity of permeated material per unit of time (Q) is given by a Fick's first law-type equation:

$$Q = -PA\frac{\Delta p}{l}, \tag{1}$$

where P is the permeability, A is the active surface area of the packaging material, Δp is the difference in pressure (or concentration) on either side of packaging material, and l is its thickness.

The permeability coefficient, P, is a function of the number of variables, which include the structure of the packaging film, the properties of the permeate, the time, the pressure, and the thickness (the temperature is constant). Depending on the sensitivity of the product to water vapor, oxygen, or loss of aroma components, the packaging material composition and barrier characteristics are determined (see **Tables 1 and 2**).

Prediction of Shelf-life

The classification of solid foods determines the choice of the model transfer equation and affects the prediction of the deteriorative reaction rate. (*See* **Storage Stability: Shelf-life Testing.**)

For nondeformable foods, the most critical transfer is that of water vapor and/or oxygen. The kinetics of oxidative degradation should be adapted to each category of product, while the model equation used for water vapor transfer may be generalized.

For deformable foods with $a_w > 0.70$, water vapor transfer generally occurs from the inside to the outside of the package and is the preponderant factor of alteration. In some cases, more than one transfer has to be taken into account, for instance, water vapor and oxygen. The mathematical model of mass transfer is then obviously more complicated.

In order to calculate the shelf-life of products sensitive to moisture, eqn (2) is applied:

$$\frac{dW}{dt} = -\frac{PA(p_e - p_i)}{l}, \tag{2}$$

where dW/dt is the flow of moisture, and p_e and p_i are, respectively, the water vapor pressure in the exterior and the interior of the package. If storage takes place at a constant temperature, p_e and p_i may be replaced by $a_{w,e}$ and $a_{w,i}$, the values of outer and inner water activities, and eqn (2) can be written in integral form as:

$$t = K \int\limits_{\text{origin}}^{\text{degradation}} f(aw)\,daw, \tag{3}$$

where t is the shelf-life of the product, and K is a constant related to the properties of the product, the permeability of the packaging material, and the conditions of storage. Eqn (3) is only valid when the degradation originates from a change in a_w due to moisture transfer. In the case of oxidative deterioration, two equations are applied, one accounting for oxygen transfer and the other for its adsorption by food. For transfer, the relation is:

$$\frac{d(O_2)}{dt} = P_{O_2} A \frac{\Delta P}{l} m_{o_2}, \tag{4}$$

where $d(O_2)/dt$ is the flow of oxygen into the package, P_{O_2} is the permeability, A is the surface of permeation, Δp is the pressure difference of oxygen between the interior and exterior of the package, m_{o_2} is the absorption capacity of food, and l, as before, is the thickness of the packaging film. It is generally supposed that oxygen is adsorbed by food, and the rate of adsorption of oxygen $d(O_2)/dt$ is given by:

$$\frac{d(O_2)}{dt} = \frac{P_{O2}}{k_1 + k_2 P_{O2}} \tag{5}$$

where P_{O_2} is the partial pressure of oxygen, and k_1 and k_2 are two constants related to the degree to which the food adsorbs oxygen. An increase in the rate of adsorption of oxygen provokes an increase of oxidation. However, many other factors affect this type of degradation (light, heavy metals, water activity, etc.), and it is difficult to find mathematical model accounting for shelf-life as regards oxidation.

Other degradations may be predicted from mathematical models, like nonenzymatic browning on model systems (reducing sugars + amino acid) in the a_w range 0.55–0.85, but a precise prediction of a real degradation in a real product is not possible.

Increase in Shelf-life

When the degradation reactions are known, it is possible to minimize their rate in order to increase the shelf-life of the product. The basis of this approach is the need to reduce the availability of the donor or acceptor of the degradation agent.

If the acceptor is water, its availability is reduced by the addition of humectants (polyols, sugars, proteins) to the formula of the baked goods. Such an addition provokes a decrease in a_w from about 0.85 to 0.75 for a sponge cake, for example, which is then screened from microbial spoilage.

Preservation against oxidation and/or microbial growth may be obtained by changing the inner atmosphere of the package, generally by using a mixture of nitrogen and carbon dioxide. A low partial pressure of oxygen, together with a carbon dioxide/nitrogen atmosphere, contributes to an increase in the shelf-life; for example, industrial pastry with a normal a_w level of about 0.90 can have its shelf-life extended to 3 months. (*See* **Chilled Storage**: Use of Modified-atmosphere Packaging; Packaging Under Vacuum; **Chill Foods**: Effect of Modified-atmosphere Packaging on Food Quality.)

The stabilization of a solid food may also be achieved by using a superficial thermal treatment using radiant energy from infrared or microwave radiation. Such a treatment applied to the packaged product can contribute to an increase in the shelf-life.

Protection of the product against degradation is also obtained by the choice of a good barrier (polymeric film or laminate), which may be transparent or opaque. The barrier properties of the packaging material change if it interacts with degradation agents like water vapor or oxygen. These interactions depend on the hydrophilic or hydrophobic character of the material. For products particularly sensitive to one or other of the degradation agents, the packaging material may include chemicals permitting the scavenging of oxygen, the adsorption of water vapor, the emission of ethanol or carbon dioxide, and so on. The packaging becomes an 'active' barrier of protection and the center of a protective chemical reaction. However, one should remember that adaptation of the packaging material to food protection must be economically feasable and that the objective is not the absolute stability of the product during storage.

It is only needed for the solid food to be accepted by the consumer.

Polymeric Packaging Materials used for Solid Foods

The selection of polymeric or laminate film for solid-food packaging involves a certain number of conditions. The material must be compatible with the food in accordance with the legislation and calculated (thickness) to fit with determined barrier properties and shelf-life of the product. Depending on the level of protection and kind of product to be packed, the packaging material will be either a monolayer or a multilayer (coextruded or laminated). It may be coated or metallized, if the aim is to prevent the transfer of water vapor or light, and the structure of a foil reinforced by orientation or plasticizing.

A wide range of films used for baked goods are listed in **Table 1**. The transfer of gases and water vapor is expressed in commonly used units rather than in SI units. The unit for the transmission rate of gases is given in cubic centimeters of gas in 1 day for 1 square meter of film at a given thickness at constant temperature and pressure. The rate of water vapor transmission is expressed in grams per day per square meter at a specific temperature and a given difference of water vapor pressure (Δp) between the two sides of the packaging material. When the barrier characteristics required are not achievable with monolayer film, laminates of multilayer material are used. The characteristics of such laminates are listed in **Table 2**. As may be seen in **Tables 1** and **2**, the basic films are polyolefins, polyethylene, and polypropylene, copolymers like ethylene-vinyl alcohol and other polymers such as polyesters, polyamides (OPA 6 and 66) and polyvinylchlorides. The laminated plastic films are coated with regenerated cellulosics like DM, which is nitrocellulose coated on one side, MXXT, which is polyvinylidene chloride coated, or paperboard or aluminum foils. Different laminating adhesives are used.

Most foods are sensitive to water vapor, and the packaging materials generally used are good barriers to water vapor; polyesters (polyvinyldene chloride, coated or metallized film) and oriented polypropylene are the major materials found. Snacks and crisps contain fats and have a large specific area. Therefore, they require protection against light, which has a catalytic effect on lipid oxidation. The laminates used to protect these products utilize aluminum foils (see **Table 2**). Metallized polyester is generally coated with vaporized aluminum.

Food powders freely flowing are packaged in polyethylene sealable films, which gives a good water

Table 1 Characteristics of some films used in the packaging of solid foods

Film (thickness 25 μm)	Gas transmission ($cm^{-3}\,m^{-2}\,day^{-1}$ (dry gas))			Water vapor transmission ($g\,m^{-2}\,day^{-1}$)	
	Oxygen 23 °C	Carbon dioxide 23 °C	Nitrogen 23 °C	ΔRH 90% 38 °C	ΔRH 75% 25 °C
LDPE (0.917)	7 400.00	40 000.00	2 800.00	12.50	4.00
HDPE (0.960)	1 600.00	11 400.00	440.00	3.70	1.45
PP-cast	3 040.00	9 760.00	690.00	8.20	3.30
OPP-coextruded	1 550.00	5 280.00	320.00	5.00	1.35
OPP-coated	15.00	88.55	4.50	5.00	2.00
OPP acrylic-coated	1 200.00	4 500.00	250.00	4.60	1.80
OPP-metallized	35.00	108.00	6.50	1.00	
PVC-rigid	120.00	320.00	20.00	32.00	12.00
PVC-oriented	27.00	68.00	20.00	17.50	7.00
PVC-plasticized	190–3 100	430–19 000	53–810	85.00	32.70
PVDC	1.25–14.5	5.0–50.0	0.4–2.5	0.6–3.20	0.25
PS-cast	4 500.00	11 000.00	640.00	170.00	70.00
SAN	900.00	2 800.00	120.00		
Polycarbonate	3 200.00	17 500.00	450.00	178.00	72.50
PET	55.00	240.00	12.40	20.00	7.00
PET–PVDC-coated	8.00	32.00	2.00	8.50	3.40
PET-metallized	0.65	3.4–10	0.20	1.00	0.40
PA6	40.00	200.00		280.00	80–110
OPA6	18.00	120.00	9.00	130.00	28.30
PA 6.6	35.00	140.00	11.00	90.00	15.00–30.00
EVAL (32% ethylene)	0.16	0.45		80.00	32.00
Cellulosic fim 445MXXT A	8.75	80.00	3.65	8.60	3.40

LDPE, low-density polyethylene; HDPE, high-density polyethylene; PP, polypropylene; PVC, polyvinyl chloride; PVDC, polyvinylidene chloride; PS, polystyrene; SAN, styrene acrylonitrile; PET, polyester; PA, polyamide; OPA, oriented polyamide; EVAL, ethyl-vinyl alcohol; MXXT, a PVDC coating.

Table 2 Characteristics of some laminates used in the packaging of solid foods

Laminate	Gas transmission ($cm^{-3}\,m^{-2}\,day^{-1}$ (dry gas))			Water vapor transmission ($g\,m^{-2}\,day^{-1}$)	
	Oxygen 23 °C	Carbon dioxide 23 °C	Nitrogen 23 °C	ΔRH 90% 38 °C	ΔRH 75% 25 °C
Cellulose film 280 XS + PE 40 μm	12.00			4.50	1.10
OPP coextruded 25 μm + OP coextruded 25 μm	650.00			2.60	0.95
PET coated PVDC 12 μm + PE 40 μm	5.00	15.00	1.00	3.70	1.40
M PET 12 μm + PE 80 μm	1.00	4.00	0.20	0.50	0.20
M PET 12 μm + M PET 12 μm + PE 80 μm	< 0.10	< 0.10	0.00	0.15	0.06
OPA6 15 μm end. PVCD + PE 60 μm	10.00	30.00	2.50	5.00	
OPA 6 20 μm + PE 80 μm	40–30				
M OPA 6 15 μm + PE 80 μm	2.00			2.50	
Kraft 45 g m^{-2} + PE 20 g m^{-2} + end. PVDC 20 g m^{-2}	34.00			1.70	0.06
Kraft 60 g m^{-2} + end. PVDC 30 g m^{-2}	15.00			1.90	0.65
PET 12 μm + 119 μm + monomer 20 μm	< 0.20			< 0.10	
Cellulosic film 320 DM + A 19 μm + PE 35 μm	7.15			0.15	0.10
Kraft 70 g m^{-2} + PE 15 g m^{-2} + A 19 μm	4.30			0.10	0.08
A 19 μm + Kraft 70 g m^{-2}	25.40			0.25	0.15
A 19 μm + TPP 20 g m^{-2} wax	28.00			0.40	0.15
30 g m^{-2} TPP 20 g m^{-2}	< 8.00	< 35.00	< 3.00		

M, metallized; Kraft, paper; DM, one-size nitrocellulose coated; A, aluminum foil; XS, cellulosic film coated with PVDC.

vapor barrier and a transparent package allowing the product to be seen. Refined sugar is generally packed in multiwall paper packages, cardboard cartons, or polyethylene coated paper. Finished packs contain 500 g to 5 kg for domestic use. Sacks of up to 50 kg are produced for industrial utilization. An intermediate stage between the bulk transport of granulated sugar and the 50 kg sacks is the big bag called

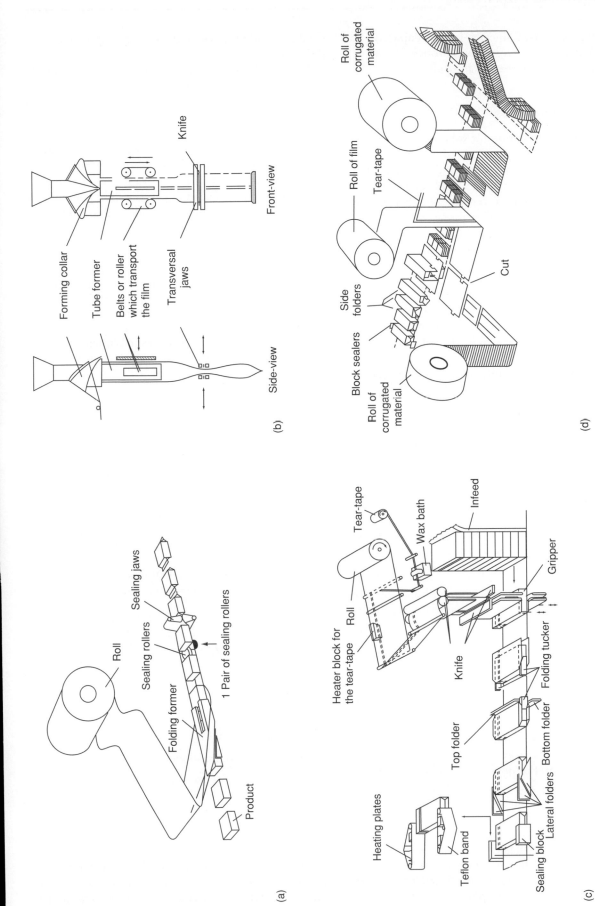

Figure 1 (a) Horizontal form–fill–seal machine; (b) vertical machine ('pillow bag'); (c) overwrapping machine; (d) overwrapping machine (with adjustment to the circumference of the product).

intermediate bulk container of 1 tonne capacity. This supersack is made of woven polypropylene and it is able to stand unsupported on the pallet.

Packaging Machines

A good packaging solution for solid foods should not ignore packaging machines and the complex interactions between the machine and the packaging material. (**Figure 1**) shows an example of a horizontal form–fill–seal machine, a vertical pouch form–fill–seal machine, and two overwrapping machines.

The suitability of a packaging machine for a certain packaging material and product can be summarized in the following key points:

- The angle at which the wrapping material comes into contact with the folding former is important, as it may cause a mark or even a cut in the material. The tightness of the packaging film depends on the shape of the former.
- The quality of heat sealing of the packaging material depends on the temperature control, pressure, and dwell time. Optimal results are obtained when the dwell time at the melting temperature of the coating or copolymer is long enough to ensure tightness without damage to the basic material.
- The wrapping material should slip easily after sealing at high temperature. In some cases, it is even necessary to cool the plate fixed above the sealing rollers.
- The quality of cutting depends on the knife, cut angle, position of the knife in relation to the sealing jaws, etc.

Overwrapping also requires compatibility between the machine and the wrapping material, and prevention of static electricity, poor slip under hot conditions, temperature control disruption, and rolling of the film. The packaging material should be sealable on both sides, the sealing area should be sufficient, and the structure (soft or rigid) of the solid food should not affect the shape and sealing surface if a good gas tightness is desired.

Packaging machines for food powders, and especially sugar, cover a large range of sizes of packages, starting with small flat pouches containing between 4 and 10 g of sugar. These are produced using simple machines that assemble sachets by bringing together two webs of polyethylene-coated paper, heat sealing three sides, filling with sugar, and sealing the top. There are also carton packing machines, especially designed for sugar cubes. For ease of handling, 'brick-packs' are preferred. These are produced by forming a polyethylene pack and dropping it into a horizontal

solid upstanding pack with similar characteristics to the paper packet.

The quality of packaged food powders should comply with the necessary regulations. Any foreign matter should be eliminated. All packages go through metal detectors to control metal contamination. Likewise, traceability of packets and even full pallets is needed for identification of the product according to EU regulations. A printed code showing the date and line of production, factory of origin, etc. is required. The spread of use of scanning at the shop point of sale imposes the printing of a bar code on each domestic package.

See also: **Browning**: Nonenzymatic; **Chilled Storage**: Use of Modified-atmosphere Packaging; Packaging Under Vacuum; **Chill Foods**: Effect of Modified-atmosphere Packaging on Food Quality; **Oxidation of Food Components**; **Spoilage**: Chemical and Enzymatic Spoilage; Bacterial Spoilage; **Storage Stability**: Shelf-life Testing; **Water Activity**: Effect on Food Stability

Further Reading

Gary JI, Harte BR and Miltz J (eds) (1987) *Food–Product Package Compatibility*, Lancaster, PA: Technomic.
Hotchkiss JH (ed.) (1988) *Food and Packaging Interactions, American Chemical Society Series*. Washington, DC: American Chemical Society.
Mathlouthi M (ed.) (1986) *Food Packaging and Preservation, Theory and Practice*. London: Elsevier.
Rockland LB and Beuchat LR (eds) (1987) *Water Activity. Theory and Application to Food*. New York: Marcel Dekker.
Rogé B and Mathlouthi M (2000) Caking of sucrose crystals: effect of water content and crystal size. *Zuckerindustrie* 125: 5.
Van der Poel PW, Schiveck H and Schwartz T (eds) (1998) *Sugar Technology, Beet and Cane Sugar Manufacture*. Berlin: Dr Albert Bartens Verlag.
Versanyi I (1985) *Food Packaging Technique [H]*. Budapest: Mezögazdasagi Kiado.

Aseptic Filling

L Mauer, Purdue University, West Lafayette, IN, USA

Background

Aseptic processing is a high-temperature–short-time thermal process to commercially sterilize a product and fill the cooled sterile product into a presterilized package in a sterile environment. Purposes for aseptic

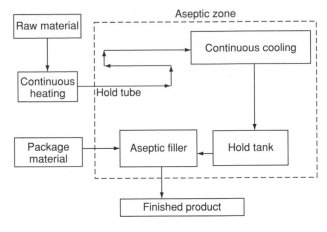

Figure 1 Aseptic process diagram.

processing include extending the storage life of food products, optimizing product quality, and reducing cost. A diagram of an aseptic processing system for consumer products is shown in **Figure 1**. In the aseptic process, the aseptic filler is designed to sterilize the package material, fill the sterile product into the package in a sterile environment, and then hermetically seal the package. Aseptic filling differs from other traditional methods of food packaging in that the food product and the package are continuously sterilized separately and then meet in the sterile environment provided by the aseptic filler. Important factors in the aseptic filling process are the type of product, type of package, obtaining and maintaining a sterile environment for filling, and the sealing process.

Types of Product

A variety of food products are aseptically packaged. Examples include milk and dairy products, fruit and vegetable juices, fruit juice concentrates, sport and dietetic beverages, tomato products, edible oils, puddings, soups, cheese and soy sauces, and products containing small or large particles of fruits, vegetables, or potatoes. The aseptic process must be designed so that the product is sterilized, the package is compatible with the product composition and storage needs, and the type of filler treats the product gently while maintaining sterility to achieve optimum quality. Products are usually sterilized by high-temperature heat treatments for short periods of time, cooled, and conveyed to the filler via aseptic pumps or nitrogen. Packaging and filling considerations for different product types include the following:

Low-acid Products (pH > 4.5)

The concept of aseptic heat treatments for low vs. high acid foods is similar to the canning principles for these food types. In low-acid foods, such as many milk and vegetable products, pH is not a barrier to pathogenic microbial growth. Therefore, aseptic processing conditions must be designed with stringent controls to achieve a 10-decimal reduction value for *Clostridium botulinum*, the most heat-resistant pathogen able to grow in low-acid foods, both in the food product and in the package materials. Targeting the most resistant pathogen ensures safety because all other pathogens are killed faster than it. The package must attain the same level of sterility as the product; therefore, package materials for low-acid foods may have stricter sterility controls than those for high acid foods. *Bacillus subtilis* is the most resistant organism to hydrogen peroxide sterilization of packages, and a four-decimal requirement for it exists to achieve the required reduction for *C. botulinum* in low acid foods.

Acid and High-acid Products (pH < 4.5)

In acid and high-acid foods, such as fruit juices and many tomato products, pH is a barrier to pathogenic microbial growth, most notably for *C. botulinum*. For these foods, spoilage microorganisms may cause more problems than pathogens. Processing temperatures and holding times may be lower than for low acid foods. Spoilage organisms are much more sensitive to hydrogen peroxide sterilization of packages than *B. subtilis*. *Aspergillus niger* is usually the target organism for dry and moist heat sterilization of packaging materials for use with acid and high-acid products.

Homogeneous vs. Heterogeneous Products

Knowledge of heat-transfer characteristics of a food, quality parameters, flow characteristics, mean residence time, type of heat exchanger or sterilizing system, and heat resistance and required death values for target microorganisms are important factors in designing an effective aseptic process. Homogeneous products contain no particles that disrupt heat transfer. Heterogeneous products contain particles with sufficient size to create a thermal gradient during processing. Therefore, heat treatments for homogeneous liquids may be less damaging than those required to insure commercial sterility of particulates in similar liquids. Pumps used to convey aseptically processed products must be able to maintain both product sterility and integrity; therefore, pumps used for heterogeneous products must limit shear forces so as not to damage particulate structures. Homogeneous liquids are often conveyed to a filler with either a centrifugal pump or sterile nitrogen; however, highly viscous liquids may require use of a positive displacement pump. Heterogeneous products are conveyed to the filler using a positive displacement pump that limits pressure and shear placed on the particulates. Opening sizes in an aseptic system designed for heterogeneous products must be sufficient to allow passage of the particulates without creating pinch points. In the filler, contamination of the seal area, especially with particulates, must be avoided.

Types of Packages

The type of aseptic package used must be suitable for a product's requirements, package surfaces must withstand sterilization by heat, irradiation, and/or chemicals prior to filling, and the package must contain, protect, and preserve the food product throughout its distribution and shelf-life. Properly aseptically processed products remain microbiologically stable as long as the package remains intact. Often, shelf-life and product quality are limited more by package performance than any other factor. Needs to consider for an aseptic packaging system include:

- Compatibility of product and package – product composition, needs for maintaining quality (barrier, mechanical), and shelf-life requirements.
- Packaging material – type (plastic, metal, glass, laminate, etc.), barrier and mechanical properties, machineability, recyclability, and cost.
- Package form – size, shape, compatible machinery, sealing properties, appeal to consumer, communication of information, and cost.
- Sterilization method (heat, irradiation, chemicals, combination of methods) – efficiency, throughput, residues, compatibility with product/package/environment, worker safety, regulations, and cost.
- Filling equipment – reliability, efficiency, capacity, and cost of installation/operation/maintenance.

Aseptic Package Materials

A significant drive in the conversion from traditional to aseptic processing is the reduction in packaging costs for both materials and transportation. Traditional thermal processes require packages to withstand high temperatures, whether for in-package sterilization or for hot-filled products, as well as vacuum forces created on cooling. Packages that meet these structural requirements include metal, glass, and plastics (rigid, semirigid, some with vacuum panels incorporated into the design, and pouches). The cold aseptic filling allows for a lighter container with a greater design flexibility (squeezable, ergonomic, etc.). By weight, most plastic materials cost significantly less than glass and metals, and less plastic is needed for a cold-filled product than a hot-filled product. Plastics are the most common material used for aseptic packaging; however, plastics are more permeable to oxygen and moisture than either glass or metals. The quality of products may suffer after extended storage in permeable packages. An example of this is oxidation of dairy products in packages permeable to oxygen. To limit permeability, plastics with different barrier characteristics are often combined using lamination or coextrusion methods to optimize barrier properties. This can maximize product quality and package function while maintaining low package costs.

The variety of plastic materials used in aseptic packaging is continuously increasing, along with the use of laminates, coextrusions, and copolymers. Adhesive and thermal lamination methods are used to create multilayer packages, often with a structure based on the diagram in **Figure 2**. More than one barrier layer may be incorporated into a package design. Commonly used plastics include polyolefins (polyethylene, polypropylene, polystyrene), polyesters (polyethylene terephthalate), vinyl plastics (polyvinyl chloride, polyvinylidene chloride, polyvinyl alcohol), polyamides (nylon), ionomers, acrylics, fluorocarbons, and polycarbonates. Polyethylene (PE) is the most widely used polymer in packages and is available in a variety of densities: low density (LDPE), linear low density (LLDPE), medium density (MDPE), and high density (HDPE). LDPE is used in flexible bags and pouches and is a good heat-sealing material. HDPE used in blow-molded bottles is more rigid and has better barrier properties than LDPE. Polypropylene (PP) provides clarity, stiffness, and heat resistance. Polyethylene terephthalate (PET) has good barrier and clarity properties and a high tensile strength, and is resistant to high temperatures. PET is often used to meet consumer demand for round and recloseable plastic bottles. Ethylene vinyl alcohol

Figure 2 General structure of a laminate aseptic package.

(EVOH) and polyvinylidene chloride (PVDC) are generally the best plastic barriers to moisture and oxygen migration. Metal and glass are used for barrier properties either alone or within a laminate package. A paperboard layer in a laminate package is used for printing and light-barrier purposes.

Package Formation

Packages formed for the aseptic process require minimum levels of contaminating microorganisms prior to the sterilization process, and the packages must pass leak-testing procedures. Packages may be preformed or formed in the filler just prior to filling. Metals are formed into cans, and metal films may be incorporated into laminate structures. Glass is blow-molded into bottles and jars. Plastics and laminated materials are blow-molded or thermoformed into various package shapes. The blow-molding process involves the melting of a glass or plastic, forming a parison (tube), and then blowing the parison into the package shape (designated by the mold) using sterile air or nitrogen. Extrusion, injection, and stretch blow-molding methods are used for different package applications. A sheet of plastic is heat-softened then molded to shape via vacuum, pressure, or matched mold forming in the thermoforming process. Cups and trays are thermoformed packages.

Aseptic Package Systems

Package forms and filler types are interwoven to the extent that discussion of one is not complete without discussion of the other, as the aseptic process hinges on the placing of a sterile product into a sterile package in the sterile filler environment. Common aseptic package forms are discussed in this section, leaving classification of filler types for a later section.

Cartons The brick-style, paperboard packaging used for cartons is recognized as the traditional aseptic package used for juice boxes. The laminate material used for carton formation generally consists of multiple polyethylene, paper, and aluminum foil layers. Outer and inner polyethylene layers provide protection and sealing properties, the paperboard is printed with product information and provides stiffness, and the aluminum foil contributes gas- and light-barrier functions. Cartons are prefabricated, or rolls of the laminate material are sterilized and formed in the filler just prior to filling and sealing. Cartons are sterilized using hydrogen peroxide and hot air.

Bottles The convenience of plastic bottles that are clear, round, resealable, recyclable, and able to be formed in a variety of shapes and sizes is a driving force in the beverage industry. Aseptic milk, creamer, sports, and nutritional beverages in PET, PE, and PP bottles are currently available to consumers. Bottles for use with high-acid products are generally sterilized using steam, a combination of heated air and steam, or heat of formation (extrusion and blow molding). Bottles for low-acid products are sterilized with hydrogen peroxide.

Cups The thermoformed cup is a common aseptic package for puddings, particulates, fruits, purees, and baby food. Package materials for cups include PS, PP, and laminates of PS, PVDC/EVOH, and LDPE materials. Cup lids are often aluminum foil coated with LDPE or other heat-sealant plastic. Cups are generally sterilized using hydrogen peroxide or saturated steam under pressure.

Pouches Pouches for institutional and fast food chain use are easier to open, take up less space for transportation, storage, and waste, and decrease product loss when compared with traditional metal cans. Form-fill-seal systems are used for converting a sheet film of plastic (LDPE, laminate) into a pouch. Plastics formed as lay-flat tubes require fewer seals to form pouches than the sheet films. Aseptic products available in pouches include tomato-based sauces, ice cream mixes, cheese sauces, and puddings. Pouches are often sterilized using hydrogen peroxide.

Bag-in-box The bag-in-box aseptic packaging system is used for bulk packaging (1–10 000 or more liters). The product is filled into a presterilized plastic bag. The bag is placed into a protective container, such as a corrugated box or metal drum, either before or after filling. A laminate of barrier (PVDC, EVOH) and heat sealant (LDPE) plastics is commonly used for bag formation. Tomato, fruit juice, and dairy products are packaged using the bag-in-box system. Preformed bags are sterilized by ionizing gamma irradiation. The exterior of bags for low-acid products is also sterilized with steam or hydrogen peroxide vapor.

Metal cans The metal can traditionally used in thermal processes also has advantages for use with aseptic products despite the added weight over plastic materials. The durability, consumer acceptance, barrier properties, and recyclability of metal cans combined with the ability to sterilize the cans with superheated steam instead of chemicals make the metal can attractive for some applications. Aseptic dietetic beverages, pudding, and cheese sauces are available in metal cans. Superheated steam is used to sterilize metal cans prior to filling.

The Sterile Environment

Obtaining a Sterile Environment

Establishing, maintaining, and validating sterility in an aseptic system are essential. Processes used for obtaining a sterile environment for products, equipment, and packages include thermal, chemical, irradiation, and mechanical treatments along with combinations of these. Thermal processes include saturated steam, superheated steam, hot air, mixtures of hot air and steam, and extrusion/heat of formation (although extrusion may not be acceptable for sterilization purposes). Heat is the most common method for product sterilization and is often used for equipment sterilization as well. The most common chemical used for sterilization is hydrogen peroxide (20–35% concentration), and a combination of hydrogen peroxide treatment followed by hot air is often used to reduce the level of hydrogen peroxide to less than 0.5 p.p.m. on packages and equipment. Peracetic acid and ethylene oxide also have applications; however, regulations for acceptable chemicals, uses, and residue limits must be checked. Chemicals are applied by dipping, spraying, or rinsing, or in vapor form. Irradiation treatments may include ultraviolet radiation, infrared radiation, and ionizing gamma radiation. Mechanical processes are generally designed to reduce initial microbial load to a level the aseptic system is designed to handle. If an initial microbial load is too high, sterility might not be achieved during aseptic processing. Therefore, water rinsing or flushing, air blasting, brushing, and ultrasound methods are used to reduce the initial load on equipment and packaging prior to other sterilization processes.

Equipment and packages must attain the same level of sterility as the products they will come into contact with. Therefore, equipment is sterilized with hydrogen peroxide or a time/temperature steam sterilization at least equivalent to what the product will receive. Thermal processes for equipment used with low-acid foods will be higher, both time and temperature, than for high-acid products. Packages, other than metal cans, may not withstand high temperature treatments; therefore, chemical and irradiation sterilization procedures are commonly used.

Maintaining a Sterile Environment

Once equipment, packages, and products are sterilized, the challenge is to maintain sterility during the filling and sealing operation. This is accomplished using either an overpressure of sterile air or continuous flow of superheated steam, the latter generally being used for the metal can aseptic process. Air is sterilized by filtration, usually using a series of high-efficiency particulate air (HEPA) filters. Since all equipment must attain commercial sterility, the HEPA air filters also must be sterilized. The positive pressure maintained by a continuous sterile air flow prevents contamination during the filling process when the product may be exposed to air. If the integrity of an aseptic zone is compromised because of line stoppages or other occurrence, it is necessary to repeat the initial sterilization process.

Validating Sterility

Validating sterility is a combination of implementing an appropriate hazard analysis critical control point program, documenting product and filler sterilization procedures, filing the process with the appropriate agency, maintaining accurate temperature, time, pressure recording devices, and testing all aseptically processed foods for sterility using appropriate microbiological sampling techniques. Challenge testing with inoculated foods is part of validating a thermal process.

The Sealing Process

At the end of the filling process, packages are often sealed using heated sealing bars, jaws, and plates. The temperature, pressure, and time of sealing must be strictly controlled for each sealing compound to maximize package integrity. Common heat sealant polymers include polyethylene, polypropylene, ethylene vinyl acetate, and polyvinyl chloride. These compounds have demonstrated a desirable mechanical strength of the seal (impact strength, tensile strength, tear strength, seal strength, puncture resistance) and product holding properties (chemical resistance, product compatibility, oxygen and moisture barrier properties). Problems occur when particulates, fibers, or other food contaminants are caught in the seal area and prevent a completely fused seal. To avoid seal contamination, especially with particulate foods, the filler must be designed to avoid or remove product from the seal area prior to fusion.

Types of Fillers

The filling method depends on both the type of product (high or low acid, homogeneous liquid, viscous product, size of particulates, etc.) and type of package (retail, institutional, cup, pouch, bottle, etc.) desired for the end product. Common filler classifications are named for the processes that occur in the filler and are outlined below.

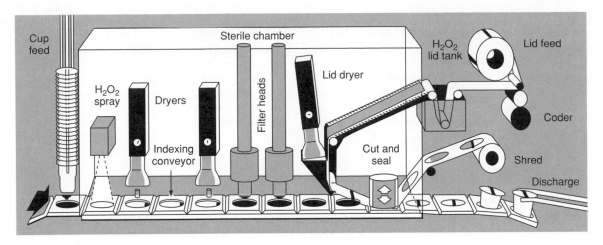

Figure 3 Aseptic filler design.

Fill and Seal

In the fill and seal process, a preformed sterile container (bottles, cups, irradiated pouches, bag-in-box) is aseptically filled and sealed. The fill and seal process for cups is shown in **Figure 3**. In bulk packaging systems (500-liter bins, large totes, million gallon storage tanks), product is filled directly into a sterilized large storage container. This system is often used for seasonal food products to extend the supply throughout the year.

Form, Fill, and Seal

Brick pack juice boxes and pouches (LDPE, laminate) enter the filler as a roll of packaging material and are sterilized, aseptically formed into the box or pouch shape, filled, and thermally sealed in the form, fill, and seal process.

Thermoform, Fill, and Seal

In the thermoform, fill, and seal process, the packaging material (PS, PP, laminates) enters the filler as roll stock and is sterilized, heated, thermoformed (usually into cups), filled, and thermally sealed with presterilized lid material (aluminum foil coated with LDPE).

Blow Mold, Fill, and Seal

Plastic bottles are formed, filled, and sealed in the blow mold, fill, and seal filling process. An extrudable material (PET, PP, PE) is blow-molded into a container that is filled and sealed in place before the mold opens. For low-acid products, a chemical sterilization after the bottles are formed is used to insure sterility prior to filling.

Advantages and Disadvantages of Aseptic Filling

Advantages of aseptic filling include a high product quality, decreased storage space for packaging materials, decreased weight of packaging materials, decreased package costs, decreased shipping costs for packaging materials, reduced energy consumption, increased flexibility for package options, increased convenience for easy open containers, feasibility of long-term bulk storage, and increased shelf-life and storage time for products. Disadvantages include the high cost of aseptic fillers, potential for product contamination in the aseptic filler owing to the complexity of the system, decreased recyclability of pouch and laminate package materials compared with metal and glass packages, and slower line speeds than traditional process methods. For many, the advantages far outweigh the disadvantages.

See also: **Canning**: Principles; **Heat Treatment**: Ultrahigh Temperature (UHT) Treatments; **Packaging**: Packaging of Liquids; **Pasteurization**: Principles; **Storage Stability**: Mechanisms of Degradation; Parameters Affecting Storage Stability; Shelf-life Testing

Further Reading

Chambers JV and Nelson PE (ed.) (1993) *Principles of Aseptic Processing and Packaging*. Washington, DC: The Food Processors Institute.

David JRD, Graves RH and Carlson VR (eds) (1996) *Aseptic Processing and Packaging of Food: A Food Industry Perspective*. New York: CRC Press.

Floros JD (1993) Aseptic packaging technology. Chambers JV and Nelson PE (eds). In: *Principles of Aseptic Processing and Packaging*, pp. 115–148. Washington, DC: The Food Processors Institute.

Holdsworth SD (ed.) (1992) *Aseptic Processing and Packaging of Food Products*. London: Elsevier Applied Science.

Moruzzi G, Garthright WE and Floros JD (2000) Aseptic packaging machine pre-sterilisation and package sterilization: statistical aspects of microbiological validation. *Food Control* 11(1): 57–66.

Reuter H (ed.) (1993) *Aseptic Processing of Foods*. Lancaster, PA: Technomic.

Robertson GL (ed.) (1993) *Food Packaging Principles and Practice*. New York: Marcel Dekker.

Sommers CH, Thayer DW, Pauli G *et al.* (2000) *New Processing and Packaging Technologies*. Washington, DC: National Food Processors Association.

Willhoft EMA (ed.) (1993) *Aseptic Processing and Packaging of Particulate Foods*. London: Blackie Academic & Professional.

Packaging Materials *See* **Chill Foods**: Effect of Modified-atmosphere Packaging on Food Quality; **Chilled Storage**: Use of Modified-atmosphere Packaging; Packaging Under Vacuum; **Packaging**: Packaging of Liquids; Packaging of Solids; Aseptic Filling

---

PALM KERNEL OIL

K G Berger, Chiswick, London, UK

Introduction

Palm kernel oil is obtained from the oil-rich seed of the oil palm (*Elaeis guineensis Jackqu*). A description of the oil palm and its development into a major crop is given in the entry on **Palm Oil**. The production process of the kernels in the oil mill is also described there.

The quantity of palm kernel oil produced is on average 12% that of palm oil. World production in 1998 was 2.9 million tons.

Production of the Oil

The palm kernels leave the oil mill at about 50% oil content and up to 8% moisture. Further processing typically consists of:

1. Comminution in hammer mills to approximately 2-mm pieces.
2. Cooking to 110–120 °C in a continuous cooker. This breaks the oil cells and reduces moisture to 2.5%.
3. Pressing in a screw press at high pressure to yield crude palm kernel oil and a meal containing 7–9% oil. The meal is used for animal feed.

An alternative process is to submit the broken nuts after step (1) to a cold pressing at low pressure, leaving a meal containing 10–12% oil. This may be further treated by solvent extraction, depending on the market prices of oil and meal.

Refining

Palm kernel oil may be refined either by an alkali or by the physical process, as described in the entry on **Palm Oil**. However, because most of the fatty acids present are of 12 carbon atoms or less, the deodorization temperatures used are lower – typically 220 °C in the alkali process and 230–235 °C in the physical process.

Further Processing

Palm kernel oil is treated by fractionation or hydrogenation and/or interesterification to obtain products with closely defined functionality. These processes are described in the entry on **Palm Oil**.)

Chemical Composition and Physical Properties

The fatty acid composition of palm kernel products is given in **Table 1**, while **Table 2** gives the glyceride composition by carbon number. Typical values for the minor components of palm kernel oil are given in **Table 3**.

Palm kernel oil has a high content of lauric and shorter-chain fatty acids and a low content of unsaturated acids. Coconut oil has a somewhat similar composition. They contrast strongly with the other common vegetable oils, which contain no shorter-chain fatty acids. In consequence the physical properties are unusual, as will be seen from **Tables 4** and **5**. Palm kernel oil has a relatively high solid fat content and a rather hard structure at 20 °C, but melts sharply at 28 °C – well below mouth temperature. These properties are accentuated in palm kernel stearin. As is clear from the last two columns of **Table 5**, hydrogenation raises the solids content further, but the products are still substantially molten at mouth temperature. Hydrogenated palm kernel olein, on the other hand, having a higher content of long-chain saturated acids, has a higher melting point. The products shown in **Table 5** are examples of the properties that can be obtained by further processing. A wide range of proprietary products based on palm kernel oil is available, giving variations in physical properties required for specific food applications.

The data shown in **Tables 1–4** are average figures taken from an extensive survey of Malaysian palm

Table 3 Minor components of palm kernel oil typical values

	Refined palm kernel oil (mg kg^{-1})
Carotenoids	Less than 8
Tocopherols and tocotrienols	Less than 33
Sterols[a]	875
Triterpene alcohols	470

[a]Main components β-sitosterol (68%), stigmasterol (14%), and campesterol (9%).

Table 4 Physical properties of refined palm kernel oils

Solid fat content % by NMR at:	Palm kernel oil	Palm kernel olein	Palm kernel stearin
5 °C	72.8	65.6	93.2
10 °C	67.6	56.9	91.6
15 °C	55.7	40.4	90.1
20 °C	40.1	20.9	82.8
25 °C	17.1	1.4	68.2
30 °C			34.6
Slip melting point °C	27.3	23.6	32.2

NMR, nuclear magnetic resonance.
Data from Palm Oil Research Institute of Malaysia, with permission.

Table 1 Fatty acid composition % of palm kernel oil products

Fatty acid	Palm kernel oil	Palm kernel olein	Palm kernel stearin
6:0	0.3	0.4	0.2
8:0	4.2	5.4	1.2–3.5
10:0	3.7	3.9	2.4–3.6
12:0	48.7	41.5	55.6–58.6
14:0	15.6	11.8	18.1–24.7
16:0	7.5	8.4	7.1–7.9
18:0	1.8	2.4	1.5–1.8
18:1	15.0	22.8	2.6–8.8
18:2	2.6	3.3	0.2–1.5

Data from Palm Oil Research Institute of Malaysia, with permission.

Table 2 Triglyceride composition of palm kernel oil products: carbon number by gas–liquid chromatography

Carbon number	Palm kernel oil (%)	Palm kernel olein (%)	Palm kernel stearin (%)
28	0.5	0.3	0.1
30	1.3	1.6	0.5
32	6.4	7.8	3.3
34	8.4	9.3	6.5
36	21.0	18.3	27.5
38	15.6	12.5	24.8
40	9.5	7.6	15.2
42	9.0	9.3	9.2
44	6.8	7.8	5.2
46	5.3	6.5	3.0
48	6.2	8.3	2.4
50	2.5	3.1	0.9
52	2.7	3.4	0.7
54	3.3	4.2	0.8

Data from Palm Oil Research Institute of Malaysia, with permission.

oil. The figures are not significantly different from those recently obtained for oils from other sources in other producing regions.

Food Uses of Palm Kernel Products

The most important applications of palm kernel oil products are in the confectionery industry, where its high solid fat content and sharp melting characteristics are important. However, palm kernel oil itself melts at too low a temperature for some applications. It is therefore hydrogenated to give a variety of products, one of which is indicated in **Table 5**. Hydrogenated palm kernel oil is suitable for chocolate-type couvertures for biscuits and sugar confectionery, and for biscuit cream fillings. Hydrogenated palm kernel oil is also used to replace butterfat in filled milk, coffee creamers, and imitation cream. The higher-melting-point grades tend to leave a waxy residue on the palate. They can be improved by fractionating to remove the highest melting point components. A superior and more expensive product is obtained by the direct fractionation of palm kernel oil. The stearin has good contraction when it solidifies and is therefore suitable for molded chocolate. (*See* **Cocoa**: Production, Products, and Use.)

The palm kernel olein produced is the lower-value fraction. It may be hydrogenated (**Table 5**) to give a range of confectionery fats of somewhat lower quality. Alternatively, palm kernel olein is used in margarine blends. It is a useful component of

Table 5 Solid fat content of further processed palm kernel oil products

	Partly hydrogenated palm kernel oil	Partly hydrogenated palm kernel olein	Fully hydrogenated palm kernel olein	Same interesterified	Palm kernel stearin	
					Partly hydrogenated	Fully hydrogenated
Melting point (°C)	34	38.6	44.5	35.6	34.5	37.5
Solid fat content % at:						
10 °C	94	85	90	82		
15 °C	92	77	82	74		
20 °C	84	61	72	60	90	90
25 °C	59	39	64	45	83	84
30 °C	27	20	38	25	38	38
35 °C	7	7	22	6	15	17
37 °C		7	18	3	3	4
40 °C	0.5	3	13		2	3

Data from Palm Oil Research Institute of Malaysia, with permission.

Table 6 Interesterified blends using palm kernel oil products

Oil	%	Solid fat content % at:					Application
		20 °C	25 °C	30 °C	35 °C	40 °C	
Palm kernel olein	75	53		15	3	0	Biscuit filling cream
Palm stearin	25						
Palm kernel olein	75}	46	24	9			Chocolate soft center
Cotton seed oil stearin	25}[a]						
Same blend	[b]	66	40	23			Couverture
Hardened palm kernel oil	70}	80	63	43	21	3	Whipping cream
Palm stearin	30}						

[a]After interesterification, hydrogenated to iodine value 17.
[b]After interesterification, hydrogenated to iodine value 8.

interesterified mixtures for various applications (**Table 6**).

Palm kernel oil without modification is used for chocolate coatings for icecream bars, usually in a blend with liquid oil or palm oil to give the right consistency without excessive brittleness.

Palm Kernel Products in Margarine

1. Palm kernel oil forms a eutectic with palm oil in a mixture containing about 30% palm kernel oil. This feature is used to improve the mouth feel (rapid melting) in the following formula:

    Palm oil   63%
    Palm stearin   7%
    Palm kernel oil   30%

2. Palm kernel olein (30 parts) interesterified with palm stearin (70 parts) forms a margarine stock. To make margarine, 60 parts are blended with 40 parts liquid oil.
3. Equal quantities of fully hydrogenated palm kernel oil and fully hydrogenated palm oil are interesterified. A margarine high in polyunsaturated fatty acids is made by blending 12% of this hard stock with 88% of a liquid oil such as sunflower oil.

See also: **Cocoa**: Production, Products, and Use; **Palm Oil**

Further Reading

Applewhite TH (1994) *Proceedings of World Conference on Lauric Oils*. Champaign, Ill: American Oil Chemists Society.

Baldwin AR (1985) Proceedings of World Conference Kuala Lumpur, Malaysia. Processing of palm, palm kernel and coconut oils. *Journal of American Oil Chemists Society* 62.

Siew WL and Berger KG (1981) *Malaysian Palm Kernel Oil Chemical and Physical Characteristics. Porim Technology 6*. Kuala Lumpur: Palm Oil Research Institute of Malaysia.

Tang TS, Chong CL and Yusoff MSA (1995) *Malaysian Palm Kernel Stearin, Palm Kernel Olein and their Hydrogenated Products. PORIM Technology 16*. Kuala Lumpur: Palm Oil Research Institute of Malaysia.

PALM OIL

K G Berger, Chiswick, London, UK

Introduction

Palm oil is obtained from the fruit flesh of the oil palm (*Elaeis guineensis* Jacqu.), a native of the equatorial region of West Africa. Unrefined oil obtained from wild palms has been a traditional food source for the indigenous population for thousands of years. However, its use for edible purposes elsewhere had to await the development of suitable refining processes to produce a bland pale-colored oil. Today, palm oil is the second most abundant edible oil after soya bean oil and is in universal use.

The Oil Palm

The oil palm flourishes best in lowland regions of high rainfall and close to the equator. Optimum conditions are a rainfall of 1700 mm or more per annum, evenly distributed through the year and a position within 10° N and S of the equator, but it also grows in isolated locations outside these limits, i.e., between 15° N and 20° S. The plant can be grown in a variety of tropical soils, regular and sufficient water being apparently more important than soil, provided that the plant nutrient requirements are supplied. Unfavorable soil factors are poor drainage and high laterite or sand content. Special techniques have been developed to enable peat soils to be used.

The oil palm is a single-stemmed plant bearing a number of fronds in a simple head. A mature palm may have up to 50 fully opened fronds. One inflorescence arises from the axil of each leaf; male and female inflorescences occur on the same palm. Both consist of a central stem carrying about 200 flower bearing spikelets. Each spikelet on the male inflorescences carries about 1000 flowers, and on the female inflorescence 15–30 flowers. Pollination is by insects, the most effective being the weevil, *Elaeidobius kamerunicus*, native to West Africa. Ripe fruit

develops in about 155 days after fertilization, but the bulk of the oil is synthesized in the final 2–4 weeks. The fruit bunch, containing 1500–2000 individual fruits weighs 20–30 kg.

Considerable variation occurs in the wild palms, and three types have been classified according to fruit type. The most common wild type is the dura, characterized by a relatively thin layer of flesh covering the seed, which consists of a hard thick shell within which lies the kernel. A small proportion of plants, called pisifera, bear fruit with a thick layer of flesh and a small kernel with a very thin or no shell. Another small proportion of plants in the wild, the tenera, has relatively thick flesh with a shell and kernel of intermediate size.

The tenera type was found to be a natural cross (D×P) of dura and pisifera palms. When plantations were first being developed in West Africa (in Belgian Congo, now Zaire), the superior economic performance of the tenera was recognized, and the heritability of shell thickness was discovered. Subsequent development of improved planting material therefore centered on selecting the best dura and pisifera plants for crossing. The continuing efforts of plant breeders have resulted in progressive increases in yield, as shown in **Table 1**.

A slow development of oil palm plantations occurred in Indonesia after World War I, soon to be followed by Malaya. By 1938, Indonesia had 90 000 ha and Malaya 30 000 ha under oil palm. After 1960, the area under oil palm increased rapidly in Malaysia, reaching 3.2 million ha by 2001. Developments in Indonesia were slower, with 2.4 million ha in 2001. These two countries are by far the largest producers of palm oil and the main suppliers to the world market. Plantation developments in Central and South America have reached about 500 000 ha in all, with Colombia, Ecuador, and Brazil having the largest areas.

A unique feature of the oil palm is that the fruit yields two distinct types of oil. Palm oil from the flesh is the major product, whereas palm kernel oil is

Table 1 Progress through breeding and selection in South-east Asia

Material	Bunches (tonnes ha^{-1})	Mesocarp to fruit (%)	Oil to bunch (%)	Oil (tonnes ha^{-1})
Bogor 1878	16.5	58.7	17.6	2.8
Elmina 1933	20.1	58.2	17.0	3.4
OPRS 1969	24.8	64.1	18.3	4.5
Commercial DXP 1978	26.0	80.1	22.0	5.7

Adapted from the Palm Oil Research Institute of Malaysia Bulletin, November 1984 with permission.

obtained from the seed. Its chemical and physical characteristics are quite distinct from palm oil, and it has different applications. (*See* **Palm Kernel Oil**.)

Commercial Importance

The importance of palm oil in the world's oils and fats economy is shown in **Table 2**, giving the total production of 17 major oils and fats, and individual figures for soya, palm, palm kernel, and rapeseed oils. The figures given are 5-year averages and include production forecasts to the year 2002. Figures in brackets are the percentage of total. Production figures for sunflower oil (not shown in **Table 2**) are similar to those of rapeseed oil, and that the other oils among the 17 included in the total show little or no growth trend.

Table 2 shows that palm and palm kernel oils together have increased their share of edible oil supplies regularly for more than 20 years, and are forecast to continue to do so. World production of palm oil in 2001 was 23.6 million tonnes of which 17.6 million tonnes were exported to world-wide destinations. Eighty-eight percent of the exports were of Malaysian and Indonesian origin.

In those African countries where the oil palm grows, it continues to hold its position as the major traditional food oil with an annual production of about 1.5 million tonnes in the region. While much of it is still eaten in the crude form, the consumer is also increasingly demanding processed products.

In a number of Latin American countries, palm oil is seen as an efficient means of reaching self-sufficiency, and current production of about 1.4 million tonnes is mainly for local consumption. In contrast, domestic consumption in Malaysia is less than 10% of production. The crop is mainly grown for export, as a successful diversification from rubber. Indonesia, the second largest producer, requires a higher proportion of its production for internal consumption, but is increasing production and exports rapidly.

The importance of palm oil in the world market is based on several factors.

1. It is comparatively cheap, often with an appreciable discount to soya bean oil, the market leader.
2. It has technical attributes, useful in food manufacture, especially its good stability to oxidation, and its natural solid fat content.
3. It is a perennial crop, planted for an economic life of 25–30 years. It mainly grows in equable climates in regions little affected by earthquakes or hurricanes, so that production fluctuates less from year to year than does that of annual crops.

Harvesting and Processing of the Fruit

Various traditional primitive methods of oil recovery are still widely used in West Africa. Typically, the fruit bunch is cut off, and allowed to ferment a few days, so that the fruit detaches easily. The fruit flesh is softened by boiling or by further fermentation, then mashed in a pestle and mortar or under foot. Hot water is added, the oil skimmed, and the water is then boiled off in a separate container. The oil is used without any refining.

In contrast, the modern plantation is usually associated with a mechanical oil mill with a processing capacity of 5–60 tonnes of fresh fruit bunches (FFB) per hour. The harvesting of FFB is still done manually, using an ax, or a sickle-type blade fixed to a pole, depending on the height of the palm. Ripeness is judged by the number of detached fruit that have fallen below the bunch. The frond and then the bunch are cut off. The bunch and any loose fruit are picked up and carried manually, by barrow, in a light animal-drawn vehicle, or with a light mechanical vehicle to the nearest estate road. It is important not to compact the soil close to the palms. Transport is then by lorry to the oil mill. Bunches are emptied into a chute, and filled into sterilizer 'cages' mounted on

Table 2 World production of oils and fats (× 1000 tonnes)

	1968–1972	1973–1977	1978–1982	1983–1987	1988–1992	1993–1997	1998–2002
Total of 17 oils and fats	40 314	45 883	56 848	67 614	80 663	92 314	105 266
Soya bean oil	6 036	8 477	12 639	14 147	16 997	19 508	23 317
	(15)	(18.4)	(22.1)	(20.8)	(21.1)	(21.1)	(21.9)
Palm oil	1 725	2 763	4 482	6 745	10 101	13 878	17 498
	(4.3)	(6.0)	(7.9)	(10.0)	(12.5)	(14.9)	(16.6)
Palm kernel oil	406	437	545	862	1 232	1 655	2 175
	(1.0)	(1.0)	(1.0)	(1.3)	(1.5)	(1.8)	(2.1)
Rapeseed oil	2 039	2 572	3 732	6 009	8 329	9 534	10 829
	(5.0)	(5.6)	(6.6)	(8.9)	(10.3)	(10.3)	(10.3)

Adapted from Oil World Annual, ISTA Mielke, Hanburg, with permission wheels, which are then moved into a horizontal cylindrical pressure vessel on rails. Typically, up to 7×2.5 tonnes of fruit are cooked in one load. Steam at 304 kPa (3 atm) is applied for about 1 h. The contents of the cage are then fed to a 'bunch stripper' consisting of a horizontal rotating drum with baffles. Bunches are repeatedly lifted and dropped, so that the fruit is shaken out. It is then transported to a 'digester', a vertical steam-jacketed cylinder fitted with rotating beater arms. The fruit is thoroughly mashed and passes directly to a single or a twin-screw press. The press extrudes a liquid at one end, consisting of about 53% oil, 7% finely divided solids, and 40% water, and a press cake containing the fruit fiber and the nuts at the other end. The liquid phase passes through a vibrating screen to a settling tank. After about 2 h, the upper layer is clarified in a sealed centrifuge; the oil is then dried under vacuum and pumped to storage. The lower layer is treated in a sludge centrifuge, or in a three-phase decanter, the oil phase being returned to the settling tank.

The cake formed by the press fiber and nuts is processed in a separate stream. The cake is broken up by rotating arms on a conveyor, and the fiber is removed in a 'polishing' drum. The nuts are partially dried by a warm air stream in a silo, before being cracked in a centrifugal cracker. Nuts drop on to a rotor, which throws them against a peripheral ring of hardened metal. Kernels are separated from shells in a pneumatic column and/or a hydrocyclone, washed, and dried with hot air to below 8% moisture.

The production of palm kernel is usually carried out in a separate factory.

Treatment of Wastes

Three aqueous waste streams arise in the oil mill:

1. condensate from the sterilizing process;
2. aqueous effluent from the centrifuges;
3. waste water from nut processing.

Typically, for every 1 tonne of oil produced, the combined aqueous waste is 2.5 tonnes containing 0.6% oil and 3.9% dissolved and suspended solids. After removal of any supernatant oil, this effluent is treated until the water is of a quality suitable for discharge or reuse in the process. A number of processes have been adopted. Typically, treatment involves:

1. 1–2 days in an open acidification pond;
2. up to 20 days in a tank for anaerobic digestion;
3. about 20 days in a lagoon with vigorous aeration for anaerobic digestion.

The press fiber, after separation of the nuts, is used as fuel in the mill boilers. Consequently, the mill is usually self-sufficient in fuel for steam and electric power. Some shell may be used in the boiler. Alternate uses are as hard-core on estate roads or as raw material for charcoal.

Recently, an alternative treatment process has been developed. The waste stream is partly dewatered in a decanter centrifuge, and then treated in a rotary drier, using heat from the boiler flue, or produced by burning empty fruit bunches. The resulting solid is used as a fertilizer in the plantation.

Refining of Palm Oil

Crude palm oil may be refined by the traditional alkali refining process. In view of its natural strong red color, a highly active bleaching earth may be required. **Table 3** gives a flow sheet for a typical process.

Crude palm oil usually contains 3–5% of free fatty acids, and so the process losses in alkali refining tend to be rather high. This has led to the development of physical or steam refining, where the use of a somewhat higher temperature in a modified deodorizer enables the free fatty acids to be removed by distillation instead of neutralization. The process has proved to be more economical and is now being adopted for other oils. A flow sheet for physical refining is given in **Table 4**.

Further Processing of Palm Oil

Palm oil is subjected to further processes in the refinery in order to make it more widely useful in food applications.

The following processes are carried out on a large scale in order to modify the physical properties:

- fractionation;
- hydrogenation;
- interesterification.

Brief descriptions of the processes will be given here, and the properties of the products will be described in a later section.

The major glyceride components of palm oil range from triolein (melting point 5 °C) to tripalmitin (melting point 66 °C) with a number of mixed glycerides of intermediate melting point.

Crystallization from the melt of this mixture at a controlled temperature, followed by separation of the liquid and solid phases, results in palm olein and palm stearin. The characteristics of these products depend on the temperature chosen for crystallization and the efficiency of the separation process. This may be by centrifugation, filtration on a rotating band filter, or in a plate and frame filter. A modification of the latter has each frame fitted with an inflatable diaphragm, which enables pressure to be applied to the solids, and results in very efficient separation of the olein. Fractionation from solution in acetone or hexane is more costly and is only used when a sophisticated midfraction is required (see later sections).

Hydrogenation

Hydrogenation is a standard process in the edible oil industry. It involves treating the oil with an activated nickel catalyst with hydrogen under pressure and at an elevated temperature. Vigorous stirring is required, since the liquid and gas phases have to contact

at the surface of the catalyst. The oil is neutralized, washed, and bleached before hydrogenation, to avoid deactivation of the catalyst. (*See* **Vegetable Oils**: Oil Production and Processing.)

Interesterification

When a glyceride oil is stirred with sodium methoxide (or other alkali catalyst) at a temperature of about 90 °C, the fatty acid radicals are detached from their original positions, and new glycerides are formed with a random distribution of fatty acids. Examples of interesterified products will be given in later sections. (*See* **Vegetable Oils**: Oil Production and Processing.)

Table 3 Flow sheet for alkali refining

Process step	Typical conditions	Main impurities removed or reduced
Pretreatment, Gum conditioning	0.1% phosphoric acid 80 °C – 20 min	Phospholipids Trace metals Pigments
Neutralization	4 N caustic soda 20% excess	Fatty acids Phospholipids Pigments
Washing		Soap
Drying		Water
Bleaching	1% active earth 80–100 °C, vacuum	Pigments Oxidation products Trace metals Soap residues
Filtration		Spent bleaching earth
Deodorization	240 °C at 133–667 Pa (1–5 torr) 90–110 min	Fatty acids Partial glycerides Oxidation products Pigment Decomposition products
Polishing filter		Traces of oil Insolubles

Table 4 Flow sheet for physical refining

Process step	Typical conditions	Main impurities removed or reduced
Gum conditioning	0.1% phosphoric acid 80 °C – 20 min	Phospholipids Trace metals Pigments
Bleaching	1–2% active earth 90 °C – 20 min Vacuum	Phospholipids Trace metals Pigments Oxidation products Spent earth
Filtration		
Deacidification and deodorization	260–265 °C at 133–667 Pa (1–5 torr)	Free fatty acids Partial glycerides Oxidation products Pigment decomposition products
Polishing filter		Traces of oil Insolubles

Chemical Composition of Palm Oil and Fractions

Average fatty acid composition data from refined palm oil and standard palm olein and stearin, as traded, are shown in **Table 5**. Tailor-made fractions are also available for specific requirements. The last column shows the composition of a midfraction obtained from palm oil under special conditions, which is suitable for use in confectionery fats. Triacylglycerol compositions for the same products are given in **Table 6**.

Physical Properties of Palm Oil

Many food products require a consistent or semisolid fat as an ingredient in order to achieve the required structure. An important feature of palm oil is its natural content of solids. The average solid fat content of palm oil products is given in **Table 7**. Palm stearins of a wide choice of composition and solid fat contents are available, for example, with solid fat contents at 20 °C ranging from 35 to 72%.

Minor Components of Palm Oil

Data for the minor components of palm oil, collected from published sources, are given in **Table 8**. The main sterols present are β-sitosterol (58%), campesterol (22%) and stigmasterol (11%). The main carotenoids are β-carotene (56%) and α-carotene (35%). The carotenoids are removed in the normal refining process. The tocol content is unusual in having a high proportion of unsaturated tocotrienols. Details for crude and refined products are given in **Table 9**. The tocols are important as vitamin E and as potent natural antioxidants. (*See* **Antioxidants**: Natural Antioxidants; **Carotenoids**: Occurrence, Properties, and Determination.)

Table 5 Mean fatty acid composition (%)

	Refined palm			Palm midfraction[a]
	Oil	Olein	Stearin	
12:0	0.24	0.27	0.18	
14:0	1.11	1.09	1.27	0.7
16:0	44.14	40.93	56.79	60.9
16:1	0.1	0.1	0.1	
18:0	4.44	4.16	4.93	4.6
18:1	39.04	41.51	29.0	31.0
18:2	10.57	11.64	7.23	2.6
18:3	0.2	0.2	0.2	0.1
20:0	0.2	0.1	0.2	0.3

[a]Data from Britannica Food Ingredients with permission.
Adapted from Siew WL, Tang TS, Oh FCH, Chong CL and Tan YA (1993) Identity characteristics of Malaysian palm oil products: fatty acid and triglyceride composition and solid fat content. *Elaeis* 6(1): 38–46, with permission.

Table 6 Mean triacylglycerol compositon of refined oils (carbon numbers by GLC)[a]

Carbon number	Palm oil	Palm olein	Palm stearin	Palm midfraction[b]
C44	0.07	0.09	0.13	
C46	1.18	0.77	3.13	
C48	8.08	3.28	23.72	2.2
C50	39.88	39.52	40.31	78.9
C52	38.77	42.74	25.28	15.1
C54	11.35	12.80	6.86	0.7
C56	0.59	0.67	0.45	
Unidentified				3.1

[a]Carbon numbers are the sum of the carbon atoms in three acyl groups.
[b]Data from Britannica Food Ingredients with permission.
Adapted from Siew WL, Tang TS, Oh FCH, Chong CL and Tan YA (1993) Identity characteristics of Malaysian palm oil products: fatty acid and triglyceride composition and solid fat content. *Elaeis* 6(1): 38–46, with permission.

Table 7 Mean solid fat content of standard refined oils

Temperature (°C)	Palm oil (mean) %	Palm olein (mean) %	Palm stearin (mean) %	Palm midfraction[a] %
10	53.6	38.27	76.04	
15	39.13	19.89	68.91	
20	26.10	5.67	60.71	89.9
25	16.28	2.05	50.55	82.6
30	10.54		40.39	50.2
35	7.85		34.30	50.2
40	4.64		28.13	0.0
45			22.38	
50			12.45	
55			0.60	

[a]Tempered at 26 °C for 40 h.
Adapted with permission from Siew WL, Tang TS, Oh FCH, Chong CL and Tan YA (1993) Identity characteristics of Malaysian palm oil products: fatty acid and triglyceride composition and solid fat content. *Elaeis* 6(1): 38–46, with permission.
Data from Britannica Food Ingredients with permission.

Table 8 Minor components of crude palm oil typical figures (p.p.m.)

Sterols	490
4-Methyl sterols	360
Triterpenic alcohols	550
Isoprenoid alcohols	80
Other alcohols	130
Tocols	830
Carotenoids	670
Squalene	350
Other hydrocarbons	40

Reproduced with permission from Palm Oil Research Institute of Malaysia.

Table 9 Tocol content of palm oil (p.p.m.)

		α-tocopherol	α-tocotrienol	γ-tocotrienol	δ-tocotrienol	Total
Crude palm oil	Mean (n=9)	162	165	324	81	774
	Range	136–241	90–205	273–439	67–94	635–890
Refined palm oil	Mean (n=3)	117	117	158	31	426
	Range	85–180	99–147	67–239	5–62	256–630
Refined palm olein	Mean (n=8)	141	152	218	49	561
	Range	107–163	131–177	113–293	28–68	478–673

Reproduced with permission from Palm Oil Research Institute of Malaysia.

Food Uses of Palm Oil

Frying

Palm oil has a good stability at the high temperatures used in frying (usually 175–185 °C), because of its content of natural antioxidants, the absence of highly unsaturated fatty acids, and the moderate content of linoleic acid. Consequently, palm oil or palm olein is widely used domestically and in industry, especially for deep fat frying whether in a batch process, as in restaurants and fast food outlets such as British 'fish and chip' shops, or in continuous fryers. Palm oil is used for doughnut frying, because the solids content assists the adhesion of the sugar coating. Palm oil is also used for the frozen 'French fry' industry and for instant noodle frying in Japan and China. For potato crisps (American 'chips'), palm olein is preferred, or a blend of palm olein with a more unsaturated oil such as sunflower or soya bean oil.

Bakery Fats

Texturized palm oil is used as such in some types of biscuits. For other biscuits and for cakes, a formulated shortening is required to obtain good aerating properties. Some formulae with a satisfactory performance are given in **Table 10**.

Vanaspati

Vanaspati may be defined as a cheaper alternative to butterfat, having a melting point of about 37 °C and a granular crystalline structure with little or no free oil. It is generally based on vegetable oils suitably hydrogenated or on a blend of oils. Vanaspati is the customary domestic cooking fat in the Indian subcontinent and the Middle East. When formulated from liquid vegetable oils, the desired structure requires a hydrogenation giving a high content of *trans* fatty acid isomers (30–60%). This is regarded as undesirable according to current nutritional advice. A number of formulae based on palm oil and having low or zero *trans* fatty acids are given in **Table 11**

Table 10 Shortening formulae containing palm oil

	1	2	3	4
Palm oil	50			
Hardened palm oil melting point (49–51 °C)	15			
Liquid oil	35			
Palm stearin		35	42	
Hardened rapeseed oil melting point 36 °C		30		
Rapeseed oil		35	40	
Hardened palm oil melting point 42 °C			18	
Palm olein (interesterified)				100

Reproduced with permission from Palm Oil Research Institute of Malaysia.

Table 11 Vanaspati formulae containing palm oil products

	1	2[a]	3	4[b]
Hardened palm olein (melting point 41 °C)	24			
Palm oil	56	80	70	80
Liquid oil	20			
Rice bran oil		20		
Palm stearin			7	
Hardened soya bean oil (melting point 34 °C)			23	20
Trans fatty acids (%)	2.7	Nil	7.5	4

[a]The blend is interesterified.
[b]Current formulae of this type in use in Pakistan.
Reproduced with permission from Palm Oil Research Institute of Malaysia.

Margarines

Margarines can be classified as being for table use in blocks or in tubs, general bakery margarines, and puff-pastry margarines.

The textual properties are the textual properties appropriate for each application, and these can be obtained by blending a variety of ingredients. **Table 12** shows some formulae based on palm oil.

Confectionery Fat

Palm oil midfraction (see **Tables 5–7**) is used as a major component of blends designed to have physical properties like cocoa butter, and compatible with it in mixtures. Palm olein, partly hydrogenated under conditions giving a high content of *trans* fatty acids, is

Table 12 Margarine formulae using palm oil products

	Block (temperate)	Block (tropical)	Tub	General bakery	Danish pastry	Puff pastry
Palm oil	50	80	50	65	40	50
Hydrogenated palm oil (melting point 44 °C)	20			10	40	50
Liquid oil	30		50	10	20	
Coconut oil				15		
Palm stearin		20				

Reproduced with permission from Palm Oil Research Institute of Malaysia.

useful as a confectionary fat with limited compatability with cocoa butter, or for use in toffee, bakery coatings, and confectionery centers.

Miscellaneous Uses

Palm oil is used instead of butterfat in icecream and in filled milk or coffee whiteners. Partly hydrogenated palm oil is used in dried soups, where its stability against oxidation is important.

Special Products

Using a modified refining process, red palm oil and palm olein, retaining about 80% of the carotenoid content of the crude oil, have become commercially available. Red olein has found uses in the manufacture of attractively golden-colored potato crisps, whereas red palm oil is being used in nutritional intervention studies in South Africa and India. This is potentially a very important use, in view of the high incidence of xerophthalmia in developing countries.

A commercial process is also in operation to obtain a 99% pure concentrate of the tocopherols and tocotrienols from the palm fatty acid distillate, the byproduct of physical refining. The concentrate is used in health supplements.

See also: **Antioxidants**: Natural Antioxidants; **Carotenoids**: Occurrence, Properties, and Determination; **Palm Kernel Oil**; **Tocopherols**: Properties and Determination

Further Reading

Berger KG (1983) Palm Oil. In: Chan HT (ed.) *Handbook of Tropical Foods*, pp. 433–468. New York: Marcel Dekker.

Berger KG (1983) Production of palm oil from fruit. *Journal of American Oil Chemists' Society* 60: 158–162.

Hartley CWS (1988) *The Oil Palm*, 3rd edn. Harlow, UK: Longman Scientific and Technical.

Kheiri SA (1987) Palm Oil. In: Gunstone FD (ed.) *Critical Reports on Applied Chemistry*, vol. 15. London: Society of Chemical Industry.

Siew WL, Tang TS, Oh FCH, Chong CL and Tan YA (1993) Identity characteristics of Malaysian palm oil products: fatty acid and triglyceride composition and solid fat content. *Elaeis* 6(1): 38–46.

Tan BK and Oh FCH (1981) *Malaysian Palm Oil Chemical and Physical Characteristics*, Porim Technology No. 3. Kuala Lumpur: Palm Oil Research Institute of Malaysia.

Palms *See* **Coconut Palm**; **Date Palms**; **Sago Palm**; **Sugar**: Sugarcane; Sugarbeet; Palms and Maples; Refining of Sugarbeet and Sugarcane

---

Pancreatic Hormones *See* **Hormones**: Adrenal Hormones; Thyroid Hormones; Gut Hormones; Pancreatic Hormones; Pituitary Hormones; Steroid Hormones

PANTOTHENIC ACID

Contents
Properties and Determination
Physiology

Properties and Determination

G F M Ball, Wembley, London, UK

Background

In 1933, a research team led by R. J. Williams isolated from a variety of biological materials an acidic substance that acted as a growth factor for yeast. Williams' team elucidated the chemical structure of the purified substance and named it pantothenic acid because of its apparently widespread occurrence (Greek *pantos*, meaning everywhere). Pantothenic acid was established as a vitamin in 1939, when it was shown to be identical to a 'filtrate factor' required by rats for normal growth, and to a chick antidermatitis factor. Sometimes referred to as vitamin B_5, pantothenic acid is a member of the water-soluble B-group vitamins.

Structure and Physicochemical Properties

The biological activity of pantothenic acid is attributable to its incorporation into the molecular structures of coenzyme A and acyl carrier protein. The molecular structures of pantothenic acid and related compounds are shown in **Figure 1**. Pantothenic acid ($C_9H_{17}O_5N$; molecular weight $= 219.23$) is composed of pantoic acid linked by an amide bond to β-alanine. The pantothenic acid molecule, having a chiral carbon atom, exhibits optical isomerism as well as being optically active. Only the D(+) isomer occurs in nature. Synthetic pantothenic acid is a racemic (DL) mixture, and, since only the D isomer is biologically active, this fact must be considered if the DL mixture is to be used therapeutically. Pantothenic acid is a pale yellow oil that is extremely hygroscopic and so is unsuitable for commercial application. For human food supplements, calcium D-pantothenate [$(C_9H_{16}O_5N)_2Ca$; molecular weight $= 476.53$] is used.

The corresponding alcohol of pantothenic acid, pantothenol (referred to commercially as panthenol), is widely used as a source of pantothenate activity for pharmaceutical vitamin products, because it is more stable than the pantothenate salts, especially in liquid multivitamin products that must be slightly acid to preserve the thiamin content. Pantothenol does not occur naturally and itself has no pantothenate activity, but it is converted quantitatively to pantothenic acid in the body.

The stability of pantothenic acid and its calcium salt in aqueous solution is highly dependent on the pH. In contrast to other B-group vitamins, pantothenic acid becomes more stable as the pH of the solution increases. Solutions of calcium pantothenate are most stable between pH 5 and 7 but, even so, are not stable to autoclaving, and therefore, sterilization by ultrafiltration is necessary for pharmaceutical preparations. Below and above these pH values, solutions of calcium pantothenate are thermolabile. Alkaline hydrolysis yields pantoic acid and β-alanine, whereas acid hydrolysis yields the γ-lactone of pantoic acid. Pantothenic acid is unaffected by atmospheric oxygen and light.

Dietary Sources

Pantothenic acid is widely distributed in foods of both animal and plant origin. In concentration units of mg per 100 g, the vitamin is particularly abundant in liver (8), kidney (3), heart (2.5), egg yolk (4.6) broad beans (4.9), and peanuts (2.7). Lesser amounts are found in beef (0.6), chicken (1.2), potatoes (0.4), broccoli (1.2), oatmeal (1.0), and milk (0.35), but these will be important food sources if consumed in sufficient quantity. Outstandingly high amounts are found in the ovaries of tuna and cod (232) and in royal jelly from the queen bee (50). In contrast, highly refined foods such as sugar, fats and oils, and cornstarch are totally devoid of the vitamin.

Coenzyme A is the major pantothenic acid-containing compound present in foods of both animal and plant origin, accompanied by small amounts of other bound forms (phosphopantothenic acid, pantetheine, and phosphopantetheine). Notable exceptions are human and bovine milk in which free (unbound) pantothenic acid constitutes around 90% of the total pantothenate content.

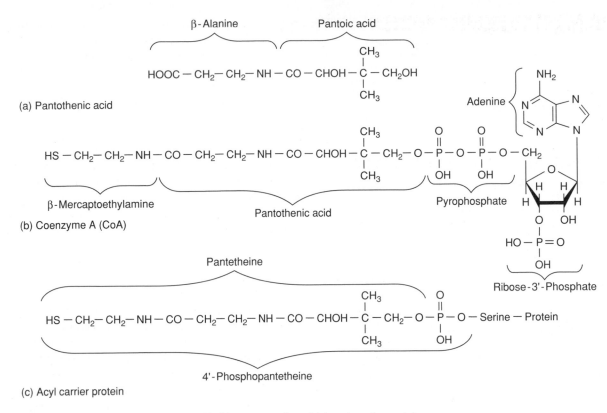

Figure 1 Structures of (a) pantothenic acid, (b) coenzyme A, and (c) acyl carrier protein.

Pantothenic acid has a good stability in most foods during home cooking but is susceptible to leaching. The roasting of meat causes degradation of less than 10%, but the meat drippings contain 20–25% of the initial vitamin content.

Estimates of dietary intakes of pantothenic acid should be based not on the raw food values, but on the cooked food values, otherwise falsely high estimates of intake will be obtained. **Table 1** presents data taken from a study in which the pantothenic acid content of 75 processed and/or cooked foods was determined by radioimmunoassay. The foods selected for analysis were those commonly consumed in the USA. Samples were brought to a ready-to-eat stage, following any package directions, and using only the edible portion. Results indicated that the canning of foods incurs large losses of pantothenic acid, as does the conversion of grains to various cereal products, and the processing of meats to produce fat- and cereal-extended products such as frankfurters and sausages.

Little information is at hand regarding the nutritional availability to humans of pantothenic acid in food commodities. In one study, based on the urinary excretion of pantothenic acid, the availability for male human subjects ingesting 'the average American diet' ranged from 40 to 61%, with a mean of 50%.

Analysis

Pantothenic acid in its various bound forms (mainly coenzyme A) is routinely determined by microbiological assay. Other published methods include radioimmunoassay, enzyme-linked immunosorbent assay, gas chromatography, and high-performance liquid chromatography. For all of these techniques, it is necessary to liberate pantothenic acid from its bound forms by enzymatic hydrolysis, because only the free vitamin can be measured. Hydrolysis is not required, however, for calcium pantothenate-supplemented foods or for milk in which the free vitamin predominates.

Extraction from Food

Neither acid nor alkaline hydrolysis is applicable for the liberation of bound pantothenic acid, since the vitamin is degraded by such treatments. The only practicable alternative is enzymatic hydrolysis, and this was successfully accomplished through the simultaneous action of intestinal phosphatase and an avian liver enzyme. This double enzyme combination liberates practically all of the pantothenic acid from coenzyme A, but it does not release the vitamin from acyl carrier protein. The phosphatase splits the coenzyme A molecule between the phosphate-containing

Table 1 Pantothenic acid content of processed and/or cooked foods purchased in Utah, USA

Food[a]	Pantothenic acid content (mg per 100 g) (mean and standard deviation)[b]
Breads, cereals, and other grain products	
Bran'ola bread (high-fiber)	0.458 ± 0.044
Rolls, hamburger	0.471 ± 0.075
Ready-to-eat cereals	
Cheerios (oats)	1.341 ± 0.198
Corn Flakes (corn)	0.284 ± 0.032
Wheat Chex (wheat)	0.502 ± 0.042
Rice, white	0.261 ± 0.036
Meat, fish, poultry, and meat products	
Beef, regular ground – pan broiled	0.671 ± 0.048
Pork loin chops – pan broiled	0.650 ± 0.051
Fish filet, frozen, breaded – baked	0.250 ± 0.016
Chicken breast – baked with skin	1.188 ± 0.049
Frankfurters	0.342 ± 0.025
Salami	0.997 ± 0.086
Fruits and vegetables	
Orange juice, frozen, reconstituted	0.197 ± 0.029
Potatoes – baked	0.318 ± 0.045
Potatoes – boiled	0.291 ± 0.018
Potatoes – canned	0.152 ± 0.026

[a]The use of brand names is for identification purposes only and does not imply endorsement of a food product.
[b]Data from Walsh JH, Wyse BW and Hansen RG (1981) Pantothenic acid content of 75 processed and cooked foods. *Journal of the American Dietetic Association* 78: 140–144.

moiety and pantetheine, while the liver enzyme breaks the link in pantetheine between the pantothenic acid and β-mercaptoethylamine moieties. Both enzymes are available commercially as powdered extracts. Liver enzyme preparations contain a relatively high amount of coenzyme A, which is converted to pantothenate during the incubation period, thus creating an unacceptably high blank value. Such preparations can be purified quite simply by treatment with Dowex 1-X4 anion exchange resin. Intestinal phosphatase preparations contain negligible amounts of coenzyme A and do not require purification.

Microbiological Assays

Microbiological methods, as applied to the determination of the B-group vitamins, are based on the absolute requirement of a particular microorganism (the assay organism) for the vitamin in question (in this case, pantothenic acid); that is, the organism can multiply only when the vitamin is present in the surrounding medium. In a typical turbidimetric microbiological assay, aliquots of a standard solution of pantothenic acid, or aliquots of the sample extract containing pantothenic acid, are added to an initially translucent basal nutrient medium, complete in all

respects except for pantothenic acid. Following inoculation with the assay organism, the organism multiplies in proportion to the pantothenic acid content of the standard or sample, and the extent of the growth is ascertained by measuring the turbidity produced. Over a defined concentration range, the measured response will be directly proportional to the amount of pantothenic acid present, and, within this range, the sample solution and standard pantothenic acid solution can be compared accurately. The usual assay organism is *Lactobacillus plantarum* (ATTC (American Type Culture Collection) No. 8014), which can also be used for assaying nicotinic acid and biotin. The basal nutrient medium can also be used for assaying nicotinic acid and biotin, with the exclusion of the relevant vitamin from the formulation. Fatty acids are stimulatory in the presence of suboptimal amounts of pantothenic acid, so a preliminary ether extraction step may be necessary.

In the standard turbidimetric procedure, the basal nutrient medium is prepared at twice its final concentration. Multiple aliquots of a standard solution of pantothenic acid and of enzyme-treated extracts of the test food are added to a series of uniform assay tubes in amounts suitable to produce gradations in growth between no growth and maximum growth. The contents of all tubes are diluted with water to the same volume, and an equal volume of the basal medium is added. The tubes are sterilized, cooled to a uniform temperature, and then inoculated with an actively growing culture of *L. plantarum*. The tubes are incubated for 6–24 h at any selected temperature between 30 and 40 °C held constant to ± 0.5 °C until growth has reached the maximum permitted by the limiting vitamin present, pantothenic acid. The growth response to standard and test extract is then determined by measuring the turbidity produced. The data obtained from the standards are used to construct a standard curve from which the pantothenic acid concentrations of the various sample aliquots are derived. The use of multiple aliquots allows a validity check to be carried out: the pantothenic acid concentration found should be directly proportional to the volume of aliquot taken. The amount of pantothenic acid present in the original sample is then calculated at the different test levels, and the results are averaged to obtain the final result.

An alternative method, the radiometric microbiological assay, is based upon the measurement of radioactive $^{14}CO_2$ generated from the metabolism of a ^{14}C-labeled nutrient by the test organism. The radioactivity is measured automatically by means of a commercially available gas flow system incorporating an ionization chamber. Sample preparation for this technique is simplified due to the fact that colored, turbid, or precipitated debris does not interfere with the $^{14}CO_2$ output or detection; furthermore, the scrupulous cleaning of glassware required for turbidimetric assays is unnecessary. An assay for pantothenic acid in human milk and blood is based on the measurement of $^{14}CO_2$ produced from the metabolism of L-[1-^{14}C]methionine or L-[1-^{14}C]-valine by the yeast *Kloeckera brevis* (ATCC 9774). Metabolic CO_2 can also be measured nonradiometrically with the aid of an infrared CO_2 analyzer, which measures automatically the infrared radiation absorbed by the CO_2 band at 4.2 μm.

Radioimmunoassay

The radioimmunoassay is based on the competition for a fixed, but limited, number of antibody binding sites by antigen (a substance capable of binding to a specific antibody) and a trace amount of radiolabeled antigen added to the sample extract. In this case, the antigen is the analyte, pantothenic acid. The presence of larger amounts of unlabeled analyte results in less radioactivity being bound to the antibody.

Wyse and colleagues raised antibodies to pantothenic acid by coupling a bromoacetyl derivative of pantothenic acid with reduced and denatured bovine serum albumin and injecting this immunogen into rabbits. For the assay, each tube contained diluted antiserum, pantothenic acid standard solution or sample extract, and radiolabeled sodium *d*-pantothenate. After incubation, neutral saturated ammonium sulfate was added to facilitate suspension of the antibody-bound pantothenic acid, which was then centrifuged. The washed precipitate, containing antibody-bound pantothenic acid, was dissolved in tissue solubilizer, and the radioactivity was counted in a scintillation counter. The amount of pantothenic acid in each unknown was determined by reference to a standard curve constructed by plotting on logit-semilog paper log concentration of nonradioactive pantothenic acid in the standard vials (ng per 0.50 ml) versus the percentage of the counts bound. Results from the radioimmunoassay and AOAC microbiological assay for pantothenic acid in 75 processed and cooked foods were highly correlated ($r = 0.94$). However, there was a statistically significant difference ($p < 0.05$) between the two assay results for all foods and for the subgroups meats, breads and cereals, and fruits and vegetables. At $p < 0.01$, only meats were significantly different. For breads and cereals, the microbiological assay results averaged 6.6% higher than those of the radioimmunoassay. For fruits and vegetables, the microbiological assay results were 11.6% higher; for meats, the results were 23.2% higher. It was postulated that bacterial enzymes in the assay organism promote further breakdown of bound pantothenic acid, or nonenzymatic breakdown occurs during the long microbiological incubation period.

Enzyme-linked Immunosorbent Assay (ELISA)

An ELISA is an enzyme-linked immunoassay in which one of the reactants is immobilized by physical adsorption on to the surface of a solid phase. In its simplest form, as used in food analysis applications, the solid phase is provided by the plastic surface of a 96-well microtitration plate. The generally preferred format for vitamin assays in food analysis is a two-site noncompetitive assay used in the indirect mode. This format employs two antibodies: a primary anti-vitamin antibody raised against an immunogen (in this case, a pantothenic acid–protein conjugate), and an enzyme-labeled, species-specific second antibody, which binds specifically to the primary antibody. To perform such an assay, a protein conjugate of pantothenic acid is immobilized to the well surface of the microtitration plate, the attached protein being different to that used for the immunogen. The protein adsorbs passively and strongly to the plastic, and, once coated, plates can usually be stored for several months. To perform the assay, the standard solution or sample extract is added to the well, followed by a limited amount of primary antibody. After incubation, the antibody becomes distributed between immobilized vitamin and free vitamin according to the amount of vitamin initially present. After phase separation, achieved by well emptying and washing, the second antibody is added in excess, and the plate is incubated for a second time. Excess unbound material is removed, and the amount of bound enzyme is determined by addition of substrate and spectrophotometric measurement of the colored product. Unknown samples are quantified by reference to the behavior of vitamin standards.

Finglas and colleagues developed a noncompetitive ELISA, which is highly specific for pantothenic acid and does not recognize coenzyme A, panthenol or pantheneine. The primary antivitamin antibody was raised in rabbits according to the method of Wyse and colleagues (*see* section Radioimmunoassay), and the enzyme-labeled, species-specific second antibody (alkaline phosphatase-labeled antirabbit IgG) was obtained commercially. Microtitration plates were coated with pantothenic acid–keyhole limpet hemocyanin as the immobilized phase of the assay. A high correlation coefficient ($r = 0.999$) was reported when ELISA values obtained for six foods were compared with corresponding values obtained by a microbiological method using *L. plantarum*. Gonthier and colleagues improved the sensitivity of the ELISA by using an immunogen composed of pantothenic acid coupled to thyroglobulin by a 6-carbon atom linker (adipoyl dichloride). By contrast, the bromoacetyl linker used in Finglas' pantothenic acid–bovine serum albumin immunogen contains two carbon atoms.

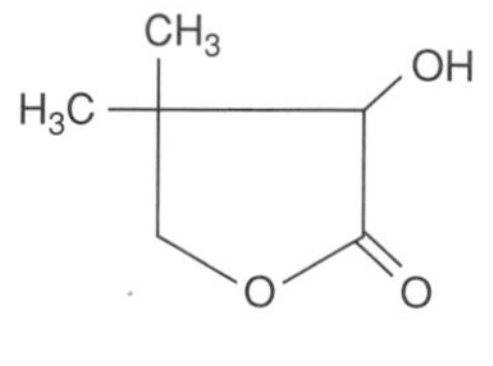

Figure 2   Pantoyl lactone.

Gas Chromatography

Salts of pantothenic acid present in pharmaceutical preparations have been analyzed by gas chromatography after conversion to volatile acetate, trifluoroacetate, or trimethylsilyl derivatives. An alternative approach to derivatization is to chromatograph the pantoyl lactone formed from pantothenic acid by acid hydrolysis (**Figure 2**). This approach is applicable to foodstuffs, because the hydrolysis reaction liberates the lactone from the free and bound pantothenic acid in the food matrix with a recovery of at least 95%. Davídek and colleagues applied this technique to the determination of pantothenic acid in fresh beef liver, spray-dried egg yolk, soybean flour, whole-grain wheat flour, and dried bakers' yeast. Samples were hydrolyzed by treatment with dilute hydrochloric acid, and the neutralized hydrolysate, after filtration, was extracted with dichloromethane. The combined extracts, to which ethyl laurate was added as an internal standard, were concentrated by rotary evaporation, and then analyzed by gas chromatography using a polar stationary phase of 10% Carbowax 20M and a flame ionization detector. Gas chromatographic results correlated with results obtained by the currently accepted microbiological method ($r = 0.975$), and no significant difference was found between the two sets of results ($p > 0.05$). Davídek and colleagues used a packed column of dimensions $2.4 \, \text{m} \times 2 \, \text{mm}$ i.d. in which the stationary phase was coated on to a porous support material of diatomaceous earth treated with dimethyldichlorosilane. Woollard and colleagues upgraded the column to a more efficient $30 \, \text{m} \times 0.25 \, \text{mm}$ open-tubular capillary column coated with stationary phase (BPX70).

High-performance Liquid Chromatography (HPLC) and Capillary Electrophoresis

The pantothenic acid molecule does not contain a characteristic chromophore, and hence it exhibits only very weak absorbance at 204 nm, owing to the presence of carbonyl groups. Detection at wavelengths below 220 nm is subject to interference from the many organic compounds present in a typical food sample extract prepared for HPLC. The problem of weak and nonspecific absorbance, coupled with the low concentrations of pantothenic acid in foods, has thwarted attempts to apply HPLC to the

determination of pantothenic acid in foodstuffs, although the technique has been successfully applied to pharmaceutical products. Chemical treatment of the pantothenic acid molecule to form a derivative that fluoresces or absorbs at a higher wavelength is a possibility, but, so far, reproducible results have not been obtained using such an approach. Refractometry as a means of detection lacks the required sensitivity and specificity.

Recognizing these problems, Woollard and Indyk in New Zealand developed an HPLC method for determining free endogenous D-pantothenic acid in milk and supplemental calcium pantothenate in infant formulas. Sample preparation simply entailed addition of acetic acid to the milk or reconstituted infant formula, followed by centrifugation and membrane filtration. This treatment resulted in a protein- and fat-free extract that could be directly injected (10-μl aliquot). The problems of poor detection specificity in the low-ultraviolet region of the spectrum were overcome by the use of a photodiode array detector that provided multiwavelength detection (selected wavelengths were 200, 205 and 240 nm) and on-line spectral analysis. The HPLC system incorporated an on-line mobile phase degasser – an important feature, as dissolved oxygen constitutes a source of interference at low ultraviolet wavelengths. Among several reversed-phase columns investigated, the column selected for routine use was of dimensions $250 \times 4.6 \, \text{mm}$ i.d. and packed with Luna (Phenomenex) $5 \, \mu\text{m} \, C_8$ (octyl) of 100 Å pore size, $400 \, \text{m}^2 \, \text{g}^{-1}$ surface area, and 13.5% carbon load. The Luna material is based on low-acidity silica, exhaustive end-capping, and shielded bonded phase ligand. The mobile phase was phosphate buffer (0.1 M, pH 2.25): acetonitrile (97:3, v/v) delivered initially at $1.4 \, \text{ml} \, \text{min}^{-1}$ and increased to $1.8 \, \text{ml} \, \text{min}^{-1}$ at 18 min. Following completion of the sample schedule, the column was purged sequentially with acetonitrile:water (50:50), water (100%), acetonitrile: water (50:50), and finally acetonitrile (100%) for column storage between runs. The retention time of the pantothenic acid was around 15 min, but the next sample was not injected until a major unknown peak with a retention time of about 35 min was removed (**Figure 3**).

Reversed-phase columns do not retain acidic solutes in the ionic state, but if the mobile phase is buffered to pH 3 or lower, the acidic solute will be nonionized and act as a neutral solute. This technique is known as ion suppression. Under these conditions, residual silanol groups on the silica support will also be nonionized. The net result is retention of undissociated acidic solutes with no peak tailing, owing to the elimination of electrostatic interactions between acidic solutes and silanol groups. A potential problem with certain reversed-phase column packings is that the siloxane ($\equiv$Si—O—Si$\equiv$) bond linking the alkyl ligand to the silica support is subject to hydrolysis at low pH, resulting in loss of bonded phase. Although longer-chain ligands such as C_{18} are relatively stable at a low pH, short-chain bonded phases, including small endcapping groups, are especially susceptible. The problem of hydrolysis and loss of bonded phase can be minimized by the use of 'shielded' stationary phases, which are sterically protected from attack by hydrolyzing protons. Results obtained by HPLC correlated with those obtained by microbiological assay utilizing *L. plantarum* ($r = 0.971$), and there was no significant difference ($p > 0.05$) between the two sets of results.

Reversed-phase HPLC with ion suppression is unable to separate the D and L enantiomers in synthetic calcium pantothenate. However, separation

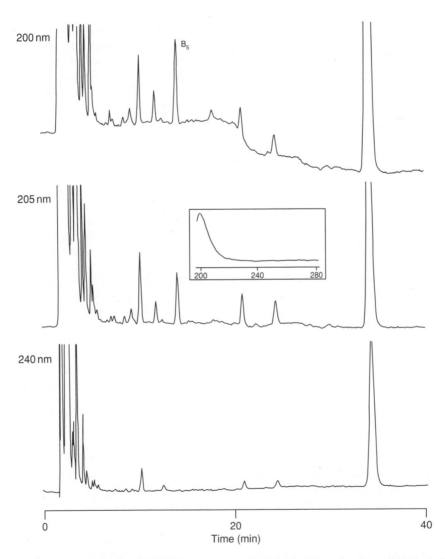

Figure 3 Multiwavelength UV chromatogram (200, 205, 240 nm) of a typical infant formula extract obtained with a Luna C_8 (octyl) column. The insert illustrates a UV spectral scan of pantothenic acid. Chromatographic parameters are given in the text. Reprinted from Woollard DC, Indyk HE and Christiansen SK (2000) The analysis of pantothenic acid in milk and infant formulas by HPLC. *Food Chemistry* 69: 201–208, with permission from Elsevier Science.

of enantiomers in DL pantothenic acid has been reported using a chiral selector in capillary electrophoresis. Optimum separation was obtained using a pH 7.0 phosphate buffer containing the chiral selector (60 mM 2-hydroxypropyl-β-cyclodextrin) and 10% (v/v) methanol. The enantiomers were unresolved when a buffer of pH 3.0 was tried, implying that dissociation of pantothenic acid was necessary for its chiral resolution under the conditions employed.

Overall Appraisal of Analytical Techniques

The 'free' (no enzyme treatment) or 'total' (after enzyme treatment) pantothenic acid content of a food has been traditionally determined by the turbidimetric microbiological assay using L. plantarum. This assay has been adopted as an official method by the AOAC on the basis of collaborative study. Once the facilities are in place, the microbiological assay can be used routinely to determine other B-group vitamins with minor changes in protocol. Inherent problems include stimulation or inhibition of bacterial growth by other compounds, and nonlinear response (drift) for various volumes of food extract analyzed. The radioimmunoassay produces results that, although correlated with microbiological assay results, are significantly lower. The use of radioisotopes would not be permitted in the vicinity of commercial food production. The ELISA is a better substitute for the microbiological assay, as results from the two techniques are in good agreement. The high technology of the ELISA is built into the reagents, so the assays are simpler to perform than microbiological assays. Routine use of the ELISA for determining pantothenic acid will depend on the commercial availability of standardized assay kits. Little interest seems to have been taken in gas chromatography, but the technique of chromatographing the lactone hydrolysis product merits further investigation using modern capillary columns. High-performance liquid chromatography is becoming increasingly popular for determining vitamins in foods, although the poor detectability of pantothenic acid limits the sensitivity of this technique. Capillary electrophoresis also suffers from poor sensitivity but, when used with a chiral selector, has the advantage of separating active d and inactive l enantiomers of racemic calcium pantothenate added to foods.

See also: **Bioavailability of Nutrients**; **Chromatography**: Principles; High-performance Liquid Chromatography; Gas Chromatography; **Immunoassays**: Principles; Radioimmunoassay and Enzyme Immunoassay

Further Reading

AOAC (1995) AOAC Official Method 992.07. Pantothenic acid in milk-based infant formula. Microbiological turbidimetric method. In: *Official Methods of Analysis of AOAC International*, Method No 50.1.22, 16th edn. Arlington VA: AOAC International.

Bell JG (1974) Microbiological assay of vitamins of the B group in foodstuffs. *Laboratory Practice* 23: 235–242, 252.

Crawley H (1993) Natural occurrence of vitamins in food. In: Berry Ottaway P (ed.) *The Technology of Vitamins in Food*. Glasgow: Blackie Academic & Professional.

Davídek J, Velíšek J, Černá J and Davídek T (1985) Gas chromatographic determination of pantothenic acid in foodstuffs. *Journal of Micronutrient Analysis* 1: 39–46.

Finglas PM, Faulks RM, Morris HC, Scott KJ and Morgan MRA (1988) The development of an enzyme-linked immunosorbent assay (ELISA) for the analysis of pantothenic acid and analogues. Part II – Determination of pantothenic acid in foods. *Journal of Micronutrient Analysis* 4: 47–59.

Goli DM and Vanderslice JT (1989) Microbiological assays of folacin using a CO_2 analyzer system. *Journal of Micronutrient Analysis* 6: 19–33.

Gonthier A, Boullanger P, Fayol V and Hartmann DJ (1998) Development of an ELISA for pantothenic acid (vitamin B_5) for application in the nutrition and biological fields. *Journal of Immunoassay* 19: 167–194.

Gonthier A, Fayol V, Viollet J and Hartmann DJ (1998) Determination of pantothenic acid in foods: influence of the extraction method. *Food Chemistry* 63: 287–294.

Guilarte TR (1989) A radiometric microbiological assay for pantothenic acid in biological fluids. *Analytical Biochemistry* 178: 63–66.

Kodama S, Yamamoto A and Matsunaga A (1998) Direct chiral resolution of pantothenic acid using 2-hydroxypropyl-β-cyclodextrin in capillary electrophoresis. *Journal of Chromatography A* 811: 269–273.

Morris HC, Finglas PM, Faulks RM and Morgan MRA (1988) The development of an enzyme-linked immunosorbent assay (ELISA) for the analysis of pantothenic acid and analogues. Part I – Production of antibodies and establishment of ELISA systems. *Journal of Micronutrient Analysis* 4: 33–45.

Walsh JH, Wyse BW and Hansen RG (1979) A comparison of microbiological and radioimmunoassay methods for the determination of pantothenic acid in foods. *Journal of Food Biochemistry* 3: 175–189.

Woollard DC, Indyk HE and Christiansen SK (2000) The analysis of pantothenic acid in milk and infant formulas by HPLC. *Food Chemistry* 69: 201–208.

Wyse BW, Wittwer C and Hansen RG (1979) Radioimmunoassay for pantothenic acid in blood and other tissues. *Clinical Chemistry* 25: 108–111.

Physiology

G F M Ball, Wembley, London, UK

Background

The biological activity of pantothenic acid is attributable to its incorporation into the molecular structures of coenzyme A (CoA) and acyl carrier protein. CoA performs multiple roles in cellular metabolism, whereas acyl carrier protein is involved in fatty acid biosynthesis. A wide variety of functional proteins are modified by the addition of acetyl, acyl, and isoprenyl groups through reactions that directly or indirectly involve CoA. These modifications allow the regulation of such important processes as gene transcription, signal transduction, and vision.

Metabolism

Intestinal Absorption

Ingested CoA, the major dietary form of pantothenic acid, is hydrolyzed in the intestinal lumen to pantetheine by the nonspecific action of pyrophosphatases and phosphatase. Pantetheine is then split into pantothenic acid and β-mercaptoethylamine by the action of pantetheinase secreted from the intestinal mucosa into the lumen. Within the alkaline medium of the luminal contents, pantothenic acid exists primarily as the pantothenate anion. Absorption of the liberated pantothenate takes place mainly in the jejunum.

At physiological intakes, pantothenate must move across the brush-border membrane of the intestinal epithelium from a region of lower concentration in the lumen to one of higher concentration in the cytoplasm of the absorptive cell (enterocyte). Such 'uphill' movement requires active transport – a mechanism that depends ultimately upon the expenditure of metabolic energy, i.e., the energy released from the hydrolysis of adenosine triphosphate produced during cellular metabolism. The precise mechanism of pantothenate absorption is secondary active transport in which a transmembrane protein (inappropriately called a carrier) mediates the sodium-coupled transfer of pantothenate across the brush-border membrane. The carrier spans the membrane in a weaving fashion and effects solute transfer through a conformational change in its molecular structure. The immediate energy source for the transport mechanism is the concentration gradient of sodium across the brush-border membrane. The gradient is maintained by the constant extrusion of sodium from the enterocyte by the action of the sodium pump at the basolateral membrane. The sodium pump is driven by metabolic energy and is the primary driving force for pantothenate absorption. As the transport process does not respond to an electrical gradient, it nust be electroneutral, indicating a 1:1 cotransport of Na^+ and $pantothenate^-$ by the same carrier. The mechanism by which pantothenic acid exits the absorptive cell at the basolateral membrane has not been established.

Intestinal microflora have been reported to synthesize pantothenic acid in mice, but the contribution of bacterial synthesis to body pantothenic acid levels or fecal loss in humans has not been quantified.

Unlike other water-soluble vitamins (ascorbic acid, biotin and thiamin) that are absorbed by specific carrier-mediated systems, the absorption of pantothenic acid is not regulated by its level of dietary intake. The absence of clear-cut deficiency symptoms in humans and the lack of toxicity at high doses could explain why a regulated absorption mechanism has not evolved for pantothenic acid.

Tissue Uptake and Metabolism

After absorption, free pantothenic acid is conveyed to the body tissues in the plasma from which it is taken up by most cells. A so-called sodium-dependent multivitamin transporter that mediates placental and intestinal uptake of pantothenate, biotin and the essential metabolite lipoate has been cloned from rat placenta and rabbit intestine. Messenger RNA transcripts of this transporter have been found in many tissues (intestine, liver, kidney, heart, lung, skeletal muscle, brain and placenta) suggesting that this carrier protein may be involved in the uptake of pantothenate, biotin and lipoate by all cell types.

In mammalian tissues (but not in red blood cells), CoA is synthesized from pantothenic acid in five enzymatic steps. Three substrates are needed to synthesize CoA: pantothenic acid, ATP, and cysteine. The rate-controlling step in the synthesis is the conversion of pantothenic acid to 4′-phosphopantothenic acid by pantothenate kinase. Tissue levels of CoA are kept in check by feedback inhibition of pantothenate kinase by CoA, acetyl-CoA, or a related metabolite.

In the event of a drastically reduced intake of pantothenic acid, such as would occur during food deprivation, the liver, and possibly other tissues, is able to maintain nearly constant CoA levels for some considerable time. In fasting rats, pantothenic acid uptake by the liver is stimulated by the natural rise in glucagon, and incorporation of pantothenic acid into CoA is stimulated by glucagon and cortisol.

In contrast to the liver, uptake of pantothenic acid by heart and skeletal muscle of fasting rats is reduced, and yet the rate of pantothenic acid conversion to CoA is increased. Evidently, myocardial and muscle CoA synthesis is not governed by the availability of pantothenic acid to these tissues, but rather is controlled intracellularly by regulation of enzymes involved in the CoA synthetic and/or degradative pathways.

Pantothenic acid derived from the degradation of CoA is excreted intact in urine. The amount excreted varies proportionally with dietary intake over a wide range of intake values. Both fasting and diabetes result in decreased excretion, thus conserving whole-body pantothenic acid under these conditions.

Biochemical Functions of Coenzyme A and Acyl Carrier Protein in Cellular Metabolism

A molecule of pantothenic acid is incorporated into the structures of CoA and acyl carrier protein. Though the functional sulfhydryl group of these co-enzymes is not part of the pantothenate moiety, the steric configuration of pantothenic acid is important for enzymatic recognition.

Acetyl-CoA and succinyl-CoA are energy-rich thioesters that play important roles in the tricarboxylic acid cycle. Acetyl-CoA is also required for the acetylation of choline to form the neurotransmitter, acetylcholine. The amino sugars D-glucosamine and D-galactosamine react with acetyl-CoA to form acetylated products, which are structural components of various mucopolysaccharides. The biosynthesis of cholesterol begins with the condensation of two molecules of acetyl-CoA to form acetoacetyl-CoA. The latter reacts with acetyl-CoA to form 3-hydroxy-3-methylglutaryl-CoA (HMG-CoA), which in turn is reduced to the key intermediate, mevalonic acid. CoA is required at two steps in each cycle of the β-oxidation of fatty acids. Acyl carrier protein, as an integral part of fatty acid synthase, is involved in the biosynthesis of fatty acids.

Physiological Roles of Coenzyme A in the Modification of Proteins

Many diverse cellular proteins are modified by acetylation and/or by the covalent attachment of lipids. The modifications fall into three main categories: acetylation, acylation, and isoprenylation. The alterations in protein structure may be relevant to the association of proteins with the plasma membrane or with subcellular membranes, protein–protein binding, or the targeting of proteins to specific intracellular locations.

In some cases, the modifications are cotranslational, i.e., they take place on the growing polypeptide chain associated with the ribosome during protein synthesis; in other cases, they are posttranslational.

Most soluble proteins are modified at their amino termini with an acetate group that is donated by CoA. Acetylation alters the protein's binding affinity for receptors or other proteins.

A wide variety of proteins are modified with long-chain fatty acids donated by CoA. The two fatty acids most commonly attached to proteins are myristic acid (14:0) and palmitic acid (16:0). The enzyme linking myristate to amino-terminal glycine residues by an amide bond is N-myristoyl transferase. For myristoylation to take place, the protein substrate must have a glycine residue at position 2, immediately following methionine, and preferably a hydroxyamino acid (typically serine) at position 6. Myristoylated proteins include G protein α subunits (signal transduction), ADP-ribosylation factors (vesicular transport), myristoylated alanine-rich C kinase substrate protein (cytoskeletal rearrangements), recoverin (vision), proteins of the immune system, and several enzymes. Palmitoyl transferases link palmitate to the side chains of cysteine residues by a thioester bond. The cysteine residues can reside at any point in the primary structure of the protein; there is little evidence for any specific sequence requirements. Unlike the highly stable amide linkages to myristate, modifications of proteins by palmitate occur in thioester or oxyester linkages that are subject to hydrolysis by esterases. Cycles of palmitoylation and depalmitoylation allow the modified protein to have a regulating function. Palmitoylated proteins include G protein α subunits, many plasma membrane-anchored receptors, cytoskeletal proteins, gap junction proteins, neuronal proteins, and the enzymes acetylcholinesterase and glutamic acid decarboxylase. Palmitate modification is also a prerequisite for the budding of transport vesicles from Golgi cisternae.

Two important isoprenoids, the 15-carbon farnesyl pyrophosphate and the 20-carbon geranylgeranyl pyrophosphate (**Figure 1**), are metabolic products of mevalonic acid. Attachment of either isoprenoid chain is the first step in the modification of proteins bearing a C-A1-A2-X motif, where C is a carboxy-terminal cysteine residue, A1 and A2 are aliphatic amino acids, and X is an undefined amino acid. The attachment is a thioester bond with the terminal cysteine. Isoprenylated proteins include Ras proteins (signal transduction), Rab proteins (vesicular transport), nuclear lamins A and B (assembly and stabilization of the nuclear membrane), G protein γ subunits, and the enzymes phosphorylase kinase and rhodopsin kinase.

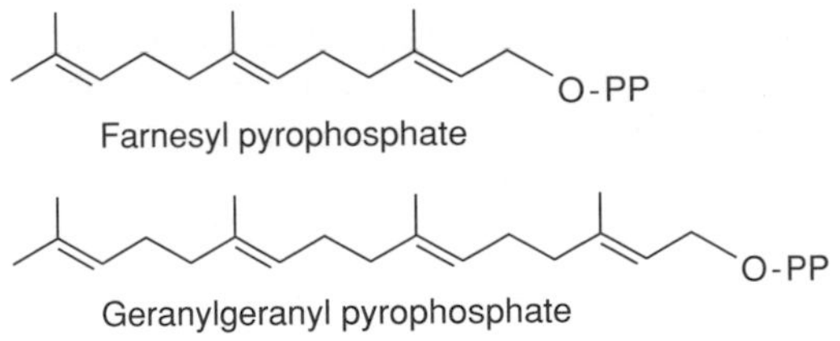

Figure 1 Structures of farnesyl pyrophosphate (C15) and geranylgeranyl pyrophosphate (C20).

The physiological implications of selected acetylation and acylation modifications of proteins are discussed below.

Acetylation of β-Endorphin

Amino-terminal acetylation plays an important role in regulating the biological activity of the brain neurotransmitter β-endorphin. This peptide has morphine-like analgesic activity and also affects sexual behavior and learning. Acetylation deactivates β-endorphin by rendering it unable to bind to specific receptors. The modification is posttranslational and occurs before or during the packaging of the peptide into the secretory granules of multineurotransmitter neurons in the pituitary gland.

Histone Acetylation

The DNA in cell nuclei does not exist in the 'naked' state – rather, it is compacted into chromatin by winding around specific DNA-binding proteins called histones. The fundamental repeating unit of chromatin is the nucleosome, which appears in electron micrographs as beads on a string. Each nucleosome consists of core histones (H2A, H2B, H3, and H4), linker histones (H1 or variants thereof) and variable lengths of linker DNA. Two molecules each of the core histones form a barrel-shaped nucleosome core particle, around which 146 base pairs of DNA are wrapped in nearly two complete turns. A model of the octamer of core histones is shown in **Figure 2**. The linker histone acts as a clamp, preventing the unwinding of DNA from the octameric complex. Each of the four types of core histone comprises a globular, hydrophobic carboxy terminus and an extended hydrophilic amino-terminal tail containing a number of positively charged amino acid residues. The tails lie on the outside of the nucleosome, where they can interact ionically with the negatively charged phosphate groups of the DNA backbone. During the periods between cell division, the 'beads on a string' chromatin filaments form higher-order structures by winding into a solenoid containing six nucleosomes per turn. In these structures, the tails of the core histones still extend outside the nucleosome.

The organization of chromatin into nucleosomes is an essential feature in the regulation of gene transcription – the step in protein synthesis in which messenger RNA is synthesized from DNA. During transcription, the enzyme RNA polymerase II combines with a host of protein transcription factors to form a multiprotein complex at a precise site on the DNA called the promoter. The polymerase moves along the DNA, temporarily unwinding and separating the two strands. As it moves along, RNA is formed by the linking of ribonucleotides under the influence of the enzyme and using one of the DNA strands as a template.

It is necessary to control gene transcription so that only those proteins needed by a particular cell for a specific purpose are synthesized. When a protein is not needed, nucleosomes prevent transcription by impeding the access of factors required to initiate and regulate this process. When protein synthesis is required, changes in cell physiology cause a partial and localized alteration of chromatin structure (chromatin remodeling) in a manner that permits the binding of initiating and regulatory factors.

One important chromatin remodeling system involves the post-translational modification of core histones by acetylation. Nuclear histone acetyltransferases (HATs) catalyze the transfer of an acetyl group from acetyl-CoA on to the ε-amino group of specific lysine residues present exclusively in the amino-terminal tails of each of the core histones. Neutralization of the positively charged lysines reduces the net positive charge of the histone tails and weakens their association with DNA. The displacement of the flexible tails permits subtle changes in nucleosomal structure and a partial unwinding or loosening of the core DNA. The result is an increase in accessibility of transcription factors to their DNA-binding sites. Acetylation does not occur randomly; multiple HATs have specificities for different lysines in the histone tails. Histone deacetylases (HDACs) counter the effects of HATs by restoring the nucleosomes to their transcriptionally repressive configurations.

In the overall scheme (**Figure 3**), the chromatin structure is transiently and reversibly altered to allow or prevent access of the transcription factors by targeting HATs or HDACs to the core promoter, thereby activating or repressing transcription, respectively. It is now clear that transcriptional activators function by recruiting coactivators, and it is the coactivators that possess HAT activity. Repressors inhibit transcription indirectly by recruiting HDACs via a bridging corepressor.

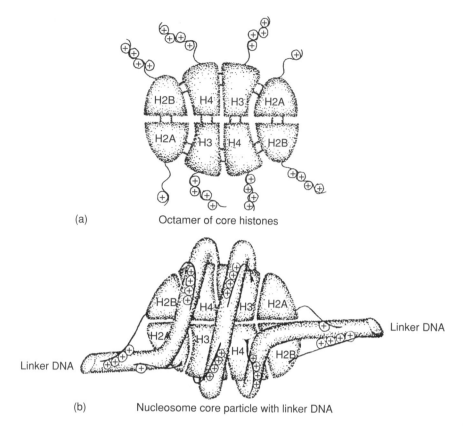

(a)   Octamer of core histones

(b)   Nucleosome core particle with linker DNA

Figure 2   Models for (a) the octamer of core histones, and (b) the nucleosome core particle with linker DNA. From Csordas A (1990) *Biochemical Journal* 265: 23–38 with permission.

α-Tubulin Acetylation

Microtubules are long, stiff, hollow cylinders composed of polymerized α- and β-tubulin dimers. As constituents of the cytoskeleton, microtubules provide structural support for the cell. They also act as lines of transport for the organized movement of mitochondria and other organelles to desired locations within the cell, facilitate delivery of transport vesicles from the Golgi complex to the apical membrane in epithelial cells, and become associated with the centrioles and chromosomes to form the spindle during mitosis and meiosis (cell division). A subset of the α-tubulin is modified, like the histones, by posttranslational acetylation of the ε-amino group of specific lysine residues. In contrast to histone acetylation, the acetylation of α-tubulin stabilizes the polymeric structure of the microtubule; deacetylation is coupled to depolymerization.

Acylation of G Proteins

Peptide hormones, being hydrophilic, cannot cross the lipid bilayer of the cell plasma membrane. To overcome this problem, the hormones bind to specific cell surface receptors, and a member of the family of guanine nucleotide-binding regulatory proteins (G proteins) acts as a signal transducer in coupling these receptors to intracellular effector proteins – enzymes that generate the second messenger (e.g., cyclic 3′,5′-adenosine monophosphate, cAMP). The second messenger mediates the biological action of the hormone through the activation of protein kinases.

The posttranslational attachment of palmitate to the G protein α subunit provides the means for reversible translocation of the subunit between the plasma membrane and the cytoplasm. The model shown in (**Figure 4**) applies to the α_s subunit responsible for stimulation of cAMP synthesis. In the unactivated state, the palmitoylated α subunit–GDP complex, α_{pal}–GDP, is associated with the β/γ subunits and the plasma membrane. Receptor activation stimulates release of GDP and binding of GTP to form active α_{pal}–GTP; the α and β/γ subunits dissociate from each other but remain at the plasma membrane by virtue of their respective palmitate and isoprenyl attachments. Palmitate is rapidly cleaved from α_{pal}–GTP by a palmitoyl esterase, and the depalmitoylated α subunit is released from the membrane into the cytoplasm. Intrinsic GTP hydrolysis converts the active GTP-bound subunits in both membrane and cytoplasm

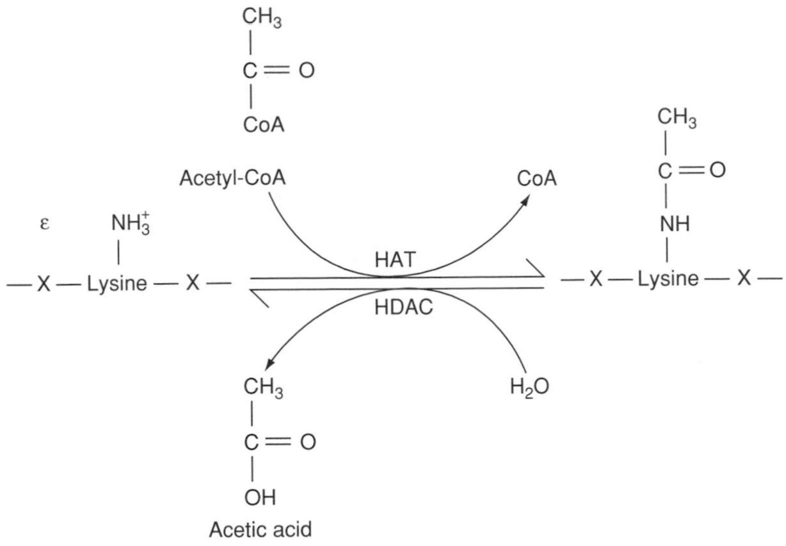

CH$_3$
|
C $=$ O
|
CoA

Acetyl-CoA

CH$_3$
|
C $=$ O
|
NH

CoA

ε NH$_3^+$
|
— X — Lysine — X —

HAT
⇌
HDAC

— X — Lysine — X —

CH$_3$
|
C $=$ O
|
OH

Acetic acid

H$_2$O

Figure 3 Opposing activities of histone acetyltransferases (HAT) and histone deacetylases (HDAC) in the control of transcription through chromatin remodeling.

into the inactive GDP-bound forms. Reattachment of palmitate to the cytoplasmic subunit by a palmitoyl transferase facilitates the return of the α_{pal}–GDP to the plasma membrane.

The G protein α subunit is further modified by the cotranslational attachment of myristic acid. This increases the affinity of the α subunit with the β/γ subunits, with the plasma membrane, and with the effector protein.

Palmitoylation of Asialoglycoprotein Receptors

The hepatic asialoglycoprotein receptor mediates the endocytosis of desialylated glycoproteins containing terminal galactose or *N*-acetylgalactosamine. (Endocytosis refers to the cellular uptake of macromolecules by entrapment within inward foldings of the plasma membrane, which then pinch off to form intracellular vesicles.) There is evidence that a cycle of palmitoylation and depalmitoylation regulates the ligand-binding activity of the asialoglycoprotein receptor. Inactivation of the receptor by depalmitoylation prevents the rebinding of dissociated ligand molecules and ensures that ligand is shuttled to lysosomes for degradation rather than nonproductively recycled back to the cell surface.

Deficiency in Animals and Humans

Pantothenic acid deficiency has been induced experimentally in many species of animals and birds by feeding diets containing low levels of the vitamin. The wide range of deficiency signs, histopathological abnormalities, and metabolic changes indicate disorders of the nervous system, reproductive system, gastrointestinal tract, and immune system. Rodents are particularly prone to necrosis and hemorrhage of the adrenal glands with consequent impairment of adrenal endocrine function. In young animals, the earliest sign of deficiency is a decline in the rate of growth. Distinctive visible signs are depigmentation of fur in rats and mice, and rough plumage and exudative lesions around the beak and eyelids of chickens. 'Goose-stepping' of the hind legs in pigs and ataxia (falling down to one side) in chicks are associated with demyelination of the motor neurons.

Human pantothenic acid deficiency has been carefully studied in healthy male volunteers given an emulsified artificial diet by stomach tube. In one study, two subjects received the basic diet devoid of pantothenic acid, a second pair received the same diet with added antagonist (ω-methyl pantothenic acid), and a third pair (the controls) received the diet supplemented with pantothenic acid. After about 4 weeks, subjects receiving the antagonist and those in the deficient group began to show similar symptoms of illness. Clinical observations were irritability, restlessness, drowsiness, insomnia, impaired motor coordination, and neurological manifestations such as numbness and 'burning feet' syndrome. The most persistent and troublesome symptoms were fatigue, headache, and the sensation of weakness. Among the laboratory tests, the loss of eosinopenic response to adrenocorticotropic hormone indicated adrenocortical insufficiency.

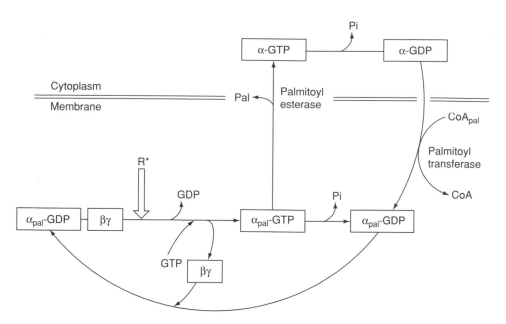

Figure 4 Model of $G\alpha_s$ palmitoylation and depalmitoylation as a means for reversible translocation of the subunit between the plasma membrane and the cytoplasm. R^* indicates receptor activation. Modified from Wedegaertner PB and Bourne HR (1994) Activation and depalmitoylation of $G_{s\alpha}$. *Cell* 77: 1063–1070, with permission.

Dietary Intake

A recommended dietary allowance (RDA) for a nutrient is derived from an estimated average requirement (EAR), which is an estimate of the intake at which the risk of inadequacy to an individual is 50%. In the case of pantothenic acid, no data have been found on which to base an EAR, and an adequate intake (AI) is used instead of an RDA by the Food and Nutrition Board of the US Institute of Medicine. The AI for infants up to 12 months old (1.7–1.8 mg day^{-1}) reflects the observed mean intake of breastfed infants. The AI for children aged 1–3 years (2 mg day^{-1}) is extrapolated from adult values. The AIs for children aged 4–13 years (3–4 mg day^{-1}), and adolescents and adults of both sexes (5 mg day^{-1}) are based on pantothenic acid intake sufficient to replace urinary excretion. AIs for women during pregnancy and lactation are 6 and 7 mg day^{-1}, respectively.

There are no known toxic effects of oral pantothenic acid in humans or animals.

See also: **Fatty Acids**: Metabolism; **Oxidative Phosphorylation**; **Tricarboxylic Acid Cycle**

Further Reading

Casey PJ (1994) Lipid modifications of G proteins. *Current Opinion in Cell Biology* 6: 219–225.

Clarke S (1992) Protein isoprenylation and methylation at carboxyl-terminal cysteine residues. *Annual Review of Biochemistry* 61: 355–386.

Food and Nutrition Board of the Institute of Medicine (1998) *Dietary Reference Intakes for Thiamin, Riboflavin, Niacin, Vitamin B6, Folate, Vitamin B12, Pantothenic Acid, Biotin, and Choline*. Washington, DC: National Academy Press.

Hassig CA and Schreiber SL (1997) Nuclear histone acetylases and deacetylases and transcriptional regulation: HATs off to HDACs. *Current Opinion in Chemical Biology* 1: 300–308.

Hodges RE, Ohlson MA and Bean WB (1958) Pantothenic acid deficiency in man. *Journal of Clinical Investigation* 37: 1642–1657.

Kadonaga JT (1998) Eukaryotic transcription: an interlaced network of transcription factors and chromatin modifying machines. *Cell* 92: 307–313.

Kuo M-H and Allis CD (1998) Roles of histone acetyltransferases and deacetylases in gene regulation. *Bioessays* 20: 615–626.

Plesofsky-Vig N (1999) Pantothenic acid. In: Shils ME, Olson JA, Shike M and Ross AC (eds) *Modern Nutrition in Health and Disease*, 9th edn, pp. 423–432. Philadelphia, PA: Lippincott Williams & Wilkins.

Robishaw JD and Neely JR (1985) Coenzyme A metabolism. *American Journal of Physiology* 248: E1–E9.

Sinensky M and Lutz RJ (1992) The prenylation of proteins. *Bioessays* 14: 25–31.

Smith CM and Song WO (1996) Comparative nutrition of pantothenic acid. *Journal of Nutritional Biochemistry* 7: 312–321.

Tahiliani AG and Beinlich CJ (1991) Pantothenic acid in health and disease. *Vitamins and Hormones* 46: 165–228.

Wedegaertner PB and Bourne HR (1994) Activation and depalmitoylation of $G_{s\alpha}$. *Cell* 77: 1063–1070.

PAPAYAS

J H Moy, University of Hawaii at Manoa, Honolulu, USA

Introduction

Papaya, a mildly sweet, melon-like tropical fruit belonging to the family Caricacae, is a native of tropical America (Central America and the Caribbean, including Mexico, Costa Rica, and the Bahamas). After the Spaniards took the fruit to Luzon Island in the Philippines in the mid sixteenth century, it reached Malacca shortly afterward, and then India. It has been widely grown throughout tropical and subtropical regions such as Australia; Hawaii, Florida, Texas, California, and Puerto Rico in the USA; Peru; Venezuela; various parts of Central and South Africa; and Bangladesh, Pakistan, and India. The Australians call it 'pawpaw' while the Venezuelans call it 'lechosa.'

Fruit size may vary from less than 0.5 kg to 3 kg. A climacteric fruit, it is mainly consumed fresh when ripened after harvest. Ripening is judged by the approximate percentage of yellowness on its skin, and more accurately by measuring its total soluble solids (TSS) contents with a refractometer. Export-grade papayas grown in Hawaii should have a minimum of 11.5% TSS. Firmness, as gaged by touch or by a texture-measuring device, is another way of judging ripening. The firmness is related to the biochemical changes in three fractions of pectins in papayas, and is not a very accurate index of ripeness. Green papayas can also be consumed as a salad or in soups, both of which are quite popular in South-east Asia.

Hawaii, part of the USA, is the world's largest exporter of papayas, with about 16 000 t in 1999, valued at US$14.15 million. Most of the outshipments are to the US mainland, with about 25% going to Japan. Brazil is the second largest exporter, with most papayas going to European countries such as Germany, France, and the UK. Kenya, the Ivory Coast, and Malaysia also ship papayas to European markets. For recent shipment statistics, (*See* **Fruits of Tropical Climates:** Commercial and Dietary Importance.)

Species and Cultivars

Papaya belongs to the family Caricaceae – a small, somewhat anomalous family with four genera and 31 species, of which three are in tropical and subtropical America and one in Africa. The genera and the number of included species are: *Carica*, 22; *Facaratia*, 6; *Farilla*, 1; and *Cylicomorpha*, 2. The edible fruits are found only in *Carica*, of which *C. papaya*, the common papaya, is extensively grown. Other species include: *C. chilensis*, *C. goudotiana*, *C. monoica*, and *C. pubesceus*. *C. candamarcensis*, known as mountain papaya, grows to a height of about 2.5 m, tolerates low temperature, and thrives well at an elevation between 1500 and 2000 m. *C. monoica* grows in the Amazon basin.

A large number of papaya cultivars are grown in different parts of the tropical and subtropical regions. However, none are true cultivars because they do not reliably transmit the parental characters to all their progeny. Some of the better known cultivars are listed below, with their average weight in kg in parentheses:

1. Washington (0.9–1.0)
2. Honey dew (1.1–1.5)
3. Coimbatore 1 (1.2–1.5) – a selection from the progenies of cv. *Ranchi*
4. Coimbatore 2 (1.0–1.5) – a high papain-yielding cultivar
5. Coimbatore 3 (0.9–1.0) – a hybrid between Co. 2 and Sunrise Solo
6. Coimbatore 4 (1.3–1.5) – a cross between Co. 1 and Washington (The coimbatore cultivars were released from the Tamil Nadu Agricultural University, Coimbatore.)
7. Pusa Delicious (1.0)
8. Pusa Majesty (0.95)
9. Pusa Giant (1.9)
10. Pusa Dwarf (1.0) (The pusa cultivars were bred at the India Agricultural Research Institute Regional Station, Pusa, Bihar.)
11. Sunrise Solo (0.4–1.0) – red-fleshed
12. Sunset Solo (0.4–1.0) – red-fleshed
13. SunUp (0.4–1.0) – a genetically transformed sunset, red-fleshed
14. Kapoho Solo (0.5–1.4) – yellow-fleshed

15. Waimanalo Solo (0.5–1.4) – yellow-fleshed
16. Rainbow (0.5–1.4) – a transgenic cross between Kapoho and SunUp, yellow-fleshed (The two transgenic cultivars have been grown on the island of Hawaii since 1998)
17. Pink-Fleshed Sweet (0.85)
18. Thailand (1.9–2.2)

Other important cultivars are: Guinea Gold, Sunnybank, and Hybrid 5 (Australia); Hortus Gold and Honey Gold (South Africa); Tai Ning No. 2, Panama red and Solo No. 1 (Taiwan); Red Rock and Cariflora (Florida, USA); Red Fleshed, (Philippines); Madhubindu, Barwani, Ranchi, and Peradeniya (India).

Botany and Horticulture

Papaya is a herbaceous, unbranched plant. Its stem is supported by phloem fibers encircling a hollow stem, which can grow up to 8–10 m high. Its leaves are large, deeply lobe-shaped with long hollow petioles, and spirally crown the stem. Flowers, male and female, can be on the same plant (monoecious) or on separate plants (dioecious), or as male and female parts on the same flower (hermaphrodite).

Papaya fruits are mostly round in the female tree and cylindrical and pear-shaped in the hermaphrodite tree. Underneath the smooth, thin skin (green when immature, orange-yellow when ripe) is a thick layer of deep yellow to orange-red pericarp with an elliptical, central cavity containing many small, round black seeds coated with jelly-like tissues (**Figures 1 and 2**).

Planting

The three major environmental factors to consider in selecting a site to grow papayas are temperature,

Figure 2 (see color plate 120) Thailand cultivar growing in Thailand, weighing 2.50 kg.

moisture (rainfall and soil drainage), and wind. The hermaphrodite papaya plant preferred for commercial orchards is more sensitive to its growing environment than the female papaya plant. Selection of a suitable site is therefore critical. Another condition to consider is the amount of sunlight the site receives to support plant growth and fruit production. Insufficient sunlight results in fruits with inadequate sugar and low yields, and encourages plant diseases that affect papaya production.

The temperature of the site is the most important factor. Optimal production is generally limited to elevation under 160 m within a temperature range of 16–31 °C. Temperature below 16 °C could cause carpeloidy, resulting in 'cat-face' deformity when floral stamens develop abnormally into fleshy, carpel-like structures. High temperature (32–35 °C) may induce female sterility, in which normally hermaphroditic papaya plants produce male flowers, resulting in poor fruit set and production.

A minimum monthly rainfall of 10 cm and an average relative humidity of 66% are considered ideal for papaya growth and production. Drip irrigation can supplement low rainfall. Good soil drainage is also essential. When soil drainage is restricted, papaya is susceptible to fungal root diseases. The plants are severely affected by waterlogging and can be killed when subjected to puddled conditions for even a few hours. The Puna area on the island of Hawaii is well-suited to commercial papaya production because of the very porous lava soils.

Papaya plants must be protected from wind. Plants exposed to constant wind develop deformed, crinkled leaves. When wind stress damage is excessive, the plant growth, fruit set, fruit quality, and productivity will be reduced. Wind-blown dust can cause sap bleeding that harms fruit appearance. In coastal

Figure 1 (see color plate 119) Rainbow cultivar growing in Hawaii, weighing 0.66 kg.

areas, salt spray carried by wind can desiccate leaves and kill papaya plants. Winds of $64\,km\,h^{-1}$ can uproot papaya trees growing in mineral soils, especially when accompanied by heavy rain. Windbreaks should be established well in advance of planting a papaya crop. On the other hand, adequate air movement is important in reducing incidence of fungal diseases such as *phytophthora* and *anthracnose*. These diseases can become severe when there is excessive free moisture and high humidity around the plant.

Papaya can either be seeded directly or transplanted into a new field. Direct seeding can be practiced in porous soil. Six-week-old seedlings can be transplanted into less porous fields. Planting in 'virgin' lands or fields in which papaya has not been grown before is preferred because of low disease and insect pressure. However, it is becoming increasingly difficult to find such fields. Other important horticultural aspects of planting papaya are plant sex selection, thinning, fertilizer applications, leaf trimming, weed control, pest management, and disease control.

Transgenic Cultivars

For more than four decades, papaya production in the Hawaiian island chain was severely affected by the papaya ringspot virus (PRV). In the 1960s, papaya planting was forced to move from the island of Oahu to the Puna area of the island of Hawaii. Since 1992, the introduction and subsequent spread of PRV in commercial orchards in Puna has resulted in a significant decrease in papaya production in the state of Hawaii.

The only solution seemed to be to produce a transgenic cultivar that would be resistant to the virus. Faced with this extremely challenging task, molecular biologists from Cornell University, the University of Hawaii, and the Upjoin Company collaborated from 1987 to 1991 to obtain the first transformed plant that appeared to have PRV resistance. It took several more years for plant breeders at the University of Hawaii to inbreed the resistant plant to produce a tree-breeding, redfleshed cultivar that was named UH sunup (also referred to as *SunUp*). Because the papaya industry wanted a yellow-fleshed fruit, plant breeder Richard Manshardt crossed *SunUp* with *Kapoho* to produce a yellow-fleshed F_1 hybrid that was named *UH Rainbow* (also called *Rainbow*).

SunUp is totally resistant to the virus, but *Rainbow*, in greenhouse studies, has shown susceptibility to PRV until about 7 weeks of age. This is referred to as young plant susceptibility. The plants become fully resistant to PRV after 3 months of age. The hybrid *Rainbow* has some differences from the *Kapoho*: (1) a lower sex-segregation ratio; (2) plants grown from seeds of F_1 hybrids do not breed true; (3) a higher sensitivity to calcium deficiency; (4) more sensitive to *phytophthora* and will require timely application of fungicides; and (5) ripens faster than *Kapoho* but slower than *Sunrise* fruits.

These two lines of transgenic cultivars have been rigorously reviewed and approved by three US agencies: the US Department of Agriculture (USDA), the Environmental Protection Agency (EPA), and the Food and Drug Administration (FDA). The USDA determined that these new plants would not contaminate other existing germ plasma, and they would not foster a new strain of virus. The USDA concluded, therefore, that they posed no hazards to agriculture in the USA. The EPA sets tolerance limits for regulated chemicals. The coat protein used in the genetic engineering process was considered a pesticide, to be regulated because its function is to eliminate diseases. The EPA subsequently agreed with University of Hawaii scientists that there was no tolerance required for coat protein. The FDA's job is to protect the public from unsafe products. The FDA was satisfied that the nutrients of the original material (fruit) were retained, and therefore did not require a toxicity test. All three agencies declared the two transgenic cultivars safe for human consumption.

Those wishing to purchase seeds of transgenic papaya cultivars must adhere to a licensing process as well as sign a sublicense agreement with the Papaya Administrative Committee in the state of Hawaii.

Nutrient Composition and Fruit Chemistry

Nutrient Composition

The papaya fruit is a good source of ascorbic acid (vitamin C) and carotenoids (provitamin A), two important nutrients for people in the subtropical and tropical regions. **Table 1** shows the nutrient composition of papaya from data published in 1965, 1990, and 1999.

Chemistry and Biochemistry

Sugars are the principal carbohydrates in papayas, with very little starch. Refractometric measurements of total soluble solids (%TSS) show some differences in various cultivars. Those grown in Florida have been reported to range from 5.6 to 7.2%; those in India vary from 6.5 to 13%, while the Hawaiian cultivars measure 11.5–13.5%. The sugars in ripe papayas in which the invertase has been inactivated were reported to be 48% sucrose, 30% glucose, and 22% fructose.

Papaya is a low-acid fruit, with the total titratable acidity of about 0.1% calculated as citric acid. The

Table 1 Nutrient composition of papaya per 100 g of edible pulp

Moisture (%)	86–89
Carbohydrate (g)	9.5–12.2
Protein (N × 6.25: g)	0.36–0.5
Fat (g)	0.06–0.1
Fiber (g)	0.5–0.6
Ash (g)	0.5–0.6
Ascorbic acid (mg)	40–84
Vitamin A (mg)	11–32[a]
Thiamin (mg)	0.027–0.04
Riboflavin (mg)	0.043–0.25
Niacin (mg)	0.20–0.33
Calcium (mg)	10–30
Phosphorus (mg)	10–12
Iron (mg)	0.2–4.0
Energy (cal)	40–48

[a]Vitamin A data assuming 12 µg of *all-trans* β-carotone = 1 µg *all-trans* retinol.

Sources of data: Wenkam NS and Miller: CD (1965) *Composition of Hawaii Fruits*. Bulletin 135. Honolulus, Hawaii: Hawaii Agricultural Experiment Station, University of Hawaii at Manoa; Muthukrishnan CR and Irulappan I (1990) In: Bose TK and Mitra SK (eds) *Fruits*: Tropical and Subtropical, pp. 303–335. Calcutta: Naya Prokash. Moy JH, Paull RE, Bian X, Chung R and Wong L (1999) Quality of tropical fruits irradiated as a quarantine treatment. In: Moy JH and Wong L (eds). *Proceedings of the workshop on the use of Irradiation as a Quarantine Treatment of Food and Agricultural Commodities* pp. 45–53. Honolulu, Hawaii: College of Tropical Agriculture and Human Resources, University of Hawaii and Dept of Hawaii, State of Hawaii.

pH of the pulp ranges from 5.5 to 5.9, far higher than the pHs of other tropical fruits at 3.2–4.5, which explains the low tartness of the papaya fruit. The organic acids in papayas are mainly equal amounts of malic and citric acid, with smaller amounts of ascorbic and α-ketoglutaric acid.

The volatile flavors of papayas were reported to consist of 124 compounds. Linalool is the major component with characteristic fresh papaya aroma and flavor. Benzyl isothiocyanate, another major component, has a pungent off-aroma. Other off-aroma, off-flavor compounds in papaya pulp have been identified as butyric, hexanoic, and octanoic acids and their corresponding methyl esters. These components were analyzed with gas chromatography and mass spectrometry. (*See* **Chromatography**: Gas Chromatography; **Flavor (Flavour) Compounds**: Structures and Characteristics; **Mass Spectrometry**: Principles and Instrumentation.)

Carotenoids are the pigments of the ripe papaya flesh, which is deep yellow. Red-fleshed papaya cultivars additionally contain lycopene, which gives the pulp an orange-red color. (*See* **Colorants (Colourants)**: Properties and Determination of Natural Pigments.)

In addition to papain being an economically important enzyme in green papaya, several other enzymes play a role in the stability and quality of processed papaya products. Pectinesterase (3.1.1.11),

the enzyme responsible for gel formation in unheated papaya purée, contributes to some increase in acidity in purée as a result of demethoxylation of the carboxyl groups in the pectins. Thioglucosidase (myrosinase) (3.2.3.1) is responsible for generating benzyl isothiocyanate, a sulfurous, pungent, odiferous compound which contributes to the off-aroma of papaya products. When fruit tissues are macerated, acid phosphatase (3.1.3.2) catalyzes the hydrolysis of the P–O bond of orthophosphoric monoesters, producing ROH and phosphoric acid, which increases the acidity of the purée. Another enzyme, β-fructofuranosidase (invertase) (3.2.1.26), is responsible for the hydrolysis of sucrose to fructose and glucose.

Harvesting, Handling, and Storage

Summer fruits become mature in 22 weeks, while 26 weeks are needed during the winter. Fruits are typically harvested weekly or twice weekly at color-break (a tinge of yellowness on a green fruit) by hand, or with the aid of a cut-off 'plumber's helper' (a rubber suction cup) attached to a pole for hard-to-reach fruits. Fruits can also be harvested with a ladder, or from a large fruit bin mounted on a tractor high-lift. Careful handling with minimal bruising and abrasion helps with postharvest sorting, packing, shipping, and marketing.

To control fungal decay of postharvested fruit, a dip in hot water at 49 °C for 20 min was found to be very effective. Hot-water spray with fungicide or a wax dip containing fungicide is also effective.

Storing papayas at temperatures below 7 °C after harvest can result in chill injury. The recommended storage temperature is 10 °C.

Quarantine Treatment

Fruit flies and other insect pests are prevalent in the tropics and the subtropics. Most soft tropical fruits are prone to infestation by fruit flies. A quarantine treatment approved by the plant quarantine authority of each country must be applied to fruits before they can be exported to noninfested areas. Up until 1984, chemical fumigants such as ethylene dibromide (EDB) were widely used, until tests showed them to be carcinogenic to laboratory animals. Since then, quarantine treatment options are thermal or cold treatment, or irradiation.

Papayas grown in Hawaii are infested with four species of fruit flies: the Oriental fruit fly (*Dacus dorsalis*), the Mediterranean fruit fly (*Ceratitis capitata*), the melon fly (*Dacus cucurbitae*), and the Malaysian fruit fly (*Bactrocera latifrons*). After EDB was banned, a double-dip hot-water treatment (42 °C for

30 min, 49 °C for 20 min) was used but was abandoned after 3 years because a high percentage of treated fruits became injured by the heat. Subsequently, two other thermal treatments were used. One is a modified vapor heat treatment, and the other is called a high-temperature forced-air (HTFA) procedure. Both are very similar. The difference is the relative humidity of the heated air admitted to the treatment chambers. The air in the vapor heat treatment has moisture added to the air throughout, with the relative humidity kept at 90% or above. In the HTFA method, ambient air is heated and admitted into the chamber, and only in the last hour of a 4-h treatment is the relative humidity required to be kept at 90–100%. Both methods require about 4 h to reach the endpoint, which is when the center of the fruit reaches 47.2 °C. Fruits are then cooled with a water spray for about 30 min. Several fruit packers in Hawaii are using these methods, which work quite well. However, both are time-consuming and commodity-specific. Total uniformity in heating is also difficult to achieve because of variation in fruit size, fruit ripeness (affecting thermal conductivity and heat transfer rates), and physical locations within a chamber. After heat treatment, it is common to find lumpy texture and lack of flavor in a few percent of the heat-treated fruits in each batch, caused by enzymes responsible for ripening being inactivated by heat.

Cold treatment requires that fruits be kept at 1 or 2 °C for 12–14 days to immobilize and inactivate fruit fly eggs and larvae before the fruits can be taken to the supermarkets. Not many tropical fruits can tolerate this time–temperature regime, papaya being one of them.

Though somewhat controversial, several decades of research in Hawaii and in other parts of the world have proven that irradiation is the most efficacious quarantine treatment procedure. The treatment is efficient (15–20 min in a commercial irradiator) and effective (all fruits are thoroughly irradiated to cause the fruit fly eggs and larvae to be sexually sterilized, regardless of fruit size and ripeness). Irradiated papayas ripen normally or slightly delayed, and their chemical, physical, nutrient, and sensory qualities are well retained. Radiation sources can be gamma-rays (from cobalt-60), an electron beam (with limited penetration), or X-rays (converted from e-beam). Fruits irradiated will not become radioactive because there is a limit on the energy level used. In April 1995, Hawaii became the first place in the world to use irradiation as a quarantine treatment of its papayas and other tropical fruits. The generic quarantine dose approved by the USDA is 0.25 kGy. Starting in 2001, more countries will be using this technology to treat their fruits for export markets.

Processed Products

Chunks (Refrigerated or Frozen)

After being washed, deseeded, and peeled, papayas can be cut into chunks, then refrigerated or frozen as a convenient food. It is available in some supermarkets in the salad section. Another niche market for this type of product is the airlines. A number of airlines prefer the convenience of serving these prepared fruits to their passengers. The frozen chunks are more suitable for serving with ice cream or for further manufacturing into dessert products.

Canned (Mixed with Other Fruits)

Papaya can be made into a canned product, or as a cocktail of several fruits. Chunks or dices of papayas or mixed fruits can be filled into a can and covered with hot (75 °C), acidified syrup (c. 40° Brix with pH adjusted to 3.6–3.8 with citric acid: 40° Brix = 40 g of sugar mixed with 60 g of water at 20 °C). After the air in the head space in the cans (c. 6.3–8.0 mm) has been exhausted by steam, the cans are sealed and pasteurized with steam, or in a boiling-water bath. (*See* **Canning**: Principles.)

Purée

Papaya purée is prepared in the form of a free-flowing paste, without seeds, skin, or unwanted fiber. The purée can be an intermediate product used for manufacturing several end products such as juices, nectars, jams, jellies, syrups, and dried fruit rolls or leathers. Papaya purée and other fruit purées in the USA can be manufactured with a high degree of mechanization. Several aspects of purée manufacturing are important to produce a quality product.

1. Ripe papaya fruits should be steamed for 2 min to coagulate the latex in the peel, preventing the latex from entering into the purée. Also, steaming increases purée yield by softening the outer layers of the fruit, and inactivates enzymes in the peel. The fruits are spray-cooled to remove the residual heat.
2. Fruits are sliced, crushed, and dropped into a centrifugal separator. All the skin and most of the seeds are separated. The remaining seeds are separated in a paddle pulper fitted with rubber paddles and a small screen (c. 0.80–0.85 mm).
3. The purée is acidified with citric acid (a 50% solution) to a pH of about 3.4–3.6 to inhibit gelatin of the purée. Acidification also helps control microbial activity and enhances the effect of subsequent heat pasteurization.

4. A smooth purée can be obtained by passing the liquid through a paddle finisher with a 0.50 mm screen to remove seed specks and undesirable fiber.
5. The purée is pasteurized in a heat exchanger at 94 °C for 2 min, then cooled to a few degrees above ambient temperature.
6. The purée can be aseptically packed in suitable flexible containers, or frozen in bulk at −20 °C or below.

Beverages

Juice, drinks, nectars, and cordials can be prepared from papaya purée through formulation by adding appropriate amounts of water, sweeteners, and acidulants to the purée. Commonly used acidulants include citric acid, malic acid, lemon and lime juice. These beverages can be pasteurized in bottles or cans after being acidified to a pH below 4.4.

Dried Products

Several forms of dried products can be prepared from papaya slices. Since papaya is a soft fruit when fully ripe, three-quarter ripe fruits are more suitable for preparing dried slices. Drying can be carried out in a hot-air drier, a vacuum drier or a freeze drier, with increasing product quality but correspondingly higher cost. Solar drying of papaya slices is also a very practical means of preserving papaya with good quality. The final moisture content of dried papaya should be around 5–8% (wet weight basis). Good packaging is important to retain quality and to prevent insects from entering the package. University of Hawaii researchers have designed and built a continuous solar drier on the island of Kauai in Hawaii, capable of drying 450 kg of papaya slices within a 24-h period. This drier uses only renewable energy – solar and biogas.

University of Hawaii researchers have also developed a vacuum-puff freeze drying process to make high-quality juice powder, including papaya. The process involves mixing a papaya purée with a small amount of sucrose, freezing the mixture, and then placing it into a freeze drier. Initially, the purée–sucrose or juice–sucrose mixture puffs into a foam when raised to its freezing point. It then refreezes under a good vacuum. The endpoint is reached when 1% of moisture is left. The freeze-dried product is a crystalline powder, and can be rehydrated into a nectar, similar to freeze-dried coffee. Freeze-dried powders also retain the color and flavor of the original purée or juice with the right degree of sweetness

Papain

Source

A milky latex in immature papaya fruit contains an enzyme called papain. Papain from the dried latex is in great demand on the international markets, especially in the UK and the USA.

Properties

Papain is a proteolytic enzyme, capable of hydrolyzing or breaking down protein materials.

Applications

In the food, pharmaceutical, textile, and tanning industries, papain is used as a meat tenderizer; for clearing beer; in the manufacture of cosmetics such as face creams and dental creams; in degumming silk and rayon; in the preshrinking of wool; and in tanning leather.

In the medical field, papain can be used to treat necrotic tissues, dyspepsia and other digestive ailments, ringworm and roundworm infections, skin lesions and ulcers, eczema, and other skin diseases and kidney disorders. Papain is used in detecting stomach and intestinal cancers and also in correcting diphtheria.

Production

The latex of a green papaya drips into a container after the skin is slit. Papain is the dried latex in powder form. The collection of latex is very labor-intensive. Papain production is influenced by several factors:

1. Fruit size and shape: oblong fruits 36 cm in length and 28 cm in diameter were found to give the highest papain yield. In general, papain yield increases with increasing fruit size.
2. Fruit maturity: unripe but fully grown fruits yield maximum papain, especially when fruits have been grown to 75–90 days after fruit set.
3. Season: the flow of latex is low if the temperature is below 10 °C.
4. Cultivars: in India and Sri Lanka, cultivars found to have high papain yield were Washington, Philippine, Botanist's Selection, Peradeniya and Coimbatore 1.

Yield

Papain yield varies from 1.23 to 7.45 g per fruit among nine cultivars, with the cultivar Washington recording the highest mean yield of 7.45 g per fruit. In a papaya tree, the total papain yield can vary from 150 g to 227 g.

Role of Papaya in Human Health

The flesh of papaya contains high levels of three carotenoids: β-cryptoxanthin, β-carotene, and lycopene. Different papaya cultivars vary in concentrations and ratios of these carotenoids. It has been reported that many individual carotenoids, including β-carotene and lycopene, have the ability to prevent transformation of cells to cancer phenotypes in a model system *in vitro*. No study has considered carotenoid combinations as they occur in fruits, such as papaya, and the possibility of synergistic effects via multiple inhibitory mechanisms. Research has been proposed to study the effectiveness of various combinations of carotenoids as inhibitors of cancer development in mammalian systems, and the cellular and molecular mechanisms if it proves to be effective. The implication is that fruits rich in carotenoids and lycopene such as papaya and tomato could be beneficial to human health.

See also: **Canning**: Principles; **Chromatography**: Gas Chromatography; **Colorants (Colourants)**: Properties and Determination of Natural Pigments; **Flavor (Flavour) Compounds**: Structures and Characteristics; **Fruits of Tropical Climates**: Commercial and Dietary Importance; **Mass Spectrometry**: Principles and Instrumentation

Further Reading

Bose TK and Mitra SK (eds) (1990) *Fruits: Tropical and Subtropical*, pp. 303–335. Calcutta: Naya Prokash.

Moy JH, Paull RE, Bian X, Chung R and Wong L (1999) Quality of tropical fruits irradiated as a quarantine treatment. In: Moy JH and Wong L (eds) *Proceedings of the Workshop on the Use of Irradiation as a Quarantine, Treatment of Food and Agricultural Commodities*, pp. 45–53. Honolulu, Hawaii: College of Tropical Agriculture and Human Resources, University of Hawaii, and Dept of Hawaii, State of Hawaii.

Nakasone HY and Paull RE (1998) *Tropical Fruits*, pp. 239–269. Oxford: CAB International.

Nishima M, Zee F, Ebesu R *et al.* (2000) *Papaya Production in Hawaii*, pp. 1–8. Honolulu, Hawaii: College of Tropical Agriculture and Human Resources, University of Hawaii at Manoa.

Shaw PE, Chan HT Jr and Nagy S (eds) (1998) *Tropical and Subtropical Fruits*, pp. 401–445. Auburndale, FL, USA: AgScience.

Wenkam NS and Miller CD (1965) *Composition of Hawaii Fruits*. Bulletin 135. Honolulu, Hawaii: Hawaii Agricultural Experiment Station, University of Hawaii at Manoa.

PARASITES

Contents

Occurrence and Detection

J Melo-Cristino, Faculdade de Medicina de Lisboa, Lisbon, Portugal
J Botas, formerly of Hospital de Santa Maria, Lisbon, Portugal

Importance of Parasites in Human Infection

Parasites are a heterogeneous group of invertebrate animals, including unicellular microrganisms (protozoa) and multicellular organisms with organ systems (helminths), which can infect a diversity of other animals, including humans.

Human infections are frequently acquired by ingestion of contaminated water or food. They are important all over the world and their prevalence can be very high, especially in tropical and subtropical regions. Many parasites have a worldwide distribution, but others occur in limited endemic areas. However, frequent and rapid travel of people from nonendemic areas (tourists, scholars, business people, military personnel, immigrants, etc.) returning from visits to endemic areas contributes to infection by such parasites.

The spectrum of human disease is extremely wide. Many parasites are not noticed, and the infection is asymptomatic, while others are major human pathogens and are responsible for high morbidity and mortality. Infections may have an acute onset with intense symptoms or may require months or years before becoming clinically evident. Recently, some parasites have become more important as they have been recognized as major pathogens in immunocompromised hosts, particularly in patients with acquired immune deficiency syndrome (AIDS). (*See* **HIV Disease and Nutrition.**)

Protozoa Involved in Foodborne and Waterborne Human Infection

The most significant foodborne and waterborne protozoa are shown in **Table 1**. They are classified into phylum Sarcomastigophora (amoebae and flagellates), phylum Ciliophora (ciliates), phylum Apicomplexa (coccidia), and phylum Microspora (microsporidia). Characteristics for differentiation include motility and stages in life cycles and replication.

Motility is accomplished by different mechanisms, namely pseudopodia in amoebae, flagella in flagellates, and movement of rows of cilia in ciliates. Coccidia and microsporidia are essentially nonmotile obligate intracellular parasites.

The life cycles of amoebae, flagellates, and ciliates include two stages, the trophozoite and the cyst. The trophozoite is the vegetative, motile, feeding stage usually found in the intestine. The cyst is the resting, resistant, thick-walled infective stage excreted in feces. Replication is accomplished by binary fission of trophozoites or by development of several trophozoites inside mature cysts.

Coccidial life cycles include stages of asexual development (trophozoite, schizont or meront and merozoite) and stages of sexual differentiation (microgamont and macrogamont) that lead to the production of oocysts, which are shed in feces. The oocyst, containing sporozoites, which may require a period of maturation outside the host, is the resistant and infective stage. Microsporidia multiply by binary fission or multiple fission, producing spores which are

excreted in urine or in feces. The spore possessing the infective agent or sporoplasm is the resistant stage.

Helminths Involved in Foodborne and Waterborne Human Infection

The most significant foodborne and waterborne helminths are shown in **Table 2**. Helminths comprise the phylum Nematoda (nematodes) and the phylum Platyhelminthes (trematodes and cestodes).

Nematodes or roundworms have cylindrical bodies, separate sexes and a complete digestive system. Life cycles can be simple and direct or complex with one or more intermediate hosts. Hosts are defined as intermediate hosts when they harbor an asexual stage of the parasite and definitive hosts when they harbor the sexual stage.

Trematodes or flukes have flattened leaf-shaped bodies, and most are hermaphroditic with an incomplete digestive system. They have complex life cycles, with snails serving as first intermediate hosts and other aquatic animals as second intermediate hosts. Some trematodes may also encyst on aquatic plants.

Cestodes or tapeworms have segmented ribbon-like bodies, are hermaphroditic, and lack a digestive system. Some have direct life cycles, while others need one or more intermediate hosts. Infective stages of helminths include eggs laid by adult worms and asexual stages encysted in animal tissues or on plants.

Occurrence of Parasites in Foods and Water and Mechanisms of Entry into the Food Chain

Parasites have access to food and water by two routes. One is by contamination with the feces of infected humans or animals passing infective resistant stages such as cysts and oocysts of protozoa and eggs of helminths. Another route exists when an infective

Table 1 Foodborne and waterborne protozoa causing human infection

Amoebae	Flagellates	Ciliates	Coccidia	Microsporidia
Endolimax nana[a]	Chilomastix mesnili[a]	Balantidium coli	Cyclospora cayetanensis	Encephalitozoon sp.
Entamoeba coli[a]	Dientamoeba fragilis		Cryptosporidium spp.[b]	Enterocytozoon sp.
Entamoeba hartmanii[a]	Enteromonas hominis[a]		Isospora belli[b]	Nosema sp.
Entamoeba histolytica/dispar[c]	Giardia intestinalis		Toxoplasma gondii[b]	Pleistophora sp.
Iodamoeba bütschlii[a]	Retortamonas intestinalis[a]		Sarcocystis spp.	
Blastocystis hominis[d]	Trichomonas hominis[a]			
	Trichomonas tenax[a]			

[a]Nonpathogenic for immunocompetent humans.
[b]Infection often serious in immunocompromised patients: most human infections are with *C. parvum*.
[c]*E. histolytica* and *E. dispar* are morphologically identical, but only *E. histolytica* causes disease.
[d]Inclusion in the amoebae group has recently been proposed by some authors.

Table 2 Foodborne and waterborne helminths causing human infection

Nematodes	Trematodes	Cestodes
Angiostrongylus spp.	*Clonorchis sinensis*	*Diphyllobothrium* spp.
Anisakis spp.	*Fasciola hepatica*	*Dipylidium caninum*
Ascaris lumbricoides	*Fasciolopsis buski*	*Echinococcus granulosus*
Bayliascaris procyonis	*Heterophyes heterophyes*	*Echinococcus multilocularis*
Capillaria hepatica	*Metagonimus yokogawai*	*Hymenolepis diminuta*
Capillaria philippinensis	*Opistorchis* spp.	*Hymenolepis nana*
Dracunculus medinensis	*Paragonimus westermani*	*Multiceps multiceps*
Enterobius vermicularis		*Spirometra* spp.
Toxocara spp.		*Taenia saginata*
Trichinella spiralis		*Taenia solium*
Trichuris trichiura		

viable stage of the life cycle of a parasite is present in animal tissues or on plants.

Environmental contamination with human or animal faeces is correlated with socioeconomic conditions and occurs more frequently in areas with poor sanitation and poor personal hygiene. However, even in developed countries, the use of animal or human feces as a fertilizer may be responsible for the contamination of natural water courses, wells or water supplies. As a consequence, fruits and vegetables may be contaminated during irrigation. Flies and other arthropods may also have an important role in passive transmission.

Prepared food may also be contaminated either by use of polluted water during preparation or by food handlers who are carriers. Asymptomatic carriage is common, and carriers may be reservoirs of the parasites and may excrete infective stages, continuously or intermittently, for long periods of time. Many excreted stages (cysts, oocysts, and eggs) are very resistant and may survive in the environment for weeks or months.

Parasites which may cause human infection transmitted by water or foods contaminated with feces include all protozoa shown in **Table 1** and the helminths *Ascaris lumbricoides, Enterobius vermicularis, Trichuris trichiura, Capillaria hepatica, Hymenolepis nana, H. diminuta, Echinococcus granulosus, E. multilocularis,* and *Multiceps multiceps.*

Cryptosporidium parvum and *Cyclospora cayetanensis*, in particular, have been responsible for several large outbreaks resulting from either contaminated drinking water or contamination on food (fruits or vegetables).

Many of them are human intestinal parasites with no known animal reservoirs, including the amoebae, *Dientamoeba fragilis, Isospora belli, A. lumbricoides, En. vermicularis,* and *T. trichiura.* Therefore, their presence implies contamination with human feces. Furthermore, the presence of human commensal or nonpathogenic amoebae also implies contamination

with human feces and reflects the possibility of transmission of pathogenic species.

The most significant helminths causing human infection by ingestion of foods containing viable asexual infective stages are shown in **Table 3**. *Toxoplasma gondii* and *Sarcocystis* spp. are protozoa which may be transmitted to humans by ingestion of infected meat.

Ingestion of raw or improperly cooked meat, as well as sausages, dried, cured, and smoked meat from animals serving as intermediate hosts, is a frequent cause of infection. Examples of parasites transmitted by beef include *To. gondii, Sarcocystis* spp., and *Taenia saginata.* Pork is implicated in the transmission of *To. gondii, Sarcocystis* spp., *Taenia solium,* and *Trichinella spiralis.* Other livestock raw meat may transmit *To. gondii.* Crustaceans involved in transmission of parasites include crabs (*Paragonimus westermani, Angiostrongylus* spp.), prawns and shrimps (*Anisakis* spp., *Angiostrongylus* spp.). (*See* **Shellfish**: Contamination and Spoilage of Molluscs and Crustaceans.)

Uncooked aquatic vegetables, where infective metacercariae encyst, such as water chestnuts and watercress may be responsible for human infection with the trematodes *Fasciolopsis buski* and *Fasciola hepatica*, respectively.

Fate on Processing/Storage

Protozoal cysts or oocysts and helminth eggs are moderately or highly resistant to chlorination at concentrations used to disinfect water, e.g., *Cryptosporidium* spp., *Giardia intestinalis, A. lumbricoides,* and *T. trichiuria.* They are, however, generally susceptible to heat and to freezing at $-20\,°C$.

Helminth-encysted metacercariae or infective larvae in meat, fish, or crustaceans are generally killed by heating and by prolonged freezing at $-20\,°C$. However, they often resist smoking, curing, or pickling, e.g., *Tr. spiralis.*

Table 3 Helminths transmitted by foods containing asexual infective stages

Food	Nematodes	Trematodes	Cestodes
Meat	Trichinella spiralis		Taenia saginata
			Taenia solium
Fish	Anisakis spp.	Clonorchis sinensis	Diphyllobothrium latum
	Capillaria philippinensis	Heterophyes heterophyes	Spirometra spp.
		Metagonimus yokogawai	
		Opistorchis viverrini	
		Paragonimus westermani	
Crustaceans	Angiostrongylus spp.	Paragonimus westermani	
	Anisakis spp.		
Vegetables		Fasciola hepatica	
		Fasciolopsis buski	

Detection in Foods and Water

Incrimination of foods and water in the transmission of parasites is almost always indirect and based on epidemiological association. Direct isolation of these parasites from foods or water is very difficult. Water-filtration techniques for detection of cysts, oocysts, or eggs may be used, but too often with poor results. Enrichment culture media are not currently available in microbiology laboratories. Thus, it is not possible to recover small numbers of parasites from foods or water by in-vitro culture methods. Furthermore, the viability and infectivity of a parasitic stage cannot be assessed.

Direct incrimination of food in human infection is obvious when ingestion of an infected food is the only possible way for the infection to occur. All parasites in Table 3 belong to this group. Encysted larvae of Tr. spiralis, Ta. solium, and Ta. saginata may be detected in raw meat after specific procedures, such as enzymatic digestion. A similar approach to raw fish may have little practical value. Other helminths, e.g., E. granulosus and F. hepatica, may be detected by the examination of viscera from livestock animals.

Detection in Humans

The diagnosis of a parasitic infection can be made directly by the finding and identification of the parasite, or indirectly by immunological methods detecting specific antibodies. DNA probes, other molecular techniques, and immunological methods using monoclonal antibodies for the detection of specific parasitic antigens in serum and in body tissues have recently been described and are promising powerful diagnostic tools for the future.

Intestinal parasites are currently diagnosed by morphological identification of trophozoites, cysts, oocysts, eggs, or adult worms in feces, whereas tissue parasites are generally diagnosed by immunological methods.

Microscopic examination of fecal specimens by direct wet mount, wet mount after concentration and/or permanent stains, is the most important method for the diagnosis of intestinal or biliary parasites. As excretion is variable, at least three specimens should be examined, the specimens being collected at 2- or 3-day intervals.

Tissue parasites may occasionally be diagnosed by direct detection and identification in biopsy material from lesions. History of traveling into endemic areas combined with appropriate clinical symptoms and radiological, ultrasound, or computed tomography examinations may be very important diagnostic clues. However, detection of rising titers of specific antibodies is the most useful way to establish the diagnosis. Current immunological methods include complement fixation indirect immunofluorescence, indirect hemmaglutination, enzyme-linked immunosorbent assay, immunoelectrophoresis, and double-diffusion tests. In immunocompromised patients, these tests are less satisfactory for the diagnosis of infection. Recently, several easy-to-use antigen detection tests have become available and are of particular value in the diagnosis of Giardia intestinalis, Cryptosporidium parvum, and Entamoeba histolytica. (See **Immunoassays**: Principles; Radioimmunoassay and Enzyme Immunoassay.)

Specific Examples

Cryptosporidium spp.

Cryptosporidium spp. have a worldwide distribution and are responsible for enteric infection in humans and many animals, especially cattle and sheep. Infection is acquired by ingestion of water or food containing oocysts, which are excreted in an infective form that does not need maturation outside the gut. Transmission can occur from animals to humans (zoonotic) as well as from humans to humans (nonzoonotic).

Oocysts are highly resistant to chlorination but are sensitive to heat and prolonged freezing. Diagnosis is by microscopic detection of oocysts in fecal specimens stained by the acid-fast technique or immunofluorescence, as well as by antigen detection assays.

Giardia intestinalis

G. intestinalis has a worldwide distribution and is responsible for the most frequently reported human protozoal infection. Giardias is an intestinal infection associated with poor socioeconomic conditions. Infection is acquired by ingestion of food or water contaminated with feces. Cysts are the infective stage excreted in feces of humans and other animals. Asymptomatic carriage and excretion are frequent. Cysts can be found in water, sewage, vegetables, fruits, and other food. They resist chlorination but are killed by heat and by prolonged freezing. Diagnosis is by microscopic identification of cysts in fecal specimens. Trophozoites may sometimes be detected in feces from patients with diarrhea.

Toxoplasma gondii

Toxoplasmosis is a worldwide zoonosis. Prevalence of human infection increases with age and is generally 30–70%. Infection is acquired by ingestion of infective oocysts from cat fecal contamination or by ingestion of improperly cooked meat from animals serving as intermediate hosts, e.g., pork, mutton, and beef. Oocysts need a period of maturation outside the cat bowel. Cysts in meat are killed by heating, by smoking or curing, and by prolonged freezing at $-20\,^{\circ}C$. Diagnosis is by demonstration of specific immunoglobulin (IgM) antibodies or increasing IgG antibody titers in blood by immunological tests. These tests are not satisfactory in immunocompromised patients with latent or reactivated infections.

Ascaris lumbricoides

A. lumbricoides has a worldwide distribution but is more prevalent in areas of poor sanitation, with 25% of the world's population estimated to be infected. It is the most common helminthic parasite causing human infection. No animal reservoir is known. Infection is acquired by ingestion of water or food contaminated with embryonated eggs from human feces. A. lumbricoides eggs need a maturation period outside the gut and can remain infectious for months. Diagnosis is by microscopic detection of eggs in fecal specimens. Occasionally, adult worms are eliminated in feces.

Taenia solium

Ta. solium is prevalent in Africa, South-east Asia, Central and South America, eastern Europe, Spain, and Portugal. The human is the definitive host and the pig the intermediate host. Human intestinal infection is acquired by ingestion of undercooked pork containing a larval stage called the cysticercus.

Humans can also harbor the larval stage of Ta. solium in tissues. This potentially severe disease, called cysticercosis, occurs after ingestion of water or foods contaminated with Ta. solium eggs (from human feces) or by autoinfection when eggs from adult worms hatch in the intestine, producing a larval form that penetrates the intestinal wall and enters the circulation to reach structures such as muscle, brain, lungs, and eyes. Diagnosis is by macroscopic detection of segments of the adult worm (proglottids) or by microscopic detection of eggs in fecal specimens. Cysticercosis may be diagnosed by radiological or scanning examinations and by detection of rising titers of specific antibodies. Sometimes, identification of cysts after surgery may be the only means of definitive diagnosis.

Trichinella spiralis

Trichinosis, the infection due to Tr. spiralis, is a disease of carnivorous animals which accidentally occurs in humans. It has a worldwide distribution but is more prevalent in temperate regions. Human infection is acquired by ingestion of raw or poorly cooked pork or pork products containing encysted larvae. Laboratory diagnosis is confirmed by the detection of spiral larvae in biopsy muscle tissue from the patient, sometimes after trypsin digestion of muscle fibers. Demonstration of rising titers of specific antibodies is also available for diagnosis.

See also: **HIV Disease and Nutrition**; **Immunoassays**: Principles; Radioimmunoassay and Enzyme Immunoassay; **Shellfish**: Contamination and Spoilage of Molluscs and Crustaceans

Further Reading

Casemore DP (1990) Foodborne protozoal infection. *Lancet* 336: 1427–1432.

Gamble HR and Murrell KD (1998) Detection of parasites in food. *Parasitology* 117 (supplement): S97–S111.

Herwaldt BL (2000) *Cyclospora cayetanensis*: a review, focusing on the outbreaks of cyclosporiasis in the 1990s. *Clinical Infectious Diseases* 31(4): 1040–1057.

Smith HV (1998) Detection of parasites in the environment. *Parasitology* 117 (supplement): S113–S141.

Todd EC (1997) Epidemiology of foodborne diseases: a worldwide review. *World Health Statistics Quarterly* 50(1–2): 30–50.

Illness and Treatment

J Botas (formerly of) **and J Melo-Cristino**, Hospital de Santo Maria, Lisboa, Portugal

This article is reproduced from *Encyclopaedia of Food Science, Food Technology and Nutrition*, Copyright 1993, Academic Press.

Protozoal Infections

Illness

Protozoa are a major cause of morbidity and mortality worldwide, either in developing or in western countries. Some of those able to infect humans by the oral route have long been known to cause disease, but others are only now being increasingly recognized as pathogenic in immunocompromised hosts, especially in patients with acquired immune deficiency syndrome (AIDS). They are very diverse in epidemiology, symptomatology, treatment, and preventive measures.

In immunocompetent hosts most infections are asymptomatic, but overt local or systemic symptoms may be present, which may be severe in some cases. However, protozoal infections are particularly troublesome, sometimes life-threatening, in immunocompromised patients.

Acute, self-limited diarrhea is the most common symptom caused by protozoa listed in **Table 1**, but sometimes more intense manifestations do occur, including abdominal cramps, flatulence, anorexia, vomiting, malabsorption, weight loss, and fever. In AIDS patients, insuperable chronic watery diarrhea with malabsorption and, eventually, wasting syndrome may be due to some of these parasites.

Blood, pus, and mucus in stools are the result of invasion of the intestinal wall and common only in intestinal infection by *Entamoeba histolytica*, which is also able to cause systemic infection, especially liver abscesses, manifested by fever, weight loss, abdominal pain, and hepatomegaly.

Humans may be either intermediate or definitive hosts for *Sarcocystis* spp. and, while intestinal infection is thought to be asymptomatic, parasites in muscle may cause swelling and pain. Microsporidia have only recently been recognized as human pathogens in AIDS and very few cases have been reported in non-AIDS patients. *Pleistophora* spp. has been found in muscle biopsy, *Encephalitozoon* spp. in liver and the central nervous system and *Nosema* spp. in disseminated infection. (*See* **HIV Disease and Nutrition.**)

Treatment

Susceptibility of protozoa to chemotherapy correlates roughly with metabolism and, accordingly, with species. Several drugs are active against different protozoa. For some of them, e.g., diloxanide furoate, the exact mechanism of action is still uncertain. Others, e.g., metronidazole, are broad-spectrum antibiotics, interfering with the DNA of susceptible infectious agents.

Some antibiotics are active directly against protozoa, but others exert their effect on enteric bacteria necessary for proliferation of protozoa, e.g., *Entamoeba histolytica*. Most of these drugs are contraindicated in pregnancy and treatment of pregnant women often must be delayed until after delivery.

Metronidazole is effective against most flagellates and amoebae and is an alternative choice in the treatment of *Balantidium coli*. Other nitroimidazoles (tinidazole and ornidazole) have similar activity and less untoward side-effects, including headache, metallic taste, nausea, vomiting, and diarrhea.

Emetine and dehydroemetine have considerable side-effects, gastrointestinal and systemic, and are

Table 1 Foodborne and waterborne protozoa causing human infection

	Amoebae	Flagellates	Ciliates	Coccidia	Microsporidia
Intestinal infection	*Endolimax nana*[a] *Entamoeba coli*[a] *Entamoeba hartmanii*[a] *Entamoeba histolytica* *Iodamoeba bütschlii*[a] *Blastocystis hominis*	*Chilomastix mesnili*[a] *Dientamoeba fragilis* *Enteromonas hominis*[a] *Giardia intestinalis* *Retortamonas intestinalis*[a] *Trichomonas hominis*[a] *Trichomonas tenax*[a]	*Balantidium coli*	*Cryptosporidium* spp.[b] *Isospora belli*[b] *Sarcocystis* spp.	*Enterocytozoon* spp.[c]
Systemic infection	*Entamoeba histolytica*			*Cryptosporidium* spp.[b] *Toxoplasma gondii*[b] *Sarcocystis* spp.	*Encephalitozoon* spp.[c] *Nosema* spp.[c] *Pleistophora* spp.[c]

[a]Nonpathogenic for immunocompetent humans.
[b]Infection often serious in immunocompromised patients. Infection usually with *C. parvum*: avian species of doubtful significance.
[c]Importance as human pathogens, particularly in patients with acquired immune deficiency syndrome (AIDS), has only recently been recognized.

reserved only for extraintestinal amoebiasis. They are not active against *E. histolytica* in the bowel and diloxanide furoate, tetracycline, paramomycin, or iodoquinol must be added in order to treat simultaneous intestinal infection.

Iodoquinol is also effective against *Dientamoeba fragilis* (sensitive also to paramomycin and tetracycline) and is used as an alternative drug against *Balantidium coli*. Infections due to *Giardia intestinalis* may be treated with quinacrine, metronidazole, or furazolidine. Tetracycline is the drug of choice against *B. coli* and *D. fragilis*. Iodoquinol must be used with great caution, as it may cause myelitis and optic atrophy. This is the reason why it is no longer available in most developed countries.

Treatment of protozoal diseases in AIDS patients is far from satisfactory. There is no effective specific therapy for microsporidia or *Cryptosporidium* spp., perhaps the most common agent of infection of the bowel in AIDS. *Isospora belli* and *Sarcocystis* spp. respond poorly, if at all, to antifolates, which include co-trimoxazole, sulfadiazine, and pyrimethamine.

Prevention

Despite high prevalence and easy fecal–oral transmission of protozoa, foodborne and waterborne infections, although surely underestimated, seem rare. Good sanitary conditions and personal hygiene are keystones in preventing these infections. Protozoal cysts and oocysts differ in sensitivity to adverse conditions (heat, desiccation, freezing, chlorination) and some are amazingly resistant.

In countries with poor sanitary conditions, water should be boiled and uncooked vegetables and unpeeled fruits avoided, as they may have been washed with contaminated water. Flies and other arthropods may also be vehicles for fecal contamination of food. But even when a visitor takes every possible precaution to avoid these infections, a local food handler,

even in the best hotel, may be a cyst passer not aware of the elementary rules of personal hygiene.

For some species there are several possible animal reservoirs, which lessens the effectiveness of preventive measures. Even for these species with humans as the only hosts, patients in some institutions, especially those who are mentally handicapped, are a difficult group to control. The occurrence of sexual transmission of some protozoa by oral–anal and oral–genital sex is also considered important and is difficult to control.

Helminthic Infections

Illness

Helminths are probably the most common infectious agents of humans. Two of those transmitted orally, *Ascaris lumbricoides* and *Trichuris trichiura*, are each thought to cause 1000 million infections worldwide. Humans may be, for different helminths, definitive or intermediate hosts and clinical manifestations, when present, depend on localization and worm burden. Unlike protozoa, helminths are not generally considered opportunistic pathogens in immunocompromised hosts.

Most intestinal infections caused by helminths listed in **Table 2** are asymptomatic. Gastrointestinal symptoms are related to the number of worms in the bowel. Diarrhea and abdominal pain are the most common complaints, but malabsorption and weight loss may occur and, if untreated, infection may even be fatal, e.g., *Capillaria philippinensis*. *Enterobius vermicularis* is frequently responsible for anal pruritus, especially nocturnal, and restless sleep in children.

Some parasites compete for absorption of specific nutrients, with symptoms resulting from their deficiency, e.g., *Diphyllobothrium latum* causes

Table 2 Foodborne and waterborne helminths causing human infection

	Nematodes	Trematodes	Cestodes
Intestinal infection	*Anisakis* spp. *Ascaris lumbricoides* *Capillaria philippinensis* *Enterobius vermicularis* *Trichuris trichiura*	*Fasciolopsis buski* *Heterophyes heterophyes* *Metagonimus yokogawai*	*Diphyllobothrium* spp. *Dipylidium caninum* *Hymenolepis diminuta* *Hymenolepis nana* *Taenia saginata* *Taenia solium*
Systemic infection	*Angiostrongylus* spp. *Bayliascaris procyonis* *Capillaria hepatica* *Dracunculus medinensis* *Toxocara* spp. *Trichinella spiralis*	*Clonorchis sinensis* *Fasciola hepatica* *Opistorchis* spp. *Paragonimus westermani*	*Echinococcus granulosus* *Echinococcus multilocularis* *Hymenolepis nana* *Multiceps multiceps* *Taenia solium* *Spirometra* spp.

deficiency of vitamin B_{12} and consequently megaloblastic anemia and neurological symptoms. (*See* **Cobalamins:** Physiology.)

Systemic infections, although sometimes asymptomatic, are more commonly associated with symptoms which depend on localization of the parasite. Liver and biliary tract are targets for some helminths, e.g., *Capillaria hepatica*, *Clonorchis sinensis*, *Fasciola hepatica*, and *Opistorchis* spp., and symptoms of hepatitis and cholangitis may occur. Furthermore, increased incidence of cholangiocarcinoma seems to be associated with these infections.

Spirometra spp. and *Dracunculus medinensis* are present in subcutaneous tissue, causing pain and swelling or chronic ulcers with protrusion of worms. Lung is infected by *Paragonimus westermani*, with resulting bronchopulmonary symptomatology, including cough, hemoptysis, bronchitis, or lung abscesses.

Infections by *Angiostrongylus* spp., *Bayliascaris procyonis*, or *Multiceps multiceps*, when symptomatic, cause central nervous system disease, especially seizures. Visceral larva migrans syndrome (fever, hepatomegaly, and eosinophilia) is caused principally by *Toxocara* spp., but may also be due to *Angiostrongylus* spp., *Anisakis* spp., and *Capillaria* spp.

Larvae of *Trichinella spiralis* may be responsible for fever, edema, myositis and, rarely, encephalitis, pneumonia, or myocarditis. Larvae may encyst in different tissues, resulting in space-occupying lesions when they are big enough. Cysts of *Echinococcus granulosus* are found more frequently in liver and lung and are often multiple, but cysts of *E. multilocularis* have an unlimited germinal membrane and spread either locally or to distant sites like neoplastic metastases.

Treatment

Susceptibility of helminths to chemotherapeutic agents correlates with species even more closely than is the case with protozoa. Some drugs are preferred because they act in the bowel, expelling adult worms, while others must be absorbed and act systemically, in order to kill parasites in tissues. For most of them, the exact biochemical mechanism of action is still unknown. Some cause spastic paralysis of adult worms, e.g., pyrantel pamoate, while others cause flaccid paralysis, e.g., piperazine, with resulting expulsion of worms by peristalsis.

Mebendazole is the drug of choice for intestinal nematodes and replaced thiabendazole, which is much more toxic. In higher doses it is also used in the treatment of infections by *T. spiralis*, *E. granulosus* and *E. multilocularis*. Liver and bone marrow toxicity are not uncommon with high doses.

Other benzimidazole drugs (albendazole and flubendazole) are new alternatives which are perhaps less toxic. Pyrantel pamoate, pyrvinium pamoate and piperazine offer no pharmacological advantages over mebendazole. Mebendazole is also the drug of choice in systemic infections by nematodes, but is much less effective. Removal of worms is the best treatment for *Anisakis* spp. and *D. medinensis*.

Praziquantel is effective against most trematodes and cestodes. It is the drug of choice for either intestinal or systemic infections due to these parasites and failures have only been reported in *F. hepatica* (bithionol is the preferred drug for fascioliasis). Untoward side-effects, including nausea, abdominal pain, and headache, although not uncommon, are mild and transient.

Niclosamide is equally effective against intestinal cestodes, is less expensive, and has fewer untoward side-effects than praziquantel. Although imidazoles might be tried, surgery is still the best treatment for *M. multiceps* and *Spirometra* spp. and should be considered in echinococcosis, in which it is the only treatment when the cyst is calcified.

Although drugs now available to treat helminthic infections are less toxic than before, some authors still prefer, particularly in intestinal infections, treating only heavy infections, or using low doses of drugs to reduce the number of parasites in order to avoid undesirable side-effects. In contrast, reinfection with *Enterobius vermicularis* is so easy that some authors prefer to treat all household contacts. As with protozoal therapy, treatment of pregnant women is better delayed until delivery.

Prevention

Foodborne and waterborne infections are common with both helminths and protozoa, although direct person-to-person, fecal–oral contamination is also responsible for transmitting some infections. Good personal hygiene and sanitary conditions are thus mandatory for prevention and control of these infections. Flies and other arthropods may transmit eggs from feces to food and water. Vegetables may be contaminated either by encysted larvae (aquatic vegetables) or after contact with contaminated water.

Veterinary inspection of meat is very important in parasites which are infectious by macroscopic cysts or larvae. Raw meat or fish may transmit several helminths and should be avoided. Larvae and eggs may be extremely resistant to adverse conditions and remain viable for prolonged periods of time in salted, smoked, or undercooked meat and fish, as well as in chlorinated water. Snails, crayfish, crabs, and prawns may harbor larvae and so should be properly cooked. (*See* **Fish:** Spoilage of Seafood; **Shellfish:**

Contamination and Spoilage of Molluscs and Crustaceans.)

Specific Examples

Cryptosporidium spp.

Cryptosporidium spp. were recently recognized as an important cause of diarrhea in humans. Infection is frequently symptomatic and in immunocompetent hosts symptoms include mild acute self-limited watery diarrhea, sometimes with abdominal cramps, flatulence, nausea, anorexia, and weight loss. In immunocompromised patients, especially AIDS, chronic diarrhea and profound weight loss may occur and are an important cause of morbidity and mortality.

Spiramicin has been used for treatment with doubtful efficacy; no effective therapy is available. Nonspecific symptomatic therapy is important to avoid fluid and electrolyte imbalance, as in acute diarrhea of other etiology. Prevention may be difficult, as oocysts are extremely resistant and fully infectious when passed in stools. Good sanitary conditions and personal hygiene are of utmost importance. Precautions with stools in hospitals and other institutions should be taken in order to avoid outbreaks.

Giardia intestinalis

Giardiasis is now the most commonly reported protozoal infection worldwide. About 20–50% of infected people present symptoms, usually self-limited acute diarrhea, but chronic diarrhea with malabsorption may follow. Treatment with quinacrine has a high degree of efficacy, but the drug is poorly tolerated (vomiting, abdominal pain). Metronidazole and other nitroimidazoles are equally effective and better tolerated. Furazolidine has no advantages over quinacrine, either in efficacy or in untoward gastrointestinal side-effects. Good water control is the most important way of prevention. When necessary, water should be filtered or boiled. Chemical treatment is of doubtful efficacy.

Toxoplasma gondii

Toxoplasmosis in immunocompetent people is generally asymptomatic or may resemble an infectious mononucleosis. Tissue cysts, however, remain viable for life. In immunocompromised patients, such as AIDS patients, transplant recipients and patients submitted to immunosuppressive therapy, either acute or reactivation of latent infection, may be responsible for severe or fatal toxoplasmosis, mostly of the central nervous system. Congenital toxoplasmosis may present several clinical manifestations, some severe

(encephalitis, retinitis), depending on the time of pregnancy when infection was acquired.

The combination of pyrimethamine plus sulfadiazine is the most effective therapy for toxoplasmosis and has been the choice in immunocompromised patients. To avoid relapses, it must be maintained for life in AIDS patients and as long as immunodeficiency persists in other cases. Bone marrow toxicity and skin rash are frequent with this combination and may cause replacement of sulfadiazine by clindamycin. Acute toxoplasmic infection in immunocompetent hosts is only to be treated when symptoms are severe or persistent.

Infection in pregnant women, either with or without symptoms, must be treated. Spiramycin can be used throughout pregnancy and after the first trimester, pyrimethamine plus sulfadiazine can be used. None of the available drugs is active on cysts, but only on vegetative forms. Oocysts are present in cat feces in large numbers and so contacts must be avoided by seronegative pregnant women and immunodeficient patients. In meat, cysts do not resist prolonged freezing ($-20\,^{\circ}$C), heating ($66\,^{\circ}$C), smoking, or curing.

Ascaris lumbricoides

Clinical manifestations of *A. lumbricoides* infection are diverse, being related to worm burden. Bronchopulmonary complaints may be present, due to migration of larvae through the lungs. If infection is slight, adult worms cause no symptoms, but heavy infections may be responsible for malabsorption or even intestinal obstruction. Other rare manifestations may be due to abnormal worm migration. Mebendazole is the treatment of choice. Pyrantel pamoate or piperazine are effective alternatives. Good sanitary conditions are keystones in prevention. Infection is also avoided by hand washing after contact with soil.

Taenia solium

Human intestinal infection by *Taenia solium* is generally asymptomatic. Symptoms due to cysticercosis depend on organ involvement, the central nervous system being the system most commonly affected (by seizures), followed by orbit, muscle, liver, and lungs. Niclosamide is widely used in the treatment of intestinal taeniasis. It is effective and has few untoward side-effects, but is not active against larvae. Praziquantel is highly active against adults and larvae, and should be preferred. Calcified cysts will not respond to drugs and may justify surgery. Symptomatic treatment of seizures may also be necessary. The most effective prevention is by good sanitary conditions, avoiding contamination of animal and human food and water by human feces. Veterinary inspection of meat decreases the risk of ingestion of cysticercus, as does thorough cooking of pork.

Trichinella spiralis

Most intestinal and tissue infections by *Trichinella spiralis* are subclinical, symptoms depending mostly on the number of viable larvae and their localization. Diarrhea is the most common symptom due to adult worms in the gut, but abdominal discomfort and vomiting may also occur. Systemic invasion by larvae is more often symptomatic, with fever, myalgia, periorbital edema, and headache. Occasionally, myocarditis, pneumonitis, or encephalitis may occur. Mebendazole and thiabendazole are the drugs of choice for treatment, along with salicylates and bed rest. The most effective method of prevention is thorough cooking of pork.

See also: **Cobalamins**: Physiology; **Fish**: Spoilage of Seafood; **HIV Disease and Nutrition**; **Shellfish**: Contamination and Spoilage of Molluscs and Crustaceans

Further Reading

Casemore DP (1990) Foodborne protozoal infection. *Lancet* 336: 1427–1432.

Mandel GL, Douglas RG Jr and Bennett JE (eds) (1990) *Principles and Practice of Infectious Diseases*, 3rd edn. New York: Churchill Livingstone.

Mehlhorn H (ed.) (1988) *Parasitology in Focus*. Berlin: Springer-Verlag.

Strickland GT (2000) *Hunter's Textbook of Tropical Medicine*. Philadelphia: WB Saunders & Co.

Thompson JH Jr (1990) Parasitology. In: Murray PR, Drew WL, Kobayashi GS and Thompson JH Jr (eds) *Medical Microbiology*, pp. 349–436. London: Wolfe.

PARENTERAL NUTRITION

C C Ashley and L Howard, Albany Medical College, Albany, NY, USA

Definition and History

The development of central venous catheters in the 1960s provided for the safe infusion of hypertonic nutritive solutions. As placement of these catheters became more widespread, so did the use of parenteral nutrition (PN). In the early 1970s, this also became an available therapy for patients with chronic intestinal insufficiency living at home. Today, a variety of vascular access options are available to patients, and portable pumps can be used to infuse outside of the home/hospital setting. PN is sometimes called total parenteral nutrition (TPN), but this is an overstatement because several hard-to-solubilize but important nutrients are left out. Another term, 'hyperalimentation,' has also gone out of vogue, as we learn the downside of overproviding substrate, especially in a stressed patient.

Indications for Therapy

Wide availability does not necessarily imply broad applicability. Only recently has the evidence for and against the use of parenteral nutrition in a variety of clinical settings become better defined. Enteral routes of nutritional support are generally cheaper, safer, and equally efficacious. PN is indicated in those patients where delivery of nourishment via the gut is either unsafe or not possible. Candidates for PN should be carefully selected because the therapy is costly, cumbersome, and risky. A systematic approach to choosing candidates for PN is detailed in **Figure 1**.

The first step in the process requires the clinician to form a clinical judgement about whether the severity and nature of the patient's illness will lead to nutritional impairment. Initial considerations are whether a lengthy course of chemotherapy will provoke nausea and reduce oral intake to negligible levels; whether severe intestinal ileus will make it impossible to use the gut; and whether a planned surgery is likely to significantly reduce the patient's absorptive surface. The second step is to assess whether the patient is at increased risk for malnutrition because of limited nutritional reserves. The physician must determine whether the patient is an alcoholic. Are they already on catabolic drugs such as steroids? Are they vulnerable by virtue of age or chronic illness such as diabetes? The third step is to determine whether the patient is already malnourished. Loss of more than 20% of lean body mass, serum albumin depression signs of vitamin or mineral deficiency independently indicate a high-risk nutritional status. The fourth step is the most critical. The physician must decide whether the quality of life and prognosis can be positively effected by the provision of specialized nutritional support and therefore whether this therapy should be recommended to the patient. The

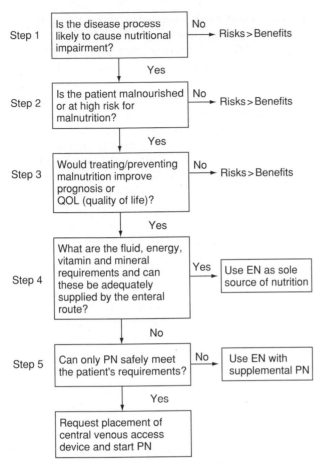

Figure 1 Systematic approach to assessing candidacy for total parenteral nutrition. Adapted from Howard L (1998) Parenteral and enteral nutrition. In: Fauci AS, Braunwald E, Isselbacher KJ and Wilson JD (eds) *Harrisons Principles of Internal Medicine*, 14th edn., chapter 78, figure 78-1, p. 473. New York: McGraw-Hill, with permission.

Table 1 Various indications for parenteral nutrition

Obstruction
   Anatomical (congenital malformation, strictures, adhesions)
   Functional (prolonged postoperative ileus, intestinal dysmotility)
Hypermetabolic state/wasting syndromes
   Critical illness
      Multisystem organ failure and bowel ischemia
      Severe pancreatitis
      Burns
   Cancer
      Bone-marrow transplantation
   HIV
Malabsorption/mucosal disease
   Insufficient surface area from short bowel syndrome
   Proximal enteric fistula
   Severe inflammatory bowel disease
   Severe chemotherapy related mucositis
   Radiation enteritis

remaining steps are concerned with choosing the route of delivery for the indicated specialized nutritional support. A summary of common clinical scenarios, where provision of PN is considered, is shown in **Table 1**.

Evidence Base for PN

Insufficient evidence exists to assess the utility of PN in many specific clinical settings, but there are several broad areas where prospective randomized controlled studies support the use of PN.

Perioperative

Strong evidence exists to support the use of PN in the perioperative setting when the patient has severe protein–calorie malnutrition. The benefit is a modest (several studies found 10%) reduction in postoperative complications.

Critical Illness

In critical illness, there is support for the use of PN in patients with severe protein–calorie malnutrition using modest calories. Complication rates are lower when lipids are not infused as a part of this specialized nutritional support. Excessive exogenous carbohydrate enhances hypermetabolism and lipogenesis, leading to increased carbon dioxide production; this makes ventilatory weaning more difficult.

Wasting Syndromes

Cancer   Like noncancer patients, cancer patients suffering from severe protein–calorie malnutrition benefit from perioperative PN when undergoing cancer surgery. There are some data to suggest that bone-marrow-transplant patients benefit from PN supplemented with glutamine and that PN during the cytoreductive phase enhances long-term survival.

HIV   HIV patients appear to benefit from specialized nutrition support if their wasting reflects poor intake or malabsorption; they do not benefit if their weight loss is principally cytokine-related wasting. The route of nutrition support is more controversial. In one well-controlled study, PN was not superior to enteral nutrition in repleting lean body mass, but it did add more weight as water and fat. One small study showed a longer survival in PN HIV patients compared to those who received diet counseling alone.

Chronic Renal Failure

PN improves survival in acute renal failure. A solution using just essential amino acids is not superior to the usual essential–nonessential combination. In chronic renal failure requiring dialysis, 15% of patients become significantly malnourished. PN may be indicated if there is severe bowel dysmotility associated with autonomic neuropathy, as may occur with diabetes. Routine use of intradialytic PN has not been shown by randomized studies to clearly alter survival.

Pancreatitis

In pancreatitis, PN can decrease mortality rate and reduce rates of infectious complications when it is used early in the management of the most severe cases.

Short Bowel Syndrome

The prognosis of acute short bowel syndrome has been dramatically improved by PN, so much so that a randomized study of PN vs. enteral nutrition (EN) would be unethical. In the adaptive phase, an open study purported to show that supplementation with growth hormone and glutamine improved bowel adaption, but this was not confirmed in a more rigorous controlled trial.

Inflammatory Bowel Disease

In inflammatory bowel disease, Crohn's disease, or ulcerative colitis, specialized nutritional support (SNS) is often necessary for severe disease associated with protein calorie malnutrition, but there are no clear data that show any advantages of PN over EN. SNS alone is not superior to steroids in reducing remission or preventing relapse; however, it may be the preferred approach in a growing child.

Technical Details

Route

Peripheral PN PN can be delivered via peripheral intravenous access, but this route is appropriate only for formulations of modest osmolarity (e.g., less than 900 mOsm or 10 g of dextrose per 100 ml), since small veins quickly sclerose and become useless for further infusion. Peripheral PN is best used in a supportive role for patients with inadequate oral intake. Patients who are unable to meet their fluid and electrolyte needs can be supported in this way without exposure to the complications of central venous access. However, peripheral PN is not a long-term solution.

Central PN Central access is a prerequisite when infusing formulations of PN that are more complex and more hyperosmolar. The catheter used to deliver PN should be a dedicated line used for nothing else in order to minimize the risk of infection. If additional intravenous therapy is needed, additional access sites should be arranged. Sometimes, multilumen catheters are used, but studies have shown higher infection rates in multilumen catheters versus single lumen catheters.

Conventional central venous line In the setting of an acute requirement for parenteral therapy or as a bridge to the placement of a longer-lasting device, an externalized central venous line (CVL) inserted by the conventional Seldinger technique in the subclavian or internal jugular vein is acceptable for the delivery of PN. This catheter may be placed at the bedside as long as full aseptic precautions are taken. This can be a considerable advantage over other devices that may take several days to arrange.

Peripherally inserted central venous catheter The peripherally inserted central venous catheter (PICC) can be placed at the bedside in the antecubital fossa by specially trained personnel. PICCs placed at the antecubital fossa are subject to mechanical stress from flexion and extension of the elbow. They are a short-term option. For long-term use, especially in patients going home on PN, PICC lines are best placed to the upper arm (brachial vein) under ultrasound and fluoroscopic guidance. Such lines are not subject to the same level of mechanical stress. One disadvantage common to all PICCs is that they are more difficult for home patients to care for without nursing support, because the patient has only one free hand.

Tunneled catheter The chief advantage of tunneled catheters is that an exit site low on the chest wall permits a patient at home to see their catheter site and manipulate the catheter with both hands. This considerably simplifies the act of catheter access and improves patient independence. Tunneled catheters have a dacron cuff just under the skin which aids catheter fixation and prevents bacteria from accessing the blood stream through the tunnel. Patients with an externalized catheter are usually advised to cover their access devices with an adhesive/occlusive dressing before showering or bathing in order to keep it dry. Swimming is not recommended.

Subcutaneous port Subcutaneous ports are often preferred by long-term patients, who have active lifestyles and want to swim. Subcutaneous ports are less obtrusive and therefore cause less disturbance to body image. The disadvantage of these catheters is that they require a needle stick for accessing, but this uncomfortable procedure is made easier by the use of topical anesthetics. Also, it is rarely possible to clear an infected part and hence removal is necessary. For this reason they are not wise in patients with a history of frequent line sepsis.

Infusion Method

Continuous   Once a route is decided upon and available, a method of delivery must be selected. Most patients begin PN as a continuous infusion. Continuous infusion is the least complicated method of delivering PN, since a metabolic steady state is quickly achieved and maintained. This is the method of choice in patients who are critically ill, hemodynamically compromised, have marginal renal function, or are brittle diabetics. However, continuous infusions are not optimal from the perspective of most patients. Long-term patients frequently seek some freedom from their parenteral infusion. If a long-term continuous infusion is necessary, portable pumps are available that allow patients to perform most activities of daily living without needing to disconnect.

Cycled   TPN infusions can usually be compressed to a period between 10 and 12 h, depending upon the volume and calories needing to be infused. Most home patients opt to infuse during night time hours. This leaves the patient free for a period of between 12 and 14 h a day to be more active. When the infusion is cycled on and off daily, the risk for dehydration rises especially toward the end of the infusion-free period and in patients with significant gastrointestinal (GI) losses. The risk of both hyper- and hypoglycemia is increased at the start and conclusion of the daily infusion. Swings in serum glucose levels demonstrate the impact of carbohydrate delivered directly to the systemic rather than the portal circulation.

Intermittent   In patients requiring less PN support, it may be possible to provide PN every second or third day. The complication of such a regimen is making sure that mineral and vitamin requirements are met.

Establishing a Nutrient Formula

Once the need for parenteral support is established and the route and method of delivery are decided on, the formulation can be determined. The first step is to estimate the total nutritional requirement of the individual patient. The requirement of an individual patient may vary from day to day, depending upon abnormal GI losses. Over time, bowel adaption occurs, so fluid and nutrient requirements may significantly decrease.

Fluid requirements
Basal   The normal daily fluid requirement is 25–35 ml kg^{-1} (for adults) and 120 ml kg^{-1} (for infants).

In a 70-kg man, normal losses are 1600 ml of urine daily, 800 ml of insensible losses, 100 ml of stool losses and 350 ml per degree Celsius per day of fever.

Extraordinary fluid losses   Fluid losses can vary considerably in patients who require TPN, often because of excessive loss from the GI tract. In the example shown in **Table 2**, the patient requires twice the usual volume of fluid because of extraordinary losses through his jejunostomy. All fluid losses must be meticulously reviewed in order to ensure adequate replacement. Commonly excessive losses from the GI tract will precipitate a greater oral intake.

Calorie requirement   The total amount of calories required is determined by weight and the patient's level of stress. Between 20 and 40 kcal kg^{-1} day^{-1} is usually provided to adults and 80–110 kcal kg^{-1} day^{-1} to infants. In patients who are simply starved but not stressed, 35–40 kcal kg^{-1} day^{-1} (100–110 kcal kg^{-1} day^{-1} for infants) can be effectively utilized. In patients who are moderately stressed and malnourished, slightly lower amounts are provided, e.g., 30–35 kcal kg^{-1} (80–100 kcal kg^{-1} day^{-1} in infants). In patients who are severely stressed, calories are kept modest (< 25 kcal kg^{-1}), since the stress induces a catabolic state that prevents efficient utilization of high levels of calories. Excessive calories precipitate hepatic steatosis and retention of carbon dioxide.

Carbohydrate content   Glucose provides 3.4 kcal g^{-1} because it is glucose monohydrate. The carbohydrate content of the formulation varies from 4 to 7 g kg^{-1} day^{-1} and usually accounts for 50–60% of total calories.

Table 2   Daily fluid requirements

Normal man (70 kg)	
35 ml day^{-1} × 70 kg = 2.5 l	
In: 1.2 l of fluid	Out: 1.6 l of urine
1.0 l of water in food	0.8 l insensible losses
0.3 l of water from metabolism of food	0.1 l stool
Same patient status post high jejunostomy	
In: 2.2 l of fluid	Out: 1.6 l of urine
1.0 l of water in food	0.8 l of insensible losses
0.3 l of water from metabolism of food	4.1 l of jejunostomy losses
3.0 l TPN	

The patient needs to have enough parenteral fluid to maintain an adequate urine output. Short-bowel patients have an increased risk of kidney stones. Adapted from Howard L (1998) Parenteral and enteral nutrition. In: Fauci AS, Braunwald E, Isselbacher KJ and Wilson JD (eds) *Harrisons Principles of Internal Medicine*, 14th edn., chapter 78, figure 78-4, p. 476. New York: McGraw-Hill.

coagulopathy. Most catheter leaks can be addressed by clamping or crimping the catheter proximal to the leak and then having the catheter repaired. Subcutaneous bleeding around a port usually responds to pressure.

Air embolism Air embolism is a feared complication of central venous access. Patients need to be reassured that dangerous volumes of air are large and rarely happen through an access device. If concern for this complication is real, and the amount of air delivered is unknown, patients crimp off their catheter and lie flat on their left side so that air is trapped in the right auricle or ventricle and does not pass on into the pulmonary artery and the lungs where the 'bubble' will disrupt gas exchange. Over 20 min, even large trapped bubbles will dissolve.

Phlebitis Phlebitis is a relatively common complication with peripherally inserted access devices. Inflammation of the small vessel, which is initially entered, can result in local pain, swelling, erythema, and heat. There may or may not be associated cellulitis. The internal surface of the catheter usually remains sterile. If conservative local therapies are insufficient to provide relief, the line should be removed, since phlebitis can progress to thrombosis and more serious complications.

Migration After initial placement, catheters can migrate in the wrong directions. For example, a line placed via the subclavian vein can end up in the internal jugular. The peripherally inserted central catheters can propagate into the heart from the superior vena cava or right atrium. The resulting endocardial irritation can result in palpitations, skipped beats, and, rarely, more serious arrhythmia. When fluoroscopy is available at the time of the placement, these complications are easily correctable. It is useful to note at the time of insertion just how much of a CVL or PICC remains externalized.

Crimping and breakage If infusion becomes difficult, and the pump signals blockage, the first consideration should be the site of central access. Crimping of the catheter is common with peripherally inserted access introduced through the veins of the antecubital fossa. Repeated flexion and extension at the elbow can introduce a kink in the catheter itself and result in breakage. The diagnosis of catheter breakage is straightforward, since it usually occurs as separation of the catheter from the external hub or as a simple breach in the external catheter wall. Rarely is a catheter sheared off internally. None the less, a central catheter should never be withdrawn

with any amount of force, as there have been incidences of catheter embolization requiring extraction under fluoroscopy.

Blockage Catheters can become blocked by thrombus, built-up fibrin and platelet deposits, crystals of infused minerals or drugs, or by lipid precipitate from intravenous fat solutions. Vigorous and frequent flushing with heparinized saline and the use of heparin-impregnated catheters has been useful in reducing the frequency of thrombotic luminal obstruction. Accumulation of fibrinous material on the outer surface of the catheter tip is relatively common and can produce a ball valve effect, allowing infusion but preventing blood withdrawal. Fluoroscopic confirmation of catheter position and clots is useful. Current treatment for thrombus or fibrin is 0.5–1 mg of tissue plasminogen activator. Crystalline material may be dissolved by 0.1 N HCL and lipid by small amounts of alcohol. Checking for blood return from access devices is discouraged because the process of drawing blood into the catheter promotes the process of fibrin and platelet precipitation and contributes to higher rates of intraluminal catheter obstruction and infection, shortening the usable life of the devices. Catheter thrombosis is frequently associated with catheter infection.

Underfeeding

Ideally, the amount of specialized nutritional support delivered to the patient is sufficient to meet the needs of the patient. Because the process of determining those needs is imperfect, it is likely that either too much or too little support will be delivered. Because the complications associated with overfeeding can be dramatic, it is now preferred to err on the side of underfeeding, especially in the patient who is under stress. In order to optimize the patient's response to specialized nutritional support, prospectively determined goals should be used to gauge the success of therapy. The clinician should assess the patient's wound healing, strength to wean from a respirator, cough effectively, and mobility. If such an evaluation is neglected, the patient may inadvertently be denied the full benefit of parenteral support.

Overfeeding

Overfeeding is the provision of one or more of the nutrient components of parenteral nutrition in an amount or at a rate that exceeds the physiologic utilization and excretion of the nutrient.

Hyperglycemia Excessive provision of carbohydrate can occur in both diabetic and nondiabetic patients and is more likely to occur in those who are critically ill. Hyperglycemia should be treated by reducing the carbohydrate calories. The use of insulin is only appropriate in patients known to be insulinopenic.

Fluid overload   Fluid overload is possible in patients who have reduced cardiac reserves and in persons with healthy hearts whose PN volume is added to an already generous regimen of crystalloid or colloid. It is essential to consider the total volume of other fluids delivered when determining a volume for PN.

Fatty liver   Delivery of carbohydrate or fat calories in excess of that which can be used by the patient creates hepatic steatosis. In the case of a patient with impaired liver function, this can occur more rapidly. Chronic hepatic steatosis leads to cholestasis and, occasionally, irreversible fibrosis and cirrhosis. This appears to occur more readily in children than adults.

Refeeding syndrome   This classic complication of starvation can occur after prolonged starvation from any cause; it happens when the nutrient supply is restored too rapidly. The underfed patient has depleted stores of adenosine triphosphate (ATP) with normal or low circulating levels of potassium, phosphorus, and magnesium. Once metabolism is restored, phosphate is drawn into cells to be coupled to adenosine monophosphate (AMP) and adenosine diphosphate (ADP). Magnesium is also drawn into cells to act as a cofactor for a number of cellular processes, for example, the conjugation of phosphate to ADP and the Na–K ATPase. Finally, potassium is drawn into cells to participate in the vigorous metabolism of newly supplied energetic compounds. This leads to depleted serum magnesium, potassium, and phosphate. This transcellular shift can have dire consequences, as a result of altered membrane conduction, resulting in arrhythmia, edema, and heart failure.

Allergic Reactions

Reactions to any component used in the delivery of PN are possible. Allergic reactions to latex are well known. Autoimmune reactions to heparin are also recognized, and even the small amount present in the lumen of an impregnated catheter is sufficient to trigger the heparin-induced thrombotic thrombocytopenia syndrome. Reactions to the actual components of the nutrient formulation are unusual.

Withdrawal of Support

Once PN is begun in a patient, when does one stop? There are three situations in which withdrawal of therapy is appropriate: first, when the patient can wean from PN to more physiologic EN; second, when desired by the patient; and third, when further treatment is futile, and the risks clearly outweigh the benefits. Assessing PN nutrition repletion and transposing to use of the GI tract is an important clinical step that often has to happen gradually. Every patient must agree to both initiation and stopping PN therapy. Occasionally, after experiencing nutritional rescue by PN, the patient or family desire continuation of therapy and are fearful of the consequences of weaning off. When continued nutritional support is felt to contribute nothing to the patient's quality of life or prognosis, it should be withdrawn like any other futile therapy. This process is ethically straightforward but emotionally charged and medicolegally complicated. The best way to avoid these difficult situations is to avoid PN in every case where its use is not clearly indicated, and the course of therapy is not prospectively defined.

See also: **Bioavailability of Nutrients**; **Body Composition**; **Burns Patients – Nutritional Management**; **Cancer**: Diet in Cancer Treatment; **Children**: Nutritional Requirements; **Dehydration**; **Diarrheal (Diarrhoeal) Diseases**; **Dietary Reference Values**; **Dietary Requirements of Adults**; **Electrolytes**: Analysis; **Enteral Nutrition**; **HIV Disease and Nutrition**; **Infants**: Nutritional Requirements; **Malnutrition**: The Problem of Malnutrition; **Minerals – Dietary Importance**; **Nutritional Assessment**: Importance of Measuring Nutritional Status; **Stress and Nutrition**; **Vitamins**: Determination; **Water**: Structures, Properties, and Determination

Further Reading

Ashley C and Howard L (2001) Evidence base for specialized nutritional support. *Nutrition Reviews* 58: 282–289.

Brooks S and Kearns P (1996) Enteral and parenteral nutrition. In: Ziegler EE and Filer LJ (eds) *Present Knowledge in Nutrition*, 7th edn. pp. 530–539. Washington, DC: ILSI Press.

Dresser RS and Boisaubin EV (1985) Ethics law and nutritional support. *Archives of Internal Medicine* 145: 122–124.

Gonzalez-Huix F, Fernandez-Banares F, Esteve-Comas M et al. (1993) Enteral versus parenteral nutrition as adjunct therapy in acute ulcerative colitis. *American Journal of Gastroenterology* 88: 227–232.

Howard L (1998) Parenteral and enteral nutrition therapy. In: Fauci AS, Braunwald E, Isselbacher KJ and Wilson JD (eds) *Harrison's Principles of Internal Medicine*, 14th edn. pp. 472–480. New York: McGraw-Hill.

Howard L, Ament M, Fleming CR et al. (1995) Current use and clinical outcome of home parenteral and enteral nutrition therapies in the United States. *Gastroenterology* 109: 355–365.

Klein S, Kinney J, Jeejeebhoy *et al.* (1997) Nutrition support in clinical practice: Review of published data and recommendations for future research directions. *Journal of Parenteral and Enteral Nutrition* 21: 133–156.

Klein S and Koretz RL (1994) Nutrition support in patients with cancer: What do the data really show? *Nutr Clin Pract* 9: 91–100.

Lipman TO (1998) Grains or veins: Is enteral nutrition really better than parenteral nutrition? A look at the evidence. *JPEN* 22: 167–182.

Semrad CE (2000) Parenteral nutrition. *Clinical Perspectives in Gastroenterology* Nov/Dec: 307–314.

Shils ME and Brown RO (1999) Parenteral Nutrition. In: Shils ME, Olson JA, Shike M and Ross AC (eds) *Modern Nutrition in Health and Disease*, 9th edn. pp. 1657–1688. Philadelphia, PA: Lippincott, Williams & Wilkins.

Viall CD (2000) Home parenteral nutrition: Finances. In: Rombeau JL and Rolandelli RH (eds) *Clinical Nutrition: Parenteral Nutrition*, 3rd edn. pp. 512–528. New York: WB Saunders.

PASSION FRUITS

D B Rodriguez-Amaya, Universidade Estadual de Campinas, Campinas, SP, Brazil

Introduction

Because of the intense, fragrant, and distinctive aroma and flavor, passion fruit has been the object of investigations and is in increasing demand worldwide. Native to Brazil, this fruit owes its name to the delicate and beautiful flowers, some features of which the early Christian missionaries to South America found symbolic of the Passion of Christ.

Species and Cultivars

Passion fruit belongs to the *Passifloraceae* family, which consists of 12 genera and over 500 species, widely distributed in tropical America, Asia, and Africa. *Passiflora*, the principal genus, has approximately 400 known species, about 50–60 of which bear edible fruits, but only a few are of any commercial importance. Many are known only in native markets in South and Central America and the West Indies. Commercial production of passion fruit is based on the purple species *Passiflora edulis* Sims and the yellow form *Passiflora edulis* f. *flavicarpa* Degener. It is still in question whether the yellow passion fruit is a mutant of the purple *P. edulis* or its hybrid with other species. These two forms are commonly called granadilla, parcha, or parchita in Spanish, maracuja in Portuguese, and lilikoi in Hawaiian.

Aside from the skin color, the purple and yellow passion fruits differ in horticultural performance and fruit properties. The purple species is more resistant to cold injury, is less acid, and is considered superior in aroma and flavor. The yellow form is faster growing, has a greater resistance to soil fungi, has more vigorous vines, bears crops over longer periods, and has a greater yield of fruit and pulp, larger fruits, and more acid juice.

P. quadrangularis L. or 'giant granadilla,' widely distributed in the tropics, is the most cultivated species after *P. edulis* in tropical America. It bears the largest fruits, which are more elongated, up to 25 cm long, and fleshy instead of hollow. The seeds are much larger, brownish, and flattened. The rind is not so hard as in the purple and yellow passion fruits. The juice content is much lower and is somewhat inferior in flavor and color. The pulp can be eaten like a melon, with or without the addition of sugar, or cooked with milk. When green, this fruit can be used as a vegetable as green papaya.

P. ligularis Juss, 'sweet granadilla' or 'water lemon,' is cultivated in the mountains of Mexico and Central America. The fruit is mostly eaten out of hand or used in drinks or icecream. Its translucent white pulp is almost a liquid, acid with sweet aroma. The peel is resistant so the fruit can be transported well, without being damaged. Colombia, where this species is cultivated in the Western Cordillera, exports this fruit to Europe.

P. mollissima, 'banana passion fruit' or 'curuba,' grows widely in the Andes and is distributed from Venezuela and Colombia to Peru and Bolivia. The flavor is more astringent and less acid than *P. edulis*. The sieved pulp is mixed with milk and sugar and served as a drink. It is also used in marmalades and desserts and for flavoring icecream

P. maliformis L. is known as 'sweet cup' in the West Indies, 'chulupa' in Colombia, and 'granadilla de hueso' in Ecuador. It is a little known species but may have a good future because of its excellent aroma and flavor.

Botany and Horticulture

The purple passion fruit crops best at higher altitudes in the tropics. It is produced commercially in Australia, South Africa, Kenya, Papua-New Guinea, New Zealand, Sri Lanka, and India. The yellow form thrives in humid tropical lowlands. It is the passion fruit most cultivated in Brazil, Peru, Venezuela, Hawaii, and Fiji. Both wild and cultivated passion fruits are also found in Bolivia, Malaysia, Indonesian, Taiwan, and the Philippines. Production of passion fruit is also possible in frost-free areas in the temperate zone, although fruiting is more limited than in the tropical regions.

The plant is a woody, perennial vine that bears a large number of purple or yellow, oval or round fruits 4–7.5 cm in diameter. It climbs with the aid of tendrils that spring from the same axils as the flowers. The alternate evergreen leaves, deeply three-lobed when mature, are finely toothed, 7.5–20 cm long, deep green and glossy above, paler and dull beneath, and, like the young stems and the tendrils, tinged with red or purple, especially in the yellow form. A single, fragrant flower, 5–7.5 cm in diameter, is borne at each node on the new growth. Clasped by three green leaf-like bracts, it consists of five greenish white sepals; five white petals; a fringed corona of straight white-tipped rays, vivid purple at the base; five stamens with large anthers; the ovary; a prominent, centrally located, triple-branched style and three stigmas. The stigmas are at the apex of the androginophore, the anthers situated below them, making self-pollination difficult.

The *P. edulis* Sims bears fruits, weighing about 35 g, with a deep purple leathery skin that wrinkles when the fruits are fully ripe. The *P. edulis* f. *flavicarpa* Degener has a yellow skin softer than that of the purple fruit and is usually larger, weighing about 80 g. It does not shrivel as badly. The rind is about 3 mm thick and has a whitish layer internally, similar to the albedo of citrus fruit. Inside the fruit is a cavity filled with a juicy mass of intensely fragrant, translucent yellow–orange pulp, with a distinctive sour-sweet flavor, in which are embedded numerous small, flattened, oval, edible, black (in purple passion fruit) or dark brown (in yellow passion fruit) seeds.

Although a hermaphrodite, the passion fruit flower is self-sterile. Cross-pollination depends on pollinators such as large bees, the carpenter bee (*Xylocopa varipuncta*) and the honey bee (*Apis melifera*) being the most important. To overcome the lack of natural pollination, hand pollination is practiced in many countries. Artificial pollination has been found to be more efficient, and fruits from hand-pollinated flowers are larger and juicier than fruits from insect-pollinated flowers.

Passion fruit species have different periods for flower opening, which are usually short, rarely passing 8 h. The flowers of *P. edulis* open at dawn and close at midday; those of the *P. edulis* f. *flavicarpa* open at noon and close in the evening. Thus, the time for possible pollination is limited.

The purple and yellow passion fruits rarely hybridize in the field but can be crossed manually. A number of successful hybrids have been produced in Queensland, Florida, and Hawaii. Commercial production in Australia, in fact, now depends almost exclusively on commercial hybrids. Hybrids are intermediate in skin color and character between the two parents and have been selected in terms of agronomic characteristics, including resistance to viral and fungal diseases, high yield, extended cropping season, and flavor quality similar to the purple form.

In countries where the fruit is grown commercially, the usual method of propagation is by seeds. However, vegetative propagation, mainly from rooted cuttings and grafting, is becoming important to obtain disease-resistant plants and high yields.

The plant can grow in different types of soil, preferably sandy-clay soil, provided it is deep, relatively fertile, with good amount of organic matter, has good drainage and a pH between 5.0 and 6.5. A temperature between 21 and 32 °C favors development of this plant, the optimum being 26–27 °C. Long exposure to sunlight (between 10 and 12 h of daylight) is required for flowering. Annual rainfall should be between 800 and 1750 mm and well distributed throughout the year. Heavy rainfall during the peak of flowering will impede pollination. Irrigation is recommended during the dry periods, and pruning is necessary for good production. Frost and strong winds are limiting factors to passion fruit cultivation. The vines are usually supported by trellises in commercial plantations or by wires strung between wooden or concrete posts in small domestic farms.

The plant starts to produce fruit a year after planting, but has a short productive life, varying from 3 to 5 years, although longer periods have been reported. The highest yield is obtained on the second or third year, decreasing thereafter.

On average, 60–70 days are required from pollination to maturation of the fruit. Thus, two or more harvesting seasons per year are possible, depending on the climatic conditions, geographic location, cultural practices, and other factors. In the tropics, production is almost uninterrupted, but in subtropical regions, vegetation ceases in winter, production being stopped during this period and in the spring months. Thus, in these regions, the productive period is reduced to 6 or 7 months per year.

Upon ripening, the passion fruits fall from the vine and are picked from the ground at least once a week, preferably more often in wet weather. The fruits are transported in lug boxes and, if not processed immediately, placed into cold storage. A temperature of 6.5 °C with a relative humidity of 85–90% is recommended. Under these conditions, the fruits can be stored for 4–5 weeks. Storage temperatures below 6.5 °C cause chill injury; at higher storage temperatures, the fruits suffer more losses due to moldiness.

Diseases and Pests

Production of passion fruit may be adversely affected by various diseases and insect pests. Common diseases are woodiness (thickening and hardening of the pericarp) or the mosaic disease, due to a virus transmissible by aphids, and several diseases of bacterial and fungal origin. Bacterial diseases include grease spot caused by *Pseudomonas passiflorae*, blast due to *Pseudomonas syringae* and bacterial spot caused by *Xanthomonas passiflorae* or *Xanthomonas campestris* pv. *passiflorae*. Diseases brought about by fungi are: brown spot by *Alternaria passiflorae*; leaf spot or septoria blotch by *Septoria passiflorae*; Fusarium wilt by *Fusarium* spp., such as *Fusarium oxysporum* f. *passiflorae*, or crown canker by *Fusarium lateritium* and *Fusarium sambucium*, which enter the plant at lesions and cracks; root rot caused by *Thielaviopsis basicola*; antrachnosis by *Colletotrichum gloeosporioides*; cladiosporioses by *Cladosporium herbarium*; and collar rot by *Phytophthora cinnamomi*.

Among the insect pests that can damage passion fruits, the most important are: (1) fruitflies, notably *Dacus dorsalis*, *D. cucurbitae*, or *D. vertebrates*, or larvae of the fruitflies *Anastrepha* spp., *Ceratitis capitata* and *Dacus tryoni*; (2) caterpillars of the butterflies *Dione juno juno* and *Agraulis vanillae*; (3) fruit mites *Tenvipalpus californicus*, *Brevipalpus papayensis*, *B. phoenicis*, *Polyphagotarsonemus latus*, *Tetranychus mexicanus*, *T. desertorum*; (4) pasion fruit or fruit bug *Diactus bilineatus* and *Holymenia clavigera*; and (5) leafhopper *Scolypopa australis*. Purple passion fruit vines can also be damaged by nematodes, such as *Meloidogyne javanica*, *M. arenaria*, *M. incognita*, *Scuttellona truncatum*, *Helicotylenchus* sp., and *Pratelenchus* sp. The yellow passion fruit is nematode-resistant. (*See* **Insect Pests**: Insects and Related Pests; Problems Caused by Insects and Mites.)

Chemical Composition

The chemical composition of the fruit and the juice is known to vary appreciably, being affected by factors such as variety, degree of ripeness, plant status, date of picking, season, climate, locality or region, soil composition, and cultural practices.

Although other tables on the composition of passion fruit are available in the literature, a list based on USDA data is presented in **Table 1** because this is constantly updated.

The purple passion fruit juice is a good source of ascorbic acid and carotene, a fair to good source of riboflavin and niacin, and a fair source of mineral matter. On an overall basis, the yellow passion fruit has less ascorbic acid, °Brix, °Brix/acid ratio, and reducing and total sugars, but a higher acid and carotene content.

Carbohydrates constitute the second largest constituents of passion fruit (**Table 1**), and three sugars, glucose, fructose, and sucrose, make up 86% of the total carbohydrates, the rest being starch. The sugar composition of yellow passion fruit consists of 29%

Table 1 Composition of passion fruit and juice[a]

Constituents	Fruit, purple	Juice, purple	Juice, yellow
Proximate			
Water (g)	72.9	85.6	84.2
Food energy (kcal)	97	51	60
Protein (g)	2.20	0.39	0.67
Total lipid (fat, g)	0.70	0.05	0.18
Carbohydrate, by difference (g)	23.4	13.6	14.4
Total dietary fiber (g)	10.4	0.20	0.20
Ash (g)	0.80	0.34	0.49
Minerals			
Calcium (mg)	12.0	4.0	4.0
Phosphorus (mg)	68.0	13.0	25.0
Iron (mg)	1.60	0.24	0.36
Sodium (mg)	28.0	6.0	6.0
Potassium (mg)	348	278	278
Magnesium (mg)	29.0	17.0	17.0
Zinc (mg)	0.10	0.05	0.06
Copper (mg)	0.09	0.05	0.05
Selenium (mcg)	0.60	0.10	0.10
Vitamins			
Vitamin A (IU)	700	717	2410
Vitamin E (mg α-tocopherol equivalents)	1.12	0.05	0.05
Riboflavin (mg)	0.13	0.13	0.10
Niacin (mg)	1.5	1.5	2.2
Vitamin B$_6$ (mg)	0.10	0.05	0.06
Folate (µg)	14.0	7.0	8.0
Vitamin C (mg)	30.0	29.8	18.2

[a]Values per 100 g of edible portion.
From USDA Department of Agriculture, Agricultural Research Service (2002) *USDA Nutrient Database for Standard References, Release 15*.

fructose, 38% glucose, and 32% sucrose. The purple passion fruit has 34% fructose, 37% glucose, and 29% sucrose.

The nitrogen content of the juice is reported to vary from 0.10 to 0.19%, and the crude protein content ($N \times 6.25$) from 0.6 to 1.2%. The free amino acids found in purple passion fruit juice are leucine, proline, threonine, valine, tyrosine glycine, aspartic acid, arginine, and lysine.

The total acid content (expressed as citric acid) ranges from 2.4 to 4.8% w/w with an average of 3.4% for the purple passion fruit juice. The yellow passion fruit juice from Hawaii has an acidity ranging from 3.0 to 5.0%, with an average of 4.0%. The predominant acid is citric acid (about 83% of the acids), followed by malic acid (16%) and lesser amounts of lactic (0.87%), malonic (0.20%), and succinic (trace) acids. The same organic acids are found in the purple fruit juice, but the relative amounts are different. Citric acid is also the most abundant (41%), followed by lactic (23%), malonic (15%), malic (12%), and succinic (7.6%) acids.

An early study identified in purple passion fruit the carotenoids phytofluene, α-carotene, β-carotene, and ζ-carotene, of which β-carotene dominated. Later, β-apo-12′-carotenal, β-apo-8′-carotenal, cryptoxanthin, auroxanthin, and mutatoxanthin were also detected. Recently, 13 carotenoids were conclusively identified in yellow passion fruit: phytoene, phytofluene, ζ-carotene, neurosporene, β-carotene, lycopene, prolycopene, monoepoxy-β-carotene, β-cryptoxanthin, β-citraurin, antheraxanthin, violaxanthin, and neoxanthin, with the yellow ζ-carotene predominating. A recent study showed cyanidin 3-glucoside to be responsible for the purple color of the rind of *P. edulis*, accounting for 97% of the total anthocyanin content. Small amounts of cyanidin 3-(6″-malonylglucoside) (2%) and pelargonidin 3-glucoside (1%) were also found.

As in other fruits, the volatile composition of passion fruit is extremely complex, consisting of numerous compounds, and the unique flavor and aroma cannot be attributed to any single constituent. A compendium in 1990 listed 84 esters, 33 aldehydes and ketones, 71 alcohols and acids, and 25 miscellaneous volatiles identified in passion fruit. The volatile constituents present in the highest concentrations are the C-2 to C-8 esters of the C-2 to C-8 fatty acids that occur in many fruits. Some volatiles are probably degradation products of carotenoids, such as linalool, β-ionone, dihydro-β-ionone and the lactone of 2-hydroxy-2,6,6-trimethylcyclohexylideneacetic acid (dihydroactinidiolide), an oxidation product of β-ionone, these compounds being detected in both purple and yellow passion fruit. Dihydro-β-ionone and 1,1,6-trimethyl-1,2-dihydronaphthalene (3,4-dehydroionene), a possible degradation product of β-ionone, have been found only in purple passion fruit. Other compounds, such as edulans, megastigmatrienes, and sulfur-containing compounds, not included in the listing cited above, have also been reported as volatile constituents of passion fruit.

Out of some 300 volatile flavoring constituents of passion fruit drinks, only 22 were found likely to contribute to the passion fruit flavor. Of these, 6-(but-2′-enylidene-1,5,5-trimethylcyclohe-1-ene and (Z)-hex-3-enyl butanoate contributed 30 and 11%, respectively, of the passion fruit flavor profile. Several esters present in the juice at relatively high concentrations had negligible flavor impact values.

The flavor and aroma of the purple passion fruit are often described as more pleasing than those of the yellow fruit; this difference is reflected in the volatile composition. The purple passion fruit has higher levels of the major esters ethyl, butyl, and hexyl butanoates, and butyl and hexyl hexanoates, and of the terpene ketones, β-ionone and dihydro-β-ionone. Some important flavor constituents of the purple passion fruit are absent in the yellow fruit: 2-heptyl acetate, butanoate, and hexanoate, the edulans, dihidroionone, 1,1,6-trimethyldihydronaphthalene, and the megastigmatrienes. There are also differences in the terpene hydrocarbons, the yellow fruit having relatively high concentrations of (E)-β-ocimene, myrcene, limonene, and 1,4-*p*-methadiene, whereas only (E)-β-ocimene is present in a significant concentration in the purple variety. (*See* **Flavor (Flavour) Compounds**: Structures and Characteristics.)

The immature fruits have lower juice, sugar, ascorbic acid, and carotene contents. These fruits are more acidic and inferior in flavor compared with the partially and fully ripe fruits. Changes in the volatile composition during maturation have also been reported, the compounds of greatest flavor significance reaching maximum concentrations in the fruit that have just fallen from the vine. This observation indicates that the common practice of harvesting passion fruit after it falls is likely to provide fruit of the best flavor, provided that gathering from the ground is not delayed. Another study, however, found the fallen fruit to be lower in soluble solids, sugar, acidity, and ascorbic acid, suggesting that the customary harvesting practice should be avoided. Anthocyanins and carotenoids also increase during ripening.

Production, Utilization, and Processing

Reliable and updated production statistics on passion fruits are not available. Although it is one of the better known so-called exotic fruits, it has not reached the point of being one of the fruits for which production is monitored by FAO. The major producing countries, listed in decreasing order of area of production in 1987, are: Brazil, Peru, Sri Lanka, Ecuador, Australia, Kenya, South Africa, Venezuela, Papua-New Guinea, Fiji, New Zealand and the USA (Hawaii). These countries account for 80–90% of the world's production of passion fruit. Taiwan and Colombia have been markedly increasing production, a good part of which is destined for export. Commercialization of the fresh fruit is limited, with Kenya being the principal exporter to Europe, especially to the UK. It is estimated that 50% of the world's production of passion fruit juice is exported. Brazil, Peru, and Colombia are the principal exporters of the juice, the major importers being the UK, Germany, France, Switzerland, the USA, and Japan.

After cutting the fruit in half, the pulp and seeds can be scooped with a spoon and eaten as is, or the pulp can be sieved to make a refreshing drink. Because of its intense flavor and high acidity, the passion fruit juice is considered a natural concentrate and is often diluted, sweetened, or blended with other fruit juices. The whole or sieved pulp is also used as a flavoring for yogurt, icecream, sherbet, meringue, or cake topping.

In Australia, there is appreciable consumption of the fresh fruit, although the bulk of passion fruit production is processed into juice, consumed locally as carbonated beverages. The juice is also used as a flavoring for icecream, confectioneries, and tropical fruit salads. Australian consumers are used to eating and drinking passion fruit products with the seeds still present, the seeds being regarded as evidence of passion fruit content. Elsewhere, the seeds must be removed.

Production of passion fruit in Fiji and Papua-New Guinea is largely for export to Australia and New Zealand as frozen, unsweetened pulp or juice. A significant part of production in Fiji is also consumed in homes and restaurants as mixed drinks. Passion fruit products in Sri Lanka include jams and sweetened and unsweetened juices. In India, this fruit is processed into passion fruit squash.

Most of the passion fruit produced in Hawaii is consumed as drinks blended with other fruit juices, such as orange and/or guava, the remainder being used as frozen juice bases. Passion fruit juice is considered too acidic for icecream, but this characteristic is advantageous in the preparation of sherbet. Minor uses are as a flavoring for syrups and as a pie filling.

The principal passion fruit products in Brazilian market are the juice and the frozen pulp, these two products serving as the base for other products such as drinks, yogurt, icecream, confectioneries (cakes, meringues, and chocolate fillings), gelatin, marmalades, and fruit cocktails. Fresh fruits are also marketed. In Venezuela, popular products include passion fruit juice, passion fruit icecream, and a bottled passion fruit and rum cocktail.

Passion fruits are preserved by freezing or thermal processing. Two characteristics of this fruit favor freezing. Firstly, the flavor is extremely sensitive to heat, so it is difficult to heat-process the juice without markedly altering the flavor. Secondly, the high starch content causes the accumulation of gelatinous deposits on the heating surfaces of the heat exchanger, lowering its efficiency, as well as causing deterioration of juice flavor. Enzymatic degradation of starch and centrifugal procedures for removing starch have been recommended.

Production of passion fruit juice can be done manually, as in many small cottage industries, consisting simply of hand-slicing the fruit, scooping out the pulp, and separating out the seeds either through sieving or expression through a cloth.

In Hawaii, the processing of passion fruit is highly mechanized, a centrifugal separator being the favored method for extraction of the pulp. A typical extractor has a capacity of $1725\,\mathrm{kg\,h^{-1}}$ of passion fruit with an extraction efficiency of 94%. Its main disadvantages are: (1) some seeds are cut in the slicing operation, which necessitates the use of a fine screen in the finishing operation; and (2) there may be some extraction of skin flavors.

In Australia, the converging cone extractor is the most commonly employed extractor. In another method, a modified apricot-pitting machine and plunger are used. Since Australians are accustomed to consuming passion fruit products with the seeds, further processing to remove the seeds from the pulp is not necessary. Elsewhere, consumers' preference for seedless passion fruit products necessitates further processing.

The extraction unit most commonly utilized in Brazil is a three-stage system, consisting of a cutter, a perforated cylinder with a series of beaters that separates the rind from the pulp and seeds, and a pulper that separates the juice from the seeds and does the finishing of the juice.

Much of the earlier canned passion fruit juice was considerably overcooked. More recent processes make use of slightly higher temperature and shorter heating times. The most successful method for thermal processing of the juice employs a spin cooker. This cooker is an inexpensive, easily constructed unit, utilizing rapid can rotation to transfer heat quickly. Pasteurization to an 88 °C center temperature is achieved in about 1¾ min, after which the cans are rapidly cooled with a cold water spray while still rotating. The color and flavor retention in this method is much better than in any alternative method of heat preservation.

To concentrate passion fruit juice, centrifugal evaporators have been successfully used in Brazil and Australia. The main advantage is the short residence time (0.2–1.0 s), which minimizes heat damage. Passion fruit concentrate can be stored at $-18\,^{\circ}C$ for 6 months, $4\,^{\circ}C$ for 3 months, and $20\,^{\circ}C$ for 1 month with good color and flavor retention and no microbial spoilage.

Passion fruit juice may be quick-frozen directly from the finisher. Preferably, a slush-type or scraped-surface freezer should be used to hasten the freezing process, aside from increasing the freezing capacity of the processing plant. The juice may also be frozen directly in containers in an air-blast freezer. This product is sold to manufacturers of juice blends and of foods with passion fruit as an ingredient or as a major flavor component.

According to folk medicine, passion fruit has sedative and muscle-relaxant properties. Medicinal utilization of the leaves has also been cited.

In the extraction of juice from the passion fruit, about two-thirds of the bulk is refuse, of which 90% is rind and about 10% is seeds. Passion fruit rind has been found to be satisfactory as a supplementing feedstuff for dairy cows, and so it is commercially utilized as feed for dairy animals in Hawaii. In Brazil, the passion fruit rind is used as a component of rations for cattle and hogs. Other possible uses of subproducts, such as pectin from the rind and oil from the seeds, have not reached the industrial scale.

See also: **Acids**: Natural Acids and Acidulants; **Amino Acids**: Properties and Occurrence; **Ascorbic Acid**: Properties and Determination; Physiology; **Canning**: Quality Changes During Canning; **Carbohydrates**: Classification and Properties; **Carotenoids**: Occurrence, Properties, and Determination; **Flavor (Flavour) Compounds**: Structures and Characteristics; **Freezing**: Structural and Flavor (Flavour) Changes; **Heat Treatment**: Chemical and Microbiological Changes; **Insect Pests**: Insects and Related Pests; Problems Caused by Insects and Mites; **Minerals – Dietary Importance**; **Vitamins**: Overview

Further Reading

Casimir DJ, Kefford JF and Whitfield FB (1981) Technology and flavor chemistry of passion fruit juices and concentrates. *Advances in Food Research* 27: 243–295.

Chan HT Jr. (1993) Passion fruit, papaya and guava juices. In: Nagy S, Chen CS, and Shaw PE (eds) *Fruit Juice Processing Technology*, pp. 334–377. Auburndale: Agscience.

Choucair K (1962) *Fruticultura Colombiana Tomo II Frutas Tropicales–Subtropicales y de Clima Templado y Frio*, pp. 795–850. Medellin: Editorial Bedout.

Escobar LK (1993) Passion fruits. In: Macrae R, Robinson RK, and Sadler MJ (eds) *Encyclopaedia of Food Science, Food Technology and Nutrition*, vol. 5, pp. 3423–3428. London: Academic Press.

ITAL (1994) *Maracuja. Cultura, matéria-prima, processamento e aspectos econômicos.* Campinas: Instituto de Tecnologia de Alimentos.

Kidoy L, Nygard AM, Andersen OM *et al.* Anthocyanins in fruits of *Passiflora edulis* and *P. suberosa*. *Journal of Food Composition and Analysis* 10: 49–54.

Mercadante AZ, Britton G and Rodriguez-Amaya DB (1998) Carotenoids from yellow passion fruit (*Pasiflora edulis*). *Journal of Agricultural and Food Chemistry* 46: 4102–4106.

Pruthi JS (1963) Physiology, chemistry, and technology of passion fruit. *Advances in Food Research* 12: 203–282.

Ruggiero C (ed.) (1980) *Cultura de Maracujazeiro.* Jaboticabal: UNESP.

Salunkhe DK and Desai BB (1984) *Postharvest Biotechnology of Fruits*, vol. II, pp. 53–58. Boca Raton, FL: CRC Press.

Shibamoto T and Tang CS (1990) 'Minor' tropical fruits – mango, papaya, passion fruit and guava. In: Morton ID, and MacLeod AJ (eds.) *Food Flavours Part C. The Flavour of Fruits*, pp. 221–280. Amsterdam: Elsevier Science.

USDA Department of Agriculture, Agricultural Research Service (2002) *USDA Nutrient Database for Standard Reference, Release 15.*

PASTA AND MACARONI

Contents
Methods of Manufacture
Dietary Importance

Methods of Manufacture

R Cubadda, University of Molise, Campobasso, Italy
M Carcea, National Institute for Food and Nutrition
Research, Rome, Italy

Background

The current range of products referred to as pasta is relatively vast, and products vary widely in terms of shape, color, composition, storage requirements, and use. However, all share a basic technology that involves the preparation of a dough made by mixing a flour with a liquid (mainly water), which is then processed (namely by extrusion) to obtain the required shape and dimension of the product itself.

It is a popular belief that pasta originated in China, where it was manufactured with soft wheat flour. From there, it is supposed to have been introduced into Europe (Italy) by Marco Polo in AD 1292 after his travels to the Far East. However, the production of dried pasta, as it is manufactured today, is comparatively recent. The first attempt to produce extruded pasta instead of cut noodles from a sheeted dough was made in Italy in the 1800s using a hand-operated mechanical wooden press. For many years, pasta production was home-based, and only at the turn of this century did it become an industry, owing to the invention of the steam engine and hydraulic presses, and to the development of artificial instead of natural drying. The first continuous press, which replaced the traditional batch method, was introduced in 1933. The final step towards a fully automated system was achieved in the early 1950s, with the introduction of the automatic weighting and packaging equipment in pasta factories.

Today, the whole process, from reception of raw materials to mixing, kneading, extrusion, drying, packaging, and dispatching, can be performed automatically by several pieces of machinery controlled by computers. Moreover, the diffusion of pasta throughout the world and the popularity of the traditional dry product have stimulated the development of production technologies leading to new convenient commodities such as fresh, precooked, and frozen pasta

Raw Materials and Their Properties

The raw material of choice for pasta production is semolina from durum wheat, although, for various reasons, the industry uses flours from soft and hard wheats, corn, and other cereals. Durum wheat semolina is the only raw material permitted for pasta production by national laws in Italy, France, and Greece. Special pasta is also produced by adding a great variety of other ingredients (i.e., fresh, frozen, or powdered eggs, powdered spinach, tomatoes, carrots and other vegetables, soy protein, wheat gluten, milk protein, vitamins, and minerals) to the two basic components, semolina (or flour) and water. This section deals with the production of pasta using the traditional raw material durum wheat semolina. However, the manufacture of pasta with flour or blends of flour and semolina is also mentioned. (*See* **Wheat**: The Crop; Grain Structure of Wheat and Wheat-based Products.)

Semolina quality requirements vary from country to country. In Germany, Austria, and Switzerland, for example, semolina color is very important, whereas in Italy, the factors affecting cooking quality are synonymous with semolina quality. In today's pasta industry, uniform semolina granulation is also a desirable factor for uniform flow in semolina feeders and for proper dough development in continuous presses. A high ash content is usually indicative of longer-extraction semolina and imparts a dull color. If the wheat is not properly cleaned, or the kernels are smudged and/or damaged by severe mildew, brown and black specks can appear in the semolina. Commercial semolina with fewer than 200 specks per dm^2 is desirable, and gives pasta with a good esthetic appearance. A low level of lipoxidase, which, during pasta processing, may destroy the yellow color given to semolina by natural carotenoid pigments (carotenes and xanthophyls), is also desirable. Nevertheless, the major quality requirement in most traditional pasta consuming areas is the ability of semolina to be processed into pasta with a good cooking quality.

According to the existing literature, gluten quality and quantity are the most important factors affecting cooking quality, but starch and minor constituents such as soluble and insoluble pentosans, lipids, lipoproteins, various enzymes, and products of enzyme reactions could also be involved. At present, little information is available on the physicochemical changes and on the behavior of starch and minor constituents during processing, cooling, and storage of pasta. It is well known that semolina quality varies widely among durum wheat varieties. Semolinas from certain durum wheat varieties grown in some locations produce pasta with a good cooking quality and/or yellow color, whereas other durum wheat varieties give semolina that produces pasta with insufficient cooking characteristics or pigment. Consequently, attention is paid to check the quality of durum varieties in order to produce the desired semolina.

Common soft wheat flour is less pigmented, translucent, and flinty than durum wheat semolina, and gives a gluten with an elastic strength suitable for making bread, but not for manufacturing a superior pasta product. However, flours milled from common hard wheats with strong gluten, a high protein content, and superior rheological characteristics are acceptable for producing pasta. The color and cooking quality of pasta from common wheat flour can be improved by adding proportional amounts of egg. As far as the water used for pasta making is concerned, it should be pure drinking water with no off-flavors.

Method of Manufacture

In a modern pasta factory, pasta products are formed by extrusion on large automatic machines that perform several processing operations. Production lines for processing short and long pasta are shown in **Figure 1**. The essential equipment includes:

- A series of metal silos that receive the various types of semolina or flour (coming either directly from a mill adjoining the pasta factory or from outside mills);
- An automatic continuous press;
- A spreader predrier for long pasta or shaker predrier for short pasta;
- A drier; and
- A storage and packing unit.

(*See* **Extrusion Cooking**: Principles and Practice.)

Mixing

The automatic press is one of the most important machines in a pasta factory. It performs three essential processing operations: mixing, kneading, and extruding.

In the mixing stage, water is added to the semolina so that the moisture content of the dough is approximately 30%. The flow of ingredients is regulated by volumetric or gravimetric dosers with constant and proportioned outputs. It is very important that the water is absorbed evenly by the semolina or flour particles to obtain a thoroughly homogeneous dough that will prevent the dried pasta from having faults (white spots, for instance). For this reason, the

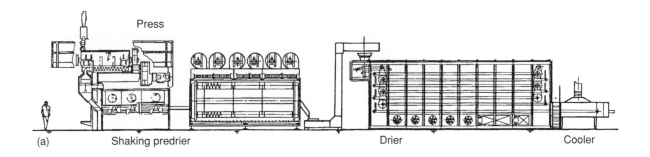

(a)   Press   Shaking predrier   Drier   Cooler

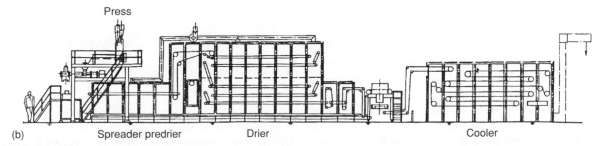

(b)   Press   Spreader predrier   Drier   Cooler

Figure 1 Modern production lines for processing (a) short and (b) long pasta. Reproduced from Pasta and Macaroni, *Encyclopaedia of Food Science, Food Technology and Nutrition*, Macrae R, Robinson RK and Sadler MJ (eds), 1993, Academic Press.

semolina particles must have the same size (for a typical semolina, a maximum of 25% of particles should pass through a 0.180-mm mesh sieve), and the time needed for the liquid to be absorbed by the particles has to be evaluated while taking into account their average size and the temperature of both flour and liquid. Steps should be taken to insure that the lower the flour temperature, the higher the temperature of the liquid in the dough.

Depending on the pasta shape and type, warm water at 40–65 °C is generally used. A uniform dough is obtained by mixing the ingredients in a special twin-shaft mixing chamber for 15–17 min for short pasta, and 16–20 and 16–18 min for long pasta and long egg pasta, respectively. The shafts of the mixer turn in opposite directions to limit the amount of dough balling that can occur.

The latest development in press technology is represented by new systems that are able to carry out the dough-making in a very short time (about 2–3 min) instead of the 15–20 min necessary for normal dough preparation. A very short dough-making time increases considerably the line production speed, since it can better match the speed of the following steps (predrying, drying, etc.). For this reason, at present, fine granulometry semolina is preferred, which best suits short dough-making times. Moreover, short dough-making times allow for a reduction in the press size (output capacity being equal) and consequently a better plant compactness and smaller dimensions.

Modern continuous presses are equipped with vacuum-producing equipment to remove any air bubbles from the pasta dough during mixing and just before extrusion. The aim of this operation is to avoid the formation of small bubbles, which can give the end product a white, chalky appearance. In addition, air bubbles can facilitate the oxidation of pigments, resulting in a pale and unattractive pasta, and can weaken the dried product.

Extruding

The extrusion screw is the core of the pasta press, and special attention is paid to its design and construction, just as to the compression cylinder. In fact, it is vital to prevent the dough from being overworked, an occurrence that would adversely affect the quality of the finished product. The dough is forced by the screw into the extrusion head. Even under the best operative conditions, a considerable amount of heat is generated during this process; therefore, the barrel and the extrusion head are equipped with a water-cooling jacket to reduce the heat and to maintain a constant extrusion temperature of no more than

40 °C. Higher temperatures could cause the deterioration of the cooking quality of the finished product.

The process in the press is completed by the extrusion of the dough through a die to make a variety of product shapes that are cut to the required length with a rotary cutter. Until recently, pasta dies were made of bronze. Now, new dies with Teflon on the extruding surface are used. Teflon extends the life of the die and improves the appearance of the product. Bronze dies are still used for small special productions characterized by the rough surface of pasta. Special equipment is required to form particular shapes such as nests, skeins and Bologna-type pasta. The cut pasta is then hung over drying rods or spread evenly, all automatically, ready for the drying stage.

Drying

Drying is the most difficult and critical step in the pasta-making process. The objective of this process is to decrease the moisture level of the extruded pasta from about 30% to less than 12.5% without causing any damage to the finished product. To achieve this, the air temperature and relative humidity must be properly controlled. If the drying is too slow, the pasta products tend to spoil or become moldy. However, if they are dried too rapidly, moisture gradients occur that may cause checking or cracking. Usually, uniform temperature and ventilation are insured in all parts of the dryer. This is achieved through a suitable hot-air circulation system using low-pressure spiral ventilators located in various parts of the dryer. Moreover, automatically controlled periods of hot-air circulation alternate with periods of rest to allow for moisture equilibration between the inner part of the pasta and its surface.

In modern plants, the drying cycle can be subdivided into three distinct stages: (1) predrying, (2) drying, and (3) cooling. Although identical basic components are used, the manufacture of different types of pasta requires machines that are technically different. The most widely used dryer for long pasta is basically made of a single complex consisting of:

- an extruded-pasta supporting structure on one or more levels;
- a separate drive unit for the section on one or more levels;
- a heat ventilation unit;
- an unloading unit; and
- a regulation and control unit.

The initial predrying takes place in the first part of the dryer, where special aluminum or steel rods loaded with pasta are normally at only one level. During this stage, the product moisture is reduced to 17–18%, as a consequence of quick evaporation of the inside water excess of pasta. Surface drying has to be avoided to prevent closing of the interstices through which the remaining water can reach the surface by capillarity and evaporates. This predrying stage lasts a short time and is followed immediately by a longer drying stage.

For the final drying, driers with more than one level are used. The product hanging on the rods is moved by means of special conveyor chains in the one-level section and with a system of racks, which alternate vertical and horizontal movements, in the section with more than one level. From a technological point of view, the drying differs from the predrying phase for a more gradual moisture reduction and particularly for the alternation of water evaporation and its even distribution inside the product.

The predrying stage of short pasta is generally carried out with vibrating trays, called shaker pre-driers, whereas there is a great diversification in the machines used for the next drying steps. Two different kinds of equipment are used: the belt dryer and the rotary dryer. The belt dryer is more suitable for large shapes, which must be treated delicately. These dryers receive the product from the shaker predriers by means of conveyor belts. An oscillating band evenly distributes the product on the whole upper belt width. The product, after having reached the far end of the belt, falls on the following ones, repeating this route as many times as the existing number of belt dryer tiers. Drying is achieved by means of aerothermal units in which hot water is kept in forced circulation by a pump. The rotary dryers can be used for both the preliminary and final drying of small- and medium-sized and Bologna-style pasta products. The rotary driers are essentially composed of a rotating drum, supported on rollers and driven by a multiple-speed gearbox. The product advances forward one cell for each revolution of the drum. Drying is achieved by means of a battery of air-circulating fans and heat radiators that are located above the drum.

At present, the drying technology varies greatly from factory to factory and from country to country, the main difference being in the degree of temperature used during the process. However, two systems can be recognized: (1) traditional or conventional drying, which uses 'low' temperatures not higher than 60 °C, and (2) 'high'-temperature drying, which uses temperatures between 70 and 85 °C, even though there is a trend to seek higher temperatures. Of course, the total drying time varies, depending on the system employed. High-temperature drying has been without doubt the most important innovation in pasta technology in recent times. This new technique has the advantage of appreciably reducing the drying time, compared with traditional methods, and of reducing microbial contamination. The former characteristic allows the construction of drying lines that maintain the same production capacity as the traditional lines, but they are smaller, more compact and constructed to facilitate maintenance operations.

With the increasing popularity of high-temperature drying among pasta manufacturers, considerable attention has been given to the effect of this treatment on the cooking quality, color, and chemical composition of the final product. The effect of high-temperature drying on the cooking quality is still controversial. The main reason is that the available information comes from investigations where the method of application of the high temperature during the drying cycle and the intrinsic characteristics of the raw materials are different. However, experimental evidence is accumulating to prove that semolina protein quality and quantity can influence cooking quality jointly or independently as a consequence of the drying technology adopted (i.e., low- or high-temperature drying). Different temperatures modify components and influence cooking quality in different ways. At low temperatures (40–50 °C), protein quantity and quality are equally important in determining the final pasta results, whereas at high temperatures (≥ 80 °C), protein quantity is more important and is strongly correlated with the resultant pasta quality. Some researchers have reported that pasta proteins are in fact polymerized by high drying temperatures, whereas starch granules appear embedded in the protein matrix. As a consequence, there may be an improvement in the cooking quality of pasta. Pasta cooking quality is determined by a physical competition between protein coagulation into a continuous network and starch swelling with spherules scattering during cooking. If the former prevails, starch particles are trapped in the network alveoli, promoting firmness in cooked pasta, whereas if the latter prevails, the protein coagulates in discrete masses lacking a continuous framework, and pasta will show softness and usually stickiness on cooking. High-temperature drying partially overcomes this competition by producing a coagulated protein framework in dry pasta without starch swelling.

Drying conditions also have a significant effect on the retention of lysine in spaghetti. As lysine is the limiting amino acid in wheat, any reduction in its availability has an important bearing on the nutritional quality of pasta. A decrease in the amount of total lysine has been reported as being directly proportional to the increase in temperature. The loss of lysine depends mainly on the blockage of the amino group as a consequence of the Maillard reaction. (*See* Browning: Nonenzymatic.)

Cooling and Packaging

The final drying is followed by a cooling treatment that must be carried out with certain precautions to avoid damage to the pasta through humidity imbalance within the shape. After this stage, the end product is unloaded, sent to storage silos, and then sent to the packaging section.

Pasta products are packaged in various ways, depending on their particular shapes and sizes. The most widely used packages are cellophane bags and packets, cartons, and special trays. All these are automatically weighed, packed, and collected by machines on continuous cycles. Sometimes, semi-automatic machines may be required.

Considerations of Quality

A good cooking quality is the most important requirement for pasta products. Aroma, taste, and, sometimes, uniformity of shape and color and breaking strength of uncooked pasta are also important to the consumer. Although it can be interpreted according to the individual taste and habits of the consumer, the cooking quality of pasta is generally regarded as the capacity of the product to maintain its shape when cooked in boiling water and to retain a good texture after cooking without becoming a thick, sticky mass. The parameters involved in the evaluation of pasta cooking quality can be defined as follows:

- stickiness is the state of surface disintegration of cooked pasta;
- firmness is the resistance of cooked pasta when sheared between the teeth;
- bulkiness is the degree of adhesion of pasta shapes after cooking.

As already mentioned, cooking quality depends essentially on the type of raw materials used and on their intrinsic characteristics, although it can be affected by the manufacturing conditions.

As far as the nutritional quality of pasta is concerned, the nutritive value appears to be completely independent of the use of durum wheat semolina or bread wheat flour. However, the addition of other ingredients such as eggs, vegetables, soy or vegetable protein concentrates and isolates, individual nutrients, etc. produces remarkable changes in the nutritional value of common pasta.

The protein quality of pasta can be influenced by a reduced bioavailability of lysine caused by severe

drying conditions. This has to be considered when adding high-quality protein integrators to the raw materials.

See also: **Browning**: Nonenzymatic; **Extrusion Cooking**: Principles and Practice; **Wheat**: The Crop; Grain Structure of Wheat and Wheat-based Products

Further Reading

Cubadda R (1988) Evaluation of durum wheat, semolina, and pasta in Europe. In: *Durum Wheat: Chemistry and Technology*, pp. 217–228. St. Paul, MN: American Association of Cereal Chemists.

Cubadda R (1989) Current research and future needs in durum wheat chemistry and technology. *Cereal Foods World* 34: 206–209.

Cubadda R (1996) Pasta quality: the relationship between raw material properties and production technologies. *Proceedings of 1st World Pasta Congress*, pp. 164–168. Pinerolo, Italy: Chiriotti Editori.

D'Egidio MG, Mariani BM, Nardi S, Novaro P and Cubadda R (1990) Chemical and technological variables and their relationships: a predictive equation for pasta cooking quality. *Cereal Chemistry* 67: 275–291.

Grzybowski RA and Donelly BJ (1979) Cooking properties of spaghetti: factors affecting cooking quality. *Journal of Agricultural and Food Chemistry* 27: 380–384.

Lirici L (1999) From the bar kneader to the modern automatic press: over 100 years of evolution. *Tecnica Molitoria* 50: 101–120.

Pavan G (1996) Pasta: processing trends and possible technological answers. *Proceedings of 1st World Pasta Congress*, pp. 79–94. Pinerolo, Italy: Chiriotti Editori.

Resmini PP and Pagani MA (1983) Ultrastructure studies of pasta. *Food Microstructure* 2: 1–12.

Dietary Importance

L J Malcolmson, Canadian International Grains Institute, Winnipeg, Manitoba, Canada

Background

The term 'pasta' is used to describe products made from a flour–water dough that fit the 'Italian' style of extruded foods. The primary ingredient used in the manufacture of pasta is a coarsely ground flour from durum wheat, called semolina. In the commercial manufacture of pasta, semolina and water are mixed into a dough, which is then extruded through a die to produce the desired size and shape of pasta. The product is then dried or sold fresh. Common pasta products include spaghetti, vermicelli, linguine, rotini, rigatoni, and macaroni.

In contrast, the term 'noodle' is used to describe an 'Oriental' style of sheeted and cut products made from grains other than durum wheat. Grains used to produce noodles include common or bread wheat, rice, barley, and buckwheat, as well as legumes such as mung bean.

The earliest written record of pasta in the Italian diet was reported to be 1279, although noodle products were most certainly part of earlier cultures of China and the later cultures of many other countries. Today, pasta products have become an important part of the diet in many countries. This has been influenced in part by changes in consumers' perception of pasta. Once thought of as starchy and fattening, pasta is now considered to be nutritious and convenient. **Table 1** lists the average per capita consumption of uncooked pasta in selected countries. The low consumption of pasta products in Asian countries such as China and Japan compared with other countries is a reflection of the much higher consumption of noodle products in these countries compared with 'Italian'-style pasta products.

The widespread consumption of pasta throughout the world is likely due to its simple formulation, relative ease of processing, extended shelf-life in the dry form, low cost, and versatility. Pasta's versatility stems from the fact that it is available in numerous shapes and sizes, and can be prepared and served with other foods as an appetizer, main dish, side dish, salad, soup, or dessert, thus allowing menu flexibility for the consumer. Because pasta is relatively bland in flavor, it lends itself to be complemented by sauces and other food ingredients.

Nutritional Value of Pasta

Although the nutritional content of pasta can vary widely, depending on the ingredients used in its preparation and added sauces, pasta provides significant quantities of complex carbohydrates, protein, B vitamins, and iron, and is low in fat. A typical 140-g serving of cooked macaroni or spaghetti provides approximately 6 g of protein, 40 g of carbohydrates, and 2 g of dietary fiber, whereas whole wheat pasta provides approximately 8 g of protein, 38 g of carbohydrates, and 3 g of total dietary fiber. The fat content of a 140-g serving of pasta is less than 1 g.

Fortification of pasta products with thiamin, riboflavin, niacin, iron, and, more recently, folic acid is permitted in many countries. The vitamin and mineral content of enriched and whole wheat pasta is provided in **Table 2**.

In recent years, studies have been undertaken to examine the beneficial role of pasta in the diet. Evidence exists indicating that a diet rich in pasta may lower the incidence of cardiovascular disease, be helpful in managing diabetes, and be useful to individuals with a specific gluten intolerance or individuals suffering from celiac disease.

According to a review published by Costantini, subjects who consumed diets rich in pasta had lower levels of serum low-density lipoproteins and total lipoproteins than those who did not consume pasta-rich diets. This finding suggests that diets rich in pasta may be responsible for lower incidents of cardiovascular disease, provided the pasta is prepared and served using other low-fat ingredients.

The exact role of pasta in the dietary management of diabetes is not fully understood. However, it has been speculated that diets rich in pasta result in reduced rates of both starch uptake and gastric emptying, thereby resulting in a lowered postprandial blood glucose and insulin response.

Table 1 Average yearly per capita consumption in selected countries

Country	Per capita consumption (kg)
Italy	28.0
Venezuela	12.7
Switzerland	9.1
USA	9.0
Russia	7.0
France	7.0
Greece	6.6
Canada	6.3
Germany	5.0
Spain	4.0
UK	2.0
Japan	1.8
China	0.8

Source: Pastavilla Pasta (2001) and Pastificio Pagani (2001).

Table 2 Vitamin and mineral content of cooked pasta (140-g serving)

Product	Thiamin (mg)	Riboflavin (mg)	Niacin (mg)	Folate (μg)	Iron (mg)	Calcium (mg)	K (mg)	P (mg)	Mg (mg)	Zinc (mg)
Enriched	0.20	0.20	2.4	10.0	2.0	10.0	44	76	26	0.8
Whole wheat	0.20	0.00	1.0	8.0	1.4	22.0	62	124	42	1.2

K, potassium; P, phosphorus; Mg, magnesium. Source: NutriQuest™ (1999).

Peptides derived from the gliadins of durum wheat have been shown to be less aggressive to celiac disease. Thus, pasta derived from durum wheat may be useful in the diets of individuals suffering from gluten intolerance, depending on the severity of the reaction.

The nutritional value of pasta can be increased by the addition of other food ingredients to increase the protein or fiber content. Research has been undertaken to investigate the effects of various protein sources on pasta quality. Studies have been published showing the successful addition of whey derivatives, nonfat dry milk, soy, corn, legumes (bean, lentil, cowpea), lupin, amaranth, buckwheat, and maize. By blending durum wheat semolina with other plant protein sources, the limiting amino acids lysine and threonine are compensated for by the other protein sources. Protein-enriched pasta products are currently available in some countries. Egg-enriched pasta is also popular in many countries. Whole grain cereals such as barley and legumes as well as the use of whole wheat have also been used to increase the mineral, vitamin, and dietary fiber content of pasta.

The addition of natural colorants such as black squid ink, puréed spinach and carrot, tomato paste, and beet juice has also received considerable attention, and many of these products are currently being sold in the marketplace. The use of herbs and chocolate has also gained popularity in the production of more flavorful product lines. These ingredient additions are made to enhance the flavor and color of the final product, since the nutritional properties are not changed substantially.

Effects of Processing and Cooking on the Nutrient Content of Pasta

The effect of drying and cooking on the nutritional content of pasta has also been studied. Drying conditions have been found to significantly affect the bioavailability of lysine retention in spaghetti. Losses as high as 47% have been found for spaghetti dried at 80 °C and as low as 22% for pasta dried at 45 °C. Other researchers have reported that the extent of lysine losses is a function of both drying temperature and length of exposure to high drying temperatures. With continued improvements in drying technologies, the deleterious effects of high-temperature drying on available lysine will likely be minimized. Furthermore, whether these losses are significant, given that pasta is combined with other foods in the diet, is also questionable. Studies that have examined the loss of

vitamins during drying reported higher losses of thiamin and riboflavin with increasing drying temperatures. These losses were smaller for thiamin under industrial drying conditions. The effect of cooking on the retention of B vitamins has suggested that a good portion, namely 50–75% of the vitamins, is retained following cooking. With fortification of B vitamins during processing, these losses are not as significant as they may be in products that are not fortified.

As a wheat-derived staple food, pasta is second only to bread in world consumption. Its worldwide acceptance can be attributed to its low cost, ease of preparation, versatility, long shelf-life, and nutritional and sensory properties. With the greater emphasis towards the consumption of cereal based foods as a result of changes in dietary nutritional recommendations, the importance of pasta in the diet of many countries will continue to remain high.

See also: **Celiac (Coeliac) Disease**; **Coronary Heart Disease**: Etiology and Risk Factor; **Pasta and Macaroni**: Methods of Manufacture; **Wheat**: Grain Structure of Wheat and Wheat-based Products

Further Reading

Costantini AM (1986) Nutritional and health significance of pasta in modern dietary patterns. In: Mercier Ch and Cantarelli C (eds) *Pasta and Extrusion Cooked Foods: Some Technological and Nutritional Aspects, Technoalimenti Food Technology and Nutrition Series, No. 1.* London: Elsevier Applied Science.

Cubadda R (1986) Effect of the drying process on the nutritional and organoleptic characteristics of pasta. A review. In: Mercier Ch and Cantarelli C (eds) *Pasta and Extrusion Cooked Foods: Some Technological and Nutritional Aspects, Technoalimenti Food Technology and Nutrition Series, No. 1.* London: Elsevier Applied Science.

Dick JW and Matsuo RR (1988) Durum wheat and pasta products. In: Pomeranz Y (ed.) *Wheat: Science and Technology*, vol. II. St. Paul, MN: American Association of Cereal Chemists.

Giese J (1992) Pasta: New twists on an old product. *Food Technology* 46(2): 118–126.

Ranhotra GS, Gelroth JA, Novak FA and Mathew RH (1985) Retention of selected B vitamins in cooked pasta products. *Cereal Chemistry* 62(6): 476–477.

Seibel W (1996) Future trends in pasta products. In: Kruger JE, Matsuo RB and Dick JW (eds) *Pasta and Noodle Technology.* St. Paul, MN: American Association of Cereal Chemists.

Pasta Filata Cheese *See* **Cheeses**: Types of Cheese; Starter Cultures Employed in Cheese-making; Chemistry and Microbiology of Maturation; Manufacture of Extra-hard Cheeses; Manufacture of Hard and Semi-hard Varieties of Cheese; Cheeses with 'Eyes'; Soft and Special Varieties; White Brined Varieties; Quarg and Fromage Frais; Processed Cheese; Dietary Importance; Mold-ripened Cheeses: Stilton and Related Varieties; Surface Mold-ripened Cheese Varieties; **Conjugated Linoleic Acid**

PASTEURIZATION

Contents
Principles
Pasteurization of Liquid Products
Pasteurization of Viscous and Particulate Products
Other Pasteurization Processes

Principles

R A Wilbey, The University of Reading, Reading, UK

Introduction

Pasteurization has become widely accepted as an effective means of destroying vegetative pathogens in food products, with the least possible damage to the sensory qualities of the product. The heat treatment also reduces the general microbial population, so that an increased shelf-life is normally obtained. Bacterial spores and some heat-resistant enzymes will survive pasteurization processes and limit the shelf-life of the product.

Historical Origins

Cooking is an age-old method of preparing many traditional foodstuffs, and for centuries it was generally appreciated that cooked products would normally take longer to putrefy than if they were raw.

In the latter part of the 18th century there was great interest in understanding the mechanism of putrefaction. Lazzaro Spallanzani demonstrated that putrefaction may not occur in a heated sealed flask of an infusion, but that aerial contamination could result in putrefaction. The presence of microorganisms was demonstrated, and it was recognized that there was a possible division into organisms that could be killed by boiling and those that would survive this heat treatment. Subsequent experiments by Franz Schutz in the early 19th century demonstrated that it was not the air itself that caused spontaneous putrefaction, but a contaminant carried in the air. Similar experiments were carried out by Theodore Schwann at the same time.

Methods for the preservation of foodstuffs were developed in parallel with this pioneering work on the basic understanding of why foods spoil. Carl Wilhelm Scheele used heat for the conservation of vinegar, and Nicholas Appert developed the preservation of foods by heating in cans. In 1824 William Dewees recommended that milk for infants be heated to near to boiling (but not boiled) then cooled as preparation for infant feeding.

The credit for the development of a mild method of processing foods, now particularly associated with milk, has been given to Louis Pasteur, after whom the pasteurization process was named. Pasteur had, amongst his many interests, an interest in fermentations. The poor hygiene conditions associated with the production of food and beverages at that time often led to unwanted fermentations, causing putrefaction and loss of product. His experiments confirmed that fermentations were not spontaneous but the result of microbial metabolism. While some of his earlier work was with lactic fermentations, most of his work in this field was based on alcoholic fermentations, brought about by yeasts. The conversion of ethanol to acetic acid was demonstrated to be brought about by bacteria, subsequently classified as *Acetobacter*. Both yeasts and *Acetobacter* could be destroyed by relatively mild heat treatments, at about 55 °C in an acid medium such as wine, in closed vessels.

While Pasteur's work on beer, wine, and vinegar laid the foundations for hygienic processing, his complementary work on the relation between specific organisms and disease also aided the recognition of the public health implications of hygiene and of heat treatments.

By the late 19th century the economic benefits from improving the shelf-life of milk and other products were appreciated, though the microbiological and public health implications of pasteurization were not fully understood. Pasteurization of wine was adopted in both France and the USA and is still used for some wines, though filtration, higher alcohol levels, and better production hygiene have largely displaced heat treatment.

Introduction to the Dairy Industry

The heat treatment of milk on a commercial scale did not develop until the end of the 19th century, with the production of commercial pasteurizers in Germany and Denmark. The earlier treatment systems were aimed at improving the storage life of milk, often using simple continuous-flow techniques to reduce costs. The realization that milk was a potential carrier for diseases such as tuberculosis led to the development of a low-temperature long-time (LTLT) batch process, the first commercial plant being installed by Charles North in New York in 1907. It was not till 1922 that pasteurization was legally recognized in the UK when the term was defined in the Milk and Dairies (Amendment) Act, using a LTLT process at 62.8–65.5 °C for a minimum of 30 min.

LTLT pasteurization was the first safe method adopted, but the processing of milk was revolutionized by the invention in the UK of the plate heat exchanger, capable of regeneration. The development of a modular heat exchanger that could be relatively easily cleaned, together with a microbiologically effective holding tube system and a flow diversion valve, enabled milk to be heat-treated with safety on a far larger scale than had been possible with the batch-based LTLT system. With better appreciation of the thermal death characteristics of pathogens, this continuous process was able to take advantage of higher process temperatures with a corresponding shorter hold, and became known as the high-temperature short-time (HTST) process. In the UK, as in the European Community, the minimum heat treatment permitted is 71.7 °C for 15 s. It has been suggested that 15 s was originally chosen as the minimum time to allow an adequate margin for the response rate of the temperature-sensing and control system at that time.

Subsequent developments in the design of process equipment have led to the construction of pasteurization plants that may be cleaned in place (CIP), with much greater thermal efficiency and with much more sensitive and responsive instrumentation and control systems. It is now technically possible to pasteurize at higher temperatures with little or no hold – the so-called 'flash' processes.

Development of safe HTST processes for milk insured that in-bottle LTLT pasteurization of milk, pioneered by Charles North in 1911, although capable of producing a high-quality product virtually free of postpasteurization contamination, was too expensive and thus unable to compete in the marketplace. The only in-bottle milk processing that continued was for sterilized milk and this product is also diminishing with the development of continuous UHT processes. (*See* **Heat Treatment**: Ultra-high Temperature (UHT) Treatments.)

The LTLT process has found an application in cream processing. With most UK dairies now producing over half of their milk as semiskimmed plus some skimmed milk, large but variable volumes of cream are produced from the separation and inline standardization processes. Carrying out separation and cream holding at 63 °C avoids problems of varying throughputs with HTST processes.

The drive to minimize energy costs has led to the widespread use of regeneration in plate heat exchangers, where the cold raw material is preheated by the hot pasteurized product. This process can enable heat recoveries in excess of 90%, giving energy savings for both heating and refrigeration. There is a risk associated with this process, however, as in simple systems with one pump, the pressure of the raw liquid will be higher than that of the pasteurized product, so if there were a leak within the heat exchanger there could be contamination of the pasteurized product. Apart from regular checks on the integrity of the heat exchanger, for many products the problem can be avoided by introducing a second pump after the preheater plus some means of maintaining a higher pressure in the downstream section of the heat exchanger, either by flow restriction within the heat exchanger or by a backpressure valve at the exit. Where such pressure drops are undesirable, double heat exchanger plates may be used. These pairs of thinner heat exchanger plates are constructed without gaskets between the pairs so that, in the event of one of the pair failing, the leaking liquid would leak out through the gap between the plates to give a failure indication without the risk of contamination of the other flow path. Maintenance of heat exchanger plates has been eased by the introduction of clip-in gaskets that do not require adhesive.

Basic Aims of Process

The basic aim of pasteurization is summarized by the definition adopted by the International Dairy Federation (IDF): 'pasteurization is a process applied to a product with the aim of avoiding public health hazards arising from pathogenic microorganisms associated with milk by heat treatment which is consistent with minimal chemical, physical and organoleptic changes in the product'.

This definition would be equally applicable to other commodities if one were to substitute the name of that product for 'milk' in the definition.

To achieve the public health objective of pasteurization in a particular product, it is essential to be aware of the pathogens associated, or potentially associated, with that product. The thermal death characteristics of the organism(s) in that product must also be known.

In most of the early work the death characteristics were expressed in terms of a temperature–time combination that would destroy the target pathogen. In the case of milk, tuberculosis was recognized as a major disease associated with milk consumption and *Mycobacterium tuberculosis* was found to be the most heat-resistant pathogen normally associated with milk. Temperature–time combinations needed to destroy *M. tuberculosis* were published by North in 1911, North and Park in 1927, Hammer in 1928 and Dahlberg in 1932; these data are included in **Figure 1**.

This knowledge enabled safer process conditions to be set up for LTLT and subsequently for HTST processes. However, with better understanding of the kinetics of thermal death rates, more quantitative data may now be obtained.

When organisms are subjected to a moist heat above their normal temperature range there is a drop in the number of survivors with time: the logarithm of the number of survivors being inversely proportional to the exposure time. Thus for a given temperature, the time taken for a 10-fold reduction in survivors, the D value, may be obtained. D values are expressed in minutes or seconds and must be accompanied by the temperature, e.g., D_{72} for 72 °C. While D values are a good approximation for the behavior of bacterial populations, there can be a 'tail' of more heat-resistant strains leading to an overestimate of the lethality of the process. The term D may also be used without a subscript to describe the number of orders that the population may be reduced by in the course of a thermal or other process, e.g. a 10^{12} reduction may be described as a $12D$ process.

The D value will decrease with increasing temperature. The rate of change is usually given as a z value,

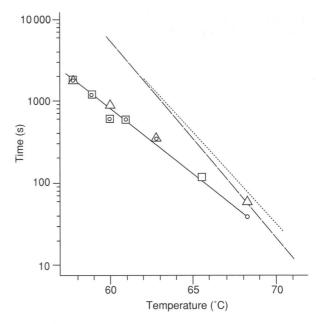

Figure 1 Time–temperature conditions for destruction of *Mycobacterium tuberculosis* in milk. Continuous line, based on data from North and Park (squares), Hammer (triangles), and Dahlberg (circles). Dashed line, virtual loss of alkaline phosphatase. Dotted line, reduction in cream line. Adapted from North and Park (1927) *The American Journal of Hygiene* 147: VII March, cited by Davis JG (1955) *A Dictionary of Dairying* p. 787. London: Leonard Hill.

the z value being the change in temperature required to give a 10-fold change in the D value. Typical z values for mesophiles are 4–6 °C in high water activity (a_w) systems, though some authorities use a value of 8 °C. The data in **Figure 1** for *M. tuberculosis* implies a z value of 6.3 °C. Some care is needed in using z values as they can vary with temperature and, as with the D values, will vary with the product being heated. Capillary tubes have been used for the estimation of thermal death data. While this method may be appropriate for low-viscosity foods, increases in viscosity during the heat treatment may give different results to those from pilot and full-scale treatments.

By using the D and z values it is possible to estimate the risks associated with a temperature–time combination, i.e., the probable level of survivors for a given level of contamination in the raw material. This is easy for a batch process such as with LTLT pasteurization, since the hold is easily measured and the contribution of the heating and cooling stages to the overall lethality of the process is small. With HTST processes, however, the temperatures are higher and the heating and cooling stages may have a significant contribution to the overall lethality of the process. Thus it is essential that the process be characterized in terms of temperature and minimum time.

Minimum time is critical as the microbiological risk (particularly the public health risk) is associated with the minimum heat treatment given to the product. Since most HTST processes are continuous, the flow characteristics of the system must be taken into account. From a microbiological viewpoint, turbulent flow in the pasteurizer will give the best results as there will be a narrower spread of flow rates and hence residence times in the equipment. However, in practice, slower flow rates may be needed to conserve desirable product characteristics or to avoid excessive pressures.

Once the plant has been characterized it is possible to analyze the process quantitatively in relation to a given risk. For instance, safe conditions for milk have been established as a minimum hold of 15 s at 72 °C. If the lethality of this process is given an arbitrary value of one unit, for instance P^*, which uses a z value of 8 °C, then we can calculate the contribution of each stage of the process to the overall lethality. An alternative unit, the pasteurization unit (PU) has also been used; in this case, taking 1 min at 60 °C ($z = 10$ °C) as the standard unit so that a safe process for milk at that temperature would need to be $\approx$60 PU if 63 °C/30 min is taken as the standard for a safe process. A comparison between PU and P^* values is made in **Figure 2**. A temperature of 10 °C is appropriate for spores but is higher than normal for vegetative pathogens – appropriate z values should be chosen to meet the substrate pathogen conditions applying to that foodstuff. Changes in the pH or a_w of the foodstuff will alter the D and z values.

It will be seen from **Figure 2** that the contribution of temperatures below 65 °C to overall lethality in an HTST process is so small that it may be ignored. However, at higher temperatures the effect of heating and cooling becomes more important. This may be illustrated by comparing two model heat treatments in heat exchangers where the rate of change of temperature is 1 °C s^{-1} (**Figure 3**).

With the HTST process A, the hold at 72 °C provides the main contribution to the lethality of the process; the heating and cooling stages contribute less than 30% to the total lethality of the process. Raising the final temperature to 80 °C as in B extends both the heating and cooling times and introduces exposure to higher temperatures; the result is that, despite there being no hold, the total heat treatment is well in excess of the minimum safe treatment.

While the primary concern in pasteurization is to obtain a safe food, this is irrelevant if the sensory quality of the food is reduced excessively, either by overcooking or due to the persistence of other less temperature-labile factors (microbiological or biochemical). Cooked flavors may be acceptable in some

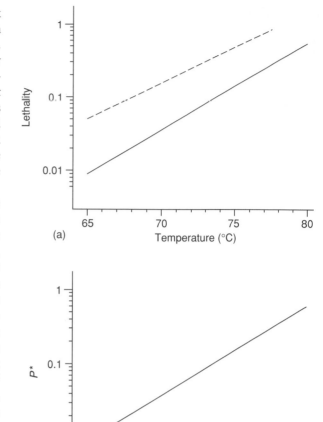

Figure 2 Lethal effect of a 1-s exposure at typical pasteurization temperatures: (a) lethality; (b) P^*. Straight line, P^* ($z = 8$ °C); dashed line, PU ($z = 10$ °C).

foods, e.g., clotted cream, but not in others such as wine.

Heat treatments may also bring about undesirable changes in the stability or functional properties of food products, often related to protein denaturation. In milk the denaturation of agglutinin (an immunoglobulin fraction) reduces the rate at which cream forms in milk on standing: the denaturation is more pronounced when more severe heat treatments are used so that less cream separates from the milk. This has been an important consideration in processing milk for bottling (**Figure 1**), where the consumer has associated cream separation with milk quality, but it is not relevant to the production of homogenized milks where cream separation would indicate a processing failure. Similarly, the protein denaturation associated with pasteurization of liquid egg white

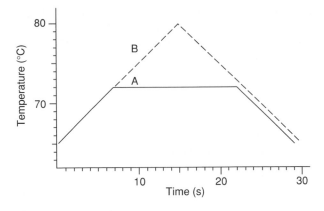

Figure 3 Example of two heat treatment profiles (A and B) with P^* values using the same heating and cooling rates. ($P^* = 1$ for a 15-s exposure at 72 °C with $z = 8$ °C.)

	A	B
Heating	0.22	2.4
Hold	1	0
Cooling	0.16	1.8
Total P^*	1.4	4.2

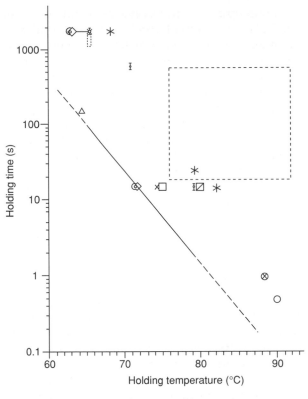

Figure 4 Examples of heat treatments used in pasteurization processes. o Milk (UK); O milk (USA); △ liquid egg (UK); ◇ cream (UK); □ cream (International Dairy Federation: IDF); ▱ cream (≥ 35%, IDF); X milk products > 10% fat or added sugar (USA); ✳ eggnog and frozen mixes (USA); ⚕ icecream (UK); --- fruit juices; $P^* = 1$ for milk. The dotted boxes indicate the range of values normally associated with low-temperature long-time (LTLT) and high-temperature short-time (HTST) pasteurization processes.

reduces the foaming properties of the product slightly, giving less volume and/or a longer whipping time than with the raw egg white.

Protein denaturation may also be used beneficially to indicate that a satisfactory heat treatment has been given, using an assay of a suitable enzyme occurring naturally in the product. The indicator enzyme should be denatured under conditions slightly more severe than those needed for microbial stability. Ideally the activity of the indicator enzyme should not be subject to wide variation with season or source.

Alkaline phosphatase is denatured under slightly more severe conditions than are required for destruction of *M. tuberculosis*, so the absence of alkaline phosphatase activity is generally used as an indicator for satisfactory pasteurization of milk. The presence of alkaline phosphatase activity may indicate either an insufficient heat treatment or contamination by raw milk. The pasteurization of liquid egg in the UK (minimum 64.4 °C for 2.5 min) is not sufficiently severe to inactivate alkaline phosphatase but will denature α-amylase, whereas the milder treatment required in the USA (minimum 60 °C for 1.75 min) aimed at achieving a 9D reduction in *Salmonella enteritidis*, will leave residual α-amylase activity.

Bacteria are more resistant to heat treatment when the a_w of the medium is lowered. Thus more severe heat treatments are normally used for pasteurization of sweetened products such as icecream and dessert products. Examples of heat treatments are given in **Figure 4**.

In fruit juices the pH is usually below 4.5 so that growth of pathogenic bacteria will not be supported, though pathogens may survive for some time. Yeasts and some lactobacilli may grow and cause spoilage of the juice, plus molds may grow at the surface. Heat treatments to eliminate yeasts and lactobacilli are more severe than for elimination of vegetative pathogens, e.g., 70 °C/60 s or 85 °C/30 s for citrus juices.

Survival of enzymes can cause problems on storage of fruit juices. In apple juice extraction, polyphenol oxidase will cause rapid browning of cold extracted juices if the juice is not immediately treated with antioxidants such as ascorbic acid or sulfur dioxide. HTST treatment at 89 °C/90 s will denature polyphenol oxidase as well as potential spoilage organisms.

In citrus juices the presence of pectinase will lead to breakdown of the cloud associated with the fresh juices. HTST treatment at 90 °C/10 s or 85 °C/4 min will denature the pectinase. For many fruit juices, including apple and orange, the juice is extracted in the country of origin, heat-treated, and concentrated.

The concentrate is then stored and transported in bulk before reconstitution and final heat treatment.

With any perishable product, the shelf-life is controlled not only by the survivors of the heat treatment but also by the posttreatment contaminants. With mild heat treatments such as the pasteurization of milk the total count may be reduced by two orders so the shelf-life will be a function of the original count, the postpasteurization contamination, and the storage temperature. Fruit juices, though subjected to more severe treatment, will also have their shelf-lives restricted by posttreatment spoilage so that the same juice may be packed as a short shelf-life chilled product or, by aseptic handling and packaging, as a long-life product stable at ambient temperatures. The advantages resulting from avoiding postprocess contamination have led to the retention of in-container pasteurization for many fruit juices and fruit drinks.

The principle of pasteurization has also been applied to the surface treatment of beef and other carcasses following slaughter. Both steam and hot water at 80 °C have been used for reduction of the surface microflora, including *Escherichia coli*. This treatment reduces both the spoilage rate and the potential public health risks associated with raw meat.

See also: **Enzymes**: Functions and Characteristics; Uses in Food Processing; **Fermented Foods**: Origins and Applications; **Heat Transfer Methods**; **Heat Treatment**: Ultra-high Temperature (UHT) Treatments; Chemical and Microbiological Changes; Electrical Process Heating; **Lactic Acid Bacteria**; **Milk**: Processing of Liquid Milk; *Mycobacteria*; **Pasteurization**: Pasteurization of Liquid Products; Pasteurization of Viscous and Particulate Products; Other Pasteurization Processes; **Spoilage**: Chemical and Enzymatic Spoilage; Bacterial Spoilage

Further Reading

Cunningham FE (1986) Egg-product pasteurisation. In: Stadelman WJ and Cotterill OJ (eds) *Egg Science and Technology*, 3rd edn. Westport: Avi Publishing.

Dahlberg AC (1932) New York Agr. Expt. Sta. Tech. Bull. 203.

Dubos RJ (1960) *Louis Pasteur, Freelance of Science*. New York: Da Capo Press.

Hammer (1928) *Dairy Microbiology*. New York: John Wiley.

IDF (1986) *Bulletin 200: Monograph on Pasteurized Milk*. Brussels: International Dairy Federation.

Kessler HG (1981) *Food Engineering and Dairy Technology*. Freising: Verlag A. Kessler.

Michalski CB, Brackett RE, Hung YC and Ezeike GOI (1999) Use of capillary tubes to validate US Department of Agriculture pasteurization protocols for elimination of *Salmonella enteriditis* from liquid egg products. *Journal of Food Protection* 62: 112–117

Rees JAG and Bettison J (1991) *Processing and Packaging of Heat Preserved Foods*. Glasgow: Blackie.

Robinson RK (ed.) (2002) *Diary Microbiology Handbook*, 3rd edn, pp. 765 New York: John Wiley.

Wilbey RA (1996) Estimating the degree of heat treatment given to milk. *Journal of the Society of Dairy Technology*. 49(4): 109–112.

Pasteurization of Liquid Products

R A Wilbey, The University of Reading, Reading, UK

Background

Liquid products are relatively easy to pasteurize, since their flow properties permit fast heat transfer by a mixture of convection and conduction. Three basic types of process have evolved to meet the requirements of the food industry: these are batch, in-container, and continuous processes.

Batch Pasteurization

Batch processing represents the simplest approach, where the bulk of the liquid is heated, and possibly cooled, within a vessel with either a jacket or a heating coil. This method was originally used for pasteurization of milk and is still used for small-scale production of icecream and yogurt mixes.

Batch processing suffers from a number of disadvantages. The product flow rate within the tank is relatively slow, and so the heat transfer rates are low, whereas increasing the severity of mixing will increase the power consumption of the plant and may damage the product. The overall heat transfer is also slow because of the relatively low surface:volume ratio, a problem that worsens exponentially with increasing vessel size. Slow heating and cooling rates contribute to a relatively high level of chemical changes in relation to the biocidal effects of the heat treatment, which can limit both the temperature and applicability of the process. Thermal efficiency of batch processing is usually low, as heat is not easily or economically recovered during the cooling process.

Against these disadvantages, batch pasteurization has the advantage of relative simplicity and lower capital cost. Thus, it has been used widely for small-scale processing, particularly where daily outputs are likely to be less than 1000 l per day.

The basic batch pasteurizing setup will usually consist of a jacketed tank with a mixer, temperature sensor and recorder, plus a source of heat and control system. The mixer should be mounted off-center to avoid excessive cavitation and swirling. A control system is essential to maintain the product at the desired holding temperature for what is usually an extended period, e.g., 30 min at 63 °C for milk or 66 °C for icecream mix. With many small batch pasteurizers, the cooling is carried out within the vessel, and the control system will actuate the changeover from heating to cooling medium.

Apart from the risk of contamination of the product towards the end of the holding period, there are two potential design/operational faults that can occur with batch pasteurizers. Inadequate heating of the head space within the pasteurizer may lead to condensation dripping back into the product with a consequent risk of contamination. The potentially more serious risk is that of cross-contamination from raw materials. The raw material feed must be kept separate from the pasteurized product exit pipe, normally by using a top feed, bottom drain setup, as illustrated in **Figure 1**. Furthermore, the drain valve must be located close to, or preferably integral with, the vessel, so that all liquid in the drain pipe between the vessel and valve will also receive an adequate heat treatment. The control system and/or the plant operating instructions must insure that the drain valve is closed before any raw materials are filled into the batch pasteurizing vessel.

The different mixing requirements and shear sensitivities have led to a wide range of types of mixers being used in batch pasteurizers. For icecream mix manufacture, an emulsification stage is required, which, on a small scale, can be carried out by using a high-shear mixer. (*See* **Emulsifiers**: Organic Emulsifiers; Uses in Processed Foods), and thus high- and low-shear mixers may be placed into the same vessel.

Pasteurizer vessels for shear-sensitive products such as cream have used large, slow-moving paddles and rocking coils, though there have been contamination problems resulting from leaks when passing coolant through the paddles.

Where outputs of about $1000 \, l \, h^{-1}$ are required, some processes use two batch pasteurizers in a semicontinuous or flip-flop mode. The vessels are equipped for mixing and heating, together with a temperature sensor, recorder, and control system. In the case of icecream, the emulsification is left to the end of the heat treatment (66 °C for 30 min or 72 °C for 10 min), the heat-treated mix being pumped to a high-pressure homogenizer then cooled by passage through a plate heat exchanger. This plant configuration is outlined in **Figure 2**. A flip-flop mode of operation is sometimes used for heat treatment of yogurt mix, where holds of 85 °C for 30 min may be applied to the mix (though it is advisable to homogenize the mix before the hold).

Although the semicontinuous mode of operation overcomes some of the objections to batch heat treatment, there remain two serious disadvantages in terms of economics and product quality. It is difficult to recover heat during the cooling phase, thus raising the cost of processing. Product quality can be compromised by the need to empty the tank, after holding, at a rate equivalent to the time taken to prepare, heat, and hold the succeeding batch. This means that the actual hold used will vary from the nominal hold to greater than twice that time when the last part of the batch is removed. Though this excess holding time can be reduced by using three tanks rather than two, there remains a significant average over-processing that can lead to cooked flavors and textural defects in the product. (*See* **Browning**: Nonenzymatic; Enzymatic – Biochemical Aspects; **Protein**: Interactions and Reactions Involved in Food Processing; **Vitamins**: Overview.)

In-container Pasteurization

Batch pasteurization of bulk liquids has a further disadvantage: the pasteurized product must then be dispensed into containers for storage, distribution, and sale. This introduces the likelihood of recontamination of the product and, without the use of aseptic handling systems, limits the shelf-life and requires refrigerated storage to minimize the growth of contaminants. Such postpasteurization contamination (PPC) may be avoided by moving the heat treatment downstream so that pasteurization is carried out in the final container.

The principle of in-container pasteurization is identical to that of batch pasteurization, but in this case, the heat-exchange surface is the container wall,

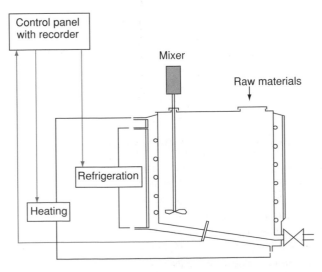

Figure 1 Outline of a simple batch pasteurizer.

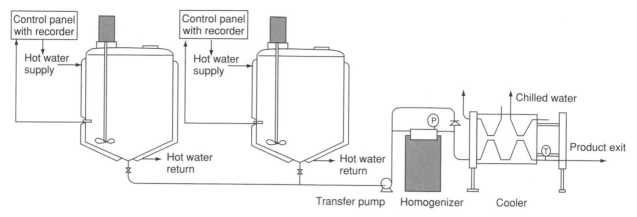

Figure 2 Outline of a semicontinuous pasteurizer.

whether plastic, glass, aluminum, or coated steel. The simplest operations may use a tank of hot water or a steam chamber for heating, followed by immersion in cold water for cooling. Adequate disinfection of the cooling water is essential to avoid contamination, either directly via pinhole leaks or indirectly via contaminated films on the surface of the pack. Unlike bulk batch pasteurization, the product temperature cannot be readily measured during the heat treatment, so test containers with the product must first be set up with temperature sensors and run through the process to establish safe processing conditions. The development of radio telemetry and solid-state logging systems now facilitates continuing checks on product temperatures, but the main control is via the temperature of the heating medium.

The majority of in-container pasteurization operations have moved to continuous processes, based on the conveyance of the containers through a tunnel where the heating and cooling operations may be carried out by spraying water over the containers. The total process time is approximately 1 h, requiring a long tunnel to give the desired holding time and capacity. The width of such tunnels increases with the capacity, as a number of containers will be treated in parallel, so that machines can be up to 25 m long and 7 m wide. Modular construction is essential.

Typical products handled in tunnel pasteurizers are soft and carbonated drinks, fruit juices, beers, sauces, and, occasionally, semisolid products such as jams and puddings. Pasteurization of acid products in this manner can yield products of a very high microbiological quality, where refrigeration is not essential for a long shelf-life.

Tunnel Pasteurizer

The product must first be filled into clean containers and sealed. A scanner or check-weigher is used to guard against under-filling. The stream of containers must then be split into groups corresponding to the number of lanes through the pasteurizer. The conveyer system must be open to permit the heat exchange liquid to drain through and must be made of corrosion resistant material, e.g., stainless steel. A combination of fixed and moving grid may be used to convey the containers on an intermittent rather than continuous basis. The conveyer speed is variable to allow for changes in throughput and holding time.

Heating is carried out by a series of zones in which increasingly hotter water, e.g., 25, 36, and 47°C, is sprayed over the containers, as illustrated in **Figure 3**. This incremental heating has two advantages: glass bottles are unlikely to suffer thermal shock, and the incremental system can allow surplus hot water (usually generated by steam injection) to cascade over into the next hottest reservoir, thus reducing the energy loss. The relatively poor heat-transfer characteristics of the containers require an extended heating time, since product flow within the container will be relatively poor, being primarily due to convection currents.

Some energy may be saved by using warm water from the cooling section to preheat the incoming containers, the cooled water then being returned to the cooling part of the tunnel. Process integrity demands that not only water temperature at the various stages be monitored but the flow checked, for instance by monitoring the feed pressure to the spray bars.

Control of In-container Pasteurization

Many tunnel pasteurizers are built for processing a specified product and packaging system and are optimized for that combination. Nevertheless, the performance of the machine must be checked in order for the safety of the process to be proved. This is done by monitoring the center temperature of test containers passing through the pasteurizer under operational conditions. An example of a temperature–time plot

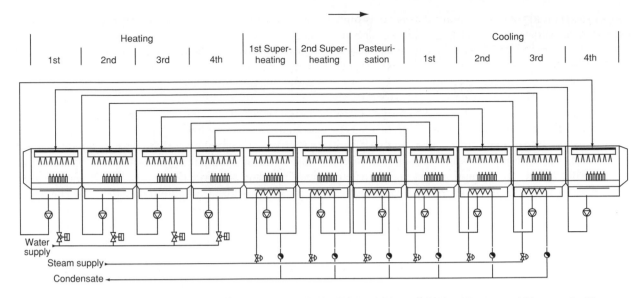

Figure 3 Schematic layout of the water circulation system in the Etna P85 tunnel pasteurizer. (Courtesy of Simonazzi srl.)

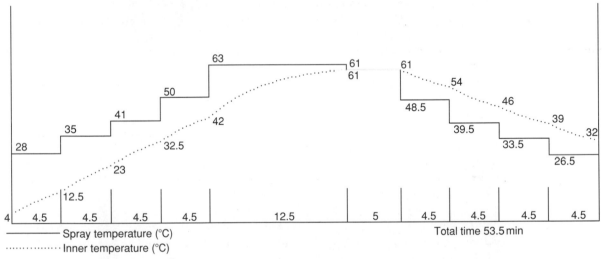

Spray temperature (°C)
Inner temperature (°C)
Times and temperatures in balanced condition.

Figure 4 Time–temperature plot for an eight-stage tunnel pasteurizer. Dotted lines indicate water temperatures. (Courtesy of Simonazzi srl.)

for a soft drink is given in **Figure 4**. The temperature–time data must then be converted into a relevant measure of the lethality of the process, such as the use of pasteurization units (PU). For beers, the PU used is a 1-min exposure at 60 °C, assuming a *z* value of 7 °C, 20 PU being accepted as a safe process with little effect on flavor, providing the dissolved oxygen has been kept low, preferably below 0.1 p.p.m. (*See* **Pasteurization:** Principles.)

Major problems can occur when the tunnel pasteurizer stops, whether through a fault in the pasteurizer itself or in the production system downstream of it. The risk then is that the containers within the system will be over-processed if no remedial action is taken. Most systems are arranged to shut off the hot water sprays in the event of a stoppage, thus avoiding further heating. Some systems may have extra cooling sprays, the best systems using a computer model of the process to model the actual process during the stoppage, predict when the held product has had sufficient heat treatment, and modify the heat treatment given to the material on start-up, including variation of the conveyer speed.

Continuous Pasteurization Processes

Continuous pasteurization processes take advantage of the shorter exposure times needed at higher temperatures, and these are often referred to as high-temperature, short-time (HTST) processes. The development of HTST processes has been dependent on the development of hygienic heat exchangers, most notably the plate heat exchanger.

The concept of the plate heat exchanger is brilliantly simple. A pair of plates are separated by a thin elastomer seal, so that the liquid flows as a thin layer, 1.5–5 mm thick, depending upon the seal design, minimizing the distance through which heat must travel. This gives an exceptionally high surface: volume ratio, typically in excess of 500:1 (compared with 5:1 for a 1000-l batch pasteurizer). The plates are rippled, e.g., as shown in **Figure 5**, so that the flow is broken up, and turbulent flow can be maintained at relatively low velocities over a larger surface area than suggested by the overall dimensions of the plate.

All the plates in the heat exchanger can be pressed with one tool, different heat exchanger configurations being achieved by the presence or absence of the ports cut at the corners of the plate and by the design of the elastomer seal. Plates are now designed so that the seals are held in place without the need for adhesive, speeding up maintenance. The range of flow rates over a plate is limited by the efficiency of heat transfer at lower flow rates, while the pressure drop and possible product damage provides the upper limit. This capacity limitation may be overcome by arranging plates in parallel so that there is an optimal flow across each plate. Similarly, the time permitted

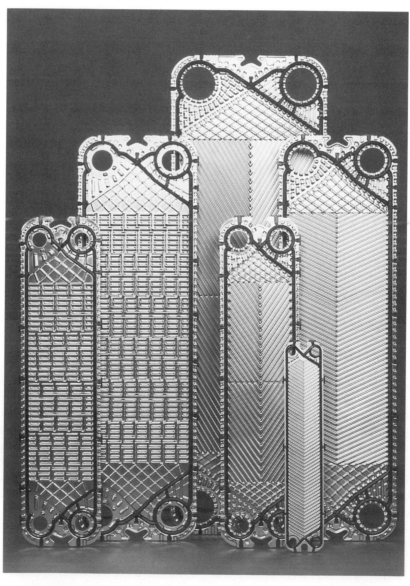

Figure 5   (see color plate 121) Heat exchanger plates. (Photo by courtesy of Tetra Pak Processing UK.)

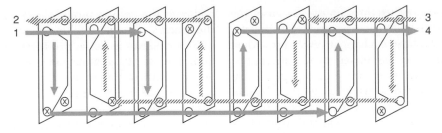

Figure 6 Arrangement of heat exchanger plates in parallel and series. 1, Cold product in; 2, Cool water out; 3, Hot water in; 4, Hot water out; x, blanked-off ports.

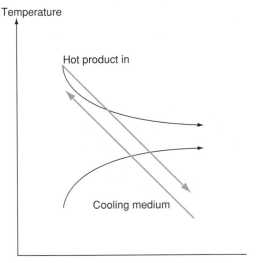

Figure 7 Illustration of co-current (——) and counter-current (——) flow through a heat exchanger using equal flow rates.

for heat exchange can be lengthened incrementally by arranging plates in series, as illustrated in **Figure 6**.

The heat-exchange medium is normally arranged to flow countercurrent to the product. This reduces the temperature differential between the two media, increasing the overall efficiency of heat transfer, as illustrated in **Figure 7**. In most cases, the heat-transfer medium will be flowing at a higher rate than the product, but with extremely heat-sensitive products such as fermented milks, similar flow rates must be used to minimize the temperature differential.

With heat-sensitive foods the build-up of a fouling layer will limit the running time of the HTST pasteurizer, owing to both pressure build-up and loss of heat exchange. The nature of the fouling layer will vary with the medium but is most commonly protein, with or without mineral deposition. Fat and carbohydrate components may become associated with fouling layers as they build up. Mineral deposits are common where hard water is used for cleaning and disinfection; additional sequestrant in the cleaning solution or periodic acid cleaning may be required. The

running time may be extended by using a wider gap between the plates, though this would require a correspondingly larger heat exchanger to compensate for lower heat-transfer rates. Larger gaps have also been used when liquids have small quantities of suspended solids, e.g., orange juice containing cells that can become lodged within the plate pack and even with back flushing are very difficult to remove. Tubular heat exchangers may be used as an alternative to plates where fouling is a serious problem but thermal efficiency would suffer.

The use of a plate heat exchanger has made it much easier to recover heat by using the hot heat-treated product to heat the incoming untreated liquid. Heat recoveries of up to 95% may be achieved by this 1:1 flow, the recovery being limited by the increasing capital cost of providing further heat exchange capacity as temperature differentials decrease and the hygiene issue of the relatively low flow rates permitting the build-up of a biofilm at temperatures that will permit microbial growth and may eventually cause significant spoilage of the product.

This heat recovery, often referred to as 'regeneration,' may also pose a hygiene risk. This risk arises because, in a simple flow-through system, the pressure of the raw liquid will be greater than that of the heat-treated liquid downstream of it. Thus, any leakage could be from the raw into the pasteurized product and could have serious public health implications. The design of heat exchanger plates seeks to avoid this problem by insuring that the two liquids are separated by either the heat exchanger plate or by two seals, one for each product with a vent to the atmosphere between. Partial failure of one seal is not a major problem as the product would leak on to the floor, but the development of a crack or pinhole crevice could lead to cross-contamination. Most heat exchanger plates are made from a corrosion and stress crack-resistant stainless alloy (e.g., AISI 316), but failure can eventually occur, and checks are needed. The risk can be avoided by using a second pump to drive the downstream part of the process at a higher pressure, raising the pressure in the downstream regeneration section either through the design of the final cooling section or by incorporating a back pressure valve into the heat exchanger discharge, when this will not damage the product. An alternative system now available is to use double plates; these plate pairs have no internal gaskets other than around the ports, so that any leakage, whether from a port or through a crack in the plate, would leak out of the heat exchanger stack and become obvious. The introduction of the air gap between the pairs reduces the heat transfer rate and would need more plates to achieve the same regeneration efficiency. It must be remembered that the savings from the use of regeneration are twofold; the first saving is from a reduction of heat input, and the second, often greater, saving comes from reduction in the refrigeration requirement.

Figure 8 illustrates a typical arrangement for the pasteurization of liquid milk. Raw milk should be taken from well-mixed bulk storage, *e.g.*, a 100 000–200 000 l silo tank, and fed to a float balance tank, which provides a constant head to the pasteurizer feed pump. Centrifugal pumps are frequently used to feed pasteurizers because of their relatively low cost, but their output is dependent on the pressure that they operate against. In the case of beer, the pump arrangement must allow an over-pressure to maintain the carbon dioxide in solution throughout the process, giving a total working pressure in excess of 8 bar. Shear-sensitive products such as cream require the use of a gentler pump such as a lobe pump, which also has the advantage of a constant output against moderate variations in pressure. Where a centrifugal pump is used, it must be accompanied either by a flow controller or by a second constant-output pump to guarantee a fixed flow rate. This requirement for a constant flow rate is set by the need to guarantee a constant residence time in the holding tube at the completion of heating. The heating is normally divided into two sections, the first preheating section using regeneration and the second heated by hot water or steam.

Sometimes, the preheating section is partially bypassed by a length of pipe with a fine control valve. This may be used to reduce the thermal efficiency during cleaning, but the main use is to control the final temperature of products such as milk for cheesemaking, where the milk will be cooled to incubation temperature (e.g., 30 °C) without the need for final cooling and without upsetting the pasteurization temperature (typically 72–73 °C). A small flow of cold raw milk past the preheater can lower the regeneration efficiency and hence raise the final milk temperature. This implies that the limit for thermal

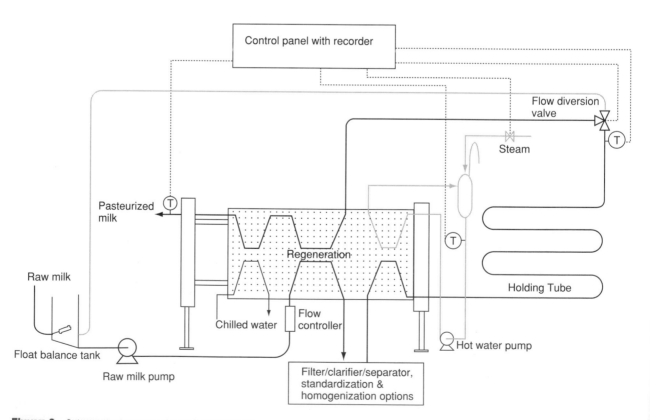

Figure 8 Schematic diagram of a milk pasteurizer.

efficiency is $\approx 63\%$ when pasteurizing milk for cheesemaking.

In the simpler plants, preheated milk may be filtered before being subjected to final heating, etc. Many modern plants, however, have replaced filtration by centrifugation, sometimes simply for clarification but more often to separate the cream. The cream may be fed back into the skimmed milk to produce standardized milks for subsequent processing. Standardization may be carried out as a continuous process, the milk then passing to a high-pressure homogenizer that will provide a constant output irrespective of back-pressure (thus requiring a pressure-relief valve). The temperature provided by preheating is usually sufficient for homogenization of milk or cream, and incorporation of homogenization at this point avoids contamination from the homogenizer. Higher temperatures (≈ 65–$80\,^\circ\text{C}$) are desirable for icecream emulsification, so the homogenizer may then be placed at the beginning of the holding tube.

Hot water is the most common heating medium, energy being provided by steam injection and the condensate being returned to the boilers. Water temperature is controlled by sensing the water temperature, the controller generating a signal to open the pneumatically operated steam control valve when the temperature drops below the preset level. Some small pasteurizers, e.g., $\leq 200\,\text{l}\,\text{h}^{-1}$, may use electrical heating only. A few larger pasteurizers may use low-pressure steam as the heating medium, though this is less easy to control and is best avoided for temperature-sensitive products.

Most pasteurizer designs require the product to be held for a fixed period of time in order to achieve the desired lethality. The most reliable method of achieving this is to use a holding tube, the design of which is critical. Any bacteria present in the liquid to be pasteurized can be treated as solid particles in suspension, and the rate at which they progress along the holding tube will depend upon the flow pattern that is induced within it. The flow pattern may be deduced by estimating the Reynolds Number (Re) for that liquid under those process conditions, i.e.,

$$Re = v\,\rho\,d/\mu,$$

where v is the velocity, ρ is the specific gravity at the holding temperature, d is the pipe diameter, and μ is the dynamic viscosity at the holding temperature. When $Re \leq 2000$ for flow in tubes, the flow is said to exhibit streamline or laminar flow, where there is a large variation in velocity across the tube, with the velocity in the center being twice the average velocity, i.e., the minimum time taken for a particle to pass along the tube would be half that of the average, as illustrated in **Figure 9**. Values of 2000–4000 are said

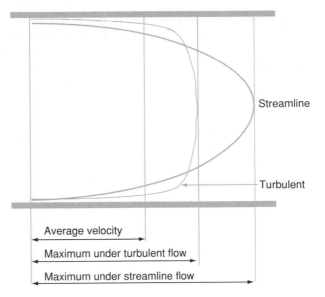

Figure 9 Illustration of the relationships between flow velocities in a tube under turbulent and laminar flow regimes.

to characterize critical or intermediate flow patterns, where the geometry of the tube, e.g., the incidence of bends, will affect the flow pattern. At $Re \geq 4000$, the flow in a tube will exhibit turbulent flow, where the distribution of velocities is much narrower, and the minimum residence time may be as high as 0.83 that of the mean. Turbulent flow is designed into most holding tubes, but the holding efficiency is assumed to be 0.75. The advantages of turbulent flow in holding tubes are tighter control of the lethality of the process and less chemical damage to the product for that lethality.

The second issue for holding tubes is heat loss. Heat is readily lost even from polished stainless surfaces, particularly with smaller plants where the surface:volume ratio is higher. Holding tubes must never be situated in draughty locations such as close to doors, since sudden draughts can lead to a drop in holding temperature that will cause the process to fail. Where cold and variable ambient conditions are expected, the holding tubes must be insulated, and if a long hold is being used, e.g., holding whole liquid egg at $65\,^\circ\text{C}$ for $150\,\text{s}$ in the UK or $60\,^\circ\text{C}$ for $210\,\text{s}$ in the USA, auxiliary heating of a box around the holding tube should be considered.

The temperature of the hot product is normally sensed by a fast responding probe close to the end of the holding tube. The signal generated is used to both provide a permanent record of the heat treatment and to generate a signal to the flow diversion valve. Should the temperature in the holding tube fall outside a preset range (e.g., 72–$78\,^\circ\text{C}$ for milk) or a failure in any of the services (steam, air, power), the valve must move into the divert position and the heated product be sent back to the balance tank. Only when the temperature is in the desired range should the valve permit product to flow forward into the downstream, cooling sections of the plant. This action protects the downstream part plant from contamination by under-processed material and hence avoids risk to the consumer. The recorder fitted to the pasteurizer must record the pasteurization temperature, final product temperature, and status of the flow diversion valve throughout the operation.

In most large-throughput plate heat exchangers, the bulk of the cooling will be achieved by regeneration, with final cooling by chilled water. Chilled water may be produced directly by refrigeration or indirectly via an ice bank. Where cooling by glycol or brine solutions is employed, more stringent controls are needed to avoid freezing. In all cases, the coolant must be of good microbiological quality and product contamination avoided. With rheologically simple liquids such as beer and milk, the cooling would use countercurrent flow to achieve the optimal exit temperature. However, with more complex liquids such as cream, where the viscosity is dependent on many factors, including the time–temperature–shear profile during cooling, a mixture of co- and counter-current cooling may be employed. For the thickest creams, there is only partial cooling in the heat exchanger, with the remaining cooling being achieved after filling, in the retail container. This approach demands the highest hygiene standards during filling and effective cooling of the containers.

Tubular Heat Exchangers

Although plate heat exchangers are excellent machines for pasteurization of low-viscosity liquids, there can be problems with those liquids either displaying higher viscosities, a tendency to foul at modest temperatures or containing suspended particles. The introduction of tubular heat exchangers enables a greater path width, e.g., 5–10 mm, so that fouling becomes less critical and particles less likely to become trapped.

In the case of fruit juices containing pulp, simple tubular heat exchangers provide an effective means of pasteurization though with higher capital and running costs. This may be attributed to lower surface: volume ratios than are possible with plate heat exchangers and to problems in heat recovery. The surface:volume ratio can be increased by adopting an annular design, and heat exchange may be improved by adopting a rippled wall profile to promote turbulence. Detail improvements in the design of tubular heat exchangers have reduced the cleaning problems,

particularly in permitting product–product heat exchange, and so direct heat regeneration has become possible. Control systems are similar to those for plate heat exchangers.

Although most fruit juices have pH values below 4.5 and are not considered high risk, pear juice, banana purée, and tomato juice may be low acid. Regardless of the public health risk, heat treatment is needed if rapid spoilage is to be avoided. Pasteurization at over 70 °C for 15 s should inactivate vegetative spoilage organisms such as yeasts, mold mycelia, and *Lactobacillus fermentum* in a high acid juice. More severe conditions such as 87 °C for 15 s would be needed for inactivation of spoilage enzymes, and for inactivation of *Byssochlamis* spp. ascospores, temperatures in excess of 100 °C would be needed. Tomato juice has been processed at 115 °C for 15 s.

Concluding Comments

Low-acid products must be kept refrigerated after pasteurization and have a relatively short shelf-life. Postpasteurization contamination poses a major threat to product quality and can only be avoided by using either in-container treatments or aseptic filling techniques. The advantages and disadvantages of the pasteurization processes are summarized in Table 1.

Table 1   Summary of the main advantages and disadvantages of pasteurization systems

Advantages	Disadvantages
Batch tank	*Batch tank*
Low capital cost	High energy cost
Low maintenance cost	Poor cooling characteristics
Very flexible	PPC risk depends on
Suitable for small quantities	downstream hygiene
In-container tunnel	*In-container tunnel*
Avoids PPC	High capital cost
High capacity possible	High space requirement
Consistent lethality	Product containers must be
More efficient than batch	heat-resistant
Low cleaning requirement	Less efficient than HTST
	Higher maintenance than
	others
Plate heat exchanger	*Plate heat exchanger*
Moderate capital cost	Viscosity limit
Low maintenance cost	Particulates limit
Low space requirement	Shear can damage product
High energy efficiency	PPC risk depends on
Consistent lethality	downstream hygiene
	Fouling problems with
	temperature-sensitive
	products

See also: **Browning**: Nonenzymatic; Enzymatic –
Biochemical Aspects; **Emulsifiers**: Organic Emulsifiers;
Uses in Processed Foods; **Homogenization**;
Pasteurization: Principles; **Vitamins**: Overview

Further Reading

Cunningham FE (1995) Egg product pasteurization. In:
Stadelman WJ and Cotterill OJ (eds) *Egg Science and
Technology*, 4th edn. Binghampton: Haworth Publish-
ing.

Hyde A (1999) Heat transfer processes in breweries – part
3. *Ferment* 12: 30–34.

Lewis MJ and Heppell NJ (2000) *Continuous thermal processing of foods: pasteurization and UHT sterilization*.
Gaithersburg: Aspen.

Zufall C and Wackerbauer K (2000) The biological impact
of flash pasteurization over a wide temperature interval.
Journal of the Institute of Brewing 106: 163–167.

Pasteurization of Viscous and Particulate Products

G Tucker, Campden & Chorleywood Food Research
Association, Chipping Campden, UK

Background

For food products of high viscosity, the choice to
pasteurize in a continuous system has distinct advantages over one where the process takes place in the
container (see **Figure 1**). This is because the heat
transfer from a heating or cooling medium to a viscous food is far from ideal, but careful design and
selection of the heat exchange system can overcome
this limitation. Of principal concern are the economics of food production, which become attractive in
continuous systems for high volumes and long
production runs. Further economic benefits can be
realized by minimizing the product's exposure to the
adverse effects of high temperatures, long processing
times and high shear preparation methods, resulting
in improved product quality, reduced processing
costs, increased safety, and increased plant throughput. This chapter considers both continuous and in-
container processing systems.

Continuous Processing Considerations

A challenge in the design of heat-transfer equipment
are the so-called prepared food products, such as

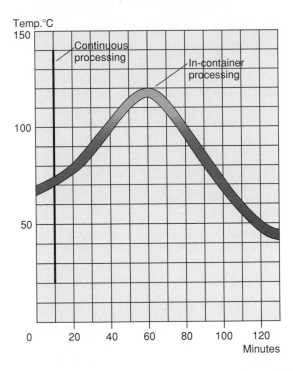

Figure 1 Time/temperature graph for comparison of in-
container and continuous processes.

tomato products, soups and sauces, and dessert products. These products are normally of high viscosity as
well as of complex composition. Also, in most cases,
the particle content is significant. With respect to the
rheological or flow behavior, the products are typically non-Newtonian, showing, in many cases, quite
extraordinary behavior.

The microbiological demands of a pasteurization
process are still to achieve commercial sterility, i.e.,
the end product must be free from pathogenic and
spoilage organisms capable of growth under normal storage and distribution conditions as well as
being free from toxins. Depending on the rheological
properties of the product and the possible presence of
particles, the design and choice of equipment can vary
significantly from case to case.

Flow Behavior

In the design and choice of heat exchangers, the flow
behavior of the product has to be taken into consideration. The flow behavior will affect, for instance, the
residence time distribution and hence the design of
heat exchangers and holding cells to obtain the sufficient thermal treatment. The basic difference between
laminar (streamline) and turbulent flow is well
known (see **Figure 2**), as is the effect on the velocity
profile from heating or cooling of the product. For
example, the maximum velocity in laminar flow is
theoretically twice the mean velocity, and in turbulent

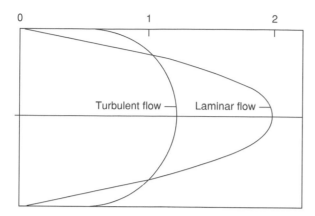

Figure 2 Velocity profiles for turbulent and laminar (streamline) flow showing the ratio of maximum velocity to average velocity.

flow around 10% higher. For viscous products, however, the flow conditions are nearly always laminar.

For liquid foods that have a complex flow behavior the velocity profiles in pipe-flow depend on the flow behaviour index. This index can range from 1.0 in simple materials such as water, milk, or fruit juice, to 0.7 in starch-based sauces, to 0.3 in tomato paste, and in theory can go to 0.0. With decreasing flow behavior index, the velocity profile increases in flatness, which means in practice that the maximum velocity decreases from twice the mean velocity. Maintaining the factor of two for the calculation of the necessary holding cell length thus creates overcooking of the product, but it is a safe assumption and one that most of the industry adopts.

A further complication to the viscous flow behavior arises with additives that give elastic properties, e.g., xanthan or gellan gum. These are sometimes used to enhance the particle-carrying properties for a carrying fluid in continuous processing. The so-called *yield value*, which normally is a measure of the product's willingness to flow by itself, e.g., from a storage tank, is also a measure of the particle carrying abilities. A significant yield value, typical of paste-like products, also adds to the flatness of the velocity profile.

Choice of Heat Exchanger

The choice of optimal heat exchanger depends to a great deal on the flow conditions. Fluids with low viscosities and no particles are preferably treated in a plate heat exchanger (see **Figure 3**). This is the most economic option. The food product and heating (or cooling) media flow in alternate channels to provide good heat-transfer characteristics. The plates are 0.5–1.25 mm thick, separated by 3–6 mm, and sealed with gaskets.

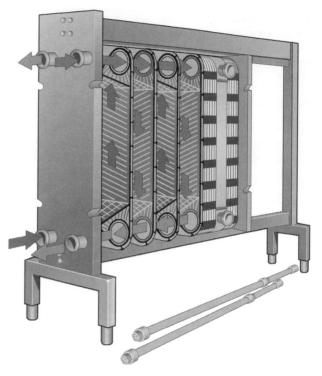

Figure 3 Plate heat exchanger.

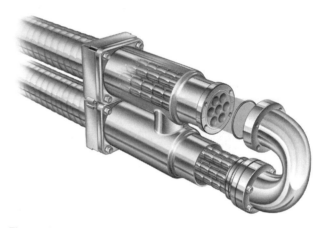

Figure 4 Multitube tubular heat exchanger.

For fruit juices with pulp and fibers up to 5 mm in length, special types of plates are available with wide gaps. Even with foods of higher viscosities, the plate heat exchanger can be utilized as long as the pressures developed are not too high, which can cause failure of the gaskets. Attempts to pasteurize foods with particles can result in blocked flow channels and increased pressure until the gaskets fail.

For fruit juices with fibers up to 15 mm in length, a multitube tubular heat exchanger (see **Figure 4**) is preferable to the plate heat exchanger. Also, foods of moderate to high viscosity with only small particles (< 5 mm) will flow through a multitube exchanger without problems. The typical inside diameter of each tube is 14 mm, and these can be grouped in bundles of four, seven, 12, 19, 27, or 37 tubes. Other combinations of tube diameters and tube bundle numbers are available, depending on the equipment supplier. Most commercial tubular heat exchangers are 6 m long.

However, if the fluid is significantly viscoelastic, i.e., exhibits a large yield value, often in combination with a high viscosity, there is a risk of maldistribution across the tubes of a multitube exchanger. In the worst case, the product flow will stop in some of the tubes, causing overcooking of parts of the product and subsequent cleaning problems. Examples of such products are hot break tomato paste or a stiff dessert pudding. In these cases, the concentric tube is the best choice, which has only one product channel and eliminates this risk. At the same time, the narrow gap and the two service medium channels surrounding the product channel provide efficient heat transfer (see **Figure 5**).

Finally, if large particles are present, the monotube is the optimal choice of tubular heat exchanger (see **Figure 6**). The drawback with a monotube compared to multi- or concentric tubes is a reduced thermal efficiency due to the thicker product layer. This is illustrated in **Figure 7**, which shows the cross-

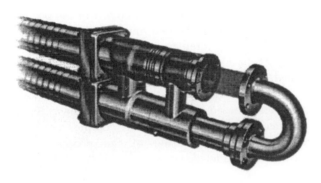

Figure 5   Concentric tubular heat exchanger.

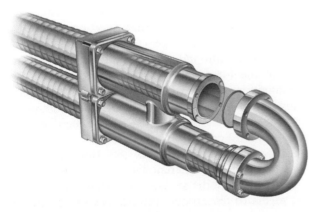

Figure 6   Monotube tubular heat exchanger.

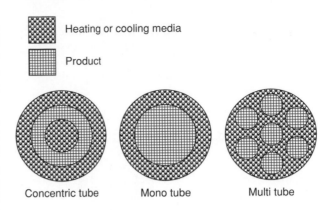

□ Heating or cooling media

□ Product

Concentric tube   Mono tube   Multi tube

Figure 7   Cross-sectional tube designs for tubular heat exchangers.

sectional flow areas of each tube type. However, to a great extent, the particles present in the product will work as 'internal mixers' and hence will promote heat transfer.

If the food viscosity is high, and there is the chance of product fouling the heated or cooled surfaces of a tubular heat exchanger, the scraped surface heat exchanger must be employed (see **Figure 8**). In principle, a scraped surface heat exchanger is a monotube equipped with a rotating internal scraper. The scraper keeps the heating surface free from any deposits and also promotes turbulence. Hence, this type of heat exchanger is ideal for products of a very high viscosity, possibly also containing large particles.

Heat-transfer Correlations

Design models for heat exchangers are normally based on empirical correlations of the dimensionless Nusselt, Prandtl, and Reynolds numbers. The derived equation is basically of the form $Nu = f(Re, Pr)$. By using dimensionless numbers, only a limited number of experiments have to be performed in which the product and heating/cooling medium flow rates and the product physical properties are varied in order to cover a large range of Reynolds as well as Prandtl numbers. The physical properties are changed preferably by changing the temperatures of the fluids involved. Decreased and increased temperatures can normally vary the viscosity of the product significantly.

The definition of the Nusselt (Nu), Prandtl (Pr) and Reynolds (Re) numbers is as follows:

$$Nu = \frac{\alpha d_h}{\lambda}$$
$$Pr = \frac{c_p \mu}{\lambda}$$
$$Re = \frac{d_h \rho}{\mu}$$

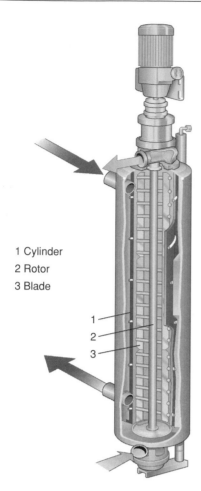

1 Cylinder
2 Rotor
3 Blade

Figure 8 Scraped surface heat exchanger.

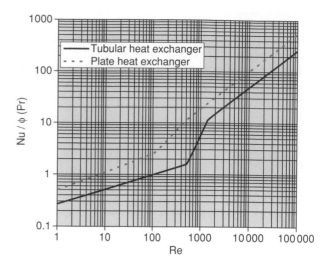

Figure 9 Comparison of *Nu–Pr–Re* correlations for plate and tubular heat exchangers.

where α is the individual heat transfer coefficient (W/m^2K^{-1}); d_h is the hydraulic diameter (m); λ is the thermal conductivity of liquid $(W/m^{-1}.K^{-1})$; c_p is the specific heat of liquid $(J/kg^{-1}.K^{-1})$; μ is the dynamic viscosity of liquid (Pa.s); v is the velocity $(m.s^{-1})$ and ρ is the density $(kg.m^{-3})$. A comparison of tubular heat exchangers with plate heat exchangers shows that the thermal performance, as read off in a typical *Nu–Pr–Re* graph, is better for the plate heat exchanger, mainly due to the complex corrugated pattern including a large number of contact points in the product channel (see **Figure 9**).

From systematic laboratory tests and from experience of commercial plants, it has been found, however, that to completely model the heat transfer of liquid food products, several more parameters have to be included. Such parameters include, for instance, particle content, particle shape, and size as well as the type of particles. In addition, the softness or hardness of the particles can effect the heat transfer by their influence on the laminar boundary layers at the exchanger surfaces.

Particulate Processes

Flow rates of particulates and carrier liquids should be balanced to give even thermal processing of the product. This relies on controlling the density of both phases to prevent sedimentation or flotation of the particulates. The limiting step will be the heat transfer to the center of the fastest and largest of the particulates, which should be determined by experimentation in the measurement of residence time in the holding tube and of thermal conductivity. Once determined, an estimation of the pasteurization level can be obtained using a mathematical procedure to solve conduction-based heat transfer equations. Before production of the food product commences, a full microbiological validation of the process safety should be conducted.

Batch Processing Considerations

Batch systems fall into two main categories, the most common being where the viscous food is pasteurized in its container, and the alternative is to carry out the processing within a large container (e.g. a closed vessel) and fill into individual containers. The advantage of the former method is that both food and container are pasteurized during the process, which allows the integral package to be commercially sterile and the potential shelf life extended. The disadvantage is that the containers need to be strong enough to withstand high temperatures (80–100°C) and the swings in pressure differential.

Irrespective of the method used, heat transfer from the media to the product thermal center is poor, because of the viscous boundary layers that develop on the inside surface of the containers. The effect is to form an insulating layer that impairs heat transfer if there is no mixing inside the containers. This results in lengthy processing times to achieve pasteurization and so few companies will process viscous foods without agitation.

In-container Systems

The traditional package is the metal can processed in a steam atmosphere, although glass jars, pouches and flexible plastic or aluminum trays can now be successfully processed in steam or hot water. These containers require an air overpressure to counteract the natural expansion of the gases present in the headspace and those released from the food as it rises in temperature. The desired effect is to push the lid or sides back to their original position and therefore minimize the stress on the seals. Examples of viscous foods pasteurized in their containers are fruit pie fillings, tomato salsas and cook-in-sauces.

The first equipment (introduced by FMC Corporation in the 1930s) to increase heat transfer to this type of food was the reel & spiral cooker-cooler. This imposes rotation of cans around their central axis and allows heating times to be reduced substantially (e.g. from 90 minutes to 15 minutes). Cans travel on a helical reel through the steam heating and water cooling sections, and in doing so rotate at high speeds. The continuous operation of the cooker-cooler allows a high throughput to be achieved and favors high volume production.

More recently, end-over-end container rotation has become popular, where the containers are constrained in baskets that rotate. Being a batch system, this offers greater production flexibility and is not restricted to cylindrical metal cans. However, heat transfer enhancement is not as great as with the cooker-cooler and so the throughput is lower and operating costs higher. Sauce products in plastic pouches are processed in steam and air mixtures, using low rotation speeds (2 to 10 rpm) to induce mixing without damaging the delicate pouch.

In-vessel Systems

The vessel acts in a similar way to a heat exchanger in that it raises the food temperature to that required for pasteurization. A typical vessel size is around 800 to 1 000 kg, and comprises a hemispherical steam-jacketed base with cylindrical sides above. A hinged lid is usually present to reduce heat loss and prevent foreign objects falling into the food. With viscous foods it is essential that the food is well-mixed, otherwise laminar boundary layers develop and the food

burns on to the heated surface. Horizontal agitators with scraped surface blades offer the most effective mixing, although recirculating pumps and vertical mixing blades are alternatives. Examples of viscous foods pasteurized this way are fruit preparations, confectionery, cook-in-sauces and tomato products.

Once pasteurized, the food can be filled either hot or cold into the containers. A hot-fill process will only require a short hold time at high temperature to ensure the inside container surfaces are pasteurized. This is usually achieved in a raining water tunnel pasteurizer, although it is possible to omit this step if the food's acidity is high (pH < 3.8), the filling temperature above 95°C and the containers pre-warmed or of low heat capacity. The shelf life of a viscous food of low pH will be many months if hot-filled, and determined by its chemistry. A cold-fill process will not guarantee commercial sterility of the container, and as such requires far greater attention to hygiene in order to minimize the introduction of microbial contamination during filling. Most cold-filled products are sold chilled and have shelf lives up to 14 days.

Heat transfer issues are similar to those for continuous processing in terms of transferring heat from steam or hot water to the core of a viscous food. Boundary layer generation on the inside walls of a food container limit the heating effect, which becomes more detrimental as the food viscosity increases. Agitation of the food container can overcome this resistance to some extent but it is important to match the rotational conditions to the product viscosity, otherwise the benefits will not be realised. Food containers can be rotated either axially in a continuous canning system or end-over-end in a batch retorting system. The continuous system offers greater efficiency for long production runs whereas the batch system offers greater flexibility.

Thermal Process Validation

Although pasteurization of the food is the desired condition, the food is referred to as commercially sterile. A pasteurization process usually operates to 6 log reductions of the target organism (further details on pasteurization treatments can be found in the CCFRA guidelines), and this differs from fully sterilized foods where the intention is to achieve at least 12 log reductions in *Clostridium botulinum* spores. The lower target log reductions for pasteurization are because of the reduced risks associated with the target microbial species when compared with the lethal botulinum toxin. The following equation is used to calculate the process or F-value from heat resistance data on the target pathogenic or spoilage organisms. When the F-value is divided by the decimal reduction time (D_T) this gives the number of log reductions of surviving spores.

$$F = D_T \ (\log N_0/N)$$

where N_{final} is the final number of organisms after a specific time-temperature history; $N_{initial}$ is the initial number of organisms; D_T is the decimal reduction time at a fixed temperature (T) to reduce the number of organisms by a factor of ten (minutes).

To prove that the pasteurization process has achieved the F-value it will be necessary to conduct validation studies using an approved method. Various methods can be selected, and their choice depends on the costs and on the nature of the food and the process type. Temperature measurements provide the cheapest method but are not appropriate for all foods, and so microbiological methods are required as alternatives.

In a validation using temperature measurements, the heat resistance data for the target organism is used to convert the measurements to log reductions. For a viscous food, the position within a container that heats slowest is usually the geometric center and so the probes are placed at the centers of several containers. To validate a process for a food containing particulates, a large food chunk or piece is usually attached to the end of a temperature probe. While this will work for incontainer processes, this method cannot be used for continuous processes.

Microbiological Methods for Process Validation

These are often referred to as direct methods, but they in fact rely on measuring the achieved log reductions for a process using a non-pathogenic organism and converting this to an F-value using the same equation. If there are no surviving organisms then it is only possible to conclude that the process achieved greater than e.g. 6 log reductions for a 10^6 initial loading. In this situation, there will be uncertainty as to whether the organisms died as a result of the process, during transportation to or from the factory, or if the spores germinated during the come-up time making them more susceptible to destruction at milder temperatures than for the heat resistant spores. Hence, controlling how these tests are performed is critical and the expertise to conduct a test using encapsulated spores or organisms tends to be restricted to a limited number of microbiology laboratories. A microbiological method can be conducted using organisms distributed evenly throughout a food product or concentrated in small beads.

The inoculated pack method is also known as the count reduction method and involves inoculating the entire food with organisms of known heat resistance. For ease of handling, the organisms are usually in the spore form. It is essential that some organisms survive the heat process in order that the containers can be incubated and the surviving organisms counted. The average thermal process received by a container can be calculated using the equation for F-value. If the product is liquid it is relatively easy to introduce the organisms but for solid products it is necessary to first mix the organisms in one of the ingredients to ensure that they are dispersed evenly throughout the container. Typical levels of the inoculum are between 10^3 and 10^5 organisms per container. An alternative is to use a gas-producing organism and estimate the severity of the process by the number of blown cans.

Encapsulating spores or organisms in an alginate bead allows the organisms to be placed at precise locations within a container or within the food particulates. The alginate bead can be made up with a high percentage of the food material so that the heating rate of the bead is similar to that of the food. This method has been used for continuous processes where the food contains particulates that require evaluation at their centers, and conventional temperature sensing methods cannot be used. Large numbers of alginate beads are used to determine the distribution of F-values that can occur in continuous processes as a result of the distribution of particle residence times. Estimating the exact number to use in a test is not straightforward because it depends on the F-value distribution, which is not known until after the test is conducted and the results analyzed. The number of organisms used will be greater than for an inoculated container test and can be of the order of 10^6 per bead. It is also important that not all are destroyed by the heat process otherwise it is not possible to estimate an F-value.

General Conclusions

When pasteurizing a viscous food, the limits to heat transfer are to overcome the development of boundary layers that impair heat transfer rates at the heated surfaces. Some degree of agitation is advantageous irrespective of whether it is a continuous heat exchange process or one processed in a container. If particulates are introduced then this imposes a further limitation to heat transfer rates. The concepts of high temperature short time (HTST) treatments do not apply easily to viscous foods either with or without particulates. Hence, most thermal processes will be considerably longer than those for foods of lower viscosity. Minimizing the thermal impact is a challenge to food processing companies operating in a market where the consumer demands products of increased viscosity or consistency, but high quality is also essential.

See also: **Heat Transfer Methods**; **Pasteurization**: Principles

Further Reading

Bolmstedt U (2000) *New Food* 3(2): 15–18.

CCFRA (1992) *Pasteurisation Heat Treatments*, CCFRA Technical Manual No. 27. Chipping Camden, UK: CCFRA.

Department of Health (1994) *Guidelines for the Safe Production of Heat Preserved Foods*. London: HMSO.

Holdsworth D (1997) *Thermal Processing of Packaged Foods*. London: Blackie Academic & Professional.

Richardson P (ed.) (2000) *Thermal Technologies in Food Processing*. Cambridge: Woodhead.

Tucker G and Bolmstedt U (1999) *Liquid Foods International* 3(3): 15–16.

Other Pasteurization Processes

G J Swart, Dairy Belle, Olifantsfontein, South Africa

C M Blignaut and P J Jooste, University of the Orange Free State, Bloemfontein, South Africa

Background

Conventional pasteurization processes usually employ continuous heat-transfer mechanisms. The primary aims of these processes are to destroy pathogenic organisms in liquid foods such as milk and to extend the shelf-life of the product for a limited period of time. (*See* **Heat Transfer Methods.**)

In addition to the conventional processes, other methods exist that employ alternative heating methods, no heating at all, or methods that are not generally regarded as true pasteurization processes. These processes may, under certain circumstances, have the same effect on foods as the conventional pasteurization techniques. It is with this in mind that the following processes have been included in this article, namely pasteurization by irradiation, microwave pasteurization, ohmic heating, 'pasteurization' by blanching, as well as lesser-known versions of the pasteurization process, namely 'cold' pasteurization, extrusion pasteurization, thermization, and chemical preservation using hydrogen peroxide.

Pasteurization by Irradiation

Foods are irradiated for different reasons. When the aim of food irradiation is the inactivation of certain spoilage microorganisms that may be present in foods without necessarily leading to sterilization, the process is called 'radurization.' The 'radicidation' of foods, however, is aimed at the inactivation of pathogenic nonspore-forming bacteria and, in some cases, will also inactivate toxigenic fungi, viruses, and parasites. Both radurization and radicidation have an effect comparable to that of heat pasteurization. To understand the process better, the following aspects are dealt with in this discussion, namely the basic theoretical principles, factors influencing the process, uses, product suitability, nutritional implications, and, finally, the acceptability and current status of the process.

Basic Theoretical Principles

The underlying principle of food irradiation (and therefore also irradiation pasteurization) is based on chemical changes caused in foods by a form of energy called 'ionizing radiation.' Different kinds of radiation are included in this concept. Only a few of these are suitable for use in the treatment of foods, namely X-rays, γ-rays (from ^{60}Co or ^{137}Cs), and electron beams (cathode rays and β-rays). Nonionizing radiation, such as ultraviolet radiation, is essentially absorbed at the food surface and does not have effective penetration properties. Although limited work has been done with ultraviolet radiation, research in this regard concentrates more on extending the shelf-life of fruit and vegetables by controlling fungal spoilage, such as fusarium rot (caused by *Fusarium solani*) charcoal rot (caused by *Marcrophomina phaseolina*) and soft rot (caused by *Rhizopus stolonifer*). Ultraviolet lamps are also used to inhibit the formation of bacterial slimes on surfaces. Of the two radiation treatments, processes using ionizing radiation are more important in the context of the present assignment. (*See* **Irradiation of Foods**: Basic Principles.)

In the case of either radurization or radicidation, an irradiation dose of less than 10 kGy will have the desired effect. For radurization the doses required vary within the 1–5 kGy range. Sometimes, even less will suffice, e.g., beer radurization at 500 Gy. An acceptable end product, however, is not necessarily guaranteed. For the radicidation of foods, the radiation doses vary according to the targeted organism. To control *Salmonella* contamination, for example, doses of between 2 and 6.5 kGy have been recommended to reduce the contamination level three- to sevenfold. An estimated sevenfold reduction of other pathogenic nonspore-forming foodborne bacteria, such as *Shigella*, *Mycobacterium*, *Escherichia*, *Staphylococcus*, *Streptococcus*, and other species, may also be obtained using doses ranging from 5 to 8 kGy. It should always be remembered that the suggested doses may be influenced by the nature of the food to be irradiated. It is, therefore, essential to take note of the factors that generally influence irradiation processes. (*See Escherichia coli*: Occurrence; *Mycobacteria*; *Shigella*; *Staphylococcus*: Properties and Occurrence.)

Factors Influencing the Process

The success of any form of irradiation depends on the influence and interaction of certain factors. These should also be taken into account when deciding on pasteurization by way of irradiation, since it could influence the postirradiation properties of the food. These factors include:

1. the resistance of some organisms to radiation;
2. the rate at which radiation is applied;
3. the water content of the food to be irradiated;
4. the influence of temperature on subsequent effects in the irradiated food;
5. the presence of oxygen and additives during irradiation;
6. the possibility of combining irradiation with other treatments to obtain the desired effect;
7. the penetration ability of the selected type of ionizing radiation within stated radiation limits;
8. the elemental composition of the food (as a whole) to be irradiated;
9. the magnitude and composition of the initial microbial population present in the food to be irradiated;
10. the susceptibility of the product (such as fruit and vegetables) to radiation damage;
11. cost factors with regard to the process.

Since the above-mentioned factors influence the end result, attention should be given to the applications of the process, product suitability, and nutritional implications that may result from irradiation.

Uses, Product Suitability, and Nutritional Implications

In order to understand the use of radurization and radicidation, it is essential to take into account the fact that these treatments should supplement normal food-processing practices. Radurization especially is used solely to prolong the shelf-life of products. As such, it has shown promise with fresh products such as fish, meat, poultry, vegetables, fruit, baked goods, etc. (*See Irradiation of Foods*: Applications.)

Radicidation is usually applied to products in which radurization is not normally able to completely eliminate pathogenic nonspore-forming bacteria. Such foods include fruits and fishery products (e.g., shellfish and fishmeal), although it is necessary to mention that radicidation of shellfish has little effect on the presence of viruses. Irradiation of meat products, such as poultry and pork at radiation levels up to 7 kGy, have, however, been successful in controlling pathogens, such as *Salmonella*, *Campylobacter*, and *Listeria* species. These pathogens cannot be controlled by good manufacturing practice alone. (*See Campylobacter*: Properties and Occurrence; *Listeria*: Properties and Occurrence.)

With regard to nutritive value, it has been shown that normal nutritional values are retained in proteins, lipids, and carbohydrates after irradiation. The nutritional availability of minerals can, however, be altered by the treatment, and a small amount of vitamins may be destroyed. These losses, however, compare favorably with those incurred during conventional processes. These aspects should have a positive influence on the acceptability and current status of radurization and radicidation. (*See* **Irradiation of Foods**: Processing Technology.)

Acceptability and Current Status of the Process

Irradiation of foods is the one process for which safety aspects have been considered very thoroughly. These include aspects such as microbiological implications, effects on nutritional value, and the possible production of toxic substances, carcinogens, and radioactivity in treated foods. It may be concluded that treated foods can be considered safe on all counts (especially at the dose levels required by radurization and radicidation), provided that:

1. approved doses are applied;
2. the foodstuffs to be irradiated do not possess abnormal levels of elements that could be rendered radioactive; and
3. acceptable packaging materials are used.

A limiting factor with regard to both radurization and radicidation processes is the possible development of objectionable changes in the properties of some foods, such as flavor, odor, color, texture, perceived freshness, etc. Negative effects of irradiation can, however, be counteracted by using the treatment in association with other processes. Lower doses of radiation (radurization) accompanied by a refrigeration process are proving to be technically and economically feasible.

It should be mentioned that irradiation can also have a product-enhancing effect, e.g., the enhanced flavor of brandy. Research with regard to this aspect of the process is continuing. **Table 1** demonstrates the current interest in low-dosage irradiation and fields

Table 1 Current interest in irradiation processes from differing perspectives

Type of interest	Author and year of publication	Title of paper	Irradiation dose applied	Source
Research	Ma CY et al. (1990)	Gamma irradiation of shell eggs; internal and sensory quality, physicochemical characteristics and functional properties	0.97, 2.37, 2.98 kGy	Canadian Institute of Food Science and Technology Journal 23(4/5): 226–232
Research	Huhtanen CN (1990)	Gamma radiation inactivation of enterococci	Varying	Journal of Food Protection 53(4): 302–305
Research	Stevens C et al. (1990)	Effect of ultraviolet radiation on mold rots and nutrients of stored sweet potatoes	Varying	Journal of Food Protection 53(3): 223–226
Research	Harris T et al. (1989)	Poultry meat irradiation – effect of temperatures on chemical changes and inactivation of microorganisms	0.5–10.0 kGy	Journal of Food Protection 52(1): 26–29
Research	Dempster JF et al. (1985)	Effect of low-dose irradiation (radurization) on the shelf-life of beefburgers stored at 3 °C	1.03, 1.54 kGy	Journal of Food Technology 20: 145–154
Research	Modi NK et al. (1990)	Effects of irradiation and temperature on the immunological activity of staphylococcal enterotoxin A	9.4, 12.2 kGy	International Journal of Food Microbiology 11: 85–92
Research	Paster N et al. (1985)	Preservation of a perishable pomegranate product by radiation pasteurization	2.0, 4.0 kGy	Journal of Food Technology 20: 367–374
Overview	Various (1989)	Food irradiation: a most versatile twentieth century technology for tomorrow		Food Technology July: 75–97
General	Lacey RW and Dealler SF (1990)	Food irradiation: unsatisfactory preservative		British Food Journal 92(1): 15–17

Reproduced from Pasteurization: Other Pasteurization Processes, *Encyclopaedia of Food Science, Food Technology and Nutrition*, Macrae R, Robinson RK and Sadler MJ (eds), 1993, Academic Press.

of application. It clearly shows the wide interest (both positive and negative) generated by the process.

In conclusion, it should be stated that the current status of the process depends on individual countries and their health standards. Before commercial exploitation of the process by a manufacturer, the government of the country involved should approve or have approved it. Although many countries permit food irradiation, doubts still exist in some quarters with regard to the safety of the process. (*See* **Irradiation of Foods: Legal and Consumer Aspects.**)

Microwave Pasteurization

Microwave processing of foods has been the focus of much research since the 1940s. As a food-processing method, however, it became of practical interest in the 1960s with the development of multikilowatt conveyorized ovens. The microwave principle only came into general use in the 1980s, when microwave heating applications started to increase. One of these applications was pasteurization using microwave energy. To understand the process better, this article concentrates on the basic theoretical principles of the process as well as its applications and acceptability.

Basic Theoretical Principles

Microwaves are electromagnetic radiant energy waves with wavelengths of between 0.025 and

0.75 m and frequencies of approximately 2450 and 915 MHz in the case of food applications. Most countries have standardized on 915 or 896 MHz for industrial use. The energy level for a specific foodstuff is chosen according to the amount of energy lost in the process due to packaging, for example, since this adversely affects the penetration capability of the microwaves. If a high penetration level is required, a microwave frequency with a lower loss factor should, therefore, be selected.

Another important aspect is the fact that microwave energy has an effect on dipolar molecules such as water molecules. Application of microwave energy causes these molecules to oscillate, causing intermolecular friction and, ultimately, producing heat. Heating is, therefore, achieved inside the foodstuff being irradiated, depending on the type of applicator and the foodstuff being treated. All in all, the treatment facilitates control and the prevention of overprocessing in the case of pasteurization.

Applications and Acceptability

Microwave pasteurization as a process has been applied especially to milk and milk products. The Bach process (using a dual-frequency application) claims great success with regard to cultured milk products. In the case of fluid milk, patents exist for a microwave heat exchanger used in the pasteurization of milk. In spite of the fact that different processes, such as the high-temperature/short-time, multitherm, and in-pack continuous process, have been developed, the feasibility of microwave pasteurization of milk on an industrial scale has not really been established.

Microwave pasteurization has the following advantages and disadvantages:

1. Amino acids may be altered during the process, and this aspect should be looked into in more detail.
2. Microwaves are successful in pasteurization because of their heat-generating action. The effect it has on microorganisms as such, however, needs to be resolved more fully.
3. Microwave-pasteurized products retain the properties of the fresh product.
4. The process shows promise as a high-efficiency, low-energy process. This limits bacterial build-up in usually restricted flow areas of the system.
5. The design of the system is of paramount importance for successful processing.
6. The effectiveness of heating is related to applicator design and can be either more or less even than conventional methods.
7. The operating costs are high.
8. Magnetrons have a relatively short lifespan and are essential in the conversion of electrical energy to microwave energy.
9. Depending on the design, less factory space may be required.
10. Safety is of great importance, since damage can be caused to the eyes and other tissues that absorb microwaves.
11. Both small static microwave ovens (home pasteurization) and complex, moving-belt microwave tunnels (industrial application) may be used. A schematic diagram of a microwave system for processing foods under pressure is given in **Figure 1**.

12. Microwaves are not necessarily effective in killing bacteria, especially in higher-density products. Manual mixing has been suggested to facilitate more uniform heat distribution. Research in this field is continuing.
13. Heat-sensitive products can be treated to great advantage using microwave technology.

In conclusion, many questions relating to microwave pasteurization and microwave technology in general still remain to be answered. Continuing research should, however, make more general applications a reality.

Blanching

Although blanching is not traditionally seen as a pasteurization process, some authors deem it to be a 'kind' or 'type' of pasteurization. While pasteurization, however, has as its main function the destruction of pathogenic microorganisms, it also destroys certain natural food enzymes in the process. Blanching, however, is applied primarily to inactivate natural enzymes in fruits and vegetables to be processed. Its secondary effect is that it reduces the microbial load. Both pasteurization and blanching are similar in that they employ temperatures below 100 °C.

A process involving acid blanching and the addition of ethylenediaminetetraacetic acid to a canning brine has been a recent topic of research. The results have shown potential with regard to the control of spoilage and botulinal toxigenesis in canned products. This draws a further parallel between blanching (a preparatory method) and pasteurization (a processing method). (*See* **Canning**: Principles.)

With this in mind, the reader is referred to the article on canning for further information.

Ohmic Heating

Ohmic heating is a new food-processing operation which offers a major advantage in the continuous processing and particulate food products.

Basic Theoretical Principle

Ohmic heating occurs when heat is internally generated by the passage of an electrical current through a food product and is in this way similar to microwave heating. In the case of ohmic heating, the penetration depth, however, is virtually unlimited, and heat penetration is more effective. The process depends on the electrical conductivity of the product. Fats, sugars, and syrups are, therefore, not suitable for this type of processing, while foods that contain dissolved ionic salts conduct electricity sufficiently and are,

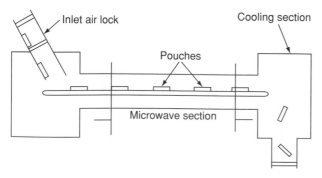

Figure 1 Schematic diagram of a microwave system for processing foods under pressure. From Decareau RV (1985) *Microwaves in the Food Processing Industry*. London: Academic Press, with permission.

therefore, suitable for ohmic heating. If both the liquid and solid phases of a product contain sufficient dissolved ionic salts, heat penetration is more rapid than with conventional heating. (*See* **Heat Treatment: Electrical Process Heating.**)

Design of Ohmic Heaters

The ohmic heater is based on a well-known principle, and the apparatus provides an alternative to tubular or scraped-surface heat exchangers for the processing of viscous and particulate food materials. The design of an ohmic heater relies on the principle that an electrical potential is generated across a moving column of product. For safety reasons, the heater column inlet and outlet electrodes are earthed.

Figure 2 illustrates the electrical configuration and construction of the electrical resistance heater.

The heater column is arranged in such a manner that an upward flow occurs in the column while the product is progressively heated to the required temperature. Back pressure is maintained at a constant

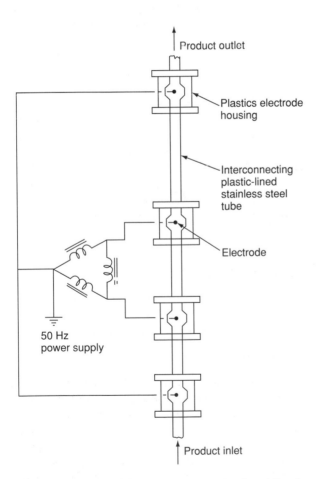

Figure 2 Electrical configuration and construction of the electrical resistance heater. From Skudder PJ, Biss CH and Coombes SA (1988) Ohmic heating: a breakthrough in profitable food manufacture. *Food Review* 15(5): 24, 27.

1 bar when heating products up to 90 °C, but a pressure of 4 bar is required up to the maximum temperature of 140 °C. The back pressure is controlled by regulation of the pressure on the surface of the product in a presterilized tank using sterile compressed air.

Acceptability and Applications

The ohmic process offers several advantages over more conventional particulate processing methods. These are:

1. the ability to handle food materials containing particles as large as 25 mm^3;
2. the absence of moving parts (feed pump excluded), which can damage shear-sensitive food products;
3. the ability to heat food products in a continuous flow without the need for any heat-transfer surfaces;
4. a considerably reduced risk of fouling;
5. the generating of heat in the product solids without reliance on thermal conductivity through liquids;
6. the absence of noise during the operation;
7. low maintenance costs;
8. the ease of control and instantaneous start up or shut down;

Examples of food products which have been processed successfully in this manner with minimal structural damage include meat chunks, prawns, diced or sliced carrots, baked beans, and sliced mushrooms. Fruits, such as diced apple and whole strawberries, have also been successfully heat-treated using the ohmic heating principle, and product quality reportedly varies from good to excellent. Ohmic heating has been developed by the APV company as a feasible practical process, and the reader is referred to the article on ohmic heating for further information.

Other Processes

To conclude this article, attention will be given to less well-known, product-specific processes, namely cold pasteurization, pasteurization by extrusion, thermization and chemical preservation by hydrogen peroxide.

Cold Pasteurization

As the name implies, the cold pasteurization technique does not make use of heat to reduce the microbial load. Instead, microporous membrane filters are used to retain the majority of bacteria and yeasts. In this respect, the technique is similar to any other pasteurization process. This process is mainly used in the case of heat-sensitive products, such as draught beer, wine with less than 17% alcohol (especially sparkling wine) and pulp-free fruit juices. (*See* **Filtration of Liquids.**)

Pasteurization by Extrusion

Although extrusion may not, in general, be comparable with conventional pasteurization, research has shown that the microbial load of thermosensitive food powders, such as egg white powder or whey powder, could be reduced by an extrusion process without any loss of functional properties of proteins and carbohydrates, for example. This advantage, however, depends on the barrel temperatures and contact times, since high temperatures combined with longer contact times lead to a loss of functional properties as a result of mechanisms such as starch gelatinization and protein denaturation. The feasibility of this process as a form of pasteurization is not very clear, and therefore requires further investigation. (*See* **Extrusion Cooking:** Principles and Practice.)

Thermization

The thermization process is a subpasteurization heat treatment of milk at 62–65 °C for 10–20 s, followed by refrigeration. It is used as a prepasteurization treatment of raw milk to safeguard milk quality during prolonged storage in insulated silos. The process is also used as a postpasteurization treatment of dairy products. Research has shown that this process effectively reduces both the total and psychrotrophic bacterial counts, enabling thermized milk to be stored for up to 3 days longer at 8 °C. In this way, detectable sensory changes and the concomitant decrease in shelf-life are limited.

Thermization is only successful if applied to raw milk and products of good quality.

Chemical Preservation by Hydrogen Peroxide

Hydrogen peroxide is undoubtedly a most effective germicidal agent. It has been proven as an effective bacteriostatic and bactericidal agent and has been used in a treatment (30 min; 49–55 °C, 0.08% by weight) that has been suggested as a substitute for pasteurization. The peroxide concentration, temperature used, exposure time, and bacterial species and number all influence the effectiveness of the process. (*See* **Preservation of Food.**)

Treatment by this method destroys coliforms, anaerobic spore-formers and also a number of pathogens. The most resistant pathogen has been found to be *Mycobacterium tuberculosis*. Treatment at levels of 0.115% (v/w) hydrogen peroxide, however, completely inactivates this pathogen. Gram-positive organisms are not inactivated to the same extent as Gram-negative organisms.

Application of hydrogen peroxide as a milk preservative has been attempted in the dairy industry. Certain countries with underdeveloped milk collection industries and high environmental temperatures have expressed interest in this preservation method. The addition of catalase after exposure or the catalase from microorganisms and leucocytes, together with heat treatment, completely destroys or decomposes the hydrogen peroxide. The breakdown products (water and oxygen) are virtually undetectable in milk.

Hydrogen peroxide treatment of milk may be applied to reduce the total bacterial count and to extend the keeping quality. This treatment should, however, be supplemented by approved processing methods to ensure complete destruction of all pathogens. The question of whether hydrogen peroxide treatment of milk may be allowed depends on the specific country and its health regulations.

See also: ***Campylobacter***: Properties and Occurrence; **Canning**: Principles; ***Escherichia coli***: Occurrence; **Extrusion Cooking**: Principles and Practice; **Filtration of Liquids**; **Heat Transfer Methods**; **Heat Treatment**: Electrical Process Heating; **Irradiation of Foods**: Basic Principles; Applications; Processing Technology; Legal and Consumer Aspects; ***Listeria***: Properties and Occurrence; ***Mycobacteria***; **Preservation of Food**; ***Shigella***; ***Staphylococcus***: Properties and Occurrence

Further Reading

Coghill DM, Mutzelburg ID and Birch SJ (1982) Effect of thermization on the bacteriological and chemical quality of milk. *Australian Journal of Dairy Technology* 37(2): 48–50.

Decareau RV (1985) *Microwaves in the Food Processing Industry*. London: Academic Press.

Jay JM (1986) *Modern Food Microbiology*, 3rd edn. New York: Van Nostrand.

Kang KH, Yoon KB and Pack MY (1983) Microbial contamination of raw milk and prevention with hydrogen peroxide. *Korean Journal of Animal Sciences* 25(4): 296–302.

Potter NN (1986) *Food Science*, 4th edn. Westport, CT: AVI.

Skudder PJ, Biss CH and Coombes SA (1988) Ohmic heating: a breakthrough in profitable food manufacture. *Food Review* 15(5): 24, 27.

Urbain WM (1986) *Food Irradiation*. Orlando, FL: Academic Press.

Young GS and Jolly PG (1990) Review: microwaves: the potential for use in dairy processing. *Australian Journal of Dairy Technology* 45(1): 34–37.

PASTRY PRODUCTS

Contents
Types and Production
Ingredient Functionality and Dough Characteristics

Types and Production

J B Harte, Michigan State University, East Lansing, MI,
USA

Background

There are many types of pastry dough. Shortcrust or
standard plain pastry contains four ingredients –
flour, fat, water, and salt – that are mixed, rolled,
and cut into shapes. Alterations in the proportions
of these ingredients, the addition of other ingredients,
and the employment of various mixing and manipulation techniques produce popular pastry doughs,
such as puff, choux, and strudel pastries. It is the
purpose of this article to outline popular pastry varieties and present some common small- and large-
scale production methods.

Plain Pastry

Pie Crusts

Shortcrust pastry or piecrusts are prepared by cutting
solid fat into the flour and salt, and gradually adding
water to obtain the consistency for rolling the dough
into the desired shape. A low-protein, pastry flour is
used to prevent the development of gluten in the pie
dough. Limiting the mixing step after the addition of
water and limiting reworking of dough will also help
to prevent gluten development that would result in a
tough crust.

Medium-flaky crust is obtained by cutting the fat
into lumps the size of rice or coarse cracked corn.
Flakier crusts are obtained by chilling the water and
sometimes the flour to 1.7–4.5°C prior to mixing,
cutting fat into larger pieces and/or keeping the
mixing step as short as possible. Chilling the dough
before rolling also helps to keep the fat firmer. A
mealy crust forms when the flour and shortening are
mixed thoroughly. The dough should be at about
18 °C after mixing and may be held there for 3–4 h
in order to distribute the moisture more evenly. The
type of fat used also affects the flakiness or mealiness
of the crust. Rolling of the dough should start from

the center and continue in the direction that will
produce the desired shape and thickness.

Many varieties of pie crust dough, with fat contents
ranging from less than 50 to 75%, are used. Pie crusts
can also be prepared with a combination of shortcrust
dough and short pastry (pâte de sucre or flan pastry)
or with short pastry alone. Pie crust dough can be
used for two-crust, fruit-filled pies, single-crust, and
soft-filled pies. The typical one- and two-crust pies
are approximately 23 cm in diameter and serve six to
eight people. Crusts are also used for baked meat or
fish pies, quiches, tarts, and turnovers. Fruit-filled
turnovers can be made from dough scraps.

In two-crust fruit pies, the bottom crust should
be slightly thicker than the top in order to support
the fruit filling. Placement of cooled fruit filling in the
prepared pie crust allows it to bake adequately before
the filling begins to boil. Two to four cuts should
be made in the upper crust to vent the steam during
baking. Crusts are baked at 281–232°C, reducing the
moisture content to a low level.

Two types of machines are used to form pies. Pie
crusts can be produced commercially on a small-scale
basis using a pie press that shapes dough balls into
pastry sheets. Alternatively, pies can be formed by
automatic sheeting and cutting. The rotary pie machine mixes the dough, reduces its thickness and cuts
it into 7.5 × 12.5-cm sections. These pieces are automatically cross-rolled into thin, round pie pastry
sheets that are manually placed into the pie plates.
The top crust is placed on top after the filling is added
to the bottom crust. The rotary pie machine then
finishes the pie by crimping, scaling, and trimming.

The straight-lined pie type is another example of a
sheeting and cutting machine that is more completely
automated than the rotary pie machine. Pie dough is
divided in two and separately reduced, cut into
7.5 × 11.5-cm pieces and cross-rolled into the top and
bottom round pastry sheets. The pastry is automatically placed into the pie plate, filled, topped with another cut pastry sheet, sealed, crimped, and trimmed.

Short Pastry

Short, pâte de sucre or flan pastry is a rich dough
containing a large amount of fat and, unlike standard

pastry, eggs and sugar. This pastry is used for biscuits, tarts, some pies, and the bottom and sides of fruit flans and cheesecakes. If a nonsweet pastry is desired for meat, vegetable, or seafood dishes, other ingredients, such as Parmesan cheese, may be substituted for the sugar. Smooth blends of shortcrust and pâte de sucre doughs are often used for pie shells and cream rolls. Chocolate or cocoa may also be added to these blends.

Puff Pastry

Puff pastry is lighter, flakier, and more delicate than standard shortcrust or flan pastry. It contains a higher fat level and has a unique preparation method that is critical for preparation of a successful product. This pastry can be used for a wide variety of products, including turnovers, shells filled with whipped cream, custards or fruit, vol-au-vents, and sausage rolls, and is considered a staple in commercial bakeries.

The pastry is prepared with a high-protein flour (bread), water, salt and shortening. A firm, plastic shortening, equal to 50% of the dough weight and containing some moisture, is optimum. Other ingredients, such as eggs and acids (cream of tartar, vinegar, or lemon juice) can be used. There are three commonly used methods for mixing and handling the puff pastry dough – the French (roll-in), English (roll-in), and the Scottish (all-in) method.

The French method involves rubbing a portion of pastry fat into the flour to lubricate the gluten. A portion of the water, acid, and eggs is placed into the center of a ring made from the mixed flour and shortening. The dough is mixed well, and the procedure is repeated until all the ingredients, with the exception of the remaining fat, have been added. The dough is rolled into a rectangle, 1.25 cm thick with a thinner outside border, and the fat is rolled to the size of the thicker dough. After the fat is placed on the thicker dough, the thinner dough is folded over on the thicker dough, and the thinner dough is folded

over, making a type of pocket. The fat and dough is then rolled into a rectangle. One half-turn of the dough is formed by folding one end of the dough two-thirds of the way down the rectangle and folding the opposite end to cover the first fold (see **Figure 1b**). The new rectangle is three layers of pastry fat separated by dough layers. The dough is relaxed, and the rolling procedure is repeated after turning the dough through 90° for a total of six half-turns. The French method of puff pastry preparation is the most popular but is not used by some, because it is complicated and difficult to produce under high-speed commercial manufacturing processes.

The English method is the most popular in the USA. The dough is sheeted into a rectangular shape, and the roll-in fat is spread uniformly over two-thirds of the area. Three-fold pastry preparation (**Figure 1**) is used to form three dough layers separated by two layers of fat.

The Scottish method is similar to pie-crust preparation, simple, and widely used commercially. All the pastry fat is broken into lumps (5 cm) and mixed loosely in the flour. The remaining ingredients are then added to form a lumpy dough. The rolling procedure is the same as the French method with the addition of one extra half-turn. After the final rolling, the dough is shaped and relaxed before baking. For maximum volume, puffed pastries should be baked at high temperatures (204–212°C). The Blitz method is a modification of the Scottish method that partially develops the gluten prior to adding the firm shortening.

Two mechanical mixers are commonly used in batch commercial pastry production. The low-speed, double-arm mixing machine with a large dough bowl and slow mixing produces a good-quality commercial puffed pastry. The Scotch pastry method must be followed when the high-speed mixer is used in large-scale, high-product-capacity operations. Careful timing is critical with these mixers to prevent overmixing that results in an intimate mixing of fat with the dough and formation of a short rather than a

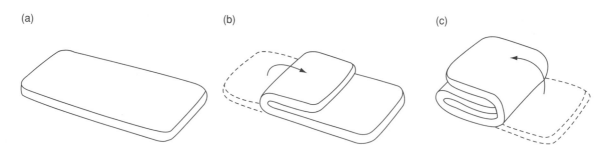

(a) (b) (c)

Figure 1 Three-fold pastry preparation: (a) the dough is rolled into a rectangle; (b) one end of the dough is folded two-thirds of the way down the rectangle; (c) the opposite end is folded to cover the first fold.

puffed pastry. After mixing the dough, any of the methods described in the commerical pastry preparation section can be applied. Continuous puff pastry dough-production equipment extrudes the dough and shortening into a double-layer continuous tube with the shortening in the center layer. This is flattened and reduced by a dough stretcher. The sheet is folded, stretched, and thinned. Pieces are cut into their final shape for pastry shells, cheese or savory sticks, napoleons (mile feuilles), turnovers, sugar crisps, etc. Products prepared with this continuous process are usually intended for wholesale distribution. Thus, their size and flakiness are limited to prevent breakage during shipping.

Choux Pastry (pâte é chou)

A wide variety of pastry products are made using choux paste: cream puffs, éclairs, French crullers, soup nuts, and decorative shapes used for pastry garnishing. Choux pastry differs from other pastries in that it is cooked before it is piped or spooned. It is composed of 150–200 parts eggs, 100 parts flour, 100 parts shortening (butter) and approximately 200 parts water. Flour is quickly stirred into a mixture of liquid, seasoning and fat, which is heated to boiling. The fat must be completely dispersed in the hot liquid to form a proper emulsion. Stirring continues until the smooth, gelatinized paste no longer sticks to the pan. The mixture is cooled, and the eggs are beaten into the choux mix one or two at a time, mixing well after each addition. Liquid may be gradually added to make the final paste the desired soft consistency. A small amount of chemical, fast-acting leavening can be added to this liquid to obtain the smooth consistency and maximum pastry volume. The paste is spooned or piped on to parchment-paper-lined pans and baked at 215–226°C until crisp. (See **Leavening Agents.**)

Choux pastry can be prepared using a continuous extrusion process. The extruder has groups of processing zones provided by varying the pitch of the screw. Throughout these zones, ingredient mixing, kneading, starch gelatinization, and comminuted/homogenization are accomplished. The final step involves extrusion of the dough through a die, forming the desired choux pastry shape. (See **Extrusion Cooking:** Principles and Practice.)

Strudel Pastry

The dough used to prepared strudel can be of the puff pastry type or can be made from water, seasoning, eggs, shortening or oil, and high-protein flour. Optional ingredients include sugar and lemon juice. The fat is mixed or cut into the flour mixture, and the liquid ingredients are added gradually. The dough is made into a ball, kneaded, and shaped into a rectangular form, relaxed, then rolled and gently stretched to obtain very thin, transparent dough. Oil is brushed on the thin layers. Home preparation by hand stretching is difficult. Commercial methods have been developed to prepare thin sheets. Strudel pastry and Fillo (phyllo) dough, which is similar to strudel, are frozen. These are sold in wholesale and retail markets and have wide applications for the home baker. Strudel filling is placed at one end of the pastry, and the dough is rolled over the filling as in the preparation of a Swiss roll. Relaxed strudel is baked at 218–233°C.

Yeast-leavened Pastry

Danish Pastry

Danish pastry is a rich pastry that combines the principles of puff pastry and fermented, sweetened dough. The dough contains high-protein flour (11.5–12%) that is often combined with up to 30% pastry flour for easier handling during lamination. Other ingredients include butter or margarine, yeast, sugar, shortening, milk solids, eggs (optional), salt, and flavoring. Fat containing moisture will help produce steam for leavening and flakiness. The total amount of fat can range up to 50% of the pastry dough. Sheeting and folding fat between layers of dough forms thick films of fat separating dough layers. Two types of Danish pastries are made in the USA. The European style is made from lean sweet dough that has a short shelf-life. The American style is more common and is made from a richer dough with a longer shelf-life.

There are two commonly accepted mixing procedures for the preparation of Danish pastry: conventional (three- or four-fold) or the short mixing time method, and the lamination or the long-mixing-time method, which is used for automated, high-speed sheeting systems. In the conventional three-fold method (**Figure 1a–c**), all the ingredients are combined with the exception of the rolled-in fat. A rectangle is formed by rolling the dough to a thickness of 1.25 cm. A large portion of roll-in pastry fat is evenly distributed over two-thirds of the dough, and the rectangle is folded in thirds, starting with the portion without fat. The dough is then rolled into a rectangle of 1.25-cm thickness and folded into four equal parts. After retarding, the dough is rerolled, and the procedure is repeated two more times. The four-fold or book-fold technique (**Figure 2a–c**) involves distributing fat over the center 50% of the dough and folding

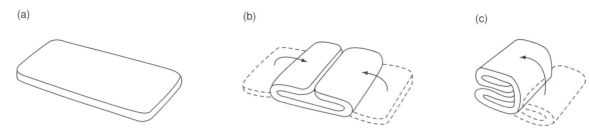

Figure 2   Four-fold pastry preparation: (a) the dough is rolled into a rectangle: (b) both ends of the dough are folded halfway down the rectangle; (c) one folded end is turned to cover the opposite fold.

the two ends of the original rectangle to the dough center. One end is folded over the other fold to line up with the far edge. The retarding and refolding steps are then repeated, similar to the procedure described for the three-fold method.

The gluten structure is developed during the laminating rather than during mixing. The commercial lamination method involves layering the shortening between two pieces of dough. This can be accomplished by extruding dough around the shortening, or high-speed pumping of the shortening between two continuous dough sheets. Automatic or overlapping of 45–60-cm dough pieces is followed by sheeting. This process is repeated two more times before cutting.

The retarded dough may be shaped into a wide variety of pastries, including crescents, sticks, rings, filled pockets, and rolls. Individual pastries are baked at 193–204 °C. Larger Danish products, such as coffee cakes, are baked at lower temperatures, 176–190 °C, to prevent over-browning before the pastry structure is set.

Croissants

The croissant is a popular, sweetened, yeast-raised product that is delicate and flaky. It has a medium-grain texture and the shape of a crescent. Dough composition and preparation techniques are similar to those of Danish pastries, with total fat contents ranging from 25 to 50%. The rolled in fat is a critical component for laminated yeast doughs that should have the same consistency as the dough. It provides the two-dimensional gluten framework that results in the characteristically flaky croissant. Additional ingredients are sugar, salt, whole eggs, yeast, and water. Dough (19°C) is mixed completely, without over-developing the gluten, for 1 min in a vertical high-speed mixer. The dough is extruded, and fat is extruded on top. It is then flattened, rolled and rerolled, and laminated for a total of 36 layers. Most of the gluten development occurs during the lamination step. Retarded dough is rolled and cut into triangles with slightly longer sides than the

base. Filling is added, and the dough is rolled with four rotations, leaving the peak touching the baking sheet. The 'rolled' pastries are proofed and then baked at 163–204°C.

Toaster Pastry

Toaster pastries are convenience pastry products that are popular in the USA. These breakfast or snack pastries are made from a filled dough, formed into a shape and size that can fit into a toaster. The pastry dough is prepared from soft wheat flour, sugar, shortening, chemical leavening, salt, flavorings, and colors. It is baked just enough to set the dough and reduce the moisture content, thus extending the product shelf-life. Although many are shelf-stable, many new varieties are frozen. Browning is completed in the toaster.

Commercial Preparation of Pastry

Most pastry products evolved after long periods of small-scale, manual preparation followed by high-speed, automated production in this century. High-capacity production of pastry is made possible by elimination of manual dough manipulation and automation of the pastry production process. Processing time is shortened by elimination of roll-in steps, and reducing resting times between fold-in steps. Variations exist in the continuous production methods for the different pastry types such as pie, Danish, and puff pastries. However, common to all are the sheeting, depositing, lamination, and cutting steps.

An example of an automated pastry production system is presented in **Figure 3**. The dough is mixed and carried to the sheeter head, or three-roll laminating head, by the dough feeder and conveyer belt. The sheeter head unit consists of three rollers with adjustable speeds for producing perfect sheeted pastry. Two are feed rollers, one with deep grooves and the other with fine triangular grooves. The third roller has a smooth surface and is located underneath the feed rollers. Fat layers can be added to the first dough to

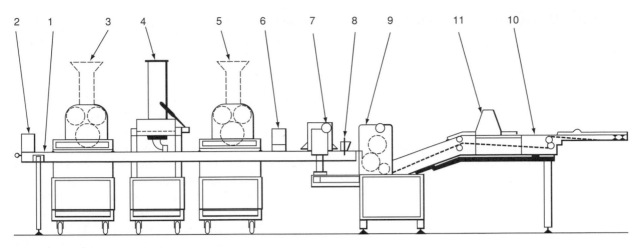

Figure 3 Automated pastry production system. (1) Supply table; (2) flour duster; (3) three-roll laminating head; (4) fat pump; (5) three-roll laminating head; (6) flour duster; (7) cross-roller; (8) guiding rollers; (9) two-roll sheeter; (10) retracting belt; (11) guillotine. From Rijkaart C (1984) Producing the Perfect Pastry. *Food Manufacture* 59: 29 with permission.

produce the flaky texture of pastries. A second sheeter head adds another dough layer over the fat. If the Scottish system is used, all the fat is incorporated into the dough, a fat extruder is not needed, and only one sheeter head unit is required.

Dusting flour boxes distribute small amounts of flour on the conveyor belt, to prevent sticking, and to the top of the dough sheets prior to cross-rolling. Conical cross-rollers travel across the belt width, thinning and spreading the dough sheet to the desired thickness. Further sheeting occurs at the two-roll sheeter, reducing dough thickness. Several systems can be utilized to obtain the desired flakiness by controlling the number of times the pastry is folded.

The book-fold method (**Figure 2a–c**) involves folding the dough sheet and then turning it through 90 °C and sheeting the dough, repeating the procedure to obtain the optimal flakiness. Another method involves continuous folding and sheeting of the dough. A third method uses a guillotine to cut the dough sheet into pieces (50 × 60 cm) and then stacks then with a retracting conveyor. The dough pieces are placed on another conveyor belt operating at a 90° angle to the first belt. The pieces are stacked five layers high by this process, after which the dough sheets are reduced in thickness. Another method utilizes a curling arm, or roll winder, which coils the dough sheet covered with extruded fat into five layers. A final technique uses a ring extruder that produces a very uniform, hollow dough cylinder, lined with fat. This method is particularly suited for long-shelf-life pastries. For all methods, the dough sheets can be reduced in size by pressing, cross-rolling, and running between a multiroller device. This multiroller allows the dough sheet to be

manipulated when in contact with the small rollers, which press the dough on to the larger roller. The small rollers revolve individually as the whole group of small rollers also rotate together in a loop. The dough is allowed to relax when it is between the rollers and is not touching the large roller. A similar stretcher system with a larger number of small rollers, but without the bottom roller, has also been used. The final pastry production steps repeat the sheeting and reducing steps, depending on the necessary dough thickness and desired flakiness. Preparation of individual pastries of varying sizes and shapes depends upon the type of cutting, depositing and shaping equipment utilized.

See also: **Extrusion Cooking**: Principles and Practice; **Leavening Agents**

Further Reading

Doerry W (1998) Formulation and production of puff pastries. *AIB Research Department Technical Bulletin* XX (2): 2–5.

Friberg BO (1990) *The Professional Pastry Chef*, 2nd edn. pp. 14–23, 29–34, 94–97. New York: Van Nostrand Reinhold.

Goodsell GR (1985) Making Danish – basics that determine quality results. *Bakers Digest* 59(3): 16–21.

Loving HJ and Brennesis LJ (1981) Soft wheat uses in the United States. In: Yamazaki WT and Greenwood CT (eds) *Soft Wheat: Production, Breeding, Milling and Uses*, pp. 169–207. St. Paul, MN: American Association of Cereal Chemists.

Matz SA (1987) *Formulas and Processes for Bakers*, pp. 1–26, 39–55, 58. McAllen, TX: Pan-Tech International.

Matz SA (1992) *Bakery Technology and Engineering*, 3rd edition. McAllen, TX: Tan-Tech International, Inc.

Nicolello LG and Dinsdale J (1991) *Basic Pastrywork Techniques*, 2nd edn, pp. 45–46, 53–58. London: Hodder & Stoughton.

Pyler EJ (1988) *Baking Science and Technology*, vol. II, pp. 1107–1183. Meriam, KS: Sosland.

Rijkaart C (1984a) Producing the perfect pastry. *Food Manufacture* 59(7): 29–31.

Rijkaart C (1984b) Producing the perfect pastry. *Food Manufacture* 59(8): 46–47.

Sultan WJ (1990) *Practical Baking*, 5th edn., pp. 217–328, 379–423, 425–485. New York: Van Nostrand Reinhold.

Ingredient Functionality and Dough Characteristics

S L Andrews and J B Harte, Michigan State University, East Lansing, MI, USA

This article is reproduced from *Encyclopaedia of Food Science, Food Technology and Nutrition*, Copyright 1993, Academic Press.

Introduction

Traditional pastry or pie crust is a simple dough made from four primary ingredients – flour, fat, salt, and water. Numerous pastry varieties have been developed by slightly altering ingredients, their amounts, mixing methods, and shaping. Common ingredient modifications include the amount or type of fat, the protein content of flour, the substitution of milk for water, use of chemical leavening agents and yeast, and the addition of other ingredients such as eggs, sugar, acid, and flavorings. It is the purpose of this article to outline pastry ingredient functionality and their proportions and desirable pastry characteristics. US and equivalent UK pastry terms are presented in **Table 1**.

Functionality of Ingredients

Flour

Flour is the main ingredient in pastry, and the type of flour used affects the characteristics of the pastry.

Table 1 US and equivalent UK pastry terms

US	UK
Standard plain pastry	Shortcrust pastry
Short pastry	Pâte de sucre or flan pastry
Cookies	Biscuits
French puff pastry	Flaky pastry
Scottish puff pastry	Rough puff pastry

Protein in flour yields gluten wherever the flour is dampened with water and manipulated. Gluten development increases as the flour's protein content increases and as the amount of added water increases. Gluten development can be minimized by incorporating fat and flour before water is added. This method insulates flour proteins from added water and limits gluten development. The extent of gluten distribution throughout pastry and the amount of gluten formed determine whether pastry is crumbly, with a tendency to be compact and tough, or tender and flaky. When dough is rolled and baked, gluten is denatured by heat, which contributes additional toughness. High-protein flours yield more gluten, which results in a more cohesive dough. In contrast, pastry flour and other low-protein flours do not yield as much gluten and tend to make a tender pastry which crumbles and cannot hold its shape. A mixture of high gluten or plain flour blended with a soft cake or pastry flour can be used to achieve the desired protein concentration.

In choux pastry a high-protein (12–13%) flour is used. During baking, egg and flour proteins in choux pastry coagulate, preventing steam and gas from leaving the pastry, and setting the pastry into the desired form. High-protein (13%) flour is also used in yeast-leavened Danish pastries. For croissants, a 12.5% protein flour is recommended. (*See* **Flour**: Analysis of Wheat Flours.)

Fat

Fat contributes tenderness, or shortness, to pastry. Depending on the type of pastry, the fat content can range from 25% to almost 75% of the dough. Fat tenderizes pastry by waterproofing flour particles. The polar groups in water have an affinity for the polar groups in both the protein and starch. Polar carbonyl groups and the double bonds in unsaturated fatty acid moieties make it possible for fat to unite with polar groups on the surface flour particles. The remaining portions of the fat molecule have no affinity for the flour or water and act as a mechanical barrier, preventing contact of the water and protein in the flour.

A fat's ability to interfere with gluten formation is known as its shortening power. Pure fats have more shortening power than do butter or margarine which contain 16% water. Even pure fats, such as lard, hydrogenated shortenings, and oils, exhibit different characteristics in a pastry product. Oil is more dense than lard, which is more dense than shortening. In addition, liquid fats have more spreading power and are able to coat flour more evenly and completely. The higher the ratio of liquid to crystals, the greater the covering power of the fat.

In addition to making pastry tender, fats also contribute desirable flakiness by separating the dough into layers or flakes. Oils, on the other hand, tend to coat each particle of flour. As a result, water contact with the flour is limited, little gluten is developed, and a tender but crumbly, or even greasy, pastry results. A flaky, tender pastry can be made with oil, but, because oil mixes so easily with flour, the mixing must be carefully monitored. Oil and flour should be mixed until the particles are the size of peas. Overmixing will cause particle size to decrease, resulting in a crumbly pastry.

Plastic fats, having the properties of both liquid and crystalline fats, tend to make a more flaky product which is tender. Lard is considered an excellent pastry fat because of its shortening action, desirable plasticity, and dispersibility which enhances flakiness. Selection of fat depends on a number of factors, including ambient temperature, method of processing, and desired crust characteristics (mealy or flaky). (*See* **Fats**: Uses in the Food Industry.)

Puff pastry dough contains up to 100 parts of a good-quality, high-protein flour, 58 parts of water, and two parts of salt. A firm, plastic, shortening, equal to 50% of the dough weight and containing some moisture, is optimum. Fat containing moisture will help to produce steam for leavening and flakiness.

Liquid

The liquid most frequently used in pastry is water, but liquid is often contributed by the addition of milk and eggs, and even fats such as butter and margarine which are 16% water. Liquid is needed to hydrate the flour, develop gluten, and provide cohesion to the dough. Without liquid, the flour particles would not adhere to form a dough. With insufficient liquid the dough crumbles and is difficult to handle. Too much liquid causes excessive gluten formation, producing a tough pastry. While too much liquid is undesirable in pastry doughs, liquid is essential in choux pastry for maintaining the desired soft consistency of the paste. Liquid in the form of water or milk is gradually added to produce the final paste consistency. If the paste is too thick, the baked volume will be low and the shell will be thick.

To promote flakiness, water used in pie dough mixtures should be chilled. This limits homogenization during dough mixing by creating hard fat particles. Ice water may also be used. Liquid is also necessary for leavening. When baked, liquid in pastry dough produces steam, leavening the product and separating the individual flakes. Fat containing moisture will also help to produce steam for leavening and flakiness.

Salt

Salt functions as a seasoning and is not required to produce a successful pastry product. A crust without salt will produce a flat-tasting crust. However, its omission does not affect the mechanical aspects of the pastry.

Other Ingredients

Eggs Eggs are incorporated into several pastries, including choux paste, short (pâte de sucre, or flan pastry), and croissant doughs. Eggs are optional ingredients in yeast-leavened Danish and puff pastry doughs. Egg yolks can function to lubricate the gluten, while the whites aid in baked pastry volume. Eggs contribute richness, structure, and increased keeping qualities. In choux pastry, for which 90–100 parts whole egg is beaten into a mixture of 60 parts high-protein flour (12–13%), 100 parts water, 40 parts shortening, and two parts seasonings to form a very thick paste, the primary functions of eggs are dough emulsification, leavening, and supplying moisture.

Acids Acids such as cream of tartar, vinegar, or lemon juice are used in some pastry formulations. Acids relax the gluten, thus improving the ease of rolling. If too much acid is added, the dough will be sticky and difficult to handle and roll. (*See* **Acids**: Natural Acids and Acidulants.)

Sugar Sugar helps to tenderize the dough, adds extra flavor, and promotes crust browning. Lactose, the simple sugar found in milk, also contributes to product browning. (*See* **Lactose**.)

Milk In some pastries, milk replaces water as the source of liquid. Milk adds crust color by providing lactose, which promotes Maillard browning during baking. Milk solids strengthen the interior cell structure, which helps to maintain pastry flakiness. Milk can also aid in producing a thin, smooth crust. (*See* **Browning**: Nonenzymatic.)

Yeast Yeast is used in croissant and Danish pastry doughs to provide leavening through the production of carbon dioxide. Danish pastry doughs may contain a high concentration of yeast to offset the growth-inhibiting effects of added sugar.

Proportions of Ingredients (by Weight)

For a traditional plain (shortcrust) pastry, 100 parts flour (average 115 g per cup) are used for each 42–59 parts fat. The optimal amount of plastic fats, such as lard, is 44 parts to 100 parts of flour. If less than 44 parts are used it may become difficult to mix in the water without overdeveloping the gluten. When liquid fat such as vegetable oil is used, 42 parts should be used to each 100 parts of flour. If more than 59 parts of fat are used, the pastry becomes crumbly and greasy. Within the recommended range, the smaller the proportion of fat, the greater the likelihood of developing excessive gluten. For each 100 parts flour, 2.5 parts salt are used for flavoring.

A minimal amount of water must be added to develop gluten (flakiness). If too little water is added, the pastry will be crumbly, but too much water will cause excessive gluten development and the pastry will be tough. A mixture of 31 parts water to 100 parts flour is sufficient to provide flakiness without promoting toughness. When a minimal amount of fat is used, careful measurement of liquid becomes even more important. A difference of as little as 2.5 parts water per 100 parts flour can cause a discernible difference in the tenderness of the pastry.

Pastry Characteristics

Tenderness and flakiness are two desirable characteristics of pastry products which can occur together or separately. Up to an optimum, tenderness increases as fat is increased, as a result of the shortening power or ability of the fat to coat the gluten strands, interfering with gluten structure. The fat can also function to coat the flour particles and prevent water from fully hydrating the flour, thus preventing gluten formation. Higher temperatures increase the ability of the fat to spread, coating the gluten and increasing tenderness. Keeping fat and water cold will serve to promote a flaky, tender pastry which is not greasy or too tender. A product which is too tender becomes crumbly and too hard to handle. Overmanipulation, including kneading of pastry dough, has the opposite effect. As the dough is kneaded, rerolled, or mechanically overmixed, the pastry becomes tough as a result of excess gluten formation. Toughness can also be caused by adding too much liquid, or by the presence of excess flour during rolling.

Flakiness results from partially hydrated flour or dough being layered between thin sheets of fat. Pastry layers are separated during baking as steam and, to some extent, air separates the layers. These fat layers must coat the flour sufficiently to produce layers, yet not enough to cause the pastry to crumble. Cutting fat into the flour in large particles promotes flakiness. Medium-flaky pastry is made by cutting fat into small lumps, while a mealy crust will form when the flour and shortening are mixed thoroughly. Use of cold water helps to promote flakiness in the final product.

For long flaky pastry the fat is cut into larger lumps and the lumpy, sticky dough is rolled three times, turning three times for each roll. The harder fats do not spread as readily as softer fats and thus are more likely to have flour particles adhering to the fat. Optimum mixing time for desirable tenderness is longer when harder fats are used. Dough rerolling and folding increase flakiness. Increased gluten development can hold steam more efficiently and separate the dough into layers. The resultant increased toughness may be reduced by rolling small pieces of fat between the dough layers.

Dough Retardation and Relaxation

During dough mixing and handling, fat must remain cold enough so that it does not soften during preparation. This is especially true for roll-in fat products, such as puff and Danish pastries. Additional fat folded into the dough creates numerous layers of dough and fat. To prevent changes in fat properties that occur quickly at elevated temperatures, dough is retarded or relaxed at refrigerator temperature between sheetings. To prevent greasiness and decreased volume during baking, cooled pastry should reach ambient temperature slowly.

Leavening

In pastry, leavening occurs when heated liquid creates steam, and individual layers (flakes) are separated. Baking temperature is important to insure sufficient leavening. Too low a temperature will not allow the optimum production of steam for aeration, while an oven temperature that is too high will result in premature setting of the proteins and low pastry volume. Leavening also occurs as air expands during baking.

In yeast-leavened Danish pastry and croissant doughs, yeast is added for additional leavening from carbon dioxide. In choux paste, a small amount of chemical, fast-acting leavening – ammonium bicarbonate and baking powder (1:1) (14 g per 454 g extra liquid) – can be added to this liquid to obtain the smooth consistency and maximum pastry volume. In choux pastry, if eggs are used to soften the paste, the liquid and chemical leavening mixture is not needed. Leavening of the choux paste during baking occurs by expansion of the eggs; carbon dioxide is released from the chemical leavening agents, and steam is formed from heated liquids in the paste. Flour and egg protein coagulate to set the shape of the aerated pastries. (*See* **Leavening Agents**.)

See also: **Acids**: Natural Acids and Acidulants; **Browning**: Nonenzymatic; **Fats**: Uses in the Food Industry; **Flour**: Analysis of Wheat Flours; **Lactose**; **Leavening Agents**

Further Reading

Charley H (1982) *Food Science*, 2nd edn, pp. 243–255. New York: John Wiley.

Daniel AR (1963) *Baking Materials and Methods*, 4th edn, pp. 362–391. Essex: Elsevier Science.

Goodsell GR (1985) Making Danish – basics that determine quality results. *Bakers Digest* 59(3): 16–21.

Loving HJ and Brenneis LJ (1981) Soft wheat uses in the United States. In: Yamazaki WT and Greenwood CT (eds) *Soft Wheat: Production, Breeding, Milling and Uses*, pp. 169–207. St Paul, Minnesota: American Association of Cereal Chemists.

Pyler EJ (1988) *Baking Science and Technology*, vol. II, pp. 1107–1183. Meriam, Kansas: Sosland.

Rijkaart C (1984) Producing the perfect pastry. *Food Manufacture* 59(7): 29–31.

Rijkaart C (1984) Producing the perfect pastry. *Food Manufacture* 59(8): 46–47.

Sultan WJ (1990) *Practical Baking*, 5th edn, pp. 217–328, 379–423, 425–485. New York: Van Nostrand Reinhold.

Pathogens   *See* **Emerging Foodborne Enteric Pathogens**; **Food Poisoning**: Classification; Tracing Origins and Testing; Statistics; Economic Implications; **Microbiology**: Classification of Microorganisms

---

PEACHES AND NECTARINES

C Blattný, Prague, Czech Republic

Introduction

Both peaches and nectarines originated from the same tree – the peach, whose botanical name is *Prunus persica* L. (rose family – Rosaceae). They are a typical stone fruit with the seed in a hard stone; soft pulpy flesh surrounds the stone and the surface is covered with a relatively thin skin. The peach and nectarine are closely related to other species of the genus *Prunus*: apricot, almond, plum, and cherry.

Origin and History

The peach is a native of China, where it was depicted in old paintings and mentioned in Chinese mythology and folklore as far back as the tenth century BC. The peach signifies long life and peach blossom has also been important in China for a very long time. There semiwild forms can be found in all native peach areas from south to north-west and north-east.

It is unclear how peach and nectarines spread to Europe. At least two or three ways are probable. First, when the Romans occupied Syria to the end of the first century BC and found the peach, they named it 'Persian apple.' It must have been important that peaches grow quickly from seed, and this permitted their rapid dispersal from China to the west. The second way was probably across Egypt to Rhodes and the third route was over Greece and the Balkans. Since the beginning of the second millennium peach and nectarines spread throughout Europe, especially in the south and south-east. France played an important role in the development of the peach – at the beginning of the seventeenth century about 30 cultivars of peach were grown there. During the following two centuries this number reached more than 350 cultivars. Also in Italy interest in peach and nectarine increased from century to century. Much archeological evidence of peach seeds has been found there along the central Donau in Hungary and Slovak republic. The Spaniards brought peaches to America (Mexico and Florida) in the sixteenth century. Peaches spread in America very quickly and, especially in California, the peach became the leading fruit grown in orchards.

Requirements of the Peach Tree

Peaches have different requirements according to their old native district. The hardiness, early or late blooming, and requirement of winter cold exposure (chilling) are most important. Most older cultivars need more than 950 h at around 6 °C. However it should also be emphasized that the peach tree is a warmth-seeking botanical species. It needs higher temperatures in summer with strong intensive sunshine and relatively mild frost in winter – attributes of a mild climate. However, cultivars from south China are suitable for subtropical regions as they need less chilling.

The peach has very similar requirements for environmental conditions as the vine. The soil for peach should be warm, not too heavy, airy, pH 6–7.5, and $CaCO_3$ 3–5%. Selection of the rootstock for given conditions is important. The following seedlings are used in various countries: peach seedling (various types), peach × almond hybrids, almond seedlings, plums seedlings, *Prunus davidiana* (China), and peach × *P. davidiana* hybrids. Peach seedlings prevail: in some countries peach × almond are very often used.

Nectarine cultivars have greater environmental demands. They need warmer summers (without rain), as the fruit's skin tends to crack after rain.

The Fruit

Peach is one of the most variable of all tree fruit species. From the pomological point of view, the cultivars are divided into five groups:

1. True peaches – freestone or semifreestone with fuzzy skin
2. Clings – clingstone with fuzzy skin
3. Nectarines (smooth-skin peaches) – with fuzz-free skin and freestone
4. Brugnons – small clingstone nectarines
5. Peento – a special group of peaches with flat fruit (almost like dried figs) and exposed stone

This division is by no means accurate. It is necessary to emphasize that in some years when there is a cold and rainy summer, a stronger fusion of flesh and stone also occurs with freestone cultivars.

Cultivars of peach and nectarine differ in their height, width, and thickness. Both peaches and nectarines are round or oval-round, oval, oblong, or somewhat flat. The suture leading from the stem pit to the top of the fruit or behind it is a typical feature. The depth of the suture also differs between cultivars. Some have a small tip on the top, some have deeper ones, and a small number of cultivars have a small knoll (like Venus). The skin of true peaches and clings is fuzzy. This fuzz differs according to the cultivars – sometimes it is very fine, sometimes it is dense and very fuzzy. The skin is orange-yellow, white with red cheeks, or red with stripes, or plain red. The flesh is soft, sometimes too firm or too fibrous, yellow, orange-yellow, nearly white, green-white, or red. In old-type cultivars, which have only been selected to a limited degree, the flesh is often farinaceous. The color of flesh around the stone and under the skin is slightly darker and sometimes red.

Nectarines usually have a somewhat smaller fruit and a smooth shiny skin (like plums); the flesh is harder, more like a plum, and not melting like true peaches. The taste differs slightly from true peaches; however it is outstanding too. Greater consumer acceptance is attained of the fruit of nectarines which are more acidic than peaches. The color of the skin is often red to purple-red. In recent years an important development of white-flesh nectarine cultivars has occurred.

The stones of peaches and nectarines are conspicuous by their surface, with different deep branch-like grooves and pits. These characteristic shape, size, grooves, and color of the stone are peculiar to every single cultivar.

Regulations

Commission regulation (EC) no. 2335/1999 lays down marketing standards for peaches and nectarines.

Definition of Produce

This standard applies to peaches and nectarines grown from cultivars of *Prunus persica* L., to be supplied fresh to the consumer; peaches and nectarines for industrial processing are excluded.

Provisions Concerning Quality

The purpose of this standard is to define the quality requirements for peaches and nectarines after preparation and packing.

Minimum requirements Peaches and nectarines must be:

> intact, sound, produce affected by rotting or deterioration such as to make it unfit for consumption is excluded, clean, practically free of any visible foreign matter, practically free from pests, practically free from damage caused by pests, free of all abnormal external moisture, free of foreign smell and/or taste

Peaches and nectarines must have been carefully hand-picked. The development and ripeness of peaches and nectarines must be such as to enable them to withstand transport and handling and to arrive in a satisfactory condition at the place of destination.

Classification

- Extra class
- Class I
- Class II

Table 1 Classification of peaches and nectarines according to size

Diameter	Size (code)	Circumference
90 mm and over	AAAA	28 cm and over
80 mm – under 90 mm	AAA	25 cm under 28 cm
73 mm – under 80 mm	AA	23 cm under 25 cm
67 mm – under 73 mm	A	21 cm under 23 cm
61 mm – under 67 mm	B	19 cm under 21 cm
56 mm – under 61 mm	C	17.5 cm under 19 cm
51 mm – under 56 mm	D	16 cm under 17.5 cm

Provisions Concerning Sizing

Sizing is determined by circumference or maximum diameter of the equatorial section (**Table 1**).

Diseases

Peaches and nectarines are subject to attack by a cosiderable number of diseases. Viruses and *Mycoplasma* are the most significant. We will discuss only a few diseases here

- Peach mosaic (vector – eriophyoid mite *Eriophyes insidiosus*)
- Peach yellows and little peach (vector – leafhopper *Macropsis trimaculata*)
- Phony (distributed through the root system)
- X-disease (vectors – more species of the leafhoppers)

These diseases have various symptoms on the leaves, twigs, and fruits. All these diseases are transmitted with buds by propagation, therefore healthy budwood and rootstocks are the most important control measures (together with the fight against vectors). Affected trees have nearly no crop and usually die within a short period.

The fungus diseases leaf curl, brown rot, and mildew are troublesome:

- Leaf curl (*Taphrina deformans*) curls and deforms the leaves early in the season; their color changes to dark red or violet. It must be prevented by applying fungicides just before bud opening.
- Peach mildew (*Sphaeroteca pannosa*) appears in the form of a white powdery substance upon the leaves and fruits. Spraying with fungicides will commonly hold it in check.
- Brown rot (*Monilia fructicola*). Infection begins during flowering but fruit rot often occurs after harvest. Control with fungicide during vegetation and then immediate cooling after harvest are important.
- Gray mold (*Bothrytis cinerea*) occurs especially after a wet spring during storage if the fruit has been infected during gathering. Gentle harvest, undamaged fruit, and temperature about 0 °C during storage and transport are effective control measures.

Harvesting of Fruit

In the 1930s the harvesting season of peaches was about 3 weeks – mid-July to mid-August. Due to intensive breeding work, especially in the USA (and in France and Italy), the harvesting season is now extended to at least 2 months; in the most favorable environmental conditions it may be more than 3 months (mid-May to 3 September). As regards ripening, the cultivar Redhaven grown worldwide is taken as the reference. Most peach cultivars ripen in the midseason 2–3 weeks after Redhaven; more nectarine cultivars ripen before Redhaven; the canning clingstone cultivars have a shorter harvesting period before and 6 weeks after Redhaven.

The harvest differs according to the purpose of using the fruit. For fresh fruit the fruit has to be harvested 2–3 days before it will reach ripeness, with a fully developed flavor. This is called tree-ripe. Therefore it is sometimes recommended to repeat harvesting at least three times according to advancing ripeness. The grade of ripeness can be determined by skin ground color changes from green to yellow in most cultivars (exept for the white-flesh cultivar). This is called minimum maturity, and is used to harvest fruits determined for distant markets and long transport. The ideal time to harvest peaches is when they are beginning their ripening process, but are still firm. At this stage they withstand picking operations and hauling quite well and complete the ripening process in good condition. If picked too soon, potential tonnage is lost because peaches and nectarines grow considerably just before reaching maturity. If the harvest delay is too long, they begin to soften, bruise easily, and shelf-life is reduced. To satisfy consumers that the fruit is ready to eat it is necessary to initiate the ripening process at the right time before the fruit is put on the market. Nowadays detailed ripening protocols for different peach and nectarine cultivars from important international export districts have been developed. For home gardens usually the best time is when the yellow-fleshed cultivars change from yellowish-green to yellow or orange-yellow. White-fleshed cultivars are ready for harvest when the color changes to cream white.

The nearly ripe fruits of the best quality should be gathered and placed in the market boxes that are provided with special niches for each individual fruit to prevent bruising during transport. Rapid transport should take place in refrigerated trucks.

Cultivars destined for processing are usually picked all at once – true peaches by hand and clings with the help of fruit-shaking machines.

Postharvest Operations

Peaches and nectarines are transported from orchard to packinghouse and cooler as soon as possible after harvest. In the packinghouse the fruits of minimum maturity are sorted to eliminate fruits with visual defects, and to select outstanding nice colored and big fruits (for the extra class). Cleaning machinery to remove fuzz from peaches is often used in all major production areas. In the defuzzing operation the peaches are scrubbed with rotating brushes to remove the fuzz and then washed and rinsed with fresh cold water. Sizing segregates fruits by size or weight. It is very important to cool the fruit as quickly as possible to near $0\,°C$. Also longer chilling at $0\,°C$ is recommend. When the fruits are stored at a higher temperature, between about $2.2\,°C$ and $7.5\,°C$, chilling injury (internal breakdown) occurs. First the flavor is lost, then the flesh becomes mealy, wooly, or harder, and brown around the pit of the stone. Cultivars of peach and nectarine differ in their susceptibility to chilling injury; nectarines are less susceptible.

Sometimes during harvest or after 1–2 days, black spots or stripes occur on the skin. This is a result of abrasion combined with touch with metal – iron or copper. Therefore it is very important to handle fruit gently during harvest and postharvest hauling operations.

Supplying Fresh Fruit to Consumers

The fruits must be intact, sound, clean, practically free of any visible foreign matter, practically free from pests, free of abnormal external moisture and free of any foreign smell or taste. The fruits must have been carefully picked and sufficiently developed that the ripening process can continue in order to reach the degree of maturity proper to their cultivar characteristic, to withstand transport and handling, and to arrive at their destinations in satisfactory conditions. Usually they are classified to four or seven classes according to the diameter of the widest part of the peach and nectarine.

Use of fresh fruits

Peaches – and, to a certain extent, nectarines – belong to the most sought-after fruits, thanks to their delicious taste and the development of global trade. Most of the peach crop (usually about 60%; in some regions much more) is marketed as fresh fruit; for nectarines, nearly all of the crop is used fresh. Cultivars of true freestone peaches with white and orange-yellow melting flesh, thin skin with not too dense fuzz, and delicious taste, fully ripened are the best for eating fresh. On the market such cultivars are sold in four to five classes. Sliced fresh peaches or nectarines are eaten alone or in combination with other fruit in salads or cocktails. Canned or frozen fruit can be used in the same way. Dried fruits are usually eaten as an energy-giving snack. In homes and confectioneries peaches and nectarines can be used for baked goods, icecream, gelatine desserts, pies, juices, and sherbets.

Some sensitive individuals can experience difficulties caused by a thicker fuzzy skin of certain peach cultivars. This unpleasant characteristic can be solved by a defuzzing operation (with rotating brushes) or by developing large nectarines, as was done in the last decades, especially in the USA.

Processing

The main method of processing is canning, usually about 30% of the whole crop, while a small amount is dried in some regions. The fruits are further used for jams, jellies, juices, brandy and liqueurs, preserves, and other miscellaneous products.

Technological Requirements

Peaches are inspected, graded, halved, pitted, and peeled before canning, freezing, and drying. The shape of the fruit should be round or flattened round and almost uniform; the diameter should be at least 45 mm. The flesh should be soft but firm enough (not farinaceous), of yellow-orange color, and have a taste full of the aroma. The smaller the stones, the better – 8–9% of total fruit weight. Tannin content should not be more than 0.08%. True peaches and clings are used in the processing industry. Cling cultivars are better, as they have firm flesh from the skin to the stone. Therefore after pitting and peeling the flesh is equally firm and does not crumble. The cling cultivars withstand better mechanical harvesting with shaking machines, longer transport, and storage. A further advantage of clings is equal ripening and slower overripening than true peaches. For processing the optimal grade of ripeness has to be measured just before processing starts. Prematurely harvested and processed fruit is too hard and lacks intense taste and flavor. Peaches are usually canned in light to heavy syrup, less so in water or juicy pack. When freezing the syrup, a small amount of ascorbic acid (to prevent browning) is used. Peeled peaches must be sulfured for drying.

Plate 111 Oats Oat panicle of cultivated *Avena sativa*.

Plate 113 Oats Oat panicle of naked (hull-less) *Avena nuda*.

Plate 112 Oats Oat panicle of *Avena* spp. From a genetic cross between wild and hull-less oats. The oat derivative can be used for decorative purposes.

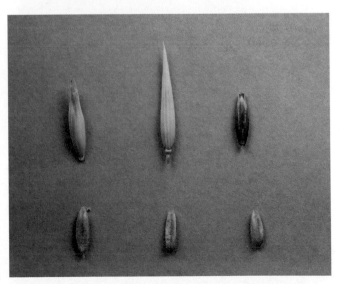

Plate 114 Oats Kernels of cultivated white oat, *Avena sativa* (left), hull-less oat, *Avena nuda* (center), and a derivative of a wild *Avena* sp. (right). Whole oat kernels are pictured on the top row and the corresponding groats on the bottom row.

Plate 115 Olive Oil Picture of a fossilized olive leaf. Reproduced from Psillakis N and Kastanas E (1999) *The civilization of olive: olive oil,* 2nd edn. Greek Academy of Taste, Karmanor, Iraklion, Crete (Greece), with permission.

Plate 117 Olive Oil Hydraulic pressure unit (courtesy of ELAIS SA, Piraeus, Greece).

Plate 116 Olive Oil Olive tree in Rethimnon, Crete (Greece), courtesy of V. Zambounis, Athens.

Plate 118 Olive Oil Pottery from Knossos Palace. Reproduced from Psillakis N and Kastanas E (1999) *The civilization of olive: olive oil,* 2nd edn. Greek Academy of Taste, Karmanor, Iraklion, Crete (Greece), with permission.

Plate 119 Papayas Rainbow cultivar growing in Hawaii, weighing 0.66kg.

Plate 120 Papayas Thailand cultivar growing in Thailand, weighing 2.50kg.

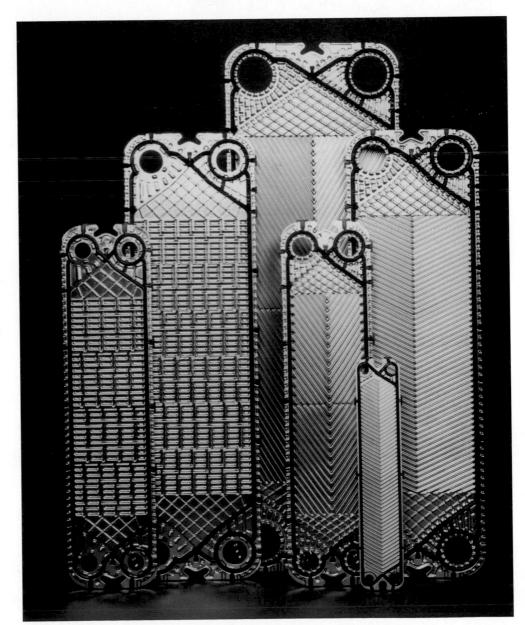

Plate 121 Pasteurization
Heat exchanger plates.
(Photo by courtesy of Tetra Pak Processing UK.)

Plate 122 Port Vila Nova de Gaia; bonded aread in Portugal.

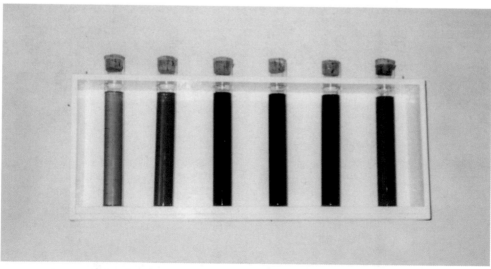

Plate 123 Port Samples of different types of port, showing the range of colors.

Recommended Cultivars

Nowadays grown cultivars have different origins. In earlier times cultivars from the Persian (Iran) region were a base for many old cultivars. But in the middle of the nineteenth century some Chinese clings were imported to America directly from China. Some from North China (with hard winters) required high chilling and some from South China required less chilling. From these Chinese imports accidental hybrids with cultivars grown in that time in the USA have been chosen. One such hybrid was introduced in the 1870s as the cultivar Elberta. It had become one of the most widespread cultivars in a short time and a parent of many new cultivars.

Some examples of worldwide cultivars (2000 AD) are:

- true peaches – freestone and semifreestone – yellow: Redhaven, O'Henry, Spring Lady, Rich Lady, Cresthaven, Maycrest, Springcrest, Elegant Lady, Royal Glory, Fayette; white: Maria Bianca, White Lady
- clingstones: Babygold 5, 7, Vivian, Harbinger, Andross, Loadel, Helford 2, Golden Queen
- nectarines – yellow: Redgold, Fantasia, Flamekist, Supercrimson, Venus, Armking, May Grand, Marycrest; white – Queen Giant, Silverking.

In many districts these has been a shift from middle and late cultivars to early-maturing ones. This resulted from the strong demand for the first peaches and nectarines harvested. This is reflected in the high prices of early peaches and nectarines. In recent years there has been an improvement in the size and color of early peach and nectarine cultivars. Many have firm, white flesh with a delicious flavor and taste. In the typical great peach district a peach cultivar now usually lasts less than 20 years before being replaced by new superior one. Elberta's (1870) longevity is a distinguishing feature. North American cultivars became popular in European districts (Italy, France, Spain, Hungary, and Yugoslavia) and also in South America and North Africa. New cultivars were selected that required less chilling in the USA around the year 1970. This allowed an important increase in peach growing in very warm regions – India, Peru, Thailand, Brasil, Israel, not just in the USA. In the last decade many new cultivars have been developed in all main peach- and nectarine-growing regions – for instance, in the USA about 250, France about 55, Italy about 90, China about 35, Mexico about 20, and in the whole world more than 500 new cultivars.

Nutritional Value

The composition of peaches and nectarines is given in **Tables 2, 3** and **4**. When ripe, most modern cultivars of peaches and nectarines exhibit a well-balanced proportion of acids and sugars. The content of dietary fiber and mineral salts, principally potassium, is important. Vitamin content is relatively low. It is influenced by the cultivar, how it is grown, the environment, and ripeness grade. There is also a rich content of delicious aromatic substances present. All this contributes to the great popularity of peaches and nectarines.

Production

Worldwide production of peaches and nectarines is increasing. This trend is mainly based on a large

Table 3 Mineral salt content of peaches and nectarines

Mineral salts	$mg\,kg^{-1}$
Calcium	50.0–70.0
Iron	1.1–12.0
Magnesium	70.0–100.0
Phosphorus	120.0–300.0
Potassium	1970.0–2050.0
Sodium	0.0–30.0
Zinc	0.2–1.4
Copper	0.5–0.7
Manganese	0.4–1.1
Selenium	Traces

Table 2 Chemical composition of peaches and nectarines

	$g\,kg^{-1}$
Water	805.0–877.0
Protein	7.0–8.0
Total lipid	0.9–2.0
Sugar	70.0–150.0
Acids	6.0–12.0
Dietary fiber	14.0–20.0
Ash	4.0–5.0

Sugar is composed on average from 13% glucose, 15% fructose, and 72% saccharose. Organic acids fluctuate according to grade of ripeness: 42% citric acid and 58% malic acid.

Table 4 Vitamin content of peaches and nectarines

Vitamins	
Beta-carotene	$0.58–4.40\,mg\,kg^{-1}$
Vitamin E	$7.00\,mg\,kg^{-1}$
Vitamin B_1	$0.17–0.50\,mg\,kg^{-1}$
Vitamin B_2	$0.40–0.53\,mg\,kg^{-1}$
Vitamin B_6	$0.18–0.30\,mg\,kg^{-1}$
Niacin	$6.00–9.90\,mg\,kg^{-1}$
Pantothenic acid	$1.40–1.70\,mg\,kg^{-1}$
Folate	$30.00–34.00\,\mu g\,kg^{-1}$
Vitamin C	$66.00–370.00\,mg\,kg^{-1}$

The content of beta-carotene (provitamin A) depends on the color of the cultivar (white, yellow, orange-yellow, red).

number of new more suitable cultivars and better cultivation methods (fertilization, irrigation, prevention of diseases and pests). These developments have led to the possibility of exploiting moderately favorable environmental conditions. Worldwide production increased over 20 years by more than 50% and in 1996–98 reached a volume of 11 146 million tonnes. The USA produced 1302 million tonnes, and Europe 3418 million tonnes, of which the most important are Italy 1447 million tonnes, Spain 894 million tonnes, Greece 648 million tonnes, and France 468 million tonnes. High production volumes have been registered in China: 2922 million tonnes, Turkey: 372 million tonnes, Argentina: 284 million tonnes, Chile: 278 million tonnes, South Africa 221 million tonnes, and Tunisia 221 million tonnes. The development of international trade was also a great contributory factor: it solved problems of rapid transport without losses, mainly due to standardization and refrigeration all the way from harvest to consumer.

See also: **Canning**: Principles; Cans and their Manufacture; **Cherries**; **Chilled Storage**: Principles; **Controlled-atmosphere Storage**: Effects on Fruit and Vegetables; **Fruits of Temperate Climates**: Commercial and Dietary Importance; Factors Affecting Quality; Improvement and Maintenance of Fruit Germplasm; **Plums and Related Fruits**; **Preservation of Food**; **Ripening of Fruit**

Further Reading

Blattný C (1986) *Koservárenské Suroviny*. Prague: SNTL.

Commission Regulation (EC) no. 2335 (1999) Laying down marketing standards for peaches and nectarines. *Official Journal of the European Communities* L 281/11: 4.11.
Crisosto CH and Kader AA (2002) Peach – postharvest quality maintenance guidelines. http://postharvest.ucdavis.edu/produce/producefacts/.
Crisosto CH and Kader AA (2002) Nectarines – postharvest quality maintenance guidelines. http://postharvest.ucdavis.edu/produce/producefacts/.
Das Grosse Lebensmittel Lexikon, 3rd edn. (1985) Innsbruck: Pinguin Verlag.
Der Kleine Souci Fachmann Kraut, 2nd edn. (1991) *Lebensmitteltabelle fuer die Praxis*, 2nd edn. Stuttgart: Wissenschaftliche Verlagsgeselchaft.
Ensminger AH, Ensminger ME, Konlande JE and Robsen JRK (1995) *The Consise Encyclopedia of Foods and Nutrition*. Boca Raton, FL: CRC Press.
Faust M and Timon B (1995) Origin and dissemination of peach. In: Janiek J (ed.) *Horticultural Reviews*, vol. 17. John Wiley.
Fideghelli C, Della Strada G, Grassi F and Morico G (1998) *The peach industry in the world: present situation and trend. Proceedings of the Fourth International Peach Symposium. Acta Horticulture* 465:
Hubbard EE, Purcell JC and Ott SL (1986) *Variety Changes in the Georgia Peach Industry, 1957–1984*. University of Georgia: Agricultural Experiment Station.
Kutina J et al. (1991) *Pomologický Atlas 1,2*. Prague: Zeměděl. Nakl. Brázda.
Kyzlink V (1990) *Principles of Food Preservation*. Amsterdam: Elsevier.
Mareček F et al. (1994) *Zahradnický Slovník Naučný*. Prague: ÚZPI.
Stehlík V et al. (1969) *Naučný Slovník Zemědělský 1,2*. Prague: UVTI-SZN.

PEANUTS

T H Sanders, North Carolina State University, Raleigh, NC, USA

Origin

Arachis hypogaea L., commonly known as peanut, groundnut, monkey nut, goober, or earth nut because the seed develop underground, is in the division Papiolionaceae of the family Leguminosae. The peanut is only one of a few hundred species of legumes that produces flowers above ground but develops the fruit below ground. Peanuts are native to South America and were cultivated in pre-Columbian native societies of Peru as early as 3000 BC. Peanuts probably originated in the region of eastern South America, where a large number of species are found growing wild. A Bolivian origin is suggested by the wide range in seed and pod morphology documented there. It has been suggested that *A. hypogaea* originated from a hybrid between *A. cardenasii* (nn) and *A. batizocoi* (K. & G.), as both parents occurred in reasonable proximity in Bolivia. Peanuts were widely distributed throughout South and Central America and the Caribbean region during the time of Columbus. Peanuts were probably brought to West Africa from Brazil in the sixteenth century and then to the African east coast and to India. Peanuts from widely separated regions of the world were brought to Africa and it is regarded as a center of genetic diversity.

Production

Worldwide production figures for peanuts are estimated at slightly less than 30×10^6 t and peanuts are produced on a significant basis in more than 30 different countries. In recent years, India, China, and the USA, the three largest producers, accounted for approximately 65% of the world production. Production figures for the USA are relatively stable at about 1.6×10^6 metric tons. However, production from China and India have been about 10.9×10^6 t and 7.4×10^6 t, although recent estimates from India at 7.4×10^6 t are probably much higher than current production. Although cultivated for many centuries, the economic importance of groundnuts and groundnut oil has increased rapidly only in the past century.

Yield per hectare is highly variable, depending on environment, variety, disease, and other factors. Important increases in peanut yield occurred between 1960 and 1990 and these increases are due to improved technology and management of factors which reduce yield. Peanuts are a high-value crop and an important source of revenue to producers throughout the world, and especially in less developed countries. The exportation of peanuts is mainly from the USA, China, and Argentina, while less developed countries utilize peanuts produced within the country, and the majority of these are processed for oil. The percentage of peanuts crushed for oil is a function of the use of peanuts as edible products and the need for readily available, cheap cooking oil. On a world basis, more than 50% of the peanuts produced are crushed for oil. In India, 75–80% of the peanuts produced are crushed for oil, while in the USA only approximately 12% of recent production is crushed.

World production of peanut oil is estimated at 4.65 million metric tons (**Table 1**). Extraction methods (hydraulic press, expeller, and/or solvent extractor) may differ widely due to technological advances in some countries; however, residue may contain 1–7% oil and with inefficient equipment the percentage may be much higher.

Peanut Plant

The leaves of peanuts are alternate and pinnate, with three to four leaflets on long petioles (**Figure 1**). The leaflets, which are obovate and softly hairy, are normally about 3–5 cm long. Peanuts have a distinct main stem and variable numbers of lateral branches. There are two major types of peanuts based on

growth habit of the plant – upright with erect central stem and spreading with numerous laterals. The upright type is suitable for mechanical production, and the spreading type is grown more often in less developed farming operations. The root system of peanuts is often expansive and may be found up to a depth of 2 m. Several strains of *Rhizobium* nitrogen-fixing bacteria are found in abundant nodules on main and lateral roots.

Flower, Fruit, and Seed

Peanuts have yellow flowers on axillary branches which are small (about 12 mm long). Depending on genotype, environment, and temperature, flowering starts at about 25 days after emergence and the most

Table 1 Area, yield, and production of peanuts: average for 1995–2000

Country	Area (million hectares)	Yield (tons per hectare)	Production (million metric tons)
India	7.78	0.95	7.39
China	3.89	2.80	10.90
USA	0.58	2.87	1.67
Senegal	0.73	0.90	0.66
Sudan	0.55	0.67	0.37
Brazil	0.09	1.77	0.16
Argentina	0.29	1.49	0.43
South Africa	0.09	1.57	0.15
Other	6.93	0.98	6.77
Total	20.94	1.36	28.51

Source: US Department of Agriculture, Foreign Agricultural Service, Cotton, Oilseeds, Tobacco, and Seeds Division. October 2001.

Figure 1 Peanut plant.

prolific flowering occurs between 5 and 11 weeks after planting (**Figure 2**).

Under the best of conditions, fewer than 20% of the flowers produce mature fruit and with drought or heat stress the percentage may decrease to fewer than 15%. Genotypes which produce most flowers during the first 2 weeks of the flowering period produce greater numbers of pods. Peanuts are generally self-pollinating, although a small amount of cross-pollination may occur. About 7–10 days after fertilization, the receptacle thickens and elongates into a long carpophore bearing the fertilized ovules to the ground. The carpophore is commonly known as the peg, and the action of burying the immature pod is known as pegging. Growth of the peg is positively geotropic until it enters the soil – approximately 5–7 cm. The tip then turns horizontal and begins to develop into a fruit, commonly referred to as a pod, containing from two to four seeds and ranging in size up to 8.0×2.7 cm. The enlargement of the pod proceeds from base to apex and reaches maximum size about 3 weeks after the penetration of the peg into the soil. As the pod matures, the inner face of the shell becomes increasingly darker in color, associated with an increased tannin content, and becomes very dark brown on maturation. Seeds show large variations in size, shape and color, all of which are fairly stable for any given cultivar. Seed size, together with the seed mass, has been used extensively in agronomic classification of peanuts. Larger seed types are generally preferred for confectionery purposes; mostly small and medium-size seeds are used to produce oil.

Peanut Composition

Peanut composition is greatly affected by variety, environment, and maturity. The interaction of these factors with harvest date and handling determines the specific composition of an individual lot. Average percentages for peanut components are presented in **Table 2**.

Oil

Peanut oil content is considered to average about 50% and the triacylglycerol content is generally in the range of 95%. Peanut oil is composed of mixed glycerides and contains a high proportion of unsaturated fatty acids, in particular, oleic (18:1) and linoleic (18:2). Additional fatty acids are palmitic (16:0), stearic (18:0), arachidic (20:0), 11-eicosenoic (20:1), behenic (22:0), and lignoceric (24:0). A comparison of fatty acid content with some other vegetable oils (**Table 3**) shows that peanuts are considered to have generally undetectable levels of linolenic acid. Peanuts contain higher levels of oleic acid than corn and soybean but lower levels than olive oil. However, peanut lines with high oleic acid concentration have recently become commercially available. The fatty acid profile from these peanuts is similar to that of olive oil.

Several years ago, research at the University of Florida examined approximately 500 peanut lines for fatty acid distribution and identified two lines with oleic-to-linoleic ratios of about 35. Using classical breeding techniques, commercial peanut varieties have been developed incorporating the high oleic acid trait. The developed lines do not have meaningful differences in oil content, flavor, color, or texture. Oxidative stability comparisons of high

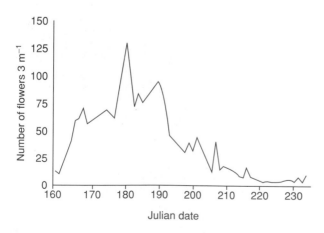

Figure 2 Number of flowers of peanut plant as a function of time.

Table 2 Typical composition of peanut kernels

Constituent	Range (%)	Average (%)
Moisture	3.9–13.2	5.0
Protein	21.0–36.4	28.5
Lipids	35.8–54.2	47.5
Crude fiber	1.2–4.3	2.8
Nitrogen-free extract	6.0–24.9	13.3
Ash	1.8–3.1	2.9
Reducing sugars	0.1–0.3	0.2
Disaccharide sugar	1.9–5.2	4.5
Starch	1.0–5.3	4.0
Pentosans	2.2–2.7	2.5

Table 3 Fatty acid profiles of peanut, olive, corn, and soybean oils

Fatty acid	16	18	18:1	18:2	18:3	20	20:1	22	24
Peanut	11.1	2.4	46.7	32.0		1.3	1.6	2.9	1.5
Olive	9.0	2.7	80.3	6.3	0.7	0.4			
Corn	10.9	2.0	25.4	59.6	1.2	0.4			
Soybean	10.6	4.0	23.2	53.7	7.6				

oleic peanut oil versus normal peanut oil were made using extracted, neutralized, and bleached oil from isogenic peanut lines varying only in fatty acid composition. Oil from the high oleic line contained 76.3% oleic and 4.7% linoleic compared to 56.6% oleic and 24.2% linoleic in the normal peanut oil. Oxidative stability was found to be as high as 14.5 times greater for high oleic peanut oil depending on the method of measurement. Details of peanut oil composition and characteristics are provided in the section on vegetable oils.

Extraction of Peanut Oil

Peanut oil is held in cells in an extremely fine emulsion. Damage to cells results in the release of oil. Peanut oil is usually extracted by hydraulic press, expeller, and solvent extractor. These may be used separately or in combination. The oil in the residue from oil recovery by the three methods is about 7, 5, and 1%, respectively. The residue (press cake) from the hydraulic press and expeller is generally used for livestock feed, and that from solvent oil extractors may be finely ground into flour for human consumption, if the nuts were of good quality.

Proteins The total protein content of whole seed peanuts ranges between 22 and 30%. Total protein can be separated into four major fractions: albumins, arachin, nonarachin, or conarachin. The arachin and conarachin fractions represent about 63% and 33% of the total, respectively. The arachin fraction is rich in threonine and proline and poor in lysine and methionine, while the conarachin fraction is poor in phylalanine and tyrosine.

The arachin and conarachin protein fractions are complex in composition and structure. Gel electrophoresis of sodium dodecyl sulfate dissociated proteins showed that these two fractions each contained five different components having molecular weights between 20 and 84 kDa. Two-dimensional polyacrylamide gel electrophoresis of seed polypeptides resulted in the detection of at least 74 major and between 100 and 125 minor peptides among various peanut cultivars and breeding lines. Variation existed among the major polypeptides of the cultivars, as noted by isoelectric points between pH 4.4 and 8.0 and molecular weights between 16 and 75 kDa. There is genetic variability in the mechanism for peanut protein synthesis, indicating that there is a potential for the development of peanut cultivars possessing nutritionally desirable proteins by manipulating protein synthesis.

Nutritional value of peanut protein depends on the amount and kind of essential amino acids available

during digestion (**Table 4**). Amino acid availability in peanuts and peanut products is enhanced by the processing they undergo before consumption. Antinutritive factors, including trypsin inhibitors, hemagglutinins, goitrogens, saponins, and phytic acid are present in raw peanuts, and heat treatment in blanching and roasting of peanuts destroys the trypsin inhibitor and hemagglutinin. Peanuts are generally considered to be deficient in lysine and methionine, thus their relative nutritive value is decreased in comparison with a reference protein. The protein efficiency ratio for peanut protein is generally considered to be 1.5–1.8 and these values are related to the limited availability of lysine and methionine.

Carbohydrates The carbohydrate content of peanut kernels is usually determined by difference from the proximate analysis. It has been found to range from 10 to 20% of the fresh weight. Defatted peanut meal contains about 38% total carbohydrate. The major components are starch, sucrose, and nonstarch polysaccharides. Defatted meal contains about 18% sugars.

The major sugars in peanut are sucrose, stachyose, raffinose, fructose, and glucose. Inositol, verbascose, arabinose, and galactose have also been reported in various studies. The total sugar content has been found to vary with genotype, maturity, growing conditions, storage conditions, and kernel size. Large kernels usually contain less sugar. Cooler growing conditions were found to produce more sugars and sugar content decreases with maturity. Sucrose is the predominant sugar, with the content ranging

Table 4 Amino acid composition of peanuts

Amino acid	% Total protein[a]
Aspartic acid	11.89
Threonine[b]	2.43
Serine	5.11
Glutamic acid	19.83
Proline	4.44
Cystine	0.81
Glycine	6.59
Alanine	4.12
Valine[b]	3.14
Methionine[b]	0.51
Isoleucine[b]	3.40
Leucine[b]	5.21
Tyrosine	4.09
Phenylalanine[b]	6.38
Histidine[b]	3.17
Lysine[b]	3.87
Arginine[b]	10.81

[a]Data represent means from several sources.
[b]Essential amino acids.

from 2.7 to 3.5%. Stachyose is the next abundant, at levels of 1.5–2.5%. Other values found are 0.02–0.06% for raffinose and 0.01–0.04% for inositol. Fructose and glucose are usually present at less than 0.02%.

The sugar content is important for peanut flavor. Free sugars are noted to be precursors to roasted peanut flavors. When roasted, there is loss of some sugar content, with the greatest losses being sustained by fructose and glucose.

Vitamins and minerals The vitamin content of peanut seed is shown in **Table 5**. Peanuts have little or no vitamin A activity. Three forms of tocopherol (α, γ, β) have been found, with γ-tocopherol the highest and β-tocopherol the lowest. Peanuts are considered to be a good source of vitamin E, niacin, and folate.

Peanuts contain several of the minerals essential for human growth and development (**Table 6**). Recommended daily intake percentages of the quantities present are significant and constitute a good source of minerals.

Production and Handling Practices that Affect Quality

Maturity

Peanut maturation is extremely important in changes of biochemical components related to quality and shelf-life potential. Those facts have been demonstrated collectively in sensory studies to evaluate the relationship of maturity and flavor quality. Mature peanuts tend to have increased intensity of positive flavor descriptors, such as roasted peanutty and sweet aromatic, while immature peanuts are more commonly associated with negative terms, such as painty and fruity fermented. The term 'painty' is related to the common term 'rancid' and indicates the propensity of oil components in more immature peanuts to undergo enzymatic or autoxidation changes that result in decreased shelf-life. The degree of dark roast color is related to maturity level, with immature peanuts roasting darker because of higher free sugars and free amino acids. The relationship of high curing temperature to decreased sensory acceptability in immature peanuts has been demonstrated. Further, decreased sensory acceptability caused by high intensities of the off-flavor term 'fruity fermented' in sized lots was related to darker roast colored immature peanuts in the lot.

Maturity in peanuts is a complex concept because peanuts of all market types are generally marketed in various commercial seed sizes based on thickness, or

Table 5 Essential vitamin composition of peanuts

	Amount[a]	%RDI
Vitamin E	3.3 IU	11
Folate	41 µg	10
Niacin	3.8 mg	19
Thiamin (B$_1$)	0.12 mg	8
Riboflavin (B$_2$)	0.03 mg	2
Vitamin B$_6$	0.07 mg	4

[a]Based on 1 oz serving of dry roasted peanuts.
RDI, recommended dietary intake.

Table 6 Composition of some minerals in peanuts

	Amount (mg)[a]	% RDI
Zinc	0.94	6
Copper	0.2	10
Potassium	186.5	6
Calcium	15.3	2
Iron	0.6	3
Phosphorus	101.5	10
Magnesium	50	13

[a]Based on 1 oz serving of dry roasted peanuts.
RDI, recommended dietary intake.
Source: US Department of Agriculture, Agricultural Research Service, 1998. USDA Nutrient Database for Standard Reference, Release 12.

on count per unit weight. The complexity arises in that all mature peanuts are not large and all immature peanuts are not small. Although size and maturity are positively related, the relationship is not absolute, and peanuts of very different maturity are found consistently in all commercial sizes. This fact is confounded even more in lots that are marketed on a count-per-weight basis wherein large and small seed may be mixed together to give the same count as seed of consistent uniform size. Peanut seed size distribution is affected by the environment in predictable ways. Soil temperature increases tend to shift the mean seed size down while cooler temperatures shift the mean size upward. Irrigation, harvest date, and soil temperature treatments generally produced significant differences among percentages of individual maturity classes in each commercial size. The distributions of maturity classes within commercial sizes are sufficiently different to suggest that flavor, roast color, shelf-life, and other quality factors would be affected in final roast products. Published data indicate that there are sensory differences in medium-grade-size peanuts harvested on a weekly basis over a 6-week period. In these studies, the percentage of immature peanuts in the seed sized lots progressively decreased with time. Roasted peanuts from earlier harvests had lower roast peanut impact and more bitter taste. Although shelf-life studies were not conducted, the increase in percentage of mature peanuts in progressive harvests predisposes those lots to longer shelf-life.

Environment

The fact that growth, production, and handling practices affect quality, flavor, and shelf-life of peanuts and peanut products is implicit in data from numerous studies, years of manufacturing use, and the wide ranges of flavor and shelf-life observed in those countless situations. However, although studies report specific biochemical constituents and the effect of certain practices and conditions on changes in those compounds, the total effect on final product shelf-life resulting from all these processes must be extrapolated using established quantitative and qualitative relationships of specific constituents to shelf-life potential.

The role of environment on peanut yield has been consistently documented, but the effect on peanut quality factors potentially related to flavor and shelf-life has received little recent research attention. Higher soil temperatures tend to result in smaller seed size distributions, but studies on biochemical comparisons among years do not relate well to specific environmental conditions. Drought and temperature stress have been shown to affect some of the protein factors of peanuts. Soil temperature increases result in decreased concentrations of the free carbohydrates fructose, glucose, sucrose, raffinose, and stachyose in peanuts.

Harvest

When peanut plants are removed from the soil, the peanuts begin to dry, and the final natural processes related to potential flavor and shelf-life are initiated. When peanuts are harvested into inverted windrows, temperatures inside pods near the ground may reach 40–50 °C when air temperatures are no higher than 32 °C. Conditions that slow the drying process with associated high moisture may lead to fungus growth with resulting high free fatty acid values due to microbial lipase activity. Peanuts affected in this way are usually visibly damaged and may be removed electronically as part of the shelling process, or in less developed countries they may be removed by hand picking. Although off-flavors may not normally be thought of as factors that reduce shelf-life, usually the physical, physiological, and biochemical mechanisms that contribute to off-flavor development so disrupt the normal processes that shorter shelf-life should be anticipated. Obviously, processors will not knowingly utilize off-flavor peanuts in formulations, but knowledge or stipulation of growing and handling practices for purchased peanuts may

contribute to the production and delivery of more stable products by the processor.

Maturity, as previously discussed, is a primary factor in total quality of peanuts, and the level of total crop maturity at harvest influences many of the handling and storage factors that follow. Moisture content decreases as peanuts mature on the plant, and at harvest, a wide range of moisture percentages are present. The range has variably been ascribed as 20–70% moisture at the time of digging. Given this, and the degree of biochemical and physical development of the various peanuts, it is readily discernible that differences in flavor and shelf-life are still being affected. Peanuts normally air-dry for some period of time before they are harvested. In mechanical harvest situations, peanuts usually dry to an average of 20–25% moisture before being picked and further dried.

The physical process of machine harvest has potential effects on the shelf-life of peanuts due to the force used to pick and transport peanuts into the machine. The physical damage in mechanical harvest, noted as cracked or broken pods, must inherently damage seed integrity, although relationships to quality reduction have not been investigated. Peanuts that are field-dried are subject to various environmental conditions which may affect quality. Harvested peanuts with a high percentage of immature pods will be difficult to dry to an overall safe storage level. High-moisture peanuts going into storage provide an increased opportunity for decrease in shelf-life potential.

Storage

Shelf-life potential may be maintained in storage of in-shell and shelled peanuts. In-shell peanuts are stored in bulk situations in many parts of the world, often at approximately 10% moisture content. In these situations, care must be given to ventilation and moisture control, because as the stored peanuts cool in response to ambient conditions that are usually cool, they dry further. The moisture that is lost must not be allowed to concentrate into the bulk lot because of obvious quality and microbial considerations. Adequate conditions of storage, especially ventilation, are critical to maintaining low levels of free fatty acids and total carbonyl compounds. Because of the removal of the shell, which is a normal protective barrier, shelled peanuts are somewhat more susceptible to quality/shelf-life deterioration. Shelled peanuts may be held in conditions of *c.* 0.5–5 °C and 55–70% relative humidity for periods of a year or more without quality loss.

Moisture content should be in the 7% range and slightly lower levels may impart additional protection. In peanuts stored at *c.* 6 and 9% moisture, the higher moisture content resulted in reduced flavor quality and changes in amino acids, carbohydrates, and the phospholipid fraction of oil.

Consumption/Uses of Peanuts

Peanut Oil

Peanut oil is used as a cooking oil, especially in deep-fat frying; peanut oil is excellent since it has a smoke point of 229.4 °C. Degradation which may occur during frying results in the increase of free fatty acids and a decrease in smoke point. Crude peanut oil has a bland, slightly beany, nut-like flavor which is removed during refining to produce an oil that is odorless. Peanut oil develops few off-flavors or odors in use as a frying oil. Peanut oil is sometimes used in salad oil and in pourable dressings because of the length of time solids are held in suspension in the oil, although it solidifies from 0 to 3 °C. Refined peanut oil is widely used to prepare pastries, shortening (hydrogenation), oleomargarine, mayonnaise, salad dressings, and other food products. It is superior to oils from corn, soybean, cotton seed, olive, or safflower for making salad dressings to be stored below −12.2 °C.

Roasted Peanuts

Peanut roasting is a common practice in much of the world since roasting brings out the unique flavor. In western countries, particularly the USA, roasted and salted peanuts are one of the popular food items, produced on a commercial scale under a highly mechanized system with excellent quality control and packaging methods. In many European countries, peanuts are roasted and then coated with various flavors. A large portion of the US peanut crop is processed into roasted product for use in peanut butter, salted peanuts, and confections. Although the salted, roasted form predominates, oil-cooked, salted, unsalted, and low-calorie forms such as whole or mixed nuts are also popular.

A typical protocol for dry-roasted peanuts is exposure to 160 °C for 20–30 min. Seed coats are generally removed before applying oil and salt. In oil-roasting, blanched peanuts are immersed for 3–5 min in heated oil. After roasting the peanuts are then subjected to air cooling and salting with about 1.8–2.2% salt. Studies conducted to determine the effect of high-oleic-oil roasting indicated slight increases in shelf-life of normal oleic peanuts. Since peanut seed coats contain phenolics, including tannins, and the germs contain bitter saponin compounds, these are often removed before consumption. The current methods

of blanching eliminate the skins and germs so that blanching always precedes oil-roasting, but often follows dry-heat roasting.

Peanuts have a high oil content and defatted peanuts have been developed to meet a specific consumer market. Partially defatted peanuts are obtained by hydraulic pressing to remove about 50% oil, followed by restoration of pressed kernels to their original size and shape by immersion in boiling water containing salt, species, or flavorings. Kernels are then dried and roasted. The pressure applied in the hydraulic press, the holding time, and the initial depth of peanuts in the compression influence the reconstitution of peanuts.

Peanut Butter

Although not a product accepted worldwide, peanut butter and products made from it are the most consumed form of peanuts in the USA, and notable consumption occurs in Canada and The Netherlands. Marketing introductions of peanut butter are still in progress in many European countries. Peanut butter is made by the fine grinding of dry roasted peanuts. Salt, hydrogenated oil, and sugar are usually added. Peanut butter may be made from any variety of peanuts but the blend of two parts Spanish or runner peanuts with one part Virginia peanuts is considered best for the most desirable consistency. Peanut quality, roasting techniques, manipulation of temperature, and grinding result in a wide range of colors, flavors, and consistencies of peanut butters on the market.

After grinding, oil separates from the solid particles unless a crystalline lattice is formed. Since crystals of fat are not present in the natural oil of peanut butter at ordinary temperatures, hydrogenated peanut oil, which is crystalline at these temperatures, is mixed uniformly with the product to provide such crystals. Generally, the incorporation is accomplished above the melting point of the hydrogenated peanut oil to insure more complete dispersion of the hydrogenated oil. In a satisfactorily stabilized peanut butter, crystallization of the added fat (hydrogenated peanut oil) takes place generally throughout the product before the oil–meal mixture can separate. *Trans*-fats infrequently associated with hydrogenated oil have recently been examined in commercial peanut butters and the work demonstrated unmeasurable quantities using current technology.

Confections

Peanuts (split, granulated, whole or ground peanut paste) are used in a wide range of other products, including confections such as candy bars, dragées, hard-coated peanuts, peanut butter cups, peanut brittle, and a variety of other specialty applications. Shelf-life is important for confectionary products and is greatly affected by the product configuration, moisture content, and storage temperature.

Aflatoxin

Aflatoxin is a potentially carcinogenic compound produced by *Aspergillus flavus* and *A. parasiticus* which can invade peanuts, corn, cottonseed, and other commodities. In technologically advanced countries, control programs and testing programs are in place which significantly reduce the concentrations of aflatoxin over those found in developing countries. Aflatoxin contamination is usually found to be the most serious when drought conditions are present during the last few weeks of growth and pod development. Preharvest contamination is a significant economic problem for the peanut industry. Preharvest aflatoxin contamination in the crop is a problem when drought-type conditions predispose peanuts to insect damage by lesser corn stalk borer and other insects which are known carriers of fungus spores. Invasion of pods and infestation of seed often result in visible damage, and high levels of aflatoxin are characteristically found in damaged kernels. Preharvest invasion can occur with no obvious kernel damage and, although extensive studies have been conducted, the exact mechanism of fungal invasion is not yet known. Aflatoxin contamination is often highest in small, immature seed. Studies have demonstrated a relationship between the water activity-related physiological ability of the seed to produce phytoalexins and the water activity at which the fungus can best invade the seed. As the ability of the seed to produce phytoalexins decreases, there appears to be a point at which the fungus can grow in the seed before decreasing water activity terminates growth of the fungus.

Although generally less frequent, postharvest contamination may be a problem, especially when moisture accumulation and proper temperatures are present in harvest, transport, or storage.

Once the pods are field-dried and stripped from plants, they must be dried to moisture content less than 9% prior to storage. Peanuts stored at moisture content greater than 10% have increased risk for growth of *A. flavus*. Once dried to less than 9% moisture, peanuts can be stored for long periods with little loss of quality if kept under controlled storage conditions. Control of aflatoxin is best addressed during production by the appropriate irrigation of the crop through harvest time. In most contaminated lots, only a relatively small number of kernels is contaminated, and in many cases physical separation is the ideal method. This is based on the assumption that contaminated kernels are either discolored or shriveled. Methods include manual sorting, mechanical sorters, or electronic scanning, which examines each kernel separately and accepts or rejects it on the basis of reflectance. Action levels of aflatoxin in peanuts and other commodities are set and rigorously enforced within producing countries, and import regulations carefully control levels which may be imported.

Peanut Allergy

In recent years, concern for food allergies in general has increased, and concern for peanut allergy is no exception. For unknown reasons, peanut allergy is associated with a higher incidence of fatal food-induced anaphylaxis than any other food allergy. Immediate hypersensitivity to foods occurs in 6–8% of children and about 1% of adults. In the USA, a recent survey suggested that 0.7% of children are allergic to peanuts in varying degrees. Avoidance is the only current method of dealing with food allergy. Research potentials in dealing with peanut allergy will be discussed later.

Five peanut allergens have been identified by different research groups. Two major allergens, Ara h 1 (a vicilin) and Ara h 2 (a conglutin) are recognized by serum immunoglobulin E (IgE) of greater than 90% of peanut allergic individuals. Ara h 3 is recognized by serum IgE of about 40–45% of allergic individuals and is thus considered to be a minor allergen. Ara h 5 (a profilin) is also a minor allergen; IgE is recognized by about 13% of allergic individuals. Refined peanut oil (heat-processed) is not allergenic; however, oils contaminated with peanut protein may indeed produce significant allergic reactions in peanut-sensitive individuals. Cold-pressed oils are more likely to contain peanut proteins than hot-pressed oils.

Current research on peanut allergy is directed at testing an approved asthma drug, anti-IgE monoclonal antibody, against peanut allergy. Development and testing of a vaccine against peanut allergy are progressing towards human trials. A third area of research involves identification of enzymes involved in mediation of the allergic reaction and development of a test kit for individuals (especially children) to determine potential for peanut allergy.

See also: **Aflatoxins**; **Amino Acids**: Properties and Occurrence; **Food Intolerance**: Food Allergies; **Sweets and Candies**: Sugar Confectionery; **Vegetable Oils**: Composition and Analysis

PEARS

C Blattný, Prague, Czech Republic

Introduction

The genus of pear *Pyrus* has about 20 species. Cultivars of the common pear (mainly European and American) belong to *Pyrus communis*. The Asian cultivars belong to the sand pear *Pyrus serotina = P. pyrifolia*. The pear is a member of the rose family (Rosaceae).

Origin and History

The common pear came from Central Asia. It was well known and popular in ancient Greece. In the year 287 BC Theofrast, in his *Historia Plantarum*, described four cultivars of pears. The Romans took over the cultivation of pears from the Greeks. In the time of Vergil (50 years BC), Cato described 35 cultivars. The number of cultivars increased very quickly and by the end of the Middle Ages over 200 cultivars were described. But most of the outstanding dessert cultivars that are now widespread were selected out in the eighteenth and nineteenth centuries in France and in Belgium. Most of these old pear cultivars arose out of the selection of seedlings that grew as a consequence of random cross-breeding. Intentional or deliberate breeding led to new cultivars during the twentieth century, but a number of excellent cultivars from the eighteenth and nineteenth century have not been replaced. By the end of the twentieth century it is roughly estimated that 3000 cultivars of the common pear had come into being.

In China the cultivation of sand pear cultivars was known some centuries before Christ. Nowadays, many cultivars of the sand pear are widespread in East Asia. By the end of the twentieth century this sand pear had spread not only to North America, but also in smaller amounts to other continents. This species is also used for hybridization with the common pear, especially in the USA, as hybrids are less susceptible to the dangerous infectious bacterial disease fireblight.

Requirements of the Pear Tree

The pear tree, depending on the species and cultivar, has moderate to high demands of temperature (in winter as well as in summer) and length of vegetation period. Winter cultivars need warmer regions with a longer vegetation period to ripen properly and to attain better quality. Some of the cultivars do not tolerate extensive winter frost, e.g., Bartlett (Williams Christ). Fairly rich, warm, loamy soil, of sufficient depth and with a well-drained subsoil are suitable for the pear. Of all temperate climate fruit species, pear quality is the most influenced by soil properties and weather.

The trees that grow naturally from seeds (or grafts on pear seedlings) reach a height of up to 9 m with trunks of 30 cm or more in diameter. The cultivars of pears can only reproduce by being grafted on a rootstock – usually a pear seedling. In order to create smaller or dwarf trees the quince can be used as the rootstock. Most widespread vegetatively propagated clones of the quince are: quince A, quince C, from East Malling Research Station in England, French Provence Quince BA 29 C, and Le Page series C. In the USA a new hybrid Old Home × Farmingdale 333 (OH × F333) is now also recommended. Certain cultivars like Bartlett (Williams Christ) and Bosc are incompatible with the quince and require an interstem, e.g., cultivar Hardy or Old Home (used in the USA).

Many pear cultivars are self-sterile. They need the pollen of other cultivars to develop fruitfully. Some cultivars have relatively often parthenocarpic fruit (without kernels). Those fruits are usually longer and slimmer.

The cultivars of the Japanese-type Asian pear will pollinate each other. Chinese types need another cultivar for pollination, and when they bloom with the cultivars of the common pear, these cultivars could be very good pollinators for them.

As regards the sand pear's requirements of the environment, these are similar to the common pear, but it seems that suitable districts for peaches will also be best for most sand pear cultivars. *Pyrus betulaefolia* is recommended as standard rootstock for sand pear cultivars. The hybrid OH × F333 is also suitable.

The Fruit

The pear is a typical pome fruit closely related to the apple. The basic shape is pear-shaped and ranges according to the cultivar from bell-shaped, to egg-shaped, conical, globular and teardrop-shaped (**Figure 1**). Further differences are in size, color, texture, flavor, taste, ripening season, and possibilities for processing and cooling.

The flesh is usually firmer than that of apples; when fully ripe it softens and finally it melts. After

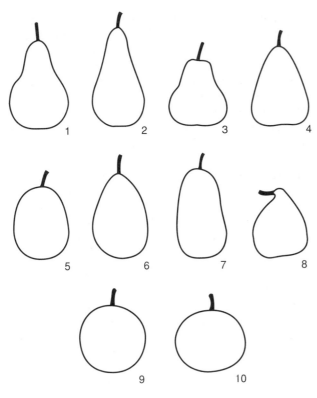

Figure 1 Different shapes of pear. 1, pear-shaped; 2, bottle-shaped; 3, bell-shaped; 4, conical; 5, egg-shaped; 6, teardrop-shaped; 7, cylinder-shaped; 8, fig-shaped; 9, globe-shaped; 10, flat globe-shaped

overripening the flesh quickly melts due to hydrolysis of pectin and total disorganization of enzymes. At optimal grades of ripeness, the fruit of the best dessert cultivars melts like butter. The skin is smooth and its color varies from yellow, yellow green, to brown (maybe russet) and also almost red. Only two cultivars are bright dark red: Max Red Bartlett and Red Anjou. The flavor and taste of cultivars differ greatly – some are spicy like muscat. There is another conspicuous difference between pears and apples. In the center of a pear around the core and along the vessels from the stem to the core, the gritty cells called stone cells are present. The more gentle the cultivar, the fewer small stone cells are present. However, very small stone cells in a very small number are always present in the best dessert cultivars, e.g. Bartlett (Williams Christ). The number of stone cells is not only influenced genetically by also by the quality of the soil, especially for winter cultivars. Most modern dessert cultivars of the common pear have large heavy fruit: medium weight 150–180 g, heavy weight 200–400 g, giant weight up to 750 g.

Asian pears are also called Oriental pears, Chinese pears, and Japanese pears. The Asian pear cultivar differs in its flesh texture – when eaten ripe they remain pleasantly firm, crisp, and juicy like apples but with a different and distinctive texture. They do not change texture after being picked and stored like common pears. Most Asian pears are sweet, flavorful, and low-acid. They are round or flat fruit with yellow or green skin, with bronze-colored and russeting skin or they are pear-shaped with either smooth or russeted skin.

Diseases and Pests

Diseases

Some virus and *Mycoplasma* diseases are relatively widespread. We will only discuss a few here. Ring pattern mosaic gives striking light-green or greenish-yellow rings and a line pattern in the leaves. Yellow veins and red mottle are typical with chlorotic banding of the tertiary and final veins. Both diseases suppress growth in a number of pear cultivars. The disease stony pit occurs wherever pears are grown. The fruits of diseased trees become pitted and deformed with many gritty cells in the tissue near pitty areas. Fruits with many pits are woody and difficult to cut. Either only a few or all of the fruits of diseased trees may develop symptoms, depending on the susceptibility of the cultivar. An affected fruit may have one or a number of pits and is not suitable for eating. The content of dry matter, ash, potassium, calcium, and mallic acid is higher in diseased fruits then in comparable healthy ones. The content of monosaccharides is less. The affected fruits cannot be used to produce sherry and brandy. Very sensitive cultivars are Bosc, Doyenné du Comice, while Bartlett in the USA is relatively resistant.

Fireblight, a bacterial disease, is the greatest problem limiting production of pears in the USA. Fireblight seldom affects Europe. Leaves and young terminal shoots infected by bacteria *Erwinia amylovora* wilt, die, and blacken. Further infection of branches and trunk results in the development of cankers and finally the death of the tree. Recommendations for decreasing the danger of fireblight are:

1. Select tolerant, better resistant cultivars. In the USA recommended hybrids between common and Asian pears include: Moonglow, Kieffer, Seckel, Waite, and Maxim.
2. Control of the vector – green apple aphids (*Aphis pomi*).
3. Avoid any practice that encourages excessive and soft terminal growth.
4. Prune out and burn the infected portion of the tree (better in winter) and avoid overfertilization.

Pear scab is a disease caused by the fungus *Venturia pirina*. The infection starts at blossom time and occurs on the blossoms, more so on the leaves than on the fruits. Infection of young pear fruits may cause the pear to be lop-sided and sometimes cracked. Then the scab spots spread on the whole fruit, especially on that of the susceptible cultivar. After very late infection, small scabs occur which then develop into so-called storage scabs. Spraying during infection time (May–July) is one of the best control measures.

During longer storage some pathological and physiological disorders can occur:

- blue mold rot (*Penicillium expansum*) – after mechanical injury
- gray mold rot (*Botrytis cinerea*) – infection during blossoming or through wounds of mature fruits
- senescent scald – dark brown skin discoloration
- superficial scald – diffuse brown skin discoloration
- freezing injury – translucent, water-soaked appearance of tissue
- bitter pit – brown, corky lesions in the flesh
- watery breakdown – rapid enzymatic flesh breakdown
- carbon dioxide injury – browning of interior walls of carpels and adjacent core tissue
- low-oxygen injury – core browning

Control measures

- effective preharvest disease control procedures
- careful handling during harvesting and postharvest preparation for market to minimize mechanical injuries
- prompt cooling to $0\,^{\circ}C$
- maintenance of proper temperature ($-1\,^{\circ}C$ to $0\,^{\circ}C$) during storage and transport
- use of postharvest chemical treatments
- use of the controlled atmosphere

Pests

Codling moth (*Laspeyresia pomonella*) larvae feed on the fruits to the core and enlarge an exit hole which they plug with frass. They can have two to three generations and they are the most dangerous pests to apples and sometimes can be pernicious to pears. Spraying when females lay their eggs (June, July) is a good control measure.

Pear midge (*Contarinia pyrivora*) can cause considerable damage to young fruits. The attacked fruitlets become deformed and enlarged, and are later black in the center and full of white maggots, and they crack, decay, and fall down in a short time. They hibernate in the soil. Spraying just before the blossoming can decrease the damage.

Pear leaf midge (*Psylla pyricola*) in some districts is an important pest, resembling a miniature cicada. The psyllas suck on the leaves which margins roll. Maggots produce a honeydew that collects in droplets on the leaves, branches, and the fruit. It is an excellent medium for sooty molds. The crop is usually lower and unsellebled. Another danger to pear psylla is in its transmission of *Mycoplasma*.

Harvesting

The pear has to be gathered from the tree before full ripeness, usually when the basic green color changes to yellowish green. It is very important to determine the correct time for the crop. Prematurely harvested fruits do not ripen but they stay hard permanently and become leathery. Therefore the best time for gathering is when the fruits attain normal size, which is optimal for the cultivar, at the time when the stem parts readily from the branch if the fruit is slightly lifted. Then it is important to get the fruit quickly to cool storage. Overripe fruits soften so quickly that they are nearly useless.

To harvest fruits of Asian pears one must hold them lightly in the palm and pluck them using an upward twisting motion to remove the fruit from the spur. A pulling motion can result in damage as the stalk may be removed from the fruit. The skin of Asian pears is much more susceptible to abrasion and friction marks than common pears. Smooth surface containers such as polystyrene trace should be used to gather fruit. Fruit should be placed into trays with the stem end upward, preferably in single layers.

Chill Storage

Only a very small number of common pear cultivars can endure longer preservation in chill storage (Bosc, Anjou, Doyenné du Comice) if they do make it to the fresh market. These cultivars can be stored in controlled atmosphere ($1–2\%\ O_2 + 0–1\%\ CO_2$) at $1\,^{\circ}C$ for up to 4 months (Bosc and Doyenné du Comice) or 6 months (Anjou) while maintaining their capacity to ripen and attain good flavor and texture. Some winter cultivars can be stored in cooler places for some months until spring, but their quality is much lower (Diel, Comtesse de Paris, Nelis). For chilling it is necessary to have optimal temperature $-1–0\,^{\circ}C$, optimal relative humidity 90–95%, and for ripening, optimal conditions are $15–22\,^{\circ}C$ and 90–95% relative humidity. This basic schedule for ripening must be introduced after each longer cool storage of all pear cultivars to help the development of flavor and optimal texture of the flesh.

Some of the Asian pear cultivars, e.g., Shinseiki, Hosui, Kosui, Nijisseiki, have an adequate storage life. If fruits are harvested at the recommended maturity, a storage life of 12–20 weeks and subsequent shelf-life of 10–15 days can be expected, depending on the cultivar.

Important World Cultivars

- Common pear
- Summer – Bartlett (= Williams Christ), Dr J. Guyot, Clapp's Favourite, Santa Maria, Max Red Bartlett
- Autumn – Doyenné du Comice, Bosc, Conference, Fondante de Charneuses, Elisabeth, Luise Bonne, Boussock, Hardy, Lucas
- Winter – Diel, Comtesse de Paris, Anjou, Red Anjou, A. Lucas, Passe Crassane, Nelis
- Sand (Asian) pear (recommended in the USA)
- Early – Shinseaki
- August – Hosui, Kosui, Nijisseiki (Twentieth Century), Yo Inashi
- Late – Shinko, Olympic (Korean Giant)
- Sand (Asian) pear (widespread in China)
- Ya Li, Tse Li

Supplying Fresh Fruit to the Consumer

The fruit must be intact, sound, clean, free of any visible foreign matter, free from pests, free of abnormal external moisture, and free from any foreign smell or taste. The fruit must have been carefully picked and sufficiently developed so that the ripening process can continue in order to reach the degree of maturity appropriate to the cultivar, to withstand transport and handling, and to arrive at the destination in a satisfactory condition. Usually fruit is classified into three or four classes according the diameter of the widest part of the pear fruit.

Use of Fresh Fruit

The time to ripeness for consumption varies according to the cultivar: summer cultivars – up to 2 weeks; autumn cultivars – 2–8 weeks; winter cultivars up to 16 weeks. It is greatly influenced by the grade of ripeness during gathering and by storage temperature.

Dessert cultivars of the common pear, when eaten at the right stage of ripeness (the flesh should melt like butter) are delicious fruits. The fruits are used in various cookies, fruit salads, desserts, and gelatine dishes.

A small warning: it should be mentioned that some sensitive people can experience flatulence after eating fruits of the common pear, especially fruits which are not fully ripened.

The Asian pear cultivars are refreshing, very juicy, crispy like apples and, as with common pears, they go well with salads.

Processing

Pears are less often used for processing, compared to apples, due to their relatively low acid content and, above all, to a rapid ripening process which leads to the already-mentioned melting disintegration and liquefaction of the fruit. Selected pear cultivars are outstanding raw material for canning. They must, however, fulfill strict conditions:

1. The shape of the fruits is one of the decisive determinants of industrial preparation, since a suitable shape facilitates basic processing – peeling, halving, and removal of the core. The most suitable is a longish pear form with a wider stem part (for instance, the parthenocarpic fruits of the Bartlett (Williams Christ)). Cultivars that are too round have their core situated too high, which causes difficulties for machine coring. Fruits that are bottle-shaped, that crack, and that are too small are unsuitable.
2. The flesh should be white or faintly yellowish.
3. The skin should be firm, not too thick, clearly demarcated from the flesh, and easily peelable.
4. For factory processing, pears must be of the prescribed size and shape. Fruits of a uniform size and shape reach most uniformly technologically required ripeness. This degree is attained when the basic green color of the skin changes to yellow. Accurate appraisal of the degree of ripeness heralded by the changing color of different cultivars calls for an experienced eye. Many manufacturers use penetrometric methods to measure the firmness of the flesh, i.e., the pressure needed to push a special needle to a certain depth of the fruit. In areas of great production of cultivars destined for canning, for instance the summer Bartlett (Williams Christ) it is possible to employ long-term storage at temperatures −1 to 0 °C for 9–12 weeks, or at temperatures +1 °C for 6–8 weeks to extend the processing season. The relative humidity must be 90–95%. The ripening process must follow this chill storage. Similar conditions can be also used for further canning of suitable cultivars like Clapps Favourite, Elisabeth, or Kieffer. The fruits are canned in syrup. To prevent browning of the flesh it is recommended to plunge the peeled halves in a solution of ascorbic or citric acid.

- Sweetening – the same requirements as for canning are applied.
- Drying of pears is relatively widespread in some regions. Pears suitable for drying should be of medium size (3 cm length), typically pear-shaped, slimmer in the stem part, the skin thinner, and the flesh with a small number of stone cells. Dried fruits should have a cinnamon color. Pears should be fully ripe with the highest content of sugar. They are dried whole, halved, or divided into small pieces, with the skin on or peeled.
- Pears are not frequently frozen. The requirements are the same as for canning.
- Brandy – due to the delicious flavor of some cultivars, e.g. Bartlett (Williams Christ), pear brandy is the mildest of the fruit brandies.
- Perry is a famous light alcoholic drink (like light wine) made from special perry cultivars. This drink was produced and beloved in Roman times. It is a drink of choice particularly in Normandy, the UK, Austria, Switzerland, and the USA. The special perry cultivars are highly fertile, with small fruit, and they should have a high content of sugar and acids and a rich content of tannins. Unfortunately, the production of those cultivars declined strikingly during the twentieth century.

Production

The pear is a popular fruit in temperate regions. Their production over the last 10 years in individual countries has developed differently. Worldwide it is increasing – in 1996–98 it reached 14 034 million tonnes, while in 1989–91 it was only 9529 million tonnes. It especially rose in China, by 245%, and amounted to 6401 million tonnes in 1996–98, that is, 45% of world output. The increase can also be observed in South America and Africa – in Chile by c. 500%, in Ecuador by 300%, in Argentina by 225%, in Morocco by 430%, in Algeria by 310%, in Tunisia by 210%, and in Egypt by 350%. Even in other warm regions production also increased, for instance, in Lebanon by 390% and in Turkey by 140%.

The changes are the result of the development of international trade which has facilitated refrigerated transport of high-quality dessert cultivars over long distances. Particularly profitable is the supply of the winter market either from southern countries of the northern hemisphere or from countries of the southern hemisphere. Among the largest exporters of pears we find now Argentina and Chile – not only Italy, France, and The Netherlands, as it was some years ago. The decrease in pear output in European countries is chiefly connected to the increase in the cultivation of apples and peaches to replace the pear.

Table 1 Chemical composition of pears

	$g\,kg^{-1}$
Water	775–869
Protein	4–5
Total lipid	3–4
Sugar	124–158
Acids	1–5
Dietary fiber	15–28
Ash	3–4

Sugar is composed on average from 54% fructose, 13% glucose, 15% saccharose, and 18% sorbit. Organic acids fluctuate according to the cultivar. In dessert cultivars malic acid, prevails citric acid is about 10% of

Table 2 Mineral salt content

Mineral salt	$mg\,kg^{-1}$
Calcium	100.0–160.0
Iron	2.0–3.0
Magnesium	60.0–80.0
Phosphorus	110.0–150.0
Potassium	1100.0–1300.0
Sodium	0.0–20.0
Zinc	1.2–2.3
Copper	0.9–1.1
Magnesium	0.4–0.8
Selenium	0.0–10.0

Table 3 Vitamin content

Vitamin	
Beta-carotene	0.15–0.30 mg kg^{-1}
Vitamin E	4.30–5.00 mg kg^{-1}
Vitamin B$_1$	0.20–0.50 mg kg^{-1}
Vitamin B$_2$	0.30–0.40 mg kg^{-1}
Vitamin B$_6$	0.15–0.18 mg kg^{-1}
Pantothenic acid	0.62–0.70 mg kg^{-1}
Folate	70.00–140.00 µg kg^{-1}
Vitamin C	40.00–50.00 mg kg^{-1}

Nutritional Value

The composition of the pear is given in **Tables 1–3**. Pears contain a good quantity of sugars (rapidly assimilated fructose, glucose, saccharose, and sorbit), as well as pectin, tannin, mineral salts, organic acids, some vitamins, and dietary fiber. There is also a rich content of delicious aromatic substances present. All this contributes to the great popularity of pears.

Further Reading

Bailey LH (1958) *The Standard Cyclopedia of Horticulture*, 17th edn. New York: MacMillan.

Blattný C (1986) *Konservárenské Suroviny*. Prague: SNTL.

Chittenden FJ (1956) *Dictionary of Gardening: the Royal Horticultural Society*, 2nd edn. Oxford: Clarendon Press.

Commission Regulation (EEC) no. 920/89 (1989) Laying down quality standards for carrots, citrus fruit and dessert apples and pears and amending Commission Regulation no. 58. *Official Journal of the European Communities* no. I.97/19.

Crisosto CH (2002) *Asian Pear – Postharvest Quality Maintenance Guidelines*. Dept. of Pomology, University of California, Davis, CA 95616. Available on line at: http: //postharvest.ucdavis.edu/producefacts.

Das Grosse Lebensmittel Lexikon (1985) 3rd edn. Innsbruck: Pinguin Verlag.

Dvořák A, Vondráček J, Kalášek J *et al.* (1978) *Atlas Odrůd Ovoce*. Prague: SZN.

Ensminger AH, Ensminger ME, Konlande JE and Robsen JRK (1995) *The Concise Encyclopedia of Food and Nutrition*. Boca Raton, FL: CRC Press.

Koch V, Ferkl FR, Blattný C *et al.* (1967) *Hrušky*. Prague: Academia.

Kutina J, Kalášek J, Blažek J *et al.* (1991) *Pomologický Atlas 1, 2*. Prague: Zeměděl. Nakl. Brázda.

Kyzlink V (1990) *Principles of Food Preservation*. Amsterdam: Elsevier.

Mareček F, Pekárková E, Blažek J *et al.* (1994) *Zahradnický Slovník Naučný*. Prague: ÚZPI.

Mitcham EJ, Crisosto C and Kader AA (2002) *Pear Anjou, Bosc and Comice. Recommendation for Maintaining Postharvest Quality*. Available on line at: http://postharvest.ucdavis.edu/producefact/fruit/pear2.html.

Senser and Scherz H (1991) *Der Kleine "Souci Fachmann Kraut". Lebensmitteltabelle fuer die Praxis*, 2nd edn. Stuttgart: Wissenchaftliche Verlagsgesellschaft.

Stehlík V, Trantírek J, Hruška L *et al.* (1969) *Naučný Slovník Zemědělský 1, 2*. Prague: ÚVTI-SZN.

PEAS AND LENTILS

G Grant and M Duncan, Rowett Research Institute, Bucksburn, Aberdeen, UK
R Alonso and F Marzo, The Public University of Navarra, Pamplona, Spain

Background

Peas and lentils are potentially rich sources of macro- and micronutrients and bioactive factors. This overview deals with the nutritional properties of the seeds, possible physiological responses (detrimental or beneficial) of animals to the constituent bioactive compounds, and the effects of processing on nutritional quality.

Peas (*Pisum* spp.) and lentils (*Lens* spp.) have been cultivated since around 6000–5000 BC, making them amongst the earliest of the developed legume crop plants. They are highly adaptable and can flourish in a wide variety of habitats and environmental conditions. As a result, they are grown in most regions of the world and provide a staple source of dietary protein and energy for a significant proportion of the population.

Much of the world output of peas and lentils continues to be produced by traditional agricultural methods, whereby they may be grown as part of a crop rotation system or as secondary crops. Large-scale intensive cultivation and industrial processing of these legumes as foodstuffs do, however, occur in some regions. Nonetheless, peas and lentils have been at a commercial disadvantage to seeds such as soyabean in that they do not contain significant amounts of oil. This has limited their commercial value. Protein and starch are the primary products of cultivation of these seeds, whereas with soyabean, they are essentially byproducts of vegetable oil production and therefore comparatively inexpensive. However, food production will need to increase greatly to meet the requirements of a rising world population. This is most likely to be met through use of food sources that grow well in a wide range of environments, in particular those that are adverse to soyabean production. Peas and lentils are good candidates for further development in this regard. Furthermore, since many legume seeds contain bioactive compounds that have been linked to health-promoting effects, peas and lentils may have significant health benefits in addition to being ready sources of protein and energy.

Composition

Although peas and lentils are highly adaptable, their actual composition varies greatly according to region, climate, cultivar, and husbandry. The carbohydrate levels may range between 525 and 740 g kg^{-1}, protein between 156 and 355 g kg^{-1}, and lipid between 6 and 61 g kg^{-1} (**Table 1**). Intensively grown peas and lentils have been developed and selected for yield and thus are less variable in composition. In general, these contain around 240–280 g protein per kilogram, 570–610 g of carbohydrate, about 100 g of lipid,

Table 1 Comparison of general composition and nutritional value of raw legume seeds

	Peas	Lentils	Soyabeans
N (g kg^{-1})	23–57	29–57	55–82
Protein (g kg^{-1})[a]	145–355	184–355	343–512
Carbohydrate (g kg^{-1})	599–700	500–697	247–370
Lipid (g kg^{-1})	15–62	15–25	181–227
Total fiber (g kg^{-1})	188–347	89–305	93–192
Crude fiber (g kg^{-1})	25–88	25–50	23–39
N digestibility (%)[b]	79–85	79–92	77–82
Biological value (%)	57–72	31–58	48–54
Net protein utilization (%)	45–61	22–46	38–44
Protein efficiency ratio	0.1–2.7	0.4–2.7	0.8–1.6

[a]N × 6.25.

[b]Protein digestibility and utilization by rats.

Data from Savage (1988), Savage and Deo (1989), Huisman and Jansman (1991) *Nutrition Abstracts and Reviews (Series B)* 61: 902–921, Lusas and Riaz (1995) *Journal of Nutrition* 125: 573S–580S, Cuadrado *et al.* (2002) *Journal of Agricultural and Food Chemistry* 50: 4371–4376, Grant *et al.* (1999), and Alonso *et al.* (2000) *Journal of Agricultural and Food Chemistry* 48: 2286–2290, Alonso *et al.* (2001) *Nutrition Research* 21: 1067–1077 .

Table 2 Comparison of the amino acid composition of cultivated raw legume seeds

	Peas	Lentils	Soyabeans	Amino acid needs Human[a]	Rat
Alanine[b]	4.4–4.9	2.4–4.2	4.7		
Arginine	8.6–8.9	3.9–7.7	7.8		5.0
Aspartic acid	4.9–11.8	9.9–11.1	12.6		
Cystine	1.1–1.5	1.3–1.5	1.6		
Glutamic acid	12.2–17.1	15.5–16.3	19.4		
Glycine	4.4–4.8	4.1–4.4	4.6		
Histidine	2.2–2.4	1.3–2.8	2.7		2.5
Isoleucine	3.1–4.1	4.0–4.3	4.9	3.5	5.0
Leucine	7.2–8.3	7.0–7.2	8.1	6.5	8.0
Lysine	7.2–8.2	4.3–7.0	6.7	5.0	6.0
Methionine	1.0–1.1	0.8–1.1	1.3	2.5[c]	4.5[c]
Phenylalanine	4.6–5.3	4.9–5.5	5.2	6.5[d]	5.0
Proline	4.1–4.9	2.6–4.2	5.9		
Serine	4.4–4.9	2.9–4.6	5.8		
Threonine	3.6–3.9	2.5–3.6	4.3	2.5	4.0
Tryptophan	0.9–1.1	0.9	1.5	1.0	1.5
Tyrosine	2.9–3.9	1.1–2.5	3.8		4.0
Valine	4.7–5.5	3.0–5.0	5.0	3.5	5.5

[a]Amino acid as g per 100 g of protein.

[b]MIT pattern of human amino acid requirements. *Journal of Nutrition* 130. 1841S–1849S.

[c]Combined sulfur amino acids.

[d]Combined aromatic amino acids.

Data from Lusas and Riaz, (1995) *Journal of Nutrition* 125: 573S–580S, Urbano *et al.* (1995) *Journal of Agricultural and Food Chemistry* 43: 1871–1877, Alonso *et al.* (2000) *Journal of Agricultural and Food Chemistry* 48: 2286–2290, Alonso *et al.* (2001) *Nutrition Research* 21: 1067–1077 and USDA database www.nal.usda.gov.

and 250–300 g of total dietary fiber. The protein and lipid levels are far lower than those in raw soyabean but are similar to the amounts found in other cultivated legumes, such as kidney beans. Carbohydrate levels in peas and lentils are about double that in soyabean, and the dietary fiber content can be two to three times greater.

The levels of individual amino acids in peas and lentils also vary greatly according to region, climate, and cultivar. Again, those in intensively cultivated varieties are more consistent (**Table 2**). The average amino acid profile of peas is close to the MIT (Massachusetts Institute of Technology) recommended amino acid pattern for humans, but that of lentils may be slightly deficient in the sulfur amino acids and possibly tryptophan (**Table 2**). However, when compared with the requirements for rats, the most widely used experimental animal, both seeds are deficient in a number of amino acids. The sulfur amino acids are limiting, being present at 47–56% of the requirements of rats. (*See* **Amino Acids**: Properties and Occurrence.)

Pea proteins are readily extractable in neutral or alkaline aqueous conditions. They are mainly of the globulin type, with the 7S globulins (vicilin) tending to predominate over the 11S globulins (legumin). Lentil proteins are also mainly of the globulin type, with legumin being the more prominent form. However, lentil proteins tend to be poorly soluble under neutral or alkaline aqueous conditions if whole seed meal is extracted. In contrast, solubility is high if the lentils have been dehulled. This may be a result of the removal of tannins present in the hulls. The globulin

proteins of both seeds generally have low contents of the sulfur amino acids. In contrast, the levels in albumin (water-soluble) fractions from these seeds are quite high, owing to the presence of sulfur amino acid-rich proteins such as the trypsin/chymotrypsin inhibitors.

Protein digestibility values for peas and lentils *in vivo* are around 79–84%. This is similar to the values obtained *in vitro* (80–84%) with activated pancreatic extracts. However, digestibility values *in vitro* are around 65–70% if purified digestive enzymes are used. This discrepancy *in vitro* is likely to be a result of slow or poor digestion of vicilins by the pure enzymes. A similar difference was noted with phaseolin from kidney bean. It was highly resistant to pure proteolytic enzymes but readily degraded when incubated with rat intestinal tissue plus contents. Enzymes or factors, in addition to pepsin, trypsin, and chymotrypsin, were shown to be necessary to facilitate digestion of this protein. The nature of these components is unknown. However, if additional factors are crucial to promote degradation of legume globulins, there is no guarantee that all animals will possess them. There could therefore be considerable species differences in ability to digest legume globulins. (*See* **Protein**: Quality.)

Purified radiolabeled pea legumin and vicilin are digested by rats almost as effectively as casein (97.4, 97.1, and 98.5%, respectively). Equally, isolated lentil globulins are readily degraded (99%) by rats. The globulins do not therefore appear to be inherently resistant to proteolysis *in vivo*. Digestibility values obtained with raw meals or crude protein extracts are much lower. This is likely to be a result of interference with proteolysis by the seed matrix or seed components, such as protease inhibitors. This would not be a problem with protein isolates prepared from peas or lentils.

Utilization of Raw Legumes

Weight gains and net protein utilization values obtained with young rats fed raw peas or lentils as the sole source of dietary protein (approximately 400 g per kilogram of diet) are low. This is mainly a result of reduced food intake and increased fecal dry matter and N and urinary N excretion. Supplementation of diets with methionine improves food intake, N retention, and weight gain, but the overall nutritional quality remains below that of high-quality protein (**Table 3**).

Inclusion of raw peas at 150–450 g per kilogram of diet severely impairs food intake and growth by weanling pigs. The effects are much less marked in pigs over 5–6 weeks old, and supplementation of diets with tryptophan or methionine greatly improves performance. The maximal inclusion levels recommended are 100 g per kilogram of diet for weaner pigs, 200 g kg^{-1} for grower pigs and 350 g kg^{-1} for finishers. The growth of chicks is severely retarded by inclusion of up to 200 g of raw pea meal in their diet. Supplementation with methionine, cystine, and tryptophan improves performance but only to a limited extent. Older birds appear more tolerant of peas, and with laying hens, up to 330 g kg^{-1} can be incorporated into the diet without any significant adverse effects. Addition of supplementary methionine and lipid to the diet allows this incorporation to be increased to around 440 g per kilogram of diet. Fecal dry matter and N outputs by human subjects are increased by intake of 130 g of lentils per day, but urinary N output is slightly decreased.

Possible Health Benefits

Peas are hypocholesterolemic for rats, pigs, and humans. The mechanisms remain unclear but may be linked to fiber and saponins in the seeds. With rats, the effect may be partly a result of reduced lipid deposition by the animals. Lentils are hypocholesterolemic for rats but not apparently for pigs or humans, at least at the dietary inclusion levels studied.

Bioactive Factors

Peas and lentils contain a number of potentially bioactive components such as protease inhibitors and lectins that may influence body metabolism in a negative or positive manner (**Table 4**).

Trypsin/chymotrypsin Inhibitors

Peas and lentils contain Bowman–Birk-like trypsin/chymotrypsin inhibitors that inhibit the activity of these digestive enzymes *in vitro* and *in vivo*. These inhibitors are of a relatively low molecular weight (around 8–10 kDa) and are double-headed in that they inhibit two enzyme molecules simultaneously. The original Bowman–Birk inhibitors (BBI) were isolated from soyabean and were found to inhibit

Table 3 Nutritive value of peas or lentils for rats after heat treatment and/or L-methionine supplementation

	Peas			Lentils	
	Raw	Cooked	Extruded	Raw	Cooked
Unsupplemented					
N digestibility (%)	80–84	84–86	80–87	80–85	82–86
Net protein utilization (%)	50–60	44–54	60–66	27–33	36–40
Biological value (%)	60–75	55–65	75–83	32–40	43–47
Protein efficiency ratio	1.8–2.4	2.0–2.2	1.9–2.3	0.8–1.6	nd
Amino acid supplemented					
Biological value (%)	nd	nd	nd	53–61	69–71
Protein efficiency ratio	2.7–3.3	3.0–3.7	3.1–3.7	2.1–3.0	nd

nd, not determined.
Data from Savage (1988), Savage and Deo (1989, and references therein), and Alonso *et al.* (2000) *Journal of Agricultural and Food Chemistry* 48: 2286–2290, Alonso *et al.* (2001) *Nutrition Research* 21: 1067–1077.

Table 4 Bioactive factors present in raw legume seeds

	Peas	Lentils	Soyabeans
Trypsin inhibitor (g kg^{-1})[a]	1–8	0.1–4	16.7–27.2
Chymotrypsin inhibitor (g kg^{-1})[a]	0.5–6	0.1–2	0.2–4.9
α-Amylase inhibitor (g kg^{-1})[b]	< 0.1	< 0.1	< 0.1
Lectin (U × 10^{-6} kg^{-1})[c]	5.1–15.1	1.8–7.7	50–350
Lectin (g kg^{-1})[c]	0.3–0.7	~0.8	2–4
Phytic acid (g kg^{-1})	2.2–8.2	1.2–8.1	10.0–14.7
Tannin (g kg^{-1})	0.2–5.0	0.6–2.7	0.2–0.9
Phenolics (mg kg^{-1})	13–27	10–30	8–15
Saponin (g kg^{-1})	1.1–2.5	1.1–4.6	2.2–6.7

[a]Trypsin and chymotrypsin inhibitor is given as the number of grams of enzyme that would be inhibited per kilogram of meal.
[b]α-Amylase inhibitor is expressed as inhibitor equivalents using pure kidney bean α-amylase inhibitor as standard.
[c]Lectin levels are given as hemagglutinating units (U) per kilogram of meal based on the ability to agglutinate rabbit blood cells or as g kg^{-1} based on recoveries obtained during purification of lectin.
Data from Trowbridge (1974) *Journal of Biological Chemistry* 249: 6004–6012, Savage (1988), Savage and Deo (1989), Liener (1991) *Journal of the American Oil Chemists Society* 56: 121–129, Anderson and Wolf (1995) *Journal of Nutrition* 125: 581–588, Huisman and Jansman (1991) *Nutrition Abstracts and Reviews (Series B)* 61: 902–921, Alonso et al. (1998) *Food Chemistry* 63: 505–512, Alonso et al. (2000) *Journal of Agricultural and Food Chemistry* 48: 2286–2290, Alonso et al. (2001) *Nutrition Research* 21: 1067–1077, Cuadrado et al. (1999), Grant (1999), and Grant et al. (1999).

only trypsin and chymotrypsin in combination. However, in other legume seeds, variants capable of inhibiting two molecules of trypsin or chymotrypsin are also present. This heterogeneity in the reactivity profiles of BBI-like inhibitors makes it difficult to quantify the amounts present in the seeds or seed products. Therefore, for comparison purposes, the amount of each enzyme inhibited by a known amount of product is usually given. Alternatively, enzyme inhibitory units are calculated from the same data.

Trypsin and chymotrypsin inhibiting activity in peas and lentils varies greatly between cultivars (**Table 4**). Assuming that the mix of inhibitor variants leads to an average profile of one molecule trypsin and one molecule chymotrypsin inhibited per molecule inhibitor, available data suggest that peas contain 0.4–2.8 g of BBI-like inhibitor per kilogram and lentils 0.15–1.4 g kg^{-1}. This is comparable with the levels found in many other legumes. Soyabean has 3.2–3.7 g of BBI per kilogram. However, it also contains 4.5–9.0 g per kilogram of Kunitz trypsin inhibitors. Thus, the total trypsin inhibiting activity in soyabean is far higher than that in peas or lentils.

BBI are highly resistant to proteolysis *in vitro*. Nonetheless, native BBI and BBI-type inhibitors seem to be readily degraded *in vivo*, although ^{125}I-BBI seems to be highly resistant to proteolysis *in vivo*. Native inhibitors have limited effects on digestion and absorption of dietary protein and growth of animals, even when they are included at high levels

(≤ 10 g kg^{-1}) in the diet. However, BBI or BBI-like inhibitors interfere with cholecystokinin-mediated control over pancreas function and trigger hypersecretion of pancreatic trypsin, chymotrypsin and α-amylase in rats, chicks, quails, and humans. In rats, chicks, and quails, this leads to pancreas enlargement, mainly as a result of hyperplasia and hypertrophy of the acinar cells and, in the long term, may lead to tissue dysfunction and disease. The effects of the inhibitors are, however, species-specific. Thus, they induce pancreas growth in young rats, mice, hamsters, guinea-pigs, and chickens but have little or no effect on the pancreas of young pigs, cattle, monkeys, or dogs.

Pancreatic enlargement is evident in rats fed pea diets, although the increase is usually much less than observed in rats given an equivalent intake of raw soyabean. Pancreatic growth was not apparent in rats fed lentil meal, although an inhibitor-enriched fraction of lentil meal did cause enlargement.

Consumption of soyabean BBI by experimental animals appears to significantly reduce the incidence and severity of liver, colon, and mammary cancers that develop as a result of treatment with chemical carcinogens or radiation. The mechanisms by which this occurs, however, remain unclear. BBI acting through localized effects on gut endocrine cells may induce the release of a number of hormones, growth factors, or peptides that interfere with tumor cell metabolism. Alternatively, bioactive fragments of BBI absorbed from the gut may be the main protective agents. The BBI-like inhibitors of field bean have also been shown to have cancer-preventing properties. It is thus possible that pea and lentil inhibitors will have a similar protective effect. (*See* **Trypsin Inhibitors**.)

α-Amylase Inhibitors

α-Amylase inhibitors inhibit the activity of salivary and pancreatic amylase *in vitro* and *in vivo*. They can impair the growth and metabolism of animals when given at high levels in the diet but may have beneficial uses in treatment of obesity or diabetes. They are generally present in very high amounts in *Phaseolus* species (kidney beans, 4.3 g of inhibitor per kilogram). However, the levels in peas and lentils are very low (**Table 4**) and unlikely to contribute significantly to the effects of these legumes on metabolism.

Lectins

Lectins are defined as carbohydrate-binding proteins/glycoproteins other than enzymes that are present in most plant materials. They are highly resistant to proteolytic degradation *in vivo* and survive passage through the gastrointestinal tract. If appropriate carbohydrate receptors are present on gut epithelial cells, lectins bind to them and may be taken up systemically. As a result, lectins can potentially interfere with and modify many aspects of gut and systemic metabolism. Individual lectins vary greatly in their effects *in vivo*, but most species appear to be responsive to dietary lectins. (*See* **Hemagglutinins (Haemagglutinins)**.)

Lectins can be separated into eight general categories on the basis of their carbohydrate-binding specificity: complex, fucose, galactose, N-acetylglucosamine, mannose, mannose/glucose, mannose/maltose, and sialic acid. Some, such as soyabean agglutinin (galactose-specific), alter gut and pancreas metabolism in rats, in particular causing rapid growth of these tissues, without affecting systemic systems. A few, including kidney bean lectin (complex specificity), have additional effects on systemic hormone balance and lipid and muscle metabolism and can be very deleterious if consumed in high amounts. Others, such as pea and lentil lectins (glucose/mannose-specific) appear to have little or no effect on the body metabolism of rats.

The sensitivity of animals to lectins may, however, vary with species, age, period of exposure, gastrointestinal bacteria, diet composition, and dietary history. Thus, the glucose/mannose-specific jack bean lectin (Con A) has no effect on mature germ-free rats but has limited effects (causes small intestine and pancreas enlargement) in specific pathogen-free rats. It is, however, highly deleterious to rats carrying a salmonella infection, to suckling guinea-pigs, and to quails. The glycoconjugates expressed on the gut surface of very young animals differ greatly from those in mature counterparts; in particular, a high proportion of mannose residues are present. This is also evident in rats with a pathogen infection. In these circumstances, lectins that would not normally affect the gut may be able to bind to it and elicit changes in body metabolism.

The levels of lectins in peas and lentil are low compared with that in soyabean (**Table 4**) and very low by comparison with kidney bean ($15-30 \, \text{g kg}^{-1}$). Furthermore, in studies with mature rats, these lectins have no significant effects on metabolism. This would suggest that they are unlikely to cause problems. However, in view of the data with Con A and the number of factors that influence the sensitivity of an animal to lectin, one cannot exclude the possibility of specific circumstances where these dietary lectins have profound effects.

Young chicks do not do well when raw peas are added to their diet in low amounts but will tolerate quite high dietary inclusions if they are a few weeks old. During this period, the gut develops from its very immature form at hatching to its adult form. The gut may be susceptible to the action of pea lectin at the early stages of the maturation period.

Antigenic Proteins

Native 11S globulins (glycinin) and 7S globulins (conglycinin) of soyabean induce very adverse immune reactions in preruminant calves and newly weaned piglets, leading to gut damage, scouring, and poor performance. Pea globulins partially survive gut passage and can trigger some immune responses in preruminant calves. However, the degree to which this occurs is very much lower than that observed in soyabean-fed animals.

Lentils have been linked to allergy problems in a small number of pediatric patients. A number of possible allergens have been identified, including subunits of vicilin. The incidence of intolerance to pea proteins seems to be low.

Phytate

Phytic acid is often the main reserve of phosphorus in legumes. However, it also chelates with minerals and metals, such as calcium, magnesium, zinc, and iron, forming insoluble salts that are not readily absorbed by animals or humans. In particular, it can severely impair availability of zinc and iron. Phytate can also complex with proteins and may thereby reduce digestibility or enzyme activity.

Mineral uptake by pigs and chickens fed with soyabean-based diets is slightly impaired. Addition of phytase, a phytate-degrading enzyme, to the diet appears to counteract this effect, leading to an improvement in mineral uptake and better overall performance by the animals. However, the efficacy of this treatment can be very variable.

Phytic acid levels in peas and lentils are lower than those in kidney beans ($11-17 \, \text{g kg}^{-1}$) and soyabean (**Table 4**). None the less, they can still affect mineral metabolism, since iron absorption from a pea protein-based infant formula is significantly enhanced after enzymatic degradation of phytate.

Pea and lentil phytate can clearly have adverse effects on mineral uptake and body metabolism. However, in many cases, their impact is likely to be minimal because the mineral content of the diets is well above the requirements. Pea and lentil phytate, however, may cause significant problems if mineral intake, particularly of zinc and iron, is close to or below requirements.

Dietary phytic acid may have health-promoting properties. It can inhibit α-amylase, limit carbohydrate digestion, and lower blood glucose. There are also indications that it is hypocholesterolemic and protective against colon cancer. However, the amounts required are quite high. It is unclear whether a normal physiological intake of peas or lentils would provide sufficient phytate to have a significant health-protective effect. (*See* **Phytic Acid**: Nutritional Impact.)

Tannins

Tannins are present in a wide array of plant crops. Legume condensed tannins are oligomers of variously substituted flavan-3-ols, and their antinutritional effects have recently been comprehensively reviewed. These compounds can reduce enzyme activity in the gut, impair gut morphology, lower nutrient (protein, carbohydrate and lipid) digestion and absorption, reduce mineral uptake, and greatly stimulate excretion of endogenous N. Thus, at high dietary intakes, they can adversely affect animal performance.

In contrast to the antinutritional effects, dietary flavan-3-ols have also been suggested to have important roles in disease prevention, particularly of cardiovascular diseases and some forms of cancer. They may act as antioxidant and free radical scavengers, inhibit tumor initiation and promotion, and have antibacterial and angioprotective properties.

The tannin contents of peas and lentils tend to be higher than that in soyabean (**Table 4**). However, the levels are comparable with those in many other legume seeds. Lentils contain moderate amounts of catechins and proanthocyanidin dimers and trimers, whereas pea samples tend to have low amounts. The compounds may have health-protective effects. However, since questions still remain as to how effectively these compounds or products derived from them are absorbed from the gut and distributed throughout the body, it is difficult to assess whether a normal intake of peas or lentils will provide sufficient flavanols to have a significant beneficial effect. (*See* **Tannins and Polyphenols**.)

Saponins

Triterpenoid saponins, found in leguminous plants, are composed of a triterpene aglycone linked to one, two, or three saccharide chains of varying size and complexity. The levels in peas and lentils are similar to, or slightly lower than, those in soyabean (**Table 4**) and other legume seeds. They may reduce weight gain if consumed at very high levels. However, at the levels present in peas, lentils, and soyabeans, they are considered to have no significant antinutritional effects. Saponins, however, may have beneficial effects. They have been found to be hypocholesterolemic in a number of species, owing in part to their ability to facilitate adsorption of bile acids to dietary fiber.

Evidence of their efficacy in humans is not conclusive. It has also been suggested that saponins may have anticarcinogenic or antioxidant properties. (*See* **Saponins**.)

Fiber

Peas and lentils contain high levels of crude and total dietary fiber (**Table 1**). The fibers have a high binding capacity for bile acids *in vitro* and also promote fermentation in the hind-gut and production of butyrate *in vivo*. As with other dietary fibers, pea and lentil fibers are generally considered to be beneficial. However, the specific effects of these dietary fibers on metabolism remain unclear. Inclusion of pea or lentil fibers in diets appears to lower postprandial blood triglyceride levels without affecting cholesterol concentrations. However, other studies suggest that pea fibers can influence cholesterol levels in humans. Equally, studies show that pea fibers reduce or delay starch absorption and reduce the blood glucose peak, whereas others show no effect on blood glucose. This variability may be a result of differences in the test meals used or the nutritional/health status of the subjects. Pea fibers are used in the treatment of hypercholesterolemic patients. However, they are not used alone but given as part of a mixture of dietary fibers. (*See* **Dietary Fiber**: Properties and Sources.)

Starches

Legume starches in general tend to be less digestible or more slowly digested than corn starches. This slow release means that their glycemic index is low. They can thus be useful in diets for those with impaired carbohydrate tolerance. Lentils have been found to have beneficial effects on blood glucose profiles of diabetic patients. This is due, at least in part, to the slow release property of the constituent starch combined possibly with the effects of the fiber. (*See* **Starch**: Structure, Properties, and Determination.)

Vitamins and Minerals

Peas and lentils contain significant amounts of important vitamins and minerals. The levels of most are fairly similar in both seeds. However, lentils appear to contain higher levels of ascorbic acid, folic acid and vitamin B_6 than peas but have considerably less vitamin A. (*See* **Vitamins**: Overview.)

Effects of Processing

Various processing methods are routinely used in the preparation of peas and lentils for consumption. These include dehulling, soaking, germination, cooking, autoclaving, extrusion, and fermentation.

In general, dehulling or soaking seems to have limited effects on the levels of bioactive factors or the nutritional quality of the seeds. Dehulling of lentil increases the solubility of lentil proteins, and soaking slightly reduces trypsin inhibitor and phytate content. Germination lowers phytate and slightly reduces trypsin inhibitor, tannin and polyphenol levels in peas, and lowers trypsin inhibitor and phytate in lentil. Nonetheless, it may also increase levels of nonprotein amino acids. Natural fermentation also reduces the levels of bioactive compounds and slightly improves the starch and protein availability from legumes. The efficacy of this treatment is, however, extremely variable, partly because the environmental factors responsible are unknown.

Cooking, autoclaving, or extrusion treatment of peas fully inactivates trypsin inhibitors and lectins, has limited effects on phytate and polyphenols, but greatly reduces tannin concentrations. Starch digestibility is significantly increased, but protein digestibility is improved only to a limited extent (**Table 3**). The effects of cooking on lentils appear similar, with the exception that the reduction in tannins and polyphenols may be more pronounced than that observed with peas. Prolonged thermal treatment (cooking, autoclaving, or extrusion) can adversely affect nutrient availability from peas and lentils.

Thermal treatment of peas and lentils clearly improves the nutritional value of the constituent proteins (**Table 3**). With rats, this is evident only after supplementation of the diet with methionine to bring the sulfur amino acid content up to the requirements for rats. The human need for sulfur amino acids is much lower than that of rats and could be met from the levels present in peas and possibly lentils (**Table 2**). For humans, the biological value of cooked peas or lentils alone may thus be close to that found for the supplemented diets in rats.

Pea-based diets are hypocholesterolemic for rats. This property is retained even after extrusion treatment of the seeds. Some lentil proteins retain their antigenicity, even after cooking.

Digestibility of Heat-treated Globulins

The moderate digestibility of raw legume proteins has mainly been attributed to the presence of antinutritional factors in the seeds and the refractory nature of the proteins. However, although heat-treatment abolishes the activity of the main antinutritional factors, it appears to have remarkably little effect on the digestibility of pea or lentil proteins. In addition, pea and lentil globulins have a very high digestibility *in vivo*, once isolated from the seeds. The proteins are thus not inherently indigestible.

It is possible that interactions between nonprotein components of the seed matrix and the constituent proteins may be preventing digestion of a proportion of the proteins. Heating may not abolish these interactions. Alternatively, globulin-like proteins from some heat-treated legumes have been shown to be less digestible than counterparts from raw seeds. This may be due to the formation of large protein multimers that are not readily accessible to digestive enzymes. Any beneficial effects of heat treatment of seeds on protein availability may be negated, in part, by the development of these poorly digestible multimers of the globulin proteins. This merits further study.

Summary

Peas and lentils are potentially good sources of dietary protein and carbohydrate and micronutrients. They contain a range of bioactive compounds that may be detrimental to consumers. However, some of the bioactive compounds may also be beneficial for health. Thermal treatment greatly improves the nutritional quality of the seeds. Both raw and processed seeds have hypocholesterolemic properties.

See also: **Carbohydrates**: Interactions with Other Food Components; **Cholesterol**: Factors Determining Blood Cholesterol Levels; **Dietary Fiber**: Physiological Effects; Effects of Fiber on Absorption; **Extrusion Cooking**: Principles and Practice; Chemical and Nutritional Changes; **Food Intolerance**: Types; **Hemagglutinins (Haemagglutinins)**; **Legumes**: Legumes in the Diet; Dietary Importance; **Phytic Acid**: Properties and Determination; Nutritional Impact; **Plant Antinutritional Factors**: Detoxification; **Protein**: Interactions and Reactions Involved in Food Processing; **Trypsin Inhibitors**

Further Reading

Bardocz S, Hajos G and Pusztai A (eds) (1999) *Cost 98 – Effects of Antinutrients on the Nutritional Value of Legume Diets*, vol VI. Luxemburg: Office for Official Publications of the European Communities.

Carbonaro M, Grant G, Cappelloni M and Pusztai A (2000) Perspectives into factors limiting *in vivo* digestion of legume proteins: antinutritional compounds or storage proteins? *Journal of Agricultural and Food Chemistry* 48: 742–749.

Castell AG, Guenter W and Igbasan FA (1996) Nutritive value of peas for nonruminant diets. *Animal Feed Science Technology* 60: 209–227.

Caygill JC and Mueller-Harvey I (eds) (1999) *Secondary Plant Products. Antinutritional and Beneficial Actions in Animal Feeding*. Nottingham, UK: Nottingham University Press.

Jansman AJM, Hill GD, Huisman J and van der Poel AFB (eds) (1998) *Recent Advances of Research in Antinutritional Factors in Legume Seeds and Rape Seeds*. Wageningen, The Netherlands: Wageningen Pers.

Martinez San Ireneo M, Ibanez Sandin MD and Fernandez-Caldas E (2000) Hypersensitivity to members of the botanical order Fabales (legumes). *Journal of Investigational Allergology and Clinical Immunology* 10: 187–199.

Rowland I (1999) Optimal nutrition: fibre and phytochemicals. *Proceedings of the Nutrition Society* 58: 415–419.

Sanchez-Monge R, Pascual CY, Diaz-Perales A *et al.* (2000) Isolation and characterisation of relevant allergens from boiled lentils. *Journal of Allergy and Clinical Immunology* 106: 955–961.

Santos-Buelga C and Scalbert A (2000) Proanthocyanidins and tannin-like compounds. Nature, occurrence, dietary intake and effects on nutrition and health (review). *Journal of the Science of Food and Agriculture* 80: 1094–1117.

Savage GP (1988) The composition and nutritive value of lentils (*Lens culinaris*). *Nutrition Abstracts and Reviews (Series A)* 58: 319–343.

Savage GP and Deo S (1989) The nutritional value of peas (*Pisum sativum*). A literature review. *Nutrition Abstracts and Reviews (Series A)* 59: 65–88.

Urbano G, Lopez-Jurado M, Aranda P *et al.* (2000) The role of phytic acid in legumes: antinutrient or beneficial function? *Journal of Physiology and Biochemistry* 56: 283–294.

van der Poel AFB, Huisman J and Saini HS (eds) (1993) *Recent Advances of Research in Antinutritional Factors in Legume Seeds*. Wageningen, The Netherlands: Wageningen Pers.

Pecans *See* **Walnuts and Pecans**

PECTIN

Contents
Properties and Determination
Food Use

Properties and Determination

L Flutto, Danisco, New Century, KS, USA

Introduction

Pectin is a high-molecular-weight carbohydrate polymer which is present in virtually all plants where it contributes to the cell structure. The term pectin covers a number of polymers which vary according to their molecular weight, chemical configuration, and content of neutral sugars, and different plant types produce pectin with different functional properties. The word 'pectin' comes from the Greek word *pektos* which means firm and hard, reflecting pectin's ability to form gels.

The gelling properties of pectin have been known for centuries, but the isolation of commercial pectin only started at the beginning of the twentieth century. In this document we highlight the chemistry, origin and production, and the functional properties of pectin.

Chemistry

The Homogalacturonic Acid Backbone

Pectin consists of a chain of galacturonic acid units which are linked by α-1,4 glycosidic bonds. The galacturonic acid chain is partly esterified as methyl esters. Pectin molecules can have a molecular weight of more than 200 000, corresponding to a degree of polymerization up to 1000 units (**Figure 1**).

Though the esters are the most significant components on the galacturonic acid backbone, other chemicals, such as acetyl, can be important in specific pectin types. Commercial pectin can also be partly

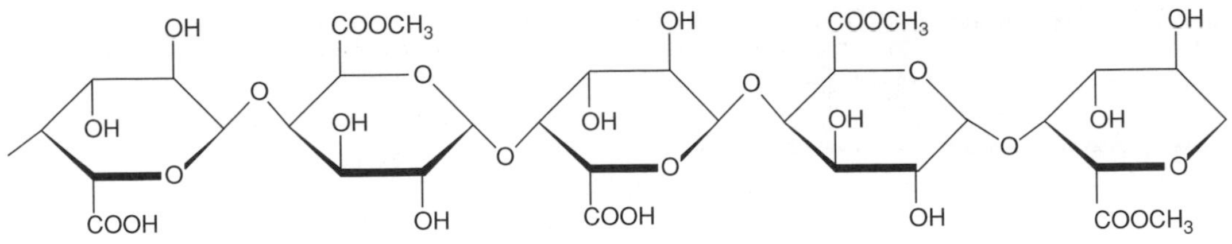

Figure 1 Pectin consists of long sequences of anhydro galacturonic acid completely or partly esterified with methanol.

Table 1 Classification of pectins

DE > 50%	DE < 50% Low-ester pectin (LE)	
High-ester pectin (HE)	No amidation Conventional LE	DA < 25% Amidated LE

DE, degree of esterification; DA, degree of amidation.

amidated with ammonia to form galacturonamide units in the molecular chain.

Some neutral sugars are also included in the homogalacturonic backbone. This is the case for rhamnose but also specifically in apple pectins for xylose.

The percentage of galacturonic acid of the whole molecule is defined as the galacturonic acid content (% GA), which is set at a minimum of 65% in the definition of pectin as a food additive.

The percentage of esterified or amidated galacturonic acid units of the total number of galacturonic acid units in the molecule are respectively defined as the degree of esterification (DE) and the degree of amidation (DA) (Table 1). Regulations limit the DA to a maximum of 25%.

Esterification Pattern

In addition to the number of components on the backbone, their position is of significant importance to the functional properties of pectin. The distribution of esters is of special importance as it affects the local electrostatic charge density of the polymer and so its interaction with other charged molecules, whether ions such as calcium, proteins, or other pectin molecules.

In apple pectin subjected to a mild extraction process, the ester distribution is reported to be almost random, while citrus pectin tends to have a somewhat blockwise distribution.

Neutral Sugars

Pectin always contains varying amounts of neutral sugars such as D-galactose, L-rhamnose, L-arabinose, and D-xylose. Some of these neutral sugars are constituents of side chains to the galacturonan backbone.

However, 1,2-linked L-rhamnose is present in the main polygalacturonic chain where it forms 'kinks' in the molecular chain. Xylose is also a very important neutral sugar in apple pectin where it is attached in the homogalacturonic backbone.

It is well documented that neutral sugar side chains are concentrated in relatively short segments of the galacturonic backbone, described as 'hairy regions.' The part of the molecule free of side chains is described as the 'smooth region.' (Figure 2).

Chemical and Physical Properties

Solubility Pectin is soluble in water, but insoluble in most organic solvents. The solubilization rate in water is related to the degree of polymerization and the number and distribution of methyl-ester groups. The pH, temperature, and ionic strength of the solution are of great importance to the rate of pectin dissolution. The calcium content of the water used to dissolve the pectin is of special relevance as it is common that a high-water hardness will translate into an incomplete dissolution of pectin.

Rheology of pectin solutions Pectin solutions are viscous but pectin is not an especially effective viscosifier compared to other gums such as guar gum.

Dilute pectin solutions are almost Newtonian, and they are only slightly affected by the presence of calcium. However, solutions with more than 1% pectin exhibit pseudoplastic behavior and are strongly affected by calcium. There is a continuum in texture starting with water through thixotropic solutions with yield value to stiff gels depending on the pectin type and concentration, level of calcium, and pH.

Stability High-ester pectins are stable at pH levels of 2.5–4.5. Above a pH level of 4.5, β-elimination will occur, depolymerizing the galacturonic acid chain. This mechanism requires an esterified carboxyl group next to the glycosidic bond to be cleaved (Figure 3), so low-ester pectins are more stable at higher pH values.

The pectin molecular structure is quite resistant to heat. At pH around 3.5 pectin is only marginally depolymerized at high temperatures. The heat-stability

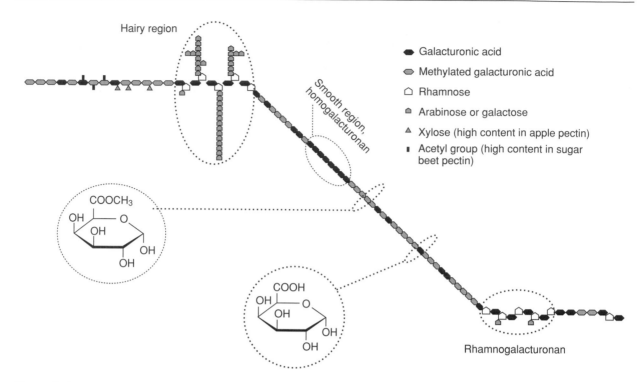

Figure 2 Primary and secondary structure of pectin.

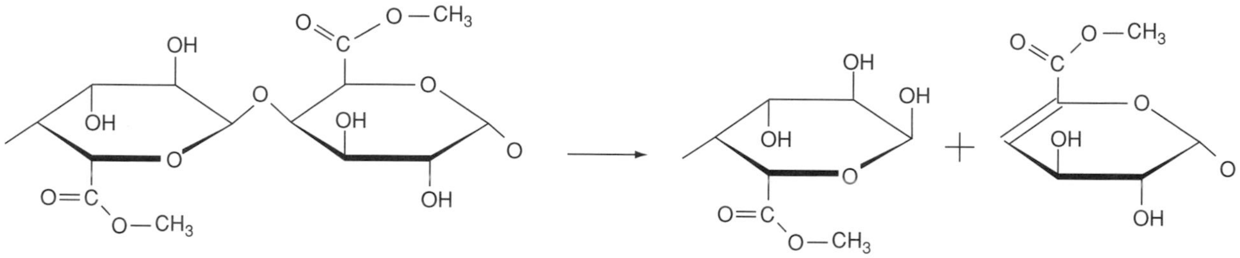

Figure 3 β-elimination mechanism.

of pectin is greatly improved when the water activity of the system is lowered through the addition of sugar.

Origin and Production

Pectin is a natural component of plants, predominantly in the form of pectic substance or protopectin, which is not soluble in water. The pectic substance is an essential part of the plant cell wall structure, acting as a cement for the cellulosic network and a hydrating agent. The exact nature of the pectic substance is not completely understood. It is, however, generally recognized that it is a complex structure in which pectin is attached to other cell wall components such as cellulose, hemicellulose, and proteins by covalent bonds, hydrogen bonds, and/or ionic interactions. In the plant the residual carboxyl groups are partly neutralized with cations of calcium, potassium, and magnesium which are present in the plant tissues.

Pectin is today commercially produced mostly from apple pomace and citrus peels by an extraction process followed by separation, purification, isolation, and then drying, milling, and standardization (**Figure 4**).

Functional Properties

Pectins are used in a broad variety of food and pharmaceutical products. Their principal properties are gel formation with both high- and low-ester pectins, viscosity build-up, and protein stabilization with high-ester pectins. As methods develop for obtaining a better understanding of the pectin molecular structure, it is likely pectin will be attributed new functional properties in the future.

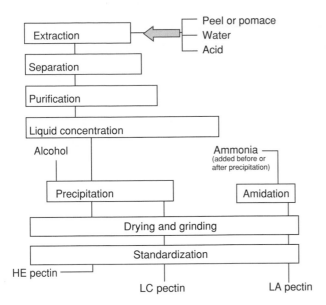

Figure 4 Pectin manufacture flow sheet. HE, high-ester; LC, low ester conventional; LA, low ester amidated.

Gelling of High-Ester Pectins

The ability of pectin to form gels in specific conditions has long been used in the production of jams and preserves. The separation of high- and low-ester pectins in accordance with the 50% esterification rule proves somewhat arbitrary when dealing with gelling mechanisms. It is clear that the dominant factor in the gelling of pectin is highly dependent on its degree of esterification but the complex network structure is most often the result of a combination of several mechanisms. The statement that high-ester pectins will gel with sugars and acid, while low-ester pectins gel with calcium, is certainly valid but certainly calcium can alter the gelling of high-ester pectins, just as pH and soluble solids will affect low-ester pectin gelling.

Gelling Mechanism

It is generally accepted that a high-ester pectin gel is formed by the cross-linking of the polymer in junction zones, in which mainly hydrogen bonds but also hydrophobic attractions between the methyl-ester groups play a part. Calcium bridges may also participate, especially if the esters are distributed in blocks, leaving large parts of the molecule as free acids.

Gelling will occur upon cooling of a media where favorable conditions are met. Cooling is necessary to decrease molecular movement and permit the formation of intermolecular interactions.

As the pectin chains carry negative charges they will tend to repel each other. This repulsion will depend on the charge density of the molecule which can be directly correlated to the pH of the medium and the frequency of free galacturonic acids of the polymer. The higher the pH and the lower the degree of esterification, the higher the charge density and hence the stronger the repulsion.

This repulsion, but more importantly the impossibility of forming hydrogen bonds between ionized pectin chains, are the reasons why a low pH is required for high-ester pectin gelling. At a low pH, typically below 3.6, the repulsion is low enough for the distance between the pectin chains to be reduced sufficiently and hydrogen bonding can occur. In order to achieve sufficient hydrophobic interactions to stabilize the molecular network, the water activity of the system also has to be decreased. Sugars are usually added for this purpose.

So, typically, high-ester pectins will only form gels when the pH is below 3.6 and soluble solids are above 55% (**Figure 5**).

Upon storage of a cooled gel, it is typical that the texture will still develop into a stronger final gel. This corresponds to a slow reorganization of the network involving the creation of new junction zones or an enlargement of the existing junctions between the pectin molecules.

Parameters Influencing Gelling

The gelling and final gel structure of high-ester pectins is influenced by a great number of parameters, the main ones being the pectin concentration, the degree of esterification, molecular weight, acetylation and branching of the pectin molecule and the pH, ionic strength, water activity, sugar type, and cooling rate of the gelling medium.

Pectin concentration The concentration of high-ester pectin will increase the final gel strength of the system due to the increase in the number of junction zones, increasing the number of chains with elastic activity. An increase in molecular weight would have the same effect. In addition to this expected effect, the gelation rate is also increased with increasing pectin concentration and a power law can be calculated between the two parameters.

Degree of esterification The degree of esterification of the galacturonic acids affects both the charge density of the polymer and the number of sites for hydrophobic interaction. As pectin molecules with a high degree of esterification are less charged, they can form gels at a higher pH and will also start gelling at a higher temperature.

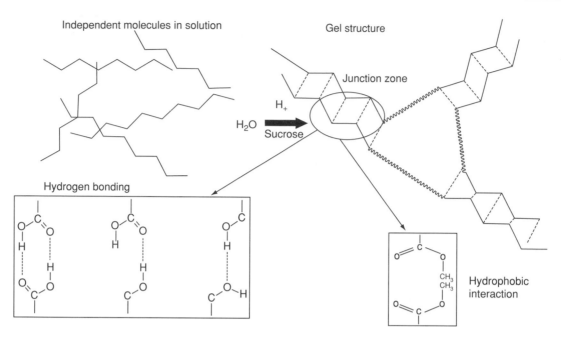

Figure 5   High-ester pectin gels through hydrogen bonding and hydrophobic interactions in an acidic water and sugar matrix.

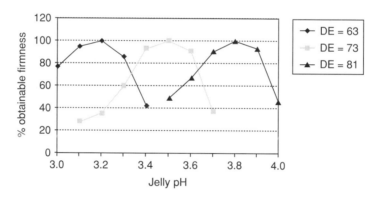

Figure 6   High-ester pectin: the degree of esterification (DE) determines the optimal pH for the gelation.

This last effect forms the basis for the classification of high-ester pectins into rapid-set, medium-rapid-set, slow-set, and extra-slow-set pectins as the degree of esterification is decreased from more than 70% to 50% (**Figure 6**).

The distribution of the ester groups on the backbone will also affect pectin gelation, as a marked blockwise distribution of esters will result in a significant contribution of calcium gelling. This contribution of calcium gelling will significantly increase the gelling temperature of the pectin (**Figure 7**).

Acetylation and branching   Acetylation sharply decreases the gelling ability of pectin as the size of the acetyl groups does not allow the pectin chains to come close enough for interaction between the molecules.

The effect of neutral sugars on high-ester pectin gelling can be twofold and would need to be studied more extensively. Neutral sugars present on the pectin molecule could result in steric hindrance of intermolecular interaction and thus decrease the ability of the pectin to form gels. However, they could also participate in gelling through hydrophobic interaction and contribute to an increased cohesion of the gel.

pH and ionic strength of the gelling system   The lower the pH, the lower the repulsion between the pectin molecules and, thus, the easier it will be for them to interact. This means that a low pH will lead to faster gelling in high-ester pectins. However, below a critical level, the gel strength will be reduced as the gelling is too fast to obtain a well organized polymer

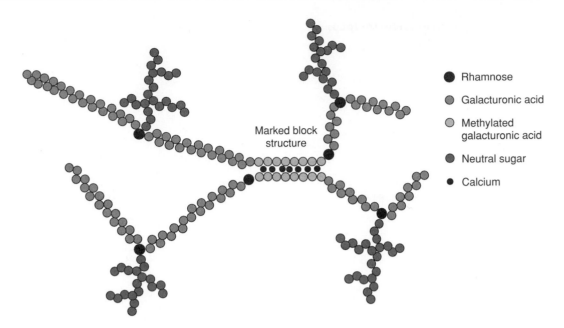

Rhamnose
Galacturonic acid
Methylated galacturonic acid
Neutral sugar
Calcium

Marked block structure

Figure 7 Participation of calcium gelling in high-ester pectin gelation.

network and precipitation can occur. The optimum pH for gelation is controlled by the degree of esterification of the pectin as well as the soluble solids content of the medium (**Figure 6**).

Through their effect on the neutralization of the pectin molecule, cations present in the system will affect gelling. High ionic strength shifts the optimum pH range towards higher values. This is particularly visible with sodium ions. However, with ions such as calcium or potassium that can bridge high-ester pectin molecules in areas with a low density of ester, there is a possible increase in junction zones. This effect can become significant for high-ester pectin with marked blockwise distribution of ester groups (**Figure 7**).

Water activity and sugar types Water activity and sugar types will both affect the way hydrophobic interactions can develop between the pectin molecules. As water activity is reduced, the hydrophobic interactions are easier to form, causing faster gelling to occur and the final gel strength to be increased.

The commonest way of reducing water activity in a food system is through the use of sugars. The effect of sugars on hydrophobic interaction and so on gel structure will be specifically linked to their molecular conformation and their interaction with the neighboring water molecules (**Figure 8**).

Influence of cooling rate and storage temperature
Cooling decreases molecular movement and allows polymer molecules to interact at close distances. As the cooling rate is increased, the gelation rate is also increased. However, during rapid cooling and with a low storage temperature, gelling can actually become very slow, reflecting the difficult development of hydrophobic interactions in these conditions. With an intermediate cooling rate and temperature range, hydrogen bonds and hydrophobic interactions together can contribute to the build-up of a network with the highest elasticity.

Properties of high-ester pectin gels Due to the nature of the molecular interactions involved in high-ester pectin gelation, gels made with these pectins will typically not be thermo-reversible or shear-reversible. When submitted to mechanical stress, the broken gel will then show a high level of syneresis.

Gelling of Low-Ester Pectin

Low-ester pectin has traditionally been used for gelling food products when the conditions required to achieve a gel with high ester pectin were not met. Recently however, low ester pectins have also found applications in high sugar and low pH systems because of their specific texture characteristics.

Though the same base mechanism applies, low ester conventional and amidated pectin differ in their gelling properties and offer a broad range of functional properties.

Gelling mechanism The gelation of low-ester pectins is the result of ionic linkage through calcium bridges between carboxylic groups from two pectin

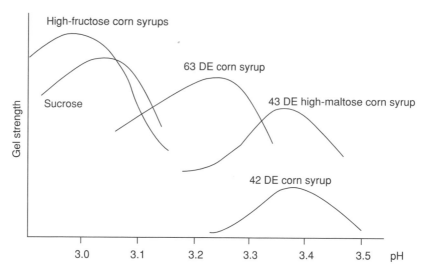

Figure 8 Effect of various sugars on high-ester pectin gel strength. DE, dextrose equivalent.

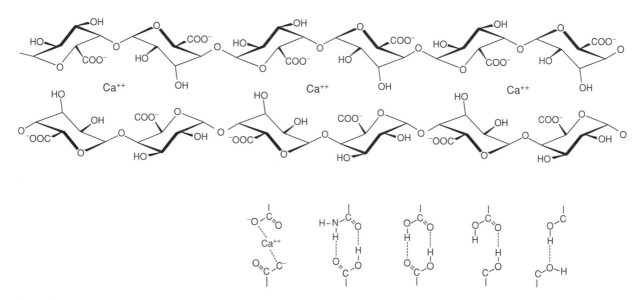

Figure 9 Eggbox model. Schematic overview of established low-ester pectin gel mechanism, showing calcium-induced junction zones. Inset shows detail of the various types of possible hydrogen bonding that participate in the junction zone, together with the calcium chelation (far left).

chains with the participation of hydrogen bonding. The linkage generally takes place upon cooling a pectin and calcium system. The most commonly accepted model of association is the eggbox model (**Figure 9**). In this model, pectin chains could be bridged by calcium ions, which incorporate in their coordination shells two polyanion oxygen atoms from one pectin molecule and three from another chain. Even though a number of positive ions can bridge pectin molecules, especially magnesium and potassium, calcium is particularly effective in complexing with carbohydrates, largely because its ionic radius (0.1 nm) is big enough to coordinate with

many oxygen atoms and because of flexibility with regard to the direction of its coordinate bonds.

The exact calcium requirements to obtain a gel highly depend on the degree of esterification of the pectin, the recipe, and process parameters such as the rate of cooling. An increase in ionic strength, increase in pH, or decrease in the degree of esterification lowers the amount of calcium required to achieve sol–gel transition. An optimum calcium level can be defined for a given pectin in specific conditions. Above this optimum level, pregelation will occur, i.e., gelling will occur at too high temperatures to obtain a coherent gel structure (**Figure 10**).

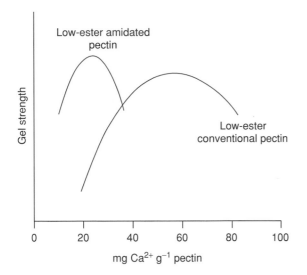

Figure 10 Influence of calcium on low-ester pectin gels.

Gelling is dependent on the length of the junction zones, that is, the number of galacturonic acid units involved in electrostatic bonds with calcium. The bonds are stable when at least seven consecutive carboxyl groups from each chain are involved. If the junction zones formed with calcium become too long, a pectin precipitate may be formed. This may occur with pectin with a very low degree of esterification when large amounts of calcium are available.

The presence of ester or amide groups prevents the formation of junction zones in the interjunction segments of the molecules, making them more flexible. Side chains also prevent aggregation of the pectin molecules through steric hindrance.

The typical high-ester pectin gelling mechanism with hydrogen bonds and hydrophobic interactions can also contribute to the final texture of low-ester pectin gels, especially at low pH and with high soluble solids concentration.

Parameters Influencing Gelling

Although the availability of calcium is a critical factor in the gelation of low-ester pectins, other parameters in relation to the pectin molecule and media have a significant influence on gelling and the final structure obtained. The main parameters are the number and distribution of ester and amide groups as well as the molecular weight of the pectin molecule and the pH, ionic strength, and water activity of the gelling system.

Influence of esterification Because calcium bonds can only occur in esterification-free zones, gel strength increases with a decreasing degree of esterification. For low-ester pectin with an average DE

above 30%, the distribution of the esters may be of significant importance as it will control the length of possible junction zones and influence the gelling temperature, the final gel strength, and texture.

Influence of amidation The amidation of pectins was developed in the 1940s as a means of modifying the functional properties of low-ester pectins in order to achieve better gelling control. The exact mechanism by which the amide groups intervene in the gelation remains to be fully explained. However, it is generally accepted that amidation increases the gelling power of low-ester pectins due to the possibility of hydrogen bonding involving amide groups. Gels made with amidated pectins are firmer and require less calcium; they are also more thermo-reversible than ones made with low-ester conventional pectins.

Influence of molecular weight As for any polymer gel, the length of the polymer governs the number of junction zones required to achieve a coherent network. Low-ester pectins with a high molecular weight will exhibit a higher gelation rate, lower calcium requirements for gelling, and an overall more cohesive and elastic gel structure with a reduced tendency towards syneresis.

Influence of pH At low sugar content, as the pH is decreased, the pectin molecules are neutralized with protons, decreasing the probability of junction zones forming with calcium. This translates into higher calcium requirements and a looser gel texture at a low pH.

On the other hand, when the water activity of the system is decreased by the addition of sugars, the high-ester pectin gelling mechanism will start to play a significant part in the gelling and the calcium requirements will then be decreased when the pH is lowered. In usual food systems, availability of natural calcium will also be increased at a lower pH and thus may reduce the need for extra calcium addition (**Figure 11**).

Influence of ionic strength An increase in gel strength can be observed at a higher ionic strength. This is usually explained by the neutralization of the polymers by the extra ions, which allows the chains to be closer, leading to a more organized and cohesive gel.

Influence of water activity As the solids level increases, calcium requirements decrease. However, for most pectins, a higher solids level accelerates gelling, and increases the setting temperature and the final gel strength. It also reduces the optimum calcium window, thus increasing the risk of pregelation.

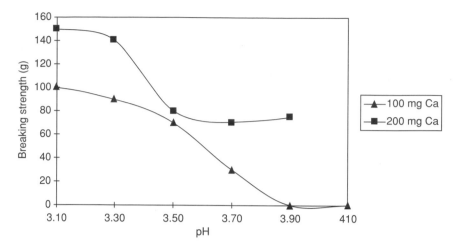

Figure 11 pH influences on Ca-curves in low-sugar system (31% soluble solids), with high reactive amidated pectin. Triangles, 100 mg Ca; squares, 200 mg Ca.

In practice, this leads to a choice of pectin with a higher degree of esterification (less calcium-reactive) at a higher solids level.

Properties of Low-Ester Pectin Gels

The properties of low-ester pectin gels are very dependent on the type of pectin used (conventional or amidated) along with the procedure and formulation employed to make the gel. Typically, they are thermo-reversible and show a high degree of thixotropy. In specific conditions it is however possible to obtain heat-resistant or very brittle gels with no shear reversibility.

Protein Stabilization with High-Ester Pectins

In acidified conditions, casein and, more generally, food proteins will tend to agglomerate and sediment if the viscosity of the system is low enough. In these conditions the proteins are also very sensitive to dehydration and can easily become sandy after heat treatment. With the rapid development of acidified dairy beverages worldwide, now expanding with other protein sources such as soy, the need for effective protein stabilizers in a low-pH environment is growing strongly. High-ester pectin has proven to be a very useful stabilizer in these conditions.

Mechanism It is generally accepted that, at sufficient pectin concentrations, the adsorption of the carboxyl blocks of the pectin molecule to the protein surface will stabilize the protein system through steric repulsion (**Figure 12**). So, the presence of blocks of free carboxyl groups on the galacturonic backbone has an important influence on the protein-stabilizing property of pectin.

Due to the lower proportion of carboxyl groups, high-ester pectin has a weaker electrostatic interaction with protein than low-ester pectin but proves to be more effective. Indeed, it seems to be important that significant parts of the pectin molecule do not interact with the protein surface in order to achieve the steric repulsion effect. It is also the key to minimizing interactions between pectin and cations available in the system.

Both the DE of the pectin and the distribution of the esters on the polymer affect its stabilization properties. Excessively large blocks of carboxyl groups will tend to interact with the ions present in the system, such as calcium, rather than with the protein – an interaction which will lead to an increase in viscosity or even gelation.

Conditions Most food proteins (isoelectric point around 5) can form complex coacervates with anionic polysaccharides such as pectin (isoelectric point around 3.5) in the intermediate-pH region, where the two macromolecules carry opposite net charges: pH above the isoelectric point of the polysaccharide but below that of the protein.

In the case of pectin–protein interaction, the strength of the complex will depend on several factors, such as the distribution of the carboxyl groups on the galacturonic acid backbone, but also on the three-dimensional protein structure and the distribution of ionizable groups on its surface. The whole interaction will also depend on several system parameters such as pH, ionic strength, presence of sugars, or fat.

Through its role in the ionization of both the protein and pectin molecules, pH is the most significant factor affecting electrostatic pectin protein interactions. It

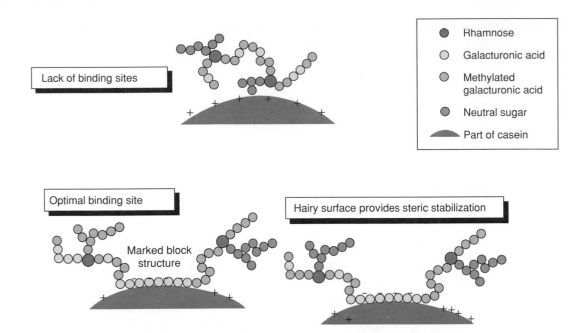

	Rhamnose
	Galacturonic acid
	Methylated galacturonic acid
	Neutral sugar
	Part of casein

Lack of binding sites

Optimal binding site

Marked block structure

Hairy surface provides steric stabilization

Figure 12   Theoretical picture of interaction between pectin and protein.

also plays a very significant role in the protein structure and how proteins interact in a complex system such as milk.

The optimum pH range for interaction between high-ester pectin and casein is 3.6–4.5. At a lower pH, the block structures of the high-ester pectin will not be sufficiently ionized for proper protein binding as the pH is too far below the pKa of the pectin. Above the isoelectrical point of the protein, the protein–polysaccharide complex is very weak or nonexistent and the electrostatic protein–protein repulsion is dominant. It is, however, clear that this repulsion is not sufficient to stabilize the proteins.

Conclusion

Pectin exhibits a wide range of functional properties and enjoys a very good public image as a natural product derived from fruit. Today it is widely used as a textural ingredient and stabilizer in a variety of food applications, and there is little doubt that its usage will grow as new functionalities are revealed.

See also: **Cholesterol**: Absorption, Function, and Metabolism; **Citrus Fruits**: Composition and Characterization; **Dietary Fiber**: Physiological Effects; **Fermented Milks**: Types of Fermented Milks; **Gums**: Properties of Individual Gums; Food Uses; Dietary Importance; **Jams and Preserves**: Methods of Manufacture; Chemistry of Manufacture; **Pectin**: Food Use; **Protein**: Interactions and Reactions Involved in Food Processing; **Rheological Properties of Food Materials**;

Rheology of Liquids; **Stabilizers**: Types and Function; Applications

Further Reading

Dickinson E (1998) Stability and rheological implications of electrostatic milk protein–polysaccharide interactions. *Trends in Food Science and Technology* 9: 347–354.

Kravtchenko TP, Voragen AGJ and Pilnik W (1994) Characterization of industrial high methoxyl pectins. *Gums and Stabilizers for the Food Industry* 7: 27–35.

May CD (1990) Industrial pectins: sources, production and application. *Carbohydrate Polymers* 12: 19–99.

Rolin C and De Vries J (1990) Pectin. In: Harris P (ed.) *Food Gels*, pp. 401–434. London: Elsevier.

Thakur BR, Singh RK and Handa AK (1997) Chemistry and uses of pectin – a review. *Critical Reviews in Food Science and Nutrition* 37: 47–73.

Walter RH (ed.) (1991) *The Chemistry and Technology of Pectin*. London: Academic Press.

Food Use

L Flutto, Danisco, New Century, KS, USA

Introduction

Pectin has been used traditionally in food ever since man started cooking fruits and vegetables. As a

Table 1 Function of pectin in food applications

Function	Food application	Typical use level (%)
Viscosity building	Juice beverage, soft drinks	0.01–0.20
	Sauces	0.10–0.50
	Sorbet (avoids crystallization)	0.10–0.40
Protein stabilization	Yogurt drinks	0.15–0.60
	Milk–Juice drinks	0.15–0.60
	Acidified soy drinks	0.20–0.60
Gelation	Jams, jellies, preserves	0.30–1.20
	Fillings, glazes	0.30–1.20
	Desserts	0.50–1.00
	Confectionery	0.50–2.50

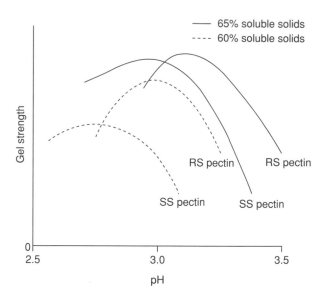

Figure 1 Choice of high-ester pectin and slow-set pectin (SS) content to determine the optimal pH for the gelation. RS, rapid set.

natural component of plants, pectin is a desirable texturizing and stabilizing agent for use in all kinds of processed foods available today. In this chapter, we take a brief look at the most important food uses of pectin and its key properties in each system.

Function in Foods

Pectin is traditionally used for its gelation properties, with more than a third of the pectin commercially available being utilized by the jam and jelly industry. Pectin is used to 'set' or gel jams, jellies, and preserves. The pectin holds fruit pieces suspended within the gel structure, trapping water between cross-linked pectin molecules. The result is a gelled food that is shelf-stable. However, in more complex applications such as fruit preparations for dairy products, pectin is used to impart specific rheological properties during the manufacturing and storage of the fruit preparation, as well as during blending with the dairy product and all during the shelf-life of the finished product. Pectin is also widely used for its unique protein stabilization properties in acidic conditions. The general acceptance of pectin as a natural material and a soluble food fiber has made it desirable for use in a variety of applications (**Table 1**).

Jams and Jellies

Jams, jellies, and preserves can be divided into two categories. Traditional products have a solids range usually above 60%. Reduced sugar products have a soluble solids content in the range of 25–55%.

High-ester pectins are used in the production of jams, jellies, and preserves with soluble solids above 60% and a pH below 3.6. As high-ester pectin gels are not shear-reversible, it is of utmost importance that the jars be filled at a temperature above the setting temperature of the pectin in order to avoid a broken gel and syneresis. Various high-ester pectins

used in recipes with different soluble solid contents will have different optimum pHs (**Figure 1**).

The choice of pectin will also be dependent on the type of product made. A fast-setting pectin is preferred for preserves where fruit pieces have to be suspended evenly throughout the container. By contrast, a slow setting pectin is used for jellies in order for air bubbles to escape prior to gelling, resulting in a fully transparent product. The size of the packaging will also influence the choice of pectin, with fast-setting pectin recommended for larger jars, as the cooling rate is slower, and a quick gelling is necessary to keep the fruit pieces in place. It is important, however, to note that for very large containers (several kilograms to several hundred kilograms), it is necessary to cool down the product before filling as the cooling rate is too slow in the center of the product, and the pectin is degraded when exposed to high temperatures for an extended time. In this case, a slow-set pectin or even a low-ester pectin is required to allow for cooling prior to filling without breaking the gel network. When filling in small packages and especially in food-service single-serve packages, the use of a slow-set pectin is also required as the filling time can be quite long. The temperature of the product in the holding tank before the packaging line will need to be reduced to avoid pectin degradation in time and insure consistent gelling from the first to the last package of the batch. Low-ester pectins are used when conditions for gelling a high-ester pectin are not met. Typically, this is the case when the sugar content is reduced below 60%. They are, however, also used in high-soluble-solids formulations when a

Table 2 Typical reduced-sugar jam formulations with a low-ester amidated pectin

Ingredients	Jam final soluble solids content					
	25%		35%		45%	
	Dosage (%)	Soluble solids (%)	Dosage (%)	Soluble solids (%)	Dosage (%)	Soluble solids (%)
High-calcium reactive low-ester amidated pectin	0.8	0.8	0.7	0.7		
Medium-calcium reactive low-ester amidated pectin					0.6	0.6
Locust bean gum	0.1	0.1				
Sugar I	3.6	3.6	2.8	2.8	2.4	2.4
Water I	18.0		14.0		12.0	
Fruit, 10% soluble solids	45.0	4.5	45.0	4.5	45.0	4.5
Sugar II	15.9	15.9	26.9	26.9	37.4	37.4
Water II	21.1		15.1		7.1	
K-sorbate, 20% w/v	0.2	0.04	0.2	0.04	0.2	0.04
Na-benzoate, 20% w/v	0.3	0.06	0.3	0.06	0.3	0.06
Citric acid, H_2O, 50% w/v	As required		As required		As required	
Flavoring	As required		As required		As required	
Total	105.1	25	105.1	35	105.1	45
Evaporation	5.1		5.1		5.1	
Yield	100.0		100.0		100.0	
pH	3.2		3.2		3.2	
Filling temperature (°C)	40–50		45–55		55–65	

Procedure (regardless of targeted soluble solids)

1. Dry-blend the low-ester pectin (the locust bean gum if present) and sugar I, and add the blend to 80 °C hot water I, agitating vigorously
2. Mix fruit, sugar II and water II and heat the blend to approx. 80 °C
3. Add the pectin solution, agitating continuously
4. Evaporate to the desired content of soluble solids
5. Add preservatives and adjust pH with citric acid
6. Add flavoring
7. Cool to filling temperature and fill

spreadable texture is desired. These pectins gel in the presence of calcium ions, which is the main factor controlling the gelling. The soluble solids level and pH are only secondary parameters, but they have an influence on the gelling temperature and the texture.

As the soluble solids are decreased, the greatest challenge for the manufacturer is to keep an even distribution of fruit pieces. Indeed, the main influence of the soluble solids level will be on the viscosity of the jam at high temperatures during the process.

The choice of a low-ester pectin is therefore mostly based on the solids level. The use of low-ester amidated pectins with increasing calcium reactivity is usually recommended when the soluble solids are decreased. This insures a faster set with an even distribution of fruits. In low-brix formulations, the addition of calcium may be required to supplement the calcium brought by the fruits and the water and to insure a fast and homogeneous gelling.

To further avoid the flotation issues, a combination of low-ester amidated pectin and conventional pectin is commonly used. The low-ester conventional pectin forms a network at a higher temperature than the amidated counterpart and helps to keep the fruit well distributed. Other gums such as locust bean gum or guar are also often used in combination with a low-ester pectin to provide viscosity in low-solids formulations (Table 2).

A rapidly growing category in jams and jellies field is products where refined sugars have been replaced by fruit-juice concentrates. These products usually have between 40 and 50% soluble solids and rely on low-ester pectin gels. The only difference with standard reduced sugar recipes is in the ionic balance, as the fruit concentrates add a significant amount of calcium, as well as magnesium and potassium, to the formulation. This influences the choice of pectin as well as the amount of calcium to be added to the product. It is important to position the product on the flat part of the calcium curve and use a pectin with a broad tolerance to calcium in order to obtain a consistent texture, despite the usual variations in calcium content of the natural raw materials (Figure 2).

Fruit Preparations

The fruit preparation industry constantly juggles between retaining the identity of the fruit, perfect fruit distribution throughout large totes, unique formulations able to accommodate fruits with a varying calcium content, and the texture and flavor of the final dairy product in which the fruit preparations are used. These challenges, combined with the pressure on manufacturers to develop new fruit preparations quickly and the need to create process-friendly formulations, call for the use of highly functional and versatile stabilizers. Pectin has found a natural application in these systems and is widely used.

Use of Pectin in Fruit Preparations

Pectin is used in fruit systems for various purposes. However, fruit preparations should only be thickened and not fully gelled. This requires pectin capable of giving fruit preparations a high degree of thixotropy.

The texture of the fruit preparation can be controlled by various parameters, as outlined in **Table 3**.

Use of a High-ester Pectin

When more than 55% soluble solids are used, a high-ester pectin with a typical dosage of 0.3% can successfully impart viscosity and control syneresis in a fruit preparation.

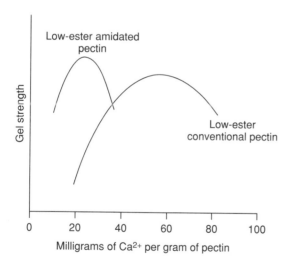

Figure 2 Low-ester-pectin calcium curves.

It is also possible to use a high-ester pectin with less than 55% soluble solids in order to yield a special texture. In these cases, a high-ester pectin typically gives the fruit preparation a high degree of shine and smoothness, although higher dosages are necessary.

Special high-ester pectins developed for their interaction with casein can also be used to obtain a high degree of texture carry-through in the white mass. At a usage level of 1–2%, the pectin gives the fruit preparation the required texture and significantly increases the viscosity of the finished stirred yogurt. It also helps maintain a constant viscosity and reduce syneresis during the shelf-life of the dairy product (**Figure 3**).

Use of a Low-ester Pectin

Conventional and amidated low ester pectins are widely used in fruit systems. While conventional pectins require a slightly higher dosage and high level of calcium, special ones have extreme processing flexibility and can be used with great success in a broad variety of formulations from 30 to 50 Brix with various pH values. It is common to include low ester conventional pectin in fruit on the bottom formulations. Low-ester amidated pectin can prove more economical with the possibility of lower dosages, although it tends to be more sensitive to calcium variations (see **Figure 2**). When the calcium level is adjusted, it is possible to achieve an excellent fruit suspension with a high level of shear reversibility. This allows low viscosity during pumping for maximum fruit identity.

Calcium Addition for Texture and Postreaction Control

When using a low-ester pectin, it is usually necessary to add calcium salts to obtain the full texture. The necessary quantity will depend on the calcium source, pH, soluble solids, fruit type, and content of the preparation. Specific calcium salts should be selected according to the speed of release and labeling considerations. Calcium chloride, -lactate, and -citrate is

Table 3 Parameters influencing the texture of fruit preparations

Choice of pectin	High-ester, low-ester conventional, and low-ester amidated pectins perform differently and allow a variety of textures
Pectin dosage	A higher pectin dosage results in a thicker texture
Calcium level	The calcium level needs to be adjusted with consideration for a possible reaction in the white mass; higher calcium levels usually increase viscosity, but pregelation can occur if too much calcium is added
Soluble solids	Higher soluble solids usually result in a shorter texture
Filling temperature	Lower filling temperatures usually result in a more fluid texture
Use of other hydrocolloids	Locust bean gum, guar gum, xanthan gum and starches can be successfully used in combination with pectin to achieve specific textures

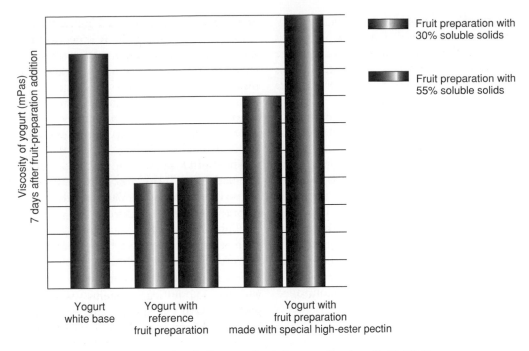

Figure 3   Use of a special high-ester pectin in a fruit preparation to increase the viscosity of yogurt.

most commonly used to supplement the calcium contained in the fruit.

If the calcium content of the fruit preparation is too low, the pectin present in the fruit preparation may react with the calcium of the yogurt, resulting in a grainy texture (stirred yogurt) or a gelled preparation (fruit on the bottom). To avoid this type of postreaction, it is highly recommended that the appropriate amount of calcium is added to the fruit preparation.

Bakery Fillings

In principle, bakery fillings are a form of fruit preparation. However, they deserve a separate paragraph as the demands on these products are fairly specific. Applied on or inside the dough, they usually need to be heat-resistant and have a level of water activity low enough to minimize water transfer from the filling to the dough. This is especially important for long-shelf-life cookies and cereal bars, which tend to be where there is greatest demand. As is the case for fruit preparations for the dairy industry, the trend is for the bakeries to outsource the production of fillings and transport them in larger containers.

Most bakery fillings have a high solids content typically between 60 and 80%. Thus, a high-ester pectin seems a natural choice also because these gels are thermally irreversible. However, high-ester pectin gels are completely nonthixotropic and lose their entire structure after pumping. So, pure high-ester pectin gels, which require cooling prior to filling and

mechanical pumping at the bakery, are not adequate for packaging in large containers. High-ester pectin gels would be destroyed by such processes and show syneresis and a very poor bake stability.

High-ester pectins may, however, be used successfully for cold gelation systems, also called delayed setting systems. These are gel systems in which all the gelling conditions for a high-ester pectin are met, with the exception of the pH, which is kept too high. This viscous syrup can then be shipped to the bakery, where the acid is then added upon depositing the filling on the dough.

For other types of bake-stable fillings, specific low-ester pectins can prove ideal. Even if low-ester pectin gels are typically characterized as being thermally reversible (a property used in glazes), they can be made bake-stable by using an optimized formulation. Low-ester conventional pectins are usually preferred, as they show a higher bake stability due to a stronger binding with calcium. The choice of the correct calcium salt as well as its dosage is the key to both the texture of the filling and its heat stability. In very high solids formulations, the use of pectin in combination with alginate can prove useful in improving the filling appearance and its bake stability.

Glazes and Bakery Decor

Glazes are widely used to cover fruit tarts and other types of cakes where they not only add consumer appeal but also extend shelf-life by forming a barrier to dehydration and microbiological contamination. The high transparency and shine of pectin gels make the use of this hydrocolloid an ideal choice for this application. In some specific applications, it is possible to use combinations of pectin with alginate or carrageenan.

Contrary to most bakery fillings, in glazes and bakery decorations, the heat reversibility of the system is usually of paramount importance. These products are most often sold to bakery shops in concentrated form and reheated and diluted with water before application. The gel must then form quickly and keep for several days, even on fruits with a high water content. In industrial applications, the gel has to be so completely reversible such that the product can be sprayed. Clearly, these requirements are best met with low-ester amidated pectins.

The use of specific calcium sources with controlled release together with calcium sequestrants is necessary to control the gelling of a low-ester pectin and secure an optimum performance after reheating.

Confectionery

Pectin is widely used in traditional confectionery items such as 'pâte de fruit,' or Zefir. However, it is also an important textural agent in a number of other confectionery articles, where its superior flavor release, high transparency, and clean bite are sought after.

Jelly candies typically have a soluble solids content of between 75 and 80% and a pH of around 3.5. Due to the high solids level, it is important to use a very slow setting high ester pectin and include buffer salts in the formulation to avoid any pre-gelling before depositing.

The standard process usually involves making a pectin solution with the buffer salts, heating the sugars and fruits when present, and mixing in the pectin solution before cooking to reduce the water level and obtain the desired solids content. Immediately before depositing, acid, color, and flavorings are added. Pectin jellies can be deposited in starch or rubber molds or in large trays for further cutting. It is also possible to use pectin on jet cooking equipment where all the ingredients are mixed together, and the pectin actually dissolves at a very high temperature (around 140 °C) with all the solids present. This procedure usually requires the use of specific pectins with particular dissolution properties.

In most cases, the pectin chosen for confectionery will be an extra-slow setting type to guarantee a low setting temperature. Even with this type of pectin, the gelling temperature of the system is usually around 80 °C, and the depositing temperature needs to be

above this level if the full gelling potential of the pectin is to be used. This is a major difference from the commonly manufactured gelatine candies, which have a much lower setting temperature. Another significant difference is the speed of gelation, as pectin jellies can be demolded quickly after depositing since a firm gel is obtained very quickly. As with all high-ester pectin gels, typical pectin jellies are not thermally reversible.

The buffer salts commonly used with pectin in confectionery include sodium citrate and sodium potassium tartrate. The latter tends to give the jellies a somewhat firmer texture.

High-ester pectin is also often used in combination with other ingredients such as gelatine, starch, or agar to give intermediate textures. In the case of gelatine candy, pectin is also used in small amounts to increase the melting point of the confectionery pieces in warm climates, thus insuring a better appearance throughout their shelf-life without significantly affecting the texture.

In specific candy pieces where a neutral flavor such as vanilla is used, the use of a low-ester pectin is necessary in order to obtain a gel at a higher pH. As with high-ester pectins, it is usually necessary to reduce the setting temperature. In these systems, calcium sequestrants such as sodium hexametaphosphate are used. Turkish delight is one such traditional candy based on a low-ester pectin gel with or without the addition of starch.

High- and low-ester pectins are also widely used for confectionery fillings, which are then either enrobed usually with chocolate or, for example, panned into jelly beans.

Fruit Beverages and Soft Drinks

Pectin is used as a viscosifier in beverages and soft drinks, and high-ester pectins may be used as a mouth-feel improver. This use has been widely developed for juice drinks with a reduced juice content or sugar-free soft drinks.

Low-concentration pectin solutions can be considered Newtonian and show a low viscosity. This is of great relevance for the use of pectin in fruit beverages and soft drinks as the concentration used rarely exceeds 0.5%. Indeed, the clean mouth feel imparted by pectin compared with the tendency towards a slimy mouth feel with some other gums could be related to the low viscosity of pectin solutions at the shear rate applied in the mouth. This property makes pectin an ideal choice when trying to replace the mouth feel lost by the reduction in sugar content.

As most juice beverages and soft drinks contain calcium, pectin with a high degree of esterification is usually recommended to minimize the calcium sensitivity of the pectin and avoid any risk of gelling. A slight gelling of the product changes the rheology of the solution, resulting in undesirable pseudoplastic behavior. For this reason, the most commonly used pectin is of the rapid setting type. Pectin manufacturers usually offer rapid-set pectins standardized to a viscosity instead of gelling properties, so as to guarantee a consistent performance in a beverage application.

Yogurt Drinks and Other Acidified Protein Beverages

From being largely restricted to certain regions or addressed to specific consumer groups, acidified protein drinks today have become a highly esteemed and popular food world-wide. Aided by the increased focus on health benefits, the market for acidified beverages with proteins is currently enjoying rapid growth.

Perfect stability and a consistent mouth feel are essential to consumers, so the use of stabilizers is necessary in order to prevent any defects such as sedimentation, whey separation, and grainy mouth feel.

At a low pH, the positive net charge of the protein clusters in an acidified drink is insufficient to prevent aggregation of the proteins, and the drink therefore suffers from severe sedimentation of protein aggregates and separation. This lack of physical stability is even more pronounced if a shelf-stable, postpasteurized product is desired as the unprotected proteins are denatured by the heat treatment. Although specific structural details vary, the same general stability issues occur with various protein sources such as casein, whey, soy, or other plant-derived proteins.

The stability problem can be avoided by adding a stabilizer such as a high-ester pectin. With its controlled block structures of nonmethylated galacturonic acid units, high-ester pectins can bind to the protein surface as the blocks are negatively charged at a pH of 3.6–4.5, a typical level in acidified beverages. Bound to the proteins, the long pectin molecules protect them from reaggregation through steric stabilization. Even with a postpasteurized drink, this stabilization can produce a shelf-stable product and prevent unwanted sedimentation and serum separation.

Yogurts and Other Dairy Products

The gelling of low-ester pectins with calcium ions is often used to set dairy products, but low-ester pectins are also used to reinforce the texture of other types of gel network such as the protein network in yogurts and fermented dairy products. Recently, the trend towards low-fat or non-fat dairy products has increased the need to use textural ingredients and stabilizers to restore a pleasant mouth feel and avoid common defects such as syneresis.

Low-ester pectins can be dissolved in neutral milk and are commonly used in the manufacture of gelled neutral dairy desserts. Giving a softer texture with more body and a high level of creaminess than widely used carrageenan, pectin is preferred in premium creamy formulations where the extra cost of pectin can be justified. It is also used when carrageenan is not desired on the label. In addition, pectin is the main component of a variety of regional home-made desserts formulations such as Fruche in Japan. Fruche is a fruit and pectin solution sold in a pouch, which the consumer can mix with milk. After half an hour in the refrigerator, the gelled product can be consumed as a refreshing dessert. In Europe, pectin is often seen in dry mix blends for home-made desserts.

Whether the effect of low-ester pectins on the texture of fermented dairy products comes through gelation with calcium or interaction with proteins is not quite clear. However, low-ester pectin is already widely used on its own or together with gelatine or starch to add texture to yogurts and other fermented products.

Owing to the high level of calcium present, the usage level of the pectin must be low in order to avoid an excessively strong reaction between the pectin and the calcium, resulting in a sandy product. The maximum recommended usage level is around 0.25%. Interestingly, high-calcium-reactive pectins usually show a better performance than their low-calcium-sensitive counterparts. Amidated pectins are usually preferred, even if conventional low-ester pectins do prove useful in specific recipes for syneresis control.

Pectin can also be used in combination with carrageenan for mousse products, where gelatine is not desired. Pectin provides a much better gelation after the mousse mix is whipped, leading to a more sliceable texture compared with carrageenan alone.

Other Applications

Pectin is used in a great number of other applications. For example, it is used in tomato-based products such as barbecue or taco sauce as an alternative to starch. Usually, low-ester conventional pectins are used to thicken the sauce and ensure a nonmelting texture when the sauce is poured over hot food.

See also: **Dairy Products – Nutritional Contribution**; **Gelatin**; **Jams and Preserves**: Methods of Manufacture; Chemistry of Manufacture; **Soft Drinks**: Chemical

Composition; **Sweets and Candies**: Sugar
Confectionery; **Yogurt**: Yogurt-based Products

Further Reading

Beli R. Thakur, Singh RK and Handa AK (1997) Chemistry and uses of pectin – a review. *Critical Reviews in Food Science and Nutrition* 37(1): 47–73.

May CD (1990) Industrial pectins: sources, production and application. *Carbohydrate Polymers* 12: 19–99.

Rolin C and De Vries J (1990) Pectin. In: Harris P (ed.) *Food Gels*, pp. 401–434. Amsterdam: Elsevier.

Walter RH (ed.) (1991) *The Chemistry and Technology of Pectin*. London: Academic Press.

PELLAGRA

D A Bender, University College London, London, UK

Introduction

Pellagra is a nutritional disease due to deficiency of the vitamin niacin and the essential amino acid tryptophan. The clinical features of pellagra are dermatitis, diarrhea, and dementia; it is commonly known as the 'disease of the four Ds,' since it is also fatal – the fourth 'D' is death.

During the first half of the twentieth century, pellagra was a major problem of public health in the southern USA, with some 87 000 deaths attributed to the disease between 1900 and 1950. Through the 1950s to 1980s, it continued to be a problem in southern Africa and parts of India, but improvements in food availability and general nutritional status have led to more or less complete eradication of pellagra as an endemic nutritional deficiency disease.

There have been a few reports of outbreaks of pellagra among refugees in southern Africa when rations have been inadequate to meet the demand, and individual cases have been reported among alcoholics and people with Crohn's disease and other gastrointestinal diseases that impair nutrient absorption.

Pellagra may also occur as a result of a variety of (relatively rare) conditions affecting tryptophan metabolism and as a side-effect of a number of drugs that inhibit tryptophan metabolism. In alcoholics, it is not clear whether pellagra is the result of an impairment of tryptophan and niacin metabolism directly attributable to alcohol, or whether it reflects general undernutrition among people who obtain a considerable proportion on their energy needs from alcohol, and hence have a low intake of (nutrient rich) foods.

The first full description of pellagra was given by Casal in 1735, working in Spain; he called it '*mal de la rosa*' – the red disease. The name pellagra was coined by Frapolli in 1771, from the Italian '*pelle*' for skin and '*agra*' for rough, thus describing the most striking feature of the disease, the roughened appearance of the skin, resembling severe sunburn in areas exposed to sunlight. Casal recognized that the underlying cause of pellagra was nutritional, associated with the limited diet of many in central Spain at the time, although he did not associate it with the then recent introduction of maize from Central America.

The spread of pellagra largely followed the introduction of maize as a dietary staple, and by the nineteenth century, it was a major problem around the Mediterranean, in the Balkans, and in the Ukraine, as well as the southern USA and southern Africa. In South Africa, pellagra became a problem following the outbreak of rinderpest in 1897, which killed most of the cattle, leading to a marked deterioration in the diet of people who had previously had ample supplies of milk and meat to supplement their maize-based diet. In the USA, it was the social and economic upheaval of the Civil War, which led to large numbers of subsistence farmers living on a diet based very largely on maize.

The other region where pellagra was a major problem is the Deccan plateau in India, where the dietary staple is jowar (a variety of millet, *Sorghum vulgare*), rather than maize.

Although Casal had described pellagra as a nutritional disease, at the beginning of the twentieth century it was generally assumed to be due to an infectious agent. The nutritional basis of the disease, and the interaction between deficiencies of both niacin and tryptophan, was established mainly by Goldberger and coworkers in the USA between 1913 and 1948. A 'pellagra-preventing factor' isolated from protein-free yeast extract was shown in 1937 to be nicotinic acid, and either nicotinic acid or its amide, nicotinamide, was shown to prevent or cure the disease in both man and experimental animals. The name niacin was coined when it was decided to enrich foods with the vitamin, since it was considered that nicotinic acid would be unacceptable as a food additive, because of its chemical (but not metabolic) relationship with nicotine. In the USA, 'niacin' is commonly used to mean nicotinic acid, and the amide is niacinamide; elsewhere, 'niacin' is used as a generic descriptor for nicotinic acid and/or nicotinamide.

Other early studies showed that feeding additional protein, and especially the essential amino acid tryptophan, would also prevent or cure pellagra, suggesting that it was a protein-deficiency disease rather than due to a lack of a vitamin. It was not until 1947 that the problem was resolved, when it was shown that tryptophan is a metabolic precursor of the nicotinamide moiety of the coenzymes nicotinamide adenine dinucleotide (NAD) and nicotinamide adenine dinucleotide phosphate (NADP). The coenzymes can be formed in the body either using preformed dietary niacin or by *de novo* synthesis from tryptophan.

The reason a maize-based diet predisposes to pellagra is that the proteins of maize are particularly poor in tryptophan, so that a diet in which there are few other sources of protein provides insufficient tryptophan for nicotinamide synthesis. Other cereals, such as wheat, barley, rye, rice, and millet, contain enough tryptophan to meet the requirement for niacin synthesis.

Although maize proteins are poor in tryptophan, the other cereal associated with pellagra, jowar, provides a minimally adequate amount of tryptophan to meet requirements. However, the proteins of jowar are considerably richer in leucine than most other proteins, and a diet based largely on jowar provides a considerable excess of leucine. There is a considerable body of evidence that this amino acid imbalance can be a precipitating factor in the development of pellagra when the dietary intake of preformed niacin is extremely low, and the intake of tryptophan is only marginally adequate. Leucine inhibits kynureninase, and hence reduces the rate of oxidative metabolism of tryptophan, resulting in reduced formation of NAD. In addition, leucine competes with tryptophan for tissue uptake, and thus has a further inhibitory effect on the rate of tryptophan oxidative metabolism and NAD synthesis.

All cereals contain preformed niacin. However, this is largely present as a variety of nicotinoyl esters, collectively known as niacytin, which are not hydrolyzed by digestive enzymes to any significant extent, so that most of the niacin present in cereals is nutritionally unavailable. A small proportion of the niacin present as niacytin (up to about 10%) may be biologically available as a result of nonenzymic hydrolysis by gastric acid.

Interestingly, pellagra has not been a problem in Central America, the original home of maize. This is because of the traditional way in which it is prepared. Rather than milling the grain, it is steeped overnight in limewater (calcium hydroxide solution), then squeezed to form the dough from which tortillas are made. This alkaline treatment results in hydrolysis of most of the nicotinoyl esters, so releasing free nicotinic acid, which is nutritionally available. Although maize spread to many countries following its discovery, it was generally milled like other grains rather than being treated in the traditional Mexican manner.

Clinical Features of Pellagra

Dermatitis

Exposure of the skin to modest amounts of sunlight results in a severe sunburn-like dermatitis in sufferers (pellagrins). Mechanical pressure can cause similar lesions, especially around the wrists and ankles. The skin in the affected areas is red and slightly swollen at first, and then becomes rough, thickened, cracked, and dry, with scaling, a shiny surface, and brown pigmentation.

The cause of this photosensitive dermatitis in pellagra is unknown and cannot be attributed to the known metabolic functions of either tryptophan or the nicotinamide nucleotide coenzymes. There is some evidence that there is increased metabolism of the amino acid histidine in the skin in pellagra, resulting in a lower than normal concentration of both histidine and an intermediate in its metabolism, urocanic acid, both of which are believed to have a role in absorbing ultraviolet light, and so minimizing damage to the skin from exposure to sunlight. Treatment of pellagrins with niacin both clears the dermatitis and also increases the concentration of histidine and urocanic acid in the dermis.

The skin lesions of pellagra may be due to secondary zinc deficiency when tryptophan intake is inadequate or its metabolism is disturbed; they resemble the lesions seen in acrodermatitis enteropathica, which is due to a failure to secrete the intestinal zinc-binding ligand, which is believed to be picolinic acid – a tryptophan metabolite. There is some evidence that some pellagrins have a poor zinc status, and the zinc depletion associated with excessive alcohol consumption may be a factor in the development of pellagra among alcoholics.

Diarrhea

Although diarrhea is common in pellagrins, it is not a constant feature of the disease, and indeed in some cases, there may be chronic constipation. The cause of both the diarrhea and the constipation is almost certainly general nutritional deficiency, resulting in atrophy of the intestinal mucosa and the intestinal musculature.

Dementia

The psychiatric disturbances of pellagra range from mild hallucinations with some psychomotor retardation, through confusion with increasing hallucinations, to severe dementia and anxiety psychosis, with melancholia, intermittent stupor and possibly epileptiform convulsions. In many ways, this resembles schizophrenia, but the dementia of pellagra can be differentiated from schizophrenia and the organic psychoses by the sudden lucid phases that alternate with the most severe mental symptoms.

Although the cause of the psychiatric disturbance in pellagra remains to be firmly established, it is likely that it is largely due to deficiency of tryptophan, as a precursor for the neurotransmitter serotonin (5-hydroxytryptamine), rather than a direct result of inadequate supply of the nicotinamide nucleotide coenzymes in the brain.

The Equivalence of Dietary Tryptophan and Niacin

The nicotinamide ring of the coenzymes NAD and NADP can arise either from preformed dietary niacin or by *de novo* synthesis from quinolinic acid, an intermediate in the oxidative metabolism of tryptophan. For an adult in nitrogen balance, the amount of tryptophan available for oxidative metabolism is about 99% of the dietary intake, since new synthesis of protein is balanced by catabolism of tissue proteins, releasing their tryptophan for metabolism. A variety of studies have shown that under normal conditions, 60 mg of tryptophan are equivalent to 1 mg of preformed dietary niacin, and it is usual to express the total niacin intake in terms of niacin equivalents – the sum of preformed niacin plus 1/60 of the tryptophan. On this basis, average Western diets provide more than enough niacin to meet requirements from tryptophan alone, ignoring preformed niacin. Indeed, reanalysis of dietary records of people who died from pellagra in the USA during the first half of the twentieth century shows that their intake of preformed niacin and tryptophan was (marginally) adequate to meet requirements, suggesting that some other factor may have been involved in precipitating the disease.

Complicating Factors in the Etiology of Pellagra

Deficiency of Vitamins B₂ and B₆

The enzyme kynurenine hydroxylase in the oxidative pathway of tryptophan metabolism is a flavoprotein, and hence its activity is impaired in riboflavin (vitamin B_2) deficiency. The enzyme kynureninase is pyridoxal phosphate-dependent, and its activity is impaired in vitamin B_6 deficiency. This means that deficiency of either vitamin B_2 or B_6 will reduce the rate of oxidative metabolism of tryptophan and hence the rate of formation of quinolinic acid and NAD. If the intake of tryptophan is anyway marginal, this could well precipitate pellagra.

Mycotoxins

A number of mycotoxins cause DNA damage, and activate poly(ADP-ribose) polymerase as part of the DNA repair mechanism. This enzyme uses NAD as the source of ADP-ribose, releasing nicotinamide. While, theoretically, this nicotinamide could be reused for synthesis of NAD, the enzymes involved (nicotinamide phosphoribosyltransferase, nicotinamide deamidase, and nicotinic acid phosphoribosyltransferase) are all more or less saturated with their substrates at normal tissue concentrations. This means that the additional nicotinamide released by poly(ADP-ribose) polymerase cannot be used for NAD synthesis but will largely be methylated to N-methylnicotinamide and excreted. Exposure to such mycotoxins may therefore be a factor in the etiology of pellagra, depleting the body of nicotinamide. The one report of an outbreak of pellagra in Mexico was associated with a shipment of maize that had suffered considerable fungal damage.

Estrogens and Progestagens

Through the first half of the twentieth century, when pellagra was a major problem in the southern USA, there was a twofold excess of women over men among those affected. In a number of reports of more recent outbreaks, there is a similar sex ratio between the menarche and menopause. There is no difference in the numbers of males and females affected among either prepubertal children or adults aged over about 40 years. This suggests that estrogens and/or progestagens may have a pellagragenic effect. Estrogen metabolites are competitive inhibitors of kynureninase, and the administration of progesterone results in reduced activity of kynurenine hydroxylase, although *in vitro*, neither progesterone nor its conjugates affect the activity of kynurenine hydroxylase.

When the intake of preformed niacin is low, and that of tryptophan is marginal, the impairment of tryptophan metabolism by estrogens and progesterone may be sufficient to precipitate pellagra more commonly in women than men.

Non-nutritional Pellagra

Pellagra can occur as a result of impairment of tryptophan metabolism due to a variety of diseases that affect the tryptophan oxidative pathway, or as a result of drugs that inhibit one or more enzymes of the pathway, despite an apparently adequate intake of tryptophan. General malabsorption associated with gastrointestinal disease; there have been a number of case reports of pellagra among people with Crohn's disease. In most cases, the condition responds well to supplements of nicotinamide.

Carcinoid Syndrome

Under normal circumstances, about 1% of the daily intake of tryptophan is metabolized by way of 5-hydroxytryptamine in the central nervous system and gut, with the remainder being oxidized by way of kynurenine, and thus available for NAD synthesis. A carcinoid is a tumor of the enterochromaffin cells of the gastrointestinal tract, which forms 5-hydroxytryptamine from tryptophan. The carcinoid syndrome occurs when the tumor has metastasized, usually to the liver, and in extreme cases as much as 60% of the daily intake of tryptophan may be metabolized by way of 5-hydroxytryptamine. The result is a considerable reduction in the rate of oxidative metabolism through kynurenine, and hence a considerable reduction in the synthesis of NAD from tryptophan, resulting in the development of pellagra in a significant proportion of patients.

Hartnup Disease

Hartnup disease is an inborn error of metabolism affecting the membrane proteins that transport the large neutral amino acids (including tryptophan). The same proteins are involved in the absorption of free tryptophan from the gastrointestinal tract into the bloodstream, from the bloodstream into tissues, and in the reabsorption of amino acids from the urine. The result of the defect is a considerable reduction in the amount of dietary tryptophan that is absorbed, as well as a considerable loss in the urine. Thus, despite an apparently adequate intake, there is a deficiency of tryptophan (and other large neutral amino acids), resulting in the development of pellagra.

Inborn Errors of Tryptophan Metabolism

A number of inborn errors of metabolism affecting enzymes of the tryptophan oxidative pathway have been reported, all of which result in the development of pellagra. Such conditions include much reduced activity of tryptophan dioxygenase, the first enzyme of the pathway; low or undetectable activity of kynureninase and kynurenine hydroxylase; and increased activity of aminocarboxymuconic semialdehyde decarboxylase, the enzyme that competes with the nonenzymatic cyclization of aminocarboxymuconic semialdehyde to quinolinic acid, the precursor for NAD synthesis.

Drug-induced Pellagra

A number of drugs can precipitate pellagra, despite an apparently adequate intake of tryptophan. The best-documented such drug is the antituberculosis drug isoniazid (isonicotinic acid hydrazide), although two antiparkinsonian drugs, Benserazide and Carbidopa, are also associated with niacin depletion. These drugs have an indirect effect on tryptophan and niacin metabolism. They are hydrazine derivatives that act as carbonyl-trapping reagents and therefore cause depletion of vitamin B_6 by forming inactive adducts with the metabolically active form of the vitamin, pyridoxal phosphate. Among other effects, this results in impaired activity of kynureninase, and hence a reduced rate of tryptophan oxidative metabolism and NAD synthesis. Although the pellagra responds to supplements of nicotinamide, it is more usual to give supplements of vitamin B_6, at least in combination with isoniazid, in order to minimize the other metabolic effects of vitamin B_6 depletion.

Pellagra has also been reported among cancer patients treated with a variety of chemotherapy agents that cause strand breaks in DNA. This activates poly(ADP-ribose) polymerase as part of the DNA repair mechanism, and so depletes tissue NAD. Indeed, this is the mechanism of action of these drugs; they deplete NAD in the tumor cells to such an extent that there is severe impairment of energy-yielding metabolism, leading to cell death. It has been suggested that patients treated with such chemotherapeutic agents should receive supplements of niacin to protect against the development of pellagra.

See also: **Cereals**: Contribution to the Diet; Dietary Importance; **Drug–Nutrient Interactions**; **Maize**; **Mycotoxins**: Toxicology; **Niacin**: Physiology; **Vitamins**: Overview; Determination

In small-fruited chilli peppers, there are two chambers, or carpels, separated by an inner wall, or septum. The seeds are attached to a spongy tissue called the placenta that develops from the septum. Large fruits are often formed from more than two carpels, and therefore have additional septa. The septa are better developed at the stalk end of the fruit, and often fail to reach the tip, so most of the seeds are borne on a hemispherical placenta at the stalk end of the single cavity in the fruit.

Special Characteristics

Quality of pepper fruits, and of the products made from them, depends on their color, aroma, and pungency. Hot peppers also have certain well-known pharmacodynamic effects. Nutritionally, both hot and sweet peppers are valuable sources of vitamins.

Color

The middle layer or mesocarp of the fruit wall is responsible for the color of both ripe and unripe fruits. The cells of the mesocarp contain plastids, which in the unripe fruit usually contain chlorophyl. The quantity of chlorophyl varies in different cultivars. Occasionally chlorophyl is absent and the unripe fruits are creamy white. Immature fruits may also contain sap-soluble, purple anthocyanin pigments. As the fruit ripens, both chlorophyl and anthocyanin usually disappear. Chloroplasts are converted to chromoplasts, and additional chromoplasts form *de novo*. There is a relationship, not yet fully understood, between chlorophyl content of the unripe fruit and carotenoid content of the ripe fruit. Immature fruits that lack chlorophyl may ripen red, but the red is deepest in fruits that are dark green when unripe. (*See* **Chlorophyl; Colorants (Colourants):** Properties and Determination of Natural Pigments; **Ripening of Fruit.**)

As a red pepper ripens, total carotenoids increase 35-fold and are sequestered within fibrils in the chromoplasts. About half of this increase represents *de novo* synthesis of red pigments. These consist of three ketocarotenoids: capsanthin (30–60% of total carotenoids present), capsorubin (5–15%), and cryptocapsin (about 5%). They are unusual in containing the cyclopentanol ring, unlike red carotenoids of fruits such as tomato (which accumulates lycopene). Lycopene is thought to be an intermediate in the biosynthesis of capsanthin and capsorubin (**Figure 1**).

Although the carotenoid pigments are formed in the chromoplasts, the quality and quantity of these pigments, which determine the different colors of the mature fruit (**Table 2**), are controlled by genes located on the chromosomes, not in the plastids. Fruits of the yellow color series cannot synthesize red carotenoids. In *C. annuum*, yellow-fruited plants carry a deletion of the DNA that codes for the enzyme capsanthincapsorubin synthase, which catalyzes the final step in the production of capsanthin and capsorubin. Recessive alleles of other nuclear gene(s) reduce the quantities of pigments in the fruits, but we do not know where or what these blocks to pigment synthesis may be. Chocolate-brown fruits, such as those used in the famous Mexican sauce mole, are homozygous recessive for an allele of another nuclear gene, which prevents chlorophyl breakdown in the ripe fruit. The combination of red carotenoids and green chlorophyl appears as brown.

Breakdown of the red pigments in harvested fruits, or in powders produced by drying and grinding these fruits, is not prevented by blanching, but is retarded by antioxidants. This suggests that pigment breakdown is not due to enzyme action, but may involve two other processes: an autoxidative degradation which is accelerated by heat, and an autocatalytic destruction in light which involves direct absorption of light energy.

Aroma

Droplets of volatile oil in the mesocarp cells produce the characteristic aroma of the pepper fruit. These droplets increase in quantity as the fruit ripens. The oil is a mixture of methoxypyrazines, aliphatic alcohols, and esters, in which the most important constituent is 2-methoxy-3-isobutylpyrazine. This has one of the lowest odor thresholds of any compound examined, being detectable at levels of 2 parts in 10^{12} parts water. Methoxypyrazines have also been reported from other vegetables such as raw potato and French bean, but at much lower levels than in *Capsicum*. (*See* **Sensory Evaluation**: Aroma.)

The aliphatic alcohols and esters seem to be responsible for the fruity and floral components of the aroma. Although aroma is a valued attribute of fresh chillies, and the different aroma and flavor properties of different chillies are usually well-known to habitual consumers, little is known about differences in composition of the volatile oil either within or between different species of *Capsicum*.

Pungency

Some botanists have used pungency of the fruits to distinguish *Capsicum* from allied genera. The biological significance of pungency is not known, but the sensory effect is produced by a group of vanillyl amides collectively known as capsaicinoids (**Table 3**), amongst which capsaicin and dihydrocapsaicin

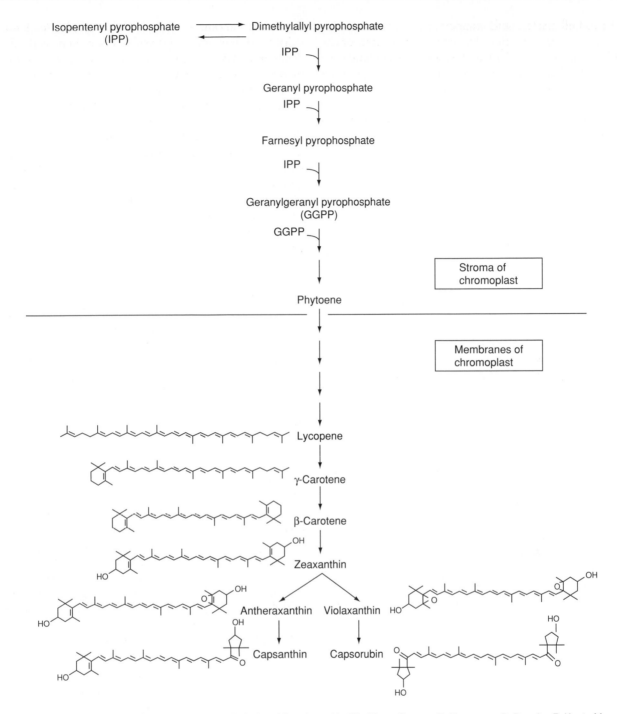

Figure 1 Biosynthesis of carotenoid pigments in fruits of *Capsicum*. Modified from Camara B, Hugueney P, Bouvier F, Kuntz M and Monéger R (1995) Biochemistry and molecular biology of chromoplast development. *International Review of Cytology* 163: 175–247, with permission.

predominate. Capsaicin is insoluble in cold water but soluble in alcohol, acetone, ether, and similar solvents. It is one of the most pungent compounds known, detectable by taste at dilutions of 1 in 15–17 million. Pungent compounds are also present in other plants, for example, ginger and black pepper (**Table 3**). The capsaicin molecule has three characteristic features: the vanillyl group, the acid–amide linkage, and the alkyl side-chain. Change in any one of these decreases pungency. Piperine from black pepper lacks the vanillyl group and has a short side-chain with a bulky substituent. Its pungency is more than two orders below that of capsaicin. Gingerols and shogaols, from ginger, have a vanillyl group and a long alkyl chain, but lack the acid–amide bond and thus are also much less pungent than capsaicin. The length of the alkyl side-chain has been varied experimentally in synthetic analogs of capsaicin. Pungency appears first in compounds with a six-carbon side-chain, peaks with the nine-carbon side-chain, then gradually declines as chain length increases. The synthetic nine-carbon analog, N-vanillylnonamide, has been used as a substitute for capsaicin.

Biosynthesis of capsaicinoids is thought to involve one pathway from phenylalanine to vanillylamine, and one from valine or leucine to their respective isoacids, followed by condensation of vanillylamine with the activated fatty acids. Two capsaicinoid-like compounds (capsiate and dihydrocapsiate) have been identified from a nonpungent Japanese cultivar of C. annuum. These compounds have an ester moiety in place of the amide moiety of capsaicin and

Table 2 Effects of pigment quality and quantity on color of the ripe fruit in Capsicum

Pigment quantity	Pigment quality	
	Red pigments present	Red pigments absent
Normal	Red	Orange-yellow
Slightly reduced	Tangerine	Lemon yellow
Trace amounts only	Pink	Cream

Table 3 Structure and pungency of capsaicinoids compared to pungent principles of ginger and black pepper

Name	Structural formula	Threshold pungency (10^5 Scoville units)
N-vanillyl alkyl amides (from Capsicum)		
Capsaicin	CH_3O—, HO—⟨⟩—CH_2–NH–CO–$(CH_2)_4$–CH=CH–$CH(CH_3)_2$	160
Dihydrocapsaicin	CH_3O—, HO—⟨⟩—CH_2–NH–CO–$(CH_2)_6$–$CH(CH_3)_2$	160
Nordihydrocapsaicin	CH_3O—, HO—⟨⟩—CH_2–NH–CO–$(CH_2)_5$–$CH(CH_3)_2$	91
Homodihydrocapsaicin	CH_3O—, HO—⟨⟩—CH_2–NH–CO–$(CH_2)_7$–$CH(CH_3)_2$	86
Homocapsaicin	CH_3O—, HO—⟨⟩—CH_2–NH–CO–$(CH_2)_5$–CH=CH–$CH(CH_3)_2$	86
N-vanillylnonamide	CH_3O—, HO—⟨⟩—CH_2–NH–CO–$(CH_2)_7$–CH_3	92
N-vanillyl alkyl ketones (from ginger)		
Gingerol	CH_3O—, HO—⟨⟩—CH_2–CH_2–CO–CH_2–$CH(OH)$–$(CH_2)_n$–CH_3	0.8
Shogaol	CH_3O—, HO—⟨⟩—CH_2–CH_2–CO–CH=CH–$(CH_2)_n$–CH_3	1.5
Substituted piperidine (from black pepper)		
Piperine	⟨⟩N–CO–CH=CH–CH=CH—⟨⟩$\begin{smallmatrix}O\\O\end{smallmatrix}$$CH_2$	1.0

dihydrocapsaicin, hence do not taste pungent. It is not known whether they are intermediates in the biosynthesis of capsaicin and dihydrocapsaicin.

Continuous light has been reported to induce formation of capsaicinoids during postharvest ripening of nonpungent *C. annuum*, and cells isolated from sweet pepper stem callus will synthesize capsaicin *in vitro* under appropriate conditions. The nonpungent mutant, at least in *C. annuum*, may therefore involve some change in a regulatory or switch mechanism affecting the pathway of capsaicin biosynthesis, rather than a simple block in this pathway.

Although many secondary metabolites of the Solanaceae (such as nicotine) are produced wholly or partly in the roots, this is not true of the capsaicinoids. Experiments in which tomatoes were grafted on to hot pepper rootstocks, and vice versa, showed that capsaicinoids are produced only in the fruits. They usually appear about 2 weeks after flowering and reach a plateau 4 weeks after flowering. Temperature, particularly night temperature, seems to affect both the formation and accumulation of capsaicinoids. Capsaicinoids are synthesized in the epidermal cells of the placenta, in the inner compartment of the endoplasmic reticulum. They are translocated across the cell membrane, through the cell wall, and accumulate in cavities beneath the cuticle covering the placenta. This may cause the cuticle to crack and the capsaicinoids to spread over the seeds and inner wall of the fruit, leading to the widespread, but erroneous, belief that pungent peppers have pungent seeds and pungent fruit walls.

For a long time, pungency of *Capsicum* was assessed organoleptically by the Scoville test, which was simple and convenient but not accurate or reproducible. A specified weight of *Capsicum* was combined with a specified volume of alcohol, diluted with sugar solution to the threshold of taste, and the reciprocal of the dilution was the measure of pungency in Scoville heat units (150 000 Scoville units = 1% capsaicinoids). The major techniques now used for separating and quantifying the various capsaicinoids are gas chromatography, high-performance liquid chromatography, and high-performance thin-layer chromatography. There is still no simple nonorganoleptic test that can be applied in the field. (*See* **Chromatography**: Thin-layer Chromatography; High-performance Liquid Chromatography; Gas Chromatography.)

Pharmacodynamic Properties

Capsaicinoids are responsible for many pharmacodynamic as well as organoleptic properties of *Capsicum*. Hot peppers stimulate saliva flow and may overcome loss of appetite. They also stimulate gastric juices, thereby aggravating stomach ulcers. They increase peristalsis, thus having a laxative action (which has benefited archeologists studying food remains in prehistoric feces). Hot peppers increase perspiration, which may partially explain their popularity in hot climates. Chilli-containing foods reputedly reduce risks of thromboembolism and significantly lower both liver and serum triglycerides. Capsaicinoids induce coughing and sneezing, inflame the skin, and have an irritating effect on mucous membranes of the eyes and nose. This was exploited in the past by some Amerindians, who punished their children by holding them over the smoke of a fire on to which dried chillies had been thrown, and exploited more recently by manufacturers of certain aerosols sold for use against 'muggers.'

The burning and pain associated with hot peppers result from pungent capsaicinoids acting on sensory neurons known as nociceptors. Repeated exposure leads to desensitization. Experimenters who desensitized their tongues to capsaicin found that their taste thresholds for other pungent compounds, such as ginger and mustard, also increased, but their ability to perceive tactile stimuli or basic tastes, such as sweet, salt, sour, or bitter, was not affected. Capsaicin apparently acts via a receptive site in the nociceptor. This site seems also to be involved in the perception of temperatures which are dangerously high (perhaps explaining why pungent foods are perceived as 'hot'). Capsaicin kills the nociceptor, or destroys its peripheral terminals. This has been exploited in the topical use of capsaicin as an analgesic to treat conditions such as shingles and rheumatoid arthritis.

Nutritional Value

Hot peppers are traditionally valued for adding interest to bland, starchy foods and for masking off-flavors in meat and other stored products. Both hot and sweet peppers are also excellent sources of vitamins, particularly vitamins A and C. Carotene (provitamin A) is deficient in many tropical diets. Green peppers have a β-carotene equivalent of 180 μg per 100 g; red peppers contain 4770 μg of carotenoids per 100 g. (*See* **Retinol**: Physiology.)

Fresh peppers are perhaps even more important as a source of vitamin C. The Hungarian chemist Szent-Györgyi won the Nobel Prize for his isolation of vitamin C, first from the adrenal gland, later from paprika and lemon. Fresh paprika fruits contain up to 340 mg per 100 g, i.e. more vitamin C per unit weight than citrus. Much of this is lost when peppers are dried (paprika powder contains only 30–60 mg per 100 g), but about two-thirds is retained in canned *Capsicum* fruit. (*See* **Ascorbic Acid**: Physiology.)

Harvesting, Handling, and Processing

Green bell peppers and fresh green chillies are picked when the fruit has reached its full size but the seeds are not yet fully mature. This is usually about 1 month after flowering, or 70 days after planting. Fruits are picked by hand at 7–14-day intervals over a harvesting period of about 3 months. They are usually picked with calyx and fruit stalk attached, since bacteria and fungi may infect the scar left by removal of the calyx and cause postharvest rots. Harvested fruits can be stored for up to 14 days, preferably under cool humid conditions (7–10 °C and 95% relative humidity). Ventilation may be advisable to remove accumulating ethylene, which would accelerate fruit ripening. Ripe peppers are harvested 2–3 weeks later than green peppers but are handled in the same way. Since the grower has to wait longer before harvest, ripe fruits usually command higher prices than green peppers.

Sweet peppers may also be marketed as dehydrated flakes. Stalks, calyces, placentas, and seeds are removed mechanically and the pericarp is diced, sprayed with sulfite-bisulfite solution and dried by hot air. Pimentos for canning have the tough outer cuticle plus the underlying epidermis removed by roasting or by treatment with lye, then are cored to remove placenta and seeds and canned either whole or diced.

The quality of *Capsicum* powders depends on their color and flavor, which are influenced by how the fruits are harvested and handled. Paprika is always made from red-fruited cultivars. The genotype influences both the quantity of pigment present and how well it is retained after harvest. Traditionally, fruits are harvested when fully ripe and dried in the open air for 3 weeks or more. The red pigments increase for the first 25 days of after-ripening, then remain constant until about 40 days after harvest, when they start to break down. Mechanical harvesting requires that 80–90% of the fruits mature simultaneously. In Hungary, plants are cut by machine at ground level and laid in the field to dry and after-ripen. Artificial drying in hot air is also widely used, but if temperatures exceed 80 °C, fruit color starts to deteriorate. Stalks, calyces, placentas, and seeds are separated from the dried fruits because they would dilute the color, but at least 5% of cleaned seeds are then re-added to facilitate grinding (the precise percentage depends on the intended grade of the final product). The seeds contain fat, and carotenoids are fat-soluble, so adding ground seeds helps to disperse the color evenly, though the fat may become rancid during storage. (*See* Drying: Theory of Air-drying.)

Pungent fruits are handled in much the same way as paprika. Grinding is done mechanically, formerly in the importing countries, but costs are high and grinding of chillies is particularly unpopular because of the cough- and sneeze-provoking effects of the dust, so grinding is increasingly left to the producing countries. The ground powder is fumigated to control microorganisms and stored under dry, cool, dark conditions to minimize breakdown of the red pigments. (*See* Fumigants.)

Oleoresins are extracted from dried and ground fruits. If the seeds are ground with the fruits, fat from the seeds dilutes both color and pungency, and hence quality, of the oleoresin. However, if the seeds are removed, the cost of production is increased and the yield of oleoresin is decreased. Acetone and ethylene dichloride are frequently used solvents. The extract is distilled to remove the solvent, leaving the concentrated oleoresin.

Global Distribution, Products, and Commercial Importance

Capsicum is adapted to tropical or subtropical climates, but can be grown in temperate climates in glasshouses or under protected cultivation. The Food and Agriculture Organization (FAO) production yearbooks do not distinguish between *C. annuum* and other species, or between pungent and nonpungent peppers. Over the last two decades, total world production of chillies and green peppers (presumably including also ripe fruits of sweet pepper and pimento) increased from about 7 million to nearly 17 million tonnes. Over half of this is produced in Asia. China alone produces over 40% of the world total.

Data in the FAO trade yearbooks combine, under the heading pimento, information for capsicum, cayenne, chillies, paprika, and red pepper. The confusion surrounding the name chilli has already been discussed. Capsicum is a similarly confusing term, which may be used for the nonpungent vegetable peppers, but may also be used for some mildly pungent dried fruits, e.g. Ancho, that are larger than most fruits known as chillies. Applied to the oleoresin, capsicum signifies the most pungent, rather than the least pungent, of the commercial grades. Cayenne is usually made from small-fruited peppers, grown in various parts of the world, and is usually very pungent. However, in the USA cayenne is made from locally grown, large-fruited cultivars and is less pungent. Paprika comes mainly from Hungary, Spain, and the USA and is usually, though not always, nonpungent. Ground red pepper and crushed red pepper have similar names, but ground red pepper is usually milder than crushed red pepper. Red pepper may also refer to ripe fruits of some sweet peppers.

The FAO trade yearbook shows that, in 1997, North America and Europe were net importers of over 100 000 t of 'pimento.' Asia and Africa were net exporters. China contributed about 40% of Asia's exports, thereby earning over US$72 million, while India contributed a further 30% and earned approximately US$40 million.

Oleoresins are valued by some commercial food processors for their greater hygiene (less contamination by microorganisms, insects, or rodent droppings) and more precise standardization of pungency. They are used mainly in the USA and the UK, but elsewhere do not seem to be replacing pepper powders. Three different oleoresins are produced commercially: oleoresin paprika, oleoresin red pepper (or oleoresin chilli), and oleoresin capsicum. Oleoresin capsicum is used mainly in the pharmaceutical industry but also in the production of some foods and beverages, such as ginger ale. One kilogram of oleoresin capsicum replaces approximately 20 kg of cayenne. Oleoresin red pepper is made from less pungent fruits, such as the long, thin peppers much grown in Asia. One kilogram of oleoresin red pepper replaces about 10 kg of red pepper. Oleoresin paprika is used for coloring salad dressings and oleomargarines, partly because it retains its color longer than annatto. One kilogram of oleoresin paprika replaces 12–15 kg of paprika powder.

Probably less than 10% of total global production of hot peppers is for export. The remainder is grown for local use. Mexico leads the world in chilli consumption at 15 g per head per day. Sweet peppers are eaten less regularly than hot peppers but are consumed in larger quantities (up to 20 g per meal). At these levels of intake, *Capsicum* adds considerably more than spice to the diet of some of the world's poorest people.

See also: **Ascorbic Acid**: Physiology; **Chlorophyl**; **Chromatography**: Thin-layer Chromatography; High-performance Liquid Chromatography; Gas Chromatography; **Colorants (Colourants)**: Properties and Determination of Natural Pigments; **Drying**: Theory of Air-drying; **Fumigants**; **Retinol**: Physiology; **Ripening of Fruit**; **Sensory Evaluation**: Aroma

Further Reading

Andrews J (1984) *Peppers: The Domesticated Capsicums.* Austin, Texas: University of Texas Press.

Camara B, Hugueney P, Bouvier F, Kuntz M and Monéger R (1995) Biochemistry and molecular biology of chromoplast development. *International Review of Cytology* 163: 175–247.

Caterina MJ, Schumacher MA, Tominaga M, Rosen TA, Levine JD and Julius D (1997) The capsaicin receptor: a heat-activated ion channel in the pain pathway. *Nature* 389: 816–824.

Govindarajan VS (1985) *Capsicum* – production, technology, chemistry and quality. Part I: history, botany, cultivation and primary processing. *CRC Critical Reviews in Food Science and Nutrition* 22: 109–176.

Govindarajan VS (1986) *Capsicum* – production, technology, chemistry and quality. Part II: processed products, standards, world production and trade. *CRC Critical Reviews in Food Science and Nutrition* 23: 207–288.

Govindarajan VS (1986) *Capsicum* – production, technology, chemistry and quality. Part III: chemistry of the color, aroma and pungency stimuli. *CRC Critical Reviews in Food Science and Nutrition* 24: 245–355.

Govindarajan VS and Sathyanarayana MN (1991) *Capsicum* – production, technology, chemistry and quality. Part V: impact on physiology, pharmacology, nutrition and metabolism; structure, pungency, pain and desensitization. *CRC Critical Reviews in Food Science and Nutrition* 29: 435–474.

Govindarajan VS, Rajalakshmi D and Chand N (1987) *Capsicum* – production, technology, chemistry and quality. Part IV. Evaluation of quality. *CRC Critical Reviews in Food Science and Nutrition* 25: 185–282.

Kobata K, Todo T, Yazawa S, Iwai K and Watanabe T (1998) Novel capsaicinoid-like substances, capsiate and dihydrocapsiate, from the fruits of a non-pungent cultivar, CH-19 Sweet, of pepper (*Capsicum annuum* L.). *Journal of Agricultural and Food Chemistry* 46: 1695–1697.

Purseglove JW, Brown EG, Green CL and Robbins SRJ (1981) *Spices*, vol. 1. London: Longman.

Somos A (1984) *The Paprika.* Budapest: Akadémiai Kiadó.

Suzuki T and Iwai K (1984) Constituents of red pepper species: chemistry, biochemistry, pharmacology and food science of the pungent principle of *Capsicum* species. In: Brossi A (ed) *The Alkaloids*, vol. 23, pp. 227–299. Orlando, Florida: Academic Press.

Peptic Ulcer *See* **Colon**: Structure and Function; Diseases and Disorders; Cancer of the Colon

PEPTIDES

D González de Llano, Centro Nacional de Biotecnologia, Cantoblanco, Madrid, Spain
C Polo Sánchez, Instituto de Fermentaciones Industriales (CSIC), Madrid, Spain

Introduction

Peptides consist of chains of amino acids linked to each other by amide bonds, also known as peptide bonds. They form a heterogeneous group of compounds, because of the large number of different naturally occurring amino acids and the high potential variability in the order and number (chain length) of residues making up the peptide.

The number of amino acids that a peptide contains is indicated by a prefix, e.g., di-, tri-, tetrapeptide; the term 'oligopeptides' or 'low-molecular-weight peptides' refers to peptides with 10 or fewer residues, while 'polypeptides' is used for peptides with higher molecular weights. Although the transition point from polypeptide to protein is not well defined, proteins are normally considered to have at least 100 residues (mol wt > 10 000). By convention, peptides are referred to by the first three letters of the amino acids making up the chain or by one-letter abbreviations. Also by convention, the amino acid with the free amino group is referred to as the N-terminus and is represented on the left, while the amino acid with the free carboxyl group is referred to as the C-terminus and is represented on the right. (*See* **Amino Acids**: Properties and Occurrence.)

The N-terminal amino group and the C-terminal carboxyl group on peptides react chemically in the same way as the α-amino and α-carboxyl groups on free amino acids, and they can be used for detection and quantification purposes.

The acid–base behavior of peptides is dependent upon the free α-amino group on the N-terminal residue, the free α-carboxyl group on the C-terminal residue, and the ionizable R groups located at intermediate positions.

In the presence of carbonyl compounds, certain peptides take part in the Maillard reaction, leading to the formation of melanoid pigments that contribute to the development of undesirable coloration in foodstuffs. Conversely, there are also lysine-containing peptides that retard browning reactions with glucose, and these may be appropriate for fortifying the lysine content of sugar-containing foods to be cooked. (*See* **Browning**: Nonenzymatic.)

Peptides also play a very important role in determining the rheological properties of foods. Hydrolysis of proteins to peptides during processing (fermentation, ripening, and cooking) of foods may modify food texture.

Peptides in Foods

Peptides are naturally present in foodstuffs. Most simple peptides are the result of partial hydrolysis of protein polypeptide chains. On the other hand, nonprotein peptides have also been recorded in foods. Such peptides usually differ in structure from the peptides derived from proteins, and these structural variations may protect them from the action of peptidases.

The tripeptide glutathione, present in the cells of all higher animals, is an example of a nonprotein peptide. It contains a glutamic acid residue linked by an unusual peptide bond involving its γ-carboxyl group. Glutathione plays a role in the active transport of amino acids, acts as an antioxidant for lipids and as an activator for certain enzymes, and is a coenzyme of glyoxalase. Other nonprotein peptides include the dipeptides carnosine, anserine, and balenine, which are present in vertebrate muscle and contain a β-amino acid (β-alanine) bound to L-histidine or 1-methyl-L-histidine. These histidine dipeptides are present in differing proportions in the muscles of different species and therefore are a mean of determining the source of the meats used in meat or of identifying marine fish species.

Peptides may also be present in foods because they have been used as additives (e.g., sweeteners, flavor enhancers, or bulking agents for light beverages) to improve food quality. Some peptides produced by the enzymatic hydrolysis of proteins possess better functional properties than the parent proteins and consequently are used by the food industry for a variety of purposes.

Peptides have lower molecular weight and less secondary structures, as well as higher number of ionizable groups and exposure of hydrophobic groups than native proteins. These facts imply that solubility, surface activity, foaming, and emulsifying properties may be different from that of the intact protein.

Enzymatic hydrolysis has been reported to enhance the emulsifying capacity of different proteins. Peptides with an inhibitory effect on autoxidation of linolenic acid and hence potentially useful as antioxidants have been found in fermented foods and in protein hydrolysates. Moderately large peptides have pronounced bulking properties and large peptides act as foam stabilizers; because of these interesting functional properties, enzymatic hydrolysates of soya bean proteins have been used in beverages. Hydrolyzed hydrophobic proteins from gluten (zeins, gliadins) have been prepared with a view to increase solubility and diversify their functional properties.

Analysis of the peptides of enzymatic hydrolysates of trypsin can be used to confirm the presence of nonmeat proteins such as soyabean proteins in meat products. Tryptic peptides of the caseins from milks of different species are distinct and may be used to detect blends of milks. The presence of the casein glycomacropeptide is also useful in detecting adulteration of powdered milks with rennet whey.

Organoleptic Properties

Peptides are tasteless or bitter, with the exception of certain dipeptides of glutamic acid and aspartic acid, which are sweet. Others, like the lower alkane members in the series of methyl esters of L-aspartyl-α-amino-cycloalkane carboxylic acid and a series of L-aspartyl-D-alanyl tripeptides, are sweet but turn bitter or tasteless when the size of the C-terminal amino acid ring increases. There is, thus, an important relationship between size and flavor in these peptides. Peptides belonging to L-aspartyl D-alanyl amides are strong sweeteners and may be used as good sugar substitutes. (See **Sensory Evaluation**: Practical Considerations.)

Foods may also contain peptides with a flavor similar to that of monosodium glutamate, termed umami taste or relish, which are able to mask the bitter taste produced in foods by bitter peptides or other bitter substances. This group includes hydrophilic peptides, in particular acid peptides that balance flavor in foods.

Warmed-over flavor in cooked–stored–recooked meat has also been related to the higher hydrophobic peptide content in such meats as compared to cooked meats. However, low-molecular-weight peptides play an important role in the flavor formation and intensity of meat, beef broth, and dry cured ham.

Bitter peptides form as a result of proteolytic reactions that take place in foods. These bitter peptides have been identified in soyabean and casein hydrolysates as well as in foods, like cheese, that undergo fermentation of ripening processes with marked proteolytic activity. The bitter taste has been related to the hydrophobic amino acid content and to chain length. An overly high proportion of hydrophobic amino acids gives rise to bitterness; however, above a given molecular weight bitter flavors are no longer perceptible, even though hydrophobic amino acids

are present. Proteins do not have a bitter taste even when they contain hydrophobic amino acids, yet hydrolysis of such proteins may indeed yield bitter peptides. Bitter flavors in di- and tripeptides have been observed to increase when the hydrophobic amino acid is located on the C-terminal residue and the basic amino acid is on the N-terminal residue. Enzymatic activity is extremely important in the development of bitter flavors in foods. Trypsin does not degrade bitter peptides, although peptidases may break down bitter peptides to nonbitter lower-molecular-weight peptides and amino acids.

Bitterness is the most frequent flavor defect limiting cheese acceptability. It has been attributed to the formation and accumulation of bitter peptides derived from casein. During cheese manufacture rennet or rennet substitutes and microbial proteases are the principal agents hydrolyzing the caseins and are thus thought to be responsible for the occasional appearance of bitter taste. The use of rennet substitutes may give rise to the appearance of bitter peptides; for instance, commercial chicken pepsin produces an objectionable level of bitterness in cheese. On the other hand, cheeses made from goats' or ewes' milk are less susceptible to bitter flavor defects than cheeses made from cows' milk.

Aged cheeses, in particular hard cheeses, require long ripening periods in which to acquire the desired organoleptic characteristics. Proteases have therefore been used to accelerate the ripening process and lower costs. However, the proteases give rise to bitter peptides that limit cheese quality. In recent years a considerable amount of work has been carried out on peptidases that break down the bitter peptides derived from the caseins to smaller, nonbitter peptides. The combined use of proteases and peptidases may be an effective means of shortening ripening times, while preventing the development of bitter flavors. (See **Cheeses**: Chemistry and Microbiology of Maturation.)

Bitter peptides that contribute to bread flavor form during the fermentation and baking of bread dough as a result of the action proteolytic enzymes on the protein fraction in the flour (gliadins and glutelins).

Analytical Procedures

Spectrophotometric Methods

The N-terminal α-amino groups on peptides react quantitatively with reagents such as ninhydrin to form colored derivatives or with o-phthalaldehyde and fluorescamine to form fluorescent derivatives. These reactions are useful in making quantitative determinations of peptides in foodstuffs and for detection during chromatographic analysis. Due to the peptide bond, peptides also take part in reactions which free amino acids do not undergo, like the classic biuret reaction. This reaction consists of the formation, in an alkaline medium, of a colored complex with a transition metal (e.g., copper or nickel) that can be measured quantitatively by spectrophotometry. This reaction is utilized in quantifying peptides. (*See* Spectroscopy: Visible Spectroscopy and Colorimetry.)

The peptide content can also he determined by assaying the aromatic amino acids tyrosine and tryptophan, which is readily achieved by measuring absorbance at 280 nm or by measuring the color reactions of these amino acids with certain reagents like Folin-Ciocalteu's reagent. Another analytical method for determining the high-molecular-weight peptide content is by precipitating the peptides with acid dyes (amino black, orange G) and measuring the excess dye remaining in solution. These methods are useful in determining the total peptide contents of foodstuffs, but their drawback is that they provide no information on how many or which peptides are present. For this, other techniques such as chromatography or electrophoresis are needed.

Chromatographic Methods

Paper chromatography and thin-layer chromatography have conventionally been used to separate peptides, particularly for the fractions isolated by gel permeation chromatography (GPC) and ion exchange chromatography (IEC), but they are lengthy, time-consuming procedures with low resolving power and low reproducibility.

Gel filtration chromatography on open Sephadex columns with different pore sizes (G-10, G-25, and G-50) using water or saline or acid buffers as solvents is used extensively in the fractionation of peptides. Detection is performed by measuring absorbance of the column eluate with an ultraviolet detector at 280 nm, the absorbance wavelength for tyrosine and tryptophan, or at 214 nm, the absorbance wavelength for most amino acids and the peptide bond. Peptides elute in fractions according to their molecular weight. Small peptides must normally be separated from the amino acids; gel filtration on columns prepared by hydroxypropylation (Sephadex LH-20), ion exchange on diethylaminoethyl (DEAE)-cellulose or DEAE-Sephadex with Cu^{2+} complexes, or ligand exchange columns with a stationary phase modified with Cu^{2+} are used for this purpose. Ion exchange chromatography on DEAE-cellulose or on Aminex A-5 is normally used for the separation of high-molecular-weight peptides, employing saline buffers as eluent. Detection is carried out at 280 nm or in the visible region after derivatization with ninhydrin.

These methods are now most often used as preparatory techniques for subsequent high-performance liquid chromatographic (HPLC) or electrophoretic analysis of the peptide fractions. (*See* Chromatography: Principles.)

High-performance Liquid Chromatography

Because of its versatility, short analysis times, high resolution, and effective separations, and because it is well suited to automation procedures, HPLC is the most widely used method of peptide analysis. Practically all known mechanisms have been employed in the chromatographic separation of peptides, e.g., separation based on molecule size (GPC), on charge (IEC), on hydrophobicity (reversed-phase and interaction chromatography), and even on combinations thereof; of these, however, reversed-phase chromatography is most commonly used to separate mixtures of peptides from foodstuffs (**Figure 1**). It affords the possibility of changing the stationary phase, the pH, the ionic strength of the aqueous buffer, or the type of organic modifier, of using different gradient shapes, and of working at a variety of temperatures, making this method suitable for analyzing highly different peptides. Moreover, the option of using microcolumns makes it possible to detect quantities on the order of picomols and has been an important step toward HPLC–mass spectrometry coupling. (*See* Chromatography: High-performance Liquid Chromatography; Combined Chromatography and Mass Spectrometry.)

Reversed-phase chromatography Reversed-phase columns with pore sizes of 6–10 nm are highly appropriate for separating small peptides. Columns with pore sizes of 30–50 nm give the best results when separating large peptides (mol wt 4000), because the molecules are able to pass freely through the pores, thus permitting greater access to the alkyl chains and increasing column efficiency and load capacity.

The mobile phase is commonly a mixture of water and an organic solvent, normally acetonitrile, methanol, or 2-propanol, although other organic solvents such as methoxyethanol, ethanol, butanol, or tetrahydrofuran are also used. Gradient elution is normally required for reversed-phase HPLC of peptides.

Because of their high polarity, peptides do not interact sufficiently with the hydrophobic chains of the stationary phase in reversed-phase chromatography. On the other hand, they do interact with any free silanol groups in these phases. Salts or strong acids must be added to the stationary phase to block the silanol groups. In order to reduce the polar nature of the peptides, chromatography should be carried

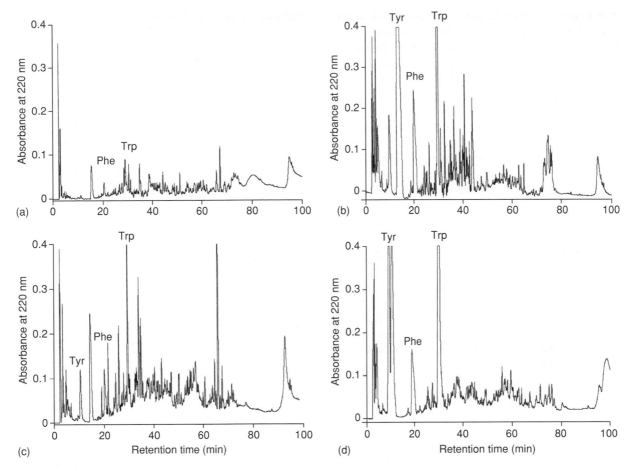

Figure 1 Chromatogram of the peptides of the water-soluble fraction from commercial (a) Vidiago, (b) Cabrales, (c) Beyos, and (d) Peral cheeses. Column: Ultraspher ODS. Solvent: TFA/acetonitrile/water. Gradient elution. Detection at 220 nm. From González de Llano D, Polo C and Ramos M (1995) Study of proteolysis in artisanal cheeses: high performance liquid chromatography of peptides. *Journal of Dairy Science* 78(5): 1018–1024 with permission.

out at a pH of less than 3, so that the carboxyl groups on the acids amino acids (aspartic acid and glutamic acid) are not in their dissociated form. However, at this pH the basic amino acids histidine, lysine, and arginine are charged and must be blocked by forming ion pairs. The selectivity of the chromatographic system can be modified depending on the pH and on the type of ion-pairing reagent used. Trifluoroacetic acid has been widely employed as the ion-pair reagent for peptides.

Ion exchange chromatography IEC has also been used in peptide separations. Two types of ion exchangers are utilized: cationic exchangers to separate neutral and basic peptides, and anionic exchangers to separate neutral and acid peptides. Peptide separations have been carried out on polymeric cationic exchange resins (polystyrene/divinylbenzene copolymers), and indeed automatic peptide analysers, similar to those used in amino acid analysis, have been designed, with postcolumn detection using ninhydrin.

Size exclusion chromatography (SEC) High-performance SEC has most commonly been applied to protein separations, though on occasion it has also been used to separate peptides from foodstuffs. The stationary phase may be inorganic, such as silica or alumina. The disadvantage of inorganic materials is that their active surfaces adsorb charged molecules and hence they must first be deactivated, for instance using organosilanes. Organic materials such as glycerol methacrylate, sulfonated styrene divinylbenzene, and polyesters with surface hydroxyl groups have also been used.

The different separation mechanisms referred to above (reversed-phase HPLC with small or large pore sizes, IEC, and SEC) are complementary, and the best separations are sometimes achieved by combining the various methods.

Detection Peptides are detected in amounts ranging from 100 to 1000 ng by absorbance at between 200 and 230 nm. Detection of peptides containing aromatic amino acids (phenylalanine, tyrosine, and trypto-phan) may be performed at 254 nm; tyrosine- or tryp-tophan-containing peptides can also be detected at 280 nm. Rapid-scanning ultraviolet/visible detectors based on diode array technology have proved an extremely useful aid in identifying peptides that con-tain aromatic residues. Wavelengths of the spectrum maxima, convexity interval, and wavelength of the second derivative spectrum maxima, all over in the 190–340 nm range, allow identification of the aro-matic amino acids that form the peptides, as well as identification of HPLC coeluted compounds such as the cinnamic derivatives detected in the peptide frac-tions of wines.

Since numerous solvents and even other sample components may absorb light in the range 200–230 nm, formation of derivatives detectable at higher wavelengths that are more specific for given chromo-phores is commonly employed. To this end, phenyl isothiocyanate and dansyl chloride are frequently used in peptide analysis as derivatizing agents. Peptides may also be detected using a fluorescence detector to detect either the natural fluorescence of certain amino acids in the peptide sequence or the fluorescence of derivatives formed artificially using such reagents as fluorescamine and o-phthalaldehyde. Other reagents used lately for derivatization of peptides are naphtha-lene-2,3-dicarboxaldehyde, 3-(4-carboxy-benzoyl) 2-quinolinecarboxaldehyde, and 6-aminoquinoyl-N-hydroxysuccimidylcarbamate.

Following the significant advances made in liquid chromatography/mass spectroscopy coupling in recent years, mass spectrometry has turned into a detection system that holds great promise for peptide analysis, since it is capable of furnishing structural information and quantitative data that are otherwise difficult to obtain.

The greatest advance in the analysis of peptides has been the coupling with spectrometric techniques, both for identification and for characterization. Their high accuracy, sensitivity, and, in some in-stances, tolerance to solvents make mass spectrom-eters ideal detectors for analyses of HPLC-separated peptides. Presently, electrospray ionization and matrix-assisted laser desorption/ionization time-of-flight (MALDI-TOF) mass spectrometry – soft ion-ization methods – provide the suitable means to determine accurate mass on peptides and proteins with high sensitivity (subpicomol range). Further-more, both methods are suitable for the creation of peptide ions for analysis using tandem mass spec-trometry and collision-induced dissociation methods. Mass spectrometry in conjunction with database searching plays an increasingly important role in the characterization of food peptides.

Electrophoretic Techniques

Despite the important advantages afforded by HPLC in the analysis of hydrophobic peptides, electrophor-esis can be useful, because insoluble peptides can be solubilized with detergents. Although a number of different supports have been employed, polyacryla-mide gel electrophoresis (PAGE) and sodium dodecyl sulfate–PAGE (SDS-PAGE) have been widely used in the study of peptide formation from proteins during food processing. Another developed method of pep-tide analysis that may be used either an alternative or a complement to HPLC is capillary electrophoresis. This technique has extremely high resolving power and allows the separation of peptides differing in only a single amino acid; it is simple, fast, and appropriate for quantitative analysis, and only small amounts of sample are required. Peptide separation by CE is now almost a routine technique in food analysis. Moreover the application of two analytical tech-niques (HPLC-CE, SEC-CE) and coupling mass spec-trometry to HPLC have acquired great importance for characterization of complex peptide samples. (See Electrophoresis: General Principles.)

Determination of Amino Acid Composition and Sequence in Peptides

The amino acid composition of peptides is generally assayed by carrying out acid hydrolysis with 6 mol l^{-1} hydrochloric acid, followed by determination of the individual amino acids by HPLC. Various types of HPLC have been employed to separate the amino acids, but reversed-phase chromatography on C_{18} columns is most commonly used. Detection of the amino acids normally involves derivatization, since maximum absorbance of amino acids is located in a region of the spectrum (214 nm) in which many other compounds also absorb. The most frequently used derivatizing agents are dansyl chloride, phenyl iso-thiocyanate and o-phthalaldehyde.

Peptide sequencing is normally performed by deg-radation of the N-terminal amino acid using phenyl isothiocyanate, Edman's reagent. The terminal amino groups react with the isothiocyanate, forming a phe-nylthiocarbamylic derivative. When treated with acid in an organic solvent, cyclization takes place and phenylthiohydantoin amino acid is formed; this can be separated from the rest of the chain, which remains intact. The process can be repeated.

The usefulness of Edman degradation in the sequencing of low-molecular-weight peptides is limited, because of the low repetitive yield in the successive sequencing cycles. For such peptides the best results are obtained by coupling sequential Edman degradation with detection by means of dansylation or by the double coupling method using 4-*N*,*N*-dimethylaminoazobenzene-4'-isothiocyanate and phenyl isothiocyanate.

Mass spectrometry has also become a powerful tool for the sequencing of small amounts of peptides and proteins. Tandem mass spectrometry and ion source collision-induced dissociation produce information specific to the amino acid sequence and the specific covalently modified amino acid. A new method, referred to as protein ladder sequencing, for the *N*-terminal sequencing of peptides utilizes multiple steps of partial Edman degradation chemistry prior to the analysis of the reaction mixture by MALDI-TOF mass spectrometry.

See also: **Amino Acids**: Properties and Occurrence; Determination; **Chromatography**: Principles; Thin-layer Chromatography; High-performance Liquid Chromatography

Further Reading

Gonzalez de Llano D, Herraiz T and Polo MC (1996) Peptides. In: Nollet L (ed.) *Handbook of Food Analysis*, vol. 1, pp. 229–276. New York: Marcel Dekker.

González de Llano D, Polo C and Ramos M (1995) Study of proteolysis in artisanal cheeses: high performance liquid chromatography of peptides. *Journal of Dairy Science* 78(5): 1018–1024.

Matsudaira PT (1989) *A Practical Guide to Protein and Peptide Purification for Microsequencing*. San Diego: Academic Press.

Polo MC, González de Llano D and Ramos M (1992) Derivatization and liquid chromatographic separation of peptides. In: Nollet L (ed.) *Food Analysis by HPLC*, pp. 117–140. New York: Marcel Dekker.

Polo MC, González de Llano D and Ramos M (2000) HPLC of peptides. In: Nollet L (ed.) *Food Analysis by HPLC*, pp. 99–125. New York: Marcel Dekker.

Persimmon *See* **Fruits of Tropical Climates**: Commercial and Dietary Importance; Fruits of the Sapindaceae; Fruits of the Sapotaceae; Lesser-known Fruits of Africa; Fruits of Central and South America; Lesser-known Fruits of Asia

---

PESTICIDES AND HERBICIDES

Contents
Types of Pesticide
Types, Uses, and Determination of Herbicides
Residue Determination
Toxicology

Types of Pesticide

R Pleština, World Health Organization, Geneva, Switzerland

Introduction

Humans have always had to cope with disease, discomfort, and great economic loss caused by pests. Some methods for pest control date back to ancient times. What may be called the 'pesticide revolution' dates back to the early 1940s when dichlorodiphenyltrichloroethane (DDT) was first used as an insecticide. Since that time, in a variety of pest-control activities around the world, both in agriculture and in public health, chemicals continue to play a significant role, and this trend will be sustained for many years to come. Although pesticides are poisons, the majority of people recognize the advantages of their use in view of the benefits they bring to the control of disease vectors and nuisance pests, and particularly to the increase in food and fiber production. Research on alternatives has produced few materials to take the place of these chemicals. Although the use of pesticides has more or less stabilized, or is even declining in the developed world, it continues to increase in the developing countries owing to development and buildup of pest resistance, because of an increase in the need for food, feed, and fibers, in addition to a limited and constantly decreasing per capita availability of arable land. (*See* **Insect Pests**: Insects and Related Pests; Problems Caused by Insects and Mites.)

There is little doubt that the principles promoted by the concept of integrated pest management will reduce excessive use of pesticides. However, it should be stressed that integrated pest management does not exclude pesticide use, and pesticide treatment should be considered as a supplement to basic sanitation; like drugs, pesticides should always be used with discretion and in conjunction with other measures in order to achieve effective pest control.

Definition and Classifications

The term 'pesticide' is defined in many different ways, each having some specific feature. Most of them, however, define pesticide as a pest-control agent aimed at killing or repelling a pest. The most widely accepted definition is that given in the *International Code of Conduct on the Distribution and Use of Pesticides* (FAO, 1990) which reads as follows:

> Pesticide means any substance or mixture of substances intended for preventing, destroying or controlling any pest, including vectors of human or animal disease, unwanted species of plants or animals causing harm during or otherwise interfering with the production, processing, storage, transport, or marketing of food, agricultural commodities, wood and wood products or animal feedstuffs, or which may be administered to animals for the control of insects, arachnids or other pests in or on their bodies. The term includes substances intended for use as a plant growth regulator, defoliant, desiccant, or agent for thinning fruit or preventing the premature fall of fruit and substances applied to crops either before or after harvest to protect the commodity from deterioration during storage and transport.

Similarly, classifications of pesticides are numerous, depending on the criteria used and their use. Thus, based on the pest they control, pesticides may be grouped as insecticides, fungicides, herbicides, rodenticides, repellents, etc., and based on the chemical structure of the active ingredient, they may be grouped as organophosphates, carbamates, organochlorines, etc. There is also a clear distinction between agricultural and public health pesticides or household pesticides. By formulation, the products may be grouped as liquids, solids, or gases. In addition, there are persistent and degradable pesticides. One classification does not exclude another; all of them are used to make a classification according to the hazard that pesticides pose to humans, and this arises from their intrinsic toxic property. Thus, the World Health Organization elaborated *The WHO Recommended Classification of Pesticides by Hazard* and *Guidelines to Classification 1992–1993* (**Table 1**). (*See* **Fungicides**.)

This document was approved by the 28th World Health Assembly in 1975 and has since gained wide recognition in a number of Member States and by pesticide registration authorities. Although the classification takes into account acute oral or dermal toxicity, whichever is higher, it also considers any irreversible effect that might be recognized. For practical reasons, a number of pesticides classified as Class III (slightly hazardous) are listed in a separate table as 'Unlikely to present acute hazard in normal use.' *Guidelines to Classification* is prone to frequent revision, based on documented scientific evidence.

Needs and Present-day Use

The costs and benefits of pesticides are currently being evaluated, on the one hand, for their ability to reduce the cost of food and feeds, and to decrease the spreading of vectorborne diseases, and, on the other hand, for the potential impact that pesticides may have on human health and the environment. All of these aspects should not, however, be considered

Table 1 WHO recommended classification of pesticides by hazard

| Class | LD50 for the rat (mg per kilogram of body weight) | | | |
| | Oral | | Dermal | |
	Solids[a]	Liquids[a]	Solids[a]	Liquids[a]
Ia (extremely hazardous)	5 or less	20 or less	10 or less	40 or less
Ib (highly hazardous)	5–50	20–200	10–100	40–400
II (moderately hazardous)	50–500	200–2000	100–1000	400–4000
III (slightly hazardous)	Over 500	Over 2000	Over 1000	Over 4000

[a]The terms 'solids' and 'liquids' refer to the physical state of the product or formulation being classified.
From WHO (1992).

separately from the costs which must be paid to ensure the correct use of pesticides.

It has been estimated that if pesticides were not used, the potential loss of food production would be 45% (30% attributed to pests, including weeds and diseases before harvest, and 15% as a postharvest loss). In addition, in over 100 countries, more than 4×10^8 clinical cases of vectorborne diseases currently exist. For some of these diseases, control of vectors remains the main control measure, and for others, pesticides are considered to be an essential part of disease control. The use of pesticides in the control of household pests should not be neglected, regardless of whether they are used in the control of vectors of diseases or to control nuisance pests. This aspect of pesticide use is of great importance, as pesticides intended for this kind of application are available to the general population.

Although the exact figures for the global use of pesticides, either according to intended use (agriculture, public health, household) or according to regional or country use, are not available, it is certain that the quantities of pesticides used are steadily growing, albeit less in the developed countries than in developing countries, and this trend may continue for some years. A few highly developed countries show a decline in the quantities used as more efficient compounds are applied, and the trend to organic farming increases. (*See* **Organically Farmed Food.**)

In western Europe and North America, herbicides are the most represented groups of pesticides, whereas in the tropics, insecticides are used more than other groups of pesticides. This fact gives rise to ever-growing concern, mainly attributable to the indiscriminate use and frequently to the misuse of pesticides. This concern is supported by the fact that, although only about 20% of pesticides are used in developing countries, they account for the majority of total deaths caused by pesticide poisonings.

Evaluation of the Risk and Testing Procedure

In general pesticides, in common with drugs, are among the most carefully studied chemicals, and this makes the toxicological evaluation of active material possible and meaningful.

According to the widely accepted *International Code of Conduct on the Distribution and Use of Pesticides*, pesticide manufacturers are expected to ensure that each pesticide is adequately and effectively tested in accordance with sound scientific procedures and good laboratory practice. Generated toxicological data and use patterns are then meticulously evaluated by competent international or national institutions, and results made available to the relevant national authorities in order to allow them to introduce the necessary legislation for their regulation, including registration. (*See* **Legislation**: Contaminants and Adulterants.)

In many countries, registration procedures are elaborated upon in great detail, offering possibilities, for preparing precise precautionary measures for their production, transport, storage, and use. Unfortunately, in many areas, the implementation of carefully prescribed safety measures is lacking, and pesticides are greatly misused.

The activities of the WHO in the safety assessment of pesticides were initiated soon after its foundation, and a number of expert committees have been devoted to the toxic hazards arising from pesticides used in public health programs. Subsequently, the experts also became concerned with the safety aspects of pesticides used in agriculture. The evaluation of pesticide residues in food started in 1963 in collaboration with the Food and Agriculture Organization (FAO). Since then, the Joint Meeting of Pesticide Residues in Food (JMPR) has evaluated a large number of pesticides. The main objectives of the WHO Expert Group on Pesticide Residues (the WHO component of JMPR) are consistent with those of the International Programme on Chemical Safety and include the formulation of guiding principles for exposure limits such as acceptable daily intake (ADI) for pesticide residues in food. Tolerances for these substances in air, water, soil, and the working environment are recommended by other WHO expert groups. (*See* **World Health Organization.**)

Although the requirements for toxicological data needed for health-risk assessment vary considerably from country to country, they comprise short- and long-term toxicity studies in several species of experimental animals, data on absorption, distribution, metabolism and excretion of a pesticide, carcinogenicity, reproduction, and genotoxicity, and other special studies for particular classes of compounds.

Evaluation of data is an extremely complicated process, and results are not always equivocal. This is particularly so because extrapolation of animal data to humans is mainly based on the introduction of a safety factor and, when possible, on pharmacokinetic extrapolation. Whenever applicable, observations on humans following occupational and/or accidental exposure are incorporated within the health-risk evaluation process.

Biological Activity of Pesticides

The inherent toxicity of a pesticide to humans is the same regardless of its use, either in agriculture or

	Type	Mode of action		Hazard
DDT	Insecticide	Disturbance of Na^+ and K^+ transport in nervous membranes	113	Moderately hazardous
Pyrethroids Permethrin	Insecticide	Disturbance of Na^+ and K^+ transport in nervous membranes	500	Moderately hazardous
Deltamethrin	Insecticide	Disturbance of Na^+ and K^+ transport in nervous membranes	135	Moderately hazardous
Substituted phenols Dinoseb	Herbicide	Electron transport inhibition	58	Highly hazardous
DNOC	Insecticide	Electron transport inhibition	25	Highly hazardous
Bipyridinium compounds Paraquat	Herbicide	Excessive production of superoxide	150	Moderately hazardous

Continued

Class	Compound	Structure	Use	Mechanism	LD₅₀	Hazard
Triazines	Atrazine	structure (see image)	Herbicide	Nonspecific (irritant)	2000	Slightly hazardous
Chlorophenoxy	2, 4-D	$OCH_2CO.OH$; Cl substituents	Herbicide	Nonspecific	375	Moderately hazardous
Dithio-carbamates	Maneb	$[-CH_2-NH-C(=S)-S-CH_2-NH-C(=S)-S-Mn-]_x$	Fungicide	Nonspecific (irritant)	6750	Slightly hazardous
Coumarins	Brodifacoum	structure (see image)	Rodenticide	Reduction of clotting power of the blood	0.3	Extremely hazardous

Pyrethrins and Pyrethroids

Pyrethrum is one of the oldest natural insecticides in use in the world today, and it has one of the best safety records of all insecticides. It is a mixture of several esters, called pyrethrins, which are extracted from flowers belonging to the genus *Chrysanthemum*. Natural pyrethrins are unstable to light and are therefore unsuitable for residual application, particularly in agriculture. This led to the development of several classes of related synthetic compounds which have a higher stability to light and a high insecticidal activity. They are known as pyrethroids. In general, pyrethroids can be toxicologically divided into two classes on the basis of signs of toxicity: those causing mainly tremor and prostration (T syndrome) and those causing choreoathetosis and salivation (CS syndrome).

For both groups, the main biological activity is mediated through the effect on sodium channels along the axon membrane of the nerves, in both insects and mammals. The effect on the sodium-exchange disturbance is fully reversible, the duration of which is different for the two classes of pyrethroids: compounds belonging to the group causing the T syndrome produce a considerably shorter effect on the sodium 'gate' openings than those belonging to the CS group.

Being highly lipophilic, pyrethroids readily pass through cell membranes and are absorbed into the body by all routes following exposure. However, the ratio of the toxic dose by the oral route to that of intravenous injection is very high; this is because of rapid detoxification, mainly by cleaving of the ester bond by esterases.

Present-day evidence indicates that doses of a pyrethroid likely to be encountered both in the workplace and through food consumption would not lead to any serious untoward effects, although reversible transient effects on the skin may be recorded and may be used as a warning response of inadvertent exposure.

Nitro- and Chlorophenols

Among dinitro compounds, the most toxic is dinitroorthocresol (DNOC), applied mostly as a winter wash for fruit trees. This and similar compounds are readily absorbed not only through the lungs and gastrointestinal tract but also through intact skin. Since DNOC is eliminated more quickly from laboratory animals than from humans, animal experiments are of limited value for the assessment of hazard. These compounds are strong metabolic stimulators, exerting a common biochemical action which affects energy at the cellular level, stimulating and metabolic

processes independently of their physiological stimulator, which is the thyroid gland.

The major effects observed with excessive overexposure to these chemicals indicate a general effect on the nervous system. As the liver and kidney accumulate high concentrations of nitro- and chlorophenols, they are often also adversely affected. The major mode of action in the acute toxic effects involves uncoupling of the energy conservation process.

In animals and humans, increased basic metabolism leads to hyperpyrexia, tachycardia, hyperventilation, dehydration, and, ultimately, depletion of carbohydrate and fat stores. In the case of a single large exposure, symptoms develop rapidly, and if death occurs, it will probably take place within 24–48 h with characteristically fast rigor mortis. A high environmental temperature might aggravate the hazards from exposure. In nonfatal cases, recovery is complete, but the skin remains stained yellow for a long period of time.

Bipyridinium Compounds

The best-known representative of this class of compounds is paraquat, a widely used herbicide of moderate acute toxicity. However, paraquat deserves special attention as it produces striking lung injury in many species regardless of its route of entrance, most probably because it accumulates in lung tissue in both experimental animals and humans.

Both its herbicidal and toxicological properties are caused by the ability of the parent cation to undergo a single electron addition, forming a free radical which reacts with molecular oxygen, producing a superoxide anion. The oxygen radical damages the cell, presumably through lipid peroxidation.

So far, a large number of paraquat poisonings have been reported, the majority being accidental or suicidal and very few resulting from occupational exposure. Acute fulminant poisoning may lead to death within a few days, and more protracted cases may last for several weeks, even months, resulting in fatal, irreversible pulmonary fibrosis.

The response to treatment of paraquat poisoning, even if started soon after the accident, is not very promising, and the mortality rate remains high.

Anticoagulants

Pesticides with hemorrhagic effects are widely used as rodenticides, exerting their action by disturbing the blood clotting mechanism through inhibition of synthesis of vitamin-K-dependent factors and by decreasing prothrombin production. Some have also been used as drugs in human medicine.

As the desired harmful effect in rodents is achieved only following repeated ingestion of small amounts of these compounds, this reduces the hazard to man, and safety records are remarkable. The existence of a very effective antidote (vitamin K_1), needed in case of poisoning, makes these compounds even safer. Owing to the development of rodent resistance to many existing anticoagulants, there is a present-day trend to develop new rodenticides of a similar or entirely different mode of action.

Fumigants

Fumigants have little in common either in their chemical structure or in their mode of action. The only common characteristic is that all have a relatively high vapor pressure, and, as a rule, they are highly toxic to both pests and humans. Handling requires great skill, and strict safety rules must be followed, including the wearing of specific personal protective clothing (gas masks with specific cartridges). Many countries have therefore posed restrictions on the availability of these fumigants. To this group belong the frequently used hydrogen cyanide, phosphine, methyl bromide, chloropicrin, and some others. (*See* **Fumigants.**)

Miscellaneous Pesticides

Diversity of pesticide structure and biological activity does not allow any further meaningful grouping according to their mode of action. Therefore, several more common classes of pesticides, which do not belong to any of the above-mentioned groups, are outlined below.

Inorganic and organometal pesticides, in particular those containing arsenic, mercury, thallium, and fluorine, were extensively used in the past. Those insecticides containing some types of arsenical active material are considered today to be the only insecticides showing sufficient evidence for carcinogenicity in humans. Therefore, their use today is restricted worldwide or banned. However, numerous pesticides with active ingredients containing zinc, copper, tin or sulfur are still widely used and are considered to be irreplaceable in the control of many plant pests and diseases. As a rule, the toxicity of most of these is rather low, and the health hazard they pose is frequently outweighed by the benefit they bring. (*See* **Heavy Metal Toxicology.**)

Those *pesticides derived from plants* which are still in use are entirely unrelated chemically and toxicologically. They range in toxicity from practically harmless (such as pyrethrins) to highly hazardous (such as nicotine). Several groups of chemicals extracted from some tropical plants (Endod, Neem) are known to be biologically active against various pests, as learned by experience. However, the evidence of safety for humans and beneficial species is still lacking.

Insect growth regulators disturb the metamorphosis of insects by interfering with juvenile hormones needed for proper functioning of this process. Their action is slow and toxicity for humans negligible as the mode of action is specific to insects. Some are even recommended for use in drinking water for control of mosquito breeding sites.

Biopesticides (e.g., *Bacillus thuringiensis* H14) are very useful tools in pests control, both in agriculture and in public health, and have been in operational use for many years. However, research into the genetic manipulation of the bacteria has been intensified, particularly regarding the incorporation of the gene responsible for toxin production into other organisms. The safety aspects of these mutants should be studied carefully.

Health Hazard Arising from Pesticides

A terminological distinction should be made between *toxicity*, which is an innate capacity of a chemical to cause damage, and *hazard*, which is a qualitative term expressing the potential that a pesticide can harm health. Thus, the most toxic pesticides may be handled with little hazard if sensible and disciplined adherence to good practice become routine, and vice versa. Therefore, toxicity and hazard are not synonyms because hazard is a function of two other variables besides toxicity: the extent of contamination and duration of exposure. This can be expressed in the form of an equation:

$$\text{Hazard} = K_1 \text{ toxicity} \times K_2 \text{ contamination} \times K_3 \text{ time}$$

(K_1, K_2, and K_3 are selected measures for a given parameter, e.g., reciprocal value of LD50 (median lethal dose), amount of active ingredient and duration of work.) If any of the three variables on the right-hand side of the equation equals zero, the hazard is also zero.

The toxicity variable can never be zero (by the definition of a pesticide), but it can be acceptably low by selecting a pesticide of favorable mammalian toxicity versus insect toxicity. The toxicity of the active material also depends on the ability of the body to detoxify and/or excrete the active material. If these mechanisms fail, hazard from exposure to an insecticide is increased. Thus, disease of the liver or kidney – the two organs which are mainly responsible for metabolism and excretion of foreign compounds – place an individual in a higher risk group.

The magnitude of the hazard will also be reduced by a reduction in contamination when the pesticide is applied by an experienced and well-trained operator who is properly equipped and properly protected. To exert a biological action, the active material must reach a sensitive tissue or organ, and it must therefore enter the body by some route. The speed of its action depends largely on the speed with which the active material reaches the blood system. Thus, exposure to volatile insecticides by the respiratory route produces an effect without much delay because the active material, which comes in through the alveolar system, is quickly absorbed into the blood circulation. Insecticide absorption is much influenced by the solvent. Regardless of the kind of compound, the speed of onset of signs of poisoning is faster if the dose is large and oral, and slower if the dose is small and dermal. There are three important routes of entry: oral (gastrointestinal), respiratory (inhalation), and dermal (through intact skin).

The longer the working time, the higher the risk, provided that the two other variables remain unchanged. It is therefore essential to limit the duration of exposure, depending upon the toxicity of the formulation used. It is preferable that work with pesticides be shared if more trained workers are available. In this way, total exposure is reduced and risk diminished.

Assessment of all of the parameters mentioned is essential for those concerned with the safe use of pesticides, as they are crucial in posing specific limitations of availability and use of a particular pesticide in agriculture, public health, or household use.

Diagnosis and Treatment of Pesticide Poisoning

In cases of acute poisoning, placing a correct diagnosis may be very difficult, even for the very experienced clinician, unless the illness is undoubtedly associated with recent exposure (occupational or accidental). This is particularly so because the signs and symptoms vary considerably both in intensity and sequence of appearance. Laboratory tests available are limited, and treatment should never be delayed pending the results of these tests.

Following excessive exposure to anticholinesterases (organophosphorus compounds or monomethyl carbamates), signs and symptoms develop very quickly, and these include nausea, headache, glandular hypersecretion (salivation, lacrimation), muscular weakness, vomiting, and cramplike abdominal pain with diarrhea. Blurred vision caused by an effect on the ciliary body is very indicative of poisoning. In severe cases of poisoning, paralysis of the diaphragm may occur, leading to respiratory depression, coma, and death.

Acute poisoning by chlorinated hydrocarbons is rare nowadays, unless they are swallowed accidentally or with suicidal intent. Signs and symptoms include apprehension and excitement, dizziness, hyperexcitability, disorientation, headache, and convulsions. In some instances, the onset of symptoms may be delayed, and remission of acute symptoms may occur in 1–3 days.

Clinical cases of poisoning with pyrethroids are extremely rare, and limited observations do not permit a detailed description of signs and symptoms in humans. However, from the studies on experimental animals and on the basis of the similarity in mode of action, one may expect similar signs and symptoms to those of organochlorine pesticides.

Acute poisoning with nitro- or chlorophenols resembles the signs and symptoms developed in a thyrotoxic crisis, with increased body temperature being the leading sign.

The peculiarity of paraquat poisoning is usually a delayed effect on the lungs with the development of fatal fibrosis. Without clear evidence of exposure, early diagnosis is virtually impossible.

As anticoagulants disturb clotting mechanisms, laboratory tests may help in clarifying obscure signs of bleeding in an otherwise healthy person.

The management of poisoning includes three essential procedures: (1) alleviation of life-threatening effects, (2) removal of nonabsorbed material, and (3) antidotal or supportive treatment. Any delay in the prompt institution of treatment can prove fatal. Rapid implementation of first-aid measures, removal of the source of contamination, and transport to a hospital, if indicated, may save life.

The sequence of procedures which should be strictly observed is as follows:

1. Check respiration, and make sure that the airway is clear.
2. Give artificial respiration if spontaneous breathing is inadequate.
3. Check the need for decontamination, and proceed if required.
4. Give antidote if available.
5. Collect evidence of exposure.
6. Transport to a medical care facility.

As a rule, emergency treatment starts in the field, continues during transport, and ends in a health center or in a hospital.

Emergency treatment in the field should be directed first towards alleviation of the life-threatening effects. All efforts should be made to maintain normal respiration, and a prerequisite for this is a clear airway.

In cases of serious poisoning with anticholinesterase insecticides, atropine is a drug of choice, to be given a soon as possible. To save life, it must be given by injection. Oral application may mask the clinical picture and is not recommended.

As soon as the diagnosis of organophosphorus compound poisoning is made, a first injection of 2–4 mg of atropine sulfate should be given intramuscularly, or intravenously in very severe cases. The effects of intravenous atropine application begin within 3–4 min and are maximal about 8 min after injection. Overatropinization is not as dangerous as underatropinization. Persons with manifest peripheral symptoms should also be given 1–2 g of a soluble salt of pralidoxime. Patients poisoned with carbamates must not be given an oxime.

For intoxication with organochlorine compounds, there is no specific antidote. Treatment is aimed at controlling the symptoms, especially hyperactivity and, in some instances, convulsions. Artificial ventilation may be required. Anticonvulsant treatment with barbiturates, diazepam, or paraldehyde should be given in sufficient dosage to calm the patient and prevent convulsions.

The principal management of intoxication by conventional or second-generation anticoagulant rodenticides is administration of phytomenadione (vitamin K_1). In addition to vitamin K_1, a seriously ill patient should initially be given a transfusion of carefully matched whole blood (as little as 50 ml may be effective); transfusions may be repeated daily until the patient's prothrombin time has returned to normal. Prolonged observation of patients affected by second-generation anticoagulants (coumarin derivatives) is required because these compounds are metabolized slowly, and repeated therapy may be indicated.

For intoxication with paraquat, no antidotes currently exist, and management essentially relies upon (1) the use of adsorbents to prevent absorption from the gut and (2) the removal of absorbed paraquat from the body, although these are rarely effective in severe poisoning. One liter of a suspension of Fuller's earth (about $300\,g\,l^{-1}$) or bentonite (about $70\,g\,l^{-1}$) should be administered orally as soon as possible. Activated charcoal may be more effective in adsorbing paraquat and should be used if it is available. In an emergency, use of ordinary soil may be beneficial if these adsorbents are not available. Administration of oxygen is contraindicated in acute poisoning because paraquat is more toxic in the oxygenated lung, and the use of oxygen should be delayed for as long as possible.

If a number of patients are found to be exhibiting symptoms suggestive of poisoning by a pesticide (or other chemical) without a history of exposure, the possibility of the cause being gross contamination of a food item or of drinking water, and being unrelated to any chemical, should be considered.

See also: **Fumigants**; **Fungicides**; **Heavy Metal Toxicology**; **Insect Pests**: Insects and Related Pests; Problems Caused by Insects and Mites; **Legislation**: Contaminants and Adulterants; **Organically Farmed Food**; **World Health Organization**

Further Reading

Aldridge WN (1990) An assessment of the toxicological properties of pyrethroids and their neurotoxicity. *Critical Reviews in Toxicology* 21: 89–104.

FAO (1990) *International Code of Conduct on the Distribution and Use of Pesticides (amended version)*. Rome: Food and Agriculture Organization.

Hayes JW, Jr. and Laws ER, Jr. (eds) (1990) *Handbook of Pesticide Toxicology*, vols. 1, 2 and 3. New York: Academic Press.

WHO (1981) *Environmental Health Criteria 39: Paraquat and Diquat*. Geneva: World Health Organization.

WHO (1986a) *Environmental Health Criteria 63: Organophosphorus Insecticides*. Geneva: World Health Organization.

WHO (1986b) *Environmental Health Criteria 64: Carbamate Pesticides*. Geneva: World Health Organization.

WHO (1990) *Environmental Health Criteria 104: Principles for the Toxicological Assessment of Pesticide Residues in Food*. Geneva: World Health Organization.

WHO (1992a) *IPCS Summary of Toxicological Evaluations Performed by the Joint FAO/WHO Meeting on Pesticide Residues (JMPR) through 1991 (WHO/PCS/92.9)*. Geneva: World Health Organization.

WHO (1992b) *The WHO Recommended Classification of Pesticides by Hazard and Guidelines to Classification 1992–1993 (WHO/PCS/92.14)*. Geneva: World Health Organization.

Worthing CR (ed.) (1990) *The Pesticide Manual, A World Compendium*, 9th edn. Farnham, UK: The British Crop Protection Council.

Types, Uses, and Determination of Herbicides

A M Au, Department of Health Services, State of California – Health and Welfare Agency, Berkeley, CA, USA

This article is reproduced from *Encyclopaedia of Food Science, Food Technology and Nutrition*, Copyright 1993, Academic Press.

Introduction

The major function of herbicides is to prevent or control weed growth in the field. Some herbicides are nonselective, that is, they kill a wide variety of plants, both desirable and undesirable. Other herbicides are selective – they kill certain kinds of plants while permitting others to survive. Selectivity of herbicides enables a desired crop to grow and produce free of competition from weeds. This article focuses on the classification of herbicides based on mode of action, the methods of application, safety implications, specific examples of uses, stability in the environment, and the analysis of residues in foods.

Classification Based on Mode of Action

Herbicides can be classified in different ways: by chemical name, by chemical characteristics of the compound, by toxicity, or by mode of action. There are two major categories of herbicides classified by mode of action: contact herbicides and translocated herbicides.

Contact herbicides affect only the part of the plant that they touch. Absorption through foliage is minimal. The application, therefore, must be made in sufficient quantity to cover the foliage thoroughly. Examples of contact herbicides are diclofop, dinoseb, diquat, and paraquat. Certain contact herbicides, like diquat and paraquat, are deactivated by soil particles. They must be mixed with clear water and applied directly to the vegetation.

At the molecular level, not all contact herbicides act in the same manner. For example, diquat and paraquat generate phytotoxic free radicals that interfere with the lipid metabolism of the plant and lead to ultimate death, whereas diphenyl ether herbicides cause chlorosis and necrosis resulting from the inhibition of the photosynthetic process. The biochemical mechanism of action of organic arsenicals such as cacodylic acid is not known.

Unlike contact herbicides, systemic herbicides can be translocated to other parts of the plant. They alter the normal biological function of the plant by interfering with certain biochemical reactions. Thus, when applied to foliage or soil, they enter the plant and translocate to their site of action. Examples of translocated herbicides are atrazine, glyphosate 2,4-dichlorophenoxyacetic acid (2,4-D) and simazine. Systemic herbicides, like contact herbicides, also have diverse modes of action at the molecular level. Chlorinated aliphatic acid herbicides, such as trichloroacetic acid (TCA), modify protein structure, causing chlorosis and necrosis in plants. Amide herbicides (e.g., alachlor, metachlor) interfere with both protein and nucleic acid synthesis. Carbamates inhibit protein synthesis only. The phytotoxic activities of thiocarbamates and dithiocarbamates appear to stem from the ability of these compounds to inhibit lipid synthesis. Phenoxy herbicides (e.g., 2,4-D), on

the other hand, apparently stimulate protein and RNA synthesis. These stimulations accelerate plant growth and, in turn, contribute to the death of the plants. Unlike the other translocated herbicides listed above, triazine herbicides block photosynthesis as the primary mode of action.

Methods of Application

Uniform and precise application of herbicides to the field is essential. Common herbicide carriers are liquid formulations containing emulsifiers. Since most herbicides are organic compounds that are not readily soluble in water, they are frequently distributed as dry granules.

Liquid herbicides are usually dispensed by sprayers. Large ground sprayers mounted on mobile units such as tractors can apply herbicide over areas as large as a hectare. Recirculating sprayers distribute the herbicide solution horizontally above the crop. Solution not lost in tall weeds is caught in a basin and pumped back to the spray tank for reuse. Most herbicides can be applied to foliage or to the soil by sprayers and granule applicators as broadcast treatments or bands.

There are three basic types of application methods for the distribution of either granular or liquid herbicides. These are ground-operated application, aerial application, and manual application. Most of the application devices can be adapted for use via ground-operated or aerial-operated methods, depending on the vehicle on which the device is mounted. Handheld sprayers, such as backpack sprayers or small plot sprayers, are examples of manual applicators.

Granular herbicide formulations can be applied by drop-type applicator, centrifuge spreaders or pneumatic systems. The drop-type applicator consists of rollers rotating over an orifice in the hopper. The flow of the herbicide granules is controlled via a metering mechanism. The maintenance of this type of applicator is cumbersome. In the centrifuge spreader system, herbicide granules are distributed into the field by two counterrotating disks. In the pneumatic system, the herbicide granules are kept in a hopper, and the flow of granules is controlled by a ground-driven fluted roller coupled to a low-volume, high-pressure airstream. The granules eventually reach the spreader nozzles via two flexible delivery chutes which provide the control necessary for the adjustment and positioning of the treatment.

Recently, because it appears to be cost-effective and environmentally beneficial, the wiper application method (first conducted in 1909 by Mahanay) was reintroduced as an application method for wipe-on herbicides. There are a wide variety of wiper applications, including the Rope-wick, carpet roller, continuous belt, rotary wiper, and Rogueing glove application. The Rope-wick device is a rig consisting of a series of short, exposed nylon ropes, each end of which is connected to a reservoir of herbicide solution. The solution passes into the ropes by both capillary action and gravitational flow. As the applicator moves through the weed-infested field, the chemical on the soaked wicks is rubbed on to the tops of the tall weeds but not on to lower-growing crop. Spray drift is eliminated. The carpet roller, as the name suggests, is a tractor mounted with a nylon carpeted roller soaked with liquid herbicide through a delivery system from a herbicide reservoir. The design is such that herbicide will only be in contact with the tall weeds without contacting the desired crop. A continuous belt consists of a V-belt, with sponges glued on to it, passing through a herbicide-containing reservoir. An adjustable pressure wheel removes excess herbicide from the sponges prior to its application on to the weeds. This helps prevent herbicide from contacting the crop. A rotary wiper applicator consists of flexible arms that allows wiper rotation around stationary objects to avoid injury to tree trunks. A glove equipped with absorbent pads that has continuous loading of herbicide solution from a reservoir is known as the Rogueing glove application.

A relatively new way of applying herbicide is through a sprinkler system. Liquid herbicide is pumped into the system and sprayed in a manner similar to a water sprinkler system.

Aerial spraying, using small aircraft or helicopters, can cover a large field in a shorter time period. Sprayers similar to those used in the ground operation can be mounted on to the aircraft for application to large areas.

Considerable research has been conducted by chemical companies to devise controlled-release formulations, a technology which has been demonstrated to be safe and effective in the drug industry. The three controlled-release systems are: (1) the chemically bound system, in which the herbicide is bound to a polymer; (2) the microencapsulation system, where the herbicide is coated by a polymer; (3) the matrix encapsulation system, where the herbicide is dispersed within a polymer matrix. This technology renders safer herbicide handling, reduces environmental pollution, and enhances herbicide selectivity.

The treatment of seeds with herbicides is another area of research currently under investigation. Certain herbicides can be applied to certain crop seeds prior to planting. Once planted in the soil, the herbicide moves swiftly away from the seed to the surrounding area to inhibit weed growth. This approach is economical and convenient to use. It also has a lesser tendency to pollute the environment.

Safety Implications

Ideally, herbicides are chemicals designed to cause injury only to undesirable weeds. Due to the biological differences between plants and animals, herbicides have low acute toxicity towards humans. However, most chemicals tend to have more than one metabolic effect on different living organisms. It is such unexpected secondary or side-effects of chemicals that cause major concern. Before the approval for marketing of any herbicide by the government, chemical companies must conduct extensive studies on the effectiveness of new herbicides. In addition, studies on metabolic fate and toxicity to animal species and plants are conducted to insure that the herbicide is safe to be used in the field. The cost of these testing procedures is tremendous. The specific requirements for herbicide registration depend upon the law of the country in which they are used but, in general, herbicides may not be used without registration by some governmental agency that carries out extensive reviews of the studies and tests performed by the chemical companies. It is also the duty of the vendor to provide users with information about all precautions and safe handling procedures of the herbicide. Adherence to those instructions by users is of utmost importance for the avoidance of harmful exposure.

The health effects of herbicides and their environmental impacts have been a major concern of the public. The health and safety issues associated with herbicide use affect not only consumers, but also farmers, formulators, applicators, and field workers, as well as the users of home and garden products. Groundwater contamination by agricultural chemicals has been an issue worldwide and will continue to be important in the years to come. Restrictions on the use and banning of certain herbicides are on the rise. Development of biodegradable herbicides with low residue levels combined with improved delivery systems can help protect the environment.

Depending on the conditions, exposure to herbicides at a high concentration can be fatal. Precautions in handling herbicides should be strictly followed to avoid unnecessary exposure. Safety training of all personnel designated to handle herbicides is imperative, and refresher safety training should also be conducted periodically to keep personnel up-to-date on all safety issues.

Equipment for application should be inspected frequently to insure that it is functioning properly. Malfunctioning equipment should be repaired prior to usage. Emergency procedures covering spillage or accidental poisoning should be established and strictly followed. There should be health surveillance programs for workers to monitor and insure worker well-being. Personal protective equipment and clothing should be checked for leaks periodically and should be kept clean. Good record-keeping of herbicide inventory and application is necessary to account for usage.

In order to avoid inhalation of herbicides during application, workers should wear respirators. There are many different kinds of respirators, namely the chemical cartridge respirators, powered air-purifying respirators, canister respirators, supplied-air respirators, and self-contained breathing apparatus. There are also certain herbicides that can be absorbed by the skin; users should handle these with due caution and should wear appropriate protective clothing, including face shields.

Care should be given in the storage and transport of herbicides to minimize spillage. In case spillage occurs, proper decontamination should be performed immediately. Maintenance of all equipment for dispensing herbicides should be done routinely. Warnings should be given in advance to alert others of possible herbicide drift during application, and warning signs and restricted-entry signs should be posted to prevent others from entering the treated areas.

Specific Examples of Uses

Weeds are usually defined as undesirable plants. Weeds are often the primary concern of farmers, because they cover many millions of productive acres that could be used to grow beneficial crops. In the past, farmers controlled weeds by manually removing them from the crops. The ancient Romans killed weeds with salt. With farms of small size, manual weed control, such as hand hoeing and pulling, mowing, burning, and machine tillage, is feasible. However, with large farms such labor is extensive and costly. As the size of the farms increased and synthetic herbicides were introduced, farmers began to rely on herbicides to control weeds.

Generally, there are three different types of treatment for the application of herbicides. They are preplanting, preemergence, and postemergence treatments. Preplanting treatment takes place prior to planting; preemergence treatment is done after planting but preceding the crop or weed emergence; and postemergence treatment is performed after the emergence of the crop or weeds.

2-4-Dichlorophenoxyacetic acid (2,4-D) was one of the first synthetic herbicides introduced to control broad-leafed weeds in cereal crops and pastures. 2,4-D is an effective systemic herbicide and is selective for broad-leafed plants but not grasses. It can be used as either a pre- or a postemergence herbicide for corn, but only as postemergence for sorghum. 2,4-D is

highly versatile and is used on a variety of crops such as wheat, barley, oats, rice, and sorghum. During the Vietnam war, Agent Orange, a 50: 50 mixture of 2,4-D and 2-4-5-Trichlorophenoxyacetic acid (2,4,5-T), was used extensively over the terrain in Vietnam as a defoliant to clear the way for US troops. 2,4,5-T is usually contaminated with dioxin, a highly toxic chemical compound and known carcinogen. It is due to this notorious contaminant that Agent Orange has been blamed for various illnesses and reproductive problems among those who came in contact with the defoliant in Vietnam.

Paraquat is used as a preemergence treatment for sugar beet. Simazine, on the other hand, is used both as a preplant and as a preemergence treatment for corn. S-ethyl dipropylthiocarbamate (EPTC) is incorporated into the soil as a preplant treatment for potatoes.

Herbicides have made it possible to grow more food on less land with less labor and at lower cost. Herbicides are also used to control aquatic weeds which impede water flow in irrigation canals and drainage systems, interfere with fishing, or promote insect-breeding grounds.

Stability in the Environment

For herbicides to exert their effects on weeds, they must be relatively stable in the treated environment. However, the stability of the chemical creates a burden on the environment, especially for those herbicides that find their way into aquifers and contaminate drinking water sources or remain on the crops at the time they reach consumers. In order to insure that the newer generation of herbicides do not linger on after accomplishing their task, research is directed towards synthesizing biodegradable compounds. The ideal herbicide is one that degrades to harmless chemicals after it performs its function and therefore does not persist in nature. Carbamates are one such class of chemicals specifically designed with that goal in mind.

Analysis of Residues in Foods

There is increasing awareness among consumers of the hazard of chemical contamination of food and drinking water. There is particular concern over the implications of food contamination by herbicide residues. Analysis for herbicide residues in food requires methods that identify not only parent structures, but also their metabolites and degradation products in a variety of food matrices. Certain food crops are perishable and therefore cannot wait for lengthy analysis to establish the suitability for consumption. Thus, rapid analytical technology is needed. Multiresidue methods, which can detect the presence of many herbicides at once, are the methods of choice for determining the presence of a multiple number of herbicides and their degradation products in a food sample.

An analytical process consists of several major steps: the sample preparation, the extraction, the clean-up, the determinations, and the confirmation. These steps are common to the determination of other agrochemical residues, including pesticides, and are discussed in detail in the following article. The basic operation of sample preparation is to separate physically food or plant parts and to chop and blend them. The essence of the extraction process is to remove the target herbicide from the other components in the sample matrix. The main function of the clean-up procedure is to remove interfering constituents, usually by selective partitioning into organic solvents followed by an adsorption or size exclusion chromatographic purification step. The determination step includes separation of the purified samples through thin-layer chromatography, gas chromatography, or liquid chromatography techniques followed by the detection procedure using a variety of specific detectors for the targeted compound. For confirmation purposes, the analyte is further subjected to mass spectrometric analysis. Recently, successful attempts have been made in using gas chromatography–mass spectroscopy (GCMS) as a primary screening method. The GCMS screening technique provides simultaneous results for both the detection and the confirmation of the targeted compound in the sample matrix. This one-step procedure will be the method of choice as it offers both rapid and definitive data. (*See* **Chromatography**: Thin-layer Chromatography; Gas Chromatography; **Mass Spectrometry**: Principles and Instrumentation.)

Improvements to existing analytical technology are well underway to reduce the time- and solvent-consuming extraction and clean-up steps. Supercritical fluid extraction, which is based on the solvent properties of gases such as carbon dioxide at its critical pressure and temperature, can selectively remove the targeted compound from the complex food matrix in a short time. Through such approaches the recovery of the compound can be easily accomplished.

The use of antibodies as analytical tools is a common practice in clinical laboratories. Antibodies have been recently developed for identifying and quantifying herbicides. Antibodies can be isolated from the plasma of an immune-challenged animal or from a hybridoma cell line. An antibody that is specifically generated from a compound will have high selectivity towards that compound even in the midst of other interfering components and can bind to it tightly to form a complex. Therefore, by attaching a tracer to the antibody molecule, one can quantify the amount of antibody complex present, which is also an indication of the amount of the compound in the sample. A variety of tracers are available, for example radioisotopes, fluorescent molecules, and so forth. One of the disadvantages of the immunoassay is that the length of time to generate the specific antibody is relatively long. Typically, it takes approximately a year to develop. However, once it is generated, the immunoassay can be performed in less than half an hour. The triazine immunoassay is now available commercially, and, like most immunoassays, it is specific, sensitive, rapid, and cost-effective. (*See* **Immunoassays**: Principles.)

One of the modes of action of herbicides is by inhibiting photosynthesis. The Hill reaction is one of the processes in the photosynthetic pathway; therefore, a screening technique based on the inhibition of the Hill reaction can be a useful tool in detecting herbicides like triazines and carbamates.

See also: **Chromatography**: Thin-layer Chromatography; Gas Chromatography; **Immunoassays**: Principles; **Mass Spectrometry**: Principles and Instrumentation; **Pesticides and Herbicides**: Residue Determination

Further Reading

Ashton FM and Crafts AS (1981) *Mode of Action of Herbicides*, 2nd edn. New York: Wiley.

FDA (1968) *Pesticide Analytical Manual*, 2nd edn, vol. 1. *Methods Which Detect Multiple Residues*. Washington, DC: US Department of Health Services, Food and Drug Administration.

Klingman GC, Ashton FM and Noordhoff LJ (1982) *Weed Science: Principles and Practices*, 2nd edn. New York: Wiley.

McWhorter CG and Gebhardt MR (eds) (1987) *Methods of Applying Herbicides. Monograph Series of The Weed Science Society of America*, no. 4. Champaign, IL: The Weed Science Society of America.

Newton M and Knight FB (1981) *Handbook of Weed and Insect Control Chemicals for Forest Resource Managers*. Beaverton: Timber Press.

Residue Determination

S Nawaz, Central Science Laboratory, Sand Hutton, York, UK

Background

Continued population growth has led to an increased demand on the world's natural resources. Pesticides are widely used to help increase the yield and improve the quality of crops. Pesticides are categorized according to their mode of action and include insecticides, herbicides, fungicides, acaricides, nematicides, and rodenticides. Pesticides are also used as plant growth regulators and for public-health purposes. Global sales of pesticides during 1996 were estimated at over US$30 billion. Herbicides, insecticides, and fungicides accounted for 48, 28, and 18%, respectively, of the total sales (see **Table 1**). There are over 900 chemicals registered for plant protection purposes in the European Union (EU) alone. In addition to the existing pesticides, there are an ever increasing number of new chemicals being granted approval.

As pesticide use can leave undesirable residues, various national and international authorities regulate the use of pesticides and set maximum residue levels (MRLs) in crops. An MRL is the maximum concentration of a pesticide and/or its toxic metabolites legally permitted in food commodities and animal feeds. If pesticides are properly applied at the recommended rates, and crops are only harvested after the appropriate time intervals have elapsed, residue levels are not expected to exceed MRLs. The residue levels in foods derived from commodities that comply with the respective MRLs are also intended to be toxicologically acceptable. In the EU, the regulation of the agrochemical industry, and the setting of MRLs is currently being harmonized across all member states by the EC. In the USA, the Environmental Protection Agency is responsible for such regulation. The Codex Alimentarius Commission (FAO/WHO) has published tables of MRLs which have official status across the world and are used to aid international trade. The monitoring of the residues in foods is often at the microgram per kilogram level or lower and has to be supported by strict analytical quality-control standards, so that the analysis produces unequivocal, precise, and accurate residue data. Before samples are analyzed, the analyst has to

demonstrate that the intended analytical method can achieve adequate specificity, sensitivity, linearity, accuracy, and precision at the relevant analyte concentration and in appropriate matrices. The calibration solutions must be prepared using certified reference standards. Residue analyses normally include the metabolites, isomers, and other related compounds included in the MRL definition. Many methods can determine a large number of residues in a single analytical run; these multiresidue methods are in common use and help reduce the total cost of analysis.

Sampling

A representative sample consists of a large number of randomly collected units. It is not always possible to collect large samples because of the cost of transportation and the practicalities of sample handling in the laboratory. Monitoring of pesticide residues for MRL compliance involves analysis of a composite sample, made up of a number of individual units. Recent research has shown that pesticide residues in individual units of fruit and vegetables can exhibit an extremely skewed distribution, and this is likely to add to the difficulty of taking a representative sample. The guidelines for obtaining composite samples for MRL compliance monitoring are published by the Codex Alimentarius Commission and are summarized in **Table 2**. In addition to checking for MRL compliance, residue analyses are also carried out to investigate other issues such as cases of misuse or the deliberate poisoning of wildlife or domestic animals. In such instances, a more targeted sampling regime may be adopted, and a qualitative analysis may suffice.

Sample Preparation/Subsampling

Samples should be analyzed without any delay, as some pesticide residues may degrade rapidly. If

Table 1 Major classes of chemicals used as pesticides

Type of chemical	Examples and their primary uses
Benzimidazole	Benomyl (F), carbendazim (F), thiabendazole (F)
Bipyridylium	Diquat (H), paraquat (H),
Carbamate	Aldicarb (I, N), carbaryl (I), carbofuran (A, I, N)
Dithiocarbamates	Mancozeb (F), ziram (H), thiram (H)
Organochlorine	DDT (I), lindane (I), endosulfan (I)
Organophosphorus	Chlorpyrifos, malathion (I, A), parathion (I)
Pyrethroids	Permethrin (I), cyfluthrin (I)
Substituted phenyl ureas	Diuron, linuron, monolinuron (H)
Triazine	Atrazine (H), simazine (H)

A, acaricide, F, fungicide, H, herbicide, I, insecticide, N, nematicide.

Table 2 Codex guidelines for collection of representative samples

Sample type	Minimum weight of sample (kg)
Small or light products (e.g., berries, peas, spinach)	1
Medium sized products (e.g., apples, carrots, potatoes)	1 (at least 10 units)
Large size products (e.g., melons)	2 (at least 10 units)
Dairy products (e.g., cheese, butter)	0.5
Meat, poultry, fish	1
Oils and fats	0.5
Cereals and cereal products	1

immediate analysis is not possible, storage of samples at $-20\,^\circ$C may help minimize the degradation process. Typically, a 20–50-g portion (subsample) is required for analysis. In order to obtain representative subsamples, it may be necessary to grind/mill and thoroughly mix the whole sample, so that any residues present are evenly distributed. This process is especially important because residue levels can exhibit a high degree of variability between individual units. Some pesticides are known to degrade during the processing of fruit and vegetable samples at ambient temperature. Milling frozen food samples in the presence of excess solid CO_2 (dry ice) has been shown to minimize the losses of most pesticide residues during the process.

Domestic mills (e.g., coffee grinder) as well as more specialized mills are used to grind samples of cereals, nuts, and pulses. Manual methods, such as cone and quarter or mechanized devices such as the riffle divider can be used to obtain representative subsamples from such samples. Samples of animal tissues are minced and mixed thoroughly before subsampling. For pesticides (e.g., organochlorines) that accumulate in the fatty tissue of animals, visible layers of fat may be removed for direct analysis. Preparation of homogenous fruit and vegetable samples prior to subsampling may be carried out using domestic food processors and blenders. However, larger specialized mechanical bowl choppers are more suitable for large samples, and heavy-duty choppers may be required to process frozen samples.

Extraction

The extraction step involves the quantitative transfer of pesticide residues from the food matrix into solvent(s). The efficiency of extraction process depends on the physicochemical properties of the solvent(s) and analytes. Important factors include pH, polarity, temperature, sample/solvent ratio, presence of water, and degree of analyte/matrix binding. Most extraction procedures employ organic solvents in the presence of water. The presence of water is critical for extraction of pesticides from cereals and cereal products, as it helps reduce the binding between residues and matrix. Samples of cereals, nuts, pulses, fruit, and vegetable samples are extracted by simple homogenization with an organic solvent. The most commonly used solvents include ethyl acetate, acetone, acetonitrile, hexane, methanol, and dichloromethane. For multiresidue extraction methods, it is not possible to establish the optimum extraction conditions for all residues with differing physical and chemical properties so the choice of solvent polarity is usually a compromise. Although the presence of water may aid the extraction process, most polar pesticides are partitioned into the water phase if the organic solvent used is not water-miscible. Addition of anhydrous sodium sulfate prior to extraction can overcome this problem. The analysis of nonpolar pesticides in fatty products involves extraction of fat using nonpolar solvents such as hexane, n-pentane, or light petroleum. After the evaporation of the solvent, the fat is redissolved in an organic solvent prior to the clean-up and determination steps.

Other methods used for residue extraction include soxhlet, where samples are exposed to solvent vapor that is condensed and vaporized repeatedly to exhaustively extract analyte(s). Supercritical fluid extraction (SFE) methods use a gas under high pressure and above the *critical* temperature to extract the residues. This technique is more widely used for samples with low moisture content (cereals) where sample/analyte binding is more common.

For residues that are not suited to multiresidue extraction methods, dedicated extraction methods may be used for single or small groups of closely related pesticides. These methods utilize physical and chemical properties of analyte and solvent to carry out selective extraction of the analyte from matrix. A number of pesticides are normally analyzed using single residue methods and include formetanate, fluazifop, 2,4-D, formetanate, propamocarb, and maleic hydrazide.

Clean-up

Sample extracts not only contain the target analyte(s) but may also contain coextractives, such as plant pigments, proteins and lipids. These coextractives may have to be removed prior to instrumental analysis to avoid possible contamination of instruments and to eliminate compounds that interfere during the determination step. To achieve low detection limits, sample extracts may also require a concentration step, which can be incorporated in the clean-up procedure. Clean-up procedures can lead to losses of residues and increases in the cost of analysis, and can reduce the sample throughput. Therefore, a number of methods utilize minimal clean-up and instead rely on the selectivity of the detector(s).

A number of analytical procedures employ liquid–liquid extraction for clean-up of sample extracts by selective partitioning of analytes between two immiscible solvents. This technique is commonly used during the analysis of fatty samples. Liquid–liquid extraction is not easy to automate and requires the use of large amounts of solvents. (*See* **Analysis of Food**.)

Adsorption chromatography is used in many residue laboratories for the clean-up of sample extracts. This process involves either:

- retention of analyte(s) on a chromatographic column while the coextractives are unretained: (the analyte(s) are then selectively eluted from adsorption medium); or
- retention of coextractives on a chromatographic column while the analyte(s) are unretained.

A number of materials are available for adsorption chromatography, including alumina, florisil, cellulose, diatomaceous earth (celite), carbon (charcoal, graphite), silver nitrate, and silica. Alumina and silica are effective for the clean-up of fatty samples for organochlorine (OC) pesticide residue analysis. Silver nitrate is used for the removal of sulfur-containing intreferences. Carbon has a high affinity for plant pigments and is particularly suitable for the clean-up of green leafy vegetable extracts with a high chlorophyl content.

Chemically modified silica sorbents are widely used for the clean-up of sample extracts. These materials are prepared by the reaction of silanol groups on silica surfaces with silane reagents to form esters containing required functional groups. These sorbents are used in cartridges, disks, membrane filters, and impregnated fiber tips. Some chemically modified silica and other sorbents used for clean-up of sample extract are given in **Table 3**.

Gel-permeation chromatography (GPC) or size-exclusion chromatography separates molecules on the basis of their molecular size. Large molecules (e.g., lipids, pigments, and polymeric coextractives) elute faster than smaller molecules such as pesticide residues. The reproducibility, suitability to automation, and compatibility with a wide range of pesticide/matrix combinations make GPC a popular clean-up method in many laboratories worldwide. The disadvantages of the technique include the use of large quantities of solvents, limited sample throughput,

and incomplete separation of high-molecular-mass pesticides from coextractives. (*See* **Chromatography: Principles.**)

Chromatography and Determination

The final stage of the pesticide residue analysis procedures involves the chromatographic separation and instrumental determination. Where chromatographic properties of some pesticides are affected by sample matrix, calibration solutions should be prepared in sample matrix. The choice of instrument depends on the physiochemical properties of the pesticide(s) and the sensitivity required. As the majority of pesticides are relatively volatile, gas chromatography (GC) has proved to be an excellent technique for pesticide determination and is by far the most widely used. A typical chromatogram from a multiresidue method is shown in **Figure 1**. (*See* **Chromatography: Gas Chromatography.**)

Most residue methods employ splitless injection of 1–3 μl of the sample extract. Cold on-column injection is employed when the pesticides are likely to breakdown in a hot injection port. A wide range of GC column types are used for residue analysis, and the choice depends on the physicochemical properties of the analytes. Fused silica capillary columns are most widely used during analysis of pesticides (see **Table 4**). Typical capillary columns are 25–30 m long with an internal diameter range of 0.1–0.5 mm and a stationary phase of 0.1–10 μm thickness. Nonpolar stationary phases are used for the separation of nonpolar pesticide residues such as OC and pyrethroids. Similarly, more polar pesticide residues (e.g., methadimophos) are separated on relatively polar columns. The conventional semiselective detectors are widely used for residue analysis. Electron-capture detectors (ECDs) are utilized for halogenated pesticides (OCs and pyrethroids). Nitrogen phosphorus detectors are used for organophosphorus (OP) and nitrogen-containing pesticides. Flame-photometric detectors (FPDs) are used for OP and sulfur-containing pesticides, while atomic emission detectors can be used for a wide range of pesticides.

Gas chromatography–mass spectrometry (GC–MS) has been the predominant technique for the confirmation of pesticide residues in the past. Relatively inexpensive bench-top instruments have made the technique more widely available for routine screening in recent years. The resolving power of GC coupled with the specificity of mass spectroscopy provides the most effective means of pesticide residue analysis. A number of ionization techniques are available for GC–MS instruments and include electron impact (EI) and chemical ionization (CI). EI impact ionization

Table 3 Examples of sorbents used for clean-up of sample extracts

Nonpolar	Polar	Ion exchange
Octadecyl (C18)	Florisil	SCX benzenesulfonylpropyl
Octyl (C8)	Diol (2OH)	PRS sulfonylpropyl
Ethyl (C2)	NH₂ aminopropyl	Water's Oasis™ divinylbenzene: vinylpyrrolidione copolymer
Polypropylene	Silica	SAX trimethylaminopropyl

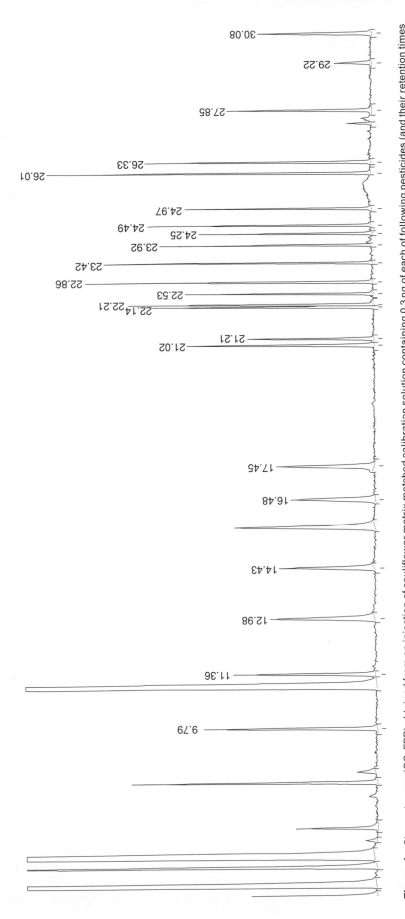

Figure 1 Chromatogram (GC–FPD) obtained from an injection of cauliflower matrix matched calibration solution containing 0.3 ng of each of following pesticides (and their retention times in minutes): heptenophos (9.79), ethoprophos (11.36), monocrotophos (12.98), dimethoate (14.43), fonofos (16.48), diazinon (17.45), parathion-methyl (21.02), malaoxon (21.21), fenitrothion (22.14), pirimiphos-methyl (22.21), malathion (22.53), parathion (22.86), pirimiphos-ethyl (23.42), mecarbam (23.92), methidathion (24.25), tetrachlorvinphos (24.49), profenofos (24.97), ethion (26.01), triazophos (26.33), pyridaphenthion (27.85), azinphos-methyl (29.22), and pyrazophos (30.08).

Table 4 GC stationary phases commonly used for pesticide analysis (listed in increasing order of polarity)

Stationary phase	Typical uses
100% methyl silicone (DB1)	Nonpolar pesticides
5% phenyl, 95% methyl silicone (DB-5)	Multiresidue screening
35% phenyl 65% methyl silicone (DB-35)	EPA method 608
50% phenyl 50% methyl silicone (DB-17)	Polar organophosphorus pesticides
14% cyanopropylphenyl 86% methyl silicone (DB1701)	Organochlorine pesticides
50% cyanopropylphenyl 50% methyl silicone (DB-225)	Polar pesticides
Poly(ethylene) glycol (DB-wax)	Polar pesticides

leads to a greater degree of fragmentation of molecules compared with CI. Hence, CI provides a greater sensitivity, but EI provides more spectral information. The detection systems most widely utilized in residues analysis are based on either the quadrupole or ion-trap principle. The quadrupole instruments have limited sensitivity in the scan mode compared with the ion-trap instruments. However, by operating in the selected ion mode, adequate sensitivity can be achieved on quadrupole detectors. A small quadrupole mass selective detector can typically detect over 100 pesticides in food extracts at the relatively high levels (typically $0.2 \, mg \, kg^{-1}$) using full scan spectra, reducing to lower levels (typically $0.01 \, mg \, kg^{-1}$) in the selected ion mode. The ion-trap detectors have a higher inherent sensitivity, and this allows screening of clean sample extracts for a wide range of pesticides at low levels (typically $0.05 \, mg \, kg^{-1}$) in full scan mode. The major advantage of the ion-trap instruments is that the characteristic ions can be selected and then further fragmented to provide added specificity (MS–MS). (*See* **Chromatography**: Combined Chromatography and Mass Spectrometry.)

High-performance liquid chromatography (HPLC) is increasingly being used for the determination of pesticide residues, as it is especially suited to the analysis of nonvolatile, polar, and thermally labile residues that are difficult to analyze using GC. The resolution achieved on HPLC can be comparatively low, and therefore, the use of selective detection systems may be necessary for reliable residue analysis. Ultraviolet (UV) spectroscopy is the most common choice for detection of (OP) residues in environmental samples (e.g., soil, water). Although UV detection is not a very selective technique, it is commonly used for screening purposes due to its low cost, simplicity, and wide application range. Elimination of interferences and optimized chromatography are essential prior to detection in order to enhance the selectivity

of UV-based methods. The use of diode array detectors can further enhance the selectivity of UV-detection procedures. Fluorescence detection offers a greater selectivity and sensitivity than UV. Pesticides with inherent fluorescence include dimethoate, ethoxyquin, azinphos methyl, phosalone, thiabendazole, and carbendazim. With the exception of thiabendazole and carbendazim, this technique is not widely used in pesticide residue analysis, as methods based on inherent fluorescence have a poor sensitivity compared with other methods available. Precolumn and postcolumn reaction systems can be employed with HPLC methods, which can help improve the chromatographic separation and detection of analytes. A number of pesticides (e.g., N-methyl carbamates, glyphosate, and phenylurea herbicides) are analyzed after derivatization to enable fluorescence detection. The electrochemical detectors are used for a number of pesticide residues (e.g., captan) in relatively clean samples.

The on-line combination of HPLC and mass spectroscopy (HPLC–MS) offers a high sensitivity and specificity, and its use in the field of pesticide residues analysis is growing. There are a number of ionization techniques used to interface HPLC with MS analyzers, of which the most widely used are electrospray and atmospheric pressure chemical ionization. The HPLC–MS methods use soft ionization techniques which typically produce protonated or deprotonated pseudomolecular ions. Therefore, the chromatographic data do not provide structural information except when the ions produced are subjected to successive fragmentation (MS^n). (*See* **Chromatography**: High-performance Liquid Chromatography.)

Derivatization

Some pesticide residues require derivitization to enhance the extractability, clean-up or subsequent chromatographic resolution and determination steps. For example, pesticides with hydroxy groups are not suited to GC analysis, and such an analysis may be possible only after derivatization to esters. Furthermore, esters of certain functional groups can enhance the detection process, e.g., pentafluorobenzyl derivatives produce a high response on the ECD.

Dithiocarbamate pesticides break down to carbon disulfide (CS_2) during analytical procedures. These residues are determined after treatment of samples with acidic tin (II) chloride. Any dithiocarbamate residues in the sample break down to produce CS_2 gas, which can be trapped in the reaction chamber. An aliquot of the gas (headspace) in the reaction chamber is analyzed for CS_2 using GC–FPD. Alternatively, the gas produced can be absorbed into a 2,2,4 trimethyl pentane, and an aliquot of the liquid layer is then analyzed using GC. This approach is more robust and more amenable to GC–MS analysis compared with the headspace procedure.

Some pesticides containing sulfur may oxidize to form sulfoxide and sulfone derivatives before or during analysis. These products are also toxic and are included in the residue definition for the monitoring purposes. These residues are analyzed after complete conversion of the pesticide and its sulfoxide to the corresponding sulfone and thus enable combined measurement of the pesticide, its sulfoxide, and sulfone residues. The conversion step involves the treatment of sample extracts with potassium permanganate in the presence of 2-methyl propan-2-ol. The sulfone formed is then extracted into an organic solvent and analyzed by GC.

Other Techniques

Enzyme-linked immunosorbent assay (ELISA) methods are used for rapid screening of an individual or a group of closely related pesticides. These methods require little or no sample clean-up, require no expensive instrumentation, and are suitable for field use. ELISA methods are especially suitable for residue analyses that are not possible using multiresidue methods. ELISA kits are available for a number of pesticides, including 2,4-D, aldicarb, carbendazim, thiabendazole, chlopyrifos, diazinon, endosulfan, and metalaxyl.

Confirmation

For regulatory purposes, it is essential that pesticide residues be unequivocally confirmed using MS. However, if an MS method is not available, the sample extract is reanalyzed using a different chromatographic column and/or a different detection system to confirm the initial results.

Emerging Techniques

There are continued improvements in the design of instruments available for residue analysis. The use of GC injectors capable of injecting large volumes can enable determination at low levels. The use of fast GC, which utilizes an improved design for heating the columns, can enable faster chromatographic runs and thus enable quicker analysis. Improvements in GC instrumentation have enabled precise control over temperature and gas flow rate. The use of electronic pressure control devices can enable more reproducible chromatographic runs, thus improving the quality of data. These advances, coupled with more sophisticated software, can enable more reproducible chromatography, with a typical retention time variation of 0.01 s. Improvements in MS instruments will continue to enhance the selectivity of methods.

See also: **Chromatography**: Principles; High-performance Liquid Chromatography; Gas Chromatography; **Mass Spectrometry**: Principles and Instrumentation; Applications; **Pesticides and Herbicides**: Types of Pesticide; Types, Uses, and Determination of Herbicides; Toxicology

Further Reading

Anon (1991) *Concerning the placing of PPPs on the market, Council Directive 91/414/EEC, Brussels: Official Journal No. L 230.* Luxembourg: Office of Official Publications of the European Communities.

Cairns T and Sherma J (eds) (1992) *Emerging Strategies for Pesticide Analysis.* Boca Raton, FL: CRC Press.

Chapman JR (1993) *Practical Organic Mass Spectroscopy, A Guide for Chemical and Biochemical Analysis.* Chichester, UK: John Wiley.

Codex Alimentarius (1993) *Codex Commission Guidelines, Recommended Method of Sampling for Determination of Pesticide Residues*, vol. 2, Section 3. Rome: Food and Agriculture Organization of the United Nations.

Codex Alimentarius (1996) *Pesticide Residues in Food, Maximum Residue Limits*, vol. 2B. Rome: Food and Agriculture Organization of the United Nations.

Fong WG (ed.) (1999) *Pesticide Residues in Foods: Methods, Techniques and Regulations.* New York: John Wiley.

Hill ARC and Reynolds SL (1999) Guidelines for in-house validation of analytical methods for pesticide residues in food and animal feeds. *Analyst* 124: 953–958.

Krause RT (1979) Resolution, sensitivity and selectivity of a high performance liquid chromatographic post-column fluorometric labeling technique for determination of carbamate insecticides. *Journal of Chromatography* 185: 615–624.

Mellon F, Self R and Startin JR (2000) *Mass Spectroscopy of Natural Substances in Food.* Cambridge: The Royal Society of Chemistry.

Sherma J (1981) *Manual of Analytical Quality Control for Pesticide Residues and Related Compounds, USA: EPA 600/2–81–059.* US Environmental Protection Agency, Research Triangle Park.

The Working Party on Pesticide Residues (1997) *Unit to Unit Variation of Pesticide Residues in Fruit and Vegetables.* York, UK: Pesticide Safety Directorate.

Tomlin C (ed.) (1997) *The Pesticide Manual, A World Compendium*, 11th edn. Bracknell, UK: The British Crop Protection Council.

Zooner P Van (1996) *Analytical Methods for Pesticide Residues in Foodstuffs*, 6th edn. The Hague, The Netherlands: Ministry for Public Health.

Toxicology

C K Winter, University of California Davis, CA, USA

Background

The use of agricultural chemicals, collectively known as pesticides, in the past several decades has led to significant reductions in crop losses resulting from insects, weeds, and plant diseases throughout the world. The toxicological properties that pesticides possess also present the potential for impacts upon human health and upon the environment. As an example, agricultural workers involved in the mixing, loading, and/or application of pesticides as well as those working in fields previously treated with pesticides face potential health risks resulting from excess exposure to the pesticides. Consumers are routinely exposed to pesticide residues in their foods, and the potential dietary risks from pesticide exposure have been the subject of considerable government study, regulation, and societal concern.

This review focuses upon the toxicology of the various types of pesticides used, how pesticide residues in foods are regulated, and the magnitude of potential risks faced by consumers from pesticide residues in the food supply.

Pesticides

Classification

The US Federal Insecticide, Fungicide, and Rodenticide Act defines a pesticide as 'any substance or mixture of substances intended for preventing, destroying, repelling, or mitigating any pest, any substance or mixture of substances intended for use as a plant regulator, defoliant, or desiccant, and any nitrogen stabilizer....' Under this broad definition, it is clear that a variety of pesticide types exist to control a wide number of different types of pests. A commonly held perception is that pesticides refer primarily to agricultural chemicals that control insects (insecticides). According to the US definition, however, pesticides also refer to chemicals that control plant diseases (fungicides) and weeds (herbicides) as well as a variety of other 'pests' (**Table 1**). For the purposes of consistency, all types of pesticides, including herbicides, will be considered under this broad umbrella in this article.

Pesticide Use

According to the US Environmental Protection Agency (EPA), approximately 2 billion kg of chemicals were used as pesticides in the USA in 1997. It should be noted that the majority of pesticide use was not for agricultural purposes. For example, 53% of pesticide use (by volume) involved chlorine or hypochlorites used for disinfection of potable and wastewater pools. 'Conventional' pesticides, defined as those developed or produced exclusively or primarily for use as pesticides, accounted for the remaining 47% of pesticide use by volume; 77% of this total was for agricultural uses, and 12% represented industry/government/commercial use, with the remaining 11% resulting from home and garden use.

Figure 1 shows the relative amounts of a variety of pesticide types used in US agriculture in 1997. Nearly half of the total volume of agricultural pesticide use came from herbicides and plant growth regulators (213 million kg), followed by sulfur/oils (65 million kg) and fumigants/nematicides (63 million kg). Agricultural insecticide use in 1997 was approximately 37 million kg, and 26 million kg of fungicides were used for agricultural purposes.

In terms of trends, total agricultural pesticide use, in terms of kg applied, has decreased slightly since 1979, with the largest drops in use representing insecticides, sulfur, and oils. The use of herbicides has been relatively steady.

Toxicity

Hundreds of different pesticide active ingredients are presently registered by the EPA, and nearly 300 pesticides are considered to have the ability to leave residues on food crops. Some of the more common classes of pesticides and some representative examples are given in **Table 2**. A comprehensive review of the toxicity of all pesticides is clearly beyond the scope of this article, and those interested in more specific and/or detailed pesticide toxicity information are encouraged to consult articles cited in the Further Reading section.

Table 1 Pesticide types and targets

Pesticide type	Pest controlled
Acaricide	Mites
Algacide	Algae
Bacteriocide	Bacteria
Defoliant	Leaves
Fungicide	Fungi
Herbicide	Weeds
Insecticide	Insects
Molluscicide	Snails
Nematicide	Nematodes
Rodenticide	Rodents

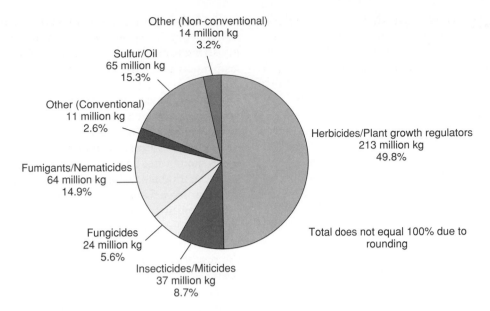

Figure 1 Agricultural use of pesticides in the USA, 1997.

Table 2 Common pesticide classes and representative examples

Pesticide	Examples
Insecticides	
Chlorinated hydrocarbons	Dicofol, methoxychlor, DDT, aldrin, chlordane
Organophosphates	Parathion, malathion, chlorpyrifos, azinphos-methyl
Carbamates	Aldicarb, carbaryl, carbofuran
Pyrethroids	Permethrin, cypermethrin
Herbicides	
Triazine	Atrazine, simazine, cyanazine
Phenoxy	2,4-D, 2,4,5-T, MCPA
Quaternary ammonium	Paraquat, diquat
Benzoic acids	Dicamba
Acetanilides	Alachlor, metolachlor
Ureas	Linuron
Fungicides	
Inorganic	Sulfur
Ethylenebisdithiocarbamates	Maneb, mancozeb, zineb
Chlorinated phenols	Pentachlorophenol

Insecticides   Insecticides are represented by a variety of different chemical classes involving a variety of mechanisms of toxic action on insects as well as mammals such as humans. Examples of mechanisms of action include metabolic interference of the nervous or muscle systems, desiccation, and sterilization. Common types of insecticides include the chlorinated hydrocarbons, organophosphates, carbamates, and pyrethroids.

The first major synthetic class of insecticides, the chlorinated hydrocarbons, was developed during the 1930s and 1940s. Representative members of this insecticide class include DDT, aldrin, dieldrin, and chlordane. The chlorinated hydrocarbons are very potent nerve toxins to insects, and their initial use led to significant improvements in insect control. Their high insect potency, combined with generally low mammalian toxicity, provided excellent selectivity and insect control that was further enhanced by their high environmental persistence. In subsequent years, however, their resistance to environmental decay, coupled with their widespread continuing use, resulted in environmental build-up and food-chain magnification that resulted in significant environmental and ecological damage. Today, very few chlorinated hydrocarbon insecticides remain registered for use in the USA, because of environmental concerns, although their use may still be significant in many areas of the world. Many recent studies have also associated chlorinated hydrocarbon insecticides with potential adverse effects on human and nontarget organism fertility and reproduction arising from enzyme-inducing or estrogenic properties of the chemicals.

Historically, many of the uses of chlorinated hydrocarbon insecticides were replaced by the organophosphate and carbamate insecticides. In contrast to the chlorinated hydrocarbon insecticides, the organophosphates and carbamates break down quite rapidly in the environment following application. They also have a far greater acute toxicity in mammals. The organophosphates and carbamates are derived as esters of two very different chemical families but share a common mechanism of toxicological action in both insects and mammals that involves inhibition of cholinesterase enzymes that can disrupt the nervous system.

A newer class of insecticides, the pyrethroids, are synthetic derivatives of pyrethrins (natural extracts from chrysanthemums) that are made to be more light-stable than their natural predecessors and thus more effective as insecticides. The pyrethroids were introduced in the 1970s and have broad-spectrum activity against many insects while possessing a much lower mammalian toxicity than the organophosphates or carbamates. The pyrethroids are frequently used in agricultural pest control, but their use is still limited by their environmental lability, relatively high cost, and their tendency to lose effectiveness through the development of insect resistance.

Herbicides Several different types of herbicides exist (**Table 2**), and each type has its own mechanism of toxic action on weeds. Since weeds and mammals differ dramatically in terms of metabolic function, herbicides targeting specific metabolic pathways in weeds frequently have little effect, and therefore a low relative toxicity, in mammals including humans.

The timing of herbicide application may determine significantly whether the potential exists for consumers to be exposed to herbicide residues in foods. Some herbicides (preplant herbicides) are applied before a crop is planted, whereas others (preemergent herbicides) are applied after planting but prior to the appearance of weeds, and others (postemergent herbicides) are used after the weeds have germinated. Some herbicides have broad-spectrum activity that makes them toxic to most types of plant life, including the crop being produced. One such example is glyphosate. Recent developments in genetic engineering have led to the development of varieties of soy, corn, and cotton that are resistant to glyphosate, meaning that glyphosate applications made while the crop is growing will control weeds without affecting the crop.

Other herbicides, such as the phenoxy herbicides (2,4-D, MCPA) are more selective in their toxicity; they are toxic to broad-leaf plants but not to narrow-leaf plants such as grasses. Some epidemiological studies have suggested links between agricultural worker exposure to phenoxy herbicides and certain types of cancer.

Fungicides A wide number of different types of fungicides are available for use in agriculture (**Table 2**). Fungicides control molds and other plant diseases by interfering with the growth and/or metabolic processes of fungal pests. In addition to improving crop

yields, fungicides may provide human health benefits by reducing the production of mycotoxins (naturally occurring toxins produced by fungi living on the food crop) such as aflatoxins and fumonisins.

Pesticides suspected as carcinogens The potential carcinogenic (cancer-causing) effects of many pesticides have generated considerable public concern and regulatory scrutiny. Although the consumption of foods containing pesticide residues has not been correlated with the development of human cancers, some epidemiological studies have linked the occupational use of specific pesticides with the development of cancers. Most pesticides considered as carcinogens, however, owe their status as suspected carcinogens to the results of long-term animal toxicology studies in rodents such as rats and mice.

The carcinogenic potential of pesticides and other chemicals is typically determined through long-term animal bioassay studies. In these studies, animals are frequently given a high dose (the maximum tolerated dose, or MTD, that may cause considerable toxicity but does not reduce life expectancy), a lower dose (commonly one-half to one-fourth of the MTD), and a control (zero) dose. Relatively high doses of chemicals are given to the animals to maximize the potential to identify toxicological effects such as tumor production.

The amounts of pesticides that cause cancer in animal studies frequently exceed the amounts that humans are exposed to from the diet by several orders of magnitude. It has been argued that many pesticides considered to be carcinogens receive such classification because of the high doses used in the long-term animal studies and that the results of such studies may not be applicable to human exposure to much lower levels of the pesticides. As an example, it is well established that high-dose toxicity may lead to increased rates of cell proliferation that in itself is linked to the development of cancer in laboratory animals. Human exposure at lower levels of exposure would not be expected to trigger such a response and, as a result, would not be expected to lead to the development of cancer.

Long-term animal studies are also prone to inconclusive and/or contradictory results among different test animals and under different conditions. As a result, the EPA has developed a weight-of-the-evidence evaluation scheme based upon results of any human data and animal testing to determine the likelihood of a pesticide to be carcinogenic. A list of several pesticides considered by the EPA to be 'probable' carcinogens is provided in **Table 3**. Some pesticides may receive classification a 'possible' carcinogens, whereas many others are considered noncarcinogens

Table 3 Some pesticides considered to be 'probable' carcinogens by the US Environmental Protection Agency

Pesticide	Pesticide type
Acetochlor	Herbicide
Aciflourfen sodium	Herbicide
Alachlor	Herbicide
Amitrol	Herbicide
Cacodylic acid	Herbicide
Chlorothalonil	Fungicide
Creosote	Wood preservative
Cyproconazole	Fungicide
Fenoxycarb	Insect growth regulator
Folpet	Fungicide
Heptachlor	Insecticide
Iprodione	Fungicide
Lactofen	Herbicide
Lindane	Insecticide
Mancozeb	Fungicide
Maneb	Fungicide
Metiram	Fungicide
Oxythioquinox	Insecticide
Pentachlorophenol	Fungicide
Pronamide	Herbicide
Propargite	Insecticide
Propoxur	Insecticide
Terrazole	Fungicide
Thiodicarb	Insecticide
TPTH	Fungicide
Vinclozolin	Fungicide

Regulating and Monitoring Pesticide Residues in Foods

The use of pesticides does not necessarily imply that food residues will occur. Many pesticides are applied to nonfood crops, whereas others such as broad-spectrum herbicides could damage or eliminate a crop if misapplied. The timing of the pesticide application is another important factor; many pesticides are applied to food crops prior to the development of edible portions of the crop, whereas other pesticides used on food crops may not result in residues, because of rapid environmental degradation between the time of application and the time of harvest.

Pesticide residue regulation In cases where the use of a pesticide on a food crop may present the potential to leave a residue in the USA, a maximum permitted allowable residue level, or tolerance, is established. Tolerances are specific to combinations of pesticides and commodities; it is possible for the same pesticide to have different tolerance levels established on different commodities, and several different pesticide tolerances for distinct pesticides may be established on the same commodity.

Although it may seem counterintuitive, pesticide tolerances are not based upon safety but rather represent the maximum expected residue of a pesticide on a particular commodity resulting from the legal use of a pesticide. The maximum levels are determined from the results of controlled field studies performed by the pesticide manufacturer using the 'worst legal case' conditions such as the maximum recommended application rate, maximum number of applications per growing season, and harvesting at the minimal anticipated time following harvest. Pesticide manufacturers typically petition the EPA to establish the tolerances at or slightly above the highest levels determined from the controlled field studies. As such, the values selected for tolerances are determined solely on the basis of agricultural practices but not as a result of human health risk assessments. Pesticide tolerances therefore should be considered to represent enforcement tools to determine whether pesticide applications may have been made in accordance to legal requirements; in cases where residues exceed the established tolerances, it is likely that such residues resulted from misapplication of the pesticides. Such a finding, however, rarely constitutes an 'unsafe' residue according to standard toxicological criteria. Pesticide tolerances, therefore, should be viewed as enforcement tools but not as standards of safety.

Before the EPA grants a tolerance, human health risk assessments are performed to determine the conditions for acceptable pesticide use. Such conditions include the listing of commodities on which a specific pesticide may be used, the target pests controlled by the pesticide, application requirements, and the acceptable interval between application and harvest. In the EPA's assessment of acceptable levels of consumer exposure to pesticides, it considers potential human exposure from all registered (and proposed) uses of the pesticide. If the resulting risk is deemed to be excessive, the EPA will not allow tolerances to be established for specific commodities. If the risks are considered to be acceptable, the tolerances are established, as discussed in the preceding paragraph.

The processes that the EPA uses to determine the acceptability of dietary pesticide risk are quite complicated and subject to ongoing evolution to meet the needs of new regulations, improved toxicology testing, and advances in computational methods.

The first step in evaluating the consumer risks from pesticide residues involves making estimates of the amount of consumer exposure. The maximum legal exposure to the pesticide is frequently calculated by assuming that (1) the pesticide is always used on all food items for which it is registered and/or proposed for registration, (2) all residues on the food items will be present at the established or proposed tolerance levels, and (3) there will be no reduction in residue levels resulting from postharvest effects such as washing, cooking, peeling, processing, and transportation. This approach leads to the calculation of the theoretical maximum residue contribution (TMRC). Although studies have indicated that the TMRC values may overestimate the actual consumer exposures by factors of 100–100 000, the TMRC provides a starting point from which to estimate consumer risks.

In practice, the TMRC is compared with established toxicological criteria such as the reference dose (RfD) or the acceptable daily intake (ADI) that represent, following analysis of animal toxicology data and extrapolations to human health, the daily exposure levels that do not constitute an appreciable level of risk. In cases where the EPA determines that exposure to a pesticide at the TMRC is below the RfD or ADI, the risks for the pesticide are typically deemed to be negligible, and the EPA allows tolerances to be established for the pesticide on specific commodities. For pesticides that are considered as potential carcinogens, the EPA also requires that the quantitative carcinogenic risk to the pesticide be below one excess cancer per million using models that calculate possible human risks from low levels of exposures to potentially carcinogenic pesticides. Such models are developed from the results from long-term studies on animals given high doses of the pesticides. In cases where the exposures at the TMRC exceed the RfD or ADI, or in cases where the carcinogenic risks at the TMRC exceed one excess cancer per million, the EPA may adopt a refined risk assessment that more accurately expresses exposures. Such refinements may include adjustments of actual pesticide use, the use of more realistic pesticide residue data, and consideration of postharvest effects that may significantly reduce residue levels prior to consumption. In cases where the refined exposure estimates are below the RfD or ADI and where the carcinogenic risks at the refined exposure estimates are below one excess cancer per million, the EPA will typically allow tolerances to be established.

The processes by which the EPA determines human risks from pesticide exposure became more complicated following the passage and adoption of the Food Quality Protection Act (FQPA) of 1996. It is clear that the new requirements of FQPA will require scientists to develop improved methods for assessing risks to pesticides and that once such methods are adopted, the regulatory requirements may be much more stringent and may lead to reductions in the amounts and types of pesticides that may be used on food crops.

Prior to FQPA, the EPA allowed tolerances to be established on a chemical-by-chemical basis and considered only exposures resulting from dietary pathways. FQPA now requires the EPA to establish tolerances only when the risks posed by pesticides

represent a 'reasonable certainty of no harm.' In determining what constitutes this 'reasonable certainty of no harm,' the EPA must now consider the *aggregate* exposure to pesticides from dietary, drinking water, and residential sources as well as the *cumulative* exposure from pesticides possessing a common mechanism of toxic action. As such, the EPA may consider the risks from entire families of chemicals rather than the risks from individual chemicals. Another important provision of FQPA is the so-called *10 × factor*, which requires the EPA to consider applying an additional 10-fold uncertainty factor in cases where infants or children may be more susceptible to the toxicological effects of pesticides than adults. The application of the full *10 × factor* would result in a subsequent reduction in the RfD by a factor of 10.

Internationally, many countries adopt the Codex Alimentarius maximum residue limits (MRLs), which, like the US tolerances, exist primarily as enforcement tools to determine if pesticide applications are made following good agricultural practices. Although many US tolerances and Codex Alimentarius MRLs are identical, there are many cases in which the US tolerances are more restrictive and many others in which the Codex Alimentarius MRLs are more restrictive.

Pesticide residue monitoring
Authority In the US, the Food and Drug Administration (FDA) has the primary responsibility for enforcing tolerances in domestic and imported foods. Domestic food samples are frequently collected near the source of production or at the wholesale level, whereas imported food samples are typically taken at the point of entry into the US. The types and quantities of samples taken by FDA are determined by a variety of factors such as regional intelligence on pesticide use, the dietary importance of specific foods, information on the amount of foods that enter interstate commerce, and pesticide use patterns. Samples are analyzed using multiresidue methods capable of detecting over 200 individual pesticides.

The US Department of Agriculture (USDA) also has a role in pesticide residue monitoring, as it is responsible for the monitoring of meat, poultry, and egg products. It also conducts the Pesticide Data Program (PDP), which collects residue data on fruits, vegetables, and processed foods. Findings from the USDA's PDP are considered to be more representative of the actual food supply than those collected from the FDA's regulatory monitoring programs and are commonly used by EPA to aid in its risk-assessment efforts.

US monitoring of pesticide residues also occurs at the state level. The largest state pesticide regulatory and monitoring program exists in California, and several other states, including Texas and Florida, have significant pesticide-monitoring programs.

Residue findings The most recent FDA pesticide monitoring residue data are available for 1999. During that year, FDA analyzed 9438 food samples for pesticide residues. More samples were taken from imported foods (6012 samples, or 63.7%) than from domestic foods (3426 samples or 36.3%).

The results of FDA's 1999 monitoring of pesticide residues in imported foods are shown in **Figure 2**. Overall, 65.0% of the samples showed no detected residues, and violations were identified in 3.1% of the samples. **Figure 3** shows the comparable results from domestic foods where 60.2% of the samples showed no detectable residues, and violations were present in 0.8% of the samples.

Pesticide residue violations commonly take one of two forms. The most common form of a violative residue results when residues of a pesticide are detected on a commodity for which no tolerance has been established. This type of violation may result from application of the pesticide to the wrong commodity, uptake from soil contaminated from a prior use of the pesticide on a different commodity, or drift of a pesticide from an adjacent field. The other type of violation results when residue levels are detected in excess of the established tolerance. In 1999, 90% of the import violations occurred where residues were detected on commodities for which tolerances were not established, and only 10% of the violations

represented residues in excess of the tolerance. For domestic samples, 69% were detected on commodities for which tolerances were not established, and the remaining 31% of violations occurred when levels exceeding tolerances were detected. Fruits and vegetables were responsible for the highest percentages of residue detections and the greatest number of violations from both imported and domestic foods.

Results obtained from California's 1997 Marketplace Surveillance Program were quite similar to those of the FDA's 1999 monitoring. This program analyzed 5660 samples in 1997, with 62.2% of those originating from California, 6.7% from other US states, and the remaining 31.1% from other countries. The majority of samples (62.1%) showed no detectable residues, whereas 36.7% of the samples contained legal residues. Violations were detected in 1.2% of the samples, and only 12% of the violations represented residues detected in excess of tolerances. The violation rate from imported foods was 2.4%, and the violation rate for domestic foods was 0.7%.

The USDA's PDP analyzed 9125 food samples for pesticide residues in 1999. Foods analyzed included apples, cantaloupe, cucumbers, grape juice, lettuce, oats, pears, spinach, strawberries, sweet bell peppers, tomatoes, winter squash, and corn syrup. Most of the samples (8637) were from fruits and vegetables, with lower numbers of samples collected for oats (332) and corn syrup (156). The majority of samples (79%) was of domestic origin. Overall, 36% of the samples contained no detectable residue, whereas

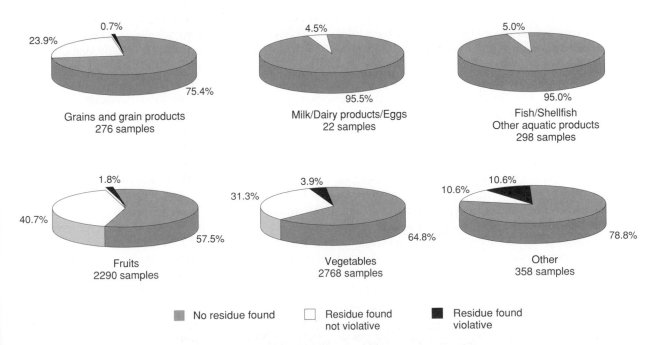

Figure 2 Results of US Food and Drug Administration monitoring of imported foods for pesticide residues, 1999.

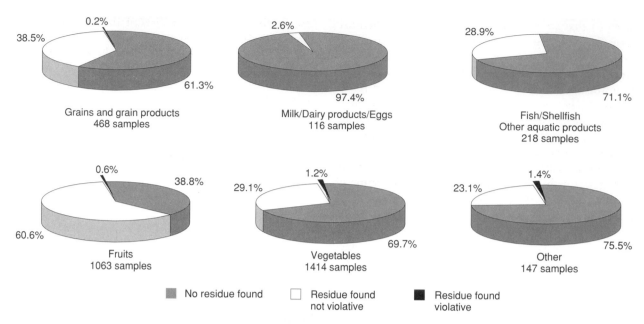

38.5% 0.2%

61.3%

Grains and grain products
468 samples

2.6%

97.4%

Milk/Dairy products/Eggs
116 samples

28.9%

71.1%

Fish/Shellfish
Other aquatic products
218 samples

0.6% 38.8%

60.6%

Fruits
1063 samples

29.1% 1.2%

69.7%

Vegetables
1414 samples

23.1% 1.4%

75.5%

Other
147 samples

■ No residue found   ☐ Residue found   ■ Residue found
                        not violative   violative

Figure 3 Results of US Food and Drug Administration monitoring of domestic foods for pesticide residues, 1999.

26% contained one residue, and 35% contained more than one residue. Residues exceeding the tolerance level were detected on 0.3% of the samples, and residues of pesticides were detected on commodities for which no tolerances of the pesticides were established on another 3.7% of the samples.

Dietary Risks from Exposure to Pesticides in Food

A common method used to discuss the potential human health risks arising from the consumption of pesticide residues in the diet is to report on the results of regulatory monitoring programs, as has been done in the previous paragraphs. Results indicate that a large percentage of food samples contain no detectable residues and that violation rates are relatively low, particularly for foods grown in the US. Frequently, the recitation of such findings is used as justification for a lack of significant human health risks resulting from pesticide residues in foods.

The problem with adopting such an approach is that it ignores the important, yet confusing, fact that pesticide tolerances are *not* safety standards but rather represent the maximum legal residues expected when pesticides have been used according to directions. Taking this a step further, violative residues should not be considered to be *unsafe* residues in most cases but merely represent cases in which pesticides have been misapplied or have been transported to commodities for which they are not registered. Although cases of pesticide misuse have historically resulted in a small number of incidents of acute poisoning of people consuming tainted foods, it can

be reasonably argued that the vast majority of violative pesticide residues do not represent significant health threats to consumers based upon common toxicological and risk assessment criteria.

A more accurate approach for estimating human dietary risks from pesticides is to use exposure data derived from market basket surveys rather than regulatory monitoring data. The FDA, for example, performs its Total Diet Study annually. This study uses a market basket approach, with each market basket containing more than 250 individual food items. Foods are collected by FDA inspectors from three cities in each of the four geographical regions of the USA and are prepared for table-ready consumption prior to analysis for pesticide residues. By combining analytical results from estimates of typical consumption rates of the various food items or their components, it is possible to estimate the typical daily exposure of members of the general population to specific pesticides as well as to break the population down further into subgroups defined by factors such as age, gender, and geographical location.

Although the FDA no longer makes its dietary pesticide exposure estimates from its Total Diet Study available to the public, results reported from the 1991 Total Diet Study indicate that, for most pesticides in a variety of different population subgroups, the typical daily exposure estimates represent only a small fraction (often less than 1%) of the corresponding RfDs or ADIs. To put this into some level of perspective, it should be noted that typical RfDs are derived by first identifying the highest level of exposure to a pesticide that causes no noticeable signs of toxicity in laboratory animals and then dividing that level by an uncertainty factor (usually 100) that presumably covers potential variability resulting from the animal to human extrapolation and from interhuman variability. Exposure at a level of 1% of the RfD represents an exposure 10 000 times below the level that does not produce noticeable effects in the animals. Such findings provide an illustration of why the majority of health professionals consider the typical human health risks from pesticides in the diet to be much lower than food safety risks posed by such factors as microbiological contamination of foods, nutritional imbalance, environmental contaminants, and naturally occurring toxins. The risks from pesticides in the human diet are clearly not zero, since consumption of tainted foods has caused documented human illnesses throughout the world, and concerns remain regarding potential long-term effects of dietary pesticide exposure.

It is also clear that the potential health benefits resulting from pesticide use should be considered. Pesticide use has resulted in increases in production of a wide variety of food crops, which translates into greater availability and lower consumer costs and thus a greater potential consumption of agricultural products. Epidemiological studies have clearly indicated that diets rich in consumption of fruits, vegetables, and grains may significantly decrease one's risk of heart disease and certain types of cancer. The US National Academy of Sciences, among other scientific bodies, has concluded that the theoretical increased risks from pesticide exposure resulting from increases in consumption of fruits, vegetables, and grains were greatly outweighed by the health benefits of these foods.

See also: **Cancer**: Epidemiology; Carcinogens in the Food Chain; **Carcinogens**: Carcinogenic Substances in Food: Mechanisms; **Food and Drug Administration**; **Food Poisoning**: Classification; **Pesticides and Herbicides**: Types of Pesticide; Types, Uses, and Determination of Herbicides

Further Reading

Fong WG, Moye HA, Seiber JR and Toth JF (1999) *Pesticide Residues in Foods: Methods, Techniques, and Regulation.* New York: John Wiley.

Hayes WJ and Laws ER (1991) *Handbook of Pesticide Toxicology.* San Diego, CA: Academic Press.

National Research Council (1987) *Regulating Pesticides in Foods: The Delaney Paradox.* Washington, DC: National Academy Press.

National Research Council (1993) *Pesticides in the Diets of Infants and Children.* Washington, DC: National Academy Press.

Tweedy BG, Dishburger HJ, Ballantine LG and McCarthy J (1991) *Pesticide Residues and Food Safety: A Harvest of Viewpoints.* Washington, DC: American Chemical Society Symposium Series No. 446.

Winter CK (1992a) Pesticide tolerances and their relevance as safety standards. *Regulatory Toxicology and Pharmacology* 15: 137–150.

Winter CK (1992b) Dietary pesticide risk assessment. *Reviews of Environmental Contamination and Toxicology* 127: 23–67.

Winter CK (2001) Contaminant regulation and management in the United States: The case of pesticides. In: Watson DH (ed.) *Food Chemical Safety, Volume 1: Contaminants,* pp. 295–313. Cambridge, UK: Woodhead.

Winter CK and Francis FJ (1997) Assessing, managing, and communicating chemical food risks. *Food Technology* 47: 85–92.

PH – PRINCIPLES AND MEASUREMENT

D Webster, Formerly of University of Hull, Hull, UK

Background

pH measurements have been, and continue to be, widely used as a rapid, accurate measure of the acidity of fluids of all sorts. There are two methods for measuring pH: colorimetric methods using indicator solutions or papers, and the more accurate electrochemical methods using electrodes and a millivoltmeter (pH meter). The development of the glass electrode, which is convenient to use in a variety of environments, and the development of the pH meter have enabled the widespread application of pH measurement and control to take place. The determination, and hence the control of pH, is of great importance in the food industry.

Basic Theory

In water, molecules (H_2O) are in equilibrium with hydrogen ions (H^+) and hydroxide ions (OH^-) (eqn (1)).

Further Reading

Eisenbrand G, Dayan AD, Elias PS, Grunow W and Schlatter J (2000) *Carcinogenic and Anticarcinogenic Factors in Food.* Weinheim: Wiley-VCH.

Harborne JB, Mabry TJ and Mabry H (eds) (1975) *The Flavonoids.* London: Chapman & Hall.

Harborne JB and Mabry TJ (1982) *The Flavonoids: Advances in Research.* London: Chapman & Hall.

Harborne JB (ed.) (1994) *The Flavonoids: Advances in Research Since 1986.* London: Chapman & Hall.

Lindsay DG and Clifford MN (eds) (2000) Nutritional enhancement of plant-based food in European trade (NEODIET). *Journal of the Science of Food and Agriculture* 80.

Macheix J-J, Fleuriet A and Billot J (1990) *Fruit Phenolics.* Boca Raton, FL: CRC Press.

Packer L and Cadenas E (eds) (1996) *Handbook of Antioxidants.* New York: Marcel Dekker.

Packer L, Hiramatsu M and Yoshikawa T (eds) (1999) *Antioxidant Food Supplements in Human Health.* San Diego, CA: Academic Press.

Rice-Evans C and Packer L (eds) (1998) *Flavonoids in Health and Disease.* New York: Marcel Dekker.

Shahidi F and Naczk M (1995) *Food Phenolics: Sources, Chemistry, Effects, Applications.* Lancaster, PA: Technomic.

Shahidi F and Ho C-T (eds) (2000) *Phytochemicals and Phytopharmaceuticals.* Champaign, IL: AOCS Press.

Shahidi F and Wanasundara PKJ (1992) Phenolic antioxidants. *Critical Reviews in Food Science and Nutrition* 32: 67–103.

PHOSPHOLIPIDS

Contents
Properties and Occurrence
Determination
Physiology

Properties and Occurrence

B F Szuhaj, Central Soya Company Inc., Fort Wayne, IN, USA

This article is reproduced from *Encyclopaedia of Food Science, Food Technology and Nutrition*, Copyright 1993, Academic Press.

Introduction

Phospholipids have been scientifically studied since the 1700s and became commercially available as lecithin in the 1930s. Their primary commercial source today is the soya bean, but phospholipids can be found in all living cells as part of the cellular membranes. This article covers the properties and occurrence of phospholipids as commercial lecithins, their chemistry, manufacture, composition, specifications, and their potential use in food systems.

Occurrence

The International Lecithin and Phospholipid Society defines lecithin as 'a complex mixture of glycerophospholipids obtained from animal, vegetable or microbial sources, containing varying amounts of substances such as triglycerides, fatty acids, glycolipids, sterols, and sphingophospholipids.' Lecithins are natural surfactants primarily derived from soya beans and eggs. They are found most abundantly in seeds and nuts, eggs, brains, and cell walls, in a concentration range of 0.5–2%. (*See* **Eggs**: Structure and Composition; **Fatty Acids**: Properties; **Soy (Soya) Beans**: Properties and Analysis; **Triglycerides**: Structures and Properties.)

Properties of Phospholipids (Lecithins)

There are three types of properties necessary to define phospholipids and lecithins: (1) chemical, (2) physical, and (3) functional.

Chemical Properties

The chemical composition of deoiled and liquid soya bean lecithin is shown in **Table 1**. There are approximately 17 different compounds in commercial lecithin, including carbohydrates, phytosterols, and minor phytoglycolipids. The three major phospholipids are phosphatidylcholine, phosphatidylethanolamine, and phosphatidylinositol.

Table 1   Chemical composition of soya bean lecithin

	Granular lecithin	Typical liquid lecithin
Phosphatides (acetone-insolubles) (%)	95 (minimum)	60 (minimum)
Soya bean oil (%)	2–3	39
Moisture (%)	1	0.7
Fat (g per 100 g product)	90	93
Monounsaturated (oleic acid) (%)	9.2	17.9
Polyunsaturated (linoleic, linolenic acids) (%)	65.9	60.7
Saturated (palmitic, stearic acids) (%)	24.9	20.3
Carbohydrates (g per 100 g of product)	8	5
Approximate composition (100-g sample)		
Fatty acid content (g)	50	66
Fatty acid content (relative composition) (%)		
Linoleic	58.9	54.0
Linolenic	7.0	6.7
Oleic	9.2	17.9
Palmitic	20.3	15.6
Stearic	4.6	4.7
Other fatty acids	0.0	1.1
Total	100.0	100.0
Primary acetone insolubles (g)		
Phosphatidylcholine	23	15
Phosphatidylethanolamine	20	12
Phosphatidylinositol	14	9
Elemental analysis (mg)		
Calcium	65	40
Iron	2	1
Magnesium	90	60
Phosphorus	3000	2000
Potassium	800	440
Sodium	30	10

Reproduced from Central Soya, Chemurgy Division (1989) *The Lecithin Book*. Fort Wayne: Central Soya.

Structure of the Major Phospholipids

The chemical backbone of the major phospholipids is a diacylglycerol molecule with the third carbon attached to a phosphate molecule. Choline, ethanolamine, serine, and inositol can be attached to the phosphate group to change the physical and functional properties, leading to the formation of phosphatidylcholine, phosphatidylethanolamine, phosphatidylserine, and phosphatidylinositol, respectively. The groups attached to positions 1 and 2 (α or β) are C_{14}–C_{18} fatty acids with double bonds associated with the lecithin source. The second carbon of the glycerol molecule, the β-position, usually contains linoleic acid.

Physical Properties

There are two major physical classes of lecithins: (1) fluid, and (2) waxy solids.

The fluid lecithins can have viscosities from 5000 to 100 000 cP, depending on processing conditions and diluents. The low-viscosity products are made through the addition of fatty acids and vegetable oil, depending on the function and stability required. Divalent metal ions like calcium can be added during drying to decrease viscosity. The moisture content can also make a difference. Water levels above 1% will increase the viscosity, eventually to a plastic state.

Deoiled lecithins are waxy solids that can be ground to various particle sizes. They are stable, free-flowing granules or powders.

Functional Properties

Lecithins are multifunctional agents. They can be used for many purposes in a food system, as shown in **Table 2**. The most popular functionalities are discussed below.

Antidusting agents   Lecithins reduce static electricity by wetting dusty particles. They can be used alone or in conjunction with vegetable oils. Oils can be selected for the degree of shelf-life required.

Crystal formation modifier   Lecithins retard nucleation in fats and even monoglycerides, reducing graininess in texture.

Emulsifiers   Lecithins are most often used as amphoteric emulsifiers. They promote stable formation of oil-in-water and water-in-oil emulsions by reducing

Table 2 Functionality of lecithins

Adhesion aid
Antibleed agent (as in fat bloom)
Anticorrosive
Antidusting agent
Antioxidant
Antispatter agent
Biodegradable additive
Biologically active agent
Catalyst
Color intensifier
Conditioning agent
Coupling agent
Dispersing agent, mixing aid
Emollient, softening agent
Emulsifier or surfactant
Flocculant
Grinding aid
Lubricant
Liposomal encapsulating agent
Machining aid
Modifier
Moisturizer
Nutritional supplement, vitamin source
Penetrating agent
Plasticizer
Promoter
Release agent, antisticking agent
Spreading agent
Stabilizer
Strengthening agent
Suspending agent
Synergist
Viscosity modifier
Water repellent
Wetting agent

Reproduced from Schmidt JC and Orthoefer FT (1985) In Szuhaj BF and List GR (eds) *Lecithins*, p. 187. Champaign: American Oil Chemists' Society, with permission.

the interfacial surface tension between immiscible liquids. (*See* **Emulsifiers**: Organic Emulsifiers.)

Mixing and blending aids Lecithins decrease time and increase efficiency of mixing of unlike ingredients such as sugar and shortening by providing lubricity as well as viscosity reduction at the contact surfaces of the incompatible solids.

Release agents Lecithins provide easy release from metallic surfaces by attaching to the metal surface during hot or cold cooking. They assist in the cleaning of hot surfaces where proteins or batters are applied. They also reduce sticking between frozen food products.

Separating agents Lecithins prevent the adhesion of products that normally stick when in contact, like cheese slices and caramel confectionery.

Viscosity modifiers Lecithins reduce viscosity by coating particles to reduce particle–matrix friction such as in chocolates.

Wetting agents Lecithins provide complete wetting of fatty or hydrophilic powders in aqueous systems. The fatty acids are attracted to the fatty portion and the hydrophilic portions of the molecules actively imbibe water and control the hydration of the powder.

Manufacture

The majority of phospholipids are solvent-extracted from their source. Usually, nonpolar hydrocarbons, like hexane, are used. Soya beans, for example, are cleaned, cracked, and dehulled before flaking and extraction. The hexane is removed from the solvent micelle and the crude soya bean oil is cooled for further refining. **Figure 1** shows a flow diagram for the degumming of crude soya bean oil and the production of lecithin. Approximately 2–3% water is added to the crude oil and agitated for at least 30 min. The phosphatides hydrate, swell, and are separated by centrifugation. The wet gums with 50% moisture are dried through a thin-film drier. The important points here are careful drying temperatures and cooling of the product below 20 °C. Lightening the color of the product can be achieved with hydrogen peroxide in the gum stage. The hydrogen peroxide is removed through drying.

Modification

The chemistry and functionality can be altered by simple chemical additions with acids and bases as well as hydrogen peroxide and acetic anhydride. These modifications increase the dispersion and hydration properties of the lecithin. Enzyme modification is also possible with lipases and phospholipases. These changes markedly affect the functionality in emulsification.

Composition of Lecithins

The composition of lecithins and their phospholipids will vary depending on their source: vegetable, animal, or bacterial. There are minor differences within a class but major differences between the sources. Vegetable lecithins are high in phosphatidylcholine, phosphatidylethanolamine, phosphatidylinositol, and phosphatidic acid, but very low in phosphatidylserine and contain no sphingomyelin. Animal lecithins are high in phosphatidylcholine, phosphatidylethanolamine, phosphatidylserine and

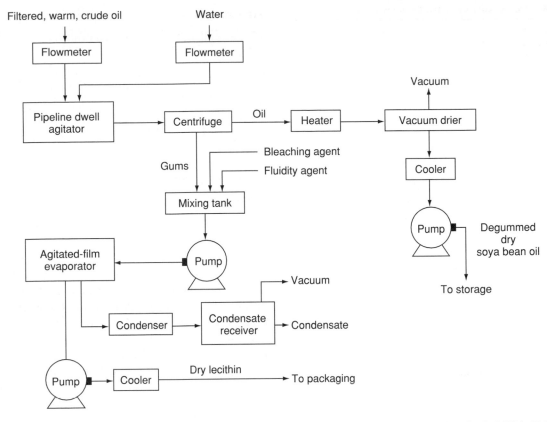

Figure 1 Flowsheet for degumming and crude lecithin production. Reproduced from List GR (1989) In Szuhaj BF (ed.) *Lecithins: Sources, Manufacture and Uses*, p. 149. Champaign: American Oil Chemists' Society, with permission.

sphingomyelin but contain no phosphatidylinositol. Microbes have phospholipids similar to the plant kingdom with high levels of phosphatidylethanolamine and phosphatidylcholine and phosphatidylserine or sphingomyelin.

Specifications of Lecithins

Lecithins may be qualified in several ways, chemically, physically, and functionally, but there are also specifications used to assess quality and purity (**Table 3**). These include acetone-insolubles (AI), acid value (AV), hexane-insoluble matter (HI), peroxide value (PV), moisture, color, free fatty acids (FFA), divalent metals (DVM), iodine value (IV), and phosphorus. Most of these analytical methods are found in the American Oil Chemists' Society (AOCS) *Official Methods and Recommended Practices*, Section J.

Acetone-Insolubles

Phospholipids are nearly insoluble in cold acetone. This quantitative method should measure the active ingredients in lecithin. Depending on the type of lecithin, the range is 35–98%.

Acid Value

The phosphorus group in lecithins have a titratable acidity that is measured with this volumetric method. FFAs are also measured in this test and should not be confused with phospholipid acidity. The AV range is 20–36 mg of potassium hydroxide per gram.

Hexane-Insolubles

In the processing of soya bean lecithin, particulate matter finds its way through some processes and gives the product a hazy appearance. The HI can be determined by dissolving the product in hexane, centrifuging and gravimetrically measuring the insolubles. The HI range for soya bean lecithins is 0.05–0.3%.

Peroxide Value

The PV is a measure of oxidation in fats and oils. In lecithin, however, the PV usually measures the residual hydrogen peroxide from processing. The range in unbleached products is 0–10 mmol kg^{-1} and in bleached products is 10–75 mmol kg^{-1}.

Table 3 Specifications range for commercial lecithins

Analysis	Typical range		
Acetone-insolubles (%)	35–98		
Acid value (mg KOH g^{-1})	20–36		
Hexane-insolubles (%)	0.05–0.3		
Peroxide value (mmol kg^{-1})			
Unbleached	0–10		
Bleached	10–75		
Moisture (%)	0.1–1.0		
Viscosity (Brookfield, 25 °C) (cP)	150–20 000		
Free fatty acids (mg KOH (g^{-1})	1–5		
Iodine value (cg I g^{-1})			
Natural	95–110		
Oil-free	80–90		
	Fatty acid composition (%)		
	Soya bean oil	Natural	Oil-free
C$_{16:0}$	10.3	15.6	20.3
C$_{18:0}$	4.4	4.7	4.6
Total saturates	14.7	20.3	24.9
C$_{18:1}$	24.5	17.9	9.2
C$_{18:2}$	53.8	54.0	58.9
C$_{18:3}$	7.0	6.7	7.0
Total unsaturates	85.3	78.6	75.1
Unsaturated: saturated ratio	5.8:1	3.9:1	3.0:1

Reproduced from Central Soya, Chemurgy Division (1990), with permission.

Moisture

The water content of lecithins is quite low, at 0.1–1.0%. It can be measured by oven drying but is more accurately determined by the Karl–Fischer method. This water content is so low that there is no measurable water activity for microbial growth. Moisture contents above 1% will change the viscosity from a fluid to a plastic state. (*See* **Water Activity: Principles and Measurement.**)

Viscosity

The Brookfield viscosity at 25 °C will have a range of 150–20 000 cP. Diluents and divalent metals can alter the viscosity to usable levels.

Free Fatty Acids

This method measures the true fatty acid levels in lecithins which range from 1 to 5 mg of potassium hydroxide per gram. Fatty acids are added as additives to adjust the viscosity. (*See* **Fatty Acids: Analysis.**)

Iodine Value

The IV is a traditional method for qualifying lecithin sources. The more unsaturated fatty acids are found in soya bean lecithins, which have a range of 95–110 in natural fluid lecithins, to 80–90 in deoiled lecithin.

Phosphorus

This wet chemical method is an indirect way of measuring the phospholipid content. The typical level of phosphorus is 2.0% in fluid lecithin and 3.0% in deoiled products. The AOCS method Ca 12–55 has an approximation for converting percent phosphorus to the phosphatides in soya bean oil. The equivalent phosphatides content is equal to percent phosphorus × 30.

Uses of Lecithin

There are many uses of phospholipids in the food industry. As seen from the functional properties, there are multiple functions for lecithins. The following is a listing of the major areas of use:

- margarines – emulsifier, stabilizer, and antispatter;
- confectionery and snack foods – crystallization control, viscosity control, antisticking;
- instant foods – wetting and dispersing agent, emulsifier;
- commercial bakery products – crystallization control, emulsifier, wetting agent, release agent;
- cheese products – emulsifier, release agent;
- meat and poultry processing – browning agent, phosphate dispersant;
- dairy and imitation products – emulsifier, wetting and dispersing agent, antispattering and release agent;
- packaging aid – release agent, sealant;
- processing equipment – internal or external release agent, lubricant.

Refer to individual foods

Applications of lecithins in foods are clearly supported by the Food Chemicals Codex, the European E322 regulations, and they are considered as generally regarded as safe substances by the US Food and Drugs Administration.

Storage and Handling

Lecithins are very stable products. They are shipped in drums or in bulk containers. They can be stored at ambient temperatures for up to 2 years without loss of activity or becoming rancid or spoiling. The water activity is so low that no microbial growth can occur. The products may be heated to 25 °C for easier application. (*See* **Water Activity: Effect on Food Stability.**)

See also: **Eggs**: Structure and Composition; **Emulsifiers**: Organic Emulsifiers; **Fatty Acids**: Properties; Analysis; **Soy (Soya) Beans**: Properties and Analysis; **Spices and Flavoring (Flavouring) Crops**: Tubers and Roots; **Triglycerides**: Structures and Properties; **Water Activity**: Principles and Measurement; Effect on Food Stability

Further Reading

Burner D (ed.) (1991) *Official Methods and Recommended Practices*. Champaign: American Oil Chemists Society.

Central Soya, Chemurgy Division (1990) *The Lecithin Book*. Fort Wayne: Central Soya.

Charalambous G and Doxastakis G (1989) *Food Emulsifiers: Chemistry, Technology, Functional Properties and Applications*. New York: Elsevier.

Hanin I and Pepeu G (1990) *Phospholipids: Biochemical, Pharmaceutical, and Analytical Considerations*. New York: Plenum Press.

Szuhaj BF (ed.) (1989) *Lecithins: Sources, Manufacture and Uses*. Champaign: American Oil Chemists' Society.

Szuhaj BF and List GL (eds) (1985) *Lecithins*. Champaign: American Oil Chemists' Society.

Determination

B F Szuhaj, Central Soya Company Inc., Fort Wayne, IN, USA

This article is reproduced from *Encyclopaedia of Food Science, Food Technology and Nutrition*, Copyright 1993, Academic Press.

Introduction

Phospholipids are a well-known class of lipids that have been thoroughly analyzed over the past three centuries. Their complete analysis was facilitated by the great advances in separation science and qualitative procedures that have occurred in the last 50 years. This article will cover the analysis of phospholipids from their structure and composition, extraction techniques, qualitative and quantitative assays, and industrial methodology. (*See* **Fats**: Classification.)

Structure of Phospholipids

There are at least a dozen compounds that fall into the class of phospholipids. They have a basic structure of a diacylglycerol backbone with a phosphate ester on the α or third carbon of the glycerol molecule. Usually another compound is attached that characterizes the phospholipid. **Figure 1** shows examples of the structures of the major phospholipids. These phospholipids (and their common abbreviations) are:

- phosphatidylcholine (PC, PtdCho)
- phosphatidylethanolamine (PE, PtdEth)
- phosphatidylserine (PS, PtdSer)
- N-acylphosphatidylethanolamine (NAPE, N-acyl-Ptd-Eth)
- phosphatidylinositol (PI, PtdIns)

- phosphatidic acid (PA, PtdA)
- phosphatidylglycerol (PG, PtdGro)
- plasmologen (PM)
- diphosphatidylglycerol (DPG, diPtdGro)
- lysophosphatidylcholine (LPC, lysoPtdCho)
- lysophosphatidylethanolamine (LPE, lysoPtdEth)

The proper nomenclature for phospholipids has been defined by the 1976 revised recommendations of the International Union of Pure and Applied Chemistry (IUPAC) and the International Union of Biochemistry (IUB) Committee on Biochemical Nomenclature. For example, the term 'lecithin' is permitted for phosphati-dylcholine but the systematic name is 1,2-diacyl-*sn*-glycero-3-phosphorylcholine. The generic name of 3-*sn*-phosphatidylcholine could be used. The abbreviation PtdCho is also allowed. This article will use the common names listed above since the literature has thousands of references with this terminology.

Composition of Phospholipids

The composition of phospholipids depends on the source of the phospholipids. Those from animal, plant, and microbial sources will have different compositions, depending on the nature of the tissue from which the lipids are extracted – for example, brain, liver, or blood. In plants, it will vary on whether they are from soya beans, corn, cotton, rapeseed, or sunflower. In microbial sources it depends upon the organism.

The phospholipid classes are similar within a species, but differ primarily in fatty acid acyl composition around the 1 and 2 positions on the glycerol backbone.

Fatty Acids

The fatty acid chain length is commonly from C_4 to C_{26}, with different degrees of unsaturation from one to six double bonds, which may be at different locations on the acyl group. However, there is a pattern that is relevant to the present discussion. (*See* **Fatty Acids**: Properties.)

In animals the primary fatty acids range from $C_{12:0}$ (lauric acid) to $C_{24:0}$ (tetracosanoic acid). Again, depending on the species and tissue extracted, the fatty acids can be variable.

In the plant kingdom, the primary diacyl groups on the phospholipids will range from $C_{12:0}$ to $C_{18:3}$, i.e., lauric to linolenic acid. There are usually no C_{20} fatty acids and higher, as in the animal kingdom. The degree of unsaturation depends on the origin of the crop, i.e., from temperate or tropical regions. Also, the climate within the zone can make a seasonal difference.

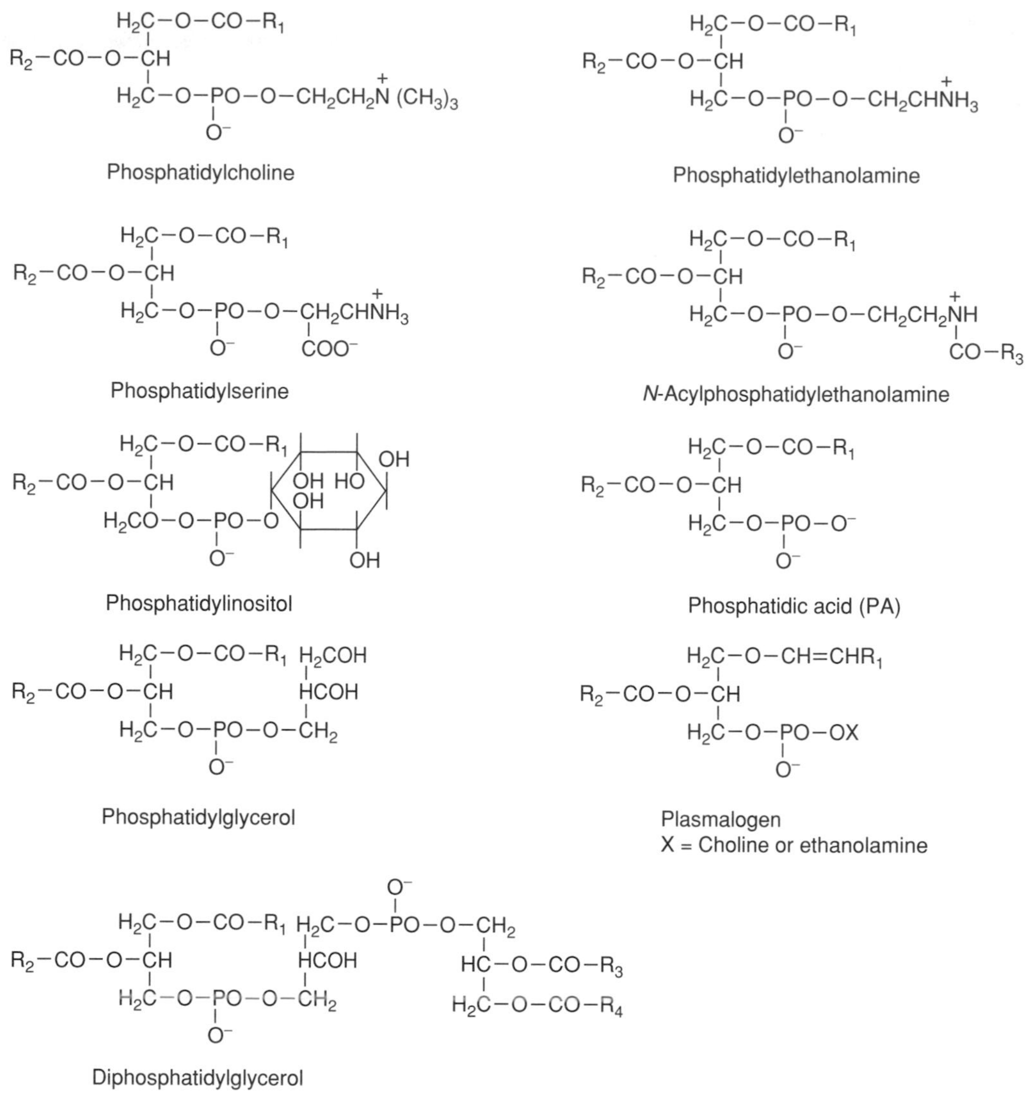

$$H_2C-O-CO-R_1$$
$$R_2-CO-O-CH$$
$$H_2C-O-PO-O-CH_2CH_2\overset{+}{N}(CH_3)_3$$
$$O^-$$

Phosphatidylcholine

$$H_2C-O-CO-R_1$$
$$R_2-CO-O-CH$$
$$H_2C-O-PO-O-CH_2\overset{+}{C}HNH_3$$
$$O^-$$

Phosphatidylethanolamine

$$H_2C-O-CO-R_1$$
$$R_2-CO-O-CH$$
$$H_2C-O-PO-O-CH_2\overset{+}{C}HNH_3$$
$$O^-\quad COO^-$$

Phosphatidylserine

$$H_2C-O-CO-R_1$$
$$R_2-CO-O-CH$$
$$H_2C-O-PO-O-CH_2CH_2\overset{+}{N}H$$
$$O^-\quad CO-R_3$$

N-Acylphosphatidylethanolamine

$$H_2C-O-CO-R_1$$
$$R_2-CO-O-CH$$
$$H_2CO-O-PO-O$$
$$O^-$$
(inositol ring: OH OH HO OH OH)

Phosphatidylinositol

$$H_2C-O-CO-R_1$$
$$R_2-CO-O-CH$$
$$H_2C-O-PO-O^-$$
$$O^-$$

Phosphatidic acid (PA)

$$H_2C-O-CO-R_1\quad H_2COH$$
$$R_2-CO-O-CH\quad HCOH$$
$$H_2C-O-PO-O-CH_2$$
$$O^-$$

Phosphatidylglycerol

$$H_2C-O-CH=CHR_1$$
$$R_2-CO-O-CH$$
$$H_2C-O-PO-OX$$
$$O^-$$

Plasmalogen
X = Choline or ethanolamine

$$O^-$$
$$H_2C-O-CO-R_1\quad H_2C-O-PO-O-CH_2$$
$$R_2-CO-O-CH\quad HCOH\quad HC-O-CO-R_3$$
$$H_2C-O-PO-O-CH_2\quad H_2C-O-CO-R_4$$
$$O^-$$

Diphosphatidylglycerol

Figure 1 Structures of the major phospholipids. Reproduced from Scholfield CR (1985) In: Szuhaj BF and List GR (eds) *Lecithins*, p.3. Champaign: American Oil Chemists' Society, with permission.

Microbial phospholipid acyl groups are more similar to those found in the plant kingdom than they are to those from the animal kingdom. The predominant acyl fatty acids range from $C_{16:0}$ to $C_{18:3}$. There are more odd-numbered carbon fatty acids in microorganisms than elsewhere.

A further factor that makes phospholipid composition more complex is the ability to have different fatty acids on the 1 and 2 positions of the molecule. Most research has shown that polyunsaturated fatty acids are usually in the 2 position.

Extraction Techniques

One of the most important factors in phospholipid analysis is the initial extraction procedure. If the analysis is on a finished commercial lecithin there is no problem, but if the analysis is from tissue samples or food samples the extraction technique will be critical in obtaining meaningful results.

Tissue Samples

For many years the Folch extraction of tissue homogenates with chloroform/methanol 2:1 (v/v) has been the method of choice by most researchers. Some found that the use of chloroform/methanol 1:1 (v/v) was preferable and some have used a biphasic system of butanol/methanol with dilute hydrochloric acid. Some have used hexane/2-propanol 3:2 (v/v). Which solvent system used depends largely on the required accuracy, but in most cases chloroform/methanol 2:1 (v/v) is the best solvent to try initially.

Food Samples

Since food samples may have lecithin or phospholipid added rather than being incorporated into the tissues of the food matrix, the dried sample can be ground and extracted with petroleum ether. If the lipids are bound in the product through processing, chloroform/methanol 2:1 (v/v) can be used. Because of environmental considerations di- or trichloroethane should be used in place of chloroform.

Drying of high-moisture products is required, but not by oven drying or air drying, as oxidation of the fatty acids can occur. Freeze drying of the product is preferred.

Qualitative Analysis

There are several ways to detect the presence of phospholipids. Traditional instrumental methods using ultraviolet and infrared are not used as often since thin-layer chromatography (TLC) can give a qualitative and semiquantitative result in one assay. Also, the use of conformational chemical sprays on the TLC plates can further identify the products.

Thin-Layer Chromatography

There are several types of silica gel plates available for TLC. Silica gel G and H are the most useful. Phospholipids may be separated on a 20×20 cm plate in one or two directions. A polar solvent and a nonpolar solvent system are used. The polar system is chloroform/methanol/water (65:25:4, v/v/v) and the nonpolar solvent is petroleum ether/diethyl ether/acetic acid (90:10:1, v/v/v). See the American Oil Chemists' Society (AOCS) recommended practice Ja 7-86 for alternative methodology. (See **Chromatography: Thin-layer Chromatography.**)

These TLC plates are air- or oven-dried after separation of 20–50 μg of sample and are sprayed with 10% sulfuric acid and heated to char the phospholipids. Alternatively, they may be sprayed with a phosphorus spray containing molybdenum blue. Phospholipids stain a deep blue on heating the TLC plate. An example is shown in **Figure 2**.

Nondestructive visualization techniques can be used if the phospholipids are to be determined or the fatty acid composition is to be run. Ultraviolet light and 2′,7′-dichlorofluorescein easily detects lipids on TLC. On prep plates, the bands are scraped off and extracted with chloroform/methanol (1:1, v/v) and the fatty acids converted to methyl esters using boron trifluoride and then determined using gas–liquid chromatography (GLC).

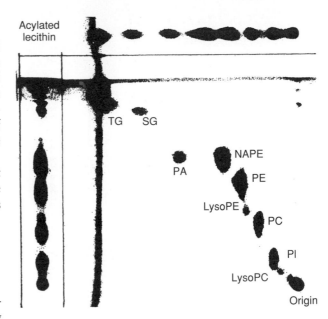

Figure 2 Thin-layer chromatogram of soya bean lecithin in two dimensions: triglycerides (TG), sterol glucosides (SG), phosphatidic acid (PA), N-acylphosphatidylethanolamine (NAPE), phosphatidylethanolamine (PE), lysophosphatidylethanolamine (Lyso PE), phosphatidylcholine (PC), phosphatidylinositol (PI), lysophosphatidylcholine (LysoPC). Silica gel plate; first dimension chloroform/methanol/acetic acid/water, 85:15:15:3 (v/v/v/v); second dimension chloroform/acetone/methanol/acetic acid/water (10:4:2:2:1) (v/v/v/v/v). Courtesy of J. Yaste, Central Soya, Food Research, Fort Wayne, Indiana, USA.

Quantitative Analysis of Phospholipids

Column Chromatography

Column chromatography precedes TLC in the separation of phospholipids. The techniques are slow and require good skill with column preparation, flow rates, and solvent removal. Commercial lecithins can be separated by dissolving the crude mixture in petroleum ether and passing it through a deactivated silica gel column with petroleum ether. The phospholipids are adsorbed and do not pass through the column, whilst triglycerides and sterol esters are eluted. The phospholipids are subsequently quantified by TLC and wet phosphorus analysis.

High-Performance Liquid Chromatography (HPLC)

Newer technologies have found that HPLC can separate and quantify phospholipids more quickly and accurately. Separation is carried out on several types of columns, including silica gel and an amino group bonded to the silica surface (μBondapak-NH₂). The columns are eluted with chloroform/methanol gradients, acetonitrile/methanol/85% phosphoric acid, or acetonitrile/methanol/ water. The eluent is measured at 205 nm or detected with flame ionization. **Figure 3**

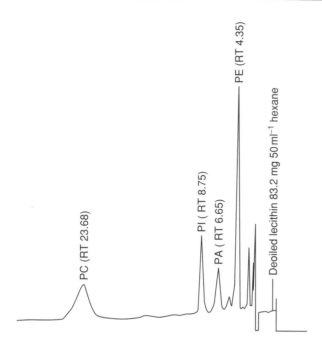

Figure 3 High-performance liquid chromatography of deoiled soya bean lecithin. PC, phosphatidylcholine; PI, phosphatidylinositol; PA, phosphatidic acid; PE, phosphatidylethanolamine. Column, μ Porasil 10 μ 3.98 × 300 mm; mobile phase, hexane/2-propanol/acetate (8:8:1, v/v/v), buffer pH 4.2; detection, ultraviolet (206 nm); injection, 10 μl; flow rate, 1 ml min^{-1}. Retenton time (RT) in minutes. Courtesy of P. Balazs, Central Soya, Food Research, Fort Wayne, Indiana, USA.

shows an HPLC separation of commercial lecithin, using ultraviolet detection. The mass detector, an evaporative analyzer, has also been successfully used for the HPLC determination of phospholipids. (*See* **Chromatography**: High-performance Liquid Chromatography.)

Densitometry

Densitometric scanning has been used as an indirect method for determining phospholipid content on TLC plates. While the method has some promise, a problem is the quanitative charring of the phospholipid spots. Each phospholipid has a different charring density and this depends on fatty acid composition. Only with proper standards can this method be useful.

Thincography

Thin-rod TLC combines TLC with quantification by flame ionization detection. Rods are used rather than plates but controversy still exists over the suitability of the technique for routine lipid analyses.

Phosphorus Analysis on Phospholipids

Phosphorus analysis is an indirect method for the quantification of phospholipids because the qualitative composition of the sample must be known, if accurate values are to be obtained. With pure phosphatides this will work well, but most separation techniques give mixed phospholipids. The preferred method for phosphorus in lecithins is the AOCS method Ja 5-55. This determines the total phosphorus content of the sample. For commercial lecithins a multiple factor of 30 is used to convert total phosphorus values to acetone-insoluble value.

There are various methods to determine phosphorus through molybdenum blue and molybdovanadophosphate yellow. To improve reproducibility many factors need to be evaluated. This includes the digestion method, chromogenesis, and sensitivity. The AOCS method is the most straightforward and should be used especially in the food area.

Industrial Methods of Analysis

Phospholipids are characterized by a different set of assays than the determination of compounds used for academic or biochemical use. Most commercially available phospholipids come from soya beans and from eggs, and the methods outlined below can be used to qualify or categorize the products.

Acetone-Insolubles (AI)

This method determines the content of phospholipids in commercial lecithins. The method employs AOCS method Ja 4-46. The AI is an approximation of the active ingredients in formulations. Cold lecithin-saturated acetone must be used in this test.

Acid Value (AV)

This method determines the phosphatide and fatty acid content of commercial phospholipids. The method utilized is AOCS method Ja 6-55. Phosphatide acidity is often confused with fatty acid addition to commercial lecithin. It is a combination of organic acids and phosphoric acid.

Peroxide Value (PV)

This is a measurement of oxidative state of commercial phospholipids. It measures the milliequivalents of peroxide per kilogram of sample which oxidize potassium iodide. It also measures the residual peroxide used in process stabilization and bleaching. AOCS method Ja 8-87 is used.

Free Fatty Acids (FFA)

This method utilizes AOCS method Ca 5a-40. When run on the acetone-soluble portion of the AI method, it gives the added fatty acids.

Phosphorus Content

The determination of total phosphorus is an indirect method for quantifying phospholipids. This method (AOCS method Ja 5-55) quantifies phospholipids through a molybdate reaction to a chromophore quantitated by phenolphthalein titration.

Gas Chromatography (GC)

This method is commonly used to measure the fatty acid composition and does not quantify the phospholipids themselves (AOCS method Ce 1-62).

High-performance Liquid Chromatography

This is gradually replacing the older techniques for qualifying and quantifying particular phospholipids in commercial mixtures. A uniform technique is being addressed by the AOCS.

Phospholipid analyses have come a long way since their study by Theuticum, *c.* 1800. Each analyst must choose the best method depending on constraints of accuracy and time.

See also: **Chromatography**: Thin-layer Chromatography; High-performance Liquid Chromatography; **Fatty Acids**: Properties; **Fats**: Classification

Further Reading

Burner D (ed.) (1991) *Official Methods and Recommended Practices.* Champaign: American Oil Chemists' Society.

Central Soya, Chemurgy Division (1990) *The Lecithin Book.* Fort Wayne: Central Soya.

Charalambous G and Doxastakis G (1989) *Food Emulsifiers: Chemistry, Technology, Functional Properties and Applications.* New York: Elsevier.

Hanin I and Pepeu G (1990) *Phospholipids: Biochemical, Pharmaceutical, and Analytical Considerations.* New York: Plenum Press.

Szuhaj BF (1989) *Lecithins: Sources, Manufacture and Uses,* Champaign: American Oil Chemists' Society.

Szuhaj BF and List GL (1985) *Lecithins.* Champaign: American Oil Chemists' Society.

Physiology

T H M Da Costa and M K Ito, Universidade de Brasília, Brasília, DF, Brazil

Introduction

Phospholipids are ubiquitous molecules that, due to their chemical properties, have important structural as well as metabolic roles in the cell. The first phospholipid to be discovered was phosphatidylcholine. It was originally named lecithin after the Greek *lekithos*, which means egg yolk. The concept of phospholipids as mere structural and metabolic inert molecules associated with methological difficulties delayed the interest in these compounds until the 1950s, when elucidation on the biosynthetic pathway of phospholipids began to be published. Today, the metabolic and physiological roles of phospholipids are an active and exciting area of research. In this article we discuss the physical and biochemical properties of phospholipids in cellular and subcellular membranes, the biosynthetic and hydrolytic pathways, and the role of phospholipids in signal transduction.

Functional Role of Membrane Phospholipids

Phospholipids are a major component of cellular membrane and play a pivotal role in the communication between extra- and intracellular space. Phospholipids represent a large class of compounds that have fatty acyl chains esterified to glycerol and a charged or zwitterionic head group (**Figure 1**). The fatty acyl chains usually have an even number of carbon atoms from 12 to 26, with 80% being 16–20 carbon atoms long. They may have up to six double bonds, commonly present as a *cis* isomer in one of the fatty acids that creates a kink in the chain. Differences in the length and saturation of the fatty acid tails are important for their influence in the ability of phospholipid molecules to pack against one another, and for this reason they affect the fluidity of the membrane.

In animal cells phosphatidylcholine (PtdCho) is the major phospholipid, whereas in bacteria phosphatidylethanolamine (PtdEtn) is the predominant species. Sphingomyelin is found in most animal cell membranes and belongs to a different group of phospholipids that fits into the overall pattern of phospholipid structure (**Figure 1**). In many mammalian cells, four major phospholipids predominate in the plasma membrane: PtdCho, phosphatidylserine (PtdSer), PtdEtn, and sphingomyelin. Only PtdSer carries a net negative charge, while the other three are electrically neutral at physiological pH with one positive and one negative charge. Sphingomyelin and PtdCho carry a molecule of choline in their head group (**Figure 1**). The phospholipid bilayer of biological membranes confers the amphipathic feature with hydrophobic and hydrophilic domains. The polar head groups of phospholipids face the aqueous exterior of the membrane while hydrophobic regions of the

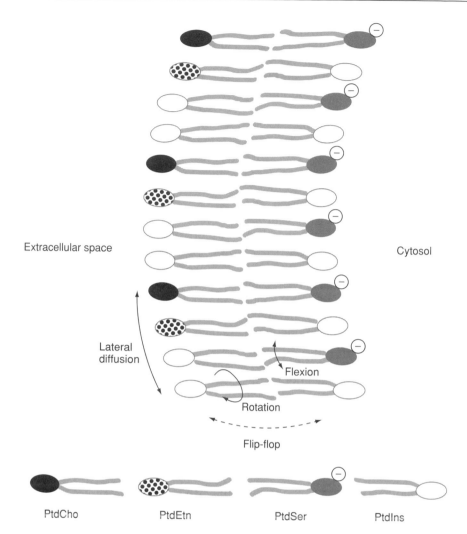

Extracellular space   Cytosol

Lateral
diffusion

Flexion

Rotation

Flip-flop

PtdCho   PtdEtn   PtdSer   PtdIns

Figure 2 Schematic representation of a biological membrane section, showing the possible phospholipid movements within each phospholipid molecule, between molecules and between the inner and outer phospholipid monolayer. The asymmetry of phospholipid species is also presented, with the predominance of phosphatidylcholine (PtdCho) and phosphatidylethanolamine (PtdEtn) facing the outer half and phosphatidylserine (PtdSer) with a negative charge and phosphatidylinositol (PtdIns) facing the inner half of the membrane.

Table 1 Phospholipid composition of cell membranes

Membrane	PtdCho	PtdEth	PtdSer	PtdIns	PtdGro	diPtdGro	PtdOH
Rat liver							
Endoplasmic reticulum	58	17	4	9			
Plasma membrane	56	15		10	2		
Mitochondrial (inner)	< 3	3	25	1	6	2	18
Mitochondrial (outer)	< 5	5	23	2	13	3	3
Nuclear	55	20	3	7			
Rat brain							
Myelin	11	14	7				
Erythrocytes							
Rat	31	15	7	2			
Human	23	20	8	3			
Sheep	1	23	8	1			
Escherichia coli plasma membrane		80			15		5

PtdCho, phosphatidylcholine; PtdEth, phosphatidylethanolamine; PtdSer, phosphatidylserine; PtdIns, phosphatidylinositol; PtdGro, phosphatidylglycerol; diPtdGro, diphosphatidylglycerol (cardiolipin), PtdOH, phosphatidic acid.

Adapted with permission from Jain MK and Wagner RC (1980) *Introduction to Biological Membranes*, p. 36. New York: John Wiley.

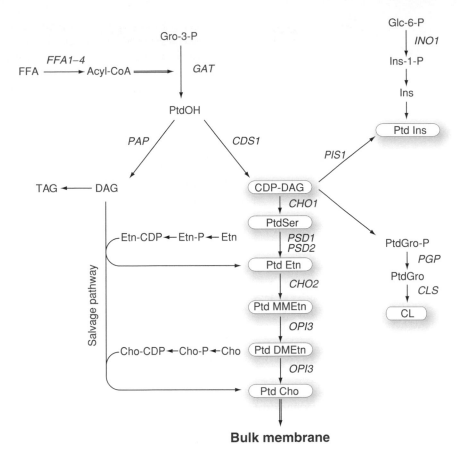

Figure 3 Phospholipid biosynthetic pathways in *Saccharomyces cerevisiae*. FFA, free fatty acid; Acyl-CoA, acyl coenzyme A; Gro-3-P, glycerol-3-phosphate; PtdOH, phosphatidic acid; DAG, diacylglycerol; TAG, triacylglycerol; CDP-DAG, cytidine diphosphate-diacylglycerol; PtdSer, phosphatidylserine; PtdEtn, phosphatidylethanolamine; PtdMMEtn, phosphatidylmonomethylethanolamine; PtdDMEtn, phosphatidyldimethylethanolamine; PtdCho, phosphatidylcholine Cho, choline; Cho-P, choline phosphate; Cho-CDP, cytidine diphosphate-choline. Etn, ethanolamine; Etn-P, ethanolamine phosphate, Etn-CDP, cytidine diphosphate-ethanolamine. Glc-6-P, glucose-6-phosphate; Ins-1-P, inositol 1-phosphate; Ins, inositol; Ptd Ins, phosphatidylinositol. PtdGro-P, phosphatidylglycerol-phosphate; PtdGro, phosphatidylglycerol; CL, cardiolipin. Enzymes abbreviations (italic): *FFA 1–4*, acyl-CoA synthetases 1–4; *GAT*, glycerol-3-phosphate acyltransferase; *PAP*, phosphatidate phosphatase; *CDSI*, CDP-diacylglycerol synthase; *CHO1*, phosphatidylserine synthase; *PSD1,2*, phosphatidylserine decarboxylase 1,2; *CHO2*, phosphatidylethanolamine *N*-methyltransferase; *OPI3*, phosphatidyl-*N*-methyltransferase; *PISI*, phosphatidylinositol synthase; *INO1*, inositol-1-phosphate synthase; *PGP*, phosphatidylglycerophosphate phosphatase; *CLS*, cardiolipin synthase.

-elongase(s) and -desaturase(s). The fatty acids are esterified to sn-1 and sn-2 to glycerol-3-phosphate by the enzyme glycerol-3-phosphate acyltransferase (GAT) (**Figure 3**). Cytidine diphosphate-diacylglycerol (CDP-DAG) synthase (CDS) converts phosphatidic acid to CDP-diacylglycerol (CDP-DAG), the major intermediate in phospholipid synthesis. The availability of CDP-DAG at specific cellular sites will direct the synthesis of different phospholipids, such as cardiolipin (CL), which occurs exclusively in mitochondria and phosphatidylinositol (PtdIns) and PtdSer to distinct subfractions of the ER or the Golgi apparatus. There is extensive transfer of intermediates and cross-compartment integration for the synthesis of phospholipids. For example, the enzyme phosphati-

dylserine synthase (CHO1) is localized in the ER, while the steps of conversion of PtdSer to PtdEtn take place in the inner mitochondrial membrane. The PtdEtn synthesized in the mitochondria must again migrate to the ER to insure synthesis of PtdCho, the most abundant phospholipid of *S. cerevisiae*.

In the absence of exogenous choline, PtdCho is synthesized by the *de novo* pathway by a three-step methylation of PtdEtn, which is catalyzed by the ER methytransferases. In the case of mutants with defects in phosphatidylserine decarboxylase (PSD) and methyltransferases, choline and ethanolamine must be provided in the medium to enter phospholipid biosynthesis via the salvage pathway (**Figure 3**). This route also insures recycling of PtdEtn and

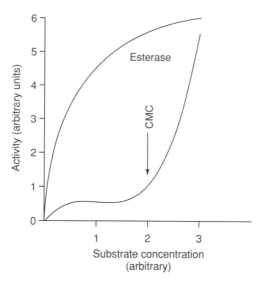

Figure 4 Esterase exhibits Michaelis–Menden kinetics, whereas phospholipase needs critical micellar concentration (CMC) of the substrate for full activity.

PtdCho degradation products and control of local levels of the potent second messenger diacylglycerol.

Degradation of Phospholipids

Undoubtedly, the most important function of phospholipids in the cell membrane relates to their breakdown by the action of phospholipases. The phospholipases are enzymes that hydrolyze phospholipids on water–lipid interfaces and are distinguished from other esterases by their relatively low activity in soluble monomeric phospholipids below their critical micellar concentrations (CMC), but become fully active in aggregated phospholipid structures above their CMC, such as in micelle, bilayers, or hexagonal structures.

As shown in **Figure 4**, whereas esterases show classical Michaelis–Menten kinetics, the phospholipases may reach more than 1000-fold increase in activity as the substrate phospholipid concentration reaches the CMC. Important factors responsible for this increased rate of hydrolysis include high local substrate concentration, amphipatic substrate orientation at the interface, enhanced diffusion of products from the enzyme to lipid or aqueous moieties, and conformational change of enzyme upon binding to the interface.

Phospholipases play a central role in the activation of various events related to phospholipid degradation. In general: (1) many phospholipases are digestive enzymes found in high concentrations in venoms, bacterial secretions, and digestive fluids of higher

animals; (2) many phospholipases have a regulatory function with their products being cellular mediators such as diacylglycerols, inositol trisphosphate, platelet-activating factor, and the eicosanoids. The actions of lipid mediators, known collectively as second messengers, are highly sensitive and under rigid control by a variety of anabolic and catabolic enzymes. Many pathological states, as in inflammation, allergic reactions, and hypertensive states, are related to activation of the cascade of events involving phospholipid degradation and the intracellular second messenger pathway.

Phospholipases are classified as type A_1, A_2, B, C, or D, according to where they act on the substrate phospholipid (**Figure 5**).

The phospholipases A are acyl hydrolases, which means they remove one acyl group, yielding one fatty acid and lysophosphatide. Phospholipase A_1 and phospholipase A_2 remove fatty acids at positions sn-1 and sn-2, respectively. Phospholipase A_1 is widely distributed in nature and in mammals the major sources are found in the pancreas and the brain. Intracellularly, phospholipase A_1 is dominant in the ER. The best known examples of phospholipase A_2 are snake venom and the pancreatic enzyme that acts mostly on PtdCho and PtdEdn. It is the major phospholipase of mitochondria. Phospholipase B hydrolyzes both acyl groups (sn-1 and sn-2) and also has high lysophospholipase activity. Phospholipase C acts on the glycerophosphate bond while phospholipase D catalyzes the removal of the polar head group. In mammalian cells, phospholipase C is an important initiator of the PtdIns cycle. Phospholipase D is the major phospholipase in many plant tissues and acts specifically on intact phospholipids to give phosphatidic acid and an alcohol. In plants, it seems to be involved in cell turnover and energy utilization. Phospholipase D is also found in bacterial and mammalian sources, where it is involved in replacement of the polar head groups of membrane phospholipids.

Phospholipids as Precursors of Cellular Signal Transduction

In the cell membrane, phospholipids store important precursors of second messengers. Transduction of signals from hormones, across a membrane, involves the coordinated actions of receptor, membrane proteins, and phospholipids that either stimulate or inhibit the synthesis of a second messenger. A number of processes inside the cell are controlled by the level of second messengers.

Among the best described second messenger systems is the G-protein and phosphatidylinositide

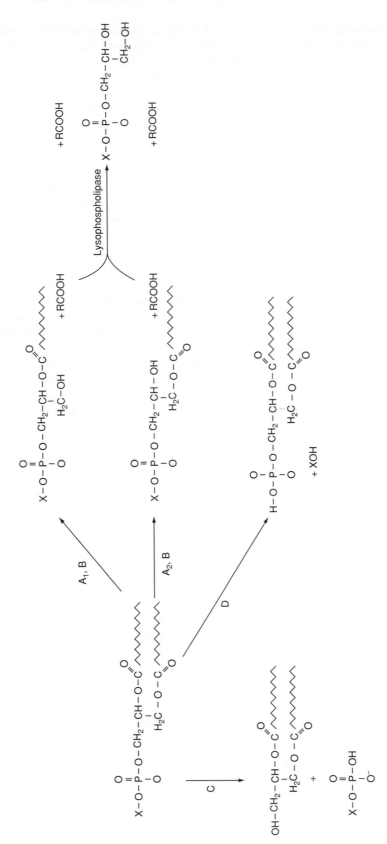

Figure 5 The action of phospholipases A_1, A_2, B, C, and D and lysophospholipase in the hydrolysis of phospholipids. X represents choline, serine, ethanolamine, etc.

system. This system is distinctive in that one stimulus activates membrane reactions that generate two second messengers. The first experimental observations noted that administration of neurotransmitter acetylcholine in pigeon pancreas led to a rapid turnover of PtdIns fraction of membrane phospholipids and release of the digestive enzyme amylase. Similar observations were made in other systems by hormones, neurotransmitters, or growth factors. Despite intensive efforts, the understanding of these mechanisms had not progressed until the early 1980s. It is now recognized that the initial events of inositol phospholipid metabolism occur within 20–30 s of binding of the agonist to the receptor.

Today, the role of specific inositol phospholipids in the phosphoinositide family, namely, phosphatidylinositol 4,5-bisphosphate (PI(P_2)), as a membrane-associated storage form of two second messengers, is quite clearly understood.

Figure 6 shows a simplified scheme of the events involved. When an agonist binds to a receptor, membrane G-protein is stimulated to bind guanosine triphosphate. The activated G-protein then acts on the membrane-bound enzyme, phospholipase C, which in turn cleaves PI(P_2) to yield two products, sn-1,2-diacylglycerol (DAG) and inositol 1,4,5-trisphosphate (IP$_3$). Both of these products are second messengers. The function of IP$_3$ is to stimulate the release of calcium ion from intracellular stores, largely from the ER. The increased calcium concentration originates a cascade of events in the intracellular metabolism, including the activation of the membrane-bound protein kinase C (C from calcium). This enzyme requires calcium and PtdSer for its activity. The specific role of the second messenger DAG is to increase the affinity of the protein kinase C for calcium ions. Activated protein kinases will then phosphorylate target proteins inside the cell, which will then be activated and proceed within the cascade. Since many metabolic events are controlled by calcium fluxes and by phosphorylation of specific proteins, the phosphoinositol system shows great ingenuity as a control mechanism. Some of the known target proteins are the insulin receptor, β-adrenergic receptor, and glucose transporter. It is now quite clear that PtdIns is not the only source of DAG which can activate protein kinase C. PtdCho and PtdEtn also seem to serve this role. PtdCho is also a major source of arachidonic acid for the biosynthesis of eicosanoids.

Another well-described example of a regulatory function involving phospholipids in membrane is the generation of another second messenger, known as arachidonate cascade. As described earlier, when an

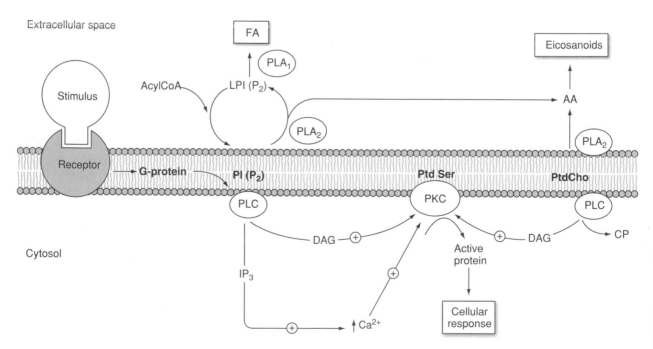

Figure 6 The role of phospholipids in signal transduction. PLA$_1$, PLA$_2$, PLC, phospholipases A$_1$, A$_2$, C; PI(P_2), phosphatidylinositol 4,5 bisphosphate; IP$_3$, inositol 1,4,5-trisphosphate; AcylCoA, acyl coenzyme A; DAG, diacylglycerol; FA, fatty acids; AA, archidonic acid; CP, choline phosphate; PKC, protein kinase C; Ptd Ser, phosphatidylserine; PtdCho, phosphatidylcholine; LPI, lysophosphatidylinositol.

agonist stimulates a membrane receptor, a series of events occurs which may lead to the release of 20 carbon polyunsaturated fatty acids from membrane phospholipids, most commonly arachidonic acid (**Figure 6**). These are tissue-specific stimulants by hormones such as bradykinin or epinephrine, or proteases such as thrombins, to name but a few. Pathological release can occur if membranes are perturbed; for example, the bee sting may stimulate the release of arachidonate from local cell membrane and cause inflammation. This release involves the action of a specific phospholipase A_2 on PtdCho or PtdEdn, yielding arachidonate, and the action of a phospholipase C on PtdIns, yielding a diacylglycerol, which in turn undergoes cleavage to give free arachidonate, which is then converted into a specific eicosanoid in the cell. Eicosanoids, such as prostaglandins, leukotrienes, and thromboxanes are compounds with potent physiological properties which are formed from 20-carbon unsaturated fatty acids into one of a series of eicosanoids according to the enzyme present in the cell and the unsaturated fatty acid released from the membrane. Long-chain polyunsaturated fatty acids of the *n*-3 series, especially from marine origin, when present in the diet will be incorporated into the cell membrane and may replace arachidonic acid in the signal pathway. All eicosanoids are metabolized very rapidly. We still know relatively little about the subsequent effects of eicosanoids at a molecular level, though recent evidence points to their function in the communication pathway with nuclear receptors.

Figure 6 presents events as a coordinated response involving membrane phospholipids and their degradation products. This figure exemplifies the events associated with arachidonic acid release (top) and the mobilization of calcium ions and protein kinase C (bottom). The action of lipid mediators is highly sensitive and under rigid control and when this regulation is not maintained, a number of pathological states, such as those mentioned above, may result.

Current knowledge on the mechanisms involved in intracellular signaling is rapidly advancing and it underscores the importance of the physiological, pathological, as well as the pharmacological properties of this diverse and ubiquitous group of compounds, known as phospholipids. The nutritional significance of phospholipid physiology may be summarized by the fact that foods are the main sources of essential elements of this system.

See also: **Choline**: Properties and Determination; **Essential Oils**: Properties and Uses; **Fatty Acids**: Properties; Metabolism; Gamma-linolenic Acid; **Fish**: Dietary Importance of Fish and Shellfish; **Fish Oils**: Dietary Importance; **Fats**: Classification; Occurrence; **Prostaglandins and Leukotrienes**; **Vegetable Oils**: Dietary Importance

Further Reading

Carman GM and Henry AS (1989) Phospholipid biosynthesis in yeast. *Annual Review of Biochemistry* 58: 635–669.

Eyster KM (1998) Introduction to signal transduction – a primer for untangling the web of intracellular messengers. *Biochemistry and Pharmacology* 55: 1927–1938.

Kohlwein SD, Daum G, Schneiter R and Paltauf F (1996) Phospholipids: synthesis, sorting, subcellular traffic – the yeast approach. *Trends in Cell Biology* 6: 260–266.

Mead JF, Alfin-Slater RB, Howton DR and Popják G (1986) *Lipids: Chemistry, Biochemistry and Nutrition.* New York: Plenum.

Sim E (1982) *Membrane Biochemistry.* Outline Studies in Biology. London: Chapman and Hall.

Vance DE (1991) Phospholipid metabolism and cell signalling in eucaryotes. In: Vance DE and Vance J (eds) *Biochemistry of Lipids, Lipoproteins and Membranes – New Comprehensive Biochemistry*, vol. 20, pp. 205–239. Netherlands: Elsevier.

Waite M (1991) Phospholipases. In: Vance DE and Vance J (eds) *Biochemistry of Lipids, Lipoproteins and Membranes – New Comprehensive Biochemistry*, vol. 20, pp. 269–295. Netherlands: Elsevier.

PHOSPHORUS

Contents
Properties and Determination
Physiology

Properties and Determination

A N Garg, Indian Institute of Technology, Roorkee, U.A., India

Introduction

The extensive and varied chemistry of phosphorus transcends the traditional boundaries of inorganic chemistry because of its vital role in the biochemistry of all living organisms as a constituent of adenosine triphosphate (ATP) and phosphoproteins. It was first isolated from urine by Hennig Brandt in 1669 as a white waxy substance. The spontaneous chemiluminescent reaction of white phosphorus with moist air was the first observed property and was also the origin of its name (Greek: '*phos*,' 'light'; '*pherein*,' 'bearing'). It was also the ancient name for the planet *Venus*, when it appeared before sunrise.

Chemical Properties

Phosphorus is a typical nonmetal placed in group VA of the periodic table. The element phosphorus (P) has an atomic number of 15 and atomic weight of 30.97 with electrons distributed as $1s^2 2s^2 2p^6 3s^2 3p_x^1 3p_y^1 3p_z^1$ and atomic energy levels as shown in **Figure 1**. Thus, three unpaired electrons, together with the availability of low-lying vacant 3d orbitals, account for the predominant oxidation states III and V in phosphorus chemistry. The most important biological form is the pentavalent oxygen compound phosphate PO_4^{3-}. Phosphorus exists in three main allotropic forms; white, red, and black, each of these being polymorphic. There are at least 11 known modifications, some amorphous, others of some indefinite identity, and all but three of unknown structure. Of these, white phosphorus is soft, waxy, most reactive, and thermodynamically least stable. It has a melting point of 44.1 °C, a boiling point of 80 °C and a specific gravity of 1.82 g cm^{-3}. Red and black forms are heavier with specific gravities of 2.2 and 2.69 g cm^{-3}, respectively. White phosphorus reacts with moist air and gives out light. It ignites

spontaneously in air at about 35 °C and is therefore stored in water to prevent combustion. It is soluble in organic solvents such as CS_2 and benzene. If white phosphorus is heated to about 250 °C, or a lower temperature in the presence of sunlight, then red phosphorus is formed. It is a polymeric solid with a melting point of 280.5 °C and sublimes at 430 °C. It is much less reactive than white phosphorus and does not phosphoresce in air. Unlike white phosphorus, red phosphorus need not be stored under water. Its structure is extremely complex, involving a crisscross packing of infinite tubular chains of P atoms. When white phosphorus is heated under high pressure, a highly polymerized form called black phosphorus is obtained. It is also obtained by heating at 220–370 °C for 8 days in the presence of mercury as a catalyst and with a seed of black P. This is thermodynamically the most stable allotrope. It is inert and has a double layer structure with P atoms being bound to three neighbors, as shown in **Figure 2**. The entire structure consists of a stacking of these double layers with the closest P–P distances within each layer being 2.17–2.20 Å and the shortest P–P distance between layers at 3.87 Å. Therefore, the crystals are flaky like graphite. All forms of phosphorus melt to give the same liquid, which consists of symmetric P_4 molecules with phosphorus atoms occupying corners of a regular tetrahedron. Each atom is directly bonded to the other three atoms P–P = 2.21 ± 0.02 Å. The same molecular form exists in the gas phase at >800 °C and low pressure. The bonding orbitals in P_4 have only 2% of 3s and 3d character. It is most likely that pure

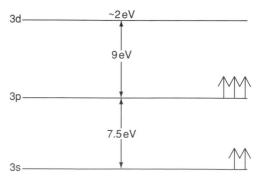

Figure 1   Atomic energy levels in phosphorus.

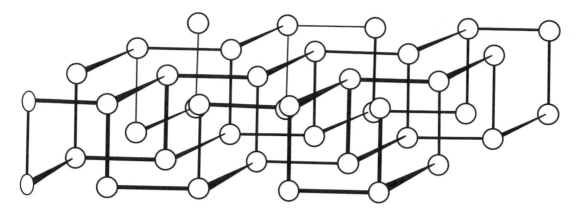

Figure 2 Arrangement of P atoms in the double layers in crystalline black phosphorus.

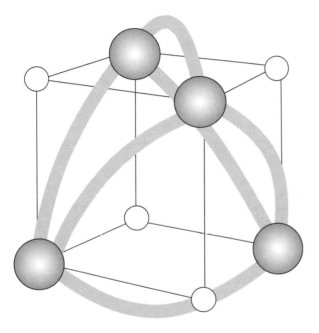

Figure 3 Three-dimensional distribution of valence shell electron density in the P_4 molecule. From Hart RR, Robin MB and Kuebler NA (1965) *Journal of Chemical Physics* 42: 3631–3638, with permission.

3p orbitals are involved, even though the bonds in P_4 are bent with $\angle$P–P–P = 60° and a strain energy of 96 kJ mol^{-1}, which accounts for its high reactivity. Electronic spectral studies suggest a resonance structure with strong π bonds. The regions of high electron density in the P_4 molecule are shown in **Figure 3**. The moment of inertia about any of its three major axes is 2.5×10^{-40} g cm^2. Thus, the P_4 molecule might be expected to have donor ability.

The only naturally occurring isotope, ^{31}P, has a nuclear spin $I = \frac{1}{2}h/2\pi$ and a large magnetic moment (1.13 NM) but no quadrupole moment. It is suitable for nuclear magnetic resonance (NMR) spectroscopy.

In a 10-kG field ^{31}P resonates at 17.24 MHz. Phosphorus forms compounds with all elements except tin, bismuth, and the inert gases. It also reacts readily with heated aqueous solutions to give a variety of products. The bonding and stereochemistry of the phosphorus atom are varied, essentially due to empty d orbitals. It is known in at least 14 coordination geometries with coordination numbers up to 9, though the most frequently met coordination numbers are 3, 4, 5, and 6. Some typical geometries along with orbitals used in the formation of bonds are given in **Table 1**. Phosphorus has a significant tendency to catenation, forming a series of cyclic compounds $(RP)_n$ where $n = 3$–6, as well as some R_2 PPR$_2$ type of compounds. (*See* **Spectroscopy**: Nuclear Magnetic Resonance.)

Forms in Foods

Plants need phosphate for healthy growth, especially for the development of roots, flowers, fruits, and seeds, though the requirement is less compared with that for nitrogen and potassium. All foods contain phosphorus in the form of the phosphate anion (PO_4^{3-}), and is consumed by living organisms as such. Very few natural compounds contain phosphorus in any other form. A summary of the types of biologically important phosphorus compounds is presented in **Table 2**. It is picked up from the soil, where it is present as organic and inorganic phosphates (soluble as well as insoluble) in nucleoproteins, nucleic acids, and the coenzymes nicotinamide adenine dinucleotide (NAD) 2-phosphate, ATP, and other high-energy phosphates. Organic phosphates include sugar phosphates such as glucose 6-phosphate (**Figure 4**), phospholipids, and pigments. Phosphate in the form of nucleotides serves as a source of a high-energy bond and performs an important function in conserving and providing bursts of metabolic energy.

deficiency in adults may occur with excessive use of alcohol, prolonged vomiting, liver disease, or hyperparathyroidism. The dietary content of P has been shown to regulate physiologically the serum concentration of PTH and thus indirectly the phosphate homeostasis. Refer to individual food.

Plants lacking phosphorus may develop necrotic areas on the leaves, petioles, or fruits; they may have a general overall stunted appearance, and the leaves may have a characteristic dark to blue–green coloration.

Properties of Phosphates

All phosphates are salts of oxyacids that contain a P = 0 group and at least one P–OH group that ionizes. Some species also have P–H group where the hydrogen atom is not ionizable. Phosphates of metal ions and other cations, mixed metal phosphates, and condensed phosphates are well known because of their commercial and technical importance. Many phosphates, especially long-chain polyphosphates, are known for their toxicity as they adversely affect the osmotic pressure of body fluids and prevent absorption of mineral nutrients. Phosphates are capable of interacting with many of the constituents of food systems, and inactivate metal ions, and are thus important in food processing.

Monosodium phosphate (NaH_2PO_4) is water-soluble and is used as a phosphatizing agent on steel surfaces. Its acidic property is used in effervescent laxative tablets and as a leavening agent in baking powder. Monopotassium phosphate (KH_2PO_4) crystals show a piezoelectric effect and are used in submarine sonar systems. Disodium and dipotassium phosphates are used as buffering agents to maintain pH. This property is used for stabilizing meat. These salts are also used as sequestering agents in the food industry. Sodium orthophosphate (Na_3PO_4) is highly alkaline and finds use in industrial hard surface cleaners. Its aqueous solution is a valuable constituent of scouring products, paint strippers, and grease saponifiers. Its complex with sodium hypochlorite [$(Na_3PO_4 \cdot 11H_2O)_4 \cdot NaOCl$] releases active chlorine when wetted; this combination of scouring, bleaching and bacterial actions makes the adduct valuable in automatic dish washing powder formulations. Potassium orthophosphate (K_3PO_4) is used to regulate the rate of polymerization of styrene--butadiene rubber. Mono- and diammonium phosphates are used as fertilizers and nutrients. An important property of ammonium phosphate is a flame-retarding agent for cellulose materials. The action depends on its dissociation to ammonia and orthophosphoric acid when heated. Acid so generated catalyzes the decomposition of cellulose to char and

smother the flame. Urea phosphate is generally used to flameproof cotton fabrics. A dilute solution of diammonium phosphate, with an initial pH of 7.85, upon boiling evolves ammonia, and the pH drops to 5.78 in 2 h. This property is used for the precipitation of colloidal dyes on wool fabrics. Dicalcium phosphates are used in pharmaceutical tablets as supplements. Natural phosphate minerals are all orthophosphates, the major ones being fluoroapatite; partly carbonated hydroxyapatite makes up the mineral part of teeth. These are important constituents of bones. Calcium orthophosphates are particularly important in fertilizer technology. (*See* **Leavening Agents; Stabilizers:** Types and Function.)

Many phosphate complexes of transition metal ions are known. Ce^{4+}, Th^{4+}, Zr^{4+}, U, and Pu form insoluble phosphates from fairly strong acid solution (3–6 M nitric acid). Condensed phosphates contain more than one phosphorus atom and P–O–P bonds with three main building units – the end unit (**Figure 7**), middle unit (**Figure 8**), and branching unit (**Figure 9**). These units can be readily distinguished by reactivity with water and ^{31}P NMR. These can also be incorporated into linear or cyclic polyphosphates. Linear polyphosphates are salts with the general formula $[P_n O_{3n + 1}]^{(n + 2) -}$ ($n = 2-10$) such as $M_4^1P_2O_7$ and

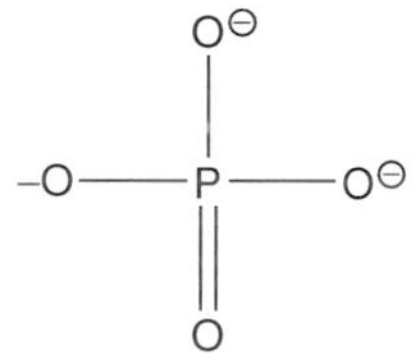

Figure 7

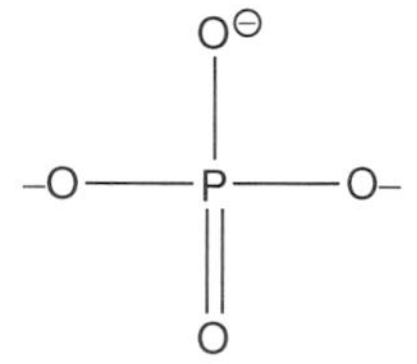

Figure 8

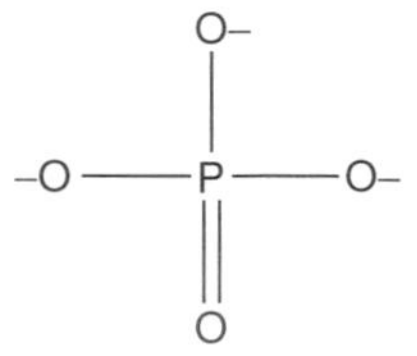

Figure 9

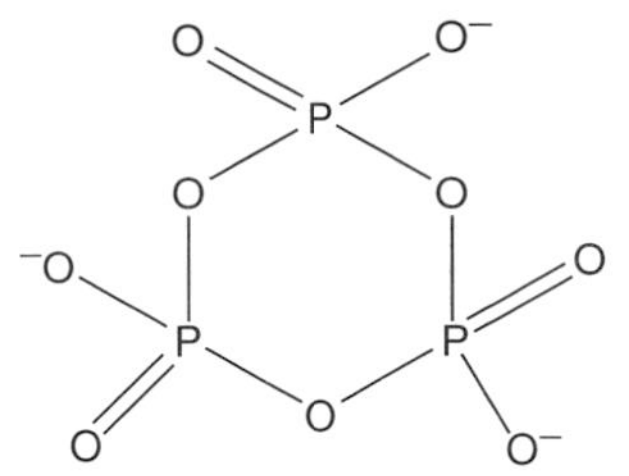

Figure 10

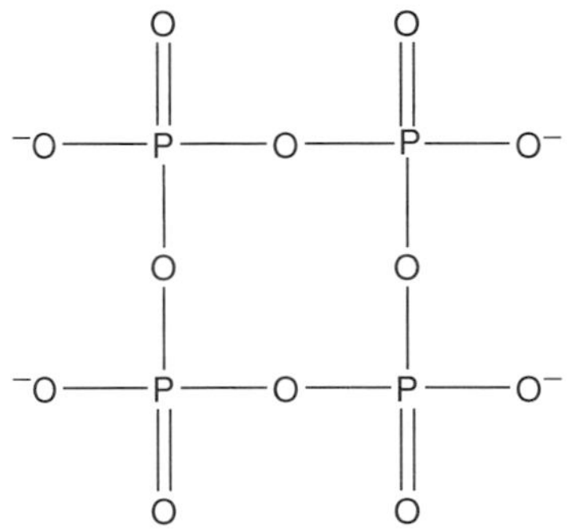

Figure 11

$M_5^I P_3 O_{10}$. Many polyphosphates with different chain lengths have been known. Disodium dihydrogen pyrophosphate, $Na_2^I H_2 P_2 O_7$ is mixed with $NaHCO_3$ and used in bread making to leaven the bread, that is to make it rise. They react and evolve CO_2 when heated together. Sodium pyrophosphate, $Na_4 P_2 O_7$, is mixed with starch and a flavoring agent to make instant pudding mixture. At one time, it was also added to soap powders and solutions as a water softener. $Ca_2 P_2 O_7$ is used as the abrasive/polishing agent in fluoride paste. Cyclic polyphosphates or metaphosphates are the salts with the general formula $[P_n O_{3n}]^{n-}$ with $n = 3-7$ such as $M_3 P_3 O_9$ (**Figure 10**) and $M_4 P_4 O_{12}$. The eight-membered ring of the $P_4 O_{12}^{4-}$ (**Figure 11**) is puckered with equal bond lengths. Condensed phosphates form soluble complexes with many metals. These are usually prepared by dehydration of orthophosphates under various conditions of temperature (300–1200 °C) and also by appropriate hydration of dehydrated species. Chain phosphates are used as water softeners in industry. Polyphosphates aid in controlling the microbiological population on the surface of poultry meat. Some metaphosphates having an infinite chain length are also known, e.g., KPO_3.

Organic phosphates contain phosphate groups linked through OH groups of organic compounds (–C–O–P linkage) such as sugars. Large numbers of phosphate esters, $RO–PO(OH)_2$ are known. These occur in the form of mono-, di-, and triphosphoesters

Each form has specific chemical properties leading to different biological functions. These are constituents of numerous highly active intracellular compounds. Release of free energy by hydrolysis of ATP provides the main source of energy for various metabolic processes and for muscle contraction. Intracellular phosphate is a regulator of enzymes in the glycolytic pathways. Some of the alkyl phosphorus compounds are of industrial importance, particularly for solvent extraction of metal ions from aqueous solutions. They extract metal ions by cation exchange and/or by solvation. Among these, di(2-ethylhexyl) phosphoric acid (DEHPA), tri-*n*-butyl phosphate (TBP), and tri-*n*-octylphosphine oxide (TOPO) have been used most extensively. Organic derivatives of fluorophosphoric acid $[FP = O(OH)_2]$ have promising properties as insecticides.

ANALYSIS

Phosphorus is generally detected on the basis of the reaction between orthophosphoric acid and the molybdate ion (MoO_4^{2-}), which gives a yellow-colored precipitate in a strongly acid solution.

Total Phosphorus

Any solution containing phosphorus is fumed with aqua regia almost to dryness, followed by heating with 1 M nitric acid, whence lower oxidation states are oxidized to the orthophosphate (PO_4^{3-}) form. The resultant solution can be used for the estimation of the total phosphorus by gravimetric, titrimetric, or spectrophotometric methods. Some of these have been recommended by the Association of Official Analytical Chemists for the analysis of total phosphorus in vegetables, fruits, cereals, and other foods. These are dried in a silica/platinum crucible and heated over a low Bunsen flame to volatilize organic matter. A 10% sodium bicarbonate solution is added and the contents evaporated to dryness. Oil may be burnt off at a lower temperature without smoking, and finally, ashing is carried out in a muffle furnace at 500 °C. The contents are dissolved in concentrated nitric acid and heated to dryness, and then dilute hydrochloric acid is added.

Gravimetric methods include the formation and weighing of phosphorus as ammonium phosphomolybdate, ammonium magnesium phosphate or pyrophosphate, and quinoline molybdophosphate. On addition of ammonium molybdate solution (12.5 g dissolved in 75 ml of water are slowly added to another solution containing 125 g of ammonium nitrate in 125 ml of water and 175 ml of nitric acid diluted to 500 ml) a yellow-colored precipitate is do contribute a small percentage to the total excretion of Pi.

Several issues of Pi homeostasis need further explanation because of their potential impact on health and disease in populations of developed nations.

Aging and Renal Function

Serum Pi concentration changes little in women with a healthy renal function, but it does increase when the renal function becomes compromised, though the percent reduction in GFR needed for this increase in serum Pi has not been established. Less is known of changes of the serum Pi concentration in men with increasing age, but the same relationship to renal function probably holds in men as in women. Because intestinal Pi absorption efficiency remains fairly constant in aging women, whereas calcium absorption efficiency declines within a decade or so following the menopause, it is possible that PTH secretion is increased in these women 10 or more years beyond the menopause without any other perturbation. Under these conditions, serum PTH concentrations rise, but they typically remain within the upper limit of normality. Severe reductions in renal function clearly elevate serum Pi concentration. Refer to Renal Secondary Hyperparathyroidism.

Nutritional Secondary Hyperparathyroidism

The major concern of nutritionally induced hyperparathyroidism, even if the PTH level remains within the upper range of normal, is the reduction of both bone mass and bone mineral density. A low Ca:P ratio from a diet that provides too little calcium and plentiful amounts of phosphorus may increase PTH in a fairly persistent manner, which in turn increases bone turnover. If bone formation cannot keep up with PTH-governed bone resorption, bone loss will follow (**Figure 1**). Although not fully established by research evidence, this scenario of a persistently elevated PTH in response to a low Ca:P ratio ($\ll 0.5{:}1$) is suspected of contributing first to osteopenia and then to osteoporotic bone that is more fragile and at increased risk of fracture (see **Table 1** for examples of foods with low Ca:P ratios). Recent reports, for example, have shown that the adverse effects of a low Ca:P diet can be improved and largely overcome by simple calcium supplementation. Because the diets of so many adolescents and adults in the USA and probably adults in other technologically advanced nations with significant use of phosphate additives and inadequate calcium intakes have low Ca:P ratios, it is expected that rates of osteoporotic fractures, especially of hip fractures, will increase in the next several decades.

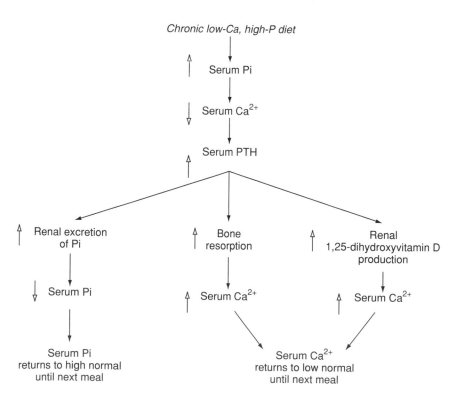

Figure 1 Illustration of steps in development of long-term nutritional secondary hyperparathyroidism: chronic low-calcium, high-phosphorus diet. Reproduced from Phosphorus, *Encyclopaedia of Food Science, Food Technology and Nutrition*, Macrae R, Robinson RK and Sadler MJ (eds), 1993, Academic Press.

Renal Secondary Hyperparathyroidism

When renal function becomes compromised to such an extent that creatinine, other nitrogenous metabolic products, and Pi are retained abnormally and excessively by the body, then several pathophysiological adaptations occur that have serious effects on health. One of the important adverse effects of the retention of Pi is the rapid and progressive loss of mineral mass. The chronic elevation of serum Pi causes a decline in serum Ca^{2+}, which triggers PTH secretion. The net effect is a constantly elevated PTH concentration that continues to act on bone tissue, i.e., resorption, to try to raise $[Ca^{2+}]$ to its homeostatic set level. Since Pi is also released from bone along with Ca^{2+} during resorption, the serum Pi concentration also increases. Because the kidneys cannot eliminate Pi adequately, $[Ca^{2+}]$ can never be raised to its set level, and bone tissue continues to be degraded as part of an unending vicious cycle.

Various dietary manipulations have been tried to control the loss of bone mass, but no one regimen has been very successful. Reductions of both dietary protein, especially animal protein, and phosphorus have been moderately successful in slowing the progress of chronic renal disease, but the diets are not very palatable or satisfying.

Conclusions

Pi metabolism is much more complex than that of calcium because of the many intracellular pathways that utilize Pi ions at one stage or another. The cytosolic utilization of Pi is closely linked with that of glucose for the formation of glucose 6-phosphate and for triglyceride synthesis through glycerol 3-phosphate formation, as well as with other molecules, during the postprandial period. Pi is utilized by cells for many diverse molecules, including regulatory peptides and phospholipids. Extracellular regulation of Pi is closely associated with that of calcium through PTH and other calcium-regulating hormones. Under typical dietary conditions of excessive phosphorus intake compared with calcium, i.e., low Ca:P ratio, nutritional secondary hyperparathyroidism and the long-term development of osteopenia are likely to result. Food fortification with calcium and calcium supplementation are common ways in which the low Ca:P ratio can be minimized, but individual behaviors aimed at selecting a diet higher in calcium will be needed to overcome the adverse ratio ($\ll 0.5$), despite calcium fortification and/or supplementation. Renal secondary hyperparathyroidism, a serious consequence of renal functional impairment, produces severe bone loss because of altered homeostatic regulation of Pi.

See also: **Aging – Nutritional Aspects**; **Bone**; **Calcium**: Physiology; **Carbohydrates**: Requirements and Dietary Importance; **Cells**; **Cholecalciferol**: Physiology; **Dietary Requirements of Adults**; **Energy**: Measurement of Food Energy; **Hormones**: Thyroid Hormones; Steroid Hormones; **Osteoporosis**

Further Reading

Akesson K, Lau K-H, Johnston P, Iperio E and Baylink DJ (1998) Effects of short-term calcium depletion and repletion on biochemical markers of bone turnover in young adult women. *Journal of Clinical Endocrinology and Metabolism* 83: 1921–1927.

Anderson JJB and Garner SC (eds) (1996) *Calcium and Phosphorus in Health and Disease*. Boca Raton, FL: CRC Press.

Anderson JJB, Sell ML, Garner SC and Calvo MS (2000) Phosphorus. In: Russell R *et al. Present Knowledge in Nutrition*, 7th edn. Washington, DC: International Life Sciences Institute.

Barger-Lux J and Heaney RP (1993) Effects of calcium restriction on metabolic characteristics of premenopausal women. *Journal of Clinical Endocrinology and Metabolism* 76: 103–107.

Bringhurst FR (1989) Calcium and phosphate distribution, turnover, and metabolic actions. In: DeGroot LJ (ed.) *Endocrinology*, 2nd edn, vol. 2. Philadelphia, PA: WB Saunders.

Brot C, Jorgensen N, Jensen LB and Sorensen OH (1999) Relationships between bone mineral density, serum vitamin D metabolites and calcium:phosphorus intake in healthy perimenopausal women. *Journal of Internal Medicine* 245: 509–516.

Calvo MS, Kumar R and Heath H III (1990) Persistently elevated parathyroid hormone secretion and action in young women after four weeks of ingesting high phosphorus, low calcium diets. *Journal of Clinical Endocrinology and Metabolism*, 70: 1340–1344.

Calvo MS and Park YM (1996) Changing phosphorus content of the U.S. diet: Potential for adverse effects on bone. *Journal of Nutrition* 126: 1168S–1180S.

Harnack L, Stang J and Story M (1999) Soft drink consumption among US children and adolescents: Nutritional consequences. *Journal of the American Dietetic Association* 99: 436–441.

Institute of Medicine, Food and Nutrition Board (1997) *Dietary Reference Intakes: Calcium, Phosphorus, Magnesium, Vitamin D and Fluoride*. Washington, DC: National Academy Press.

Karkkainen M and Lamberg-Allardt C (1996) An acute intake of phosphate increases parathyroid hormone secretion and inhibits bone formation in young women. *Journal of Bone and Mineral Research* 11: 1905–1912.

McKane WR, Khosla S, Egan KS *et al.* (1996) Role of calcium intake in modulating age-related increases in parathyroid function and bone resorption. *Journal of Clinical Endocrinology and Metabolism* 81: 1699–1703.

National Research Council (1989) *Recommended Dietary Allowances*, 10th edn, pp. 184–187. Washington, DC: National Research Council, National Academy Press.

Slatopolsky E, Dusso A and Brown A (1999) The role of phosphorus in the development of secondary hyperparathyroidism and parathyroid cell proliferation in chronic renal failure. *American Journal of Medical Sciences* 317: 370–376.

USDA (1978) *Nutritive Value of Foods. Home and Garden Bulletin* No. 72. Washington, DC: US Department of Agriculture.

Phylloquinone *See* **Vitamin K**: Properties and Determination; Physiology

---

Physical Properties of Food *See* **Rheological Properties of Food Materials**

PHYTIC ACID

Contents
Properties and Determination
Nutritional Impact

Properties and Determination

U Konietzny, Karlsruhe, Dettenheim-Liedolsheim, Germany
R Greiner, Centre for Molecular Biology, Federal Research Centre for Nutrition, Karlsruhe, Germany

Introduction

The proper chemical designation for phytic acid is *myo*-inositol(1,2,3,4,5,6)hexakisphosphoric acid. Salts of this acid, designated as phytates, are found in plants, animals and soil. Phytate has been considered as an antinutrient due to its inhibitory effect on the bioavailability of essential dietary minerals. During food processing and digestion, phytate can be dephosphorylated to produce degradation products, such as *myo*-inositol pentakis-, tetrakis-, tris-, bis-, and monophosphates. Besides the adverse effects of phytate and other highly phosphorylated *myo*-inositol phosphates on mineral bioavailability, some novel metabolic effects of phytate and some of its degradation products have been recognized. Certain *myo*-inositol phosphates have been suggested to have positive effects on heart disease by controlling hypercholesterolemia and atherosclerosis and to prevent renal stone formation. The most extensively studied positive aspect of *myo*-inositol phosphates is their potential for reducing the risk of colon cancer. Furthermore, much attention has been focused on *myo*-inositol with fewer than six phosphate residues, since some of these compounds have been shown to play an important part as intracellular second messengers and some have shown important pharmacological effects, such as the prevention of diabetes complications and antiinflammatory effects. The position of the phosphate groups on the *myo*-inositol ring is therefore of great significance for their physiological function. Thus, it is important to have reliable techniques available to determine qualitatively and quantitatively *myo*-inositol phosphates not only by the number of phosphate groups, but also by the position of the phosphate groups on the *myo*-inositol ring. (*See* **Plant Antinutritional Factors: Characteristics.**)

Structure, Occurrence, and Biological Significance

Phytate is a *meso* compound and consequently possesses a plane of symmetry with either five equatorial and one axial phosphate groups (5-eq/1-ax) or five axial and one equatorial phosphate groups (5-ax/1-eq: **Figure 1**). The carbon bearing the single axial or equatorial phosphate group is numbered C2 and the other ring carbons can be numbered C1–C6 from a C1 atom either side of C2, proceeding around the ring in a clockwise or counterclockwise fashion. Some less phosphorylated *myo*-inositol derivatives are optically active (**Table 1**). Their absolute configuration must be clearly defined. According to convention, a counterclockwise numbering gives rise to *myo*-inositol phosphates with a D-prefix and a clockwise numbering to *myo*-inositol phosphates with an L-prefix. The choice of prefix is normally determined by giving preference to that which results in the lowest numbering of substituents (**Figure 2**). The predominant conformation of the *myo*-inositol phosphates depends on the specific *myo*-inositol phosphate, pH value, type of cations present, and ionic strength. At pH values above 9.5, phytate exists exclusively in the 5-ax/1-eq conformation, whereas with *myo*-inositol pentakis-phosphates a small amount of the 5-eq/1-ax conformer is found in equilibrium with the predominant 5-ax/1-eq conformer. In *myo*-inositol phosphates with fewer than five phosphate residues, the *myo*-inositol ring appears to have a conformation in which only the phosphate group at C2 is axially oriented. Below pH 9.5 the 5-eq/1-ax conformer is predominant with all *myo*-inositol phosphates. The binding of some cations such as Cu^{2+} and Ca^{2+} is proposed to occur mainly via phosphate groups at the equatorial position of phytate; other cations such as

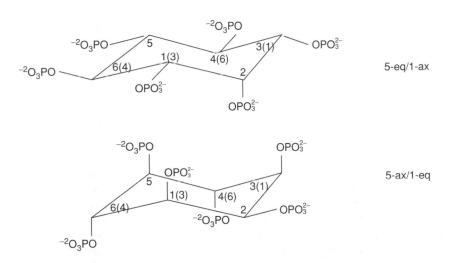

Figure 1 Possible chair conformations of phytate.

Table 1 *myo*-inositol phosphate isomers

No. of phosphate residues	No. of isomers	No. of enantiomeric pairs	myo-*inositol phosphate isomers*
6	1	0	$I(1,2,3,4,5,6)P_6$
5	6	2	D-$I(1,2,3,4,5)P_5$/L-$I(1,2,3,4,5)P_5$ D-$I(1,2,4,5,6)P_5$/L-$I(1,2,4,5,6)P_5$, $I(1,2,3,4,6)P_5$, *I(1,3,4,5,6)P_5*
4	15	6	D-$I(1,2,3,4)P_4$/L-$I(1,2,3,4)P_4$, D-$I(1,2,4,5)P_4$/L-$I(1,2,4,5)P_4$, D-*I(1,2,5,6)P_4*/L-$I(1,2,5,6)P_4$, D-$I(1,2,4,6)P_4$/L-$I(1,2,5,6)P_4$, D-*I(1,3,4,5)P_4*/L-$I(1,3,4,5)P_4$, D-*I(1,4,5,6)P_4*/L-*I(1,4,5,6)P_4*, $I(1,2,3,5)P_4$, *I(1,3,4,6)P_4*, $I(2,4,5,6)P_4$
3	20	8	D-$I(1,2,4)P_3$/L-$I(1,2,4)P_3$, D-$I(1,2,5)P_3$/L-$I(1,2,5)P_3$, D-*I(1,2,6)P_3*/L-$I(1,2,6)P_3$, D-$I(1,3,4)P_3$/L-$I(1,3,4)P_3$, D-$I(1,3,6)P_3$/L-$I(1,3,6)P_3$, D-*I(1,4,5)P_3*/L-$I(1,4,5)P_3$, D-$I(1,4,6)P_3$/L-*I(1,4,6)P_3*, D-*I(2,4,5)P_3*/L-$I(2,4,5)P_3$, $I(1,2,3)P_3$, $I(1,3,5)P_3$, $I(2,4,6)P_3$, $I(4,5,6)P_3$
2	15	6	D-$I(1,2)P_2$/L-$I(1,2)P_2$, D-*I(1,4)P_2*/L-$I(1,4)P_2$, D-$I(1,5)P_2$/L-$I(1,5)P_2$, D-$I(1,6)P_2$/L-*I(1,6)P_2*, D-$I(2,4)P_2$/L-$I(2,4)P_2$, D-$I(4,5)P_2$/L-$I(4,5)P_2$, *I(1,3)P_2*, $I(2,5)P_2$, $I(4,6)P_2$
1	6	2	D-*I(1)P*/L-*I(1)P*, D-*I(4)P*/L-$I(4)P$, *I(2)P*, $I(5)P$

Myo-inositol phosphate isomers found in nature are indicated in italics.

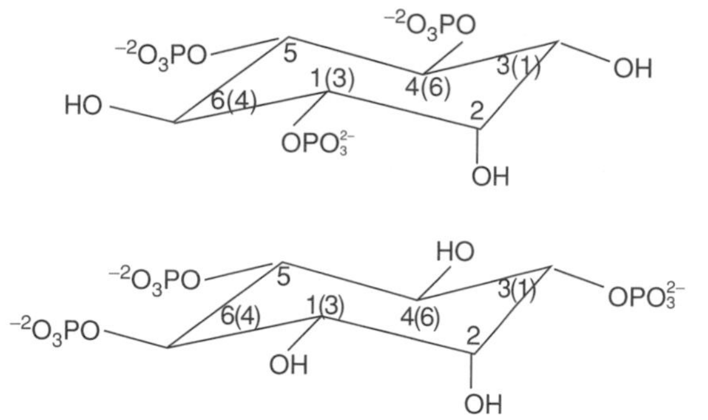

Figure 2 Absolute configuration of D- and L-*myo*-inositol(1,4,5)trisphosphate.

Zn^{2+}, Mn^{2+}, and the alkali metal ions seem to have preference for phosphate groups at the axial position. A single-crystal X-ray analysis of the sodium salt showed the 5-ax/1-eq conformer. In solution, the inositol ring also occurs over a wide pH range in a chair conformation in which five phosphate residues are arranged axially and only the phosphate at C2 is equatorially-oriented. Raman data indicate that alkali metal ions preferentially bind to, and thus stabilize, the 5-ax/1-eq phytate conformer in the order $Li^+ \sim Na^+ > Cs^+$. In contrast, the Raman spectrum of solid Ca_6-phytate is characterized by the 1-ax/5-eq conformer.

Plants

Phytate is ubiquitous among plant seeds and/or grains, comprising 0.5–5% (w/w). It is primarily present as a salt of the mono- and divalent cations K^+, Mg^{2+}, and Ca^{2+} and accumulates in the seeds during the ripening period. In dormant seeds phytate represents 60–90% of the total phosphate. Only a very small part of the *myo*-inositol phosphates exists as *myo*-inositol penta- and tetrakisphosphate of unknown isomeric state. The function of these high phytate concentrations in plant seeds is unclear. It has been suggested that phytate may serve as a store of phosphate, of cations, of the cell wall glucuronate precursor, of high-energy phosphoryl groups and, by chelating free iron, as a potent natural antioxidant. In amoeba, two diphospho-*myo*-inositol pentakisphosphate isomers and one bis(diphospho)-*myo*-inositol tetrakisphosphate isomer are present in concentrations exceeding that of phytate and thus may in fact represent a compact store of high-energy phosphate.

Until now only little is known of the pathway of phytate synthesis in either the plant or animal kingdom. A study in the slime mould *Dictyostelium discoideum* established that phytate synthesis from *myo*-inositol proceeds via Ins(3)P, Ins(3,6)P$_2$, Ins(3,4,6)P$_3$, Ins(1,3,4,6)P$_4$, and Ins(1,3,4,5,6)P$_5$. Early studies

of phytate synthesis in plants led to the proposal that phytate synthesis from Ins(3)P was mediated by phosphoinositol kinase(s) via a series of undefined *myo*-inositol phosphates. Recently the first description of the synthetic sequence to phytate in the plant kingdom was given. From the identities of *myo*-inositol phosphates found in duckweed (*Spirodela polyrhiza* L.), at a development stage associated with massive accumulation of phytate, it was concluded, that synthesis of phytate from *myo*-inositol proceeds according to the sequence D-I(3)P, D-I(3,4)P$_2$, D-I(3,4,6)P$_3$, D-I(3,4,5,6)P$_4$, I(1,3,4,5,6)P$_5$. An unanswered question that relates to the pathway of phytate synthesis in plants concerns the source of D-I(3)P. Two enzyme activities that are capable of synthesizing D-I(3)P have been identified. These are *myo*-inositol phosphate synthase (EC 5.5.1.4), which converts glucose-6-phosphate, the ultimate source of *myo*-inositol in plants, to D-I(3)P, and *myo*-inositol kinase (EC 2.7.1.64), which converts *myo*-inositol to D-I(3)P. The spatial and temporal distribution and the relative contribution of these two enzymes to phytate synthesis via D-I(3)P are unclear.

During germination, phytate is rapidly hydrolyzed in a stepwise manner by phytate-specific phosphohydrolases (phytases, EC 3.1.3.8, EC 3.1.3.26) or a concerted action of phytases and other phosphatases to supply the nutritional needs of the plant without an accumulation of less phosphorylated *myo*-inositol intermediates. Neither the isomer structure of these intermediates nor the final product of phytate degradation is known to date. From *in vitro* investigations on the stereospecificity of phytate hydrolysis by purified phytases from cereals it was established that these enzymes dephosphorylate phytate in a stereospecific way by sequential removal of phosphate groups via D-I(1,2,3,5,6)P$_5$, D-I(1,2,5,6)P$_4$, D-I(1,2,6)P$_3$, and D-I(2,6)P$_2$ to finally I(2)P. Moreover, the phytases from bacteria and fungi investigated for phytate degradation release five of the six phosphate groups, and the end product was identified as I(2)P. Thus, the

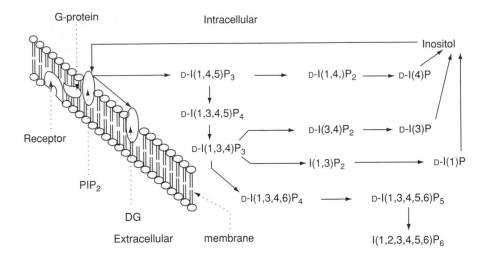

Figure 3 The phosphatidylinositol pathway. PIP$_2$, phosphatidylinositol 4,5-bisphosphate.

phosphate at C2 seems to be particularly resistant to enzymatic cleavage. (*See* **Enzymes:** Functions and Characteristics.)

Soil

In the soil, phytate as well as less phosphorylated *myo*-inositols are found. Their biological sources are unknown.

Animals

In animal tissue, a considerable number of *myo*-inositol phosphates containing one to six phosphate residues have been found. In most tissues stimulation of the phosphatidylinositol pathway (**Figure 3**) causes release of D-*myo*-inositol(1,4,5)trisphosphate, which is subsequently metabolized to a wide range of *myo*-inositol phosphate isomers. For D-I(1,3,4)P$_3$, metabolism is complex, involving both dephosphorylation via D-I(1,4)P$_2$ to D-I(4)P and phosphorylation to D-I(1,3,4,5)P$_4$, this latter compound being eventually degraded to D-I(1)P or D-I(3)P. The full significance of this complex metabolism is not clear, but there is now evidence that certain products of the phosphoinositide metabolism play second messenger roles in most cells. D-I(1,4,5)P$_3$ and D-I(1,3,4,5)P$_4$ bind to specific receptors and regulate Ca^{2+} release from or movement between intracellular Ca^{2+} stores. D-I(1,3,4,5)P$_4$ is also the starting point for metabolic pathways generating other *myo*-inositol tetrakisphosphate isomers as well as higher phosphorylated *myo*-inositols. There are no known functions for these higher phosphorylated *myo*-inositols; these metabolites comprise the bulk of *myo*-inositol phosphate content in mammalian cells, but evidence for their association with cell signaling was recently suggested.

D-*myo*-inositol(1,3,4,5,6)pentakisphosphate was also found in the erythrocytes of birds, turtles, and frogs. The functional importance of this compound as a key regulator of oxygen affinity becomes evident with the discovery that erythrocytes of adult birds contain virtually no 2,3-bisphosphogylcerate, the potent allosteric regulator of hemoglobin in mammalian erythrocytes.

Chemical Properties

Numerous studies have been made on the protonation constants of phytate, but the results are often conflicting. This could be due to the fact that the protonation constants of phytate to a large extent are dependent on the ionic strength of the medium. Phytic acid contains six strong acid groups which are completely dissociated in solution (pK_a 1.1–3.2), three weak acid protons (pK_a 5.2–8.0), and three very weak acid protons (pK_a 9.2–12). There unusually high pK_a values for the second protonization step seem to be due to intramolecular hydrogen bonding between the *syn*-axial phosphate residues at C1 and C3 as well as C4 and C6. These pK_a values imply that phytic acid will be strongly negatively charged over a wide pH range and have immense potential for binding positively charged species, such as cations or proteins. Free phytic acid is an unstable compound and decomposes to yield lower *myo*-inositol phosphates and orthophosphate. It is generally isolated as a sodium or calcium salt. In its free form, phytic acid is a light-yellow to light-brown syrupy liquid, soluble in polar solvents (water, methanol, ethanol, 2-propanol, acetone tetrahydrofuran, dimethyl sulfoxide, dimethyl formamide), but insoluble in nonpolar solvents (benzene, toluene, hexane, chloroform). In contrast to free phytic acid, their salts are very stable compounds. The phosphate groups can be removed hydrolytically by enzymes or acid/heat to yield a large number of homologs and positional isomers ranging from *myo*-inositol mono- to pentakisphosphates (**Table 1**). In spite of the considerable number of isomers identified *in vivo*, they still represent only a small percentage of the number possible in theory. Above pH 5, there is almost no decomposition of phytate at 100 °C within 10 h. The rate of acid hydrolysis is low – as in other orthophosphoric esters it reaches a maximum at pH 4, e.g., 27% of phytate is cleaved after 6 h at 100 °C, and even the use of strong acids leads to an only moderate increase in the rate of hydrolysis. In 5 mol l^{-1} HCl 47% of phytate is cleaved after 6 h at 100 °C. Final products of acid hydrolysis are *myo*-inositol and *myo*-inositol(2) monophosphate. As with enzymatic cleavage, the axial phosphate at C2 seems to be particularly resistant to hydrolysis. Complete decomposition of phytate was achieved with 3 mol l^{-1} H$_2$SO$_4$ at 165 °C for 4 h. Under acidic conditions and higher temperatures in particular, monophosphorylated *cis*-diol groups of the inositols, in competition with hydrolysis, exhibit the phenomenon of phosphate migration. Thus, for example, D-*myo*-inositol(1)phosphate can yield a mixture of D-*myo*-inositol(1)phosphate, *myo*-inositol(2) phosphate and L-*myo*-inositol(1)phosphate. For this migration, an intermediate formation of cyclic phosphodiesters is essential which is only sterically favored in *cis*-diol groups.

Phytate–Cation Interaction

Phytate forms complexes with numerous divalent and trivalent cations. The stability and solubility of the cation–phytate complexes depend on the specific cation, pH value, phytate-to-cation molar ratio, and the presence of other compounds in the solution. Phytate has six reactive phosphates and meets the criterion of a chelating agent. In fact, a cation can complex not only within one phosphate group or between two or more phosphate groups of one phytate, but also between two or more phytate molecules (**Figure 4**).

Studying the solubility and relative stability of various phytate–metal complexes by potentiometric titration, the following order of stability at pH 7.4 was found: Cu^{2+} > Zn^{2+} > Ni^{2+} > Co^{2+} > Mn^{2+} > Fe^{3+} > Ca^{2+}. Most phytates tend to be more soluble at lower than at higher pH values. The pH value below which the solubilities increase is about 5.5–6.0 for calcium, 7.2–8.0 for magnesium, and 4.3–4.5 for zinc phytate. In contrast, ferric phytate is insoluble at pH values in the 1–3.5 range at equimolar Fe^{3+}-to-phytate molar ratios. Solubility increases above pH 4, reaching 50% at pH 10. When Fe^{3+}-to-phytate molar ratio is increased to 3.5:1, there is increased solubility below pH 2, reaching a maximum of 90% at pH 1.5 and lower solubility at pH values above pH 4. By forming a complex with Fe^{3+} that lacks iron-coordinated water and thus is unable to catalyze the formation of hydroxyl radicals in the Fenton reaction, phytate is a good antioxidant.

Another important fact is the synergistic effect of secondary cations, among which Ca^{2+} has been most prominently mentioned. Two cations may, when present simultaneously, act together to increase the quantity of phytate precipitation. For example, Ca^{2+} enhanced the incorporation or adsorption of Zn^{2+} into phytate by formation of a Ca-Zn phytate. The effect of Ca^{2+} on the amount of Zn^{2+} coprecipitated with phytate is dependent on Zn^{2+}-to-phytate molar ratios. For high Zn^{2+}-to-phytate molar ratios, Ca^{2+} displaces Zn^{2+} from phytate-binding sites and increases its solubility. The amount of free Zn^{2+} is directly proportional to the Ca^{2+} concentration. For low Zn^{2+}-to-phytate molar ratios, Ca^{2+} potentiates the precipitation of Zn^{2+} as phytate. The higher the Ca^{2+} level, the more extensive the precipitation of the

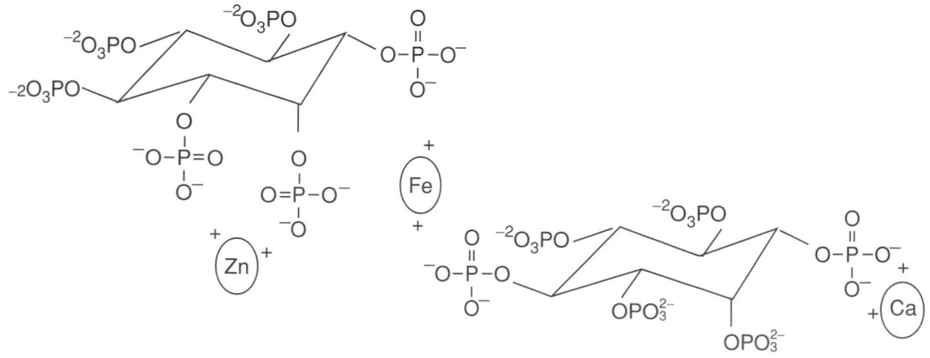

Figure 4 Phytate–cation interaction. DG, diacylglycerol.

ions. Mg^{2+} also has been shown *in vitro* to potentiate the precipitation of Zn^{2+} in the presence of phytate; however, Mg^{2+} has been found to exert a less pronounced effect on Zn^{2+} solubility than Ca^{2+}.

The knowledge about the interaction of the lower *myo*-inositol phosphates with different cations is limited. Recent studies have shown that *myo*-inositol pentakis-, tetrakis-, and trisphosphates have a lower capacity to bind cations (Ca^{2+}, Cu^{2+}, Zn^{2+}, Fe^{2+}, Fe^{3+}) at pH values in the 5–7 range. The capacity to bind cation was found to be a function of the number of phosphate groups on the molecule. The cation-*myo*-inositol phosphate complexes seem to become more soluble as the number of phosphate groups decreases. There is also some evidence for weaker complexes when phosphate groups are removed from phytate. Furthermore, the binding affinity of cations to *myo*-inositol phosphates has been shown to be affected by the orientation of the phosphate groups.

Phytate–Protein Interaction

Phytate interactions with proteins are pH-dependent. Phytate is known to form complexes with proteins at both acidic and alkaline pH (**Figure 5**). At pH values below the isoelectric point of the protein, the anionic phosphate groups of phytate bind strongly to the cationic groups of the protein to form insoluble complexes that dissolve only below pH 3.5. The α-NH_2 terminal group, the ε-NH_2 of lysine, the imidazole group of histidine, and guanidyl group of arginine have been implicated as protein-binding sites for phytate at low pH values. These low-pH protein–phytate complexes are disrupted by the competitive action of multivalent cations.

Above the isoelectric point of the protein, both protein and phytate have a negative charge, but in the presence of multivalent cations soluble protein–cation–phytate complexes occur. The major protein-binding site for the ternary complex appears to be the unprotonated imidazole group of histidine. The ionized carboxyl group of the protein are also suggested sites. These complexes may be disrupted by high ionic strength, high pH (> 10), and high concentrations of the chelating agents.

Protein–phytate complexation may effect changes in protein structure that can decrease enzymatic activity, solubility, and vulnerability to attack by proteolytic enzymes. Phytate has been shown to reduce the activity of lipase, α-amylase, pepsin, trypsin, and chymotrypsin *in vitro*. The inhibitory effect increases with the number of phosphate groups per *myo*-inositol molecule and the *myo*-inositol phosphate concentration. (*See* **Protein**: Interactions and Reactions Involved in Food Processing.)

Application

Phytate has found industrial application, including uses in the food industry (**Table 2**). The focus of research on phytates includes occurrence and functions in plant seeds, nutritional significance, preservative applications in food technology, and potential medical and industrial uses. (*See* **Preservatives**: Food Uses.)

Determination

The measurement of *myo*-inositol phosphates in any material requires an initial extraction. The reagents most commonly used to extract *myo*-inositol phosphates from foodstuff and biological samples include 3% trichloroacetic acid and 2.4% hydrochloric acid. Since *myo*-inositol phosphates do not have a

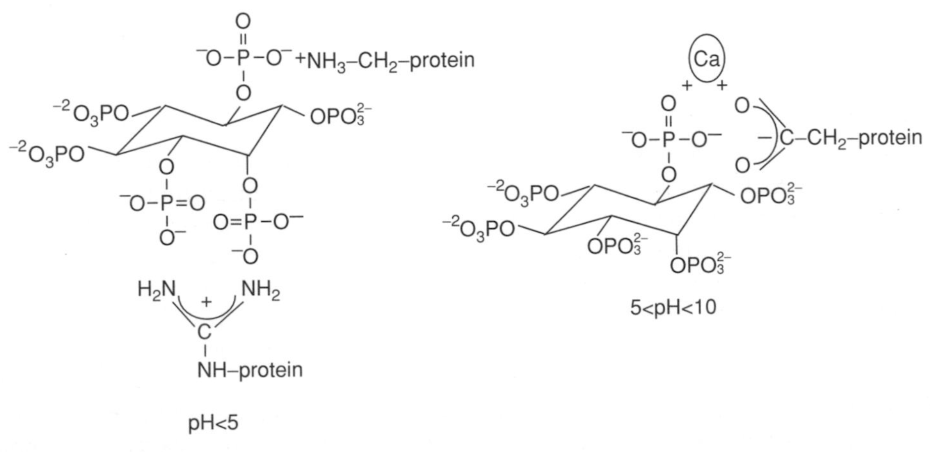

Figure 5 Phytate–protein interaction.

Table 2 Application of phytate

Action	Application
Metal chelation	Prevention of color and quality changes in processed agricultural (chestnut, bean sprouts, pickles, asparagus, etc.) and fishery (tuna, clams, shrimps, crabs, etc.) products
	Removing metal ions from wine
	Rust-proofing and dissolving-out prevention inside cans
	Prevention of oxidation in oil/water emulsion-type food such as cream, dressings, butter, chesses, soups
	Additive for etching solution for offset printing
	Anticorrosion agent for paints, antifreezes, and metal surfaces (steel, tin, aluminum, iron)
	Stabilizer for perfumes and cosmetics
	Antioxidant for industrial oils and greases
pH control	Prevention of quality changes by controlling pH value
Fermentation promoter	Improvement of product yield and quality by promoting the growth of microorganisms such as lactic acid bacteria and yeasts (fermented food, antibiotics, methanol, etc.)

characteristic absorption spectrum, nor can they be identified using specific colorimetric reagents, the determination of these compounds has remained a persistent problem.

Qualitative Separation Methods

Qualitative separation and detection of *myo*-inositol phosphates have been developed in the 1950s and 1960s. Paper chromatography has been shown to be useful for separating *myo*-inositol phosphates by the number of phosphate groups. *Myo*-inositol mono- to hexakisphosphate could also be resolved relatively rapidly by electrophoresis. Thin-layer chromatography, even if successfully applicable, has not been widely adopted for the separation of *myo*-inositol phosphates.

Quantitative Separation Methods

Precipitative methods Quantitative methods for determining phytate often employ the addition of a controlled amount of Fe^{3+} to an acidic sample extract to precipitate the phytate. Phytate is subsequently estimated either by determining the phosphate, inositol, or iron content of the precipitate (direct method), or by measuring the excess iron in the supernatant (indirect method). The indirect methods are generally more convenient and reproducible, because the stoichiometric ratio of phosphate to iron in Fe^{3+}-*myo*-inositol phosphate precipitates is affected by several variables, including the way in which the precipitate is washed. These methods are not specific for phytate due to coprecipitation of less phosphorylated *myo*-inositols and should therefore be limited to the analysis of material which contains negligible amounts of these *myo*-inositol phosphates.

Nonprecipitative methods Nonprecipitative methods for *myo*-inositol phosphate determination include ^{31}P-Fourier transform nuclear magnetic resonance

(^{31}P-FT NMR) spectroscopy, near-infrared reflectance spectroscopy, low-pressure anion-exchange chromatography, several high-performance liquid chromatographic (HPLC) separation systems, and capillary electrophoresis. The main limitation of ^{31}P-FT NMR and near-infrared spectroscopy is that these methods are specific for phytate only when the sample contains negligible amounts of less phosphorylated *myo*-inositols. Additionally, sophisticated instruments which are not available in most laboratories are required.

Low-pressure anion-exchange chromatography
Low-pressure anion-exchange chromatography is widely used in the determination of *myo*-inositol phosphates. The method currently accepted by the Association of Official Analytical Chemists (AOAC) for measuring phytate in foods and feeds is based on a step gradient (0.7 mol l^{-1} NaCl) anion-exchange method (Dowex AG1-X8). Unfortunately, the anion-exchange resin also retains less phosphorylated *myo*-inositols. The method should therefore be limited to the analysis of material with negligible amounts of these *myo*-inositol phosphates. *Myo*-inositol mono-to hexakisphosphate and even some positional isomers could be resolved using anion-exchange chromatography with a linear eluting gradient of hydrochloric acid or a stepwise elution with either hydrochloric acid or ammonium formate/formic acid solutions of increasing concentrations. Unfortunately, these methods require long elution times (up to 24 h) and a large number of eluate fractions must be hydrolyzed for quantitation as phosphate or inositol, since these systems preclude the use of refractive index and conductivity detection methods. Methods designed by those studying calcium metabolism are dependent on the use of radiolabeled *myo*-inositol phosphates to facilitate detection and quantitation, but it is not feasible to label existing *myo*-inositol phosphates in dietary constituents.

High-performance liquid chromatoraphy More recently, HPLC techniques have been introduced into *myo*-inositol phosphate determination. Purification of crude acid extracts of biological samples is usually required prior to injection on to the analytical HPLC system. The techniques used for detection and quantitation of the *myo*-inositol phosphates is heavily dependent on the system employed for their separation. The *myo*-inositol phosphates may be separated using anion-exchange, reverse-phase, micellar and ion chromatography, and detected/quantified by a variety of techniques, including refractive index, conductivity, indirect photometry, online postcolumn spectrophotometric detection, and offline phosphate or inositol assay. Among these, ion-pair reverse-phase and anion-exchange chromatography are largely used. (*See* **Chromatography:** High-performance Liquid Chromatography.)

Ion-pair reverse-phase chromatography Ion-pair reverse-phase chromatography with refractive index detection has been successfully applied to analysis of *myo*-inositol phosphates. The retention of *myo*-inositol phosphates on reverse-phase packings is markedly increased through the use of ion-pair reagents, allowing the simultaneous separation of *myo*-inositol tris- to hexakisphosphates, but neither *myo*-inositol mono- or bisphosphates nor the individual positional isomers are resolved (**Figure 6**). However, sample extracts must be passed through anion-exchange resin to remove orthophosphate and concentrate the *myo*-inositol phosphates. Acidic column eluent is then evaporated to dryness to remove hydrochloric acid and reconstituted in water prior to injection on to a silica-based C18 reverse-phase HPLC column. The mobile phase consisted of formic acid/ methanol and tetrabutylammonium hydroxide. The affinity of *myo*-inositol phosphates for the stationary phase increases with the increasing number of phosphate groups on the inositol ring and with increasing pH. (*See* **Chromatography:** High-performance Liquid Chromatography.)

Anion-exchange chromatography To date, published procedures involving anion-exchange HPLC fall into two categories: isocratic and gradient ionchromatographic techniques. The capability of resolving the different *myo*-inositol phosphates depends on the stationary phase used and the chromatography conditions. *Myo*-inositol mono- to hexakisphosphates have been successfully resolved by isocratic elution from low-capacity weak anion-exchange columns. These single eluent systems are compatible with refractive index, indirect photometric, thermospray mass spectrometric, and conductivity detection.

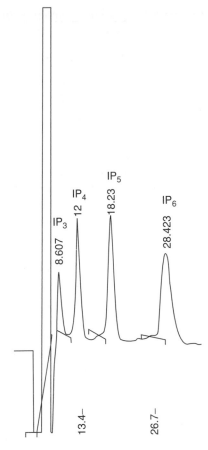

Figure 6 Chromatographic profile of a *myo*-inositol phosphate standard by high-performance liquid chromatography (HPLC) ion-pair chromatography on Ultrasep ES 100 RP18 (2 × 250 mm). The column was run at 45°C and 0.2 ml min⁻¹ of an eluent consisting of formic acid:methanol:water: tetrabutylammonium hydroxide (TBAH: 44:56:5:1.5 v/v), pH 4.25. *Myo*-inositol phosphates were detected by refractive index a standard. Peaks (8.607) IP₃; (12) IP₄; (18.23) IP₅; (28.423) IP₆.

However, in the case of conductivity detection, sensitivity is low unless counterions in the eluent are continuously removed using a suppressor column or membrane suppressor system.

In the last few years a number of isomer-specific ion-exchange chromatography methods with gradient elution for separation and quantitation of *myo*-inositol phosphates in the picomolar range have been developed. Eluents with high ionic strength, such as formate, acetate, citrate, phosphate, nitrate, sulfate, sodium chloride, or hydrochloric acid, have been used. The most commonly used detection method with gradient elution is online postcolumn derivatization or complexation reactions followed by spectrophotometric detection. Three approaches have been employed in the postcolumn detection and quantitation of *myo*-inositol phosphates. The first is based on the direct reaction of *myo*-inositol phosphates with a reagent to form a fluorescent complex or one which has an absorbance in the ultraviolet or visible part of the spectrum. For example, the eluate from the column was mixed, online, with 0.1% $Fe(NO_3)_3$ in 2% $HClO_4$ to form ultraviolet-absorbing phytate–Fe^{3+}–ClO_4 complexes. The use of postcolumn derivatization through ligand-exchange reaction between the iron(III)-sulfosalicylate complex and eluted myo-inositol phosphates has been described as an alternative. Furthermore, a complexometric technique based on competition between myo-inositol phosphates and the cation-specific reporter dye 4-[2-pyridylazo]resorcinol for the transition metal yttrium has been described. The second approach is based on the online enzymatic hydrolysis of myo-inositol phosphates which is then mixed with a molybdate solution in the reaction coil. The colored phosphomolybdate complex may be quantified spectrophotometrically. Finally, myo-inositol phosphates may be quantified by online thermospray mass spectrometric techniques.

A remaining problem, however, is to separate isomers from the whole spectrum of myo-inositol phosphates in the same run. Separation is generally performed on HPLC columns with gradient elution in two combined systems. Myo-inositol mono- to trisphosphates have been acidic gradient-eluted, postcolumn-derivatized, and ultraviolet-detected and myo-inositol bis- to hexakisphosphates have been alkali gradient-eluted and detected using chemically suppressed conductivity detection. The sensitivity of the analysis of myo-inositol mono- and bisphosphates was improved 10–100 times by using sodium acetate gradient elution in a sodium hydroxide environment and pulsed amperometric detection.

Capillary electrophoresis Recently capillary electrophoresis has been applied to the determination of myo-inositol phosphates. Capillary electrophoresis is attractive, since only a few nanoliters of sample are used in each analysis, there is the potential for concurrent separation of mono- to hexakisphosphate species in the same analysis, and run times are usually short due to the intrinsically high efficiency of the technique. Indirect ultraviolet detection was used to allow the detection of the nonchromophoric myo-inositol phosphates. Thus, no derivatization of the compounds is needed. Separations of all six myo-inositol phosphate groups in deionized water has been achieved in about 13 min. The position isomers myo-inositol(1)phosphate and myo-inositol(2)phosphate are easily separated in a phthalate electrolyte system, demonstrating the potential for separating myo-inositol phosphate isomers. However, further work is required on the development of capillary electrophoretic methods for the separation and quantiation of the different myo-inositol phosphate isomers.

All in all, efficient analytical systems for separation and quantitation of myo-inositol phosphate isomers are available. One problem in developing methods to determine myo-inositol phosphate isomers is their availability as reference compounds. They may be produced by chemical or enzymatic hydrolysis of phytate. Then identification of these isomers is needed. This requires sophisticated methods. The earliest developed technique is chemical analysis by oxidation with periodate, reduction, dephosphorylation, and subsequent identification of the polyols found. Further information can be obtained from the above-mentioned cis phosphate migration. In the past few years, high-resolution nuclear magnetic resonance (NMR) spectroscopy has evolved as a much simpler technique in the identification of the isomeric nature of myo-inositol phosphates. Thirty-nine of the 63 theoretically possible myo-inositol phosphate isomers can be identified by NMR. Only for the 24 enantiomeric pairs among these isomers (**Table 1**) absolute configurations are indistinguishable from NMR, but these enantiomers are also not separated on the achiral columns in use. Separation techniques using chiral columns have to be developed to resolve the enantiomers. The absolute configuration of such enantiomers may be determined using high-affinity binding proteins or enzymatic assays.

See also: **Chromatography**: High-performance Liquid Chromatography; **Electrophoresis**: General Principles; **Enzymes**: Functions and Characteristics; **Phosphorus**: Properties and Determination; Physiology; **Plant Antinutritional Factors**: Characteristics; **Preservatives**: Food Uses

Further Reading

Buscher BAP, Irth H, Anderson E, Tjaden UR and van der Greef J (1994) Determination of inositol phosphates in fermentation broth using capillary zone electrophoresis with indirect UV detection. *Journal of Chromatography* A678: 145–150.

Cheryan M (1980) Phytic interactions in food systems. *CRC Critical Reviews in Food Science and Nutrition* 13: 297–335.

Cosgrove DJ (ed.) (1980) *Inositol Phosphates: Their Chemistry, Biochemistry and Physiology.* Amsterdam: Elsevier.

Dean NM and Beaven MA (1989) Methods for the analysis of inositol phosphates. *Analytical Biochemistry* 183: 199–209.

Graf E (ed.) (1986) *Phytic Acid: Chemistry and Application.* Minneapolis: Pilatus Press.

Irvine RF (ed.) (1990) *Methods in Inositide Research*. New York: Raven Press.

Morré DJ, Boss WF and Loewus FA (eds) (1990) *Inositol Metabolism in Plants*. New York: Wiley.

Oberleas D (1971) The determination of phytate and inositol phosphates. In: Glick D (ed.) *Methods of Biochemical Analysis*, vol. 20, pp. 87–101. New York: Wiley.

Potter BVL (1990) Recent advances in the chemistry and biochemistry of inositol phosphates of biological interest. *Natural Products Report* 7: 1–24.

Reddy NR, Pierson MD, Sathe SK and Salunkhe DK (eds) (1989) *Phytates in Cereals and Legumes*. Boca Raton: CRC Press.

Skoglund E, Carlsson N-G and Sandberg A-S (1997) Determination of isomers of inositol mono- to hexaphosphates in selected foods and intestinal contents using High Performance Ion Chromatography. *Journal of Agricultural and Food Chemistry* 45: 431–436.

Turk M, Sandberg A-S, Carlsson N-G and Andlid T (2000) Inositol hexaphosphate hydrolysis by Baker's yeast. Capacity, kinetics, and degradation products. *Journal of Agricultural and Food Chemistry* 48: 100–104.

Xu P, Price J and Aggett PJ (1992) Recent advances in methodology for analysis of phytate and inositol phosphates in foods. *Progress in Food and Nutrition Science* 16: 245–262.

Nutritional Impact

U Konietzny, Karlsruhe, Dettenheim-Liedolsheim, Germany
R Greiner, Centre for Molecular Biology, Federal Research Centre for Nutrition, Karlsruhe, Germany

Introduction

Mixed salts of phytic acid [*myo*-inositol(1,2,3,4,5,6)-hexakisphosphoric acid) are common constituents of foods and feed, since phytate is a naturally occurring compound formed during the maturation of seeds and cereal grains. Depending on the amount of plant foods in the diet and the grade of food processing, the daily intake of phytate can be as high as 4500 mg. On average, the daily intake of phytate was estimated to be 2000–2600 mg for vegetarian diets as well as diets of inhabitants of rural areas of developing countries and 300–1300 mg for mixed diets (Table 1).

Phytate behaves in a broad pH region as a highly negatively charged ion and therefore has a tremendous affinity for food components with positive charge(s) (*See* **Phytic Acid**: Properties and Determination). There is a large body of evidence that minerals are less available from foods of plant origin as compared to animal-based foods. Minerals of concern in

this regard include zinc, iron, calcium, magnesium, manganese, and copper. The formation of insoluble cation–phytate complexes at physiological pH values is regarded as the major reason for the poor mineral bioavailability, since these complexes are essentially nonabsorbable from the gastrointestinal tract. Furthermore, phytate phosphorus may not be nutritionally available, since phytate is not hydrolyzable quantitatively in the human gut. Consumption of phytate, however, seems not to have only negative aspects on human health. In the last few years, results of epidemiological and animal studies also suggest beneficial effects, such as decreasing the risk of heart disease and colon cancer, but data from human studies are still lacking.

Table 1 Phytate content of foods of plant origin

Food	Phytate (mg g^{-1} dry matter)
Cereal-based	
French bread	0.3–0.4
Mixed-flour bread (70% wheat, 30% rye)	0.2–0.7
Mixed-flour bread (70% rye, 30% wheat)	0–0.3
Sourdough rye bread	0–0.3
Wholewheat bread	4.3–6.8
Whole rye bread	2.5–4.8
Unleavened wheat bread	9.2–19.5
Corn bread	5.2–7.1
Unleavened corn bread	11.4–16.3
Oat bran	12.4–29.6
Oat flakes	8.2–10.3
Oat porridge	7.7–10.6
Pasta	2.2–8.6
Maize	11.5–14.2
Cornflakes	0.8–1.3
Rice (polished, cooked)	1.4–2.9
Wild rice (cooked)	16.4–20.1
Sorghum	5.6–9.8
Legume-based	
Chickpea (cooked)	4.9–6.1
Cowpea (cooked)	5.8–10.3
Black beans (cooked)	8.5–11.3
White beans (cooked)	9.1–10.9
Lima beans (cooked)	6.2–9.8
Faba beans (cooked)	10.1–13.7
Kidney beans (cooked)	8.9–11.2
Navy beans (cooked)	7.4–10.6
Soybeans	9.9–14.9
Tempeh	9.1–10.3
Tofu	8.2–9.3
Lentils (cooked)	6.5–9.3
Green peas (cooked)	5.7–7.8
Peanuts	16.5–19.1
Others	
Sesame seeds (toasted)	43.2–55.1
Soy protein isolate	4.3–11.7
Soy protein concentrate	12.4–21.7
Buckwheat	10.3–14.1
Amaranth grain	12.6–14.3

Phytate as an Antinutrient

The main concern about the presence of phytate in the diet is its negative effect on mineral uptake. Most studies have shown an inverse relationship between phytate content and mineral availability, although there are great differences in the behavior of individual minerals (**Table 2**).

Difficulties in Experimental Approaches

The effect of phytate on mineral absorption is highly controversial, since many investigations have shown a negative effect, but some studies have also shown no effect or even enhancement of mineral uptake. This controversial result gives an idea of the complexity of mineral absorption in the intestine (**Figure 1**). The differing types of experimental design may explain much of this controversy. *In vitro* studies can only incompletely simulate the physiological factors and physicochemical conditions affecting mineral availability. Also, *in vivo* approaches, widely used in mineral bioavailability studies, are not easily comparable due to the existence of many factors that cannot be reproduced in the different experiments. In addition, part of the variability may arise from differences in the method of phytate analysis and experimental techniques for measuring mineral bioavailability. (*See* **Minerals – Dietary Importance.**)

The solubility of phytate complexes is a critical and perhaps overriding issue, because complexes that are insoluble in the upper small intestine, where maximum mineral absorption normally occurs, are highly unlikely to provide absorbable essential elements. Thus, chemical interactions of phytate in the upper gastrointestinal tract are of particular concern. The form in which many minerals occur in foodstuffs is largely unknown, as is also the form in which they occur in the gut. Thus, predicting the specific interactions of phytate in the gastrointestinal tract and the nutritional implications of these interactions is very difficult. As foods are ingested and the digesta travels through the gastrointestinal tract, phytate may continue to maintain associations developed during ripening or food processing or phytate complexes may dissociate and other chelates form, since binding of phytate with minerals or proteins depends upon the pH value, which changes from low pH in the stomach to about neutral in the upper intestine.

The total composition of experimental diets has a great importance on mineral bioavailability, since the reduced availability of essential minerals depends on several dietary factors, such as the total concentration and composition of minerals or the phytate concentration. Phytate *per se* seems not to have a direct adverse affect on mineral absorption, since the physiological concentration of a single mineral is generally not sufficient for the formation of insoluble phytate complexes in the small intestine. Since calcium concentration in the diet is high enough for complete precipitation of phytate, leading to a coprecipitation of other minerals, the calcium content of the diets is of vital importance to the negative impact of phytate on mineral bioavailability. Calcium clearly augments the adverse effects of phytate on mineral absorption, and numerous other dietary components have lesser effects, both beneficial and adverse. For example, the intake of organic acids such as ascorbic acid and/or meat effectively counteracted the inhibitory effect of phytate, whereas dietary fiber and polyphenols intensified it. Both phytate and fiber have a high potential binding capacity for minerals and, because they are generally presented together in many foods, it is very difficult to separate completely the effects of these two in studies with typical human diets.

Phytate degradation during digestion in the gastrointestinal tract may have a positive effect on mineral absorption, since results of animal and human studies indicated that besides phytate, only the first degradation product of phytate, *myo*-inositol pentakisphosphate, showed negative effects on zinc, iron, and calcium absorption in its isolated form, while *myo*-inositol tetrakis- and trisphosphates had no effect in the concentrations under investigation. Furthermore, phytate phosphorus becomes available for utilization after release from the *myo*-inositol ring. The hydrolysis of phytate in the gastrointestinal tract of monogastric animals, including humans, may be carried out by the action of phytate-degrading enzymes from three sources: dietary phytases, phytases from the bacterial flora in the gut, and intestinal mucosal phytases. The level of dietary calcium is thereby of particular interest. A low calcium concentration may favor increased phytate hydrolysis in the gut, whereas elevated dietary calcium decreases its hydrolysis. It was recently reported that mucosal phytase, if present in the human small intestine, does not seem to play a significant role in phytate digestion, whereas the dietary phytases are an important factor, since these enzymes are active in the stomach for a certain time. Phytate degradation by phytases from the bacterial flora in the colon does not contribute to mineral and phytate phosphorus absorption, since, with the exception of calcium, absorption is negligible there.

Furthermore, the history of food processing appears to be of great interest insofar as it can affect the availability of phytate phosphorus and phytate-associated cations. Depending on the manufacturing process, reduction in phytate content of the foods,

Table 2 Effect of phytate on mineral and phosphorus availability

Zinc	Iron	Calcium	Copper, manganese, magnesium	Phosphorus
Zinc was reported to be the essential mineral most adversely affected by phytate. Zinc deficiency in humans was first reported in 1963 in Egyptian boys whose diet consisted in the main of bread and beans. These patients, who were characterized by dwarfism and hypogonadism, showed a response to zinc supplementation of their diet. It became accepted that the presence of phytate in plant products was an important factor in the reduction of zinc absorption from foodstuff. Zinc absorption was shown to be inversely correlated with phytate content of the meal. There is, however, some lack of agreement among studies, particularly with respect to specific foods and their components. In addition, it was shown that phytate not only depresses the bioavailability of dietary zinc, but also substantially reduces the reabsorption of endogenous intestinal zinc	A great deal of controversy exists regarding the effect of phytate on the availability of dietary iron. Much of this controversy may be due to the low absorption of iron in general, the presence of different iron phytates with different solubility, and the existence of two types of food iron, heme and nonheme iron. Heme iron is better absorbed and is little influenced by dietary factors; nonheme iron is less easily absorbed, and its absorption is affected by other dietary factors. Since many human studies indicated that phytate had a very strong inhibitory effect on iron absorption, it is well accepted today that phytate appeared to be the major contributor to the reduction in iron availability in humans, but some other factor(s) also contribute	Human studies indicated that phytate inhibited calcium absorption, but the effect of phytate on calcium availability seems not to be as extreme as on that of iron and particularly zinc. This may be due to the relatively high calcium content of plant-based foods, the ability of phytate degradation by microbial phytases of the gut flora, and the absorption of calcium in the large intestine	Relatively few studies have dealt with the effects of phytate on dietary copper, manganese, and magnesium utilization. Phytate has been shown to decrease their bioavailability in *in vivo* studies. It appears that the effect of phytate on copper, manganese, and magnesium bioavailability is less marked than those for some other essential elements	The fact that phytate phosphorus is poorly available to monogastric animals, including humans, was demonstrated several years ago. Phosphorus is absorbed as the orthophosphate. No conclusive evidence for absorption of phytate or other *myo*-inositol phosphates is available at present. Thus, the utilization of phytate phosphorus by monogastric animals will largely depend on their ability to hydrolyze phytate

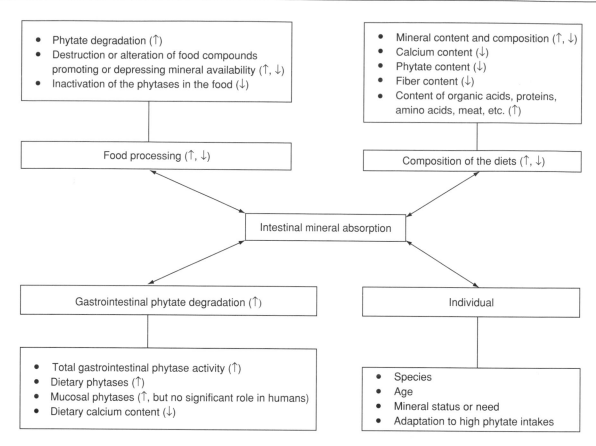

Figure 1 Factors interacting with intestinal mineral absorption. ↑ promoting mineral uptake; ↓ depressing mineral uptake.

destruction or alteration of food compounds promoting or depressing mineral availability, and inactivation of phytate-degrading enzymes may occur. Inactivating the endogenous dietary phytase leads to a limited degradation of phytate in the stomach and therefore to a lower bioavailability of minerals and phytate phosphorus. During food processing and food digestion, phytate can be partially dephosphorylated to yield a large number of positional isomers of *myo*-inositol pentakis-, tetrakis-, tris-, bis-, and monophosphates (*See* **Phytic Acid:** Properties and Determination). Especially with respect to the isomers, their chelating potential has not been carefully studied. This is of great importance, since processed foods and digesta may contain lower *myo*-inositol phosphates in great amounts and human studies have shown that *myo*-inositol tetrakis- and tris-phosphates contribute in high concentrations to the negative effect on mineral absorption. On the other hand, recent evidence has suggested that, besides the adverse effects of phytate and other highly phosphorylated *myo*-inositol phosphates on mineral absorption, lower phosphorylated *myo*-inositol phosphates may improve mineral absorption. Therefore, a reliable and accurate determination of the individual *myo*-inositol phosphates in diets must be utilized

before any useful evaluation can be made of their physiological effect on mineral availability (*See* **Phytic Acid:** Properties and Determination).

The source of phytate and minerals also influences the results of experimental studies. Endogenous phytate and added phytate are not necessarily equal. Usually sodium phytate is added to experimental diets; this is capable of chelating the multivalent cations therein, whereas endogenous phytate is already bound to those cations in the seeds and grains. Furthermore, the availability of minerals has mostly been examined after addition of ionic salts to the experimental diets. Results obtained under such conditions may not represent the level of absorption of minerals present in natural sources.

The effect of phytate on mineral absorption also seems to be dependent on the experimental species used. Phytate degradation in the stomach and small intestine varies with the species. For example, it was suggested that rats may not be a good model for assessing mineral absorption from phytate-containing foods due to the existence of rat intestinal phytase activity. In addition, the age of the individuals is of importance, since it seems that phytate digestion decreases with age. The ability of endogenous carriers in the intestinal mucosa to absorb essential minerals bound to phytate or other dietary substances as well as the mineral status and need of the individual also have to be considered. A possible adaptation to a high-phytate intake is controversially discussed. A long-term study did not show any human adaptation to a high phytate diet with respect to iron absorption, whereas the normal bone and teeth calcification throughout the world in several populations who depend almost exclusively on cereal diets suggests human adaptability towards high phytate consumption with respect to calcium absorption.

Protein

Phytate also interacts with proteins, which may affect protein digestibility negatively. Strong evidence exists that phytate–protein interactions negatively affect protein digestibility *in vitro*. The extent of this effect depends on the protein source. The inhibition of digestive enzymes such as alpha-amylase, lipase, or proteinase by phytate may also be of significance, as shown in *in vitro* studies. This may be due to the nonspecific nature of phytate–protein interactions, the chelation of calcium ions which are essential for the activity of trypsin and alpha-amylase, or the interaction with the substrates of these enzymes. The inhibition of proteases may be partly responsible for the reduced protein digestibility. *In vivo*, phytate has also been considered to inhibit alpha-amylase, as indicated by a negative relationship between phytate intake and blood glucose response. A negative effect of phytate on the nutritive value of protein, however, was not clearly confirmed *in vivo*. While some have suggested that phytate does not affect protein digestibility, others have found improvement in protein utilization with decreasing levels of phytate. This difference may be at least partly due to the use of different protein sources. Thus, the significance of protein–phytate complexes in nutrition is still under scrutiny. (*See* **Protein**: Digestion and Absorption of Protein and Nitrogen Balance.)

Beneficial Effects of Phytate

In view of the above results, the evidence seems overwhelming that high intakes of phytate can have adverse effects on mineral uptake in humans. In the last few years, however, some novel metabolic effects of phytate or some of its degradation products have been recognized. Results of epidemiological and animal studies suggest possible protective effects of certain *myo*-inositol phosphates on heart disease, renal stone formation, and colon cancer in humans. Moreover, the potential beneficial effects of phytate in the prevention of severe poisoning should also be

considered. Calcium phytate 1–2% in the diet has been found to protect against dietary lead in experimental animals and in human volunteers. Furthermore, calcium phytate was capable of lowering blood lead levels. Thus, phytate seems to be a helpful means of counteracting acute oral lead toxicity. The effect of calcium phytate on acute cadmium toxicity is still controversially discussed.

Phytate and Diabetes Mellitus

Diabetes mellitus is one of the commonest nutrition-dependent diseases in western society. It may be caused by hypercaloric diets with a high percentage of quickly available carbohydrates. Foods that result in low blood glucose response have been shown to have great nutritional significance in the prevention and management of diabetes mellitus. In this regard phytate-rich foods are of interest, since a negative relationship between phytate intake and blood glucose response was reported. For example, phytate-enriched unleavened bread based on white flour reduced *in vitro* starch digestibility as well as flattening the glycemic response in five healthy volunteers compared to bread without the addition of phytate. The *in vitro* reduction of starch digestion was positively correlated with the *myo*-inositol phosphate concentration and negatively with the number of phosphate groups on the *myo*-inositol ring. It has to be noted that there are also studies which have not found an inhibition of alpha-amylase and starch digestion by phytate. (*See* **Diabetes Mellitus**: Etiology.)

Phytate and Coronary Heart Disease

Heart disease is a leading cause of death in western countries, yet it is low in Japan and developing countries. Elevated plasma cholesterol or, more specifically, elevated low-density lipoprotein (LDL)-cholesterol concentrations have been shown to be one of the risk factors. It has been proposed that dietary fiber or, more specifically, phytate as a component of fiber may influence the etiology of heart disease. Animal studies have demonstrated that dietary phytate supplementation resulted in significantly lowered serum cholesterol and triglyceride levels. This effect was accompanied by a decrease in serum zinc level and in the ratio of zinc to copper. Thus, the hypothesis was put forward that coronary heart disease is predominantly a disease of imbalance in regard to zinc and copper metabolism. The hypothesis is also based on the production of hypercholesterolemia, which is a major factor in the etiology of coronary heart disease, in rats fed a diet with a high ratio of zinc and copper. It was thought that excess zinc in the diet resulted in decreased copper uptake from the small intestine, since both minerals compete for common mucosal carrier systems. As phytate preferentially binds zinc rather than copper it is presumed that phytate exerts its effect probably by decreasing zinc without affecting copper absorption. It should be pointed out that the support for the preventive role of phytate in heart disease is only based on a few animal and *in vitro* studies. Results from human studies are still lacking. (*See* **Coronary Heart Disease**: Etiology and Risk Factor.)

Phytate and Renal Calculi

The increase of renal stone incidence in northern Europe, North America, and Japan has been reported to be coincident with the industrial development of these countries, making dietary intake suspect. Epidemiological investigations found that there were substantial differences in renal stone incidences between white and black residents of South Africa. Whereas renal stone occurred in the white population with no less frequency than in other Western communities, it was seldom seen in the black South Africans. The major dietary difference is that, compared to the white population, blacks consumed large amounts of foods containing high levels of fiber and phytate. Furthermore, a high-phytate diet has been used effectively to treat hypercalciuria and renal stones in humans. Thus, there is evidence to support the role of phytate in the prevention of renal stone formation.

Experimental evidence indicates that *myo*-inositol bis- and trisphosphates are effective in preventing the formation of hydroxyapatite crystals *in vivo*, which can function as nuclei for stone formation. Thus, the hypothesis was put forward that a high dietary phytate intake results in an increased urinary content of lower *myo*-inosital phosphates, which may act as very effective inhibitors of stone formation. (*See* **Renal Function and Disorders**: Kidney: Structure and Function.)

Phytate and Caries

The higher incidence of caries in industrialized compared to developing countries was suggested to be nutrition-dependent. Phytate lowers the solubility of calcium, fluoride, and phosphate, the major components of enamel. Thus, teeth are more protected against the leading cause of caries, the attack of acids and bacteria. Furthermore, the very high affinity of phytate for hydroxyapatite may prevent the formation of plaque and tartar. (*See* **Dental Disease**: Etiology of Dental Caries.)

Phytate and Cancer

The frequency of colonic cancer varies widely among human populations. It is a major cause of morbidity

and mortality in western society. The incidence of cancer, especially large intestinal cancer, has been associated principally with dietary fat intake and is inversely related to the intake of dietary fiber. It was further suggested that the apparent relationship between fiber intake and rate of colonic cancer might arise from the fact that many fiber-rich foods contain large amounts of phytate and that this latter might be the critical protective element, since an inverse correlation between colon cancer and the intake of phytate-rich fiber foods, but not phytate-poor fiber foods, has been shown. A high phytate intake may also be an important factor in reducing the breast and prostatatic cancer mortality in humans. (*See* **Cancer**: Epidemiology.)

Both *in vivo* and *in vitro* experiments have shown striking anticancer potential for phytate. It was demonstrated that a treatment regimen of 1–2% sodium phytate in the drinking water of growing rats significantly decreased the number of colon tumors and tumor volume when treatment was commenced prior to carcinogenic induction or when administered up to 8 months postinitiation. The inhibitory effect of phytate was dose-dependent and a mixture of 1% *myo*-inositol and 1% phytate was more effective in suppressing cell proliferation. Unfortunately, these investigations failed to study and/or report the mineral status of the animals.

A recent study demonstrated that phytate also has antineoplastic effects on another tumor model, the murine fibrosarcoma. Further rat studies indicated that phytate may also reduce the risk of breast cancer.

Not only pure phytate dissolved in the drinking water, but also phytate in experimental diets such as wheat bran has a protective effect against colon cancer, as shown in animal studies. It was concluded that, while having a role, endogenous phytate is not the sole active component in wheat bran. This result clearly emphasizes the fact that dietary components should not be studied individually for their antineoplastic effect. Though phytate at the 2% level significantly reduced the weight gain of the animals, it does not seem to be of major consequence.

How phytate exerts its antineoplastic and antiproliferative action is not understood. It was proposed that the mechanism by which dietary phytate reduced colon cancer was via chelation of iron and suppression of iron-related initiation and promotion of carcinogenesis, since *in vitro* experiments demonstrated that phytate is a powerful inhibitor of the Haber–Weiss reaction, in which iron catalyzes hydroxyl radical formation (**Figure 2**). These radicals are mediators of several tissue damages related to tumor initiation and promotion. Phytate not only suppresses

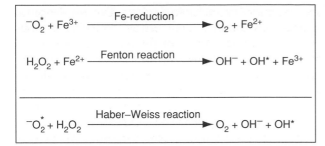

$$^{-}O_2^* + Fe^{3+} \xrightarrow{\text{Fe-reduction}} O_2 + Fe^{2+}$$

$$H_2O_2 + Fe^{2+} \xrightarrow{\text{Fenton reaction}} OH^- + OH^* + Fe^{3+}$$

$$^{-}O_2^* + H_2O_2 \xrightarrow{\text{Haber–Weiss reaction}} O_2 + OH^- + OH^*$$

Figure 2 Iron-catalyzed radical formation.

iron-catalyzed hydroxyl radical generation, but also acts to inhibit almost totally iron-catalyzed lipid peroxidation. *Myo*-inositol tris-, tetrakis-, and pentakisphosphate are also shown to be effective. The effects of phytate on cancers distant from the intestine, such as breast cancer, might be simply explained by limitation of iron absorption.

According to a further hypothesis, suppression of tumor incidence in experimental animals may be mediated in part by natural killer cells, since phytate treatment enhanced the activity of natural killer cells involved in the destruction and growth inhibition of tumor cells. It was also suggested that phytate acts by influencing cell regulation. Decreased tumor cell growth may be due to the complexation of zinc and magnesium, which play an important role in cell regulation and gene expression, or due to an influence of extracellular phytate on the intracellular phosphatidylinositol phosphate system, which is important in regulating a variety of physiological and biochemical processes, including cell growth via the second messengers D-Ins(1,4,5)P$_3$ and D-Ins(1,3,4,5)P$_4$ (*See* **Phytic Acid**: Properties and Determination). Theoretically, dietary phytate could serve as a precursor for second messengers *in situ* or as a negative-feedback inhibitor for their formation. As a prerequisite, phytate or its hydrolysis products have to be taken up by tumor cells. It was shown that, after exposure of human-transformed cells to [^{3}H]-phytate, a rapid uptake of radioactivity occurred, whereas nontransformed cells showed only a limited uptake. The uptake of radioactivity seems to correspond with tumor cell growth rate. Furthermore, intragastric-administered [^{3}H]-phytate is rapidly absorbed through the upper gastrointestinal tract and distributed to various organs in the rat. Analysis demonstrated that in both studies most of the radioactivity was due to *myo*-inositol and *myo*-inositol monophosphate. The authors suggested that phytate could be absorbed as such or after a variable degree of dephosphorylation, but it cannot be ruled out that phytate was taken up after complete dephosphorylation to *myo*-inositol. From the extracellular area, phytate or

its degradation products may be internalized by endocytosis involving fluid-phase pinocytosis, phagocytosis, or receptor-mediated endocytosis, as well as by partitioning into and/or through the plasma membrane. The possibility of receptor-mediated endocytosis was supported by identification of a receptor protein for phytate in rat brain.

Conclusion

Numerous reports suggest that a high phytate intake may suppress absorption of minerals such as zinc, iron, and calcium and that removal of dietary phytate significantly improves the bioavailability of those minerals in humans and animals. Thus, maintaining a diet exceptionally high in phytate may not be entirely without risk. However, there is no uniform agreement on this score, and several studies indicate that phytate may have minimal or no adverse effects on mineral bioavailability in humans. Furthermore, it has been suggested that phytate may exhibit some beneficial health effects, such as reducing the risk of heart disease, renal stone formation, and certain types of cancer.

The most severe effects attributable to phytate in humans have occurred in populations with unrefined cereals and/or pulses as a major dietary component. In particular, zinc and iron deficiency, but not calcium, magnesium, manganese, or copper deficiency, were observed as a consequence of high phytate intake. Therefore, the anticalcifying and rachitogenic effect of phytate has been questioned; it appears that marginal vitamin D status is more important in the etiology of rickets and osteomalacia. Furthermore, marked mineral deficiency syndromes attributed to phytate have not been identified in highly developed countries. Thus, phytate intake does not necessarily result in mineral deficiency. The absorption of minerals depends on the total composition of the meal and in a balanced diet containing animal protein, a high phytate intake does not imply a risk of inadequate mineral supply. Therefore, the recommendation for increasing dietary fiber in western diets would not be expected to have any adverse effect on mineral absorption. The higher phytate intake with whole-grain products will undoubtedly lead to a percentage decrease in mineral absorption, but the absolute amount of absorbed minerals may remain unchanged, because of the large amounts of minerals in these products. In addition, the impact of phytate on phosphorus availability can be considered to be of little consequence in humans, since the phosphorus intakes are usually high, and phytate phosphorus represents only a small portion of the total phosphorus in the diets.

In population groups where phytate-containing foods contribute to poor or marginal mineral status, fortification of some such foods with the corresponding minerals and/or the reduction in phytate content are possibilities to reduce the risk of mineral deficiencies. Since phytate and certain phytate degradation products have been proposed to exhibit health benefits in humans, complete degradation or removal of phytate during food processing should not be followed up, but a controlled degradation to lower *myo*-inositol phosphates. This seems to be a way of avoiding the adverse effects of phytate on intestinal mineral absorption and of utilizing their potential beneficial properties. Controlled degradation is most easily feasible by using phytases. The addition of exogenous or activation of endogenous phytases during food processing has already been shown to result in extensive phytate degradation without affecting other food components.

Phytases are also of interest in areas of intensive animal agriculture. There has been widespread concern in recent years about possible phosphorus pollution from animal manures. The accumulation of phosphorus over time from excessive excretion in manure increases the potential for contaminating water sources from runoff and erosion. The effect of increased content of phosphorus in water is eutrophication, which decreases the available oxygen within the water, thus posing a hazard to fresh-water animal life. Effluent control is therefore a high priority in areas of intensive animal production and, in this context, phytase can become an important waste management tool, since phytase is active in the stomach for a certain time and thus increases phytate hydrolysis during digestion. Phytase not only has the potential to reduce environmental pollution by minimizing the excretion of phosphorus and nitrogen in manure, but also reduces the need for mineral supplementation by increasing the availability of cations bound to phytate.

Because suggestions for the beneficial aspects of *myo*-inositol phosphates are only derived from *in vitro*, animal, or epidemiological investigations, carefully controlled human and animal studies should be carried out to evaluate simultaneously the potential benefits and adverse effects of dietary *myo*-inositol phosphates. If certain phytate degradation products induce physiological effects (the position of the phosphate groups on the *myo*-inositol ring is of great significance for the physiological function), phytase may also find application in food processing to produce foods with improved nutritional value, health benefits, and maintained sensory properties (functional foods). By adding the phytase to the raw material, the antinutritional factor phytate will be degraded to physiologically active *myo*-inositol phosphates during food processing. Thus, foods with a reduced content of phytate and a regulated content and composition of lower *myo*-inositol phosphate with beneficial health effects could be designed.

See also: **Cancer**: Epidemiology; **Coronary Heart Disease**: Etiology and Risk Factor; **Dental Disease**: Etiology of Dental Caries; **Diabetes Mellitus**: Etiology; **Minerals – Dietary Importance**; **Phytic Acid**: Properties and Determination; **Protein**: Digestion and Absorption of Protein and Nitrogen Balance; **Renal Function and Disorders**: Kidney: Structure and Function

Further Reading

Challa A, Ramkishan R and Reddy BS (1997) Interactive suppression of aberrant crypt foci induced by azoxymethane in rat colon by phytic acid and green tea. *Carcinogenesis* 18: 2023–2026.

Erdman JW Jr and Poneros-Schneier A (1989) Phytic acid interactions with divalent cations in foods and the gastrointestinal tract. *Advances in Experimental Biology* 249: 161–171.

Fox MRS and Tao TH (1989) Antinutritive effects of phytate and other phosphorylated derivatives. *Nutritional Toxicology* 3: 59–96.

Harland BF and Morris ER (1995) Phytate: a good or a bad food component? *Nutrition Research* 5: 733–754.

Jariwalla RJ, Sabin R, Lawson S and Herman ZS (1990) Lowering of serum cholesterol and triglycerides and modulation of divalent cations by dietary phytate. *Journal of Applied Nutrition* 42: 18–28.

Jenab M and Thompson LU (1998) The influence of phytic acid in wheat bran on early biomarkers of colon carcinogenesis. *Carcinogenesis* 19: 1087–1092.

Ohkawa T, Ebisuno S, Kitagawa M *et al.* (1984) Rice bran treatment for patients with hypercalciuric stones: experimental and clinical studies. *Journal of Urology* 132: 1140–1145.

Porres JM, Stahl CH, Cheng WH *et al.* (1999) Dietary intrinsic phytate protects colon from lipid peroxidation in pigs with a moderately high dietary iron intake. *Proceedings of the Society for Experimental Biology and Medicine* 221: 80–86.

Sandberg AS, Brune M, Carlsson NG, Hallberg L, Skoglund E and Rossander-Hulthen L (1999) Inositol phosphates with different numbers of phosphate groups influence iron absorption in humans. *American Journal of Clinical Nutrition* 70: 240–246.

Shamsuddin AM, Vuvenik I and Cole KE (1997) IP-6: a novel anti-cancer agent. *Life Science* 61: 343–354.

Shen X, Weaver CM, Kempa-Steczko A, Martin BR, Phillippy BQ and Heany RP (1998) An inositol phosphate as a calcium absorption enhancer in rats. *Nutritional Biochemistry* 9: 298–301.

Thompson LU (1988) Antinutrients and blood glucose. *Food Technology* 42: 123–132.

Vucenik I and Shamsuddin AM (1994) [^{3}H]Inositol hexaphosphate (phytic acid) is rapidly absorbed and metabolized by murine and human malignant cells in vitro. *Journal of Nutrition* 124: 861–868.

Zhou JR and Erdman JW Jr (1995) Phytic acid in health and disease. *Critical Reviews in Food Science and Nutrition* 35: 495–508.

Phytochemicals *See* **Functional Foods**

PICKLING

D M Barrett, University of California Davis, CA, USA

Introduction

Pickling is one of the oldest, and most successful, methods of food preservation known to humans. This article reviews the origins of pickling, the various methods of pickling employed commercially, and the nature of preservative action. The optimization of pickle quality depends on maintenance of proper acidity, salt concentration, temperature, and sanitary conditions.

History and Tradition

It is difficult to suggest a date for production of the first pickled foods, but it is known that both vinegar and spices were being used during biblical times. Fermentation of plant and animal foods was known to the early Egyptians, and fish were preserved by brining in prehistoric times. By the third century BC, Chinese laborers were recorded to be consuming acid-fermented mixed vegetables while working on the Great Wall. In about 2030 BC, northern Indians brought the seed of the cucumber to the Tigris Valley. The Koreans created kimchi from acid-fermented Chinese cabbage, radish, and other ingredients many centuries ago. Corn, cassava, and sorghum were fermented and became staples of the African diet. In the west, acid fermentation of cabbage and cucumbers produced sauerkraut and pickles, products that are still popular today. Early explorers carried kegs of sauerkraut and pickles that prevented scurvy on their voyages.

Peterson in 1977 defined pickling, in a broad sense, as 'the use of brine, vinegar or a spicy solution to preserve and give a unique flavour to a food.' In a 1936 document, the US Department of Agriculture described cucumber pickles as:

> immature cucumbers properly prepared, without taking up any metallic compounds other than salt, and preserved in any kind of vinegar, with or without spices. Pickled onions, pickled beets, pickled beans and other pickled vegetables are vegetables, prepared as described above, and conform in name to the vegetables used.

Literature references concerning the technology of acid fermentation began to appear in the western press in the early 1900s. In 1919, Orla-Jensen isolated strains of *Betacoccus arabinosaceus* from sour potatoes, sour cabbage, and sour dough. Pederson, in a number of classic studies in the 1930s, enumerated and identified the sequence of microorganisms involved in sauerkraut fermentation. *Leuconostoc mesenteroides* was identified as being one of the most important microorganisms for initiation of vegetable fermentation. Numerous investigators have carried out studies on acid-fermented vegetables over the past century. (*See* **Fermented Foods**: Beverages from Sorghum and Millet.)

Advent of New Pickling Methods

Brining vegetables in salt, and the resultant lactic acid fermentation, is an ancient form of preservation. Two new methods of pickling cucumbers, which represent the largest volume of a single vegetable preserved by pickling, have been developed during the 1900s. Both methods use lower salt concentrations and result in a milder product. The first new method, pasteurization and direct acidification, was developed and began commercial production in the 1940s. The second,

Table 1 Selected pickling problems

Problems	Cause
Soft, slippery slimy pickles (discard pickles, spoilage is occurring)	Hard water
	Acid level too low
	Cooked too long or at too high a temperature
	Water bath too short, bacteria not destroyed
	Jars not airtight
	Jars in too warm a resting place
Shriveled, tough pickles	Pickles overcooked
	Syrup too heavy
	Too strong a brine or vinegar solution
	Pickles not fresh enough at outset
	Fruit cooked too harshly in vinegar/sugar mixture
Dark, discolored pickles	Iron utensils used
	Copper, brass, iron, or zinc cookware used
	Hard water
	Metal lid corrosion
	Too great a quantity of powdered and dried spices used
	Iodized salt used

Reproduced from McNair JK (ed.) (1975) *All About Pickling*. San Francisco: Ortho Books, with permission.

Table 2 Nutritive analysis of pickles: the composition of 100 g of edible portion (approx. 1 large dill pickle or 1/2 cup of fresh cucumber pickle slices)

	Fermented dill pickles	Sweet pickles	Sour pickles	Fresh pack cucumber pickles
Water (%)	93	60.7	94.8	78.7
Food energy (J)	46.2	613.2	42	306.6
Protein (g)	0.7	0.7	0.5	0.9
Fat (g)	0.2	0.4	0.2	0.2
Carbohydrate (g)	2.2	36.5	2.0	17.9
Ash (g)	3.6	1.7	2.5	2.3
Calcium (mg)	26.0	12.0	17.0	32.0
Iron (mg)	1.0	1.2	3.2	1.8
Vitamin A (IU)[a]	100	90	100	140
Thiamin	Trace	Trace	Trace	Trace
Riboflavin (mg)	0.02	0.02	0.02	0.03
Vitamin C (mg)	6.0	6.0	7.0	9.0
Phosphorus (mg)	21.0	16.0	15.0	27.0
Potassium (mg)	200.0			
Sodium (mg)	1428.0		1353.0	673.0

Reproduced from McNair JK (ed.) (1975) *All About Pickling*. San Francisco: Ortho Books, with permission. 1 IU = 0.6 μg β-carotene.

See also: **Acids**: Natural Acids and Acidulants; **Fermented Foods**: Beverages from Sorghum and Millet; **Lactic Acid Bacteria**; **Pasteurization**: Principles; **pH – Principles and Measurement**; **Vinegar**; **Sensory Evaluation**: Sensory Characteristics of Human Foods

Specific Examples

Pickled cucumbers are by far the most abundant pickled product available in the western world today. Other common pickled products include sauerkraut, pickled pears, peaches and plums, pickled nuts, relishes, cured meats, fish and poultry, and specialty items such as pickled mushrooms and cherries.

Salt stock is used to prepare sour cucumber pickles, which typically have a final acidity not lower than 2.5%. Sweet pickles are prepared in a similar fashion, except that a sweet, spiced vinegar solution is added to the salt stock.

Sauerkraut is produced through salt-controlled bacterial fermentation. Cabbage selected for sauerkraut should have at least 3.5% sugar to insure an adequate carbohydrate source for bacteria. Shredded cabbage is mixed with salt (2.25% by weight) and the final product contains an average of 1.5–2.0% lactic acid.

Further Reading

Binsted R, Dewey JD and Dakin JC (1939) *Pickle and Sauce Making*. London: Food Trade Press.

Fleming HP (1984) Developments in cucumber fermentations. *Journal of Chemical Technology and Biotechnology* 34B: 241–252.

Fleming HP and McFeeters RF (1981) Use of microbial cultures: vegetable products. *Food Technology* 1: 84–88.

McNair JK (ed.) (1975) *All About Pickling*. San Francisco: Ortho Books.

Pederson CS (1930) Floral changes in the fermentation of sauerkraut. *NYSAES Technical Bulletin* 168.

Pederson CS (1979) *Microbiology of Food Fermentations*, 2nd edn. Westport: AVI.

Peterson MS (1977) Pickles. In: Desrosier NW (ed) *Elements of Food Technology*, pp. 690–691. Westport: AVI.

Steinkraus KH (1983) *Handbook of Indigenous Fermented Foods*. New York: Marcel Dekker.

US Department of Agriculture (1936) *Service and Regulatory Announcements*, no. 2, rev. 5. Washington, DC: Food and Drug Association.

Pigments *See* **Colorants (Colourants)**: Properties and Determination of Natural Pigments; Properties and Determinants of Synthetic Pigments

Pilchards *See* **Fish**: Introduction; Catching and Handling; Fish as Food; Demersal Species of Temperate Climates; Pelagic Species of Temperate Climates; Tuna and Tuna-like Fish of Tropical Climates; Demersal Species of Tropical Climates; Pelagic Species of Tropical Climates; Important Elasmobranch Species; Processing; Miscellaneous Fish Products; Spoilage of Seafood; Dietary Importance of Fish and Shellfish; **Fish Farming**; **Fish Meal**

PINEAPPLES

L G Smith, Maroochy Research Station, Nambour, Queensland, Australia

This article is reproduced from *Encyclopaedia of Food Science, Food Technology and Nutrition*, Copyright 1993, Academic Press.

Introduction

The pineapple (*Ananas comosus* (L.) Merr.), botanically a member of the ornamental Bromeliaceae family, originated in tropical South America but is now widely grown in all tropical and subtropical areas of the world. In Spanish-speaking countries the fruit is known as pina; in Portuguese-speaking countries as abacaxi; as ananas in Dutch- and French-speaking former colonies, and as nanas in southern Asia. More than 4.5×10^6 t, both fresh and canned, is marketed worldwide each year from at least nine major countries (**Tables 1** and **2**), with the major cultivar by far being the large juicy-fruited smooth-leafed cultivar Smooth Cayenne. Wild pineapple varieties still grow in the tropical savanna of South America but most have small, seedy, fibrous fruits. (*See* **Fruits of Tropical Climates**: Commercial and Dietary Importance.)

The Plant, its Appearance, and Physiology

The pineapple is a perennial, monocotyledonous, xerophytic plant, up to 1.5 m high, of herbaceous, lily-like habit, but with tough, spiny-tipped leaves that are waxy on the upper surface and possess a fragile dusty bloom on the underside. The leaves in all but a few cultivars, such as the important Smooth Cayenne, also have numerous formidable barbs along the edges, which make cultivation hazardous. In all varieties the concave leaves channel any precipitation into the plant center for absorption by spongy leaf tissue and roots. Other features which enhance the shallow-rooted plant's adaption to low rainfall include leaves that do not wilt, and its crassulacean acid metabolism (CAM), in which the stomata open at night to take up carbon dioxide rather than in the day, greatly reducing water loss. Malic acid is accumulated during the night and is decarboxylated during the day.

The fruit of the pineapple, botanically a sorosis or a syncarp, comprises spirals of fused fleshy fruitlets radiating from a fibrous but succulent core and topped with a leafy crown or top, an extension of the peduncle or central plant stem of the plant (**Figure 1**). The crown is planted for reproduction, when available. The commercial fruit, grown as a monoculture of genetically identical plants, is normally seedless due to a genetic self-incompatibility. If cross-pollinated, by wind, natural vectors, or deliberately by humans, as in breeding trials, viable small hard-coated seeds 1–2 mm across will form, capable of producing a new plant and fruit within 2–3 years. The commercial plant produces one fruit 14–24 months after planting, but two or more vegetative shoots (suckers) subsequently produce additional rattoon crops. Fruits become smaller with successive rattoons and usually a maximum of two rattoon crops are commercially viable.

Because pineapple plants cannot tolerate frost or prolonged cold, production is restricted to coastal or near-coastal areas of low or moderate elevation. They tend to thrive on tropical and subtropical islands, where the surrounding water mass maintains a more ideal constant temperature and moderate humidity.

Varieties

The classification of existing pineapple varieties is currently under complete review. The Cayenne variety was previously seen as but one of five recognized groups of *A. comosus*, the other four being Spanish, Queen, Pernambuco, and Mordilona. The five groups are compared in **Table 3**. Most likely these will soon be reclassified to be just varieties of the one species *A. comosus*, better to reflect the true genetic

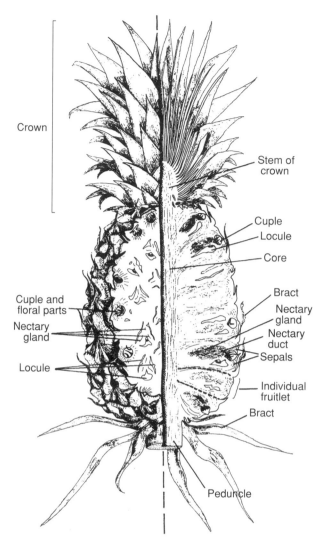

Figure 1 Mature pineapple: on the right, median longitudinal section; on the left, tangential longitudinal section. Adapted from Py C, Lacoeuilhe JJ and Teisson C (1987) *The Pineapple: Cultivation and Uses.* Paris: GP Maisonneuve et Larose, with permission. Reproduced from Pineapples, *Encyclopaedia of Food Science, Food Technology and Nutrition*, Macrae R, Robinson RK and Sadler MJ (eds), 1993, Academic Press.

many unwanted genotypes. Genetic engineering is being investigated as a way around the problem of extreme heterozygosity, but to date no new genetically modified variety has been commercially released.

Composition and Nutritional Value

The pineapple fruit is typically a supplementary food rather than a staple. The pineapple's very sweet and sour taste, its mild aromatic flavor, and firm succulent flesh in a large and attractive form make it somewhat unique among foods. While very palatable by itself, it is equally used as a flavoring component, both cold and hot, making many other less attractive but

nutritionally sound foods more edible. The dietary composition of Smooth Cayenne, both fresh and canned, is shown in **Table 4**. The nutrient profile of the pineapple is, in general, similar to many other fruits, containing high levels of carbohydrate and low levels of fat and protein. Dietary fiber constitutes about 14% of the dry matter and, as with most fruits, can be incorporated into a cholesterol-lowering diet. Its vitamin C content is about one-tenth that in citrus, and the level of retinol equivalents (3 per 100 g) is low compared to those of pawpaw (papaya) (153) and mango (356). Refer to individual nutrients.

The total sugar content is the major quality determinant of the fruit for both fresh and canned markets and in commercial practice sugar content is often regularly monitored, using a refractometer calibrated in degrees Brix which, in pineapples, is virtually identical to percentage total sugar. Fresh Smooth Cayenne fruit above 15° Brix would be considered of excellent eating quality by most tasters and fruit of 9–12° Brix of marginal quality; however, taste preferences towards Smooth Cayenne of a particular Brix level vary considerably between individuals and also between ethnic groups. In canned pineapple the sugar content of the fresh fruit makes less difference to consumer quality.

Variability of Organoleptic Parameters

Wide variations occur in the total sugar content and titratable acidity in fresh pineapples. Not only is the bottom of the fruit always about 3° Brix higher than the top but, more importantly, there is wide variation between samples. For example, the 8% total sugars from Queensland fruit in **Table 4** is very low compared to fruit from Hawaii or Ivory Coast and is either from immature summer fruit (average mature fruit is 15–18° Brix) or is from winter-grown fruit (6–11° Brix). Fruit grown under cooler conditions is lower in sugar and higher in acid, although the sour taste of winter-grown pineapples is caused by the low sugar content rather than the high acidity.

Postharvest Changes

The postharvest changes in various parameters in fruit held near 20°C during 15 days after harvest are illustrated in **Figure 2**. Total sugar concentration and eating quality remain relatively constant after harvest; acidity and carotene content increase moderately, while the shell color (degree of yellowness of the skin) and ester concentration increase substantially.

Skin Color and Ripeness

Although the fruit normally changes from green to yellow as the fruit ripens on the plant, skin color is a

Table 3 Characteristics of the major pineapple varieties[a]

	Cayenne[b]	Spanish	Queen	Pernambuco[c]	Perolera
Main production	Worldwide	Caribbean Malaysia	South Africa Australia	Brazil Venezuela	Colombia Ecuador, Peru
Weight (kg)	1.8–3.5	1–1.8	0.5–1.2	1–1.8	1.8–2.5
Shape	Cylindrical	Spherical	Conical	Conical	Cylindrical
Ripe skin color	Orange-yellow	Reddish yellow	Bright yellow	Green-yellow	Bright red-yellow
Ripe flesh	Near translucent	Near translucent	Opaque	Opaque	Opaque
Flesh fibrosity	Nonfibrous	Fibrous	Crisp	Nonfibrous	Crisp
Flesh color	Pale yellow	Near white	Bright yellow	White-yellow	Bright and pale yellow
Flavor	Sweet and acid	Spicy	Aromatic	Low acidity	Low sugar and acidity
Wilt resistance	*****	*	**	**	**
Nematodes	*****	*	**	**	**
Uses	Fresh, canned Major export	Fresh Minor export	Fresh Minor export	Fresh Minor export	Fresh Minor export

[a]The term 'varieties' is currently under review: see text.
[b]Includes Smooth Cayenne, which is but one variety of the Cayenne group.
[c]Other name abacaxi, but this should be avoided as it means 'pineapple' in Portuguese.
* resistant, ***** very susceptible.
Based on data from Py C, Lacoeuilhe JJ and Teisson C (1987) *The Pineapple: Cultivation and Uses*. Paris: GP Maisonneuve et Larose.

Table 4 Composition of 100 g of pineapple flesh

	Canned[a] heavy syrup, drained	Canned[a] pineapple juice, drained	Fresh[a], peeled	Fresh[b] peeled
Proximate				
Water (g)	74.8	84.6	86.0	80–86
Energy (kJ)	350	188	158	[c]
Protein (g)	0.8	0.7	1.0	0.2
Fat (g)	0.0	0.0	0.1	0
Carbohydrate, total (g)	20.4	10.2	8.0	
Sugars, total (g)	20.4	10.2	8.0	10–18
Starch (g)	0.0	0.0	0.0	
Ash (g)	0.2	0.3	0.5	0.3–0.6
Cholesterol (mg)	0	0	0	
Acids, total (g)				0.5–1.6
Total nitrogen (mg)				45–120
Pigments, xanthophylls (mainly carotenoids) (g)				0.2–0.3
Dietary fiber (g)	1	2	2	
Minerals (mg)				
Sodium	1	4	2	14
Potassium	76	140	180	11–330
Calcium	5	6	27	3–16
Magnesium	10	14	11	10–19
Iron	0.3	0.5	0.3	0.05–0.3
Zinc	0.2	0.3	0.2	0
Vitamins				
Retinol equivalents (μg)	3	3	4	
Retinol (μg)	0	0	0	
β-Carotene equivalents (μg)	18	18	25	
Thiamin (μg)	40	30	40	69–125
Riboflavin (μg)	30	30	30	20–88
Nicotinic acid equivalents (mg)	0.3	0.3	0.3	
Nicotinic acid (mg)	0.2	0.2	0.1	0.2–0.3
Vitamin C (mg)	15	14	21	3–25

[a]Queensland fruit; data from *Composition of Foods Australia*. Australian Food Publishing Service, Canberra. 1989.
[b]From Py C, Lacoeuilhe JJ and Teisson C (1987) *The Pineapple: Cultivation and Uses*, Paris: GP Maisonneuve et Larose; data presumably based on fruit from Ivory Coast.
[c]Data not given.

developing a cannonball shape, or lacking a crown), but because of its good blackheart resistance and lo \w acidity, it is slowly developing wider acceptance in Australia. Dipping the fruit in a carnauba-based wax, or applying a lipid-carboxymethyl cellulose coating, can control blackheart. These skin coatings function by reducing the oxygen concentration in the flesh, necessary for the melanin pigment to develop, but the anaerobic flesh can develop an alcoholic flavor. Heating the fruit to 40 °C for 24 h either immediately before or after cool storage also reduces blackheart and improves pineapple eating quality, but is not commercially used because of logistical limitation.

Pineapples that are cool-stored for prolonged periods can also develop other symptoms of chilling injury such as flaccid watery flesh, skin necrosis, crown wilting, and crown necrosis, but fruit with such injuries are often already damaged by internal chilling. Overcoming chilling injury in fresh-market pineapples is currently under investigation.

Controlled-Atmosphere Storage

Controlled-atmosphere storage does not prolong the life of the cool stored pineapples and is not used for international shipping. (*See* **Controlled-atmosphere Storage**: Applications for Bulk Storage of Foodstuffs.)

Field Blackheart

Pineapples from cooler production areas are particularly susceptible to blackheart, where it can develop in the field following periods of cool overcast weather. Because the disorder shows no external symptoms, affected fruit cannot be identified. In Queensland and Hawaii pineapple canneries close or restrict intake during such periods and thus avoid excessive losses.

Pineapple Products

Canned pineapple is a commodity of major international trade; figures showing exports and imports are shown in **Table 1**. Pineapples are also processed into a number of other products other than canned fruit. Pineapples are often used as a component of commercially marketed fresh fruit salad, either fresh or frozen. Less ripe fruits freeze best as they have firmer flesh. Soft ripe fruits can develop off-flavors when frozen, as well as turning out very flaccid. Fried pineapple pieces are commonly used in Asian cuisine. Dried pineapple pieces are easily prepared and are being sold in increasing quantities as they retain a more attractive appearance than other fruits that strongly darken when dried. Sliced thin and dried, pineapple 'chips' are also sold as a snack food.

Glacé pieces are readily prepared from either the fresh product or, more easily, from canned rings. Pineapple juice is a major item of commerce, as fresh or frozen, full strength, or as a concentrate. Reasonably high in carbohydrates, fermented juice is used to make vinegar. It does not make a good-quality wine but, fermented and distilled, it makes a potent, if somewhat disreputable, alcoholic spirit, available in several Pacific island countries. Cannery waste (crushed skins, etc.) makes a good cattle feed supplement, while old plantation fields are often let out to cattle. Pineapple leaves are used to make cloth or rope in both Taiwan and the Philippines, while use of the whole plant as a source of energy is under current investigation. (*See* **Canning**: Quality Changes During Canning; **Freezing**: Structural and Flavor (Flavour) Changes.)

Bromelain

Pineapple fruits contain a proteolytic enzyme, bromelain, which can constitute nearly half the measured protein in the fruit. Much higher concentrations of the basic isoenzyme (the fruit form is acidic) exist in the stem. Stem bromelin is commercially extracted and sold, as a buff-colored powder, slightly soluble in water and practically insoluble in alcohol, chloroform, and ether. Bromelin is used in the pharmaceutical industry for digestive and antiinflammatory products, in the manufacture of cattle feed and to 'chill-proof' beer (to prevent formation of a haze of proteinaceous material which can form when brewed products are refrigerated). Some recent pharmacological reports indicate that bromelin may have value in modulating tumor growth and blood coagulation, as well as debridement of third-degree burns.

Commercial Farm Production

Modern commercial culture of pineapples is generally capital-intensive, requiring substantial land preparation, machinery for planting, boom (fertilizer) spray and harvesting, plus bulk equipment for cannery fruit and/or harvester-packers for fresh market fruit.

By using a combination of plant hormones and appropriate planting practices, the harvest date and the fruit size and quality can be preset and regulated as required. This makes modern pineapple production function with almost clockwork precision. On modern fresh-fruit farms a year-round cycle is maintained of weekly, if not daily, planting and harvesting. Combined with modern planting machinery, modern herbicides, and large mobile harvesting equipment, production of 1000–2000 t of fruit per staff year is common.

Being parthenocarpic (seedless, but otherwise normal), pineapple propagation is vegetative and can be problematical as only limited planting material is available from any one plant. The best planting material, that producing the most uniform plants and fruits, is the fruit crown, but if the fruit is sold as fresh fruit (as opposed to canning), tops are not available, so that suckers and other vegetative material must be collected and used. Up to 80 000 pieces of planting material (tops, suckers or slips) are required per hectare, planted in double rows. The time of planting and the size and type of planting material affect the date of harvest and fruit size, and are predetermined. Weeds can be readily controlled as the plant is resistant to several effective herbicides (e.g., diuron, fluazifopbutyl, ametryne) which are applied over the rows of growing plants with a boom spray. Ten to fourteen months after planting, a flower-inducing hormone (commonly the ethylene-releasing compound Ethrel, 2-chloroethyl-phosphonic acid) is applied, and the treated plants uniformly produce a single, blue-petalled inflorescence about 8 weeks later. Fruit development occurs without further interruption, but as sunburn can be a serious injury, fruit may be individually covered (bagged) or sprayed with a reflective bentonite clay suspension.

Harvesting pineapples at the correct stage of internal ripeness is important as it is the major factor that determines the fruit quality at consumption. As skin color is an inconsistent guide to eating quality, pickers have to use past experience and market demands.

Sodium naphthaleneacetic acid (NAA) is sometimes applied as a spray on the fruit to increase fruit size by delaying degreening (becoming yellow). Adverse effects on fruit quality from using NAA are reportedly minor but are under current debate. Ethrel is also sometimes sprayed on both fresh and cannery crops to increase flesh yellowness and to 'accelerate ripening.' While very effective and cost-saving on cannery fruit, its use on fresh-market fruit has been shown to be of questionable value as the variability of the eating quality is greatly increased.

See also: **Canning**: Quality Changes During Canning; **Controlled-atmosphere Storage**: Applications for Bulk Storage of Foodstuffs; **Freezing**: Structural and Flavor (Flavour) Changes; **Fruits of Tropical Climates**: Commercial and Dietary Importance; **Fungicides**; **Ripening of Fruit**

Further Reading

Bartholomew DP and Paull RE (1986) Pineapple. In: Monselise SP (ed.) *Handbook of Fruit Set and Development*, pp. 371–388. Boca Raton, Florida: CRC Press.

Dull GG (1970) The pineapple: general. In: Hulme AC (ed.) *The Biochemistry of Fruits and their Products*, vol. 2, pp. 303–324. London: Academic Press.

Py C, Lacoeuilhe JJ and Teisson C (1987) *The Pineapple: Cultivation and Uses*. Paris: GP Maisonneuve et Larose.

Smith LG (1988) Indices of physiological maturity and eating quality in Smooth Cayenne pineapples. *Queensland Journal of Agricultural and Animal Sciences* 45(2): 213–228.

PINE KERNELS

B T Styles[†], University of Oxford, Oxford, UK

Background

Pine nuts, also known as Indian nuts, piñons, or pignolias, have been an important food crop in some areas of the world since prehistoric times. The 'nut' is, in fact, the seed of different species of pine (*Pinus*), nearly all of which belong to a group known as 'soft' or 'white' pines or their relatives. These species are evergreen, coniferous trees whose cones are softly woody and have few scales and two large seeds per scale that lack a wing (see **Figure 1**). The kernel consists of the endosperm tissue of the seed containing the stored food material and the developing embryo (germ). The 'shell' surrounding it is the testa. This must be removed before the kernel can be eaten.

Sources

Pine nut kernels are almost all obtained from wild forest trees. Piñons from the drier areas of southwest USA and Mexico form the largest wild source, but the Italian pignolia tree has been cultivated and protected in the Mediterranean region for centuries. Almost all the other species grow in upland mountainous areas, but information about them is scanty (**Table 1**). The harvest of pine nuts is only a fraction of the production of cultivated nuts such as pecan, macadamia, walnut, or filbert, and the crop is generally irregular, with bumper harvests occurring approximately every

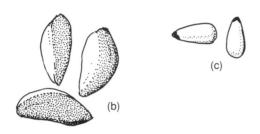

Figure 1 Pignolia (*Pinus pinea*): (a) cone; (b) seeds; (c) polished kernels (all natural size). Reproduced from Pine Kernels, *Encyclopaedia of Food Science, Food Technology and Nutrition*, Macrae R, Robinson RK and Sadler MJ (eds), 1993, Academic Press.

Table 1 Sources of pine kernels (*Pinus* species and their distribution)

Group	Latin name	Distribution
Piñon pines	Pinus edulis	Southwestern states of USA and Mexico
	P. monophylla	
	P. quadrifolia	
	P. maximartinezii	Mexico
	P. cembroides	
Stone pines		
Italian (pignolia)	P. pinea	Mediterranean basin and Turkey
Japanese	P. pumila	Northeast Asia, Siberia to Korea and Japan
Korean	P. koraiensis	Korea, southeast Siberia and Japan
Siberian	P. siberica	Western Russia to Siberia and Mongolia
Swiss	P. cembra	Alps and Carpathian Mountains
Chilgoza pine	P. gerardiana	Eastern Afghanistan to northern India and Pakistan

5 years. Harvesting is chiefly by an itinerant labor force that moves into the pine forests during the autumn. Green cones are cut from the trees, dried in the sun and opened to release the seeds, or the seeds are collected from beneath the trees. The collection and later preparation can at best be considered as a cottage industry. The kernels of only two or three of the piñon pines and the pignolia pine are important in commercial trading on a large scale. They may be bought in health food shops or delicatessens but are considered expensive, £12.50 per kilogram in the UK.

Kernels of most of the other pines may be found on sale in local markets near the site of collection. There are no reliable statistics on yields of these wild crops, but it has been estimated that about a million kilograms of Colorado pine nuts were collected annually in southwest USA. This quantity has probably not changed a great deal since then.

Storage and Preparation

After harvesting, nuts should be kept unshelled, in a dry, cool, well-ventilated place in cloth or paper bags. Tannins in the shell and in the seed coat of the kernel may function as antioxidants in preserving the fat and oil in the nuts. Fresh kernels stored in closed containers soon go moldy or rancid, but after drying, they can be safely preserved and have excellent keeping qualities for up to 3 years. They freeze well when fresh. Shelling of the nuts can be done domestically by lightly crushing the nuts on a cloth with a rolling pin. Those of the thin-shelled, single-leaf pine can be easily removed with the fingers. The thicker Colorado piñon, stone pine, and pignolia nuts must be cracked mechanically to release the kernels. (*See* **Antioxidants**: Natural Antioxidants; **Oxidation of Food Components**.)

Although the kernels can be eaten raw, roasting is necessary to bring out their full flavor. They may be roasted with or without the shell, the amount of time required depending on shell thickness and moisture content.

Composition and Nutritional Value

The richness and flavor of pine kernels are well known, and they rank highly among all other nuts in food value. There are, however, considerable differences in the properties of the kernels of different species. A comparison of the nutritional value of pine kernels and some commercial nuts is provided in **Table 2**.

Refuse and Wastage

When compared with other nuts, piñons have low percentages of shell waste or refuse. Shell thickness varies considerably: 30–35% of seed weight of the very-thin-shelled single-leaf piñon is made up of shell, whereas the thicker-shelled Colorado pine has a waste factor of 42%. Mediterranean pignolia has a particularly thick shell that must always be removed before it is sold. Thus, pine kernels of piñons constitute some 58–70% of the edible portion. The average is lower than that found in all other commercial nut crops except peanuts.

Protein

The average protein content is about 15% for piñon nut kernels and about 34% for pignolias. It is higher than for pecan nuts and about the same as that for English walnut and Brazil nut. It is only greatly exceeded by peanuts. A study in the former Yugoslavia has shown that kernels of pignolias are richer in proteins than pork and goose meat. Kernel protein has a digestibility value almost as good as that of beef and is distinctly better than almost all other nuts. It is near the proportion required for a balanced diet. Colorado and single-leaf kernels are especially rich in tryptophan and cystine. (*See* **Protein**: Requirements.)

Fat (Lipid) and Oil

Piñon pine kernels average about 60% of fatty materials. This is lower than pecan, English walnut, and Brazil nut, which approximate to 70%. Pignolia kernels in the former Yugoslavia are reported to contain 48% of lipid, which is again higher than for fat pork (37%) and goose meat (44%). As **Table 2** indicates, the Siberian stone pine is also very rich in fats

and oil. This is processed commercially for the production of cooking oil, but no published details of production methods or quantities are available. The fats of piñons, particularly Colorado and single-leaved pine, contain monounsaturated oleic, and polyunsaturated linolenic and linoleic acids. It has been reported that pignolia kernels from the Mediterranean region contain up to 50% of linoleic acid. Commercially produced pine kernels studied contained 46.40 g of fat by weight, of which 6.12 g is saturated (fatty acid not named). (*See* **Fats**: Requirements.)

Carbohydrate

The average oily piñon kernel of Colorado pine contains 19% carbohydrate, but this can increase to as high as 54% in the more starchy kernels of the single-leaf pine. In the Parry pine, it is 44%, but the former is much preferred. (*See* **Carbohydrates**: Requirements and Dietary Importance.)

Other Substances

Pine kernels are extremely rich in phosphorus (6040 mg kg^{-1}) which is about the same as for soya bean, and iron (53 mg kg^{-1}). They also contain significant amounts of vitamin A, thiamin, riboflavin, and niacin. Refer to individual nutrients.

Uses

In the past, pine kernels were a staple component in the diet as well as a subsistence food, but because of their enormous versatility, they now hold a high profile in modern cuisine among people of the developed nations, particularly in the USA, Europe, Asia, and North Africa. As roasted nuts, they are very widely used in the preparation of soups, sauces, and dressings, and may be served as a garnish for fish meals, in mixtures with cooked meat and as part of side-dishes with rice. They form suitable ingredients in cakes, puddings, and biscuits, and as a garnish for icecream. They are also highly appreciated in fruit desserts and vegetable salads. The kernels are so nutritious that a number of recipes have recently appeared under the label of 'Backpackers' Friends.' These products bear such names as Piñon Pemmican and Granola, and may be carried as bars in the rucksack as emergency rations or snacks. Pine kernels will probably become even more popular as the rise in vegetarianism continues.

Table 2 Comparison of the nutritional value of pine kernels and some commercial nuts

Type of nut	Food content		
	Protein (%)	Fat (%)	Carbohydrate (%)
Colorado pion, Pinus edulis	14	62–71	18
Single leaf piñon, P. monophylla	10	23	54
Mexican piñon, P. cembroides	19	60	14
Parry piñon, P. × quadrifolia	11	37	44
Digger pine, P. sabiniana	30	60	9
Pignolia pine, P. pinea	34	48	7
Siberian stone pine, P. siberica	19	51–75	12
Chilgoza pine, P. gerardiana	14	51	23
Pecan, Carya illinoensis	10	73	11
Peanut, Arachis hypogaea	26	39	24
English walnut, Juglans regia	15	68	12
Almond, Prunus dulcis	21	54	7
Brazil nut, Bertholletia excelsa	16	69	8

Typical energy value per 100 g = 556 kcal.
Data from Lanner RM (1981) *The Piñon Pine*. Reno, Nevada: University of Nevada Press and Botkin CW and Shires LB (1948) The Composition and Value of Piñon Nuts. Bulletin 344, New Mexico Experiment Station.

See also: **Antioxidants**: Natural Antioxidants; **Carbohydrates**: Requirements and Dietary Importance; **Fats**: Requirements; **Oxidation of Food Components**; **Protein**: Requirements

Further Reading

Botkin CW and Shires LB (1948) *The Composition and Value of Piñon Nuts*. Bulletin 344, New Mexico Experiment Station.

Budzynska J (1964) Note sur la composition en acides gras de l'huile des noix de *Pinus pinea*. *Revue Française des Corps Gras* 11: 143–145.

Kaic M (1985) Prehrambenotenoloske osobine sjemenki oraha (*Juglans regia*), crnog oraha (*Juglans nigra*) i pinije (*Pinus pinea*). *Sumarski List* 109: 325–328.

Lanner RM (1981) *The Piñon Pine*. Reno, Nevada: University of Nevada Press.

Little EL (1938) *Food Analyses of Piñon Nuts*. Research notes, Southwestern Forest and Range Experiment Station, Arizona.

Rowe DB, Blazich FA and Weir RJ (1999) Mineral nutrient and carbohydrate status of loblolly pine during mist propagation as influenced by stock plant nitrogen fertility. *Hortscience* 34: 1279–1285.

Siew WL (2001) Crystallisation and melting behaviour of palm kernel oil and related products by differential scanning calorimetry. *European Journal of Lipid Science and Technology* 103: 729–734.

Zakaria S, Hamzah H, Murshidi JA and Deraman M (2001) Chemical modification on lignocellulosic polymeric oil palm empty fruit bunch for advanced material. *Advances in Polymer Technology* 20: 289–295.

Pituitary Hormones   *See* **Hormones**: Adrenal Hormones; Thyroid Hormones; Gut Hormones; Pancreatic Hormones; Pituitary Hormones; Steroid Hormones

---

PLANT ANTINUTRITIONAL FACTORS

Contents
Characteristics
Detoxification

Characteristics

G D Hill, Lincoln University, Canterbury, New Zealand

Background

The growing of plants as a major food source for humans only started about 10 000 years ago with the birth of agriculture. At that stage, human intervention disrupted the chain seed – plant – seed. Because the seeds, tubers, and fruits of a number of plant species are rich in carbohydrates, protein, and fats, plant materials now provide a substantial proportion of the food energy that is consumed by humans and their domestic animals. When that process happened, a number of chemical compounds occurring naturally in plants, and often with a role in protecting the plant from attack by animals – be they insect, ruminant herbivore, or monogastric – became antinutritional factors (ANFs).

Other articles will deal in detail with most of these ANFs, which are sometimes called secondary metabolites. This chapter will consider the plant species involved and their interaction with animals and humans when they are eaten. The chapter will also consider briefly how the effects of ANFs can be overcome. It will not cover their exact chemical structure, their determination, or their biosynthesis.

Plant Antinutritional Factors

Antigens

A number of legume species, but particularly soybean (*Glycine max*) and peanut (*Arachis hypogaea*), produce severe antigenic reactions in both people and animals. Affected persons can go into anaphylactic shock, and unless treatment is quickly applied, death can follow. In a recent case in the USA, a person who was allergic to peanuts died after licking his fingers having touched a chicken leg coated with peanut sauce. Similarly, some airlines now have peanut-free flights. Other plant foodstuffs that are reported to induce allergenic reactions include rice (*Oryza sativa*), hazelnuts (*Corylus avellana*), apples (*Malus sylvestris*), and celery (*Apium graveolens*)

Compounds responsible The proteins that are thought to be responsible for the development of allergenic reactions are generally water-soluble glycoproteins that are resistant to breakdown by heat, acid, and digestive enzymes. In soybean and peanuts, the glycoproteins are thought to be the storage proteins, glycinin and β-conglycincin. These proteins resist digestion and are absorbed into the intestinal mucosa. The body responds by producing immunoglobulins or antibodies in the Peyer's patches of the gut and their associated lymphoid tissue. These antibodies are then released into the gut and react with the antigens to prevent their absorption. Antigens that escape absorption cause inflammation of the intestinal mucosa.

Three mechanisms are involved. An immediate mechanism that involves IgE antibodies and mast cells (type I hypersensitivity), semidelayed, which is brought about by antigen–antibody complexes (type III), and delayed caused by specific activated T-lymphocytes (type IV). The last is so called because there can be a delay of several days between the antigen challenge and the onset of observable symptoms. (*See* **Food Intolerance:** Food Allergies.)

Symptoms Affected animals show poor growth and diarrhea, and may die. In the gut, there are changes in the structure of the mucosa, increased gut motility, and reduced absorption of electrolytes (Na^+, K^+, and Cl^-). Once animals are sensitized, they can remain so for a considerable period. In laboratory animals, chronic stimulation by antigens has given reduced disacchiridase activity, reduced surface area of the brush borders of the microvilli, reduced microvilli height, crypt hyperplasia, cell degranulation, and increased numbers of mast cells.

Minimization of problems In humans, the development of food allergies seems to be related to age. Among young children, up to 8% can suffer from food allergies. In adults, the rate of occurrence is considered to be <1%. Thus, as people get older, they seem to be less prone to the development of food allergies. Where products such as soybean are used as milk substitutes, it is possible to treat the material by procedures such as steam heating to reduce the level of antigenicity.

α-Galactosides

It is probably not strictly correct to refer to the α-galactosides as ANFs, as they are not toxic. However, their presence in plant foods can cause severe digestive upsets in monagastrics that are embarrassing in humans and can lead to severely reduced growth rates in animals such as pigs.

Compounds responsible The α-galactosides or oligosaccharides comprise the sugars raffinose, stachyose, verbascose, and ajugose. They are simple sugars comprising a sucrose molecule to which are attached α-D-galactopyranosyl units in α-1,6-galactosidic linkages. They are a common carbohydrate present in the seed of grain legumes. It is thought that on seed germination, they provide energy for the emerging plant embryo.

Symptoms When members of the raffinose group of sugars are consumed by monogastrics, the animals do not have the necessary gut enzymes to digest these sugars in the intestinal tract. Consequently, they pass undigested to the large intestine where they are anaerobically fermented by gut bacteria. The products of this fermentation are, among other things, the gases carbon dioxide, methane, and hydrogen. Thus, consumption of legumes rich in these sugars leads to a considerable increase in flatus production. Some legumes can produce up to $137\,ml\,h^{-1}$ of flatus, causing severe abdominal discomfort, bloating, belching, flatulence, constipation, and diarrhea. In pigs, in particular, high levels of oligosaccharides in their rations can cause severe digestive upsets and reduced growth.

Minimization of problems There is considerable interspecific variation in the level of oligosaccharides in legume seed. Soybean, chickpea (*Cicer arietinum*), and *Phaseolus* spp. tend to contain high levels, whereas lentil (*Lens culinaris*) and mung bean (*Vigna radiata*) contain lower levels. There is also some evidence of intraspecific variation among cultivars within species.

Traditional methods of processing grain legumes prior to their consumption significantly reduced the level of oligosaccharides. In the production of bean sprouts, the oligosaccharide level falls as germination continues. *Tofu*, which is a traditional protein precipitate produced from soybean, has low levels of oligosaccharide. In the production of *tempe*, which is a traditional Indonesian soy-based food, cooked beans are fermented with *Rhizopus oligosporus*. During fermentation, the oligosaccharides are utilized by fungus as an energy source, and the resulting *tempe* contains low levels of oligosaccharides.

Traditional legume-processing methods cannot be used for bulk quantities, however, because of the cost of processing. In the stock food industry, the problem is approached by the plant breeders trying to reduce the level of these sugars in the seed using selection within the species.

Alkaloids

A number of plant species contain significant amounts of alkaloids. Of the alkaloids, the most universally known is probably nicotine, which is a component of tobacco (*Nicotiana tabacum*) and is ingested on a daily basis by millions of people. Edible members of the plant family Solanaceae, such as potato (*Solanum tuberosum*), tomato (*Lycopersicon esculentum*), and and eggplant (*Solanum melongena*), may contain high levels of glycoalkaloids. (*See* **Alkaloids: Toxicology.**)

Among legumes that are eaten by humans, or used in stock rations, plants of the Genus *Lupinus* contain high levels of alkaloid in both their foliage and their seed. In most lupin species, the alkaloids present are quiniolizidine alkaloids. However, *Lupinus luteus*, which is from the Iberian Peninsula, contains the indol alkaloid gramine.

Compounds responsible The alkaloids present in food plants of the family Solanaceae are glycoalkaloids. They have a stereoalkaloid nucleus, which is attached to a sugar side-chain. Their role is to protect the plant from insect attack. They are also implicated in both fungal and nematode resistance. The alkaloids tend to be concentrated near the epidermis and in the leaves. In green tomatoes, the fruits are protected while ripening by the alkaloid tomatine. As the fruit matures, the tomatine level falls.

In potato, the major alkaloids present are α-solanine and α-chaconine. They are present throughout the plant. Apart from a case in Scandinavia, where plant breeders increased the alkaloid levels in a new cultivar to the level where they produced toxic symptoms in humans, potato alkaloids are not a problem, provided that the tubers have been properly grown and stored. If potato tubers form too close to the soil surface, or are exposed to light after harvest, there is a rapid increase in alkaloid level. Harvested potatoes should be stored in the dark. In modern shopping systems, potatoes are often exposed to light in supermarkets when they are packed in clear plastic bags.

Among lupin species that are grown for their seed, there are a large number of quinolizidine alkaloids (**Table 1**). However, only eight of these occur in amounts of more than 4% of total alkaloids. The main alkaloids are sparteine, lupanine, 13-hydroxylupanine, albine, multiflorine, angustifioline, tetrahydrorhombifoline, and lupinine. The distribution of the alkaloids tends to vary with plant species. Most of the alkaloid in *Lupinus albus, L. angustifolius*, and *L. mutabilis* is lupanine. In *L. luteus*, the major alkaloid present is lupinine. The gramine in *L. luteus* (an indol alkaloid) can comprise up to 0.12% of total seed dry matter.

Symptoms Both potato and lupin alkaloids affect the nervous system. An excess of potato alkaloids can produce symptoms similar to a bad hangover. There have been deaths reported from the ingestion of excess potato alkaloids. Ingestion of lupin alkaloid leachate by a human has been known to produce blurred vision, an irregular heart beat rate, and urine retention. In animal feeding, the response to foods high in alkaloids is a loss of appetite, because of their bitter taste. This induces a reduction in feed intake and reduces growth rate.

Minimization of problems Fortunately, the alkaloids in both potatoes and lupins are water-soluble. Further, in potatoes, they are concentrated under the skin. Peeling the potato removes most of the alkaloid in the tuber. Soaking and boiling in water further reduce alkaloid levels. In both the altiplano of South America and southern Europe, *L. mutabilis* and *L. albus* are processed to reduce alkaloid levels. The seed is soaked and boiled and then soaked in running water for up to three days, by which time the alkaloid levels are significantly reduced. As noted above, with potatoes, correct storage in the dark, after harvest and prior to use, is also important.

Table 1 Distribution of total alkaloids present in smooth seeded lupin species used in human diets and animal feeding

Alkaloid	Lupin species			
	L. albus	L. angustifolius	L. luteus	L. mutabilis
Sparteine	Trace	Trace	30–50%	5–20%
Lupanine	50–80%	50–80%	Trace	50–70%
17-Oxylupanineo	Trace	Trace	a	
4-Hydroxylupanine				7%
13-Hydroxylupanine	5–15%	10–20%		10–20%
Albine	5–15%			
Multiflorine	3–10%			Trace
Angustifoline	Trace	5–20%		Trace
Tetrahydrorhombifoline	Trace	Trace	Trace	4%
Lupinine			40–70%	Trace
Esteralkaloids	1–5%	1–3%	Trace	1–5%

[a]Not detected.

In lupin, since the early 1920s, an alternative approach has been to select for low-alkaloid genotypes. Since then, because of work by plant breeders in Germany, Australia, and South America there are now low-alkaloid genotypes of *Lupinus albus*, *L. angustifolius*, *L. luteus*, and *L. mutabilis*. In Australia, low-alkaloid genotypes of rough-seeded Mediterranean lupins have now been bred. Thus, lupin seed is a safe food, and alkaloids are not likely to be a problem, provided the seed is processed before consumption or only low-alkaloid types are used.

Cyanogenic Glycosides

A number of plant species produce hydrogen cyanide (HCN) from cyanogenic glycosides when they are consumed. These cyanogens are glycosides of a sugar, often glucose, which is combined with a cyanide containing aglycone.

Plant species of major importance in human and animal feeding are cassava (*Manihot esculenta*), linseed (*Linum usitatissmium*), various sorghums (*Sorghum* spp), white clover (*Trifolium repens*), and some species of *Lotus*. Lesser quantities are found in the kernels of such plants as almonds (*Amygdalus communis*), apricots (*Prunus armeniaca*), peaches (*Prunus persica*), and apples (*Malus sylvestris*).

When plant material containing the glycoside is consumed, it is broken down by a β-glucosidase to produce a sugar and an aglycone. The aglycone is then acted upon by a hydroxynitrile lyase to produce cyanide and an aldehyde or a ketone.

Compounds responsible The compounds responsible vary with plant species. However, the major cyanogens are amygdalin and prunasin, which are found in fruit kernels. The latter also occurs in bracken fern (*Pteridium aquilinum*), dhurrin, found in members of the genus *Sorghum* and linamarin, found in clovers, linseed, cassava, and lima beans (*Phaseolus lunatus*). Cassava root contains relatively low levels at 53 mg of CN per 100 g of plant tissue. Sorghum forage contains 250 mg of CN per 100 g of plant tissue and lima beans up to 300 mg of CN per 100 g of plant tissue.

Symptoms As cyanide is extremely toxic, one of the most obvious symptoms is death. In the body, cyanide acts by inhibiting cytochrome oxidase, the final step in electron transport, and thus blocks ATP synthesis. Prior to death, symptoms include faster and deeper respiration, a faster irregular and weaker pulse, salivation and frothing at the mouth, muscular spasms, dilation of the pupils, and bright red mucous membranes.

In Africa, where many people consume cassava on a regular basis, many members of the human population are regularly exposed to low levels of cyanide in their diet. This is associated with a condition called tropical ataxic neuropathy. Symptoms include neurological disturbances, which affect vision, hearing, and the peripheral nervous system. There are also raised levels of blood thiocyanate and goiter. Cassava consumption, combined with protein deficiency, which is often common in societies that consume large amounts of starchy tubers, can lead to reduced glucose tolerance and diabetes. There is some evidence that the symptoms can be partially alleviated by the administration of vitamin B_{12} and methionine.

Minimization of problems As with many other problems of plant ANFs, one method of reducing the problem is via plant breeding. Older varieties of white clover used to produce high levels of cyanide. Newer cultivars have much lower concentrations. Similarly, breeding work has been carried out with lima beans to reduce their potential toxicity.

To produce cyanide, plant enzyme systems need to be active, and so heating plant material, and denaturing the enzymes, will generally render it safe. Thus, in the production of linseed meal, which is the residue left over after oil extraction, the material is generally heated sufficiently, while passing through an oil press, to render the enzyme system inactive. The use of entire linseed seeds in bakery products usually leads to their being exposed to high temperatures before they are eaten. In home consumption of cassava, cooking is used to inactivate the enzymes. Roots are also washed, grated, and soaked in water, and further washed to eliminate the cyanide. As with clover and lima beans, plant breeders are also trying to reduce the level of the cyanogen in cassava roots.

Glucosinolates

Glucosinolates are a common component in plants of the family Brassicacea. They have become of more interest in recent years following the considerable increase in the production of rape oil (Canola) from the seed of spring and winter forms of *Brassica campestris* and *B. napus*. Following oil extraction in these, and other *Brassica* spp., a high-protein meal is left, which contains high levels of sulfur amino acids. However, the presence of glucosinolates in the resulting meal has limited the use of the meal in animal feeding because of its tendency to produce goiter and other symptoms associated with low iodine availability. (*See* **Glucosinolates.**)

Compounds responsible Glucosinolates are glycosides of β-D-thioglucose. They contain an aglycone, which, on hydrolysis, can yield an isothiocyanate, a nitrile, or a thiocyanate. Glucosinolates are responsible for the sharp flavor of such plants as horseradish (*Armoracia rusticana*) various mustards (*Brassica* spp), watercress (*Nasturtium officinale*), and radish (*Raphanus sativus*).

In plant tissues, the glucosinolates are present at the same time as the enzyme myrosinase, which is a thioglucosidase. In intact plants, the enzyme and the substrate occur in different parts of the plant. However, when plant tissues are disrupted, as may occur in chewing, or in pressing and grinding for oil extraction, the two components come together. As a result, the myrosinase breaks the thioglucoside bond on the glucosinolate. The result is the production of glucose and an aglucone intermediate product. The latter breaks down to produce a number of possible toxic components. The most common are isothiocyanates and nitriles. However, this depends on the conditions at the time of breakdown, and other products, such as goitrin, may be produced.

Symptoms   The major effect of the breakdown products of the glucosinolates is interference with thyroid function. Goiter can occur in both humans and animals, but it mainly appears in animals when they are fed on rations in which rapeseed meal provides a considerable amount of the protein in the ration. Besides obvious morphological effects on the thyroid gland of animals, differences can be shown in biochemical parameters such as decreased serum thyroxin levels. In pigs, when sows are fed on high-glucosinolate rations, it can lead to abortion of piglets and increased death of piglets after birth.

Minimization of problems   As with many other ANFs, moist heating the meal prior to feeding it to the animals or heat extrusion can reduce problems associated with the feeding of glucosinolates in *Brassica* meals. Bacterial fermentation and enzymes have also been used. However, the most significant recent achievement in reducing the problem was wrought by Canadian plant breeders who produced new cultivars of both major oils seed rape species, which contain considerably reduced levels of glucosinolates. These were originally known as '00' rapes but have since been renamed Canola. This major plant breeding advance has meant that considerably greater amounts of *Brassica* meal can now be incorporated into monogastric rations without any deleterious effects on thyroid function and iodine metabolism.

Lectins

Lectins, or phytohemagglutinins, are present in the seed of a number of legume species that are consumed by humans. The most important legume species, which is high in lectin, is *Phaseolus vulgaris*, because of its major role in the diet of central and northern South American populations. *Phaseolus vulgaris* lectin is highly toxic, as is the lectin from scarlet runner bean (*Phaseolus coccineus*) and tepary bean (*P. acutifolius*). At the other extreme, legume species with low to zero lectin activity include chick pea (*Cicer arietinum*), cow peas (*Vigna unguiculata*), faba bean (*Vicia faba*), lentil (*Lens culinaris*), mung bean (*Vigna radiata*), peas (*Pisum sativum*), pigeon pea (*Cajanas cajan*), and soybean (*Glycine max*). (*See* **Hemagglutinins (Haemagglutinins)**.)

In the plant, it appears that lectins are involved in the chemical recognition of the legume root by *Rhizobium* bacteria. Thus, although plant breeding may be used to reduce their absolute concentration in seed, it is unlikely that they could be entirely eliminated from legume seeds because of their importance in biological nitrogen fixation.

Compounds responsible   Lectins are either carbohydrate-binding proteins or glycoproteins. They are capable of recognizing and binding carbohydrates in complex glycoconjugates. More than 70 lectins have now been isolated from legumes. They usually contain two or four subunits, each with a single carbohydrate-binding site.

Symptoms   The main effect of lectins, when they are included in the diet, is through their interference with digestion in the small intestine. Lectins are resistant to breakdown in the digestive tract and bind to surface receptors. These bound lectins then cause changes in the metabolism of the epithelial cells. These changes include cell hypertrophy and hyperplasia. Indeed, animals fed high levels of toxic lectin can suffer rapid body weight loss, while the weight of the small intestine is significantly increased. Other changes that may occur are changed gut endocrine systems and hormone production, changed gut immune systems and disturbances in gut bacterial ecology, which, combined, can cause a greater sensitivity to bacterial infection. At the gross level, these effects are mediated via reduced growth, diarrhea, reduced nutrient absorption, and increased incidence of bacterial infections. In highly toxic species such as the red kidney bean (*Phaseolus vulgaris*), these symptoms, when combined, cause death.

Minimization of problems   As indicated above, it is unlikely that it will be possible to breed the lectins out of legume seed entirely because of their role in the establishment of biological nitrogen fixation. However, it may be possible to select for reduced levels and/or to select for less toxic lectins. Fortunately, most lectins are heat-labile. Dry *Phaseolus* beans require long soaking and cooking times before they are soft enough to eat, and therefore, for human consumption, most seed is rendered safe before it is consumed. However, there are reports that if beans are heated at 80 °C, lectin levels may be increased, and thus the beans are more, not less, toxic.

For animal feeding, however, lectin-containing seed needs to be processed to reduce lectin levels before the seed can be fed to monogastric animals. Generally, seed can be rendered safe by heat treatment prior to consumption. Other techniques that have been employed include chemical treatment, production of protein isolates, germination, and moist extrusion.

Saponins

Saponins are a diverse group of chemicals, which derive their name from their ability to form soap-like foams in aqueous solutions. They occur in a considerable number of plant species ranging from asparagus (*Asparagus officinalis*) to cucumber (*Cucumis sativus*). They are also present in a number of commonly used herbs and spices such as fenugreek (*Trigonella foenumgraecum*), ginseng (*Panax* spp.), liquorice (*Glycyrrhiza glabra*), and nutmeg (*Myristica fragrans*). (*See* **Saponins**.)

Saponins fall in a gray area between being ANFs or beneficial plant constituents. When monogastric animals such as pigs and poultry are fed rations containing such products as alfalfa (*Medicago sativa*) meal, there is a reduction in growth. However, experiments with a number of animal species have failed to show any significant effect of saponins on a number of growth parameters. As saponins are very bitter, one hypothesis is that their effects are mediated through reduced food intake.

On the positive side, it has been shown that saponins can form complexes with cholesterol in the gastrointestinal tract. This leads to increased excretion of cholesterol and a reduction in blood cholesterol level. With the high level of coronary artery disease in most developed countries, this is seen to have potential medical benefits. The most common form of ingestion by humans is through alfalfa sprouts. Recent publicity would suggest that the sprouts are more dangerous because of bacterial contamination, leading to gastroenteritis, than because of their saponin content.

Compounds responsible   All saponins contain an aglycone, which is either sapogenol or sapogenein. The aglycone is linked to one or more sugars or oligosaccharides. However, the saponins fall into two groups, depending on whether the aglycone is triterpenoid or steroidal. Most cultivated plants have a triterpenoid aglycone, whereas in many herbs,

it is steroidal. Saponin's structure is complex and depends on the variation in the aglycone structure and the position and nature of attachment of the glycosides to the molecule. Aglycones can be linked to D-glactose, L-arabinose, L-rhamnose, D-glucose, D-xylose, D-mannose, and D-glucuronic acid. The chain lengths are generally linear and comprise two to five saccharide units. This diversity in structure leads to a diversity of biological activity.

Symptoms   The major physiological effect of saponins is on cell membrane permeability. It has been proposed that the saponins combine with membrane cholesterol to form permeable micelles in the plane of the membrane. However, as indicated above, the major effect of saponins in reducing intake appears to be the result of their bitter flavor-reducing feed intake, as experiments on a range of animal species have not shown any pathogenic effects of saponin intake.

Minimization of problems   Saponins that are present in herbs and spices are likely to be present in such small quantities in the diet that they are unlikely to have any deleterious effect. Alfalfa has been selected by plant breeders to contain low levels of saponins. Similarly, quinoa (*Chenopodium quinoa*) seed, which is a minor Andean food crop, has reported saponin levels ranging from 0.14 to 2.3%. This has allowed for the selection of low saponin genotypes. In the Andes, in households, saponins are removed by soaking, washing, and rubbing to reduce the saponin levels in quinoa grain. Industrially, the seed is milled, or washed and milled.

Tannins

Many plant species contain tannins. However, the seed of a number of legume species and some sorghum lines contain appreciable quantities of tannins. These have dietary impacts on animals that consume them. Tannin is a generic name for polyphenolic compounds derived from plants. They have molecular weights of between 500 and 3000 and are divided into two classes, hydrolyzable and condensed. The former can be broken down by acid, alkali, and some hydrolytic enzymes. The latter are resistant to being broken down.

Faba bean (*Vicia faba*) tannins are involved with the human disease favism. The inclusion of high levels of tannin in rations for a range of monogastric animals reduces weight gain and feed-conversion efficiency. When fed to hens, tannins reduce egg production. However, the consumption of forage with high tannin levels can be advantageous in ruminant animal production. Tannins prevent the formation of the stabilized foams that produce bloat. Further, tannins form complexes with free protein in the rumen and prevent protein degradation in the rumen. (*See* **Tannins and Polyphenols**.)

Compounds responsible Hydrolyzable tannins have a polyhdric alcohol at their core, the hydroxyl groups of which are partially, or fully, esterified with either gallic or hexahydroxydiphenic acid. They may have long chains of gallic acid coming from the central glucose core. On hydrolysis with acid or enzymes, the hydrolyzable tannins break down into their constituent phenolic acids and carbohydrates.

Condensed tannins are dimers, the simplest of which is procyanidin, or higher oligomers of substituted flavan-3-ols. Monomeric tannins are usually linked by carbon–carbon bonds between carbon-4 and carbon-8 of two flavan-3-ols. The bonds that are formed are highly stable and require heating with strong acids to break them down.

Symptoms In humans, favism produces acute hemolytic anemia. After susceptible subjects eat the beans, symptoms can occur in 5–24 h. The symptoms include headache, vomiting, nausea, yawning, stomach pains, and a raised temperature. The symptoms may subside naturally or, in severe cases, lead to hemolytic anemia, followed by hemoglobinuria. Previously, deaths in children could reach 8%. The disease is due to the lack of an enzyme in red blood cells, glucose-6-phosphate dehydrogenase. It is thought that over 100 million people, of mainly eastern Mediterranean origin, are susceptible to the disease. There has been some suggestion that the high levels of tannins in the diet of people who use sorghum as their main staple may be associated with an increased incidence of esophageal cancer.

The condensed tannins provide an important plant defense against herbivory. In particular, bird damage can be reduced by high tannin levels in crops such as sorghum and insect damage in seed of species like faba beans. The adverse nutritional effects of tannins are due to their reaction with proteins to form indigestible complexes. They can also form complexes with digestive enzymes and reduce the overall digestibility of food. They can irritate the gut lining and stimulate the secretion of mucus. This increases endogenous protein secretion and therefore increases protein demand. They also form complexes with divalent metals and reduce mineral absorption. A common feature of diets high in tannin is weight loss.

Minimization of problems In most legume species, there is a strong relationship between the level of condensed tannin in the seed and their flower color. Plants that have white flowers have low tannin levels

in their seed. Thus, in a number of the legume species that have high tannin levels, the development of white-flowered genotypes gives pale-colored seed with low tannin levels.

In both legume seed and sorghum seed, the tannins are located mainly in the seed coat. Thus, dehulling the seed and discarding the hulls reduce the tannin level. Heat treatments do not destroy tannins in sorghum. However, sorghum tannins are unstable under alkaline conditions. Soaking in an alkaline solution followed by washing reduces the level of condensed tannins. The tannins in sorghum can also be neutralized by treatment with anhydrous ammonia. In contrast, in faba beans, the principal thermolabile antinutritional factor is the condensed tannins.

Toxic Amino Acids

In the legume family, there are two toxic amino acids that have a major negative effect on animals or humans that consume them. The tropical tree species *Leucaena leucocephala* contains the amino acid mimosine, which is an amitotic agent. Seed of the hardy annual legume species *Lathyrus sativus* contains a neurotoxic amino acid, generally referred to as ODAP or BOAA. The latter can cause paralysis of the lower limbs when eaten in excess.

Compounds responsible The full chemical name of mimosine is β-[1-(3-hydroxy-4-pyridone)]-α-amino propionic acid. The amino acid ODAP is β-*N*-oxalyl-L-α-β-diamino-propionic acid.

Symptoms Mimosine is an amitotic agent, and one of its major effects is to stop hair growth. It has been tested for a possible role in the chemical shearing of sheep. *Leucaena* seed, which is high in mimosine, is usually only used to a limited extent in human diets as a curry sambal in Indonesia.

The effects of *Lathyrus sativus* seed consumption are somewhat more problematic. Consumption of large amounts of *Lathyrus* seed in the diet often occurs in countries such as India and Bangladesh at times of food shortages as a result of drought. The condition produced is neurolathyrism, which is paralysis of the lower limbs caused by damage to nerve cells in the spinal cord. There is some evidence that the response to *Lathyrus* consumption may be sex-linked, as it seems to have a major effect on young males.

Minimization of problems With regard to mimosine, there have been some attempts to breed genotypes of *Leucaena leucocephala* that are low in mimosine. As it is such a minor component in human diets, it is unlikes to be a major problem. During

World War II in Hawaii, *Leucaena* leaf meal was used in both pig and poultry rations as a substitute for alfalfa meal. Provided that the levels that were fed were less than 10% there were no ill effects. Finally, ruminants can modify the spectrum of their rumen microflora to convert mimosine to 3,4-dihydroxypyrididine, which appears to have no biological activity.

With regard to *Lathyrus sativus* consumption, three approaches have been taken to reduce the problems associated with its consumption. In India, the crop has been banned. However, it apparently has a pleasant flavor and is a common adulterant of red gram (pigeon pea dhal). There is some suggestion that since the ban, the total area sown to the crop has increased. Therefore, the two following methods of dealing with the problem would appear to have a greater potential.

As with many other legume species that contain ANFs, they can be removed by traditional processing methods such as soaking, boiling, and fermentation. The major problem is that often in times of famine, it is not only food that is in short supply. Reduced amounts of wood for cooking often mean that food is cooked for less time, and the ANFs are therefore not destroyed to the same extent. At the same time, because of the drought resistance of the mother plant, it may become the only grain legume that is available, so the total amount consumed, and thus the dose of ODAP, is increased.

In Bangladesh, *Lathyrus sativus* is the most important single pulse crop and is grown on about 82 000 ha. To this end, plant breeders in Bangladesh, in association with workers from Belgium and Canada, have selected lines of *Lathyrus* with low levels of ODAP. The aim is to produce cultivars that are safe to eat with no, or a reduced, requirement for processing.

Trypsin Inhibitors

Trypsin inhibitors that inhibit the activity of the enzymes trypsin and chymotrypsin in the gut, thus preventing protein digestion, are found in many plant species. These species include a range of grain legumes such as common bean (*Phaseolus vulgaris*), cowpea (*Vigna unguiculata*), Lima bean (*Phaseolus lunatus*), peanut (*Arachis hypogaea*), peas (*Pisum sativum*), soybean (*Glycine max*) and winged bean (*Psophocarpus tetragonolobus*). However, they are also found in cereals such as wheat (*Triticum aestivum*) and barley (*Hordeum vulgare*), potatoes (*Solanum tuberosum*), and a number of species in the genus *Cucurbita*. Because of the extensive use of soybean meal in monogastric feeding, the trypsin inhibitors associated with this plant species have been studied most extensively. (*See* **Trypsin Inhibitors**.)

When monogastric animals are fed rations, which contain large amounts of raw soybean meal, there is poor growth, poor hair and feather production, and digestive disturbances. However, there is recent evidence to suggest that consumption of food containing trypsin inhibitors may have a role in combating breast cancer in humans.

Compounds responsible There are many plant-derived trypsin inhibitors. Most of these inhibitors differ in their specificity. Many can inhibit one or two enzymes. Different forms of inhibitor may be present in the same seed, and most of the inhibitors can inhibit trypsin, but they may also inhibit chymotrypsin. However, in legume seed, the two most important inhibitor families are the Kunitz trypsin inhibitor family and the Bowman–Birk trypsin inhibitor family.

The Kunitz inhibitor family was the first family to be isolated. It is a peptide comprising 181 amino acids containing two disulfide bridges with a molecular weight of about 21 000 Da. As this inhibitor inhibits trypsin stoichiometrically to form a stable complex, it is known as a single-headed inhibitor. It primarily inhibits trypsin, but it can weakly inhibit chymotrypsin. It is inactivated by heat and by gastric juices.

The Bowman–Birk inhibitor family is widely distributed in legume seed. It is a smaller peptide molecule and contains 71 amino acids. It contains a high level of cystine and has seven disulfide bridges. The molecular weight is about 8 000 Da. It is a double-headed molecule and inhibits both trypsin and chymotrypsin at two different binding sites. Bowman–Birk inhibitors are resistant to gastric juices and to proteolytic enzymes. There is also a suggestion that they may be resistant to breakdown by heat.

Symptoms Animal response to trypsin inhibitors in the diet varies with animal species. In chicks, rats, and mice, the inhibitors cause pancreatic hypertrophy and increased pancreatic secretion. However, they do not have these effects on pigs, dogs, or preruminant calves. Most monogastrics have reduced growth when fed on rations with high levels of raw soybean meal. Protein digestibility may be reduced, and dietary protein is excreted in the feces. There is also reduced nitrogen and sulfur absorption. One of the effects of the inactivation of digestive enzymes in the intestine is the stimulation of trypsin and chymotrypsin secretion from the pancreas, which can create an increased demand for the sulfur amino acids methionine and cystine. In turn, this leads to increased endogenous loss of both nitrogen and sulfur. Finally, the inhibitors can stimulate the release of cholecystokinin into the blood stream, which further increases pancreatic secretion.

A number of factors can modify the effect of trypsin inhibitors in the diet. These include the animal species, the age of the animal, other ANFs present in the diet, and the type and level of protein in the diet. It appears that human trypsins are not inactivated by either family of the soybean trypsins.

Minimization of problems   As with many other ANFs, the level of trypsin inhibitors can be reduced by moist heat treatments. Loss of activity is associated with temperature, duration of heat treatment, moisture conditions, and particle size. Up to 95% of activity can be destroyed by heating at 100 °C for 15 min. Autoclaving for 15–30 min has also been shown to destroy most trypsin inhibitor activity in a range of grain legume species. However, the results are not universally applicable. In the industrial production of soybean oil and meal, the heating caused during the process substantially reduces the level of trypsin inhibitor activity.

Plant breeders have been able to show that there is considerable variation in the level of inhibitor activity among genotypes of the same legume species. Recently, in peas, breeders have produced isogenic lines that differed only in their level of seed trypsin inhibitor. Similarly, there are also inhibitor-free lines of soybean. Thus, the breeding of plants with reduced levels of trypsin inhibitors in their seed is possible. However, as with a number of other ANFs, it is not clear what role the inhibitors serve in the possible protection of the plant. Until this is resolved, given that they are mainly rendered inactive by the processing of the seed, it may be better to leave the inhibitors in the seed.

Conclusions

It appears that our ancestors, and people in many societies in the developing world today, have been well aware of the ANFs that are present in the plant material eaten. They have long established methods of reducing the risks involved by one or more of the processes of soaking, testa removal, heating, fermenting, germinating, or protein extraction to produce isolates. However, these systems do not always work, and in many developing countries, the problems of already-poor nutrition are exacerbated by the presence of ANFs in the diet.

In the developed countries, in recent years, there has been a swing away from meat-based diets towards the consumption of more legume seed. A result has been an increased number of cases reported where people have become ill after eating legume seed that had not been fully processed. It has been suggested that this is because the need to remove the ANFs is no longer common knowledge.

In the processing of some plant materials to produce products like Canola or soybean oil, the heat that is produced during crushing and milling may be sufficient to render the resulting press cake safe to feed to nonruminants, including humans. Industrial producers prefer to use products that require as little extra processing as possible. To try to meet this demand, plant breeders have been actively trying to reduce the levels of ANFs in plant products. However, the downside is that this often renders plants more susceptible to attack by birds, insects, and fungi while they are growing, and to increased depredations of stored products by insects, particularly weevils.

It is possible that, if the controversy over the use of GM-modified plants for human food can be resolved, plant breeders could remove the ANFs from food plants. The partial protection that the ANFs currently provide could be obtained by the insertion of one or a few specific genes.

See also: **Bioavailability of Nutrients**; **Carbohydrates**: Classification and Properties; **Galactose**; **Glucosinolates**; **Goitrogens and Antithyroid Compounds**; **Heat Treatment**: Chemical and Microbiological Changes; **Legumes**: Legumes in the Diet; **Phytic Acid**: Nutritional Impact; **Hemagglutinins (Haemagglutinins)**; **Plant Antinutritional Factors**: Detoxification; **Sucrose**: Properties and Determination; **Tannins and Polyphenols**; **Trypsin Inhibitors**

Further Reading

Bruneton J (1999) *Toxic Plants Dangerous to Humans and Animals*. Paris: Éditions TEC & DOC.

Cheeke PR (1998) *Natural Toxicants in Feeds, Forages and Poisonous Plants*, 2nd edn. Danville: Interstate Publishers.

D'Mello JPF, Duffus CM and Duffus JH (eds) (1991) *Toxic Substances in Crop Plants*. Cambridge: Royal Society of Chemistry.

Huisman J (1990) *Antinutritional Effects of Legume Seeds in Piglets, Rats and Chickens*. PhD thesis, Wageningen Agricultural University, The Netherlands.

Jansman AJM, Hill GD, Huisman J and Poel AFB van der (eds) (1998) Recent advances of research in antinutritional factors in legume seeds and rapeseed. In: *Proceedings of the 3rd International Workshop on 'Antinutritional Factors in Legume Seed and Rapeseed'. Wageningen, July 1998*. Wageningen: Wageningen Pers.

Liener IE (ed.) (1980) *Toxic Constituents of Plant Foodstuffs*. New York: Academic Press.

Nartey F (1977) Manihot esculenta *(Cassava): Cyanogenesis, Ultrastructure and Seed Germination*. DSc thesis, University of Copenhagen, Denmark.

Raffauf RF (1996) *Plant Alkaloids: A Guide to their Discovery and Distribution*. New York: Food Products Press.

Roberts MF and Wink M (1998) *Alkaloids: Biochemistry, Ecology, and Medicinal Applications*. New York: Plenum Publishing.

Shahidi F (ed.) (1997) Antinutrients and phytochemicals in food. In: *ACS Symposium Series 662*. Washington, DC: American Chemical Society.

Waller GR and Nowacki EK (1978) *Alkaloid Biology and Metabolism in Plants*. Newyork: Plenum Press.

Wink M (ed.) (1999a) Biochemistry of plant secondary metabolism. In: *Annual Plant Reviews*, vol. 2. Sheffield: Sheffield Academic Press.

Wink M (ed.) (1999b) Functions of plant secondary metabolites and their exploitation in biotechnology. In: *Annual Plant Reviews*, vol. 3. Sheffield: Sheffield Academic Press.

Detoxification

I E Liener, University of Minnesota, St. Paul, MN, USA

Introduction

Proteins of plant origin, particularly oil seeds and legumes, provide a valuable source of protein for humans and animals, despite the fact that they contain substances that can adversely affect the nutritional quality of the protein (**Table 1**). Fortunately, in many cases, these substances are heat-labile and can be readily eliminated or inactivated by the heat treatment involved in domestic cooking and commercial processing, or by treatment with chemicals. In other cases, potential toxicants remain innocuous until they are acted upon by enzymes of endogenous origin. Paradoxically, advantage can sometimes be taken of the action of these enzymes to detoxify certain plants employing such traditional modes of food preparation as germination or fermentation.

Protease Inhibitors

Proteins capable of inhibiting mammalian digestive enzymes, such as trypsin and chymotrypsin, are widely distributed in nature and have been shown to retard the growth of animals by virtue of their ability to interfere with the digestion of dietary protein. The animal tends to adapt to this situation by stimulating the secretory activity of the pancreas as well as its size. In extreme cases, such as the long-term ingestion of these inhibitors, the pancreas may exhibit precancerous lesions.

Because of its economic importance, the soya bean, which is very rich in protease inhibitors, has received the most attention with respect to the means whereby these inhibitors can be inactivated. Because they are proteins, they can be denatured by heat treatment, and this is accompanied by a loss in inhibitory activity. In general, destruction by heat treatment is a function of the temperature, duration of heating, particle size, and moisture content, conditions that are carefully controlled in order to minimize thermal damage to the nutritive value of the protein, which may occur if excessive heat is applied. An example of the relationship between the destruction of the trypsin inhibitor and the concomitant improvement in the nutritional value of the protein in shown in **Figure 1**. It is reassuring to note that most commercially available soybean products intended for human consumption such as tofu, soy protein isolates and concentrates, and textured meat analogs have received sufficient heat treatment to reduce the trypsin inhibitor activity to nontoxic levels.

Although the treatment of soya beans and other legumes by boiling in water or exposure to live steam (sometimes referred to as 'toasting') are the most commonly employed methods for inactivating the trypsin inhibitor, other modes of heat treatment or processing have proved to be equally effective. These

Table 1 Examples of naturally occurring toxicants of plant origin, their distribution and physiological effects

Toxicant	Distribution	Physiological effect
Protease inhibitors	Most legumes	Impaired growth, enlarged pancreas
Lectins	Most legumes	Impaired growth
Goitrogens	Cabbage family	Hyperthyroidism
Cyanogens	Lima beans, cassava	Respiratory failure
Phytate	Most plants	Interference with mineral availability
Tannins	Most legumes	Interference with protein digestion
Oligosaccharides	Most legumes	Flatulence
Mimosine	*Leucaena*	Goiter
β-N-oxalyl-α,β-diamino-propionic acid	*Lathyrus sativus*	Lathyrism

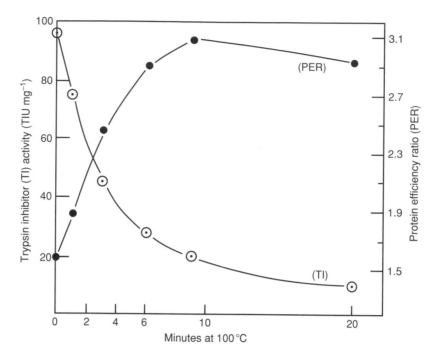

Figure 1   Effect of heat treatment on the trypsin inhibitor (TI) activity and protein efficiency (PER) of soya beans. PER=gain in weight (g)/protein intake (g). Courtesy of Academic Press, New York.

include dry roasting, microwave cooking, γ irradiation and infrared radiation.

Traditional methods of preparing various foods derived from beans generally result in products that are quite low in protease inhibitor activity. For example, tofu comprises mainly protein that has been precipitated from a hot-water extract of soya beans with calcium–magnesium salts, and soya milk is simply a hot-water extract of soya beans that may have been clarified by filtration. In both cases, the boiling or steaming of the soya beans prior to extraction with water serves to inactivate the inhibitors. The same holds true for fermented preparations of soya beans and other legumes such as tempeh, miso, and natto in which the beans have been subjected to boiling prior to the fermentation step.

The trypsin inhibitors are protein molecules whose activity is retained only if the disulfide bonds of the cystine residues remain intact. Thus, cleavage of these disulfide bridges by treatment of soya beans with reducing agents (cysteine, N-acetylcysteine, mercaptoethanol, or reduced glutathione) or with sodium metasulfite causes inactivation of the inhibitors.

A search for varieties of soybeans that might be devoid of trypsin inhibitors has led to the indentification of a cultivar that has only about half of the inhibitory activity of most commercial varieties

of soya beans. This is due to the absence of the Kunitz trypsin inhibitor, the Bowman–Birk inhibitor, presumably accounting for the remainder of the activity. This particular variety of soya bean has the economic advantage of requiring less heat than conventional soya beans in order to produce a product of comparable nutritional value for animal feeding. (*See* **Plant Antinutritional Factors**: Characteristics.)

Lectins

Paralleling the distribution of protease inhibitors present in oil seeds and legumes is a class of proteins referred to as 'lectins'. These are proteins that exhibit the unique property of being able to bind to specific sugars that are part of the structure of so-called glycoproteins. Their antinutritional effect lies in the fact that, by binding to the glycoproteins located on the surface of the cells lining the small intestines, they interfere with the absorption of nutrients; the result is a failure in growth and eventual death.

The toxic effects of the lectins can be effectively eliminated by heat treatment, essentially under the same conditions as those that inactivate the protease inhibitors. However, there are documented cases where beans that have been eaten raw or only partially cooked have led to serious outbreaks of gastrointestinal illness. For example, bean-containing stews and casseroles, when cooked in a slow cooker where a relatively low temperature is maintained for a long period of time, retain sufficiently high levels of lectin to cause illness.

With the availability of the soya bean variant devoid of the Kunitz trypsin inhibitor (see above) as well as one that was found to be free of the lectin, it became possible to compare the relative contribution of these two antinutritional factors to the poor nutritive value of raw soya beans. In comparison with conventional raw soya beans, a greater improvement in the growth performance of chicks was noted for the Kunitz inhibitor-free soya beans than that obtained with the lectin-free soya beans. These results may be taken to indicate that the Kunitz trypsin inhibitor is a more important antinutritional factor than the lectins. It is important to note, however, that heat treatment is necessary for the inactivation of both of these factors in order to achieve a maximum nutritive value.

Goitrogens

Goiter-causing agents in the form of glycosides (also referred to as 'glucosinolates') are found in members of the cabbage family, which includes not only cabbage but also broccoli, Brussels sprouts, cauliflower, turnips, rapeseed, and mustard seed. These compounds are biologically inactive as long as they remain bound to glucose ((1) **Figure 2**), but an enzyme, myrosinase, present in the same plant, serves

to release the active goitrogenic principle, goitrin ((2) **Figure 2**). The glucosinolates found in the cabbage family appear to pose little risk to human health since the enzyme responsible for the release of goitrin is inactivated by household cooking. The glucosinolates may also be removed to a great extent by leaching out into the cooking water. The common usage of rapeseed meal in the feeding of livestock may prove to be toxic unless it has been treated with moist heat. Alternative methods of detoxification of rapeseed may involve prior extraction of the glucosinolates with water or acetone or by decomposition with iron salts or soda ash. These procedures, however, do not preclude the possibility that some goitrin produced enzymatically prior to processing may still remain in the meal. A more effective means of detoxification is one in which an aqueous slurry of the meal is deliberately allowed to undergo autolysis, which serves to liberate virtually all of the goitrin; the latter is then removed by extraction with water or acetone. Lactic acid fermentation or treatment with a specific fungus (*Geotrichum candidum*) has also been reported to be an effective means for the biological destruction of the glucosinolates in rapeseed meal. The immobilization of myrosinase on a solid matrix offers a promising approach for the hydrolysis of the glucosinolates, provided the goitrogenic end products are subsequently removed by extraction, as described above. (*See* **Plant Antinutritional Factors**: Characteristics.)

Cyanogens

Many plants are potentially toxic because they contain glycosides, principally linamarin ((3) **Figure 3**), from which hydrogen cyanide may be released by the action of an endogenous enzyme, linamarinase, when the plant tissue is macerated. Among the cyanogenic

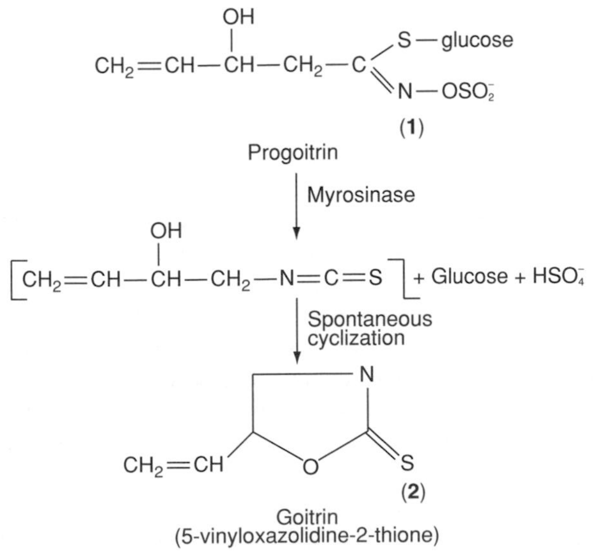

Figure 2 Structure of one of the goitrogenic compounds present in the cabbage family. Progoitrin (1) is biologically inactive, but, upon hydrolysis by the enzyme myrosinase, the active principle goitrin (2) is released.

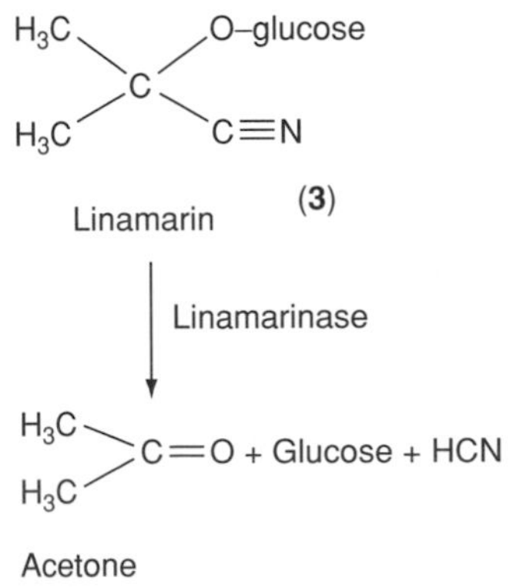

Figure 3 Structure of the major cyanogen present in lima beans. Enzymatic hydrolysis of linamarin (3) releases acetone, glucose and hydrogen cyanide.

plants most likely to be consumed by humans are the lima bean and cassava. Although some tropical varieities of the lima bean may contain toxic levels of cyanide, those varieties consumed in the USA and Europe generally have levels of cyanide below the dosage known to be toxic to humans. Cassava is a staple food item in the tropics and often contains toxic levels of cyanide unless properly processed. The traditional method of preparing cassava involves the removal of the peel, which is particularly rich in linamarin, followed by a thorough washing of the pulp with running water. A further reduction in toxicity is achieved by the application of heat (boiling, roasting, or sun drying), which serves to inactivate linamarinase and to volatilize any hydrogen cyanide that may have been produced. The cyanide content of casava can also be reduced by a lactic acid fermentation to produce a native Nigerian dish called gari. The hydrogen cyanide released by the enzymes elaborated by the fermenting agent is then eliminated by frying. The use of an exogenous source of linamarinase derived from one of several species of fungi and bacteria has also been proposed for enhancing the detoxification of gari. An appreciable degree of detoxification of cassava pulp for use in the feeding of livestock can be achieved by ensilage.

Phytate

Phytic acid ((**4**) **Figure 4**) or its salt, phytate, is inositol combined with six phosphate groups and is a common constituent of most plants. Its antinutritional effect lies in the fact that it forms a chelate with metal ions such as calcium, magnesium, zinc, and iron to form poorly soluble compounds that are not readily absorbed from the intestine, thus interfering with the bioavailability of these essential minerals. The ability of phytate to bind metal ions is

lost when the phosphate groups are removed by hydrolysis through the action of phytase. Heat alone is relatively ineffective in reducing the phytate content of plant materials, but the phytate content can be reduced by taking advantage of the endogenous phytase that accompanies the phytate in separate compartments of the plant tissue. For example, by the simple expedient of allowing ground soya beans to undergo autolysis under conditions that are optimal for phytase activity, one can obtain a significant reduction in phytate content. An exogenous source of phytase is commercially available as a feed additive to livestock and poultry diets. This not only serves to eliminate the mineral binding ability of the phytate that may be present from other plant ingredients of the diet, but also makes more phosphate available to the animal.

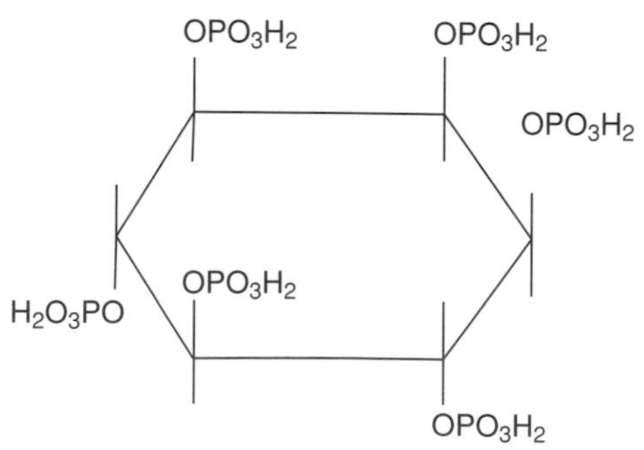

Traditional fermented dishes derived from such plants as soya beans, cassava, rice, maize, cowpeas, or yams have reduced levels of phytate, presumably due to the action of phytase produced by various molds, bacteria or yeasts involved in the fermentation. Even the use of yeast in breadmaking serves to reduce the phytate content of wheat flour. The germination of mature seeds of various legumes results in a great increase in phytase activity with a concomitant reduction in phytate. Other techniques for removing phytate such as ultrafiltration and ion exchange chromatography have been proposed, but these are unlikely to replace the simpler methods already described. (*See* **Phytic Acid**: Properties and Determination; Nutritional Impact.)

Tannins

Oilseeds and legumes contain appreciable levels of polyphenol compounds, broadly referred to as 'tannins' ((**5**) **Figure 5**). The negative nutritional effects of tannins are diverse and incompletely understood, but the major effect is to interfere with the digestibility of dietary protein. This may be due to the binding of the tannins to the protein to form substrates that are resistant to digestive enzymes or to a direct binding to these enzymes themselves.

Since tannins are located primarily in the seed coat of dry seeds, the physical removal of the seed coat by dehulling or milling markedly reduces the tannin content with a resultant improvement in the nutritional quality of the protein. Soaking in water or salt solution prior to household cooking also causes a significant reduction in tannin, provided the cooking broth is discarded. Treatment with a variety of chemicals such as ammonia, hydrogen peroxide, formaldehyde, polyethylene glycol, or polyvinylpyrrolidone has also proved to be an effective means for reducing the

Phytic acid (**4**)

Figure 4 Structure of phytic acid (**4**). Enzymatic removal of the phosphate groups by phytase serves to eliminate its metal-binding property.

tannin content of plant sources of protein. Germination leads to more than a 50% loss in tannins in legumes such as the chick pea, green gram, mung bean, and black gram, presumably due to the action of polyphenol oxidase of endogenous origin. (*See* **Tannins and Polyphenols**.)

Flatulence-producing Factors

One of the important factors limiting the use of legumes in the human diet is the production of intestinal gas (flatulence), which is generally attributed to the inability of the gut enzymes to hydrolyze the α-1,6-galactosidic linkages of such oligosaccharides as raffinose, stachyose and verbascose ((**6**), (**7**) and (**8**) respectively **Figure 6**). In their unhydrolyzed forms, these oligosaccharides become metabolized by the microflora of the large intestine into carbon dioxide, hydrogen, and methane, gases that are responsible for the characteristic features of flatulence, namely nausea, cramps, diarrhea, and the social discomfort associated with the release of rectal gases. Advantage may be taken of the endogenous galactosidases present in the plant tissue by allowing a slurry of the raw ground beans to undergo autolysis under suitable conditions or by treatment with an exogenous source of these enzymes derived from molds or bacteria. Such traditional dishes as tofu and tempeh, which have undergone fermentation by microorganisms, likewise have little flatus activity. Protein isolates from which these oligosaccharides have been eliminated during the course of their isolation are essentially devoid of the offending oligosaccharides. Preparations of galactosidases from microbial sources that have been immobilized as continuous flow reactors or in the form of a hollow-fiber dialyzer have also been employed for removing flatulence-producing factors.

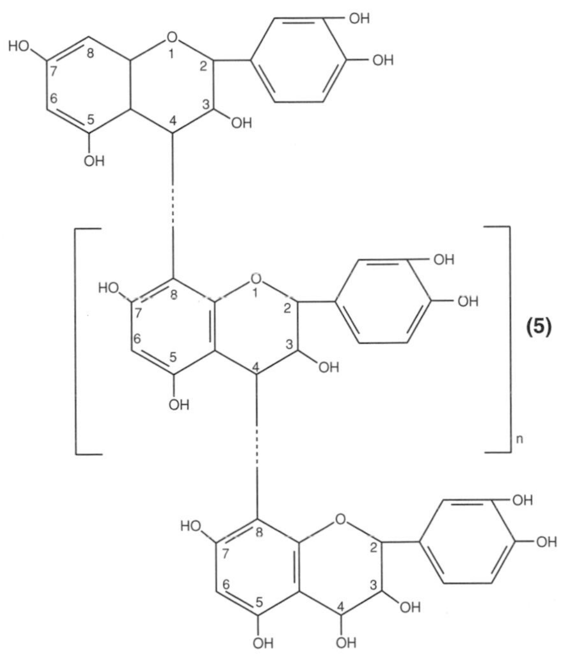

Saponins

Saponins comprise a large family of structurally related compounds containing a steroid or triterpenoid aglycone (sapogenin) linked to one or more oligosaccharide moieties. Despite the fact that saponins are widely distributed in the plant kingdom, only a small number of such plants are actually toxic to mammals. It has long been recognized, however, that saponins, such as the glycoside of medicagenic acid ((**9**) **Figure 7**), which is found in alfalfa, is considered to have an adverse effect on the productive performance of nonruminant animals such as swine and poultry.

Figure 5 Structure of condensed tannin (5), which, by forming a complex with dietary protein, interferes with the digestion of protein.

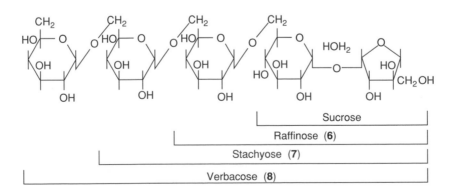

Figure 6 Structure of the oligosaccharides (6, 7, and 8) responsible for flatulence. In the absence of enzymes in the small intestine of humans that are capable of hydrolyzing α-1,6-galactosidic likages, these compounds are metabolized by the microflora of the gut into gaseous end products.

The negative effect of the saponins can be reversed, however, by the inclusion of dietary cholesterol, which interferes with the absorption of saponins by forming an insoluble complex with saponins. By the same token, dietary saponins may exert a positive effect by reducing cholesterol levels in the tissue and serum of experimental animals. (*See* **Saponins**.)

Mimosine

Leucaena leucephela is a tropical legume used primarily as a forage crop for feeding livestock, but its use is limited by the fact that it contains an unusual amino acid, mimosine ((**10**) **Figure 8**). This compound has an adverse effect on the growth of ruminants because bacteria can convert mimosine to 3,4-dihydroxypyridine ((**11**) **Figure 8**) which acts as a goitrogenic agent. In Hawaii, ruminants can convert greater amounts of the *Leucaena* before it becomes toxic than ruminants in Australia because of the presence of bacteria in the rumen capable of detoxifying mimosine. In fact, when Australian ruminants were inoculated with these organisms, they became more resistant to the toxic effects of mimosine.

Lathyrism

Human consumption of the legume chickling vetch (*Lathyrus sativus*) causes a neurological condition known as 'lathyrism,' a disease that is common in India and Bangladesh. The causative factor of this disease has been identified as β-*N*-oxalyl-α, β-diaminopropionic acid ((**12**) **Figure 9**), but over 90% of this toxin can be eliminated by the simple expedient of soaking the seeds overnight in an excess of water followed by steaming, roasting, or sun drying.

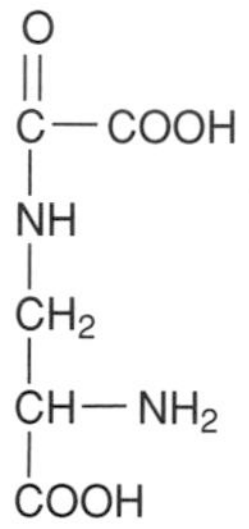

Figure 9 The structure of β-*N*-oxalyl-α,β-diaminopropionic (12), the causative factor of lathyrism in humans.

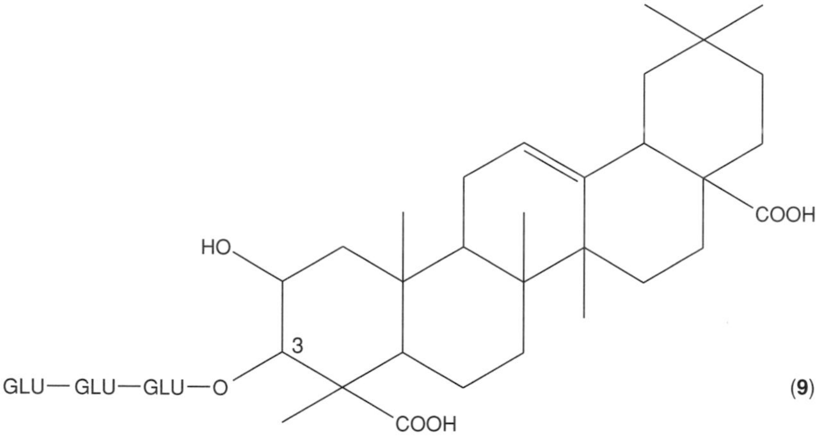

Figure 7 Structure of medicagenic acid (9), a saponin that is present in alfalfa and is responsible for toxic effects in livestock.

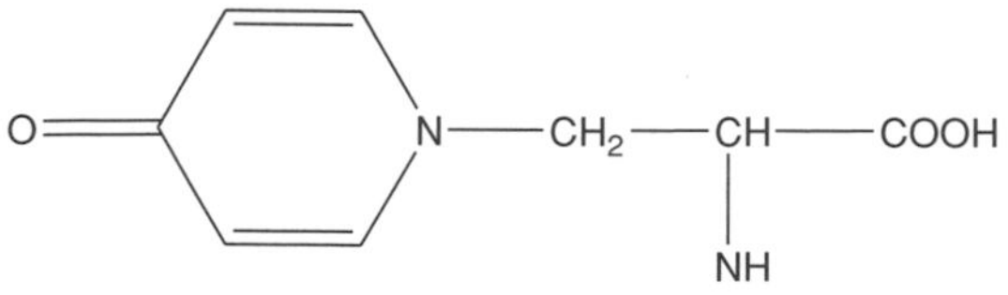

Mimosine
(10)

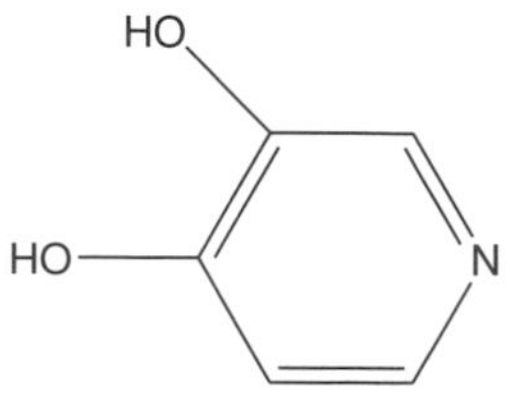

3,4-Dihydroxypyridine
(11)

Figure 8 Structure of mimosine (10) and its goitrogenic metabolite, 3,4-dihydroxypyridine (11).

See also: **Glucosinolates**; **Phospholipids**: Properties and Occurrence; **Phytic Acid**: Properties and Determination; Nutritional Impact; **Saponins**; **Soy (Soya) Beans**: Properties and Analysis; Dietary Importance; **Tannins and Polyphenols**; **Plant Antinutritional Factors**: Characteristics

Further Reading

Cheeke PR (ed.) (1989) *Toxicants of Plant Origin*, vols. I–IV. Boca Raton, FL: CRC Press.

Douglas MW, Parsons CM and Hymowitz T (1999) Nutritional evaluation of lectin-free soybeans for poultry. *Poultry Science* 78: 91–95.

Jones RJ and Megarrity RG (1986) Successful transfer of DHP degrading bacteria from Hawaiian goats to Australian ruminants to overcome the toxicity of *Leucaena*. *Australian Veterinary Research Journal* 63: 259–261.

Legras JL, Jory M, Arnaud A and Galzy P (1990) Detoxification of cassava pulp using *Brevibacterium* sp. R312. *Applied Microbiology and Biotechnology* 33: 529–533.

Liener IE (ed.) (1980) *Toxic Constituents of Plant Foodstuffs*, 2nd edn. New York: Academic Press.

Liener IE (1987) Detoxifying enzymes. In: King RD, Cheetham PSJ (eds) *Food Biotechnology*, pp. 249–271, vol. 1. London: Elsevier.

Liener IE (1989) Control of antinutritional and toxic factors in oilseeds and legumes. In: Lucas EW, Erikson DR and Nip W-K (eds) *Food Uses of Whole Oil and Protein Seeds*, pp. 249–271. Champaign, IL: American Oil Chemists Society.

Liener IE (1997) Plant lectins: properties, nutritional significance, and function. In: Shahidi F (ed.) *Antinutrients and Phytochemicals in Food*, ACS Symposium Series 662, pp. 31–43. Washington, DC, American Chemical Society.

Majak W (1992) Mammalian metabolism of toxic glycosides from plants. *Journal of Toxicology* 11: 1–40.

Marfo EK, Simpson BK, Idou JS and Oke OL (1990) Effect of local food processing on phytate levels in cassava, cocoyam, maize, sorghum, rice, cowpea and soybean. *Journal of Agricultural and Food Chemistry* 38: 1580–1585.

Moneam NMA (1990) Effect of presoaking on faba bean enzyme inhibitors and polyphenols after cooking. *Journal of Agricultural and Food Chemistry* 36: 1479–1482.

Sessa DJ and Ghoutous PE (1987) Chemical inactivation of soybean trypsin inhibitors. *Journal of American Oil Chemists Society* 64: 1682–1687.

Watson D (ed.) (1998) *Natural Toxicants in Food*. Boca Raton, FL: CRC Press.

PLANT DESIGN

Contents
Basic Principles
Designing for Hygienic Operation
Process Control and Automation

Basic Principles

H L M Lelieveld and G J Curiel, Unilever Research and Development Vlaardingen, Vlaardingen, The Netherlands

Background

If a plant is not of a proper hygienic design, it will be difficult to clean, and if not clean, the inactivation of microorganisms can be extremely difficult. After inadequate cleaning, harmful microorganisms may grow and affect product quality and safety. This chapter summarizes the basic principles of hygienic plant design to control the microbiological quality and safety of the final product.

Raw materials usually contain microorganisms. When given time and the right environment, they can grow out and increase to very high numbers. In food-processing equipment, the temperature and availability of nutrients are often ideal for the growth of many types of microorganisms. Consequently, when given enough time, such as in areas where the product is stagnant or residing for a long time for any other reason, microorganisms will cause problems. Product that is free from (viable) microorganisms may become recontaminated by microorganisms in the environment (air, people, insects, and other pests), by the processing and packaging equipment and by the packaging materials. Equipment must be hygienically designed and fabricated, installed in a sufficiently hygienic environment, and be operated by staff who consistently observe the critical hygiene rules that apply.

Requirements for Hygienic Food Plants

Location and Buildings Exterior

Preferably, food plants should not be located near areas where levels of microorganisms, insects, or rodents can be exceptionally high. Therefore, a food plant should not be near a sewage treatment plant or in the middle of an animal farm area. Also, although decorative, there should be no trees, bushes, and plants near the factory. Such flora attracts insects and birds, which are sources of all kinds of pathogenic microorganisms. Also, there should be no breeding places for insects, so there should be no decorative ponds and generally no stagnant water at all in the vicinity. To avoid attracting insects, external lighting should never be mounted to, or near to, the wall of the plant. If positioned at a suitable distance, illumination will be adequate while insects move away from the plant. The outer walls of factories should not provide landing or breeding places for birds or other animals. The lower part of the walls must be rodent-resistant. Also, walls should be deep enough to prevent rodents from digging tunnels underneath to gain access to the plant. Measures must be taken to prevent animals gaining access to any area between the ceiling and roof. Roofs should not retain water (breeding of insects and bacteria) and thus slope such that draining is adequate. Doors must be self-closing; windows and doors that are opened infrequently, including emergency exits, should be such that there is little space between door and frame.

Layout

As the requirements for raw materials differ from those applying to intermediate or final products, the design of food plants should start with defining the hygiene requirements for the product in the subsequent stages of the production process. Based on that inventory, the layout of the plant should be decided, taking into account the various flows: of product, packing material, people, and air. Air, unless freed from microorganisms, should always move away from the exposed final product. Primary packing materials should be clean (complying with the microbiological requirements for the product to be packed). Neither people nor any material should move from the raw materials area to the finished exposed product area unless measures have been taken to prevent the carryover of microbial contamination. To protect

microbiologically sensitive products, high-care areas may be needed. The entrance to such an area should have facilities for (compulsory) change of footwear and coats. All areas should have well-designed and well-maintained handwash facilities.

It is essential too, that, if the layout is correct, staff do not nullify the effect of correct measures by not knowing or not following the rules. For example, one is easily tempted to open doors and windows on a hot summer day, thereby completely destroying the airflows needed to prevent product contamination. Also, if possible and unobserved, people are likely to take 'shortcuts' to move from one place to another and thereby entering high-care areas from raw material areas. It is part of correct plant design to ensure that the rules are correct and sensible and that everybody is aware of their importance – and perhaps of the sanctions in case of breach: the health of consumers and the reputation of the factory are at stake.

Construction

Floors Floors are one of the great challenges: on the one hand, for obvious reasons, they should not be slippery. On the other hand, because they must be cleaned, floors must be smooth. In the case of dry floors, the problem is relatively easy to control, but any spilled liquid (water or product) may present an occupational hazard. The hazard selfevidently is related to both floor and footwear, and their interaction must be taken into account. Clearly, a 'dry-floor policy' will help to reduce both hygiene and occupational problems. Smooth concrete floors are acceptable for reception and many raw material storage areas. In areas where product is not protected, ceramic tiles or seamless mortar resin floors are preferred. In high-care areas, often, seamless resin floors are the best choice. Floors must have rounded corners where the floor meets the wall and should slope approximately 2% towards drains or gutters.

Walls Walls should be smooth and cleanable and should be covered by ceramic tiles or a good-quality coating or paint. If sandwich panels are used, all sides should comply with the chemical resistance required. It is important that all seams are reliably sealed. In some cases, seams can be sealed by welding with the parent material. In other cases, and where walls meet the floor, sealants should be used. Also, anything mounted to the wall should be sealed all around to prevent access to insects and dirt. The surface of choice should be strong enough to meet factory conditions, and repairs to any damage should be carried out within a short time. Where walls separate various hygiene areas, care must be taken that any holes for service lines are properly sealed.

Ceilings Panels similar to those used for walls (see above) may be used, and the same requirements apply. Lighting should be preferably built into the ceiling, with the underside flush with the ceiling. All suspended items, including lights at lower levels and exposed beams must be such that the upper side is sloped sufficiently to avoid the accumulation of dust and dirt. Ceilings and anything suspended from them must be easy to clean. Where acoustic ceilings are needed, it should be noted that sound-absorbing panels often cannot be cleaned and that panels that are easily cleanable usually are excellent sound reflectors. Consequently, a compromise will have to be found. Whenever possible, try to avoid noisy equipment in the plant, and when unavoidable, try to contain the noise near its location, where it may be easier to inspect for cleanliness.

Service lines and air ducts Service lines (hot and cold water, drains, electricity, compressed air) may harbor insects and other pests. Preferably, they should be combined in suitable, thus externally, smooth ducts, with all ends effectively sealed against vermin. Where mounted to a flat surface, sealant may be needed to close recesses.

Air ducts as well as cold-water pipes and drains, owing to their lower temperature, often may collect moisture, resulting in drops falling down on to product and food contact surfaces, and these may also support microbial growth. Hence, in areas where this will happen, cold surfaces must be insulated. The insulation should be sealed, sealing requirements being the same as for the ducts mentioned above. If the conditioned air is too humid, water may collect within the air ducts, allowing microbial growth, and so, the air becomes a significant source of contamination. This should be taken into account when designing the air system in the plant.

Equipment

Cleaning Cleanability is very important. Equipment that is not cleanable cannot be decontaminated other than with great difficulty or not at all.

Inactivation of microorganisms It must be possible to treat all product contact surfaces with whatever means has been chosen to inactivate the relevant microorganisms. Where the equipment is intended for pasteurized products, the equipment should preferably be 'pasteurizable.' In other words, it must be ascertained that all surfaces come in contact with water of the correct temperature for the required time. Entrapped air, e.g., in the top of tanks, will hamper pasteurization. Equipment for sterilized products must be made sterile. Steam sterilization is preferable, but again, there are pitfalls: in this case, lower parts of equipment may collect condense and thereby not be treated with steam. The considerably lower degree of heat transfer of water may reduce the temperature too much at such a spot. Where equipment is not sufficiently heat-resistant, chemical means may be used. Also, here, air entrapment must be prevented to ensure treatment of all surfaces.

Migration of microorganisms – aseptic processing and packaging A hygienic plant should protect the product from contamination. The ingress of microorganisms should be limited and, in the case of aseptic processing, fully prevented. For so-called closed plants, microorganisms may migrate to the product through crevices, leaking connections and moving parts of machinery, such as valves and pumps. Prevention is achieved by correct mounting and sealing of parts and by the application of barriers where movable parts are unavoidable. When choosing equipment, it should be realized that reciprocating shafts facilitate the ingress of microorganisms: the shaft transports small amounts of product to the outside, providing nutrients to any microbes present there, and every other stroke, microorganisms are transported to the inside. Rotary shafts display a similar effect, but to a greatly reduced extent. Where ingress must be prevented completely, such as for aseptic processing and packaging, diaphragms should be used to separate the product from the environment. Where this is not possible, such as with rotating shafts, double seals should be used, with an antimicrobial liquid circulating between them.

Growth Where, for functional reasons, such as blending (nonsterile), food products reside in equipment for a long time, measures must be taken to prevent excessive growth. This can be done by keeping the product below or above temperatures suitable for growth of the relevant microorganisms. Where this is done, it must be taken into account that there will be warm spots, e.g., where a shaft enters the equipment (centrifugal pump, rotary stirrer). Where the residence time is too long, because of the presence of avoidable dead areas in equipment, the equipment must be considered unhygienic. Such equipment should not be used for food processing, unless it can be cleaned easily and is used for production runs that are too short to cause microbial problems.

Principle hygienic design requirements for equipment Although hygiene is essential, the measures taken to ensure hygiene should not generate other health problems. Therefore, the principle requirements for the design of hygienic equipment also must take other aspects into account.

The basic principles are as follows:

- To cope with the aforementioned issues, food-processing equipment should meet the following requirements to be acceptable: all product contact surfaces must be noncorrosive, i.e., resistant to the product and chemicals under the operating and cleaning conditions. Furthermore, materials used (basic construction materials, but also auxiliary materials such as coatings, lubricants, sealants, adhesives, and fillers) should not have the potential of contaminating the food product such that the food becomes toxic or allergic.

- All product contact surfaces must be free from crevices and other irregularities that would be difficult to clean. Thus, no metal-to-metal connections (other than by welding or soldering) are allowed, and screw threads must be avoided or otherwise sealed.

- Product contact surfaces must be smooth enough for easy cleaning. Both the Hygienic Engineering and Design Group and 3-A prescribe a surface roughness for product contact surfaces of $R_a = 0.8 \, \mu m$ or less.

- All product contact surfaces must be drainable to make it possible to remove all liquids from process equipment. Thus, surfaces and pipes should not be completely horizontal but slope towards drain points, and there should be no ridges, which may hamper draining. Where it is not possible to build equipment in such a way that proper draining is possible, procedures must be designed to ensure that residues of cleaning and disinfection liquids can be removed in another way. Even if no chemicals are used for cleaning, draining is important because many microorganisms can grow easily in residual water, needing only tiny amounts of nutrients.

- Dead legs and product stagnant areas must be avoided or, where unavoidable, should be designed to minimize the residence time. The presence of such areas should be taken into account when deciding on the frequency and procedure of cleaning. Dead legs of tees, if not too long, may be reasonably easy to clean if the direction of the flow of cleaning liquid is towards the dead leg. Special attention should be given to provisions for temporary connections, such as those used for filling transportable containers, road tankers, etc. If not used for some time and not cleaned, they present a large 'dead' area, very much alive with microorganisms.

Fabrication and installation   The fabrication must be such that product contact surfaces are not damaged to adversely affect the cleanability. Welds should not be rough and, where necessary, should be polished to obtain a smooth finish. Welding should be done skillfully, as welds otherwise may soon corrode. This is particularly important for welds that cannot be seen anymore after assembly, such as with pipelines. Pipe-bending can be applied to preclude the need for too many welds and too many piping components. In general, the risk of contamination of food product is proportional to the number of microorganisms in the surrounding, and so the nonproduct side of equipment also must be accessible for cleaning. External features should not collect water, product, or dirt in areas that cannot be inspected and cleaned and thus provide places for harboring pests. That means that there must be either no space at all or an insufficiently large space between parts of machinery. The same applies to the space between machinery and the supporting structure, i.e., the floor, wall, ceiling, or any other support.

The installation of components should be such that the assembly is also hygienic. Hygienic components may be built together in such a way that the completed installation is not hygienic because the assembly cannot be drained or cleaned, and creates dead areas. Connections may result in crevices, components may cause obstructions for draining, and in a closed process line, a hygienic diaphragm pressure sensor on an easily cleanable tee may create a large dead area. The food-processing manager should not only decide on hygienic equipment but also keep a close eye on how the contractors intend to build the entire process line. In addition, care should be taken that overhead lines and components are not a source of contamination. Vertical piping is preferred, and horizontal pipework should slope to drains.

Sources of Further Information

There are a number of professional organizations that provide detailed information on the hygienic aspects of plant design. The EHEDG is an organization with about 25 subgroups specializing in the many different aspects of hygienic design and processing. The organization is supported financially by the European Commission to support the development of guidelines and to ensure their availability in many European languages. Currently, over 20 guidelines are available, covering subjects such as hygienic design criteria, the design of hygienic equipment for open and closed plants, designing hygienic pumps and valves, microbiologically safe pasteurization and sterilization of liquid products, hygienic as well as aseptic packaging of food products, and welding to meet hygienic requirements. In addition, the EHEDG has developed methods to determine whether equipment meets the hygienic design criteria. Equipment complying with the EHEDG criteria may be certified. Much of EHEDG's guidelines can be found in the standards for machinery developed by the European Standardization Organization (CEN, Commission Européenne de Normalisation). Details can be found at www.ehedg.org. In the USA, the 3-A organization develops standards for food-processing equipment. Currently, 3-A provides a self-certification scheme for equipment complying with the 3-A standards, which may result in authorization to use the 3-A symbol on equipment. Information on 3-A can be found at www.3-A.org. Also, the USA-based NSF (National Sanitation Foundation) International produces standards for hygienic equipment, but traditionally focused on the catering area. NSF's website can be found at www.nsf.org. The three organizations have recognized the importance of harmonization between the various regions in the world and increasingly work together. NSF International has an office in Brussels, and one of EHEDG's regional sections is in Japan. EHEDG guidelines are produced with the assistance of 3-A and NSF, and 3-A standards are produced with assistance from EHEDG. Recently, NSF and 3-A have started to produce joint standards. The organizations also all work closely together with national, regional, and global authorities.

Concluding Remarks

It is important to ensure that the plant or process line will deliver safe products. Validation of the cleanability and the disinfectability and, where relevant, imperviousness to microorganisms are essential in the selection and commissioning steps. Not only must the food be safe just after commissioning the plant, but the plant or installations must continue to comply with the hygienic requirements. Therefore, an appropriate scheme for preventive maintenance should be devised.

See also: **Cleaning Procedures in the Factory**: Types of Detergent; Types of Disinfectant; Overall Approach; Modern Systems; **Plant Design**: Designing for Hygienic Operation; Process Control and Automation

Further Reading

3-A Sanitary Standards Committees (1999) *Model Document for Preparing 3-A Sanitary Standards and 3-A Accepted Practices*, 2nd edn. McLean, VA: Dairy & Food Industries Supply Association.

Curiel GJ, Hauser G, Peschel P *et al.* (1993) *Hygienic Equipment Design Criteria, EHEDG Document No. 8.* Chipping Campden, UK: Campden & Chorleywood Food Research Association.

Imholte TJ (ed.) (1984) *Engineering for Food Safety and Sanitation.* Crystal, MN: Technical Institute of Food Safety.

Lelieveld HLM, Mostert MA, White B and Holah JT (eds) (2002) *Hygiene in Food Processing* Cambridge, UK: Woodhead.

Shapton DA and Shapton NF (eds) (1991) *Principles and Practices for the Safe Processing of Foods*, Cambridge, UK: Woodhead.

Designing for Hygienic Operation

J P Clark, A. Epstein and Sons International Inc., Chicago, IL, USA

This article is reproduced from *Encyclopedia of Food Science, Food Technology and Nutrition*, Copyright 1993, Academic Press.

Background

As with facility design, discussed in the previous article, process and equipment design has many elements specific to the food being processed and others that represent common principles. This article discusses a number of those common principles, types of equipment, and common processes.

Overall Considerations

Equipment used in food-processing plants must be appropriate, effective, noncontaminating, easy to clean, and easy to inspect. It must also be safe for the workers using it. In the USA, equipment used in meat, poultry, and fish processing must be specifically approved in advance by the US Department of Agriculture (USDA). Equipment used in dairy processing should be listed in 3A standards (a voluntary industry standard-setting group), and equipment used in baking should satisfy Baking Industry Sanitation Standards Committee requirements. There are other standard-setting groups in other countries and for other industries.

In general, such standards address the aforementioned general requirements in light of the specific conditions. For example, the USDA emphasizes cleaning and inspection, because most meat processing equipment is cleaned daily. Dairy equipment is often cleaned in place, and so surface finish and absence of crevices are important. Baking equipment is less often cleaned with water, but is vulnerable to insect infestation. (*See* **Cleaning Procedures in the Factory**: Overall Approach; **Insect Pests**: Insects and Related Pests.)

As a result of nearly universal sanitation requirements, much food equipment is constructed of stainless steel, usually type 304, but often 316 and occasionally some other alloy. The alloys differ in their weldability and resistance to stress corrosion cracking, which is accelerated by chloride ions found in many foods and cleaning materials. Stainless steels are nonmagnetic, and so fragments that may occur as a result of wear or misuse are not removed by magnets. Metal detectors, which function by measuring changes in inductance, can detect stainless steels and are often used on final packages of foods. (*See* **Corrosion Chemistry.**)

Many polymers are acceptable for food use, especially Teflon, Neoprene, polyethylene, and polypropylene. Carbon steel is acceptable where conditions will not promote corrosion and contamination. Copper is used occasionally, for special purposes such as candy cooking.

Equipment designed for food processing can usually be disassembled with few tools and uses wide screw-threads or flanged connections to avoid creation of places where residues can be caught. Wherever possible, pipelines are welded, taking care to make inside surfaces of welds smooth and flush. Long runs of pipe or tubing must have inspection access and must be installed carefully so that drainage is complete. Dead spots are avoided in pipes so that systems can be cleaned thoroughly.

Clean-in-place

The concept of cleaning in place by flowing cleaning solutions and rinse water through a system in place of food liquids has significantly increased the productivity of food plants. Previous to its wide use, dairies and other food plants were disassembled and washed completely by hand. This consumed so much time and labor that it limited the practical size of plants.

The elements of a typical clean-in-place (CIP) system are chemical storage, solution supply and recovery tanks, supply and return pumps (usually centrifugal), spray devices in tanks, air-controlled automatic valves, manual connection stations (flowverters), additional piping to complete circuits, and automatic control devices such as programmable logic controllers. Empirical design rules have evolved from experience for flow velocities, spray intensities and solution strengths for effective cleaning of typical soils from food plants.

Most experience is with dairy and other liquid-product plants, such as soft drink, fruit juice, and syrup plants. Minimum velocities of $1.5 \, \mathrm{m \, s^{-1}}$ are generally required in pipelines to achieve sufficient turbulence. Tees and dead legs should not exceed

three pipe diameters' length from the flowing stream so as to ensure cleaning. Tanks are sprayed at about 37 liters per minute per meter of circumference. Often, such flow rates surpass the normal process flow rates, and so special pumps may be needed just for CIP. Sometimes, it may be possible to use dual-speed process pumps, with a low speed for process flow and a higher speed for CIP. Such a technique saves investment cost and space.

It is common to design CIP systems so that cleaning solutions and final rinse water are recovered for reuse, to save costs and reduce discharges. Caustic soda solutions (1.5%) and nitric, phosphoric, and citric acids are common cleaning agents. Sometimes, detergents are used, alone or with acids or bases. In addition, sanitizing agents, such as iodophores, quaternary ammonium compounds, and chlorine-releasing agents, may be used, usually as a final rinse that is allowed to sit overnight. In the morning, the sanitizing agent is displaced with fresh water and then with product. (*See* **Cleaning Procedures in the Factory**: Types of Detergent; Types of Disinfectant; Modern Systems.)

With a properly designed CIP system, most of the plant waste water should be discharged during cleaning from the CIP unit. This permits monitoring and pretreatment, if necessary.

Some manual labor is involved in cleaning even highly automated systems. Positive-displacement pumps must be disassembled, bypassed, and cleaned by hand because their construction prevents complete cleaning by flushing. It is common practice to require manual connections at flowverters to prevent accidental contamination of food with cleaning solutions. Connections are commonly verified by electronic signals to the controller.

It is difficult to separate CIP design from process system design in many cases; they are best developed together. Physical arrangements, details of nozzles on tanks, valve placement, pump utilization, pipe runs, and other details are significantly affected by CIP.

Food Plant Layout

Overall process flow is dictated in part by location of shipping and receiving and also by constraints on equipment. For example, cookie and cracker baking ovens are as long as 100 m and must run in a straight line, because they use metal bands as baking surfaces, and partially baked pieces cannot be transferred easily. Hydrostatic sterilizers for cans and jars are usually located on outside walls for ease of access and maintenance. Cooling tunnels for chocolate enrobing are usually long and straight.

Dry equipment and processes should not be physically adjacent to wet ones; cold process areas should be separated from hot ones; 'dirty' areas should be separated from clean ones, and so forth. Such separations are not merely physical, in the sense of walls and doors, but also include separate air-handling units, and perhaps limited access by people from other areas.

Vehicle traffic should be separate from people traffic. Often, it is desirable to keep visitors, who may be customers, off the production floor, in part for their own safety, but also to reduce exposure of food to humans. At the same time, if visitors are likely, it is desirable to offer a decent and controlled view of the process. One solution is an elevated, enclosed walkway with viewing windows. If the process side of the windows is refrigerated, as many food plants are, the windows need to be heated to prevent condensation.

Plant layout can encourage or inhibit communication among workers; with fewer people running modern food processes, good communication is essential. A serpentine layout may be helpful, compared to a long, straight arrangement. Clever use of closed-circuit television can overcome distance and obstructions.

Typical Unit Operations

Some unit operations are so common and important to food processing that they deserve individual discussion.

Mixing

Processes of mixing solids with solids, solids with liquids, and liquids with liquids are all quite common in food processing. For hygienic operation, mixers must be designed to empty completely, to be easily cleaned, and to be easily inspected. Access for addition of minor ingredients and for sampling may also be necessary. Doughs and pastes, such as bread dough, cookie batter, and confection fondants are prepared in specially designed machines with relatively high-powered motors.

Incorporation of gas, especially of air, may be desirable or undesirable for a given case; the mixer design changes accordingly. (High speed, giving high shear, leads to gas incorporation.)

Dispersion of powders into liquids, such as starch or gums into water or syrup, requires high-speed agitation to avoid the formation of partially wet agglomerates, called 'fish eyes.' One technique is to pour solids into a deliberately inefficient centrifugal pump, which is circulating liquid from and to a tank. The inefficiency means that energy is applied to dispersing the solids rather than pumping the liquid.

Dissimilar liquids, such as oil and water, do not form solutions but can be made to form stable suspensions or emulsions by formation of very small droplets of oil and by using surfactants. The small droplets are formed by high-shear agitation and by pumping under high pressure through small orifices or special valves in a homogenizer. Many foods contain natural surfactants, such as proteins and polysaccharides, which reduce surface tension at the interfaces between liquids and so help maintain an emulsion. Milk, icecream and salad dressings are examples of foods containing oils and aqueous solutions in stable suspension. (*See* **Colloids and Emulsions.**)

Heat Transfer

Heating and cooling of foods is critical to many preservation and processing techniques, including pasteurization, freezing, baking, and other types of cooking. One of the more common heat exchangers used in food processing is the plate type. This has dimpled or otherwise embossed plates separated by gaskets and held together by mechanical pressure on a sturdy frame. Hot and cool fluids pass on opposite sides of the plates, providing a large heat transfer surface in a rather compact space. The major advantage of the plate exchanger for food service is the ease with which it can be disassembled for inspection and maintenance. It can also be reconfigured easily by adding or removing plates on the same frame (up to the limit of the frame). (*See* **Heat Transfer Methods.**)

Shell and tube, concentric tube, and spiral tube heat exchangers are also found in food plants. Direct steam injection is also common.

Cooling and freezing are performed in several ways. Wiped-surface heat exchangers are common for freezing flowable fluids, such as icecreams, and for cooling fluids that have been heated for sterilization. Refrigerated plates are used for contact freezing of packages of food, such as vegetables or preplated dinners. Direct contact with very cold vapors or snow from liquid nitrogen or carbon dioxide is used in cryogenic freezing, usually on continuous belts passing through tunnels. Air cooled by refrigeration can also be used in similar tunnels. Cryogenic equipment is less expensive than mechanical refrigeration equipment but is more expensive to operate. The exact balance depends on local costs, but cryogenics are a good choice when minimizing capital is key; mechanical freezing is more common for established, ongoing processes, where operating costs are more critical. (*See* **Freezing**: Principles; Operations; Blast and Plate Freezing; Cryogenic Freezing.)

Steam heating of food in sealed containers, such as cans, jars, and pouches, is commonly performed in pressure vessels known as retorts. Batch retorts may be horizontal or vertical, and may have water in addition to steam or steam alone for heating. For containers such as jars and pouches, which cannot tolerate internal pressure, air is added during the cooling phase, which uses water, to counter the internal pressure built up during cooking.

There are several approaches to continuous cooking of foods in containers. One of the more efficient is the hydrostatic retort, in which containers are transported in carriers on a chain through a 'U'-shaped tower in which the vertical leg of water serves as a seal to contain a high-pressure chamber. (In both batch and continuous retorts, pressures to reach 120 °C are used so as to shorten the time required to sterilize food without giving it overcooked taste and texture.) (*See* **Sterilization of Foods.**)

Another approach uses a helical track inside a horizontal cylindrical shell. The track rotates, transporting metal cans. Cans enter and leave through a star wheel valve. A similar shell and track is used to cool the cans under pressure.

Conveying

Many foods are solids or are made from solids, and so solids material handling is critical to food processing. Pneumatic conveying, by pressurized air or vacuum, is commonly used to move flour, sugar, and grains in food plants. Pressure conveying permits movement from one source to several delivery points, such as flour to several dough mixers. Vacuum conveying works well for several sources to one delivery point, as in dust collecting. It is also often used to unload bulk material delivery vehicles, such as rail cars or trucks. Dense-phase pneumatic conveying is a variation used for fragile materials, such as sugar, which might be damaged by the high velocities encountered in pressure or vacuum conveying. Typically, for dense-phase conveying, powders are blown from a holding vessel in slugs rather than being entrained in conveying gas.

Screw conveyors are used to move pasty or granular materials, often up inclines. They can be difficult to clean and are very dangerous, but for some purposes, they are uniquely appropriate.

Belt conveyors can be used to move loose food materials, but are best for packages and containers. They can also pose safety hazards, with numerous pinch points and motors, which start and stop under remote control.

Vibratory conveyors are well suited to many food applications, especially for fragile materials such as fried snacks, because they can be quite gentle and are easily cleaned.

No matter what solids conveying system is used, it can rarely be properly designed without specific measurement of the physical properties of the actual materials involved. Solids used in food plants vary widely in the properties that influence conveying and handling, and these properties are difficult to predict from theory or correlations. Examples of key properties include bulk density, particle size distribution, angle of repose, cohesiveness, abrasiveness and moisture content.

Pumping

Many foods are quite viscous and non-Newtonian, so proper design of pumps is challenging. Positive-displacement pumps, especially those using close-fitting lobes and screw-shaped rotors, are common. Such pumps are usually made of stainless steel and are designed for easy disassembly. Centrifugal pumps are more conventional but also likely to be stainless steel and built to be taken apart by hand with few tools.

A special case of pumping, which combines the functions of mixing and heating, is screw extrusion. Extruders may be single or double screw, with heating or cooling in jackets, and with a wide variety of die designs. When starchy materials, such as corn, are extruded under pressure through a die, an expanded foam is formed, which has a desirable texture and density. Pet foods, cereals, and snacks are made in large quantities in this way. (*See* **Extrusion Cooking:** Principles and Practice.)

Typical Food Processes

Many food processes have a number of steps in common, including mixing of ingredients, forming of pieces or shapes, filling of containers or packages, cooking and cooling, packaging for shipment, and unitizing. The specific details obviously depend on the product and circumstances. However, some issues are almost universal.

Formulation

In order to be mixed, a formula must be weighed or measured; this may involved feeding and scaling of particulate solids, metering of liquids and addition of small quantities of key ingredients. Most formulas for foods contain three categories of ingredients: major (percentages in the tens), minor (single-digit percentages), and micro (less than 1%). (Even these distinctions may vary from case to case.) It is common to prepare formulas in batches of several thousand kilograms, even if subsequent processing is continuous. Issues that arise include whether to use one scale, which must then be capable of weighing the entire batch, or multiple scales, sized for accuracy of each category. The cycle time of batching is longer if fewer scales are used, but the costs are lower. Liquid ingredients can be metered using any of several devices.

Sometimes, ingredients are simply added by units, such as entire bags or drums, but this is less precise than scaling, because usually, such units are slightly heavier than their label weight.

Critical issues in any automatic formulating system are reliable feeding and flow of highly variable solids; even such common commodities as flour and sugar can clump, cake, and bridge in hoppers and chutes, disrupting flow and upsetting operations.

Forming

Forming operations are highly specific, but can include such examples as flaking, extrusion, sheeting (rolling into thin layers), cutting, laminating, decorating, dicing, and many others. Usually, these operations are continuous and so need careful metering of a mix, which may have difficult flow properties (doughs or pastes, for instance). Forming operations usually work best when operated at a smooth and constant rate, so the surge capacity ahead of the operation and reliable removal of output are important.

Filling

Primary containers may be filled before or after such processing as cooking, baking, or sterilization. If the material is liquid or a suspension, it can be filled volumetrically; if it is free flowing solid, it may be filled volumetrically, or it may be scaled. Weighing of valuable particulate solids is often done with multi-compartment digital scales, in which a computer scans compartments to select several whose summed weight most closely approaches the target. Such scales are very accurate and justify their considerable cost by reducing overweight packages.

It is common for food packages, such as cans of soup, to be filled in several successive steps, with the final 'mixing' occurring in the package. Chicken noodle soup, for instance, has noodles, chicken, and broth each added separately using specialized fillers.

Cooking

The meaning of cooking varies from food to food but almost always involves some heating, directly or indirectly, to cause desirable chemical reactions, remove excess moisture, develop color (usually browning), and reduce microbial populations. Heat may be applied by flames, steam, hot air, oil, microwaves, or hot water. After heating, cooling often occurs by exposure to ambient air but may be accelerated by refrigeration. Cooking almost always involves more than one reaction, some of which may be undesirable, so control is complex. In frying, for example, water is removed and largely replaced by fat, the volume increases, and browning occurs.

(*See* **Browning**: Nonenzymatic; **Cooking**: Domestic Techniques.)

Packaging

In addition to a primary package, which may be filled before or after cooking, there are usually additional layers of package to provide additional protection during storage and shipping. These may include cartons and cases. Cartons are usually made of heavy paper and are part of the consumer package, so usually have multicolor printing of brand names, ingredients, use instructions, weight, and nutritional information. Shipping cases are usually removed in a store, are rarely seen by consumers, and so can be made of sturdy corrugated cardboard with minimum printing and graphics.

Plastics are increasingly important as components of food packaging, because they are lighter in weight than metal or glass, can be used in microwave ovens to reheat contents, and do not break as easily as glass. Plastic containers usually cost more than the metal or glass containers they replace but are perceived to add value to the consumer. Recycling of food containers is an increasingly important concern to consumers and complicates the selection of packaging material. Some of the most effective plastic containers are composed of several types of plastic, each providing particular properties, such as barriers to oxygen or strength. Such multicomponent materials are less easily recycled than single-component materials.

Unitizing

Shipping cases of food are usually assembled into larger units, often of about 450 kg on wooden pallets, for storage and shipment. These pallet loads may be wrapped in stretch plastic, wrapped with heat-shrinkable plastic, strapped, or stabilized by glueing boxes or bundles together. Alternatively, large cardboard sheets known as slip sheets may be used to hold a unit load. Special equipment is needed to pick up and move loads on slip sheets, in contrast to the common forklift truck for which wooden pallets are designed, but the slip sheets are less expensive than pallets and take less space in storage and shipping.

In automatic storage systems and racks, used to achieve higher volumetric density in warehouses, special captive or slave pallets may be used, which are not shipped with the product load. This may require an additional transfer operation.

Some short-shelf-life foods are unitized on special carts, baskets, and trays suited for storage on small delivery trucks for direct shipment to stores. Fluid milk, cultured dairy products, and baked goods are often handled this way.

Analysis of the best packaging and unitizing procedure and equipment is a specialized area of engineering and requires balancing investment costs, operating costs and delivery system costs to achieve the best overall arrangement.

Conclusion

Design of food plants for hygienic and effective operations requires consideration of the interaction of the building with the process, selection of the proper equipment, development of an efficient layout, and consideration of the roles of people, materials, and processes. The goals are safety, quality and cost efficiency. Many special pieces of equipment have been developed to aid in reaching these goals.

See also: **Browning**: Nonenzymatic; **Cleaning Procedures in the Factory**: Types of Detergent; Types of Disinfectant; Overall Approach; Modern Systems; **Colloids and Emulsions**; **Cooking**: Domestic Techniques; **Corrosion Chemistry**; **Extrusion Cooking**: Principles and Practice; **Freezing**: Principles; Operations; Blast and Plate Freezing; Cryogenic Freezing; **Heat Transfer Methods**; **Insect Pests**: Insects and Related Pests

Further Reading

Matz SA (1976) *Snack Food Technology*. Westport, CT: AVI.

Matz SA (1988) *Equipment for Bakers*. McAllen, TX: Pantech International.

Mercier C, Linko P and Harper JM (eds) (1989) *Extrusion Cooking*. St. Paul, MN: American Association of Cereal Chemists.

Sacharow S and Griffin RC (1980) *Principles of Food Packaging*. Westport, Ct: AVI.

Woodroof JG and Phillips GF (1981) *Beverages: Carbonated and Noncarbonated*. Westport, Ct: AVI.

Process Control and Automation

J P Clark, A. Epstein and Sons International Inc., Chicago, IL, USA

This article is reproduced from *Encyclopaedia of Food Science, Food Technology and Nutrition*, Copyright 1993, Academic Press.

Background

Food processes, like other industrial operations, are measured and controlled in order to obtain consistent, safe, and efficient results. Wherever possible, response to measurements is automated, to reduce human labor and the risk of error. Modern processes rely heavily on computers to monitor operations, record data, and generate reports. Increasingly, computing power is in the form of microprocessors, which are dedicated components of instruments, relieving control computers of functions, simplifying installation, and increasing versatility. This article discusses techniques for measuring typical parameters, especially those found most often in food processes, and the ways in which measurements are used to control processes. The food industry has adopted measurement and control techniques from other industrial processes, including computer-integrated manufacturing (CIM), which is discussed briefly at the end of the article.

Typical Parameters

Weight

Weight, usually of solids, but also of mixes, packages, and storage vessels, is one of the most important and common measurements made in a food plant. Mechanical lever arrangements with counterweights and indicating dials are still widely used, but more common is the use of electronic strain gauges or load cells. These have a quick response and give a signal that is converted to a digital signal that can be read directly by a computer. Load cells need to be calibrated and their limitations understood; they can be sensitive to temperature, moisture, and vibration, and they must be selected for the total weight range they will experience.

The combination of an electronic scale with a variable-speed screw or vibrating cone creates a feeder that can be programmed to deliver a given rate or a given total amount of material. Such a feeder may operate by tracking the loss in weight from a hopper or by weighing the receiving bin (gain in weight). Loss in weight is most common. Periodically, the feed bin must be refilled; during this time, the feeder continues to run at the last speed it had. To minimize the time in which this 'volumetric' feeding occurs, refilling is kept to 5–10% of the cycle time. This means that the refill flow rate is 10–20 times the controlled feed rate; for high flow rates, this can be quite substantial and represents a significant shock to the receiving bin. Refilling can also aerate solids and make them hard to control or force them out of the feeder. Thus, bin sizing, delivery system, and flow control are all integral to successful weighing and delivery of solids.

Checkweighers are special scales placed in line with a package conveyor to weigh individual packages. Usually, a checkweigher controls a kick out device to remove over- or underweight packages. Data are also accumulated to calculate averages, standard deviations, and other quality statistics, which are used for statistical process control. This involves identifying process capabilities by measuring performance, then using such data to detect abnormalities and gradually to improve performance. Occasionally, a checkweigher is used to control a filler or other process affecting package weight.

When storage or process tanks and bins are to be weighed, they must be carefully isolated from other supports so that the load cells bear the true weight. This requires flexible connections from delivery pipes and other gear, which may be attached to a typical tank. Sometimes, load cells are installed to permit calibration of other instruments, such as meters, which are regularly used to control formulations.

Flow of Fluids

The traditional method of measuring flow rates of liquids and gases is by measuring pressure drop across an orifice (a small hole in a plate inserted in a pipe). For food fluids, an orifice plate may pose an unacceptable obstruction to suspended solids and may also be hard to clean in place. For homogeneous, clean fluids, this approach is reliable and inexpensive, but it provides a limited turn-down ratio of about 3:1.

Better for measuring homogeneous fluids (liquids or gases) is the vortex meter, which detects small changes in capacitance created as vortices are shed from a bluff body inserted in the fluid path. The vortices, or swirls of fluid, are created and shed in direct proportion to the flow rate. The turn-down ratio can be 10:1 or greater, and the measurement is independent of temperature, density, and viscosity. Installation is simple, and obstruction of the flow path is slight.

An increasingly common technique, which has no flow obstruction and does not require long runs of straight pipe before measurement, is the mass flow meter. A tube is inserted into a pipe as part of the flow path. The tube is vibrated at a resonant frequency. As fluid flows through the vibrating tube, it changes the phase difference of vibration at two points on the tube (or it imparts a twist to a bent tube), which can be detected and converted to a signal proportional to the flow rate. Magnetic or optical detectors measure the location of the tube, and electronic devices convert the signals to flow rates.

Other flow-rate-measuring devices count turns of rotors, vanes, or some other signal of flow rate. For electrically conducting fluids, changes in a magnetic field due to flow can be measured without any obstruction to the flow path. Positive-displacement meters use the flow directly to generate a signal by driving a lobe or piston connected to a shaft; rotations are proportional

to flow. If a measurement is used to control flow, it usually signals an air or electrically operated valve, but it could signal a variable-speed drive on a pump.

Temperature

Temperature is a critical measurement in many food processes because it controls sterilization and cooking operations; too high or too low a temperature may be harmful. Thermocouples are inexpensive and reliable. They generate a small but consistent voltage difference at the junction of dissimilar metals, which can be correlated with temperature.

Most common for accurate measurement over larger ranges are resistance temperature detectors, which use the change in resistance of platinum with temperature to measure temperature. A common standard is to set the resistance at 100 Ω for 0 °C.

Bimetallic thermometers are commonly used as temperature indicators – the familiar dial thermometer – but they do not provide an electronic signal. Various solid-state devices, such as thermistors, are used to measure temperature because they give a signal directly compatible with electronic equipment, are compact, and do not require special wiring, as do thermocouples; however, they are not very accurate.

When temperature must be controlled, the signal usually drives a steam, gas, or water valve. Control of temperature in multizone equipment, such as baking ovens, can become very complex, as there are many temperatures to measure and many valves to control, most of which interact with each other. Safety interlocks are also required, to detect temperature excursions, to trigger alarms, to activate fire-suppression equipment, to control exhaust blowers (preventing buildup of gas), and to indicate restoration of operating conditions.

Mercury-in-glass and alcohol-in-glass thermometers are still found in canning operations, as calibration standards for retorts, but are rarely used in other food plants, because of the risk of contamination. (See **Canning**: Principles; **Mercury**: Toxicology.)

Temperature can be measured at a distance by infrared instruments. These are useful for checking insulation on overhead lines, roofs, and tall tanks.

A difficult temperature measurement challenge is to follow the temperature history of a container in a continuous sterilizer, such as a hydrostatic retort. The solution is a solid-state measuring and recording device, which fits within or on a container and is plugged to a computer when it is recovered. The data are recorded in semiconductor memory, which is read by computer or special instruments. Similar devices are used to track the temperature history of refrigerated and frozen foods in the distribution chain.

Traditionally, sterilization and pasteurization processes were required to keep ink tracings of temperature history continuously, usually on familiar circular instruments charts. Such records are still seen, but electronic data collection is increasingly accepted. It is critical to ensure the integrity of such data, if it is to displace the ink recorder. (*See* **Pasteurization:** Principles; **Sterilization of Foods.**)

Pressure

Solid-state strain gauges, similar to those at the heart of electronic scales, can also be used to measure pressure, since both parameters involve force. Sanitary pressure gauges involve flush-mounted diaphragms whose deflection generates a signal. The same device can serve as a level indicator in a tank.

Traditional Bourdon tube pressure indicators are found in food plants, especially on utilities such as steam. If used in direct contact with food, they need to be designed to prevent contamination by the food or of the food by any filling material.

Level

The level of fluids is measured by weight or pressure at the bottom, but also can be detected by floats, position detectors, or conductivity switches. The level of solids is more difficult to measure because hydrostatic pressure is not directly proportional to solids height in a bin. A variety of devices, using plum bobs, ultrasonics, and light are used to measure the distance to the top surface of solids, but these suffer from the tendency of solids to remain in a pile with an uneven surface. Bin weight is the best way to control inventory of solids.

Probes can be used to measure capacitance change, as liquids or solids cover more or less of the length, thus giving level indication around the location of the probe. Vibrating reeds or 'tuning forks' generate a change in signal when they are covered, thus indicating the presence of solids or liquid at their location in a bin. Capacitance detectors can also be used as point indicators of level, as can devices that detect change in resistance to rotation of a small paddle.

Composition

The refractive index correlates with the solids content of many liquid food materials, such as syrups, milk, and juices. Thus, instruments to measure refractive index can report results in terms of composition and can be used to control blending operations. Other properties found to correlate with solids content include viscosity and density. On-line instruments exist to measure these properties also.

Direct measurement of composition is quite difficult except by versions of analytical instruments, such as gas chromatographs, mass spectrometers, and pH meters. Ion content can be detected and quantified by ion-specific electrodes, and salt content can be measured by conductivity. Relatively few of these techniques are found directly connected to processes; more often, they are used as off-line quality control devices. (*See* **Chromatography:** Gas Chromatography; **Mass Spectrometry:** Principles and Instrumentation; **pH – Principles and Measurement.**)

Moisture

Moisture is very important in foods and is usually measured by drying a sample. Instruments exist to give moisture determination in minutes, but this is still too slow for direct process control. Water activity can be detected by a humidity sensor over a sample in a closed container, but again it requires minutes. Changes in radio-frequency signals or of microwaves can be used for noncontact, on-line sensing of moisture in some foods.

Moisture, fat, and protein content can be measured quickly for meats and other foods using infrared spectrometery. This has been an off-line technique, but it can be done on the process floor, giving results in time to modify a batch, if necessary. Reflected near infrared is used on-line to measure some composition changes, including moisture and oil content. (*See* **Protein:** Determination and Characterization; **Spectroscopy:** Near-infrared; **Water Activity:** Principles and Measurement.)

Color can be detected and quantified on-line and has been used to control baking ovens, and also to help sort out burnt potato chips (crisps). (*See* **Colorants (Colourants):** Properties and Determination of Natural Pigments; Properties and Determinants of Synthetic Pigments.)

Other Parameters

Counting, most often of packages and cases, can easily be done simply by breaking a light beam to a photo detector or by using proximity switches, which detect changes in capacitance or magnetic fields due to the presence of an object. Such switches are also used to detect position and control starting and stopping of machinery, such as accumulation conveyors.

Seal integrity, most often of heat-sealed polymer packages, such as snack foods, is difficult to detect quickly. Visual inspection is the most frequently used technique. Often, such packages have a small amount of inert gas, such as nitrogen, injected to displace oxygen; the 'pillowing' effect that this creates can be used to detect presence of leaks in seals by lightly compressing the package and measuring its resistance to deformation.

The integrity of cans and jars is measured by detecting the deflection of lids due to the internal vacuum. The vacuum is created when hot-filled contents cool or by steam-flushing of the head space. The deflection can be measured physically or by tapping to create a distinctive tone; 'duds' are rejected as likely to have leaked.

Vision systems are complex combinations of optical devices and computers to observe, quantify, and report on shapes. Such systems are becoming less expensive and more useful as the cost of computing power continues to decline. A vision system can check pizzas for the proper count and placement of garnish, such as pepperoni slices; it can look for and discard 'doubles' in a candy bar process; and it can detect the presence and proper placement of labels on containers. Special devices can read bar codes on cases and report data to an inventory management system.

Vision systems will become more complex and useful with time and will assume many duties now performed by humans, such as sorting, grading, trimming, and assembling.

Integration of Processes

Integration of processes means the linking of data from various sources to maintain smooth operation and the use of such data to improve operations. At a primitive level, such techniques as signaling downstream or upstream operations about stoppages can avoid waste and prevent damage to equipment. On a more sophisticated level would be recording data about stoppages for later analysis. A computer can be used to coordinate the signaling and to do the analysis. It can even write the report, recommending process improvements!

CIM encompasses both levels cited plus more. The basic idea is to avoid ever having to reenter data or information once it exists in a computer. For example, direct process data, such as case counts or batch weights, are available in a process control computer. It should not be necessary to obtain a report and reenter that data in another computer in order to calculate daily, weekly, or monthly production results. Nor should it be necessary to correlate data from two or more different sources to identify production problems and detect improvements. In a CIM environment, all data are available on a common data base shared among various computers linked electronically.

Computers included in such a network range from mainframes and minicomputers through desktop or microcomputers to factory-floor devices such as programmable logic controllers. The power found in each of these devices is increasing as the cost decreases, so the capabilities available to those running food plants are constantly improving. Current practice in modern food plants is to control nearly all equipment (pumps, conveyors, ovens, sterilizers, feeders, etc.) through operator-interface terminals, which have color graphic displays and keyboards, keypads, or touch screens.

The displays, of which there may be several dozen, usually show flow diagrams of the process and status of the equipment (on or off, temperature, etc.). Time histories of any measured parameter can be plotted on command, deviations from desired ranges flagged, and correlations tested among measured parameters.

Standard reports are generated, and data can be summarized for use at the next level of management. For example, on the operational level, the instantaneous rate of production is a concern (cans per minute, kilograms per hour, etc.); at the next level, the interest might be cases or pallets per day (calculated by summing all the actual cans, kilograms, cases, or pallets counted); at the plant and corporate levels, totals for weeks, months, quarters, and years are the only concern. If all the results, at whatever level, are calculated from the same basic data, with little human intervention, they will be consistent and available quickly, and people will be relieved of tedious and unnecessary time-consuming labor.

Barriers to this ideal situation have included incompatibilities among computers, languages, programs, and communications media. Additional barriers have been low computer literacy among workers involved and institutional resistance to sharing of information. Evolution and cooperation among vendors have led to the emergence of common protocols for exchanging information electronically, so that many of the previous technical barriers have fallen. Training and the use of computers in early school grades is reducing the literacy problem. The institutional resistance may be the most difficult, but as the significant benefits of computer integration are appreciated, this too will disappear.

Conclusion

The overwhelming importance of computers in process control and business systems, which are increasingly linked together, has a significant impact on food plant design. The need to have data compatible with computers influences the selection of instruments and measurement techniques. Instruments are being made 'smarter' by incorporating computing power in them with microprocessors, leaving only simple connections to control computers. Such smart instruments calibrate themselves, control drift, correct for changing conditions, and are inexpensive to install. The ability of fewer people to manage and control large processes reduces the labor requirements while raising the education and training requirements of the employees.

See also: **Canning**: Principles; **Chromatography**: Gas Chromatography; **Colorants (Colourants)**: Properties and Determination of Natural Pigments; Properties and Determinants of Synthetic Pigments; **Mass Spectrometry**: Principles and Instrumentation; **Mercury**: Toxicology; **Pasteurization**: Principles; **pH – Principles and Measurement**; **Protein**: Determination and Characterization; **Spectroscopy**: Near-infrared; **Sterilization of Foods**; **Water Activity**: Principles and Measurement

Further Reading

Bernard JW (1989) *CIM in the Process Industries.* Research Triangle Park, NC: Instrument Society of America.
Food and Process Engineering Institute (1990) *Food Processing Automation Conference Proceedings.* St. Joseph, MI: American Society of Agricultural Engineers.
Schwartzberg H and Rao MA (1990) *Biotechnology and Food Process Engineering.* New York: Marcel Dekker.

Plantains *See* **Bananas and Plantains**

PLUMS AND RELATED FRUITS

C F Hansmann and J C Combrink, ARC Infruitec-Nietvoorbij, Stellenbosch, South Africa

Background

The European plum (*Prunus domestica*) probably originated in eastern Europe or western Asia around the Caucasus and the Caspian Sea. It has been known in Europe for more than 2000 years. Prunes are the most important subgroup. Others are the greengage, yellow egg, imperatrice, and lombard groups. The Japanese plum (*P. salicina*) originated in China and was domesticated in Japan. An American fruit breeder, Luther Burbank, developed many commercial cultivars from it after its introduction into the USA around 1870. Native plum species occur in some countries, such as North America (e.g., *P. americana, P. hortulana, P. subcordata*) and the UK (*P. spinosa*, blackthorn sloe; *P. institia*, damson or bullace), but these are not extensively grown commercially. Some are used in breeding programs with Japanese plums. Other plum types, used as rootstocks or in plum breeding programs, or occurring in home orchards, are *P. cerasifera* (myrobalan, cherry plum), *P. simoni* (apricot plum), and *P. institia*.

Plums are cultivated over a wide range of climatic conditions. In 2000, world production of plums was estimated at 8 830 000 tonnes (**Table 1**), representing about 8.5% of the world production of deciduous fruit. Fifteen countries produced more than 100 000 tonnes each, or more than 85% of the total production. The ratio of Japanese plum to European plum is approximately 99:1 for Japan, 60:40 for South Africa and Israel, 40:60 for Canada (Ontario) and Italy, and 0:100 for Yugoslavia (Montenegro, Serbia).

Anatomy and Morphology of the Fruit

The plum is a succulent fruit, called a drupe, and is formed by thickening of the ovary wall following fertilization. The floral remnants become senescent and drop off. The skin (epicarp) consists of a layer of elongated living cells (epidermis) covered by a thin film of cutin (cuticle). It protects the underlying tissue and permits exchange of metabolites or gases with the external environment through openings (lenticels). Wax deposited on the cuticle forms a light gray bloom that gives the skin a dull appearance and makes it impermeable to water. The thick, fleshy, edible portion (mesocarp) consists of parenchyma cells that have an active protoplast where all metabolic reactions occur. The protoplast is enclosed by a pectinaceous cell wall. Middle lamellae, which occur between adjacent cell walls, have a high calcium content and play a vital role in maintaining cohesion between cells, thus providing structural rigidity. The single seed is enclosed by a hard stone (endocarp) consisting of isodiametric cells with thick, lignified cell walls (sclerenchyma). The mesocarp usually clings to the endocarp, although some cultivars are freestone (i.e., the mesocarp is free from the endocarp).

Japanese plums are large and attractive, round or heart-shaped, with or without a prominent apex. The surface of some cultivars is covered by a heavy bloom. The flesh and skin color of ripe fruit may vary from

Table 1 World production of plums during 2000

Country	Production ($\times 10^3$ tonnes)
Countries producing more than 20 000 tonnes	
Africa	
Algeria	26
Egypt	21
Libya	33
Morocco	39
South Africa	55
Asia	
China	4192
India	74
Iran	130
Iraq	27
Israel	23
Georgia	49
Japan	119
Korea	52
Lebanon	41
Pakistan	81
Russia	120
Syria	26
Turkey	190
Uzbekistan	76
Europe	
Austria	57
Belarus	43
Bosnia and Herzegovina	90
Bulgaria	62
Croatia	40
France	214
Germany	316
Hungary	91
Italy	188
Moldova	27
Poland	107
Romania	345
Spain	156
Ukraine	85
Yugoslavia (Montenegro, Serbia)	370
North and Central America	
Mexico	61
USA	668
Oceania	
Australia	24
South America	
Argentina	69
Chile	158
Countries producing less than 20 000 tonnes	250
Total	8830

Data from Food and Agriculture Organization (2002) *FAOSTAT Agricultural Database* (http://apps.fao.org/). Rome: United Nations.

yellow to blood red. The skin of the red-colored cultivars changes during ripening from green to red, usually starting at the apex. Fruit texture varies from firm and nonmelting to soft and melting. European plums are usually oval, with bulging of the central side, and compressed bilaterally. Their skin color is blue or purple, and they have a thick, meaty, freestone flesh. Fruit texture is usually firm and semimelting.

Physiology of Plum Fruit

Cultivar, climatic, environmental, and cultural factors significantly affect physiology. Cell division, multiplication, and differentiation of tissue occur during the first few weeks after fertilization. Cell enlargement and maturation follow. Numerous biochemical changes occur in plum fruits during the last few weeks prior to harvest, whilst the fruits are still attached to the tree. Many of these physiological changes are consistent and predictable, and can be used as indices for determining harvesting maturity. Substances synthesized in the leaves through photosynthesis are translocated to the fruit and transformed into products, which eventually determine the quality and nutritional value of the fruit.

The sugar and total soluble solids content increase throughout the period of growth. Organic acids accumulate in the fruit during the early stages of growth and then gradually decrease. The phenolic content of the fruit is high during the early stages, decreases, and then remains contant until harvest. Volatiles, which determine the flavor or aroma, are produced. Wax develops on the skin of the fruit. The fruit attains its full size and optimum maturity, although it is still unripe. (*See* **Phenolic Compounds; Sensory Evaluation:** Aroma; Taste.)

During ripening, pectic substances in the cell walls change from an insoluble to a soluble form, resulting in softening of the fruit. The chlorophyl (green pigments) content of the skin decreases, and the carotenoid (yellow pigments) and anthocyanin (red pigments) content increases. Based on its respiration pattern, the plum is a climacteric fruit. The respiration rate is high during and immediately after cell division. As maturity approaches, it decreases to the preclimacteric minimum and then increases irreversibly to a maximum (climacterium) during ripening. At the climacterium, the fruit are soft and sweet, with a characteristic flavor, and are ideal for eating. Subsequently, senescence sets in, whereupon the respiration rate decreases, and the fruits become overripe and decayed. (*See* **Ripening of Fruit; Phenolic Compounds.**)

Chemical Composition and Nutritional Value

Food composition tables generally do not specify the cultivar analyzed, nor whether a sample was fresh or cold-stored before analysis. Nevertheless, they are useful guides to the composition and nutritional value of fruit. A typical composition table for plums and related fruits is given in **Table 2**. The anthocyanins present in plums are 3-glucosides and rutinosides of cyanidin and peonidin. Refer to individual nutrients.

Table 2 Nutrient and mineral content of fresh plums, damsons, and prunes

Nutrient	Content (per 100 g edible portion)		
	Plum[a]	Damson[b]	Prune[a]
Moisture (%)	85.2	69.7	32.4
Energy value (kJ)	230.0	146	1000.0
Carbohydrate (g)	11.0	8.6	55.5
Dietary fiber (g)	2.0	1.6	7.2
Protein (g)	0.8	0.5	2.6
Total fat (g)	0.6	Trace	0.5
Nicotinic acid (mg)	0.5	0.3	2.0
Pantothenic acid (mg)	0.18	0.24	0.46
Riboflavin (mg)	0.10	0.03	0.16
Thiamin (mg)	0.04	0.09	0.08
Folic acid (μg)	2.0	3.0	4.0
Carotene (μg)	295[b]	265	140[b]
Vitamin B_6 (mg)	0.08	0.05	0.26
Vitamin C (mg)	10.0	5	3.0
Vitamin E (mg)	0.65	0.60	[c]
Sodium (mg)	0	2.0	4.0
Potassium (mg)	172	260	745
Calcium (mg)	4.0	22	51.0
Magnesium (mg)	7.0	10	45.0
Phosphorus (mg)	10.0	14	79.0
Iron (mg)	0.1	0.4	2.5
Copper (mg)	0.04	0.07	0.43
Zinc (mg)	0.10	0.10	0.53

[a]Data from Langenhoven ML, Kruger M, Gouws E and Faber M (1991) *MRC Food Composition Tables*, 3rd edn. Parow, South Africa: South African Medical Research Council.
[b]Data from Holland B, Welch AA, Unwin ID, Buss DH, Paul AA and Southgate DAT (1991) *McCance and Widdowson's -The Composition of Foods*, 5th edn. Cambridge: Royal Society of Chemistry and Ministry of Agriculture Fisheries and Food.
[c]No data available.

Taste is largely determined by a balance between the sugar and acid contents. Low acid and high sugar contents result in a bland taste, and high acid and low sugar contents in a sour taste. The contents of individual sugars and acids vary between cultivars, but citric acid as well as malic acid are present in the fruit. For example, in Australia, the cultivar Santa Rosa contains approximately twice as much malic acid as Mariposa, but has a lower soluble solids content. (*See* **Acids**: Natural Acids and Acidulants.)

Harvesting, Handling, and Storage

Plums are harvested in summer, from about December to February in the southern hemisphere and from about July to October in the northern hemisphere. It is important to harvest at the correct stage of maturity. If plums are harvested while still immature, the characteristic flavor will not develop, and fruit quality will be inferior. Plums harvested too ripe are very susceptible to injuries and fungal infection during harvest and postharvest handling, and their storage life will be short.

Dessert Plums

Dessert plums are harvested when they are physiologically mature but unripe. Ripening to a stage at which the fruit has a pleasant taste and aroma occurs during and, mostly, after storage. Plums that are to be sold soon after harvest must be picked at a riper stage than usual, or cold-stored for a few days to promote proper ripening. They require extra care during handling.

Total soluble solids content, estimated by means of a refractometer, and mesocarp firmness, measured by means of a probe (penetrometer) forced into the flesh, are good indicators of maturity. However, these tests are destructive, and pickers use skin color, in combination with these tests, as a maturity index. Fruit size is an additional criterion used to determine optimum maturity stage. Fruits are picked by hand into picking bags and carefully transferred to containers with a capacity of about 500 kg. Plums bruise easily and must be handled carefully during and after harvest. Filled containers are transported to a packhouse. Bruised, cut, or decayed fruit and culls are removed. The plums are sorted according to size and either hand- or mechanically packed into cardboard or wooden containers. These usually have a capacity ranging from about 5 to 20 kg. Trays or padding material immobilize the fruit in the container to prevent transit injuries. Packed containers are unitized (palletized) to facilitate handling and protect the fruit.

To extend the storage life of the fruit, its temperature must be lowered as soon after harvest as possible. Precooling, which removes field heat from the fruit, commences immediately after packing and palletizing. Cold air is distributed through the pallets in such a way that it comes into contact with all fruit. The ideal storage conditions for plums are a temperature of -0.6 to $0\,°C$ and a relative humidity of 85–90%. The storage period depends on the cultivar. Most cultivars cannot be stored for longer than 3–4 weeks, but others can be stored for up to 3 months. The fruit of some cultivars are susceptible to chilling injury. To prevent chilling injury, the temperature is raised from $-0.6\,°C$ to about $7\,°C$ after 7–10 days, and this temperature is maintained for the rest of the storage period. This is called dual-temperature storage. Controlled- or modified-atmosphere storage increases the storage life of some cultivars. (*See* **Controlled-atmosphere Storage**: Effects on Fruit and Vegetables; **Storage Stability**: Mechanisms of Degradation.)

Prunes

Prunes are harvested when fully ripe and in an ideal condition for fresh consumption. Overmature fruit discolor and break down during drying. The fruit of some cultivars drop from the trees naturally and must be gathered promptly to prevent losses. Soluble solids content and flesh firmness are important maturity indices. As in the case of fresh plums, these indices are related to skin color.

Most prunes are dried rather than sold fresh. Fresh prunes can be cold-stored, but their storage life is shorter than that of dessert plums. Sorted prunes are dipped into a hot caustic solution (generally 0.5% sodium hydroxide at boiling point) for sufficient time to remove the wax layer and cause minute cracks in the skin. This facilitates moisture loss and speeds up drying. The prunes are then packed on to trays and dried in the sun. Prunes can also be dried in mechanical dehydrators, usually tunnel dehydrators. The fruits are washed and placed on trays in the dehydrator without application of a lye treatment. Hot air circulates through the trays, and drying takes place within 24 h. Dehydrators used for drying prunes are frequently operated in the parallel-flow mode (concurrent movement of fruit and heated air through the drying section), with the fruit entering the dehydrator at air temperatures of up to 90 °C. Prunes are gathered when dry but still pliable, with a moisture content of about 20%. Considerable degradation of the red pigments present in the fresh fruit occurs during drying. (*See* **Drying**: Drying Using Natural Radiation; Theory of Air-drying.)

Market Disorders and Diseases of Plum Fruit

Disorders

Disorders are transit- or storage-related. Transit-related disorders include bruising as a result of rough handling, freezing injury owing to temperatures below the freezing point of plums, and shriveling owing to moisture loss caused by low relative humidities in the storage atmosphere. These disorders can be controlled by effective postharvest management.

Storage-related disorders are similar to transit-related disorders but include internal breakdown or internal browning. When a fruit is cut open, internal browning can be seen as a reddish brown discoloration, which is more intense around or near the endocarp. It is responsible for extensive losses during cold storage. It usually occurs at low temperatures and can be prevented by storing fruit at slightly elevated temperatures. Some cultivars are more susceptible than others. Dual-temperature storage is used where plums have to be shipped over long distances.

Diseases

Fungi cause market diseases. The most important postharvest diseases of plums and prunes are brown rot (*Monilinia laxa*, *M. fructicola*), blue mold rot (*Penicilliun expansum*), gray rot (*Botrytis cinerea*), mucor rot (*Mucor piriformis*), and rhizopus rot (*Rhizopus* spp.). These fungi cause decay in a wide range of commodities, and their occurrence is not limited to plums. Fungal spores present in the orchard on plant debris or in the air are dispersed to fruit by wind or insects. The spores of *B. cinerea* and *Monilinia* spp. can infect uninjured fruit in the orchard, especially during the 1–2 weeks prior to harvest, when the susceptibility of fruit to infection increases. However, disease symptoms are only expressed during cold storage. The other fungi are wound pathogens that penetrate fruit through skin breaks caused by rough handling or insects. Decay symptoms are related to the fungus involved. Mucor and rhizopus rots develop rapidly, and affected plums become soft and watery, and are covered by black spore masses. Brown rot starts as a small, water-soaked spot, which enlarges rapidly and becomes brown or black. The skin becomes leathery but remains intact. Fruit infected by *B. cinerea* are firm, spongy, and only slightly moist. The skin covering lesions readily slips away when slight pressure is applied. Decay development is slower, and, in advanced stages of decay, gray spores cover the affected areas. Blue mold rot lesions are wet and soft and covered with blue–green mold growth. (*See* **Spoilage**: Molds in Spoilage.)

Control of postharvest decay involves the integration of preharvest factors (soil preparation, spray programs, orchard hygiene, etc.) with sound postharvest crop management. Fruit produced under optimal conditions possess a high natural resistance to infection. Proper sanitation in the orchard and packshed keeps the number of spores to a minimum and is an important precautionary measure. All fungi causing postharvest decay, except *Rhizopus* spp., can grow at 0 °C, and cold storage does not prevent decay development. Chemicals can be used to control decay, but only a few are registered for postharvest use. (*See* **Fungicides**.)

Industrial Uses

Plums have a wide variety of industrial uses. Japanese plums are mostly eaten fresh, although a small percentage are dried. European plums are mostly dried (prunes), cooked (greengages), canned (greengages, yellow egg, damsons, other European plums), or used in jams (plums, greengages, damsons) or jellies (American plums). Prune juice is used as a laxative and plum purée as a baby food. Some plums are used as rootstocks (e.g., marianna, myrobalan) or in breeding programs (*P. simoni*) and are planted as ornamentals in home gardens. In some parts of Europe, plums are fermented and distilled into a 'brandy.' In Romania, Hungary, and Yugoslavia (Montenegro, Serbia) it is called slivovitz, and in other parts of Europe the name depends on the cultivar used (e.g. Quetsch or Mirabelle).

See also: **Acids**: Natural Acids and Acidulants; **Controlled-atmosphere Storage**: Effects on Fruit and Vegetables; **Drying**: Theory of Air-drying; Drying Using Natural Radiation; **Fungicides**; **Phenolic Compounds**; **Ripening of Fruit**; **Sensory Evaluation**: Aroma; Taste; **Spoilage**: Molds in Spoilage; **Storage Stability**: Mechanisms of Degradation

Further Reading

Chandler WH (1965) *Deciduous Orchards*, 3rd edn. Philadelphia, PA: Lea & Febiger.

Eksteen GJ and Combrink JC (1987) *Manual for the Identification of Post-harvest Disorders of Pome and Stone Fruit.* Stellenbosch, South Africa: Van der Stel Printers.

Harvey JM, Smith WL, Jr. and Kaufman J (1972) *Market Diseases of Stone Fruits. US Department of Agriculture, Agriculture Handbook 414.* Washington, DC: US Government Printing Office.

Holland B, Welch AA, Unwin ID, Buss DH, Paul AA and Southgate DAT (1991) *McCance and Widdowson's The Composition of Foods*, 5th edn. Cambridge: Royal Society of Chemistry and Ministry of Agriculture Fisheries and Food.

Langenhoven ML, Kruger M, Gouws E and Faber M (1991) *MRC Food Composition Tables*, 3rd edn. Parow, South Africa: South African Medical Research Council.

Macheix JJ, Fleuriet A and Billot J (1990) *Fruit Phenolics.* Boca Raton, FL: CRC Press.

Ramming DW and Cociu V (1990) Plums (*Prunus*). In: Moore JN and Ballington JR, Jr. (eds) *Genetic Resources of Temperate Fruit and Nut Crops*, vol. 1, pp. 235–281. Wageningen, The Netherlands: International Society for Horticultural Science.

Salunkhe DK, Bolin HR and Reddy NR (1991) *Storage, Processing, and Nutritional Quality of Fruits and Vegetables*, 2nd edn, vol. 1. Boston, MA: CRC Press.

Teskey BJE (1978) *Tree Fruit Production*, 3rd edn. Westport, CT: AVI.

Wills RBH, Scriven F and Greenfield H (1983) Nutrient composition of stone fruit (*Prunus* spp.) cultivars: apricot, cherry, nectarine, peach and plum. *Journal of the Science of Food and Agriculture* 34: 1383–1389.

POLITICS AND NUTRITION

C Geissler, King's College London, London, UK

Introduction

Politics can be defined as both the art and science of government, as well as the play of competing economic and ideological interests. This article reviews some aspects of government nutrition policies and the effects of disparate interests on nutrition in developing and developed countries.

Nutrition Policies

Governments in both developed and developing countries have over the last two decades increasingly focused attention on nutrition through explicit national nutritional policies. (*See* **Nutrition Policies in WHO European Member States.**)

Hunger and malnutrition were put on the international agenda by the League of Nations in the 1930s and remained an important focus of the United Nations (UN) technical agencies, the Food and Agricultural Organization (FAO), World Health Organization (WHO), and United Nations Children's Emergency Fund (UNICEF), which were created immediately after World War II. The types of nutrition interventions and development policies supported were shaped by the World Food Surveys conducted every decade by the FAO (these estimated the extent and defined the causes of malnutrition) and by postcolonial theories of economic development. During the 1950s and 1960s, the prevailing policy encouraged industrialization and large-scale agriculture in order to increase economic wealth, from which improved nutrition would 'trickle down' rather than be achieved by improved distribution. Nutrition continued to be approached mainly through piecemeal interventions, including child-feeding programs and nutrition education, but also via some integrated village-level 'Applied Nutrition Programs', that addressed the economic, educational, and food resource constraints on good nutrition. (*See* **World Health Organization.**)

By the early 1970s, it had become clear that rapid economic growth had led to increased inequalities, as well as increased absolute poverty in some countries such as Pakistan, Nigeria, and Brazil, resulting in greater levels of undernutrition in the impoverished,

Dissatisfaction with the effectiveness of past approaches in improving malnutrition led to the concept of National Nutrition Planning, based on the coordination of sectoral policies to focus on nutrition. The aim was to remove the barrier to economic growth of malnutrition that results in increased morbidity and mortality, poor educational achievement, low work capacity, increased absenteeism, and low productivity. This integrated, systematic, intersectoral approach was endorsed and promoted by the FAO and the United States Agency for International Development (USAID), that assisted several countries to prepare explicit integrated nutrition policies. (*See* **Malnutrition**: The Problem of Malnutrition.)

In this international context, the need for integrated national food and nutrition policies was also expressed in the developed countries, where the malnutrition of poverty had been largely eliminated but where malnutrition of affluence, reflected in increased rates of coronary heart disease, diabetes, obesity, and cancer, is the current concern. Several countries developed national dietary goals in the 1960s and 1970s to encourage their populations to consume a more healthy diet, but Norway was the first to pass an intersectoral Food and Nutrition Policy in 1975. (*See* **Malnutrition**: Malnutrition in Developed Countries.)

During the 1980s, this extent of integration fell out of favor because of practical difficulties, and the international community concentrated on improving the nutritional impact of agricultural and other sectoral development programs. In the 1990s, the focus moved to micronutrient interventions to eradicate the most prevalent vitamin and mineral deficiencies. However, in 2000, the pendulum of opinion is swinging away from specific 'magic bullet' interventions and back to calls for more integrated approaches to national policy.

Types of Nutrition Interventions

The types of interventions that form part of national nutrition policies tend to be limited to palliative measures such as vitamin supplementation, nutrition education, and child-feeding programs, because the underlying political issues that lead to malnutrition, whether undernutrition or overnutrition, involve fundamental economic and political interests. These are much more difficult and contentious issues and rarely features overtly in nutrition policies. (*See* **Community Nutrition.**)

Famine

One extreme example of the political aspects of nutrition is famine, with mass starvation and death, often attributed to climatic causes such as drought or flooding. Such factors may precipitate famine, but it is only the most vulnerable members of society who starve in this type of famine. To quote Amartya Sen, 'starvation is the characteristic of some people not *having* enough food to eat. It is not the characteristic of there *being* not enough food to eat.' He argues that poverty and famines are related to 'entitlement' or the ability of individuals to obtain food through the legal means available in the society, including production possibilities, trade opportunities, state provisions, and other methods of acquiring food. When individuals can no longer obtain enough food in exchange for their products or services, they suffer from famine, as happens with food price rises owing to local shortages and hoarding. Several examples exist where aggregate food supplies were adequate but certain groups could not afford to buy them, such as the Irish famine of 1846, the Bengal famine of 1943, and more recent famines in the Sahelian countries.

War

War is an extreme example of political crisis and is an increasingly common contributor to the causes of famine through the blockade or diversion of food supplies, the destruction of cropland and the migration of farmers away from their lands in the war zones. War contributed to the famines in Bengal, The Netherlands and Warsaw (World War II) and more recently in Nigeria, Timor, Kampuchea, Laos, Vietnam, Afghanistan, Chad, Uganda, and the Sudan. War and its aftermath also have an strong impact on nutrition, short of famine. One example is the 1990–1991 Gulf War, which resulted in extensive destruction of Iraq's infrastructure and economy so that widespread malnutrition, particularly of children, has occurred.

Refugees

War and political crises also create refugees. The number of people in the world who have either fled their country as refugees, or have been displaced internally, mainly as a result of civil war increased greatly during the 1980s and 1990s and at the beginning of the new millenium stands at around 50 million. Refugees suffer from the same type of diseases as other vulnerable groups in developing countries, the main killers being measles and diarrhea, to which they are more susceptible because of malnutrition. They often receive food inadequate in quantity and quality. If prolonged, this deprivation leads to starvation, and debilitating outbreaks of scurvy, pellagra, beriberi, and other deficiency diseases are now common. This is attributable to the dependency of refugees on the food provided by governments and aid donors, which has little diversity and is inadequately planned. (*See* **Refugees**; **Scurvy**.)

Poverty

Famine is only the tip of the iceberg of the larger problem of poverty and access to resources, which affects nutrition on a longer-term basis. Various economic models or political views exist on the causes of poverty and wealth creation, ranging from simplistic models of idleness and ignorance to systematic models such as the Marxist model, which views poverty as an inevitable and even necessary part of the capitalist economic system. These models form the framework of what policymakers believe can be done to break out of poverty and malnutrition, what the implications are for the rest of the economy, and therefore the type of interventions they support.

Poverty reduction

Poverty reduction became a specific target of the international community: an International Development Goal was set in 1990 to halve poverty and was expected to be met within 20 years in most areas of the world. However, the financial crisis later in the decade led to a rise in levels of poverty and inequalities, setting back progress and leading to renewed calls for debt relief.

Debt relief

Some countries are unable to tackle poverty adequately because past loans have led to a level of debt that is now too high for them to repay. For example, in Africa as a whole, governments spend four times as much paying back their debts to developed countries in repayments as they do on health and education. In 1996, the International Monetary Fund (IMF), the World Bank, Regional Development Banks and other external creditors launched the Heavily Indebted Poor Countries Initiative (HIPCI) to reduce the debts of the poorest countries who are deemed eligible by their economic track record over the previous 5–6 years. However, few countries qualified (Uganda and Bolivia), and in 1998, a fundamental review was agreed upon to speed up the process and increase the level of relief.

Equity

Poverty is a major, but not the only, determinant of undernutrition. At the same level of family or national income, some individuals or countries do better than others in terms of nutrition. Internationally, there is a clear relationship between gross national product (GNP) and several indicators of health and nutrition,

including food available for consumption, infant mortality, etc. Countries that have done best in improving nutrition in recent decades are those where there is a greater equity or where policies have concentrated on insuring the satisfaction of basic needs, including adequate food. These include a wide spectrum of political ideologies from communist China to capitalist South Korea. China has been a classic example of a country that was still poor but conquered millennial problems of malnutrition and famine – except during the disastrous Great Leap Forward policy of rural industrialization at the end of the 1950s – through effective organization of food production and distribution. Other examples are Kerala state in India, Sri Lanka, Vietnam, Indonesia, Malaysia, and Thailand, which have had better nutrition conditions than other countries with similar GNPs. In contrast, some countries have extensive chronic malnutrition, e.g., Brazil and some other Latin American countries (Bolivia, Ecuador, Gwatemala), most countries in Sub-Saharan Africa, and Bangladesh despite massive aid.

Since the early 1980s, there has been a widespread rise in income inequalities with an acceleration in the 1990s in several countries, including all economies in transition such as China. This has challenged the poverty reduction targets set by the international community.

International Relations

Poverty and malnutrition are controlled not only by national policies and resource distribution but also by the power play of international economic relations. Again, opposing political views shed very different lights on their effects.

On one side are those who believe that the rich industrialized countries and the international agencies do much to assist poor developing countries through technical advice and monetary or food aid, while international corporations provide them with employment opportunities.

On the other side are those who believe that the odds are stacked against poor countries breaking out of poverty, as the countries that became industrialized first now hold economic power internationally, dominating trading and banking systems. In this view, national elites of many countries assist the world powers to dominate their economies through concessions to multinational companies, the acceptance of aid, and requests for finance from the international banking system.

During the 1980s, the latter view received wider acceptance and led to calls for a 'new economic order' by such groups as the Brandt Commission, which warned that the inequalities of opportunity between the North (the rich industrialized countries) and the South (the poor developing countries) could lead to political and environmental instability, and it was therefore in the interests of both to allow the South greater control of its destiny.

The questions of international food trade, food aid, and the role of international agencies in relation to poverty reduction and nutrition are therefore highly contentious political issues that can be touched on only briefly here.

Food Trade

The economies of the developing countries are affected by the policies of the major producers of food and other essential commodities. For example, the 'world food crisis' in 1973, which was marked by soaring prices of grain and other food commodities on the world market, was the result of several factors. These included climate, the USA policy of deliberately reducing its grain stocks by taking land out of production to maintain producer prices, and Russia's sudden increase in its grain purchases from the world market to support its policy of livestock production. These, and the oil crisis, caused great economic hardship for many developing countries that depended on imports for their food supply, and led to a spate of policies to increase food self-sufficiency.

Similarly, the Common Agricultural Policy (CAP) of the European Union (EU) has had a controversial impact on the Third World through price fluctuations caused by periodic dumping of surplus food on the world market and through tariff barriers that restrict imports. Some trade concessions were made by the EU to the developing countries through the Lome Convention, but this provided little protection in practice. (See **European Union**: European Food Law Harmonization.)

These protectionist policies are the subject of international bargaining for free trade that is carried out in periodic rounds of talks in the council on the General Agreement on Trades and Tariffs (GATT). However, these mainly reflect the interests of the seven most economically powerful nations, the USA, Canada, Germany, France, Italy, the UK, and Japan (the G7) and only in the 1986–1992 round included food commodities, prompted by the food price policies that are detrimental to the food trade of the USA and other countries.

GATT was succeeded by the World Trade Organization (WTO) from 1995, which, by the end of the decade, had 134 members, accounting for 90% of world trade and 30 others, including China, negotiating for membership. The costs of participation are high for the poorest countries, which means that the agenda is still dominated by the large economies.

Food Aid

Food and other international aid has often been used as a political weapon, given to politically friendly countries, such as South-east Asia in the late 60s and to Egypt and Syria in 1974, and withdrawn from those that do not conform to the policies of the donor countries, such as Chile in 1971 and Mozambique in 1981.

Food aid has also received much criticism on the grounds that it has been used by the donor countries to dispose of surpluses and to penetrate food markets of developing countries in order to create a long-term trade demand, that it is unreliable, involves excessive opportunity costs, reinforces expensive subsidy programs, provides opportunities for corruption, and is used by governments to keep urban prices down, thereby acting as a disincentive to agriculture and maintaining rural poverty and malnutrition.

Food aid and capital intensive agricultural policies are also seen to contribute to the rapid rural urban migration seen in so many developing countries, where cities are unable to cope with the rate of influx to provide adequate housing, sanitation, water supply, and employment. As a result, the migrants can often only scrape a living through petty trading, begging, and low wage employment. This influx undermines minimum wage legislation or enforcement and keeps the cost of labor, and therefore incomes, low. The consequent inability of the urban poor to purchase adequate diets, and the insanitary living conditions, result in extensive urban malnutrition in many developing countries. Such effects are seen to negate the short-term benefits of food aid used in famine situations, and of child-feeding and food-for-work projects that act as welfare benefits or resource transfers to the poor.

The structure of food aid has altered since it became a permanent international transfer mechanism in the 1954 when the USA enacted the public law PL480, referring to sales to 'friendly nations,' to be paid initially with local money deposited as counterpart funds for use by the USA or with their approval (Title 1), and donations for famine relief and projects (Title II). The latter represents a small proportion contrary to the popular belief that all aid is a gift. Initially, the USA supplied nearly all aid, but now the EU, Japan, Canada, Australia, and other countries contribute about 50%. About 35% of food aid is channeled through the multilateral World Food Program for feeding projects; this reduces political bias. However, 40% is still bilateral, and several countries, such as Bangladesh, remain dependent on food aid and, therefore, politically dependent on the donor. There is widespread recognition of the need for reform of food aid and its international infrastructure to better handle hunger and meet targets in poverty reduction, but consensus is difficult to reach.

Structural Adjustment

Over the last decades of the millennium, many developing countries experienced severe economic crises owing to rapid changes in oil prices, falling and unstable prices of export commodities, rapidly increasing rates of interest, and increasing dependence on foreign borrowing, resulting in reduced foreign exchange reserves, and inability to service debts.

To deal with the crisis, countries had to implement a variety of 'macroeconomic adjustment policies,' including reductions in government spending. These conditionalities have been rigidly imposed by the International Monetary Fund (IMF) and the World Bank (International Bank for Reconstruction and Development (IBRD)), to obtain new financial loans. These institutions are funded by quotas from members who have voting rights in proportion to their contribution, assessed according to economic status. The USA has the majority of the votes, followed by the European countries, and others very small percentages, so that decisions are effectively in the hands of the major industrialized countries, especially the USA. This banking structure means that the policies of borrowing countries are dictated by the industrialized nations.

The primary aim of macroeconomic adjustment policies is to improve the balance of payments. Therefore, short-term effects on the poor have often been ignored unless they threaten political stability, e.g. through urban riots. Adjustment has frequently included changes that are of particular concern to the poor, such as increased food prices and decreased expenditure on social programs. The effects of these policies on health care, food consumption, incomes, and prices have led to a serious deterioration in indicators of nutrition, health status, and school achievement in several countries. Efforts have been made by UNICEF and other bodies to buffer vulnerable groups from these effects.

Affluence

In most developed countries, the problems of poverty-related malnutrition have been largely overcome by higher incomes and a safety net of welfare benefits, and replaced by the problems associated with affluent diets. The politics of nutrition in this case are very different from those associated with poverty. Only a small proportion of the population is involved in primary food production (2–3% in the US and UK), and the majority of the population depends on the food industry, which is a major industry in economic terms in Europe and the USA: in the UK agriculture,

food processing and distribution contribute more than 7% to the GDP. Therefore, the agricultural and food industry lobbies are potentially important in food and nutrition policy formulation. In food policy, nutrition generally has a very low priority. For example, the EU CAP has no explicit nutritional component, its main aims being to support farm incomes and food security for Europe in terms of adequate supplies and stable prices. The emphasis is therefore on quantity, not on nutritional quality. The concept of food quality has lagged behind current values and has generally referred to the purity, hygiene, and richness of nutrients. These are all extremely important, but now that nutritional deficiencies have been largely overcome, the concept of a healthy diet can no longer be related only to purity and abundance. Government policies have also lagged behind current needs such that food legislation and producer incentives continued to favor a high fat diet, including premium producer prices for high-fat milk and inappropriate grading systems for livestock.

Efforts during the 1980s to translate general nutritional advice on healthy eating into explicit policies were initially met with considerable concern and opposition from some sections of the food industry that were threatened with change. The debate included the following arguments: the evidence for the relationship between diet and disease is not adequately conclusive; the level of risk to the population as a whole for certain dietary components such as salt does not warrant blanket nutritional policies, only targeted interventions to those at risk; and in a democratic society, the individual must be free to make his or her own choice of food consumption without coercive pressures from a 'nanny state.'

This initial resistance of the food industry and of government was viewed in some quarters as a conspiracy of vested interests between food producers, government policymakers and advisory committees. However, others viewed it as a natural cautious attitude that is necessary because of the widespread industrial and economic changes entailed. Despite this resistance, the food industry has responded to pressures for change in various ways including research into leaner livestock, the formulation of a variety of low-fat and low-energy products, and voluntary nutritional labeling. Also, since 1993, UK health policy explicitly addresses nutrition and associated 'diseases of affluence,' including obesity and heart disease.

Much stronger political support exists for interventions based only on informed choice, and little for coercive measures to improve the diet such as price manipulation through consumer subsidies and taxation, even though these instruments are used for other purposes. There is therefore much effort put into legislation on food labeling.

The food industry has become progressively more vertically integrated, from production through processing to retailing, by the development of multinational agribusiness and large chain food stores, that increasingly order food directly from the producers. This means that the buyers for these food chains become exceedingly important arbiters of the national diet in deciding the choice available to the consumer.

Emerging Issues

The creation of the WTO and the increasing power of large food corporations have highlighted several issues of control over food, from consumer pressures to national and international governance.

New Technologies

Consumer and environmental lobby groups have put a brake on the use of genetically modified (GM) foods: they disrupted the WTO meeting in Seattle in early 2000 by protests against the trade in GM crops; they caused agribusiness to discontinue the development of terminator genes to prevent plant replication; and consumer pressure in developed countries such as the UK has curtailed the sale of GM products as well as the conduct of field trials.

Resistance to the use of the new technology in agriculture and food production is based on several fears: the safety for consumption; the control over world food supply of the large corporations that develop the technology; the potential detrimental effects on the environment from cross-pollination; and the lack of benefits to developed country consumers as opposed to the seed and pesticide manufacturers. These arguments currently outweigh the arguments for the potential benefits, within appropriate socioeconomic structures, to farmers and consumers in other parts of the world by the development of crops that are resistant to adverse climatic conditions and pests and that have higher yields and content of specific nutrients, factors that could expand the areas of crop production, increase total food production, and hence price in line with the expected population growth, and improve nutritional value.

Governance

Such consumer pressures have also led to other issues of governance in relation to national and international political bodies that control economics, including food. For example, consumer and professional pressures following a series of food safety scares in the UK during the 1980s have led to the separation within government of the structures that regulate the food supply from those that regulate agriculture as there are potentially conflicting interests between consumer and producer. A Food Standards Agency was established in 2000 to regulate food 'from farm to fork.' The EU has subsequently proposed a similar structure at European level. This brings to the fore the tensions in governance between national and regional control over food and the extent to which these are dominated by WTO rules in the interests of free trade. Similar issues arise in global governance, underlining the need for new arrangements to make more effective the relationships between the UN, WTO, G7, IMF, and other bodies.

Conclusion

Food and nutrition involve highly political issues that determine both access to sufficient food and the commercial and consumer pressures on a balanced diet.

See also: **Community Nutrition**; **European Union**: European Food Law Harmonization; **Malnutrition**: The Problem of Malnutrition; Malnutrition in Developed Countries; **Nutrition Policies in WHO European Member States**; **Refugees**; **Scurvy**; **World Health Organization**

Further Reading

ACC/SCN (1997) *Nutrition and Poverty. Nutrition Policy Paper 16*. Geneva: ACC/SCN.

ACC/SCN (2000) *Fourth Report on the World Nutrition Situation*. Geneva: ACC/SCN.

Atkinson J (1996) *Farmers, Food and the WTO. Oxfam Policy Briefing Paper*.

Berg A (1973) *The Nutrition Factor. Its Role in National Development*. Washington, DC. The Brookings Institution.

British Nutrition Society (1991) Nutrition and Development: symposium & workshop. *Proceedings of the Nutrition Society* 51(1): 71–107.

Food Industry, Nutrition and Public Health (1997) Symposium. *Proceedings of the Nutrition Society* 56(3): 807–888.

George S (1984) *Ill Fares the Land. Essays on Food, Hunger and Power*. London: Writers and Readers Publishing Cooperative Society.

Jolly R and Cornia GA (eds) (1984) *The Impact of World Recession on Children. UNICEF Report*. New York: Pergamon Press.

Overseas Development Institute (2000) *Reforming Food Aid: Time to Grasp the Nettle? ODI Briefing Paper, January*. London: ODI.

Royal Society (1998) *Genetically Modified Plants for Food Use*. London: Royal Society.

Sen A (1982) *Poverty and Famines: an Essay on Entitlement and Deprivation*. Oxford: Oxford University Press.

Tansey G and Worsley T (1995) *The Food System. A Guide*. London: Earthscan.

UN (1997) *Renewing the United Nations: A Programme for Reform. Report by the Secretary General*. New York: UN.

World Health Organization (1988) *Nutrition Policy Experiences in Northern Europe. Report on a WHO Consultation*. Copenhagen: WHO EUP/ICP/NUT.

products. The same limit was also set in Germany for smoked cheese and cheese products. In the European Union, the maximum BaP concentration permitted in foodstuffs as a result of using flavorings, including smoke flavorings, is 0.03 μg kg^{-1} (Council directive 88/388/CEE of 22-06-1988). In addition, another Council directive (91/493/CEE of 22-07-1991) concerning fishery products lays down some health conditions under which these products must be smoked, and lists wood materials which cannot be burned in smoking foods. (*See* **Vegetale Oils**: Types and Properties; Composition and Analysis.)

Waters   The following limits were set in the European Union to PAH presence in water intended for human consumption (Council directive 98/83/EC of 3-11-1998): 0.010 μg l^{-1} for BaP, and 0.10 μg l^{-1} for the sum of benz[*e*]acephenanthrylene, benzo[*k*]fluoranthene, benzo[*ghi*]perylene, and indeno[1,2,3-*cd*]pyrene. BbFA, BkFA, BghiP, and IP.

Refined oils and fats   Some European oil industries set their own guideline values for refined oils and fats: 5 μg kg^{-1} for the sum of seven higher-molecular PAHs (BaP, benzo[*e*]pyrene (BeP), BghiP, dibenz[*a,h*]anthracene (DBahA), perylene, anthanthrene, coronene), and 25 μg kg^{-1} for the sum of 13 PAHs, including the previous seven, plus another six lower-molecular PAHs. (*See* **Chromatography**: High-performance Liquid Chromatography; Gas Chromatography; Supercritical Fluid Chromatography; **Spectroscopy**: Visible Spectroscopy and Colorimetry.)

Analysis

The determination of PAHs in food is complex, time-consuming, and requires experienced personnel. Nowadays, it is accepted that it has to be performed under a program of quality control/quality assurance, with the aim of providing fully reliable results.

PAH Selection

There is no standard list of PAHs to be determined and the selection performed by each investigator is commonly arbitrary, based on the available instrumentation and reference standards. Almost invariably, BaP has been determined in every investigation, due to its well-known carcinogenicity. Nowadays, special attention is generally given to the determination of other carcinogenic PAHs, especially in health-related studies. Also when other PAH concentrations are available, BaP is usually taken as a reference compound to compare the contamination and the toxicological potency of different food items. This is supported by the fact that BaP is by far the

species of higher toxicological concern, due to the combination of its carcinogenic potency – relatively to the other PAHs – with its concentration levels.

Sample Preparation

Most foods are not homogeneous and must be carefully homogenized prior to analysis. Extraction and clean-up are crucial steps in the determination because of the very low amounts of PAHs and the need to separate them from substances which may interfere during analysis, especially the lipids. The general scheme shown in **Figure 1** has been widely used, as such or with some modifications which were later introduced in its application to a wide range of foods and to total diet samples. Modifications to the scheme of **Figure 1** most commonly involve the extraction and partitioning solvents (e.g., dimethyl sulfoxide replacing *N,N*-dimethylformamide), the chromatographic sorbent (e.g., XAD-2 resin combined with other sorbents or florisil or alumina combined with silica gel), and thin-layer chromatography as the clean-up step. The sonication extraction (e.g., of plants and smoked foods) has also been used efficiently; its advantages are the reduced time of extraction and possibly superior recovery efficiencies and reproducibilities, but the results depend on matrix, solvent, and experimental conditions. Conventional chromatographic columns are often substituted by prepacked commercial cartridges (solid-phase extraction), which give advantages in terms of time and solvents consumed, and of reproducibility performance.

Recently, supercritical fluid extraction (SFE: e.g., of smoked and broiled fish, and toast) and accelerated solvent extraction (ASE: e.g., of smoked meat) have gained attention as rapid alternatives to conventional liquid extraction.

Sample Analysis

Identification and quantification are performed by capillary gas chromatography (GC) or by high-performance liquid chromatography (HPLC). In GC analysis, the most widely used stationary phases are the methylpolysiloxanes, especially SE-54 (5% phenyl, 1% vinyl-substituted) and SE-52 (5% phenyl-substituted) or equivalent phases. A flame ionization detector is almost universally employed: it has an excellent response linearity and, coupled with cold on-column injections, gives an accurate and precise quantification; because of its nonselectivity, however, samples need to be highly purified. Mass-spectrometric detectors are powerful tools in identifying and confirming compounds, and have gained wide acceptance. In HPLC analysis, the most used packed material consists of silica particles chemically bonded to

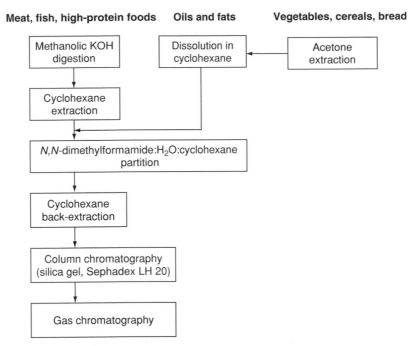

Figure 1 A classical general scheme for the determination of polycyclic aromatic hydrocarbons (PAHs) in food. Adapted from Grimmer G and Böhnke H (1979) Method 4 – Gas chromatographic profile analysis of polycyclic aromatic hydrocarbons in (I) high protein foods; (II) fats and vegetable oils; and (III) plants, soils and sewage sludge. In: Egan H, Castegnaro M, Kunte H, Bogovski P and Walker EA (eds) *Environmental Carcinogens: Selected Methods of Analysis*, vol. 3. *Analysis of Polycyclic Aromatic Hydrocarbons in Environmental Samples*, pp. 163–173. Lyon: International Agency for Research on Cancer.

linear C18 hydrocarbon chains. Ultraviolet (UV) and fluorescence detectors are used, usually arranged in series. The fluorescence detector is more sensitive, and its specificity allows for PAH determination in the presence of nonfluorescing interferences. For confirmation purposes, much information on the isomeric structure may be obtained from UV spectra acquired during the elution of chromatographic peaks by the UV diode-array detector.

Standard Methods

Standard methods for BaP in fats and oils have been prepared by International Standards Organization (ISO), American Oil Chemists' Society (AOCS), and IUPAC. General methods for PAHs in food have been published by IARC and Association of Official Analytical Chemists (AOAC).

Occurrence in Foods

Table 2 summarizes the levels of individual PAHs in various foods, as reported in the literature. The investigations for which the results are included in the table were selected on the basis of their representativeness of typical or (conversely) extreme situations, their recency (since the 1980s), and the wide spectrum of PAHs reported. While not exhaustive, the table aims to give general information on the contamination

levels which may be found and to allow them to be compared in different foods. A number of investigations concerning only BaP determination are not included in the table but are discussed in the following sections.

Levels of individual PAHs generally range from below $1 \, \mu g \, kg^{-1}$ to a few $\mu g \, kg^{-1}$. Occasionally, they are in the order of ten, and even hundreds of $\mu g \, kg^{-1}$.

The highest PAH levels were found in grilled food, smoked and broiled fish, mussels from polluted waters, and leafy vegetables grown in areas heavily exposed to air pollution. Shellfish such as crustaceans and bivalves do not metabolize PAHs appreciably and may accumulate high PAH amounts.

Food Smoking

There are considerable variations in the PAH contents of smoked foods, even within the same type of food. These are due to the variations in the smoke generation conditions, which include smoke production temperature, type and comminution of wood material, type of generators, time of smoking, and fat content of products. In general, the higher the smoking temperature, the more PAHs are formed.

High levels of individuals PAHs, and BaP in particular, were found in smoked fish (BaP, in the order of $10 \, \mu g \, kg^{-1}$); this may be locally of major health concern for populations who consume large

Table 2 Concentrations (μg kg^{-1}) of selected polycyclic aromatic hydrocarbons (PAHs) in various foods

	n	PHE	FA	PY	BaA	CHR	BbFA	BkFA	BaP	BeP	BghiP	IP	DBahA
Meat and meat products													
Meat and meat products[a]	4	3.0	0.5–1.1	0.5	0.02–0.5	0.1–0.6	0.04–1.0	0.01–0.2	0.01–0.6	0.03	0.03–0.6	0.01–0.7	0.01
Poultry and eggs[a]	1		0.9		0.5		0.3	0.2	0.1		0.2	0.2	
Eggs	1		0.1	ND	0.03	ND	0.4	0.01	0.01		ND		
Eggs, after Gulf War[b]	1	18(295)	1.2(3.9)	5.5(27)	4.5(7.8)	5.6(32)	3.5(13)	4.5(6.5)	7.5(19)		1.2(1.5)	8.7(20)	4.7(5.8)
Bacon	1		7.8				0.3	0.05	0.05		3.70	2.5	
Sausages	1				0.04–0.1				0.03–0.3		0.05–0.2	0.05–0.1	
Smoked meat	1				Tr–0.3[k]				0.01–0.1		Tr–0.1	Tr–0.1	
Smoked beef	1				0.02–0.6				0.02–0.4		0.03–0.3	0.04–0.4	
Smoked chicken	1	32	16	20	2.1	6.3	4.6	2.6	3.2			1.7	
Smoked sausages	2				0.04–0.4				<0.01–0.3		0.06–0.3	0.04–1.4	
Grilled sausages	2	3.5–618	1.1–376	1.2–452	0.2–144	0.3–140[k]	ND–92	ND–172[n]	ND–212	ND–81	ND–153	ND–171	ND–8.8
Grilled duck[c]	1	25	20	24	2.4	31	10	5.8	9.2		ND	5.2	ND
Fish and marine products													
Fish	4	3.5	0.8–1.8	0.8	Tr–7.5	0.6–2.9	0.1–2.0	0.04–0.7	Tr–4.5	0.1	Tr–0.9	ND–1.6	0.03
Fish, after an oil spill[d]	1	<2–101	<2–123	<2–145	<2	<2	0.1–5.8		<2–7.6				
Fish and shrimps, after Gulf War[e]	1	5.8–87	ND–33	ND–68	0.1–5.3	ND–16	0.1–5.8		ND–5.3		0.2–31	0.3–29	0.2–39
Fish, oil-contaminated sea[f]	1	13	1.5	1.7	0.3	1.7	0.4	0.4	0.8		0.3	0.1	0.1
Oysters	1	2.1–4.2	5.1–17	3.1–12	0.8–3	3.2–8.8[k]	4.5–12[m]	see BbFA	0.4–1.0	2.4–6.3	0.4–0.8	0.3–0.6	0.1–0.2[o]
Trout[g]	1	ND–0.07	0.1–0.4	0.8–2.9	0.2–1.5	0.07–0.3	ND		ND	ND		ND	ND
Smoked trout[g]	1	7.6–8.3	27–39	82–175	9.6–16	1.5–2.9	ND–0.08		5.1–8.4			2.0–4.2	3.2–4.0
Smoked fish	3	5–330	1.4–210	1.3–68	ND–86	ND–50	ND–3.9	ND–6.7[n]	ND–40	ND–2.8	ND–25	ND–37	
Smoked fish, traditional kilns[h]	1	65	26	20	2.5	2.5	1.2	0.5	1.2		0.7	1.1	<0.1
Smoked fish, modern kilns[h]	1	32	9.1	5.3	0.6	0.6	0.1	0.07	0.1		0.03	ND	ND
Smoked or canned oysters and mussels	1	1.9–20	4.5–19	2.6–11	0.8–21	3.9–31[k]	1.2–24[m]	see BbFA	0.2–12	0.7–7.6	0.3–5.7	0.2–6.4	<0.1–0.5[o]
Broiled fish	1	320	1080	390		390			400				
Vegetables													
Kale	1		117	70	15	62			4.2	7.9	7.7	7.9	1.0
Lettuce	4	0.5–12	0.09–28	0.4–18	0.05–4.6	0.8–4.0	0.1–7.3	ND–17	0.007–6.2	0.07–6.7	0.02–11	0.1–8	
Lettuce, industrial area	1	28					6.1	3.7	5.6		10	2.4	
Potatoes[a]	1		ND		0.4	0.8	0.2	0.1	ND		0.1	ND	
Potato products[a]	1	3.0	0.6		0.1	ND	0.6	0.3	0.3		0.9	0.6	
Tomatoes	2		0.09	0.4	0.006–0.3	0.1–0.5	0.008	0.003	0.003–0.2	0.2	ND	ND	0.04
Soups[a]	1		ND		ND	0.4	0.05	0.08	ND		ND	ND	
Fruits and confectionery													
Fresh fruit[a]	1	7.8	3.6		0.5	0.5	0.1	0.1	ND		ND	ND	
Apples	1	ND	2.4	3.5	0.3	ND	0.3	0.08	0.5		0.02		
Canned fruit and juices[a]	1	ND	1.0		ND	ND	0.1	0.1	0.1		0.1	ND	
Nuts	1	10(17)	1.3(3.0)		0.1(4.2)	4.0(69)	0.2(0.4)	ND(0.1)	ND(0.2)		ND(0.4)	ND(0.4)	
Biscuits, pudding, and cakes	2	2.9	0.5–3.6	0.6–2.4	0.08–2.7	0.09–2.8	0.03–1.3	0.04–1.4	0.04–2.2	0.08–2.9	0.1–2.5	0.1–3.2	
Sugar and sweets	1	ND(3.2)	0.7(2.3)		0.2(4.2)	0.7(36)	0.2(3.5)	0.1(0.5)	0.1(0.4)		ND(0.2)	ND(0.2)	
Cereals													
Breakfast cereal	1		0.2–0.6	0.3–1.2	0.06–0.2		0.02–0.05	0.02–0.07	0.03–0.05	0.06–0.2[f]	0.06–0.08	0.08–0.15	<0.01

	n												
Bread	3	3.3	0.2–2.8	ND–0.9	0.1–0.8	0.1–1.9	0.04–1.2	0.02–0.06	0.02–0.8	0.06–0.1	0.01–0.5	0.11–0.6	<0.01–0.01
Pasta, macaroni, and rice	2	2.1	2.5–3.9	ND	0.03–0.4	1.3–1.9	0.04–1.0	0.02–0.5	0.017–0.8		ND–0.6	0.5	0.5
Noodles and pizza	1	4.2	3.7		0.5	2.0	0.5	0.1	0.2		0.5	0.3	0.3
Pizza in a wood-burning oven	1		0.2	0.6	9.1	3.0	0.04	0.02	0.02		0.06		
Oils and fats													
Olive oils	3	0.4–38	Tr–15	ND–14	0.03–0.9	0.4–1.5	0.3–1.4	0.06–0.5	0.1–1.3	<1	ND–1.9	0.3–2.0	ND–0.1
Extra virgin olive oils	1	1.7–53	Tr–53	0.4–28	Tr–10	Tr–2.3	Tr–1.3	Tr–0.5	Tr–1.2		Tr–0.7	Tr–0.6	Tr–0.3
Various unrefined oils[i]	1	ND–69	0.2–18	0.1–14	ND–6.1	0.1–8.6[k]	ND–8.9[m]	see BbFA	ND–4.1	ND–3.8	ND–4.2	ND–4.3	ND–0.2[o]
Coconut oil, deodorized	1	8.7	23	29	5.7[l]	see BaA	1.0[m]	see BbFA	0.2	0.4	<0.1		ND
Corn oil	1	0.1	ND	ND	ND	ND	0.09[m]	see BbFA	0.02	0.04	0.1	ND	0.03[o]
Butter	1		0.1	1.8	0.6	ND	0.02	0.03	0.02		ND		
Margarine	3	<0.2–6.0	0.09–9.0	<0.1–15	<0.1–5.2	<0.2–7.5	<0.2–9.2	<0.1–11	0.05–6.0	0.09–6.1	0.02–11	0.03–9.7	0.05–9.2
Beverages													
Wine	2		0.08	ND	0.003	ND	Tr–0.04	Tr–0.01	Tr–0.009		ND–0.03	ND	ND
Beer	2		0.04	ND	0.1	ND	Tr–0.2	ND–0.1	ND–0.07		ND–0.2	ND–0.1	ND–Tr
Coffee	1		1.2	ND	0.1	ND	0.08	0.02	0.01		0.01		
Tea infusion[j]	1	0.2–0.7	0.007–0.03	0.02–0.2	0.003–0.043	ND–1.5	0.01–0.06	0.003–0.03	0.002–0.02	0.007–0.056	0.007–0.03		0.005–0.02
Milk	2	ND	0.1–0.2	0.04–0.2	0.01			0.003–0.03	0.01–0.02	ND	0.01–0.03	ND	ND
Milk, after Gulf War[b]		3.0(9.8)	3.4(12)	35(145)	2.4(3.7)	8.6(32)	3.1(4.4)	ND	1.5(1.5)	ND	ND	ND	ND

A large part of data was adapted from International Programme on Chemical Safety (1998). *Selected Non-heterocyclic Polycyclic Aromatic Hydrocarbons. Environmental Health Criteria 202.* Geneva: World Health Organization. Unless otherwise specified, an individual value is the mean concentration (if $n = 1$) or the only available mean concentration (if $n > 1$). Concentrations in parentheses are maximum values.

n, number of investigations; ND, not detected; Tr, traces; for PAH abbreviations, see **Table 1**.

a Maximum concentrations (mean concentrations were mostly ND).

b Kuwait, from locally reared animals.

c Grilling time: 0.5 h.

d Arabian Gulf.

e Kuwait.

f Red Sea coast; oil operations and heavy ship traffic.

g 'Trout' and 'smoked trout' were investigated within the same study.

h 'Traditional kilns' (27 samples, three kilns) and 'modern kilns' (35 samples, five kilns) were investigated within the same study; mean concentrations are reported.

i Olive, safflower, sunflower, maize germ, sesame, linseed, wheat germ oils.

j Concentrations in µg l^{-1}.

k In sum with triphenylene.

l In sum with CHR and triphenylene.

m In sum with BjFA and BkFA.

n In sum with BjFA.

o In sum with dibenz[a,c]anthracene.

quantities of these products. Considerable amounts of PAHs were found in home-smoked meat in Iceland, with most of BaP detected in the superficial layers of the meat, and in smoked cheese too. The highest BaP concentrations were found in samples of smoked cereals (up to $160\,\mu g\,kg^{-1}$) and smoked teas (up to $110\,\mu g\,kg^{-1}$).

Traditional smoking techniques, in which smoke from incomplete wood burning comes into direct contact with the product, can lead to extensive contamination with PAHs. Hence, smoke flavorings are used as an alternative. Current commercial smoking practices appear to be effective in controlling the deposition of PAHs. In a recent Spanish survey of commercial smoked products, BaP was found at low levels in almost all samples: up to $0.3\,\mu g\,kg^{-1}$ in meat products and up to $0.9\,\mu g\,kg^{-1}$ in cheese and in fish, with the exception of $2.5\,\mu g\,kg^{-1}$ in a sample of smoked sardine. The oil associated with the fish, if any, may be more contaminated than the fish itself, possibly due to the PAH leaching from the fish.

In a German survey conducted in the 1990s, PAH concentrations were compared between 27 smoked fishery products from traditional smoking kilns and 35 from modern smoking kilns with external smoke generation. The BaP levels from the modern kilns were one order of magnitude lower (on average, $0.1\,\mu g\,kg^{-1}$) than by the traditional systems ($1.2\,\mu g\,kg^{-1}$).

Meat and Meat Products

The amount of PAHs formed during cooking depends markedly on the method of cooking and the conditions. In particular, PAH formation during charcoal grilling is dependent upon the fat content of the meat, the time of cooking, and the temperature.

This formation may be due to various causes: the incomplete combustion of charcoal, the transformation of some food components such as triglyceride and cholesterol, or – the most likely source of high PAH levels – the melted fat of the meat. Actually, during charcoal grilling at high temperatures, the fat drippings fall on the hot coals where they are pyrolyzed, producing PAHs which are then volatilized and deposited onto the meat surface.

An investigation of the effect of the method, including broiling on electric and gas heat, charcoal broiling, and broiling over charcoal with a no-drip pan, showed that PAH formation may be minimized by avoiding direct contact of the food with the cooking flame, cooking meat at lower temperatures for longer periods, and using meat with a low fat content.

The total concentration of six carcinogenic PAHs in a lamb sausage increased from $1.9\,\mu g\,kg^{-1}$ when grilled over charcoal under standard barbecue practices (distance between fire and meat 10 cm; temperature at the surface of the meat 200–250 °C) to $13.2\,\mu g\,kg^{-1}$ when heavily grilled for a prolonged time.

An experiment on PAH formation in duck meat during various processing methods showed the highest concentrations of carcinogenic PAHs were due to smoking (1 m distance between smoke generation source and meat), followed by charcoal grilling (1 m distance between charcoal and meat) and roasting (at 200 °C); no carcinogenic PAHs were detected in samples subjected to steaming or to flavoring by liquid smoke. The BaP concentration during smoking increased from $6.9\,\mu g\,kg^{-1}$ after 0.5 h to $13.9\,\mu g\,kg^{-1}$ after 3 h.

Fishes and Marine Products

Processing and cooking may increase PAH levels by the same routes described above for meats. The highest levels of BaP and other PAHs were generally found in smoked fishes, and especially in the smoked skin. In a study of commercial Baltic herrings, BaP was found at about $40\,\mu g\,kg^{-1}$ in a smoked specimen, but at a level one order of magnitude higher in a broiled one (about $400\,\mu g\,kg^{-1}$, which is the highest level of contamination reported). High concentrations are also found in marine products from contaminated sea.

Vegetables

The differences in PAH content may be due to variations in the ratio between the surface area and the weight, to location (rural or industrialized) or to growing season.

It is recognized that broad-leaved vegetables may have particularly high levels of PAHs, due to the deposition of airborne particulate matter. The importance of atmospheric pollution was shown by the high levels detected in lettuce grown close to a highway, with levels decreasing with distance from the road. The PAH profiles in lettuce were found to be similar to those in ambient air, confirming the deposition of airborne particles as the main source of contamination. In an investigation near a coking plant, PAH levels in vegetables with large, rough leaf surfaces (spinach and lettuce) were found to be 10 times higher than in other vegetables (carrots and beans), and this was likely due to deposition from ambient air.

The effect of washing vegetables on reducing PAH contamination due to vehicle exhaust appears controversial: in an investigation on kale, PAH content was not reduced other than at levels lower than about 10% (BaP, 10%). Conversely, in an experiment with lettuce, the higher-molecular PAHs were considerably reduced (BaP, 67–95%). In no study, however, were lower-molecular PAHs significantly affected by washing.

In an experimental comparison of growth of terrestrial plants in a 'clean-air' chamber and in the open field, the contamination was shown to be due almost exclusively to airborne PAHs and not to synthesis by plants.

Cereals and Dried Foods

Growing crops (wheat, corn, oat, barley) in industrialized areas increases PAH levels in comparison to more remote areas. Grain samples from a heavily industrialized area contained 10 times more PAHs than samples from areas remote from industry. The growth of rye near a high-traffic highway resulted in PAH contamination, which decreased slightly 7–25 m away from the road.

Drying by combustion gases was also found to increase contamination by three- to 10-fold. In an experiment with wheat, drying over a light fuel oil flame increased the BaP deposition from six- to 130-fold depending on the degree of exposure.

Oils

The presence of PAHs in vegetable oils is due to contamination from technological processes (namely, smoke drying of oil seeds or contaminated extraction solvents) or to environmental contamination (e.g., traffic exhaust or industrial emissions).

BaP in olive oils is commonly within the $1 \, \mu g \, kg^{-1}$ level; higher levels may be present if the oils are from plants exposed to industrial emissions or if they are blended with previously contaminated vegetable oils.

The refining of olive oils (especially the deodorization step) has a marked reducing effect on the light PAHs but not on the heavy ones. The latter may be reduced by treatment with activated charcoal.

High PAH levels may be found in oil seeds. They are usually due to the process of direct drying the seeds, using wood or oil as a fuel. BaP at levels up to $25 \, \mu g \, kg^{-1}$ were found in corn oils. The PAH content of oil seeds is drastically reduced during refining, particularly by treatment with activated charcoal. This refining method is now widely used.

Dietary PAH Intake

Table 3 shows the dietary intake of PAHs, as estimated from different studies in three European countries. The intakes were calculated on the basis of food consumption surveys, except in one Dutch study where portions of a 24-h intake of food were provided by the participants and analyzed. There are numerous differences between the diets of northern and southern Europe and the criteria adopted in

Table 3 Daily intake of selected polycyclic aromatic hydrocarbons (PAHs) in total diet, as estimated in various countries (mean and, in parentheses, where available, maximum values)

PAH	Intake ($\mu g \, person^{-1}$)			
	UK[a]	The Netherlands[b,c]	Italy[c,d]	Range of mean values
PHE	NG	2.69 (5.13)	NG	
FA	0.99	1.32 (2.11)–2.7 (10.4)	4.22	1.0–4.2
PY	1.09	1.6 (5.1)	1.73	1.1–1.7
BaA	0.22	0.16 (0.48)–0.28 (0.65)	0.41–1.29	0.2–1.3
CHR	0.5	1.19 (3.90)–1.2 (5.0)	1.46 (1.70)–2.20	0.5–2.2
BbFA	0.18	0.33 (0.59)	0.65	0.2–0.6
BkFA	0.06	0.12 (0.24)	0.17	0.06–0.2
B[b+j+k]FA	> 0.24	> 0.45	1.10	> 0.2–1.1
BaP	0.25	0.08 (0.35)–0.20 (0.42)	0.17 (0.32)–0.37	0.08–0.4
BeP	0.17	0.14 (1.2)	NG	0.1–0.2
BghiP	0.21	0.16 (0.58)–0.28 (1.03)	0.02	0.02–0.3
IP	ND	0.16 (1.2)–0.27 (0.55)	0.16 (0.20)	ND–0.3
DBahA	0.03	0.04 (0.53)	0.08 (0.17)	0.03–0.1

ND, not detected in any food group; NG, not given; PAH abbreviations: see **Table 1**. B[b+j+k]FA: sum of BbFA, BjFA and BkFA; BjFA was not determined in any study.

[a]From Dennis MJ, Massey RC, McWeeny DJ, Knowles ME and Watson D (1983) Analysis of polycyclic aromatic hydrocarbons in UK total diets. *Food and Chemical Toxicology* 21: 569–574.

[b]From De Vos RH, Van Dokkum W, Schouten A and De Jong-Berkhout P (1990) Polycyclic aromatic hydrocarbons in Dutch total diet samples (1984–1986) *Food and Chemical Toxicology* 28: 263–268; Vaessen HAMG, Schuller PL, Jekel AA and Wilbers AAMM (1984) Polycyclic aromatic hydrocarbons in selected foods: Analysis and occurrence. *Toxicological and Environmental Chemistry* 7: 297–324.

[c]When a range is reported, it refers to two different studies.

[d]From Turrio-Baldassarri L, Di Domenico A, La Rocca C, Iacovella N and Rodriguez F (1996) Polycyclic aromatic hydrocarbons in Italian national and regional diets. *Polycyclic Aromatic Compounds* 10: 343–349; Lodovici M, Dolara P, Casalini C, Ciappellano S and Testolin G (1995) Polycyclic aromatic hydrocarbon contamination in the Italian diet. *Food Additives and Contaminants* 12: 703–713.

the surveys are not homogeneous. Consequently, the intakes estimated in different studies cannot be directly compared and have to be considered as a rough estimate of the actual intakes. Notwithstanding these limitations, it is worth noticing that they are generally of the same order of magnitude and the differences are relatively small compared with the variation in cancer potency estimates.

The Contribution of the Different Food Groups

The contribution of different food groups to the total dietary intake of PAHs was estimated in the UK (Table 4).

The major contributors were the oil-and-fats group and the cereals group. The former has the highest individual PAH levels, but the latter, although never showing high individual PAH levels, is a major contributor due to its relative weight in the total diet. Although several studies indicated quite large amounts of PAHs in smoked meat and smoked fish, these products made a very small contribution to the pertinent food groups and these in turn were not major components of the diet. Consequently, at least in areas where eating barbecued food is an infrequent activity, this provides a very small part of the dietary intake of PAHs. The third major contributor was vegetables, and this was likely due to the atmospheric fallout of particle-bound PAHs.

These results were substantially confirmed in a subsequent study of the Dutch diet: cereal products were found to contribute most to the daily intake of PAHs (again attributed to the high consumption share), followed by sugar-and-sweets and by oils and fats. It was not possible to explain the surprisingly large share of the sugar-and-sweets group because none of the constituents (sugar, chocolate products, jellies, licorice) was suspicious with regard to high PAH levels. The relatively high contribution

of the oils-and-fats group was at least partly attributed to the well-known elevated PAH concentrations possibly present in vegetable oils. Nuts, which were included in this group, contributed 6% to the total PAH intake, and this was most likely due to the roasted peanuts.

Cereals were also found to be the highest contributors in a Swedish study (where they were followed by vegetables, and by oils and fats), and in an Italian one.

A more accurate comparison of the results of all these studies is not possible due to the differences in the reporting criteria adopted.

See also: **Analysis of Food**; **Cancer**: Carcinogens in the Food Chain; **Carcinogens**: Carcinogenicity Tests; **Chromatography**: High-performance Liquid Chromatography; Gas Chromatography; **Dietary Surveys**: Surveys of National Food Intake; **Flavor (Flavour) Compounds**: Production Methods; **Legislation**: International Standards; **Smoked Foods**: Applications of Smoking; Production; **Smoking, Diet, and Health**

Further Reading

Dennis MJ, Massey RC, McWeeny DJ, Knowles ME and Watson D (1983) Analysis of polycyclic aromatic hydrocarbons in UK total diets. *Food and Chemical Toxicology* 21: 569–574.

De Vos RH, Van Dokkum W, Schouten A and De Jong-Berkhout P (1990) Polycyclic aromatic hydrocarbons in Dutch total diet samples (1984–1986). *Food and Chemical Toxicology* 28: 263–268.

Fazio T and Howard JW (1983) Polycyclic aromatic hydrocarbons in foods. In: Bjørseth A (ed.) *Handbook of Polycyclic Aromatic Hydrocarbons*, pp. 461–505. New York: Marcel Dekker.

Grimmer G and Böhnke H (1979) Method 4 – Gas chromatographic profile analysis of polycyclic aromatic hydrocarbons in (I) high protein foods, (II) fats and vegetable oils and (III) plants, soils and sewage sludge. In: Egan H, Castegnaro M, Kunte H, Bogovski P and Walker EA (eds) *Environmental Carcinogens Selected Methods of Analysis*, vol. 3, pp. 163–173. *Analysis of Polycyclic Aromatic Hydrocarbons in Environmental Samples*. Lyon: International Agency for Research on Cancer.

Guillén MD (1994) Polycyclic aromatic compounds: extraction and determination in food. *Food Additives and Contaminants* 11: 669–684.

IARC (1983) *Polynuclear Aromatic Compounds, Part 1, Chemical, Environmental and Experimental Data. IARC Monographs on the Evaluation of the Carcinogenic Risk of Chemicals to Humans*, vol. 32. Lyon: International Agency for Research on Cancer.

IARC (1987) *Overall Evaluations of Carcinogenicity: An Updating of IARC Monographs vols. 1–42. IARC Monographs on the Evaluation of Carcinogenic Risks*

Table 4 Contribution from individual food groups to the total UK dietary intake of polycyclic aromatic hydrocarbons (PAHs)

	% of total dietary intake	
	BaP	Sum of six PAHs[a]
Cereals	30	35
Meat	3	4
Fish	1	2
Oils and fats	50	34
Fruit, sugar and sweets	5	7
Vegetables	8	18
Beverages	0	0
Milk	2	1

Modified from Dennis MJ, Massey RC, McWeeny DJ, Knowles ME and Watson D (1983) Analysis of polycyclic aromatic hydrocarbons in UK total diets. *Food and Chemical Toxicology* 21: 569–574.
[a]Olin, DuA, DeF, BaP, DBahA and IP: for abbreviations, see Table 1.

to *Humans*, Supplement 7. Lyon: International Agency for Research on Cancer.

International Programme on Chemical Safety (1998) *Selected Non-heterocyclic Polycyclic Aromatic Hydrocarbons. Environmental Health Criteria 202.* Geneva: World Health Organization.

Lijinsky W (1991) The formation and occurrence of polynuclear aromatic hydrocarbons associated with food. *Mutation Research* 259: 251–261.

Lodovici M, Dolara P, Casalini C, Ciappellano S and Testolin G (1995) Polycyclic aromatic hydrocarbon contamination in the Italian diet. *Food Additives and Contaminant* 12: 703–713.

Phillips DH (1999) Polycyclic aromatic hydrocarbons in the diet. *Mutation Research* 443: 139–147.

Thomson B and Muller P (1998) Approaches to the estimation of cancer risk from ingested PAH. *Polycyclic Aromatic Compounds* 12: 249–260.

Turrio-Baldassarri L, Di Domenico A, La Rocca C, Iacovella N and Rodriguez F (1996) Polycyclic aromatic hydrocarbons in Italian national and regional diets. *Polycyclic Aromatic Compounds* 10: 343–349.

Vaessen HAMG, Schuller PL, Jekel AA and Wilbers AAMM (1984) Polycyclic aromatic hydrocarbons in selected foods: Analysis and occurrence. *Toxicological and Environmental Chemistry* 7: 297–324.

Polyphenols *See* **Tannins and Polyphenols**

Polysaccharides *See* **Carbohydrates**: Classification and Properties; Interactions with Other Food Components; Digestion, Absorption, and Metabolism; Requirements and Dietary Importance; Metabolism of Sugars; Determination; Sensory Properties

---

PORK

R W Mandigo, University of Nebraska, Lincoln, NE, USA

This article is reproduced from *Encyclopaedia of Food Science, Food Technology and Nutrition*, Copyright 1993, Academic Press.

Development of the Swine Industry

The earliest known records of swine domestication are from China and date to 4900 BC. Christopher Columbus brought the first pigs to the USA via the Canary Islands in 1493. The early colonists of the USA brought livestock with them from England throughout the 1600s. Gradually, as more grain was grown, larger herds of pigs developed and swine production became a true industry.

By the early 1800s, swine-slaughtering plants had emerged at several large population centers in the USA. The largest of these was at Cincinnati (also known as Porkopolis at that time) because of its prime location on the Ohio river. Early slaughtering plants were located near rivers for four reasons: rivers provided ice to keep the meat cold, transportation for products to be shipped, a source of water, and a place to deposit plant waste materials such as blood.

The advent of mechanical refrigeration led to industry expansion as pork could be processed year-round and kept fresh longer. Development of rail systems and the use of refrigerated rail carriages boosted industry growth as both livestock and meat could be more widely distributed. In the 1800s, pigs were often allowed to roam free in pastures and were fed garbage or what little grain was available. Today, most pigs are raised in large numbers in environmentally controlled buildings with a very specific diet designed to maximize growth.

The first slaughter operations were separate from packing operations. By the 1850s, the two had started to become integrated. Today, many plants have reverted back to specializing in either slaughter and fabrication or pork processing. This specialization has allowed modern-day slaughtering facilities to process around 1000 head per hour.

The pigs in the USA today are thought to have descended from two wild stocks, the European wild boar (*Sus scrofa*) and the East Indian pig (*Sus vittatus*). From these two genetic bases, modern pigs have been refined into several breeds which each have distinguishing characteristics. In addition to hair color and marking differences, breeds have been developed for specific purposes such as prolificacy, bacon production, ham production, and others.

In general, pigs have become leaner as the demand for lard has decreased and the demand for leaner meats has increased. The modern market pig is approximately 71.5% carcass while the remaining 28.5% is mostly byproducts used in the medical and animal feed industries.

Slaughter, Fabrication, and Processing

Procurement

Although pigs of all sizes can be slaughtered, most swine-slaughtering plants have equipment designed to handle pigs marketed at about 105 kg. Swine producers can sell their pigs to slaughter plants in various ways. Pigs can be marketed at a central location such as an auction or a packer buying station or they can be purchased from the farm by a packer buyer. The latter method is the most popular today. A new way to market pigs that has some popularity is contract producing for a specific packer.

After pigs arrive at the slaughtering plant, they are slaughtered within 2–3 h of arrival. If a substantial delay is expected, then they will usually be held in pens for 12–24 h. It is important to allow pigs time to become acquainted with their environment as high-stress conditions can cause undesirable changes in muscle color and juiciness. In pigs with a certain genetic make-up, severe stress can cause the meat to be pale and watery (known as pale, soft, and exudative, PSE). Long-term stress, on the other hand, can lead to a condition where the muscle is dark and very dry-appearing (known as dark, firm, and dry, DFD). The physiological basis for these conditions is an abnormal muscle pH change between the time of death and the completion of rigor mortis. Therefore, it is necessary to minimize stress by proper handling of pigs prior to slaughter. Feed is usually kept from the animals during this period to decrease stomach fill. This improves percentage of carcass dressing, aids evisceration during slaughter, helps minimize contamination, and improves color. (*See* **Meat**: Slaughter.)

Slaughter

All meat sold interstate in the USA must be inspected by the US Department of Agriculture, Food Safety and Inspection Service (USDA-FSIS). Inspection begins *antemortem* and continues throughout the entire slaughter, fabrication, and processing system. This helps ensure the wholesomeness of meat products offered to the consumer. If the meat will be sold intrastate only, some states allow inspection by state-employed inspectors. This inspection, by law, must be equal to federal inspection. Similar inspection regimes apply in Europe and other developed regions of the world.

At the time of slaughter, pigs are herded into a chute where they are stunned to render them unconscious. Stunning can be accomplished by one of several methods, but is mandatory to comply with the US Humane Slaughter Act of 1958. If stunning is not performed correctly, the incidence of poor-meat-quality problems could increase. An example of this is a condition referred to as blood splashing. Blood splashing occurs when small capillaries in muscle tissue burst during stunning, leaving spots of blood in the muscle. The most common method of stunning is accomplished using an electrical current. Other methods include a blow from a captive bolt to the head, moving pigs through a carbon dioxide gas chamber to cause suffocation, or cardiac arrest stunning. The goal of stunning is to render the animals unconscious while allowing the heart to continue to beat and pump blood out of the body.

After stunning, the pig is immediately shackled, hoisted by the hind leg, and the jugular vein is severed with a knife to allow blood to exit the body. Bleeding will remove around 50% of the blood in the body. Much of the remaining blood is removed with the organs during evisceration.

Pigs can either be skinned or they can be scalded and dehaired. To dehair carcasses, the pig is put in a scalding vat of hot water (60 °C) with lye to loosen the hair follicles. The pig is removed from the scalding tank and dehaired mechanically and/or by hand using blades to scrape away the hair. Whole skins from skinned pigs can be saved and made into pigskin for use in garments and for gelatin production.

After dehairing or skinning, the abdominal and thoracic cavities are opened and the internal organs removed. Certain organs are washed, inspected, and saved as edible byproducts. These include the liver, heart, tongue, and brain. Other organs may also be saved as inedible byproducts, and will be discussed later in this article.

The carcass is washed, weighed, and placed in a cooler for chilling. The cooler temperature is maintained at about −2 to 5 °C so that internal areas of the carcass will be adequately chilled in 24 h. While in the cooler, most of the remaining stored energy in the muscle will be depleted and the carcass will undergo rigor mortis. Some carcasses are processed before they go through rigor mortis. This is called hot boning. Although fabrication is more difficult, the cuts can be chilled faster than whole carcasses, so product that has been hot-boned can be moved through the system faster.

Fabrication

Pig carcasses are fabricated into wholesale cuts, using knives, saws, and other mechanical equipment. The wholesale cuts include the ham, loin, Boston shoulder, picnic shoulder, belly, and spare ribs. The wholesale cuts may be shipped fresh to further processors, food service operations, or to the retail market. Some plants will continue to fabricate the wholesale cuts into retail cuts and trimmings and sell these items. Many plants that slaughter will carry out further processing such as curing and smoking bellies and hams in the same plant. (*See* **Curing.**)

Processing

Further processing of pork can involve many different processes. Processing encompasses many steps, including particle size reduction, the addition of nonmeat ingredients, cooking, smoking, and a variety of packaging procedures. Patties can be made by grinding pork trimmings, with a specified fat level, with salt and other seasonings. Restructured products often have salt and phosphates added to flaked or chunked meat. These ingredients are then mixed and formed into steaks, chops, or roasts. Sausage is another type of further processed meat. Sausages can be cured, smoked, dried, or cooked. Some sausages are coarsely ground while others are finely chopped, forming what is sometimes called an emulsion or batter. Bratwurst, frankfurters, and bologna are sausages that typically contain pork. (*See* **Meat**: Sausages and Comminuted Products.)

The hams and bellies from pork carcasses are usually cured and smoked. The curing may be done by packing in a dry cure or, more commonly, by pumping or injecting a solution containing the cured ingredients into the meat. The curing solution may contain any or all of the following ingredients: water, salt, sugar, phosphate, nitrite, erythorbate, or spices and seasonings. Water is used to add moisture to the product as well as acting as the solvent in a curing solution. Salt solubilizes proteins in meat so the meat particles bind together better. Salt also adds flavor and, at high levels, retards microbial spoilage. Sugar can be used for flavor, to promote browning, and to act as a food source for desirable bacteria in some fermented meat products. Phosphates are added to help retain water in the product

and to prevent oxidative rancidity from occurring. Nitrite is the curing agent. Nitrite binds to meat pigments and causes the meat to develop a stable pink color. It also lends flavor, acts as an antimicrobial agent, and helps to prevent oxidative rancidity. Erythorbate is used to speed up the curing process, which is economically important to pork processors. Spices and seasonings are added to develop flavors unique to a particular product. (*See* **Smoked Foods**: Principles.)

Some sausages are cured and may also contain acidulants, starter cultures, or extenders. Acidulants and starter cultures of bacteria can be used to lower the pH of products in order to develop certain flavors and protect the product from spoilage. Nonmeat extenders, such as soya proteins or nonfat dry milk, may be used as inexpensive protein sources.

The US Department of Agriculture (USDA) has regulations regarding the composition of processed meat products and the proper use of some ingredients used in processing. Hams must be labeled according to the amount of protein they contain on a fat-free basis. The amount of phosphates, nitrites, erythorbate, and nonmeat ingredients added are also limited. These regulations, and similar ones in other countries, are designed to provide consumers with high-quality, safe meat products.

Pork is also regulated by the USDA in an attempt to control the disease trichinosis. Trichinosis is caused by the organism *Trichinella spiralis* and can be transmitted from pig to humans by ingestion of infected muscle tissue. Simply put, the regulation states that pork for use in products that are not fully cooked in the home must be certified as 'trichina-free.' Pork may become certified trichina-free by being subjected to specific heat or freezing to kill the organism. (*See* **Zoonoses.**)

Chemical Composition

Fresh pork muscle is 70–75% water, and the protein content ranges from 18 to 22%. There are three main types of protein in the pig: myofibrillar (skeletal), stromal (connective tissue), and sarcoplasmic (pigments). Lipid or fat is another major constituent of fresh pork. It makes up between 5 and 7% of the muscle tissue. Lipids include phospholipids and triacylglycerols (also known as triglycerides). The carbohydrate content of meat is negligible, generally less than 1%. Vitamin and mineral content of fresh pork is usually about 1–2%. (*See* **Fats**: Classification; **Meat**: Structure; Nutritional Value; **Protein**: Chemistry.)

Different cuts in the carcass will have different compositions. This is largely due to the varying fat

Table 1 Composition of various raw pork cuts (g kg^{-1} raw meat)

	Moisture	Fat	Protein	Ash
Ham slice	730	55	205	10
Bacon	370	530	95	5
Loin chop	710	70	210	10
Spare ribs	575	245	170	10

level in different areas of the carcass. **Table 1** contains some example cuts and their typical raw composition.

Nutrient Value and Dietary Significance of Pork

Pork supplies many nutrients essential for maintenance and growth.

As with other meat, pork is an excellent source of protein. A single 85-g serving of pork contributes 41% of the daily protein requirement for a normal adult male. Not only does pork contain a large amount of protein, this protein is of good quality. (*See* **Protein**: Requirements.)

Pork also contains lipids and fats. About 34% of pork fatty acids are saturated and 66% are unsaturated. Cholesterol is another lipid found in pork. Cholesterol is found in cell membranes in the animal body and is synthesized in the liver of humans and animals. Consumption of animal products, therefore, provides a dietary source of cholesterol which can be used in the body. Cholesterol, like saturated fats, has been associated with increased risk of developing heart disease. The relationship is not well understood, but the American Heart Association recommends keeping dietary intake of cholesterol to less than 300 mg day^{-1}. One 85-g serving of pork provides about 79 mg of cholesterol or about 26% of the recommended 300 mg. (*See* **Cholesterol**: Role of Cholesterol in Heart Disease; **Fatty Acids**: Properties.)

Pork is an excellent food source for several vitamins and minerals. It supplies large amounts of thiamin, vitamin B$_{12}$, niacin, riboflavin, and zinc. Pork is also a good source of vitamin B$_6$, phosphorus, and iron. Dietary iron can be classified into two types, heme and nonheme. Heme iron, which is the major type found in pork, is absorbed more easily and better utilized by the body. Iron is a component of the molecule hemoglobin, which is the major carrier of oxygen in our blood stream. Intake of heme iron is especially important in warding off anemia, which may result from a low level of hemoglobin in the blood. Refer to individual vitamins and minerals.

Pork, when consumed in moderation, is an excellent source of many important dietary nutrients.

Microbiological and Other Hazards

Muscle is essentially sterile prior to death. However, meat destined for human consumption is cross-contaminated with microorganisms by equipment and handling at the time of slaughter and processing. Just as pork is an excellent source of nutrients for our bodies, muscle or meat is also an excellent growth medium for microorganisms. Controlling the growth of microorganisms on pork by acidifying, curing, salting, modified-atmosphere packaging, drying, cooking, or refrigerating is essential. (*See* **Meat**: Preservation.)

Food poisoning can result from consuming pork that has been mishandled, allowing certain microorganisms to grow. Causative organisms of food poisoning may include *Staphylococcus aureus*, *Bacillus cereus*, *Salmonella* spp., *Listeria monocytogenes*, *Yersinia enterocolitica*, *Clostridium botulinum*, *C. perfringens*, and *Campylobacter fetus* ssp. *jejuni*. Food poisoning is quite common and, in its mildest forms, is often mistaken for influenza because the symptoms are very similar. Generally, foodborne illnesses are relatively short-lived and more uncomfortable than harmful. However, food poisoning can be a very serious matter. It can be debilitating or even fatal for those with poor immunological defenses such as infants or the elderly. Like other microorganisms, pathogens which cause food poisoning are well controlled by heat, refrigeration, chemicals, or other means mentioned earlier. However, undercooking, improper cooling, or recontamination of cooked food by raw food are common ways that pathogens appear in the food supply. (*See* **Bacillus**: Food Poisoning; **Campylobacter**: Properties and Occurrence **Clostridium**: Food Poisoning by *Clostridium perfringens*; Botulism; **Listeria**: Properties and Occurrence; **Staphylococcus**: Food Poisoning.)

Of particular concern in pork is the parasitic nematode *T. spiralis*. This organism forms a cyst in porcine muscle. The organism can be transmitted to humans who consume the contaminated pork, but is readily destroyed by heating the muscle to 62 °C. Processing plants that sell pork which is not likely to be cooked again are required to heat or freeze the meat to certify that it is trichina-free.

Pork-slaughtering and processing plants have rigid sanitation programs that allow production of safe food. Good sanitation at the plant and proper handling throughout the food chain help keep microbial growth under control. Plants producing pork must keep processing temperatures below 10 °C or stop production and sanitize the equipment every 8 h. Most plants keep their working temperature low enough to require cleaning and sanitizing only once every 24 h. Cleaning and sanitizing are not the same thing, but they are most effective when carried out together. Generally speaking, a plant is cleaned with soap and water to remove residual meat particles. Then the soap is rinsed off and sanitizer is applied to kill microorganisms that survived the cleaning. Common sanitizers used can be chlorine, ammonia, or iodine-based. (*See* **Sanitization.**)

A great deal of effort is spent to clean and sanitize the plant thoroughly, as well as educate employees about the importance of good hygiene in reducing food contamination. Microbiological status of the processing area is monitored daily. In addition, the plant must pass a sanitation inspection by a USDA inspector before the day's production can begin.

Despite all the in-plant efforts to control microorganisms, pork can still be contaminated or growth of microorganisms already present can occur as a result of product abuse in the warehouse, on the delivery truck, in the retail outlet, or in the home. Perishable foods should always be frozen or refrigerated at temperatures below 4 °C. Once cooked, pork should be kept above 60 °C or quickly cooled to under 4 °C. Many microorganisms grow rapidly in the temperature range 4–60 °C. Two very common mishandling problems that occur in the home are failing to refrigerate leftovers promptly and recontaminating cooked product by using the same utensils used with the raw product.

To minimize microbiological hazards and maximize eating quality, the recommendations in **Table 2** have been devised as maximum limits for storage of pork. Molds and yeasts are of little concern in fresh pork because the high water activity allows bacteria to dominate. In dried pork items such as pepperoni, molds may grow on the surface. However, mold growth is retarded by a potassium sorbate dip applied by the manufacturer or by vacuum packaging. (*See* **Spoilage**: Bacterial Spoilage; Molds in Spoilage; Yeasts in Spoilage; **Water Activity**: Effect on Food Stability.)

Very few other hazards exist with the consumption of pork. Muscle from pigs is regularly monitored for drug and pesticide residues by the USDA. Incidences of contaminated meat have been isolated and total far less than 1% of the pork supply.

Table 2 Pork storage recommendations[a]

	Refrigerator (2–4 °C)	Freezer (−18 °C)
Fresh pork	4 days[b]	3–6 months[c]
Cured pork	7 days	2 months

[a]Packaging and handling prior to the consumer will greatly impact shelf-life of pork.
[b]Ground meat, 2 days.
[c]Ground meat, 1–2 months.

Food Uses and Products

Pork is a very versatile food that can be prepared in many ways. Popular entrée items include pork roast, ham steaks, barbecued spare ribs, and grilled pork chops. Pork is also a common component of many foods such as bacon on pizza or ham dices in a chef's salad.

Pork trimmings can be ground and made into patties or sausages of all types, including frankfurters, pepperoni, and Italian sausage.

Waste or Byproduct Utilization

People in the swine industry say that no part of the pig is wasted. This statement is not far from the truth. The major waste product from the live pig is manure, which is a good nitrogenous fertilizer.

Waste products from swine slaughter and processing are perhaps better referred to as byproducts because they are seldom wasted. Blood, bones, and inedible viscera from pig slaughter are usually dried and ground for use in animal feeds. Gelatin is made from collagenous proteins found in pig skin. Industrial lubricants, plastics, and rubber are made from fat trimmed off pig carcasses. Pig skin fabric can be produced from cured pork skins. Even the pig's hair is sometimes used for brushes or insulation.

Perhaps the most important byproducts of swine slaughter and processing are those used in medicine. Hormones such as pig insulin or heparin can be prepared for human use. Heart valves from pigs have been used to replace damaged or diseased human heart valves and are often used in heart disease research. Skin grafts from pig skin have been successfully used on human burn victims. These are only a few examples of many medicinal uses of byproducts from the swine industry.

Acknowledgments

Appreciation is expressed to Doreen Blackmer for research and editorial assistance.

See also: **Bacillus**: Food Poisoning; **Campylobacter**: Properties and Occurrence; **Cholesterol**: Role of Cholesterol in Heart Disease; **Clostridium**: Food Poisoning by *Clostridium perfringens*; Botulism; **Curing**; **Fats**: Classification; **Listeria**: Properties and Occurrence; **Meat**: Structure; Slaughter; Preservation; Sausages and Comminuted Products; Nutritional Value; **Staphylococcus**: Food Poisoning; **Zoonoses**

Further Reading

Mandigo R (1989) *Pork Operations in the Meat Industry.* Washington, DC: American Meat Institute.

National Academy of Sciences, National Research Council, Food and Nutrition Board (1986) *Recommended Daily Dietary Allowances*, 10th edn. Washington, DC: National Academy of Sciences.

National Live Stock and Meat Board (1973) *Hog is Man's Best Friend*. Chicago: National Live Stock and Meat Board.

National Live Stock and Meat Board (1977) *Lessons on Meat*. Chicago: National Live Stock and Meat Board.

National Live Stock and Meat Board (1987) *Exploring Diet and Health*. Chicago: National Live Stock and Meat Board.

Troller J (1983) *Sanitation in Food Processing*. Orlando: Academic Press.

US Department of Agriculture (1983) Composition of Foods; pork products, raw, processed, prepared. *USDA Handbook*, nos. 8–10. Washington, DC: Human Nutrition Information Service.

PORT

Contents

The Product and its Manufacture

E Cristovam, University of Adelaide, Adelaide, Australia
A Paterson, University of Strathclyde, Glasgow, UK

Background

Ports are fortified wines based upon a long tradition with expert blending employed for style, create defined characters and produce the quality factor that makes port highly prized. Production methods differ from table wines primarily in that the fermentation – in some case still from spontaneous endogeneous microfloora of the must – takes place only during the 2–3 days of maceration. During this period, there is extraction of soluble compounds responsible for characteristic color and flavor characters of ports. Addition of brandy terminates the fermentation, leaving a young wine with high amounts of residual sugars. Such fortified wines have been produced in Portugal since the seventeenth century. Originally exported to Britain, since the 1960s France has been the largest importer. Many port houses are still British-owned and blenders follow house styles, believed to yield wines suited to British markets. Port is also used in elegant cocktails and receptions ('Porto de honra'). Three major product styles are recognized – Portuguese, English, and French based on house cultures. Ports have many myths and traditions associated with Christmas, gout, the British upper classes, and formal dinner parties. Consumer rituals abound, Port should be passed to the left from the host, and ladies should not touch a Port bottle. These wines are established products with an image of luxury, with seasonal peaks in demand dominated by that from the older generation and males. Port wine is a key product for a national economy of Portugal and a symbolic asset representing the country world-wide: unfortunately, demand is decreasing, probably as a result of social and lifestyle changes. However, port is now drunk on formal occasions, and such changes may enhance the demand for higher-quality wines.

History and Origin

The designation 'port' in wines appeared during the second half of the seventeenth century when this fortified product increased in popularity in Britain, replacing Bordeaux wines in English trade. This was enhanced by the Methuen Treaty between England and Portugal in 1703, guaranteeing preferential customs tariffs for port wines to balance similar tariffs for imports of British textiles by Portugal.

At that time such wines, from the Douro in Northern Portugal, were red, dry, and coarse, with an alcohol content of 12–13%. Subsequently, brandy was added to stabilize wines, to reduce spoilage during the long sea journey around Western Europe. However, this fortification practice became central to the production process – removing excess acidity, rounding the flavor character and creating a characteristic set of attributes that became unique to port wine. In 1756, the Marquis of Pombal set legal boundaries for the Douro Valley vineyards, annual production quotas, and strict rules for cultivation, transportation, and prices. Even a tasting was required before wine was sold. The aim was to safeguard port wine's authenticity, and effectively, a new product was created. In 1921, a further addition was made to the definition of port wines in that these must be *lodged* in the country for a given period before being deemed fit for export.

In the 1860s, phylloxera destroyed most vineyards in the demarcated region. This forced a restructuring of viticulture and winemaking techniques. At the same time, imitations of port started to appear from France and Spain. At the beginning of the twentieth century, in order to counteract such fraud and maintain quality, regulations were imposed for the production, sale, and export of Ports. Denomination of origin was reserved for fortified wines from the Douro region, in Northeast Portugal, with a minimum of 16.5% alcohol. Distillation of Douro wines was prohibited, and fortifying spirit had to be obtained from other regions.

In 1926, a bonded area was created in Gaia – Vila Nova de Gaia – a city opposing Porto on the Douro river mouth (**Figure 1**). Gaia became part of the demarcated region, and here, port trade companies were forced to build lodges to age wines. Now, shipments can be made from both Gaia and the demarcated

Douro region. Lodges maturing in the Douro produce wines with characteristic baked flavor notes through the high temperatures experienced upriver.

By the 1930s, exports were at a peak, but prices soon fell as quality was neglected, and production exceeded market demands. To counter this, legislation was introduced to control the two facets of the port industry – production and trade. In 1932, winegrowers were organized into guilds, now Casa do Douro (CD), responsible for monitoring vineyards, wine quality, and grape brandy production. A year later, a guild of port wine shippers was established, and the Instituto do Vinho do Porto (IVP) was formed to coordinate activities. The IVP regulates the description of ports with respect to dates and indication of age, controls the seal of guarantee, and still coordinates Casa do Douro activities.

Prior to entering or leaving Gaia, all ports are submitted to a chamber of tasters, constituted by experienced specialists. This is overseen by a consultative committee of tasters from individual port firms, chosen on the basis of reputation and expertise, but confirmed and formally appointed by the Ministry of Economy. This gives expert tasting opinions and has the final decision.

In theory, any must produced within the Douro demarcated region can be used for port. However, the IVP sets limits to annual quantities of wine to be 'beneficiado' (improved or benefited). Growers must apply to the CD for the right to produce within this allowance. Applications must list the name of the

Figure 1   (see color plate 122) Vila Nova de Gaia; bonded area in Portugal.

petitioner, the municipality and parish of each property, property names, number of vines over 4 years old, estimated production, and amount of wine to be used for port. Applications are checked against a register that makes reference to a classification system. Class A vineyards are, on average, granted a *benefit* of 600 l per thousand vines and Class E 400 l per thousand vines. Production is quantified in *pipes* – the volume of a traditional 550-l oak cask. Allowances are varied according to vineyard classification, total allowance for the year, and prospects for wine production in that year.

Today, port is a product that requires careful quality management strategies. To protect the industry, there is an upper limit on the number of vintages that can be declared and maxima to the fractions of wine in houses that can be traded in any year.

Production Area

The boundaries of the demarcated Douro region – the oldest in the world – have remained largely unchanged since 1756 (although revised in 1907, with minor alterations in 1921). This region consists of four districts: Vila Real, Bragança, Viseu, and Guarda covering *c.* 250 000 acres, with three subregions: Baixo Corgo, Alto Corgo, and Douro Superior (**Figure 2**), each with individual characters. Vines cover approximately 15% of the region, with the Baixo Corgo dominating. Small vineyards can be found throughout the region, with the larger vineyards mostly located in the Douro Superior. Before planting on the steep mountainous slopes, terraces must be formed. This was done in the past with hammers and dynamite but is now done with bulldozers when possible. Grape production is difficult because of the nature of the soil, schistose with granite in certain areas, and narrow terraces. The climate is harsh with frosty winters, hot (> 37 °C) dry summers, and in certain areas wet springs.

Grape Varieties

Only approximately 10 500 hectares (26 000 acres) are authorized for vineyards that can be used for port production. Most vine varieties are regional, grafted on to rootstock types chosen for grape variety and soil compatibility. The rootstocks used are Rupestris du Lot, Berlandier hybrids, either Riparia strains (420-A, SO4, 161/49) or the Rupestris varieties (R.99 and R.110).

The varied microclimates in the Douro region have resulted in 48 grape varieties being used in port production. However, there has been a rationalization to increase quality and optimize vine planting and grape production. Currently, an official EC recommendation quotes 29 varieties for port: 15 red and 14 white. Desirable red grape varieties produce tannic, intensely colored wines. Until recently, one of the most popular was *Tinta Francisca*, derived from Pinot Noir. Other important varieties are *Touriga Nacional*, *Tinta Roriz* (*Tempranillo* in Spain), *Touriga Francesa*, *Tinta Barroca*, and *Tinto Cão*. The dominant white grape varieties are *Malvasia Fina*, *Viosinho*, *Donzelinho*, and *Gouveio*.

Of the Douro red grapes, *Touriga Nacional* is considered to produce the finest and most complex wine, although the yield is low. It is also a vigorous and adaptable variety, growing in a wide range of conditions. Wines from *Touriga Nacional* have an aroma character suggesting red fruit, often with floral notes, notably of violets. Flavors are complex, elegant, balanced, and rich in nonaggressive tannins, and finishes are long and smooth.

Another red grape, *Tinta Roriz*, grows best on fertile soils, producing large, tight clusters of thick-skinned grapes of a high quality. Wines, less colored than those from *Touriga Nacional*, also have rich aroma characters dominated by red berries, with strong herbal overtones. Flavor characters are strong and astringent from harsh tannins, and lengthy fruity finishes contribute structure in blending.

The most fragrant wines are from *Touriga Francesa*. This variety is adaptable but requires a warm growing season to completely ripen grapes. Strong floral characters, notably of rose, dominate the aroma notes. Flavor, with fruity and floral attributes and earthy overtones, is thought to be well balanced with a long, concentrated finish. These wines contribute aroma and structure to products in blending.

Tinta Barroca is a high-yielding variety with large, loose clusters of thin-skinned, oversized berries. Wines are soft, fragrant, and delicate with a lingering red fruit aftertaste, but are often lacking in complexity and light in color. These wines contribute little to blend structure. *Tinto Cão* produces a smaller grape than other varieties.

A central problem in the Douro is finding a balance between the desirable character attributes in varietal grapes and production yield. The overall must color

Figure 2   Região Demarcada do Douro.

in young port wines, central to production, is influenced more by cultivar than by season.

Production Method

Harvesting of port grapes is difficult. Grapes are picked entirely by hand, between 15 September and 15 October depending on the weather and region, in vineyards that slope at an average of 45°. After grape destemming, traditionally grape treading was carried out to extract must in a 'lagar' (a large open, shallow stone tank). However, this quaint procedure has generally been replaced by more readily controlled and standardized practices. Currently, most port is produced in closed fermentation tanks often equipped to turn over the wine automatically. Alternatives include open tanks, in which must is pumped around, and concrete or stainless steel tanks, fitted with mechanical stirrers. Grapes are generally broken in centrifugal crushers with pumps. In certain production sites, for total or partial removal of skins, crushed grapes fall into a blender that shreds skins completely. The fermenting must is run off when most sugars have been converted into alcohol after 2–3 days of fermentation. After separation from solids, fermentation is inhibited by *aguardente* (grape brandy) addition. Wine officially becomes port at 19–20% (abv).

Brandy fortification, in addition to quenching the spontaneous fermentation, contributes specific sensory attributes to wine, improves chemical stability, and controls the final sweetness intensity. The sweetness in ports comes only from unfermented sugars. Fermentation must therefore be terminated when the residual wine sugars contribute to the desired sweetness. The quality of brandy added to the fermenting must is also a key factor in determining the chemical composition and final flavor character potential of quality ports.

The nascent port wines are stored in tanks, vats, or casks. It is not clear whether container capacity influences final wine character, but container material certainly does. More important seems to be the storage site, i.e. whether upriver at Douro or by the sea at Gaia – concrete tanks with a tartaric acid coating or stainless steel are often used. Once transported to Gaia, wines must be submitted to the IVP tasters for classification. Group 1 wines are considered 'of immaculate quality, carrying special privileges'. Group 2 wines have slight imperfections but are fit for usage. The final Group, 3, is for wines rejected either by tasters or by analysts.

Traditional Vintage

The traditional process, predominant until 1960, consisted of handpicking of grapes by groups of workers with transport in large baskets either on the shoulders or by cart. At the winery, workers crushed grapes barefoot in the *lagar*, marching back and forth in a line, until the juice floated on top of a pulpy residue. This initiated the fermentation with must darkening as the color was extracted from skins. To maintain the fermentation, paddles stirred the must, and when sufficient sugar was converted to alcohol, the liquid was run off into wooden vats and 110 l of *Aguardente* added for each 440 l of wine.

The nascent port was held in 'toneis' and tasted regularly until it was ready to be transported by expert boatmen down the river Douro in 'barcos rabelos,' in the spring following the harvest, or vintage. Gaia was originally chosen as the bonded area because port pipes could be easily off-loaded from riverboats to lodges and transferred to seagoing vessels.

Modern Vintage

With decreased manpower through migration, major changes have been, and are, taking place in port production. The vinification process, blending policies, maturation practices, and transport to lodges in Vila Nova de Gaia have changed considerably.

The widespread introduction of electricity in the early 1960s led to the replacement of human contributions with the autovinification process. Such mechanization led to establishment of wine centers at larger *quintas*. The result was that farmers sold more grapes, and wine production was centralized to facilitate improved quality control. Closed autovinification plants and open tanks came into use, but certain producers still believe that open-tank fermentations with automatic paddles optimize the color extractions from must.

Since the 1970s, production has moved from individual *quintas* to wineries with a greater plant capacity: currently, both fermentation and storage tanks are considerably larger. By the 1980s, over 80% of port was produced in wineries, with cooperatives producing approximately 40%.

Grapes arrive by lorry in open steel containers and are weighed, tipped into reception tanks, and immediately carried by screw feeders into a crushing and stemming machine. This is typically an upright cylinder, approximately 1 m in diameter, which centrifugally removes stalks to an extent depending on the tannin requirement. The must is pumped into square concrete or triangular stainless steel tanks, autovinification vats, an autovinificator cylinder is clamped on, and a water escape valve is filled. Provided that the temperature is *c.* 16 °C, the must starts fermenting almost immediately. Fermentations last 2–3 days, with sugar contents assessed continuously from specific gravity. The total sweetness of the port is derived from the original grape sugars.

Young ports stay in wooden vats or cement holding tanks in the Douro until they are ready for transport to Gaia in the spring by the road tankers. The narrow and winding roads of the Douro retain some of the hazards of bygone eras.

White port production uses white grapes, often with limited maceration, and wines are aged under oxidizing conditions. Grape skins acquire a golden-brown hue at harvest, and in order to produce a light gold, rather than a yellow–white, port, it is necessary to minimize skin contact during fermentation. Consequently grape skins are removed after crushing, and the maceration time is also kept to a minimum. Oxidation is not encouraged if complex floral aroma notes and pale yellow color are desired in the final wine.

Maturation and Blending

Young ports are stored in vats or casks made of oak or mahogany in the Douro. During the first winter, wine is racked (pumped off) from the lees into fresh casks. The first tasting in November uses a classification based on sensory attributes. The finest wines, produced in an exceptional grape quality year, are usually set aside with a view to *declaring a vintage*; most are used for blending. Vintages represent highpoints in port wine quality. During the initial 2 years, wines are racked frequently with aeration as often and vigorously as required for the attributes considered desirable after aging, as determined by the winemaker. Winter conditions help precipitate colored matter together with yeast and potassium salts of tartaric acid. Young ports are then observed to 'close up' or show a color intensification. Ports are transported to Gaia for classification in early spring.

Port was traditionally stored in lodges or cellars in Gaia but can now be retained in the Douro. Ruby wines, matured in the Douro, are subjected to its climatic extremes and are considered to lose vigor and freshness. A common view is that older tawny ports should mature in the Douro, and younger ruby ports in Gaia, where temperature and humidity are less variable.

On arrival at Gaia, casks are stored by vineyard and vintage in the cellars of individual port houses. Large cylindrical stainless steel vats and lined concrete are often used for the temporary storage of wines on arrival, but maturation, for at least 2 and usually for 3–40 years or longer, is in wood. Cellar-masters decide between the 550-l 'pipas' (pipes or oak casks), and larger casks (25 000–35 000 l) or vats (550 000 to 750 000 l) (**Figure 3**).

Figure 3   Wooden casks and vats used for wine maturation.

On arrival in Gaia, a second tasting takes place: wines are classified, and the first *lotas* (blends) are married in large vats fitted with stirrers or pumping. This is on the basis of decisions on the potential of each wine and amalgamating similar wines. There is then separation of the top-quality wines i.e. those to be used when declaring a vintage. Such wines are kept separate for two winters and a summer, and if they do not develop to predefined standards, a vintage will not be declared that year, and the wines will be used for blending.

The purpose of blending is to obtain specific final sets of sensory attributes, port types, and styles. Consequently, young port is always stored in well-seasoned vats or pipes to retain the fresh aroma characters and deep red color with purple hues. As production is based on traditions, wooded ports are created to ensure continuity of style; new blends contain a percentage of previous wines. Previously, individual pipes were blended, but currently, concrete or large wood vats are filled from maturation casks.

Ports are blended from wines varying in age and character. Cellar-masters choose young wines on the basis of fruit flavor notes and color, and older wines for body and 'smoothness.' The objective is a character and quality typical of the style of that port house.

Regulations stipulate aging of ports at least 3 years in cask or bottle prior to release.

Aging

Port wine storage leads to varied changes in the wine character and composition. The type and capacity of containers are selected for the character evolution required, but those from grapes also contribute. Ports are aged under oxidizing conditions except for vintage and late bottled vintage (LBV) products. Young ports are blue–purple in color – gradually turning red, then tawny and progressively paler with age, eventually achieving amber and golden tones (**Figure 4**). The aging method varies. All red ports are aged for at least 2 years in oak casks and continue maturation either in casks (ruby, tawny, aged tawnies) or in bottles (vintage, LBV). Ruby ports retain an intense dark red color with the sensory characteristics of young wines.

Red ports are aged in wooden vats and pipes, stacked four high to minimize evaporation. With white ports, in order to avoid color gain through prolonged wood maturation, concrete vats are used.

With labor costs increasing, larger vats (5–400 pipes in capacity) are used more often for aging wines. The rate of maturation is inversely related to volume. As a consequence, wines are selected early according to style requirements and then stored in appropriately sized containers. Traditional lodges in the Douro exhibit wine liquid temperature changes of up to 20 °C. The atmosphere is also dry, leading to faster evaporation and maturation rates and baked or maderized flavor notes. In Gaia, the relative humidity is constant (approximately 54%), evaporation lower, and temperature fluctuations under normal storage conditions limited to 14–16 °C. Maturation is slower, producing fresher wines than the Douro.

Wood

Although the character of the grape variety is central to wine flavor development, wood has been considered to play important roles in port maturation. Staves are from local and imported oak (including some used for Scotch whisky casks), chestnut, and Brazilian mahogany. Dried fruits and spicy aroma notes, and smooth mouthfeel (texture) characters, attributed to lengthy wood maturations, characterize premium tawny ports. Traditional guidelines for production include the use of small oak pipes with frequent racking to induce sufficient aeration. Contact between the wood surfaces and the wine, regular racking, and oxygen migration through the staves have been reported to be important variables in port maturation.

Unlike table wines, port is aged in a varied range of container sizes. Wood is often reused many times, depending on the style and cellar practice. New wood is rarely used and, if required, is seasoned thoroughly with a second pressing wine, table wine, and/or young red port before deployment. Consequently, some experts believe that little woody character should be detectable in ports, through the use of extractive-depleted staves, except in the older tawnies matured in the Douro.

LBV ports are stored in large vats, and tawny ports are stored in pipes. Frequent racking with aeration of wines influences maturation. Wood aging of such wines is believed to be primarily an oxidative process with limited parallel extraction of nonflavonoid phenolics from stave lignins and tannin degradations.

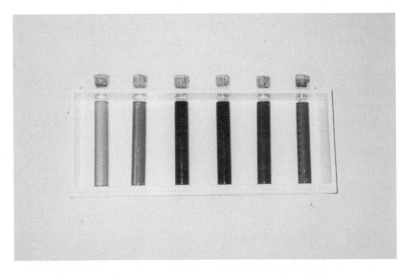

Figure 4   (see color plate 123) Samples of different types of port, showing the range of colors.

Wine maturation is thought to occur mainly from evaporation and oxidation.

Fining and Stabilizing

Fining, traditional for clarifying wines, removes a certain amount of color and tannins to produce softer and 'older' wines. Port is usually filtered: modern methods have replaced the cloth bag method of separating wine from lees. Modern filters consist of a series of woven nylon pads that allow lees to circulate until the pads are evenly coated with solid material. The resulting solid cake then acts as a filter that introduces clarity in the wine. Bentonite may be used with white wines to remove protein.

Refrigeration is now used to stabilize wines: chilling precipitates unstable coloring matter and the potassium salts of tartaric acid. Chilled wine is passed through a vessel that encourages deposition of potassium hydrate tartrate. Chilling is followed by further filtration (continuous refrigeration). There may be wine retention in an insulated tank for up to a week before filtering (Gasquet process). Finally, the clarified product is passed through a plate heat exchanger separated by thin stainless steel plates from incoming wine at ambient temperature.

Refrigeration and sulfur dioxide addition, prior to bottling, allow wines to remain bright in bottle for up to 18 months under normal storage conditions. Brief pasteurization may also be used after cold treatments. In addition, Kieselguhr with pad filtration may be a final processing after fining or refrigeration.

Bottling

In earlier years port was shipped in pipes to England for transfer to squat and bulbous bottles, designed for transporting wine from cask to table. Oxidation was a problem, since stored on their side such bottles did not allow cork contact. In 1775, bottles with longer bodies and shorter necks, were introduced to allow better storage. This change initiated the development of bottle aged vintage ports.

By the 1930s, bottling had become more centralized as suppliers, instead of merchants, could now bottle under customs and excise supervision. Duty was not payable until port was declared from bond. In the 1970s, more shippers began bottling under their own labels instead of shipping in bulk. The result was that it became possible to develop the international brands that currently dominate the consumer's attention.

Port Styles

The IVP governs how producers may use labels to describe port wines. There are four major premium categories of port: vintage, LBV, quinta, and aged tawny (**Figure 5**).

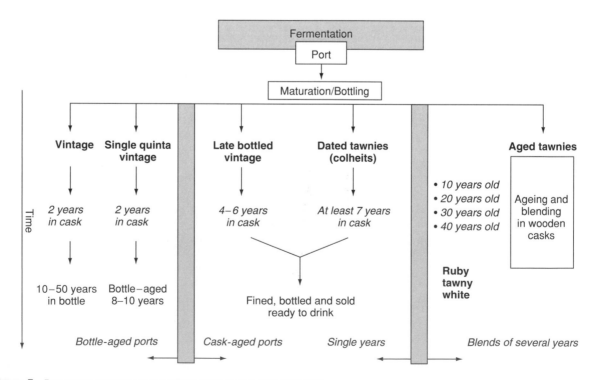

Figure 5 Processes involved in the different types of port production.

Vintage ports are wines of a single harvest, produced in a specifically recognized year, and have sensory characters regarded as exceptional. Such port wines spend 2–3 years in wooden casks or vats prior to bottling. The IVP has rights to descriptions of vintage and corresponding dates. Subsequently, maturation reactions occur in the bottle for 10–20 years and in some cases up to 50 years, with little possibility of oxidation. Such ports are never fined or filtered; the wines generate precipitates, or 'throw a crust,' and require decanting prior to consumption.

LBV ports are wines of a single harvest, aged in wood for 4–6 years. Bottling is between 1 July of the fourth year and 31 December of the sixth year. Bottle maturation reactions may continue for 5 years or more with continuing improvements. Traditional, now rare, LBVs are not fined or filtered, and throw a crust with improvements on bottle aging. However, modern LBVs are treated to remove solids.

Quinta are wines that come from grapes from a single property, named on the label. Such ports are made only in years not deemed good enough to declare a vintage. Single Quinta vintage ports have been available for some time, but currently, the market is expanding to include this designation for both ruby and tawny ports.

Aged tawnies are typically matured for 10, 20, 30, or 40 years, the average of ports blended. Old tawnies have a characteristic gold color. *Dated tawnies* are wines from single harvests, matured in oak casks for at least 7 years with the harvest and bottling date specified on the labels. A special reserve tawny contains wines that have spent 3–4 years in the cask prior to blending.

Wines Matured in Wood

Wood-matured ports, such as tawnies, are treated to encourage color lightening. Blend components are racked at least once a year to remove deposits and ensure oxygen dissolution. During the initial years of aging, container size is thought to be of little importance. Frequent use of wooden containers of considerable size such as 650-l oak casks, or vats, for short periods, stainless steel or tartaric acid-lined tanks.

Racking and the resulting wine aeration lead to oxidation, essential for the maturation of these port wines. During racking, more alcohol may be added, or maturing wine may be 'refreshed' by adding small quantities of younger wine. On creation of the blend, to assist maturation, wines are usually transferred to smaller vats or casks. When the desired degree of maturity has been reached, the final standard blends, with sensory attributes specific to each company, are prepared for bottling and export.

As noted previously, no port may be sold for consumption in Portugal or abroad until it is at least 3 years old.

Wines Matured in Bottle

Vintage and LBV ports, bottled after only a few years in casks, have sensory attributes that are more dependent on must characters than on wine treatment. Vintage ports are racked infrequently to prevent excessive aeration, using smaller casks that are considered to provide a more controlled and uniform aeration. Bottles are stored in dark cellars of constant temperature with controlled ventilation and humidity. Color loss occurs slowly compared with wooded wines, and bottled ports retain their characteristic fruitiness, retaining a full flavor and purple–red colors for many years.

See also: **Grapes**; **Port**: Composition and Analysis

Further Reading

Bakker J (1986) HPLC of anthocyanins in port wines: determination of aging rates. *Vitis* 25: 203–214.

Bakker J, Bellworthy SJ, Hogg TA *et al.* (1996) Two methods of port vinification: a comparison of changes during fermentation and of characteristics of the wines. *American Journal of Enology and Viticulture* 47: 37–42.

Bakker J, Bridle P, Timberlake CF and Arnold GM (1986) The colors, pigment and phenol contents of young port wines: Effects of cultivar, season and site. *Vitis* 25: 40–52.

Bakker J and Timberlake CF (1985) An analytical method for defining a Tawny port wine. *American Journal of Enology and Viticulture* 36: 252–253.

Bakker J and Timberlake CF (1986) The mechanism of color changes in aging port wine. *American Journal of Enology and Viticulture* 37: 288–292.

Burnett JK (1985) Port. In: Birch GC and Lindley MG (eds) *Alcoholic Beverages*, pp. 161–169. London: Elsevier Applied Science.

Cristovam E, Paterson A and Piggott JR (in press a) Development of a vocabulary of terms for sensory evaluation of dessert port wines. *Italian Journal of Food Science.*

Cristovam E, Paterson A and Piggott JR (in press b) Differentiation of port wines by appearance using a sensory panel: comparing free choice and conventional profiling. *European Food Research and Technology.*

Fletcher W (1978) *Port: An Introduction to its History and Delights*. London: Philip Wilson.

Fonseca AM, Galhano A, Pimentel ES and Rosas JR-P (1984) *Port Wine. Notes on its History, Production and Technology*. Oporto: Instituto do Vinho do Porto.

Garcia-Viguera C, Bakker J, Bellworthy SJ *et al.* (1997) The effect of some processing variables on non-coloured phenolic compounds in port wines. *Zeitschrift für Lebensmitteluntersuchung und -forschung* 205: 321–324.

Gowell RW (1972) The manufacture of port. *Process Biochemistry* Oct.: 27–29.

Howkins B (1982) *Rich, Red and Rare. The International Wine & Food Society's Guide to Port.* London: William Heinemann.

Reader HP and Dominguez M (1995) Fortified wines: sherry, port and Madeira. In: Lea AGH and Piggott JR (eds) *Fermented Beverage Production*, pp. 159–203. London: Blackie Academic & Professional.

Robertson G (1992) *Port.* London: Faber and Faber.

Williams AA, Langron SP, Timberlake CF and Bakker J (1984) Effect of colour on the assessment of ports. *Journal of Food Technology* 19: 659–671.

Composition and Analysis

E Cristovam, University of Adelaide, Adelaide, Australia
A Paterson, University of Strathclyde, Scotland, UK

Introduction

The demarcated Douro region – established in 1756 – consists of three subregions: Baixo Corgo, Alto Corgo, and Douro Superior (**Figure 1**), each with individual characters. Vines cover approximately 15% of this steep mountainous region, with the Baixo Corgo dominating. Small vineyards can be found throughout the region, with the larger mostly located in the Douro Superior. Vineyards producing grapes for port wines are graded on a point system, biased towards environment – *c.* 75% for vine productivity, altitude, soil, locality, and vine training, 14% from grape cultivars and degree of site slope, and >10% from local features. Grape varieties favored (around 7% points) include Tinta Francisca, Tinta Cão, Touriga Nacional, and Mourisco. Low-scoring musts are utilized in table wines.

Oxidative Aging – Tawny Ports

Tawny ports, matured from 10 to 40 years and particularly important, have maturations dominated by oxidative aging in wooden casks allowing oxygen migration, supplemented by oxygenation during frequent initial rackings. Small 'pipes,' as opposed to large vats, insure high cask surface area-to-volume ratios.

Reactions between wine congeners, together with cask-derived components, are important. Congener oxidations facilitate flavor evolution to characteristic mellowness with dried fruits, caramel, and vanilla notes. Most obvious, and dominating, are relationships between aging and color, with oxidations of red and purple anthocyanins, and losses as insoluble precipitates that include tannins. Tawny ports have yellow, amber, and tawny hues at intensities related to maturation period and cask choice, at rates dependent on original varietal choice and wine composition.

Reactions between acetaldehyde, anthocyanins and catechins yield reddish-blue condensation pigments that darken ports, 'closing up,' over the initial 6 months. Progressive lightening follows, with hue changes and gradual browning from competing aldehyde-induced reactions and direct condensations of anthocyanins with phenolics (**Figure 2**). Less is known about wood contributions to tawny ports spending over 50 years in old and recycled casks than character in red table wines. For tawny ports, color loss is essential and style-defining; in red wines it is detrimental to quality. (*See* **Antioxidants**: Natural Antioxidants.)

Nonoxidative Aging – Vintage Ports

Vintage ports age in bottles for up to 30 years. Initial oxygen, absorbed during bottling, is slowly reduced by reactions with anthocyanins and other wine components. Color is lost more slowly than in cask

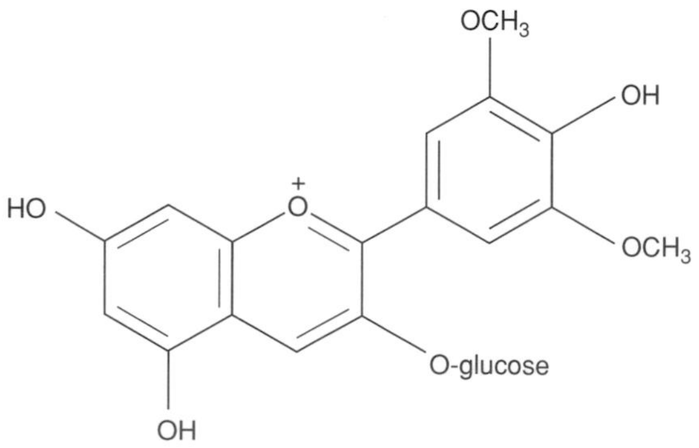

Figure 1 Região Demarcada do Douro – demarcated Douro region.

Figure 2 Oxidation/polymerization of phenolic components in ports.

maturations and the initially full-bodied wines retain flavor intensity, fruity characters, purple-red coloration, and freshness of youth for many years. Limited oxidation reduces aldehyde formation, precipitation of insoluble phenolic products, and leucoanthocyanins loss, yielding more condensed products.

Vintage ports show direct relationships between age and volatile acidity, and contents of ethyl acetate, pentyl alcohols, and sodium. This is from either maturation or modern wine-making strategies, including reductions in air contact of grapes; introduction of coating equipment; decanting without aeration; replacement of sodium metabisulfite by SO_2; and quality control in fortifying grape spirit. Clear aging effects include decreases in tartaric acid through deposition of potassium hydrogen tartrate, in citric acid content and total acidity, and increases in glycerol content and changes in alcohols.

Oxidative Reactions

Formation of Esters

In acidic, high-ethanol port wines, esterifications, slow in table wines, proceed at rates influenced by temperature and pH (pH 3.5–4.0), yielding ethyl esters of organic acids – lactic, succinic, malic and tartaric acids. Nonvolatile phenolics influence ester solubilities in the aqueous ethanol, removing such aroma-active congeners from head spaces, influencing perceived character.

Esters of higher (*iso*amyl, hexyl and 2-phenylethyl) alcohols show inverse maturation relationships through hydrolyses with equilibria functions of initial concentrations, those of parent acids and storage temperature. In aging, ethanol and acetaldehyde oxidations yield acetic acid, influencing pH, ester equilibria, and forming ethyl acetate. Such ester formation appears central to desirable flavor changes in fortified wines. (*See* **Oxidation of Food Components.**)

Formation of Carbonyl Compounds: Aldehydes, Acetals

In maturations, important aldehydes are converted to alcohols or combine with sulfur dioxide. Acetaldehyde is most abundant, but other aldehydes are derived from carbohydrate (furfural, 5-methylfurfural, and 5-hydroxymethylfurfural) or lignin degradations (vanillin, syringaldehyde, coniferaldehyde, and sinapaldehyde, cinnamaldehyde). Residual carbohydrates (approximately $100 \, g \, l^{-1}$) in port contribute to such reactions.

Furfural, acetaldehyde, and other higher aldehyde contents can be directly related to maturation period,

especially at higher temperatures. Woody flavor characters have been linked to aldehyde polymerization.

In acidic wines, acetals are principally formed from reactions between monohydric alcohols, and aldehyde carbonyl groups, notably of acetaldehyde. Oxidations yield flavor-active aldehydes, notably benzaldehydes, also contributing to acetal formation.

Other Reactions

Phenols derived from wood maturations include guaiacol, phenol, *m*-cresol, eugenol, 4-vinylphenol, and 4-vinylguaiacol. Lactones include β-methyl-γ-octactone. Wood-extracted aldehydes include vanillin, syringaldehyde, coniferaldehyde, and sinapaldehyde.

Classic Enological Analyses

Port Acidity

Total titratable acidity values of port wines at pH 7.0 range from 5.86 to $3.45 \, g \, l^{-1}$ as tartaric acid. Older wines have increased volatile and total acidities through acetic acid formation. In young ports, total acidity is reduced by tartrate precipitation.

Volatile Acidity

Determined by official Office International de la Vigne et du Vin (OIV) methods, volatile acidity is dominated by acetic acid (around 90%) but propionic, formic, and butyric acids also contribute. Such acids are from yeast and lactic acid bacterial fermentations and wood maturation reactions. Port wines generally have volatile acidity values $<0.35 \, g \, l^{-1}$ acetic acid, although $1.02 \, g \, l^{-1}$ was recorded for a 40-year-old tawny and a 1960 Colheita tawny.

Organic acids

Grapes contribute tartaric and malic acids to wines. Tartaric acid concentration varies during grape ripening, and is more related to climatic conditions than varietal character. Tartrate precipitation during maturation decreases wine contents to 0.445–$1.688 \, g \, l^{-1}$, independent of age or wine type.

In contrast, malic acid contents, more dependent on grape variety than climate, fall during ripening. Contents are lower in younger port wines, than older, possibly through later picking in recent vintages.

Lactic acid, formed by strain-dependent anaerobic yeast metabolism, is present in port wines in the range 0.3–$1.68 \, g \, l^{-1}$, independent of wine age.

Type differentiation was achieved on the basis of contents of citric acid and alcoholic strength (**Table 1**) in 21 red and four white ports, and six Spanish port-style wines. Australian port-style wines have higher ethanol contents. (*See* **Acids**: Natural Acids and Acidulants.)

Alcoholic Strength

Alcoholic strength (% v/v ethanol) in red ports ranges between 19 and 22%, with white ports as low as 16.5% (light, dry white port). Spirit contributes about 20% initial volume to port wines; subsequent alcoholic strength adjustments dilute wine congeners, and increase precipitation of tartrates and extraction of wood substances. Alcoholic strength changes with water and ethanol evaporation through staves; it is dependent on storage conditions and wine sugar content.

Sulfur Dioxide

Total SO_2 is from summations of iodometric titrations of free and bound components in acid conditions after a standard pretreatment by the Portuguese official method (Portaria 985/82 and pr NP-2220). Values between $9 \, mg \, l^{-1}$ (Colheita tawny, cask wine) and $99 \, mg \, l^{-1}$ (10-year-old tawny, commercial wine) have been reported. In vinifications, SO_2 is added for microbial stability and antioxidant properties. Free SO_2 rapidly binds to ethanol but also hydrogen peroxide, produced during phenol oxidations, reducing further oxidation.

Reducing Sugars and Glycerol

Must origin and quality, shipper requirements, and house style, all influence residual sugar content, typically $80–120 \, g \, l^{-1}$ from the Lane-Eynon method.

Certain sweet red wine musts – geropigas – can be $150 \, g \, l^{-1}$ residual sugar; drier, more fermented musts have $20–50 \, g \, l^{-1}$. White geropigas are made for blending or specialty products; dry white aperitif ports contain $< 50 \, g \, l^{-1}$ sugar. Glucose and fructose concentrations range from $161 \, g \, l^{-1}$ (Colheita tawny, cask wine) to $63 \, g \, l^{-1}$ (30-year-old tawny, commercial wine). Fructose at $60–70 \, g \, l^{-1}$ is more abundant than glucose ($40–50 \, g \, l^{-1}$).

Typical must specific gravities are 1090–1100 for red and 1085–1095 for white ports; fermentations are terminated around 1045. In Australian port-style wines, specific gravities (20/20 °C) of 1036 and 1080 make these even sweeter than the sweetest port wines (**Table 2**).

The IVP (Instituto do Vinho do Porto) recognizes classes of port wines, varying in sugar content (**Table 3**).

Glycerol, the second most abundant alcohol, contents are from 3 to $8 \, g \, l^{-1}$. Glycerol acetals are found as in other aged fortified wines, and *trans* 5 hydroxy-methyl-2-methyl-1, 3-dioxolane in port.

Dry Extract, and Folin-Ciocalteu Index

Dry extract values are determined indirectly as density of the residue without alcohol, following replacement of alcohol with an equivalent volume of water in the Portuguese norm (Portaria 985/82 and pr NP-2222) and OIV methods. Values in Colheita tawnies range from 120 to $188 \, g \, l^{-1}$. Maturation changes include: alcohol and water evaporations; extractive increases; and precipitation of salts,

Table 1 Enological analyses of red and white ports and similar Spanish and Australian wine styles

	Portuguese red ports	Portuguese white ports	Spanish port-styles	Australian port-styles
Alcoholic strength (% vol)	18.6–21.5	19.5–19.9	16.9–19.0	22–23
pH	3.54–3.99	3.54–3.57	3.10–3.72	3.69
Total acidity ($g \, l^{-1}$)	1.81–3.04	2.06–2.63	2.26–3.09	NA
Volatile acidity ($g \, l^{-1}$)	0.12–0.49	0.16–0.25	0.26–0.41	NA
Phosphoric acid ($g \, l^{-1}$)	0.23–0.33	0.23–0.29	0.18–0.28	NA
Tartaric acid ($g \, l^{-1}$)	0.94–1.45	1.19–1.69	0.97–1.85	NA
Citric acid ($g \, l^{-1}$)	0.08–0.26	0.11–0.17	0.12–1.03	NA
Malic acid ($g \, l^{-1}$)	0.06–2.07	0.94–1.39	0.35–0.70	1.66–1.71
Lactic acid ($g \, l^{-1}$)	0.34–2.42	0.33–0.94	0.34–1.19	0.37–1.20
Succinic acid ($g \, l^{-1}$)	0.18–0.33		0.17–0.31	NA

NA, not analysed.

Table 2 Sugars in red and white ports, Spanish and Australian port-style wines

	Portuguese red ports	Portuguese white ports	Spanish port-styles	Australian port-styles
Glucose ($g \, l^{-1}$)	30–48	41–44	24–49	76–118
Fructose ($g \, l^{-1}$)	48–67	53–59	32–52	93–148
Reduced sugars ($g \, l^{-1}$)	101–119		69–115	NA

NA, not analysed.

Table 3 Sweetness in port wines

Sweetness level	Gravity	°Baumé	Sugars (g l⁻¹)
Very sweet (Lágrima)	> 1034	> 5.0	> 130
Sweet	1018–1033	2.8–5.0	90–130
Semidry	1008–1017	1.4–2.7	65–90
Dry	998–1007	0.0–1.3	40–65
Extra dry	< 998	0.0	< 40

phenolic compounds, proteins, polysaccharides, and other components.

The Folin-Ciocalteu index, from European Economic Community (EEC) recommended method (EEC no. 2676/90), expresses wine total phenols (anthocyanins, tannins, flavones, flavonols, catechins, cinnamic acids, and related compounds).

Tannins and Phenolic Compounds

Tannin levels in ports have been reported to between 400 and 600 mg l⁻¹.

Color in ports, principally from soluble grape-skin anthocyanins (**Figure 3**), is from malvidin (3-glucoside, 3-*p*-coumarylglucoside, and 3-acetylglucoside, in decreasing order). Varietal differences in such compounds are important. Aldehyde-bridged polymers of anthocyanins and other phenolics dominate over direct condensation of anthocyanins with phenolics, through free aldehyde contents (50–100 mg l⁻¹) of young ports. With aging, anthocyanins rapidly disappear, replaced by polymers of progressively decreasing absorbance, at rates depending on racking and cask choice. Prolonged oxygen diffusion yields wine of tawny coloration influenced by wood extractives and composition.

Wine aeration during filling and racking seems more important for color changes than subsequent oxygen diffusion from studies of varietal port wines aged in miniature recycled-stave casks. The reduced

soluble extracts, including ellagic tannins, of recycled wood suggest limited contribution to wine changes, dominated by an evolution in phenolic components. Port maturations are likely primarily driven by oxidation, influenced by grape varietal differences, with inert but permeable casks making little direct contribution. (*See* **Tannins and Polyphenols.**)

Color Measurements

Spectrophotometric measurements at $A_{520\,nm}$ and $A_{420\,nm}$, yield tint ($A_{420\,nm}/A_{520\,nm}$), and color density ($A_{520\,nm} + A_{420\,nm}$) values and at natural pH specify redness and brownness in color, respectively. CIELAB (Commission Internationale de l'Éclairage) 76 L*a*b values (hue, angle, and chroma), are high L* for lightness and low for darkness; a* measures redness and b* brownness. Total pigments and total phenols can also be determined.

Metal Ions

Trace metal contents differentiate between Portuguese ports and similar styles in wine from Spain. Rubidium and manganese contents (**Table 4**) mirror soil differences, although rubidium varies with maceration intensity, as can lithium, and manganese from vineyard treatments. Iron and aluminum concentrations in Spanish wines may also reflect soil differences.

Volatile Compounds in Port

Thirty-five volatile aroma components were quantified in 20-year and 100-year Australian port-style wines with higher ethyl lactate (110–370 mg l⁻¹) and diethyl succinate (61–130 mg l⁻¹) than in younger wines. Only the former contributes to aroma. *Trans*-β-methyl-γ-octalactone (oak lactone) contents were 0.8–2.2 mg l⁻¹.

Furfural, 2-acetylfuran, 5-methyl furfural, ethyl lavulinate, ethyl furoate, and 5-ethoxymethyl furfural are derived from carbohydrate degradation, heating, and prolonged storage. The older wine had more volatile components from wood and carbohydrate degradation.

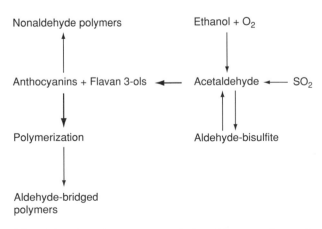

Figure 3 Structure of malvidin 3-glucoside, an anthocyanin occurring in grapes used for port wine production.

Table 4 Metal ions in red and white ports and Spanish port-style wines

	Portuguese red ports	Portuguese white ports	Spanish port-styles
Rb (mg l⁻¹)	0.66–1.43	NA	0.00–0.44
Mn (mg l⁻¹)	0.77–1.32	NA	0.20–0.58
Al (mg l⁻¹)	0.09–1.81	NA	1.91–4.01
Fe (mg l⁻¹)	1.0–3.4	NA	3.6–7.4
Li (μg l⁻¹)	17–54	31–37	45–64

NA, not analysed.

In two commercial Portuguese ports, a ruby (young wines retaining an intense dark-red color) and a tawny, 141 volatile flavor components were identified: 81 esters, 14 alcohols, 9 dioxolanes, 6 hydrocarbons, 5 acids, 5 carbonyls, 4 nitrogen components, 3 hydroxy carbonyls, 2 phenols, 2 alkoxyalcohols, 2 alkoxyphenols, 2 halogen compounds, 2 lactones, 2 other oxygen heterocyclics, 1 sulfur component, and 1 diol.

Alcohols were most abundant, followed by esters, dominated by ethyl esters and lower-molecular-weight acetates, the latter increasing with maturation. The tawny port was rich in flavor-active succinates (approximately $100 \mu g\,g^{-1}$) from slow esterification and *trans*-esterification reactions at wine pH and maturation stave contact. Ethyl lactate was present at 15 and $30 \mu g\,g^{-1}$ in ruby and tawny ports, respectively, and together with diethyl malate and succinate may indicate age, as can acidic degradation of fructose to 5-hydroxymethylfurfural. Other furan derivatives contribute baked and caramelized flavor notes to older wines. The flavor contribution of β-methyl-γ-octactone, important in oak-aging, is unclear in port wines, as are relationships between maturation and concentration of vitispirane (abundant in the ruby; traces in the tawny).

Reserva ports (single harvest without blending), matured in wood under oxidative aging, show maturation increases in diethyl succinate, ethyl acetate, ethanal, octanal, glyoxal, and methyl glyoxal (**Table 5**) and also benzaldehyde (oxidation of the alcohol), *trans*-2-octenal, 2-decanone, and 2-tridecanone, but inverse relationships for acetoin and methanol.

Differentiation of red ports (21) from similar Spanish wine (6) on the basis of volatiles concentrations are shown in **Table 6**.

Characteristic port volatile components are esters and alcohols, the former dominated by ethyl esters,

Table 5 Concentration of volatile compounds in two Portuguese Reserva ports

	Reserva ports	
	1968	1991
Glyoxal ($\mu g\,l^{-1}$)	1302.9	669.7
Methyl glyoxal ($\mu g\,l^{-1}$)	1981.2	898.1
Diethyl succinate (mg l^{-1})	16.3	2.2
Methanol (mg l^{-1})	202	213
Ethanal (mg l^{-1})	78.3	94
Ethyl acetate (mg l^{-1})	40.4	48
Acetoin (mg l^{-1})	27.6	64.3
Octanal ($\mu g\,l^{-1}$)	128	79.7
Benzaldehyde ($\mu g\,l^{-1}$)	561.9	83.4
Trans-2-octenal ($\mu g\,l^{-1}$)	291.8	90.2
2-Decanone ($\mu g\,l^{-1}$)	71.2	13.7
2-Tridecanone ($\mu g\,l^{-1}$)	91.5	65.8

Table 6 Volatile congeners in red ports and Spanish port-style wines

	Portuguese red ports	Spanish port-styles
Methanol (mg l^{-1})	259–360.3	113.6–275.8
Propanol (mg l^{-1})	46.9–66.4	15.9–43.9
Isobutanol (mg l^{-1})	93.3–146.7	21.7–30.0
2-methylbutanol (mg l^{-1})	62.7–97.1	14.9–23.2
3-methylbutanol (mg l^{-1})	237.4–344.1	54.0–92.8
Hexanol (mg l^{-1})	3.5–6.6	1.0–2.3
Other higher alcohols (mg l^{-1})	477.6–701.3	137.7–220.6
Propanol/isobutanol (mg l^{-1})	0.40–0.57	0.72–1.83
Isoamylacetate (mg l^{-1})	0.33–0.56	0.13–0.33
Octanoic acid (mg l^{-1})	0.56–1.34	0.24–0.66
Ethyl hexanoate (mg l^{-1})	0.17–0.49	0.11–0.26
Ethyl octanoate (mg l^{-1})	0.33–0.69	0.09–0.27
Other esters (mg l^{-1})	0.53–1.30	0.22–0.53

acetates, and succinates. Older tawny ports are rich in, and age-differentiated by, contents of 2-propenoic acid, diethyl succinate, butanedioic acid, oak lactone, ethyl dodecanoate, *iso*amylbutyrate, 1-butanol, decanoic acid, and ethyl lactate, amongst other components. Vintage ports have been differentiated from late-bottled vintage (wines of a single harvest, aged in wood for 4–6 years) from 3-butanoic acid contents.

Flavor in Port Wines

Port wine flavors originate in the volatile and nonvolatile components in grapes, from fermentation or maturation. Although ports have traditionally been vinified using endogenous microflora, including a range of yeasts and bacteria, there have been moves towards encouragement of yeast mono- or at least mixed cultures, for example *Saccharomyces cerevisiae* UCD 522. Butanol, isobutanol, and 1-hexanol originate from yeast fermentation of carbohydrate and amino acid sources; hexyl acetate and 2-phenethyl acetate are also fermentation products. Ethyl lactate and diethyl succinate, resulting from esterification reactions, increase in the wines due to storage and long maturations. Oxidation reactions originate mainly from benzaldehyde and ethyl-2-phenylacetate, while 2-phenethanol and oak lactone result from wood extraction.

Few published studies describe flavor components, as opposed to volatiles. Young ports have notably astringent flavor characters through high contents of total and individual phenols with a *harshness* mellowed by wood maturations. Older ports have 'softer' characters with increased flavor complexity.

Flavor characters evolve through oxidation and polymerization of solution nonvolatile phenolic

Table 7 Variables used to model vintage age prediction

Variable	Code
Alcoholic grade	A
Volatile acidity	B
Total acidity	C
Glycerol	D
Ethyl acetate	E
Pentyl alcohols	F
Tartaric acid	G
Citric acid	H
Na	I
DO420	J
DO520	K
Sudraud's index (IC)	L
Tone (TON)	M
Garoglio's aging index (IE)	N
Anthocyanins	O
Tone (H*)	P
Chroma (C*)	Q
Clarity (L*)	R
Saturation (S)	S

Tone (H*), chroma (C*), clarity (L*), and saturation (S) are four CieLab parameters.

compounds, changing partitioning of aroma-active components between wine head spaces and liquid. The products of oxidations of congeners and ethanol, depolymerizations and extraction of wood components all influence partitioning. Grape varietal character is central to flavor development, but wood aging provides oxidative processes and nonflavonoid phenolics from lignin and tannin degradations determining final character.

Relating key sensory attributes to concentrations of volatile flavor components would assist understanding of port wine character. Partial least-squares (PLS) regression has facilitated modeling relationships between compositional factors and age in vintage ports. Certain variables (A-I, **Table 7**) showed a low modeling power (32–49%) in predicting age, while others (J–S) contributed greatly (above 70%). Wine aging is generally understood as polyphenol evolution, using Sudraud, Garoglio, and other indices including color evolution. The importance of carbonyl compounds in vintage port wine aroma has also been recognized.

In further PLS modeling of ports, clear relationships between sensory data and concentrations of head space components, and between sensory and nonvolatile component data, support hypotheses that changes in nonvolatile phenolics influence the perception of flavor character through changing head space congener concentrations. Relationships were also established between such components and total maturation age, through oxidation and polymerization of wine compounds. Prediction of intensity of certain flavor notes has also been possible from solution concentrations of congeners, notably for caramel and smooth characters. (*See* **Flavor (Flavour) Compounds**: Structures and Characteristics; **Sensory Evaluation**: Aroma.)

Port and Replica Wines Production

With the success of the Portuguese product, there have been a number of attempts to produce similar wines, including Starboard, in other countries, notably Australia, California, and South Africa. In 2001 port sales were healthy, suggesting a good future. However, non-Portuguese producers have been required, with changes in international trading, to stop the use of the port descriptor in the labeling of their fortified wines and future exports look problematic.

See also: **Alcohol**: Properties and Determination; **Grapes**; **Port**: The Product and its Manufacture; **Tannins and Polyphenols**

Further Reading

Bakker J and Timberlake CF (1986) The mechanism of color changes in aging port wine. *American Journal of Enology and Viticulture* 37: 288–292.

Bakker J, Bridle P, Timberlake CF and Arnold GM (1986) The colors, pigment and phenol contents of young port wines: Effects of cultivar, season and site. *Vitis* 25: 40–52.

Bakker J, Bellworthy SJ, Hogg TA et al. (1996) Two methods of port vinification: a comparison of changes during fermentation and of characteristics of the wines. *American Journal of Enology and Viticulture* 47: 37–42.

Burnett JK (1985) Port. In: Birch GC and Lindley MG (eds) *Alcoholic Beverages*, pp. 161–169. London: Elsevier Applied Science.

Cristovam E, Paterson A and Piggott JR (2000) Development of a vocabulary of terms for sensory evaluation of dessert port wines. *Italian Journal of Food Science* 2: 129–142.

Cristovam E, Paterson A and Piggott JR (2000) Differentiation of port wines by appearance using a sensory panel: comparing free choice and conventional profiling. *European Food Research Technology* 211: 65–71.

Curvelo-Garcia A (1988) *Controlo de Qualidade dos Vinhos*. Portugal: Odivelas.

Fonseca AM, Galhano A, Pimentel ES and Rosas JR-P (1984) *Port Wine. Notes on its History, Production and Technology*. Porto: Instituto do Vinho.

Garcia-Viguera C, Bakker J, Bellworthy SJ et al. (1997) The effect of some processing variables on non-coloured phenolic compounds in port wines. *Zeitschrift für Lebesmittel Untersudnung und Forschung A*, 205: 321–324.

Gowell RW (1972) The manufacture of port. *Process Biochemistry* Oct.: 27–29.

Howkins B (1982) *Rich, Red and Rare. The International Wine & Food Society's Guide to Port.* London: William Heinemann.

Ortiz MC, Sarabia LA, Symington C, Santamaria F and Iniguez M (1996) Analysis of ageing and typification of vintage ports by partial least squares and independent modeling class analogy. *The Analyst* 121: 1009–1013.

Reader HP and Dominguez M (1995) Fortified wines: sherry, port and Madeira. In: Lea AGH and Piggott JR (eds), *Fermented Beverage Production*, pp. 159–203. London: Blackie Academic and Professional.

Williams AA, Lewis MJ and May HV (1983) The volatile flavour components of commercial port wines. *Journal of Science and Food Agriculture* 34: 311–319.

Williams AA, Langron SP, Timberlake CF and Bakker J (1984) Effect of color on the assessment of ports. *Journal of Food Technology* 19: 659–671.

Postharvest Deterioration *See* **Spoilage**: Chemical and Enzymatic Spoilage; Bacterial Spoilage; Fungi in Food – An Overview; Molds in Spoilage; Yeasts in Spoilage

---

POTASSIUM

Contents
Properties and Determination
Physiology

Properties and Determination

M A Amaro López and R Moreno Rojas, University of Córdoba, Córdoba, Spain

Background

Potassium is a light, soft metal and a strong reducing agent, discovered by Davy in 1807 in caustic potassium (KOH). It is widely distributed in minerals such as sylvin (KCl), carnallite ($KMgCl_3$ or $MgCL_2 \cdot KCl$), langbeinite ($K_2Mg_2(SO_4)_3$) or polyhalite ($K_2Ca_2Mg_2(SO_4)_4 \cdot 2H_2O$). Potassium is a very abundant metal and is the seventh most abundant element in the earth's crust (2.59% corresponds to potassium in the combined form). Sea water contains 380 p.p.m. of potassium, making this mineral the sixth most abundant in solution.

Potassium is used extensively, mainly in the form of potassium salts, in various fields such as medicine (iodide as a disinfectant), photographic processing (carbonate), explosives (chlorate and nitrate as powder), defreezing agents, poisons (potassium cyanide), metallurgy and basic chemistry (hydroxide in, for example, strong bases, oil sweetening, CO_2 absorbent, chromate and dichromates as oxidants, carbonate in glass industry), fertilizers (nitrate, carbonate, chloride, bromide, sulfate), detergents and soaps (hydroxide and carbonate), coolants in nuclear reactors (NaK alloy), etc.

Potassium (atomic number = 19) forms part of the group of alkali metals classified between sodium (atomic number = 11) and rubidium (atomic number = 37). It has an atomic weight of 39.098, oxidation state 1 and an electronic orbital structure of $[Ar] 4s^1$. This electronic configuration (one electron in the 4s orbital) enables potassium to ionize easily to the cation K^+ owing to the loss of its most external electron. It does not appear as a native metal in nature. Potassium exists in nature in three isotopes: ^{39}K (93.26%), ^{40}K (0.0117%) and ^{41}K (6.73%). ^{40}K is radioactive and responsible for most of the naturally occurring internal radioactivity in the body. This property enables researchers to monitor total body potassium values as a function of age and disease.

The behavior of potassium in its metallic form is very similar to that of sodium, with which it is closely related, both being alkali metals essential for life owing to their involvement in cellular physiological development and the regulation of body fluids, together with chlorine, thus explaining why it is common for both alkali metals to appear in the form of their corresponding chloride salts. Given this relationship with cellular liquids, potassium, sodium, and chlorine are also known as electrolytes (ions with greater proportions in the composition of organic fluids).

The K^+ ion is the main intracellular cation, with an approximate $[K^+]$ of $5.6 \, gl^{-1}$ in cellular fluids, approximately 30 times more concentrated than in plasma or interstitial liquid $(0.15–0.20 \, gl^{-1})$. Its high intracellular concentration is regulated by the cell membrane through the sodium–potassium pump. The intracellular K^+ regulates the catalytic action of numerous enzymes, through the attachment of the cation to active locations with negative charges of the enzymatic proteins, modifying the conformation of the molecule and its activity, as well as participating in cellular division processes.

The small percentage of extracellular potassium (2% of body potassium) is of great physiological importance, since it is a critical determinant of neuromuscular excitability (nervous impulse and contraction of bone muscles). At the cellular membrane level, the transport and permeability of energy-dependent potassium, with simultaneous excretion of sodium linked to the Na/K enzyme ATPase, is essential for generating the potentials of membranes required for the proper functioning of nervous and muscular cells. It also helps to maintain the acid–base equilibrium and blood pressure.

Homeostasis of potassium is still the subject of research, although it is known that 90% of dietetic potassium is absorbed in the small intestine and that body potassium (1.6–2 g per kilogram of body weight) is regulated by renal glomerular filtration and tubular secretion, potassium being lost on a daily basis through urine, gastrointestinal secretions (ileum and colon) and, to a lesser degree, sweat. Provided that renal function is normal, it is practically impossible to reach an excessive level of potassium with normal dietary intake, since the kidney is capable of excreting more potassium than it can filter. (See Potassium: Physiology; Sodium: Physiology.)

The effect of potassium on blood pressure has been discussed in recent reports on metaanalysis; these have confirmed that increases in doses of potassium from 60 to 80 mmol per day (2.3–3.1 g per day) may prompt a decrease of 4 mmHg in systolic blood pressure and possibly reduce the number of deaths related to high blood pressure by 25%.

The use of potassium chloride salts as substitutes for sodium chloride, in individuals in whom the intake of sodium is restricted because of problems relating to hypertension, is a questionable alternative, since cardiac arrythmias have been reported in association with the excessive intake of salt substitutes containing potassium chloride. In this connection, hyperkalemia is recommended in healthy individuals for intakes exceeding 17.5 g per day (acute toxicity limit), highly unlikely in normal diets, and certain risks for individuals with renal dysfunctions not detected with intakes of potassium exceeding 5.9 g per day (safest maximum dose). Cardiovascular or neuromuscular complications arising from situations of hypo- and hyperkalemias are resolved favorably by correcting the plasmatic potassium levels. (See **Hypertension**: Hypertension and Diet.)

The mean potassium intake in Western populations ranges between 1.6–5.9 g per day and the required dietary intakes of this are met without any problem, thanks to the ubiquitous presence of potassium in both vegetable- and animal-based foods. The Scientific Committee for Food of the European Community recognizes that the Minimum Recommended Intake of potassium is 1600 mg d^{-1}, and the Reference Intake for the population 3100 g d^{-1}.

Analysis of Potassium

Potassium can be analyzed using a number of different methods, although many of these are not commonly used for routine analysis of potassium in food and biological samples. In the case of potassium, there has been a shift from the old gravimetric methods, based on the precipitation of potassium using chloroplatinate or tetraphenylboron, to the spectroscopic methods, mainly flame emission and flame absorbance spectroscopy and which, according to the scientific community, are the most commonly used.

However, there are other techniques for analyzing potassium that are not as 'popular' as those mentioned above, and equally applicable to other mineral elements. These include selective electrodes, nuclear magnetic resonance spectroscopy, X-ray analysis, helium glow photometry, inductively coupled plasma optical emission, inductively coupled plasma atomic fluorescence, ion-scattering spectrometry, and other methods. Another procedure can be used – namely, radioactive dilution of potassium isotopes – when determining both potassium content and its distribution in extra- and intracellular compartments.

Emission and absorbance spectrometry have been the most widely used techniques for analyzing trace elements in biological and food samples. Their widespread use is justified by their analytical specificity, good detection limits, excellent accuracy, and relatively low cost. Both techniques are based on energy modifications of the electronic orbital structure of the atoms of mineral elements in response to certain stimuli (flame emission spectrometry and incidence of a beam of light of specific wavelength for absorption). (*See* **Sodium**: Properties and Determination.)

The determination of potassium by spectroscopy requires a stage of preparation of samples for analysis that entails the drying, grinding, and destruction of the organic matter of the sample. There are two alternatives for destroying the organic matter of biological and food samples, both with different variants: wet oxidation and dry ashing. Both techniques yield reasonably comparable results, and it is often the analyst who decides on which technique to employ. (*See* **Cadmium**: Properties and Determination; **Spectroscopy**: Atomic Emission and Absorption.)

Flame Emission Spectroscopy

The determination of potassium in biological and food samples, as in the case of sodium, may be performed using flame photometers and atomic absorption spectrophotometers (with the option of working in emission conditions). A specific wavelength of 766.5 nm is used in order to determine potassium, and the detector is adjusted to the response given to the pattern of greatest concentration used. Efforts must be made to avoid possible interference problems owing to the matrix characteristics of the sample in flame emission photometry; this can be achieved by separating the element from the object of study, eliminating the ions responsible for the interferences, or using a compensatory technique. For this purpose, recommended practice for correcting interferences includes the addition of lithium (as an internal standard) both to the samples and to the standards. It is important to obtain a suitable calibration curve, either linear by diluting the samples, or by logarithmic adjustment between emission values and potassium concentration. (*See* **Sodium**: Properties and Determination.)

Flame Absorption Spectroscopy

The determination of potassium using an atomic absorption spectrophotometer requires a light source (wavelength for $K = 766.5$ nm) and an atomization source (flame). For the light source, arc discharge lamps may be used for K, although hollow cathodes are preferred. Standard working conditions for analyzing potassium by atomic absorption spectroscopy are listed in **Table 1**.

Comparing flame emission photometry and atomic absorption spectroscopy, the detection limits for potassium ($\mu g\,ml^{-1}$) are 0.0005 and 0.005, respectively. Using an atomic absorption spectrophotometer, the accuracy and precision of potassium analysis by flame emission can be obtained (**Table 2**).

Table 1 Standard conditions for potassium analysis by atomic absorption spectrometry

Spectrometers setting	Potassium
Wavelength/slit (nm)	766.5/0.2
Nebulizer	Spoiler
Oxidant	Air
Fuel	C_2H_2
Flame condition	Oxidizing
Optimum concentration range in solution ($\mu g\,ml^{-1}$)	1–10
Detection limit ($\mu g\,ml^{-1}$)	0.005
Sensitivity 1% absorption ($\mu g\,ml^{-1}$)	0.05
Interferences	Ionization

Table 3 shows a series of official methods for determining potassium, based on flame emission and absorption spectrophotometry.

Inductively Coupled Plasma-atomic Emission Spectrometry (ICP-AES)

There are numerous references to the use of this powerful technique in multielement analysis by emission spectroscopy. ICP-AES enables samples with a high degree of variability and minimum interferences to be used. Samples can be prepared by either wet oxidation or dry ashing, and a wide range of concentrations can be used without the need to dilute or concentrate the sample. Also, many of the other mineral elements can be quantified, thus reducing analysis time. These advantages, together with the high speed and excellent instrument stability, make ICP-AES highly attractive.

For potassium determined by ICP-AES in plant tissues, a solution detection limit of 0.06 ($\mu g\,ml^{-1}$) and a sample detection limit of 2.0 ($\mu g\,g^{-1}$, based on a 2-g sample diluted to 50 ml) are recommended. There is also an AOAC Official Method (984.27) for the determination of potassium in infant formula by means of ICP-AES, which recommends the following ICP emission spectrometer operating parameters for potassium; wavelength (nm) = 766.5, no background correction and low standard = 0 and high standard = 200 ($\mu g\,ml^{-1}$). The AOAC also recommends an Official Method (985.01) for the analysis of potassium in plants and pet foods by ICP spectroscopy.

Potassium-selective Electrodes

Potassium-selective electrodes have developed from the first ion-selective glass electrodes, with little ion selectivity, to potassium-selective electrodes based on valinomycin, a highly specific neutral carrier compound for analyzing potassium, to potassium-sensitive electrodes with ion-exchange compounds dissolved in organic solvents that are not very specific

Table 2 Accuracy and precision of the potassium analysis by flame emission

Accuracy	Certified (g kg^{-1})	Found (g kg^{-1})	IC (95%)^a	rec (RSD)^b
Citrus leaves SRM-1572	18.2 ± 0.6	17.7 ± 0.5	16.8–18.7	97 (3)
Nonfat milk powder	16.9 ± 0.30	16.3 ± 0.41	15.5–17.1	96 (5)
NIST-1549				
Precision	RSD			
Plant food	1.35			
Dairy food	2.55			

^{a}Intervals of confidence (95%).
^{b}Recovery percentages and relative standard deviation (RSD).

Table 3 Official methods recognized for potassium determination by flame emission and absorption spectrophotometry

Official methods	
Sodium and Potassium in Seafood	Flame Photometric Method, AOAC Official Method 969.23
Potassium in Fruits and Fruit Products	Rapid Flame Photometric Method, AOAC Official Method 965.30
Potassium and Sodium in Wines	Flame Spectrophotometric Method, AOAC Official Method 963.13
Potassium in Distilled Liquors	Flame Photometric Method, AOAC Official Method 963.08
Potassium in Beer	Atomic Absorption Spectrophotometric Method, AOAC Official Method 987.02
Potassium in Water	Atomic Absorption Spectrophotometric Method, AOAC Official Method 973.53
Potassium in Infant Formula, Enteral Products and Pet Foods	Atomic Absorption Spectrophotometric Method, AOAC Official Method 985.35

because of significant interference induced by the presence of sodium and other cations in organic solutions. Ion-exchange electrodes have faster response times than those of organic solvent solution membranes of valinomycin, although the latter show fewer interferences from other ions. However, neither type of electrode is effective in organic solvents. Furthermore, electrodes measure ion activity and therefore do not detect ions linked to other molecules or potassium that is not in solution.

Ion-selective electrodes are also incorporated into gasometric equipment for measuring electrolytes (Na, K, Cl, and ionized Ca), in oxygenation, ventilation, basic-acidic state, and electrolytic metabolism of patients with respiratory problems. In this context, it is useful to quantify the exchange of gases and electrolytes through cellular membranes; for this purpose, the so-called *anion gap* is calculated, which interrelates with sodium (*anion gap Na*), potassium (*anion gap K*), bicarbonate, and chlorine. This parameter provides information on electrolytic alterations and the presence of toxins in blood, and also helps to control quality in laboratories by interrelating informed values (in normal individuals, not very high values).

Some developments in the field of potassium-selective electrodes have focused on the appearance of novel potassium-selective valinomycins in the development of ion-selective sensors based on new technologies and on the combination of these ion-selective electrodes with other analytical methodologies, as shown in **Table 4**.

Nuclear Magnetic Resonance (NMR)

NMR is used to study the relationship between metabolism and function in living systems as simple as cells in culture and as complex as human subjects (e.g., the effect of temperature on the sodium/potassium pump in red blood cells). As a noninvasive, nondestructive spectroscopic technique, NMR offers a powerful approach to the study of ion balance in intact biological systems. The aims are to validate new NMR methods, to elucidate tissue-specific mechanisms, and to draw conclusions with respect to basic concepts and design in tissue metabolism and function.

The possibility of determining potassium by NMR is due, on the one hand, to the relatively high concentrations of the cation in cellular tissue and, on the other hand, to the natural abundance of ^{39}K nuclide (NMR sensitivity is about 2000 times less than that of one hydrogen atom) and the short longitudinal relaxation time. The concentrations of intracellular and extracellular potassium in tissues may be quantified by applying impermeable cell-membrane chemical shift reagents.

However, this quantification by NMR measurement depends on the NMR invisibility of the potassium actually present, in accordance with other techniques. Several previous studies have shown that only approximately 20% of cardiac intracellular potassium is visible with current NMR techniques. This NMR sensitivity of potassium (about 60%) represents a serious obstacle to its proper quantification in biological tissues. (*See* **Spectroscopy:** Nuclear Magnetic Resonance.)

Table 4 Analytical methodologies with ion-selective electrodes for potassium analysis

Investigation	Reference
A potassium-ion selective electrode with valinomycin-based poly (vinyl chloride) membrane and a poly(vinyl ferrocene) solid contact	Hauser PC, Chiang DWL. and Wright GA (1995) *Analytica Chimica Acta* 302(2–3): 241–248.
Miniaturized ion-selective sensor chip for potassium measurement in a biomedical application	Uhlig A, Dietrich F, Schnakenberg U, Hintsche R and Lindner E (1996) *Sensors and Actuators B: Chemical Abstract Export* 34(1–3): 252–257.
Synthesis of novel potassium selective valinomycins	Dawson JR, Dory YL, Mellor JM and McAleer JF *et al.* (1996) *Tetrahedron* 52(4): 1361–1378.
Optical sensors for sodium, potassium, and ammonium ions based on lipophilic fluorescein anionic dye and neutral carriers	Wang E, Zhu L, Ma L and Patel H (1997) *Analytica Chimica Acta* 357: 85–90.
Recording of neuronal network properties with near-infrared dark-field microscopy and microelectrodes	Holthoff K and Witte OW (1997) *Electrochimica Acta* 42(20–22): 3241–3246.
Flow injection analysis of potassium using an all-solid-state potassium-selective electrode as a detector	Komaba S, Arakawa J, Seyama M, Osaka T, Satoh I and Nakamura S (1998) *Talanta* 46: 1293–1297.
Transferability of results obtained for sodium, potassium and chloride ions with different analyzers	Rodriguez-Garcia J, Sogo T, Otero S and Paz JM (1998) *Clinica Chimica Acta* 275: 151–162.
Plasticizer-free all-solid-state potassium-selective electrode based on poly(3-octylthiophene) and valinomycin	Bobacka J, Ivaska A and Lewenstam A (1999) *Analytica Chimica Acta* 385: 195–202.
Simultaneous detection of monovalent anions and cations using all solid-state contact PVC membrane anion and cation-selective electrodes as detectors in single column ion chromatography	Isildak I and Asan A (1999) *Talanta* 48: 967–978.
Novel sensors for potassium, calcium, and magnesium ions based on a silicon transducer as a light-addressable potentiometric sensor	Seki A, Motoya K, Watanabe S and Kubo I (1999) *Analytica Chimica Acta* 382: 131–136.
Reference ranges of electrolyte and anion gap on the Beckman E4A, Beckman Synchron CX5, Nova CRT, and Nova Stat Profile Ultra	Lolekha PH, Vanavanan S, Teerakarnjana N and Chaichanajarernkul U (2001) *Clinica Chimica Acta* 307: 87–93.

A study of factors affecting ^{39}K NMR detectability in rat thigh muscle showed that the signal may be substantially higher (up to 100% of total tissue potassium) than values previously reported of around 40%, these signals presenting two superimposed components – one broad and one narrow – and involving improvements in spectral parameters (signal-to-noise ratio and baseline roll), together with computer simulations of spectra that enable a spectrum quality with a major effect on the amount of signal detected, largely owing to the loss of detectability of the broad signal component.

Biological and biochemical cellular research in connection with the quantification of ions in intracellular and extracellular compartments, transport and ionic metabolism at the level of different cells and biological tissues, are based on the use of ^{39}K NMR spectroscopy, together with ^{23}Na and ^{31}P NMR studies.

Proper ion equilibrium between intra- and extracellular compartments is required for normal physiological function. Conversely, alterations in membrane ion transport occur in numerous pathological states. By introducing an anionic paramagnetic shift reagent into the medium, NMR signals of intra- and extracellular Na$^+$ and K$^+$ can be resolved, enabling ion transport

processes to be studied by NMR. Unfortunately, rare NMR active nuclides that are isotopes of the 100% naturally abundant ^{23}Na$^+$ and ^{39}K$^+$ are not available for tracer kinetic studies of Na$^+$ and K$^+$ transport. However, Cs is a biologically active analog of K$^+$, and the 100% naturally abundant NMR active ^{133}Cs$^+$ nuclide can be employed to examine K$^+$ transport. Other studies have shown the potential of ^{39}K NMR as a useful tool in the study of protein–cation interactions and the binding of K$^+$ to double-helical DNA.

X-ray Analysis

X-ray analysis is another nondestructive, noninvasive, *in-vivo* technique for determining mineral elements – including potassium – in foods and biological samples. X-ray analysis is based on bombarding a small area of the powdered sample with high-energy X-rays, measuring the Kα line of the element using one of several different detectors according to the wavelength of the emitted radiation. The limitations of this technique are related to scope, in terms of detected elements and their concentration range. X-ray analysis is combined with the use of an electronic microscope for quantifying and locating the mineral element in the sample, even at the subnuclear compartmental level. Another limitation of the X-ray technique relates to the complexity of the sample preparation process for electronic microscopy.

The bibliography describes different X-ray analysis techniques for measuring potassium in food and biological samples, as indicated in **Table 5**.

X-ray fluorescence (XRF) can be successfully used for the qualitative and quantitative elemental analysis of various agricultural products. Its simplicity, high throughput and automation possibilities make it useful for screening large numbers of samples. The K content of tea samples has been determined by XRF analysis (an uncommon method for mineral analysis in food), and the findings have been compared with the results obtained using atomic emission techniques, with the conclusion that the XRF system can be used effectively for quantitative analysis of the K content of tea samples.

The methods for preparing the specimens of liquid media of the organism and XRF analysis are simple and fast, entail no disintegration of the sample, and allow measurements of elements at sigma $1 = 0.02$ at concentrations of $12\,\mu g\,ml^{-1}$ to be obtained. The method is preferable and promising for several basic elements (P, S, Cl, K, Ca), which are difficult to measure using other methods.

The use of total reflection X-ray fluorescence analysis (TXRF) in life sciences is considered a powerful analytical tool for simultaneous multielement determination. TXRF is basically an energy-dispersive technique, with the sample being excited in total reflection geometry. This technique only requires minute samples with simple preparation and involves a large dynamic measuring scale. Simultaneous detection of almost all chemical elements and lower limits of detection are achievable in optimized excitation conditions. The preferred sample types are aqueous and acidic solutions, particularly samples digested with HNO_3. Special preparation techniques are required for solids and other samples.

The application of the proton-induced X-ray-emission (PIXE) method has enabled us to determine not only the concentrations of elemental composition, including K, but also their localization in different artery-wall regions. Furthermore, the usefulness of the micro-PIXE method for studying biomedical materials has also been considered.

Radioactive Isotopes

The dilution of radioactive isotopes of potassium is applied to establish the concentration and distribution of the amounts of potassium in the different sample compartments. The dilution of radioactive isotopes enables the volumes of the organic tissue compartments to be determined. The differences in potassium concentrations between the extra- and intracellular spaces may be studied by diluting the radioactive isotopes of potassium.

For this purpose, it is important to take into account that animal and plant tissues present different potassium compartments in which the ion is accumulated, and this conditions the radioactive isotope dilution process. In animal tissue, three cellular compartments (plasma + extracellular fluid, cells, and spaces where ions are closely fixed or form complexes) are described, whereas in plant tissue, only two spaces (cytoplasm and vacuoles) must be considered, since the extracellular space is small in size. This compartmental distribution conditions the process of radioactive isotope dilution.

In animal tissue, the method involves incorporating a known amount of the tracer isotope of potassium into the extracellular compartment and monitoring the changes in plasma isotope concentration until equilibrium is reached with all the compartments. This entails estimating the distribution of potassium (1) in extracellular fluid, using tracer polymers (inulin or polyethylene glycols) that cannot penetrate cell membranes, (2) in cellular volume, by subtracting the extracellular volume from the total distribution space (determined using labeled water, tritiated or deuterated), and (3) in the space where the ion is strongly fixed or forms complexes (this is calculated by subtracting the amount of extracellular potassium + cellular potassium from the amount of potassium isotope injected).

In plant tissue, the dilution of radioactive isotopes of potassium presents limitations owing to the difficulty involved in terms of separating vacuolar volume and cytoplasmic volume. For this reason, more appropriate methods than radioactive dilution methods have been

Table 5 X-ray techniques for potassium analysis in food and biological samples

Samples	X-ray analysis
Meat and meat products	X-ray fluorescent analysis
Serum and peripheral blood cells	
Fresh green tea, black tea, and tea residue	
Leaves of lettuce	Electron microprobe X-ray analysis
Walls of the mosaic virus-infected wheat leaf cells	Energy-dispersive X-ray analysis
Perisperm tissues of seeds	
Cancerous and normal tissues	Total reflection X-ray fluorescence analysis
In vegetable foodstuffs and their respective cell fractions	
Cancerous breast tissue	
Human brains	Proton-induced X-ray emission analysis
Human atherosclerotic artery wall	

developed for determining cytoplasmic and vacuolar potassium concentrations, namely ion-selective microelectrodes or X-ray microprobe analysis.

See also: **Cadmium**: Properties and Determination; **Hypertension**: Hypertension and Diet; **Potassium**: Physiology; **Sodium**: Properties and Determination; Physiology; **Spectroscopy**: Atomic Emission and Absorption; Nuclear Magnetic Resonance

Further Reading

Ammon D (1986) *Ion Selective Microelectrodes*. New York: Springer-Verlag.

Benton Jones J, Jr. (1984) Developments in the measurement of trace metal constituents in foods. In: Gilbert J (ed.) *Analysis of Food Contaminants*, pp. 157–202, Barking, UK: Elsevier.

Cappuccio FP and MacGregor GA (1991) Does potassium supplementation lower blood pressure? A meta-analysis of published trials. *Journal of Hypertension* 9: 465–473.

Commission of the European Communities (1993) *Nutrient and Energy Intakes for the European Community. Reports of the Scientific Committee for Food, Thirty-first Series*, pp. 174–178.

Gunther K and von Bohlen A (1990) Simultaneous multi-element determination in vegetable foodstuffs and their respective cell fractions by total-reflection X-ray fluorescence (TXRF). *Zeitschrift für Lebensmitteluntersuchung und -forschung* 190(4): 331–335.

Intersalt Cooperative Research Group (1988) Intersalt: a international study of electrolyte excretion and blood pressure. Results for 24 hour urinary sodium and potassium excretion. *British Medical Journal* 297: 319–328.

Luft FC (1990) Sodium, chloride and potassium. In: Brown M (ed.) *Present Knowledge in Nutrition*, 6th edn, pp. 233–240. Washington, DC: International Life Sciences Institute Nutrition Foundation.

Springer C (1987) Measurement of metal cation compartmentalization in tissue by high-resolution metal cation NMR. *Annual Review of Biophysics and Biochemistry* 375–399.

Physiology

M P Navarro, Unidad de Nutrición, Estación Experimental del Zaidin, CSIS Granada, Spain
M P Vaquero, Ciudad Universitaria, Madrid, Spain

Introduction

Potassium is the most abundant intracellular cation in the human body and plays an important role in a variety of cell functions. This review summarizes the main aspects of potassium physiology and nutrition, including absorption, transport, distribution, storage, and excretion, as well as homeostatic mechanisms of K balance.

Role in the Body

Potassium is essential for the muscular, cardiovascular, nervous, endocrine, respiratory, digestive, and renal systems. Cell metabolism depends on the maintenance of a high intracellular K^+ concentration. Many of the body functions of potassium are due to its ionic character: it generates gradients of concentration, potential, and pressure. Potassium is the predominant osmotically active species inside the cell. Together with other ions such as sodium and chloride, which are characteristic of the extracellular fluid, potassium determines osmolarity and plays a major role in the distribution of fluids inside and outside the cell and hence in the maintenance of cellular volume. In addition, potassium participates in the regulation of the acid–base balance and is involved in cellular growth and division, energy transduction, glycogenesis, protein synthesis, hormone secretion, etc. Consequently, its deficiency causes growth retardation, with a pronounced decrease in circulating levels of growth hormone and somatomedin C and inhibition of protein synthesis.

Cell excitability and muscle contraction depend on potassium. The transmembrane electrical potential is determined by the ratio of the intracellular to extracellular potassium and sodium concentrations, in particular that of potassium. Differences in K^+ and Na^+ concentrations across cell membranes are maintained by the specific permeability to each of these ions and by K^+/Na^+-ATPase activity (K^+/Na^+-pump). Thus, K^+ is critical for the excitability of nerve and muscle cells.

During vigorous exercise, potassium is released from muscle cells, leading to an increase in extracellular potassium concentrations which facilitates ongoing muscle contraction and induces vasodilatation, increasing local blood flow. However, liberation of potassium also leads to muscular fatigue. Training reduces the exercise-induced rise in plasma K^+ concentration and also increases the total activity of Na-K pumps in muscle. The potassium internal balance helps to delay the onset of fatigue during exercise and to restore homeostasis during recovery.

Potassium can be defined as a cardioprotector nutrient. It acts directly on the heart, regulating its mechanical and electrical properties. Epidemiological data suggest that potassium intake and blood pressure are inversely correlated. The greatest hypotensive effect of potassium occurs in hypertensive patients and in subjects with a high sodium intake. The mechanisms involved include: enhancing natriuresis, baroreflex sensitivity, direct vasodilatation, catecholaminergic functions, improvement of glucose tolerance, and effects on the central nervous and the renin–angiotensin–aldosterone systems.

Potassium also protects against stroke, which is the third leading cause of death worldwide (after coronary artery disease and cancer). This electrolyte may decrease ischemic as well as hemorrhagic stroke risk through its effect on blood pressure on one hand, and by inhibiting the formation of free radicals at the endothelial cell level, thus affecting vasomotion, on the other hand. Moreover, potassium inhibits the proliferation of smooth muscle cells, platelet aggregation, and arterial thrombosis. Potassium may also prevent the death of cerebellar neurons.

Interrelationships exist between potassium and other nutrients. Potassium and sodium are strongly metabolically interrelated, principally due to K^+/Na^+-ATPase. This enzyme also provides the driving force for the transport of other solutes, such as amino acids, phosphate, vitamins, and glucose. Potassium depletion, which is more intense if there is a simultaneous excess of sodium, enhances urinary loss of calcium. This interaction may have adverse effects on bone and blood pressure.

Requirements and Daily Intakes

To date, no recommended potassium intake has been unanimously established. The UK report on dietary reference values established a reference nutrient intake of $3500\,mg\,day^{-1}$ for adults. Lower amounts were recommended for infants ($mg\,day^{-1}$): 800 and 850, 0–3 months and 4–6 months, respectively; 700, 7–12 months; 800, 1–3 years; 1100, 4–6 years; 2000, 7–10 years; and 3100, 11–14 years of age. No specific levels were recommended during pregnancy, lactation, or for the elderly. The Nutrition Working Group of the European International Life Sciences Institute (ILSI) suggested a higher recommended daily intake (3900 mg). This recommendation was aimed at promoting a potassium intake similar to that of sodium, in molar terms, i.e., an intake of $100\,mmol\,day^{-1}$ of each cation (2300 mg and 3900 mg for Na and K, respectively). However, more recently a Na:K ratio below 1 (1600 mg and 3500 mg for Na and K, respectively) has been recommended to prevent hypertension, stroke, and cardiovascular disease.

During exercise potassium needs may be increased owing to higher losses in sweat, especially in hot climates and in unaccustomed individuals. These needs should be satisfied by eating food rich in potassium and exceptionally by potassium supplements.

Published values of nutrient consumption demonstrate that in some populations, sodium intake is higher than recommended, while potassium intake should increase. In the UK, the 1991 report of the National Food Survey Committee showed that the average sodium intake was 170% of the reference intake, while that of potassium was 80% of the reference value. In Spain, data from the same year revealed that the estimated potassium intake was satisfactory but that of sodium was 140% above the reference value.

Potassium is ubiquitous in all kinds of food, but the best sources are vegetables and fruits because they combine a high potassium content with low sodium concentration (Table 1). Potassium is an essential element of all forms of life, whereas sodium and

Table 1 Potassium and sodium content of various foods (mg per 100 g of edible portion)

	Potassium	Sodium
Vegetables		
Potatoes	320	11
Cabbage	275–310	7–12
Carrots	235	60–90
Spinach	500–633	65–140
Legumes		
Red kidney beans	1370	18
Soya beans	1730	5
Lentils	940	12
Fruits		
Banana	400	1
Melon	100–310	5–32
Avocado pear	400–600	2–4
Orange	195	2–3
Apple	120	2–3
Dried fruits		
Raisins	1020	60
Figs	970	62
Almonds	780	14
Walnuts	450	7
Cereals		
White bread	100	460–540
Brown bread	175–240	540–636
Rice	110	6
Meat and fish		
Beef, veal, lamb	230–360	52–110
Chicken	320	81
Bacon	350	1860
Herring	320	120
Halibut	410	60
Tuna	400	47
Mussels	320	290
Milk		
Whole cows' milk	140	55
Various		
Chocolate	300	11
Tomato ketchup	590	1120

chlorine are essential for animals but not for many plants. Meat and fish contain important amounts of potassium because they are highly cellular tissues, but they also contain large amounts of sodium chloride, due to the extracellular fluid present in animal food. Raw foods are preferable to processed ones because during food preparation salt is usually added to enhance flavor and retard bacterial growth (e.g., bread, bacon, tomato ketchup, and cookies). Potassium salts, however, are not palatable and are not added to food during processing.

Legumes represent a good source of potassium because they provide at least 1 g potassium per 100 g portion and very little sodium. Dried fruits, for example, apricots, figs, and prunes, should also be mentioned, since they often exceed 0.5 and even 1 g potassium per 100 g portion. Nuts such as pistachios and almonds contain a considerable quantity of potassium, but the addition of salt during roasting upsets the ideal sodium-to-potassium ratio normally present in these foods. Bananas are known to be one of the fruits richest in potassium, together with coconuts, melons, avocados, and kiwi fruit. Milk is the largest dietary source of potassium for infants.

All these data should be considered, taking into account the amount of each food item consumed. For example, a ration of legumes is approximately 80 g while those of meat and milk can easily reach 200 g. Although cereals are low in potassium, they account for almost 75% of potassium ingestion in western diets. There is a wide range of potassium intake by western populations (1560–5850 mg). It is higher in vegetarians and people who consume more food from plant than animal sources. The Mediterranean diet provides a mean daily intake of potassium equivalent to the recommended value (3500 mg).

Although thermal treatments have no effect on dietary potassium, during processes such as boiling, soaking, or canning it can be reduced as a result of leaching, particularly if food is to be cut into pieces and the liquid is not consumed.

In order to protect against hypertension and stroke, except in clinical hypokalemia, potassium from good dietary sources should be encouraged rather than potassium from supplements, because such supplements may have severe adverse effects. Fruits and vegetables also contain magnesium and fiber which seem to have protective effects.

Absorption

Normal subjects absorb about 90% of ingested potassium in the intestinal tract. The net absorption is the difference between fluxes from lumen to blood and from blood to lumen. In the human small

intestine, K^+ permeability is high and potassium absorption is carried out across the epithelium of duodenum, jejunum, and ileum by passive mechanisms in response to electrochemical gradients and solvent drag. In the proximal small bowel K^+ is concentrated through the absorption of water, providing a driving force for the movement of this cation across the intestinal mucosa, preferably through the tight junctions between enterocytes. Duodenum and jejunum absorb this ion even more rapidly than water. Indeed, shortly after a meal, the K^+ concentration $[K^+]$ in jejunum rapidly reaches plasma levels. In the ileum, the transepithelial electrical potential difference strongly influences its movement. There is no evidence of active potassium absorption in the small intestine, but the existence of an apical membrane H^+/K^+-ATPase could suggest active K^+ transport.

In the colon K^+ may be secreted or absorbed in response to variations in potassium status: net secretion occurs when the luminal concentration is less than 20–25 $mmol\,l^{-1}$, while net K^+ absorption takes place when levels are above 25 $mmol\,l^{-1}$. There are two mechanisms for K^+ secretion; the major mechanism involves potential-dependent, passive flux, mostly via the tight union, or by facilitated transcellular diffusion. The second is an active serosa-to-mucosa secretory mechanism in the proximal and distal colon which is the result of uptake across the basolateral membrane mediated by Na^+/K^+-ATPase and Na^+-K^+-Cl^- cotransport and movement across the apical membrane through potassium channels. Active K^+ absorption also occurs in the distal colon, in which the K^+ uptake appears to be an exchange for H^+ across the apical membrane. This process is energized by adenosine triphosphatases (ATPases) located in that membrane. Passive potassium absorption may also take place in the proximal colon.

Bioavailability

The availability of dietary potassium is high because potassium salts are entirely soluble and few dietary factors modify its digestive utilization. Olive oil favors potassium uptake, while some fibers and certain ion exchange resins may decrease K^+ absorption.

Physiological status also affects the nutritive utilization of potassium. Infants absorb a greater proportion of salts than adults. This may be due to the increased permeability of the immature small intestinal epithelium of newborns, and especially to a higher activity of the K^+-absorptive pumps in the colonic apical membrane, while K^+-secreting channels are less relevant. During the second half of pregnancy, fecal potassium decreases. Urinary K^+ output increases in parallel with Na^+ retention, nevertheless, enough potassium is retained to cover gestational needs. Glomerular filtration rates fall with age, accompanied by limitations of K^+ secretion and Na^+ conservation. Many drugs used in the elderly alter potassium homeostasis.

Transport

Potassium, being an electrolyte, is transported mainly under ionic form in the body fluids. Intracellular potassium concentration is $110–160\ mmol\ l^{-1}$ of cells while plasma concentration averages $4–5\ mmol\ l^{-1}$, of which only 10–20% is bound to proteins.

Figure 1 schematically illustrates the intracellular and paracellular mechanisms of potassium transport. Different potassium transport mechanisms can operate simultaneously. They are usually interrelated and are also linked to other ion transport systems. Each name, except the last one, corresponds to an entire family of transporters.

Na^+/K^+-ATPase activity is responsible for the maintenance of the extra- and intracellular Na^+ and K^+ concentrations against electrochemical gradients. This ATPase is found in the plasma membrane of virtually all animal cells and represents an enormous metabolic energy cost which increases the entropy of the system. It is a carrier protein that pumps $2K^+$ in and $3Na^+$ out of the cell during each cycle of conformational changes driven by adenosine triphosphate (ATP) hydrolysis. This process is electrogenic, meaning that one net positive charge is removed from the cell every pump cycle. Many hormones and various neurotransmitters can modulate K^+/Na^+-ATPase activity.

Another active K^+ transport is controlled by the H^+/K^+-ATPase which ejects H^+ in exchange for K^+. This pump has an important role in some gastrointestinal cells and in the renal tubules.

Other potassium transport systems are driven by the force of ion gradients. The free energy gained during the movement of an inorganic ion down an electrochemical gradient (sum of the concentration gradient and the electrical potential difference) is used as the driving force to pump other solutes against their electrochemical gradient. Thus, the carrier protein acts as a coupled transporter and electroneutrality is maintained. Cotransport depends on ATP only indirectly. Various Na^+-K^+-Cl^- cotransporters, which carry $1Na^+$, $1K^+$ and $2Cl^-$ inside the cell, have been identified in salivary glands, gastrointestinal tract, and renal tubules. The K^+-Cl^- cotransporter, related to the former ones, plays an important role in erythrocytic volume. The first cotransport mediates K^+ influx while the second mediates efflux.

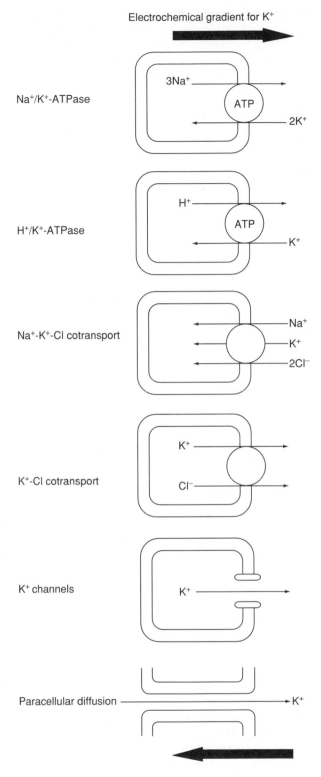

Figure 1   Potassium transport via intracellular and paracellular pathways. ATP, adenosine triphosphate.

Passive transport of K^+ occurs via intracellular and paracellular pathways. The intracellular mechanism involves K^+ channels, which are pores in the plasma membrane made by a specific protein. Channels have 'gates' which open and close in response to specific stimuli: voltage, stretch, ATP, Ca^{2+}, hormones, neurotransmitters, etc. Various stimuli sometimes act together on a channel. K^+ channels exhibit great diversity and participate in many cell functions.

Distribution and Storage

Body potassium concentration is $45-55\,mmol\,kg^{-1}$ of body weight, thus a 70-kg adult man contains approximately 135 g potassium. Women contain $35-40\,mmol\,kg^{-1}$ of body weight, and during childhood and in the elderly, body potassium concentration is also lower than that for young men. These differences are due to the lower muscle mass in these groups. Almost 98% of the potassium in the body is inside the cells, making it the major intracellular cation. More than 95% of total potassium is exchangeable.

Total potassium is an index of lean body mass because potassium is only present in the fat-free compartments of the body. It can be measured with K^{40}, a natural isotope present as a small fraction of the total potassium, which emits γ-rays and can be detected by a sensitive whole-body counter, although other isotopes and other techniques that are not based on total potassium are also available.

Studies of body composition in healthy humans have demonstrated that the total body potassium/total body nitrogen ratio is constant. However, in chronic protein-energy malnutrition this ratio can decrease owing to the reduction of muscle proteins and loss of intracellular potassium. The cellular exchange of sodium and potassium is altered, leading to potassium loss and increased intracellular sodium and water. In contrast, exercise and training increase total lean body mass, thereby causing a rise in potassium body content. Increase in muscular mass is only possible if proteins, as well as potassium, phosphorus and magnesium, which are the main intracellular elements, are present in the diet in adequate amounts.

In the extracellular fluid (ECF) an increases of potassium equivalent to 1% of total body potassium may double the concentration of potassium in plasma, resulting in muscle hypopolarization. However, if such an increase is stored in the cells, only minor intracellular changes would result, with little change to the potassium concentration difference across the cell membranes. To accomplish this buffer function, the major reservoir of potassium is the muscle, followed by the liver and erythrocytes, although each cell possesses the capacity for accumulating potassium.

Excretion

About 90% of daily potassium intake is excreted in the urine and 10% in the stool, although this last percentage can be somewhat higher in cases of diarrhea. When dietary potassium is severely restricted, its fecal loss decreases to approximately $3.5\,mmol\,day^{-1}$. This presumably represents obligatory potassium losses related to K^+ digestive secretions (salivary, gastric, biliary, and pancreatic), cell desquamation, and mucus secretion. Potassium concentration in sweat is about $10\,mmol\,l^{-1}$, so loss through perspiration is small, provided that the climate and exercise conditions are not extreme.

Renal Potassium Handling

The kidney plays the main role in potassium excretion. Renal potassium excretion shows a circadian rhythm, characterized by peak output during the subject's activity.

K^+ is freely filtered by the glomerulus. Usually, urinary potassium excretion is 5–15% of the amount filtered, which indicates the existence of tubular potassium reabsorption. The renal tubules are capable of reabsorbing and secreting potassium in response to various stimuli.

Proximal tubule The proximal tubule is responsible for the reabsorption of approximately 60% of previously filtered potassium. Solvent drag and diffusion have been proposed as the major driving forces of this process, and K^+ is reabsorbed largely via the paracellular pathway. However, some observations suggest participation of a transcellular route of potassium reabsorption by an active mechanism. The fractional rates of K^+ reabsorption are similar to those of sodium and water along the proximal tubule and changes in fluid and potassium transport are closely coupled.

In this region the basolateral K^+ channels have higher conductance than the apical ones, and this is essential to maintain a negative intracellular potential and hence promote reabsorption of positively charged carriers (Na^+-glucose cotransport, etc.)

Loop of Henle The concentration of K^+ increases as the filtrate passes through the loop of Henle, where permeability for potassium is very high. At the end of the descending limb the amount of K^+ present usually exceeds that of the glomerular filtrate. There is a net passive secretory entry of potassium into the proximal straight tubule and the descending thin limb of Henle, which arises from reabsorption in the collecting tubule and partly in the ascending limb. These pathways of K^+ transport constitute potassium recycling in the renal medulla.

In the thick ascending limb potassium is mainly reabsorbed. The mechanism of active transport includes apical uptake by Na^+/K^+-Cl^- cotransport and passive paracellular exit across the basolateral membrane by diffusion. Driven by transepithelial potential difference, reabsorption also occurs through the paracellular pathway.

The role of this limb in potassium secretion has also been recognized. The K^+ secretion through apical K^+ channels is important for net Na^+ and Cl^- reabsorption through Na^+/K^+-Cl^- cotransport, since luminal Na^+ and Cl^- concentrations are higher than luminal K^+ levels.

Distal tubule and collecting duct These tubules are the major determinants of urinary K^+, since they are able either to reabsorb or secrete potassium at rates which depend on K^+ intake and other factors.

Potassium secretion occurs in 'principal cells' by active uptake across the basolateral membrane by Na^+/K^+-ATPase and passive diffusion into the lumen across the apical membrane by K^+ channels or using a K^+-Cl^- cotransport (**Figure 2**). K^+ transport follows this route towards the lumen thanks to a favorable electrochemical gradient and the greater K^+ permeability of the apical membrane in comparison with the basolateral one. High Na^+ concentration in the lumen favors K^+ secretion, as the entrance of Na^+ into the cell generates a potential difference that tends to make the interior positive, so K^+ diffuses in the opposite direction. Simultaneously, the enhanced delivery of Na^+ into principal cells facilitates the action of Na^+/K^+-ATPase, and the accelerated potassium uptake favors its secretion towards the lumen.

Potassium reabsorption mechanisms have been detected in the intercalated cells of the distal tubule and collecting duct. It appears that there is a component of potassium reabsorption even during net secretion. Several isoforms of H^+/K^+-ATPases are able to reabsorb K^+ in exchange for H^+. The Na^+-K^+-Cl^- contransport may also introduce K^+ into the cell while the basolateral K^+ channels enable potassium to pass from the cell into the interstitial space. Proton secretion by H^+/K^+-ATPase, together with that by H^+-ATPase, contributes to urine acidification. These enzymes participate in the acid–base balance and potassium status.

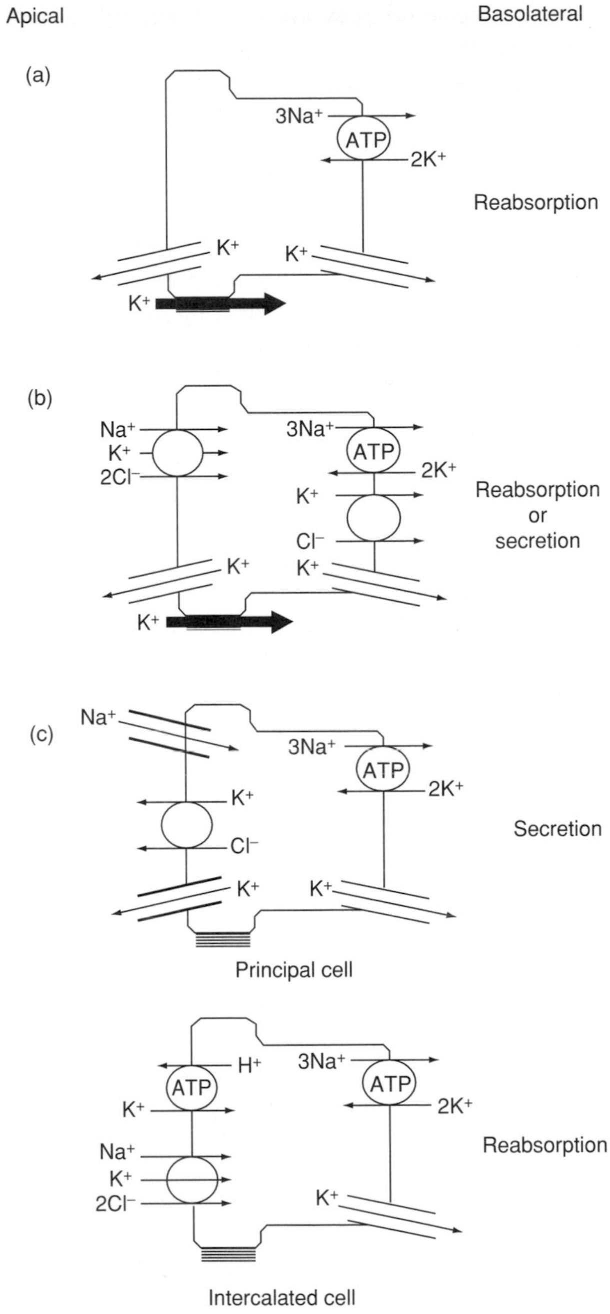

Potassium Balance and Homeostasis

Intra- and extracellular potassium concentrations must be maintained within narrow limits, despite wide fluctuations in dietary intake. This is achieved, on the one hand, by 'internal balance': distribution of K^+ between the intracellular fluid and ECF carried out mainly by

Figure 2 Principal mechanisms of potassium transport in the major tubular segments of the nephron: (a) proximal tubule; (b) thick ascending limb; (c) distal tubule and collecting duct. ATP, adenosine triphosphate.

Na^+ K^+-ATPase, and on the other hand, by 'external balance': maintenance of the amount of K^+ in the body through equilibrium between K^+ intake and excretion, which is achieved principally in the kidney but also in the intestine. After a meal, the absorbed K^+ rapidly enters the ECF. The subsequent rise in plasma $[K^+]$ is attenuated by a rapid cellular K^+ uptake (which occurs within minutes). To maintain external K^+ balance, all of the K^+ absorbed in excess must be excreted slowly by the kidneys.

Potassium Homeostasis

To maintain the equilibrium previously described, several factors which act at different levels contribute to potassium homeostasis (**Figure 3**).

Internal balance A rise in plasma $[K^+]$, especially with hyperkalemia, and the hormones insulin, epinephrine (adrenaline: by activating β_2-receptors), and aldosterone promote K^+ uptake by muscle, livers, bone, and red blood cells. In contrast, a decrease in plasma $[K^+]$, such as with hypokalemia, and stimulation of α-adrenergic receptors induce K^+ transport from cells to the ECF. Moreover, hyperkalemia stimulates insulin, aldosterone, and epinephrine secretions, while hypokalemia has the opposite effect. These are the major physiological factors involved in internal K^+ balance homeostasis.

Furthermore, other factors which are not normal homeostatic mechanisms influence K^+ movements across the plasma membrane: metabolic acidosis promotes exit of K^+ from cells, whereas metabolic alkalosis favors its uptake. Increased osmolality of the ECF enhances K^+ release by cells. Cell lysis (severe trauma, burns, etc.) induces K^+ release and may produce hyperkalemia. During intense exercise, K^+ is also released from skeletal muscle cells.

External balance There are two hypotheses concerning the homeostatic mechanisms which regulate external K^+ balance:

1. The first involves a peripheral mechanism without central nervous system (CNS) intervention. In this case $[K^+]$ would stimulate urinary K^+ excretion and aldosterone secretion directly.
2. The second attributes K^+ regulation to a reflex mechanism initiated by K^+ receptors in portal vein and liver stimulated by high $[K^+]$. Afferent fibers would transmit signals to CNS – 'K control center' – and efferent limbs, mediated by several hormones and other factors, would modulate urinary K^+ excretion in response.

In any case, plasma $[K^+]$ and aldosterone are the major physiological regulators of K^+ excretion. High $[K^+]$ increases aldosterone secretion and both act synergistically, stimulating K^+ elimination by the intestine (particularly increasing colonic secretion) and, above all, by the kidney (promoting secretion in the distal tubule and collecting duct). Renal secretion is due to stimulation of the Na^+/K^+-ATPase pump, increased permeability of the apical membrane to K^+, and greater K^+ uptake across the basolateral membrane. Hypokalemia decreases K^+ secretion by antagonist action. Under repletion conditions, potassium transported by Na^+/K^+-ATPase and H^+/K^+-ATPase recycles back into the lumen through K channels; however, when potassium levels are low, little potassium is wasted through luminal channels and passage of K through basolateral channels predominates.

Aldosterone secretion is favored by angiotensin II. On the other hand, aldosterone stimulates Na^+ and

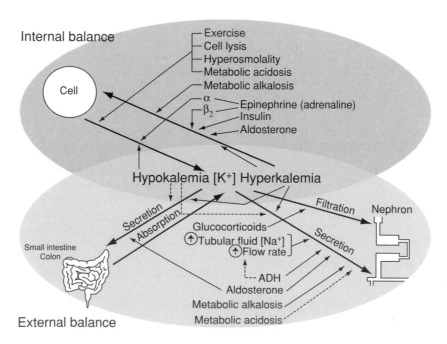

Figure 3 Potassium homeostasis. ADH, antidiuretic hormone. Continuous arrow, increase, dashed arrow, decrease.

water reabsorption and consequently decreases tubular flow at first. However, flux is restored in time, permitting aldosterone to increase K^+ renal excretion.

Antidiuretic hormone (ADH) increases K^+ secretion in the apical membrane of the principal cells in exchange for Na^+ absorption. However, because ADH also decreases tubular flow, changes in ADH levels do not substantially alter urinary K. K^+ secretion rises along the distal tubule in a flow-dependent manner and it is influenced by dietary potassium intake. This means that diuretics may enhance urinary K^+ excretion. However, K^+ excretion and distal flow rate are not completely coupled and, the K^+ balance is maintained through compensatory mechanisms when necessary.

Physiological $[Na^+]$ does not alter K^+ secretion but high $[Na^+]$ in tubular fluid increases K^+ secretion, whereas a fall in concentration has the opposite effect. Acid–base balance is another factor that modulates K^+ secretion: alkalosis increases secretion, whereas acidosis decreases it. Glucocorticoids are kaliuretic, but their effects appear to be a consequence of an increase in the glomerular filtration rate.

K^+ itself exerts a feedback regulation over most of its own control factors. Together, all these mechanisms allow the human body to maintain normal plasma $[K^+]$. However, the equilibrium can sometimes be altered, causing either hyperkalemia (extracellular $[K^+]$ over $5.5 \, mmol \, l^{-1}$) or hypokalemia (extracellular $[K^+]$ below $3.5 \, mmol \, l^{-1}$).

Hyperkalemia may be the consequence of a K^+ shift from cells to ECF or the effect of excessive K^+ retention. The first signs are flaccid paralysis, natriuresis, and other minor effects. The most important consequences are ventricular fibrillation and cardiac arrest. Hyperkalemia may be:

• Without potassium excess: caused by K^+ shift from cells (acute acidosis, hyperosmolality, insulin deficiency, β-adrenergic blockers, vigorous exercise, etc.)
• With potassium excess: impaired renal excretion (renal failure, aldosterone deficiency, primary tubule dysfunction). Inappropriately high oral or parenteral K^+ administration.

Hypokalemia may be the result of either intracellular shift or K^+ depletion, or both. Among its clinical sequelae are hyperpolarization of membrane, which affects nerves and muscle activity, and produces renal, cardiac, and metabolic alterations. The hypokalemia may be caused by:

• Intracellular shift: alkalosis, insulin excess, familial periodic paralysis, catecholamine increase, intoxications, etc.
• Extrarenal causes: insufficient K^+ intake, diarrhea, vomiting, skin loss, etc.
• Excessive renal losses: renal tubular acidosis, hypermineralocorticoidism, syndrome of choride depletion, diuretic condition.

See also: **Body Composition**; **Colon**: Structure and Function; **Diarrheal (Diarrhoeal) Diseases**; **Dietary Requirements of Adults**; **Electrolytes**: Water–Electrolyte Balance; Acid–Base Balance; **Hormones**: Adrenal Hormones; Pituitary Hormones; **Hypertension**: Hypertension and Diet; **Legumes**: Dietary Importance; **Potassium**: Properties and Determination; **Renal Function and Disorders**: Kidney: Structure and Function; Nutritional Management of Renal Disorders; **Vegetarian Diets**; **Water**: Physiology

Further Reading

Aizman R, Grahnquist L and Celsi G (1998) Potassium homeostasis: ontogenic aspects. *Acta Pediatric* 87: 609–617.

Berne RM and Levy MN (1998) *Physiology*, 4th edn, pp. 699–754. St Louis: Mosby.

Eaton SB, Eaton III SB, Konner MJ and Shostak M (1996) An evolutionary perspective enhances understanding of human nutritional requirements. *Journal of Nutrition* 126: 1732–1740.

Giebisch G (1998) Renal potassium transport: mechanisms and regulation. *American Journal of Physiology* 274 (*Renal Physiology* 43): F817–F833.

Johnson LR, Alpers DH, Christensen J, Jacobson ED and Walsh JH (1994) *Physiology of the Gastrointestinal Tract*, 3rd edn, pp. 2027–2171. New York: Raven Press.

Navarro MP and Vaquero MP (1998) Potassium. Physiology, dietary sources and requirements. In: Sadler M, Caballero B and Strain S (eds) *Encyclopedia of Human Nutrition*, pp. 1573–1580. London: Academic Press.

Nicholes CG and Lopatin AN (1997) Inward rectifier potassium channels. *Annual Reviews of Physiology* 59: 171–191.

Nilius B, Viana F and Droogmans G (1997) Ion channels in vascular endothelium. *Annual Reviews of Physiology* 59: 145–170.

Shils ME, Olson JA, Shilke M and Ross AC (1999) *Modern Nutrition in Health and Disease*, 9th edn, pp. 105–139, 1217–1225. Philadelphia: Lippincott/Williams & Wilkins.

Suter PM (1998) Potassium and hypertension. *Nutrition Reviews* 56: 151–153.

Suter PM (1999) The effects of potassium, magnesium, calcium, and fiber on risk of stroke. *Nutrition Reviews* 57: 84–91.

Wang W, Hebert SC and Giebisch G (1997) Renal K^+ channels: structure and function. *Annual Reviews of Physiology* 59: 413–436.

Yokoshiki H, Sunagawa M, Seki T and Sperelakis N (1998) ATP-sensitive K^+ channels in pancreatic, cardiac, and vascular smooth muscle cells. *American Journal of Physiology* 174 (*Cell Physiology* 43): C25–C37.

POTATOES AND RELATED CROPS

Contents
The Root Crop and its Uses
Fruits of the Solanaceae
Processing Potato Tubers

The Root Crop and its Uses

M V Rama and P Narasimham, Formerly of Central Food Technological Research Institute, Mysore, India

Introduction

The potato (Irish potato, white potato, *Solanum tuberosum* L.) is of ancient origin. It has originated and was first domesticated in South America, even before the appearance of maize. It was initially introduced to Spain and the UK, and gradually to parts of Asia and North America. It has spread around the world over the past 400 years, and gained recognition as an inexpensive and nutritive food in the eighteenth century. It is now among the 10 major food crops of the world and grown in over 140 countries. This article reviews the geographical distribution, varieties, commercial importance, morphology, and anatomy of the tuber, chemical composition, nutritive value, and postharvest handling and storage of potatoes.

Geographical Distribution

The potato (*S. tuberosum* L.) belongs to the Solanaceae family and is largely grown in cool regions where the mean temperature during the growing season does not exceed 18 °C. World production of potato accounts for 47% from 37% acreage of total root and tuber crops. It is the fourth most important food crop in the world, next to wheat, maize, and rice in global tonnage (**Table 1**). The principal producers are the European and Asian countries, North America, and the Andean countries of Latin America (**Table 2**). Production shows a yearly increase in both the developing and developed countries, more significantly in the former. Russia, China, USA, Poland, India, Germany, UK, France, Turkey, and Canada are some of the highest potato-producing countries of the world (**Table 3**). The USA and Canada together account for 8% of the world production from 4% of acreage.

Commercial Varieties and Transgenic Potatoes

There are more than 150 wild species of potato found in Central America, Mexico, and the USA. *S. tuberosum* L. (tetraploid) represents the cultivated species and there are seven other cultivated species, including *S. ajanhuiri*, *S. goniocalys*, *S. phureja*, *S. stenotomum* (diploids), *S. × juzepczukii*, *S. × chaucha* (triploids), *S. × curtilobum* (pentaploid), which are grown in different parts of Peru, Bolivia, Ecuador, and Venezuela. Many improved local varieties have been developed and introduced in several countries and these may show increased yield, disease resistance, better tuber shape, texture, and quality for processing as well as flavor.

Table 1 World production of potatoes compared to major tuber crops and cereals

Crop	Area harvested (× 1000 ha)	Yield (kg ha^{-1})	Production (× 1000 Mt)
Potato	17 586	164 063	288 522
Cassava	16 601	99 983	165 986
Yam	3802	96 122	36 545
Sweet potato	9002	133 215	199 919
Wheat	213 790	27 313	583 918
Rice	513 177	38 437	588 766
Maize	139 878	43 209	604 400

Data from www.fao.org – Food and Agriculture Organization Statistical database, provisional 1999 production and production indices data.

Table 2 Potato production in parts of the world

Parts of the world	Area harvested (× 1000 ha)	Yield (kg ha⁻¹)	Production (× 1000 Mt)	Percent production
World	17 586	164 063	288 522	100
Africa	811	111 790	9072	3.1
Asia	9171	148 288	92 058	31.9
Europe	6208	156 305	143 349	49.7
North Central America	306	918 324	28 134	9.7
South America	1030	135 924	14 008	4.8
Oceania	58	326 278	1899	0.6
Developing countries	7585	141 777	107 550	37.3
Developed countries	10 000	180 969	180 972	62.7

Data from www.fao.org – Food and Agriculture Organization Statistical Database, provisional 1999 production and production indices data.

Table 3 Major potato-producing countries of the world (production – × 1000 Mt)

Country	Year 1995	1997	1999
Former Soviet Union	39 909	37 039	30 300
China	45 754	45 534	43 477
USA	20 122	20 861	21 840
Poland	24 891	20 776	26 000
India	17 401	19 240	22 100
Germany	10 888	12 438	12 074
UK	6396	7154	7000
France	5882	6500	6500
Turkey	4750	5000	5315
Canada	3374	4050	4260

Data from www.fao.org – Food and Agriculture Organization Statistical Database, provisional 1999 production and production indices data.

In recent years, attempts have also been made worldwide to generate transgenic potato plants and assess the impact of transgenic expressions on diverse parameters such as yield, quality, carbohydrate metabolism, stress physiology, and pest and disease resistance. The outcome of the transgenic approach is promising and already insect- and herbicide-resistant potato plants are in commercial use.

Commercial Importance

The potato is grown as an early or late crop for seed or ware purpose. Even in comparatively harsher climates, it produces more dry matter (DM) and protein per hectare than major cereal crops of the world. It is a major food crop in the temperate zone and is either a staple food or merely a vegetable in tropical and subtropical countries. Tubers are consumed in various forms after cooking or processing. In the UK, USA, and India 65–90% of the production is consumed as food, and a small quantity is used for industrial purposes. In The Netherlands, a large percentage is used for starch production. Starch, alcohol, glucose, and dextrin are the industrial products of potato. World production of potato starch is about 2×10^6 t. Potatoes are generally boiled, cooked in oil, or baked. Dehydrated, frozen, and canned products are also popular. In the USA, 50% of the potato crop is consumed in processed form; the most relished is potato chips. Varieties with high DM content are preferred for potato chips. Frozen products include hash browns (patties), French fries, croquettes (puffs), and mashed potato, while dehydrated products include granules, flakes, diced chunks, and julienne strips. The common canned products are new potatoes, hash, stew, and soups. Papa seca and chuno are the traditional dried potatoes, which form a vital part of the diet in the highland areas of Peru and Bolivia. Dough-like sliceable, storage-stable potato products have also been developed. (*See* **Starch: Sources and Processing.**)

The waste water collected from peeling, cutting, and blanching in the processing industries has been used to recover highly nutritive protein of 80–85% purity. The use of this recovered protein of high biological value is limited as human food, due to the presence of antinutritional factors, but the residual water can be used in preparing a growth medium for single-cell protein production. Potato peel itself is superior to wheat bran, containing high amounts of

Table 4   Advantages of using true potato seeds (TPS) for potato production

Parameters	Advantages
Seed rate	25 g of TPS against 2 t seed tubers per hectare
Profit	Planting material cost is very low; saves cost on transport, cold storage, and needs less storage room
Productivity	More productive due to higher yield and more dry matter
Technical feasibility	Can be effectively used in different agroclimatic zones
Less risky/environment-friendly	Built in broad-based resistance to late blight: reduced usage of pesticides

Adapted from Upadhya MD (1994) True potato seed: propagule for potato production for the 21st century. In: Shekhawat GS, Khurana P, Pandey SK and Chandla VK (eds), *Potato: Present and Future*, pp. 15–22. Shimla, India: Indian Potato Association.

minerals and fiber. The water extract of the peel is a nonmutagenic antioxidant with potential antimicrobial activity. Snakin-1 (SN-1), a peptide isolated from the potato tuber, is shown to be active against potato bacterial and fungal pathogens. Further, potato plant having tuber as the sink for carbon and nitrogen is said to serve as a model species for various experiments involving bacteria-mediated gene transfer systems. (*See* **Alkaloids**: Properties and Determination.)

Tuber is also used for propagation. Tuber seed is the costliest input, accounting for nearly 40% of the cost in potato production; the availability of an adequate quantity of virus-free seed stock is still a constraint for increased production, especially in developing countries. Hence, in recent years commercial crops are being raised using true potato seed (TPS), which has several advantages over traditional tuber seeds (**Table 4**). However, usage of TPS is still limited in developed countries, due to erratic germination and low yield. A host of hybrid TPS families with desirable attributes has been developed by the International Potato Research Center (CIP), Peru, and Central Potato Research Institute (CPRI), India, using *S. andigena* and *S. tuberosum* as parental lines.

Morphology and Physiology of the Tuber

The potato is an underground, modified, fleshy stem, with a shortened axis, developing from the subapical region of a diageotrophic rhizome or lateral shoot. The tuber end that is attached to the rhizome is called the stem end or heel end, and the other end is the bud end or rose end. There are as many as 20 eyes in the axils of leaf scars (eye brows) arranged spirally around a tuber. Each eye has one main bud and several small lateral buds. Physiologically, the youngest bud is the last formed apical bud (**Figure 1a**). Tubers sprout after completing the inherent dormancy period, which lasts from 6 to 16 weeks after harvest, depending upon the variety. Tubers exhibit apical dominance. Studies have shown that the balance between the two hormones, abscisic acid

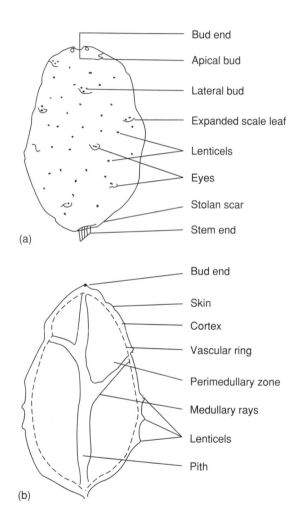

Figure 1   A potato tuber: (a) morphological details; (b) longitudinal section.

(ABA) and gibberellic acid (GA$_3$), in the tubers determines the extent of the dormancy period. Dormancy breaking is also related to protein degradation and mobilization of nitrogen reserves for sprout formation. Morphological characters of the tuber such as shape (round, oval, or elongated), size (small, medium, or big), skin color (light yellow to black or purple), and depth of eye (deep or shallow) determine its acceptability for various end uses.

Anatomy of the Tuber

The potato tuber is protected by the outermost skin or periderm, consisting of six to 10 layers of suberized cells. Skin color depends on the anthocyanin concentration in the periderm and peripheral cortex. The active periderm of a young tuber can be easily removed, and tuber enlargement is associated with sloughing off the periderm, which is then replaced by a new 'cork' layer, formed from beneath. Similarly, wound healing takes place in damaged tuber tissue by the formation of a wound periderm, which is more impervious than normal skin. A number of lenticels are found in the periderm, and these pores also facilitate the exchange of gases and the entry of pathogens. Inside the periderm is the parenchymatous cortex (0.3–1.0 cm thick) in which food material is stored in the form of starch granules. Cortex (potato flesh) may be white or various shades of yellow depending upon the variety. Yellow-fleshed varieties are sometimes highly prized. The parenchymatous perimedullary zone is seen between the vascular ring and the medullary ray (**Figure 1b**). Xylem is visible as a ring, while phloem forms many bundles in the cortex and perimedullary zones. Medullary rays run from the stem end to the eyes. The pith present at the center of the tuber is translucent, as it has less starch.

Chemical Composition

The chemical composition of potato varies with variety, soil type, cultural practice, maturity stage, disease, and storage conditions. The distribution of DM and chemical compounds (**Figure 2**) is extremely heterogeneous within a tuber. The data in **Tables 5** and **6** shows potato to be a good source of carbohydrates, vitamins, and minerals. The small fraction of protein present in the tuber consists of many essential amino acids.

Carbohydrates

Starch, sugars, and nonstarch polysaccharides constitute the carbohydrate fractions of potato tuber.

Table 5   Chemical composition of the dry matter (22.5%) of the potato tuber

Chemical composition	Content[a]
Crude fiber	2.2
Starch	74.2
Total sugar	1.3
Reducing sugar	0.6
Fat	1.0
Total nitrogen (N)	1.2
Protein N	1.0
Protein fractions (% of total protein N)	
Albumin	48.9
Globulin	25.9
Prolamin	4.3
Glutelin	8.3
Minerals	
Calcium	0.02
Magnesium	0.08
Potassium	1.47
Sodium	0.02
Phosphorous	1.87
Iron (p.p.m.)	15.70
Vitamins (mg 100 g^{-1})	
Thiamin	0.73
Ascorbic acid	92.08
Nicotinic acid	10.08
Riboflavin	0.12

[a]Values given as percentages, unless otherwise stated.
Data adapted from Rastovski A, Van Es A *et al.* (1987) *Storage of Potatoes. Post Harvest Behaviour, Store Design, Storage Practice, Handling.* Wageningen, The Netherlands: Center for Agricultural Publishing and Documentation.

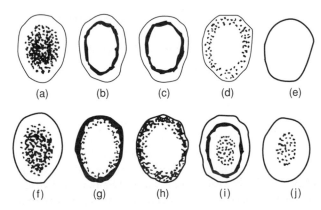

Figure 2 Pattern of distribution of chemical compounds (shaded areas) within a potato tuber. (a) Starch; (b) sugars; (c) proteins and amino acids; (d) fat; (e) crude fiber; (f) vitamins; (g) minerals; (h) organic acids; (i) phenol compounds; (j) alkaloids. Adapted from Rastovski A, Van Es A *et al.* (1987) *Storage of Potatoes. Past Harvest Behaviour. Store Design, Storage Practice, Handling.* Wageningen, The Netherlands: Center for Agricultural Publishing and Documentation.

Table 6   Essential amino acid composition of potato tuber (dry-weight basis)

Amino acid	Mean concentration[a] (mg g^{-1} tuber)
Histidine	2.5
Isoleucine	3.4
Leucine	5.0
Lysine	4.9
Methionone + cysteine	1.9
Phenylalanine + tyrosine	7.0
Threonine	3.1
Tryptophan	1.3
Valine	5.7

[a]Average of lowest and highest values reported.
Reproduced from Jadhav SJ and Kadam SS (1998) Potato. In: Salunkhe DK and Kadam SS (eds) *Handbook of Vegetable Science and Technology: Production, Composition, Storage and Processing*, pp. 11–69. New York: Marcel Dekker.

Starch, being the major carbohydrate (16–20% on a fresh-weight basis or fwb), constitutes 60–80% of the dry matter, and it is composed of amylose and amylopectin in a 3:1 ratio. Potato starch gelatinizes above 70 °C. The total sugar content ranges from 0.1% to 0.7% (fwb), and this is chiefly associated with maturity, senescence, and sprouting. The major sugars of potato are glucose, fructose, and sucrose. Trace amounts of melibiose, raffinose, stachyose, glycerol, galactinol, and glucosyl have also been identified. Tubers containing more than 2% (fwb) reducing sugars give dark-colored chips owing to Maillard reactions and are not suitable for processing. In mature stored tubers, starch and sugars exist in a state of dynamic equilibrium. The nonstarchy polysaccharides, such as cellulose, hemicellulose, and pectic substances, constitute about 1.2% (fwb) and are present in the cell walls and middle lamellae. They contribute to the final texture of cooked potato, and act as a source of dietary fiber. (*See* **Browning**: Nonenzymatic; **Carbohydrates**: Requirements and Dietary Importance; Classification and Properties; **Starch**: Structure, Properties, and Determination.)

Proteins and Amino acids

Potato is considered to be low in protein (2% fwb), but is rich in lysine compared to cereal proteins, while the concentration of sulfur amino acids is less than in cereals. The protein is concentrated more in the cortex and pith. The levels of proteins such as albumin, globulin, prolamine, and glutelin are 48.9%, 25.9%, 4.3%, and 8.3% of total protein respectively. Nearly 75% of the nonprotein nitrogen (NPN) occurs as free amino acids and amides. Two-thirds of the NPN fraction is composed of free amino acids and 21 of the amino acids have been identified. The essential amino acids and their concentrations are given in **Table 6**. Sprouting, storage, diseases, and fertilizer applications influence the concentration of free amino acids in the tuber. Amides like glutamine and asparagine occur in almost equal amounts. (*See* **Protein**: Chemistry; Requirements.)

Vitamins and Minerals

Potato contains substantial quantities of vitamins B and C (**Table 5**). Vitamin C is present in both oxidized and reduced forms. Freshly harvested tubers may contain 20 mg of ascorbic acid per 100 g; losses of vitamin C occur during long-term storage (40–60%), cooking, and processing (20%). Ascorbic acid content of potato is higher than that of several other vegetables like carrots, pumpkins, onions, and green beans. The vitamin B group comprises thiamin, riboflavin, nicotinic acid, and pyridoxine; folic acid along

with pantothenic acid is also present in potato. The ash content is about 1% (fwb), which is equivalent to 4–6% of the DM content. Fat-soluble vitamins occur in traces or are absent. This necessitates supplementation of other food sources rich in vitamin A in a potato diet. The major elements present are phosphorus, potassium, magnesium, sodium, and calcium, with wide ranges in their amounts. Potato is a poor source of calcium and sodium. A small percentage of phosphorus (25%) occurs as insoluble phytic acid. Others, such as boron, copper, zinc, iodine, aluminum, arsenic, nickel, and molybdenum, are found in trace amounts. (Refer to individual minerals and vitamins.)

Lipids and Organic Acids

Approximately 0.1% (fwb) lipid is found in potato; it is concentrated in the periderm. Linoleic, linolenic, and palmitic acids are the major fatty acids. A number of organic acids are present in potato in varying quantities; they contribute to the flavor and buffering of the potato sap. The major organic acids are citric, oxalic, fumaric, and malic acids. Other than phytic acid, nicotinic and chlorogenic acids have also been reported. Chlorogenic acid reacts with ferric iron, forming a complex, which causes darkening after cooking. Enzymatic browning in cut and homogenized potato tissue is caused by the oxidation of tyrosine. Both enzymatic and nonenzymatic browning can be inhibited by treatments with sulfur dioxide and sulfites. Other phenolic compounds found in potato are polyphenols, flavones, anthocyanins, and tannins. Phenols are again associated with after-cooking discoloration of tubers, particularly at the stem end. Tannins being localized in the periderm impart tan coloration to the skin. (*See* **Acids**: Natural Acids and Acidulants; **Fats**: Classification; **Fatty Acids**: Properties; **Phenolic Compounds**; **Tannins and Polyphenols**.)

Enzymes and Pigments

The enzymes reported in potato include amylase, glyoxalase, phosphorylase, tyrosinase, peroxidase, catalase, aldehydrase, phosphatase, sistoamylase, and zymohexase. The phosphorylase and amylase systems form sugars at low temperatures. The probable role of D-enzyme (EC 2.4.1.25; 4-alpha-glucanotransferase) in starch metabolism has been suggested. Transgenic potato plants with reduced D-enzyme activity have been obtained. Polyphenol oxidase, peroxidase, and catalase enzymes involved in the oxidation of phenols bring about browning of the freshly cut surface of the potato. The yellow color of potato flesh is attributable to carotenoids such as α-carotene, auroxanthin, violaxanthin, lutein, isolutein, and neo-β-carotene. Flavonols, flavones, flavin, and neoxanthin are also found in small quantities. Red-skinned potatoes have anthocyanin in the periderm and outer cortical cells. Chlorophyl is present in tubers exposed to light, and these green potatoes lose their market value. (*See* **Carotenoids**: Occurrence, Properties, and Determination; **Chlorophyl; Colorants (Colourants)**: Properties and Determination of Natural Pigments; **Enzymes**: Functions and Characteristics.)

Antinutritional Factors

The glycoalkaloids, proteinase inhibitors, and the lectins are the major antinutritional factors present in the tubers. The synthesis of toxic glycoalkaloids such as α-solanine and α-chaconine, which are believed to be a part of a disease-resistance mechanism, takes place in damaged and light-exposed tubers; a normal tuber contains an insignificant amount (5–10 mg $100\,g^{-1}$) of these alkaloids. They are recognized as natural products with insect deterrent and antifungal activity, concentrated more in the peel, sprouts, and around the eyes, and absent in the pith. Tubers develop bitterness and off-flavors when the alkaloid content is 25–80 mg $100\,g^{-1}$, and consumption of tubers with more than 20 mg of glycoalkaloid causes fatal illness. It can be eliminated to an extent of 60%, by peeling the tuber, while its accumulation can be controlled by waxing, oil and water dipping, and storing in the dark. The nutritional significance of two other toxic substances, proteinase inhibitors and lectins, has not been much studied, but they are reported to be heat-labile. (*See* **Alkaloids**: Toxicology.)

Nutritive Value

Potato is one of the richest sources of energy and has a greater capacity to supply energy than any other food crop. In some developing countries, potato is considered merely a vegetable rather than a staple food because of the belief that excessive consumption causes flatulence in humans. Experiments have shown that cooked potato starch is easily digestible, and hence forms a valuable food, even for infants. Potato has a slightly lower energy content (335 kJ $100\,g^{-1}$) than other roots, tubers, cereals, and legumes (**Table 7**), but this is advantageous in overcoming the problem of obesity in the developed world. On the other hand, large quantities of potato have to be consumed in developing countries to meet their populations' daily energy needs. It has been calculated that 100 g of potato can supply 5–7% of the daily energy and 10–12% of the daily protein need of children aged 1–5 years.

Potato protein has a biological value equal to that of soya bean protein and the ratio of total essential amino acids to total amino acids is so balanced that it can meet the needs of infants and small children. It has been calculated that 100 g of potato can supply 3–6% of proteins, the recommended daily allowance (RDA) for adults (depending upon sex and body weight), and one medium-sized potato provides 15 mg of vitamin C, which is about 20% of the recommended allowance (75 mg) per person per day. Eating the tuber with the skin on increases the dietary fiber intake. It is also a most valuable food for those who suffer from excess acidity of the stomach as it has an alkaline reaction. Potato fat is too low to have any nutritional significance, but it does contribute towards palatability. The tuber also provides most of the trace elements needed to maintain good health, and although potato is not a primary source of iron, 100 g of cooked potato can supply between 6% and 12% of the daily iron requirement of children or adult men. About 7% of the US RDA of phosphorus for both children and adults is available from 100 g potato. A low percent of phytic acid present in potato tuber, compared to other staple cereals and vegetables, is an advantage as it allows greater availability of free calcium, iron, zinc, and phosphorus for absorption by the human intestine. (*See* **Amino Acids**: Metabolism; **Ascorbic Acid**: Physiology; **Dietary Fiber**: Physiological Effects; **Energy**: Measurement of Food Energy; **Iron**: Physiology; **Protein**: Quality.)

Table 7 Composition of potato (raw and dried) and other major cereals and root crops (per 100 g edible portion)

Crop	Energy (kJ)	Moisture (%)	Crude protein (g)	Fat (g)	Total carbohydrate (g)	Crude fiber (g)	Ash (g)
Potato (raw)	335	78.0	2.1	0.1	18.5	2.1	1.0
Potato (dried)	1343	11.7	8.4	0.4	74.3	8.4	4.0
Rice	1523	12.0	6.8	0.5	80.2	2.4	0.6
Wheat	1389	12.3	13.3	2.0	70.9	12.1	1.7
Sweet potato	485	70.2	1.4	0.4	27.4	2.5	1.1
Yam	444	72.0	2.2	0.2	24.2	4.1	1.0
Cassava	609	62.6	1.1	0.3	35.2	5.2	0.9

Adapted from Woolfe JA (1987) The Potato in the Human Diet. Cambridge: Cambridge University Press.

Handling and Storage

In most countries, potatoes are only harvested at certain times of the year, and harvested tubers must be stored for at least a few months. Harvesting and subsequent processes such as assembling, grading, bagging, transport, and marketing, may cause damage to the tubers; such tubers lose water quickly and are highly susceptible to microbial spoilage. Bruising also causes a physiological disorder called black spot which is as the result of oxidative biochemical reactions involving chlorogenic acid, polypohenol oxidase, amino acids, and tyrosine, resulting in melanin formation (enzymatic browning); bruised black-spotted tubers have less market value and shorter shelf-life. Degree of dormancy, rate of tuber respiration, chemical composition, and ability of wound healing influence the spread of disease-causing organisms through the tuber tissue. Thus curing, a process which promotes wound healing, is necessary before potato storage. Complete curing takes 3–6 days at 20 °C and 85–90% relative humidity (RH), but is faster at higher temperature. The physiological disorders, diseases, and pests of the tubers in the field and in storage are listed in **Table 8**.

Good storage conditions should prevent excessive sprouting, root development, moisture loss, sugar accumulation, greening, and temperature damage. The postharvest losses in potatoes are estimated to vary from 5 to 40%. For short periods, the tubers can be stored in clamps, in specially designed simple sheds, pits, cellars, or buildings with controlled temperature and humidity. Modern potato storages are equipped with computer-based control systems to monitor and maintain optimum levels of temperature and humidity with air envelop facilities. Long-term storage without sprouting and minimum loss of moisture is possible at 4–5 °C, but temperatures below 6 °C lead to the formation of reducing sugars, and this is not a desirable feature for industrial processing. Tubers can be reconditioned by transfer to a higher temperature, but the sweetness developed by long-term, low-temperature storage (senescent sweetening) cannot be eliminated completely; nearly 30% of the starch is lost as sugars, reducing tuber quality, particularly for processing.

The freezing of tubers takes place at −1 °C to −2 °C, and frozen tubers soon become soft and unusable. Potato stores are artificially heated using oil or gas heaters when the outside air temperature is lower than the required temperature. On the other hand, when potatoes are stored above 30 °C, accumulation of carbon dioxide leads to death of cells, causing black heart. In temperate regions, where the ambient temperature falls below 4 °C, potatoes are

Table 8   Potato tuber diseases and pests in the field and during storage

Disease	Causal agent/organism
Nonparasitic diseases	
Soft rot, mahogany browning, and net necrosis	Low-temperature storage (chilling injuries)
Black heart and hallow heart	Low oxygen supply
Black spot	Bruising
Pathological diseases	
Bacterial	
Bacterial soft rot	*Erwinia carotovora*
Brown rot	*Pseudomonas solanacearum*
Bacterial ring rot	*Corynebacterium sepedonicum*
Common scab	*Streptomyces scabies*
Fungal	
Powdery scab	*Spongophora subterranea*
Dry rot	*Fusarium coeruleum*
Late blight	*Phytophthora infestans*
Early blight	*Alternaria solani*
Charcoal rot	*Macrophomina phaseoli*
Wart	*Synchytrium endobioticum*
Gangrene	*Phoma exigua*
Pink rot	*Phytophthora erythroseptica*
Insect (pest)	
Potato tuber moth	*Gnoremoschema operculella*
Viral	
Tobacco rattle, mop top, yellow viruses dwarf diseases	

Adapted from Pushkarnath (1976) *Potato in Sub-Tropics*. New Delhi: Orient Longman; and Rich AE (1983) *Potato Diseases*. New York: Academic Press.

cooled with outside air. In the tropics and subtropics, refrigeration is used. The recommended temperatures for tubers to be used for different purposes are as follows:

- Seed potato 2–4 °C
- Fresh consumption 4–5 °C
- Chipping 7–10 °C
- French frying 5–8 °C
- Granulation 5–7 °C

(*See* **Storage Stability**: Mechanisms of Degradation.)

Sprouting increases weight loss, softens the tuber, and favors accumulation of glycoalkaloids, thus decreasing marketability and nutritive value. Apart from low-temperature storage, other methods of achieving sprout inhibition include the use of maleic hydrazide (MH), isopropyl-N-phenylcarbamate (IPC) and isopropyl-N-(3-chlorophenyl)-carbamate (CIPC), tetrachloronitrobenzene (TCNB), naphthalene acetic acid (NAA), methyl ester of naphthalene acetic acid (MENA), and irradiation. Time, method of application, and concentration of the chemical are important factors, as treatment may have adverse effects such as

Table 9   Sprout yield in naphthalene acetic acid (NAA) and vapor heat (VH)-treated tubers at the end of 3 months of storage

Storage temperature (°C)	Sprout weight (fresh – g per 100 tubers)			
	Control	NAA-treated	Control	VH-treated
2 ± 1	17.2	6.2		
10 ± 1	13.9	4.8		
25 ± 5	26.0	2.5	58.7	10.1

Adapted from Rama MV and Narasimham P (1985) *Studies on the effect of various post-harvest treatments for controlling shriveling, sprouting and spoilage of potatoes during storage.* PhD thesis. India: University of Mysore.

Table 10   Physiological loss in weight (PLW), sprouting, and spoilage of potatoes during evaporative cooling storage for 30 days

Parameter	Ambient storage (25–35 °C)		Evaporative cooling storage	
	Untreated	NAA-treated	Untreated	NAA-treated
PLW (%)	2.6	2.4	1.1	0.9
Fresh sprout weight (g per 100 tubers)	86.1	21.6	70.1	17.1
Spoiled tubers (%)	4.5	0.0	0.0	0.0

Adapted from Rama MV and Narasimham P (1989) *Studies on the effect of various post-harvest treatments for controlling shrveling, sprouting and spoilage of potatoes during storage.* PhD thesis. India: University of Mysore.

inhibition of wound healing, increasing the reducing sugar content, and the problem of toxic residues. TCNB, CIPC, and IPC treatments are to be coupled with cold storage, as they are ineffective at tropical ambient temperatures (> 20 °C). The sodium salt of NAA (1000 p.p.m.) is an effective and economical postharvest sprout-retardant for storing potatoes under tropical ambient conditions (**Table 9**).

Several naturally occurring volatile and aromatic compounds have proved to be potent sprout inhibitors, but their acceptability for commercial application needs further investigation. Ware potatoes (those potatoes intended for human consumption in contrast to seed potatoes) can be kept sprout-free for a long time by irradiation, but its economic feasibility remains to be ascertained. Irradiated tubers need to be stored below 20 °C, and irradiation often lead to increased *Fusarium* attack and discoloration after cooking. Controlled-atmosphere (CA) storage of potatoes with as low as 5% carbon-dioxide concentration causes black heart and low oxygen concentrations inhibit wound healing; therefore CA storage appears not to be beneficial. Sprout inhibition beyond 3 months at tropical temperatures (22–35 °C) is possible by periodical vapor heat treatment at 60 °C, RH 95% for 60 min (**Table 9**). This method is superior to manual desprouting, and avoids the use of toxic chemicals. Hot-water dip treatment at 57.5 °C for 20 or 30 min prior to storage at 18 °C enhances storability, by inhibiting sprouting and spoilage by *Fusarium* and *Erwinia*. (*See* **Irradiation of Foods**: Applications.)

Evaporative cooling systems can be used effectively in tropical conditions to store potatoes. The containers needed for this storage can be fabricated locally using brick, sand, bamboo, and metal sheets; it is a cheaper method of storing potatoes, with reduced water loss, at farm level (**Table 10**). These containers maintain a high RH (85–95%) in the atmosphere and the tuber temperature, which is close to the wet bulb temperature, is the lowest possible for evaporative cooling. Seed potatoes are often exposed to natural or artificial light during storage, as light retards physiological aging, inhibits sprout growth, and increases resistance to fungal infections. On the other hand, exposure of ware potatoes to light should be avoided, as it leads to the formation of the toxic α-solanine and green pigment, chlorophyl; the market value of green potatoes is also low.

Avoiding damage to potatoes and dry, cool storage are necessary to prevent bacterial and fungal attack. Benzimidazole compounds are used to combat fungal diseases. Efficient ventilation in any store is necessary to eliminate excess moisture, which otherwise may enhance rotting. (*See* **Fungicides**; **Spoilage**: Bacterial Spoilage.)

See also: **Acids**: Natural Acids and Acidulants; **Ascorbic Acid**: Physiology; **Carbohydrates**: Classification and Properties; Requirements and Dietary Importance; **Carotenoids**: Occurrence, Properties, and Determination; **Dietary Fiber**: Physiological Effects; **Energy**: Measurement of Food Energy; **Enzymes**: Functions and Characteristics; **Fungicides**; **Protein**: Chemistry; Requirements; **Spoilage**: Bacterial Spoilage; **Starch**: Structure, Properties, and Determination; Sources and Processing; **Storage Stability**: Mechanisms of Degradation

Further Reading

Anonymous (1984) *A Potato Store Run on Passive Evaporative Cooling*. Technical bulletin no. 11, pp. 1–14. Simla, India: Central Potato Research Institute.

Davis HV (1996) Recent developments in our knowledge of potato transgenic biology. *Potato Research* 39: 411–427.

Harris PM (1978) *The Potato Crop*. London: Chapman and Hall.

Jadhav SJ and Kadam SS (1998) In: Salunkhe DK and Kadam SS (eds) *Handbook of Vegetable Science and Technology: Production, Composition, Storage and Processing*, pp. 11–69. New York: Marcel Dekker.

Pushkarnath (1976) *Potato in Sub-Tropics*. New Delhi: Orient Longman.

Rama MV (1985) *Studies on the effect of various postharvest treatments for controlling shriveling, sprouting and spoilage of potatoes during storage*. PhD thesis. India: University of Mysore.

Rama MV and Narasimham P (1986) Heat treatments for the control of sprouting of potatoes during storage. *Annals of Applied Biology* 108: 597–603.

Rama MV and Narasimham P (1989) Control of potato (*Solanum tuberosum* L. cv Kufri Jyoti) sprouting by sodium naphthyl acetate during ambient storage. *Journal of Food Science and Technology* 26(2): 83–96.

Rama MV and Narasimham P (1991) Evaporative cooling of potatoes in small naturally ventilated chambers. *Journal of Food Science and Technology* 28(3): 145–148.

Rastovski A, Van Es A, Van Der Zaag DE *et al.* (1987) *Storage of Potatoes. Post Harvest Behaviour, Store Design, Storage Practice, Handling*. Wageningen, The Netherlands: Center for Agricultural Publishing and Documentation.

Rich AE (1983) *Potato Diseases*. New York: Academic Press.

Rodriguez De Sotillo D, Hadley M and Wolf-Hall C (1998) Potato peel extract: a nonmutagenic antioxidant with potential antimicrobial activity. *Journal of Food Science* 63(5): 907–909.

Shashirekha MN (1988) *Studies on the extension of storage life of seed potato tubers stored under different temperature conditions*. PhD thesis. India: University of Mysore.

Singh G, Kapoor IPS and Kumar Pandey S (1997) Studies on essential oils VII. Natural sprout inhibitors for potatoes. *Pesticide Research Journal* 9(1): 121–124.

Smith O (1977) *Potatoes, Production, Storing, Processing*. Westport, Connecticut: AVI Publishing.

Upadhya MD (1994) True potato seed; propagule for potato production for the 21st century. In: Shekhawat GS, Khurana P, Pandey SK and Chandla VK (eds) *Potato: Present and Future*, pp. 15–22. Shimla, India: Indian Potato Association.

Van Der Zaag DE (1990) Recent trends in development, production and utilization of potato crop in the world. *Asian Potato Journal* 1: 1–11.

Woolfe JA (1987) *The Potato in the Human Diet*. Cambridge: Cambridge University Press.

Yamaguchi M (1983) *World Vegetables: Principles, Production and Nutritive Value*, p. 415. Westport, Connecticut: AVI Publishing.

Fruits of the Solanaceae

M V Rama and P Narasimham, Formerly of Central Food Technological Research Institute, Mysore, India

Background

The family Solanaceae includes about 75 genera and 2000 species of herbs, shrubs, and small trees distributed in the tropical and temperate regions of the world. Important vegetable crops such as tomato, brinjal, pepper, tree tomato, and husk tomato are included in this family. Potato and tobacco are also from this family. This article reviews the geographical distribution, morphology, anatomy, chemical composition, nutritive value, and uses of some solanaceous fruits.

Geographical Distribution, Morphology, and Anatomy

The important fruits of the family solanaceae are shown in **Table 1**. Tomato (*Lycopersicon esculentum*) is the most popular and widely cultivated vegetable and has been consumed by the inhabitants of Central and South America since prehistoric times. It is considered to be a native of the Peruvian and Mexican regions. It is an indispensable fruit today, grown all over the world (outdoors in temperate regions and in green houses in the colder regions), and constitutes about 15% of the world total vegetable production (**Table 2**). World production of tomato for processing is over 20 million tonnes, and in the USA, it ranks second to potato in dollar value, among all the other vegetables grown. Major tomato-producing countries

Table 1   Some important solanaceous fruits

Scientific name	Common name
Lycopersicon esculentum	Tomato
L. pimpinellifolium	Redcurrant or grape tomato
Solanum melongena	Eggplant, brinjal, berenjana, aubergine, guinea squash
S. macrocarpon	African eggplant
S. nigrum	Garden huckleberry, wonderberry
S. muricatum	Pepino, melon pear
S. quitoense	Naranjillo, lulo
S. gilo	Jilo
Capsicum annuam	Bell pepper, sweet pepper
C. frutescens	Pimiento, chili, aji, hot pepper, bird chili
Physalis peruviana	Cape gooseberry, uchida
P. ixocarpa	Tomatillo, ground cherry
P. purinosa	Husk tomato
Cyphomandra betacea	Tree tomato, tamarillo

Table 2 World production of major solanaceous fruits (× 1000 tonnes)

Regions	Tomato	Eggplant	Capsicum
World	91 663	21 165	18 024
Africa	10 799	799	2 078
Asia	41 246	19 543	10 715
Europe	19 789	709	2 705
North Central America	13 423	106	2 191
South America	5 990	59	301
Oceania	416		33
Developing countries	56 930	19 852	14 323
Developed countries	34 733	1 313	3 701

Source: Food and Agricultural Organization Statistical Database – Provisional 1999 Production and Production Indices Data www.fao.org
World production of vegetable + melon = 622 428 × 1000 tonnes.

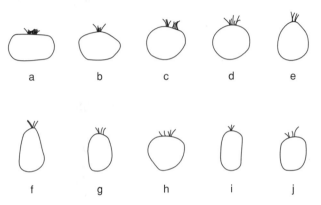

Figure 1 Shapes of tomato cultivars. a, flat; b, oblate; c, square round; d, round; e, pear; f, elongated pear; g, egg; h, oxheart; i, blocky elongated; j, blocky round. Adapted from Gould WA (1992) *Tomato Production, Processing and Technology*, 3rd edn. Baltimore, MD: CTI.

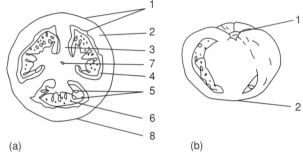

Figure 2 Structure of the tomato fruit: (a) transverse section; (b) longitudinal section. (a) 1, pericarp; 2, outer wall; 3, radial wall; 4, locule; 5, seeds; 6, gel; 7, axile fleshy core; 8, skin. (b) 1, stem scar; 2, blossom end scar.

are China, USA, Spain, Italy, Turkey, and Russia. The fruit is normally round, lobed or pear-shaped, and the diameter ranges from 1 to 12 cm. New cultivars with differently shaped fruits have been developed to suit mechanical harvesting and handling systems (**Figure 1**). It is a two- to many-loculed berry with a fleshy placenta and many small, kidney-shaped seeds, covered with short, stiff hairs and a jelly-like parenchyma (**Figure 2**). The tough skin or pericarp thickness varies with the cultivar.

The genus *Lycopersicon* includes red-fruited edible species with carotenoid pigmentation (subgenus *Eulycopersicon*), and green-fruited species with anthocyanin pigmentation (subgenus *Eriopersicon*). *Lycopersicon esculentum* and *L. pimpinellifolium* are red-fruited cultivars, while *L. pissisi, L. peruvianum, L. hirsutum, L. glandulosum,* and *L. cheesmanii* are green-fruited. Varieties evolved on pure-line selection, hybridization, and genetic engineering techniques for market and processing needs are also cultivated in many countries. The Flavr-savr™ is currently a genetically engineered approved

tomato variety in the US market. Cultivars suited for home, market, shipping, and processing are listed in **Table 3**.

The edible *Physalis* species are grown widely in the warmer parts of the world. The fruit of *Physalis purinosa* (husk tomato) is round, yellowish, and acidic-sweet. *Physalis peruviana* (cape gooseberry) is a native of the Andes, and grows from Venezuela to Chile; the fruit is a spherical or ellipsoidal, smooth berry, measuring about 4 cm long and 3 cm wide. The skin color is greenish yellow. *Physalis ixocarpa* (tomatillo or ground cherry) is of Mexican origin; the fruit is a round, green or purplish berry, with a high ascorbic acid content.

Cyphomandra betacea (tree tomato or tamarillo) is a native of Peru and is also grown in India, Sri Lanka, and New Zealand. The fruit is greenish or purple in the early stages and turns reddish at maturity. It has a musky acid taste and tomato-like flavor. The fruit rind is rough with a disagreeable flavor. Several species in the genus *Solanum* are prominent vegetable crops. *Solanum melongena* (eggplant or aubergine) is a popular and staple vegetable crop in Asia, accounting for 92% of the total world production; in Africa

Table 3 Tomato varieties and end uses

Cultivar suitability	
Home use	Shipping
Better Boy F$_1$	Burpees' Big Boy F$_1$
Red Cherry	Campbell 1327
Market	Empire F$_1$
Earypark 707	Glamour
Empire F$_1$	Heinz 1350, 1439
Mountain price F$_1$	Jet Star F$_1$
Pik Red F$_1$	Jubilee
Traveler 76	Morton Hybrid F$_1$
Processing	Pik Red F$_1$
Heinz 722, 2653	Rutgers 39
Roma VF	Spring set F$_1$
VF 134-1-2	Supersonic F$_1$
UC 82, 204	

Adapted from Madhavi DL and Salunke DK (1998) Tomato In: Salunkhe DK and Kadam SS (eds), pp. *Handbook of Vegetable Science and Technology – Production, Composition, Storage and Processing*, pp. 171–201. New York: Marcel Dekker.

and Europe, its production is 3.7 and 3.3%, respectively (**Table 2**). The leading country of its cultivation is China, followed by India, Turkey, Japan, Indonesia, and Philippines. It is an extensively grown perennial crop in the tropics and an annual in the temperate zones. The fruit can be round, globose, long, or pear-shaped. Most cultivars have purple to blackish skin, while some have a white–green or mottled green skin, with white flesh; in the white–green types, the mature fruit has yellow skin. There are regional preferences for the color and shape of the fruit; purple, large, more or less round fruits are preferred in North America. Wild eggplants with spiny and bitter fruits are found in India; of the several improved Indian cultivars, black beauty, long black, round purple, Pusa purple long, Pusa purple round, Pusa purple cluster, long purple and hybrids such as Vijay, S-1, S-4, S-5, and S-8 are popular. *Solanum melongena* has three main varieties; see **Table 4**.

The African eggplant (*S. macrocarpon*) is a perennial crop grown in the Ivory Coast. The fruit resembles the eggplant. Garden huckleberry (*S. nigrum*) is native to North America and has a wide distribution in temperate and tropical regions; the fruit is a berry about 6 mm in diameter, the color being green when

unripe and red, yellow, or black when ripe. Jilo (*S. jilo*) is a major crop of Nigeria and a minor crop in Central and Southern Brazil. Immature green fruits with a spherical to oval shape (of 4 cm in diameter and 6 cm long) are harvested, and on ripening, the fruit develops an organish red color.

Naranjillo or Lulo (*S. quitoense*) is native to Ecuador and is cultivated in Ecuador, Peru, Colombia, and some Central American countries. The fruit is spherical, 3–5 cm in diameter, and with a yellow, rough skin when ripe. The pulp containing many white seeds is green and acidic in taste.

Pepino or melon pear (*S. muricatum*) is an ancient cultivated crop of the Andes, chiefly grown along the central coast of Peru and at elevations of 1000–3000 m from Colombia to Bolivia. It is also cultivated to a limited extent in northern Argentina, Chile, New Zealand, and Australia. The fruit is long, ovoid, or ellipsoidal, and the color varies from light green to pale yellow.

Capsicum (hot, chili, bell, or sweet pepper) is the second most important fruit of the Solanaceae, grown as a vegetable and condiment crop in tropical and subtropical regions of the world. The world production of this pepper is shown in **Table 2**. There are five major recognized cultivated species of *Capsicum* (**Table 5**) and seven botanical varieties of *C. annuum* (**Table 6**). The important species are *Capsicum annuam* (annual with flowers borne singly in leaf axils) and *C. frutescens* (perennial with flower clusters in leaf axils). The bell pepper or sweet pepper (*C. annuum*), which is either mildly pungent or nonpungent, has a thick pericarp and is used in flavoring vegetable preparations. The highly pungent fruit of *C. frutescens* has a thin, smooth pericarp and is used as a condiment. Paprika is a nonpungent type of pepper with a thick pericarp, selectively bred for color and flavor.

China, Korea, Indonesia, Sri Lanka, Pakistan, Japan, Mexico, Ethiopia, Spain, Italy, and Hungary are some of the principal chili-producing countries. The world demand for chilies is increasing. The main exporters of chili are India, China, Indonesia, Hungary, Singapore, Malaysia, Mexico, and Japan. The bell pepper, *C. annuum* var. *grossum* (syn. Simla

Table 4 Common cultivated varieties of *Solanum melongena* and their fruit characteristics

Variety	Fruit characteristics
Var. *esculentum* (Wees)	Large, pendent, ovoid, oblong berries, 5–30 cm long, shiny
Var. *depressum* (Baily)	Small, pear-shaped fruits, purple in color, 10–13 cm long
Var. *serpentium* (Deeft)	Slender, greatly elongated, 30 cm long, 2–5 cm in diameter, end-curled

Adapted from Lawande KE and Chavan JK (1998) Eggplant (brinjal), In: Salunkhe DK and Kadam SS (eds), *Handbook of Vegetable Science and Technology – Production, Composition, Storage and Processing*, pp. 225–244. New York: Marcel Dekker. *Solanum melongena* var. *incanum* (Lhm) is a nonedible variety.

Table 5 Cultivated *Capsicum* species and their distribution

Species	Synonyms	Distribution
C. annuum L.	C. purpureum C. grossum C. cerasiformae	Columbia to southern USA, thoughout Asia and America
C. baccatum L.	C. pendulum C. microcarpum C. angulosum	Argentina, Bolivia, Brazil, Columbia, Equador, Peru, Paraguay, etc.
C. frutescens (Tabasco pepper)	C. minimum	Columbia, Costa Rica, Guatemala, Mexico, Puerto Rico, Venezuela
C. chinese L.	C. luteum C. umbelicatum C. sinense	Bolivia to Brazil, Costa Rica, Mexico, Nicaragua, West Indies
C. pubescence	C. eximium C. tovari C. cardenasii	Bolivia to Columbia, Costa Rica, Guatemala, Honduras, Mexico

Adapted from Rajput JC and Parulekar YR (1998) Capsicum In: Salunkhe DK and Kadam SS (eds) *Handbook of Vegetable Science and Technology – Production, Composition, Storage and Processing*, pp. 203–224. New York: Marcel Dekker.

Table 6 Botanical varieties of *C. annuum* and their fruit characters

Variety	Common names	Fruit characteristics
Var. abbreviatum, Fingerh	Wrinkled peppers	Ovate wrinkled fruits, about 5 cm long or less
Var. accuminatum, Fingerh		Linear, oblong, pungent, pointed fruits about 9 cm long
Var. cerasiformae (Miller) Irish	Cherry peppers	Globose, pungent fruits red, yellow, or purple in color, with firm flesh and 1.2–2.5 cm in diameter
Var. conoides (Miller)	Cone peppers (Tabasco type)	Erect, conical, pungent fruit, about 3 cm long
Var. fasciculatum (Stuart) Irish	Cluster peppers	Clustered, erect, slender, very pungent fruits, about 7.5 cm long
Var. grossum (L.) Sendt.	Sweet peppers, paprika	Large inflated fruit with a basal depression, red or yellow thick flesh with mild pungency
Var. longum (DC) Sendt.	Long peppers	Mostly dropping fruits with tapering apex

Adapted from Rajput JC and Parulekar YR (1998) Capsicum In: Salunkhe DK and Kadam SS (eds) *Handbook of Vegetable Science and Technology – Production, Composition, Storage and Processing*, pp. 203–224. New York: Marcel Dekker.

mirch, sweet pepper, bullnose capsicum) is cultivated widely in India, Central and South America, Bolivia, Peru, Costa Rica, Mexico, Hong Kong, and almost all European countries. Based on characteristics such as fruit size, color, texture, flavor, pungency, and uses, a 13-group classification has been proposed for cultivars of this species (**Table 7**).

The *Capsicum* fruit is a pod-like berry with a short, thick peduncle, developing from a bi-carpellary ovary with axile placentation. Many seeds are present in the cavity between the placenta and fruit wall. The unripe fruit is commonly green, but fruits with cream, greenish yellow, orange, purple, and purplish black colors also exist. Ripe fruits are usually red, but sometimes yellow or orange.

Chemical Composition and Nutritive Value

There is considerable variation in the chemical composition of the fruits of this family (**Tables 8** and **9**). The stage of maturity, cultivar, environmental, and cultural practices influences the chemical composition.

Tomato

Glucose and fructose are the principle sugars in tomato (representing more than 60% of the solids) with small amounts of sucrose. Ripe fruits also contain raffinose; the glucose concentration increases with ripening. Tomato being a climacteric fruit, ripening is associated with increase in respiration and ethylene production: these in turn are integrated with the disappearance of starch, degradation of chlorophyll, synthesis of lycopene, flavor components, and polygalacturonase. The pectic constituents mainly control the texture and firmness of the fruit. In the ripening process, the predominant protopectin in the green fruit decreases, and the pectin increases, making the fruit soft. The protein content is about 1.0% in ripe fruit, and all the essential amino acids except tryptophan are present; other amino acids identified are tyrosine, aspartic acid, glutamic acid, serine, glysine, α-aminobutyric acid, and pipecolic acid; glutamic and aspartic acids occur in greater concentrations.

Citric acid is the principle organic acid of tomato which gives the fruit its typical taste. Along with a

Table 7 Bell pepper groups, fruit character, and uses

Group	Cultivars	Uses	Fruit characteristics
Bell	California wonder Yellow wonder	Fresh market, salads, pitza, meatloaf, and canning	Large, nonpungent with thick flesh (7–12 cm × 5–10 cm)
Pimento	Pimento select	Fresh market, salads, soups, processed meat, and canning	Large, nonpungent, thick-walled (5–10 cm × 5–7 cm)
Squash or cheese	Cheese, antibois and gambo	Processing, canning, freezing, pickling, salad, and culinary purposes	Small to large, thick-walled, nonpungent, green or yellow to red (2.5–5 cm × 5–10 cm)
Ancho	Mild California, big jam, New Mexican chili	Dried, powdered, and culinary purposes	Large, heartshaped, thin walled, less pungent fruits (10–12 cm × 5–7 cm)
Anahein chili	California chili, paprika	Sauces and canning, processed into powder dehydrated pods	Slender, medium-thick flesh, medium to dark green color turning red, sweet to moderate pungency (12–15 cm × 2–3 cm)
Cayenne		For market, pickling, dry powder, sauces culinary purposes	Irregular, wrinkled, highly pungent fruits turning green to red (12–22 × 1.5–6 cm)
Cuben		Fresh market, salads, pickling, and frying	Long, thin walled, irregular fruits of mild pungency, turning yellowish green to red (10–20 cm × 1.5–5 cm)
Jalapeno	Jalapeno	Fresh green pods, dried powder, canning, and sauces	Elongated, round, cylindrical shape, highly pungent, thin-walled, red at maturity (5–7 cm × 2–5 cm)
Small hot	Red chili, sontaka	Dried powder and sauces	Slender, medium- to thick-walled, highly pungent, turning red (4–7 cm × 1–2 cm)
Cherry	Red cherry, large Red cherry, small	Pickling	Small, spherical, pungent fruit turning red (2–5 cm in diameter)
Short wax	Floraljam, caloro, cascabella	Pickle, cooking, sauce, and processing	Medium- to thick-walled, tapering, turning yellow to orange red (5–7 cm × 2–5 cm)
Long wax	Hungarian yellow wax, sweet banana	Fresh market, pickle, sauce, and canning	Pointed or blunt fruits turn yellow to red (8–12 cm × 1–4 cm)
Tabasco	Green leaf Tabasco, Tabasco	Vinegar, sauce, and pickles	Slender, highly pungent fruits, turning yellow to red (2–5 cm × 0.5 cm)

Adapted from Rajput JC and Parulekar YR (1998) Capsicum In: Salunkhe DK and Kadam SS (eds) *Handbook of Vegetable Science and Technology – Production, Composition, Storage and Processing*, pp. 203–224. New York: Marcel Dekker.

small amount of malic acid, traces of acetic, formic, lactic, and succinic acids have also been detected. The lipid fraction of the tomatoes is composed of triglycerides, sterols, sterol esters, free fatty acids, and hydrocarbons. The phenolics reported are caffeic, ferulic, chlorogenic, and *p*-coumaric acids. The chief coloring materials of tomato at the mature green stage are chlorophyll a and b, while at the ripe stage, carotene and lycopene contents dominate, contributing 7 and 87%, respectively. The final color of tomato depends upon the ratio of the carotene and lycopene. The various volatile compounds in ripe fruit are alcohols, aldehydes, carbonyls, and sulfur compounds, contributing to the flavor quality.

Tomato is regarded as an essential protective food. It is a rich source of ascorbic acid: On a fresh-weight basis, the vitamin C content averages about 25 mg per 100 g, varying with the variety, maturity, and season.

Table 8 Macronutrient content (per 100-g edible portion) of some solanaceous fruits

Crop	Energy (J)	Water (g)	Protein (g)	Fat (g)	Carbohydrates (g)
Tomato (green)	97	93.0	1.9	0.1	3.6
Tomato (ripe)	84	93.8	1.2	0.3	4.2
Eggplant	109	92.0	1.6	0.3	4.0
Chili pepper	487	65.4	6.3	1.4	24.8
Bell pepper	109	92.0	1.3	0.2	6.0
Tree tomato	202	85.9	1.5	0.3	11.3
Naranjillo	118	92.0	0.7	0.1	6.8
Tomatillo	134	92.0	0.4	1.0	6.3
Cape gooseberry	223	83.0	1.8	0.2	11.1
Husk tomato	105	92.0	0.7	0.6	5.8
Pepino	134	92.0	0.4	1.0	6.3

From FAO (1972) *Food Composition Table for Use in East Asia*. Rome: Food and Agriculture Organization; Gopalan C, Ramasastri BV and Balasubramanian SC (1993) *Nutritive Value of Indian Foods*. Hyderabad, India: Indian Council of Medical Research; Yamaguchi M (1983) *World Vegetables, Principles, Production and Nutritive Value*, pp. 291–310. Westport, CT: AVI.

Table 9 Vitamin and mineral content (per 100-g edible portion) of some solanaceous fruits

Crop	Vitamin (mg)[a]					Mineral (mg)			
	A (IU)	B₁	B₂	Nicotinic acid	C	Calcium	Iron	Magnesium	Phosphorus
Tomato (green)	320	0.07	0.01	0.4	31	20	1.8	15	36
Tomato (ripe)	385	0.06	0.04	0.6	23	7	0.6	12	30
Eggplant	124	0.08	0.07	0.7	6	22	0.9	16	37
Chili pepper	576	0.37	0.51	2.5	96	86	3.6	24	120
Bell pepper	530	0.07	0.08	0.8	103	12	0.9	13	34
Tree tomato		0.04	0.04	1.4	17	13	0.8	34	24
Naranjillo	170	0.06	0.04	1.5	65	8	0.4		14
Tomatillo	80	0.05	0.05	2.1	2	7	0.4	23	40
Cape gooseberry	2380	0.05	0.05	0.1	180	10	2.0	31	67
Husk tomato	380	0.05	0.02	2.1	0.2	7	0.4	23	40
Pepino	200	0.08	0.04	0.5	32	18	0.8		14

[a]Vitamin A content is given in IU.
From FAO (1972) *Food Composition Table for Use in East Asia* Rome: Food and Agriculture Organization; Gopalan C, Ramasastri BV and Balasubramanian SC (1993) *Nutritive Value of Indian Foods*. Hyderabad, India: Indian Council of Medical Research; Yamaguchi M (1983) *World Vegetables, Principles, Production and Nutritive Value*, pp. 291–310. Westport, CT: AVI.

Research has shown that the inclusion of tomato in the diet can prevent deficiencies of vitamins and minerals. One medium-size, raw tomato provides 47% of the B vitamin, 33% of vitamin A, and 1% of the energy of the daily dietary requirement of an average person. The mineral content varies between 0.3 and 0.6%; apart from calcium, potassium, magnesium and iron, the tomato contains zinc, boron, iodine, cobalt, and aluminum in trace amounts.

The tomato contains a glycoalkaloid, tomatine, and traces of solanine. Narcotine is present in unripe fruit. The tomatine content is lowest in the pink stage of ripeness and increases slightly in the fully ripe tomato.

Brinjal or Eggplant

This contains about 4% carbohydrate, composed of sucrose, glucose, and fructose. Brinjal protein has a high biological value and contains amino acids such as arginine, histidine, lysine, tryptophan, leucine, isoleucine, and valine, although the protein is relatively poor in lysine and tryptophan, isoleucine, and methionine. Brinjal contains a higher percentage of vitamins than many other vegetables. However, it is not a rich source of vitamin B₂. The physical characteristics such as the shape, color, and presence of spines on the calyx have also been found to influence the composition; hence the green, white, and purple varieties are often compared for chemical composition. Dark-purple-skinned varieties contain more vitamins than those with white skins. The white cultivar contains twice as much crude fiber as the purple and green cultivars. The amino acid content is higher in purple cultivars and lowest in the white. Purple cultivars are poor in potassium and chloride content, while green cultivars are rich in these minerals. The polyphenol oxidase activity is highest in the purple type, thus making the fruit turn dark faster when cut surfaces are exposed to air. Variations in the activities of other enzymes have been observed in differently colored fruits. The main pigment of the fruit is an anthocyanine, nasunin; lycopene and lycoxanthin are also present. The seed oil is reported to be rich in linoleic acid.

The fruit peel contains a bitter principle, solasonine. The phenolic compounds present in the fruit are chlorogenic acid, neochlorogenic acid, scopoletin, and caffeic acid; a trace amount of hydrocyanic acid is also present. Higher concentrations of glycoalkaloids (20 mg per 100 g fresh weight) result in a bitter taste and off-flavor.

Pepper

Pepper is a good source of vitamins A and C, and is superior to tomato and eggplant in this respect. The vitamin C content varies with the variety and maturity stage (160–210 mg in green chilies and 113–160 mg in red ripe fruits). The pungent principle of chili is capsaicin (a crystalline, colorless compound), which is concentrated more in the placenta (90%) that connects the seeds with pericarp. A highly irritating vapor is liberated on heating capsaicin. Yellow chilies are more suitable for capsaicin extraction. An African chili variety, 'Mombusa,' is known for its high capsaicin content. The green fruits contain chlorophyll a and b. The coloring matter of ripe fruit includes capsanthin, capsorubin, zeacanthin, cryptoxanthin, and α- and β-carotene.

Other Solanaceous Fruits

Husk tomato, tomatillo, and cape gooseberry are also important sources of vitamins and minerals (**Table 9**). Tree tomato contains substantial amounts of carbohydrates and minerals. It is rich in vitamin C and is also a good source of provitamin A.

Handling and Storage

Tomato

Tomato fruits are harvested at different stages of maturity (**Table 10**), depending upon the use, and are often harvested at the mature green stage for the market and ripened either in transit or during storage. For canning or juice extraction, the fruit is harvested at the ripe stage. The postharvest losses in tomato may range from 5 to 50%; the highest loss is attributable to mechanical damage, and other causes may be physiological disorder, diseases, and heat and chilling injury.

Packing-house operations include cleaning, grading, waxing, wrapping, and packing. Fruits are transported to the market in bamboo baskets, corrugated board cartons, or wooden boxes. In general, the optimum storage temperature and relative humidity (RH) for tomato storage is 12 °C and 86–95%. However, depending upon the stage of fruit maturity at harvest, storage conditions vary. Mature green tomatoes can be stored at 13–18 °C and 85–90% RH for 3–4 weeks as the fruits ripen at this temperature without chilling injury. Holding them below 10 °C for more than 24 h should be avoided to overcome the symptoms of chilling injury. However, storing green fruits above 18 °C results in early ripening, which leads to rapid deterioration in the quality of the fruits, while at about 30 °C, the red pigment formation is hindered, and the ripe fruit develops an orange to yellow color. Fruits at the breaker stage are less sensitive to chilling injury than immature green fruits. Ripe tomatoes can be preserved for 7–10 days at 3–5 °C and 85–90% RH. Fully ripened fruits are held between 1.7 and 4.4 °C until they are ready for use. Prolonged storage of fully ripe tomato causes softening and loss of color of the fruits.

Controlled-atmosphere and hypobaric storage can also prolong the shelf-life of the tomato. The packaging of tomatoes in polyethylene bags provides a modified atmosphere, which extends the shelf-life. Postharvest treatment with gibberellic acid markedly retards ripening, while 2,4-dichlorophenoxyacetic acid, naphthalene acetic acid, and ethylene hasten ripening, and this facilitates the marketing of mature green tomatoes. Ethephon, an ethylene-releasing compound, has been used commercially to accelerate

Table 10 Maturity stages of tomato

Stage	Characteristics
Immature green	Seeds and jelly-like substance surrounding the seeds not yet formed
Mature green	Fully grown fruit shows a brownish ring on the stem scar, and blossom end turns yellowish green; seeds are surrounded by a jelly-like substance
Turning (breaker stage)	One-quarter of the surface at blossom end appears pink
Pink	Three-quarters of the surface appears pink
Hard ripe	Nearly all red or pink, but flesh is firm
Over-ripe	Fully red and soft

Adapted from Gould WA (1992) *Tomato Production, Processing and Technology*, 3rd edn. Baltimore, MD: OTI.

ripening. Chemicals such as captan, dithane, and thiram have the greatest potential as fungicidal treatments on tomatoes.

Brinjal

Brinjal is a heavy vegetable and requires care in handling, even though it is not as perishable like the tomato. The fruit becomes edible from the time it attains one-third or the full growth size. Over-mature fruits are dull, seedy, and fibrous, sometimes tasting bitter. After harvest, the graded fruits are packed in special crates or bushel baskets. In tropical countries, chilling injury and certain pests and diseases cause high losses of eggplants. Temperatures below 10 °C cause chilling injury, the symptoms of which are pitting, surface bronzing, and browning of the seeds and pulp. Fruit rot and anthracnose caused by *Phomopsis parasitica* and *Gleosporium melongense*, respectively, are the most serious postharvest diseases. Hot water (50 °C) and a low concentration of dithiocarbamate provide good disease control. The fruit can be stored successfully for 2–3 weeks at 10.0–12.8 °C and 92% RH. Controlled-atmosphere storage has not been found to be beneficial in extending the storage life of the eggplant. Small fruits are not well suited for long-term storage, and chilling injury is more common in tender fruits. Prepackaging of brinjal in adequately ventilated polyethylene bags (100-gauge) increases the shelf-life, and treating brinjal with a fungicidal wax emulsion extends the shelf-life by 30–40%.

Pepper

Chili, used as a raw vegetable, is harvested green, but for processing, only red ripe chilies are harvested. The important postharvest losses in chili are weight loss, chilling injury, and microbial spoilage. At the recommended storage conditions of 7.2 °C and 85–90% RH, green peppers can be stored for 2–3 weeks, and ripe peppers at 5.6–7.2 °C and 90–95% RH can be stored for about 2 weeks. Storage in paper bags rather than in polyethylene bags can prolong the shelf-life of hot peppers. Bell peppers are harvested when they attain their full size and are still green. Storage below 7 °C results in chilling injury and, subsequently, *Alternaria* rot. Controlled-atmosphere storage at various temperatures has been recommended to extend the shelf-life of bell peppers, while hypobaric storage is not beneficial as it greatly increases the weight loss. Fruits are commonly wax-coated to reduce the moisture loss before shipment. Ripe fruit can be preserved at room temperature (20 °C) for only a week, but for 12–14 weeks at 3–4 °C after dipping in hot water at 50 °C for 10 min.

A common storage disease is bacterial decay by *Erwinia carotovora*.

Uses of Solanaceous Fruits

Fresh, ripe tomato fruits are refreshing and appetizing, and are consumed raw in salads or after cooking. Unripe fruits are usually cooked and eaten. Large quantities are processed in southern European countries and in California. Tomatoes can be processed into juice, purée, paste, ketchup, sauce, soup, and powder. Special varieties have been evolved to meet these processing requirements. The tomato cultivars for juice should have a bright color, rich flavor, and high total acidity. They should also be juicy and not meaty. Tomatine, a glycoalkaloid present in the fruit, yields, on hydrolysis, tomatidine, which can be transformed into hormones such as progesterone and testosterone. The oil of tomato seeds is used in the soap and paint industry. The pressed cake can be used as a feed for livestock or a fertilizer. Tomato also has several medicinal applications and is used in the preparation of traditional medicine in Japan, Greece, and Peru. The hot water extract of dried fruits has been used in the treatment of ulcer wounds and burns, and tomato has been reported to reduce the risk of prostrate and digestive tract cancers.

Brinjal is valued more as a vegetable during the autumn when other vegetables are scarce. It is a fairly good source of calcium, iron, phosphorus, and vitamin B. Sun-dried or hot-air-dried eggplant powder with good organoleptic and nutritive qualities has been prepared. Treatment with brinjal is recommended in liver complaints. It has been reported to stimulate the intrahepatic metabolism of cholesterol and produce a marked drop in blood cholesterol level.

The garden huckleberry (*S. nigrum*) leaf has great medicinal value. The tender leaves are boiled like spinach and eaten in many parts of India. Ripe fruits are used as a substitute for raisins in plum puddings. The berry is considered to possess tonic, diuretic, and cathartic properties, and is useful in the treatment of heart disease. The berries also find use as a domestic remedy for fever, diarrhea, ulcers, and eye problems.

Peppers are used green, or after ripening, in a great variety of ways. Chilies, when taken with food, stimulate the taste buds and increase the flow of saliva. The commercial red pepper(s) consists of the fruits of a small, pungent variety, which is dried and ground to a fine powder. Dry chili powder and green chilies are commonly used in various culinary preparations and in the preparation of sauces, soup, ketchup, and salads. Pepper sauce, prepared in a variety of ways, consists of the fruit of pungent varieties preserved in brine or strong vinegar. Both red and green bell peppers are sold frozen and are also canned, along with other vegetables. Tabasco sauce is prepared from the juice of a pungent variety, expressed by applying pressure. Paprika, a Hungarian condiment, is made from fruit ground after removing the seeds. Pepper is used in pickles of various kinds. The sweet varieties are stuffed and baked. *Capsicum* preparations are used as a counterirritant in neuralgia and rheumatic disorders. Green chilies are a good source of L-asparaginase, which has an antitumor effect and is used in the treatment of acute lymphocytic leukemia.

The oleoresins have varying color properties and pungency, are prepared from the dried fruits of *Capsicum*, and are used in the pharmaceutical, food, and cosmetic industries. The three types of oleoresins are oleoresin capsicum, oleoresin red pepper, and oleoresin paprika; oleoresin capsicum is used in making pain-balms and vaporubs. The nonpungent type of oleoresin is chiefly used as a coloring agent, while the pungent type is used for flavoring and coloring snack foods. Chili seed cake, a byproduct of seed oil, has a protein content of 27–29% and can be used as a fertilizer or as an animal feed. Chili color is a new product, which can replace artificial food colors.

The melon pear fruit is cooked and eaten when immature, while the mature fruit has a fleshy pulp with an aroma and flavor similar to those the cucumber. Lulo is cultivated mainly for the juice, especially in Ecuador, Peru, Columbia, and some Central American countries. The young shoots of jilo are finely chopped and used in soups in Nigeria. Tomatillo can be used to make chili sauce and is used for dressing meats in Mexico. The ground cherry, which is acidic-sweet, is used for preserves and sometimes for sauces, and the Cape gooseberry is eaten raw or preserved as a pickle.

See also: **Peppers and Chillies**; **Potatoes and Related Crops**: The Root Crop and its Uses; Processing Potato Tubers; **Tomatoes**

Further Reading

Bajaj KL, Bansal KD, Chadha ML and Kaur PP (1990) Chemical composition of some important varieties of eggplant (*Solanum melongena* L.). *Tropical Science* 30: 255.

Chadha YR (1972) *The Wealth of India, A Dictionary of Indian Raw Materials and Industrial* products, vol. IX, pp. 383–393. New Delhi: CSIR.

Dighe AH (1995) *Studies on Biochemical and Nutritional Composition of Promising Brinjal* (Solanum melongena L.) *Cultivars*. M.Sc. thesis, Mahathma Phule Agricultural University, Rahuri, India.

Gopalan C, Ramasastri BV and Balasubramanian SC (1993) *Nutritive Value of Indian Foods*. Hyderabad, India: Indian Council of Medical Research.

Gould WA (1992) *Tomato Production, Processing and Technology*, 3rd edn. Baltimore, MD: CTI.

Habson G and Greirson D (1993) Tomato. In: Seymour GB, Taylor JE and Tucker GA (eds) *Biochemistry of Fruit Ripening*, p. 405. London: Chapman & Hall.

Lawande KE and Chavan JK (1998) Eggplant (brinjal). In: Salunkhe DK and Kadam SS (eds) *Handbook of Vegetable Science and Technology – Production, Composition, Storage and Processing*, pp. 225–244. New York: Marcel Dekker.

Madhavi DL and Salunke DK (1998) Tomato. In: Salunkhe DK and Kadam SS (eds) *Handbook of Vegetable Science and Technology – Production, Composition, Storage and Processing*, pp. 171–201. New York: Marcel Dekker.

Pruti JS (1998) *Major Spices of India – Crop Management and Post-harvest Technology*, pp. 180–238. New Delhi, India: Indian Council of Agricultural Research.

Rajput JC and Parulekar YR (1998) Capsicum. In: Salunkhe DK and Kadam SS (eds) *Handbook of Vegetable Science and Technology – Production, Composition, Storage and Processing*, pp. 203–224. New York: Marcel Dekker.

Salunkhe DK and Desai BB (1984) *Postharvest Biotechnology of Vegetables*, vol. 1, pp. 39–82. Boca Raton, FL: CRC Press.

Wall MM, Waddell CA and Bosland PW (2001) Variation in β-carotene and total carotenoid content in fruits of capsicum. *Hortscience* 36(4): 746–749.

Yamaguchi M (1983) *World Vegetables, Principles, Production and Nutritive Value*, pp. 291–310. Westport, CT: AVI.

Processing Potato Tubers

M G Lindhauer, N U Haase and B Putz, Federal Centre for Cereal, Potato and Lipid Research, Detmold, Germany

Background

The convenience aspect of processed potatoes (*Solanum tuberosum* ssp. *tuberosum* L.) has become of significant importance in the total use of potatoes by humans. Especially in some industrialized countries, the percentage of potato products in relation to total use of potatoes as foodstuff has reached a value of up to 65% (e.g., in the USA). To get an overview, the main directions of potato processing are discussed in this article.

Raw Material for Processing

Potato processing requires special quality profiles of the raw material. Therefore, the number of varieties suitable for processing is limited. Plant breeders make a great effort to select new varieties with better agronomical behavior and with higher internal quality. Today, only a few varieties with a known processing quality are grown worldwide.

Several specific aspects must be observed:

- The oxidative potential of the harvested tubers should be low to prevent discoloration during the first processing steps.
- Key constituents like protein, minerals, and vitamins should be high because of inevitable losses during processing. Also the total content of solids should be high to make processing more worthwhile, by achieving an higher input-to-output ratio.

Typical nutrient contents of selected potato products are listed in **Table 1**.

Storage Aspects

Storage has to provide high-quality potato tubers to the processing plant for several months. Good storage management should prevent excessive loss of moisture, development of rot, and excessive sprout growth. Also the loss of potato constituents (e.g., ascorbic acid) should be low. Finally, the accumulation of reducing sugars must be controlled to prevent dark-colored products.

After harvest, potatoes pass through a period of several months without any sprouting. The following sprout growth is little or absent at temperatures below 5 °C. However, reducing sugars accumulate in excessive quantities, resulting in undesirable dark-brown colored products. Therefore, potatoes for processing are stored at 8–10 °C. At that temperature sprouting must be controlled. Experiments have been conducted with gamma-irradiation of tubers or treatment of vines with maleic hydrazide. Also several chemical compounds are known to inhibit sprouting (e.g., ethylene, volatile monoterpenes, ethanol, abscisic acid, and naphthalene), but in practice

isopropyl-N-(3-chlorophenyl) carbamate (CIPC) is preferred. To prevent the use of antisprouting agents, potato breeders have created cold storable varieties without cold sweetening to avoid the problem described.

Immature tubers with considerable amounts of free sugars will not reduce that level during storage. Also storehouses with little gas exchange capacity lead to high contents of reducing sugars inside tubers. This is because of the increasing carbon dioxide concentration in the storage atmosphere. An increase of carbon dioxide towards 0.5% in the atmosphere already shows a significant deterioration of baking color. Depending on the geographical region, storehouse equipment must include a mechanical ventilation and cooler. The relative humidity should be 90% minimum.

Potato Processing for Food

A schematic overview of several processing techniques is given in **Figure 1**.

Common Techniques

Washing   Washing systems remove dirt and attached soil from the tubers. Stones can be removed through stone catchers or through exploitation of the different specific density in comparison to tubers. Often the washing water is taken from the overflow of waste water decanters within processing.

Peeling   With few exceptions, all tubers are peeled before processing. Abrasive and steam peelers are common, while in the past lye (caustic) peelers were also used. Because of concern about environmental effects, the importance of that type of peeler has decreased. In abrasive peelers, designed with carborund layers or knives or combinations thereof, the amount of peeling waste can be up to 60% of tuber weight, if the tuber shape is irregular or deep potato eyes exist. Most single cells are disrupted and oxidative enzymatic discoloration must be avoided (often by additives, such as sulfur). Steam peelers remove much less of the tuber cortex. Most of the tissue with high nutritive value achieves further processing.

Table 1   Typical nutrient concentrations of processed potatoes

	Per 100 g edible material					
	Energy (kJ) (kcal)	Main constituents				
		Protein (g)	Fat (g)	Carbohydrates (g)	Potassium (mg)	Ascorbic acid (mg)
Cooked potatoes	298 (71)	2.0	traces	14.8	443	14
Potato chips (crisps)	2241 (536)	5.5	39.4	40.6	1000	8
French fries	1215 (290)	4.2	14.5	35.7	926	28
Mashed potatoes	314 (75)	2.0	1.9	12.2	259	9

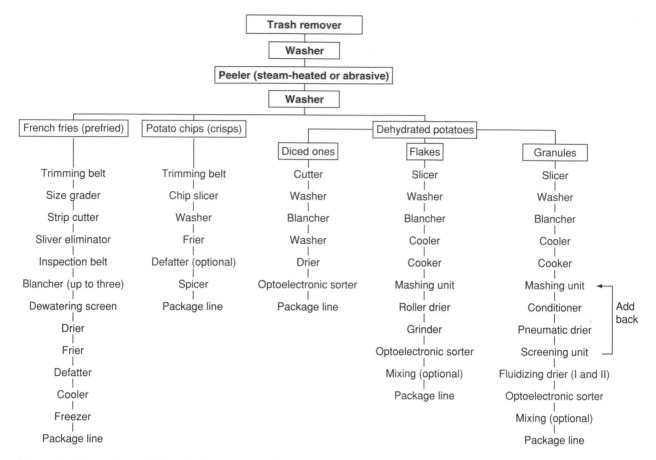

Figure 1 Schematic description of main processing lines.

Tubers come into contact with high-pressure steam (up to 15 bar) for a few seconds. The outer cell layers are softened, and the cohesion of the tissue is loosened. Through fast relaxation of the vessel the peel is separated from the tubers and can be removed by brushes or by a hard water jet. The main disadvantage is the layer of cooked tissue. Therefore, the preferred application is when making French fries and dehydrated products, whereas potato chips (crisps) are predominantly produced using abrasive peelers.

Cutting   Several types of cutting exist depending on final use. The stream of tubers is directed to specially designed knives with the help of centrifugal power or water pressure. High-pressure water knives need precise positioning of tubers. Except for when slicing potato chips (crisps), using smooth and crinkled knives, cutting is carried out in several steps. Disks are cut first. Then, stripes are cut with rotating knives. If cubes are to be manufactured, a third cutting is done using chopping knives. Stripes for French fries production can be cut using water knives. High-pressure water leads tubers through a pipe with a cutting fence at its end.

Sorting   Irregular cuts must be removed, if the shape of single pieces is a quality criterion, e.g., when manufacturing French fries, and thus is done before and after the main processing. Separation takes place by screening or by roller sorters. Optoelectronic sorters allow an additional sorting according to discoloration. Single pieces are visually inspected by charge-coupled device (CCD) cameras. Next to light and dark differentiation, state-of-the-art systems also offer the possibility of color differentiation. Discolored pieces are removed from the main stream by an ejection system using compressed air.

Blanching   Blanching means a short-term heat treatment (70–100 °C) of raw potatoes or potato pieces. Several reactions occur. Enzymes are inactivated (e.g., polyphenoloxidase) to prevent discoloration. Often the peroxidase system is regarded as a key enzyme mirroring the efficiency of enzyme inhibition. Reducing sugars are leached off, to minimize the Maillard reaction, as well as starch pastes, to reduce fat uptake during frying. Blanching for about 20 min at 70–75 °C results in a firm structure. Native pectin methyl esterase is activated to reduce cross-linking of pectin, and free carboxyl groups may react with calcium or magnesium liberated from starch granules after gelatinization to form a thermostable pectin network. The addition of calcium to the blanching water intensifies that reaction.

Blanching takes place predominantly in a continuous screw motion through a water bath or through a steam chamber. Depending on the raw material and the shape of single pieces, blanching parameters vary widely.

Dehydrated Potato Products

Dehydrated diced potatoes Dehydrated diced potatoes have been produced since the beginning of the twentieth century. Their first use was as a staple food. Today, snack industry and other food manufacturers (e.g., for dumplings, hash brown) are the main users of dehydrated diced potatoes. Because of different outlets, no uniform quality level can be addressed, but discoloration is to be prevented, flavor should be fairly typical of the varietal characteristics, and texture of the reconstituted product should be similar to that of fresh prepared potatoes.

Processing Preliminary inspection by hand or by optoelectronic sorters immediately after washing removes potatoes that are unfit for processing.

Standard peeling equipment uses steam or abrasive peeling. The expense of the inspection and trimming of peeled tubers can be reduced by optoelectronic inspection belts.

Mechanical dicers which can be adjusted to various cutting sizes (e.g., cubes, disks) cut the tubers. Irregular cuts depend on the sharpness of the cutter blades and on the mass flow of potatoes. Released solids (up to 15%) cannot be avoided but should be reduced as far as possible. Often, free starch is separated from the wash water and is used for special purposes.

Blanching for 2–10 min at 90–100 °C reduces adverse enzymatic reactions. Under- and overblanching negatively affects the texture and appearance of the finished product. Therefore it is to be controlled periodically. After blanching, gelatinized starch must be removed from the surface by a further water bath or a rinsing spray.

Sulfiting during or just after blanching protects against nonenzymatic browning reaction (darkening after cooking) during drying and reduces adverse oxidation during storage. Chemicals to be used are sodium sulfite, sodium bisulfite, sodium metabisulfite or combinations thereof. The European Community tolerates up to 400 mg sulfite as sulfur dioxide per kg ground dehydrated potatoes.

Potato drying is a two-stage process by migration of water to the surface and by removal of surface water by evaporation. Slow dehydration at a low temperature (e.g., 30 °C) gives a hard dense product, while a higher temperature results in a more porous material with better rehydration capacity. Hot air driers, constructed as hordes drier with a batch charging, or as a continuous-belt drier (conveyor drier) are used in most cases. During the first stage of drying moisture is readily removed and the drying rate is high. Therefore, the temperature can be high (70–90 °C). Within the second stage of drying, progress is much less and temperature is reduced to 50–70 °C. Potato size, shape, and raw material characteristics influence the efficiency of water removal. Microwave-assisted drying reduces the drying time drastically. The rapid diffusion of moisture through the surface will tend to eliminate surface hardening. Rehydration potential rises. It is most suitable at moisture levels between 6 and 30%, but often energetic and economic disadvantages prevent utilization.

Screening and inspection by pneumatic separators follows drying, to remove small pieces, discolored pieces or peeling left on potato dices, and black spots.

Dehydrated mashed potatoes

Dehydrated flakes Potato flakes are dehydrated mashed potatoes made by drum drying a thin layer of cooked and mashed potatoes, and breaking the sheet of dehydrated solids into a suitable size for packaging. Next to retail consumption in a pure state or prepared with milk solids and other constituents, flakes are taken as a food ingredient for potato snacks (e.g., potato chips made from a dough with uniform size and curvature, and for dehydrated French fries). A finely ground potato flour for soups and baked goods is also available.

Processing Raw potatoes should be washed to remove adhering soil, and sorted (by hand or by optoelectronic sorters) to remove defective tubers.

Peeling may be done by any process described above. In most cases steam peeling is used. Trimming may ignore some peel and defects, because of a final product control by optoelectronic sorters.

The tubers are sliced into disks of 1 cm thickness to obtain a more uniform heat treatment during precooking and cooling. Thinner slicing is possible, but flavor deteriorates and loss of solids clearly increases.

Precooking leads to gelatinization of starch and enzyme inactivation. Cooling below 20 °C for about 20 min reduces the level of free starch in the final flakes. For a long time that effect was described as retrogradation of gelatinized starch, but that reaction needs several hours to become visible. More evidence is given in toughing cell wall components (e.g., reduced pectin esterification) for an increased resistance against shear forces during drying.

For cooking, steam of atmospheric pressure is injected into a screw conveyor. To determine the proper degree of cooking, time must be evaluated periodically. Overcooking results in a poorer texture of the finished product, whereas undercooking causes excessive loss of mash from the applicator rolls and a high level of defects.

Mashing immediately follows cooking to avoid cell rupture of cooled and firmed potatoes. Also mixing with additives is easier in the hot stage. Next to emulsifiers (0.5% monoglyceride, to complex free gelatinized starch), chelating agents like sodium acid pyrophosphate (which reacts with iron, preventing darkening after cooking) and antioxidants like butylated hydroxyanisole (which extends storage life and provides reduced rancidity) are used widely. Also sulfur can be added to prevent nonenzymatic browning. The proper dosage of additives is difficult because of high losses during drying (up to 75% of antioxidants is lost and 50% of sulfur dioxide).

Drying takes place on a single-drum drier fitted out with four to six applicator rolls. The drum (e.g., 1.5 m diameter and 5 m length) is steam-heated. The temperature on the surface is between 140 and 160 °C. The mash is applied above the top roll while the direction of drum rotation is downward. Therefore, the distance between the top roll and the drum surface determines the rate of flow of mash to the rolls below. Each applicator roll applies a layer of potato cells to the drum. The distance of the applicator rolls to the surface of the drum must be controlled very carefully, because cell rupture is influenced by distance. A wider distance may result in a nonuniform deposition of potato cells and lumps from larger agglomerates can be in the product. Finally, a knife removes the dry sheet. It drops down into a screw conveyor which transfers the roughly broken sheet to a collecting tank. Then it is ground with hammer mills, creating a product with acceptable texture. High amounts of fines is necessary to prevent a sticky reconstituted product. The shelf-life depends on the atmosphere in the final products packages. Autoxidation of free fatty acids remains a problem.

Dehydrated granules Potato granules are dehydrated single cells or aggregates of cells dried to about 6–7% moisture after prior cooking. The principal process used is the 'add-back' process. Potatoes are partially dried by adding back previously dried granules to give a moist mix, which is air-dried after conditioning. The result is potato flakes, but the product is much less sticky, because most cells remain intact

Processing The first steps of processing are as for flake processing. Raw potatoes are washed and sorted (by hand or by optoelectronic sorters) before steam peeling. Trimming removes the rest of peel and the defects, and slicing results in disks with 1–2 cm thickness to obtain a uniform heat treatment.

During precooking sulfur dioxide is added to prevent nonenzymatic browning.

Water cooling below room temperature for about 20 min reduces the level of broken cells in the final product.

Steam cooking at atmospheric pressure in a screw conveyor for about 30–50 min gives fully cooked potatoes ready for mashing and mixing with dry add-back granules. Repeated mild shearing and pressing of the added granules against the cooked mash causes separation of the latter into single cells with a low content of broken cells.

The moist mix is cooled to approximately 15–25 °C and held for about 1 h at that temperature (conditioning) by moving the moist mix on rubber belts with a depth of product of 10–20 cm, to avoid excessive compression. Since some agglomeration does occur during conditioning, gentle mixing is advised after that. Very large agglomerates and bruised tissue can be separated by a scalping reed. Poor granulation is a self-perpetuating process. Therefore, the amount of large granules will increase progressively with continued recycling. Because the larger particles do not absorb moisture rapidly enough, the process is to be controlled periodically.

Flash drying with a heated air stream, an inlet temperature of 170 °C, and an outlet temperature of 80 °C reduces the moisture content to about 12%. Material coarser than 60–80 mesh is screened off and returns to the process as add-back. Further drying takes place in a two-chamber fluidizing drier with alternate hot- and cold-air streams towards a final moisture content of about 6%.

For retail purposes the granules can be agglomerated with fat, milk solids, and flavors to give a ready-to-eat product. In that case only rehydration with hot water is necessary.

French Fries

French fries are deep-fat-fried potato strips. Commercially distributed French fries (UK: chips) are semifinished, and must be fried or heated before consumption (par-fried = partially fried). Shelf-life depends on the kind of distribution, either fresh (cooled) or deep-frozen. With respect to the desirable quality profile, large and long oval-shaped tubers (50 mm up) are preferred.

Processing Sorted tubers are cleaned. Subsequent peeling is with steam peelers. After removal of the skin, strips are cut either by mechanical knife systems or by water knives (e.g., 6×6, 9×9, 11×11, 13×13 mm). In an inspection unit small pieces and defects are removed. The total amount of waste can be up to 50%, but most rejected material is processed to other products, e.g., pommes croquettes. In a washing unit free starch and other solids are washed off. Blanching reduces enzyme activity. A layer of pasted starch reduces fat uptake during frying. To prevent nonenzymatic discoloration, sulfur is often added to the blanching water. After washing, sometimes a second blancher with a low concentration of glucose is implemented to adjust the content of reducing sugars for a more stable frying color. Also dipping in a water bath containing a coating substance occurs, to enhance the texture stability for about 5–10 min after final frying. Predriers remove free water from the surface and the outer layer. Fat uptake will be reduced by the higher dry-matter content. Frying units can have a capacity of $4 \, t \, h^{-1}$ of product, or more. The fat content of such units can be more than 5000 l of fat. Temperature is between 140 and 180 °C. The potato strips are moved through the oil bath by conveyor belts. To enhance texture behavior and capacity, two frying units can work in a line. Normally, the total frying time is between 3 and 5 min. After frying a defatting unit removes free fat from the surface by vibration or by hot-air streams. Mechanical cooling reduces the temperature of prefried strips to 4 °C. That product can be packaged and distributed as par-fried fresh French fries. Alternatively, the French fries are frozen inside tunnel freezers, packaged, and distributed as par-fried deep-frozen French fries.

Potato Chips

Potato chips (UK: crisps) are thin slices (1.1–1.6 mm) of fresh potatoes which have been normally peeled and fried in hot oil to reduce the water content to less than 2%. For processing, special varieties with a very low content of reducing sugars (< 0.15% fresh weight) are required.

Processing After cleaning and visual inspection, the tubers are peeled abrasively with losses up to 25%. Thin slices are cut with smooth or crinkled knives. In a washing unit free starch and other solids of disrupted cells are washed off. Optionally, blanching may reduce the content of reducing sugars. Predriers remove free water from the surface and the outer layer. Frying units can have a capacity of $4 \, t \, h^{-1}$ of product or greater. The oil content of such units can be more than 5000 l of oil. Temperature is up to 190 °C. The potato slices are of lower specific density

than the oil. Therefore specially designed rotating paddles dip and move the slices forward. Total frying time 2–4 min. The slices absorb up to 50% oil. That high amount of oil must be continuously replaced. After frying, a defatting unit removes free fat from the surface by vibration or by hot-air streams. The hot slices are seasoned and packaged.

Others

Peeled potatoes Peeled table potatoes are produced with abrasive peelers, either carborund or knives or combinations thereof. Shelf-life depends on the hygienic status. Preservation with chemicals (e.g., sulfur) leads to a shelf-life of 8 days, whereas a substitution of air by nitrogen or carbon dioxide or combinations thereof leads to a shelf-life of 10 days.

Potato dough Several potato products prepared of dough are available, either fresh or deep-frozen (e.g. pommes croquettes, pommes duchesses). Dough preparation of cooked potatoes and several other ingredients can be from fresh potatoes (e.g., small pieces from French fries production) or from dehydrated potatoes. A final frying or heating is necessary.

Sterilized potato products Peeled potatoes are available as canned potatoes. After peeling, diced or whole tubers are blanched with water at 85 °C, filled in cans and covered with brine, sealed, and sterilized at 121 °C. In addition to table potatoes, other potato products are also sealed in plastic bags and sterilized.

Potato Processing for Industrial Purposes

Ethanol

Ethanol originates from sugar fermentation of yeast (*Saccharomyces cerevisiae*). Processing of ethanol production integrates these steps: first starch is broken down by steaming the potatoes. After mashing the potatoes, yeast fermentation runs in a batch process for up to 3 days. Finally the crude spirit is distilled and rectified.

Starch

Starch is the main component of solids in potato tubers. In comparision with other types of starch, isolated potato starch offers several technological advantages for applications in the food and nonfood industry. Also plastic substitutes (e.g., shopping bags) have become of strong interest. The main disadvantage of potato starch separation is the campaign separation outside the winter season: destroying the cells by rasping liberates the starch granules. After fruit water separation, starch is washed out in jet extractors of various types. In addition to welded sieves, plate sieves improve the efficiency of this step. After desanding, crude starch milk is purified by countercurrent washing with demineralized fresh water, dewatered, and dried. Optimum engineering results in starch recovery rates of at least 97–98%.

Fruit water is separated and concentrated by decanting for subsequent protein recovery by iso-electric precipitation and heat coagulation. The remaining deproteinized fruit water can be concentrated by ultrafiltration or reverse osmosis. Further evaporation results in a potato protein liquid.

See also: **Browning**: Nonenzymatic; Enzymatic – Biochemical Aspects; Toxicology of Nonenzymatic Browning; Enzymatic – Technical Aspects and Assays; **Chilled Storage**: Use of Modified-atmosphere Packaging; Packaging Under Vacuum; **Controlled-atmosphere Storage**: Applications for Bulk Storage of Foodstuffs; **Cooking**: Domestic Techniques; **Drying**: Theory of Air-drying; **Potatoes and Related Crops**: The Root Crop and its Uses

Further Reading

Agblor A and Scanlon MG (1998) Effects of blanching conditions on the mechanical properties of French fry strips. *American Journal of Potato Research* 75: 245–255.

Bergthaller W, Witt W and Goldau H-P (1999) Potato starch technology. *Starch/Stärke* 51: 235–242.

Boskou D and Elmadfa I (1999) *Frying of Food*. Lancaster, PN: Technomic.

Lisinska G and Leszczynski W (1989) *Potato Science and Technology*. London: Elsevier Applied Science.

McMinn WAM and Magee TRA (1997) Physical characteristics of dehydrated potatoes – part I and part II. *Journal of Food Engineering* 33: 37–48, 49–55.

Mensah-Wilson M and Gierschner K (2000) An improved method for the production of canned potatoes. *Fruit, Vegetable and Potato Processing* 85: 132–139.

Moreira RG, Castell-Perez ME and Barrufet MA (1999) *Deep-Fat Frying*. Gaithersberg, MD: Aspen, USA.

Talburt WF and Smith O (eds) (1987) *Potato Processing*, 4th edn. New York: AVI Book.

Willard M (1993) Potato processing: past, present and future. *American Potato Journal* 70: 405–418.

POULTRY

Contents
Chicken
Ducks and Geese
Turkey

Chicken

A J Maurer[†], Formerly of University of Wisconsin-Madison, Madison, MI, USA

Introduction

Chicken is one of the most widely accepted muscle foods in the world. Its high-quality protein, relatively low fat content, new products, and generally low selling price because of favorable feed conversion make chicken a high-demand food in the market-place. Furthermore, the absence of cultural or religious taboos allows increased chicken production and consumption worldwide.

This article reviews specific characteristics of chicken, its chemical composition and nutritive importance, production advances, slaughter and further processing, food uses and products, and micro-biological problems.

The USA leads the world in poultry meat production, followed by the former Soviet Union, China, and Brazil. Consumption of chicken in the USA has increased dramatically over the last 20 years. **Table 1** shows the changes in chicken consumption and

Table 1   Meat consumption per capita (kg) in the USA

	1994	1995	1996	1997	1998
Beef	30.3	30.5	30.6	30.3	30.5
Pork	24.0	23.8	22.2	22.1	23.6
Poultry	40.4	40.3	40.8	41.1	41.9

Source: http://meat.tamu.edu/consum.html.

[†]Deceased

compares those changes with other meats. Broiler consumption in the USA continues to increase.

Chemical Composition, Nutrition, and Dietary Significance

Chicken is a very digestible source of high-quality proteins. Its muscles are differentiated into light meat (primarily the breast) and dark meat (the legs). These muscles vary in myoglobin content as well as in fat content. **Table 2** lists the chemical composition of selected chicken parts, and shows some of the proximate changes that occur on cooking. (*See* **Protein**: Food Sources.)

Chicken meat is low in total fat and saturated fat, and light meat is lower in fat than dark meat (**Table 2**). Most of the fat in poultry is deposited under the skin and is easily removed by pulling the skin from carcasses or parts. Because chicken fat has a high ration of unsaturated to saturated fatty acid (approximately 30% saturated), the fat melting point is relatively low and liquefies easily. Chicken fat is also prone to oxidative rancidity. Although most consumers believe otherwise, sodium and cholesterol contents in chicken are similar to those in most meats. Light meat contains more protein, and dark meat contains higher levels of fat and cholesterol. Older chickens generally contain more fat and less moisture. Cooking tends to concentrate the level of cholesterol and protein because water is lost in the heating process. (*See* **Cholesterol**: Properties and Determination; **Fatty Acids**: Properties; **Sodium**: Properties and Determination.)

Chicken contains all the essential amino acids, and is a good source of B vitamins and minerals such as iron and phosphorus. Poultry meat also provides potassium, calcium, magnesium, and copper, but carbohydrates and fiber are negligible. Light meat contains more nicotinic acid than dark meat, although both are good sources. Conversely, dark meat contains more riboflavin, iron, and zinc than light meat. (*See* **Amino Acids**: Properties and Occurrence; Refer to individual vitamins and minerals.)

Economics of Fast-Growing Broiler Chickens

Chicken broilers have been bred to a uniform size specifically for meat yield. A common broiler strain combines Cornish genes for conformation and fleshing with the White Plymouth Rock for white feathers and faster growth. White-feathered birds are the norm for commercial broiler production because of their cleaner-looking defeathered carcasses.

Many factors have contributed to the efficiency and economics of broiler production. Advances in breeding, nutrition, disease control, and management practices have enabled the broiler industry to produce a chicken weighing 1.8 kg in 6–7 weeks. Vaccines, antibiotics, confinement rearing, and computer-balanced rations assist in producing broilers with a feed conversion ratio of less than 2 kg of feed per kg gain. Vertical integration in the broiler industry has increased efficiency even more by combining most of the activities in producing, processing, and marketing broilers under the same ownership and management.

Chicken Processing

Chicken slaughter and processing have evolved into a sophisticated and automated procedure whereby a processing plant can handle as many as 20 000 broilers per h. Transforming live chickens into a ready-to-cook form involves live bird catching, crating and hauling, unloading, hanging on a conveyor line, stunning, slaughtering and bleeding, scalding,

Table 2 Proximate analysis of chicken broilers (per 100 g edible portion, wet-weight basis)

Nutrient	Breast meat with skin (raw)	Breast meat without skin (raw)	Breast meat with skin (roasted)	Leg meat with skin (raw)	Leg meat with skin (roasted)
Water (g)	69.46	74.76	62.44	69.91	60.92
Protein (g)	20.85	23.09	29.80	18.15	25.96
Energy (kJ)	772	462	827	785	974
Lipid					
Total (g)	9.25	1.24	7.78	12.12	13.46
Saturated (g)	2.66	0.33	2.19	3.41	3.72
Monounsaturated (g)	3.82	0.30	3.03	4.89	5.24
Polyunsaturated (g)	1.96	0.28	1.66	2.65	3.00
Cholesterol (mg)	64	58	84	83	92
Ash (g)	1.01	1.02	0.99	0.85	0.92
Sodium (mg)	63	65	71	79	87

Reproduced from Posati LP (1979) *Composition of Foods. Poultry Products, Raw, Processed, Prepared.* Agriculture Handbook 8–5. Washington, DC: Science and Education Administration, US Department of Agriculture.

defeathering, eviscerating, inspecting, chilling, grading, packing, and shipping. (*See* **Meat**: Slaughter.)

Assembly of Live Birds

Broilers should be taken off feed 8–12 h before the time of slaughter to reduce fecal contamination during processing. Bruising in chickens can be minimized by removing feeders and waterers prior to the arrival of the catching crew, and by careful handling of the birds as they are caught and placed into coops. Novel mechanical methods (e.g., herding, sweeping, and vacuum systems) for catching chickens are also available and commercially used by a few companies in Europe.

Hauling and Unloading

After the birds are loaded into wooden or plastic crates or specially built compartments, they are transported via open-sided trucks to the slaughter plant. Large-volume, slow-speed fans or evaporative cooling provides good ventilation and comfort for the birds in the holding shed.

Trailer loads move into the plant unloading area as needed. Various methods (e.g., manual bird removal, or dumping the birds through side-doors) unload the crates. Birds are hung by their shanks on shackles attached to an overhead monorail conveyor for transfer through the slaughtering operations. The unloading area should be dimly lit, and a breast rub bar can be used to calm the birds. According to UK regulations, birds should be slaughtered within 3 min after they are suspended from shackles.

Stunning, Killing, and Bleeding

Stunning is used to enhance bleeding and feather release. The heads of the birds are usually dragged through an electrically charged water bath. For consistent stunning, the water should contain from 0.1 to 1.0% sodium chloride. A fine mist of water spray should be directed at the bird's feet to provide a positive electrical contact prior to entering the stunner. The shackles complete the circuit and cause an electric current to run through the body of the bird as a result of the applied voltage (approximately 50 V). Overstunning can rupture blood vessels and break carcass bones due to excess muscle contraction. It is essential that the electric shock does not kill the bird; bleeding must be the cause of death.

Killing can be either manual or mechanical. In the manual operation, a skilled worker with a sharp knife severs the jugular vein and carotid artery by cutting across the side of the neck at the base of the bird's head. In mechanical neck cutting, a guide bar with grooved rollers holds the head rigid and extends the neck in preparation for the cuts. The bird's head is guided across a single, revolving, circular blade or between a pair of revolving blades.

Bleed time should be 1.5–2 min, and the blood lost accounts for approximately 4% of the live weight. The shackles convey the birds through a blood tunnel, so that the blood can be collected and disposed of properly.

Scalding

After bleeding, the birds are scalded to loosen the feathers by immersion in agitated hot water. Wing flapping and struggling should have ceased by the time birds enter the scalder. In general, a soft- or semiscald temperature of 50–54 °C is used, with an immersion time of 1.5–2.5 min. For the fresh chilled broiler market, low-temperature scalding permits retention of the yellow cuticle on the chicken skin and retards skin drying.

Chickens are subscalded at 59–60 °C for cuticle removal if there is a preference for lighter-skinned birds, if the carcasses are to be frozen, or if the parts are to be battered and breaded. Higher temperatures greatly facilitate feather removal, and necks and wings of semiscalded birds are often scalded at the subscald conditions to achieve cleaner carcasses. In the USA, the US Department of Agriculture (USDA) recommends a minimum overflow from the scalder of 11 per bird to reduce build-up of contamination. Various chemicals added to scald tanks assist feather removal by reducing surface tension and enhancing wetting of the feathers. An alternative to immersion scalding with separate defeathering machines is a combined spray-scalding and defeathering system. This type of scalding may reduce contamination spread from bird to bird.

Defeathering

After the carcasses leave the scalder, they enter a series of online defeathering machines. The picking machines consist of banks of counter rotating stainless-steel disks or drums, with rubber 'fingers' mounted on them. As the birds are conveyed through the rotating picker fingers, the feathers are rubbed or plucked from the carcass. Continuous water sprays flush away the feathers.

Remaining pinfeathers are removed by hand. In the USA, the birds pass through a gas flame to singe the fine hairs. A final step in the defeathering area is an outside spray wash.

The head removal operation may take place in the defeathering area between the picker and the outside bird washer, or in the eviscerating room to facilitate crop removal. The bird's neck passes through a device which restrains the head as the overhead conveyor pulls the body forward, separating the neck vertebrae at the base of the skull. Some head removal devices can include a set of rollers to separate the neck bones and a rotating knife to sever the neck skin instead of tearing it; yield is improved with this system.

The birds pass through an automatic hock cutter which severs the shanks at the hock joint and causes the carcass to drop onto a conveyor for transfer to the evisceration area. In the UK, the feet of the broilers are cut off just above the spur by means of a rotating knife. The severed feet remain on the shackles and are removed mechanically on the return line.

The scalding, defeathering, and other operations occur in a separate portion of the plant, apart from the evisceration and final processed bird area. The transfer conveyor takes the defeathered carcasses into the evisceration area, where the carcasses are rehung on the clean evisceration line.

Evisceration

The first cutting operation in the evisceration area involves removing the preen or oil gland, located on the top side of the tail next to its base. A sharp, short-bladed knife is used to remove the gland manually, or a machine with a cutting blade can be used online.

An opening cut into the body cavity is then made using a knife, a special drill-like vent cutter with a rotary blade, or an online automated machine. Making an incision in the abdomen and removing the vent can be performed manually or by machine to complete the opening process for evisceration.

Removing or drawing the viscera from inside the bird can be performed by hand or by machine. To retain the identity of the viscera within the carcass and be in a position for easy inspection, the viscera are draped uniformly to one side of the tail but remain attached to the carcass until the carcasses have passed the inspection for wholesomeness.

Mechanized removal of the viscera is very common in large processing plants (**Figure 1**). An eviscerating spoon is mechanically inserted into the cavity and withdraws the viscera. Back-up workers may be necessary to remove viscera missed by the machine.

Mandatory USDA inspection of every bird for wholesomeness is performed under the supervision of an inspector employed and trained by the government. If evidence of unwholesomeness is detected, the carcass or affected parts are trimmed or condemned.

Workers remove the giblets (heart, liver, and gizzard) from the viscera of inspected birds and clean and chill them. Then the lungs are removed with a vacuum lung gun, or they can be withdrawn with the viscera on a mechanical eviscerator. The neck is generally cracked with shears, the crop and windpipe are

Figure 1 Automatic chicken eviscerator. Reproduced from Poultry: Chicken, *Encyclopaedia of Food Science, Food Technology and Nutrition*, Macrae R, Robinson RK and Sadler MJ (eds), 1993, Academic Press.

removed, and the neck is then pulled off by stationary guide bars.

Finally, the shackles convey the carcasses through a bird washer where spray nozzles rinse the inside and outside of the carcasses.

Chilling

The most common chilling operations in the USA immerse carcasses in long flowthrough tanks containing agitated chlorinated water or slush ice. The USDA regulations require that a chiller overflow rate be maintained at 2 l of water per chilled bird, in order to minimize microbial build-up in the chill water.

The chickens are first placed in a prechiller, containing water at 10–18 °C, and then into a slush ice chiller at 0–1 °C. Chickens must be chilled to 4 °C or lower within 4 h to meet USDA requirements; the chilling time generally needed to obtain 4 °C carcasses is about 40–60 min.

After chilling, carcasses are hung by one leg on a drip line for 2.5–4 min for draining, and are then conveyed to the packing area. Although the meat absorbs some moisture during washing and chilling, the drained moisture level is strictly regulated. The drip line shackles are usually weighing devices which drop the sized birds into appropriate bins.

Air chilling of eviscerated carcasses is used extensively in Europe. Spraying the carcasses with water at intervals avoids weight loss during evaporative chilling. (*See* **Meat**: Preservation.)

Grading

Carcass grading in the USA is a voluntary practice, but is often required by purchasers. Government graders sort the birds into A, B, and C categories according to body conformation, fleshing, fat cover, deformities, bruises, and defects such as pinfeathers, disjointed or broken bones, and missing parts. Europe uses similar grading standards.

Packing

Giblet packs containing a heart, liver, gizzard, and neck are stuffed into the body cavity of chilled, sized carcasses. The carcasses then go into overwrapped trays which are heat-sealed, or into polyethylene bags which are clipped or taped shut ready for market. Another common method in the USA for distributing broiler carcasses from the processing plant is to put them into corrugated, wax-impregnated boxes. After boxing, ice covers the birds to keep the skin moist during shipment. Nearly half of the broilers in the USA leave the processing plant as ice-packed birds. An increasingly popular packing method is chill-packing, whereby a −6 °C air blast lowers carcass temperatures to between −2 °C and −1 °C. Refrigeration then maintains that temperature during marketing of the birds.

Waste Products

Poultry processing results in large amounts of waste waters, semisolids, and solids, which require separation and treatment before being discharged into the environment. Waste material that can be reclaimed and, for example, used for animal feeds or fertilizer helps to reduce the overall load for disposal.

There are several ways to handle poultry offal (heads, feet, viscera, inedible parts, condemned whole birds, feathers, and blood). Usually, all offal except the blood and condemned birds is floated in water from the processing areas to an accumulation area for removal by trucks. Screens and rotating drums remove water and separate offal components. Because of its high oxygen demand for decomposition, blood is usually handled separately.

Most of the offal solids goes to a rendering plant, while liquid generally goes through a primary and secondary lagoon treatment system. Some plants can discharge the liquid effluent into a municipal water system after removing the fat and as many solids as possible.

Another method of waste treatment is spray irrigation or spreading the processing effluent on the land. This type of waste disposal requires large fields to prevent overloading. Runoff into streams is a major concern. (*See* **Effluents from Food Processing**: Composition and Analysis; Disposal of Waste Water.)

Food Uses and Products

Most broilers in the USA are sold as fresh (unfrozen) carcasses, but frozen deboned meat is used in further processing. A decline in the sale and consumption of whole chickens in the USA has been counterbalanced by a steady increase in the sale of cut-up carcasses and parts. More than half (53%) of the processed carcasses in 1990 were marketed as cut-up broilers and parts.

Cut-Up

The cut-up operation is usually mechanized with motor-driven equipment and shielded circular blades. Individual cut-up stations or online automated machines can be used. A popular method of packaging is to place the cut-up parts in a tray containing an absorbent pad to collect seepage and then overwrap the tray. The bagged whole carcasses and tray packs may be prepriced before delivery to the retail market.

Deboning

The spectacular growth in the sale of value-added, further-processed poultry products in recent years has placed a heavy demand on the production of deboned meat. Poultry meat has traditionally been removed from the carcass by hand with a sharp knife, while the carcasses hang from special shackles on a slow-moving line or are positioned on static or moving cones.

Several automatic deboning machines are also available for removing breast fillets and thigh and drumstick meat. One system operates by holding a particular chicken part in position above a contoured recess in a base plate. The machine forces the meat from the bone into the depression. Another meat-deboning system pushes the bone lengthwise out of the carcass part (e.g., drumstick) and strips off the meat in the process.

Currently, deboning is primarily done on aged chilled carcasses. Hot-stripping the muscles and skin from defeathered but uneviscerated carcasses is being studied. Some researchers are also testing hot deboning, the removal of meat after evisceration but prior to chilling. Generally, hot deboned meat is tough, which greatly limits its potential uses. Some poultry meat, especially that of mature hens, is cooked prior to hand deboning.

After the major muscles are deboned from the carcass, the remaining frames (also necks, backs, and low-value parts) can be more completely deboned by special machines. In the first stage of mechanical deboning, a grinder reduces bone and meat particle size. In the second step, a pressure system squeezes the ground meat and bone against a perforated screen or microgrooved cylinder. Advances in mechanical deboning have eliminated pregrinding and produced a more texturally attractive product. An alternative system is a batch operation. Pressure from a ram forces meat to flow from the bones through a sieve screen; the bone cake is ejected, and a new batch is introduced.

Further-Processed Products

Much of the hand- and mechanically deboned chicken meat is directed into value-added products. Deboned meat can be marketed 'as is' fresh, as whole breast or split breast fillets, strips, and chunks. Alternatively, deboned meat can be tumbled, marinated, chopped, formed, ground, emulsified, or prepared in a number of ways for sale.

Many nonmeat ingredients are used in further processing. Salt helps to extract proteins for improved binding and texture, acts as a preservative, and enhances flavor. Up to 0.5% phosphate increases water-holding capacity and final product yield. Other ingredients, such as sweeteners, spices, binders, and curing salts, are used in the wide array of poultry products on the market. (See **Curing**.)

The driving force behind poultry product development is the consumer. Fat content in meats is a current concern; many people want healthier diets, and the low fat in poultry is a major reason for its popularity. Fat pads are removed from many whole carcasses in response to consumer desires. As new poultry products are designed, diet and health, microwave ovens, and an aging population cannot be ignored.

The fast-food restaurants have had a tremendous influence on new poultry product popularity for items such as nuggets, tenders, marinated breast fillets, and frozen–fried parts. Ethnic foods are popular in the USA, especially Italian, Mexican, and Chinese. Different flavors in poultry products are in demand, including cajun, barbecue, and honey. Spicy chicken wings are an example of a popular seasoned finger-food appetizer. In addition to less fat, smaller portions, and different flavors, consumers want less salt and cholesterol in their chicken products. The food must be of high quality, convenient, safe, and nutritious.

In the retail supermarket, low-fat chicken products such as patties, rolls, sausages, and sliced luncheon meats are selling well. Poultry frankfurters and bologna are generally lower in fat than similar red-meat products and are a healthy alternative. (See **Meat: Sausages and Comminuted Products**.)

Most recent entries among the new poultry products are boneless chicken breasts stuffed with ingredients such as cheese and broccoli or wild rice and mushrooms. These products are sold in a variety of packages and combinations, but are almost always ready for cooking in the microwave. Other gourmet items include chicken Kiev (breaded chicken breasts filled with garlic butter and parsley), chicken cordon bleu (filled with ham and cheese), and herb-roasted chicken dinner (low-fat, low-cholesterol, low-sodium meal). Even the luncheon market has new chicken 'lunchable' products, which are individual packages containing items such as cooked sliced poultry, cheese slices, and crackers.

Microbiological Concerns

Modern poultry husbandry and processing techniques have greatly improved the quality of poultry meat. However, chicken can carry many kinds of organisms. The two major concerns are control of spoilage organisms which cause consumers to reject the product due to odor or flavor, and the minimization of pathogenic organisms which may (under faulty handling) lead to a health hazard.

Spoilage Microorganisms

Although poultry is refrigerated, and even frozen, for shelf-life extension, spoilage will invariably occur owing to the growth and metabolic activities of specific types of bacteria. The psychrophiles, such as *Pseudomonas*, can generally grow at refrigerator temperatures and are responsible for most of the spoilage. Spoilage organisms are present in high numbers when an off-odor becomes apparent (10^7 cells per cm^2) and even greater when slime formation occurs from coalescence of colonies (c. 10^8 cells per cm^2). (See **Spoilage: Bacterial Spoilage**.)

Pathogenic Microorganisms

Poultry is considered to be a major source of *Salmonella*, which can spread during processing, and even after cooking if raw meat preparation surfaces were not cleaned before placing a cooked product on the same surface. However, normal cooking procedures destroy *Salmonella* and it will not grow well under refrigeration. The most common problems are undercooked poultry or poultry contaminated after cooking. According to data on the vehicle of foodborne salmonellosis outbreaks in the USA between 1973 and 1987, only 3.8% (30 of 790) of the cases were caused by chicken.

In addition to *Salmonella*, other pathogens sometimes found in poultry include staphylococci, *Campylobacter, Listeria* spp., clostridia, and coliforms. Careful product handling, proper refrigeration, and adequate cooking will almost always insure product safety. (*See Campylobacter*: Campylobacteriosis; *Clostridium*: Food Poisoning by *Clostridium perfringens*; Botulism; *Listeria*: Listeriosis; *Staphylococcus*: Food Poisoning.)

Sources of Bacteria

Microbes in poultry-processing plants are ultimately found on the product and come from three main sources: the birds (feet, feathers, intestinal contents), the environment (water, air, supplies), and the workers. Careful management at the production site and adherence to good manufacturing practices at the processing plant (e.g., filtered air, cool temperatures, thorough cleaning and sanitation, and good worker hygiene) will minimize final product contamination with bacteria.

See also: **Amino Acids**: Properties and Occurrence; **Campylobacter**: Campylobacteriosis; **Cholesterol**: Properties and Determination; **Clostridium**: Food Poisoning by *Clostridium perfringens*; Botulism; **Curing**; **Effluents from Food Processing**: Disposal of Waste Water; Composition and Analysis; **Fatty Acids**: Properties; **Listeria**: Listeriosis; **Meat**: Slaughter; Preservation; Sausages and Comminuted Products; **Sodium**: Properties and Determination; **Spoilage**: Bacterial Spoilage; **Staphylococcus**: Food Poisoning

Further Reading

Anonymous (1990) *Broiler Industry Marketing Practices, Calendar Year 1989*. Washington, DC: National Broiler Council.

Bean NH and Griffin PM (1990) Foodborne disease outbreaks in the United States, 1973–1987: pathogens, vehicles, and trends. *Journal of Food Protection* 53(9): 804–817.

Brant AW, Goble JW, Hamann JA, Waback CA and Walters RE (1982) *Guidelines for Establishing and Operating Broiler Processing Plants*. Agriculture Handbook 581. Washington, DC: Agricultural Research Service, US Department of Agriculture.

Cunningham FE and Cox NA (1987) *The Microbiology of Poultry Meat Products*. Orlando: Academic Press.

Mead GC (1989) *Processing of Poultry*. Essex: Elsevier Science.

Mountney GW (1976) *Poultry Products Technology*, 2nd edn. Westport, CT: AVI.

North MO (1984) *Commercial Chicken Production Manual*, 3rd edn. Westport, CT: AVI.

Posati LP (1979) *Composition of Foods. Poultry Products, Raw, Processed, Prepared*. Agriculture Handbook 8–5. Washington, DC: Science and Education Administration, US Department of Agriculture.

Putnam JJ (1990) *Food Consumption, Prices, and Expenditures, 1967–88*. Statistical bulletin no. 804. Washington, DC: Economic Research Service, US Department of Agriculture.

Stadelman WJ, Olson VM, Shemwell GA and Pasch S (1988) *Egg and Poultry-Meat Processing*. Cambridge: VCH.

Ducks and Geese

A J Maurer[†], Formerly of University of Wisconsin-Madison, Madison, WI, USA

This article is reproduced from *Encyclopaedia of Food Science, Food Technology and Nutrition*, Copyright 1993, Academic Press.

Introduction

Ducks and geese (waterfowl) are a delicacy to many people, while others object to the higher amounts of fat in the carcasses compared to broilers and turkeys. However, breeding and mass selection programs are improving the meat-to-bone and meat-to-fat ratios in waterfowl. Annual duck consumption on a ready-to-cook basis in the USA is only about 0.2 kg per person. Goose consumption is somewhat less, at approximately 0.01 kg per person per year. Although chickens and turkeys dominate the world poultry industry, in parts of Asia ducks are commercially more important than broilers (chickens), and there are more geese than turkeys in areas of Europe. This article reviews duck and goose processing and preparation, meat composition and nutrition, food uses and products, waste products, and microbiological problems.

Processing and Preparation

With a few exceptions, duck and goose processing is similar to that of broilers and turkeys. For a more complete slaughter and evisceration procedure, (*See* **Poultry**: Chicken.)

In the USA, White Pekin-type ducks are generally slaughtered at 7 weeks of age and weigh about 3.2 kg live. White Muscovy ducks are also raised for meat, but they are a slower-growing breed and commercially

[†]Deceased.

less popular than the White Pekin. Geese are 3–5 months old at slaughter, and their average live weight is 4.5–6.8 kg. The average ready-to-cook weight of geese varies, but is generally about 4.5–5 kg. Feather maturity (lack of pinfeathers) is an important factor in determining the best time to slaughter waterfowl. (*See* **Meat**: Slaughter.)

Processors transport ducks and geese to slaughter plants on open trucks or trailers. The loading and unloading process is unique as the birds are not usually caught by their legs; they are herded on to the conveyance, and off into holding pens at the processing plant. Processors use turkey-sized crates when there are relatively few geese to transport.

Ducks and geese are driven from holding pens through chutes on to scales for weighing and then into the shackle hanging area. Care in herding the waterfowl is essential to avoid pile-ups, skin scratches, damaged legs, and smothered birds.

It is important to minimize struggling by the birds when hanging them on motorized conveyor line shackles as excessive flapping can result in bruises. In smaller processing plants, killing funnels can be used to restrain and position the birds for slaughter.

The birds are usually stunned with an electric current and then bled. Sometimes a cut inside the mouth is used, but a common method of killing the birds is to cut the outside of the throat on the left side at the base of the jaw, severing the left jugular vein and carotid artery.

Following bleeding, both ducks and geese can be scalded or dry-picked. However, the latter method is slow and laborious. Generally, ducklings proceed through immersion scalders containing agitated water at 58–63 °C. Geese can be scalded in a similar commercial scalder, or they can be hand-scalded for a smallscale operation. Scald water temperature for geese should be 63–66 °C. The duration of the scald varies from 1.5 to 3 min, usually longer for geese than for ducks. The time and temperature will also change depending on the age of the bird, time of year (season), and density of feathering. The lower the temperature, the longer the scald. A little detergent or an alkaline defeathering agent can be added to facilitate thorough wetting of the feathers. Processors can hand-scald waterfowl by pulling the bird repeatedly through the water against the lay of the feathers.

After scalding, the birds may be rough-picked by hand, picked by a conventional rubber-fingered picking machine (online), or placed in a spinner-type picker. Because pins and down remaining on the carcass are difficult to remove, it is common practice to finish rough-picked birds by dipping each bird into melted wax specially formulated for this purpose.

The defeathered birds should be surface-dried (a jet of compressed air can be used) just long enough to allow the wax to adhere. The waxing operation is usually mechanized and often includes conveying the ducks in and out of two wax tanks. A resin-based microcrystalline wax is used for greater resilience and improved stripping.

Good wax penetration and adhesion are achieved in the first tank with immersion in 90 °C wax for 15 s. The second wax dip at 71 °C puts a heavier coating on the birds to thicken the wax for good pulling power and cleaner stripping. After waxing (and sometimes between wax dips), a cold-water spray over the birds or a dip into a tank of cold water will cool and harden the wax to a tacky state. The wax is then removed by hand or by a rubber-fingered wax stripper. Some processing plants dip the wax-picked birds into water at 82 °C to tighten the skin for easier manual removal of any remaining pinfeathers. Grasping pinfeathers between the thumb and a dull knife will assist in this operation.

The wax is reclaimed by remelting and straining out the pins, down, and feathers. Occasionally, the used wax should be 'cooked down' to remove all water that may have been mixed with it by emulsion.

After completely removing feathers, pins, and down, processors eviscerate the carcasses by making an opening cut in the abdomen and pulling out the viscera. Waterfowl viscera are somewhat difficult to withdraw, so that evisceration is most often done manually. However, processing plants in Europe are using automatic duck evisceration and automatic head and trachea pulling with processing rates of 4000 ducklings per h.

Each carcass on the eviscerating line is inspected for wholesomeness before the withdrawn viscera are detached. After the inedible viscera and lungs are discarded, the heads, tracheas, and feet are removed; the giblets are cleaned and wrapped for later placement in the body cavity. In many processing plants, only the two lobes are cut from gizzards because the inner lining is difficult to peel.

Following evisceration, birds receive an inside and outside spray wash before proceeding through a chilling treatment of cold tap water or ice and water. Airagitated cooling vats, a continuous immersion chiller, or air-spray chilling can be used.

After carcass cooling and draining in commercial waterfowl plants, each carcass is graded for conformation, fleshing, fat covering, and defects such as pinfeathers and exposed flesh. Processors then vacuum-package carcasses in barrier bags, heat-shrink the bags, and quick-freeze the packaged carcasses. Young ducks and geese may also be sold fresh, unfrozen. (*See* **Meat**: Preservation.)

Composition and Nutrition

Dressing percentages and meat yield of ducks and geese will vary with breed, age, sex, weight, and grade. The processing loss from live to ready-to-cook (carcass, neck, and giblets) for ducks is approximately 30%, 25–32% for geese, and 24% for broilers.

As the data in **Table 1** show, raw ducks and geese contain more fat, less water, and less protein than are found in broiler chickens. Roasting causes fat and some water loss, thereby concentrating (increasing) the protein content. Because of the relatively high fat content of waterfowl, the carcasses become rancid more easily, and frozen shelf-life is shorter than that of broilers or turkeys. (*See* **Fats**: Digestion, Absorption, and Transport; **Protein**: Food Sources.)

Skinning the carcass can remove much of the fat in poultry. For example, the fat content of raw duckling meat with skin is 39.3%; without skin, it is 6.0%. Values for goose are 33.6% with skin and 7.1% without skin. The fat in ducks and geese is highly unsaturated, as is the case in all poultry. Duck and goose skin (including separable fat) makes up 34–38% of the ready-to-cook carcass, but broiler skin is about 15% of the carcass. Conversely, the meat yield from ready-to-cook carcasses for ducks and geese is 34–47%, compared to 52% for broilers.

Duck is a good source of thiamin, and goose is an excellent source of phosphorus. The other vitamins and minerals are also present in ample amounts. Iron content is especially high in waterfowl, contributing to the darker color of the breast meat. (*See* **Phosphorus**: Properties and Determination; **Thiamin**: Properties and Determination.)

Food Uses and Products

Most duck and goose meat is consumed in whole carcass form. The carcasses are roasted in the same way as other poultry, but they yield less cooked edible meat. Because waterfowl contain more fat than broilers or turkeys, they do not require basting during roasting. If the carcasses are excessively fat, it helps to puncture or scratch the skin to permit some fat to cook out during roasting. Thorough cooking of waterfowl (to an internal temperature of at least 85 °C) is important to attain a crispy skin and for complete customer satisfaction.

Food manufacturers sell some further-processed waterfowl products. Duckling breast portions, semi-boneless halves, marinated breasts, fully cooked roast half duckling, and other gourmet items are available to the institutional trade. A few shops market specialty duck and goose products, such as smoked duck, boneless breast of duckling, smoked goose leg, or goose liver sausage.

Goose livers are popular in delicacies such as pâté de foie gras; the product must contain at least 30% goose liver. For a period of time in Europe and in parts of the USA, many people force-fed, or noodled, their geese. The starchy noodling ingredients (corn, wheat, barley, and rye) used in this frequent forced feeding produced an extra large liver (weighing nearly 1 kg each), which was sold at premium prices.

Goose fat is sometimes used in place of butter. It can also be used in frying, baking, cooking, and preparation of gravies, broths, and soups.

Waste Products

Duck and goose slaughtering produces the usual waste products of blood, feathers, feet, heads, inedible viscera, grease, debris, and cleaning water. The control and handling of these waste materials depend on the standards set by individual countries and local authorities. The methods of handling can include disposal to a public sewer after the fat and as many solids as possible have been removed, or treatment

Table 1 Proximate analysis of ducks, geese, and broilers (100 g edible portion, wet-weight basis)

Nutrient	Duck meat with skin (raw)	Duck meat with skin (roasted)	Goose meat with skin (raw)	Goose meat with skin (roasted)	Broiler meat with skin (raw)
Water (g)	48.50	51.84	49.66	51.95	65.99
Protein (g)	11.49	18.99	15.86	25.16	18.60
Energy (kJ)	1697	1415	1558	1281	903
Lipid					
Total (g)	39.34	28.35	33.62	21.92	15.06
Saturated (g)	13.22	9.67	9.78	6.87	4.31
Unsaturated (g)	23.77	16.55	21.53	12.77	9.47
Cholesterol (mg)	76	84	80	91	75
Ash (g)	0.68	0.82	0.87	0.97	0.79
Sodium (mg)	63	59	73	70	70

Reproduced from Posati LP (1979) *Composition of Foods. Poultry Products, Raw, Processed, Prepared.* Agriculture Handbook 8–5. Washington, DC: Science and Education Administration, US Department of Agriculture.

on-site and disposal of the filtered effluent to a lagoon or spray irrigation system.

In general, the first stage of any effluent system is the removal of coarse solids by screening. A fat trap or dissolved air flotation can then remove fine solids, fat, and grease. The effluent is further cleaned by either anaerobic digestion or aeration. (*See* **Effluents from Food Processing**: Composition and Analysis; Disposal of Waste Water.)

In contrast to other poultry-processing waste treatment problems, ducks and geese provide several valuable byproducts. The Far East is a good market for frozen duck feet, where they are stuffed with pork and considered a delicacy. Duck tongues are also valued by orientals as hors d'oeuvres. Some parts (heads and offal) go into mink or other animal food.

Other important byproducts are duck and goose down and feathers, used chiefly by the bedding and clothing industries. The poultry slaughter plant rinses and centrifuges the wet feathers to decrease the water content prior to shipping them to a feather-processing plant. A machine separates the down (15–25% of the feather mixture) and feathers, and washes and dries them. About five ducklings or three goslings are needed to produce 0.45 kg of dry feathers.

On a smaller scale, home processors can wash feathers in soft, lukewarm water which includes either a mild detergent or a little borax and washing soda. After rinsing, feathers are spread out to dry.

Microbiological Problems

In general, microbiological problems in ducks and geese are similar to those in other poultry. Pseudomonads are the main spoilage organisms. However, because of the very low duck and goose meat consumption, few illnesses have been attributed to waterfowl. The combination of scalding at 60 °C, followed by immersion in molten wax at *c*. 90 °C to aid final removal of feathers, appears to have a beneficial effect on the microbial quality of the finished product. (*See* **Spoilage**: Bacterial Spoilage.)

See also: **Effluents from Food Processing**: Disposal of Waste Water; Composition and Analysis; **Fats**: Digestion, Absorption, and Transport; **Meat**: Slaughter; Preservation; **Phosphorus**: Properties and Determination; **Poultry**: Chicken; **Protein**: Food Sources; **Spoilage**: Bacterial Spoilage; **Thiamin**: Properties and Determination

Further Reading

Baeza E, Dessay C, Wacrenier N, Marche G and Listrat A (2002) Effect of selection for improved body weight and composition on muscle and meat characteristics in Muscovy duck. *British Poultry Science* 43(4): 560–568.

Cunningham FE and Cox NA (1987) *The Microbiology of Poultry Meat Products*. Orlando: Academic Press.

Franson JC, Hoffman DJ and Schmutz JA (2002) Blood selenium concentrations and enzyme activities related to glutathione metabolism in wild emperor geese. *Environmental Toxicology and Chemistry* 21(10): 2179–2184.

Mead GC (1989) *Processing of Poultry*. Essex: Elsevier Science.

Mountney GW (1976) *Poultry Products Technology*, 2nd edn. Westport, CT: AVI.

Orr HL (1978) *Duck and Goose Raising*. Publication 532. Ontario: Ministry of Agriculture and Food.

Posati LP (1979) *Composition of Foods. Poultry Products, Raw, Processed, Prepared*. Agriculture Handbook 8–5. Washington, DC: Science and Education Administration, US Department of Agriculture.

Stadelman WJ, Olson VM, Shemwell GA and Pasch S (1988) *Egg and Poultry-Meat Processing*. Cambridge: VCH.

Turkey

A J Maurer[†], formerly of University of Wisconsin-Madison, Madison, WI, USA

This article is reproduced from *Encyclopaedia of Food Science, Food Technology and Nutrition*, Copyright 1993, Academic Press.

Introduction

Chickens and turkeys dominate the world poultry industry. The increasing popularity of poultry meat, including turkey, comes from low cost (value for money), a healthy nutritious image, and availability in a variety of convenient forms. Efficiencies in integrated turkey production, processing, and marketing have helped maintain favorable retail prices. This article reviews specific characteristics of turkey, processing of turkeys, food uses and products, microbiological concerns, waste products, and the nutritional profile of turkey meats and products.

Specific Characteristics

Turkey consumption has increased in many countries throughout the world, with Israel as the leader (over 9 kg per person in 1986). Less than one-third of this per capital consumption (2.7 kg) is of whole-carcass birds. Clearly, the growth in turkey popularity has come in cut-up parts and more fully prepared products.

Turkey hens in the USA, sold at about 16 weeks of age, provide almost 6 kg of ready-to-cook carcass. Toms (male turkeys) are slightly older, at 20–24

[†]Deceased.

weeks, and yield 10–12 kg of marketable carcasses. Desirable characteristics of turkeys include heavy bird weights without excess fat, a high dressing yield, and an ample proportion of valuable parts. The breast (white meat) is 35–40% of the ready-to-cook carcass. Leg meat, 25–30% of the carcass, contains more myoglobin and is darker than breast meat.

Other desirable turkey meat attributes include tenderness, bland flavor (allows further processing and seasoning), and good functional properties, such as protein extraction, protein gelation, water holding, meat binding, and emulsification. Turkey meat is similar to broiler meat in its composition and qualities, easily digestible, high in protein, and low in fat.

Processing of Turkeys

Turkey processing refers to slaughtering, feather removal, evisceration, and chilling. Other operations, such as inspection and packaging, are also important steps in processing turkeys for market. Most of the processing procedures are similar to those used for broilers.

Procurement

Special trailers or trucks with built-in cages haul turkeys to the slaughter plant. If crates are used to transport turkeys, they are considerably higher than those used for chickens. Handlers must catch, load, and unload turkeys very carefully to minimize bruises and broken bones. Escalator-type loaders are sometimes used to elevate turkeys to the built-in truck cages. The turkeys should be off feed 8–12 h prior to slaughter to minimize fecal contamination in the processing plant. Large-volume, slow-moving fans or evaporative cooling provides good ventilation and comfort for the birds in holding sheds at the processing plant.

Because turkeys are large and heavy, shackle line height in the unloading area and truck cages must be adjusted to the same level to minimize lifting required by the hangers. Workers hang the turkeys by both feet to the overhead moving shackles.

A dimly lit hanging area discourages the birds from struggling and flapping. Some processors install a smooth plastic bar parallel to and slightly below the overhead conveyor line so that the turkey's breast rubs against it to provide a soothing effect. The turkeys should move from the hanging area to the stunner within 6 min.

Stunning

The shackle line drags the heads of the birds through 0.1–1.0% saline water which contains a submerged electrode. Careful control of the voltage and current is vital. Too little current will not immobilize the birds, and too much current can cause a violent muscle contraction, often resulting in broken clavicle fragments in the muscle tissue.

Slaughter

Killing is usually performed manually by cutting across the side of the neck at the base of the bird's head, severing a jugular vein and carotid artery. Bleed time should be at least 2 min in a blood tunnel to collect the blood for proper disposal. Workers in some turkey slaughter plants cut both sides of the neck for more complete bleeding. (See **Meat**: Slaughter.)

Scalding

After bleeding, processors scald turkeys by immersion in agitated hot water to facilitate feather release. For some fresh chilled markets, they semiscald turkeys at 50–52 °C to retain the epidermal cuticle, which reduces skin dehydration. Most processors, however, scald turkeys at approximately 60 °C for 2–2.5 min. The USA has a requirement for an overflow from the scald tank of approximately 1 l of water per bird to help float debris from the water.

Defeathering

After scalding, the shackle line carries the birds through a series of picking machines containing rotating rubber fingers on disks or drums which rub or pluck the feathers from the carcasses. Workers remove the remaining pinfeathers by hand.

The birds pass through a gas flame to singe off the filoplumes or hairs protruding from the skin surface. The final process on the kill line is a thorough washing of the external surface of the carcass with pressurized water jets.

The carcasses then pass through a shank-cutting station. In some plants, an automatic tendon puller removes the shanks and pulls up to nine of the main sinews from the drumstick. At this point, a conveyor transfers the carcasses from the killing, scalding, and defeathering area into another room for evisceration.

Evisceration

For evisceration, turkeys hang from both legs with their heads also placed in a center slot of the shackle, creating a three-point suspension. This presents the birds horizontally, breast-up, for easier cutting and eviscerating. Plants equipped with mechanical eviscerating equipment use the two-point leg suspension.

Evisceration begins with an incision made through the abdominal wall. The cut continues around the vent and enlarges in the abdomen for easier removal of the viscera. In some plants, a bar (transverse) cut in the abdomen allows evisceration and later trussing of the legs. A mechanical vent cutter with a revolving cylindrical blade is another method to make an opening cut and remove the vent.

Processors withdraw the viscera through the abdominal opening, taking special care to avoid damaging the intestines or spilling the contents. The viscera are left attached to the carcass and draped over the outside of the bird for inspection. An inspector then examines each bird (inside, outside, and viscera) for wholesomeness. Carcasses can be trimmed to remove damaged parts or dressing defects, or condemned if unfit for food.

Workers harvest, clean, and save the giblets (heart, liver, and gizzard). They clean the gizzard by splitting it, washing away the contents, peeling off the hard lining, and washing. They discard the inedible viscera, and remove the lungs with a special tool or vacuum lung gun.

As one of the last operations, the head is cut or pulled off, and the esophagus and crop are removed from the front of the carcass. Finally, the skinless neck is cut off, washed, and retained for packing. Following evisceration, the birds must be washed inside and out before chilling.

It is possible to perform some of the slaughter and evisceration operations mechanically, but the variability in turkey carcass sizes presents unique problems. Fully automated equipment for turkey processing is still being developed.

Chilling

Most processors chill poultry meat in cold water or water and ice, but they use air chilling extensively in Europe, where carcasses are more often soft-scalded and sold fresh. The most-used chillers in the USA drop the birds from the evisceration line into a prechiller which also serves as a very effective washer. The agitated water has a temperature of less than 18 °C. The birds travel to a second chiller with a water temperature of less than 2 °C. Inspectors carefully monitor chiller water overflow, and carcass exit temperature and water uptake.

After chilling, graders evaluate the carcasses according to conformation, fleshing, fat covering, and defects such as pinfeathers and exposed flesh. (*See* **Meat**: Preservation.)

Packaging

In some plants, turkey carcasses move directly to a cut-up or deboning line. If turkeys are sold as whole carcasses, workers place a giblet packet in the crop cavity and a neck in the body cavity. Processors truss the legs into the bar cut, or use a metal or plastic hock lock. These turkeys will probably be frozen, in which case they are placed in an oxygen-impermeable, shrink-film bag. After the air is evacuated, the bags are clipped shut and passed through a hot-water shrink tunnel prior to freezing.

The freezing process is very rapid, with the first step being a brine or blast freezer to set the carcass surface color. Final freezing and storage occur in a holding freezer, where the birds can be stored for relatively long periods of time until they are moved into market channels.

Food Uses and Products

About 75% of the turkeys produced in the USA are cut up or further processed. In some countries, such as Israel, France, and Italy, the proportion of cut-up or further-processed turkey is 90% or more. However, whole birds are still popular at holiday times. Some of the whole carcasses are injected with various flavored solutions to create a self-basting, juicier carcass when roasted.

Deboning

Deboning of chilled turkey carcasses usually occurs as they hang from special shackles on a slow-moving line, or sit on deboning cones which may be static or moving. Precise cuts are made to remove parts, breast meat, thigh meat, and trimmings from the skeleton. The drumsticks (and also the thighs) are often deboned on automated machines, whereby the drumstick bone may be pushed lengthwise out of the meat; thigh pieces may be compressed in a specially designed mold, squeezing the meat away from the bone. More deboning automation will occur in the near future.

After removal of intact muscles using knives or semiautomated procedures, the remaining frames, necks, and backs are usually mechanically deboned. One system uses an auger principle, pressing the meat and bone against a perforated cylinder screen or microgrooved cylinder. The alternative system is a hydraulic-press-type batch design which squeezes the meat through a series of stationary filter rings. Depending on the incoming meat materials (meat-to-bone ratio), screen perforation size, and machine adjustments, the mechanically deboned meat can be a very fine purée or have a particle size of 5 mm, which is ideal for some sausages.

Further Processing

Hand- and mechanically deboned meats are useful in many further-processed products. The term 'further processing' encompasses procedures such as deboning, size reduction, injection, tumbling, massaging, reforming, and emulsifying. The principles of water-holding, protein extraction, protein gelation, and meat binding are very important in making further-processed products. Other processes, such as battering, breading, cooking, and freezing, may take place. Further processing also refers to whole carcasses that are basted, marinated, or smoked.

The types of products now available are increasing, ranging from cut-up portions to reformed roasts, breasts, rolls, steaks, hams, burgers, frankfurters, bolognas, coarse-ground sausages, salamis and bacons. More recently, ready-prepared meals have utilized an increasing amount of turkey meat. (*See* **Meat**: Sausages and Comminuted Products.)

Consumers with changing careers and households are the driving force behind trends in the consumption of poultry products. Smaller families often lack time but can afford the convenience of further-processed poultry. Diet and health, microwave ovens, and an aging population are important factors in new product development. Ethnic foods, especially Italian, Mexican, and Chinese, are popular in the USA. Different flavors and marinades, such as honey and barbecue, are well liked in turkey products. Smaller portions, less fat, less salt, and low cholesterol are in vogue. Poultry appetizers, finger foods, sliced luncheon meats, and center-of-the-plate, cooked, vacuum-packaged, chilled boneless breasts are very popular. New turkey products could soon include sous-vide (vacuum-sealed, cooked, refrigerated, ready-to-heat) entrées, or possibly surimi (minced washed gelled protein, usually from fish) seafood-style favorites.

Microbiological Concerns

Turkey meat quality is highest immediately after processing. The maintenance of acceptable quality depends on initial microbial levels and measures taken to minimize organism growth and prevent further contamination. The two major concerns are spoilage organisms which cause odors or off-flavors, and pathogenic organisms which may, under faulty handling such as undercooking or temperature abuse, lead to a health hazard. The cutting, deboning, handling, mixing, and packaging of turkey meat also increase possible microbial contamination and growth. (*See* **Spoilage**: Bacterial Spoilage.)

Although turkey and turkey products are refrigerated or frozen for shelf-life extension, spoilage can occasionally occur as a result of the growth and metabolic activities of specific types of bacteria. Psychrophiles, such as *Pseudomonas*, can grow at refrigerator temperatures and cause problems. Turkey has reached spoilage conditions when an off-odor becomes apparent (10^7 cells per cm^2) or when slime formation occurs (10^8 organisms per cm^2). It is important to maintain refrigeration temperatures at 0–4 °C to minimize microbial growth in turkey and turkey products.

Food pathogens are more serious than spoilage organisms because the food product may not look or smell spoiled. Some of the pathogens of concern in turkey are *Salmonella*, *Staphylococcus*, *Campylobacter*, *Listeria*, and coliforms. Recently, *Salmonella* has been a major worry for consumers, and turkey carcasses do harbor the organism. Surveys in the USA indicate that in food-borne salmonellosis outbreaks only 4.5% are caused by turkey. To prevent food-borne illness from poultry, it must be kept refrigerated and cooked properly, and cross-contamination or postcooking contamination from unclean utensils or equipment must be avoided. (*See* **Campylobacter**: Campylobacteriosis; **Listeria**: Listeriosis; **Staphylococcus**: Food Poisoning.)

Bacteria can come from many sources. At the processing plant, bacteria arrive on the feet and feathers of the birds; they are present in the intestinal contents, and can also come from the workers and the environment (air and water supplies). Bacterial problems can be minimized by following good production and manufacturing practices such as feeding clean feed and keeping the litter dry at the production site, using clean hauling equipment, filtering incoming air at the processing plant, monitoring the water supply, eviscerating carefully, chlorinating chiller water, insisting on good worker hygiene, and using an approved plant clean-up and sanitation program.

Utilization of Waste Products

Poultry processing results in large amounts of highly polluting waste waters, semisolids, and solids, which must be separated and treated before being discharged into the environment. Where practical, the use of waste products for livestock food or fertilizer reduces the overall load for disposal. Rendered poultry byproducts as an animal feed ingredient can provide 50–60% protein. Feather meal and dried blood also have value as a feedstuff. Such products must be carefully processed before being recycled as animal feed, to avoid microbial contamination.

The types of waste and byproducts differ at varying stages of processing. Manure, feathers, blood, viscera, flesh debris, grease, and cleaning water are examples of the pollutants to be treated and either used or discarded.

The methods of disposal are to a public sewer, or treatment on-site followed by disposal to a water course or to fields. Preliminary treatment of turkey-processing wastes using a coarse-solids screen separator as well as a fatty-matter trap, or chemical flocculation combined with a dissolved air flotation system, will significantly reduce the pollution potential of the effluent before discharge or further biological treatment. Based on the strength of the effluent, a secondary biological treatment uses a mixed culture of microorganisms for anaerobic digestion or aerobic treatment. The cheapest and most cost-effective options for the disposal of stabilized processing sludges arising from the biological treatment of poultry processing effluents are to spread them on the land or to discharge them at land-fill sites. (*See* **Effluents from Food Processing**: Composition and Analysis; Disposal of Waste Water.)

Nutritional Significance

Table 1 shows the nutrient composition of selected raw and cooked turkey meats, and **Table 2** lists the composition of several further-processed turkey products.

Turkey and its products have a favorable reputation as nourishing and healthy foods. The composition of raw turkey meat depends on factors such as diet, age, sex, and growth environment. Processed product nutrition (**Table 2**) is a result of incoming

meat and nonmeat ingredients as well as cooking processes and formula variations by different manufacturers. Salt and fat can vary markedly, depending on whether salt was needed for protein extraction and binding, or whether skin (with adhered fat) was added for juiciness and flavor.

Consumers recognize turkey as a good protein source. Turkey meat is easily digestible, contains all the essential amino acids, and is a good source of the B vitamins and iron. (*See* **Amino Acids**: Metabolism; **Vitamins**: Overview.)

Turkey has a relatively low fat content, and the fat is only about 30% saturated. Because the fat is highly unsaturated (more than beef and pork), it is a softer type of fat and prone to oxidation. As turkeys grow, they deposit more fat under the skin. The fat content is higher in dark meat, and protein content is greater in light meat. Moisture and some fat are lost during heating, and protein is concentrated in cooked turkey.

Turkey products are somewhat similar to other processed meat products in protein content but are higher in moisture and lower in fat and energy. This favorable nutritional profile for turkey is one of the major reasons for its increasing popularity. (*See* **Protein**: Food Sources.)

Table 1 Proximate analysis of turkey (per 100 g edible portion, wet-weight basis)

Nutrient	Breast with skin (raw)	Breast with skin (roasted)	Leg with skin (raw)	Leg with skin (roasted)	Light meat only (raw)
Water (g)	70.05	63.22	72.69	61.19	73.82
Protein (g)	21.89	28.71	19.54	27.87	23.56
Energy (kJ)	659	794	605	874	483
Lipids					
Total (g)	7.02	7.41	6.72	9.82	1.56
Saturated (g)	1.91	2.10	2.06	3.06	0.50
Unsaturated (g)	4.32	4.25	3.89	5.59	0.69
Cholesterol (mg)	65	74	71	85	60
Ash (g)	0.91	1.03	0.89	0.99	1.00
Sodium (mg)	59	63	74	77	63

Reproduced from Posati LP (1979) *Composition of Foods. Poultry Products, Raw, Processed, Prepared.* Agriculture Handbook 8-5. Washington, DC: Science and Education Administration, US Department of Agriculture.

Table 2 Proximate analysis of selected turkey products (per 100 g edible portion, wet-weight basis)

Nutrient	Ham	Roll (light)	Salami	Frankfurter	Loaf (breast)
Water (g)	71.38	71.55	65.86	62.99	71.85
Protein (g)	18.93	18.70	16.37	14.28	22.50
Energy (kJ)	538	617	823	949	462
Lipids					
Total (g)	5.08	7.22	13.80	17.70	1.58
Saturated (g)	1.70	2.02			0.48
Unsaturated (g)	2.67	4.24			0.73
Cholesterol (mg)		43	82	107	41
Ash (g)	4.23	2.00	3.42	3.53	4.18
Sodium (mg)	996	489	1004	1426	1431

Reproduced from Posati LP (1979) *Composition of Foods. Poultry Products, Raw, Processed, Prepared.* Agriculture Handbook 8-5. Washington, DC: Science and Education Administration, US Department of Agriculture.

See also: **Amino Acids**: Metabolism; *Campylobacter*: Campylobacteriosis; **Effluents from Food Processing**: Disposal of Waste Water; Composition and Analysis; *Listeria*: Listeriosis; **Meat**: Slaughter; Preservation; Sausages and Comminuted Products; **Protein**: Food Sources; **Spoilage**: Bacterial Spoilage; *Staphylococcus*: Food Poisoning; **Vitamins**: Overview

Further Reading

Bean NH and Griffin PM (1990) Foodborne disease outbreaks in the United States, 1973–1987: pathogens, vehicles, and trends. *Journal of Food Protection* 53(9): 804–817.

Cunningham FE and Cox NA (1987) *The Microbiology of Poultry Meat Products*. Orlando: Academic Press.

Mead GC (1989) *Processing of Poultry*. Essex: Elsevier Science.

Mountney GW (1976) *Poultry Products Technology*, 2nd edn. Westport, CT: AVI.

Nixey C and Grey TC (1989) *Recent Advances in Turkey Science*. Oxford: Butterworth-Heinemann.

Posati LP (1979) *Composition of Foods. Poultry Products, Raw, Processed, Prepared*. Agriculture Handbook 8–5. Washington, DC: Science and Education Administration, US Department of Agriculture.

Stadelman WJ, Olson VM, Shemwell GA and Pasch S (1988) *Egg and Poultry-Meat Processing*. Cambridge: VCH.

POWDERED MILK

Contents

Milk Powders in the Marketplace

M A Augustin and C L Margetts, Food Science Australia, Werribee, Victoria, Australia

Background

The conversion of milk into milk powder enables milk to be preserved in a convenient form for transport and for later use. The drying of milk was used initially as a means of dealing with surplus milk as it preserves the nutrients in milk. This allowed consumers to obtain a liquid milk product similar to fresh milk by reconstituting the milk powder with water. The transformation of liquid dairy products into powder has also paved the way for the effective utilization of byproducts such as skim milk and buttermilk from cream and butter processing and whey from cheese manufacture. Although skim and buttermilk powders were originally considered as byproducts of the dairy industry, these are now regarded as valuable food ingredients.

Commercial manufacture of milk powders began in the second half of the nineteenth century. Through the latter half of the nineteenth century and the early twentieth century, advances in drying technology made possible by improvements to drum and spray-drying equipment led to the growth of the milk powder industry. Drum drying was the major method used for the production of milk powders in the first half of the twentieth century. Today, spray drying is the method most widely used in the commercial manufacture of milk powders. However, drum drying is still used for the production of some specialized milk powders.

The export of milk powders from dairying countries such as Australia, New Zealand, and Europe has made possible the establishment of the recombined dairy products industry in countries that are not self-sufficient in milk. Recombining operations are increasing in South-East Asia, North Asia, Middle East, Africa, and South America, and this has led to a greater demand for milk powders.

A range of powders can be derived from milk (**Figure 1**). Today, dried milk and dairy-based powders represent a significant portion of the trade of dairy products. Milk powders are used directly by consumers and as ingredients in a range of manufactured dairy and food products. This article discusses the production and applications of traditional dried milk products and new tailor-made formulated milk powders on the market. Brief descriptions of casein and whey-based powders, although not strictly milk powders, have been included for completeness.

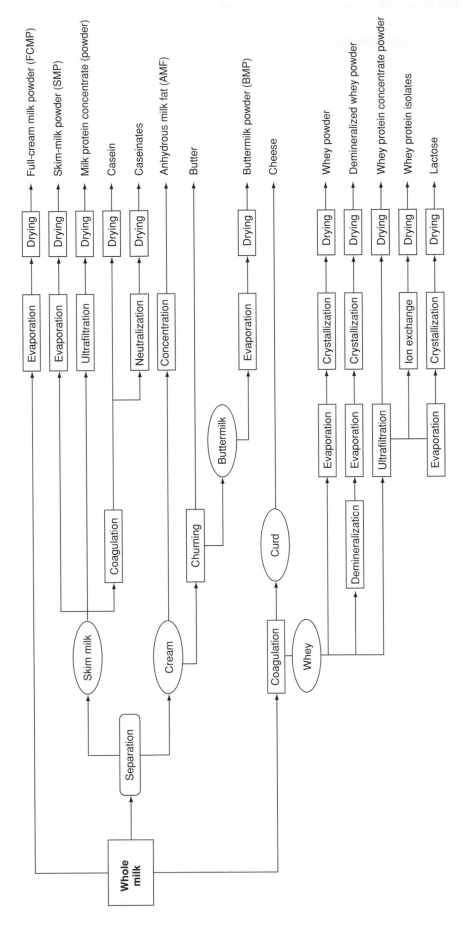

Figure 1 Milk ingredients. Reproduced with permission of the Australian Dairy Corporation.

Principal Spray-Dried Milk Powder Products

Skim, full-cream, and buttermilk powders are the traditional milk powders and are the major milk powders in the marketplace. The typical compositions of these powders are given in **Table 1**. The powders are dried to achieve specified standards for various grades of milk powders (**Table 2**).

Production of Skim, Full-Cream, and Buttermilk Powders

The main steps in the commercial manufacture of milk powders are: the preparation of the milk, heat treatment of the milk, concentration by evaporation, and spray drying. During the production of full-cream milk powder, there is an additional step, involving the homogenization of the concentrate prior to drying. **Figure 2** shows the main steps in the manufacture of milk powders. Where a powder with instant properties is required, an agglomeration step is included during powder manufacture.

Preparation of milk This step defines the final composition of the powder. For the production of skim-milk powder, the milk is first separated into skim milk and cream by centrifugal separation. The skim milk should have a fat content of < 0.1%.

For full-cream milk powders, the milk is normally standardized, generally by blending cream or skim to obtain a fat content of 25–28% in the final powder. Supplements such as vitamins and minerals may be added to the skim or full-cream milk to enhance the nutritional value of these powders. Buttermilk powders are made from the aqueous fraction remaining after churning of cream during buttermaking or during manufacture of anhydrous milkfat.

Heat treatment of milk The milk has to receive a heat treatment prior to concentration. A range of heat treatments may be used. These range from low-heat treatment (72 °C for 15 s) for pasteurization to high-heat treatments (e.g., 85 °C for 30 min or ultrahigh-temperature (UHT) at 140 °C for a few seconds). A primary purpose of the heat treatment is to achieve a required microbiological specification. The heat treatment also reduces the enzyme activity in the milk and increases the shelf-life of the powder products. Another consequence of the heat treatment is the denaturation of the whey proteins in milk. Appropriate heat treatments can be used to develop desired physical, chemical, and functional properties in milk powder products.

Skim-milk powders are generally classified on the basis of the heat treatment received during powder manufacture. Although the heat treatment classification is not used in the grading of milk powders, they are used as a general guide to the selection of powders for specific applications. The most commonly used heat classification is that of the American Dried Product Institute which is based on the amount of undenatured whey protein in the powder (**Table 3**). (*See* **Powdered Milk**: Characteristics of Milk Powders.)

Concentration of milk This is the first stage for removal of water from milk. The milk is concentrated at a low temperature by vacuum evaporation. The temperature of the milk reaches a maximum of 70 °C in modern evaporators. Generally, about 90% of the water is removed by the evaporator, as removal of

Table 1 Typical composition of milk powders

Constituent	Skim-milk powder	Full-cream milk powder	Buttermilk powder
Moisture (%)	3.0	2.25	3.0
Fat (%)	0.7	26.75	5.0
Protein (N × 6.38: %)	36.0	26.0	34.0
Lactose (%)	51.0	38.0	48.0
Ash (%)	8.2	6.0	7.9
Calcium (%)	1.31	0.97	1.3
Phosphorus (%)	1.02	0.75	1.0

Adapted from *Standards for Grades of Dairy Milks Including Methods of Analysis* (1990). Bulletin 916. American Dairy Products Institute.

Table 2 Specifications for milk powders

Constituent	Skim-milk powder[a]	Full-cream milk powder[a]	Buttermilk powder[a]
Moisture (%)	Max. 4.0	Max. 4.5	Max. 4.0
Fat (%)	Max. 1.25	Min. 26.0, max. 40.0	Min. 4.5
Protein (%)			Min. 30.0[b]
Titratable acidity (%)	Max. 0.15		0.10–0.18
Insolubility index	Max. 1.25 ml	Max. 1.0 ml	Max. 1.25 ml
Scorched particles	Not more than disc B (15.0 mg)	Not more than disc B (15.0 mg)	Not more than disc B (15.0 mg)
Bacterial estimate	Not more than 50 000 per g	Not more than 50 000 per g	Not more than 50 000 per g

[a]Extra-grade spray-dried powder.
[b]Label should specify the minimum protein content.
Adapted from *Standards for Grades of Dry Milks Including Methods of Analysis* (1990). Bulletin 916. American Dairy Products Institute.

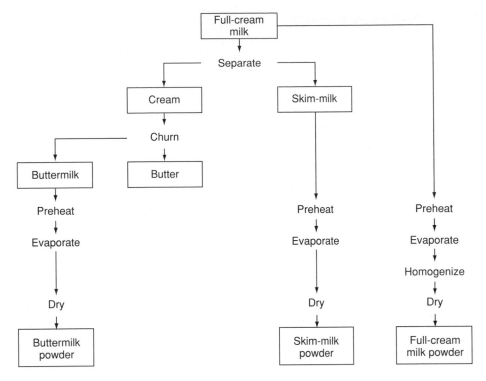

Figure 2 Schematic diagram of manufacture of milk powders.

Table 3 Heat classification of skim-milk powder

Heat class	Whey protein nitrogen index (mg undenatured whey protein N g^{-1} powder)
Low heat	Not less than 6
Medium heat	1.51–5.99
High heat	Not more than 1.5

Adapted from *Standards for Grades of Dry Milks Including Methods of Analysis* (1990). Bulletin 916. American Dairy Products Institute.

water by evaporation requires 16–20 times less energy per kg water than spray drying. The milk concentrate of about 48–50% total solids is then fed into the spray drier.

Homogenization of milk concentrate Homogenization of full-cream milk concentrates is carried out to reduce the 'free-fat' content (i.e., solvent-extractable fat) of the powder. During homogenization, the fat globule size is reduced, proteins are transferred to the surface of the globule, and the globule is stabilized. Homogenization of the milk concentrate results in powders with improved flow properties and resistance to caking and clumping.

Spray drying Spray drying transforms the fluid milk concentrate that is fed into the hot air of the drier to a powder. The moisture content of the spray-dried powder is usually 3–4% (w/w). The main stages of

spray drying are atomization of the concentrate into a spray through pressure nozzle atomizers or rotary atomizers, the mixing of the atomized particles with the hot air, the evaporation of moisture from the surface of the particles, and separation of the powder from the air. Single- and double-stage spray driers are used in milk powder manufacture. Commercial milk powders are typically dried to 4% moisture or less.

Agglomeration of milk powders For the production of milk powders with instant properties, the milk powders are agglomerated. This process yields large powder agglomerates. The wettability and dispersibility of powders are improved by agglomeration. Surfactants (e.g., lecithin) may be used during agglomeration of full-cream milk powders to improve their instant properties in cold water. (*See* **Agglomeration; Drying:** Spray Drying; **Evaporation:** Basic Principles; **Homogenization.**)

Applications of Skim, Full-cream Milk, and Buttermilk Powders

Milk powders are used in many applications. Instant milk powders, which dissolve readily in water, are used by consumers as a substitute for fresh milk and in beverage mixes. Also available in the market are a range of nutritionally enriched milk powder products that have been tailored to meet the needs of consumers at various stages of life. These

Table 4 World use of skim-milk powder in products

Product	% Used
Condensed milk	30%
Ultrahigh-temperature (UHT) fluid	26%
Icecream	18%
Cultured products and yogurts	9%
Bakery	5%
Cheese	4%
Other products	3%

Based on figures from *Dairy Foods* Jan 1999, 100(1): 15.

Table 5 Typical compositions of acid casein, caseinates, and coprecipitate

Constituent	Acid casein	Sodium caseinate	Calcium caseinate	Coprecipitate
Moisture (%)	9.0	3.5	3.5	4.0
Fat (%)	1.0	1.0	1.0	1.0
Protein (N × 6.38: %)	88.0	91.4	90.9	89.0
Lactose (%)	0.1	0.1	0.1	1.5
Ash (%)	2.2	4.0	4.5	4.5

Adapted from Chandan R (1997) *Dairy-Based Ingredients*. St Paul, Minnesota: Eagan Press.

include powders fortified with various nutrients. Most common in the market are milk powders enriched with calcium, iron, and folate.

Milk powders have major applications as ingredients in manufactured dairy and processed food products. A significant amount of milk powder is used in the manufacture of traditional recombined dairy products such as evaporated milk, sweetened condensed milk, and UHT milk in countries which do not have an adequate supply of fresh milk. Milk powders are also used as ingredients in a range of food products, including icecream, cultured milks and yogurts, chocolate, confectionery, bakery products, soups, and sauces. Buttermilk powders are used as replacers for skim-milk powder in applications where enhanced dairy flavors are desired. **Table 4** gives the proportion of skim-milk powders used in various applications. Their ability to bind water, thicken and gel, and their emulsifying and foaming properties make milk powders valuable food ingredients. These properties of milk powders can be modulated by the amount of heat treatment received by the powder during manufacture. The characteristics of milk powders that make them useful in food applications are discussed in more detail elsewhere. (*See* **Powdered Milk:** Characteristics of Milk Powders.)

Dried Casein Products

Caseins are the principal milk proteins, accounting for ∼ 80% of the proteins in milk. They can be separated or precipitated from milk. Among the major products in this class of dairy products are acid and rennet casein, caseinates, and coprecipitates.

Production of Casein Products

Acid casein is made by direct acidification of skim milk to pH 4.6 by mineral or organic acids. Rennet casein is made by adding rennet to skim milk. The casein is separated, washed, and dried. Acid casein is insoluble in water. This limits the applications of acid casein in some applications. To obtain a casein product that is soluble, the acid casein is neutralized with

an alkali and spray-dried to obtain caseinates. Various types of alkali have been used. Sodium, potassium, and calcium caseinates are the most common types of caseinates on the market. Coprecipitates are total milk protein powders; they contain both casein and whey proteins. The proteins are precipitated from skim milk under controlled conditions of heat, acid, and calcium concentration. The composition of casein products is given in **Table 5**.

Applications of Casein Products

Casein products have good emulsifying, whipping, and water-binding properties which make them suitable in applications such as desserts, confectionery, bakery products, salad dressings, processed meats, soups, and sauces. They have applications in dairy-based products such as coffee whiteners, yogurt, icecream, and processed cheese and imitation cheese products. As they are high-protein products, they are also used as supplements in dietetic foods. (*See* **Casein and Caseinates:** Uses in the Food Industry.)

Dried Whey and Whey Protein Concentrates

Whey, the serum remaining after the manufacture of cheese or casein, can be converted into powder. Sweet whey (pH ∼ 6.3) is obtained during rennet-coagulated cheese manufacture whereas acid whey (pH < 5.1) is produced during cottage cheese or casein manufacture. Acid whey has a higher mineral content than sweet whey.

Production of Whey-Based Powders

Whey is preheated gently, evaporated, and dried for production of whey powders. To facilitate drying, the lactose in the whey concentrate is precrystallized for the production of nonhygroscopic whey powders. For the production of reduced-lactose or reduced-mineral whey powders, a portion of these components can be selectively removed from the whey. Whey can also be concentrated by membrane separation (ultrafiltration, with or without diafiltration). These

Table 6 Typical composition of whey-based powders

Constituent	Sweet whey powder	Acid whey powder	Dried whey protein concentrates
Moisture (%)	3.5–5.0	3.5–5.0	3.0–4.0
Fat (%)	1.0–1.5	0.5–1.5	1–10
Protein (N × 6.38: %)	11.0–14.5	11.0–13.5	34–80
Lactose (%)	63.0–75.0	61.0–70.0	10–55
Ash (%)	9.8–12.3	9.8–12.3	4–8

Adapted from Chandan R (1997) *Dairy-Based Ingredients*. St Paul, Minnesota: Eagan Press.

processes separate the proteins from the water, lactose, and minerals. The concentrated protein solutions are then spray-dried to obtain dried whey protein concentrates with a range of protein contents, typically between 34 and 80% protein. **Table 6** gives the typical compositions of dried whey and whey protein concentrates. Whey protein concentrates with a lower protein content (25% protein) and whey protein isolates (> 90% protein) can also be produced and are on the market.

Applications of Whey-Based Powders

Dried whey, reduced-lactose and reduced-minerals whey and whey protein concentrates are used as ingredients in many food applications, including confectionery, bakery products, snack foods, yogurts, dips, desserts, meat products, pasta products, icecream, soups, sauces, beverages, and processed cheese products. Whey protein concentrates have been also used as economic egg-white replacers in food formulations. With increasing protein content, the whey protein concentrates provide greater nutritional value as well as improved functional properties such as emulsification, foaming, water binding, viscosity building, and gelling to foods. Whey protein isolates are widely used as nutritional supplements in sports drinks and health foods such as nutritional bars and protein supplements. (*See* **Whey and Whey Powders**: Production and Uses; Protein Concentrates and Fractions.)

Other Milk Powder Products

Nowadays, a range of other types of milk powders may be formulated to achieve a desired fat, protein, lactose, or mineral content, as well as to obtain target functional attributes in the powder. These include a range of cream powders, high 'free-fat' milk powders for the chocolate industry, high-heat 'heatstable' milk powders for evaporated milk, milk protein concentrate powders, skim milk/whey powder blends, lactose-hydrolyzed powders, yogurt powders, and a variety of customized formulations for use in

target food applications. These tailor-made ingredients are finding their way into the marketplace, as food manufacturers increasingly demand ingredients with enhanced performance in their applications.

The major steps in the manufacture of skim milk/whey powder blends and specialized milk powders generally include the same essential unit processes of heating, concentration, and drying that are used in conventional milk powder manufacture, except that sometimes additional processing steps are required during the preparation of the milk or milk concentrate prior to drying.

Cream and High-Fat Powders

Cream and high-fat powders, containing 40–75% fat, may be produced by spray drying of cream or milk with an increased fat content. A typical cream powder obtained from drying of cream contains ~70% fat. Cream powders were developed more than 40 years ago. The early powders were difficult to dry and handle because of their high fat content. Improvements in formulation science and drying technology have enabled cream powders with superior properties to be made. These powders are becoming more prevalent in the marketplace and can be used as an alternative to fresh cream or in formulae where milk fat is required. Some of their applications include chocolate, confectionery, icecream, desserts, soups, and sauces.

High Free-Fat Milk Powders

High 'free-fat' powders for use in chocolate manufacture have traditionally been produced by rollerdrying. In roller-dried powder, greater than 90% of the fat in the powder is readily extractable. In addition, roller-dried powders are flaky and do not contain much air. These properties of roller-dried powders make them more suitable for chocolate manufacture than conventional spray-dried powders that have a high level of entrapped air and a low freefat content. The use of high free-fat powders in chocolate reduces the requirement for cocoa butter in chocolate making. As roller-drying is more expensive than spray drying, there have been attempts to increase the level of free-fat in milk powders intended as ingredients for chocolate. Nowadays, some high free-fat powders are produced by spray drying and have improved performance compared to conventional milk powders in chocolate manufacture. (*See* **Fats**: Uses in the Food Industry.)

High-Heat Heat-Stable Milk Powders

High-heat heat-stable milk powders have been processed to insure that they are suitable for use in the manufacture of recombined evaporated milk. For the manufacture of this product, reconstituted milk powder is recombined with fat to produce a concentrate containing typically 26% total solids (18% skim-milk solids and 8% fat) and the concentrate is sterilized. A heat-stable milk powder is required in these applications, as the milk concentrate that is prepared from the powder has to withstand in-can sterilization (120 °C for 12 min) without coagulation or excessive thickening.

Milk Protein Concentrate Powders

In the production of milk concentrate powders, skim-milk concentrates obtained using a membrane process that concentrates the casein and whey proteins without precipitation are spray-dried. A range of milk protein concentrates with varying protein, lactose, and mineral contents may be made. The degree of concentration and process conditions used during membrane processing dictates the composition of the final powder. These powders are used in various applications where traditional milk powders and milk protein products have been used previously. As the protein content of these powders is higher than that of skim-milk powder, they have better functional properties.

Skim Milk/Whey Powder Blends

Skim milk/whey powder blends are made by replacing a portion of the skim-milk solids by whey-based solids. These replacers can be dry blends of skim-milk powders and whey-based solids or, alternatively, skim milk and whey may be blended prior to heat treatment, concentration, and drying. There are currently no specifications for the composition of these powders which contain a higher ratio of whey protein to casein than conventional milk powders. Blends are formulated to obtain the desired composition and functionality in the target applications, such as confectionery, bakery, icecream, and yogurts. Skim milk/whey powder blends are used as economic alternatives to skim-milk powder in many recombined dairy products. (*See* **Whey and Whey Powders: Production and Uses; Protein Concentrates and Fractions.**)

Lactose-Hydrolyzed Milk Powder

Lactose-hydrolyzed milk powders are niche products that have been developed for people who are lactose-intolerant. For the production of lactose-hydrolyzed milk powders, an enzyme, β-galactosidase (EC 3.2.1.23), that hydrolyzes lactose to glucose and galactose is added to the milk and allowed to act until the desired degree of hydrolysis is obtained; the enzyme is then inactivated by heat treatment prior to further processing. These powders may be used as alternatives to milk powders.

Yogurt Powder

A small amount of yogurt powder is available in the market. Milk is fermented prior to spraying to yield powder with a yogurt flavor. The powder has applications in confectionary as fillings and pastes, bakery products, and in a range of soups, sauces, and dips.

Milk Protein Hydrolysates

A number of milk protein, casein, and whey protein hydrolysates have been developed. These products are aimed at the nutritional supplement and sports nutrition markets. The hydrolyzed products are soluble and are quickly absorbed by the body.

Milk Powder Production Worldwide

Milk powder is the third largest category of dairy products, after cheese and fresh dairy and liquid milk products. World production of milk has continued to grow at the rate of about 1% a year through both increased cow numbers and better yields from herds. Stagnant sales of drinking milk in high-milk-production countries have meant that there has been a steady increase in manufacturing milk available for drying and milk powder production, particularly in Argentina, New Zealand, and Australia (**Tables 7** and **8**), with the last two countries exporting the majority of their production (**Tables 9** and **10**).

China and India have also shown steady increases in their production of full-cream milk powder, which has been used mostly for home consumption to shift production of milk products internally to cover areas and times of low production (**Table 8**). The

Table 7   Skim-milk powder production ('000 tonnes)

Country	1995	1996	1997	1998	1999
Argentina	36.6	36.6	40.0	38.0	81.0
Australia	229.6	237.9	230.8	272.8	275.4
Belarus	31.4	30.7			
Brazil	60.0	50.0	55.0	58.0	
Canada	71.1	64.6	66.2	69.8	77.4
Czech Republic	65.6	51.6	33.6	31.9	34.6
Estonia	19.2	21.6	16.1	14.1	
European Union (15 countries)	1187.7	1186.0	1130.4	1074.0	1113.9
Japan	189.4	200.3	199.9	201.8	191.1
Lithuania			30.1	26.4	17.4
New Zealand	127.1	172.0	177.0	160.0	227.0
Poland	118.0	120.6	119.9	131.3	110.9
Russia		107.0		79.0	83.0
South Africa	18.7	9.0	10.5	22.1	9.3
Switzerland	28.9	26.9	27.2	29.1	29.6
Ukraine	34.6	28.0	19.8	22.4	19.2
USA	564.0	477.6	548.0	515.0	625.1

Adapted from *The World Dairy Situation 2000* (2000) Bulletin 355, Table 14. International Dairy Federation.

Table 8 Full-cream and semi-skimmed milk powder production ('000 tonnes)

Country	1995	1996	1997	1998	1999
Argentina	146.5	162.3	166.0	207.0	269.0
Australia	113.0	133.7	126.5	144.8	172.0
China	352.0	358.0	391.0	422.0	552.0
Czech Republic	29.6	29.2	22.5	25.9	21.9
European Union (15 countries)	942.6	877.5	898.8	927.4	896.0
India	82.0	103.8	115.2	120.0	130.0
Japan	30.6	23.7	18.9	18.7	17.8
Lithuania	15.4	13.4	13.2	16.4	16.0
New Zealand	342.0	337.2	396.0	375.0	373.0
Poland	40.0	35.2	39.8	39.3	32.2
Russia	124.0	107.0	89.0	79.0	83.0
South Africa	11.6	9.7	13.0	10.8	11.3
Switzerland	10.3	9.9	9.3	8.0	8.0
Ukraine	11.5	6.3	4.8	3.2	3.1
USA	74.8	58.8	55.4	64.6	53.5

Adapted from *The World Dairy Situation 2000* (2000) Bulletin 355, Table 13. International Dairy Federation.

Table 9 World trade in skim-milk powder – exports ('000 tonnes)

Country	1995	1996	1997	1998	1999
World	958	1074	975	1130	1200
Argentina	22	17	15	29	
Australia	168	205	199	238	240
Canada	45	30	31	41	
European Union	227	282	175	272	370
New Zealand	127	183	166	174	157
Poland	77	112	101	83	
Other countries	270	152	177	114	

Adapted from *The World Dairy Situation 2000* (2000) Bulletin 355, Table 15. International Dairy Federation.

Table 10 World trade in full-cream milk powder – exports ('000 tonnes)

Country	1995	1996	1997	1998	1999
World	1154	1302	1392	1420	1430
Argentina	55	62	97	149	
Australia	93	109	110	139	
European Union	540	571	588	571	580
New Zealand	278	341	359	362	370
USA	16	27	20	17	20
Other countries	227	254	315	331	

Adapted from *The World Dairy Situation 2000* (2000) Bulletin 355, Table 15. International Dairy Federation.

USA and the European Union continue to be major producers and exporters of all forms of milk powder (Tables 7–10).

During the 1990s, skim-milk production increased an average of only 1.7% compared with the average annual increase in full-cream milk powder of 7%. Over this period, conventional skim-milk powder

lost market share as a livestock feed component. It also faced increased competition in the food ingredient market from full-cream milk powders and specialist dairy powders, such as whey powders and other milk components, for the production of both dairy and general food products. The demand for dairy products made from preserved milks and manufactured food items with dairy ingredients such as bakery goods, confectionery, processed meats, and beverages continued to grow in all markets.

Factors which have influenced the increased consumption and demand for a greater variety of dairy products and manufactured foods containing dairy products throughout the period include:

- an increased middle class with more discretionary income
- changes in consumption patterns as diet becomes increasingly international in style
- increased identification of dairy products as a healthy food
- newly identified applications for dairy ingredients and components
- growth in food service establishments worldwide
- growth in food-manufacturing capacity in key markets
- extended availability of dairy technology as an increasingly concentrated dairy industry moves into new markets and expands its processing capability in developing countries
- better distribution systems to support and extend the availability of dairy products in developing countries

Whilst milk production has outstripped demand for dairy products in high-production countries, the demand for dairy products and components, including powders, has continued to grow in developing areas. Key markets for imported milk powders in the 1990s included South-East Asia, Africa, and Latin America. Algeria, Japan, the Philippines, and Mexico were the world's largest importers of skim-milk powders throughout the latter parts of the 1990s, with Algeria, Brazil, Venezuela, and Saudi Arabia taking large amounts of the world's production of full-cream milk powders (Tables 11 and 12).

Economic factors, such as the financial crises suffered by Brazil, some South-East Asian and Eastern bloc countries and changing agreements on world trade policies, affected trade in milk powders through the 1990s, with downturns occurring in affected economies. Import trends (Tables 10 and 11) show the influences of these factors on changes in the sales of milk powders throughout this period. Other factors which affect the demand for milk powders and prices obtained on international markets include:

Table 11 World trade in skim-milk powder – imports ('000 tonnes)

Country	1995	1996	1997	1998	1999
World	1190	958	1074	975	1130
Algeria	108	53	40	43	
Brazil	54	34	29	24	
European Union	43	61	74	65	75
Japan	87	75	73	57	57
Mexico	107	127	130	149	
Philippines	104	79	98	90	
Russia				31	109

Adapted from *The World Dairy Situation 2000* (2000) Bulletin 355, Table 16. International Dairy Federation.

Table 12 World trade in full-cream milk powder – imports ('000 tonnes)

Country	1995	1996	1997	1998	1999
World	1165	1154	1300	1390	1420
Algeria	75	78	91	120	
Brazil	217	116	100	100	
Malaysia	60	62	65	50	
Mexico	30	30	30	20	
Philippines	36	42	52	47	
Russia				35	35
Saudi Arabia	63	69	63	64	
Singapore	30	26	26	20	
Venezuela	66	66	56	80	

Adapted from *The World Dairy Situation 2000* (2000) Bulletin 355, Table 16. International Dairy Federation.

- protectionist policies, such as subsidies and market price supports
- quotas and stockpiles in storage in producing countries
- domestic consumption by producing countries
- policies on self-sufficiency and import reduction by importing countries
- competitiveness of major exporting countries
- currency fluctuations

Future of Milk Powders in the Marketplace

Consumers and food product manufacturers recognize that milk powders have many desirable attributes. The nutritional, physical, and physiological functionality of milk powders, coupled with the ease with which they can be handled and stored, make them sought-after ingredients in the market place. This is evidenced by the growing demand for milk and dairy-based powders. As users of ingredients are placing more stringent requirements for consistent performance and enhanced functional performance in food, the trend towards tailored milk powder ingredients that are matched to specific food application sectors is expected to grow.

See also: **Agglomeration**; **Casein and Caseinates**: Uses in the Food Industry; **Drying**: Spray Drying; **Evaporation**: Basic Principles; **Fats**: Uses in the Food Industry; **Homogenization**; **Powdered Milk**: Characteristics of Milk Powders; **Whey and Whey Powders**: Production and Uses; Protein Concentrates and Fractions

Further Reading

American Dairy Products Institute (1990) *Standards for Grades of Dry Milks Including Methods of Analysis.* Bulletin 916. Chicago: ADPI.

Anonymous (1996) Milk materials multiply. *Ingredients and Analysis International* 18: 30–35.

Ashton D, Brittle S and Shaw I (2000) Dairy outlook to 2004–05. *Australian Commodities* 7: 66–72.

Chandan R (1997) *Dairy-Based Ingredients.* St Paul: Eagan Press.

Early R (1998) Milk concentrates and milk powders. In: Early R (ed.) *The Technology of Dairy Products*, 2nd edn, pp. 228–300. London: Blackie Academic and Professional.

Hall CW and Hendrick TI (1971) *Drying of Milk and Milk Products*, pp. 1–16. Westport: AVI.

Hardcastle S, Gleeson T and Topp V (1999) Dairy outlook to 2003–04. *Australian Commodities* 6: 69–78.

Hunziker OT (1935) *Condensed Milk and Milk Powder: Prepared for Factory, School and Laboratory*, 5th edn, pp. 457–527. La Grange, IL: Hunziker.

International Dairy Federation (1990) *Recombination of Milk and Milk Products*, special issue no. 9001. Brussels: International Dairy Federation.

International Dairy Federation (1999) *Third International Symposium on Recombined Milk and Milk Products.* Special issue no. 9902. Brussels: International Dairy Federation.

International Dairy Federation (2000) *The World Dairy Situation 2000.* Bulletin no. 355. Brussels: International Dairy Federation.

Knipschildt ME and Andersen GG (1994) Drying of milk and milk products. In: Robinson KR (ed.) *Modern Dairy Technology*, vol. 1, 2nd edn, pp. 159–254. London: Chapman & Hall.

Masters K (1985) *Spray Drying Handbook*, 4th edn. London: Godwin.

Pisecky J (1986) Standards, specifications, and test methods for dry milk products. In: MacCarthy D (ed.) *Concentration and Drying of Foods: Proceedings of the Kellogg Foundation 2nd International Food Research Symposium (Cork: 1985)*, pp. 203–220. London: Elsevier Applied Science.

Rasmussen KW (2000) World market for milk powder in for a boom. *Scandinavian Dairy Industry* 2/00: 46–47.

Singh H and Newstead DF (1992) Aspects of proteins in milk powder manufacture. In: Fox PF (ed.) *Advanced Dairy Chemistry – 1: Proteins*, pp. 735–765. London: Elsevier Applied Science.

Wong NP, Jenness R, Keeney M and Marth EH (1988) *Fundamentals of Dairy Chemistry*, 3rd edn. New York: Van Nostrand Reinhold.

Characteristics of Milk Powders

M A Augustin, P T Clarke and H Craven, Food Science Australia, Weribee, Victoria, Australia

Background

Milk powders are used by consumers as a substitute for fresh milk and as ingredients for the manufacture of a range of processed food products. In order to be acceptable to consumers and users of ingredients, it is essential that milk powders are of a good quality. Milk powders are manufactured to meet certain specifications and standards for composition. These have been developed for milk powders by authorities such as the American Dairy Products Institute, the International Dairy Federation, the Food and Agricultural Organization of the United Nations and national food authorities in individual countries. In addition, a range of other technical specifications have been developed for the characterization of milk powders to ensure that they have the required functional performance in specific target applications. Milk powders may be similar in composition but have different functional properties.

There are many types of milk powders in the market place. This article focuses on the characteristics of skim and full-cream milk powders, which are the major types of milk powders produced. The microbiological quality, physical and chemical attributes of these milk powders, and their functional properties are discussed. Aspects of deteriorative changes that may occur in milk powders during transport and distribution that have an impact on the sensory properties of powders and their performance as food ingredients are included. The production, composition, and applications of various types of milk powders have been discussed elsewhere. (*See* **Powdered Milk**: Milk Powders in the Marketplace.)

Microbiological Aspects

Standards for Quality and Safety

Milk powder is a microbiologically stable product. It has a water activity of 0.3–0.4, which is too low to support the growth of microorganisms. However, after milk powder has been reconstituted, it is susceptible to microbial growth and spoilage in a similar manner to pasteurized milk. Provided milk powder is protected from moisture contamination before use, the numbers of microorganisms present generally decrease during storage, although the numbers of spores may remain constant.

Although milk powder does not support the growth of microorganisms, the microbiological content is an important consideration in the subsequent use of the powder. For this reason, government bodies and customer groups have developed microbiological limits or specifications that apply to certain groups of microorganisms that may be present in milk powder. These specifications may relate to expectations of raw milk quality, hygiene during manufacture, microbial safety, or compatibility with the intended use of the milk powder.

Common end-product standards relate to the total number of bacteria (mesophilic aerobes), coliforms, *Salmonella*, and *Staphylococcus aureus*. Criteria may also be applied for *Bacillus cereus*, *Listeria*, thermophiles, Enterobacteriaceae and spore-forming bacteria. The standards developed by the International Dairy Federation, for example, are shown in **Table 1**. Many countries have either adopted these standards or developed their own local specifications based on the principles of the International Commission on Microbiological Specifications for Foods (ICMSF).

The microbiological count of milk powder is influenced by both the numbers and types of microorganisms in the raw milk and the processing conditions under which the milk powder is produced. In powders subjected to a high heat treatment, the microorganisms present will be predominantly spore-formers, belonging to the genus *Bacillus*. When heat treatment is less severe, vegetative cells

Table 1 Microbiological specifications for milk powder, as recommended by the International Dairy Federation

Criteria[a]	Total count (per gram)	Salmonella (per 25 g)	Coliforms (per gram)	Staphylococcus aureus (per gram)
m	50 000	0	10	10
M	200 000	na	100	100
n	5	15	5	5
c	2	0	1	1

[a]For a production batch, $n =$ number of samples that must be tested, $c =$ number of samples that may exceed the microbiological limit specified as m, and M is the maximum allowable microbiological limit specified for any of the samples examined.
na = not applicable.

of thermoduric bacteria will be present, with their proportion to spore-formers decreasing with the intensity of the heat treatment applied. Vegetative cells of pathogenic bacteria and Gram-negative milk spoilage bacteria are destroyed during the heat treatment.

Coliforms, *Salmonella*, and other Enterobacteriaceae are killed when the milk is heated prior to evaporation; however, they may contaminate milk powder if conditions are not sufficiently hygienic during drying. These bacteria can enter the dryer through the intake air from the factory environment, or the equipment used to dry or transport the milk powder. Cracks in dryers have been shown to be a particularly significant source of *Salmonella*. Here, *Salmonella* are harbored in the insulation material. Although dryers operate at a high temperature, the concentrated milk offers protection to the bacteria, and they will survive heating at dryer air inlet and outlet temperatures. *Salmonella* spp. are significant pathogens, and several notable outbreaks of illness have been attributed to the presence of this organism in milk powder.

Staphylococcus aureus is significant, as certain strains can produce a heat-stable toxin that is not destroyed during powder manufacture. Although *Staphylococcus aureus* is common in raw milk, it does not normally grow to produce toxin unless the milk is stored at a high temperature prior to processing. The risk of toxin production increases with temperature and storage time. Although the bacteria will be killed during the process, the toxin remains and can be detected only through specific tests. Large outbreaks of illness have been attributed to the presence of *Staphylococcus aureus* toxin in milk powder.

Another bacterium of potential significance in milk powder is *Bacillus cereus*. This is commonly found in milk, and its spores may survive heat processing. Specialty powders such as infant formula often have specifications for this bacterium, owing to the potential risk of the growth of this organism in warmed milk and sensitivity of the target group of consumers.

Sometimes, yeasts and molds or their toxins, and *Listeria* are included in powder specifications. Yeasts and molds may be significant spoilage organisms if powder is contaminated with moisture, and *Listeria* may contaminate powder from the factory environment, especially if the environment is not kept dry.

In the milk powder process, milk is subjected to heat whilst concentrated under vacuum. Such conditions are conductive to the growth of thermophilic *Bacillus* species that may form biofilms in the process lines. When this occurs, the product may be contaminated with thermophiles that can reach more than 10^6 per gram in long production runs. Thermophiles may sporulate in the process, leading to the presence of large numbers of thermophilic spores in the

powder. The spores can be extremely heat-resistant and may not be completely destroyed when the reconstituted powder is used in ultrahigh-temperature (UHT) processes. They are significant because they may cause sterility failures or spoilage in other heated products. If not properly cleaned from the plant between production runs, residues of thermophiles will seed subsequent batches of milk powder.

Although milk powder is a microbiologically stable product, the microbial quality of the raw milk may influence the shelf stability of the powder. Some bacteria present in raw milk, particularly *Pseudomonas* species, produce heat-stable spoilage enzymes, including proteases and lipases, that remain active in milk powder over many months. Experience has shown that lipase can act in full-cream milk powder to degrade milk fat to cause rancidity and other objectionable flavors. Proteases retain activity in milk powder and degrade milk proteins to cause objectionable flavors after the milk powder has been reconstituted. Proteases and lipases may be particularly detrimental in recombined milk products, or if milk powder is used to prepare UHT milk. Here, very low levels of protease and lipase may cause spoilage during long storage periods. (*See* **Bacillus**: Occurrence; Detection; Food Poisoning; **Biofilms**; **Listeria**: Properties and Occurrence; **Pasteurization**: Principles; **Salmonella**: Properties and Occurrence; **Spoilage**: Bacterial Spoilage; Fungi in Food – An Overview; Molds in Spoilage; Yeasts in Spoilage; **Staphylococcus**: Properties and Occurrence.)

Control of Microorganisms

The manufacture of microbiologically sound milk powder is dependent upon processing good-quality raw milk under hygienic conditions. To ensure the supply of good-quality milk, farm milk should be tested regularly for microbial quality. Many countries now use the total count test to monitor levels of bacteria in farm supplies. Thermoduric counts are sometimes used also. Raw milk ideally should be stored at less than 5 °C and used within 72 h of collection to minimize bacterial growth.

The pasteurization of milk is important and is normally identified as a critical control point. An example of process criteria for pasteurization would be heating of the milk for at least 15 s at 72 °C or 5 s at 80 °C.

Within the factory, application of good manufacturing practice is essential to minimize the risk of milk powder contamination with undesirable types or levels of microorganisms. To achieve this, consideration must be given to the design of the premises and control of staff or vehicular movement to separate raw materials from drying areas. Manufacturing equipment and the processing environment must be maintained, cleaned, and sanitized to ensure that microbial build-up and spread are prevented. Staff must be trained in practices to maintain high standards of hygiene. A supply of good-quality water and air for the process is also essential. Many factories now have ongoing monitoring systems for *Salmonella* and *Listeria* in place. If these bacteria are detected in the processing environment, special clean-up regimes and extra product testing are implemented.

The modern approach to ensuring that milk powder is microbiologically safe involves preventative management to ensure manufacture under appropriate conditions of hygiene. Many factories now either have in place, or are moving towards, the hazard analysis critical control point (HACCP) system. Although end-product testing is still used to verify compliance and to detect gross process failures, it cannot be relied upon to ensure the safety of a batch of product. Testing can be labor-intensive and time-consuming, taking up to 7 days to obtain final results. To overcome these problems, samples may be composited and rapid techniques for detection of pathogens based on ELISA or DNA methods applied. These methods have advanced efficiencies in testing, and product can now be cleared in 24–48 h. (*See* **Hazard Analysis Critical Control Point; Quality Assurance and Quality Control.**)

Physical Properties

The physical properties of milk powders are governed by process variables, the type of dryer, and the composition of the milk. The physical properties of milk powders play an important role in their use as food ingredients. Their ability to be readily incorporated into products and to perform specific functions in a food formulation can be influenced by many physical properties. There are many physical attributes that must be taken into account when either evaluating a current product, setting specifications for new products or designing or modifying a drying system. Tighter and more demanding specifications have meant that powders are now often manufactured on specialist dryers designed specifically to produce the best possible product of defined specification.

Moisture

The final moisture content is critical for several reasons and is therefore defined in all powder specifications. It can affect functionality and microbiological quality, and is an economic consideration in the manufacture of powders. There are several factors during manufacture that can influence the moisture content of powders. These include the characteristics of the concentrate fed into the dryer, the type of atomization used, and the operating conditions during drying.

Insolubility Index

The insolubility index of a powder is a measure of the degree to which it can be readily solubilized in water prior to use. It is related to the amount of sediment obtained under defined conditions of mixing milk powders. The main reason for loss of solubility is the temperature of the particles during the primary stage of the drying process where the majority of the moisture is removed. During this stage, an impermeable crust can form on the particle surface that severely restricts water removal, leading to the production of case-hardened particles and subsequent loss of solubility.

Bulk Density and Particle Density

Bulk density is the amount of powder by weight that is present in a defined volume. It is usually expressed as $g\ ml^{-1}$ and is obtained by measuring the volume of a fixed weight of powder after it has been tapped for a defined number of times. A high bulk density is very important in packaging and transportation, and is desirable as it can significantly reduce costs. The bulk density is influenced by a range of factors. These include the amount of air entrapped in the powder particles (occluded air), the overall density of the particle (determined by the composition), the air between the individual powder particles (interstitial air), the particle size distribution and the particle shape. The bulk density of powders is influenced by dryer design and configuration (**Tables 2** and **3**). Particle density is the density of the solids (determined by the composition), which determines the particle density, together with the amount of occluded air.

Particle Size Distribution

The individual particles produced during drying can vary greatly in size. The distribution of particle size then can be further altered by the degree of agglomeration or after grinding. An indication of the range of particle sizes obtained from different dryer configurations is given in **Table 4**.

Interstitial Air and Occluded Air

Interstitial air is the amount of air that exists between particles or agglomerates as well as the air inside porous agglomerates. The sphericity of the particles, the particle size distribution, and the degree of agglomeration determine the amount of interstitial air. To obtain minimum interstitial air, the particles need to be smooth, have a range of particle sizes, and be in compact agglomerates.

Table 2 Ability of various spray dryers to manufacture nonagglomerated and agglomerated skim milk powder with low or high bulk densities

Type of drying process	Chamber configuration	Postprimary treatment	Atomizer type	Nonagglomerated skim milk powder		Agglomerated skim milk powder	
				Low BD[a]	High BD[a]	Low BD[a]	High BD[a]
Single stage	Conventional	None	Rotary/nozzle	Yes	Yes	No	No
	Tall form	None	Nozzle	Yes	Yes	No	No
	Conventional	Cooling bed	Rotary/nozzle	Yes	Yes	No	Yes
	Tall form	Cooling bed	Nozzle	Yes	Ideal	No	Yes
Two stage	Conventional	External bed	Rotary/nozzle	Yes	Yes	No	Yes
	Tall form	External bed	Nozzle	Yes	Ideal	No	Yes
	Compact	Integrated bed	Rotary/nozzle	Ideal	Ideal	No	No
Three stage	Compact	Integrated bed + external bed	Rotary/nozzle	Yes	Yes	No	Yes
	Multistage	Integrated bed + external bed	Rotary/nozzle	Yes	Yes	Ideal	Ideal
	Integrated belt	Integrated belt	Nozzle	Ideal	No	Yes	Yes

[a]BD = bulk density; nonagglomerated skim milk powder: low bulk density ≤ 0.72 g ml^{-1}; high bulk density ≥ 0.72 g ml^{-1}; agglomerated skim milk powder: low bulk density 0.30–0.50 g ml^{-1}; high bulk density 0.45–0.55 g ml^{-1}.
Adapted from Pisecky J (1997) *Handbook of Milk Powder Manufacture*, p. 79. Copenhagen: Niro A/S.

Table 3 Ability of various spray dryers to manufacture nonagglomerated and agglomerated full-cream milk powder with low or high bulk densities

Type of drying process	Chamber configuration	Postprimary treatment	Atomizer type	Nonagglomerated full-cream milk powder		Agglomerated full-cream milk powder	
				Low BD[a]	High BD[a]	Low BD[a]	High BD[a]
Single stage	Conventional	None	Rotary/nozzle	Yes	Yes	No	No
	Tall form	None	Nozzle	Yes	Yes	No	No
	Conventional	Cooling Bed	Rotary/nozzle	Yes	Yes	No	Yes
	Tall form	Cooling Bed	Nozzle	Yes	Yes	No	Yes
Two stage	Conventional	External bed	Rotary/nozzle	Yes	Yes	Yes	No
	Tall form	External bed	Nozzle	Yes	No	Ideal	No
	Compact	Integrated bed	Rotary/nozzle	No	Ideal	Ideal	No
Three stage	Compact	Integrated bed + external bed	Rotary/nozzle	Yes	Yes	Yes	No
	Multistage	Integrated bed + external bed	Rotary/nozzle	Ideal	Yes	No	Ideal
	Integrated belt	Integrated belt	Nozzle	Yes	No	No	Yes

[a]BD = bulk density; nonagglomerated full-cream milk powder: low bulk density ≤ 0.63 g ml^{-1}; high bulk density ≥ 0.63 g ml^{-1}; agglomerated full-cream milk powder: low bulk density 0.30–0.50 g ml^{-1}; high bulk density 0.45–0.55 g ml^{-1}.
Adapted from Pisecky J (1997) *Handbook of Milk Powder Manufacture*, p. 79. Copenhagen: Niro A/S.

Table 4 Mean particle size obtained from dryers of different configuration

Powder characteristics	Dryer configuration	Particle size (μm)
Individual particles	Concurrent with pneumatic conveying	20–200
	Tall form – tower	30–250
Flakes	Roller dryer	200–5000
Loose agglomerate	Mixed flow with integrated fluid bed	100–400
– open structure	Concurrent with integrated fluid bed	100–200
Compact agglomerate	Concurrent spray dryer with integrated belt	300–2000
– porous structure	Mixed flow with integrated fluid bed	100–400

From personal communication (E. Refstrup), Niro A/S, Denmark.

Occluded air is the amount of air entrapped within the powder particles. It is affected by the preheat treatment of the original milk, with a higher pretreatment of milk resulting in less occluded air, and the amount of air incorporated in the concentrate. Higher total solids generally result in lower occluded air. Powders atomized by a nozzle contain less air than rotary atomized powders, despite improvements to the modern rotary atomizers. Gentle drying also reduces the level of occluded air, and therefore, the use of multistage dryers is recommended for the production of powders with low occluded air.

Flowability

With the ever-increasing diversity of use of milk powders today, the need for properties such as flowability is increasing. Powders are used in applications ranging from dispensing machines through to the large-scale recombining operations that utilize mechanical handling and dosing. For both agglomerated and nonagglomerated powders, a better flowability can be obtained by producing larger powder particles with smooth and rounded particle surfaces within a narrow particle size distribution. Flowability is also influenced by other factors such as total fat in the powder and the amount of 'free fat.'

'Free Fat'

'Free fat' in powder is defined as the fat fraction that is extractable by organic solvents under specific conditions of solvent type, time, and temperature of extraction. In most instances, 'free fat' is considered a defect. The exception is where 'free fat' is required for a specific application, e.g., chocolate manufacture. One of the most critical influences of 'free fat' is the moisture content of the powder. If the moisture is too low (< 2.5%), 'free fat' increases and then decreases as the moisture content is raised from 2.5 to 4–5% but increases again if the moisture content is > 6–7%.

Instant Properties

Very fine powder particles are difficult to handle and have poor reconstitution properties. Agglomeration of powders allows water to permeate the powder particles more readily, breaking up the agglomerate and allowing the individual powder particles to dissolve. Instant milk powder is highly soluble and designed to reconstitute completely in water at both hot and cold temperatures. The other properties required in instant powders are wettability and dispersibility. The wettability of a powder is measured by determining the time taken for a given amount of powder to pass through the surface of water. Wettability may be enhanced by lecithination. The dispersibility of a powder is a measure of how completely a powder dissolves under controlled conditions. Other tests carried out on milk powders related to their instant properties include slowly dispersible particles, coffee test, white flecks number (minute particles that are seen on the surface of reconstituted milk), and the sludge test.

Color

The color of a powder is determined by composition, preheat treatment, drying conditions, and particle size distribution. Scorched particles can be a visual defect that will often show up as deposits on the bottom of mixing vats and in strainers.

Other Properties

Apart from the properties described above, there are others that influence a powder's acceptability. These include the mechanical stability of the powder, which influences the degree of agglomeration breakdown during transport and storage, hygroscopicity, which is related to the degree of water attraction a powder exhibits, and cakiness, an attribute that is a measure of the extent to which a powder adheres to itself, especially under compression. (*See* **Agglomeration**; **Drying:** Spray Drying; **Rheological Properties of Food Materials.**)

Chemical Characteristics

The chemical properties of milk powders are determined by the composition of the milk and the heat treatment applied during powder manufacture.

Chemical Composition

Skim and full-cream milk powders are obtained by dehydration of skim milk and full-cream milk to ∼ 4% moisture. Full-cream milk is usually standardized to a fat:solids-nonfat ratio of 1:2.67 to meet the 26% legal minimum fat content for this powder. The protein content of skim milk powders may be standardized also. Variations in milk composition owing to factors such as cow breed, feed, stage of lactation, and season are reflected in the composition of milk powders. The American Dairy Products Institute standards for skim and full-cream milk powder compositions are as follows: skim milk powder should have a maximum fat content of 1.25% and a maximum water content of 4.0%, whereas full-cream milk powder should have a minimum fat content of 26% and a maximum water content of 4.0%. Control of the moisture content of milk powders to a maximum of 4% is essential for good shelf-life stability. **Table 5** shows the range of values observed in milk powder.

Table 5 Composition of milk powders[a]

Constituent	Skim milk powder	Full-cream milk powder
Moisture (g per 100 g)	3–5	2–4
Fat (g per 100 g)	0.7–1.3	25–28
Crude protein (g per 100 g)	35–37	25–27
Lactose (g per 100 g)	49–52	36–38
Citric acid (g per 100 g)	1.8–2.1	1.3–1.4
Ash (g per 100 g)	7.5–8.0	6.0–7.0
Sodium (mg per 100 g)	400–550	370–420
Potassium (mg per 100 g)	1550–1750	1150–1350
Calcium (mg per 100 g)	1200–1300	900–1000
Magnesium (mg per 100 g)	110–140	85–100
Phosphorus (mg per 100 g)	950–1050	700–770
Chloride (mg per 100 g)	∼1100	750–800

[a]Adapted from Walstra P and Jenness R (1984) *Dairy Chemistry and Physics*, pp. 418–419. New York: John Wiley.

Another important indicator of milk powder quality is the titratable acidity of the reconstituted powder. This is an indicator of the microbiological quality of the milk. The American Dairy Products Institute sets a maximum of 0.15% for titratable acidity of skim milk powder.

Heat-treatment Classification

The characteristics of milk powder can be influenced by the heat treatment received by the milk powder during manufacture. The time and temperature of the preheat treatment affects the level of whey-protein denaturation. The whey-protein nitrogen index, which is a measure of the undenatured whey-protein nitrogen in the powder and was developed by the American Dairy Products Institute, is commonly used to classify powders into low-heat, medium-heat and high-heat milk powders. Typical preheat treatments used for the manufacture of these powders are listed in **Table 6**. As the composition of milk, including the initial level of whey proteins in milk, can vary with season, the same heat treatment can result in a different whey-protein nitrogen index. Other methods for heat classification of milk powders, such as the heat number, cystine number, and thiol number also may be used as a measure of the heat treatment given to the milk during powder manufacture. (*See* **Heat Treatment**: Ultra-high Temperature (UHT) Treatments.)

Functional Properties of Milk Powders

When milk powders are used as ingredients in food applications, they contribute to the physical attributes of the food. The ability of milk powders to impart desirable properties to food is related primarily to functional properties of milk-protein components in the powders. These functional properties include solubility, hydration, heat stability, viscosity, gelling, foaming, and emulsifying. In milk powders, the functional properties of the milk proteins may be modulated by heat, ions, and other components. Heat treatment of milk prior to concentration and drying is the most common method used to alter the functional properties of milk powders. Milk powders with the same composition given different preheat treatments prior to concentration and drying have different functional attributes when used as ingredients.

Solubility

Solubility is a fundamental functional property that is a prerequisite for most other desired functionalities. The solubility of milk powders is dependent on pH. Proteins have a minimum solubility at the isoelectric pH, and solubility is increased on the acid and alkaline side of this pH. Caseins, the major proteins in milk, are least soluble at pH 4.6.

Hydration

Hydration is related to the ability of the milk proteins to bind or entrap water. Caseins hold about 3.3 g of water per gram, whereas undenatured whey proteins hold ~ 0.4 g of water per gram. Heat denaturation of whey proteins increases the water holding to 2.5 g of water per gram. Milk powder contains other components, such as lactose, that bind water in addition to the protein. Skim milk powders have a water sorption of 0.96–1.28 g water per gram, depending on the conditions used during powder manufacture.

Heat Stability

Heat stability is an important property in certain applications such as the manufacture of recombined evaporated milk. Single-strength milks made from low-, medium- or high-heat milk powders have a similar heat stability to fresh milk. They are heat-stable at the pH of milk (pH 6.7), being able to withstand coagulation for up to ~ 20 min at 140 °C. However, for adequate heat stability of evaporated milks under sterilization conditions (e.g., 120 °C for 12 min), high-heat milk powders are necessary. Heat stability is affected by the pH of the milk, mineral content, and other components in the milk (e.g., lecithin, urea).

Viscosity

The viscosity of milks reconstituted from milk powders is dependent on their state of dispersion, concentration of solids, and temperature. Increasing the concentration of milk solids increases the viscosity. Decreasing the temperature increases the viscosity, but heating milk to a temperature that results in denaturation of whey proteins also increases the viscosity.

Table 6 Heat classification of skim milk powder

Heat class	Whey protein nitrogen index[a] (milligrams of undenatured whey protein N per gram of powder)	Preheat treatment of milk[b]
Low heat	Not less than 6	72 °C for 15 s
Medium heat	1.51–5.99	75 °C for 3 min
High heat	Not more than 1.5	90 °C for 10 min
		120 °C for 2 min

[a]From American Dairy Products Institute (1990) *Standards for Grades of Dry Milks including Methods of Analysis, Bulletin 916.*
[b]A range of other preheating conditions may be used to achieve a desired whey protein nitrogen index.

Gelation

Milks reconstituted from milk powders have the ability to form gels under similar conditions to those required for the formation of gels from fresh milk, i.e., by rennet action for formation of rennet gels and by acidification of milk under quiescent conditions.

Foaming and Emulsifying

Milk powders can be used in applications where foaming and emulsifying properties are required. The ability of milk proteins in the milk powders to stabilize foams and emulsions may be exploited when these properties are required. (*See* Aerated Foods; Emulsifiers: Uses in Processed Foods; Mixing of Powders.)

Functional Requirements of Milk Powders in Major Food Applications

For milk powders to have the desired performance in food applications, the functional characteristics of the powders have to be matched to the application. This requires an understanding of the required functional properties of the milk powder ingredients in the target application.

Milk Powders for Recombined Dairy Products

A significant amount of milk powder is used in the manufacture of reconstituted and recombined dairy products. In these applications, the milk powders are combined with water and milkfat to reestablish the fat:solids-nonfat:water ratio of milk or other dairy products. Some of the major applications of milk powders in the recombination industry are for the preparation of pasteurized fluid milk, UHT milk, cream, evaporated milk, sweetened condensed milk, yogurt and cultured dairy products, recombined cheese, and icecream. Different functionalities of the milk powder ingredients are needed in these various recombined dairy products. **Table 7** lists the major functional requirements of milk powders for recombined dairy products.

Pasteurized milks and UHT milks These products have a similar composition to fresh milk. For pasteurized milks, low-heat or medium-heat powders are used to obtain a flavor similar to milk and to minimize heat-induced flavors. In the case of UHT milks, any type of powder can be used, as single-strength milks made from low-, medium-, or high-heat powders are stable to UHT conditions.

Evaporated milks It is essential to use high-heat powders for this application to obtain evaporated

Table 7 Functional requirements of milk powders in recombined dairy products and selected processed foods

Product	Functional properties required in milk powder	Heat treatment of milk powder
Pasteurized milk	Good flavor Emulsifying	Low–medium heat
UHT milk	Good flavor Heat stability Emulsifying	Low–medium–high heat
Cream	Good flavor Emulsifying	Low–medium heat
Evaporated milk	Heat stability Viscosity	High heat
Sweetened condensed milk	Viscosity	Low–medium heat
Yogurt	Water-binding Viscosity Gelling	Low heat[a]
Cheese	Rennetability	Low heat
Icecream	Foaming/whipping Emulsifying	Low–medium–high heat
Confectionery	Water-binding Foaming/whipping Emulsifying Heat stability	High heat
Bakery	Water-binding Foaming/whipping Emulsifying Gelling	High heat

[a]If a low-heat milk powder is used, the yogurt milk has to be given a high-heat treatment during yogurt manufacture. Alternatively, a high-heat milk powder may be used, in which case, the yogurt milk requires only a low-heat treatment to pasteurize the milk during yogurt manufacture.

milk with the desired viscosity. A high-preheat treatment improves the heat stability of a recombined milk concentrate (typically 26% total solids; 18% solids-nonfat: 8% fat) to in-can sterilization conditions used in its manufacture. Additionally, high-heat powders are screened using heat-stability tests to ensure that they withstand sterilization without excessive thickening or coagulation.

Sweetened condensed milk This is a traditional dairy product containing 74% total solids (20% milk solids nonfat: 8% fat: 46% sucrose). The most important physical attribute of this product is its viscosity. Low- and medium-heat powders are used in this application. Milk powders given a high-heat treatment (e.g., 85 °C for 30 min) cannot be easily processed, because the high viscosity of concentrates made from these powders also results in rapid age thickening during storage of the product. There are a number of viscosity tests that may be used as indicators of suitability of powders for sweetened condensed milk manufacture.

Yogurt Milk powders may be used as a partial or total replacement for fresh milk in this application. Viscosity development, gelling, and good water-binding properties are necessary for the production of high-quality yogurts. These properties are obtained in yogurt by preheating the yogurt milk at a temperature that causes significant denaturation of whey proteins (e.g., 90 °C for 10 min). Low-heat milk powder may be used if a high-heat treatment is given during yogurt manufacture. If a high-heat milk powder is used, the yogurt milk requires only a low-heat pasteurization treatment during yogurt manufacture.

Cheese Only low-heat milk powders are suitable for recombined cheese manufacture. This ensures good rennetability of the reconstituted milk. With a high-heat treatment of milk, there is association of the denatured whey proteins with the casein, which hinders the reaction of the rennet.

Ice cream Milk powders contribute to the flavor and texture of ice cream. The milk powder aids in the emulsification of the ice cream mix and has a role in the development of an aerated matrix. (*See* **Condensed Milk**; **Evaporation**: Basic Principles; Uses in the Food Industry; **Recombined and Filled Milks.**)

Milk Powders for Selected Food Applications

Milk powders are used as functional ingredients in a number of processed foods.

Chocolate and confectionery products Milk powders contribute to the flavor, color, and texture development in chocolate and confectionery applications. The emulsifying properties of the milk proteins influence the miscibility of the ingredients used in chocolate and confectionery, hence influencing flow properties and texture. In confectionery products such as toffee, good water-binding properties of milk proteins contribute to the texture of these products. The Maillard reaction, which is the reaction of the amino groups of the proteins with reducing sugars in the formulation, is responsible for color development and for the production of caramelized flavors; the lactose in milk powders participates in the Maillard browning reactions.

Bakery products High-heat milk powders are useful in bakery applications. In addition to enhancing the nutritive value of cereal-based baked goods, milk powders contribute to the texture and flavor of these products. Their emulsification and foam-stabilization properties and their ability to participate in the Maillard browning reaction are important requirements in bakery applications.

Other applications The functional properties of milk powders also make them useful in a number of other applications, such as processed meat products, soups, gravies, and dips.

Characteristics of Stored Milk Powders

The characteristics of milk powder are dependent on the quality and composition of the raw milk and the manufacturing process used during its manufacture. However, even if milk powders are manufactured to meet the desired standards and specifications, changes in the properties of milk powders may occur during storage and distribution. The composition of the powder, the type of packaging material used, and the conditions of handling and storage influence the shelf-life of the powder.

Deterioration of milk powders resulting from Maillard browning, lactose crystallization, and oxidation of fat may lead to flavor and physical defects in the powder. It may also affect the functionality of the milk powder when it is used in a food product. Some of the changes that may occur during storage include the development of a brown color, a reduction in pH, reduced solubility, development of off-flavors, and reduced heat stability of powders.

See also: **Biofilms**; **Condensed Milk**; **Emulsifiers**: Uses in Processed Foods; **Evaporation**: Basic Principles; Uses in the Food Industry; **Heat Treatment**: Ultra-high Temperature (UHT) Treatments; **Mixing of Powders**; **Pasteurization**: Principles; **Rheological Properties of Food Materials**; **Powdered Milk**: Milk Powders in the Marketplace; **Quality Assurance and Quality Control**; **Recombined and Filled Milks**

Further Reading

American Dairy Products Institute (1990) *Standards for Grades of Dry Milks Including Methods of Analysis, Bulletin 916*, rev ed. Chicago, IL: ADPI.

Early R (1998) Milk concentrates and milk powders. In: Early R (ed.) *The Technology of Dairy Products*, 2nd edn, pp. 228–300. London: Blackie Academic & Professional.

International Commission on Microbiological Specifications for Foods (1998) *Microorganisms in Foods 6, Microbial Ecology of Food Commodities*. London: Blackie Academic & Professional.

International Dairy Federation (1984) *General Code of Hygienic Practice for the Dairy Industry and Advisory Microbiological Criteria for Dried Milk, Edible Rennet*

Casein and Food Grade Whey Powders, Bulletin No. 178. Brussels: International Dairy Federation.

International Dairy Federation (1990) *Recombination of Milk and Milk Products, Special Issue No. 9001.* Brussels: International Dairy Federation.

International Dairy Federation (1991) *IDF Recommendations for the Hygienic Manufacture of Spray Dried Milk Powders, Bulletin No. 267.* Brussels: International Dairy Federation.

International Dairy Federation (1999) *3rd International Symposium on Recombined Milk and Milk Products, Special Issue No. 9902.* Brussels: International Dairy Federation.

Knipschildt ME and Andersen GG (1994) Drying of milk and milk products. In: Robinson KR (ed.) *Modern Dairy Technology*, 2nd edn., vol. 1, pp. 159–254. London: Chapman & Hall.

Masters K (1997) Spray dryers. In: Baker CGJ (ed.) *Industrial Drying of Foods*, pp. 90–112. London: Blackie Academic & Professional.

Mettler AE (1994) Present day requirements for effective pathogen control in spray dried milk powder production. *Journal of the Society of Dairy Technology* 47: 95–107.

Pisecky J (1986) Standards, specifications, and test methods for dry milk products. In: MacCarthy D (ed.) *Concentration and Drying of Foods: Proceedings of the Kellogg Foundation 2nd International Food Research Symposium (Cork: 1985)*, pp. 203–220. London: Elsevier Applied Science.

Pisecky J (1997) *Handbook of Milk Powder Manufacture.* Copenhagen: Niro A/S.

Walstra P and Jenness R (1984) *Dairy Chemistry and Physics*, pp. 418–419. New York: John Wiley.

POWER SUPPLIES

Use of Electricity in Food Technology

D Graham, R and D Enterprises, Walnut Creek, CA, USA

Introduction

The food industry in the USA consumed 17.4 billion kcal in 1999. Fifty-eight percent of total food industry energy costs are from electricity. The industry purchases 64.0 billion kWh of electricity annually at a cost of $3.36 billion. Purchased energy is 1.3% of the value of product shipments for the food industry compared with 1.7% for the total US industry.

Major uses of electricity include freezer and refrigeration compressors, conveyors, air handling, pumping, lighting, process controllers and monitors, and packaging forming and sealing. Natural gas is the primary source of thermal energy for ovens, fryers, dryers, evaporators, and boilers, with relative costs being the main reason. Closely related uses of electricity in agricultural food production include crop irrigation, pest control, produce disinfection, grain harvesting and storage, weed reduction, livestock waste management, and fish farming.

Irradiation, ohmic heating, microwave processing, ozonation, freeze concentration, nonthermal pasteurization, and the use of heat pumps are examples of relatively new electrotechnologies that the food industry may apply increasingly in the future. Application of these technologies provides opportunities for food processors to improve operating efficiencies and helps insure the quality and safety of processed food products.

Electricity Use by Food Industry Sectors

Food energy costs rose steadily over the past two decades. In 1996, purchased electricity comprised over half of the total energy used in food processing. Electrical consumption by grain milling, meat processing, preserved fruits and vegetables accounts for 23, 19, and 13%, respectively, of the total food industry use of electricity. Some food processors operate cogeneration facilities using surplus heat to produce electricity, but approximately 92% of total electricity used by the food industry in the USA is purchased from utility companies. Approximately 87% of the electricity is used by motor-driven equipment such as compressors, pumps, mixers, grinders, fans, etc. **Table 1** compares the cost of electricity with total energy costs in food processing. Electricity consumption has increased approximately 3% annually from 48.9 billion kWh in 1986 to 69.1 billion kWh in 1999. **Table 2** shows the trends in electric energy cost from 1980 to 1996.

New Electrotechnologies for Food Processing and Preservation

Electron beam irradiation, X-ray, microwave processing, membrane separation technology, ozonation, ohmic heating, high pressure pasteurization, infrared

Table 1 Electricity vs total energy used in food processing

	Electricity ($ million)	Total energy ($ million)	Electricity as a percentage of the total
Total food industry	3364	5799	58.0
Meat products	614	884	69.5
Dairy products	401	632	62.4
Preserved fruit/vegetables	459	804	57.1
Grain mill products	683	1142	59.8
Bakery products	241	426	56.6
Sugar/confectionery	177	424	41.7
Fats and oils	180	456	30.8
Beverages	345	584	59.1
Miscellaneous	464	447	59.1

Source: 1996 Annual Survey of Manufactures, US Dept. of Commerce, with permission.

Table 2 Growth of electricity use in food processing as a percentage of the total energy cost

	1980	1986	1994	1996
Total food	41.0	51.8	57.4	58.0
Meat products	51.4	59.6	69.2	68.5
Dairy products	47.3	58.8	66.1	64.3
Preserved fruit/vegetables	37.3	52.1	55.8	57.1
Grain mill products	44.0	53.0	57.5	59.8
Bakery products	43.3	47.9	55.7	56.6
Sugar/confectionery	n/a	35.9	34.9	41.7
Fats and oils	26.8	40.1	42.5	39.4
Beverages	43.1	52.1	61.1	59.1
Miscellaneous	48.0	54.8	53.8	59.1

Source: Annual Survey of Manufactures. US Dept. of Commerce, with permission.

radiation, laser, ultraviolet light, freeze concentration, and pulsed electric fields are developing technologies that offer future potential for improvements in food safety, quality, and more efficient processing.

Irradiation of Food Products

Treatment of a food product with a controlled source of irradiation can be very effective for improving the safety of the food product, since the irradiation can destroy any viable food pathogens that may be present. Food products are not made radioactive when irradiated, and no toxic byproducts are formed. Irradiation has been used successfully on a variety of products, including cereals, fruits, vegetables, prepared foods, dairy products, meat and poultry, beverages, spices, and seasonings. Products can be irradiated after being sealed in packaging material, which reduces the possibility of contamination by handling after treatment.

Safety of irradiated foods Controversy has surrounded the use of food irradiation since its discovery in the late 1940s. Much concern has resulted from fear that the irradiation process might make the food radioactive. Treatment of food with gamma irradiation does not inject radioactivity into the food. High-energy electrons pass through the food product and, if sufficient energy is applied, kill microorganisms that they impact. In approving food irradiation, the Food and Drug Administration (FDA) reviewed over 400 toxicity studies, including animal feeding studies before issuing any approvals. The FDA concluded that food irradiated up to 1 kGy is "wholesome and safe for human consumption, even where the food that is irradiated may constitute a substantial portion of the diet . . . "

Electron beam irradiation may provide an effective way to control harmful E. coli 0157:H7 and other emerging pathogens in ground beef and other processed meat products such as pork sausage and refrigerated poultry. Electron beams (gamma rays) from machine sources up to 10 million electron volts (10 MeV) and X-rays generated from machine sources up to 5 MeV are permitted commercially. Similar energetic rays also can be produced by decay from sealed units of radioactive sources such as cobalt-60 or cesium-137. Gamma rays are used effectively for sterilizing a variety of products including food packaging materials, syringes, bandages, and other heat-sensitive pharmaceutical products. X-rays also are effective in some applications. One kilogray is 1 J of energy absorbed per kilogram of the substance being irradiated. The FDA has approved electron beam radiation providing absorbed doses of radiation from 0.2 to 7.0 kGy, depending of the particular food product. **Table 3** lists the radiation levels approved by the FDA for food products.

Microwave Heating

Microwave ovens found extensive use in home food preparation and in commercial food service during the past two decades. Industrially, microwaves are used primarily for tempering, thawing large blocks of frozen meat, and preheating. Studies of microwave application for sterilization of food products have shown limited success.

Microwaves are electromagnetic waves similar to TV, radio, light, radar, or infrared waves, with the main difference being the frequency of the wave motion. Food products contain molecules such as water, salts, and proteins that have dipolar properties. Microwave energy passing through a food product causes these dipolar molecules to vibrate. The resulting internal friction produces heat. Thus, the food product is heated from within, and heating time is greatly reduced compared with heating by external application of heat. This rapid heating provides improved food quality and energy savings in many applications.

Table 3 Food irradiation processes approved by the FDA

Year approved	Food product	Dose (kGy)	Purpose
1963	Wheat and wheat flour	0.2–0.5	Disinfestation of Insects
1964/5	White potatoes	0.05–0.15	Inhibit sprouting
1983	Spices and seasonings	30 max.	Disinfestation of insects
1985	Pork, fresh unheat-processed	0.3–1.0	Control of *Trichinella spiralis*
1985/6	Dry of dehydrated enzymes	10 max.	Control of insects and pathogens
1986	Fruit	1 max.	Disinfestation and delay ripening
1986	Fresh vegetables	1 max.	Disinfestation of insects
1986	Herbs and spices	30 max.	Control of microorganisms
1986	Vegetable seasonings	30 max.	Control of microorganisms
1990	Poultry, fresh or frozen	3 max.	Control of microorganisms
1995	Meat, frozen and packaged (limited to use by NASA)	44 minimum	Sterlization
1995	Animal feed and pet food	2–25	Control of *Salmonella*
1997	Red meat, uncooked, chilled	4.5 max.	Control of microorganisms
1997	Red meat, frozen	7.0 max.	Control of microorganisms

Table 4 Microwave energy penetration

	Material temperature (°C)	Half power depth (cm)	
		915 MHz	2450 MHz
Ice	−12	1600	780
Water	2	4	0.6
Water	45	14	2
Water	75	21	3
Water	95	30	5
Beef, lean	−51	70	46
Beef, lean	−18	10	8
Beef, lean	4	2	2
Beef, lean (freeze-dried)	−18	550	190

Microwaves are in the radio frequency portion of the electromagnetic spectrum between 300 and 300 000 MHz. Microwaves are generated by a magnetron that converts electrical energy into an electromagnetic field with centers of positive and negative charges that change direction billions of times a second. In the USA, the Federal Communication Commission limits commercial microwave sources to 915 and 2450 MHz.

Microwave heating is used for cooking or partial cooking of meat, chicken, other prepared foods, rapid thawing of frozen meat, poultry, and seafood, drying of pasta, fruits, and cereal products, freeze drying of meat and juices, and, pasteurization of milk, yogurt, beer, and wine. **Table 4** shows the penetration of microwave energy into a variety of food substances at different temperatures.

Ohmic Heating

Ohmic heating describes an innovative process in which an electric current is passed directly through a food product to generate heat by internal resistance without any need for intermediate heat exchangers. The electric energy is applied from electrodes inserted into the product and is converted directly to thermal energy within the food product with an overall conversion efficiency of 90%.

Food products sterilized by ohmic heating undergo less damage to flavor and color as compared with conventional heating methods. Electrical conductivity of the food product influences the flow of electric current through the food and can cause uneven heat distribution. This has been troublesome with some prepared foods that contain ingredients of varied composition, size, and shape. Ohmic heating can process a wide variety of pumpable products including stews, dairy sauces, eggs, custards, soups, diced fruits, pie fillings, etc. Microorganisms present in the food are killed by heat, just as in conventional heat processing. The temperatures required to assure commercial sterility are identical to the temperatures required in conventional heat processing. The ohmic advantage stems from the rapid temperature increase due to internally generated heat, making conductance of heat through the food to the center of the mass unnecessary, and the 90% conversion of electrical energy to heat. Conventional heat exchangers achieve only about 50% transfer of heat to the food product. **Figure 1** shows a typical electrode assembly for ohmic heating.

A typical 75-kW commercial unit capable of heating 750 kg h^{-1} uses a 5-cm-diameter electrode housing. A 300-kW, 3000 kg h^{-1} unit requires a 10-mm-diameter housing. A basic ohmic heating system typically consists of four electrodes in series connected by short spacer tubes. The top and bottom electrodes are connected to one phase of a three-phase transformer and are grounded. Each of the two middle electrodes is connected to one of the other phases of the transformer. The voltage required for heating the product is varied by adjusting the primary side of the transformer.

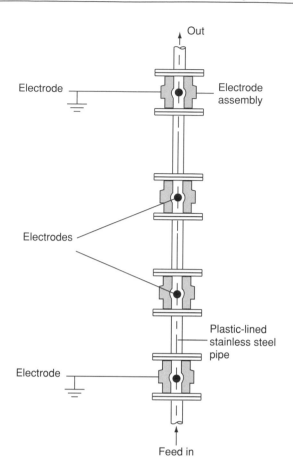

Out

Electrode

Electrode assembly

Electrodes

Plastic-lined stainless steel pipe

Electrode

Feed in

Figure 1 Typical electrode assembly. Drawing courtesy of APV.

Membrane Technology

Membrane technologies separate fluid mixtures in many different applications. They can separate materials based on size, electrical nature, physical properties, and are used widely in the food processing and chemical industries, as well as in water treatment. All membrane processes are electrically driven, using high-velocity/high-pressure pumps.

This technology originated in the early 1930s with cellulose acetate membranes, and has developed rapidly as new and improved media have been created. The arsenal of membranes now includes a wide range of polymeric membranes, ceramic, and sintered stainless steel. Among many applications are: concentration of fluid products; sterilization by removal of bacteria or viruses; removal of constituents such as proteins, acids, and fats from process streams; desalination; purification; demineralization; refining of oils; and clarification.

Membrane separation could be considered as a modern, highly efficient, versatile form of the traditional filtering process. The objective is still the same,

i.e., to remove one component such as small solid particles from a second component such as a liquid or to separate a given molecular component from a solution.

The efficiency with which separations can be made and the wide range of components that can be processed with membranes are noteworthy. Separations can be based on particulate size differences of a less than 1 μm down to a molecular level. Separation depends on the passage of specific molecules through a semipermeable membrane while the membrane retains other molecules.

In traditional simple filtration, the osmotic pressure due to differences in concentration on the two sides of a membrane, a differential pressure, and/or gravity were the only driving forces. In membrane processes, pumps apply high pressure and high cross-flow velocity to the solution being filtered to speed up the separation process. High-velocity turbulent flow at the membrane/liquid interface continuously cleans the membrane surfaces and extends their useful life. In some instances, when strongly highly ionic systems are being filtered, an electrical potential will be applied to accelerate the process and increase selectivity of the separation process.

Membrane separations typically are classified into subtechnologies based on the pore size of the filter media. Common classifications include microfiltration (coarsest), ultrafiltration, nanofiltration, and reverse osmosis (finest). **Figure 2** illustrates the tremendous range and flexibility of the filtration spectrum. Particle sizes are measured in microns and alternatively stated as Daltons. (MWCO, molecular weight cutoff).

The ability to concentrate delicately flavored solutions such as fruit juices or dairy products without the application of heat provides the food processor with the potential to significantly improve product quality. The ability to separate components selectively from a product stream or effluent stream can lead to the creation of new products, allow recycling or reclaiming of valuable ingredients, or simplify and reduce the cost of effluent disposal.

Membrane filtration has been commercially used in the USA since about 1930, primarily in the water and beverage industries. Widespread applications and use were initially limited due to the lack of practical membranes. Early membranes tended to clog easily, were hard to clean, and had short process lives. Development of new polymeric membranes, or ceramic membranes in the 1980s provided the food processor with filtering systems with greater durability, greater efficiency, far more versatility and lower costs. These membranes have a higher capacity, are less prone to clog, are easier to clean, and withstand multiple

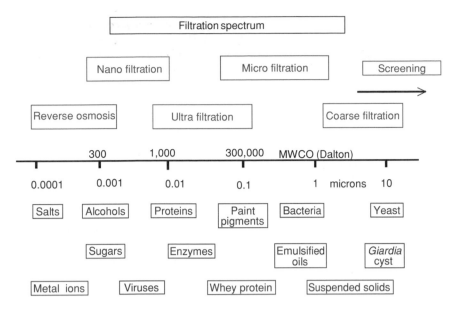

Figure 2 Membrane filtration spectrum. Drawing courtesy of J. Strasser, Process & Equipment Technology.

filtration/cleaning cycles. Many of today's routine commercial applications of membrane separation technology were only dreams two decades ago.

The cost for many types of membranes has dropped significantly in the past 10 years, e.g., from $800 to $350 for similarly sized spiral wound modules. Energy efficiencies have improved partially due to lower pressure differential required across the membranes, higher throughput and more sophisticated computer controls. These efficiencies, coupled with lower cost membranes, have made membrane separation processes not only more cost-effective but also more dependable and easier to maintain and operate.

Membrane separation technology is finding multiple industrial applications not only in food processing but also in process water and plant effluent treatment. Most food processors are very aware that water management has become a major consideration to an efficient operation. Effluent discharge requirements have become more stringent, dependable supplies of process water can no longer be taken for granted, and both effluent disposal and water supply costs have increased.

Ozone: Antimicrobial Agent

Ozone (CAS No. 10028-15-6) is a gas at ambient and refrigerated temperatures and is partially soluble in water. In the stratosphere, UV and lightening generate ozone naturally from atmospheric oxygen; ozone is generated industrially by passing oxygen through a high-voltage electric corona discharge field.

Ozone (O_3) is a molecule composed of three oxygen atoms, in contrast to ordinary oxygen (O_2), which has only two oxygen atoms. Because of its atomic structure, ozone is an unstable gas that quickly decomposes into ordinary oxygen, especially in water. This tendency to decompose naturally makes ozone an important chemical. As ozone changes into oxygen, the extra oxygen atom splits off from each ozone molecule. These free oxygen atoms have two important characteristics: they are toxic to microorganisms, and they oxidize many chemical compounds, usually changing them into nontoxic substances.

Ozonation is used extensively for sanitizing bottled water, and may become widely used as an antimicrobial agent in processing foods for human consumption. Ozone is a powerful oxidant, can be dissolved in water, and can be used effectively to kill harmful microorganisms such as *Listeria*, *E. coli*, *Salmonella*, and other pathogenic bacteria, viruses, fungi, and cyst organisms such as *Giardia*. Ozone degrades to oxygen in a few minutes and thus leaves no chemical residue like chlorine, iodine, or other common disinfectants. Current food uses in the USA include aqueous ozone for cleaning sausage curing racks, washing apples, garlic cloves, strawberries, and fresh-cut produce including celery and lettuce; gaseous ozone is used in dry storage of onions, garlic, potatoes, and citrus products.

A schematic of a typical corona discharge system used to generate ozonated water from dry air or oxygen is shown in **Figure 3**. **Figure 4** shows an ozone system installed in a food processing plant. In the foreground is a pressure swing absorption (PSA) air dryer. The sealed cabinet in the corner houses the ozone generator; the cabinet to the right contains the electronic operating controls for the system.

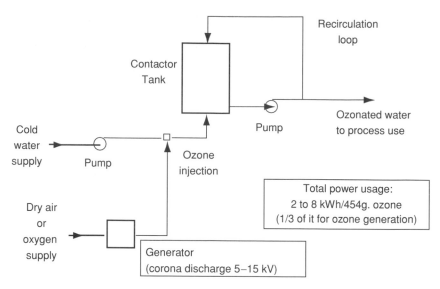

Figure 3 Typical system for generating ozonated water.

Figure 4 Ozone system installed in a grain-processing plant. Photo courtesy of RGF Environmental Systems.

Freeze Concentration

Separation is an important step in every food processing operation. Evaporators heated with steam produced from coal or gas fired boilers is the dominant separation process used in the food industry. Freeze concentration (FC) technology is an electrically driven alternative for conventional evaporation technologies. The dairy industry is the food industry's largest user of energy for evaporation. FC offers both energy and economic savings to dairy processors using this innovative freeze concentrate technology. In addition, the food processing industry and the environment benefit by shifting from fossil fuels to more environmentally clean electrical energy.

FC is a process for removing water from food by crystallizing the water and mechanically separating ice crystals from the unfrozen liquid medium. The FC concentration process has been installed in more than 100 locations throughout the world. It has helped improve the quality and reduce the operating costs for concentrating orange juice, grapefruit juice, mandarin juice, coffee extract, vinegar, and ice beer, and to remove toxic organic residues from problematic process waste streams in the chemical and refining industries.

FC stemmed from fundamental research at the University of Eindhoven in 1960, and the first prototypes were developed by Grasso/Grenco in The Netherlands from 1970 to 1975. Early commercial applications were mainly for the production of soluble coffee in 1975 and orange juice in 1980. Pioneering work by the Electric Power Research Institute (EPRI), the Dairy Research Foundation (DRF), and Niro Process Technology successfully applied the freeze concentration process to milk and milk products. Freeze concentration yields milk products with a remarkably improved product quality, no heat degradation, lower operating costs, reduced waste burden, and a sustained continuous operation.

Freeze concentration provides several unique advantages over traditional thermal concentration technologies. FC products have a better product quality with no heat degradation and are clearly superior in taste, color, and functionality compared with other commercial concentrates. In addition, the improved operating efficiencies reduce the processors' bottom line cost. Freezing consumes 80 kcal (335 kJ) per kilogram of water made into ice versus 526 kcal (2200 kJ) per kilogram of water converted to steam. The bottom-line additional cost of FC concentrate is no more than 6 cents per kilogram of milk solids higher than the cost of the best thermal evaporators of $2.05 to $2.20 kg^{-1}.

How freeze concentration works The process works by pumping the milk through a scraped surface heat exchanger to form crystals. It is then fed to the recrystallizer, where the small crystals are mixed with a population of larger ice crystals. The larger crystals are grown through a ripening process. A continuous supply of small crystals from the scraped surface heat exchangers provides the fuel for this ripening growth. The ice crystals need to be removed to complete the concentration step. The wash column provides the most efficient method to remove these ice crystals and minimize the loss of product. **Figure 5** provides a simplified schematic of the process.

The ripening process takes advantage of the equilibrium temperature at which crystals form. The equilibrium temperature of the solution is when crystals neither grow nor melt. When placed in the same concentration solution, small crystals have a slightly lower equilibrium temperature than larger crystals. The bulk solution temperature of a mixture of small and large crystals will be somewhere between the

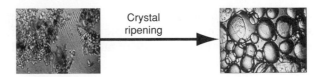

Figure 6   Ripening of crystals.

equilibrium temperature of the small and large crystals. The smaller crystals will be in a warmer environment and will melt, and the larger crystals will be in a colder environment and will grow. The heat of crystallization is exactly balanced by melting and growing crystals. The driving forces are therefore very small, producing spherical crystals, as shown in **Figure 6**.

Once crystals are formed in the solution, the remaining liquid is concentrated in the solute(s) contained in the original feed because a portion of the original water is now in the form of ice crystals. The ice crystals need to be removed to complete the concentration step. The wash column provides the most efficient method to remove these ice crystals and minimize the loss of product.

Other Developing Electrotechnologies

Ultra-high-pressure processing Pressure can kill microorganisms in food products at room temperature with little or no damage to the food product. The most promising results to date have been with acidic liquid foods such as orange juice. Ultra-high-pressure processing, also called high-pressure processing (HPP) can extend the shelf-life of food products without the application of heat or chemical treatment.

In HPP, food products are exposed to pressures in the range of 103 422–620 000 kPa for a few minutes. Spores and common food enzymes appear to tolerate high pressures and are not destroyed.

Pulsed electric fields Pulsed Electric Field (PEF) is a unique nonthermal method of inactivating microorganisms, including many of the common food pathogens, without heating the product to the usual pasteurization temperatures. The destruction or inactivation of the microorganism is achieved by the breakdown of the microorganism's cell membranes during exposure to electric fields.

PEF applies multiple short pulses (1 µs each) of high-intensity electric fields (20–50 kV cm^{-1}) between two electrodes. The food product being treated flows between the two electrodes and is exposed to the electric field. The degree of treatment is adjusted to the characteristics of the food product being processed by:

- exposure time (related to flow rate and fluid volume of the electrode chamber);

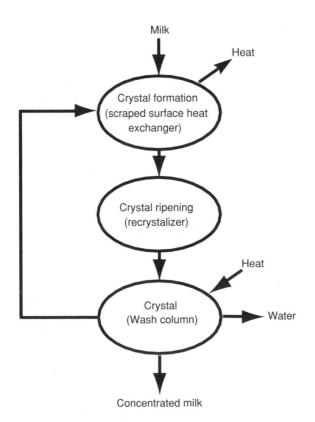

Figure 5   Overview of the freeze concentration process.

- frequency of the pulses;
- intensity (kV cm^{-1}) of the electric field; and
- shape of the pulse wave (rate of build and decay of the electrical intensity).

Even though this technology has been applied only recently in the USA, it was described in German Patent 1,237,541 issued in 1960. PEF can be an effective pasteurization process, but the FDA has expressed concerns about the consistency of some of the microbial destruction data. These apparent inconsistencies may be due to the effect of the growth phase, type of organism, media, and pH. PEF presently is *not* recommended as a food sterilization method. Promising test results have been obtained with liquid eggs, milk, emulsions, juices, rice wine, some sauces, fermented products, and water.

See also: **Convenience Foods**; **Crystallization**: Basic Principles; **Effluents from Food Processing**: On-Site Processing of Waste; **Filtration of Liquids**; **Freezing**: Operations; **Irradiation of Foods**: Applications; **Membrane Techniques**: Applications of Reverse Osmosis; **Pasteurization**: Pasteurization of Liquid Products; **Radioactivity in Food**

Further Reading

A Century of Food Science (2000) Chicago, IL: Institute of Food Technologists.

Castellari M, Arfelli G, Riponi C and Amati A (2000) High hydrostatic pressure treatments for beer stabilization. *Journal of Food Science* 65(6): 975–977.

Gerber J (1998) The electric farm. *EPRI Journal* 23(3): 8–17.

Graham DM (1997) Use of ozone for food processing. *Food Technology* 51(6): 72–75.

EPRI (1993) *Electroconductive (Ohmic) Heating for Continuous Sterilization of Solid–Liquid Food Mixtures.* Palo Alto, CA: EPRI.

EPRI (1997) *Pulsed Electric Field Technology.* Palo Alto, CA: EPRI.

EPRI (1999) *Resource Guide for Food Irradiation.* Palo Alto, CA: EPRI.

EPRI (2000) *Ozone and UV for Grain Milling Systems.* Palo Alto, CA: EPRI.

EPRI (2000) *Food Industry 2000: Food Processing Opportunities, Challenges, New Technology Applications.* Palo Alto, CA: EPRI.

EPRI (2001) *Freeze Concentration of Dairy Products: Phase 3 and Phase 4.* Palo Alto, CA: EPRI.

Kim JG and Yousef AE (2000) Inactivation kinetics of foodborne spoilage and pathogenic bacteria by ozone. *Journal of Food Science* 65(3): 521–528.

McDonald CJ, Lloyd SW, Vitale MA, Petersson K and Innings F (2000) Effect of pulsed electric fields on microorganisms in orange juice using electric field strengths of 30 and 50 kV/cm. *Journal of Food Science* 65(6): 984–989.

Marcotte M, Trigui M, Tatiboquet J and Ramaswamy HS (2000) An ultrasonic method for assessing the residence time distribution of particulate foods during ohmic heating. *Journal of Food Science* 65(7): 1181–1186.

Marquez VO, Mittal GS and Griffiths MW (1997) *Journal of Food Science* 62(2): 399–401.

Moore T (1994) A separable feast. *EPRI Journal* 19(6): 17–23.

Saldo J, Sendra E and Guamis B (2000) High hydrostatic pressure for accelerating ripening of goat's milk cheese: proteolysis and texture. *Journal of Food Science* 65(4): 636–640.

Vega-Mercado JR, Powers GV, Barbosa-Canovas G and Swanson BG (1995) Plasmin inactivation with pulsed electric fields. *Journal of Food Science* 60(5): 1143–1146.

Prader–Willi Syndrome　*See* **Developmental Disabilities and Nutritional Aspects**: Down Syndrome; Prader–Willi Syndrome

---

Prawns　*See* **Shellfish**: Characteristics of Crustacea; Commercially Important Crustacea; Characteristics of Molluscs; Commercially Important Molluscs; Contamination and Spoilage of Molluscs and Crustaceans; Aquaculture of Commercially Important Molluscs and Crustaceans

PREBIOTICS

M B Roberfroid, Université Catholique de Louvain, Louvain-la-Neuve, Belgium

Background

The microbial flora that live symbiotically within the large intestine represent an essential element of the physiology of most mammals. The flora constitute a complex ecosystem that needs to be correctly fed in order to maintain a well-balanced composition in which the health-promoting bacterial species quantitatively predominate over the potentially harmful species. Foods that resist hydrolysis by the digestive enzymes and are not absorbed in the upper part of the gastrointestinal tract, including the small intestine, become substrates for the colonic microflora. These dietary components pass into the large bowel, where most of the indigenous intestinal microflora are located in a healthy human individual. A wide variety of dietary carbohydrates, especially resistant starch, dietary fiber, some polyols and nondigestible oligosaccharides, have such characteristics, and they provide quantitatively the majority of colonic food, i.e., dietary components available for bacterial fermentation in the colon. The colonic fermentation of such 'malabsorbed,' 'non-digestible,' or 'resistant' carbohydrates (oligosaccharides and polysaccharides) plays a role in salvaging part of the energy of these dietary components, in controlling transit time, stool bulking, and stool frequency, in influencing nutrient, especially mineral, bioavailability, in producing short-chain fatty acids that are known to play physiological roles such as control of mucosal motility and epithelial cell proliferation, or in modulating immune activity and endocrine functions. Amongst colonic foods, it has been shown recently that some components are especially beneficial to health. These compounds have been called 'prebiotics,' defined as 'nondigestible food ingredients that beneficially affect the host by selectively stimulating the growth and/or activity of one or a limited number of bacteria in the colon.' Prebiotics are thus colonic foods, but they are more than simply malabsorbed, nondigestible, or resistant carbohydrates; when they reach the large bowel, they have specific effects that promote the growth of advantageous species of bacteria to the detriment of adverse species. The inclusion of prebiotics in the diet can lead ultimately to a marked change in the composition of the colonic microflora, e.g., by selectively stimulating the growth of bacteria that are generally recognized as being beneficial for health and, at the same time, reducing the number of potentially harmful bacteria. The most efficient prebiotics identified today stimulate the growth of bifidobacteria, and sometimes lactobacilli, whilst reducing the numbers and activities of potentially pathogenic organisms. (*See* **Bifidobacteria** in Foods; Lactic Acid Bacteria.)

The contents of the human gut are not readily accessible for microbiological analysis; therefore, demonstration of changes in the composition of the fecal microflora is often used as a surrogate marker for the prebiotic effect. However, for such a demonstration to be convincing, it is critical that as many components of the fecal microbiota as possible are measured. These should include at least bacteroides, bifidobacteria, clostridia, eubacteria, Gram-positive cocci, coliforms, lactobacilli, total aerobes, and total anaerobes. Simple stimulation of growth of bifidobacteria and/or lactobacilli is insufficient to substantiate a prebiotic property without determining the effects on other fecal microorganisms, since it is the selectivity of effect that determines classification as a prebiotic. Clearly, studies using pure bacterial cultures are of very limited, if any, value in this respect, unless they are supported by mixed culture work in a well-validated set-up. But the ultimate proof must come from human studies in which correctly collected and stored fecal samples are analyzed for their composition in terms of bacterial species that are further well characterized using classical techniques combined with a conventional microbiological approach towards identification and/or modern molecular genotyping methods. Indeed, it is the effect of the prebiotic in a competitive ecological environment that is important.

The Prebiotics

At present, the food components for which a prebiotic effect has been reported are nondigestible carbohydrates that consist of mixtures of oligomers of different chain lengths and are characterized by the average number of osyl moieties, referred to as the 'degree of polymerization' (DP). They have been classified as 'nondigestible oligosaccharides,' or NDOs, the osidic bond of which is in a spatial configuration that confers resistance to the hydrolytic activities of the digestive enzymes of the upper gut. But once they reach the large bowel, the NDOs are fermented by at least some of the colonic bacteria. This fermentation produces gases which are excreted, and short-chain fatty acids, which may be absorbed by the host and thus represent a source of energy. The NDOs affect the growth and, ultimately, the selective proliferation of these bacteria. Currently, there are few data on the relative efficiencies of different prebiotic oligosaccharides, or on their selectivity at a species, or even genus, level; however, the newly developed methodologies based on hybridization with specific rRNA probes will help us to make progress in that direction. It is often the case that a prebiotic effect is claimed for certain NDOs, or other dietary carbohydrates, without a full and careful investigation of their fermentation profile. Such data can never be accepted as a proof for a prebiotic attribute. (*See* **Carbohydrates:** Classification and Properties.)

Among the nondigestible oligosaccharides, those composed primarily of fructose units occupy a leading position in food science. Fructooligosaccharide is used as a generic name for all nondigestible oligosaccharides composed mainly of fructose. Strictly, however, the linear β-(2–1)-fructans are inulin-type fructans, which are different from levans, which are β-(2–6), often branched, fructans. The inulin-type fructans are by far the most extensively studied compounds and are clear market-leader prebiotics. They are composed of β-D-fructofuranose moities attached by β-2–1 linkages. The first monomer of the chain is either an α-D-glucopyranosyl or a β-D-fructopyranosyl residue. They constitute a series of homologous oligosaccharides derived from sucrose and may be represented by the formula GF_n or FF_n (where G = glucose and F = fructose). The naturally occurring NDOs, which are extracted from the roots of the chicory plant (*Cichorium intybus*), are a mixture of either GF_n (α-D-glucopyranosyl-[β-D-furanosyl]$_{n-1}$-D-fructofuranoside) or GF_n + FF_n (β-D-fructopyranosyl-[β-D-fructofuranosyl]$_{n-1}$-D-fructofuranoside) molecules, with the number of fructose units varying from 2 to ~60–65 units. As food ingredients, they are available as native inulin (inulin ST, average DP 10) and high-molecular-mass inulin (inulin HP, average DP 20), enzymatically hydrolyzed inulin (oligofructose, average DP 4), and a mixture of inulin HP and oligofructose (synergy 1), all of which occur naturally in a variety of food plants such as garlic, onion, asparagus, artichoke, banana, and wheat. Oligofructose can be produced by enzymatic conversion of sucrose to give a mixture composed of only GF_2, GF_3, GF_4 (average DP 3.8), sucrose, glucose, and fructose. The galactooligosaccharides, or transgalactooligosaccharides, are a potentially important class of prebiotics; they are produced industrially from lactose by transglycosylation reactions and consist of galactosyl derivatives of lactose with β1–3 and β1–6 linkages. The purported bifidogenic (i.e., stimulation of growth of bifidobacteria) nature

of fructooligosaccharides and galactooligosaccharides is explained, at least in part, by the linkage specificity of the bifidobacterium β-fructosidase and β-galactosidase enzymes, respectively. These enzymes are cell-bound.

The glucose-based maltooligosaccharides and xylooligosaccharides are candidate prebiotics; however, specific bacterial enzymes for the degradation of these molecules have not been identified.

Malabsorption of the Nondigestible Oligosaccharides

The β configuration of the anomeric C-2 in the fructose moieties of the inulin-type fructans as well as in the galactose monomer subunits of galactooligosaccharides explains why they are resistant to hydrolysis by human digestive enzymes (α-glucosidase, maltase-isomaltase, and sucrase), which are mostly specific for α-osidic linkages. In normal physiological conditions, the inulin-type fructans and the galactooligosaccharides are resistant to acid hydrolysis in the stomach. The most convincing data have been obtained from human intervention studies with ileostomy volunteers. These studies show that 86–88% of the ingested dose (10–30 g) of inulin or oligofructose is recovered in the ileostomy effluent, supporting the claim that these carbohydrates are practically nondigestible in the small intestine of man. Using an intubation technique in human volunteers, it has been concluded that fructooligosaccharides are malabsorbed in the human small intestine (89% recovery). The small, but still significant, loss of fructooligosaccharides in the upper part of the gastrointestinal tract could be due to fermentation by the microbial population colonizing the ileum, especially in ileostomy patients, and/or to hydrolysis of the lowest-molecular-mass oligomers such as the enzymatically synthesized oligomers. In a recent review of the malabsorption characteristics of inulin-type fructans, the authors concluded that 'inulin and oligofructose pass through the small bowel without degradation and (furthermore) without influencing the absorption of nutrients and minerals especially calcium, magnesium and iron.'

There are fewer data on the resistance of other oligosaccharides to digestion in the human upper gastrointestinal tract than for the inulin-type fructans. The available evidence comes predominantly from *in vitro* experiments or is based on hydrogen production and exhalation as an indirect marker of colonic fermentation, or on stimulation of growth of specific fecal microorganisms in animal models. No *in vivo* human data are available. Thus, the nondigestibility of isomaltooligosaccharides, soybean oligosaccharides, galactooligosaccharides, palatinose condensates, or xylooligosacchairdes *in vivo* remains to be demonstrated.

Fermentation in the Large Bowel: the Prebiotic Effect

The large bowel is by far the most heavily colonized segment of the human gastrointestinal tract, with up to 10^{12} (mostly anaerobic) bacteria per gram of gut content. These bacteria belong to a wide variety of genera, species, and strains. Through the process of fermentation, these colonic bacteria produce a wide variety of metabolites, among which the short-chain fatty acids represent salvage of part of the energy of malabsorbed food components, especially malabsorbed carbohydrates, and they play important systemic physiological roles.

Evidence for the fermentation of inulin-type fructans by bacteria colonizing the large bowel has come from *in vitro* and *in vivo* studies. At nutritional doses (up to 20–40 g per day), these malabsorbed carbohydrates are fermented quantitatively and are not excreted in the feces; the products of fermentation are gases and short-chain fatty acids, mainly acetate, butyrate, and propionate. Compared with most other malabsorbed carbohydrates (e.g., resistant starch and dietary fiber), the colonic fermentation of inulin-type fructans is accompanied by a significant change in the composition of the colonic microbiota due to selective proliferation of bifidobacteria and a concomitant reduction in the number of other bacteria, like bacteroides, fusobacteria, and pathogenic clostridia. On the basis of the results of well-designed human studies that have shown significant changes in the composition of human fecal flora, it can be concluded that inulin-type fructans (5–15 g per day for a few weeks) are prebiotic. But even though some studies showed a significant reduction in the number of pathogenic clostridia, the health benefits (e.g., reducing the risk of intestinal infections) of such a change in the composition of the colonic microbiota have yet to be established. A recent report has shown that oligofructose (daily dose of 6 g (3×2 g)) had no therapeutic value in patients with irritable bowel syndrome. But in an experimental model of necrotizing enterocolitis in quails, data have been reported that support the hypothesis that oligofructose might prevent overgrowth of the bacteria known to play a role in this pathology in preterm neonates.

For the other NDOs, studies *in vivo* have been performed with doses ranging from 3 to 15 g per day, given for periods of 1, 2, or 3 weeks. For soybean oligosaccharides, a dose of 10 g given twice daily for 3 weeks significantly increased the number of bifidobacteria, whilst slightly decreasing clostridia counts. A dose of 3 g per day increased bifidobacteria, bacteroides, and eubacteria. For the galactooligosaccharides, an increase in bifidobacteria and lactobacilli in response to doses ranging from 3 to 10 g per day has been reported. Similarly, a daily dose (5 or 10 g) of galactosylsucrose stimulated the growth of bifidobacteria after 1 and 2 weeks of ingestion. A dose of isomaltooligosaccharide of 13.5 g per day for 2 weeks significantly increased the number of bifidobacteria in adult and elderly volunteers. Early results suggest that platinose condensate may stimulate the growth of bifidobacteria. For all these NDOs, except the galactooligosaccharides, only a single human intervention study has been performed, and this will need to be repeated before any prebiotic effect can be substantiated. Moreover, as discussed above, great care should be taken to quantify the component species of the fecal microflora and to identify changes in its composition.

Physiological Effects in the Gastrointestinal Tract

The fermentation of prebiotics in the colon has a series of consequences that affect large-bowel physiology. Firstly, it contributes to the production of short-chain fatty acids, thus creating a more acidic environment, which is favorable for the development of bacteria like bifidobacteria or lactobacilli but unfavorable for the growth of potentially pathogenic species like *Clostridium* spp. or *Escherichia coli*. Furthermore, in this acidic environment, ammonia and amines become protonated and are thus much less absorbable and hence more readily excreted. Secondly, it leads to a proliferation of colonic bacteria, which increases fecal mass and thus contributes to a beneficial bulking effect and a regularization of stool production. As a consequence of these two large-bowel processes (i.e., production of acids and proliferation of bacteria), only part of the metabolizable energy of the NDOs is salvaged. As compared with absorbed carbohydrates (e.g., nonresistant starch or sucrose), NDOs represent, weight for weight, a lower energy value for the host. Caloric values of $1–2.1 \, \text{kcal g}^{-1}$ ($4.1–8.8 \, \text{kJ g}^{-1}$), i.e., 25–50% of the caloric value of sucrose, have been reported, especially for inulin-type fructans. But, as stated recently by a group of European experts, 'all carbohydrates that are more or less completely fermented in the human colon should be given a caloric value of $1.5 \, \text{kcal g}^{-1}$ ($6.3 \, \text{kJ g}^{-1}$).' In fact, the daily intake of these carbohydrates is likely to remain small, probably often not representing more than $\sim 5\%$ of total daily energy intake. Therefore, it is not justifiable to spend a great deal of effort in trying to give, for such carbohydrates, a precise calorie value, the determination of which will depend on the protocol used and probably also on the diet in which they are included.

Another physiological consequence of the consumption of NDOs that has been reported recently for the inulin-type fructans is an increased bioavailability of Ca^{2+}. Such an effect has been studied extensively in rat and hamster. These studies have led to the conclusion that, most probably because of their malabsorption and colonic fermentation, these food ingredients facilitate Ca^{2+} absorption from the large bowel compartment, thus complementing the process that takes place in the small bowel. A change in colonic pH, production of short-chain fatty acids, and increase in mucosal concentration of the protein calbindin in the colon have been proposed as hypotheses to explain that effect. Besides increased Ca^{2+} bioavailability, it has been shown both in rat and in hamster that feeding fructooligosaccharides increases Ca^{2+} concentration and improves the structure and density of the bones. More recently, three human trials (two in adolescents and one in adults) have shown that supplementing the diet every day with 16.8 g of oligofructose, 8 g of a mixture of oligofructose and high-molecular-mass inulin, or 40 g of inulin significantly increases the apparent absorption of Ca^{2+} by 10–12%. One of these studies used measurement of Ca^{2+} balance, and two used a double stable isotope technique.

NDOs, because they are malabsorbed and fermented in the colon, may be considered to be part of the dietary fiber complex. In particular, it has been shown that inulin-type fructans have a fecal bulking effect that is comparable on a weight-for-weight basis with that of soluble fiber such as pectin. Moreover, an internationally validated method derived from the AOAC method for dietary fiber analysis has been developed to quantify inulin and oligofructose in plants and food products. For the purpose of food labeling, the NDOs are thus classified as dietary fiber in most countries. (*See* **Dietary Fiber**: Properties and Sources; Determination; Physiological Effects.)

Prebiotics and the Risk of Colon Cancer

Over the last two years, experimental reports have been published that repeatedly demonstrate that feeding inulin-type fructans to rats previously treated with a colon carcinogen (i.e., dimethylhydrazine or azoxymethane) reduces the incidence of the so-called aberrant crypt foci in the colon. In one of these studies, the synbiotic approach that combines oligofructose (prebiotic) and bifidobacteria (probiotic) was reported to be more active than either the prebiotic

or the probiotic alone. Even though still experimental, these data suggest that inulin-type fructans might play a role in reducing the risk of developing preneoplastic lesions and possibly cancer in the colon. Moreover, such an effect might not be limited to the inulin-type fructans. Indeed nondigestible oligosaccharides may exert anti-carcinogenic effects first because they have been shown to beneficially affect certain biomarkers known to be associated with cancer risk (e.g. increase in apoptotis in colonic mucosa, reduction of bacterial β-glucuronidase and nitrate reductase, pH, and conversion of a dietary carcinogen to its genotoxic metabolite in caecal contents or feces of NDOs fed rats and human volunteers respectively), and second because they stimulate the growth of lactic acid bacteria for which evidence of anti-genotoxic and anti-carcinogenic effects have been reported. (*See* **Colon**: Cancer of the Colon.)

Conclusion: Prebiotics, what Benefit(s) for Human Health?

Prebiotics have nutritional properties that, according to the present state of knowledge, derive mainly from resistance to the hydrolytic activities in the upper part of the gastrointestinal tract of monogastric organisms followed by extensive fermentation in the large bowel, which leads to significant and possibly health-beneficial changes in the composition of the colonic microbiota. The gastrointestinal target functions that are associated with a balanced microflora together with an optimal gut-associated lymphoid tissue (GALT) are relevant to the state of well-being and health and to the reduction of the risk of diseases. The colonic microflora is a complex ecosystem, the functions of which are a consequence of the combined action of the microbes that, besides interacting with the GALT, contribute to salvage of nutrient energy and yield metabolic end products like the short-chain fatty acids (SCFAs) that play a role in cell differentiation, cell proliferation, and metabolic regulatory processes. It is generally assumed that the group of potentially health-promoting bacteria includes principally bifidobacteria, lactobacilli, eubacteria, and bacteroides, which are, and possibly should remain, the most important genera in humans. Changes in the composition of the fecal flora, a recognized surrogate marker for the residual colonic microbiota, can be considered as a marker, both indicator and factor, of large-bowel functions. They might play a role in gastrointestinal infections and diarrhea, constipation, irritable bowel syndrome, inflammatory bowel diseases, and colorectal cancer. Probiotics (e.g., lactobacilli or bifidobacteria), prebiotics (like chicory inulin and its hydrolysate oligofructose), and synbiotics (a combination of probiotics and prebiotics) are recent concepts in nutrition that are being used to support the development of functional foods targeted towards gut functions. Their effects may include:

- stimulation of the activity of the GALT (e.g., increased IgA response, production of cytokines, etc.);
- reduction of the duration of episodes of rotavirus infection;
- change in the composition of the fecal flora to reach/maintain a composition in which bifidobacteria and/or lactobacilli become predominant in number, a situation that is considered optimal;
- increase in fecal mass (stool bulking) and stool frequency;
- increase in calcium bioavailability via colonic absorption (e.g., inulin).

By reference to the recently published European consensus on scientific concepts of functional foods, prebiotics, especially inulin-type fructans (but also synbiotics) are good candidates to be recognized as functional food ingredients for which claims shall become authorized. Such claims should relate to enhanced gastrointestinal functions (composition of colonic flora, bulking effect and bowel habit, Ca^{2+} bioavailability) or risk of developing a disease like colon cancer.

See also: **Carbohydrates**: Digestion, Absorption, and Metabolism; **Colon**: Cancer of the Colon

Further Reading

Gibson GR and Roberfroid MB (1995) Dietary modulation of the human colonic microbiota: introducing the concept of prebiotics. *Journal of Nutrition* 125: 1401–1412.

Gibson GR and Roberfroid MB (eds) (1999) *Colonic Microbiota, Nutrition and Health*. Dordrecht: Kluwer Academic.

Roberfroid MB (1999) Concepts in functional foods: the case of inulin and oligofructose. *Journal of Nutrition* 129 (supplement): 1398S–1401S.

Roberfroid MB and Delzenne N (1999) Dietary fructans. *Annual Review of Nutrition* 18: 117–143.

Van Loo J, Cummings J, Delzenne N *et al.* (1998) Functional food properties of non-digestible oligosaccharides. A consensus report from the ENDO project. *British Journal of Nutrition* 81: 121–132.

PREGNANCY

Contents

Metabolic Adaptations and Nutritional Requirements

D J Naismith, University of London, London, UK

Predicted Energy and Nutrient Costs

Women during pregnancy are generally believed to be nutritionally 'at risk.' There are obvious energy and nutrient costs arising not only from the growth of the fetus and placenta, but also from enlargement of the maternal reproductive tissues and the deposition of a substantial energy reserve in the form of fat. These have been determined by direct chemical analyses and by indirect measurements (**Table 1**) and have been used as the basis for predicting additional nutritional needs for a successful outcome in pregnancy.

Importance of Body Fat

The relationship between nutrition and reproduction, however, begins at an earlier stage. In late childhood, body fat accounts for about 12% of body weight in both boys and girls. With the onset of puberty, the proportion of body fat rises in girls to reach around

Table 1 Calculated energy and nutrient costs of pregnancy

Energy/nutrient	Cost
Energy (fat reserve, new tissue synthesis increase in basal metabolic rate)	335 MJ (80 000 kcal)
Protein (maternal reproductive tissues, conceptus)	910 g
Calcium (fetal skeleton)	28 g
Iron	300–400 mg

17% at menarche, and 24% in the physiologically mature woman some 12 months later. The attainment of this additional fat deposit, amounting on average to 226 MJ (54 000 kcal) of stored energy, is believed to be the major determinant of fertility in normal healthy women. Young women who, for professional reasons, must maintain a low proportion of body fat, such as track athletes and ballet dancers, are frequently infertile, and the cessation of menstruation is an early symptom of the wasting disease anorexia nervosa. Should conception occur, total body fat is calculated to be sufficient to meet the energy costs of pregnancy (excluding further deposition of fat during gestation) and to sustain a 3-month period of lactation. (*See* **Anorexia Nervosa**; **Fats**: Requirements; **Lactation**: Human Milk: Composition and Nutritional Value.)

Effects of Undernutrition

The validity of this argument was revealed in a study of the effects of acute severe malnutrition that occurred in the winter of 1944 and in the following spring in cities of the western regions of The Netherlands. Immediately before the famine, the nutritional status of the population had been generally satisfactory. Thereafter, however, the situation deteriorated rapidly, and food from all sources provided no more than 2.1–2.5 MJ (500–600 kcal) per day. The incidence of amenorrhea rose dramatically in the young female population. At the height of the famine, the number of infants conceived fell to 50% of the previous conception rate. Nevertheless, in the most vulnerable women, exposed to famine in the third trimester, the average birthweight declined to a maximum of about 300 g below the prefamine level. Since the body of a healthy newborn infant contains more than 500 g of fat, most if not all of the deficit in birthweight may have been attributable to an inability to accumulate fat, priority being given to lean tissue growth.

Acute severe malnutrition, however, occurs rarely among normally well-nourished populations, and might give a misleading picture of the interaction of undernutrition and reproductive performance. In most developing countries, chronic moderate malnutrition is commonplace, and nutritional surveys invariably report mean birthweights significantly below the average for Europe and North America. Although this has commonly been thought to reflect the poor nutritional status of the mothers, it has been impossible to isolate the putative influence of maternal diet on fetal growth from that of other environmental factors associated with endemic malnutrition, perhaps the most important being the smaller stature of the mothers. Even within an affluent society, an association between birthweight and maternal height and weight-for-height is found. (*See* **Malnutrition**: Malnutrition in Developed Countries.)

Nutrition Intervention and Pregnancy Outcome

A number of well-controlled nutrition intervention studies were carried out in developing countries in the early 1970s, no doubt with the aim of demonstrating the need for, and benefit to be derived from, diet supplementation during pregnancy. The subjects lived in countries in which protein–energy malnutrition was endemic. Energy intakes ranged from 5.86 to 7.53 MJ (1400–1800 kcal) per day. Given the small stature of the mothers, such a marginal plane of nutrition would just satisfy the needs of a nonpregnant woman. The dietary supplements provided energy alone or energy with protein, and were administered on a scale that often exceeded the estimated additional costs for energy and protein during pregnancy. The findings were as surprising as they were informative.

Only in one study, conducted in rural Guatemala, was a significant increase in mean birthweight achieved, amounting to 117 g. Although most women showed no improvement in reproductive performance, the response was very variable, those with the lowest customary energy intakes experiencing the greatest benefit, and the proportion of low-birthweight infants (< 2500 g) was consequently reduced.

It was evident from these studies that there exists an 'energy threshold,' around 6.7–7.1 MJ (1600–1700 kcal) per day, above which diet supplementation has no effect, but below which fetal growth may be stimulated, the improvement being directly related to the degree of energy deprivation. It was also revealed that the major determinant of fetal growth was the maternal energy supply; the inclusion of protein in the supplements had no effect independent of the energy it provided.

Dietary Habits in Well-nourished Populations

The results of the various intervention studies clearly indicate that if a woman is adequately nourished, then pregnancy entails no appreciable additional energy (food) cost. It could be argued that the fetus is remarkably protected from the influence of maternal malnutrition, and that women exposed to generations of deprivation have simply adapted to a limited food supply. Small stature is often cited (incorrectly) as an example of adaptation. Studies of the dietary habits of women living in affluent circumstances, however, have shown that this is not so.

As long ago as 1953, analyses of the diets of 2300 well-nourished women in Nashville, USA failed to show any association between maternal food intake and birthweight. Furthermore, on average, no increase in food consumption during pregnancy was apparent (**Figure 1**). During the second trimester, an increase of around 251 kJ (60 kcal) per day was noted, but during the last trimester, when a marked acceleration in fetal growth occurs, energy intake fell by 753 kJ (180 kcal) per day. This surprising discovery, ignored for 30 years, has been amply confirmed by studies in the 1980s in Paris, Cambridge, and Glasgow.

How, then, is the obvious requirement for increased nutrition during pregnancy (**Table 1**) to be reconciled with the mother's failure, even under the most favourable environmental circumstances, to increase food intake? Only a small fraction of the fat stored in the first two trimesters of pregnancy is mobilized

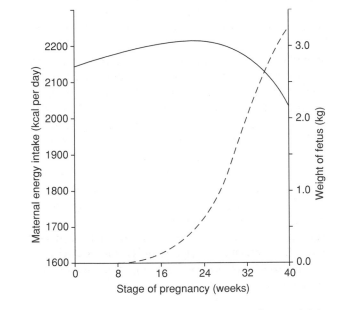

Figure 1 Diagram showing changes in maternal energy intake (—) and weight of the fetus (– – –) at different stages of pregnancy. Reproduced from Pregnancy: Metabolic Adaptations and Nutritional Requirements, *Encyclopaedia of Food Science, Food Technology and Nutrition*, Macrae R, Robinson RK and Sadler MJ (eds), 1993, Academic Press.

before term, and there exists no *ad-hoc* store of minerals and water-soluble vitamins. The answer to this question is to be found in the complex mechanisms whereby nutrients are transported from the maternal circulation to the fetus, and in the adjustments that occur in the maternal physiology in response to the gravid state.

Role of the Placenta

The role of the placenta is discussed in **Pregnancy**: Role of Placenta in Nutrient Transfer. It is primarily an organ for nutrition of the fetus, but functions also as an endocrine gland throughout the greater part of gestation, regulating nutrient utilization by the mother.

Glucose provides the energy for fetal growth, the fetus being unable to oxidize fatty acids. Transfer of glucose across the placenta is by 'facilitated diffusion,' and is therefore influenced by the maternal blood glucose concentration. A prolonged low blood glucose concentration, as may occur in mothers in the developing world engaged in hard physical work, or an abnormally high concentration associated with gestational diabetes can therefore lead to growth retardation, or to excessive growth and fat deposition (macrosomia). From early pregnancy, the fasting blood glucose concentration is reduced, a change that may favor transfer to the fetus since the placenta is particularly effective at extracting nutrients from low concentrations in the blood. Glucose tolerance following a glucose load is also progressively relaxed so that the glucose concentration remains elevated for a longer period, again facilitating placental uptake. Whether or not the metabolic response to carbohydrate consumed as part of a meal is similarly affected is unclear. (*See* **Glucose**: Maintenance of Blood Glucose Level; Glucose Tolerance and the Glycemic (Glycaemic) Index.)

Fat-soluble vitamins also cross the placenta by diffusion, maternal blood levels being maintained by dietary input and mobilization of tissue stores. In contrast, amino acids, minerals, and the water-soluble vitamins are actively transported, the concentrations in the fetal circulation being much higher than in the maternal circulation. Furthermore, there is evidence that some vitamins are chemically altered by the placenta to a form that is unable to diffuse back into the maternal bloodstream, so reinforcing the one-way transport mechanism. Thus, the fetus has prior claim on the nutrient supply, much of which is maintained, by homeostatic regulation, within narrow concentration limits in the maternal plasma. In adverse nutritional circumstances, it is the mother, not the fetus, who is 'at risk.' (*See* **Amino**

Acids: Metabolism; Minerals – Dietary Importance; Vitamins: Overview.)

Metabolic Adjustments in Pregnancy

Under normal circumstances, the integrity of the tissues is assured by homeostatic regulation of the absorption, excretion and catabolism of the nutrients. During reproduction, these control mechanisms are adjusted, under the influence of hormones secreted by the placenta, in order to make available an additional nutrient supply for growth of the conceptus.

Absorption of the Nutrients

The macronutrients, fat, carbohydrate, and protein, are almost completely digested and absorbed in the small intestine, so that no appreciable change occurs in pregnancy. The absorption of minerals, however, is carefully regulated. Most Western diets provide a considerable excess of calcium, and no more than around 30% is normally absorbed. From early in pregnancy, the blood concentration of the physiologically active form of vitamin D (1,25-dihydroxycholecalciferol) is elevated, leading to an increased uptake of calcium from the gut, thus allowing the mother to satisfy the needs of her developing fetus without recourse to her own skeletal calcium or the need to increase her dietary intake. Only when her customary calcium intake or the biosynthesis of vitamin D is severely restricted is calcification of the fetal skeleton compromised. (*See* **Calcium:** Physiology; **Cholecalciferol:** Physiology.)

As with calcium, the proportional absorption of iron from a diet providing 12–14 mg per day is very small (about 10%), but as gestation advances, iron uptake increases progressively and may exceed 40%. This adjustment in absorption, in association with a reduction in iron loss resulting from the cessation of menstruation, is sufficient to satisfy the iron cost of pregnancy. Judged by normal standards, the characteristic fall in hemoglobin concentration that is noted in pregnancy would indicate a state of anemia. Blood volume, however, is considerably increased so that the total amount of hemoglobin in circulation, and consequently the oxygen-carrying capacity of the blood, is actually raised. Supplementation of the diet of women who begin pregnancy in an iron-sufficient state may therefore be injurious as it has been shown to increase red cell size and could adversely affect blood flow in the small capillaries. (*See* **Iron:** Physiology.)

Other minerals, such as copper and zinc, may also show improved absorption. (*See* **Copper:** Physiology; **Zinc:** Physiology.)

Urinary Excretion of Nutrients

The glomerular filtration rate is increased during pregnancy. It is suggested that the presence of traces of glucose in the urine at some stages of pregnancy in most women merely indicates that more glucose has been filtered from the plasma than can be reabsorbed by the kidney tubules at that time. Some women, however, may lose large amounts of glucose in their urine, particularly those who develop gestational diabetes.

The urinary excretion of most nutrients is altered, but whether or not this reflects specific modifications in kidney function is unclear since the picture is obscured by withdrawal of nutrients from circulation by the placenta, which changes with time. The pattern of excretion of amino acids is most consistent. In nonpregnant women, about 1% of total nitrogen excretion is accounted for by free amino acids, but urinary loss rises in early pregnancy, reaching a plateau in midpregnancy at a level approximately double that found in the nongravid state. Most of the increase is accounted for by the nonessential amino acids, possibly reflecting selective uptake by the placenta. Since the plasma amino acid concentrations show little change, the large increase in urinary excretion cannot be attributed to kidney overload and is most likely the result of altered hormonal status affecting kidney function. The return to normal levels of excretion during lactation supports this conclusion.

One example of altered kidney function that favors conservation has clearly been identified. The amino acid taurine, a derivative of the nonessential amino acid cystine, shows a dramatic and sustained reduction in excretion from early pregnancy. It has been suggested that taurine may act as a membrane stabilizer, as an inhibitory neurotransmitter or neuromodulator, and as a growth modulator in fetal tissues. Tissue concentrations are particularly high in the fetus, and the suppression of urinary excretion of taurine during pregnancy is seen as a means of satisfying the needs of the developing fetus, which lacks the ability to synthesize taurine *de novo*.

Metabolism of Protein

The rate of accretion of nitrogen by the fetus and maternal reproductive tissues is known to rise 10-fold between the first and last quarters of gestation. Since, on average, no appreciable change in food intake, and thus protein intake, occurs in pregnancy, one would expect to find a positive nitrogen balance, rising continuously throughout pregnancy. In the few nitrogen-balance studies carried out on pregnant women, however, no differences have been found between the values for nitrogen retention measured at different stages of gestation. This apparent paradox was resolved by the study of reproduction in the rat. (*See* **Protein**: Digestion and Absorption of Protein and Nitrogen Balance.)

During the first 2 weeks of pregnancy, when competition between the dam and her fetuses for nutrients is minimal, a substantial reserve of protein is built up in the muscles (the 'anabolic phase'). During the final week, the period of rapid fetal growth, the protein reserve is broken down, the amino acids released being taken up by the placenta for growth of the conceptus. This 'catabolic phase' is not influenced by the protein content of the maternal diet, and is controlled by the hormone progesterone. (*See* **Hormones**: Steroid Hormones.)

Evidence for a similar biphasic system of protein metabolism in women was provided by measurement of the excretion of the amino acid 3-methylhistidine. This amino acid is found predominantly in muscle. Histidine undergoes methylation after its incorporation into the peptide chains of the contractile proteins actin and myosin, and, in the course of muscle protein turnover, the 3-methylhistidine released is not reutilized, but is quantitatively excreted in the urine. In pregnancy, the excretion of 3-methylhistidine rises sharply during the last trimester, indicating the hydrolysis of an amount of protein approximating the estimated protein cost of pregnancy.

Such a redistribution of amino acids from maternal to fetal tissues would not be detected in measurements of nitrogen balance.

This cyclic course of protein metabolism has important nutritional implications. The protein cost of pregnancy is distributed over the whole gestational period, and the influence of acute or chronic maternal undernutrition on fetal growth is minimized.

Nitrogen retention occurs when the intake of protein-nitrogen exceeds nitrogen losses from the body, largely the products of amino acid catabolism in the urine (mostly urea) and undigested dietary protein in the feces along with cells shed from the gut epithelium. If, during pregnancy, the catabolism of amino acids were in part suppressed, the fraction spared could be used for fetal protein synthesis. In the case of a woman existing on 40 g of protein per day, as in many developing world populations, a reduction of less than 9% in amino acid oxidation would spare enough protein to cover the entire estimated requirement for pregnancy.

The hypothesis was tested in the rat. The activities of two rate-limiting enzymes (alanine aminotransferase and argininosuccinatesynthetase), which regulate the oxidation of amino acids and the conversion of amino-nitrogen to urea, were measured in the livers at different stages of gestation. The activities of both

enzymes were markedly depressed by the end of the first week, and declined even further (by around 50%) as pregnancy advanced. A parallel change in plasma urea was also noted.

Evidence for a similar adjustment in amino acid catabolism in pregnant women was later obtained in a metabolic study using urea labeled with a stable isotope of nitrogen. Measurements made in the last trimester and in the postpartum postlactational period showed a reduction of 30% in the rate of urea synthesis in late pregnancy and a similar fall in the plasma urea concentration.

The mechanism for protein sparing, which operates progressively throughout pregnancy, combined with the biphasic system of early storage and later breakdown of stored protein, ensures a supply of amino acids commensurate with the demands of the growing fetus. The suppression of hepatic amino acid oxidation was also shown to be induced by the placental hormone progesterone. The fetoplacental unit thus indirectly controls its supply of amino acids as well as its energy needs.

Metabolism of Energy

As stated earlier, the dietary energy supply is the major determinant of fetal growth. In healthy pregnant women, energy balance becomes positive during the first trimester, probably in response to the rising secretion of progesterone. The purpose of the augmented fat reserve, amounting, on average, to some 4 kg of fat, is primarily to subsidize the high energy cost of lactation, but a small amount is mobilized in late pregnancy to provide an alternative fuel for use by the maternal tissues and enhance the availability of glucose for use by the fetoplacental unit. The human fetus derives its energy almost exclusively from the oxidation of glucose. (*See* **Energy**: Energy Expenditure and Energy Balance.)

The discrepancy between the measured energy intakes of healthy pregnant women and the values predicted for energy expenditure has yet to be explained. There is no doubt that some saving is made in energy expenditure from a reduced level of physical activity, but this is difficult to quantify.

The alterations in carbohydrate metabolism that are characteristic of pregnancy could also lead to the sparing of energy. In pregnancy, there is an increased output of insulin in response to a glucose stimulus, and reduced glucose uptake by the peripheral tissues (muscle and adipose tissue). Insulin antagonism has been attributed to the action of placental lactogen. Consequently, more of the ingested carbohydrate is directed to the liver, the organ that maintains the blood glucose concentration. The conversion of carbohydrate to fat is a very costly process. Approximately 20% of the energy that is available from the direct oxidation of glucose is lost as heat if the glucose is first converted to fat, and the fat is later oxidized to produce energy. This adjustment in carbohydrate utilization, therefore, not only conserves energy for anabolic purposes, but also safeguards the fetal energy supply. (*See* **Carbohydrates**: Digestion, Absorption, and Metabolism.)

Dietary Recommendations for Pregnancy

Growth of the fetus is little affected by transient or prolonged moderate undernutrition of the mother. Relatively small deficits in birthweight may be accounted for by a lower proportion of body fat resulting from maternal dietary energy restriction in late pregnancy, from the diversion of glucose (the fetal fuel) to the muscles should the mother be engaged in hard physical work throughout pregnancy, or from malfunction of the placenta.

The security of the fetus is provided by the placenta, an organ designed not only to ensure a priority claim on all nutrients present in the maternal circulation, but also, by the secretion of hormones, to modulate the homeostatic regulation of nutrient utilization at all levels – absorption, excretion, and catabolism, in order to augment the nutrient supplies. No government committee responsible for devising dietary guidelines would be so incautious as to suggest that women during pregnancy require no more food than when in the nongravid state, although all evidence points very clearly to that conclusion. One obvious *a priori* condition would be that nutritional status should be satisfactory before conception and throughout pregnancy.

Attention has been focused on the latter half of pregnancy, when fetal growth is most rapid, the need for nutritional input is at its greatest, and the effects of maternal food deprivation most apparent. These considerations led a Joint FAO/WHO (Food and Agriculture Organization, World Health Organization) Ad Hoc Expert Committee in 1973 to propose an additional 628 kJ (150 kcal) per day for the first trimester, rising to 1464 kJ (350 kcal) per day for the second and third trimesters, the stage of pregnancy when women, unencumbered by professional advice, would spontaneously reduce their food consumption. In the light of continuing research, however, estimates of energy and nutrient requirements are being revised in a downward direction. In 1974, the US report on Recommended Daily Allowances proposed an additional daily supplement of 30 g of protein, 400 mg of calcium and 30–60 mg of iron for the pregnant woman, acknowledging that such a high

intake of iron could not be met by the iron content of habitual US diets. The implication was that pregnancy was a clinical condition that required therapeutic intervention. One decade later, the proposed increments in energy and calcium were little changed, but the supplement of iron was reduced to 15 mg, and the protein allowance was changed to anticipate the pattern of accretion by the fetus, rising from 1.2 g per day in the first trimester to 10.7 g in the final trimester of pregnancy. (*See* **Dietary Requirements of Adults.**)

The UK dietary recommendations for pregnancy have consistently been on a less generous scale. Over a similar period, the daily allowance for energy has fallen from 1004 kJ (240 kcal) in the second and third trimesters to 837 kJ (200 kcal) during the third trimester only. Protein is unchanged at 6 g per day throughout pregnancy, but the recommendation for calcium has fallen from 700 mg per day in the third trimester to zero. Likewise, no recommendation was made for an increase in iron intake in the 1991 report on Dietary Reference Values, compared with the small increase of 1 mg per day in the earlier 1979 edition.

There is no doubt that as scientific opinion changes, other values will also be reduced, and nutritional guidelines for pregnancy ultimately will correspond to the dietary practices of healthy women satisfying their natural appetites on a well-balanced diet.

See also: **Anorexia Nervosa**; **Calcium**: Physiology; **Carbohydrates**: Digestion, Absorption, and Metabolism; **Dietary Requirements of Adults**; **Energy**: Energy Expenditure and Energy Balance; **Fats**: Requirements; **Hormones**: Steroid Hormones; **Iron**: Physiology; **Lactation**: Human Milk: Composition and Nutritional Value; **Malnutrition**: Malnutrition in Developed Countries; **Minerals – Dietary Importance**; **Premenstrual Syndrome: Nutritional Aspects**; **Protein**: Digestion and Absorption of Protein and Nitrogen Balance; **Vitamins**: Overview

Further Reading

Campbell DM and Gilmer DG (eds) (1983) *Nutrition in Pregnancy*. London: Royal College of Obstetricians and Gynaecologists.

Department of Health (1991) *Dietary Reference Values for Food Energy and Nutrients for the United Kingdom*. London: The Stationery Office.

Hytten F and Chamberlain GVP (eds) (1980) *Clinical Physiology in Obstetrics*. Oxford: Blackwell Scientific.

Naismith DJ (1983) Maternal nutrition and fetal health. In: Chiswick ML (ed.) *Recent Advances in Perinatal Medicine*, vol. I, pp. 21–39. Edinburgh: Churchill Livingstone.

National Research Council (1989) *Recommended Daily Allowances* 10th edn. Washington, DC: National Academy Press.

Role of Placenta in Nutrient Transfer

Y Kudo[†] and C A R Boyd, University of Oxford, Oxford, UK

Introduction

During intrauterine life, the developing fetus effectively receives all of its nutrition across the placenta, which, in the human, has a chorionic villus structure in which maternal blood directly superfuses the external surface, the trophoblast of the placenta. This trophoblast forms an unusual epithelium separating the maternal plasma from the fetal extracellular space of the villus core through which the fetal capillary circulation flows. For the fetus, the trophoblast is thus equivalent to the epithelium lining the small intestine of the newborn, since, from a nutritional perspective, it must absorb those molecules required for both growth and maintenance of the organism. The trophoblast also has an important role in acting as the lung for the fetus (since all gas exchange between mother and baby must occur across this surface), as the fetus's kidney (since excretion from the conceptus to the mother occurs across this structure), as well as being an important endocrine tissue secreting peptide and steroid hormones into the mother. The placenta itself also plays a very substantial role in intermediary metabolism, so that it cannot be assumed that there is no metabolism of absorbed nutrients during solute movement from mother to fetus.

As an epithelium, the trophoblast has two surfaces, the one facing the mother and the other facing the fetus. These surfaces differ structurally; for example, microvilli are found projecting into the maternal bloodstream at the apical surface, where they form a brush border, whereas the basal surface facing the fetus does not have this surface specialization. From the functional point of view, it is the differences in the distribution of transport membrane proteins (channels, carriers, and pumps) that determine the overall transport of nutrients from mother to baby, at least as far as water-soluble molecules are concerned. In the placenta, all transport appears to be across the trophoblast since, uniquely, the cells that compose this structure are fused to form a syncytium; this is in contrast to other epithelia where transport between adjacent epithelial cells allows a functional paracellular route to lie in parallel to the route through the epithelial cells themselves (the transcellular route). The question in the human as to whether there is a special route available for the transport of large molecules (MW 5000) is not yet resolved: certainly, at term, the human placenta does appear to be able to permit the transport of larger molecules at a slow rate from baby to mother. This route is unlikely to be of nutritional significance, since the rate of transport is low, but it may be important with regard, for example, to immunological sensitization.

The placenta, together with the growing fetus, has a very substantial metabolic energy requirement, and in the human, ATP synthesis appears to be met by the very substantial rate of glucose delivery from the mother. The glucose transporter systems that are found in both the apical and basal surfaces of the human trophoblast appear to be the type named GLUT1; in other words, they are sodium-independent facilitated transporters that are not regulated by insulin. The Michaelis constant (K_m) for glucose transport at both surfaces is relatively high (approximately 30 mM), and the maximal transport rate (V_{max}) is very substantial. The result of this is that glucose delivery across the brush-border membrane of the placenta will be in a direction and rate that are dependent solely on the chemical driving force from mother to baby (maternal–fetal plasma glucose concentration). It seems likely that this fundamental property is the basis for the macrosomia ('large-for-dates') found in the babies of mothers with elevated plasma glucose concentration, as typically found in diabetes mellitus. Transport of glucose, which is stereospecific, may be inhibited by glucose analogs that share the chemical structure of D-glucose at carbon 1; for example, both 3-0-methylglucose and 2-deoxyglucose are transported, whereas α-methylglucoside (with a methyl group on carbon 1) is not. The transport of other monosaccharides has been studied rather little; in the human, fructose is transported much more slowly than glucose itself (in contrast to other nonprimate species). The question of regulation of carbohydrate delivery across the placenta is not fully resolved since some glucose transporters may be regulated by phosphorylation, and the gradient for transplacental glucose delivery will also depend upon factors regulating glucose utilization and production by the placenta itself; little is known of the physiological regulation of either of these processes. (*See* **Carbohydrates**: Digestion, Absorption, and Metabolism; **Glucose**: Maintenance of Blood Glucose Level.)

In contrast to glucose transport, amino acid transport can be powered. The overall gradient of amino acid between maternal plasma and fetal plasma varies

---

[†]Deceased.

for individual amino acids. Typically, individual amino acid concentrations are twice as high in fetal as in maternal plasma. Current understanding of the mechanisms responsible for this relates to the distribution of the membrane transport proteins between the two faces of the trophoblast and in particular to the distribution of sodium-coupled transporters that are found predominantly (although not exclusively) in the brush border. Recent work using isolated membranes that reseal to form artificial structures (vesicles) has been useful in establishing the numbers and properties of such transporters. In addition to the direct effect of sodium ions in moving amino acids into the trophoblast across the brush-border membrane against a concentration gradient, these transporters are often electrogenic and are thus also driven physiologically by the membrane potential. One example of such a process is the transport system called 'A,' which uses alanine, serine, and proline as transported substrates and accumulates these amino acids in the trophoblast against a concentration gradient. These amino acids then leave the trophoblast across the basal membrane by a different transport system. Other amino acids may be transported via tertiary active transport; for example, leucine is found in higher concentrations in fetal plasma than in maternal plasma, but it is not itself a substrate for sodium-coupled transport; rather, it appears to exchange with amino acids, such as alanine, that have been accumulated in the trophoblast as just described.

The cationic and anionic amino acids are unusual in that, having their own charge, they will be accelerated or retarded by the membrane potential in crossing each of the plasma membrane surfaces of the trophoblast. For cationic amino acids (lysine, arginine, histidine), entry into the placenta appears to be largely by system y^+ (Na^+-independent), whereas exit into the fetus involves an electroneutral system (y^+L), which exchanges the cationic amino acid for a neutral amino acid (e.g., leucine) and a sodium ion, thus effectively solving the problem of permitting positively charged amino acids to exit against an inside-negative membrane potential. For anionic amino acids (glutamate and aspartate), very high intratrophoblast concentrations are achieved by a transport system that is coupled to K^+ efflux as well as Na^+ entry. Essential amino acid requirements for the fetus are different from those of adults. However, it is not clear whether transport of specific amino acids across the placenta ever becomes rate-limiting for fetal growth. In the human, intrauterine growth retardation not associated with other disease has been shown to be associated with reduced placental delivery of amino acids through specific systems, e.g.,

associated with decreased function of system A. IGF (insulin-like growth factor) and IGFBP (insulin-like growth factor binding protein) are now recognized as having an important role in either normal or abnormal fetal growth via controlling placental amino acids and glucose transport (e.g., IGF-I (insulin-like growth factor-1) selectively enhances system A activity). In certain unusual metabolic disorders (e.g., maternal phenylketonuria (PKU)) maternal levels of one particular amino acid may be elevated; this results in competition between this amino acid and others that share the same transporters. Some of the abnormalities found in the developing babies of such mothers may be a consequence of nutritional deprivation of tyrosine, for example, owing to competition by raised maternal phenylalanine levels for the delivery of this amino acid across the placenta. The fact that amino acid transport across the placenta involves a family of transport proteins with overlapping substrate (amino acid) specificities means that the nutritional consequences for the fetus of changing the level of one amino acid in the mother will be complex. This follows because, in contrast to placental glucose transporters, the K_m and V_{max} of the amino acid transporters are relatively low.

Lipid transport across the placenta in relation to human nutrition has been studied less rigorously, in part because it is likely to be perfusion- rather than membrane-limited, since the lipid-soluble nature of such a substrate allows ready transmembrane transport. Nutritionally, the nervous system of the developing fetus requires substrate delivery of precursors for myelin synthesis. Studies suggest that placental binding proteins may provide a pool of essential fatty acids for fetal utilization. (*See* **Amino Acids**: Metabolism; **Fatty Acids**: Metabolism.)

Transport of the inorganic cations of sodium and potassium involves both channels and transporters. Sodium transport into the trophoblast is coupled to the entry of those solutes (which include both organic and inorganic molecules) powered by secondary active transport. The extrusion of sodium across the basal surface of the trophoblast is likely to be a result of sodium pumping by Na^+/K^+ ATPase activity. In contrast, potassium, accumulated in the trophoblast, as in other epithelia by the sodium pump, requires channel-mediated release to account for its movements between mother and baby. Potassium channels have recently been shown to be sensitive to modulation in this tissue (e.g., by G proteins, by arachidonic acid, and by pH). These regulatory factors may themselves be controlled by circulating factors in both mother and fetus. It is clear that fetal plasma potassium is carefully regulated by control of placental transport of this cation.

The divalent cation of calcium is found in higher concentrations in fetal than in maternal plasma. Active transport processes are therefore involved in placental calcium metabolism, and these are unusual in that regulation of transport is clearly precisely controlled. It seems likely that calcium extrusion across the basal surface of the trophoblast is ATP-driven and regulated by calmodulin. Entry of calcium across the brush border is now known to be via ECaC (epithelial calcium channels) that are regulated: membrane potential hyperpolarization activates, and intracellular calcium inhibits, these channels. Fetal parathyroid hormone and vitamin D are likely to regulate all of these events.

Iron is also transported from mother to fetus in a regulated fashion, and again, there is a greater concentration of iron in the fetal circulation. In some species, transport of iron has been shown to be active in that it is inhibited by anoxia. The role of the transferrin receptor found in the human placental microvillus membrane is now known to be related to the mechanisms of iron entry into the cytosol. From this compartment, iron has to leave, but the mechanisms responsible for this are not fully understood. (*See* **Calcium**: Physiology; **Hormones**: Thyroid Hormones; **Iron**: Physiology; **Potassium**: Physiology; **Sodium**: Physiology.)

Anion entry into the placenta has been studied using isolated membrane preparations. These studies show that for monovalent ions (chloride and other halides), two routes are available, an exchange and a conductive route, the latter likely to be via channels. The anion exchange system appears to be functionally linked to the transport of organic anions (bicarbonate, lactate) from placenta into the maternal circulation. Phosphate transport is also regulated and appears to involve a sodium-dependent cotransporter at the maternal-facing surface and an efflux mechanism (possibly driven by the membrane potential) at the basal surface of the tissue. As for calcium transport, phosphate delivery is regulated by parathyroid hormone concentration in the fetus.

Trace-element delivery across the placenta also involves specific placental binding proteins analogous to those found in adult liver; however, membrane transport is also required to allow such ions to gain access to and from the trophoblast. The nature of such transporters varies greatly, although the role of the divalent cation transporter (DCT-1) in the placenta may be more generally important. For zinc, there is evidence that the histidine amino acid transporter is responsible for delivery of a histidine–zinc complex, whereas for transition-metal oxides, it appears that anion exchange is important.

The transport of iodide across the human placenta is also likely to be by anion exchange since SCN can inhibit it, but the mechanisms responsible for the concentration of this element in the fetal compartment are not clear. Selenate appears to share a pathway with sulfate for entry across the brush border, and the transport of both of these ions is inhibited by blockers of anion exchange. This pathway appears to be shared with those available for transport of the trace elements chromium and molybdenum as chromate and molybdate. (*See* **Trace Elements.**)

Vitamin transport also is highly specific for individual substrates; thus, for ascorbic acid, a sodium-independent transporter for the reduced from of this nutrient has been described in the brush-border membrane; this transporter appears to be functionally coupled to a placental system that maintains ascorbate in this chemical form. The Na^+-dependent multivitamin transporter that transports pantothenate, biotin, and lipoate is expressed in human placenta.

See also: **Amino Acids**: Metabolism; **Calcium**: Physiology; **Carbohydrates**: Digestion, Absorption, and Metabolism; **Fatty Acids**: Metabolism; **Glucose**: Maintenance of Blood Glucose Level; **Hormones**: Thyroid Hormones; **Iron**: Physiology; **Potassium**: Physiology; **Sodium**: Physiology; **Trace Elements**

Further Reading

Bain MD, Copas DK, Taylor A *et al.* (1990) Permeability of human placenta *in vivo* to four non-electrolytes. *Journal of Physiology* 431: 505–513.

Boyd CAR and Kudo Y (1990) Placental amino acid transport. *Biochimica et Biophysica Acta* 121: 169–174.

Hoenderop JGJ, van der Kemp AW, Hartog A *et al.* (1999) Molecular identification of the apical Ca^{2+} channel in 1,25-dihydroxyvitamin D_3-responsive epithelia. *Journal of Biological Chemistry* 274: 8375–8378.

Kudo Y and Boyd CAR (1996) Placental tyrosine transport and maternal phenylketonuria. *Acta Paediatrica* 85: 109–110.

Parkkila S, Waheed A, Britton RS *et al.* (1997) Association of the transferrin receptor in human placenta with HFE, the protein defective in hereditary hemochromatosis. *Proceedings of the National Academy of Sciences, USA* 94: 13198–13202.

Shennan DB and Boyd CAR (1988) Trace element transport in placenta. *Placenta* 9: 333–343.

Sibley CP, Birdsey TJ, Brownbill P *et al.* (1998) Mechanisms of maternofetal exchange across the human placenta. *Biochemical Society Transactions* 26: 86–91.

Wang H, Huang W, Fei YJ *et al.* (1999) Human placental Na^+-dependent multivitamin transporter. *Journal of Biological Chemistry* 274: 14875–14883.

Safe Diet

R B Fraser and F A Ford, University of Sheffield, Sheffield, UK

Introduction

Would-be pregnant and pregnant women have heightened concerns about their diet, which can be summarized by the common question, 'Will it harm my baby?' Because the question is so common, many sources of advice are available in the lay press, not all as scrupulously researched and scientifically based as they might be. Another source of potentially misleading advice is the food industry and written media, the agents of which possess a lack of objectivity that may not be obvious to the casual reader. Articles proliferate with titles such as 'How to have a beautiful baby' and 'How to have a perfect baby'. These tend to contain a mixture of dietary and lifestyle advice with the implication that, if followed, the undesirable outcomes of pregnancy such as miscarriage, congenital malformation, and fetal death are avoidable. Professionals in the field of nutrition must be aware of this for the following reason. If a woman learns of such advice only after her pregnancy has failed in some way there is a potential for a lifelong burden of guilt. For this reason purveyors of advice should restrict themselves to that which has been demonstrated in scientifically valid experiments. In particular, they should be guarded in their extrapolations from observational studies and animal experiments.

This article highlights the principal areas in which women have dietary concerns. Some of these are dealt with in more detail elsewhere and are cross-referenced.

Periconceptional Nutrition

In an ideal world, pregnancies would be planned but in the real world a high proportion are not. It is important to remember that many women who have not planned their pregnancies and have not subsequently followed the advice outlined below may need reassuring that they have probably not caused harm to their baby.

The work of Frisch in the USA identified the link between body composition and ovulation in the human female. She suggested that fat must comprise at least 22% of body weight for the maintenance of ovulatory cycles and also observed that in normal postpuberty women fat is about 28% of body weight. It is well recognized that the relationship is to the fat content of the body rather than absolute body weight, as trained athletes of average or above-average body weight may have very low body fat content and may be oligo- or amenorrheic. Frisch observed that amongst trained athletes who became fit after a normal menarche, 60% continued regular cycles, but 40% had irregular cycles and presumably associated subfertility. Other causes of secondary amenorrhea leading to infertility include psychological stress, thyrotoxicosis, and the various malabsorption syndromes. Eating disorders such as anorexia nervosa and bulimia nervosa which may have been underdiagnosed in the past are now been seen as contributing to the problem. Particularly for bulimics, a remission in the disorder may allow the body fat to build up to a stage where ovulation and conception can take place, but the disorder may relapse during the pregnancy, leading to serious nutritional deficiencies for the mother and possibly for the fetus.

The Nurses Health Study in the USA has also produced some interesting information about the relative risk of menstrual cycle irregularity, not only in underweight but also in overweight women. For women with a body mass index (BMI) below 20, at the age of 18 years, ovulatory infertility was found with a relative risk of about 1.2 compared to women with a BMI between 20 and 25. Interestingly, however, the relative risk of ovulatory infertility was 1.5 in those with a BMI of 28 and more than 2 in the obese group with a BMI above 30. About half of the risk is associated with polycystic ovarian syndrome (PCOS) in which ovulatory infertility and obesity coexist but there is still a doubling in relative risk of ovulatory infertility in women with a BMI above 30 who do not have ultrasonically detectable polycystic ovaries.

Vitamin Intakes and Congenital Malformations

Folic Acid

In 1991, the Medical Research Council (MRC) vitamin trial reported that a significant reduction in the recurrence of neural tube defects (NTD) had been obtained by a daily dietary supplement of 4 mg of folic acid in women who were at high risk. Women at risk of a recurrence of NTD should be advised to take a folic acid supplement of 5 mg (5000 µg) per day when planning a pregnancy and continue with it until 12 weeks' gestation. A later study reported that first occurrence of NTD could also be prevented by daily supplements containing 800 µg of folic acid. In the UK, in line with many other nations, women are advised to take a prepregnancy supplement of 0.4 mg (400 µg) of folic acid as a daily medicinal supplement from when they begin trying to conceive until 12 weeks' gestation, in addition to eating a folate-rich diet and breads and breakfast cereals fortified with folic acid.

The mechanism for the effect of folic acid in reducing the risk of NTD is not clear. The most likely explanation is that it overcomes genetically determined defects in folate metabolism that interfere with normal neural tube development. One other hypothesis is that it might decrease the likelihood of an affected pregnancy surviving, and there is some concern that folic acid supplementation could be associated with a higher risk of fetal death. However, the generally accepted explanation for this is that folic acid might initially permit the survival of affected or nonviable pregnancies to a point where they are recognized as spontaneous abortions.

Educational campaigns conducted since 1995 have had a substantial effect in increasing the number of women in the UK who are aware of the link between folic acid and NTD. However, such education can only have a limited effect because it is estimated that only 50% of all pregnancies are planned. NTD arise very early in gestation and by the time a woman realizes that she is pregnant it is usually too late to prevent an NTD by taking supplements. For this reason it has been suggested that the most effective way of reducing NTD in the UK would be to fortify the food supply with folic acid so that even women with unplanned pregnancies would be less likely than at present to have offspring with NTD. The Department of Health Committee on Medical Aspects of Food and Nutrition Policy (COMA) has recently recommended universal fortification of flour at $240 \mu g\ 100 g^{-1}$, which would reduce the risk of NTD by 41% without resulting in unacceptably high intakes in any group of the population.

Retinol

It has been known for a long time from animal studies that high levels of vitamin A in the form of retinol are teratogenic in the periconceptional period. This led to the Department of Health in 1990 advising women to avoid food sources containing high levels of the retinal form of vitamin A in the periconceptional period and indeed during pregnancy. The Department of Health advises that very high intakes of retinol, i.e., more than 10 times the recommended dietary allowance either in liver/liver products or vitamin/fish liver oil supplements, can damage the developing embryo. This advice remains current and all women of child-bearing age should avoid liver and its products and should not take supplements containing more than four times the recommended daily amount of retinol.

The same risks do not seem to be present for vitamin A derived from β-carotene. Manufacturers of multivitamin supplements recommended for pregnancy have generally recognized this and switched the source of vitamin A to β-carotene.

Nutritional Management of Common Symptoms in Pregnancy

Heartburn

Heartburn is thought to be caused by gastroesophageal reflux. Although occasionally experienced in the first trimester, it is generally more common in the last trimester and occurs in 30–50% of women.

Small, frequent meals or snacks are usually tolerated better than large, well-spaced meals. Common foods cited as causing heartburn as spicy and fatty foods, fizzy drinks, citrus fruits, fruit juices, and cucumber. Milk and milk products can help to relieve symptoms but antacids are frequently used.

Nausea and Vomiting

Psychological factors, changing hormone levels, hunger, altered carbohydrate metabolism, and vitamin deficiencies have all been proposed as possible causes for nausea and vomiting, but none has been confirmed.

Symptoms may start before the woman knows she is pregnant or in later pregnancy but commonly they are worst between weeks 6 and 10, and subside by about 13 weeks. Nausea is experienced at any time of day or night and can be either slight or severe. It often becomes worse when the stomach is empty, and eating small, frequent meals based on starchy carbohydrates may relieve it. Morning sickness is common and consuming dry biscuits or toast before getting up can relieve this.

Nausea can also be triggered by traveling, fried and spicy foods, and smells such as coffee, perfume, and cigarette smoke.

Some women just feel nauseated while others actually vomit as well; this may cause minor weight loss but rarely causes nutrient deficiency. Women need reassurance that not eating proper cooked meals or losing some weight and their taste alterations will cause no problems for their developing fetus. The more severe cases of pregnancy vomiting (hyperemesis gravidarum) require hospital admission, intravenous fluids and, sometimes, parenteral nutrition.

Constipation

Constipation is common at all stages of pregnancy. It may be related to a general reduction in motility in the gastrointestinal tract, with prolonged transit times and increased water resorption from the stool. General advice about constipation is also suitable for pregnancy, i.e., increased intake of fiber, particularly cereal fiber, and increased fluid intake.

Constipation may be aggravated by the consumption of iron tablets; if it is not appropriate to reduce or stop them, bulking agents may be prescribed.

Qualitative Aspects of Diet

The following section refers to common questions of dietary safety; some arise because of suspicion of harm when items are included in the diet, others for the paradoxical reason that their omission from the diet might be dangerous.

Alcohol

As far as alcohol is concerned, the picture is confusing. There is no doubt that a heavy intake of alcohol can damage the unborn baby and cause miscarriage. Many women choose to give up alcohol and this seems to be a sensible practice but there is no evidence of harm from occasional drinking or the consumption of less than 2 units per day (**Table 1**). Despite this, many women do give up alcohol when trying to conceive or whilst pregnant, and this seems a sensible but not mandatory practice. (One unit is 15 g of absolute alcohol, e.g., 0.28 l of beer, one glass of wine, or one measure of spirits.)

A well-defined group of anomalies referred to as the fetal alcohol syndrome is now recognized. The major defects of affected infants are weight and length below the 10th centile for gestational age, and microcephaly. Microcephaly is a condition of small head size associated with an underdeveloped brain. Such children are likely to be mentally retarded and their physical growth is stunted. The syndrome has been reported in up to 40% of the infants of women drinking more than 6 units of alcohol per day.

Definite harm has been recorded to the offspring of women drinking between two and six units of alcohol per day and long-term growth retardation and mental retardation to offspring of those drinking more than 6 units a day (fetal alcohol syndrome).

Table 1 Congenital malformation rates and maternal alcohol consumption

Drinks per day	Malformation rates (per 1000)[a]
None	78
Less than one	77
One to two	83

[a]No statistically significant difference.
From Mills JL and Graubard BI (1987) Is moderate drinking during pregnancy associated with an increased risk for malformations? *Pediatrics* 80: 309–314

A recent study showed that a woman's alcohol intake is associated with decreased fertility, even among women with a weekly alcohol intake corresponding to five or fewer drinks. This finding needs further corroboration, but it seems reasonable for the moment to encourage women to avoid alcohol if they are having trouble conceiving.

Caffeine

Concern about the detrimental effects of caffeine during pregnancy is not new but to some extent the literature is conflicting. Caffeine is present in tea, coffee, cola drinks, and many over-the-counter remedies for colds and allergic symptoms. Animal experiments with high doses have shown that some congenital malformations may be induced but there is no evidence of harm in humans. Current recommendations are that pregnant and breast-feeding women should limit their caffeine intake to $300 \, mg \, day^{-1}$, which is equivalent to 4 cups ordinary-strength coffee per day (or 6 cups of tea or 7 cans of cola). A recent review from the UK Committee on Toxicity of Chemicals in Food, Consumer Products and the Environment (DoH 2001) issued a statement on the reproductive effects of caffeine in October 2001. A meta-analysis of studies of maternal caffeine intake during pregnancy and the risk of spontaneous abortion or low birth weight, compared maternal caffeine intakes during pregnancy of more than 150 mg/day with less than 150 mg/day. Calculated odds ratios were significantly increased for spontaneous abortion (miscarriage) (odds ratio = 1.36; 95% confidence interval, 1.29–1.45) and low birth weight (odds ratio = 1.51; 95% confidence interval, 1.39–1.63), for low birth weight (< 2.5 kg).

Two studies, which examined the links between caffeine intake and pregnancy in SIDS, received a huge amount of publicity and caused a lot of worry for women. However, the results were conflicting.

Peanuts

Peanut allergy is increasing in children, although the cause is unclear. The use of peanuts and peanut oil in the British diet has increased rapidly over the last few years and it is thought that being exposed to peanuts at a young age may cause the allergy. However, it is not known if this can happen as a result of a mother eating peanuts when she is pregnant or breast-feeding. Current advice is that women, their partners, and existing children who have a family history of atopic disease should avoid peanuts.

Over-the-Counter Remedies

As a general rule, self-prescribing in pregnancy should be kept to a minimum. Those drugs that have been used most widely and have been shown to be safe should be preferred to anything new or untested. Small doses of simple analgesic drugs such as aspirin and acetaminophen (paracetamol) appear to be safe, as do common cold remedies, which often have these drugs as their active contents.

No problems have been reported with the use of homeopathic remedies in pregnancy. There has been much interest in possible therapeutic benefit from the consumption of oil of evening primrose in pregnancy. It has been hypothesized that the linoleic acid and γ-linoleic acid in the compound may stimulate production of vasodilatory prostaglandins and be of benefit in lowering blood pressure in abnormal states of pregnancy such as preeclampsia. The limited studies so far reported do not support this hypothesis.

Recent reports have compared women who are on diets rich in natural fish oil, the Faroese, with women on the Danish mainland whose diets contain much less fish oil. The Faroese had fewer premature babies, fewer problems with blood pressure, and heavier babies than the mainlanders. Unfortunately, these studies cannot be interpreted as showing uniform benefit from fish oil supplements as the Faroese also lost more babies as stillbirths. A further report, comparing women in Aberdeen with women in the North Sea Orkney Islands, showed the latter to have average birth weights 250 g higher. After correction, a small but significant proportion of the difference was found to be related to genetic or environmental factors. The latter included a diet in Orkney containing 30% more fish than in Aberdeen.

Ginger root may act as an antiemetic by a local effect in the stomach. Evidence of benefit in pregnancy nausea exists.

Garlic is widespread in the human diet. There has always been a recognition of possible benefits of garlic supplements on a wide range of disorders, including hypertension, hyperlipidemia, and thrombosis. No studies of supplementation in pregnancy have been reported.

Vitamin and Mineral Supplementation

Some pregnant women benefit from iron and/or folic acid supplements when dietary deficiency provokes anemia. Although self-medication with vitamin and mineral supplements is widespread, there is a good case during pregnancy to restrict such supplements to those prescribed by doctors or preparations containing less than 100% of the recommended dietary allowance. Vitamin A in the form of retinol and its analogs can cause congenital malformations in megadosage, and vitamin D supplementation is thought to have been the cause of idiopathic hypercalcemia. This

was a congenital syndrome consisting of abnormal facial features, mental retardation, and abnormal calcium metabolism. It was common in the UK in the late 1950s and early 1960s when vitamin D supplementation of a variety of foods was practiced. (The relationship of periconceptional multivitamins to NTDs is discussed elsewhere.)

It seems clear that some members of dark-skinned races are at risk of vitamin D deficiency when resident in temperate climates. Their newborns may be at risk of hypocalcemia, a cause of convulsions in the newborn period. These children may also have delayed growth in the first year of life. Both problems may be corrected by vitamin D supplementation in pregnancy.

At various times, maternal zinc deficiency has been proposed as a cause of both congenital malformation and retarded fetal growth. No valid evidence has yet emerged of any benefit derived from maternal zinc supplementation with regard to either outcome in humans.

Food Safety in Pregnancy

Pregnant women share the concerns of the general public about food poisoning but there is a great deal of confusion in the mind of the public and the clinicians treating them about which foods are safe to eat during pregnancy.

Listeria monocytogenes

This organism is widely distributed in the environment and the low incidence of infection suggests that infection requires the ingestion of a large dose by a susceptible individual. Listeriosis is a significant problem in pregnancy because infection of the mother can cause fetal death following transmission of the organism across the placenta. It is not certain at the present time whether the apparent increase in fetal loss from this infection represents improved ascertainment or a genuine increase in disease frequency.

Most women do not realize that the symptoms of *Listeria* infection are not the same as those for some other foodborne pathogenic organisms: *Listeria* symptoms are often described as mild 'flu-like symptoms' rather than diarrhea and vomiting. The perceived risk of *Listeria* infection in pregnancy for many women appears to be much greater than the actual risk.

During pregnancy the following foods should be avoided to reduce the risk of *Listeria* infection:

- Mold-ripened cheese and some types of sheep and goats' milk cheese
- Pâté – meat, fish, and vegetable (unless canned or ultra heat-treated)

- Cook–chill foods that have not been thoroughly reheated

However, many health professionals advise women to avoid 'unpasteurized soft cheeses' which, to most pregnant women, means avoiding all types of cream and cottage cheese together with yogurt and fromage frais. In fact, it is only mold-ripened cheeses with either a skin such as that found on Brie and Camembert or a mold inside such as Danish blue that should be avoided, together with sheep and goats' milk cheese that has not been made with pasteurized milk. Unnecessary avoidance of these dairy foods will substantially decrease calcium intake.

Toxoplasmosis

Toxoplasma is a protozoan organism which can cross the placenta when primary infection occurs in pregnancy. It can lead to fetal death, mental handicap, and/or blindness in an estimated 30% of offspring of infected mothers. To reduce the risk of infection during pregnancy women are advised to:

- Avoid eating raw and undercooked meat.
- Always wash vegetables and salads well to remove soil.
- Wear gloves when gardening.
- Avoid unpasteurized sheep and goats' milk.
- Wash hands after contact with cats and kittens and avoid contact with stray cats.
- Wear rubber gloves when cleaning out cat litter trays or get somebody else to do this job if possible.

(*See* **Vegetarian Diets.**)

Salmonellosis

Salmonella bacteria are one of the most common causes of food poisoning in the UK, giving rise to sickness and diarrhea. Steering clear of raw eggs and products containing uncooked egg and exercising appropriate food hygiene can reduce this risk.

Campylobacteriosis

The main source of *Campylobacter* infection is raw milk and poultry but a large number of foods have been implicated in outbreaks. Again, campylobacter may not have any direct effect on the fetus, but the Department of Health advises women in the UK to avoid this distressing illness.

Shellfish

Pregnant women are advised not to eat oysters and other shelled seafood, such as prawns, mussels, and crab, unless they are part of a hot meal and have been thoroughly cooked. When raw these foods may be contaminated with harmful bacteria and viruses.

Bovine spongiform encephalopathy

In 1996 evidence of a new variant of Creutzfeldt–Jacob disease (CJD) began to be observed, which is most probably the result of eating beef infected with bovine spongiform encephalopathy (BSE). A particular source of concern at the moment is vertical transmission which, although common in cattle, has not been shown to occur in humans. However, there has recently been a possible case reported in the media of mother to child infection.

Other concerns of pregnant women in the area of food safety include the hazards of fungicide and pesticide residues. and possible harmful effects of natural food toxicants. At present it seems unlikely that there are toxic residues on foods which can be associated with harm in pregnancy, but this observation should not be taken as grounds for complacency. One of the main difficulties for scientists wishing to explore this matter is the limited availability of reproducible assays for the minute quantities to be studied. Similarly, the study of natural food toxicants for possible harmful effects on the fetus is in its infancy.

Conclusions

It is important to monitor continuously any effects our dietary habits may be having on child-bearing and child development, but we should be able to achieve this without causing unnecessary scares. Most women who consume a good general diet, and are not addicted to tobacco, alcohol, or other drugs, can contemplate pregnancy without concern that their dietary habits can 'harm the baby'.

See also: **Caffeine**; *Campylobacter*: Campylobacteriosis; **Cholecalciferol**: Properties and Determination; Physiology; **Coffee**: Physiological Effects; **Folic Acid**: Properties and Determination; Physiology; **Food Intolerance**: Food Allergies; **Food Poisoning**: Classification; **Food Safety**; **Garlic**; *Listeria*: Properties and Occurrence; Detection; Listeriosis; **Retinol**: Properties and Determination; Physiology

Further Reading

Department of Health (2001) Committee on Toxicity of Chemicals in Food, Consumer Products and the Environment. Statement on the Reproductive Effects of Caffeine 2001/06. London: HMSO.

Department of Health and Social Security (1990) *Vitamin A and Pregnancy.* PL/CMO (90) 11, PL/CNO (90). London: Stationery Office.

Department of Health and Social Security (2000) *Folic Acid and the Prevention of Disease.* London: Stationery Office.

Frisch RE and McArthur JW (1974) Menstrual cycles: fatness as a determinant of minimum weight for height necessary for their maintenance and onset. *Science* 185: 949–951.

Jensen KT, Hjollund NH and Henriksen TB (1998) Does moderate alcohol consumption affect fertility? Follow up study among couples planning first pregnancy. *British Medical Journal* 317: 505–510.

Karen G (ed.) (1990) *Maternal–Fetal Toxicology. A Clinician's Guide*. New York: Marcel Dekker.

natural history of untreated children, a lack of randomized trials, and a lack of diagnostic criteria for neurologic complications, there are controversies about optimal care. The economic costs of spina bifida are huge. The role and effectiveness of intra-uterine myelomeningocele repair is controversial; clinical trials have been recommended.

NTDs affect at least 300 000 newborns worldwide each year. NTD rates vary considerably around the world, with northern China reporting some of the

Maternal Diet, Vitamins, and Neural Tube Defects

M L Watkins, National Center on Birth Defects and Developmental Disabilities, Atlanta, GA, USA
S L Carmichael, March of Dimes/California Birth Defects Monitoring Program, Oakland, CA, USA

Background

The most interesting work relative to neural tube defect (NTD) etiology and prevention in the last few years has focused on nutritional factors. This chapter summarizes what is known about the relation between NTDs and various nutritional factors, with particular emphasis on folate, and includes a discussion of the methodologic challenges inherent in identifying such associations.

Neural Tube Defects

NTDs result from failure of normal closure of the embryonic neural tube at one or more of five separate closure sites. Complex forces are involved, with varying closure mechanisms at different sites within the neural tube. Of the several NTD types (**Figure 1**), spina bifida is the most common. The extent of physical disability (e.g., paralysis, bowel and bladder dysfunction) associated with spina bifida depends on the lesion level, size, and extent of spinal nerve involvement. Developmental, learning, and social problems are common. Most children born with spina bifida in developed countries now survive because of extensive medical and surgical treatment. Many of these children, however, face lifelong physical and developmental disabilities. The death rate for affected infants in the USA is about 10%, but the rate is much higher in countries with less accessible medical care. Clinical deterioration and secondary conditions occur in many affected children in a variety of body systems, resulting in complex medical and surgical care. Because of a lack of information about the

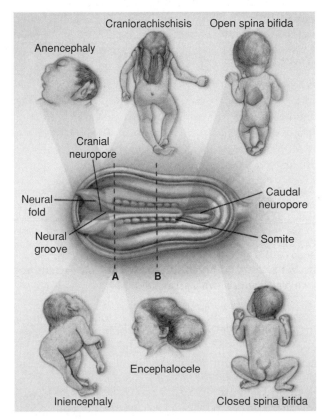

Figure 1 Neural tube defects. Schematic view (from above) of a neural tube. Main clinical types of neural tube defects. Anencephaly (lateral view): the absence of the brain and skull can be total or partial, always fatal. Craniorachischisis (posterior view): anencephaly occurs with contiguous bony defect of the spine and exposed neural tissue. Open spina bifida (posterior view): bony defect of the vertebrae (in this case, of the lower thoracic vertebrae), accompanied by exposure of neural tissue and meninges that are not covered by skin. Closed spina bifida (posterior view): bony defect of the vertebrae and, if present, the herniated meninges and neural tissue, are covered by skin. Extent of the disability associated with spina bifida related to the level and size of the lesion and the extent of spinal nerve involvement. Encephalocele (lateral view): the brain and meninges herniate through a skull defect. Iniencephaly (lateral view): the dysraphic process in the occipital region is accompanied by severe retroflexion of the neck and trunk. Spina bifida and anencephaly are more common than encephalocele, craniorachischisis, or iniencephaly. Adapted from Botto LD, Moore CA, Khoury MJ and Erickson JD (1999) Neural-tube defects. *New England Journal of Medicine* 341: 1509–1519, with permission.

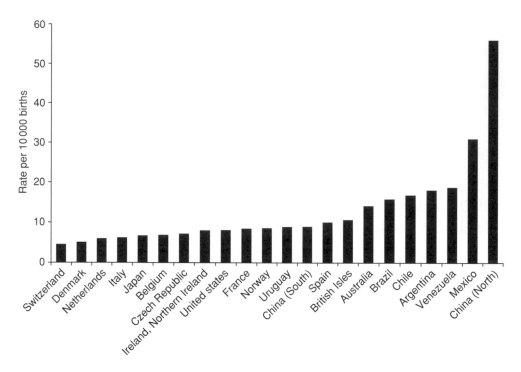

Figure 2 Rates of spina bifida and anencephaly by country. Cases reported in pregnancy terminations are included, where available. Rates for Latin-American countries and Japan are estimated from hospital-based registries. Rates for other locations are based on data from birth defects registries that usually monitor only part of a country's birth population. The following countries and registries are included: Switzerland (selected counties), Denmark (Odense), The Netherlands (North), Italy (Campania, Emilia-Romagna, Toscana), Belgium (Hainaut-Namur), Czech Republic, Ireland (Dublin, Belfast), United States (Atlanta, Hawaii), France (Bouches-du-Rhone, Paris, Strasbourg, Central East France), Norway, Spain (Basque country), UK (Glasgow, North Thames), Australia (Victoria, South Australia), and China (selected counties). Data from the International Center for Birth Defects and the European Registration of Congenital Anomalies (1998) *World Atlas of Birth Defects*, pp. 20–31. Geneva: World Health organization.

highest rates (**Figure 2**). Both environmental and genetic factors likely explain the variation in rates. Of NTDs, 80% are thought to be 'multifactorial,' meaning that they are influenced by both genetic and environmental factors. A minority of NTDs are caused by chromosomal abnormalities, single gene mutations and teratogens. Risk factors include maternal diabetes, use of folic acid antagonists, fever or hyperthermia in early pregnancy, low socioeconomic status, and obesity. In the USA, NTD rates are higher among Hispanic whites than non-Hispanic whites; blacks have the lowest rates. NTD-rate monitoring is complicated by the increasing use of prenatal diagnosis and selective termination, which results in lower birth prevalence. Because the neural tube develops very early in pregnancy (closure by 28 days postconception) before most women are aware they are pregnant, prevention efforts are challenging, especially in populations with high unplanned pregnancy rates. Approximately 95% of all NTD-affected pregnancies occur in women with no previous history, which is referred to as 'occurrence.' However, women who have already had an NTD-affected pregnancy have a 10-fold greater risk for having an NTD-affected

subsequent pregnancy than women who have not. Having more than one NTD-affected pregnancy is referred to as 'recurrence.'

The increased prevalence of NTDs in lower socioeconomic groups, decreasing NTD rates, and seasonal patterns of occurrence are consistent with a dietary etiology for NTDs. Certain nutrients (specifically folic acid) were thought to be involved because of their role in human growth and because birth defects occurred with maternal use of folic acid antagonists (e.g., aminopterin, when used as an abortifacient, induced fetal NTDs). Also, animal data have demonstrated NTD inducement resulting from various nutritional deficiencies and excesses, suggesting the plausibility of nutritional causes of NTDs.

Methodologic Issues in the Study of Nutrition and NTDs

However plausible the link between nutrition and NTDs, studying these associations presents inherent challenges. Approaches to measuring nutritional status can be broadly classified as dietary, biochemical, anthropometric, or genetic. Many of the difficulties in studying nutrition and birth defects stem from the fact that most studies are retrospective in nature, resulting from the relatively low prevalence of NTDs. This limitation has varied implications, depending on the particular approach used to measure nutritional status.

The food-frequency questionnaire is a tool commonly used to assess dietary intake retrospectively. This method produces reasonably valid and reliable, comprehensive, and semiquantitative data on usual or average intake of foods and nutrients. The validity of the data depends, however, on several factors, including the appropriateness of the food list for the study population; the accuracy of the nutrient database; and variability in individual nutrient requirements and in the bioavailability and absorption of nutrients from foods (which are affected by, for example, food processing, fiber intake, alcohol consumption, cigarette smoking, and infection). Challenges related to the analysis of dietary data include whether to analyze intake from foods and supplements separately or together; foods versus nutrients; or single versus multiple nutrients simultaneously. The high correlation that exists between nutrients makes it particularly difficult to isolate the independent effects of individual nutrients.

Depending on the particular nutrient, biochemical measures of nutritional status may vary by time of day, by the tissue from which the measurement is made (serum and toenail zinc, for instance, reflect more short-term versus long-term zinc status, respectively), and by various host factors (e.g., genes, behaviors such as smoking and alcohol consumption, health status, physical activity, stress, and medications). Furthermore, pregnancy itself is a time of changing nutritional requirements and absorption. These factors must be taken into consideration when interpreting results from individual studies and when comparing results across studies.

Anthropometric measurements (e.g., height and prepregnancy weight) can be recalled reasonably well. Whether retrospectively collected data on more complex measures (e.g., skinfold thickness and waist-to-hip ratio) can serve effectively as proxies for these parameters around the time of conception is questionable.

Genetic material is most commonly obtained from blood samples or buccal smears. The challenge is in discovering functionally relevant polymorphisms and in having sample sizes large enough to detect meaningful differences in risk. Recent advances in technology are enabling more rapid identification and analysis of polymorphisms, and innovative approaches to epidemiologic study design are being developed that enable smaller numbers of subjects for examination of certain types of genetic hypotheses.

Folic Acid and NTDs

In 1930, Dr. Lucy Wills discovered a factor that cured the nutritional deficiency anemia of pregnancy among women in India. This factor was later isolated from spinach and named 'folic acid' (*folium* is Latin for 'leaf'). Because humans are unable to synthesize folate, they must depend solely upon dietary sources. Folate-rich food sources include green leafy vegetables, grains, legumes, certain fruits, and liver. Because heat, ultraviolet light, and air inactivate food folate, food processing, preparation, and cooking can reduce the amount of food folate ingested by an estimated 50–95%. *Bioavailability*, the extent to which folates are available for use at the cellular level, varies widely across foods. The bioavailability of synthetic folic acid (monoglutamate form), which is used in cereal grain fortification and multivitamin/mineral supplements, is estimated to be about twice that of food folate (polyglutamate form).

Studies in the mid-twentieth century linked nutrition and NTDs. Lower vitamin C and folate levels in one study and poorer diet quality (in terms of macronutrient and fresh fruit and vegetable intake) in another were found among women with NTD-affected children. Intervention and observational studies followed (**Figure 3**). The most convincing and solid evidence for a preventive effect of folate is provided by two intervention trials. The international Medical Research Council (MRC) trial, a large UK-sponsored *recurrence* prevention trial, was conducted at 33 sites in seven countries. Women who had a previous NTD-affected pregnancy were randomized to one of four vitamin-use groups. Among 1195 pregnancies with known outcomes, folic acid (alone or with other vitamins) was associated with a 72% reduction in risk.

In 1992, a Hungarian randomized trial provided strong evidence for the efficacy of folic acid-containing multivitamin supplementation to prevent NTD *occurrence*. Results from this trial, combined with results from several observational studies (**Figure 3**) provided convincing evidence that occurrence is preventable. As a result, several countries recommended in 1992 that women of childbearing potential consume periconceptional folic acid daily (most commonly 400 µg) to prevent NTDs. The recommendation was made for all women capable of becoming pregnant (not just those planning a pregnancy), because these birth defects occur before many women are aware that they are pregnant, and because many pregnancies are unplanned. Since then, a large, nonrandomized community intervention in China demonstrated significant NTD reductions associated with the use of a 400-µg supplement containing only folic acid. Risk reductions were greater (85%) in the

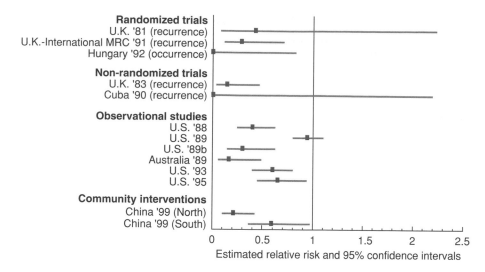

Figure 3 Risk of neural tube defects and use of folic acid or multivitamin supplements: summary of studies 1980–95, by country and year of publication. The figure shows risk estimates (boxes) and 95% confidence intervals. Risk estimates < 1 indicate a reduction in risk among the 'exposed' mothers (periconceptional use of folic acid or multivitamins). If the upper limit of confidence intervals is > 1, the risk estimate is not statistically significant. Some studies used folic acid (varying doses); some used multivitamins. Two randomized controlled trials (UK'81 and UK-international-MRC'91) and two nonrandomized trials (UK'83 and Cuba'90) examined the efficacy of folic acid with or without other vitamins among women with a previously affected pregnancy (recurrence). One randomized trial (Hungary'92) assessed the efficacy of a multivitamin containing folic acid among women without a previously affected pregnancy (occurrence). Of the observational studies, five (US'88, US'89, Australia'89, US'93, US'95,) were case-control studies, whereas one (US'89b,) was an observational cohort. The community intervention study (China'99) was conducted in two areas, one in Northern China and one in Southern China; it assessed the effectiveness of 0.4-mg folic acid supplements (without other vitamins) among women without a previously affected pregnancy in areas of high (North) and low (South) occurrence rates of neural tube defects. From Botto LD, Moore CA, Khoury MJ, and Erickson JD (1999) Neural-tube defects. *New England Journal of Medicine* 341: 1509–1519, with permission.

high-prevalence northern countries than in the lower-prevalence southern counties (40%). Additional evidence for the efficacy of folic acid comes from another study that found that periconceptional exposure to folic acid antagonist medications more than doubles the risk for NTDs and that folic acid supplementation reduced the risk elevation associated with certain folic acid antagonists.

Efficacy Issues

Based on the risk reduction observed in several studies, estimates are that at least 50% of NTDs could be prevented by the use of folic acid. The proportion of NTDs that are preventable may be less in low-prevalence areas than in high-prevalence areas. Some studies also suggest a smaller reduction in risk for NTDs associated with folic acid use for Hispanics than for non-Hispanic whites or blacks, and for obese women compared with average weight women. These observations require more study. The minimum daily effective dose of folic acid for NTD prevention is unknown; it may be < 400 μg for occurrence prevention, but determining the effectiveness of lower dosages is difficult because of the size and cost of studies necessary to assess the relative efficacy of various dosages.

The reasons for variability in folate efficacy are unknown, largely because the biologic mechanism by which periconceptional folic acid use protects against NTDs is unknown. Folic acid plays a role in transfer of methyl groups in the amino acid methylation cycles, a process essential for recycling homocysteine back to methionine (**Figure 4**). It is also a cofactor for enzymes involved in DNA and RNA synthesis, providing single carbon units for the synthesis of nucleotide bases, a process that is essential to the rapid cell division that occurs in early fetal development. Extrinsic folic acid may work by increasing tissue folate levels enough to override a failure in folate metabolism, or it may compensate for deficiency or metabolic defects related to other nutrients closely related to folate metabolism. Recent research has focused on the role of mutations in the genes that code for enzymes involved in folic acid metabolism.

Approaches to Increasing Folic Acid Consumption

Although the mechanism is unknown, a significant proportion of NTDs can be prevented with folic acid. The challenge is in how to translate science into public health policy and practice. **Table 1** summarizes three approaches for increasing folic acid consumption.

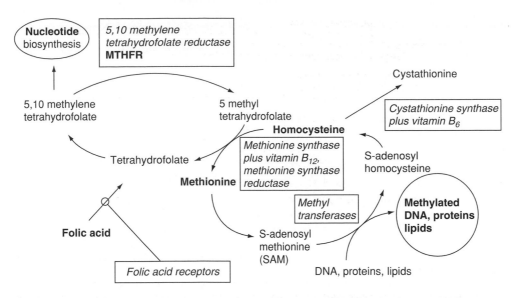

Figure 4 Metabolic roles of folic acid and related dietary factors. Simplified schematic of some of the metabolic processes that involve folic acid and other related factors. Folic acid (bottom left), after entering the cell (aided by folate receptors), participates in the transfer of carbon units used for the synthesis of nucleotides (top left) or, through the conversion of homocysteine to methionine (center), for methylation purposes. These processes are regulated by many factors including enzymes (in rectangular boxes) and vitamins other than folic acid (e.g., vitamins B_6 and B_{12}). The activity of some enzymes (e.g., methionine synthase) may be influenced by other enzymes (e.g., methionine synthase reductase). Adapted from Botto LD, Moore CA, Khoury MJ, and Erickson JD (1999) Neural-tube defects. *New England Journal of Medicine* 341: 1509–1519, with permission.

Table 1 Ways to increase folic acid/folate consumption to prevent neural tube defects

Approach	Advantages	Disadvantages	Status
Fortify staple foods	Proven to increase blood folate level Reaches almost all women Behavior change not necessary unless higher intakes needed Inexpensive	All consumers (not just target population) increase intake; concern that some will have excess intake Variable staple consumption results in variable folic acid intake Sustained behavior change required for consumption of higher amounts	Mandatory fortification in the USA, Canada, and several Central and South American and eastern European countries; proposed in the UK
Increase use of supplements	Efficacy proven by clinical trials Proven to increase blood folate level Relatively inexpensive	Requires sustained behavior change Reservations of some medical and nutritional professionals	Varies by country and survey question (e.g., in the USA, 30% of reproductive-age women report regular use; in The Netherlands, 54% of women report 'any periconceptional use' and 21% 'appropriate use')
Increase consumption of foods with high levels of natural folates (e.g., fruits and vegetables)	Other benefits of healthy diet that includes fruits and vegetables	Efficacy not proven Much smaller increases in blood folate levels than supplements or fortified foods Requires significant and sustained behavior change May be expensive	Various healthy diet campaigns under way

Fortification of a food staple provides wide population coverage at low cost without requiring behavior change. The optimal fortification level is controversial. Some experts had concerns that daily folic acid intakes in excess of 1 mg could adversely affect some persons, particularly persons with untreated cobalamin deficiency. Folic acid can ameliorate the anemia associated with cobalamin deficiency, possibly leading to a failure to detect and treat the cobalamin deficiency, which could result in the initiation or progression of neuropathy. Although some experts in the study of birth defects advocated a higher fortification level, the US Food and Drug Administration ruled that, effective January 1, 1998, all flour, corn meal, pasta, and rice labeled as 'enriched' be fortified with folic acid at 140 mg per 100 g of cereal grain product. Canada and several Central and South American and eastern European countries also have mandated fortification of wheat flour. The fortification level in some countries (e.g. Chile) was established at a higher level, with the goal of achieving a higher average daily intake of folic acid. In 2000, the UK Committee on Medical Aspects of Food and Nutrition Policy recommended folic acid fortification at 240 µg per 100 g of flour. At the time of writing, this proposal is under review but not yet approved, although voluntary optional fortification is allowed in the UK. Serum and red blood cell folate levels among reproductive aged women in the USA have more than doubled since fortification of cereal grain products. In addition, preliminary analyses suggest an approximate 20% decline in NTD rates postfortification.

Although multivitamin supplements containing folic acid are of proven efficacy in preventing NTDs, their use requires a sustained behavior change by most women. Educational campaigns in the USA have increased awareness and knowledge about folic acid more than they have increased actual supplement use. In 2001, although 80% of childbearing aged women in the USA had heard of folic acid, only 29% took a daily vitamin containing folic acid, up only slightly from 25% in 1995. In the UK and in The Netherlands, public-awareness campaigns significantly increased both knowledge about, and use of, folic acid. For example, after a 1995 campaign in The Netherlands, use of folic acid for 'any period around conception' increased from 25% before the campaign to 54% after, and 'appropriate' use increased from 5 to 21%. Whether increases in supplement use can be sustained and whether levels of use plateau at a certain level is unknown. Supplement use is less common among younger women, less educated women, and those with lower incomes – populations that are often harder to reach through health education efforts.

Attempts to increase population consumption of folate-rich foods (e.g., fruits and vegetables) through programs like the 'Five-a-Day' campaign in the USA have generally resulted in only modest increases. One small study found that neither dietary advice nor folate-rich foods significantly increased women's red cell folate levels, whereas supplements and fortified foods did. This could result from the lower bioavailability of food folate and/or the challenges of making dietary behavior changes.

Folate-related Nutrients

As noted previously, the preventive effects of folate supplementation are more likely to result from its compensation for metabolic errors closely related to folate metabolism, rather than overt folate deficiency. Folate is involved in an intricate network of reactions that include many enzymes, nutrients, and other substrates that could affect NTD risk through a variety of mechanisms (**Figure 4**). The conversion of 5-methyl tetrahydrofolate to tetrahydrofolate produces the substrate necessary for the transmethylation of homocysteine to methionine; the enzyme methionine synthase (MS) catalyzes the reaction, and vitamin B_{12} is a cofactor for this enzyme (**Figure 4**). Homocysteine may also be metabolized to cystathionine via cystathionine β synthase (CBS), of which vitamin B_6 is a cofactor.

Hyperhomocysteinemia, which has been shown to cause NTDs in animal models, may reflect impaired metabolism of folate, B_{12}, B_6, or homocysteine, may reflect a defect in the enzymes 5,10-methylenetetrahydrofolate reductase (MTHFR), MS, or CBS, or may be harmful on its own. Several studies have reported an increased NTD risk associated with high homocysteine levels in humans, in the amniotic fluid of mothers carrying affected fetuses, in maternal serum during pregnancy and postpartum, and among children with NTDs, although results have not been entirely consistent.

Animal models indicate that methionine is also important to neural tube closure. Although the mechanism is unknown, methionine is necessary for protein synthesis and transmethylation reactions. Two population-based, case-control studies have reported a 30–50% reduction in risk for NTDs with increasing quartiles of methionine intake. Other studies have been inconsistent: some have demonstrated higher levels of methionine in the amniotic fluid of mothers carrying NTD-affected fetuses, whereas another study reported no difference in maternal serum methionine post-partum.

Several studies have indicated lower levels of vitamin B_{12} in the amniotic fluid and serum of women with affected pregnancies. Studies of methyl malonic acid (an indicator of B_{12} deficiency or a defect in B_{12} metabolism) and B_{12} carrier proteins also suggest a defect in B_{12} metabolism among case mothers. Other studies have not supported these differences or have been inconsistent in their findings depending on the timing of and tissues used for measurement. Although the results are mixed, they do not seem to provide strong evidence for a relation between B_{12} deficiency and NTD risk. Few studies have examined the association between vitamin B_6 and risk for NTDs;

significant differences have not been observed, but studies have involved only small sample sizes.

There is some evidence that polymorphisms in the genes for MTHFR, MS, and CBS contribute to altered levels of these nutrients and to NTD risk, but again, results have been mixed.

The individual contributions of each of these indicators of metabolic disturbances to NTD risk, and their interdependence, are unclear at this point. Folic acid supplementation might play a role in overcoming the negative effects of many of these disturbances. However, until the exact causes of NTDs are determined, the understanding of the implications of these findings and the way in which they should be used for prevention purposes remains limited.

Other Nutrients

At least 30–50% of NTDs are folate-resistant; therefore, other nutrients have been implicated as possible causes of these cases. The evidence for the role of vitamin A, zinc, and inositol in NTD etiology is reviewed here. Other nutrients – including vitamin C, riboflavin, iron, calcium, magnesium, selenium, lead, and copper – also may be associated with increased NTD risk, but evidence is very limited.

The finding that isotretinoin, a synthetic derivative of vitamin A, is teratogenic led to investigations of whether naturally derived vitamin A is associated with increased risk for NTD-affected pregnancy. Most studies have found no evidence of an association of high vitamin A levels with NTD risk, based on intake from foods and supplements and on serum samples taken during the first trimester of pregnancy or postpartum. The safe level of vitamin A intake is still debatable, in part because of its potential association with other types of malformations and its association with other negative health outcomes. The effects of deficient vitamin A intake on NTD have not been explored.

Zinc is important to cellular growth and differentiation and to folate absorption. Various experimental studies in animals and case reports of women with specific disturbances of zinc metabolism have suggested an association between zinc and increased risk for NTD-affected pregnancy. Results for analytic studies of zinc and NTDs in humans are inconsistent. Postpartum studies have reported NTD case mothers to have levels of zinc in the hair that were higher, lower, or the same as those in control mothers, higher levels in toenails, and the same as control mothers in serum levels. Studies of pregnant women have reported that mothers with NTD-affected pregnancies have higher levels of zinc in serum, lower levels in white blood cells, and lower dietary intakes, whereas

other studies have reported no difference in serum or amniotic fluid. Despite the number of studies, inference is limited by the wide variability in zinc measurement and by the limited understanding of the relationships between these measures.

Inositol is a 6-carbon polycyclic alcohol that is present in all cells, with especially high levels in nerve cells. It protects against NTDs in the curly-tailed mouse, a folate-resistant NTD model, leading some researchers to recommend supplementation with a combination of folate and inositol. However, the applicability of these findings to humans has not been studied.

Limitations

In general, the proposed associations reviewed here are biologically plausible and supported by animal models. Unfortunately, systematically testing these associations in humans is difficult, and most existing evidence is clinical or experimental and speculative. Sample sizes for many of the studies have been small (often involving < 20 case-patients). Study population characteristics and selection methods have been highly variable; very few have been population-based. Few studies have adjusted for potential confounding or effect modification by other maternal characteristics, including status with respect to other nutritional factors or individual nutrients.

The ability to measure maternal nutritional status is also limited. Because pregnancy is a time of tremendous metabolic change, the ability of measurements taken later in pregnancy to represent the status during the time of neural tube closure is questionable. For example, the fetal contribution to levels of nutrients in amniotic fluid is uncertain and may vary by timing of measurement and by nutrient. Also, many women begin taking multivitamin/mineral supplements soon after the time of neural tube closure.

The cause of observed variability in nutrient levels is usually unknown, limiting the understanding of how to use such information for preventive measures. Nutrient metabolism is highly complex, and many reactions are interdependent. For example, depending on their cause, high homocysteine levels could be remedied with supplementation of B_{12}, methionine, folate, or B_6; low calcium intake may result in increased bioavailability of lead, reduced bioavailability of vitamin B_{12}, and, accordingly, reduced methionine synthase activity; and calcium, fiber, protein, and methionine are all important to zinc bioavailability. Therefore, it is critical that future studies examine interrelationships, including nutrient–nutrient interactions as well as gene–nutrient interactions, whenever possible.

Other Nutritional Factors

Evidence for a link between tea and NTDs is weak. A study in England and Wales showed an excess of anencephaly among women drinking three or more cups of tea per day, but no dose–response effect was observed, and the relation was observed only in regions with higher NTD prevalence rates. Furthermore, a study in Boston, Philadelphia, and Toronto found no elevated risk for NTDs associated with tea drinking. Nor is the link between coffee and NTDs strong. In the Boston study, no significant differences in coffee consumption were observed between the mothers of NTD-affected infants and control mothers. Also, no association with decaffeinated coffee or cola drinks has been found. A Finnish study found no association between central nervous system defects and coffee intake. Although there are case reports of NTDs occurring among infants of mothers who consumed significant amounts of alcohol during pregnancy, epidemiologic studies have provided little evidence for a link between NTDs and alcohol, especially for mothers consuming minimal or moderate amounts.

Some food contaminants have been proposed to be NTD teratogens. Potato blight, caused by a fungus (*Phytophthora infestans*) that rots potatoes, was hypothesized to cause NTDs in the early 1970s, although most evidence does not support this hypothesis. Fumonisin, a potent mycotoxin prevalent in corn and a carcinogen and cause of various animal diseases has been suggested as a possible cause of high NTD rates among persons living along the Texas–Mexico border who commonly consume corn products. Recent research suggests that fumonism may be involved in NTD formation by inhibiting folate uptake by cells. Also, fumonisins have been shown to produce NTDs in mouse whole embryo cultures; folic acid has ameliorated this teratogenic effect. More research is needed to establish the link between fumonisin consumption and NTD risk in humans.

About one-fourth of childbearing age women in the USA and almost as many such women in Canada, the UK, and Continental Europe, are obese, and obesity prevalence is increasing rapidly. Studies consistently find a twofold elevated risk for NTDs among children of obese women compared with mothers of average weight, although risk estimates vary with the body mass index cutoffs and referent groups used (**Figure 5**). Most studies found a higher risk estimate for spina bifida than anencephaly and found no risk elevation for encephalocele. Reasons for the increased risk are unknown. Potential explanations include teratogenic metabolic abnormalities associated with obesity including elevated levels of glucose, insulin,

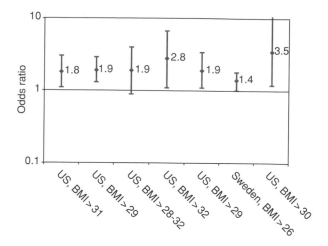

Figure 5 Risk of neural tube defects in offspring of obese women, odds ratios, and 95% confidence intervals. All studies except one (Sweden) were performed in the USA. The body mass index (BMI) used to define obesity varied (see x-axis). The referent group was average-weight or nonobese women.

uric acid, and endogenous estrogens. Obese women may be more likely to be nutritionally deficient because of a self-imposed diet restriction, poor choice of foods, or increased requirement for folate or other nutrients. More studies are needed to help elucidate the mechanism. The population burden of NTDs associated with obesity is substantial and increasing.

Conclusion

There is still much to be learned about the complex link between nutrition and birth defects. Evidence is emerging regarding the role of nutritional factors in the etiology of birth defects other than NTDs (e.g., heart defects and orofacial clefts). Despite the challenges in studying nutritional causes of birth defects, the prevention potential is substantial and worthy of pursuit.

See also: **Cobalamins**: Properties and Determination; **Folic Acid**: Properties and Determination; Physiology; **Food Fortification**; **Nutritional Assessment**: Importance of Measuring Nutritional Status; Anthropometry and Clinical Examination; Biochemical Tests for Vitamins and Minerals; Functional Tests; **Obesity**: Epidemiology; **Retinol**: Properties and Determination; Physiology; **Vitamins**: Determination; **Zinc**: Properties and Determination; Physiology; Deficiency

Further Reading

Anonymous (2000) Folic acid and the prevention of disease. Report of the Committee on Medical Aspects of Food and Nutrition Policy. *Reports on Health & Social Subjects* 50: i–xv, 1–101.

Berry RJ, Li Z, Erickson JD *et al.* (1999) Prevention of neural-tube defects with folic acid in China. *New England Journal of Medicine* 341: 1485–1490.

Botto LD, Moore CA, Khoury MJ and Erickson JD (1999) Neural-tube defects. *New England Journal of Medicine* 341: 1509–1519.

Centers for Disease Control and Prevention (1992) *Recommendations for the Use of Folic Acid to Reduce the Number of Cases of Spina Bifida and Other Neural Tube Defects. MMWR 41 (No. RR-14),* pp. 1–7.

Cogswell ME, Perry GS, Schieve LA and Dietz WH (2001) Obesity in women of childbearing age: risks, prevention, and treatment. *Primary Care Update Ob/Gyns* 8: 89–105.

Elwood M, Elwood H and Little J (1992) Diet. In: Elwood JM, Little J and Elwood JH (eds) *Epidemiology and Control of Neural Tube Defects,* pp. 521–603. Oxford: Oxford University Press.

Institute of Medicine (1998) *Dietary Reference Intakes for Thiamin, Riboflavin, Niacin, Vitamin B_6, Folate, Vitamin B_{12}, Pantothenic Acid, Biotin, and Choline: A Report of the Standing Committee on the Scientific Evaluation of Dietary Reference Intakes and its Panel on Folate, Other B Vitamins, and Choline and Subcommittee on Upper Reference Levels of Nutrients.* Washington, DC: National Academy Press (see http://www.nap.edu/openbook/ 0309065542/html).

International Centre for Birth Defects and European Registration of Congenital Anomalies (1998) *World Atlas of Birth Defects,* pp. 20–31. Geneva: World Health Organization.

International Clearinghouse for Birth Defects Monitoring Systems (1998) *Annual Report 1998 with Data for 1996,* pp. 100–101. Rome: International Centre for Birth Defects.

Kirke PN, Molloy AM, Daly LE *et al.* (1993) Maternal plasma folate and vitamin B_{12} are independent risk factors for neural tube defects. *Quarterly Journal of Medicine* 86: 703–708.

Mills JL, Scott JM, Kirke PN *et al.* (1996) Homocysteine and neural tube defects. *Journal of Nutrition* 126: 756S–760S.

Refsum H (2001) Folate, vitamin B_{12} and homocysteine in relation to birth defects and pregnancy outcome. *British Journal of Nutrition* 85 (supplement 2): S109–S113.

Rosenquist TH and Finnell RH (2001) Genes, folate and homocysteine in embryonic development. *Proceedings of the Nutrition Society* 60: 53–61.

Scott JM, Weir DG and Kirke PN (1995) Folate and neural tube defects. In: Bailey LB (ed.) *Folate in Health and Disease,* pp. 329–361. New York: Marcel Dekker.

Sells CJ and Hall JG (eds) (1998) *Neural Tube Defects. Mental Retardation and Developmental Disabilities Research Reviews* 4(4). New York: Wiley-Liss.

Wald NJ (1994) Folic acid and neural tube defects: The current evidence and implications for prevention. CIBA. *Foundation Symposium* 181: 192–208.

Preeclampsia and Diet

E Abalos, G Carroli and L Campodónico, Centro Rosarino de Estudios Perinatales, Rosario, Argentina
J Villar, World Health Organization, Geneva, Switzerland

Background

Hypertensive disorders during pregnancy are one of the main causes of maternal death worldwide, most of these deaths being attributed to eclampsia. Hypertensive disorders occur in 6–8% of all pregnancies, contributing significantly to stillbirths and neonatal morbidity and mortality. Babies are also at increased risk of intrauterine growth restriction, low birth weight and preterm delivery. Pregnant women with hypertension, either newly diagnosed or pre-existing, are prone to the development of potentially lethal complications, notably abruptio placentae, disseminated intravascular coagulation, cerebral hemorrhage, pulmonary edema, hepatic failure, and acute renal failure. The etiology of hypertensive disorders related to pregnancy, particularly preeclampsia, remains unknown. The most important consideration in the classification of the disease is differentiating hypertensive disorders that antedate pregnancy from those that are pregnancy-specific, from which the more ominous are preeclampsia and eclampsia. Preeclampsia is a pregnancy-specific syndrome of reduced organ perfusion secondary to vasospasm and activation of the coagulation cascade. Although our understanding of this syndrome has increased, the criteria used to identify the disorder remain a subject of confusion and controversy. In chronic hypertension, elevated blood pressure is the primary pathophysiologic feature, whereas in preeclampsia, increased blood pressure is important primarily as a sign of the underlying disorder. As might be expected, the impact of the two conditions on mother and fetus is different, as is their management **Table 1**.

Classification

There are controversies about the definition of hypertensive disorders during pregnancy, and several classifications have been suggested. Recently, the National High Blood Pressure Education Program Working Group on High Blood Pressure in Pregnancy updated the 1990 report, and classified the hypertensive disorders during pregnancy

Table 1 Effectiveness of nutritional interventions in hypertension during pregnancy and preeclampsia

Intervention	Hypertension during pregnancy		Preeclampsia	
	Practice	Research	Practice	Research
Nutritional advice	No evidence		No effect RR = 0.89 (0.42–1.88)	
Balanced protein (<25%) / energy	No evidence		No effect RR = 1.20 (0.77–1.89)	
Isocaloric balanced protein (<25% of total energy)	No evidence		No effect RR = 1.00 (0.57–1.75)	
Energy/protein restriction for high PI or high weight gain	No effect RR = 0.97 (0.75–1.26)		No effect RR = 1.13 (0.59–2.18)	
Salt restriction	No effect RR = 0.97 (0.49–1.94)		No effect RR = 1.11 (0.46–2.66)	
Calcium	Possibly beneficial for women at high risk RR = 0.45 (0.31–0.66) and with low baseline intake RR = 0.49 (0.38–0.62)	RCT in progress	Possibly beneficial for women at high risk RR = 0.21 (0.11–0.39) and with low baseline intake RR = 0.32 (0.21–0.49)	RCT in progress
Iron and folate	No effect RR = 1.15 (0.41–3.18)		No evidence	
Folate	No effect RR = 1.26 (0.90–1.76)		No evidence	
Magnesium	No evidence	Needed	No effect RR = 0.87 (0.57–1.32)	Needed
Fish oil	No effect RR = 0.96 (0.86–1.07)		Possibly beneficial (data from low quality studies) RR = 0.70 (0.55–0.90)	
Zinc	No effect RR = 0.87 (0.65–1.15)	Needed	No evidence	Needed
Antioxidants	No evidence	Needed	Possibly beneficial for vitamins C and E (data from one RCT) RR = 0.46 (0.24–0.91)	Needed

as (1) chronic hypertension defined as hypertension observable before pregnancy, or diagnosed before the 20th week of gestation; (2) preeclampsia, which is a pregnancy-specific syndrome usually occurring after 20 weeks' gestation, determined by hypertension with proteinuria; (3) preeclampsia superimposed on chronic hypertension; and (4) pregnancy-induced hypertension or gestational hypertension, which is transient hypertension detected for the first time after midpregnancy if preeclampsia is not present at the time of delivery and blood pressure returns to normal by 12 weeks post partum (a retrospective diagnosis). The system suggested by the International Society for the Study of Hypertension in Pregnancy (ISSHP) defines hypertension as a diastolic blood pressure of 90 mmHg or above on two consecutive occasions at least 4 h apart, or a single diastolic blood pressure of 110 mmHg or more. The definition of preeclampsia has the same criteria for high blood pressure, but with the addition of significant proteinuria, usually at least 300 mg per 24 h or 1+ on dipsticks.

Pathophysiology of Preeclampsia

Preeclampsia is a syndrome with both fetal and maternal manifestations. The maternal disease is characterized by vasospasm, activation of the coagulation system, and perturbations in many humoral and autacoid systems related to volume and blood pressure control. The pathologic changes in this disorder are primarily ischemic in nature and affect the placenta, kidney, liver, and brain. Of importance, and distinguishing preeclampsia from chronic or gestational hypertension, is that preeclampsia is more than hypertension; it is a systemic syndrome, and several of its 'nonhypertensive' complications can be life-threatening, even when blood pressure elevations are quite mild.

The cause of preeclampsia is not known. Many consider the placenta as the pathogenic focus for all manifestations of preeclampsia because the delivery of both the baby and the placenta is the only definitive cure of this disease. There is no disease without the placenta. Thus, research has focused on the changes in the maternal blood vessels that supply blood to the placenta. Failure of the spiral arteries to remodel is postulated as the morphologic basis for decreased placental perfusion in preeclampsia, which ultimately may lead to early placental hypoxia. Oxidative stress and inflammatory-like responses may also be important in the pathophysiology of preeclampsia.

Research on how alterations in the immune response at the maternal interface might lead to preeclampsia addresses the link between placenta and maternal disease. A nonclassical human leukocyte antigen (HLA), HLA G, is expressed in normal placental tissue and may play a role in modulating the maternal immune response to the immunologically foreign placenta. Placental tissue from preeclamptic pregnancies may express fewer or different HLA G proteins, resulting in a breakdown of maternal tolerance to the placenta. Additional evidence for alterations in immunity in pathogenesis includes the disease prominence in nulliparous gestations with subsequent normal pregnancies, a decreased prevalence after heterologous blood transfusions, long cohabitation before successful conception, and observed pathologic changes in the placental vasculature in preeclampsia that resemble allograft rejection. Finally, there are increased levels of inflammatory cytokines in the placenta and maternal circulation, as well as evidence of increased 'natural killer' cells and neutrophil activation in pre-eclampsia.

The mechanisms underlying vasoconstriction and altered vascular reactivity in preeclampsia remain obscure. Research has focused on changes in the ratio of vasodilative and vasoconstrictive prostanoids, since there is evidence suggesting decrements and increments in the production of prostacyclin and thromboxane, respectively. More recently, investigators have postulated that the vasoconstrictive potential of pressor substances (e.g., Angiotensin II and endothelin) is magnified in preeclampsia as a consequence of a decreased activity of nitric oxide (NO) syntheses and decreased production of NO-dependent or -independent endothelium relaxing factor (EDRF). Also under investigation is the role of endothelial cells (the site of prostanoid, endothelin, and EDRF production), which, in preeclampsia, may be dysfunctional, owing perhaps to inflammatory cytokines (e.g., tumor necrosis factor α) and increased oxidative stress. Other systems postulated to play a role in pre-eclamptic hypertension are the sympathetic nervous system, calciotrophic hormones, insulin, and magnesium metabolism.

Finally, some nutritional factor deficiencies have been postulated as playing a role in the pathogenesis of preeclampsia. This chapter discusses their possible role on the hypertensive disorders of pregnancy.

Possible Role of Nutrition in the Pathophysiology of Preeclampsia

Nutrient supplements in addition to food sources of nutrients are provided to populations either to increase the intake among those with a deficiency (in order to prevent or treat functional outcomes related to such a deficit) or to obtain a pharmacological, perhaps nonnutritional effect, among individuals with an adequate intake of the nutrient in question.

Epidemiological observations have long suggested a role for nutritional deficiencies in preeclampsia (i.e., calcium, proteins, vitamins, etc.). However, intervention evaluations have failed to confirm such promising observations. We will describe here the evidence from randomized-controlled trials that support the relationship between different nutrients and preeclampsia.

Calcium

There is considerable evidence linking calcium intake and hypertension during pregnancy from observational and experimental studies. However, there is still no satisfactory explanation for the mechanisms involved in the calcium-mediated effect on blood pressure prevention. It was postulated that parathyroid hormone could be involved in this relationship. Demonstrated alterations in extracellular calcium homeostasis in preeclampsia include hypocalciuria and decrease serum levels of calcitriol. Increased parathyroid hormone (PTH) and decreased plasma ionized calcium concentration have not been consistently observed. Also, consistent abnormalities of intracellular calcium metabolism have been described in pre-eclamptic women, such as increased intracellular free calcium concentration in platelets and lymphocytes. Increases in intracellular free calcium concentration in circulating cells are hypothesized to result from fluctuation in hormones or vasoactive substances that cause similar alteration in vascular smooth muscle. Pregnancy is a state of high calcium requirements as a result of fetal demands while maternal adaptive mechanisms are partially inhibited. These phenomena lead to the hyperparathyroid state of pregnancy. An increase in parathyroid hormone serum levels would involve an increase in free intracellular calcium. Thus, the concentration of intracellular free calcium in vascular smooth muscle cells determines the degree of tension and is the trigger for muscular contraction. So, the vasoconstrictive effect, with a rise in blood pressure, results from an increase in vascular smooth muscle tension.

Antioxidant Agents

An additional role for nutrition in the genesis of pre-eclampsia could be nutritional factors that strengthen oxidative stress, leading to preeclampsia. A nutritional factor could be the deficiency of antioxidant intake, specifically vitamins C and E. Vitamin C is central for the neutralization of both water-soluble and lipid-soluble free radicals. As a water-soluble molecule, its ability to neutralize free radicals in the aqueous compartment is clear. Also, ascorbate is not made in humans and must come from the diet. Vitamin E, a potent antioxidant, has been suggested to play a role in preventing preeclampsia. Vitamin E is not usually reduced by dietary deficiencies, and reductions are more likely due to consumption.

Other Nutrients

Nutritional factors other than antioxidants can also contribute to oxidative stress. Hyperhomocysteinemia can be a result of dietary deficiencies. Hyperhomocysteinemia is a risk factor for preeclampsia and atherosclerosis, and endothelial function is said to be altered, at least in part, by the genesis of oxidative stress. Vitamins B_6 and B_{12} and folic acid are involved at different steps in the metabolic pathway for removing or recycling homocysteine to methionine. Dietary deficiencies of any of these micronutrients can increase circulating homocysteine. Preeclampsia is characterized by increased triglycerides that favor the formation of small, dense low-density lipoprotein (LDL). This lipoprotein variant has increased access to the subendothelial space, where it is sequestered from bloodborne antioxidants. The relevant role of triglycerides in the genesis of preeclampsia is indicated by the fact that they are increased long before clinically evident disease. Similarly, free fatty acids are increased in preeclampsia, and this increment was observed months before the diagnosis. Recent studies indicate that this effect may be secondary to altered copper binding by albumin to which large amounts of free fatty acids are bound. Unbound copper is a potent stimulator of free radical formation. Normally, this effect of copper is prevented by protein binding (quantitatively, primarily to albumin). However, with fatty acid binding, albumin binds copper differently. In this configuration, copper bound to albumin maintains its ability to participate in redox reactions. Thus, it appears that increased free fatty acids can also contribute to oxidative stress.

All of these nutritional alterations may be agreeable to dietary modification raising the possibility of nutritional prophylaxis.

Nutritional Interventions and Hypertensive Disorders of Pregnancy

Prevention

The ability to prevent hypertensive disorders of pregnancy is limited by a lack of knowledge of its underlying etiology. Prevention is focused on identifying women at higher risk of developing pregnancy-induced hypertension or preeclampsia during pregnancy, followed by close clinical and laboratory monitoring to recognize the clinical symptoms of the disease in its early stages. These women and their pregnancies can then be selected for more intensive monitoring or delivery. Although these measures do not prevent the disease, they may be helpful for preventing some adverse maternal and fetal sequelae.

As part of many other nonpharmacological interventions, some dietary interventions were proposed to prevent the development of pregnancy-induced hypertension and preeclampsia.

Nutritional advice in pregnancy Literature was reviewed in order to assess the effects of advising pregnant women to increase their energy and protein intakes on the outcome of pregnancy, and maternal and fetal/infant morbidity and mortality. Nutritional advice appears to be effective in increasing pregnant women's energy and protein intake, but the implications for fetal, infant, or maternal health cannot be judged from the available evidence. Preeclampsia prevention was assessed only in one small trial involving 136 women with no beneficial effects [relative risk (RR): 0.89, 95% confidence intervals (CI) 0.42–1.88].

Protein/energy supplementation The effect of balanced protein/energy supplements for pregnant women on gestational weight gain and pregnancy outcomes was assessed on a Cochrane systematic review. Preeclampsia prevention was evaluated in three trials involving 516 women, with no significant beneficial effects (RR: 1.20; 95% CI: 0.77–1.89). However, these trials had methodological flaws, alternate allocation, and large lost-to-follow-up for the main outcomes, so results should be taken cautiously. In another Cochrane systematic review, only one trial involving 782 women evaluated preeclampsia prevention when isocaloric balanced protein/energy supplements were given to underweight pregnant women, showing no effect (RR: 1.00; 95% CI: 0.57–1.75).

Energy/protein restriction for obese pregnant women Excessive weight gain during pregnancy has long been recognized as a clinical sign of edema and impending preeclampsia. Epidemiological studies suggest that a high maternal weight is positively associated with the risk of preeclampsia.

Energy/protein restriction for high weight-for-height or weight gain during pregnancy was assessed in a Cochrane systematic review. Preeclampsia was evaluated in two trials (284 women) showing no reduction in the risk of occurrence (RR: 1.13; 95% CI: 0.59–2.18), and the same as for pregnancy-induced hypertension (three trials, 384 women; RR: 0.97; 95% CI: 0.75–1.26). The limited evidence available suggests that protein/energy restriction of pregnant women who are overweight or show a high weight gain is unlikely to be beneficial and may be harmful to the developing fetus. Although weight reduction may be helpful in reducing or preventing high blood pressure in nonpregnant women, it is not recommended during pregnancy, because there is no effect on preventing preeclampsia, even in obese women. Clinicians frequently ask pregnant women to restrict their food intake in an attempt to prevent preeclampsia, despite the absence of evidence that such advice is beneficial.

Salt restriction   Even in an early phase of pregnancy, marked hemodynamic changes occur, including a fall in vascular resistance and blood pressure and a rise in cardiac output. To compensate for the increased intravascular capacity, the kidney retains more sodium and water. Apparently, the set point of sodium homeostasis shifts to a higher level at the expense of an expansion of extracellular volume. In nonpregnant women, a strong positive association of sodium intake with blood pressure has been established, but the relationship between sodium intake and blood pressure in human pregnancy remains obscure up to date. For decades, a low salt diet has been often recommended as treatment for edema, with the idea that restricting salt intake would treat, and also prevent, preeclampsia. Recently, this practice was questioned, and even a high sodium intake was proposed for preeclampsia treatment and prevention.

Concerns about the effect of a low sodium diet during pregnancy on maternal nutritional status led researchers to investigate whether such changes could alter other nutrient intakes. It was shown that the reduction in sodium intake also caused a significant reduction in the intake of energy, protein, carbohydrates, fat, calcium, zinc, magnesium, iron, and cholesterol. Even though women are no longer advised by many clinicians to alter their salt intake during pregnancy, this is still current practice in many other settings around the world.

A recently published Cochrane systematic review evaluates the effect of the advice about low dietary salt intake during pregnancy. The review includes two trials with data reported for 603 women. Both compared nutritional advice to restrict dietary salt with advice to continue a normal diet. Women with established preeclampsia were not enrolled, so this review provides no information about the effects of advice to restrict salt intake for the treatment of preeclampsia. No effect was found in preventing pre-eclampsia (RR: 1.11; 95% CI: 0.46–2.66) or pregnancy-induced hypertension (RR: 0.97; 95% CI: 0.49–1.94). Women's preferences were not reported, but authors presumed that a low-salt diet was not very palatable and was therefore difficult to follow.

Calcium supplementation   A role for altered calcium metabolism in the pathogenesis of preeclampsia is suggested by epidemiological evidence linking low dietary levels of calcium with increased incidence of the disease. In agreement with these observations, several modifications in calcium metabolism have been observed in pre-eclamptic women and in calcium-supplemented mothers.

A Cochrane systematic review of calcium supplementation during pregnancy has been published. Authors pre-specified comparison groups taking into account women's risk of hypertensive disorders of pregnancy (low versus increased) and women's baseline dietary calcium intake (low, < 900 mg per day, versus adequate, ≥ 900 mg per day).

High blood pressure with or without proteinuria was evaluated in nine trials, involving 6604 women. Overall, there is less high blood pressure with calcium supplementation (RR 0.81; 95% CI: 0.74–0.89), but there is a variation in the magnitude of the effect across the subgroups. The effect was considerably greater in women at high risk of developing hypertension (four trials, 327 women: RR: 0.45; 95% CI: 0.31–0.66) than in those at low risk (11 trials, 6894 women: RR: 0.68; 95% CI: 0.57–0.81). Taking into account women's calcium intake, the effect was also greater in those with low baseline dietary calcium (five trials, 1582 women: RR: 0.49; 95% CI: 0.38–0.62) than in those with adequate calcium intake (four trials, 5022 women: RR: 0.90; 95% CI: 0.81–0.99).

There is a reduction in the risk of preeclampsia, evaluated on 10 trials involving 6864 women (RR: 0.70; 95% CI: 0.58–0.83). When predefined subgroups are considered, there is a significant reduction in women with low baseline dietary calcium intake (six trials, 1842 women: RR: 0.32; 95% CI: 0.21–0.49), but not in those with adequate calcium intake (four trials, 5022 women: RR: 0.86; 95% CI: 0.71–1.05).

Preeclampsia was considerably reduced in women at high risk of hypertension (four trials, 557 women: RR: 0.22; 95% CI: 0.11–0.43), and less consistently in those at low risk of hypertension (six trials, 6307 women: RR: 0.79; 95% CI: 0.65–0.94).

The results from the largest trial conducted by NIH, which studied low-risk women with adequate baseline calcium diet, and in whom all women in both groups received low-dose calcium supplementation as part of their routine antenatal care, showed no significant effect on hypertension and preeclampsia. Based on this, authorities from developed countries where adequate dietary calcium intake is common discourage the use of routine calcium supplementation during pregnancy. Evidence from this review supports the idea that calcium supplementation might benefit women at high risk of gestational hypertension and women with a low dietary calcium intake at risk of developing preeclampsia. However, further methodologically sound randomized controlled trials with adequate sample size are needed to confirm or reject this hypothesis.

Iron and folate supplementation Numerous trials involving various populations of pregnant women with normal hemoglobin levels have evaluated the effects of iron and/or folate supplementation on several outcomes, some of them including hypertensive disorders of pregnancy. A Cochrane systematic review of two trials involving 87 women with normal hemoglobin levels, in which iron and folic acid were compared with no treatment, showed no effect on the occurrence of gestational hypertension (RR: 1.15; 95% CI: 0.41–3.18). Preeclampsia was not evaluated. Another Cochrane review that included two trials involving 696 pregnant women already receiving iron, in which women were allocated to receive folic acid or no treatment/placebo, showed no effect on prevention of gestational hypertension either (RR: 1.26; 95% CI: 0.90–1.76).

Although evidence shows that iron and folate supplementation is not effective in preventing hypertensive disorders during pregnancy, they should be prescribed for other established beneficial effects on pregnancy such as prevention of anemia.

Magnesium supplementation Magnesium is one of the essential minerals needed by humans in relatively large amounts. Magnesium works with many enzymes in regulating body temperature and synthesizing proteins as well as maintaining electrical potentials in nerves and muscle membranes. Magnesium occurs widely in many foods; dairy products, breads and cereals, vegetables, and meats are all good sources. It is therefore not surprising that genuine clinical magnesium deficiency has never been reported to occur in healthy individuals who eat standard diets. However, dietary intake studies during pregnancy consistently demonstrate that many women, especially those from disadvantaged backgrounds, have intakes of magnesium below recommended levels. Observational studies based on medical records have reported that magnesium supplementation during pregnancy was associated with a reduced risk of fetal growth retardation and preeclampsia, and that magnesium intake was associated with increased birth weight. Stimulated by these encouraging epidemiological studies, randomized clinical trials have been undertaken to evaluate the potential benefits of magnesium supplementation during pregnancy on pregnancy and neonatal outcomes.

A Cochrane systematic review of these randomized-controlled trials was carried out in order to assess the effects of magnesium supplementation during normal or high-risk pregnancies on maternal, neonatal, and pediatric outcomes. Results from two trials (474 women) showed no apparent effect of magnesium treatment on prevention of preeclampsia (RR: 0.87; 95% CI: 0.57–1.32). However, these results may have been confounded by the fact that in the largest trial all women (both magnesium-supplemented and placebo groups) received a multivitamin and mineral preparation containing low doses of magnesium. Trials also have a poor methodological quality, especially related to concealment of allocation, which could give biased results. Authors conclude that dietary magnesium supplementation of pregnant women cannot be recommended for routine clinical practice because of the poor methodological quality of the current evidence.

Fish oil supplementation Studies of nonpregnant subjects suggest that fish oil, rich in long-chain n-3 fatty acids, has a moderate effect on blood pressure in normotensive as well as hypertensive individuals. A meta-analysis of controlled clinical trials of the effect of fish oil on blood pressure has demonstrated a significant reduction in systolic and diastolic blood pressure in untreated hypertensive nonpregnant individuals, but found no other significant effect on normotensives. Fish oil has been shown to interfere with prostaglandin metabolism, and its effect on blood pressure has often been assumed to be due to such interference. Epidemiological studies suggested that marine diets could have a preventive effect on early delivery and hypertensive disorders of pregnancy.

Fish oil supplementation during pregnancy was evaluated in a systematic review of two trials (5135 women), showing no effect on pregnancy-induced hypertension (two trials, 5135 women; RR: 0.96;

95% CI: 0.86–1.07). There was a statistically significant, but modest, reduction in the rate of preeclampsia (RR: 0.70; 95% CI: 0.55–0.90). However, this reduction is strongly influenced by a large nonrandomized trial conducted in 1942, in which vitamins and minerals were given to women in addition to fish oil. Another seven trials of fish oil supplementation involving more than 2000 women have been published recently, none of which demonstrates any differences in the incidence of hypertension and preeclampsia between groups. Based on current evidence, fish oil supplementation is not recommended during pregnancy.

Zinc supplementation Zinc is thought to play an important role in many biological functions, including protein synthesis and nucleic acid metabolism. There are controversies in the literature in demonstrating the relationship between low serum zinc levels with abnormalities on pregnancy outcomes such as pregnancy-induced hypertension, prolonged labor, post partum hemorrhage, preterm or postterm pregnancies, small-for-gestational-age babies, or poor perinatal outcomes.

The role of routine zinc supplementation during pregnancy on outcomes for both mother and newborn was assessed in a Cochrane systematic review. Routine zinc supplementation in pregnancy had no detectable effect on gestational hypertension (four trials, 1962 women; RR: 0.87; 95% CI: 0.65–1.15). However, there appears to be inconsistency between trials regarding the effects from other pregnancy outcomes. This may be related to varied population characteristics of women recruited in the various trials, as some included normal pregnant women with no systemic illness, other studies specifically selected women at high risk of low zinc status, and in one study, participants were selected on the basis of proven low plasma zinc levels. There is at present no evidence of overall benefit from routine as opposed to selective zinc supplementation in pregnancy on pregnancy-induced hypertension or preeclampsia.

Vitamin supplementation In terms of vitamins A, E, and C, an oxidant/antioxidant imbalance has been suggested among the pathogenic factors involved in preeclampsia. As vitamin E is one of the most important antioxidants in body components, its levels and their relation with circulating levels of lipids peroxides on pre-eclamptic women have been intensively studied in recent years. As with other antioxidants, several studies found decreased vitamin E levels in serum from women with gestational hypertension and preeclampsia compared with controls. However, these findings could not be demonstrated in other studies. Increased ascorbate radical formation and ascorbate depletion were also found in plasma from women with preeclampsia. Recently, a randomized-controlled trial involving 283 women at high risk of developing preeclampsia was conducted. Women were randomly assigned to vitamin E and vitamin C supplements, or placebo at 16–22 weeks of gestation. Authors found a significant reduction in the risk of developing preeclampsia in the vitamin-supplemented group compared with controls (RR: 0.46; 95% CI: 0.24–0.91). The authors concluded that supplementation with vitamins C and E may be beneficial for preventing preeclampsia in women at increased risk of the disease. However, these findings need to be evaluated further in different settings and populations, as well as in low-risk women.

The role of vitamin A in pregnancy-induced hypertension and preeclampsia is a subject of controversy. It was proposed as a chain-breaking antioxidant on free radical cascade. Some studies found significantly reduced serum vitamin A levels in pre-eclamptic and eclamptic women when compared with those found in healthy women in the third trimester. To date, no trials have been published that assess the effect of vitamin A supplementation on pregnancy-induced hypertension or preeclampsia. A double-blind, cluster randomized trial of low dose supplementation with vitamin A or β-carotene carried out in Nepal in 44 646 married women showed a 40% reduction in maternal mortality related to pregnancy in vitamin A-supplemented women. However, differences in cause of deaths, including preeclampsia/eclampsia, could not be reliably distinguished between supplemented and placebo groups. The future prospect of giving vitamin A supplements for the prophylaxis/management of pregnancy-induced hypertension and preeclampsia needs to be evaluated further before it is recommended.

Treatment

The objectives of treatment for established preeclampsia, or pregnancy-induced hypertension are to prevent eclampsia as well as other severe maternal complications. Close maternal evaluation is aimed at observing the progression of the condition, both to prevent maternal complications and to determine whether fetal well-being can be assessed. As this disorder is often completely reversible and usually begins to abate with delivery, an imbalance between the mother's condition and the chance of the fetus surviving without any significant neonatal complications *in utero* or in the nursery must be continuously evaluated. Even though the only definitive treatment of preeclampsia is delivery, several nonpharmacological approaches were proposed as part of an overall strategy of management of the disease to achieve these goals.

Unfortunately, there is no information from randomized controlled trials related to dietary approaches on the management of the disease in its mild to moderate stage, when conservative management is decided on.

Pre-existing (Chronic) Hypertension

Mild and uncomplicated chronic hypertension during pregnancy has a better prognosis than preeclampsia. However, there is an increased risk of superimposed preeclampsia and possible complications if pre-existing renal disease or systemic illness is present. The primary aim of therapy, if necessary, is to prevent cerebrovascular complications and to avoid the progression to superimposed preeclampsia with worse prognosis. Nonpharmacological management of this condition during pregnancy remains controversial.

In a published review of management of mild-chronic hypertension during pregnancy, no trials were found that compared non-pharmacological interventions with either pharmacological agents or no intervention in pregnant women. This comprehensive search identified 50 randomized controlled trials, but they involved either normotensive women or women with a history of preeclampsia. For the management of established chronic hypertension during pregnancy, no relevant evidence could be located to assess the effects of non-pharmacological interventions, such as limiting activity, diet modifications or stress reduction.

Weight reduction during pregnancy, even in obese women, is generally not recommended to improve pregnancy outcomes. As weight reduction may be helpful in reducing blood pressure in nonpregnant women, for obese hypertensive women planning a pregnancy, weight reduction before conception is advisable. Even though obesity may be a risk factor for superimposed preeclampsia, there is no evidence that limiting weight gain during pregnancy reduces its occurrence.

Pregnant women with hypertension have a lower plasma volume than do normotensive women, and some studies suggest that the severity of hypertension correlates with the degree of plasma volume concentration. For this reason, sodium restriction is generally not recommended during pregnancy for the reduction of blood pressure. In addition, an increase in plasma volume concentration is a risk factor for intrauterine growth restriction. If, however, a pregnant woman with chronic hypertension is known to have salt-sensitive hypertension and has been treated successfully with low salt diet before pregnancy, it is reasonable to continue some sodium restriction for

blood pressure control during pregnancy, but not for preventing superimposed preeclampsia.

High alcohol intake is related to hypertension in nonpregnant women but is not associated with an increased risk for gestational hypertension, preeclampsia, or eclampsia. There is no conclusive evidence of adverse effects on pregnancy outcomes, including fetal growth, at levels of consumption below 120 g of alcohol per week. However, there are suggestions that excessive consumption of alcohol can cause or aggravate maternal hypertension.

There is no reliable information from well-designed, randomized controlled trials assessing the best dietary approach for the management of pre-existing hypertension during pregnancy. Recommendations come from experts' consensus and authorities' statements. It seems that mild-to-moderate pre-existing (chronic, essential) hypertension without any risk factor should be managed in the same way as in the nonpregnant state. However, additional concerns include effects on fetal well-being (mainly intrauterine growth restriction) and worsening of hypertension, particularly as a result of superimposed pre-eclampsia.

Conclusions

In short, based on the available data from systematic reviews (presented in the Appendix), we can conclude that there is some evidence that calcium supplementation in populations with a low calcium intake and/or at risk of developing pregnancy-induced hypertension could be beneficial. Antioxidants (particularly vitamins E and C) are promising, but there is a need for adequately sized and designed randomized controlled trials to confirm these findings before any widespread recommendation. Although pregnant women living in developing countries could be exposed to several other nutrient deficiencies, no evidence precludes recommending other nutrient supplementations as part of their routine antenatal care in order to prevent the occurrence of pregnancy-induced hypertension or preeclampsia.

See also: **Antioxidants**: Natural Antioxidants; **Calcium**: Properties and Determination; **Copper**: Properties and Determination; **Hypertension**: Physiology; **Obesity**: Etiology and Diagnosis; **Pregnancy**: Metabolic Adaptations and Nutritional Requirements; **Vitamins**: Overview

Further Reading

Atallah AN, Hofmeyr GJ and Duley L (2002) Calcium supplementation during pregnancy for preventing hypertensive disorders and related problems (Cochrane

Review). In: *The Cochrane Library, Issue 3*. Oxford: Update Software.

Belizán JM, Villar J and Repke J (1988) The relationship between calcium intake and pregnancy-induced hypertension: up-to-date evidence. *American Journal of Obstetrics and Gynecology* 158: 898–902.

Duley L and Henderson-Smart D (2002) Reduced salt intake compared to normal dietary salt, or high intake, in pregnancy (Cochrane Review). In: *The Cochrane Library, Issue 3*. Oxford: Update Software.

Ferrer RL, Sibai BM, Murlow CD *et al.* (2002) Management of mild chronic hypertension during pregnancy: a review. *Obstetrics and Gynecology* 96: 849–860.

Kramer MS (2002a) Balanced protein/energy supplementation in pregnancy (Cochrane Review). In: *The Cochrane Library, Issue 3*. Oxford: Update Software.

Kramer MS (2002b) Energy/protein restriction for high weight-for-height or weight gain during pregnancy (Cochrane Review). In: *The Cochrane Library, Issue 3*. Oxford: Update Software.

Kramer MS (2002c) Isocaloric balanced protein supplementation in pregnancy (Cochrane Review). In: *The Cochrane Library, Issue 3*. Oxford: Update Software.

Kramer MS (2002d) Nutritional advice in pregnancy (Cochrane Review). In: *The Cochrane Library, Issue 3*. Oxford: Update Software.

Kulier R, de Onis M, Gülmezoglu AM and Villar J (1998) Nutritional interventions for the prevention of maternal morbidity. *International Journal of Gynaecology and Obstetrics* 63: 231–246.

Mahomed K (2002a) Folate supplementation in pregnancy (Cochrane Review). In: *The Cochrane Library, Issue 3*. Oxford: Update Software.

Mahomed K (2002b) Iron and folate supplementation in pregnancy (Cochrane Review). In: *The Cochrane Library, Issue 3*. Oxford: Update Software.

Mahomed K (2002c) Zinc supplementation in pregnancy (Cochrane Review). In: *The Cochrane Library, Issue 3*. Oxford: Update Software.

Makrides M and Crowther CA (2002) Magnesium supplementation in pregnancy (Cochrane Review). In: *The Cochrane Library, Issue 3*. Oxford: Update Software.

Moutquin JM, Garner PR, Burrows RF *et al.* (1997) Report of the Canadian Hypertension Society Consensus Conference: 2. Nonpharmacologic management and prevention of hypertensive disorders in pregnancy. *Canadian Medical Association Journal* 157: 907–919.

National High Blood Pressure Education Program (1990) Working group report on high blood pressure in pregnancy. *American Journal of Obstetrics and Gynecology* 163: 1689–1712.

Olsen SF, Secher NJ, Tabor A *et al.* (2000) Randomised clinical trials of fish oil supplementation in high risk pregnancies. *British Journal of Obstetrics and Gynaecology* 107: 382–395.

Report of the National High Blood Pressure Education Program Working Group on High Blood Pressure in Pregnancy (2000) *American Journal of Obstetrics and Gynecology* 183: S1–S22.

Roberts JM and Cooper DW (2001) Pathogenesis and genetics of preeclampsia. *Lancet* 357: 53–56.

Villar J and Belizan JM (2000) Same nutrient, different hypotheses: disparities in trials of calcium supplementation during pregnancy. *American Journal of Clinical Nutrition* 71 (supplement): 1375S–1379S.

Appendix

The Cochrane Collaboration

The Cochrane Collaboration is an international non-for-profit organization that aims to help people make well-informed decisions about healthcare by preparing, maintaining, and promoting the accessibility of systematic reviews of the effects of healthcare interventions. The Cochrane Collaboration's work is based on principles of collaborative efforts in order to avoid duplication by good management and coordination to maximize economy of effort. A variety of approaches such as scientific rigour, insuring broad participation, and avoiding conflicts of interest are promoted to minimize bias. Published reviews are keeping up to date by a commitment to insure that Cochrane Reviews are maintained through identification and incorporation of new evidence.

Randomized Controlled Trial

This is an epidemiologic experiment in which the investigators randomly allocate eligible subjects into intervention groups, usually called 'study' and 'control' groups, to receive or not to receive one or more interventions that are being compared. The results are assessed by comparing outcomes in the treatment and control groups.

Systematic Review

A systematic review is a systematic search and critical evaluation of all primary studies answering the same question. Statistical methods may or may not be used to analyze and summarize the results of the studies included in the review. The metaanalysis involves the use of statistical techniques within a systematic review to integrate the results of the included studies. Systematic reviews are important tools that help clinicians, health providers, researchers, and policymakers to summarize the existing information in order to make evidence-based decisions.

Relative Risk (RR)

The ratio of risk in the intervention (exposed) group to the risk in the control (unexposed) group. A RR = 1 indicates no difference between comparison groups. For harmful or undesirable outcomes, a RR that is less than 1 indicates that the intervention was effective in reducing the risk of that outcome.

Confidence Interval (CI)

This is the range within which the 'true' value is expected to lie with a given degree of certainty (e.g., 95 or 99%). Confidence intervals represent the probability of random errors.

Nutrition in Diabetic Pregnancy

A Dornhorst and G Frost, Nutrition and Dietetic Research Group, Imperial College of Science and Medicine, London, UK

Introduction

Gestational diabetes mellitus (GDM) is glucose intolerance first recognized in pregnancy. For the majority of women this is a transient deterioration in glucose tolerance which is due to the hormonal changes of pregnancy. In a minority of women the glucose intolerance continues after pregnancy and these women have either had previously unrecognized type 2 diabetes or are in the early stages of type 1 diabetes. Older, obese women of high parity are most at risk of GDM, especially when from an ethnic background with a high prevalence of type 2 diabetes.

Gestational diabetes predisposes to accelerated fetal growth which is the main contributor to the obstetric complications of GDM. Both cesarean section rates and birth trauma are increased in GDM pregnancies. The stillbirth rate is increased when GDM remains either unrecognized or untreated. Many women with GDM are obese, with a prepregnancy body mass index (BMI) $> 29 \, \mathrm{kg \, m^{-2}}$; this itself is associated with an increased risk of a large-for-gestational-age (LGA) infant, operative delivery, pregnancy-induced hypertension, and thromboembolic disease. Neonatal complications of GDM include transient hypoglycemia and hypocalcemia. The long-term sequelae of GDM are future type 2 diabetes in the mother and future obesity, insulin resistance and type 2 diabetes in the child.

Dietary management of GDM can alone, or in combination with insulin, reduce maternal hyperglycemia sufficiently to improve pregnancy outcome. Providing dietary education and advice that extends beyond the pregnancy is also important and this has

the potential to lessen the future risk of GDM in subsequent pregnancies as well as type 2 diabetes in the mother and her child.

Despite the recognition of the importance of dietary management for GDM, there is little consensus regarding the optimal diet to prescribe. There are too few controlled dietary studies during a GDM pregnancy with sufficient nutritional detail to provide the necessary guidelines. There are no long-term prospective studies on whether the dietary advice given during pregnancy reduces future recurrence of GDM or diabetes in either the mother or her child in later life. Different and sometimes conflicting dietary advice is advocated for the management of GDM. Debate surrounds the energy content of the diet and optimal proportions and type of carbohydrate and fat that should be prescribed. The minimum weight gain that is safe in the obese GDM woman still needs to be defined. Below we attempt to provide a rational framework on which we believe dietetic advice should be given for the management of GDM. We discuss the potential short- and long-term benefits to the mother and her child which may be gained from dietary intervention while emphasizing the need for future dietary studies.

Background: General Dietetic Principles Applicable to Pregnancy

Energy Requirements in Pregnancy

Pregnancy is an anabolic state. During pregnancy, sufficient energy sources need to be acquired to cover the products of conception, the fetal–placental unit, and the changes in maternal tissues. This last includes mammary, uterine, adipose deposits, and the expansion in blood volume. The basal metabolic rate increases by 15–26% during pregnancy to meet the metabolic demands of these newly synthesized tissues. The total energy cost for pregnancy was theoretically calculated by Hytten and Leitch in the 1960s to be 355 640 kJ (85 000 kcal), equivalent to an extra 1191.3 kJ (285 kcal) a day. Subsequent longitudinal studies during pregnancy, using the latest physiological techniques, in well-nourished nonobese women confirm these original calculations. These early calculations have formed the basis of the dietary recommendations for pregnancy. Cross-sectional and longitudinal nutritional studies during pregnancy have repeatedly shown that these energy costs are seldom met by an equivalent increase in food intake. The increase in dietary energy appears to provide only 20% of the total energy costs of pregnancy.

How any one women meets the energy needs of pregnancy is highly variable and it is this variability and adaptability to environmental pressures that allow pregnancies to occur under adverse circumstances, such as food deprivation or the need to perform physical labor during pregnancy. These highly adaptive processes include limiting adipose deposition, reducing diet-induced thermogenesis, and limiting physical energy expenditure. Any reduction in diet-induced thermogenesis is either small or of marginal significance. It has been estimated that a 20% reduction in physical activity during pregnancy alone would save the necessary energy to complete pregnancy. Recent work, however, suggests that significant energy savings from reducing physical activity only occur in women with high physical expenditures before pregnancy and those involved in high physical work loads during pregnancy. Under conditions of extreme calorie restriction, a fall in basal metabolic rate has been observed in the first half of pregnancy.

Rigid guidelines for increasing dietary energy consumption in pregnancy in well-nourished and obese pregnant women are probably misguided. Overall, there is a poor correlation between dietary intake and gestational weight gains. In nonobese and underweight women, pregnancy outcome improves when the minimal weight gain targets are achieved. In obese mothers minimal or no weight gain in pregnancy does not appear to jeopardize the pregnancy. As most women with GDM are obese, these women should not be expected to meet their energy costs of pregnancy through increased dietary intake alone. Despite firm evidence, it is generally proposed that all obese women should have weight gains in pregnancy to account for the products of conception and significant calorie restriction should be avoided.

Postabsorptive and Postprandial Metabolism in Normal Pregnancy

The metabolic changes in pregnancy insure optimal maternal fetal fuel transfer. Glucose is the primary fetal oxidative substrate and by late pregnancy the fetus utilizes an estimated 17–26 g glucose per day. This high obligatory fetal oxidative metabolism results in an increase in maternal carbohydrate oxidation and an accompanying rise in the 24-h respiratory quotient. Glucose is preferentially diverted from the mother to the fetus by the fall in maternal insulin sensitivity. This reduction in insulin sensitivity is due to the antagonistic actions of placental hormones, many of which are also lipolytic. The rise in circulating free fatty acids in late pregnancy also contributes directly to the increase in peripheral insulin resistance. Increased maternal insulin resistance helps maximize the transfer of glucose postprandially to

the fetus as glucose uptake into maternal peripheral tissues is impaired due to the insulin resistance. Postprandial glucose and insulin concentrations rise during pregnancy in well-nourished women consuming western-style diets and insulin resistance falls. By contrast, little or no rise in postprandial glucose concentrations or insulin occurs in rural populations consuming more traditional low glycemic index diets and in these women insulin resistance does not fall. This observation strongly suggests that habitual diets and lifestyles can influence maternal glucose tolerance and insulin sensitivity.

The high postprandial insulin levels associated with decreased maternal insulin sensitivity facilitate maternal fat deposition. The metabolic milieu of pregnancy is very permissive for fat accumulation and even undernourished women have a 2-kg increase in their adipose stores. In well-nourished women body fat increases by approximately 9% with 2–5 kg of tissue being accumulated by the end of pregnancy. Fatty acid oxidation falls in late pregnancy and this too will further favor adipose accumulation.

Metabolic changes in maternal fat metabolism insure a constant supply of fetal glucose in the postabsorptive state. This is achieved by a faster switchover from lipogenesis to lipolysis. Human placental lactogen and other placental hormones increase maternal lipolysis which generates ketone bodies and glycerol – two important gluconeogenic substrates that help spare maternal glucose and amino acids for the fetus. Ketone bodies, but not nonesterified acids, can cross the placenta and be used as fetal fuels. Despite lower fasting glucose values, hepatic glucose output increases in pregnancy due to the glucogneogenic nature of many of the pregnancy-related hormones and the increase in hepatic insulin resistance. This increase in hepatic glucose output under fasting conditions also helps maintain a steady supply of glucose to the fetus.

Metabolic Changes Associated with GDM

A degree of deterioration in glucose tolerance in pregnancy is characteristic of well-nourished women consuming western-style diets. However, only in a minority of women is the deterioration sufficient to warrant the diagnosis of GDM. Most women remain glucose-tolerant through their ability to treble their insulin secretion in the face of increasing insulin resistance. Women who do develop GDM have subtle abnormalities of β-cell function and, despite being able to maintain glucose tolerance outside pregnancy, are unable to when faced with the high insulin resistance levels encountered in pregnancy. A degree of β-cell dysfunction is universal in women who develop GDM, both during and following pregnancy. By the time GDM has developed insulin resistance, both peripheral and hepatic insulin resistance are higher than for glucose-tolerant women. There appears to be a relatively greater β-cell defect in the nonobese women who develop GDM than the obese women in whom insulin resistance is a greater contributing factor. The combination of inadequate β-cell secretion and increased insulin resistance results in inadequate nonesterified free fatty acid (NEFFA) suppression postprandially in GDM. Postprandial increases in NEFFA levels would be anticipated to reduce insulin sensitivity further and compromise β-cell function.

The Principle of Dietary Management GDMs

The dietary advice given in GDM needs to lessen the metabolic abnormalities while insuring adequate nourishment for the mother and fetus. Dietary advice should prevent unnecessary weight gain in all pregnancies, especially in obese women. Dietetic advice should in addition provide general guidelines for healthy eating beyond the pregnancy.

General Dietary Recommendations for GDM

The diet should include the recommended vitamins and minerals for pregnancy and potentially harmful foods such as uncooked meats and soft cheese should be avoided. Alcohol is potentially harmful to pregnancy and should be actively discouraged in the first trimester and severely limited for the rest of pregnancy.

When diabetes predates the pregnancy, 5 mg of folate acid should be started before conception and continued throughout the period of organogenesis (12th gestational week/4th week from last menstrual period). Higher doses of folic acid than 400 μg are recommended for nondiabetic pregnancies for the prevention of neural tube defects despite any direct evidence that folate protects against glucose-mediated neural tube defects. All women with GDM diagnosed before the 12th gestational week should start 5 mg of folic acid, as should all previous GDM women in future pregnancies. A recent large nutritional analysis of over 1000 British pregnant women showed relatively low folate intakes – fewer than 10% of women were taking folate supplements in early pregnancy.

Dietary antioxidants protect diabetic rodents against the teratogenic effects of hyperglycemia. Although similar evidence is not available for human pregnancies, insuring adequate antioxidants in the diets of diabetic mothers during organogenesis seems reasonable.

Calcium and vitamin D supplements during both pregnancy and lactation should be considered for Asian Indian women and others with poor sunlight exposure and/or low calcium intakes. Recently a relationship between insulin resistance and certain vitamin D receptor polymorphism has been reported, providing a potential association between the risk of vitamin D deficiency and diabetes.

Recommended Maternal Weight Gains for Normal Pregnancies

The published guidelines on recommended maternal weight gains are based on large obstetric surveys in the USA (**Table 1**). The maternal weight gain required to minimize the frequency of small-for-gestational-age (SGA) infants is higher for underweight (BMI $< 19.8 \, \mathrm{kg \, m^{-2}}$) than overweight or obese women. When the prepregnancy BMI is $> 35 \, \mathrm{kg \, m^{-2}}$ with little or no maternal weight gain, there is no added risk of an SGA infant. Overweight (BMI $26.1–29 \, \mathrm{kg \, m^{-2}}$) and obese (BMI $> 29 \, \mathrm{kg \, m^{-2}}$) women are more susceptible to giving birth to an LGA infant and this risk increases further as maternal weight gain increases. The US obstetric recommendation for a 7-kg minimum weight gain for all obese women may not be universally appropriate, especially for morbidly obese women (BMI $> 35 \, \mathrm{kg \, m^{-2}}$).

The dietary management of GDM women includes the management of obese pregnant women. Unnecessary weight gain in pregnancy contributes to postpartum obesity, which will increase the likelihood of future GDM recurrence, diabetes, and other obesity-related comorbidities. The American Diabetic Association (ADA) has endorsed dietary guidelines for management of diabetes in pregnancy recommending a daily calorie allowance based on prepregnancy ideal body weight (IBW) and current pregnancy weight. These guidelines recommend 36–40 $\mathrm{kcal \, kg^{-1}}$ when IBW $< 90\%$, 30 $\mathrm{kcal \, kg^{-1}}$ for IBW of 90–120% and 24 $\mathrm{kcal \, day^{-1}}$ for IBW 121–150% and as little as 12–18 $\mathrm{kcal \, kg^{-1}}$ during pregnancy

Table 1 The 1990 guidelines of the US Institute of Medicine on maternal weight gain targets according to prepregnancy body mass index

	Underweight ($< 19.8 \, kg \, m^{-2}$)	Normal weight ($19.8–26 \, kg \, m^{-2}$)	Overweight ($> 26 \, kg \, m^{-2}$)
Weight gain term target	12.5–18 kg	11.5–16 kg	7.0–11.5 kg

when the prepregnancy IBW > 150% IBW. The ADA currently endorses the minimum recommended weight gain of 7.0 kg for obese (BMI > 29 kg m^{-2}) women, as published by the American College of Obstetricians and Gynecologists. No equivalent weight or daily calorie guidelines exist for the UK.

Calorie Restriction in the Obese Woman with GDM

To date there are no long-term studies on the psychological or physical development of infants whose obese mothers are mildly calorie-restricted in late pregnancies for GDM. Long-term follow-up studies are available from infants born to well-nourished Dutch women with severe calorie-restricted daily allowance of 800 kcal, in late pregnancy during the 5½-month famine of 1944–45. These infants were thinner at birth but had normal subsequent childhood development and were reported to be less obese than other recruits on entry into the army at 18 years old. However, subsequent follow-up has shown that among this cohort of children born to calorie-restricted mothers in late pregnancy, the incidence of glucose intolerance and diabetes was increased in middle age. It is these long-term studies combined with genuine concerns relating to the harmful effect of maternal ketosis, caused by calorie restriction, on fetal neurophysiological and cognitive development that have resulted in a general reluctance to advocate calorie restriction in pregnancy.

While no one recommends severe calorie restriction in pregnancy, there is probably a place for modest calorie constraint for obese women with GDM. Women with GDM are relatively more ketosis-resistant to mild calorie restriction than glucose-tolerant women; this may be attributable to their higher hepatic glucose outputs. Theoretically, modest calorie-restricted diets that are given as small frequent meals containing slowly absorbed carbohydrates will help minimize maternal ketosis as these diets are associated with attenuated insulin responses which delay lipolysis and therefore ketogenesis.

We have previously reported that in obese GDM women modest calorie restriction of 20–25 kcal kg^{-1} day^{-1} from the 24th gestational week reduces the frequency of LGA infants compared with obese glucose-tolerant women not dieted. In this study the GDM women gained only half the weight of the controls from 28 weeks to term (1.7 ± 1.6 versus 4.1 ± 3.1 kg) and this was associated with fewer LGA infants than for obese non-GDM controls. Other studies using hypocaloric diets have also shown improvements in glycemic control with mild calorie restriction in pregnancy.

On today's evidence it would seem appropriate to limit weight gain in GDM pregnancies to the bottom rather than the top for those recommended for average, overweight, and obese women, while setting no minimum weight gain for the grossly obese (> 34 kg m^{-2}). This would insure that following pregnancy there would be no overall weight gain in the overweight women and potentially allow weight loss in the morbidly obese woman.

The Optimal Amount and Type of Dietary Carbohydrate and Fat for GDM

Glucose crosses the placenta in both a concentration-dependent manner and by specific glucose transporters. The reduction of postprandial peak glucose concentrations has a greater impact on limiting accelerated fetal growth than lowering postabsorptive levels. Therefore diets that prevent excessive postprandial placental uptake of glucose should be the basis of our dietary advice to all women with GDM.

Controversy surrounds the optimal dietary proportion that should be taken as carbohydrate. Recent advice from the USA has advocated limiting the overall percentage of dietary carbohydrate to < 45%, as in short-term studies this has been shown to improve glycemic control. The current guidelines approved by the ADA suggest limiting carbohydrate to 40% of the total energy content while increasing dietary fat to 40%. Like others, we would argue that it is the type rather than the absolute amount of carbohydrate that dictates postprandial glycemia. The glycemic index of different ingested carbohydrates quantifies their different glycemic responses. Starches, fiber, and refined sugars are the main dietary carbohydrates and their rates of intestinal absorption are different. The highest postprandial glucose and insulin values occur with refined sugars, the most rapidly absorbed, while the lowest values are with the soluble fibers that have the slowest absorption rates. In pregnancy, if the carbohydrate is predominantly refined, any increase above 45% of the total dietary energy content is associated with a deterioration in glycemic control, especially if given first thing in the morning. However, 60% of the total energy content of the diet can be consumed as low glycemic index carbohydrates with no change in glucose tolerance or an actual improvement. Both during and outside pregnancy the introduction of low-glycemic-index diets has been shown to reduce insulin sensitivity. There is a degree of gastric stasis in pregnancy and this itself can further lower the glycemic index of many carbohydrates.

The basis of the dietary management of GDM is to reduce postprandial glycemia. This theoretically can be achieved either by limiting the total amount of carbohydrate ingested or insuring that the dietary carbohydrates consumed have a low glycemic index. Reducing carbohydrates for breakfast has also been advocated by some as the postprandial glucose values tend to be highest mid-morning. However, limiting carbohydrates will result in a higher proportion of the diet consumed as fat, which is more energy-dense and is likely therefore to result in weight gain. In addition, diets that emphasize limiting the proportion of carbohydrate over fat are, we believe, sending out the wrong educational message. In practice it is often difficult to achieve compliance when greater than 50% of the energy content of the diet is given as carbohydrate. This is especially so with the low-glycemic-index carbohydrates, which produce greater satiety than refined sugars or fats.

Dietary Fat

We believe the current American recommendations, endorsed by the ADA, that 40% of the GDM diet should be derived from dietary fat, are misguided. The short-term dietary studies which demonstrated a benefit of high-fat diets on postprandial blood glucose values may, as discussed above, have been accounted for by the high glycemic index of the carbohydrates used in these studies. In addition, promoting high-fat diets in women destined to develop diabetes and therefore cardiovascular disease is highly questionable. High-fat diets have also been linked with both β-cell toxicity and increased insulin resistance. There are currently insufficient data to know if chain length and degree of saturation of fatty acids are important dietary factors in determining glucose tolerance during pregnancy.

Diet and Insulin Therapy for GDM

The aim of medical management of women with GDM is to insure maternal glycemia can be kept at a safe level. The aim of management is not to delay insulin treatment if required. The fasting and postprandial glucose levels at which insulin therapy should be introduced in addition to diet have been specified as a fasting value $> 5.5\,\mathrm{mmol\,l^{-1}}$ and 1-h postprandial value $> 7\,\mathrm{mmol\,l^{-1}}$. However, the time that dietary management should be tried before starting insulin is less well-specified. Even when very tight glycemic targets are being achieved, most women can be managed with diet alone. The women who require insulin are the most metabolically compromised and have the highest perinatal complications and the fastest deterioration postpartum for the development of diabetes. It is important

to recognize these women early in pregnancy and not to lay any sense of failure or noncompliance on them or to delay the introduction of early insulin treatment. Insulin is also occasionally introduced later in pregnancy for obstetric rather than pure glycemic reasons; these would include evidence of accelerated fetal growth and unexplained polyhydramnios.

The principles of dietary management for GDM change little with the introduction of insulin. One should continue to limit weight gain in obese women while providing sufficient carbohydrate snacks throughout the day to prevent hypoglycemia.

Whenever insulin is prescribed with the intention of achieving euglycemia, the frequency of hypoglycemia increases. Although short periods of hypoglycemia are not detrimental to the fetus, it is both frightening and unpleasant for the woman. In addition, after an episode of hypoglycemia the blood glucose rises acutely, both through the action of the counterregulatory hormones and the usual subsequent consumption of a sugary drink. Frequent episodes of hypoglycemia often result in women chasing these high-rebound glucose levels by increasing their insulin dosage and this can result in further hypoglycemic attacks.

Women on insulin should have a regular intake of carbohydrate throughout the day. This needs be taken with each meal, as well as in a snack between meals and at bedtime. To avoid unnecessary calorie intake, snacks that are low in fat should be advised: fruit is ideal for this purpose. Meanwhile, low-glycemic-index carbohydrate should be advised for meals. This strategy will increase the absorption period postprandially, thereby minimizing the risk of hypoglycemia. The use of the soluble fiber supplement guar gum can be introduced if blood glucose levels cannot be controlled adequately without increasing insulin to levels that are causing hypoglycemia.

Long-Term Dietary Advice for the Mother and her Child

Women with GDM are at greatly increased risk of future diabetes. Many of these women are obese and all have at least one child at increased risk of adolescent obesity and diabetes. Given this, it is obvious that dietary advice, education, and reinforcement should continue beyond the pregnancy. Women should be made fully aware of their added risk of both future GDM pregnancies and type 2 diabetes and the importance of not gaining weight postpartum. Low-fat diets have been shown to minimize the recurrence of subsequent GDM pregnancies. Suitable healthy living advice should be given to all women following a GDM pregnancy, emphasizing the benefits of such a diet for the whole family. Ideally, all these women would receive annual long-term dietetic input within the community. The case for continual follow-up of these women will become more apparent with the result of the ongoing diabetic prevention studies currently being carried out among high-risk groups, of which previous GDM women are included.

The Need and Feasibility of Future Dietary Studies in Pregnancy

There remains a lack of good randomized studies on the dietary management of GDM. Such studies are required for both short-term pregnancy outcomes and long-term outcomes for the mother and her child. One of the main difficulties in conducting such studies is the control arm: even when no dietary advice is given, once diagnosed with GDM women make lifestyle changes based on family beliefs of information gathered from a variety of sources. Also if the health care providers are aware of the diagnosis they too unintentionally are likely to influence lifestyle factors. The need to blind both the women and the health care staff of the diagnosis is difficult and often considered unethical, as, if ignored, GDM does carry a risk to that pregnancy.

Summary

Gestational diabetes is a common complication of pregnancy. Controlling maternal hyperglycemia with diet alone and diet and insulin can reduce the risk of inappropriate accelerated fetal growth. The prescribed diet for GDM needs to include regular meals and snacks. These should contain a large component of slowly absorbed carbohydrate. As many women with GDM are obese, the diet must not allow excessive maternal weight gain as this further compromises pregnancy outcome and adds to the mother's risk of future diabetes. Dietary advice in pregnancy should extend beyond pregnancy as a woman with GDM has a lifetime risk of future diabetes, as does her child. There remains a need for future dietary studies in the management of GDM.

See also: **Diabetes Mellitus**: Etiology; **Obesity**: Etiology and Diagnosis; **Pregnancy**: Metabolic Adaptations and Nutritional Requirements

Further Reading

Åberg A, Rydhström H, Kallen B *et al.* (1997) Impaired glucose tolerance during pregnancy is associated with increased fetal mortality in preceding sibs. *Acta Obstetrica Gynecologica Scandinavica* 76: 212–217.

Adams KM, Li H, Nelson RL *et al.* (1998) Sequelae of unrecognised gestational diabetes. *American Journal of Obstetrics and Gynecology* 178: 1321–1332.

American College of Obstetricians and Gynecologists (1993) Nutrition during pregnancy. *ACOG Technical Bulletin* 1–7.

American Diabetes Association (1995) Nutritional management. In: *Medical Management of Pregnancy Complicated by Diabetes*, 2nd edn. Virginia: American Diabetes Association. pp. 47–56.

Berkowitz GS, Lapinski RH, Wein R *et al.* (1991) Race/ethnicity and other risk factors for gestational diabetes. *American Journal of Epidemiology* 135: 965–973.

Bianco AT, Smilen SW, Davis Y *et al.* (1998) Pregnancy outcome and weight gain recommendations for morbidly obese women. *Obstetrics and Gynecology* 91: 97–102.

Buchanan TA, Kjos SL, Montoro MN *et al.* (1994) Use of fetal ultrasound to select metabolic therapy for pregnancies complicated by mild gestational diabetes. *Diabetes Care* 17: 275–283.

Butte NF, Hopkinson JM, Mehta N *et al.* (1999) Adjustments in energy expenditure and substrate utilisation during late pregnancy and lactation. *American Journal of Clinical Nutrition* 69: 299–307.

Casey BM, Lucas MJ, McIntire DD *et al.* (1997) Pregnancy outcomes in women with gestational diabetes compared with the general obstetric population. *Obstetrics and Gynecology* 90: 869–873.

Catalano PM, Huston L, Amini SB *et al.* (1999) Longitudinal changes in glucose metabolism during pregnancy in obese women with normal glucose tolerance and gestational diabetes mellitus. *American Journal of Obstetrics and Gynecology* 180: 903–916.

Dabelea D, Pettitt DJ, Hanson RL *et al.* (1999) Birth weight, type 2 diabetes, and insulin resistance in Pima Indian children and young adults. *Diabetes Care* 22: 944–950.

Dornhorst A and Frost G (1997) The potential for dietary intervention postpartum in women with gestational diabetes. *Diabetes Care* 20: 1635–1637.

Dornhorst A and Rossi M (1998) Risk and prevention of type 2 diabetes in women with gestational diabetes. *Diabetes Care* 21 (suppl. 1): B43–B49.

Durnin JVGA (1987) Energy requirements of pregnancy: an integrated study in five countries. Background and methods. *Lancet* ii: 895–897.

Felig DS and Naylor CD (1998) Eating for two: are guidelines for weight gain during pregnancy too liberal? *Lancet* 351: 1054–1055.

Freinkel N (1980) Banting lecture 1980: pregnancy and progeny. *Diabetes* 29: 1023–1035.

Freinkel N (1985) Effects of the conceptus on maternal metabolism during pregnancy. In: Leibel BS and Wrenshall GA (eds) *On the Nature and Treatment of Diabetes*, pp. 679–691. Amsterdam: Exerpta Medica.

Harris MI (1988) Gestational diabetes may represent discovery of pre-existing glucose intolerance. *Diabetes Care* 11: 402–411.

Hawdon JM and Aynsley-Green A (1996) Neonatal complications, including hypoglycaemia. In: Dornhorst A and Hadden D (eds) *Diabetes and Pregnancy: An Inter-*

national Approach to Management, pp. 303–318. Chichester, UK: John Wiley.

Hay WWJ (1991) The role of placental–fetal interaction in fetal nutrition. *Seminars in Perinatology* 15: 424–433.

Hay WWJ (1994) Placental supply of energy and protein substrate to the fetus. *Acta Paediatrica Supplement* 405: 13–19.

Hytten FE (1980) Nutrition. In: Hytten FE and Chamberlain G (eds) *Clinical Physiology in Obstetrics*, pp. 163–192. Oxford, UK: Blackwell Scientific.

Hytten FE and Leitch I (1964) *The Physiology of Human Pregnancy*. Oxford, UK: Blackwell Scientific.

Institute of Medicine (1990) *Nutrition During Pregnancy: Weight Gain and Nutritional Supplements*. Washington, DC: National Academy Press.

Jacobsen JD and Cousin L (1989) A population-based study of maternal and perinatal outcome in patients with gestational diabetes. *American Journal of Obstetrics and Gynecology* 161: 981–986.

Janerich DT (1990) Alcohol and pregnancy. An epidemiologic perspective. *Annals of Epidemiology* 1: 179–185.

Jovanovic L (1998) American Diabetes Association's fourth international workshop-conference on gestational diabetes mellitus: summary and discussion: therapeutic interventions. *Diabetes Care* 21 (suppl. 2): 131–137.

Kalhan S, Rossi K, Gruca L *et al.* (1997) Glucose turnover and gluconeogenesis in human pregnancy. *Journal of Clinical Investigation* 100: 1775–1781.

Kühl C (1991) Insulin secretion and insulin resistance in pregnancy and GDM. *Diabetes* 40(suppl. 2): 18–24.

Major CA, Henry MJ, Veciana M *et al.* (1998) The effects of carbohydrate restriction in patients with diet-controlled gestational diabetes. *Obstetrics and Gynecology* 91: 600–604.

McFarland MB, Langer O, Conway DL *et al.* (1999) Dietary therapy for gestational diabetes: how long is long enough? *Obstetrics Gynecology* 93: 978–982.

Metzger BE, Coustan DR and the Organizing Committee (1998) Summary and recommendations of the fourth international workshop-conference on gestational diabetes mellitus. *Diabetes Care* 21 (suppl. 2): 161–167.

Moses RG (1996) The recurrence rate of gestational diabetes in subsequent pregnancies. *Diabetes Care* 19: 1349–1356.

Moses RG, Shand JL, Tapsell LC *et al.* (1997) The recurrence of gestational diabetes: could dietary differences in fat intake be an explanation? *Diabetes Care* 20: 1647–1650.

National Research Council (1989) *Recommended Dietary Allowance*, pp. 33–34. Washington, DC: National Academic Press.

O'Sullivan JB (1982) Body weight and subsequent diabetes mellitus. *Journal of the American Medical Association* 248: 949–952.

Peterson CM and Jovanovic-Peterson L (1991) Percentage of carbohydrate and glycaemic response to breakfast, lunch, and dinner in women with gestational diabetes. *Diabetes* 40 (suppl. 2): 172–174.

Pettitt D, Aleck K, Baird HR *et al.* (1988) Congenital susceptibility to NIDDM: role of intrauterine environment. *Diabetes* 37: 622–628.

Silverman B, Metzger B, Cho NH *et al.* (1995) Impaired glucose tolerance in adolescent offspring of diabetic mothers. Relationship to fetal hyperinsulinism. *Diabetes Care* 18: 611–617.

Spellacy WN and Goetz FC (1963) Plasma insulin in normal late pregnancy. *New England Journal of Medicine* 268: 988–991.

The MRC Vitamin Research Study Group (1991) Prevention of neural tube defects: the results of the Medical Research Council Vitamin Study.

Wolever TMS, Bentum-Williams A, Jenkins DJ *et al.* (1995) Physiological modulation of plasma free fatty acid concentrations by diet. *Diabetes Care* 18: 962–970.

PREMENSTRUAL SYNDROME: NUTRITIONAL ASPECTS

F Ford, University of Sheffield, Sheffield, UK

Background

Many women seek medical help for symptoms related to premenstrual syndrome (PMS). Physicians, dietitians, and nutritionists often recommend some form of nutrition advice, including incorporating extra vitamin B_6, vitamin B_1, vitamin E, calcium, magnesium, essential and ω-3 fatty acids, herbal remedies, carbohydrates, low-fat and vegetarian diets, and chocolate. Although much (but not all) of this advice is part of a healthful diet, there is relatively little sound scientific evidence that they relieve or prevent symptoms.

PMS is a condition characterized by a number of physical, emotional, and psychological symptoms that appear in the second half of the menstrual cycle, after ovulation. It can affect women in the reproductive age from adolescence until the menopause. There is a list of menstrual symptoms identifying over 150 different symptoms linked to the menstrual cycle. The main symptoms reported are irritability, depression, anxiety, weight gain, edema, breast pain, fatigue, and headache. Greene and Dalton first used the term premenstrual syndrome in 1953. The first author to describe the syndrome known as premenstrual tension was Frank, in 1931. The intensity of symptoms may vary considerably between women and for each cycle. Severe cyclical symptoms cause suffering to many women in their daily activity and personal relationships.

PMS is perceived to be a common complaint, but data on prevalence vary according to different measurement scales and cultural variations. It is estimated that up to 95% of women suffer mild symptoms, and 5–10% of women have symptoms severe enough to disrupt their lives in the 2 weeks before the onset of menstruation. PMS can be defined as a regular pattern of symptoms occurring just before the start of menstruation and which gradually abate soon after the start of bleeding. Severe PMS can be defined as causing functional impairment in work, relationship, or usual activities. The social consequences of premenstrual syndrome are important and have been the focus of much research. It has been reported that PMS is responsible for a high incidence of crimes, alcoholism, school absence, and admittance to hospital for accidental injury.

The etiology of PMS is unknown, although many theories have been suggested. Factors such as estrogen levels, deficiency of progesterone and progesterone metabolites, increase of adrenal activity, subclinical hipoglucosemy, increase of prolactin and monoaminoxidases, imbalance of renin–angiotensin–aldosterone, deficiency of pyridoxine and other vitamins, and neuroendocrine dysfunction have all been cited.

The diagnosis of the PMS is essentially based on clinical symptoms. A prospective evaluation of three consecutive months that show these clinical symptoms occurring during the luteal phase of the menstrual cycles characterize PMS. However, diagnosis of the condition may be difficult, as there is no standardized diagnostic tool. Women's experiences of PMS are so varied that it is difficult to fit them into categories or even subgroups of symptoms. In the past, some doctors have regarded PMS as a psychological syndrome, so women report finding it difficult to receive satisfactory treatment and attention. This might be one of the reasons why it is a condition that women often diagnose and treat themselves.

Treatment of PMS has been largely empirical. The following dietary components have been manipulated for the treatment of PMS.

Vitamin B₆ (Pyridoxine)

Vitamin B_6 is involved in the production of prostaglandin E2 (which contributes to myometrial relaxation) and in the utilization of magnesium, so higher levels of vitamin B_6 could also influence dysmenorrhea cramps. Vitamin B_6 has been used in the doses of 50–500 mg daily in the second half of the menstrual cycle. It has been suggested that vitamin B_6 may act by correcting a deficiency at the hypothalamic end of the complex psycho-endocrine reproductive pathways. Pyridoxal 5′ phosphate, the active form of vitamin B_6, serves as the coenzyme of a wide variety of enzymes of amino acid metabolism. For example, it serves as a cofactor in the metabolism of tryptophan (the precursor of serotonin) and also in the metabolism of tyrosine (leading to dopamine and noradrenaline) and glutamate (leading to γ-aminobutyric acid). Low levels of dopamine and serotonin lead to high levels of prolactin and aldosterone, thus explaining the fluid retention, and the effect on the neurotransmitters could explain the psychological symptoms in PMS.

In recent years, a systematic review on the efficacy of vitamin B_6 in the treatment of PMS concluded that:

- Randomized placebo controlled studies of vitamin B_6 treatment for PMS were of insufficient quality to draw definitive conclusions.
- Limited evidence exists to suggest that 100 mg of vitamin B_6 daily (and possibly 50 mg) is likely to be beneficial in the management of PMS.
- Vitamin B_6 was significantly better than placebo in relieving overall premenstrual symptoms and in relieving depression associated with PMS, but the response was not dose-dependent.
- No conclusive evidence was found of neurological side-effects with these doses.
- A randomized controlled trial of sufficient power and quality is needed to compare vitamin B_6 with placebo to establish definitive recommendations for treatment.

A systematic review published in the Cochrane Database concluded that no definite results could be reported regarding the efficacy of vitamin B_6 in improving dysmenorrhea.

Calcium

Previous reports have suggested that disturbances in calcium regulation may underlie the pathophysiologic characteristics of PMS and that calcium supplementation may be an effective therapeutic approach. In one study, each participant received 6 months of treatment, involving 3 months of daily calcium supplementation

(1000 mg of calcium carbonate) and 3 months of placebo. Three premenstrual factors (negative affect ($p = 0.05$); water retention ($p = 0.005$); pain ($p = 0.05$)) and one menstrual factor (pain ($p = 0.05$)) were significantly alleviated by calcium.

Another study evaluated the effect of 1200 mg of elemental calcium on the luteal and menstrual phases of the menstrual cycle in 497 women with PMS, in a prospective, randomized, double-blind, placebo-controlled, multicenter clinical trial. During the luteal phase of the treatment cycle, a significantly lower mean symptom complex score was observed in the calcium-treated group for both the second ($p = 0.005$) and third ($p < 0.001$) treatment cycles. By the third treatment cycle, calcium effectively resulted in an overall 48% reduction in total symptom scores from baseline compared with a 30% reduction in placebo. All four-symptom factors were significantly reduced by the third treatment cycle.

Magnesium

Results from a recent metaanalysis have suggested that magnesium supplements of 500 mg per day may be a promising treatment for one of the major symptoms of PMS, i.e., dysmenorrhea. Magnesium treatment has been shown to be more effective than placebo, with those taking magnesium requiring significantly less absence from work and less use of additional medication. Minimal adverse effects have been experienced in both magnesium and placebo groups with no significant differences shown. However, overall, there has been little conclusive evidence for the effectiveness of magnesium for symptom relief. Although all the trials have been randomized and double blind with adequate methodological quality, they were all small trials, and there was poor measurement and reporting of pain outcomes. Two of the included trials had withdrawal rates of over 30%. Many of these participants may have failed to complete the trial due to a lack of efficacy or adverse effects from the treatments, and so the high withdrawal rates have an impact on the strength of the evidence. It is also unclear which treatment regimen was more effective, as the trials both administered treatment in quite different ways, one daily and the other during menses only, and at different doses. Therefore, no strong recommendation can be made about the efficacy of magnesium until a further evaluation is carried out. A randomized controlled trial of magnesium with a larger number of participants and adequate outcome measurement is needed.

The largest of the included trials also reported data on the levels of prostaglandin F2 α in menstrual blood. Overproduction of this prostaglandin has been shown to be a substantial contributing factor to the painful cramps associated with dysmenorrhea. Women taking the magnesium therapy had substantially lower levels of PGF2 α in their menstrual blood than those on placebo ($p < 0.05$), which mirrored the therapeutic decrease in pain experienced by the participants. This highlights the possible biological rationale behind magnesium therapy for dysmenorrhea; it inhibits the biosynthesis of PGF2 α as well as having a role in muscle relaxation and vasodilation.

Vitamin B$_6$ and Magnesium

Magnesium was shown to be no different in pain outcomes from both vitamin B$_6$ and a combination of vitamin B$_6$ and magnesium by one small trial. The same trial also showed that a combination of magnesium and vitamin B$_6$ was no different from placebo in reducing pain. However, vitamin B$_6$ alone was more effective at reducing pain than both placebo and a combination of magnesium and vitamin B$_6$. In a recent systematic review, no definite conclusions could be made about the efficacy of the combination of vitamin B$_6$ and magnesium supplements.

Vitamin B$_1$ (Thiamin)

Vitamin B$_1$ plays an important role in metabolism, and vitamin B$_1$ deficiency can be characterized by fatigue, muscle cramps, various pains, and a reduced tolerance to pain, all factors that could be associated with dysmenorrhea and PMS. Vitamin B$_1$ was shown to be an effective treatment for dysmenorrhea, taken at 100 mg daily, although this conclusion is tempered slightly because it is based on only one large randomized controlled trial (RCT). This RCT trial demonstrated a significant effect of vitamin B$_1$ taken daily compared with placebo in reducing pain. Results from phase one of the trial only, before participants were crossed over to the other treatment group, were used when calculating odds ratios. After crossover, those swapped from vitamin B$_1$ treatment to placebo maintained pain relief. The authors of the trial interpreted this as a curative effect of vitamin B$_1$ treatment. The length of the trial (2 months of either treatment) makes this interpretation difficult to confirm, as no longer-term follow-up was made, and the strong placebo effect that is typical in dysmenorrhea trials could account for some of the maintenance of effect. Vitamin B$_1$ needs to be assessed further by a randomized trial, preferably a large multicentered trial with different study populations that would generate generalizable results that could confirm the positive results of the above trial.

Vitamin E

It has been suggested that vitamin E has analgesic and antiinflammatory properties. A randomized trial of vitamin E for rheumatoid arthritis has shown a significant reduction in pain parameters, which lends further support to this theory. One trial comparing a combination of 100 mg per day of vitamin E and ibuprofen with ibuprofen alone on dysmenorrhea showed no significant difference between the two groups for the outcome of pain relief after 1 month of treatment. Both treatment groups experienced excellent treatment responsiveness, although this conclusion is based on a comparison of baseline scores with post-treatment scores, which is more subject to bias than if a placebo control group had been used. Overall, the addition of vitamin E supplements to the diet has no further effect on pain over that of standard treatment with ibuprofen. The treatment time of 1 month of each treatment was a limitation of the trial, as the efficacy of a dietary therapy cannot really be assessed adequately over such a short period of time.

Essential Fatty Acids

Findings from uncontrolled studies have indicated that essential fatty acid (EFA) metabolism may be abnormal in women with PMS. EFAs are the precursors of prostaglandins. Low levels of prostaglandins are thought to lead to increased sensitivity to prolactin. Women with PMS may be abnormally sensitive to normal amounts of prolactin. This hypothesis – that there is a deficiency in EFAs, leading to low levels of prostaglandin E1 that may attenuate the effects of prolactin – is one rationale for using EFAs as a treatment for women with PMS.

Evening primrose oil (EPO) contains two EFAs – linoleic acid and γ-linolenic acid (GLA). Linolenic acid is needed for the synthesis of prostaglandin E, and GLA is needed for the synthesis of prostaglandin E1. Whilst EPO is generally accepted as a safe product, reported side-effects include occasional nausea, indigestion, and headache. A small number of less common side-effects have also been documented. A potential risk of inflammation, thrombosis, and immunosuppression with prolonged use of GLA has been described.

EPO is available in most countries without a prescription and is heavily promoted for women as being an effective treatment for a range of conditions including PMS. It is, however, an expensive treatment that many women have to pay for themselves. In a systematic literature search of clinical trials of EPO for the treatment of PMS with a view to performing a meta-analysis, only seven placebo-controlled trials have been found, and randomization was clearly indicated in only five trials. Inconsistent scoring and response criteria made statistical pooling and hence a rigorous metaanalysis inappropriate. The two most well-controlled studies failed to show any beneficial effects for EPO, although, because the trials were relatively small, modest effects cannot be excluded. From the current evidence, EPO is of little value in the management of PMS.

ω-3 Fatty Acids

One trial has compared ω-3 fatty acids (fish oil) and placebo, and data have shown that the treatment is significantly more effective than placebo after 2 months' administration. A meta-analysis has examined the outcome of the use of additional medication and showed that the fish oil treatment group consumed significantly fewer tablets than the placebo group. Minimal adverse effects were significantly more likely in the fish oil group compared with the placebo group. None of the adverse effects were particularly serious (nausea, acne exacerbation, and difficulty swallowing capsules), but they were acute enough to cause some women to discontinue treatment. One methodological limitation of this trial is the poor reporting of data. Pain relief was reported, as the average of the two groups after the treatments allocated was crossed-over, and so it is not an adequate assessment of efficacy. The trial was of short duration, only 2 months in each treatment arm, which may not be enough time to properly assess the effect of a dietary intervention.

Levels of polyunsaturated fatty acids (PUFAs) have been correlated with menstrual pain, with higher levels of the ω-3 fatty acids associated with milder menstrual symptoms. The PUFA ω-6 is metabolized into the specific prostaglandins associated with dysmenorrhea, and it appears that the ratio of ω-3 to ω-6 is associated with menstrual symptoms, therefore a diet higher in ω-3 fatty acids is possibly associated with less dysmenorrhea.

Herbal Remedies

It is advisable that women consult a qualified medical practitioner who specializes in complementary medicine before taking unproven herbal remedies. There is a Practitioner Directory, which lists a selection of the best doctors and practitioners who specialize in treatment using complementary medicine.

Japanese Herbal Combination (Toki-Shakuyaku-San)

The combination herbal remedy Toki-shakuyaku-san (TSS) has been compared with placebo in one trial. Pain was assessed using a visual analog scale (VAS). TSS was significantly more effective in reducing VAS scores, after 2 months of treatment. This difference was maintained after a 2-month untreated followup period. The use of additional medication in the treatment group was significantly less than in the placebo group. One major limitation of this trial was that participants were included in the trial using a particular traditional Chinese medicine diagnosis of a complex set of symptoms as well as the typical diagnosis of PMS. It is not clear how this would translate into Western therapeutics, or if this remedy would help women without this particular pattern of diagnosis.

Agnus Castus

Agnus Castus has been reported to be a hormone regulator and tonic for the nervous system, being used for PMS to treat mood swings, depression, water retention, and breast pain. The efficacy and tolerability of agnus castus fruit (*Vitex agnus castus* L extract Ze 440) with placebo has been investigated in women with PMS in a prospective, randomized, placebo controlled study. The main efficacy variable was a change from baseline to end point (end of the third cycle) in women's self-assessment of irritability, mood alteration, anger, headache, breast fullness, and other menstrual symptoms including bloating. Results have shown a significant improvement in the active group compared with the placebo group ($p < 0.001$). Some women have reported mild adverse events, none of which caused discontinuation of treatment. It has been concluded that a dry extract of agnus castus fruit is an effective and well-tolerated treatment for the relief of symptoms of PMS.

Hypericum perforatum

A pilot observational study has been carried out to investigate whether *Hypericum perforatum* could relieve the symptoms of PMS in a small group of women. Participants took hypericum tablets for two complete menstrual cycles (1×300 mg of hypericum extract per day standardized to 900 µg of hypericin). The degree of improvement in overall PMS scores between baseline and the end of the trial was 51%, with over two-thirds of the sample demonstrating at least a 50% decrease in symptom severity. The results of this pilot study suggest that there is scope for conducting a RCT to investigate the value of hypericum as a treatment for PMS.

St John's Wort

St John's Wort has been reported to raise serotonin levels, which helps to alleviate the mild to moderate depression associated with PMS. It should not be taken with prescription antidepressants, progesterone, or other hormone treatments.

Dong Quai – (*Angelica sinensis*) or Chinese Angelica

Dong Quai (*Angelica sinensis*), or Chinese Angelica, is sometimes known as 'women's ginseng' and is used for PMS.

Natural Progesterone

Natural progesterone is sometimes prescribed where low progesterone levels are indicated in PMS. Overall, there is insufficient evidence to recommend the use of any herbal remedies apart from *Agnus castus* for the treatment of PMS. The Complete German Commission E Monographs list a number of herbal products as being used for dysmenorrhea or menstrual disorders, yet no evidence from clinical trials to support the use of any of these herbal products has been found to date.

Chocolate

Chocolate's ingredients satisfy many aspects of PMS. Its high sugar content induces the brain to make new serotonin and make the chocolate eater feel calm and relaxed. Moreover, chocolate is usually eaten by itself rather than as part of a meal. Thus, there is little chance that protein foods will interfere with the body's ability to make new serotonin. The high butterfat content of chocolate may also change premenstrual mood. High-fat foods can have a numbing effect on mood and on mental and physical energy, and the combination of the fat and sugar content of the chocolate is reported to soothe and tranquilize women. However, chocolate has other ingredients – like phenylethylamine – that are also claimed to have positive mood effects. There are no reliable studies proving this, but anecdotal evidence supports the view that chocolate may have some ingredients that women with PMS crave.

Carbohydrates

The effect of a carbohydrate-rich beverage on mood, appetite, and cognitive function was investigated in a small group of women with PMS. Twenty-four women with confirmed PMS took a beverage or placebo for three menstrual cycles after a 1-month placebo run-in period, in a double-blind placebo-controlled study. Patients were tested for mood, cognition, and food cravings, using an interactive computer-telephone system, during the luteal phase of the menstrual cycle. The beverage, but not the placebos, significantly decreased self-reported depression, anger, confusion, and carbohydrate craving, and improved memory word recognition.

Isoflavones (Plant Estrogens)

These are often derived from soya and are suggested as a natural hormone replacement for postmenopausal women. However, many are also advocating their use for PMS. There is little evidence for a beneficial effect on PMS, and in fact supplementing with soya protein and other sources of plant estrogens is now controversial, many experts believing that they can also have a harmful effect. They should certainly not be taken if a woman has had breast cancer or a family history of the disease.

L-Tryptophan

Serotonin reuptake inhibitors have been shown to be beneficial in the treatment of PMS. The efficacy of L-tryptophan, which acts specifically on serotonergic neurons, was investigated in a RCT. Results suggested a significant ($p = 0.005$) therapeutic effect of L-tryptophan relative to placebo for the cluster of mood symptoms comprising the items of dysphoria, mood swings, tension, and irritability. These results suggest that increasing serotonin synthesis during the late luteal phase of the menstrual cycle has a beneficial effect in patients with PMS.

Low-fat Vegetarian Diets

A low-fat vegetarian diet has been associated with increased serum sex-hormone binding globulin concentration and reductions in body weight, dysmenorrhea duration and intensity, and PMS duration. The symptom effects might be mediated by dietary influences on estrogen activity.

The UK PMS society states, 'a healthy diet, particularly one which is low in fat and high in fibre can relieve PMS,' and they recommend the following:

- Eating three main meals per day with three smaller snacks in between, all of which should contain starchy foods. This is particularly important in the luteal phase of the cycle.
- Cutting down on sugar and sugary food, as these provide sudden bursts of energy rather than a steady release, which will help combat fatigue and mood swings.

- Eating fiber helps avoid premenstrual constipation, but it is important to drink plenty of water or sugar-free drinks (six to eight glasses a day) to prevent bloating associated with PMS.
- Having one pint of milk a day of either skimmed or semi-skimmed, or an alternative source of calcium such as low-fat cheese or yogurt,
- Reducing salt intake.
- Eating five portions of fruit and vegetables per day.
- Reducing alcohol intake to no more than 2 units per day (1 unit = half a pint (284 ml) of lager or beer, or one small glass of wine, or one measure of spirits).
- Limiting tea and/or coffee to no more thtrun an 5 cups per day, as excess caffeine can make premenstrual symptoms worse.

See also: **Calcium**: Properties and Determination; **Carbohydrates**: Metabolism of Sugars; **Cocoa**: Chemistry of Processing; Production, Products, and Use; **Essential Fatty Acids**; **Herbs**: Herbs and Their Uses; **Magnesium**; **Thiamin**: Properties and Determination; Physiology; **Tocopherols**: Properties and Determination; Physiology; **Vegetarian Diets**; **Vitamin B$_6$**: Properties and Determination; Physiology

Further Reading

Barr SI, Janelle KC and Prior JC (1995) Energy intakes are higher during the luteal phase of ovulatory menstrual cycles. *American Journal of Clinical Nutrition* 61: 39–43.

Budeiri D, Li Wan Po A and Dornan JC (1996) Is Evening Primrose Oil of value in the treatment of PMS? *Controlled Clinical Trials* 17: 60–68.

Dalton K (1984) *The Premenstrual Syndrome and Progesterone Therapy*. London: William Heineman.

Freeman EW (1997) Premenstrual syndrome: current perspectives on treatment and aetiology. *Current Opinions in Obstetrics and Gynecology* 9: 147–153.

Horrobin DF and Manku MS (1989) Premenstrual syndrome and premenstrual breast pain: disorders of essential fatty acid (EFA) metabolism. *Prostaglandins, Leukotrienes and Essential Fatty Acids Review* 37: 255–261.

O'Brien PMS (1993) Helping women with premenstrual syndrome. *British Medical Journal* 307: 1471–1475.

Proctor ML and Murphy PA (2002) Herbal and dietary therapies for primary and secondary dysmenorrhoea. Cochrane Menstrual Disorders and Subfertility Group. *Cochrane Database of Systematic Reviews. Issue 1.*

Schellenberg P (2001) Treatment for the premenstrual syndrome with agnus castus fruit extract: prospective, randomised, placebo controlled study. *British Medical Journal* 322: 134–137.

Speroff L, Glass RH and Kase NG (1994) *Clinical Gynaecologic Endocrinology and Infertility*. Baltimore, MD: Williams & Wilkins.

Wyatt KM, Dimmock PW, Jones PW and O'Brien PMS (1999) Efficacy of vitamin B-6 in the treatment of premenstrual syndrome: systematic review. *British Medical Journal* 318: 1375–1381.

PRESERVATION OF FOOD

M F Sancho-Madriz, California State Polytechnic University, Pomona, CA, USA

Introduction

Food preservation consists of the application of science-based knowledge through a variety of available technologies and procedures, to prevent deterioration and spoilage of food products and extend their shelf-life, while assuring consumers a product free of pathogenic microorganisms. Shelf-life may be defined as the time it takes a product to decline to an unacceptable level. Deterioration of foods will result in loss of quality attributes, including flavor, texture, color, and other sensory properties. Nutritional quality is also affected during food deterioration. Physical, biological, microbiological, chemical, and biochemical factors may cause food deterioration. Preservation methods should be applied as early as possible in the food production pipeline and therefore include appropriate postharvest handling before processing of both plant and animal foods (**Figure 1**). Processing techniques usually rely on appropriate packaging methods and materials to assure continuity of preservation. Handling of processed foods during storage, transportation, retail, and by the consumer also influences the preservation of processed foods.

Selection of technology and procedures for food preservation depends on factors inherent to the product, common pathogenic and spoilage microorganisms, and cost. Product-inherent factors include customary ways of consuming the particular food, sensitivity to heat or other principle used to inactivate microorganisms, and other physical and chemical characteristics of the food.

Rationale and Goals of Food Preservation

Many food products are seasonal, and from year to year the yields of agricultural production vary depending on weather, pests, and other factors. During the production season, especially when the yields are high, the offer of fresh products often exceeds the demand. The products that are not consumed in the fresh market can be used in processed form for long periods of time and by people in regions located far from the production areas. Processed foods not only reduce the waste of food but also offer the consumer many convenient features that make preparation at home easier and faster. Food

preservation plays an important role in special circumstances like wars or space missions far from earth. In the nineteenth and twentieth centuries many new developments in food preservation occurred during war times, in both industry and government-owned laboratories. More recently, the development of foods suitable for space missions that spend a significant amount of time away from earth has posed new challenges to food scientists.

If the food production system is pictured as a long pipeline that carries food from producers to consumers, one will discover a number of leaks in the pipeline that prevent large amounts of food reaching their final destination. Those leaks or losses represent different forms of food deterioration and spoilage, and are likely to happen from the beginning of food production (**Figure 2**).

Food preservation aims to prevent and reduce the loss of food in the production system and extend its shelf-life. While developed nations have been very successful in reducing food losses and increasing efficiency of the production system, food deterioration remains an important issue in developing nations and very poor countries where undernutrition of the population is common. In many instances the food problems of those countries could be significantly reduced by the implementation of adequate postharvest handling and food preservation methods without increasing food production. (*See* **Food Security**.)

Appropriate postharvest handling of food products is essential in food preservation, especially for products sold in the fresh market or as minimally processed items. Exposure to sunlight and heat, high

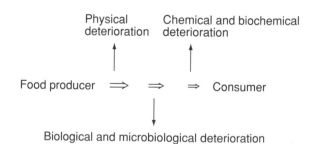

Figure 1 Losses in the food production pipeline.

Food preservation = Adequate postharvest handling
+ food-processing technology + packaging technology
+ adequate transport and storage

Figure 2 Main components of food preservation.

relative humidity, rain, dirt, insects, and rodents increases food deterioration by accelerating chemical reactions in the food and favoring microbial growth. Fast cooling of produce after harvest, storage under temperature- and humidity-controlled conditions, protection of the products by physical barriers, and minimizing the time between harvest and processing are all forms of reducing postharvest losses.

Factors that Cause Deterioration and Spoilage of Food

The factors that cause food deterioration may be grouped under physical, biological and microbiological, and chemical and biochemical factors. Most of the time more than one factor is responsible for the spoilage of a food product and some factors may favor or enhance the effect of others. Light and other forms of radiation, heat, cold, moisture loss, or gain, and the application of force that may alter the structure of the food are examples of physical factors that cause food deterioration. Chemical and biochemical factors include reactions of food components with oxygen or with each other, and reactions catalyzed by enzymatic activity. Maillard browning is caused by a chemical reaction between proteins and carbohydrates and it results in changes in color, flavor, and odor of the food product. This type of browning is desirable in baked products and roasted coffee amongst others, but it is detrimental in foods where browning is not desired. Oxidative rancidity is a chemical change in unsaturated fats and oils and it produces off-flavors and odors. These and other chemical reactions affect sensory qualities of food and can significantly alter the nutritional value of products. (*See* **Browning**: Enzymatic – Biochemical Aspects; **Oxidation of Food Components; Spoilage**: Chemical and Enzymatic Spoilage.)

Biological factors include birds, rodents, insects, and parasites. (*See* **Insect Pests**: Insects and Related Pests; Problems Caused by Insects and Mites.) These may consume or destroy the food and contaminate it with pathogenic or spoilage microorganisms. Bacteria, yeasts, and molds are different types of microorganisms that may cause food deterioration and foodborne illness. (*See* **Aflatoxins; Microbiology**: Classification of Microorganisms; **Mycotoxins**: Classifications; Occurrence and Determination; Toxicology; **Spoilage**: Bacterial Spoilage; Fungi in Food – An Overview; Molds in Spoilage; Yeasts in Spoilage.) Some viruses may also be transmitted through food but they are not associated with the deterioration of products. Not all microorganisms are detrimental to food; some are indeed very useful and have been utilized since ancient times for the production of fermented products like wine, yogurt, and cheese. Fermented cassava and corn

are typical in some African and Latin American countries. (*See* **Fermented Foods**: Origins and Applications; Dietary Importance.)

Because consumers assume that processed foods are safe, the goals of food preservation must include not only measures for the prevention of quality deterioration, but also destruction of pathogenic microorganisms that may cause foodborne illness. A food product that does not show evidence of deterioration may not necessarily be free of pathogenic microorganisms. (*See* **Food Poisoning**: Classification; **Food Safety**.) A preventive strategy for assuring safe food products is the Hazard Analysis Critical Control Point (HACCP) system. HACCP considers biological, physical, and chemical hazards. (*See* **Hazard Analysis Critical Control Point**.)

Factors that Affect Microbial Growth

In addition to the common nutrients found in most foods, microorganisms require certain conditions to survive and grow. The control and optimization of those factors is a fundamental principle of food preservation. The main factors are discussed below.

pH   The degree of alkalinity or acidity of a water solution is based on the concentration of hydrogen (H+) or hydroxyl (OH−) ions. The term pH is by definition the negative logarithm of the hydrogen ion concentration. The pH scale ranges from 0 to 14, and pure water has a pH of 7. Each microorganism will grow only within an optimum pH range. (*See* **pH – Principles and Measurement**.)

Water activity   Water activity is a measure of the availability of water in a food product. Solutes dissolved in water bind the water molecules and make them not available for microorganisms to use. Water activity may be defined as the ratio of the vapor pressure of water in food to the vapor pressure of pure water at the same temperature. Microorganisms require a minimum level of water activity in order to grow and thus water activity reduction can be used to control microbial growth. (*See* **Water Activity**: Principles and Measurement; Effect on Food Stability.)

Oxygen   Microorganisms that require the presence of free oxygen for their growth are known as aerobes, whereas oxygen inhibits the growth of anaerobes. When they can tolerate the presence of oxygen to a certain level they are known as facultative anaerobes.

Temperature   An optimum temperature range is required for microbial growth. Bacteria have been classified into groups depending on their required temperature range. The temperature range for psychrotrophs is 14–20 °C, for mesophiles is 30–37 °C, and for thermophiles is 50–66 °C. Some thermophiles can grow at temperatures up to 77 °C.

Traditional Methods of Food Preservation

The following section provides a brief description of traditional food preservation methods currently used by the food-processing industry. For more details on each method see the indicated articles within this encyclopedia or the suggested publications for further reading. (*See* **Traditional Food Technology**.)

Acidification

The use of acid to preserve foods, whether by fermentation of the product or by addition of acid foods or acids, is a very common method to preserve foods that does not require a high level of technology or special processing equipment. The acid acts as a preservative by controlling microbial growth and pH meters are used to determine if the food has reached the target pH level. (*See* **Acids**: Properties and Determination; Natural Acids and Acidulants; **pH – Principles and Measurement**.) Acidified foods, sometimes called pickled foods (*see* **Pickling**), have been defined by the US Food and Drug Administration (FDA) as low-acid foods to which acid(s) or acid food(s) is (are) added to produce a product that has a finished equilibrium pH of 4.6 or less and a water activity greater than 0.85. The pH value under 4.6 is necessary to prevent the growth of *Clostridium botulinum*; however, an additional preservation method such as pasteurization, refrigeration, or chemical preservatives is used to destroy or control the growth of deterioration microorganisms. (*See* **Clostridium**: Botulism.) Foods have different degrees of buffering capacity against acids and therefore different foods require different amounts of acid to reach a desired pH level; the higher the buffering capacity, the higher the amount of acid needed to acidify the product. Other changes in fermented foods such as metabolites produced by some microorganisms also have an effect on microbial inactivation. (*See* **Beers**: History and Types; **Cheeses**: Types of Cheese; **Fermented Foods**: Origins and Applications; **Fermented Milks**: Dietary Importance; **Wines**: Types of Table Wine; **Yogurt**: The Product and its Manufacture.)

Thermal Processing

The development of canning as a preservation method for foods can be traced back to Nicolas Appert (1749–1841), a French chef who published his procedures in 1810. Appert was awarded 12 000 francs for his proposal for preserving foods by heating them in hermetically sealed containers. Not only did Appert not have scientific training but also microorganisms were not known at the time he developed his method, long before Louis Pasteur laid the foundations for the science of bacteriology.

The degree of microorganism destruction achieved by thermal processing varies, depending on the specific temperature and time of the thermal treatment. Thermal process design is based on a target microorganism and desired number of logarithmic reductions in microorganism concentration. Blanching, pasteurization, commercial sterilization, and sterilization are different kinds of thermal processing. Blanching is a mild heat treatment utilized mostly for enzyme inactivation but it also serves other functions. Although it reduces the number of containing microorganisms on the surface of foods, it is not intended as a sole preservation method. Pasteurization involves heating the product under atmospheric pressure without exceeding the boiling point of water (100 °C). Milk and other low-acid foods (pH > 4.6) are pasteurized to eliminate pathogenic microorganisms and extend their shelf-life in combination with refrigeration or other preservation methods. In acid or acidified foods pasteurization is used to destroy spoilage microorganisms and sometimes to inactivate enzymes.

The US FDA defined commercial sterility of thermally processed food as the condition that renders the food free of microorganisms capable of reproducing under normal nonrefrigerated conditions of storage and distribution, and viable microorganisms (including spores) of public health significance. According to the FDA definition, commercial sterility may be achieved by the application of heat alone or combined with water activity control. (*See* **Canning**: Principles; **Heat Treatment**: Ultrahigh Temperature (UHT) Treatments; **Pasteurization**: Principles; Pasteurization of Liquid Products; Pasteurization of Viscous and Particulate Products.)

Concentration by Evaporation

Concentration by evaporation consists of partially removing water from liquid foods by the application of heat. Native Americans used natural evaporation in wooden or bark vessels to manufacture maple syrup using sap. Water removal causes a reduction in water activity. The rate of heat transfer into the food and the rate of mass transfer of vapor from the food are the factors that determine the rate of evaporation. This method of food preservation has a high level of energy consumption and is therefore more expensive. It offers the convenience of a concentrated product that the consumer can dilute at home and it reduces the cost of transportation and packaging. For example, a can of concentrated orange juice weighs less than the single-strength juice and it requires a smaller package. Destruction of heat-sensitive vitamins and loss of aroma are important issues in evaporation; however, the addition of vitamins and the use of aroma-recovery systems reduce the effect of those factors. (*See* **Evaporation**: Basic Principles; Uses in the Food Industry.)

Dehydration

In drying or dehydration, water is removed from the food by hot air or heated surface driers. Examples of the former include cabinet, tunnel, conveyor, and fluidized bed driers, and the latter include drum and vacuum shelf driers. As water in the surface evaporates it is replaced by water from inside the food by several mass transfer mechanisms, resulting in a reduction of the water content and water activity of the product. In addition to preserving the food, dehydration reduces the weight and bulk of the food, lowering transportation and packaging costs. Despite added convenience, dehydration also has a significant effect on the sensory properties of food. Some dehydrated products such as prunes or raisins are consumed in dehydrated form or used as ingredients in recipes; however, other dehydrated products such as dry milk or vegetables in a soup mix are reconstituted with water before consumption. (*See* **Drying**: Drying Using Natural Radiation; Fluidized-bed Drying; Spray Drying; Dielectric and Osmotic Drying; Hygiene.)

Freeze-drying or lyophilization is a method that accomplishes dehydration of the food by sublimating the water. When water sublimates it goes directly from a solid to a gas without passing through the liquid phase. Freeze-dried foods exhibit superior sensory and nutritive qualities when compared with products dehydrated by other methods. This method can be used to dehydrate high-value solid and liquid foods such as shrimp, strawberries, coffee, and juices, and it is also used by the pharmaceutical industry. (*See* **Freeze-drying**: The Basic Process.)

Refrigeration or Chilled Storage

Chilled storage in refrigerated chambers at temperatures above freezing is a widely used food preservation method. Refrigeration temperatures usually range from 0 to 7 °C in commercial and household refrigerators. The low temperatures lower the rate of metabolic reactions in unprocessed fruits and vegetables and other chemical reactions in foods. Microbial growth is usually slowed at refrigeration temperatures because metabolic reactions of microorganisms are enzyme-catalyzed and their rate depends on temperature. Refrigeration will preserve perishable foods for days or weeks, depending on the food. This method of preservation has very mild effects on sensory and nutritive attributes of products; however, it does not prevent food deterioration in the same degree and for as long as freezing or most other preservation methods. Refrigerated foods require low storage temperatures during transportation, retail, and home storage and their use is limited in rural areas of developing nations. (*See* **Chilled Storage**: Principles; Attainment of Chilled Conditions; Quality and Economic Considerations; Microbiological Considerations.)

Freezing

Foods were frozen in ancient times using ice and snow. Frozen storage cabinets were developed in the nineteenth century. Nowadays, food is frozen using several types of industrial mechanical refrigerators through cooled surfaces, and cooled liquid or air. Cryogenic freezing uses liquid nitrogen or carbon dioxide in solid or liquid form in direct contact with the food. Frozen food storage requires temperatures that maintain the food in frozen condition, usually −18 °C or less, and will preserve foods for months or years if properly packaged. A proportion of the water in the food is frozen and the concentration of solutes in unfrozen water increases, lowering the water activity. Freezing usually stops microbial growth but it does not destroy bacteria and molds. The parasite *Trichinella spiralis* and fish parasites are killed during frozen storage. Freezing temperatures significantly reduce the rate of chemical reactions in foods. Freezing has a low effect on nutritive quality of the food but sensory qualities, especially texture, may be affected by the formation of ice crystals. Fish and seafood, meats, fruits, and vegetables have been sold in frozen form for a long time. Baked goods and other prepared foods have become popular for their convenience, especially because of the use of microwave ovens in households. (*See* **Freezing**: Principles.)

Salting, Sugaring, and Curing

The addition of large amounts of salt or sugar to food is an old method of food preservation. Fish preservation by salting was used in Mesopotamia, Egypt, China, and the Mediterranean basin since ancient times. Meat was also preserved in old Mesopotamia by the addition of salt. Salt has also been used to preserve butter, cheese, and milk curds. Jams and preserves are the most common products preserved by adding sugar. When salt or sugar is added water moves from inside the cells to the outside solutes by osmosis, causing a partial dehydration of the cell, known as plasmolysis, that interferes with microorganism multiplication.

Both salting and sugaring have a significant effect in lowering water activity in food.

Curing is a method to preserve meats that also changes the flavor, color, and tenderness of the product. With more effective preservation methods available, the main purpose of curing is to produce characteristic products with unique flavor and to preserve the red color of meat after cooking. The main ingredients for curing or picking meat are sodium chloride, sodium nitrate and/or sodium nitrite, sugar, and spices. Sodium nitrate and nitrite have been linked with the formation of nitrosamines, compounds shown to be carcinogenic in animal studies. Nitrates and nitrites are naturally present in some vegetables and other foods and therefore cured meats may not be the main source of these compounds in human diets. (*See* **Curing; Jams and Preserves**: Methods of Manufacture; Chemistry of Manufacture.)

Smoking

Smoking is an old method of food preservation and it continues to be used today for fish and meats. The smoke is obtained by burning hickory or a similar wood and it contains formaldehyde and phenolic compounds that have antimicrobial properties. The heat also dries the food, increasing preservation. Smoking of foods is currently used more for its unique flavor properties than for its preservative action.

Chemical Preservatives

A number of chemical substances are used to inhibit the growth of microorganisms in foods. Sodium benzoate, sodium and calcium propionate, sorbic acid, ethyl formate, and sulfur dioxide are examples of commercially used food preservatives. Preservatives must undergo a review process that includes data on toxicological aspects prior to their approval. The allowed amount of a preservative in a product varies depending on the substance and the food. The pH level is also an important consideration as preservatives have an optimum pH range for their antimicrobial activity. Antioxidants prevent oxidative rancidity of fats and oils. Examples of commonly used antioxidants are butylated hydroxyanisole (BHA), butylated hydroxytoluene (BHT), tertiary butylated hydroquinone (TBHQ), and propyl gallate. Ascorbic acid (vitamin C) and tocopherols (vitamin E) are also used as antioxidants. (*See* **Antioxidants**: Natural Antioxidants; Synthetic Antioxidants; Synthetic Antioxidants, Characterization and Analysis; Role of Antioxidant Nutrients in Defense Systems; **Preservatives**: Classifications and Properties; Food Uses.)

Increased interest in natural antimicrobials has been noted in recent years due to toxicological concerns about synthetic preservatives and market trends that show increased popularity of foods with no artificial substances. Several naturally occurring substances have been identified and studied for their antimicrobial properties.

Role of Packaging in the Preservation of Foods

Adequate food packaging is essential and it must be used in combination with all food preservation methods. Packaging must protect the processed food against chemical attack, physical damage and contamination with microorganisms, insects, and rodents. Hermetic packaging refers to containers that are completely sealed against the ingress of vapors and gases and therefore also resistant to bacteria, yeasts, molds, and dust. Glass and metal cans are the most common hermetic containers. Nonhermetic containers may allow gas exchange but should provide protection against microbial contamination. Other packaging materials are plastics, paper, paperboard, and combinations of more than one material by lamination. Combinations are also achieved by simultaneous extrusion of two or more layers of different polymers to form a single film. Packaging must also be nontoxic, act as a barrier to moisture loss or gain, protect against oxygen and odor absorption, protect the food against ultraviolet light, be tamper-resistant or tamper-evident, and be compatible with the food. Food packages have other important functions in marketing the product and educating consumers on product use and handling. Food packages often carry the date of manufacturing or an expiration date, and some include a code that allows companies to trace the product in case of a recall. Expiration dates inform the consumer of the approximate shelf-life of the product. Shelf-life depends on the type of food, the processing method, packaging, and storage conditions. (*See* **Packaging**: Packaging of Liquids; Packaging of Solids.)

Active packaging is an innovative concept that has received a lot of attention in recent years. Active packaging may be defined as a type of packaging that changes the condition of the packaging to extend the shelf-life or improve safety or sensory properties while maintaining the quality of the food. Major active packaging techniques involve the absorption of oxygen, ethylene, moisture, odors, or carbon dioxide; they may also involve the release of antimicrobial agents, carbon dioxide, antioxidants, and flavors.

Controlled- or Modified-Atmosphere Storage and Packaging

Modification of the gas composition in storage rooms or inside the food packaging reduces the rate of respiration of fresh fruits and vegetables and also inhibits microbial and insect growth. Lowering respiration rates reduces biochemical and enzymatic activities that promote ripening and senescence. Gas modifications include oxygen reduction and increase in carbon monoxide or nitrogen. This method may also be combined with chilled storage, resulting in extended shelf-life and high quality of the products. (See **Chilled Storage**: Use of Modified-atmosphere Packaging; Packaging Under Vacuum.)

Aseptic Packaging

Aseptic processing and packaging is a sophisticated food preservation method where the food is sterilized or commercially sterilized outside the container and then placed in previously sterilized containers and sealed in an aseptic environment. Aseptic systems use ultrahigh-temperature (UHT) sterilization, a fast heating treatment at temperatures higher than pasteurization temperatures. Paper and plastic packaging materials are sterilized, formed, filled, and sealed in a continuous operation at the end of the processing line. Aseptic packaging is also used with metal cans, large plastic or metal drums, or large flexible pouches. Packages for aseptic processing may be sterilized using heat, chemicals, irradiation, or a combination of these methods.

Nontraditional or Alternative Methods of Food Preservation

Several alternative methods for food preservation are currently used or considered for future use. Some methods have been approved by regulatory angencies only for certain products and after a review of available scientific information on kinetics of microbial inactivation, toxicological aspects, and the effects of the treatment on the food products. The US FDA has identified several research needs applicable to alternative technologies. The following are brief descriptions of alternative food-processing methods evaluated by the FDA.

Irradiation

Irradiation is a preservation method where food is exposed to radiation. Irradiated food does not become radioactive. Machine or radionuclide radiation sources are used. Machine sources include electron acelerators and X-ray generators, and radionuclide sources include radioactive materials that give off ionizing gamma-rays. Bacteria, molds, yeasts, and insects are inactivated by irradiation. In 1963 the US FDA first approved irradiation for use on wheat and wheat flour. FDA requires that foods that have been irradiated bear both a logo and a statement that the food has been irradiated. The safety of irradiated foods has been studied extensively. (See **Irradiation of Foods**: Basic Principles.)

Microwave and Radiofrequency Processing

Electromagnetic waves of certain frequencies generate heat in foods by dielectric and ionic mechanisms. Microwave and radiofrequency heating have the advantage that they require less time than conventional heating, particularly for solid and semisolid foods. Industrial microwave pasteurization has been used for at least 30 years, but not radiofrequency processing systems. Food does not heat uniformly during microwave processing, and this remains an important issue when considering this technology. Several methods have been used to improve the uniformity of heating but equipment design can significantly influence processing parameters and establishing general conclusions has proved difficult.

Ohmic and Inductive Heating

In ohmic heating electric currents are passed through the food in order to heat it. This technology is also known as Joule heating, electrical resistance heating, electroheating, and electroconductive heating. Inductive heating is a process that induces electric currents within the food by the use of oscillating electromagnetic fields generated by electric coils. Microbial death kinetics data have been published for ohmic heating but not for inductive heating. The main advantage of ohmic heating is its ability to heat materials in a fast and uniform fashion, including products with particulates. Potential future uses of ohmic heating are in dehydration, evaporation, blanching, and extraction.

High-Pressure Processing (HPP)

HPP is also known as high hydrostatic pressure (HHP) or ultrahigh pressure (UHP) processing. During HPP, liquid or solid foods are subject to pressures between 100 and 800 MPa, at temperatures below 0 °C to above 100 °C and for times ranging from a millisecond pulse to more than 20 min. Temperatures ranging from 45 to 50 °C during treatment appear to increase the inactivation of pathogens and spoilage microorganisms. Temperatures in the range of 90–100 °C, together with pressures from 500 to 700 MPa, have been used to inactivate spore-forming bacteria. The effect of HPP works instantaneously and uniformly throughout the mass of food, independent of size, shape, and food composition. The temperature of the food increases by approximately 3 °C per 100 MPa of applied pressure. Water activity and pH are critical product factors in microbial inactivation by HPP. Process factors that affect HPP include pressure, time at pressure, time to achieve treatment pressure, decompression time, treatment temperature, product initial temperature, vessel temperature distribution at pressure, packaging material integrity, and concurrent processing aids.

Pulsed Electric Fields (PEF)

This method involves the application of pulses of high voltage (20–80 kV cm^{-1}) to foods placed between two electrodes. Most PEF systems are at the laboratory phase with not many commercial systems in use. Application of PEF is restricted to foods that can withstand high electric fields, have low electrical conductivity, and do not contain or form bubbles. Although various theories have been proposed to explain the microbial inactivation by PEF, the most studied are electrical breakdown and electroporation. Process factors that affect microbial destruction by PEF include electric field intensity, pulse width, treatment time, and temperature.

Ultraviolet (UV) Light

This type of processing involves the application of radiation from the UV region of the electromagnetic spectrum. Microbial inactivation occurs by DNA mutations upon absorption of the UV light and exposure must be at least 400 J m^{-2} in all parts of the product. Critical factors during UV light processing include the transmissivity of the product, the radiation path length, the geometric configuration of the reactor and the power, the wavelength, and physical arrangement of the UV source(s). UV light is used for bottled-water processing and for sanitizing food contact surfaces. There has been an increased interest in using UV light as a preservation method for fruit juices to replace pasteurization and other thermal treatments that have more impact on sensory attributes. (*See* **Ultraviolet Light**.)

Other Alternative Technologies

High-voltage arc discharge, pulsed light technology, oscillating magnetic fields (OMF), ultrasound, and pulsed X-rays are other types of technology that have been explored for their potential to inactivate microorganisms and may show promise for use in food preservation; however, more extensive research is needed at this time.

See also: **Antioxidants**: Natural Antioxidants; **Browning**: Enzymatic – Biochemical Aspects; **Canning**: Principles; **Chilled Storage**: Principles; Microbiological Considerations; Use of Modified-atmosphere Packaging; Packaging Under Vacuum; *Clostridium*: Botulism; **Curing**; **Drying**: Drying Using Natural Radiation; Fluidized-bed Drying; Spray Drying; Dielectric and Osmotic Drying; Hygiene; **Evaporation**: Basic Principles; Uses in the Food Industry; **Fermented Foods**: Origins and Applications; **Food Poisoning**: Classification; **Food Safety**; **Food Security**; **Freeze-drying**: The Basic Process; **Freezing**: Principles; **Hazard Analysis Critical Control Point**; **Heat Treatment**: Ultra-high Temperature (UHT) Treatments; **Irradiation of Foods**: Basic Principles; **Mycotoxins**: Classifications; **Oxidation of Food Components**; **Pasteurization**: Principles; **Pickling**; **Preservatives**: Classifications and Properties; **Spoilage**: Bacterial Spoilage; Molds in Spoilage; Yeasts in Spoilage; **Storage Stability**: Mechanisms of Degradation

Further Reading

Borgstrom G (1968) *Principles of Food Science*, vol. 1. *Food Technology*. London: Collier-MacMillan.

Buffler CR (1993) *Microwave Cooking and Processing*. New York: Van Nostrand Reinhold.

Connor JM and Schiek WA (1997) *Food Processing, An Industrial Powerhouse in Transition*, 2nd edn. New York: John Wiley.

deMan JM (1999) *Principles of Food Chemistry*, 3rd edn. Gaithersburg, MD: Aspen.

Fellows PJ (2000) *Food Processing Technology*. Cambridge, UK: Woodhead Publishing.

Food and Drug Administration (2000) *Code of Federal Regulations, Title 21*. Washington, DC: US government.

Food and Drug Administration (2000) *Kinetics of Microbial Inactivation for Alternative Food Processing Technologies: Executive Summary*. Washington, DC: US government.

Heldman DR and Hartel RW (1997) *Principles of Food Processing*. New York: Chapman and Hall.

Jay JM (1998) *Modern Food Microbiology*, 5th edn. Gaithersburg, MD: Aspen Publishers.

Potter NN and Hotchkiss JA (1995) *Food Science*. New York: Chapman and Hall.

Rhaman MS (ed.) (1999) *Handbook of Food Preservation*. New York: Marcel Dekker.

VanGarde SJ and Woodburn M (1994) *Food Preservation and Safety: Principles and Practice*. New York: Chapman and Hall.

ISBN 0-12-227055-X

9 780122 270550